Mechanics of Solids and Structures

David W. A. Rees
Brunel University, UK

ICP
Imperial College Press

Published by

Imperial College Press
57 Shelton Street
Covent Garden
London WC2H 9HE

Distributed by

World Scientific Publishing Co. Pte. Ltd.
5 Toh Tuck Link, Singapore 596224
USA office: Suite 202, 1060 Main Street, River Edge, NJ 07661
UK office: 57 Shelton Street, Covent Garden, London WC2H 9HE

British Library Cataloguing-in-Publication Data
A catalogue record for this book is available from the British Library.

First published 2000
Reprinted 2003

MECHANICS OF SOLIDS AND STRUCTURES

ISBN 1-86094-217-2
ISBN 1-86094-218-0 (pbk)

Printed in Singapore by Uto-Print

CONTENTS

CHAPTER 4

BENDING OF BEAMS AND PLATES **137**

CHAPTER 5

THEORIES OF TORSION **197**

CHAPTER 6

MOMENT DISTRIBUTION **253**

CHAPTER 7

FLEXURAL SHEAR FLOW 281

CHAPTER 8

ENERGY METHODS 331

CHAPTER 9

INSTABILITY OF COLUMNS AND PLATES 389

CHAPTER 10

FINITE ELEMENTS 435

CHAPTER 11

YIELD AND STRENGTH CRITERIA 497

CHAPTER 12

PLASTICITY AND COLLAPSE 539

CHAPTER 13

CREEP AND VISCO-ELASTICITY 591

CHAPTER 14

HIGH AND LOW CYCLE FATIGUE 631

CHAPTER 15

FRACTURE MECHANICS 663

APPENDIX I

APPENDIX II

APPENDIX III

PREFACE

This book has been developed from subject matter and examples that I have used in my teaching of solid mechanics, structures and strength of materials in Universities over the last two decades. It is intended for engineering degree and diploma courses in which solid mechanics and structures form a part. Postgraduates and those preparing for the membership of professional institutions by examination in these subjects will also find this book useful. The contents illustrate where overlapping topics in civil, aeronautical and materials engineering employ common principles and thereby should serve engineering students of all disciplines. In the author's experience this broadening of the subject base is also aligned to the teaching of applied mechanics within engineering science degree courses.

A concise approach has been employed for the theoretical developments in order to provide the space for many illustrative examples. It should become obvious that these calculations are all related to the load carrying capacity of materials used in engineering design. Amongst the requirements are the choice of material, its physical shape, the assessment of the nature of imposed loading and its effect on life expectancy. The text illustrates where and how the necessary techniques are to be employed in each case. The reader will soon recognise, for example, that under elastic loading, the solution to the stress and strain suffered by a material invariably becomes that of satisfying three requirements: equilibrium, compatibility and the boundary conditions. The style adopted has been to provide mostly self-contained chapters with a logical and clear presentation of the subject matter. Earlier material underpins the analyses given in later chapters. This allows occasional reference to other chapters without detracting from the main argument. The choice of general chapter titles, that contain many specific topics, emphasise the more wide ranging principles of the subject.

The first three chapters of the text are arranged to cover the necessary fundamental material on stress and strain analyses and plane elasticity theory. A structures theme follows with the full treatment of theories of bending and torsion. This theme continues with coverage of the moment distribution method, shear flow and strut buckling. The chapter on energy methods and virtual work precedes chapters on finite elements, yield and strength criteria. Thereafter, the mechanics of inelastic solids appears with chapter on plasticity and collapse, creep and visco-elasticity. The final two chapters on high and low cycle fatigue and fracture mechanics reflect some of the more recent developments in solid mechanics.

Each topic, as it appears, is illustrated by worked examples throughout. Many exercises on these topics appear at the end of each chapter. The interested reader and user of the book may, at a later date, wish to consult a solution manual to the exercise sections, which is now in preparation.

Acknowledgements are made to Imperial College, London, Kingston University, Trinity College, Dublin and to the C.E.I. for granting permission to include questions from their past examination papers as worked examples and exercises. The author also thanks Mrs M. E. J. Williams for proof reading the manuscript and his past teachers, colleagues and students who have all helped to shape this work.

D.W.A. REES

CHAPTER 1

STRESS AND STRAIN TRANSFORMATION

1.1 Three-Dimensional Stress Analysis

In dealing with the state of stress at a point we consider the element in Fig. 1.1a, set in a Cartesian co-ordinate frame *x*, *y* and *z*. Nine stress components σ_x, σ_y, σ_z, τ_{xy}, τ_{yx}, τ_{xz}, τ_{zx}, τ_{yz} and τ_{zy} act on the six rectangular faces, as shown. The three stress components existing on any one face might arise as the components of an oblique force applied to that face. In general, these component stresses are a consequence of any manner of combined loading, consisting of moments, torques and forces. In the double subscript engineering notation used to identify shear stress, the first subscript denotes the direction of the normal to the plane on which that stress acts. The second subscript denotes the direction of the shear stress. Some authors reverse the order of these subscripts but this does not alter the analysis because of the complementary nature of the shear stress. This means that $\tau_{xy} = \tau_{yx}$, $\tau_{xz} = \tau_{zx}$ and $\tau_{yz} = \tau_{zy}$, which is a moment equilibrium requirement. As a consequence there are six independent stress components: three normal σ_x, σ_y and σ_z and three independent shear components.

A *stress tensor* contains the nine Cartesian stress components shown in Fig. 1.1a and is conveniently represented within a 3 × 3 matrix. We shall also represent these components later with a single mathematical symbol σ_{ij} where *i* and $j = 1, 2$ and 3, so that:

$$\sigma_{ij} \equiv \begin{bmatrix} \sigma_x & \tau_{xy} & \tau_{xz} \\ \tau_{yx} & \sigma_y & \tau_{yz} \\ \tau_{zx} & \tau_{zy} & \sigma_z \end{bmatrix} \equiv \begin{bmatrix} \sigma_{11} & \sigma_{12} & \sigma_{13} \\ \sigma_{21} & \sigma_{22} & \sigma_{23} \\ \sigma_{31} & \sigma_{32} & \sigma_{33} \end{bmatrix}$$

1.1.1 Direction Cosines

In Fig. 1.1b ABC is an oblique plane cutting the volume element to produce the tetrahedron OABC. On the front three faces of the element in Fig. 1.1a the stress components act in the positive co-ordinate directions. Negative directions apply to the back three orthogonal faces in Fig. 1.1b. It is required to find, for Fig. 1.1b, the stress state (σ, τ) on the front triangular face ABC in both magnitude and direction. To do this it becomes necessary to find the areas of each back face. In Fig. 1.2a, we let the area ABC be unity and construct CD perpendicular to AB and join OD. The normal vector **N** to plane ABC is defined by the direction cosines *l*, *m* and *n* with respect to *x*, *y* and *z*,

$$l = \cos\alpha,\ m = \cos\beta \text{ and } n = \cos\gamma \qquad (1.1\text{a,b,c})$$

Then, as area ABC = ½AB × CD and area OAB = ½AB × OD, it follows that

$$\text{Area OAB/Area ABC} = \text{OD/CD} = \cos\gamma = n$$

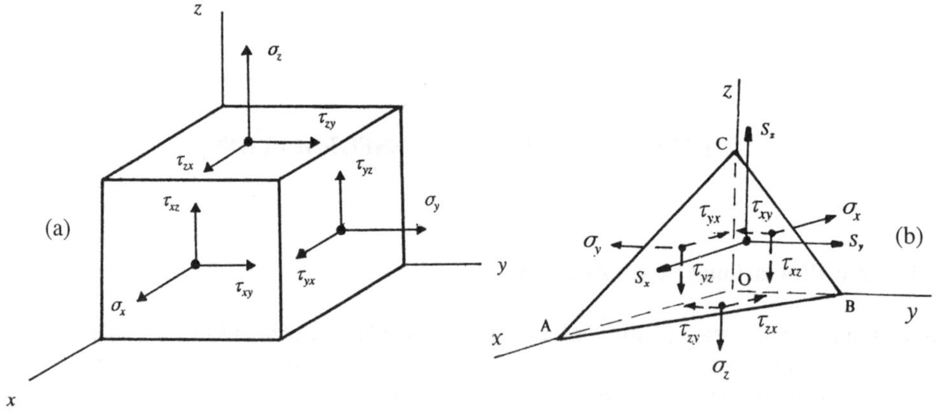

Figure 1.1 3-D stress state for an oblique plane ABC

Hence area OAB = n and, similarly, area OBC = l and area OAC = m. The direction cosines l, m and n are not independent. The relationship between them is found from the vector equation for **N**:

$$\mathbf{N} = N_x \mathbf{u}_x + N_y \mathbf{u}_y + N_z \mathbf{u}_z \tag{1.2a}$$

where $\mathbf{u}_x$, $\mathbf{u}_y$ and $\mathbf{u}_z$ are unit vectors and N_x, N_y and N_z are scalar intercepts with the co-ordinates x, y and z respectively, as seen in Fig. 1.2b.

The unit vector $\mathbf{u}_N$ for the normal direction is found by dividing eq(1.2a) by the magnitude $|\mathbf{N}|$ of **N**:

$$\mathbf{u}_N = (N_x/|\mathbf{N}|)\,\mathbf{u}_x + (N_y/|\mathbf{N}|)\,\mathbf{u}_y + (N_z/|\mathbf{N}|)\,\mathbf{u}_z \tag{1.2b}$$

but from eqs(1.1a-c) $l = \cos\alpha = N_x/|\mathbf{N}|$, $m = \cos\beta = N_y/|\mathbf{N}|$, $n = \cos\gamma = N_z/|\mathbf{N}|$. Hence eq(1.2b) becomes

$$\mathbf{u}_N = l\,\mathbf{u}_x + m\,\mathbf{u}_y + n\,\mathbf{u}_z \tag{1.2c}$$

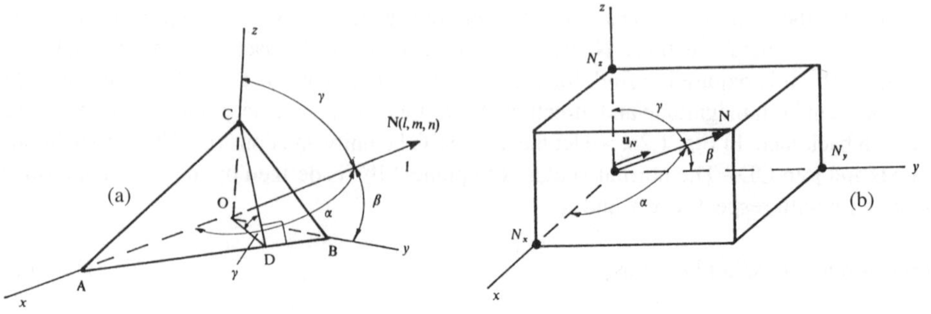

Figure 1.2 Normal to an oblique plane

Clearly l, m and n are the intercepts that the unit normal vector $\mathbf{u}_N$ makes with x, y and z respectively. Furthermore, the magnitude $|\mathbf{N}|$ reveals the relationship between l, m and n:

$$(N_x)^2 + (N_y)^2 + (N_z)^2 = |\mathbf{N}|^2$$
$$[N_x / |\mathbf{N}|\,]^2 + [N_y / |\mathbf{N}|\,]^2 + [N_z / |\mathbf{N}|\,]^2 = 1$$
$$l^2 + m^2 + n^2 = 1 \quad (1.3)$$

1.1.2 Normal and Shear Stress on Plane ABC

(a) Magnitudes

The normal and shear force (stress) components σ and τ respectively act upon plane ABC. Their resultant is **S**, with the co-ordinate components S_x, S_y and S_z, as shown in Fig. 1.3a.

Now **S** is also the equilibrant of the forces produced by the stress components acting on the back faces. Hence the following three equilibrium equations apply to its components:

$$S_x = l\sigma_x + m\tau_{yx} + n\tau_{zx} \quad (1.4a)$$
$$S_y = m\sigma_y + n\tau_{zy} + l\tau_{xy} \quad (1.4b)$$
$$S_z = n\sigma_z + l\tau_{xz} + m\tau_{yz} \quad (1.4c)$$

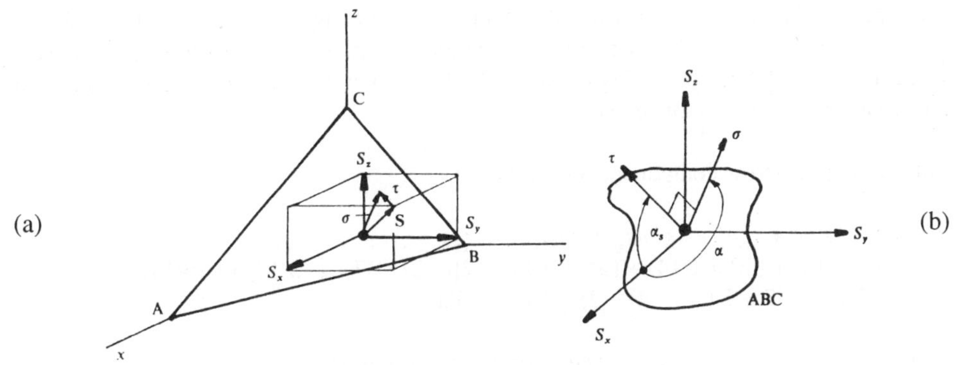

Figure 1.3 Stress state for the oblique plane ABC

As the area of ABC is unity, then σ is the sum of the S_x, S_y and S_z force components resolved into the normal direction. This gives

$$\sigma = S_x \cos\alpha + S_y \cos\beta + S_z \cos\gamma = S_x l + S_y m + S_z n$$

where, from eqs(1.4a,b,c), this becomes

$$\sigma = \sigma_x l^2 + \sigma_y m^2 + \sigma_z n^2 + 2\,(lm\tau_{xy} + mn\tau_{yz} + ln\tau_{zx}) \quad (1.5a)$$

The resultant force on plane ABC is expressed in two ways:

$$S^2 = S_x^2 + S_y^2 + S_z^2 = \sigma^2 + \tau^2$$

$$\tau^2 = S^2 - \sigma^2 = S_x^2 + S_y^2 + S_z^2 - \sigma^2 \tag{1.5b}$$

Substituting eqs(1.4a,b,c) and (1.5a) into (1.5b) enables τ to be found.

1.1.2 Directions of Shear Stress

The direction of σ is defined by l, m and n since σ is aligned with the normal vector **N**. The direction of τ, lying tangential to plane ABC, is defined by $l_s = \cos\alpha_s$, $m_s = \cos\beta_s$ and $n_s = \cos\gamma_s$ (see Fig. 1.3b). Because S_x S_y and S_z are the x, y and z components of σ and τ, it follows that

$$S_x = \sigma\cos\alpha + \tau\cos\alpha_s = l\sigma + l_s\tau$$
$$S_y = \sigma\cos\beta + \tau\cos\beta_s = m\sigma + m_s\tau$$
$$S_z = \sigma\cos\gamma + \tau\cos\gamma_s = n\sigma + n_s\tau$$

from which the direction cosines of τ are

$$l_s = (S_x - l\sigma)/\tau \tag{1.6a}$$
$$m_s = (S_y - m\sigma)/\tau \tag{1.6b}$$
$$n_s = (S_z - n\sigma)/\tau \tag{1.6c}$$

Example 1.1 The state of stress (in MPa) at a point is given by $\sigma_x = 14$, $\sigma_y = 10$, $\sigma_z = 35$, $\tau_{xy} = \tau_{yx} = 7$, $\tau_{xz} = \tau_{zx} = -7$ and $\tau_{yz} = \tau_{zy} = 0$. Determine the normal and shear stresses for a plane whose normal is defined by $l = 2/\surd 14$, $m = -1/\surd 14$ and $n = 3/\surd 14$. What is the direction of the shear stress acting in this plane?

Substituting the stress components into eq(1.5a),

$$\tau = \sigma_x l^2 + \sigma_y m^2 + \sigma_z n^2 + 2\,(lm\tau_{xy} + mn\tau_{yz} + ln\tau_{zx})$$
$$= (14 \times 4/14) + (10 \times 1/14) + (35 \times 9/14) + 2[(-2/14)7 + (-3/14)0 + (6/14)(-7)]$$
$$= 4 + 5/7 + 45/2 + 2\,(-1 - 0 - 3) = 19.21 \text{ MPa}$$

Now, from eqs(1.4a,b,c), the x, y and z component forces are

$$S_x = l\sigma_x + m\tau_{xy} + n\tau_{xz}$$
$$= (2/\surd 14)14 - (1/\surd 14)7 - (3/\surd 14)7 = 0$$
$$S_y = m\sigma_y + n\tau_{yz} + l\tau_{yx}$$
$$= -(1/\surd 14)10 + (3/\surd 14)0 + (2/\surd 14)7 = 4/\surd 14$$
$$S_z = n\sigma_z + l\tau_{zx} + m\tau_{zy}$$
$$= (3/\surd 14)35 - (2/\surd 14)7 - (1/\surd 14)0 = 91/\surd 14$$

Then, from eq(1.5b), the shear stress acting along the plane is

$$\tau^2 = S_x^2 + S_y^2 + S_z^2 - \sigma^2 = 0 + 16/14 + 91^2/14 - 19.21^2 = 223.62$$

from which $\tau = 14.95$ MPa. The direction cosines for this shear stress are, from eqs(1.6a,b,c),

$$l_s = (S_x - l\sigma)/\tau = [0 - (2/\surd 14) \times 19.21]/14.95 = -0.678$$
$$m_s = (S_y - m\sigma)/\tau = [4/\surd 14 - (-1/\surd 14) \times 19.21]/14.95 = 0.415$$

$n_s = (S_z - n\sigma)/\tau = [91/\sqrt{14} - (3/\sqrt{14}) \times 19.21]/14.95 = 0.597$

The corresponding inclinations of τ to x, y and z are

$$\alpha_s = \cos^{-1}(-0.687) = 133.4°$$
$$\beta_s = \cos^{-1}(0.415) = 65.5°$$
$$\gamma_s = \cos^{-1}(0.597) = 53.37°$$

1.2 Principal Stresses and Invariants

1.2.1 Magnitudes of the Principal Stresses

When, for the plane ABC in Fig. 1.1a, the shear stress τ is absent, then by definition, σ becomes a *principal stress*. Resolving forces in the x, y and z directions allows eqs(1.4a,b,c) to be written as

$$S_x = l\sigma = l\sigma_x + m\tau_{yx} + n\tau_{zx}$$
$$S_y = m\sigma = m\sigma_y + n\tau_{zy} + l\tau_{xy}$$
$$S_z = n\sigma = n\sigma_z + l\tau_{xz} + m\tau_{yz}$$

Rearranging,

$$\begin{aligned} l(\sigma_x - \sigma) + m\tau_{xy} + n\tau_{xz} &= 0 \\ l\tau_{yx} + m(\sigma_y - \sigma) + n\tau_{yz} &= 0 \\ l\tau_{zx} + m\tau_{zy} + n(\sigma_z - \sigma) &= 0 \end{aligned} \qquad (1.7a)$$

By Cramar's rule, the solution to σ from eq(1.7a) can be found from the determinant:

$$\begin{vmatrix} (\sigma_x - \sigma) & \tau_{xy} & \tau_{xz} \\ \tau_{yx} & (\sigma_y - \sigma) & \tau_{yz} \\ \tau_{zx} & \tau_{zy} & (\sigma_z - \sigma) \end{vmatrix} = 0 \qquad (1.7b)$$

Expanding eq(1.7b) leads to the *principal stress cubic equation*:

$$(\sigma_x - \sigma)[(\sigma_y - \sigma)(\sigma_z - \sigma) - \tau_{yz}\tau_{zy}] - \tau_{xy}[\tau_{yx}(\sigma_z - \sigma) - \tau_{yz}\tau_{zx}]$$
$$+ \tau_{xz}[\tau_{yx}\tau_{zy} - \tau_{zx}(\sigma_y - \sigma)] = 0$$

$$\sigma^3 - (\sigma_x + \sigma_y + \sigma_z)\sigma^2 + (\sigma_x\sigma_y + \sigma_y\sigma_z + \sigma_z\sigma_x - \tau_{xy}^2 - \tau_{yz}^2 - \tau_{zx}^2)\sigma$$
$$- (\sigma_x\sigma_y\sigma_z + 2\tau_{xy}\tau_{yz}\tau_{zx} - \sigma_x\tau_{yz}^2 - \sigma_y\tau_{zx}^2 - \sigma_z\tau_{xy}^2) = 0 \qquad (1.8a)$$

The three roots (the eigen values) σ_1, σ_2 and σ_3 to this cubic equation give the principal stress magnitudes. Equation (1.8a) is simplified to

$$\sigma^3 - J_1\sigma^2 + J_2\sigma - J_3 = 0 \qquad (1.8b)$$

There is a unique set of principal stresses for any given applied stress system. It follows that the coefficients J_1, J_2 and J_3 will be independent of the co-ordinate frame (x, y, z) chosen to define the applied stress components. Thus J_1, J_2 and J_3 are called the *invariants of the stress tensor*. Equation (1.8a) must include the case where the x, y, z frame coincides with the principal stress directions 1, 2 and 3. Hence the invariants may be written in both general (subscripts x, y, z) or principal (subscripts 1, 2, 3) forms:

$$J_1 = \sigma_1 + \sigma_2 + \sigma_3 = \sigma_x + \sigma_y + \sigma_z = \sigma_{ii} \quad (1.9a)$$

$$J_2 = \sigma_1\sigma_2 + \sigma_2\sigma_3 + \sigma_1\sigma_3 = \sigma_x\sigma_y + \sigma_y\sigma_z + \sigma_x\sigma_z - \tau_{xy}^2 - \tau_{yz}^2 - \tau_{zx}^2 = \tfrac{1}{2}(\sigma_{ii}\sigma_{jj} - \sigma_{ij}\sigma_{ji}) \quad (1.9b)$$

$$J_3 = \sigma_1\,\sigma_2\,\sigma_3 = \sigma_x\sigma_y\sigma_z + 2\tau_{xy}\,\tau_{yz}\,\tau_{zx} - \sigma_x\tau_{yz}^2 - \sigma_y\tau_{zx}^2 - \sigma_z\,\tau_{xy}^2 = \det(\sigma_{ij}) \quad (1.9c)$$

For the shorthand tensor notation used in eqs(1.9a-c), repeated subscripts on a single symbol, or within a term, denote summation for i and j = 1, 2 and 3. Where there are exact roots, the principal stresses are more conveniently found by expanding the determinant in eq(1.7b), with numerical values having been substituted. Otherwise, the major (σ_1), intermediate (σ_2) and minor (σ_3) principal stresses ($\sigma_1 > \sigma_2 > \sigma_3$) must be found from the solution to the characteristic cubic equation(1.8b).

Example 1.2 At a point in a loaded material, a resultant stress of magnitude 216 MPa makes angles of $\alpha_r = 43°$, $\beta_r = 75°$ and $\gamma_r = 50.88°$ with the co-ordinates x, y and z respectively. Find the normal and shear stress on a plane whose direction cosines are $l = 0.387$, $m = 0.866$ and $n = 0.3167$. Given that the applied shear stresses are $\tau_{xy} = 23$, $\tau_{yz} = -3.1$ and $\tau_{xz} = 57$ (MPa), determine σ_x, σ_y, σ_z, the invariants and the principal stresses.

Resolve the resultant force S (stress S acting on unit area ABC) in the x, y and z directions (Fig. 1.3a) to give

$$S_x = S\,l_r = S\cos\alpha_r,\; S_y = S\,m_r = S\cos\beta_r \text{ and } S_z = S\,n_r = S\cos\gamma_r$$

The normal stress in eq(1.5a) is then

$$\sigma = S_x l + S_y m + S_z n = S\,(l\cos\alpha_r + m\cos\beta_r + n\cos\gamma_r)$$
$$= 216\,(0.387\cos 43° + 0.866\cos 75° + 0.3167\cos 50.88°) = 152.71 \text{ MPa}$$

Now, from eq(1.5b), the shear stress is

$$\tau^2 = S^2 - \sigma^2 \;\Rightarrow\; \tau = \sqrt{(216^2 - 152.71^2)} = 152.76 \text{ MPa}$$

Substituting into eqs(1.4a,b,c),

$$S_x = l\sigma_x + m\,\tau_{yx} + n\,\tau_{zx}$$
$$216\cos 43° = 0.387\sigma_x + (0.866 \times 23) + (0.3167 \times 57)$$
$$\sigma_x = 310.1 \text{ MPa}$$

$$S_y = m\sigma_y + n\,\tau_{zy} + l\,\tau_{xy}$$
$$216\cos 75° = 0.866\sigma_y - (0.3167 \times 3.1) + (0.387 \times 23)$$
$$\sigma_y = 55.41 \text{ MPa}$$

$$S_z = n\sigma_z + l\,\tau_{xz} + m\,\tau_{yz}$$
$$216\cos 50.88° = 0.3167\sigma_z + (0.387 \times 57) - (0.866 \times 3.1)$$
$$\sigma_z = 369.15 \text{ MPa}$$

Substituting into eqs(1.9a,b,c), the invariants are

$$J_1 = \sigma_x + \sigma_y + \sigma_z = 310.1 + 55.41 + 369.15 = 734.66$$
$$J_2 = \sigma_x \sigma_y + \sigma_y \sigma_z + \sigma_x \sigma_z - \tau_{xy}^2 - \tau_{yz}^2 - \tau_{zx}^2$$
$$= (310.1 \times 55.41) + (55.41 \times 369.15) + (369.15 \times 310.1) - (23)^2 - (-3.1)^2 - (57)^2$$
$$= 152110.66 - 3787.61 = 148323.05$$

$$J_3 = \det \begin{vmatrix} 310.1 & 23 & 57 \\ 23 & 55.41 & -3.1 \\ 57 & -3.1 & 369.15 \end{vmatrix}$$

$$= 310.1\,[(55.41 \times 369.15) - (3.1)^2] - 23[(23 \times 369.15) - (-3.1 \times 57)]$$
$$+ 57[-(23 \times 3.1) - (57 \times 55.41) = 5956556.22$$

The principal stress cubic (eq1.8b) becomes

$$\sigma^3 - 734.66\sigma^2 + 148323.05\sigma - 5956556.22 = 0 \qquad \text{(i)}$$

Using Newton's approximation to find the roots of eq(i),

$$f(\sigma) = \sigma^3 - 734.66\sigma^2 + 148323.05\sigma - 5956556.22 \qquad \text{(ii)}$$
$$f'(\sigma) = 3\sigma^2 - 1469.32\sigma + 148323.05 \qquad \text{(iii)}$$

One root lies between 50 and 60. Take an approximation $\sigma = 55$ MPa so that the numerical values of eqs(ii) and (iii) are $f(\sigma) = 145240.03$ and $f'(\sigma) = 76585.45$. A closer approximation is then given by

$$\sigma - f(\sigma)/f'(\sigma) = 55 - (145240.03 / 76585.45) = 53.1$$

Again, from eqs(ii) and (iii), $f(\sigma) = -2335.66$, $f'(\sigma) = 78761$, giving an even closer approximation to the root:

$$\sigma = 53.1 - (-2335.66 / 78761) = 53.13 \text{ MPa}$$

The remaining roots are found from the quadratic $a\sigma^2 + b\sigma + c = 0$ where

$$(\sigma - 53.13)(a\sigma^2 + b\sigma + c) = \sigma^3 - 734.66\sigma^2 + 148323.05\sigma - 5956556.22$$
$$a\sigma^3 - (53.13\,a - b)\sigma^2 + (c - 53.13\,b)\sigma - 53.13\,c$$
$$= \sigma^3 - 734.66\sigma^2 + 148323.05\sigma - 5956556.22$$

Equating coefficients of σ^3, σ^2, σ gives

$a = 1$
$53.13\,a - b = 734.66, \therefore b = -681.53$
$c - 53.13\,b = 148323.05, \therefore c = 112113.36$
Thus
$\sigma^2 - 681.53\sigma + 112113.36 = 0$

for which the roots are 404.07 or 277.46. Hence, the principal stresses are $\sigma_1 = 404.07$, $\sigma_2 = 277.46$ and $\sigma_3 = 53.13$ MPa.

1.2.2 Directions of Principal Stresses

Substituting for the applied stresses into eq(1.7a) with $\sigma = \sigma_1$ (l_1, m_1, n_1) leads to three simultaneous equations in l_1, m_1 and n_1, of which only two equations are independent, together with the relationship $l_1^2 + m_1^2 + n_1^2 = 1$ from eq(1.3). A similar deduction is made for the separate substitutions of σ_2 (l_2, m_2, n_2) and σ_3 (l_3, m_3, n_3). Now from eq(1.2c) the principal sets of direction cosines (l_1, m_1, n_1), (l_2, m_2, n_2) and (l_3, m_3, n_3) will define unit vectors aligned with the principal directions. They are

$$\mathbf{u}_1 = l_1\mathbf{u}_x + m_1\mathbf{u}_y + n_1\mathbf{u}_z \qquad (1.10a)$$
$$\mathbf{u}_2 = l_2\mathbf{u}_x + m_2\mathbf{u}_y + n_2\mathbf{u}_z \qquad (1.10b)$$
$$\mathbf{u}_3 = l_3\mathbf{u}_x + m_3\mathbf{u}_y + n_3\mathbf{u}_z \qquad (1.10c)$$

Since these are orthogonal, the dot product of any two is zero. For the 1 and 2 directions

$$\mathbf{u}_1 \bullet \mathbf{u}_2 = (l_1\mathbf{u}_x + m_1\mathbf{u}_y + n_1\mathbf{u}_z) \bullet (l_2\mathbf{u}_x + m_2\mathbf{u}_y + n_2\mathbf{u}_z) = 0$$

Now $\mathbf{u}_x \bullet \mathbf{u}_x = \mathbf{u}_y \bullet \mathbf{u}_y = \mathbf{u}_z \bullet \mathbf{u}_z = 1$ and $\mathbf{u}_x \bullet \mathbf{u}_y = \mathbf{u}_x \bullet \mathbf{u}_z = \mathbf{u}_y \bullet \mathbf{u}_z = 0$. Similar dot products between the 1, 3 and 2, 3 directions leads to three orthogonality conditions:

$$l_1 l_2 + m_1 m_2 + n_1 n_2 = 0 \qquad (1.11a)$$
$$l_2 l_3 + m_2 m_3 + n_2 n_3 = 0 \qquad (1.11b)$$
$$l_1 l_3 + m_1 m_3 + n_1 n_3 = 0 \qquad (1.11c)$$

The following example will show that eqs(1.11a,b,c) are the only conditions that satisfy eq(1.7a). This confirms that the principal stress directions and their planes are orthogonal.

Example 1.3 Find the principal stresses and their directions, given the following stress components: $\sigma_x = 3$, $\sigma_y = 0$, $\sigma_z = 0$, $\tau_{xy} = \tau_{yx} = 1$, $\tau_{xz} = \tau_{zx} = 1$ and $\tau_{yz} = \tau_{zy} = 2$ (kN/m²). Show that the principal directions are orthogonal.

The determinant in eq(1.7b) becomes

$$\det \begin{vmatrix} 3-\sigma & 1 & 1 \\ 1 & 0-\sigma & 2 \\ 1 & 2 & 0-\sigma \end{vmatrix} = 0 \qquad \text{(i)}$$

Principal stresses are found from the expansion of eq(i),

$$(1 - \sigma)(\sigma - 4)(\sigma + 2) = 0$$

The roots are principal stresses $\sigma_1 = 4$, $\sigma_2 = 1$ and $\sigma_3 = -2$ kN/m². Equation (1.7a) becomes

$$(3 - \sigma)\,l + m + n = 0 \qquad \text{(i)}$$
$$l - m\sigma + 2n = 0 \qquad \text{(ii)}$$
$$l + 2m - n\sigma = 0 \qquad \text{(iii)}$$

Equations (i)-(iii) supply direction cosines for each principal stress. Substituting $\sigma = \sigma_1 = 4$

$$- l_1 + m_1 + n_1 = 0 \qquad \text{(iv)}$$
$$l_1 - 4m_1 + 2n_1 = 0 \qquad \text{(v)}$$
$$l_1 + 2m_1 - 4n_1 = 0 \qquad \text{(vi)}$$

Because only two of eqs(iv)-(vi) are independent, we can let any vector $\mathbf{A} = A_x\mathbf{u}_x + A_y\mathbf{u}_y + A_z\mathbf{u}_z$, lie in the 1 - direction. Thus the unit vector for the 1 - direction becomes

$$\mathbf{u}_1 = l_1\,\mathbf{u}_x + m_1\mathbf{u}_y + n_1\mathbf{u}_z = (A_x/|\mathbf{A}|)\mathbf{u}_x + (A_y/|\mathbf{A}|)\mathbf{u}_y + (A_z/|\mathbf{A}|)\mathbf{u}_z$$

Hence $l_1 = (A_x/|\mathbf{A}|)$, $m_1 = (A_y/|\mathbf{A}|)$, $n_1 = (A_z/|\mathbf{A}|)$ and eqs(iv)-(vi) become

$$- A_x + A_y + A_z = 0$$
$$A_x - 4\,A_y + 2\,A_z = 0$$
$$A_x + 2\,A_y - 4\,A_z = 0$$

We may solve these by setting $A_x = 1$ (say) to give $A_y = A_z = ½$. Then $|\mathbf{A}| = \surd(3/2)$ thus giving $l_1 = \surd 2/3$ and $m_1 = n_1 = 1/\surd 6$. The unit vector aligned with the 1 - direction becomes

$$\mathbf{u}_1 = \surd(2/3)\mathbf{u}_x + (1/\surd 6)\mathbf{u}_y + (1/\surd 6)\mathbf{u}_z$$

The direction cosines for the 2 and 3 - directions are similarly found by substituting, stresses in turn, $\sigma = \sigma_2 = 1$ and $\sigma = \sigma_3 = -2$ in eqs(i) - (iii). The unit vectors for these directions are

$$\mathbf{u}_2 = (1/\surd 3)\mathbf{u}_x - (1/\surd 3)\mathbf{u}_y - (1/\surd 3)\mathbf{u}_z$$
$$\mathbf{u}_3 = (1/\surd 2)\mathbf{u}_y - (1/\surd 2)\mathbf{u}_z$$

The directions are orthogonal as $\mathbf{u}_1 \bullet \mathbf{u}_2 = \mathbf{u}_1 \bullet \mathbf{u}_3 = \mathbf{u}_2 \bullet \mathbf{u}_3 = 0$.

1.3 Principal Directions as Co-ordinates

When the applied stresses are the principal stresses $\sigma_1 > \sigma_2 > \sigma_3$, the co-ordinate axes become aligned with the orthogonal principal directions 1, 2 and 3 as shown in Fig. 1.4. Because shear stress is then absent on faces ACO, ABO and BCO, the expressions for the normal and shear stress acting on the oblique plane ABC are simplified.

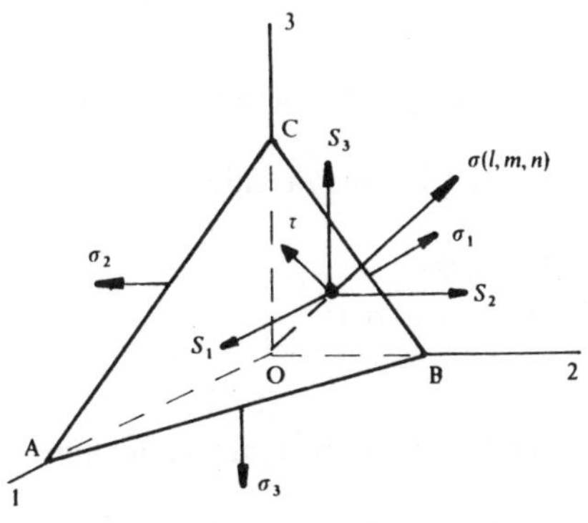

Figure 1.4 Principal stress axes

Replacing x, y and z in eqs(1.4a,b,c) and (1.5a,b) by 1, 2 and 3 respectively and setting $\tau_{xy} = \tau_{yz} = \tau_{xz} = 0$ gives the reduced forms to the co-ordinate forces and oblique plane stresses,

$$S_1 = l\sigma_1,\ S_2 = m\sigma_2,\ S_3 = n\sigma_3 \tag{1.12a,b,c}$$

$$\therefore \sigma = S_1 l + S_2 m + S_3 n$$
$$= \sigma_1 l^2 + \sigma_2 m^2 + \sigma_3 n^2 \tag{1.13a}$$

$$\tau^2 = S^2 - \sigma^2 = (S_1^2 + S_2^2 + S_3^2) - \sigma^2$$
$$= (l\sigma_1)^2 + (m\sigma_2)^2 + (n\sigma_3)^2 - (\sigma_1 l^2 + \sigma_2 m^2 + \sigma_3 n^2)^2$$
$$\tau = \sqrt{[(l\sigma_1)^2 + (m\sigma_2)^2 + (n\sigma_3)^2 - (\sigma_1 l^2 + \sigma_2 m^2 + \sigma_3 n^2)^2]} \tag{1.13b}$$

with associated directions, from eqs(1.6a,b,c),

$$l_s = (S_1 - l\sigma)/\tau = l\,(\sigma_1 - \sigma)/\tau \tag{1.14a}$$
$$m_s = (S_2 - m\sigma)/\tau = m\,(\sigma_2 - \sigma)/\tau \tag{1.14b}$$
$$n_s = (S_3 - n\sigma)/\tau = n\,(\sigma_3 - \sigma)/\tau \tag{1.14c}$$

1.3.1 Maximum Shear Stress

It can be shown that the maximum shear stresses act on planes inclined at 45° to two principal planes and perpendicular to the remaining plane. For the 1 - 2 plane in Fig. 1.5, for example, normal **N** to the 45° plane shown has directions $l = m = \cos 45° = 1/\sqrt{2}$ and $n = \cos 90° = 0$.

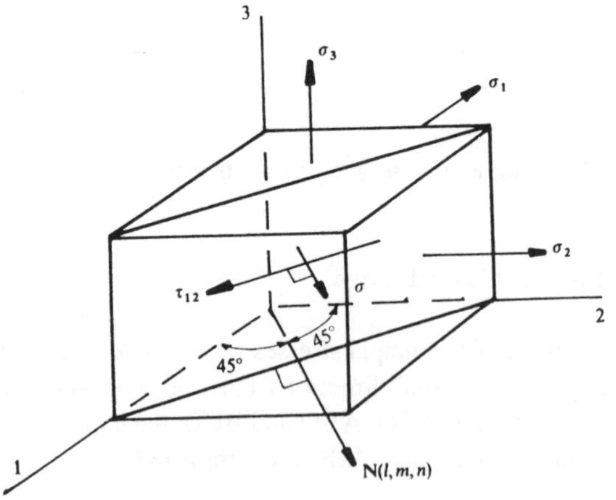

Figure 1.5 Maximum 45° shear plane

Substituting l, m and n into eq(1.13b), the magnitude is

$$\tau_{12}^2 = \sigma_1^2/2 + \sigma_2^2/2 - (\sigma_1/2 + \sigma_2/2)^2$$
$$= \tfrac{1}{4}(\sigma_1^2 + \sigma_2^2 - 2\sigma_1\sigma_2) = \tfrac{1}{4}(\sigma_1 - \sigma_2)^2$$
$$\tau_{12} = \pm \tfrac{1}{2}(\sigma_1 - \sigma_2) \tag{1.15a}$$

Similarly, for the plane inclined at 45° to the 1 and 3 directions ($l = n = 1/\sqrt{2}$ and $m = 0$), the shear stress is

$$\tau_{13} = \pm \tfrac{1}{2} (\sigma_1 - \sigma_3) \qquad (1.15b)$$

and, for the plane inclined at 45° to the 2 and 3 directions, where $n = m = 1/\sqrt{2}$ and $l = 0$,

$$\tau_{23} = \pm \tfrac{1}{2} (\sigma_2 - \sigma_3) \qquad (1.15c)$$

The greatest shear stress for the system $\sigma_1 > \sigma_2 > \sigma_3$ is $\tau_{max} = \tau_{13}$. When the 45° shear planes are joined from all four quadrants they form a rhombic dodecahedron. The normal stress acting on the planes of maximum shear stress is found from eq(1.13a). For example, with $l = m = 1/\sqrt{2}$ and $n = 0$ for the 45° plane in Fig. 1.5, $\sigma = \tfrac{1}{2} (\sigma_1 + \sigma_2)$.

1.3.2 Octahedral Planes

It follows from eq(1.3) that the direction cosines for the normal to the plane equally inclined to the principal directions are $l = m = n = 1/\sqrt{3}$ ($\alpha = \beta = \gamma = 54.8°$). Substituting these into eq(1.1a) gives the octahedral normal stress (σ_o)

$$\sigma_o = \sigma_1 l^2 + \sigma_2 m^2 + \sigma_3 n^2 = \sigma_1 (1/\sqrt{3})^2 + \sigma_2 (1/\sqrt{3})^2 + \sigma_3 (1/\sqrt{3})^2$$
$$\sigma_o = (\sigma_1 + \sigma_2 + \sigma_3)/3 \qquad (1.16a)$$

Since σ_o is the average of the principal stresses it is also called the mean or hydrostatic stress σ_m (see Section 1.4.3). The octahedral shear stress τ_o is found by substituting $l = m = n = 1/\sqrt{3}$ into eq(1.13b)

$$\begin{aligned}\tau_o^2 &= (\sigma_1/\sqrt{3})^2 + (\sigma_2/\sqrt{3})^2 + (\sigma_3/\sqrt{3})^2 - [\sigma_1 (1/\sqrt{3})^2 + \sigma_2 (1/\sqrt{3})^2 + \sigma_3 (1/\sqrt{3})^2]^2 \\ &= (\sigma_1^2 + \sigma_2^2 + \sigma_3^2)/3 - [(\sigma_1 + \sigma_2 + \sigma_3)/3]^2 \\ &= (2/9)(\sigma_1^2 + \sigma_2^2 + \sigma_3^2 - \sigma_1\sigma_2 - \sigma_1\sigma_3 - \sigma_2\sigma_3) \\ &= (1/9)[(\sigma_1 - \sigma_2)^2 + (\sigma_2 - \sigma_3)^2 + (\sigma_1 - \sigma_3)^2]\end{aligned}$$
$$\tau_o = \tfrac{1}{3} \sqrt{[(\sigma_1 - \sigma_2)^2 + (\sigma_2 - \sigma_3)^2 + (\sigma_1 - \sigma_3)^2]} \qquad (1.16b)$$

or, from eqs(1.15a,b,c)

$$\tau_o = (2/3) \sqrt{(\tau_{12}^2 + \tau_{23}^2 + \tau_{13}^2)} \qquad (1.16c)$$

Equations (1.14a,b,c) supply the direction cosines for τ_o

$$l_o = (\sigma_1 - \sigma_o)/\sqrt{3}\tau_o \qquad (1.17a)$$
$$m_o = (\sigma_2 - \sigma_o)/\sqrt{3}\tau_o \qquad (1.17b)$$
$$n_o = (\sigma_3 - \sigma_o)/\sqrt{3}\tau_o \qquad (1.17c)$$

When the eight octahedral planes are joined they form the faces of the regular octahedron as shown in Fig. 1.6. Here σ_o and τ_o act on each plane while τ_{12}, τ_{23} and τ_{13} act along the edges. The deformation occurring under any stress state may be examined from its octahedral plane. Since σ_o acts with equal inclination and intensity it causes an elastic volume change which is recoverable irrespective of the principal stress magnitudes. Superimposed on this is the distortion produced by τ_o. As the magnitude of τ_o, given in eq(1.16b), depends upon differences between the principal stresses, a critical value of τ_o will determine whether the deformation will be elastic or elastic-plastic. In Chapter 11 it is shown that a yield criterion may be formulated on this basis.

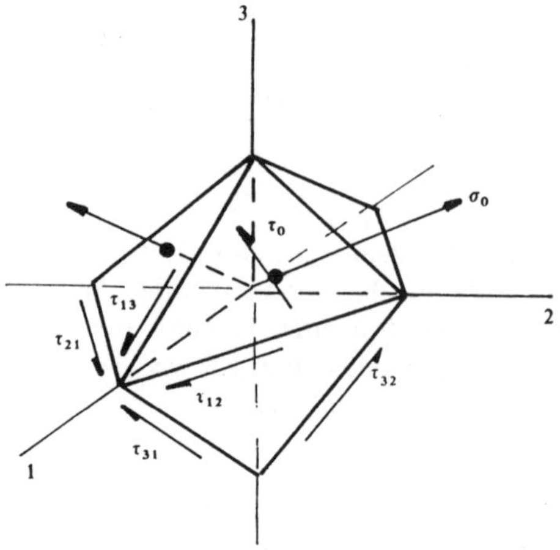

Figure 1.6 Regular octahedron formed from octahedral planes

Example 1.4 The given matrix of stress components σ_{ij} (MPa) describes the stress state at a point. Find, in magnitude and direction, the normal and shear stress on a plane whose unit normal vector equation is $\mathbf{u}_N = 0.53\mathbf{u}_x + 0.35\mathbf{u}_y + 0.77\mathbf{u}_z$. Determine the principal stresses, the greatest shear stress and the stress state on the octahedral plane.

$$\sigma_{ij} = \begin{bmatrix} 6 & 2 & 2 \\ 2 & 0 & 4 \\ 2 & 4 & 0 \end{bmatrix}$$

Substituting $\sigma_x = 6$, $\sigma_y = 0$, $\sigma_z = 0$, $\tau_{xy} = 2$, $\tau_{xz} = 2$ and $\tau_{yz} = 4$ MPa in eq(1.5a) with $l = 0.53$, $m = 0.35$ and $n = 0.77$ gives

$$\sigma = \sigma_x l^2 + \sigma_y m^2 + \sigma_z n^2 + 2\,(lm\,\tau_{xy} + mn\,\tau_{yz} + ln\,\tau_{zx})$$
$$= 6(0.53)^2 + 2[(0.53 \times 0.35 \times 2) + (0.35 \times 0.77 \times 4) + (0.53 \times 0.77 \times 2)] = 6.216 \text{ MPa}$$

Now, from eqs(1.4a,b,c), the x, y and z stress resultants are

$$S_x = l\sigma_x + m\,\tau_{yx} + n\,\tau_{zx}$$
$$= (0.53 \times 6) + (0.35 \times 2) + (0.77 \times 2) = 5.42 \text{ MPa}$$
$$S_y = m\sigma_y + l\,\tau_{xy} + n\,\tau_{zy}$$
$$= (0.35 \times 0) + (0.77 \times 4) + (0.53 \times 2) = 4.14 \text{ MPa}$$
$$S_z = n\sigma_z + l\,\tau_{xz} + m\,\tau_{yz}$$
$$= (0.77 \times 0) + (0.53 \times 2) + (0.35 \times 4) = 2.46 \text{ MPa}$$

Then, from eq(1.5b), the shear stress is

$$\tau^2 = S_x^2 + S_y^2 + S_z^2 - \sigma^2$$
$$= 5.42^2 + 4.14^2 + 2.46^2 - 6.216^2 = 13.93, \quad \Rightarrow \quad \tau = 3.732 \text{ MPa}$$

The direction of τ is defined by directions supplied from eqs(1.6a,b,c),

$$l_s = (S_x - l\sigma)/\tau = [5.42 - (0.53 \times 6.216)] / 3.732 = 0.57$$
$$m_s = (S_y - m\sigma)/\tau = [4.14 - (0.35 \times 6.216)] / 3.732 = 0.526$$
$$n_s = (S_z - n\sigma)/\tau = [2.46 - (0.77 \times 6.216)] / 3.732 = -0.623$$

These cosines may be checked from eq(1.3) when $l_s^2 + m_s^2 + n_s^2 = 1$. The unit vector in the direction of τ with respect to x, y and z becomes

$$\mathbf{u}_s = l_s\mathbf{u}_x + m_s\mathbf{u}_y + n_s\mathbf{u}_z = 0.57\mathbf{u}_x + 0.526\mathbf{u}_y - 0.623\mathbf{u}_z$$

The principal stresses are found from the determinant:

$$\det \begin{vmatrix} 6-\sigma & 2 & 2 \\ 2 & 0-\sigma & 4 \\ 2 & 4 & 0-\sigma \end{vmatrix} = 0 \quad \Rightarrow \quad \begin{aligned} &(\sigma + 4)(2 - \sigma)(\sigma - 8) = 0 \\ &\sigma_1 = 8,\ \sigma_2 = 2 \text{ and } \sigma_3 = -4 \text{ MPa} \end{aligned}$$

The greatest shear stress is found from eq(1.15b)

$$\tau_{max} = \tfrac{1}{2}(\sigma_1 - \sigma_3) = \tfrac{1}{2}[8 - (-4)] = 6 \text{ MPa}$$

which acts along the plane defined by the normal $l = 1/\sqrt{2}$, $m = 0$, $n = 1/\sqrt{2}$, relative to the principal directions 1, 2 and 3 (see Fig. 1.7a).
The normal stress acting on the octahedral plane is found from eq(1.16a)

$$\sigma_o = (\sigma_1 + \sigma_2 + \sigma_3)/3 = (8 + 2 - 4)/3 = 2 \text{ MPa}$$

which acts in the direction of the normal $l = m = n = 1/\sqrt{3}$. The octahedral shear stress is found from eq(1.16b)

$$\begin{aligned} \tau_o &= \tfrac{1}{3}\sqrt{[(\sigma_1 - \sigma_2)^2 + (\sigma_2 - \sigma_3)^2 + (\sigma_1 - \sigma_3)^2]} \\ &= \tfrac{1}{3}\sqrt{[(8 - 2)^2 + (2 + 4)^2 + (8 + 4)^2]} = 4.9 \text{ MPa} \end{aligned}$$

with direction cosines, from eqs(1.17a,b,c)

$$\begin{aligned} l_o &= (\sigma_1 - \sigma_o)/(\sqrt{3}\tau_o) = (8 - 2)/(\sqrt{3} \times 4.9) = 0.707 \quad \text{(i.e. 45° to 1)} \\ m_o &= (\sigma_2 - \sigma_o)/(\sqrt{3}\tau_o) = (2 - 2)/(\sqrt{3} \times 4.9) = 0 \quad \text{(i.e. 90° to 2)} \\ n_o &= (\sigma_3 - \sigma_o)/(\sqrt{3}\tau_o) = (-4 - 2)/(\sqrt{3} \times 4.9) = -0.707 \quad \text{(i.e. 135° to 3)} \end{aligned}$$

The stresses σ_o and τ_o are shown in Fig. 1.7b.

(a)
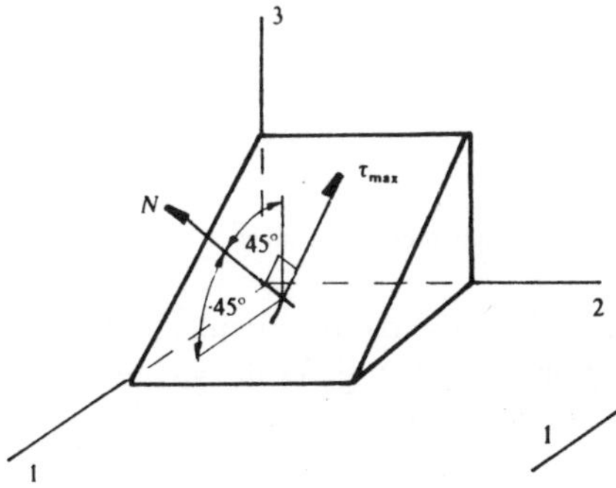

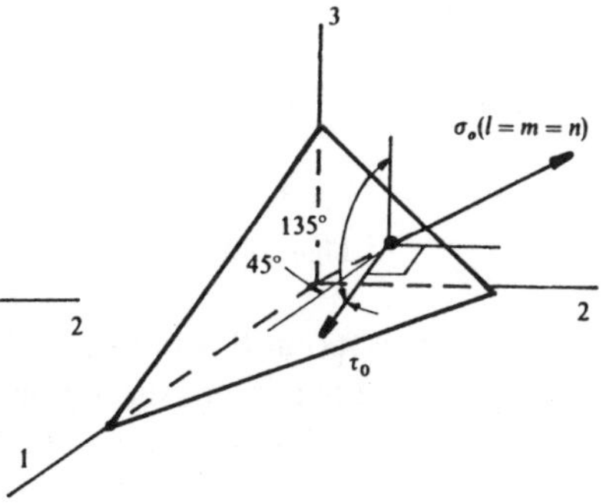

(b)

Figure 1.7 Maximum shear and octahedral planes

1.3.4 Geometric Representation

When the applied stresses are the principal stresses σ_1, σ_2 and σ_3, a construction due to Otto Mohr (1914) enables σ and τ to be found for a plane whose normals are defined by $\alpha = \cos^{-1} l$, $\beta = \cos^{-1} m$ and $\gamma = \cos^{-1} n$ relative to the 1, 2 and 3 directions. Assuming $\sigma_1 > \sigma_2 > \sigma_3$ are all tensile, eqs(1.3 and 1.5a,b) combine graphically, as shown in Fig. 1.8. The Mohr's circle is constructed as follows:

(i) Erect perpendicular σ and τ axes
(ii) Fix points σ_1, σ_2 and σ_3 to scale from the origin along the σ axis
(iii) With centres C_{12}, C_{23} and C_{13} draw circles of radii τ_{12}, τ_{23} and τ_{13} respectively
(iv) Draw the line σ_1AB inclined at α from the vertical through σ_1
(v) Draw lines σ_2D and σ_2C with inclinations of β on each side of the vertical through σ_2
(vi) With centres C_{13} and C_{23} draw the arcs DC and AB
(vii) The intersection point P has co-ordinates σ and τ as shown

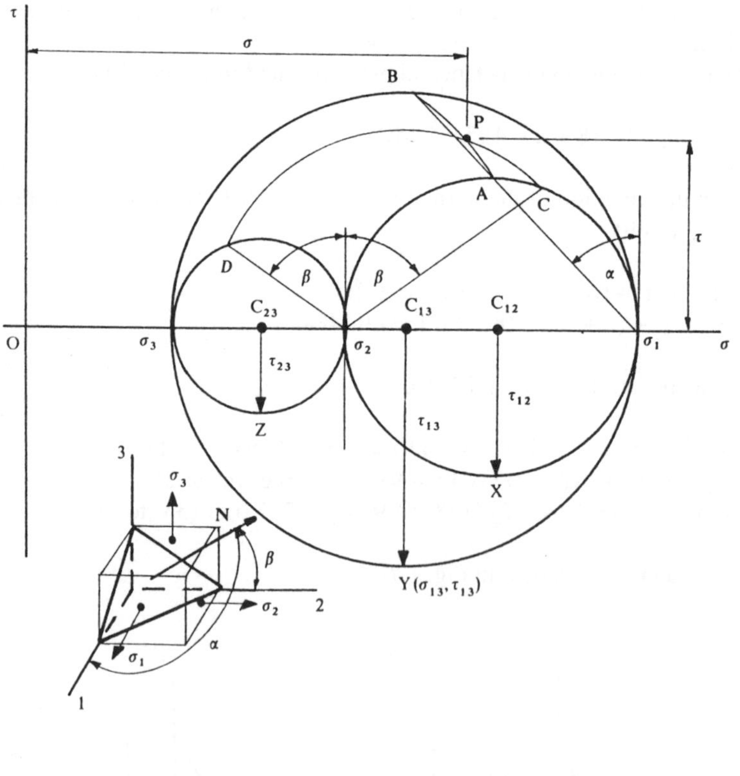

Figure 1.8 Mohr's circle for applied principal stresses

Note that for all planes, point P is mirrored about the σ - axis. Points X, Y and Z represent the state of stress for the maximum shear planes. Point Y gives the greatest shear stress and its associated normal stress as

$$\tau_{13} = ½ (\sigma_1 - \sigma_3) \text{ and } \sigma_{13} = ½ (\sigma_1 + \sigma_3)$$

Example 1.5 Given the principal applied stress system σ_1 = 15.4, σ_2 = 11.3 and σ_3 = 6.8 (MPa), determine graphically the normal and shear stresses for a plane whose direction cosines are l = 0.732 and m = 0.521. Find also the stress state for the octahedral plane. Check the answers numerically.

The Mohr's circle construction Fig. 1.9 gives σ = 12.65, τ = 3.38 MPa. The resultant stress $S = \sqrt{(\sigma^2 + \tau^2)}$ = 13.15 MPa becomes the length of OP_1. The octahedral plane is defined by $\alpha = \beta$ = 54.73° which, when applied to Fig. 1.9, gives another intersection point P_2 with co-ordinates σ_o = 11.17, τ_o = 3.51 MPa. The greatest shear stress τ_{13} = 4.3 MPa is the radius of the largest circle.

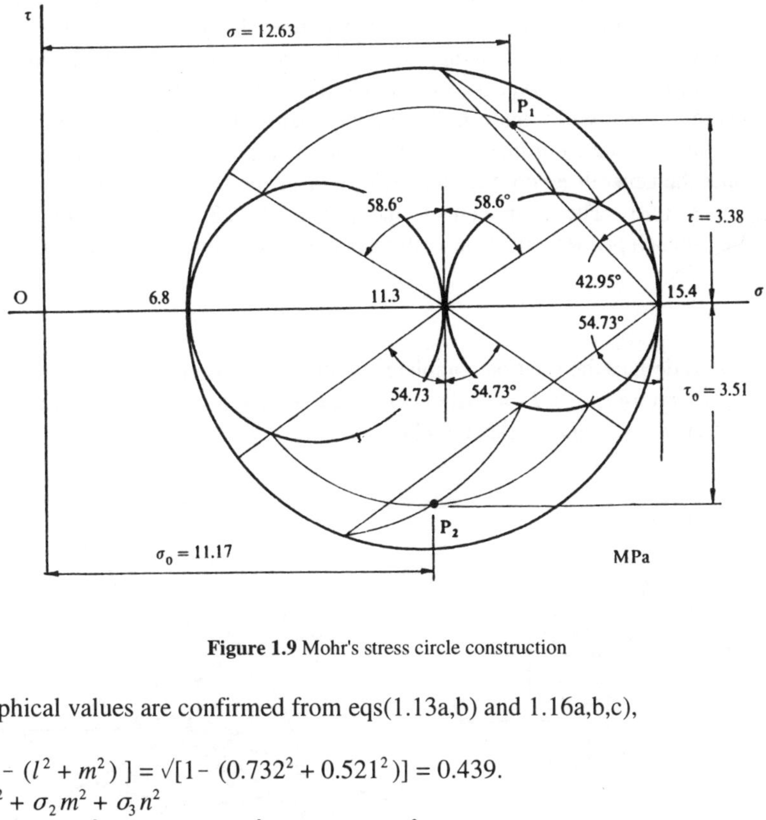

Figure 1.9 Mohr's stress circle construction

The graphical values are confirmed from eqs(1.13a,b) and 1.16a,b,c),

$n = \sqrt{[1 - (l^2 + m^2)]} = \sqrt{[1 - (0.732^2 + 0.521^2)]} = 0.439.$

$\sigma = \sigma_1 l^2 + \sigma_2 m^2 + \sigma_3 n^2$

$= 15.4\,(0.732)^2 + 11.3\,(0.521)^2 + 6.8\,(0.439)^2 = 12.63$ MPa

$\tau^2 = (\sigma_1 l)^2 + (\sigma_2 m)^2 + (\sigma_3 n)^2 - \sigma^2$

$= (15.4 \times 0.732)^2 + (11.3 \times 0.521)^2 + (6.8 \times 0.439)^2 - 12.63^2 \Rightarrow \tau = 3.38$ MPa

$\sigma_o = (\sigma_1 + \sigma_2 + \sigma_3)/\,3$

$= (15.4 + 11.3 + 6.8)/\,3 = 11.17$ MPa

$\tau_o = \tfrac{1}{3}\sqrt{[(\sigma_1 - \sigma_2)^2 + (\sigma_2 - \sigma_3)^2 + (\sigma_1 - \sigma_3)^2]}$

$= \tfrac{1}{3}\sqrt{[(15.4 - 11.3)^2 + (11.3 - 6.8)^2 + (15.4 - 6.8)^2]} = 3.51$ MPa

$\tau_{max} = \tau_{13} = \tfrac{1}{2}(\sigma_1 - \sigma_3) = \tfrac{1}{2}(15.4 - 6.8) = 4.3$ MPa

1.4 Matrix and Tensor Transformations of Stress

1.4.1 Tensor Subscript Notation

Equations (1.5a,b) and (1.13a,b) obey a more general transformation law defining any symmetrical tensor of second rank, of which stress is a member. In double subscript tensor notation, this law appears as

$$\sigma_{i'j'} = l_{ip}\, l_{jq}\, \sigma_{pq} \tag{1.18a}$$

where σ_{pq} are the six independent components of stress lying in co-ordinate axes x_1, x_2 and x_3 (see Fig. 1.10a). We now abandon x, y and z in favour of these equivalent mathematical co-ordinates. The shear and normal stress components are each identified with two subscripts. The first subscript denotes the normal direction and the second the stress direction. Hence the normal stress components are σ_{11}, σ_{22} and σ_{33} and the shear stress components are σ_{12}, σ_{13}, σ_{23} etc. Equation(1.18a) will transform these components of stress to those lying in axes $x_{1'}$, $x_{2'}$ and $x_{3'}$ following the co-ordinate rotation shown in Fig. 1.10b. Note that here x_1, x_2 and x_3 can be either the generalised co-ordinates x, y and z or the principal stress co-ordinates 1, 2 and 3 used previously. The components l_{ip} and l_{jq} ($i, p = 1, 2, 3$) in eq(1.8a) are the direction cosines defining each primed direction with respect to each unprimed direction. That is

$$l_{ip} = \cos\,(x_{i'}, x_p) \tag{1.18b}$$

Equation (1.18a) defines the rotation completely within 9 direction cosines. For example, $l_{11} = \cos\,(x_{1'}, x_1)$, $l_{12} = \cos(x_{1'}, x_2)$, $l_{13} = \cos(x_{1'}, x_3)$ are the direction cosines of the rotated axis 1' relative to directions 1, 2 and 3 respectively.

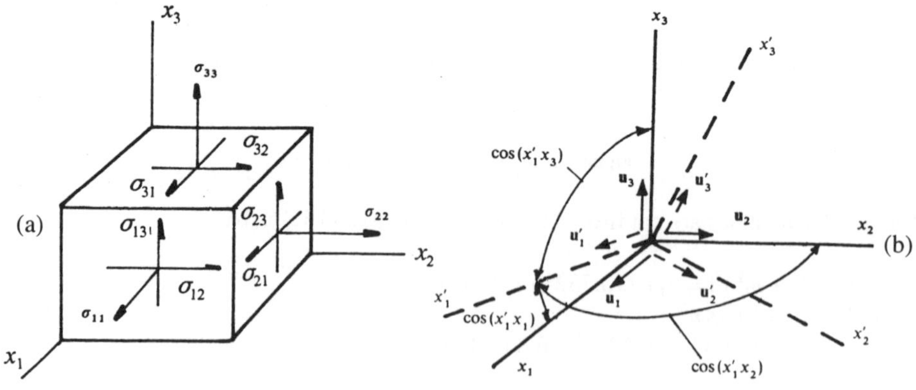

Figure 1.10 Generalised stress components with rotation in orthogonal axes

Equations (1.2b,c) show how the nine direction cosines appear within the unit vectors' expressions for the 1', 2' and 3' directions. They are

$$\mathbf{u}_{1'} = l_{11}\,\mathbf{u}_1 + l_{12}\,\mathbf{u}_2 + l_{13}\,\mathbf{u}_3 \tag{1.19a}$$
$$\mathbf{u}_{2'} = l_{21}\,\mathbf{u}_1 + l_{22}\,\mathbf{u}_2 + l_{23}\,\mathbf{u}_3 \tag{1.19b}$$
$$\mathbf{u}_{3'} = l_{31}\,\mathbf{u}_1 + l_{32}\,\mathbf{u}_2 + l_{33}\,\mathbf{u}_3 \tag{1.19c}$$

1.4.2 Matrix Notation

Equation (1.18a) may be converted to an equivalent matrix equation. Let the 3×3 matrices **S**, **S** ' and **L** represent the nine components σ_{pq}, $\sigma_{i'j'}$ and l_{ip} (and l_{jq}) respectively. Before the conversion of eq(1.18a) to matrix form can be made, similar subscripts must appear adjacent within each term. This leads to the correct order of matrix multiplication as follows:

$$\sigma_{i'j'} = l_{ip} l_{jq} \sigma_{pq} = l_{ip} \sigma_{pq} l_{jq} = l_{ip} \sigma_{pq} l_{qj}^{\mathrm{T}}$$

$$\therefore \mathbf{S}' = \mathbf{L}\mathbf{S}\mathbf{L}^{\mathrm{T}} \tag{1.20a}$$

in which $\mathbf{L}^{\mathrm{T}}$ denotes the transpose of **L**. In full, eq(1.20a) becomes

$$\begin{bmatrix} \sigma_{1'1'} & \sigma_{1'2'} & \sigma_{1'3'} \\ \sigma_{2'1'} & \sigma_{2'2'} & \sigma_{2'3'} \\ \sigma_{3'1'} & \sigma_{3'2'} & \sigma_{3'3'} \end{bmatrix} = \begin{bmatrix} l_{11} & l_{12} & l_{13} \\ l_{21} & l_{22} & l_{23} \\ l_{31} & l_{32} & l_{33} \end{bmatrix} \begin{bmatrix} \sigma_{11} & \sigma_{12} & \sigma_{13} \\ \sigma_{21} & \sigma_{22} & \sigma_{23} \\ \sigma_{31} & \sigma_{32} & \sigma_{33} \end{bmatrix} \begin{bmatrix} l_{11} & l_{21} & l_{31} \\ l_{12} & l_{22} & l_{32} \\ l_{13} & l_{23} & l_{33} \end{bmatrix} \tag{1.20b}$$

Equations (1.20a,b) provide the normal and shear stresses for three orthogonal planes within the $x_{i'}$ $(i = 1, 2, 3)$ frame. In the analytical method the stress state for a single oblique plane ABC (Fig. 1.2a) was found. We can identify ABC with a plane lying normal to $x_{1'}$ (say) with directions l_{11}, l_{12} and l_{13} (previously l, m and n in the engineering notation). The stress components associated with this plane are $\sigma_{1'1'}$, $\sigma_{1'2'}$ and $\sigma_{1'3'}$. In the analytical solution, the normal and shear stress referred to in eq(1.5a,b) becomes $\sigma = \sigma_{1'1'}$ and $\tau = \sqrt{[(\sigma_{1'2'})^2 + (\sigma_{1'3'})^2]}$. Clearly τ is the resultant shear stress acting on plane ABC and $\sigma_{1'2'}$, $\sigma_{1'3'}$ are its components in the 2' and 3' directions. It follows from eqs(1.19a-c) and (1.20b) that the individual components are given as the dot products

$$\sigma_{1'1'} = \mathbf{u}_1 \bullet \mathbf{S} \bullet \mathbf{u}_1$$
$$\sigma_{1'2'} = \mathbf{u}_1 \bullet \mathbf{S} \bullet \mathbf{u}_2$$
$$\sigma_{1'3'} = \mathbf{u}_1 \bullet \mathbf{S} \bullet \mathbf{u}_3$$

When the co-ordinates x_i become aligned with the principal directions, eq(1.20b) reduces to

$$\begin{bmatrix} \sigma_{1'1'} & \sigma_{1'2'} & \sigma_{1'3'} \\ \sigma_{2'1'} & \sigma_{2'2'} & \sigma_{2'3'} \\ \sigma_{3'1'} & \sigma_{3'2'} & \sigma_{3'3'} \end{bmatrix} = \begin{bmatrix} l_{11} & l_{12} & l_{13} \\ l_{21} & l_{22} & l_{23} \\ l_{31} & l_{32} & l_{33} \end{bmatrix} \begin{bmatrix} \sigma_1 & 0 & 0 \\ 0 & \sigma_2 & 0 \\ 0 & 0 & \sigma_3 \end{bmatrix} \begin{bmatrix} l_{11} & l_{21} & l_{31} \\ l_{12} & l_{22} & l_{32} \\ l_{13} & l_{23} & l_{33} \end{bmatrix} \tag{1.20c}$$

The elements of eq(1.20c) will contain eqs(1.13a,b). For example, to confirm eq(1.13a) we multiply rows into columns to yield $\sigma_{1'1'} = \sigma_1 l_{11}^2 + \sigma_2 l_{12}^2 + \sigma_3 l_{13}^2$, which is equivalent to eq(1.13a). Similarly, the shear stress expression(1.13b) may be identified with the result of matrix multiplication for the $\sigma_{1'2'}$ component.

Example 1.6 Given the following matrix **S** of stress components, determine **S**' when the coordinates $x_{1'}$ and $x_{2'}$ are aligned with the vectors $\mathbf{A} = \mathbf{u}_1 + 2\mathbf{u}_2 + 3\mathbf{u}_3$ and $\mathbf{B} = \mathbf{u}_1 + \mathbf{u}_2 - \mathbf{u}_3$.

$$\mathbf{S} = \begin{bmatrix} 1 & 5 & -5 \\ 5 & 0 & 0 \\ -5 & 0 & -1 \end{bmatrix}$$

Firstly, divide the vector equations by their respective magnitudes $|\mathbf{A}| = \sqrt{14}$, $|\mathbf{B}| = \sqrt{3}$, to give the unit vectors:

$$\mathbf{u}_{1'} = (1/\sqrt{14})\mathbf{u}_1 + (2/\sqrt{14})\mathbf{u}_2 + (3/\sqrt{14})\mathbf{u}_3$$
$$\mathbf{u}_{2'} = (1/\sqrt{3})\mathbf{u}_1 + (1/\sqrt{3})\mathbf{u}_2 - (1/\sqrt{3})\mathbf{u}_3$$

The coefficients are the direction cosines in eqs(1.19a,b),

$$l_{11} = 1/\sqrt{14},\ l_{12} = 2/\sqrt{14},\ l_{13} = 3/\sqrt{14}$$
$$l_{21} = 1/\sqrt{3},\ l_{22} = 1/\sqrt{3},\ l_{23} = -1/\sqrt{3}$$

The cosines for the third orthogonal direction $x_{3'}$ are found from the cross product. Thus, if a vector **C** lies in $x_3{}'$ then, by definition

$$\mathbf{C} = \mathbf{A} \times \mathbf{B} = \begin{vmatrix} \mathbf{u}_1 & \mathbf{u}_2 & \mathbf{u}_3 \\ A_1 & A_2 & A_3 \\ B_1 & B_2 & B_3 \end{vmatrix} = \begin{vmatrix} \mathbf{u}_1 & \mathbf{u}_2 & \mathbf{u}_3 \\ 1 & 2 & 3 \\ 1 & 1 & -1 \end{vmatrix}$$

$$\mathbf{C} = (-2 - 3)\mathbf{u}_1 - (-1 - 3)\mathbf{u}_2 + (1 - 2)\mathbf{u}_3 = -5\mathbf{u}_1 + 4\mathbf{u}_2 - \mathbf{u}_3$$

$$\mathbf{u}_{3'} = \mathbf{C}/|\mathbf{C}| = (-5/\sqrt{42})\mathbf{u}_1 + (4/\sqrt{42})\mathbf{u}_2 - (1/\sqrt{42})\mathbf{u}_3$$

and from eq(1.19c): $l_{31} = -5/\sqrt{42}$, $l_{32} = 4/\sqrt{42}$, $l_{33} = -1/\sqrt{42}$. Substituting into eq(1.20b),

$$\mathbf{S}' = \begin{bmatrix} 1/\sqrt{14} & 2/\sqrt{14} & 3/\sqrt{14} \\ 1/\sqrt{3} & 1/\sqrt{3} & -1/\sqrt{3} \\ -5/\sqrt{42} & 4/\sqrt{42} & -1/\sqrt{42} \end{bmatrix} \begin{bmatrix} 1 & 5 & -5 \\ 5 & 0 & 0 \\ -5 & 0 & -1 \end{bmatrix} \begin{bmatrix} 1/\sqrt{14} & 1/\sqrt{3} & -5/\sqrt{42} \\ 2/\sqrt{14} & 1/\sqrt{3} & 4/\sqrt{42} \\ 3/\sqrt{14} & -1/\sqrt{3} & -1/\sqrt{42} \end{bmatrix}$$

$$= \begin{bmatrix} 1/\sqrt{14} & 2/\sqrt{14} & 3/\sqrt{14} \\ 1/\sqrt{3} & 1/\sqrt{3} & -1/\sqrt{3} \\ -5/\sqrt{42} & 4/\sqrt{42} & -1/\sqrt{42} \end{bmatrix} \begin{bmatrix} -4/\sqrt{14} & 11/\sqrt{3} & 20/\sqrt{42} \\ 5/\sqrt{14} & 5/\sqrt{3} & -25/\sqrt{42} \\ -8/\sqrt{14} & -4/\sqrt{3} & 26/\sqrt{42} \end{bmatrix} = \begin{bmatrix} -1.286 & 1.389 & 1.980 \\ 1.389 & 6.667 & -2.762 \\ 1.980 & -2.762 & -5.381 \end{bmatrix}$$

1.4.3 Deviatoric Stress Tensor

A deviatoric stress tensor σ_{ij}' is what remains of the absolute stress tensor σ_{ij} after the mean (hydrostatic) stress σ_m has been subtracted. Figures 1.11a-c show how the principal triaxial stress system in (a) is decomposed into (b) its mean and (c) its deviatoric stress components. The mean stress (Fig. 1.11b) is $\sigma_m = (\sigma_1 + \sigma_2 + \sigma_3)/3$ and the deviatoric stresses (Fig. 1.11c) are $\sigma_1' = \sigma_1 - \sigma_m$, $\sigma_2' = \sigma_2 - \sigma_m$ and $\sigma_3' = \sigma_3 - \sigma_m$. The subtraction of σ_m will apply only to the normal stress components σ_{11}, σ_{22} and σ_{33} and does not alter the shear stresses when they are present. To ensure this, the deviatoric stress tensor is written as

$$\sigma_{ij}' = \sigma_{ij} - \delta_{ij}\,\sigma_m = \sigma_{ij} - \tfrac{1}{3}\,\delta_{ij}\,\sigma_{kk} \qquad (1.21a)$$

Similar subscripts together (kk) denote summation over 1, 2 and 3, i.e. $\sigma_{kk} = (\sigma_{11} + \sigma_{22} + \sigma_{33})$.

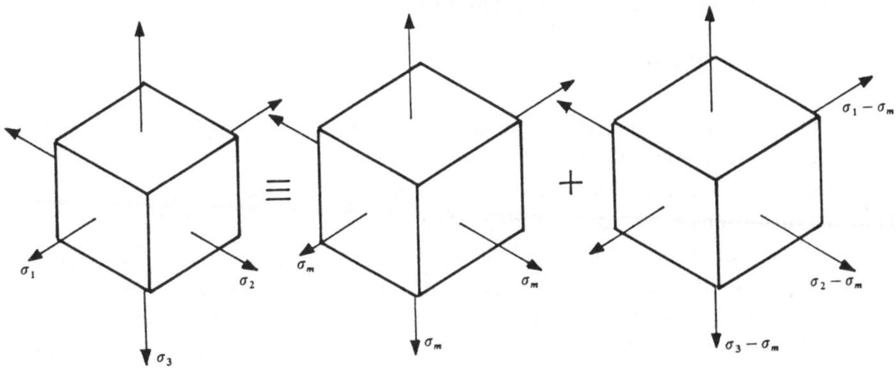

(a) (b) (c)

Figure 1.11 Hydrostatic and deviatoric components of principal stress

Thus, in eq(1.21a) the shear stress components remain unaffected since the Kronecker delta δ_{ij} is unity for $i = j$ and zero for $i \neq j$. For example, eq(1.21a) gives, with $i = 1$ and $j = 1, 2$ and 3, a normal stress deviator

$$\sigma_{11}' = \sigma_{11} - (1)\,\sigma_m = \sigma_{11} - (\sigma_{11} + \sigma_{22} + \sigma_{33})/3$$

but the shear stresses from eq(1.21a) are unaltered

$$\sigma_{12}' = \sigma_{12} - (0)\,\sigma_m = \sigma_{12}$$
$$\sigma_{13}' = \sigma_{13} - (0)\,\sigma_m = \sigma_{13}$$

It follows that the deviatoric stress tensor σ_{ij}' in eq(1.21a) is composed of σ_{11}', σ_{22}' and σ_{33}' and the original shear stresses σ_{12}, σ_{13} and σ_{23}. When the invariants of σ_{ij}' are expressed in a scalar functional form $f(\sigma_{ij}')$ = constant, this is known as a yield criterion. We shall show in Chapter 11 how the yield criterion governs the inception of plasticity under any multiaxial stress state. This reveals that the one important deviatoric invariant, identified with the von Mises yield criterion, is proportional to the expression (1.16b or c) for τ_o. The matrix equivalent of eq(1.21a) is written as

$$\mathbf{S}^D = \mathbf{S} - \tfrac{1}{3}\,(\mathrm{tr}\,\mathbf{S})\,\mathbf{I} \tag{1.21b}$$

where $\mathbf{S}^D$ is a 3 × 3 matrix of the deviatoric stress components, **I** is the unit matrix and tr **S** is the trace of **S** (the sum of its diagonal components). In full, eq(1.21b) appears as

$$\begin{bmatrix} \sigma_{11}' & \sigma_{12}' & \sigma_{13}' \\ \sigma_{21}' & \sigma_{22}' & \sigma_{23}' \\ \sigma_{31}' & \sigma_{32}' & \sigma_{33}' \end{bmatrix} = \begin{bmatrix} \sigma_{11} & \sigma_{12} & \sigma_{13} \\ \sigma_{21} & \sigma_{22} & \sigma_{23} \\ \sigma_{31} & \sigma_{32} & \sigma_{33} \end{bmatrix} - \frac{(\sigma_{11} + \sigma_{22} + \sigma_{33})}{3} \begin{bmatrix} 1 & 0 & 0 \\ 0 & 1 & 0 \\ 0 & 0 & 1 \end{bmatrix} \tag{1.21c}$$

The components of deviatoric stress agree with those found from eq(1.21a). A summary of the important equations used for transforming stress and strain is given in Table 1.1.

1.5 Three-Dimensional Strain Transformation

1.5.1 The Strain and Rotation Matrices

The analysis of the distortion produced by the stress components σ_x, σ_y, σ_z, τ_{xy}, τ_{xz} and τ_{yz} in Fig. 1.1a reveals two types of strain:
(i) direct strain ε_x, ε_y and ε_z arising from direct stress σ_x, σ_y and σ_z
(ii) angular distortions e_{xy}, e_{xz} and e_{yz} arising from shear stress τ_{xy}, τ_{xz} and τ_{yz}
The distortions in (ii) are composed of shear strain due to τ_{xy}, τ_{xz} and τ_{yz} and rigid body rotations due to the differences in direct strains. It is necessary to subtract rotations from the angular distortions to establish a shear strain component responsible for shape change, i.e. the angular change in the right angle. In Fig. 1.12a, for example, the angular distortions e_{xy} and e_{yx} for one corner of an element $\delta x \times \delta y \times \delta z$ in the x, y plane are shown.

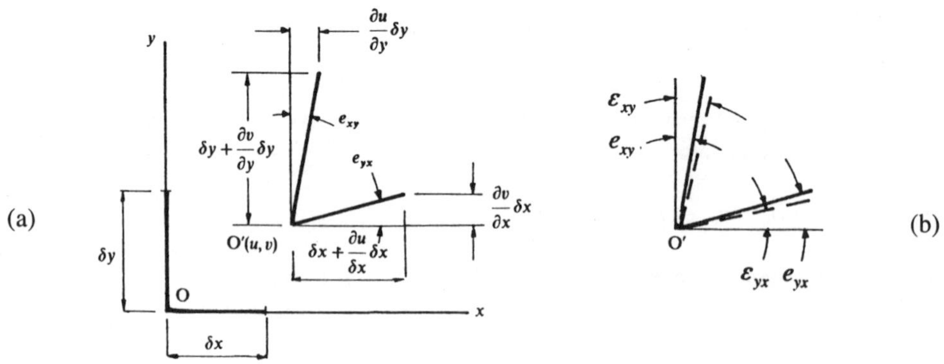

Figure 1.12 Distortion of one corner in the x - y plane

These are given as

$$e_{xy} = \partial u/\partial y \text{ and } e_{yx} = \partial v/\partial x \qquad (1.22a,b)$$

Because $\delta x \neq \delta y$ and $\varepsilon_x \neq \varepsilon_y$ then $e_{xy} \neq e_{yx}$. The direct strains ε_x and ε_y are associated with the displacement of the corner point from O to O'. Let the respective components of this displacement in the x and y directions be u and v, which are each functions of the co-ordinates $u = u\ (x, y, z)$ and $v = v\ (x, y, z)$. Now δx and δy change their lengths in proportion to the displacement gradients, i.e. the direct strains

$$\varepsilon_x = \partial u/\partial x \text{ and } \varepsilon_y = \partial v/\partial y \qquad (1.23a,b)$$

The engineering shear strain is defined as the total change in the right angle

$$\gamma_{xy} = \partial u/\partial y + \partial v/\partial x \qquad (1.24a)$$

Rotating the distorted corner so that it becomes equally inclined to the x and y directions (see Fig. 1.12b) reveals two, complementary tensor shear strains:

$$\varepsilon_{xy} = \varepsilon_{yx} = \tfrac{1}{2}\gamma_{xy} = \tfrac{1}{2}(\partial u/\partial y + \partial v/\partial x) \qquad (1.24b)$$

The corresponding rotations are

$$\omega_{xy} = e_{xy} - \varepsilon_{xy} = \partial u/\partial y - ½(\partial u/\partial y + \partial v/\partial x) = ½(\partial u/\partial y - \partial v/\partial x)$$
$$\omega_{yx} = e_{yx} - \varepsilon_{yx} = \partial v/\partial x - ½(\partial u/\partial y + \partial v/\partial x) = -½(\partial u/\partial y - \partial v/\partial x)$$

With an additional displacement function $w = w(x, y, z)$, for the z - direction and further displacement - gradient relationships, similar decompositions apply to normal and shear distortion in the x, z and y, z planes. Consequently, the complete distortion of an element $\delta x \times \delta y \times \delta z$ may be expressed as the sum of the corresponding strain and rotation matrices

$$e_{ij} = \varepsilon_{ij} + \omega_{ij} \tag{1.25}$$

Equations (1.21)-(1.23) then appear as particular components within each expanded matrix in eq(1.25) as follows

$$\begin{bmatrix} e_x & e_{xy} & e_{xz} \\ e_{yx} & e_y & e_{yz} \\ e_{zx} & e_{zy} & e_z \end{bmatrix} = \begin{bmatrix} \frac{\partial u}{\partial x} & \frac{\partial u}{\partial y} & \frac{\partial u}{\partial z} \\ \frac{\partial v}{\partial x} & \frac{\partial v}{\partial y} & \frac{\partial v}{\partial z} \\ \frac{\partial w}{\partial x} & \frac{\partial w}{\partial y} & \frac{\partial w}{\partial z} \end{bmatrix}$$

$$\begin{bmatrix} \varepsilon_x & \varepsilon_{xy} & \varepsilon_{xz} \\ \varepsilon_{yx} & \varepsilon_y & \varepsilon_{yz} \\ \varepsilon_{zx} & \varepsilon_{zy} & \varepsilon_z \end{bmatrix} = \begin{bmatrix} \frac{\partial u}{\partial x} & \frac{1}{2}\left(\frac{\partial u}{\partial y} + \frac{\partial v}{\partial x}\right) & \frac{1}{2}\left(\frac{\partial u}{\partial z} + \frac{\partial w}{\partial x}\right) \\ \frac{1}{2}\left(\frac{\partial u}{\partial y} + \frac{\partial v}{\partial x}\right) & \frac{\partial v}{\partial y} & \frac{1}{2}\left(\frac{\partial v}{\partial z} + \frac{\partial w}{\partial y}\right) \\ \frac{1}{2}\left(\frac{\partial u}{\partial z} + \frac{\partial w}{\partial x}\right) & \frac{1}{2}\left(\frac{\partial v}{\partial z} + \frac{\partial w}{\partial y}\right) & \frac{\partial w}{\partial z} \end{bmatrix}$$

$$\begin{bmatrix} \omega_x & \omega_{xy} & \omega_{xz} \\ \omega_{yx} & \omega_y & \omega_{yz} \\ \omega_{zx} & \omega_{zy} & \omega_z \end{bmatrix} = \begin{bmatrix} 0 & \frac{1}{2}\left(\frac{\partial u}{\partial y} - \frac{\partial v}{\partial x}\right) & \frac{1}{2}\left(\frac{\partial u}{\partial z} - \frac{\partial w}{\partial x}\right) \\ -\frac{1}{2}\left(\frac{\partial u}{\partial y} - \frac{\partial v}{\partial x}\right) & 0 & \frac{1}{2}\left(\frac{\partial v}{\partial z} - \frac{\partial w}{\partial y}\right) \\ -\frac{1}{2}\left(\frac{\partial u}{\partial z} - \frac{\partial w}{\partial x}\right) & -\frac{1}{2}\left(\frac{\partial v}{\partial z} - \frac{\partial w}{\partial y}\right) & 0 \end{bmatrix}$$

This shows that the strain matrix is symmetric $\varepsilon_{ij} = \varepsilon_{ji}$ and that the rotation matrix is skew symmetric $\omega_{ij} = -\omega_{ji}$. Recall that the 3 × 3 stress matrix is also symmetric ($\sigma_{ij} = \sigma_{ji}$) and it may therefore be deduced that the transformation properties of strain ε_{ij} will be identical to those of stress. Referring again to Fig. 1.1b, let us assume that the strain states for the co-ordinate planes OAC, OAB and OBC are known and that we wish to find the normal strain for the oblique plane ABC. Given the direction cosines for plane ABC, the expression for the normal strain is identical in form to the corresponding stress expression (1.5a). However, the conversion requires τ to be associated with the tensor component of shear strain, which is defined as $½\gamma$. The normal strain on plane ABC will then appear in terms of co-ordinate strains ε_{ij} as

$$\varepsilon = l^2\varepsilon_x + m^2\varepsilon_y + n^2\varepsilon_z + 2(lm\varepsilon_{xy} + mn\varepsilon_{yz} + ln\varepsilon_{xz}) \tag{1.26a}$$
$$= l^2\varepsilon_x + m^2\varepsilon_y + n^2\varepsilon_z + lm\gamma_{xy} + mn\gamma_{yz} + ln\gamma_{zx} \tag{1.26b}$$

Similarly, the principal strain cubic may be deduced from eq(1.8b)

$$\varepsilon^3 - I_1\varepsilon^2 + I_2\varepsilon - I_3 = 0 \tag{1.27}$$

The strain invariants in eq(1.27) follow from the conversion of eqs(1.9a,b,c)

$$I_1 = \varepsilon_x + \varepsilon_y + \varepsilon_z = \varepsilon_{ii} \tag{1.28a}$$
$$I_2 = \varepsilon_x\varepsilon_y + \varepsilon_y\varepsilon_z + \varepsilon_z\varepsilon_x - \varepsilon_{xy}^{\;2} - \varepsilon_{yz}^{\;2} - \varepsilon_{zx}^{\;2} = ½(\varepsilon_{ii}\varepsilon_{jj} - \varepsilon_{ij}\varepsilon_{ji}) \tag{1.28b}$$
$$I_3 = \varepsilon_x\varepsilon_y\varepsilon_z + 2\varepsilon_{xy}\varepsilon_{yz}\varepsilon_{zx} - \varepsilon_x\varepsilon_{yz}^{\;2} - \varepsilon_y\varepsilon_{zx}^{\;2} - \varepsilon_z\varepsilon_{xy}^{\;2}$$
$$= \varepsilon_x\varepsilon_y\varepsilon_z + ¼\gamma_{xy}\gamma_{yz}\gamma_{zx} - ¼\varepsilon_x\gamma_{yz}^{\;2} - ¼\varepsilon_y\gamma_{zx}^{\;2} - ¼\varepsilon_z\gamma_{xy}^{\;2} = \det(\varepsilon_{ij}) \tag{1.28c}$$

Example 1.7 Find the principal strains and their directions for the following plane strain state. Show that the principal strain directions are orthogonal.

$$\varepsilon_{ij} = \begin{bmatrix} 65 & 33 & 0 \\ 33 & -73 & 0 \\ 0 & 0 & 4 \end{bmatrix} \times 10^{-4}$$

Substituting the strains $\varepsilon_x = 65 \times 10^{-4}$, $\varepsilon_y = -73 \times 10^{-4}$, $\varepsilon_z = 4 \times 10^{-4}$, $\varepsilon_{xy} = \varepsilon_{yx} = 33 \times 10^{-4}$ and $\varepsilon_{xz} = \varepsilon_{yz} = 0$ into eq(1.27) leads to the principal strain cubic

$$\varepsilon^3 + (4 \times 10^{-4})\varepsilon^2 - (5049.3 \times 10^{-8})\varepsilon + (20069.2 \times 10^{-12}) = 0 \tag{i}$$

For the plane strain condition, the absence of shear strains ε_{xz} and ε_{yz} is a consequence of the absence of associated shear stresses τ_{xz} and τ_{yz}. Thus, the normal strain $\varepsilon_z = 4 \times 10^{-4}$ is a principal strain and a root of eq(i). It follows from eq(i) that

$$[\varepsilon - (4 \times 10^{-4})](a\varepsilon^2 + b\varepsilon + c) = \varepsilon^3 + (4 \times 10^{-4})\varepsilon^2 - (5049.3 \times 10^{-8})\varepsilon + (20069.2 \times 10^{-12})$$

Equating coefficients,

$[\varepsilon^3]$: $a = 1$
$[\varepsilon^2]$: $b - (4 \times 10^{-4}) = 4 \times 10^{-4}$, $\Rightarrow b = 8 \times 10^{-4}$
$[\varepsilon]$: $c - (4 \times 10^{-4}\, b) = -5049.3 \times 10^{-8}$, $\Rightarrow c = -5017.3 \times 10^{-8}$.

The remaining two principal strains become the roots to the quadratic:

$$\varepsilon^2 + (8 \times 10^{-4})\varepsilon - (5017.3 \times 10^{-8}) = 0$$

giving $\varepsilon = -75 \times 10^{-4}$ and 67×10^{-4}. The principal strains $\varepsilon_1 > \varepsilon_2 > \varepsilon_3$ are then

$$\varepsilon_1 = 67 \times 10^{-4},\ \varepsilon_2 = 4 \times 10^{-4} \text{ and } \varepsilon_3 = -75 \times 10^{-4}$$

As the direction of ε_2 (= ε_z) is parallel to the z - direction, it follows that the cosines are $l_2 = 0$, $m_2 = 0$ and $n_2 = 1$. The strain equivalent to eq(1.7a), together with eq(1.3), enables a calculation of the cosines for the remaining two directions. That is,

$$2l\,(\varepsilon_x - \varepsilon) + m\varepsilon_{xy} + n\varepsilon_{xz} = 0 \tag{ii}$$
$$l\varepsilon_{xy} + 2m(\varepsilon_y - \varepsilon) + n\varepsilon_{yz} = 0 \tag{iii}$$
$$l\varepsilon_{zx} + m\varepsilon_{zy} + 2n\,(\varepsilon_z - \varepsilon) = 0 \tag{iv}$$

Only two of equations (ii) - (iv) are independent. Substituting $\varepsilon = 67 \times 10^{-4}$ and the given strain components leads to the three simultaneous equations:

$$-4\,l_1 + 33\,m_1 = 0,\quad 33l_1 - 280m_1 = 0 \text{ and } -126\,n_1 = 0$$

Thus $n_1 = 0$ and $l_1 = 8.25m_1$. As $l_1^2 + m_1^2 + n_1^2 = 1$, this gives the direction cosines for the major principal strain as

$$l_1 = 0.993,\ m_1 = 0.120 \text{ and } n_1 = 0 \tag{iii}$$

For $\varepsilon_3 = -75 \times 10^{-4}$, eqs(ii) - (iv) give

$$280\,l_3 + 33\,m_3 = 0,\quad 33\,l_3 + 4\,m_3 = 0 \text{ and } 158\,n_3 = 0$$

Thus $n_3 = 0$ and $m_3 = -(8.2513)\,l_3$. Substituting into $l_3^2 + m_3^2 + n_3^2 = 1$ gives cosines for the minor principal strain as

$$l_3 = 0.120,\ m_3 = -0.993 \text{ and } n_3 = 0 \tag{iv}$$

Substituting eqs(iii) and (iv) into eqs(1.10a,b,c) provides the unit vectors aligned with the principal directions:

$$\mathbf{u}_1 = 0.993\mathbf{u}_x + 0.120\mathbf{u}_y,\ \mathbf{u}_2 = \mathbf{u}_z \text{ and } \mathbf{u}_3 = 0.120\mathbf{u}_x - 0.993\mathbf{u}_y$$

These are orthogonal when the dot product of any two unit vectors in the 1, 2 and 3 directions are zero. Clearly $\mathbf{u}_1 \bullet \mathbf{u}_2 = \mathbf{u}_1 \bullet \mathbf{u}_3 = \mathbf{u}_2 \bullet \mathbf{u}_3 = 0$

1.5.2 Strain Tensor Transformation

The strain tensor transformations, equivalent to eqs(1.20a) are

$$\varepsilon_{i'j'} = l_{ip}\,l_{jq}\,\varepsilon_{pq} \;\Rightarrow\; \mathbf{E}' = \mathbf{L}\mathbf{E}\mathbf{L}^{\mathrm{T}} \tag{1.29a,b}$$

Equations (1.29a,b) provide the nine components of strain following a rotation in the co-ordinate axes (see Fig. 1.10b). Upon the plane lying perpendicular to a rotated axis there is one normal strain and two shear strains. Equation (1.29a) provides the strain state for this plane and two further orthogonal planes. Of the three such transformed normal strain components, one is the normal strain given in eq(1.26b). To show this we must refer eq(1.29a) to the mathematical notation and set $i = j = 1$. This gives

$$\varepsilon_{1'1'} = l_{1p}\,l_{1q}\,\varepsilon_{pq} = l_{ip}\,\varepsilon_{pq}\,(l_{1q})^{\mathrm{T}} \tag{1.30a}$$

The second expression in eq(1.30a) is equivalent to a shortened matrix multiplication

$$\varepsilon_{1'1'} = \begin{bmatrix} l_{11} & l_{12} & l_{13} \end{bmatrix} \begin{bmatrix} \varepsilon_{11} & \varepsilon_{12} & \varepsilon_{13} \\ \varepsilon_{21} & \varepsilon_{22} & \varepsilon_{23} \\ \varepsilon_{31} & \varepsilon_{32} & \varepsilon_{33} \end{bmatrix} \begin{bmatrix} l_{11} \\ l_{12} \\ l_{13} \end{bmatrix}$$

$$\varepsilon_{1'1'} = l_{11}^{\,2}\varepsilon_{11} + l_{12}^{\,2}\varepsilon_{22} + l_{13}^{\,2}\varepsilon_{33} + 2(l_{11}\,l_{12}\,\varepsilon_{12} + l_{12}\,l_{13}\,\varepsilon_{23} + l_{11}\,l_{13}\,\varepsilon_{13}) \tag{1.30b}$$

Clearly, eqs(1.26b) and (1.30b) are equivalent between the two notations. The six components of tensor shear strain appear in eq(1.29a) with different valued subscripts i and j. For example, setting $i = 1$ and $j = 2$ in eq(1.29a) will yield the expression for a transformed shear strain component

$$\varepsilon_{1'2'} = l_{1p}\,l_{2q}\,\varepsilon_{pq} = l_{1p}\,\varepsilon_{pq}\,(l_{2q})^{\mathrm{T}} \tag{1.31a}$$

Equation (1.31a) is equivalent to the matrix multiplication

$$\varepsilon_{1'2'} = \begin{bmatrix} l_{11} & l_{12} & l_{13} \end{bmatrix} \begin{bmatrix} \varepsilon_{11} & \varepsilon_{12} & \varepsilon_{13} \\ \varepsilon_{21} & \varepsilon_{22} & \varepsilon_{23} \\ \varepsilon_{31} & \varepsilon_{32} & \varepsilon_{33} \end{bmatrix} \begin{bmatrix} l_{21} \\ l_{22} \\ l_{23} \end{bmatrix}$$

$$\begin{aligned} \varepsilon_{1'2'} &= l_{11}\,l_{21}\,\varepsilon_{11} + l_{12}\,l_{22}\,\varepsilon_{22} + l_{13}\,l_{23}\,\varepsilon_{33} + (l_{11}\,l_{22} + l_{21}\,l_{12})\varepsilon_{12} \\ &\quad + (l_{12}\,l_{23} + l_{22}\,l_{13})\varepsilon_{23} + (l_{11}\,l_{23} + l_{21}\,l_{13})\varepsilon_{13} \end{aligned} \tag{1.31b}$$

In the engineering notation eq(1.31b) appears as

$$\begin{aligned} \tfrac{1}{2}\gamma &= l_1\,l_2\,\varepsilon_x + m_1\,m_2\,\varepsilon_y + n_1\,n_2\,\varepsilon_z + \tfrac{1}{2}(l_1\,m_2 + l_2\,m_1)\,\gamma_{xy} \\ &\quad + \tfrac{1}{2}(m_1\,n_2 + m_2\,n_1)\,\gamma_{yz} + \tfrac{1}{2}(l_1\,n_2 + l_2\,n_1)\,\gamma_{xz} \end{aligned} \tag{1.31c}$$

where (l_1, m_1, n_1) and (l_2, m_2, n_2) define the respective rotations of the axes 1 and 2. Note that eqs(1.30a) and (1.31a) may also be written as dot products

$$\varepsilon_{1'1'} = \mathbf{u}_{1'} \bullet \mathbf{E} \bullet \mathbf{u}_{1'} \tag{1.32a}$$
$$\varepsilon_{1'2'} = \mathbf{u}_{1'} \bullet \mathbf{E} \bullet \mathbf{u}_{2'} \tag{1.32b}$$

where **E** is the strain matrix in eq(1.30b) and $\mathbf{u}_{1'}$ and $\mathbf{u}_{2'}$ are the unit vectors for the pair of perpendicular directions $x_{1'}$ and $x_{2'}$. These vectors are expressed from eqs(1.19a,b). We see that the direction cosines of $x_{1'}$ and $x_{2'}$ relative to x_1, x_2 and x_3 appear as the coefficients (scalar intercepts) for each unit vector. Equations (1.32a,b) also provide a generalised shear strain expression for when the angular change in a right angle, whose initial perpendicular directions are specified, is required. This shear strain is associated with shear stress τ in eq(1.5b). We use eq(1.32b) where $\mathbf{u}_{1'}$ is the unit vector normal to plane ABC and $\mathbf{u}_{2'}$ is the unit vector aligned with τ. The vector equation $\mathbf{u}_{1'}$ employs the direction cosines for ABC and $\mathbf{u}_{2'}$ employs the direction cosines given in eqs(1.6a,b,c).

The reader will now recognise the similarities in stress and strain transformation equations. These are summarised in Table 1.1). To complete this comparison, use is made of the octahedral and deviatoric strains in Example 1.8.

Table 1.1 Stress and strain transformation equations

SYSTEM	STRESS	STRAIN
General	$\mathbf{S}' = \mathbf{LSL}^{\mathrm{T}}$ $\sigma_{i'j'} = l_{ip}\, l_{jq}\, \sigma_{pq}$	$\mathbf{E}' = \mathbf{LEL}^{\mathrm{T}}$ $\varepsilon_{i'j'} = l_{ip}\, l_{jq}\, \varepsilon_{pq}$
Oblique Plane	$\sigma_{1'1'} = l_{1p}\, l_{iq}\, \sigma_{pq} = \mathbf{u}_{1'} \bullet \mathbf{S} \bullet \mathbf{u}_{1'}$ $\sigma_{1'2'} = l_{1p}\, l_{2q}\, \sigma_{pq} = \mathbf{u}_{1'} \bullet \mathbf{S} \bullet \mathbf{u}_{2'}$	$\varepsilon_{1'1'} = l_{1p}\, l_{1q} \varepsilon_{pq} = \mathbf{u}_{1'} \bullet \mathbf{E} \bullet \mathbf{u}_{1'}$ $\varepsilon_{1'2'} = l_{1p}\, l_{2q}\, \varepsilon_{pq} = \mathbf{u}_{1'} \bullet \mathbf{E} \bullet \mathbf{u}_{2'}$
Max Shear Plane	$\sigma = \frac{1}{2}(\sigma_1 + \sigma_3)$ $\tau_{\max} = \pm \frac{1}{2}(\sigma_1 - \sigma_3)$	$\varepsilon = \frac{1}{2}(\varepsilon_1 + \varepsilon_3)$ $\gamma_{\max} = \pm(\varepsilon_1 - \varepsilon_3)$
Octahedral Plane	$\sigma_m = \sigma_o = \frac{1}{3}\sigma_{kk}$ $= \frac{1}{3}(\sigma_1 + \sigma_2 + \sigma_3)$ $\tau_o = \pm \frac{1}{3}\sqrt{[\sum(\sigma_i - \sigma_j)^2]}$	$\varepsilon_o = \varepsilon_m = \frac{1}{3}\varepsilon_{kk}$ $= \frac{1}{3}(\varepsilon_1 + \varepsilon_2 + \varepsilon_3)$ $\frac{1}{2}\gamma_o = \pm \frac{1}{3}\sqrt{[\sum(\varepsilon_i - \varepsilon_j)^2]}$
Deviatoric	$\mathbf{S}^{\mathrm{D}} = \mathbf{S} - \frac{1}{3}(\mathrm{tr}\,\mathbf{S})\,\mathbf{I}$ $\sigma_{ij}' = \sigma_{ij} - \frac{1}{3}\delta_{ij}\,\sigma_{kk}$	$\mathbf{E}^{\mathrm{D}} = \mathbf{E} - \frac{1}{3}(\mathrm{tr}\,\mathbf{E})\,\mathbf{I}$ $\varepsilon_{ij}' = \varepsilon_{ij} - \frac{1}{3}\,\delta_{ij}\,\varepsilon_{kk}$
Principal	$\sigma^3 - J_1\sigma^2 + J_2\sigma - J_3 = 0$	$\varepsilon^3 - I_1\varepsilon^2 + I_2\,\varepsilon - I_3 = 0$
Invariants	$J_1 = \sigma_{kk}$ $J_2 = \frac{1}{2}(\sigma_{ii}\,\sigma_{jj} - \sigma_{ij}\sigma_{ji})$ $J_3 = \det\,\sigma_{ij}$	$I_1 = \varepsilon_{kk}$ $I_2 = \frac{1}{2}(\varepsilon_{ii}\varepsilon_{jj} - \varepsilon_{ij}\,\varepsilon_{ji})$ $I_3 = \det\,\varepsilon_{ij}$
Plane	$\sigma = \sigma_y \sin^2\alpha + \sigma_x \cos^2\alpha + \tau_{xy}\sin 2\alpha$ $= \frac{1}{2}(\sigma_x + \sigma_y) + \frac{1}{2}(\sigma_x - \sigma_y)\cos 2\alpha + \tau_{xy}\sin 2\alpha$	$\varepsilon = \varepsilon_y \sin^2\alpha + \varepsilon_x \cos^2\alpha + \frac{1}{2}\gamma_{xy}\sin 2\alpha$ $= \frac{1}{2}(\varepsilon_x + \varepsilon_y) + \frac{1}{2}(\varepsilon_x - \varepsilon_y)\cos 2\alpha + \frac{1}{2}\gamma_{xy}\sin 2\alpha$
Stress	$\tau = \frac{1}{2}(\sigma_x - \sigma_y)\sin 2\alpha - \tau_{xy}\cos 2\alpha$	$\frac{1}{2}\gamma = \frac{1}{2}(\varepsilon_x - \varepsilon_y)\sin 2\alpha - \frac{1}{2}\gamma_{xy}\cos 2\alpha$
&	$\sigma_{1,2} = \frac{1}{2}(\sigma_x + \sigma_y) \pm \frac{1}{2}\sqrt{[(\sigma_x - \sigma_y)^2 + 4\tau_{xy}^2]}$	$\varepsilon_{1,2} = \frac{1}{2}(\varepsilon_x + \varepsilon_y) \pm \frac{1}{2}\sqrt{[(\varepsilon_x - \varepsilon_y)^2 + \gamma_{xy}^2]}$
Strain	$\tan 2\alpha = 2\tau_{xy}/(\sigma_x - \sigma_y)$ $\tau_{\max} = \frac{1}{2}(\sigma_1 - \sigma_2)$ $= \sqrt{[(\varepsilon_x - \varepsilon_y)^2 + \gamma_{xy}^2]}$	$\tan 2\alpha = \gamma_{xy}/(\varepsilon_x - \varepsilon_y)$ $\gamma_{\max} = \varepsilon_1 - \varepsilon_2$ $= \frac{1}{2}\sqrt{[(\sigma_x - \sigma_y)^2 + 4\tau_{xy}^2]}$

Example 1.8 Find, for the given tensor of micro-strains ($1\mu = 1 \times 10^{-6}$),(a) the normal strain in a direction defined by the unit vector: $\mathbf{u}_{1'} = (2/3)\mathbf{u}_1 + (1/3)\mathbf{u}_2 + (2/3)\mathbf{u}_3$, (b) the shear strain between the normal vector in (a) and a perpendicular direction defined by a unit vector: $\mathbf{u}_{2'} = -0.25\mathbf{u}_1 + 0.942\mathbf{u}_2 - 0.221\mathbf{u}_3$, (c) the principal strains, (d) the state of strain on the maximum shear plane, (e) the octahedral normal and shear strains, (f) the elastic dilation and mean strain and (g) the deviatoric strains.

$$\mu\varepsilon_{ij} = \begin{bmatrix} 100 & 100 & -100 \\ 100 & 200 & 200 \\ -100 & 200 & 200 \end{bmatrix}$$

(a) Substituting $\varepsilon_{11} = 100\mu$, $\varepsilon_{22} = 200\mu$, $\varepsilon_{33} = 200\mu$, $\varepsilon_{12} = 100\mu$, $\varepsilon_{13} = -100\mu$ and $\varepsilon_{23} = 200\mu$ into eq(1.30b) gives the normal strain

$$\varepsilon_{1'1'} = (2/3)^2 100 + (1/3)^2 200 + (2/3)^2 200 + 2[(2/3)(1/3)100 + (1/3)(2/3)200 + (2/3)^2(-100)] = 196.7\mu$$

(b) The unit vectors' equations yield the direction cosines (see eq 1.19a,b) $l_{11} = 2/3$, $l_{12} = 1/3$, $l_{13} = 2/3$, $l_{21} = -0.25$, $l_{22} = 0.942$ and $l_{23} = -0.221$. Substituting these into eq(1.31b) with the given strain components,

$$\begin{aligned}\varepsilon_{1'2'} &= (2/3)(-0.25)100 + (1/3)(0.942)200 + (2/3)(-0.221)200 \\ &+ [(2/3)(0.942) + (-0.25)(1/3)]100 + [(1/3)(-0.221) + (0.942)(2/3)]200 \\ &+ [(2/3)(-0.221) + (-0.25)(2/3)](-100) = 213.05\mu\end{aligned}$$

(c) The principal strains may be found from the cubic eq(1.27). When the roots are exact it is easier to find them by expanding the following determinant and factoring:

$$\begin{vmatrix} 100-\varepsilon & 100 & -100 \\ 100 & 200-\varepsilon & 200 \\ -100 & 200 & 200-\varepsilon \end{vmatrix} = 0$$

$$(\varepsilon - 400)(\varepsilon + 100)(\varepsilon - 200) = 0$$

$$\varepsilon_1 = 400\mu,\ \varepsilon_2 = 200\mu \text{ and } \varepsilon_3 = -100\mu$$

(d) The maximum shear strain expression (see Table 1.1) applies when $\varepsilon_1 > \varepsilon_2 > \varepsilon_3$

$$\tfrac{1}{2}\gamma_{max} = \tfrac{1}{2}(\varepsilon_1 - \varepsilon_3)$$
$$\tfrac{1}{2}\gamma_{max} = \varepsilon_1 - \varepsilon_3 = 400 + 100 = 500\mu$$

The norml strain on this plane is

$$\varepsilon = \tfrac{1}{2}(\varepsilon_1 + \varepsilon_3) = 100\mu$$

(e) Table 1.1 gives the expressions for the octahedral strains. The mean or hydrostatic strain in the normal direction is

$$\begin{aligned}\varepsilon_m &= (\varepsilon_1 + \varepsilon_2 + \varepsilon_3)/3 \\ &= (400 + 200 - 100)/3 = 166.7\mu\end{aligned}$$

The octahedral shear strain is

$$\gamma_o = (2/3)\ \surd[(\varepsilon_1 - \varepsilon_2)^2 + (\varepsilon_2 - \varepsilon_3)^2 + (\varepsilon_1 - \varepsilon_3)^2]$$
$$= (2/3)\ \surd[(400 - 200)^2 + (200 + 100)^2 + (400 + 100)^2] = 411\mu$$

This is a radian measure of the angular change between two perpendicular directions: one aligned with the normal to the octahedral plane and the other aligned with the direction of the shear stress τ_o in that plane (direction cosines being given by eqs 1.17a,b,c).

(f) The elastic dilatation (volumetric strain) is

$$\delta V/V = \varepsilon_1 + \varepsilon_2 + \varepsilon_3 = 3\varepsilon_m = 500\mu$$

which equals the strain invariant I_1 and from which the mean strain $\varepsilon_m = 166.7\mu$.

(g) Deviatoric strains will remain when the mean or hydrostatic strain ε_m has been subtracted from the normal strain components. Correspondence with eq(1.21a) gives the strain deviator tensor

$$\varepsilon_{ij}' = \varepsilon_{ij} - \delta_{ij}\,\varepsilon_m$$
$$\varepsilon_x' = \varepsilon_x - \varepsilon_m = 100 - 166.7 = -66.7\mu$$
$$\varepsilon_y' = \varepsilon_y - \varepsilon_m = 200 - 166.7 = 33.3\mu = \varepsilon_z$$

These normal deviators, together with the given tensor shear strains, (recall $\delta_{ij} = 0$ for $i \neq j$), constitute the deviatoric strain tensor ε_{ij}'. The six independent deviatoric components of strain define unsymmetrical distortion.

1.6 Plane Stress

1.6.1 Analytical Method

The previous analytical expressions (1.5a,b) may be reduced to plane stress states. Say we wish to find expressions for the normal and shear stresses on each of the oblique planes in Figs 1.13a,b.

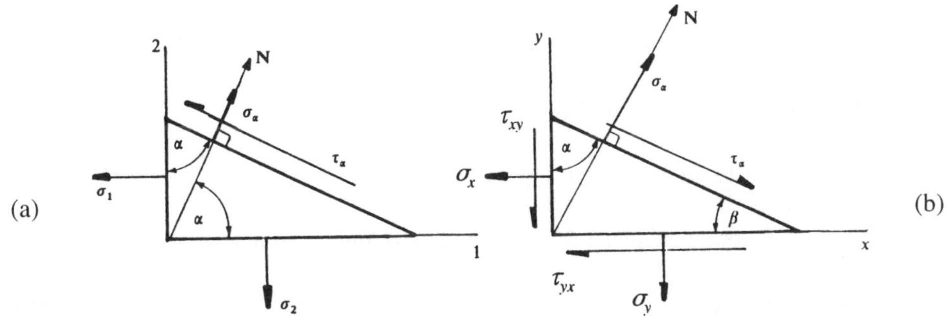

Figure 1.13 Reduction to plane stress states

With principal stresses σ_1 and σ_2 in Fig. 1.13a, the oblique plane has normal direction cosines $l = \cos\alpha$, $m = \cos(90° - \alpha) = \sin\alpha$, $n = 0$, relative to 1, 2 and 3. Substituting into eq(1.13a,b)

$$\begin{aligned}\sigma &= \sigma_1 \cos^2\alpha + \sigma_2 \sin^2\alpha \\ &= \tfrac{1}{2}(\sigma_1 + \sigma_2) + \tfrac{1}{2}(\sigma_1 - \sigma_2)\cos 2\alpha \end{aligned} \tag{1.33a}$$

$$\begin{aligned}\tau^2 &= \sigma_1^{\,2}\cos^2\alpha + \sigma_2^{\,2}\sin^2\alpha - (\sigma_1\cos^2\alpha + \sigma_2\sin^2\alpha)^2 \\ &= (\sigma_1 - \sigma_2)^2\sin^2\alpha\cos^2\alpha \\ \tau &= \tfrac{1}{2}(\sigma_1 - \sigma_2)\sin 2\alpha \end{aligned} \tag{1.33b}$$

When $\alpha = 45°$, the shear stress (1.33b) has its maximum value

$$\tau_{max} = \tfrac{1}{2}(\sigma_1 - \sigma_2) \tag{1.33c}$$

Figure 1.13b shows a general plane stress state σ_x, σ_y and $\tau_{xy} = \tau_{xy}$. The cosines are again $l = \cos\alpha$, $m = \sin\alpha$ and $n = 0$, relative to x, y and z. Substituting l, m and n into eq(1.5a) provides the normal stress on the oblique plane as

$$\begin{aligned}\sigma &= \sigma_x\cos^2\alpha + \sigma_y\sin^2\alpha + \tau_{xy}\sin 2\alpha \\ &= \tfrac{1}{2}(\sigma_x + \sigma_y) + \tfrac{1}{2}(\sigma_x - \sigma_y)\cos 2\alpha + \tau_{xy}\sin 2\alpha \end{aligned} \tag{1.34a}$$

The shear stress on this plane is found from eqs(1.4a,b,c) and (1.5b)

$$S_x = \sigma_x\cos\alpha + \tau_{yx}\sin\alpha, \quad S_y = \sigma_y\sin\alpha + \tau_{xy}\cos\alpha \ \text{ and } \ S_z = 0$$

$$\begin{aligned}\tau^2 &= S_x^{\,2} + S_y^{\,2} + S_z^{\,2} - \sigma^2 \\ &= (\sigma_x\cos\alpha + \tau_{yx}\sin\alpha)^2 + (\sigma_y\sin\alpha + \tau_{xy}\cos\alpha)^2 - [\sigma_x\cos^2\alpha + \sigma_y\sin^2\alpha + \tau_{xy}\sin 2\alpha]^2 \\ &= \tfrac{1}{4}(\sigma_x - \sigma_y)^2\sin^2 2\alpha - \tau_{xy}(\sigma_x - \sigma_y)\sin 2\alpha\cos 2\alpha + \tau_{xy}^{\,2}\cos^2 2\alpha \\ &= [\tfrac{1}{2}(\sigma_x - \sigma_y)\sin 2\alpha - \tau_{xy}\cos 2\alpha]^2 \\ \tau &= \tfrac{1}{2}(\sigma_x - \sigma_y)\sin 2\alpha - \tau_{xy}\cos 2\alpha \end{aligned} \tag{1.34b}$$

Now from eqs(1.9a,b,c), the invariants become

$$J_1 = \sigma_x + \sigma_y, \quad J_2 = \sigma_x\sigma_y - \tau_{xy}^{\,2} \ \text{ and } \ J_3 = 0$$

The principal stress cubic eq(1.8b) reduces to a quadratic

$$\sigma^2 - (\sigma_x + \sigma_y)\sigma + (\sigma_x\sigma_y - \tau_{xy}^{\,2}) = 0$$

giving the principal stresses as its roots

$$\sigma_{1,2} = \tfrac{1}{2}(\sigma_x + \sigma_y) \pm \tfrac{1}{2}\sqrt{[(\sigma_x - \sigma_y)^2 + 4\tau_{xy}^{\,2}]} \tag{1.34c}$$

The directions of a principal plane AC are found by setting $\tau = 0$ in eq(1.34b),

$$\tan 2\alpha = 2\tau_{xy} / (\sigma_x - \sigma_y) \tag{1.34d}$$

Equation (1.34d) supplies the directions of two perpendicular principal planes upon which σ_1 and σ_2 act. The maximum shear stress is again given by eq(1.33c) and this lies at 45° to the principal directions. Equations (1.34a-d) may be reduced to simpler plane systems by setting one or more of the applied stress components to zero. For example, eqs(1.33a,b) will reappear when $\tau_{xy} = 0$ in eqs(1.34a,b) because, in the absence of shear stress, the normal stresses σ_x and σ_y become the principal stresses σ_1 and σ_2.

1.6.2 Mohr's Circle

The plane stress state (σ, τ) on the inclined plane (Fig. 1.13a,b) may also be found from a simpler Mohr's circle construction. Given the stress states for two perpendicular planes AB and BC (Fig. 1.14a), normal stresses σ_x and σ_y are plotted to the right if tensile and to the left if compressive. Clockwise shear stresses τ_{yx} on BC are plotted upward and anti-clockwise shear stress τ_{xy} on AB is plotted downward. These locate two co-ordinate points for planes AB and BC lying on opposite ends of a diameter, thus enabling the circle to be drawn (see Fig. 1.14b). A focus point F is found by intersection with the projection of the normal to plane AB through the corresponding co-ordinate point AB in the circle. F may also be located by projecting the normal to plane BC in a similar manner. Indeed F is the single focus point of intersections between all such normal projections. Thus the stress state on plane AC is found by projecting the normal to AC through the focus F. The point of intersection with the circle shows that both σ_θ and τ_θ are positive, so their directions are as shown in Fig. 1.14a.

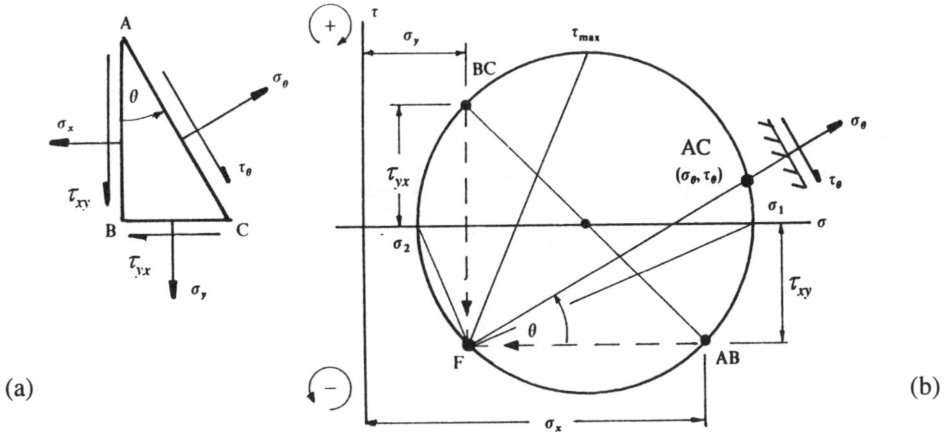

Figure 1.14 Mohr's circle construction for a general plane stress state

The normal to other planes of interest may be projected from F. In particular, the major and minor principal planes lie normal to Fσ_1 and Fσ_2 where σ_1 and σ_2 are the principal stresses (points of zero shear stress on the circle). The line joining F to the top point (τ_{max}) is the normal to the maximum shear plane.

Example 1.9 Given $\sigma_x = 100$ MPa, $\sigma_y = 50$ MPa and $\tau_{xy} = 65$ MPa in Fig. 1.14a determine graphically the principal stresses, the orientation of the principal planes and the maximum shear stress. Check your answers analytically.

A construction similar to Fig. 1.14b has co-ordinates AB (100,− 65) and BC (50, 65). The circle yields major and minor principal stresses of σ_1 = 144.6 MPa and σ_2 = 5.36 MPa respectively. The orientations of their planes are $\alpha = 44.48°$ and $\alpha + 90 = 134.48°$, relative to AC. The vertical radius identifies with τ_{max} = 69.62 MPa, this being inclined at α + 45° = 89.48° to AC. Checking analytically, eq(1.34c) yields the major and minor principal stresses:

σ_1, σ_2 = ½ (100 + 50) ± ½ √[(100 - 50)² + 4(65)²] = 75 ± 69.64
σ_1 = 144.6 MPa and σ_2 = 5.36 MPa

Equation (1.34d) gives the respective orientations of their principal planes as

α = ½ tan⁻¹ [2 × 65 / (100 - 50)] = 33.48° and 124.48°

Equation (1.33c) gives the maximum shear stress

τ_{max} = ½(144.6 - 5.36) = 69.62 MPa

1.6.3 Matrix Method

Figure 1.15 shows a plane element with normal and shear stresses applied to its sides. The mathematical notation is now used to define the co-ordinate directions x_1 and x_2 and the stress components, as shown.

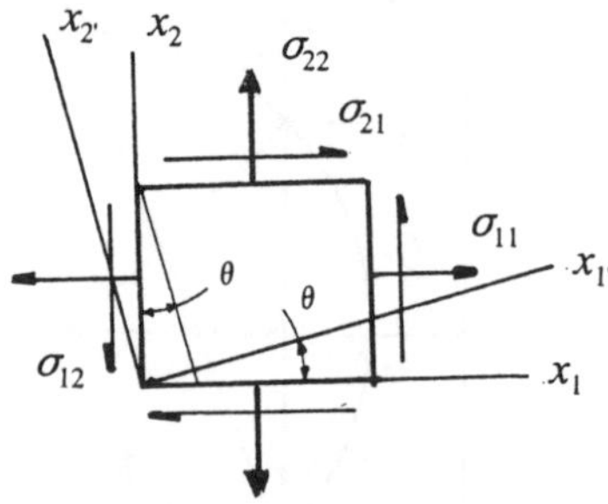

Figure 1.15 Plane rotation in co-ordinate axes

We may again use eq(1.20a) to transform stress components following a rotation from axes x_1, x_2 into axes $x_{1'}$, $x_{2'}$ when it is reduced to a plane transformation. That is, **S** and **L** are multiplied as the 2 × 2 matrices

$$\begin{bmatrix} \sigma_{1'1'} & \sigma_{1'2'} \\ \sigma_{2'1'} & \sigma_{2'2'} \end{bmatrix} = \begin{bmatrix} \cos\theta & \sin\theta \\ -\sin\theta & \cos\theta \end{bmatrix} \begin{bmatrix} \sigma_{11} & \sigma_{12} \\ \sigma_{21} & \sigma_{22} \end{bmatrix} \begin{bmatrix} \cos\theta & -\sin\theta \\ \sin\theta & \cos\theta \end{bmatrix} \qquad (1.35)$$

The components of the 2 × 2 matrix of direction cosines **L** are $l_{ij} = \cos x_{i'} x_j$. Subscript i refers to the primed axis and j to the unprimed axis. This gives

$l_{11} = \cos\theta$, $l_{12} = \cos(90° - \theta) = \sin\theta$,
$l_{21} = \cos(\theta + 90°) = -\sin\theta$ and $l_{22} = \cos\theta$

The matrix multiplication in eq(1.35) yields the transformed stress state in which the complementary shear conditions $\sigma_{12} = \sigma_{21}$ and $\sigma_{1'2'} = \sigma_{2'1'}$, have been applied,

$$\sigma_{1'1'} = \sigma_{11}\cos^2\theta + \sigma_{22}\sin^2\theta + \sigma_{12}\sin 2\theta$$
$$\sigma_{2'2'} = \sigma_{11}\sin^2\theta + \sigma_{22}\cos^2\theta - \sigma_{12}\sin 2\theta$$
$$\sigma_{1'2'} = -\tfrac{1}{2}(\sigma_{11} - \sigma_{22})\sin 2\theta + \sigma_{12}\cos 2\theta$$

These agree with the form of eqs(1.34a,b). We may interpret the change in sign of the shear stress by reversing their directions along the x_1 and x_2 axes in Fig. 1.15. This will restore the sign found by alternative analyses for an element with shear directions shown in Fig. 1.14a. Principal stresses follow from reducing the determinant in eq(1.7b) to

$$\det\begin{vmatrix} \sigma_{11} - \sigma & \sigma_{12} \\ \sigma_{21} & \sigma_{22} - \sigma \end{vmatrix} = 0 \qquad (1.36)$$

Equation (1.36) expands to agree with eq(1.34c)

$$\sigma_{1,2} = \tfrac{1}{2}(\sigma_{11} + \sigma_{22}) \pm \tfrac{1}{2}\sqrt{[(\sigma_{11} - \sigma_{22})^2 + 4\sigma_{12}^2]}$$

The principal stress directions follow from eq(1.7a)

$$l(\sigma_{11} - \sigma) + m\sigma_{12} = 0 \qquad (1.37a)$$
$$l\sigma_{21} + m(\sigma_{22} - \sigma) = 0 \qquad (1.37b)$$

where l and m are the direction cosines for a given principal plane. Either eq(1.37a) or (1.37b) define the unit vectors aligned with the two principal directions, as the following example shows.

Example 1.10 Determine the state of stress for an element inclined at 30° to the given element in Fig. 1.16a. What are the principal stress values and their directions?

(a)

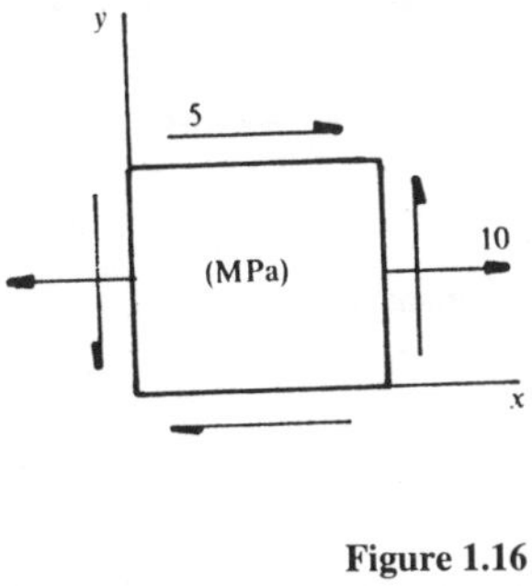

(b)

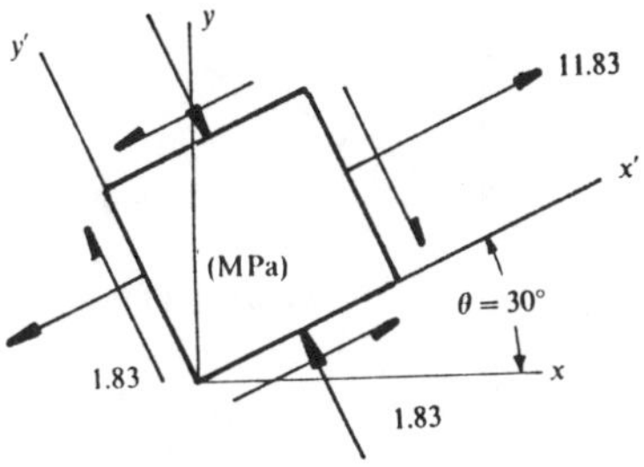

Figure 1.16 Plane stress transformation

Substituting into eq(1.35) for $\theta = 30°$, $\sigma_{11} = 10$, $\sigma_{12} = \sigma_{21} = 5$, $\sigma_{22} = 0$ gives

$$\mathbf{S}' = \begin{bmatrix} \sqrt{3}/2 & 1/2 \\ -1/2 & \sqrt{3}/2 \end{bmatrix}\begin{bmatrix} 10 & 5 \\ 5 & 0 \end{bmatrix}\begin{bmatrix} \sqrt{3}/2 & -1/2 \\ 1/2 & \sqrt{3}/2 \end{bmatrix}$$

$$= \begin{bmatrix} \sqrt{3}/2 & 1/2 \\ -1/2 & \sqrt{3}/2 \end{bmatrix}\begin{bmatrix} 11.16 & -0.67 \\ 4.33 & -2.5 \end{bmatrix} = \begin{bmatrix} 11.83 & -1.83 \\ -1.83 & -1.83 \end{bmatrix}$$

which act as shown in Fig. 1.16b. Note that the matrix supplies the complete stress state for the rotated element. The principal stresses follow from expanding the determinant in eq(1.36)

$$\begin{vmatrix} 10-\sigma & 5 \\ 5 & 0-\sigma \end{vmatrix} = 0 \quad \Rightarrow \quad \begin{aligned} & -10\sigma + \sigma^2 - 25 = 0 \\ & \sigma = 10/2 \pm \tfrac{1}{2}\sqrt{[(-10)^2 - 4(-25)]} \\ & \sigma_1 = 12.07,\ \sigma_2 = -2.07 \text{ MPa} \end{aligned}$$

Substituting σ_1 and σ_2 in turn for σ in eq(1.37a) will provide l and m for the principal directions. The following unit vectors $\mathbf{n}_1$ and $\mathbf{n}_2$, aligned with principal directions, may be derived from an arbitrary vector lying in each direction in the manner of Example 1.3.

$$\mathbf{n}_1 = 0.9239\mathbf{u}_1 + 0.3827\mathbf{u}_2$$
$$\mathbf{n}_2 = 0.3827\mathbf{u}_1 - 0.9239\mathbf{u}_2$$

These directions are perpendicular as $\mathbf{n}_1 \bullet \mathbf{n}_2 = 0$. Alternatively, for the major plane eq(1.34d) gives $\tan 2\theta = 1$, from which $\theta = 22.5°$ and $l = \cos\theta = 0.9239$.

1.7 Plane Strain

1.7.1 Reductions to 2D Equations

Putting $l = \cos\theta$, $m = \cos(90° - \theta) = \sin\theta$ and $n = \cos 90° = 0$ in the normal strain expression (1.26b) leads to the normal strain in the direction of x' in Fig. 1.17a.

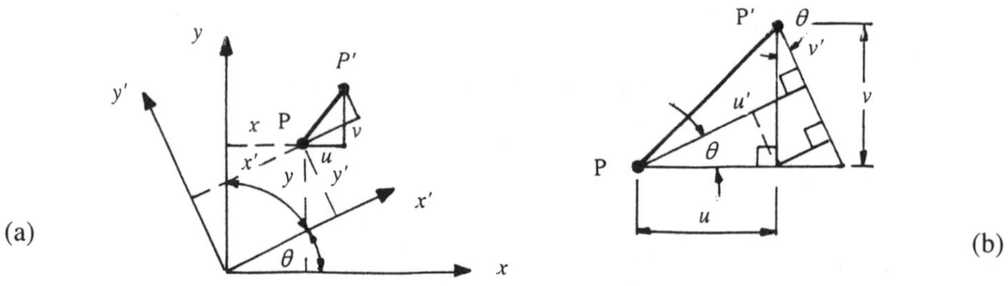

Figure 1.17 Plane strain transformation

$$\begin{aligned} \varepsilon_x' &= l^2\varepsilon_x + m^2\varepsilon_y + n^2\varepsilon_z + lm\varepsilon_{xy} + mn\varepsilon_{yz} + ln\varepsilon_{zx} \\ &= \varepsilon_x\cos^2\theta + \varepsilon_y\sin^2\theta + \tfrac{1}{2}\gamma_{xy}\sin 2\theta \end{aligned} \qquad (1.38a)$$

The normal strain in the direction of y' can be found from setting $l = \cos(90 + \theta) = -\sin\theta$, $m = \cos\theta$ and $n = 0$ in eq(1.38a). This gives

$$\varepsilon_{y'} = \varepsilon_x\sin^2\theta + \varepsilon_y\cos^2\theta - \tfrac{1}{2}\gamma_{xy}\sin 2\theta \qquad (1.38b)$$

The shear strain between the primed directions x' and y' is found from eq(1.31c). The direction cosines of x' and y' are respectively

$$l_1 = \cos\theta,\ \ m_1 = \cos(90° - \theta) = \sin\theta \text{ and } n_1 = \cos 90° = 0$$
$$l_2 = \cos(90 + \theta) = -\sin\theta,\ \ m_2 = \cos\theta \text{ and } n_2 = 0$$

$$\therefore \tfrac{1}{2}\gamma_{x'y'} = -\varepsilon_x \sin\theta\cos\theta + \varepsilon_y \sin\theta\cos\theta + \tfrac{1}{2}\gamma_{xy}(\cos^2\theta - \sin^2\theta)$$
$$= -\tfrac{1}{2}(\varepsilon_x - \varepsilon_y)\sin 2\theta + \tfrac{1}{2}\gamma_{xy}\cos 2\theta \qquad (1.38c)$$

Equations (1.38a and b) may be reduced to simpler strain systems by setting ε_x, ε_y and γ_{xy} to zero. Principal strains ε_1 and ε_2 exist along two perpendicular directions for which shear strain is absent. Setting $\gamma_{x'y'} = 0$ in eq(1.38c) gives the orientation of the major principal plane as

$$\tan 2\theta = \gamma_{xy} / (\varepsilon_x - \varepsilon_y) \qquad (1.39a)$$

The magnitudes of the two principal strains are found by eliminating θ between eq(1.39a) and eqs(1.38a,b)

$$\varepsilon_{1,2} = \tfrac{1}{2}(\varepsilon_x + \varepsilon_y) \pm \tfrac{1}{2}\surd[(\varepsilon_x - \varepsilon_y)^2 + \gamma_{xy}^2] \qquad (1.39b)$$

1.7.2 Analytical Method

The strains given in eqs(1.38a-c) may be confirmed by direct differentiation of the displacements according to the infinitesimal strain definitions in eqs(1.23a,b) and (1.24a). The geometry in Fig. 1.17b reveals the following relationships between the co-ordinates:

$$x = x'\cos\theta - y'\sin\theta$$
$$y = x'\sin\theta + y'\cos\theta$$

The displacements of a point P as it moves to P' are given by

$$u' = u\cos\theta + v\sin\theta$$
$$v' = -u\sin\theta + v\cos\theta$$

where u and v are the displacements of P' along x and y, and u' and v' are the displacements of P' along x' and y'. The normal strain and shear strains in the x', y' plane are found from eqs(1.23a,b) and (1.24a)

$$\varepsilon_{x'} = \partial u'/\partial x' = (\partial u'/\partial x)(\partial x/\partial x') + (\partial u'/\partial y)(\partial y/\partial x')$$
$$= [(\partial u/\partial x)\cos\theta + (\partial v/\partial x)\sin\theta]\cos\theta + [(\partial u/\partial y)\cos\theta + (\partial v/\partial y)\sin\theta]\sin\theta$$
$$= \varepsilon_x \cos^2\theta + \varepsilon_y \sin^2\theta + \tfrac{1}{2}\gamma_{xy}\sin 2\theta$$

$$\varepsilon_{y'} = \partial v'/\partial y' = (\partial v'/\partial y)(\partial y/\partial y') + (\partial v'/\partial x)(\partial x/\partial y')$$
$$= [-(\partial u/\partial y)\sin\theta + (\partial v/\partial y)\cos\theta]\cos\theta + [-(\partial u/\partial x)\sin\theta + (\partial v/\partial x)\cos\theta](-\sin\theta)$$
$$= \varepsilon_x \sin^2\theta + \varepsilon_y \cos^2\theta - \tfrac{1}{2}\gamma_{xy}\sin 2\theta$$

$$\gamma_{x'y'} = \partial' v/\partial x' + \partial u'/\partial y'$$
$$= (\partial v'/\partial x)(\partial x/\partial x') + (\partial v'/\partial y)(\partial y/\partial x') + (\partial u'/\partial x)(\partial x/\partial y') + (\partial u'/\partial y)(\partial y/\partial y')$$
$$= [-(\partial u/\partial x)\sin\theta + (\partial v/\partial x)\cos\theta]\cos\theta + [-(\partial u/\partial y)\sin\theta + (\partial v/\partial y)\cos\theta]\sin\theta$$
$$+ [(\partial u/\partial x)\cos\theta + (\partial v/\partial x)\sin\theta](-\sin\theta) + [(\partial u/\partial y)\cos\theta + (\partial v/\partial y)\sin\theta]\cos\theta$$

$$= - (\partial u/\partial x)2\sin\theta\cos\theta + (\partial v/\partial y)2\sin\theta\cos\theta + [(\partial v/\partial x) + (\partial u/\partial y)](\cos^2\theta - \sin^2\theta)$$

$$\tfrac{1}{2}\gamma_{x'y'} = - \tfrac{1}{2}(\varepsilon_x - \varepsilon_y)\sin 2\theta + \tfrac{1}{2}\gamma_{xy}\cos 2\theta$$

1.7.3 Mohr's Circle

Because of the identical nature of stress and strain transformations, the directions of the principal stress and strain will be coincident. Moreover, a geometric similarity exists between the Mohr's stress and strain circles. A Mohr's circle can therefore be drawn for strain and a focus located within axes of ε and $\tfrac{1}{2}\gamma$, as shown in Fig. 1.18.

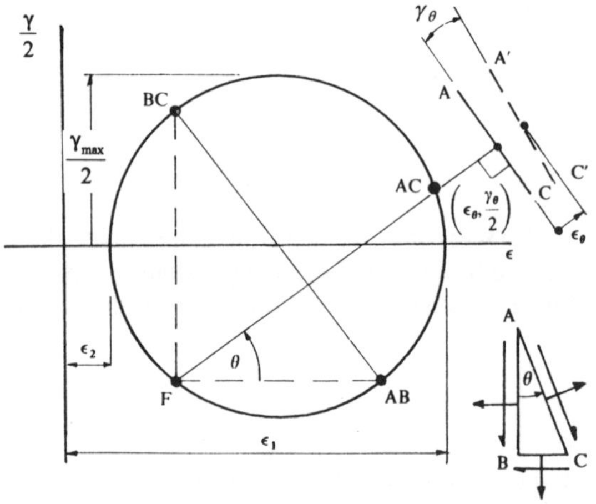

Figure 1.18 Mohr's strain circle construction showing focus point F

For the stressed element (inset) ε_x is positive, γ_{xy} is negative for plane AB while both ε_y and γ_{xy} are positive for the plane BC. These locate two diametrically opposite points on the circle thus enabling its construction. The focus F is then found by projecting the normal either to the plane AB or to BC through the corresponding point on the strain circle. For any plane AC the state of strain (ε_θ, $\tfrac{1}{2}\gamma_\theta$) is found by projecting the normal to AC through F as shown. The intercept point shows tensile strain (ε_θ) in a direction normal to AC, accompanied by a net clockwise angular change of γ_θ between AC and its normal. The inset diagram shows how to interpret these in terms of a translation and a rotation of plane AC. Points on the circle along the horizontal diameter provide the principal strains ε_1, ε_2, and points along the vertical diameter supply the maximum shear strains $\pm\tfrac{1}{2}\gamma_{max}$. Joining each of these extreme points to F will again provide the associated planes. Since the normals to planes AB, BC and AC are the respective directions in which ε_x, ε_y and ε_θ act, we see that all normal strain directions converge upon F.

In practice, the magnitude and direction of the major and minor principal strains at the surface of a structure may be determined by the measurement of surface strain in any three orientations. A three-element, electrical resistance strain gauge rosette, when bonded to a point on the surface, will measure the direct strains along each gauge axis. Knowing the inclination between the gauges (usually 45° or 60°) allows the principal strains and their orientations to be found. Example 1.11 shows how this very useful experimental technique does not require the shear strains associated with non-principal directions to be known.

Example 1.11 A strain gauge rosette is arranged with the axes of gauges 2 and 3 lying respectively at 60° and 120° anti-clockwise to gauge 1, as shown in Fig. 1.19a. These gauges record direct tensile micro-strains along these axes of 700μ, - 250μ and 300μ respectively. Determine: (i) the maximum shear strain and (ii) the principal strains and their orientations to the 1 - direction. Use both analytical and graphical methods.

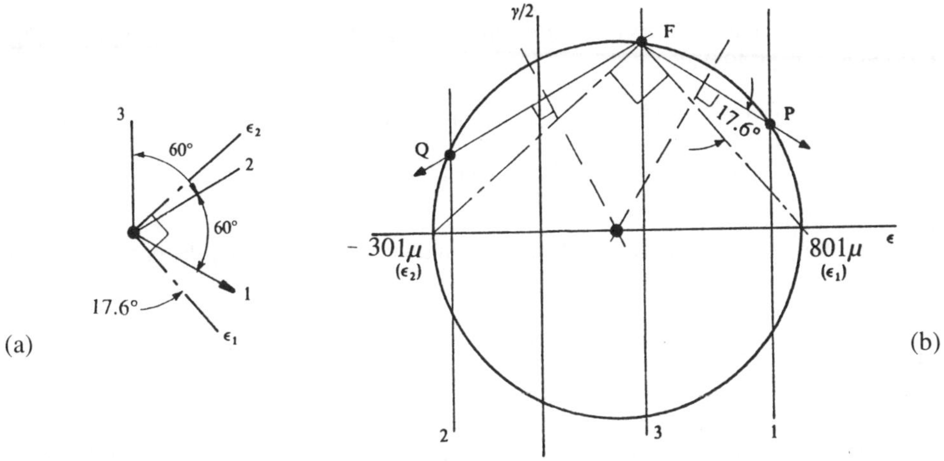

Figure 1.19 Mohr's strain construction for a 3-element rosette

For the analytical solutions, substitute for the three values ε for θ = 0°, 60° and 120° in eq(1.38a) to give the three simultaneous equations (micro-strain units):

$$\varepsilon_x = 700 \qquad \text{(i)}$$

$$\varepsilon_y \sin^2 60° + \varepsilon_x \cos^2 60° + (\gamma_{xy}/2) \sin 120° = -250$$
$$0.75\varepsilon_y + 0.433\gamma_{xy} = -425 \qquad \text{(ii)}$$

$$\varepsilon_y \sin^2 120° + \varepsilon_x \cos^2 120° + (\gamma_{xy}/2) \sin 240° = 300$$
$$0.75\varepsilon_y - 0.433\gamma_{xy} = 125 \qquad \text{(iii)}$$

The solution to eqs(i), (ii) and (iii) provides the x and y components ε_x = 700μ, ε_y = - 200μ and γ_{xy} = - 635μ. Substitution into the principal strain expression (1.39b) gives

$$\varepsilon_{1,2} = \tfrac{1}{2}(700 - 200) \pm \tfrac{1}{2}\sqrt{[(700 + 200)^2 + (635)^2]}$$
$$\varepsilon_1 = 800.8\mu \text{ and } \varepsilon_2 = -300.8\mu$$

Equation (1.39b) provides the inclination of the major principal direction to the gauge 1- axis, as shown,

$$\tan 2\theta = -635 / (700 + 200)$$
$$\theta = -17.6° \text{ and } 72.4°$$

1.7.4 Matrix Method

Equation (1.29b) contains a matrix representation of a plane strain transformation when the mathematical shear strains $\varepsilon_{12} = \varepsilon_{21} = \tfrac{1}{2}\gamma_{12}$, appear as matrix components. That is, $\mathbf{E}' = \mathbf{LEL}^T$

becomes, in full

$$\begin{bmatrix} \varepsilon_{1'1'} & \varepsilon_{1'2'} \\ \varepsilon_{2'1'} & \varepsilon_{2'2'} \end{bmatrix} = \begin{bmatrix} l_{11} & l_{12} \\ l_{21} & l_{22} \end{bmatrix} \begin{bmatrix} \varepsilon_{11} & \varepsilon_{12} \\ \varepsilon_{21} & \varepsilon_{22} \end{bmatrix} \begin{bmatrix} l_{11} & l_{21} \\ l_{12} & l_{22} \end{bmatrix} \quad (1.40a)$$

The direction cosines $l_{ij} = \cos(x_i' x_j)$ are applied to directions (x_1, x_2) and $(x_{1'}, x_{2'})$ (see Fig. 1.15) to give eq(1.40a) as

$$\begin{bmatrix} \varepsilon_{1'1'} & \varepsilon_{1'2'} \\ \varepsilon_{2'1'} & \varepsilon_{2'2'} \end{bmatrix} = \begin{bmatrix} \cos\theta & \sin\theta \\ -\sin\theta & \cos\theta \end{bmatrix} \begin{bmatrix} \varepsilon_{11} & \varepsilon_{12} \\ \varepsilon_{21} & \varepsilon_{22} \end{bmatrix} \begin{bmatrix} \cos\theta & -\sin\theta \\ \sin\theta & \cos\theta \end{bmatrix} \quad (1.40b)$$

Matrix multiplication gives three independent strain components along the primed axes

$$\varepsilon_{1'1'} = \varepsilon_{11}\cos^2\theta + \varepsilon_{22}\sin^2\theta + \varepsilon_{12}\sin 2\theta$$
$$\varepsilon_{2'2'} = \varepsilon_{11}\sin^2\theta + \varepsilon_{22}\cos^2\theta - \varepsilon_{12}\sin 2\theta$$
$$\varepsilon_{1'2'} = \varepsilon_{2'1'} = -\tfrac{1}{2}(\varepsilon_{11} - \varepsilon_{22}) + \varepsilon_{12}\cos 2\theta$$

When these are converted into an engineering notation we confirm eqs(1.38a-c). We have seen that the sign of the shear strain term can differ from the corresponding stress expression. Figures 1.20a,b show how the distortion, which accompanies an element in each notation, will differ between their corresponding reference states.

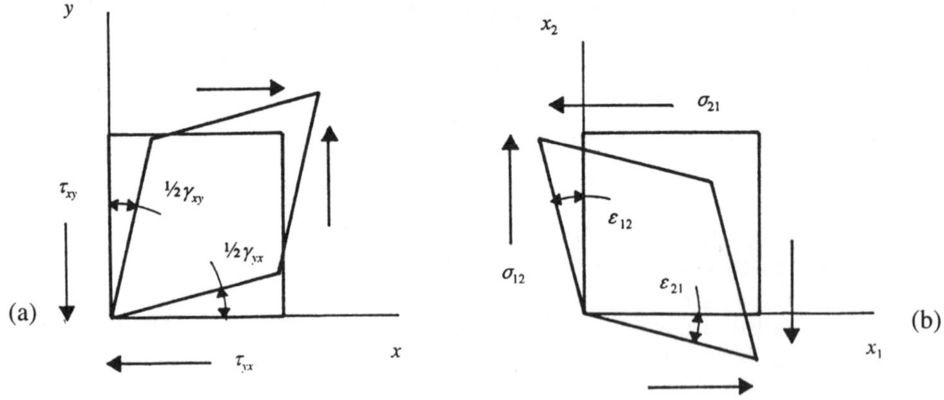

Figure 1.20 Shear distortion in the engineering and mathematical notations

EXERCISES

Plane Stress Transformation

1.1 Working from first principles, find the magnitude and direction of the principal stresses and the maximum shear stress for each element in Fig. 1.21a-d.
Answer (MPa): (a) 342.8,-3.1, 31.75°, 173, 76.75°; (b) 171.7,-48.17, 38.5°, 64.9, 83.5°(c) -102.1, 86.6, 27.6°, 94.4, 72.6°; (d) 197.6,-135.9, 11.32°, 166.8, 56.32°

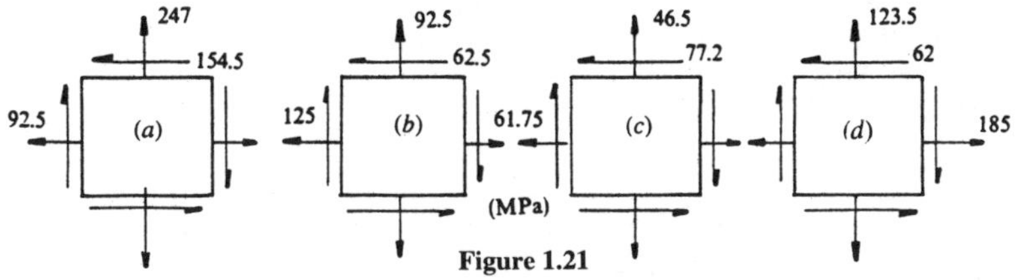

Figure 1.21

1.2 At a certain point a tensile stress of 123.5 MPa acts at 90° to a compressive stress of 154.4 MPa in the same plane. If the major principal stress is limited to 208.5 MPa, determine the maximum permissible shear stress that may act in these directions. What is the maximum shear stress and the orientation of the major principal plane?
Answer: 133.9 MPa, 193 MPa, 67°, 22°

1.3 A block of brittle material, 6.5 mm thick, is subjected to the forces in Fig. 1.22. Find the normal and shear stresses acting on the 30° plane AC. If the ultimate compressive and shear strengths are: u.c.s. = 30 MPa and u.s.s. = 17 MPa respectively, find the planes along which the material would fail when the applied stresses are increased in proportion. State the condition for compressive failure.
Answer: 11 MPa, 9.65 MPa, 45°, 135°, u.s.s. > ½ u.c.s.

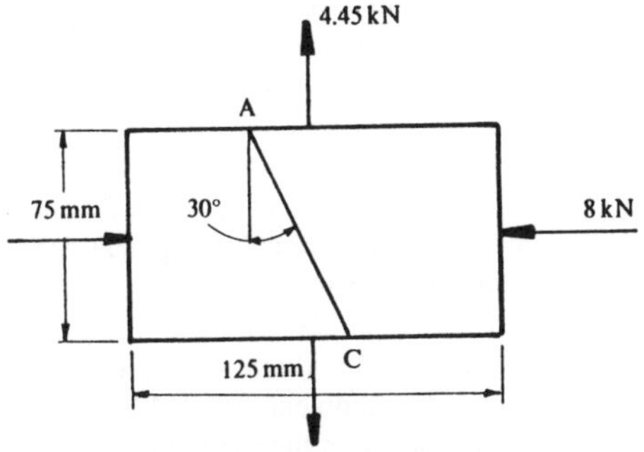

Figure 1.22

1.4 Determine the principal stresses, the maximum shear stress and their orientations for each of the following general plane stress systems (MPa): (a) $\sigma_x = 50$, $\sigma_y = 10$, $\tau_{xy} = 20$; (b) $\sigma_x = -20$, $\sigma_y = 30$, $\tau_{xy} = 15$; (c) $\sigma_x = 45$, $\sigma_y = 30$, $\tau_{xy} = 20$; (d) $\sigma_x = -30$, $\sigma_y = 50$, $\tau_{xy} = 40$. What for (a) is the state of stress on a plane making 75° to the plane on which σ_x acts and what in (b) is the angle between the σ_y plane and the plane on which the shear stress is 25 MPa?

1.5 The states of stress in MPa for two elements with a common plane are shown in Fig. 1.23. Determine the stress state for this plane and comment on the result.
Answer: 1 and 5 MPa, a uniform stress exists for all parallel planes

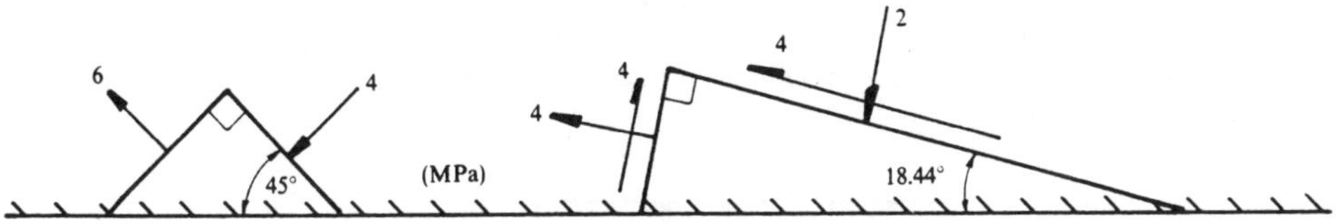

Fig.ure 1.23

1.6 A flat brass plate, stressed in plane perpendicular directions, gave respective extensions of 0.036 mm and 0.0085 mm over a 50 mm length. Find the inclination θ to the length for a plane whose normal stress is 60.4 MPa. Take $E = 82.7$ GPa and $\nu = 0.3$.
Answer: 30°

1.7 Calculate θ for the cantilevered arc in Fig. 1.24 in order that the maximum bending and torsional stresses are equal. If, for this condition $W = 10$ N and $d = 10$ mm, find R when the maximum shear stress is limited to 80 MPa.
Answer: 53°, 880 mm

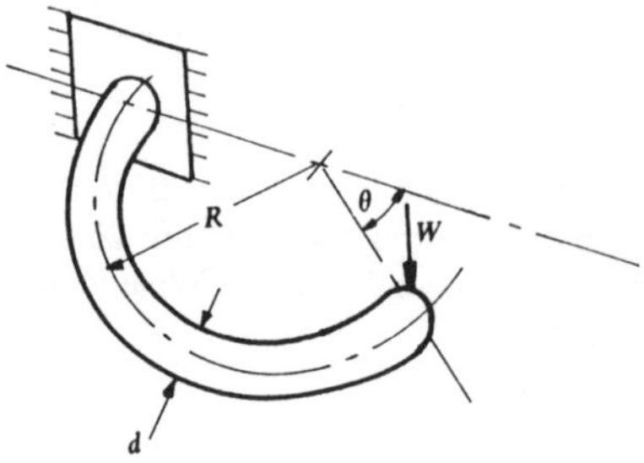

Figure 1.24

1.8 A cylindrical vessel is formed by welding steel plate along a helical seam inclined at 30° to the cross-section. Determine the stress state on the seam at a test pressure of 27.5 bar when the mean diameter and thickness are 1.5 m and 20 mm respectively.
Answer: 68.9 MPa, 23.87 MPa

1.9 At a point on the surface of a shaft the axial tensile stress due to bending is 77 MPa and the maximum shear stress due to torsion is 31 MPa. Determine graphically the magnitude and direction of the maximum shear stress and the principal stresses.

1.10 A steel shaft is to transmit 225 kW at 150 rev/min without the major principal stress exceeding 123.5 MPa. If the shaft also carries a bending moment of 3.04 kNm, find the shaft diameter. What is the maximum shear stress and its plane relative to the shaft axis?

1.11 The bending stress in a tube with diameter ratio of 5:1 is not to exceed 90 MPa under a moment of 80 kNm. Determine the tube diameters. If the moment is replaced by a torque of 80 kNm, find the major principal stress.
Answer: 130 mm, 650 mm, 45 MPa

1.12 A steel bar withstands simultaneously a 30 kN compressive force, a 5 kNm bending moment and a 1 kNm torque. Determine, for a point at the position of the greatest compressive bending stress (a) the normal stress on a plane inclined at 30° to the shaft axis and (b) the shear stress on a plane inclined at 60° to the shaft axis.

1.13 A fibre-reinforced pipe, 240 mm inside diameter and 300 mm outside diameter, has a density of 1 kg/m^3. The pipe is simply supported when carrying fluid of density 12 kg/m^3, as shown in Fig. 1.25. If the allowable axial and bending stresses in the pipe are 150 and 8 MPa respectively, check that these are nowhere exceeded.

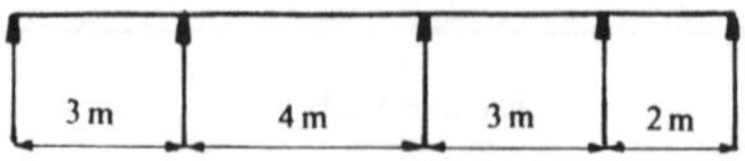

Figure 1.25

1.14 Calculate the diameter of a solid steel shaft required to transmit 89.5 kW at 200 rev/min if the angle of twist is not to exceed 0.13°/m. If the maximum bending stress is 155 MPa, calculate the principal stresses and the orientation of their planes on the compressive side of the shaft. What is the maximum shear stress induced in the shaft? Take G = 77.2 GPa.
Answer: 70 mm, 177.6 MPa, - 23.2 MPa, 20°, 110°, 100.4 MPa

1.15 The fixed end of an I-section cantilever is subjected to a bending moment of 190 kNm. What is the value of the vertical shear force at this position, given that the major principal tensile stress is 92.3 MPa at the web-flange interface? The flanges are each 152.5 mm wide × 50 mm deep and the web is 25 mm thick × 200 mm deep.
Answer: 312 kN

1.16 An I-section cantilever is 150 mm long. If the flanges are 50 mm wide × 5 mm thick and the web is 65 mm deep × 5 mm thick, what is the maximum uniformly distributed loading that the beam can carry when the bending moment is not to exceed 95 MPa? Determine, for the section at which the shear force is a maximum, the principal stresses and planes for (a) the flange top, (b) the web top and (c) at the neutral axis of bending.
Answer: 164.7 kN/m, 95 MPa, 114 MPa, - 31 MPa, 27.5°, ± 77 MPa at ± 45°

2D Strain Transformation

1.17 In plane strain, line elements in the x and y directions increased by 17% and 22% respectively and change their right angle by 7°. What is the percentage increase in an element of line originally at 45° to x and y and by how much has this line rotated?
Answer: 27%, 2.4°

1.18 A 25 mm diameter solid shaft is placed under torsion. A strain gauge records 800 × 10^{-6} when it is bonded to the shaft surface in the direction of the major principal strain. What will the change in this strain be when a 150 mm length of this bar is simultaneously subjected to an axial force causing a 0.25 mm length change and a 0.0125 mm diameter reduction?
Answer: 0.0013

1.19. Three strain gauges A, B and C, with A and C at 30° on either side of B, form a rosette which is bonded to a structural member made from an aluminium alloy having a modulus of rigidity G = 25 GN/m^2. When the structure is placed under load gauges A, B and C read 5 × 10^{-4}, 1 × 10^{-4} and - 4 × 10^{-4} respectively. Find the maximum shear stress in the material at the point of application of the rosette. (CEI)

1.20 A strain gauge rosette is bonded to the surface of a 20 mm thick plate. For the three gauges A, B and C, gauge B is at 60° to gauge A and gauge C is at 120° to gauge A, both measured in the same direction. When the plate is loaded, the strain values from A, B and C are 50μ, 700μ and 375μ respectively. Assuming that the stress and strain do not vary with thickness of plate, find the principal stresses and the change in plate thickness at the rosette point. Take E = 200 GPa and ν = 0.25. (CEI)

1.21 A strain gauge rosette is bonded to the cylindrical surface of a 40 mm drive shaft of a motor. There are three gauges A, B and C and the centre line of gauge A lies at 45° to the centre line of the shaft. The angles between gauges A and B, B and C are both 60°, all angles measured anti-clockwise. When the motor runs and transmits constant torque T, the readings of gauges A, B and C are found to be 625μ, - 312.5μ and - 312.5μ respectively. If E = 208 GN/m^2 and ν = 0.3, calculate T, the principal stresses and the maximum shear stress in the shaft. (CEI)

1.22 The right-angled cantilever steel pipe in Fig. 1.26a is 100 mm o.d. with 6 mm wall thickness. When forces P and W are applied at the free end in the directions shown, a 60° strain gauge rosette (see Fig. 1.26b) fixed to the centre point D of the horizontal portion gave the following strains: ε_a = - 275μ,

ε_b = 198μ and ε_c = - 208μ when gauge *a* is aligned with the pipe axis. Calculate *P* and *W*, given *E* = 200 GN/m^2, ν = 0.3. Plane strain transformations in Table 1.1 may be assumed. (CEI)

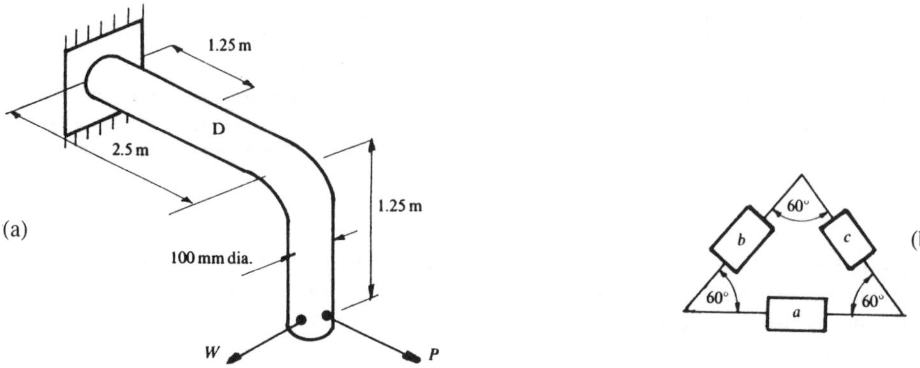

Figure 1.26

3 D Stress Transformation

1.23 The state of stress (MPa) at a point is given by σ_x = 13.78, σ_y = 26.2, σ_z = 41.34, τ_{xy} = 27.56, τ_{yz} = 6.89 and τ_{xz} = 10.34. Determine the stress resultants and the normal and shear stresses for a plane whose normal makes angles of 50° 70' and 60° with the directions of *x* and *y* respectively.
Answer (MPa): 23.45, 30.12, 29.67, 40.14 and 48.35

1.24 At a point in a material, the state of stress in MPa is given by the components σ_x = 12.8, σ_y = 27, σ_z = 51.3, τ_{xy} = 23.4, τ_{xz} = 11 and τ_{yz} = - 6.24. Determine the normal and shear stresses on a plane whose normal makes angles of 48° and 71° to the *x* and *y* axes respectively. Also find the direction of the shear stress relative to *x* and *y*.
Answer: 48.8 MPa, 12.31 MPa, 137.8°, 69.1°, 55.34°)

1.25 A stress resultant of 140 MPa makes angles of 43°,75° and 50°53' with the *x* and *y* axes. Determine the normal and shear stresses on an oblique plane whose normal makes respective angles of 67°13', 30° and 71°34' with these axes.
Answer: 98.9 MPa, 99.1 MPa

1.26 A resultant stress of 170 MPa is inclined at 23° and 72° to the *x* and *y* directions respectively. This resultant stress acts on a plane with normal direction cosines 0.656 and 0.259 relative to *x* and *y* respectively. Determine the normal and shear stresses on this plane. Given component shear stresses τ_{xy} = 35.5 MPa and τ_{xz} = - 47.86 MPa, find the normal stresses when invariant J_1 = 926.4 MPa.
Answer: 144.7, 89.9, 276.4, 497.5, 152.6 MPa

1.27 Find the invariants and principal stresses for the following stress components (MPa): σ_x = 9.68, σ_y = 14.32, σ_z = 17.28, τ_{xy} = 2.44, τ_{xz} = 10.21, τ_{yz} = 7.38. If τ_{xy} = τ_{yz} = 0, what will the principal stresses then be? Show that the principal stress directions are orthogonal.
Answer: (MPa) 14.28, 388.65, 640.15, 28.43, 10.7, 2.15; 24.38, 14.31, 2.59

1.28 Given that the partial state of stress at a point is τ_{xy} = 30, τ_{xz} = 10, τ_{yz} = 30, σ_x = 20 MPa, σ_y = 20 MPa and that J_1 = 50 MPa, determine the remaining normal stresses, the principal stresses and the maximum shear stress.
Answer: (MPa) 10, 65.22, 5.25, - 20.47, ± 42.87

1.29 The stress components at a point are $\sigma_x = 5$, $\sigma_y = 7$, $\sigma_z = 6$, $\tau_{xy} = 10$, $\tau_{xz} = 8$ and $\tau_{yz} = 12$ MPa. Find the magnitudes of the principal stresses and the maximum shear stress. What are the direction cosines for the normal to the major principal plane?
Answer: 26.2, - 2.37, - 5.83, 16.01 MPa, 0.516, 0.631, 0.579

1.30 Principal stresses of 77, 31 and - 46 MPa act in directions 1, 2 and 3. Determine the normal and shear stresses for a plane whose normal lies at 30° to the 3 - axis. The normal projection on the 1, 2 plane is inclined at 55° to the 1, 3 plane.
Answer (MPa) - 23.2, 41.4

1.31 Given the principal stresses $\sigma_1 = 7.5$, $\sigma_2 = 3.1$ and $\sigma_3 = 1.4$ MPa, find graphically and analytically the maximum shear stresses, the normal and octahedral shear stresses.
Answer: (MPa) 2.2, 3.05, 0.85, 4.0, 2.57

1.32 Given principal stresses $\sigma_1 = 130$, $\sigma_2 = 40$ and $\sigma_3 = 30$ MPa, determine the normal and shear stresses on a plane whose normal makes an angle of 60° with the 3 - axis. The projection of the normal in the 1, 2 plane is inclined at 45° to the 1 - axis.
Answer: (MPa) 71.25, 91.25

1.33 If $\sigma_1 = 6$ MPa, $\sigma_2 = 2$ MPa and $\tau_{max} = 2.5$ MPa, find the remaining principal stress and the resultant, normal and shear stresses on a plane whose normal is inclined at 45°, 60° and 60° to the 1, 2 and 3 principal stress directions.
Answer: (MPa) 1, 4.39, 3.75, 2.28

1.34 The major principal stress in a material is 6 MPa. If $J_1 = 4$ MPa and $J_3 = -48$ (MPa)^3 find (a) the remaining principal stresses, (b) the normal and shear stresses on a plane whose normal direction cosines are 0.6, 0.3 and 0.742, (c) the state of stress existing on the maximum shear and octahedral planes and (d) the unit vectors which define the planes in (c) relative to the principal directions.
Answer: 2, - 4; 0.138, 4.7; 5, 1, 4.11, 1.33, $\mathbf{u}_N = (1/\sqrt{2})(\mathbf{u}_1 + \mathbf{u}_3)$, $\mathbf{u}_N = (1/\sqrt{3})(\mathbf{u}_1 + \mathbf{u}_2 + \mathbf{u}_3)$

1.35 Find the principal stresses and directions for the components of the stress tensor (kPa): $\sigma_x = 2$, $\sigma_y = 2$, $\sigma_z = 1$, $\tau_{xy} = 2$, $\tau_{xz} = \tau_{yz} = 0$.
Answer: 4, 1, 0; 1/√2, 1/√2, 0; 0, 0, 1; - 1/√2, 1/√2, 0

1.36 Find the principal stresses in magnitude and direction and the invariants for the given stress matrix (MPa). Determine the direction cosines for the two greatest principal stresses. Show that their directions are orthogonal.
Answer: 4, 1,- 2, 3,- 6,- 8, √(2/3), 1/√6, 1/√6, 1/√3, - 1/√3, - 1/√3

$$\sigma_{ij} = \begin{bmatrix} 3 & 1 & 1 \\ 1 & 0 & 2 \\ 1 & 2 & 0 \end{bmatrix}$$

1.37 At a point in a stressed material, the state of stress in MPa is defined for the given matrix components. Find (a) the magnitude and direction of the normal and shear stresses acting on a plane whose direction cosines are 0.67 and 0.326 relative to x and y, (b) the principal stresses, (c) the maximum shear stress and (d) the normal and shear stresses acting on the octahedral plane. Sketch the planes on which (c) and (d) act with respect to the principal planes.
Answer: (a) 1.967, 2.245, - 0.441, 0.898, 0.0018; (b) 4, 2, - 1;
(c) 2.5; (d) 1.667, 2.055 MPa

$$\sigma_{ij} = \begin{bmatrix} 1 & 1 & -1 \\ 1 & 2 & 2 \\ -1 & 2 & 2 \end{bmatrix}$$

1.38 Complete the stress matrix (MPa) given that the first and second invariants are 6 and - 24 respectively. Find the principal stresses, then determine graphically the normal and shear stresses on the octahedral and maximum shear planes.
Answer: $\sigma_z = 0$, $\tau_{xy} = 2$; 8, 2, -4; 2, 4.8 ; 6, 2

$$\sigma_{ij} = \begin{bmatrix} 6 & ? & 2 \\ ? & 0 & 4 \\ 2 & 4 & ? \end{bmatrix}$$

1.39 At a point in a material the stress state (kN/m^2) is given by the following matrix of components. Determine the normal and shear stresses on a plane whose normal direction is given by the unit vector $\mathbf{u}_N = 0.67\mathbf{u}_x + 0.326\mathbf{u}_y + 0.668\mathbf{u}_z$. Also find the unit vector that describes the direction of the shear stress upon this plane.

$$\sigma_{ij} = \begin{bmatrix} 1860 & 3400 & 1600 \\ 3400 & 3920 & -905 \\ 1600 & -905 & 7445 \end{bmatrix}$$

Answer: 7097, 1768.7 kN/m^2, $\mathbf{u}_N = -0.753\mathbf{u}_x + 0.362\mathbf{u}_y + 0.573\mathbf{u}_z$

3 D Strain Transformation

1.40 Given a major principal strain of 600μ and strain invariants $I_1 = 400\mu$ and $I_2 = -4800\mu$, find the remaining principal strains. Find, in magnitude and direction, the octahedral normal and shear strains.
Answer: ($1\mu = 1 \times 10^{-6}$) 200μ, − 400μ, 133μ, 822μ

1.41 Determine graphically the normal and shear strains for a plane whose direction cosines are: $l = 0.732$, $m = 0.521$ and n (positive) with respect to the principal directions 1, 2 and 3, when the principal strains are 6800×10^{-6}, 1540×10^{-6} and 1130×10^{-6} respectively. Check analytically.
Answer: 1264×10^{-6}, 662×10^{-6}

1.42 Determine the principal strains and the corresponding transformation matrix **L** for the principal directions relative to orthogonal axes, in which the micro-strain components are $\varepsilon_{11} = 10$, $\varepsilon_{22} = 10$, $\varepsilon_{33} = 60$, $\varepsilon_{12} = 30$, $\varepsilon_{13} = -20$, $\varepsilon_{23} = -20$.
Answer: μstrain $\varepsilon_1 = 80$, $\varepsilon_2 = 20$, $\varepsilon_3 = -20$; direction cosines $l_{11} = 1/\sqrt{2}$, $l_{12} = -1/\sqrt{2}$, $l_{13} = 0$, $l_{21} = 1/\sqrt{3}$, $l_{22} = 1/\sqrt{3}$, $l_{23} = 1/\sqrt{3}$, $l_{31} = -1/\sqrt{6}$, $l_{32} = -1/\sqrt{6}$, $l_{33} = 2/\sqrt{6}$

1.43 Determine the principal strains and the deviatoric strain tensor in respect of the absolute micro-strain tensor components $\varepsilon_x = 400$, $\varepsilon_y = 700$, $\varepsilon_z = 400$, $\varepsilon_{yz} = 200$, $\varepsilon_{xy} = \varepsilon_{xz} = 0$.
Answer: $\varepsilon_1 = 800$, $\varepsilon_2 = 400$, $\varepsilon_3 = 300$; $\varepsilon_x' = -100$, $\varepsilon_y' = 200$, $\varepsilon_z' = -100$, $\varepsilon_{yz}' = 200$, $\varepsilon_{xy}' = \varepsilon_{xz}' = 0$

1.44 The strain state at a point is given by the following matrix of its components. Determine the normal strain in the direction defined by a unit vector $\mathbf{u} = (1/2)\mathbf{u}_1 - (1/2)\mathbf{u}_2 + (1/\sqrt{2})\mathbf{u}_3$ and the shear strain between this direction and a perpendicular direction $\mathbf{u} = -(1/2)\mathbf{u}_1 + (1/2)\mathbf{u}_2 + (1/\sqrt{2})\mathbf{u}_3$. Comment on the result.

$$\varepsilon_{ij} = \begin{bmatrix} 1 & -3 & \sqrt{2} \\ -3 & 1 & -\sqrt{2} \\ \sqrt{2} & -\sqrt{2} & 4 \end{bmatrix} \times 10^{-4}$$

Answer: 600×10^{-6}, 0

1.45 The given micro-strain tensor applies to a point in an elastic solid. Find (a) the principal strains, (b) the magnitudes of the maximum and octahedral shear strains and (c) the normal strain for a direction defined by the unit vector $\mathbf{u}_N = 0.53\mathbf{u}_x + 0.35\mathbf{u}_y + 0.77\mathbf{u}_z$.

$$\varepsilon_{ij} = \begin{bmatrix} 600 & 200 & 200 \\ 200 & 0 & 400 \\ 200 & 400 & 0 \end{bmatrix} \times 10^{-6}$$

Answer: 800, 200, − 400; 600, 9.8; 622 ($\times 10^{-6}$)

1.46 Transform the given micro-strain tensor from reference axes x_1, x_2 and x_3 to axes $x_{1'}$, $x_{2'}$ and $x_{3'}$ given the respective vector equations for directions $x_{1'}$ and $x_{2'}$ as: $\mathbf{A}_{1'} = \mathbf{u}_1 + 2\mathbf{u}_2 + 3\mathbf{u}_3$ and $\mathbf{A}_{2'} = \mathbf{u}_1 + \mathbf{u}_2 - \mathbf{u}_3$.

$$\varepsilon_{ij} = \begin{bmatrix} 100 & 500 & -500 \\ 500 & 0 & 0 \\ -500 & 0 & -100 \end{bmatrix} \times 10^{-6}$$

Answer $\varepsilon_{1'1'} = -128.6$, $\varepsilon_{2'2'} = 666.7$, $\varepsilon_{3'3'} = 538.1$, $\varepsilon_{1'2'} = 138.9$, $\varepsilon_{1'3'} = 198$, $\varepsilon_{2'3'} = -276.2$ ($\times 10^{-6}$)

CHAPTER 2

PLANE ELASTICITY THEORY

2.1 Elastic Constants

Four elastic constants can be defined when isotropic materials are stressed elastically. The four constants apply to both metallic and non-metallic materials provided the stresses produce reversible, proportional strains. Because elastic strains are normally small, three different elastic moduli may be calculated from the ratio between engineering stress and strain depending upon the manner of the applied loading. These moduli define the stiffness of a given material structure and are related to the strength of the interatomic forces bonding that material. Since these forces are controlled by interatomic spacing, the elastic modulus of a material is relatively insensitive to changes in its microstructure arising from alloying and heat treatment. However, the moduli will fall with increasing temperature as the forces between atoms decrease with the increase in their spacing due to thermal expansion.

2.1.1 The Modulus of Elasticity (Young's Modulus)

Thomas Young (1773-1829) defined his elastic modulus E under uniaxial stress (see Fig. 2.1). Most metallic materials are linear in their load-displacement relationship. That is, when the axial load W is elastic, the axial displacement, x, will fully recover upon unloading.

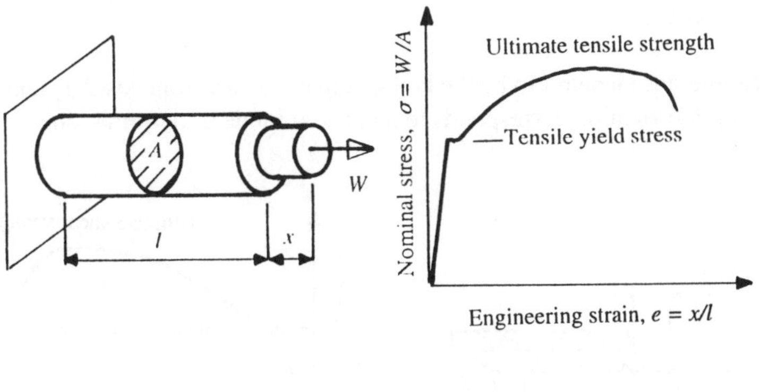

Figure 2.1 Uniaxial stress

For a bar of length l and section area A, it follows that the ratio between the engineering stress, $\sigma = W/A$, and strain, $\varepsilon = x/l$, is a constant for a particular material. This ratio is the modulus of elasticity and is defined as

$$E = \sigma / \varepsilon = (W\,l) / (Ax) \tag{2.1}$$

2.1.2 Poisson's Ratio

Simon-Denis Poisson (1781-1840) identified a further elastic constant ν under the uniaxial stress state in Fig. 2.1. Poisson's ratio refers to the constant ratio between either of the lateral strains, ε_2 or ε_3, and the axial strain ε_1. Poisson's ratio is a therefore a dimensionless elastic constant and normally lies in the range ¼ to ⅓ for metallic materials. To make the ratio positive, a minus sign must accompany the lateral strain within the definition. The sign of the lateral strain is always opposite to that of the axial strain. Each of the lateral and axial strains is a diagonal component of the infinitesimal strain tensor, ε_{ij}. Thus, in Fig. 2.1, if x is the increase in length l under tension and y is the contraction in the width dimension (or diameter) w, then

$$\nu = -\varepsilon_2/\varepsilon_1 = -(l\,y)/(wx) \tag{2.2}$$

Equation (2.2) is also valid for compression when x and y are respectively the decrease in length and the increase in width. It follows that the two lateral strains may also be found from E for a given uniaxial stress σ. That is, from combining eqs(2.1) and (2.2),

$$\varepsilon_2 = \varepsilon_3 = -\nu\varepsilon_1 = -\nu\sigma/E \tag{2.3}$$

2.1.3 The Modulus of Rigidity (Shear Modulus)

In the elastic region, the shear displacement x (see Fig. 2.2) increases linearly with the tangential shear force F. The distortion produced is referred to the angular change in the right angle ϕ rad. The modulus of rigidity, G, is identified with the constant ratio between the shear stress $\tau = F/A$ and the shear strain $\gamma = \tan\phi$. Provided ϕ is small we may write $\tan\phi \approx x/l$ (rad) and hence G becomes

$$G = \tau/\gamma = (Fl)/(Ax) \tag{2.4}$$

In taking the reference length l to be the block height, we see from eqs(2.1) and (2.4) how a shear mode of deformation corresponds to a uniaxial mode of deformation.

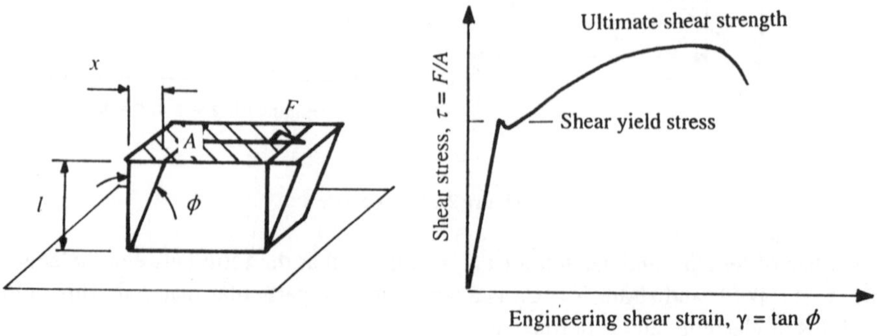

Figure 2.2 Shear deformation

2.1.4 Bulk Modulus

Stress and strain are also linearly related when deformation is produced by hydrostatic stress. The bulk modulus, K, defines the ratio between the mean or hydrostatic stress, σ_m, and volumetric strain, $\delta V/V$. That is,

$$K = \sigma_m /(\delta V/V) \tag{2.5a}$$

Hydrostatic stress arises directly when a solid is subjected to mutually perpendicular equal stresses, typical of fluid pressure (see Fig. 2.3).

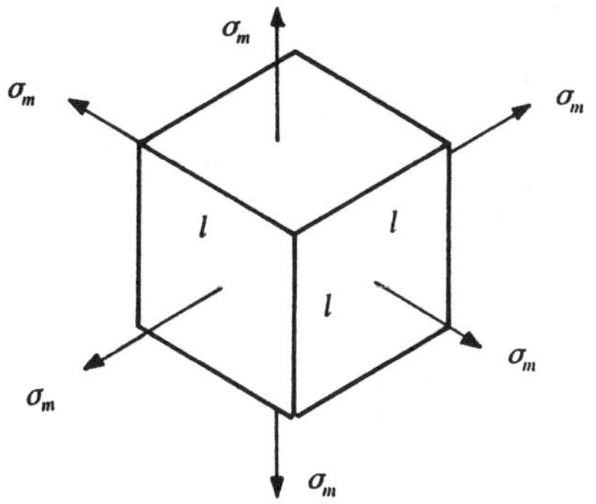

Figure 2.3 Hydrostatic stress

Equation (2.5a) remains positive irrespective of whether σ_m is tensile or compressive because δV will change sign accordingly. Assuming hydrostatic tension of a unit cube (l = 1), the increase in length of each side is $x = \varepsilon \times l = \varepsilon$, and the volume change is δV = strained volume - initial volume. This gives, ignoring ε^2 and ε^3 terms,

$$\delta V/V = (1 + \varepsilon)^3 - 1 \approx 3\varepsilon$$

Then, from eq(2.5a),

$$K = \sigma_m / (3\varepsilon) \tag{2.5b}$$

When the normal stresses are unequal the mean stress is defined from eqs(1.21a) as

$$\sigma_m = ⅓\, \sigma_{kk} = ⅓\, (\sigma_1 + \sigma_2 + \sigma_3) \tag{2.6a}$$

which causes elastic compressibility only. Note that the remaining deviatoric stress components (see Fig. 1.11c) $\sigma_1' = \sigma_1 - \sigma_m$, $\sigma_2' = \sigma_2 - \sigma_m$ and $\sigma_3' = \sigma_3 - \sigma_m$ are responsible for distortion without volume change. With associated length changes (strains), ε_1, ε_2 and ε_3, in the sides of a unit cube (V = 1),

$$\delta V/V = (1 + \varepsilon_1)(1 + \varepsilon_2)(1 + \varepsilon_3) - 1$$
$$\delta V/V \approx \varepsilon_1 + \varepsilon_2 + \varepsilon_3 = \varepsilon_{kk} \tag{2.6b}$$

Using the summation convention, eq(2.5a) becomes

$$K = \tfrac{1}{3}(\sigma_1 + \sigma_2 + \sigma_3) / (\varepsilon_1 + \varepsilon_2 + \varepsilon_3)$$
$$= \tfrac{1}{3}\sigma_{kk} / \varepsilon_{kk} = \sigma_m / (3\varepsilon_m) \qquad (2.6c)$$

Typical room-temperature elastic constants for engineering materials are given in Table 2.1.

Table 2.1 Elastic Constants at 20°C

Material	E (GPa)	G (GPa)	ν	K (GPa)
Aluminium	70.3	26.1	0.345	75.6
Al-Cu Alloy	75	28.5	0.31	65.8
Brass	103.5	39	0.33	101.5
Bronze	117	44.8	0.31	102.6
Carbon Steel	207	81	0.28	157
Cast Iron	113.4	45	0.26	50.3
Chromium	279	115.5	0.21	160.3
Copper	115.2	43	0.34	107.5
Iron	211	82	0.29	167.5
Nickel	200	76	0.31	175.4
Stainless (18/8)	200	77	0.30	167
Ni-Cr Steel	206	82	0.26	143
Titanium	112.6	42	0.34	104
Tungsten	400	157	0.27	290
Nimonics	210	80	0.31	184
Concrete	10	4.2	0.20	13.5
Glass	72.1	29.3	0.23	58
Quartz	73	31	0.17	37
Tungsten Carbide	534.4	219	0.22	318.1
Timber	4	1.3	0.50	_
Rubber	0.003	0.001	0.50	_

2.2 Relationship Between Elastic Constants

Engineers have now become used to working with four elastic constants E, G, K and ν, but they are not independent constants. Gabriel Lamé (1795-1870) first showed that, provided a solid is isotropic, three independent elastic constants were sufficient to describe all possible modes of deformation. Thus, we should expect a relationship to exist between any three of our four engineering constants.

2.2.1 E, G and ν

This relationship is established from the stress state under pure shear. Figure 2.4a shows that shear stresses, τ_{AD} and τ_{BC}, resulting from the shear force on opposite faces, cannot exist without complementary shear upon each adjacent face.

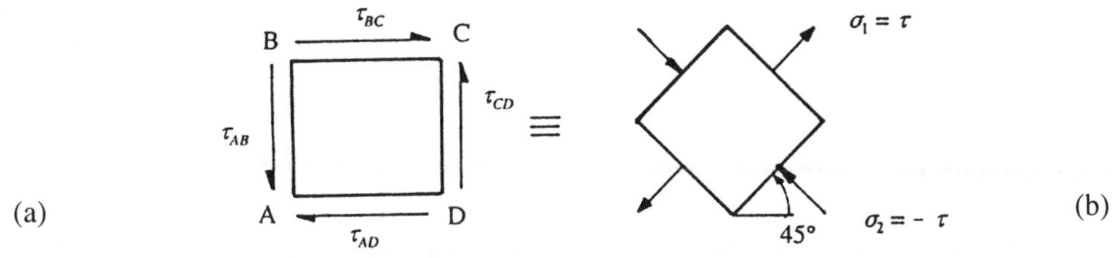

Figure 2.4 Analysis of stress and forces under pure shear

The complementary shear stresses, τ_{AB} and τ_{CD}, are found from taking moments about point A. For a thickness, t, moment equilibrium gives

$$\tau_{BC} \times (\mathrm{BC} \times t) \times \mathrm{AB} = \tau_{CD} \times (\mathrm{CD} \times t) \times AD$$
$$\therefore \; \tau_{CD} = \tau_{BC}$$

and since $\tau_{AD} = \tau_{BC}$ then $\tau_{AD} = \tau_{CD}$. Similarly, taking moments about C gives $\tau_{AB} = \tau_{AD}$. This shows that all shear stresses are the same, which will be simply labelled τ hereafter. The corresponding shear strain is $\gamma = \tau / G$. To determine the states of stress and strain following a rotation in the axes, we apply eqs(1.35) and (1.40) to give

$$\begin{bmatrix} \sigma_{1'1'} & \sigma_{1'2'} \\ \sigma_{2'1'} & \sigma_{2'2'} \end{bmatrix} = \begin{bmatrix} \cos\theta & \sin\theta \\ -\sin\theta & \cos\theta \end{bmatrix} \begin{bmatrix} 0 & \tau \\ \tau & 0 \end{bmatrix} \begin{bmatrix} \cos\theta & -\sin\theta \\ \sin\theta & \cos\theta \end{bmatrix} \tag{2.7a}$$

$$\begin{bmatrix} \varepsilon_{1'1'} & \varepsilon_{1'2'} \\ \varepsilon_{2'1'} & \varepsilon_{2'2'} \end{bmatrix} = \begin{bmatrix} \cos\theta & \sin\theta \\ -\sin\theta & \cos\theta \end{bmatrix} \begin{bmatrix} 0 & \gamma/2 \\ \gamma/2 & 0 \end{bmatrix} \begin{bmatrix} \cos\theta & -\sin\theta \\ \sin\theta & \cos\theta \end{bmatrix} \tag{2.7b}$$

where the primed axes define the rotation shown in Fig. 1.15. In particular, when $\theta = 45°$, eq(2.7a,b) yields coincident principal stress and strain systems (see Fig. 2.4b)

$$\begin{bmatrix} \sigma_1 & 0 \\ 0 & \sigma_2 \end{bmatrix} = \begin{bmatrix} \tau & 0 \\ 0 & -\tau \end{bmatrix} \quad \text{also} \quad \begin{bmatrix} \varepsilon_1 & 0 \\ 0 & \varepsilon_2 \end{bmatrix} = \begin{bmatrix} \gamma/2 & 0 \\ 0 & -\gamma/2 \end{bmatrix} \tag{2.8a,b}$$

Thus pure shear is equivalent to a major principal tensile stress σ_1 and a minor principal compressive stress σ_2, each having a magnitude equal to τ. Correspondingly, one principal strain ε_1 is tensile and the other ε_2 is compressive, each with magnitude $\gamma/2$. Both stresses contribute to the strain in each principal direction. One gives a direct strain σ/E and the other a lateral strain $-\nu\sigma/E$:

$$\varepsilon_1 = (1/E)(\sigma_1 - \nu\sigma_2) \tag{2.9a}$$
$$\varepsilon_2 = (1/E)(\sigma_2 - \nu\sigma_1) \tag{2.9b}$$

Setting $\sigma_1 = \tau$, $\sigma_2 = -\tau$, with $\varepsilon_1 = \gamma/2$ and $\varepsilon_2 = -\gamma/2$ reduces both eqs(2.9a,b) to

$$\gamma/2 = (1 + \nu)\tau/E \tag{2.9c}$$

and, since $G = \tau/\gamma$ from eq(2.4), it follows from eq(2.9c) that

$$E = 2G(1 + \nu) \tag{2.10}$$

2.2.2 E, K and ν

Consider the total strain ε, for any one direction under hydrostatic stressing in Fig. 2.3. This mean strain is the sum of three contributions: the direct tensile strain, σ_m/E, and two lateral compressive strains produced by the remaining stresses, which are each $-\nu\sigma_m/E$, from eq(2.3). The total strain is

$$\varepsilon = \sigma_m/E - \nu\,\sigma_m/E - \nu\,\sigma_m/E = (\sigma_m/E)(1 - 2\nu) \tag{2.11}$$

Substituting eq(2.11) into eq(2.6c) for when $\varepsilon = \varepsilon_1 = \varepsilon_2 = \varepsilon_3$

$$K = \sigma_m/(3\varepsilon_m) = \tfrac{1}{3}\,\sigma_m/[(\sigma_m/E)(1 - 2\nu)]$$

from which,

$$E = 3K(1 - 2\nu) \tag{2.12a}$$

To connect the mean stress and strain we combine eqs(2.6c) and (2.12a)

$$\varepsilon_{kk} = (1 - 2\nu)\sigma_{kk}/E \tag{2.12b}$$

2.2.3 Relationships Between G, K, E and G, K, ν

Further relationships between G, K and ν are found by eliminating E and ν between eqs(2.10 and 2.12a). These are

$$G = 3KE/(9K - E) = 3K(1 - 2\nu)/[2(1 + \nu)] \tag{2.13a}$$

From eq(2.6c) and (2.13a), the mean stress and strains become connected through G and ν

$$\varepsilon_m = \sigma_m(1 - 2\nu)/[2G(1 + \nu)] \tag{2.13b}$$

where $\varepsilon_m = \tfrac{1}{3}\varepsilon_{kk}$ and $\sigma_m = \tfrac{1}{3}\sigma_{kk}$

2.3 Cartesian Plane Stress and Plane Strain

Many problems which are plane in nature may be analysed using a reduced form of the general theory of elasticity. Specifically, in Cartesian co-ordinates x,y,z, plane stress refers to a body with small thickness in the z-direction. For the thin plate in Fig. 2.5a, the stress in the plane of the plate is two dimensional, while the stress through the thickness (σ_z) is zero. Plane strain, on the other hand, refers to a body with a large dimension in the z - axial direction (see Fig. 2.5b) and this gives either zero axial strain if the ends are constrained or some constant value (say ε_o).

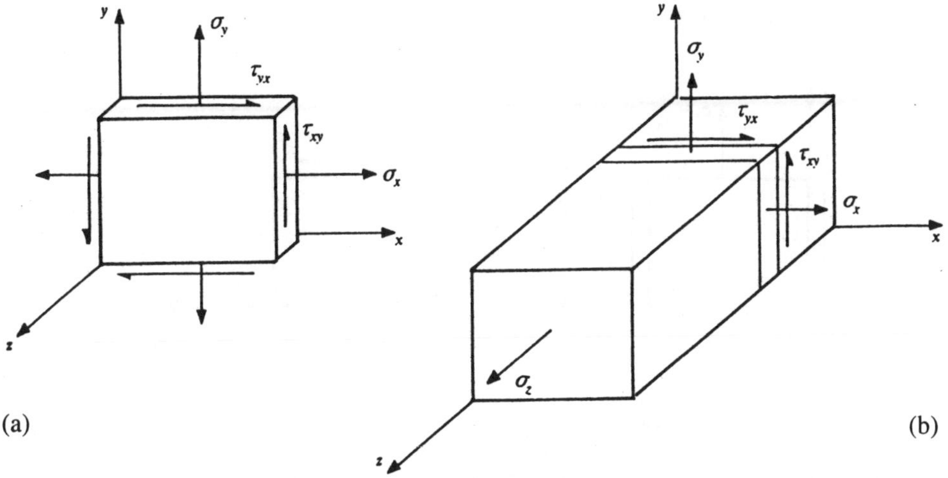

Figure 2.5 Plane stress and plane strain in Cartesian co-ordinates

2.3.1 Stress-Strain (Constitutive) Relations

The non-zero stresses σ_x, σ_y and τ_{xy} may be functions of x and y but not z. In *plane stress* $\sigma_z = 0$ and the strains become

$$\varepsilon_x = (\sigma_x - \nu\sigma_y)/E \quad (2.14a)$$
$$\varepsilon_y = (\sigma_y - \nu\sigma_x)/E \quad (2.14b)$$
$$\varepsilon_z = -\nu(\sigma_x + \sigma_y)/E \quad (2.14c)$$
$$\gamma_{xy} = \tau_{xy} / G = 2(1 + \nu)\tau_{xy}/E \quad (2.14d)$$

where the through-thickness strain ε_z is not zero. The *plane strain* constitutive relations are

$$\varepsilon_x = [\sigma_x - \nu(\sigma_y + \sigma_z)]/E \quad (2.15a)$$
$$\varepsilon_y = [\sigma_y - \nu(\sigma_x + \sigma_z)]/E \quad (2.15b)$$
$$\varepsilon_z = [\sigma_z - \nu(\sigma_x + \sigma_y)]/E \quad (2.15c)$$
$$\gamma_{xy} = \tau_{xy}/G = 2(1 + \nu)\tau_{xy}/E \quad (2.15d)$$

Since $\varepsilon_z = \varepsilon_o$ is constant, then from eq(2.15c), $\sigma_z = \varepsilon_o E + \nu(\sigma_x + \sigma_y)$. It follows from this that plane sections will remain plane when the stress sum $(\sigma_x + \sigma_y)$ is a constant. Moreover, since σ_x, σ_y and τ_{xy} are each functions of x and y only, the following analyses of equilibrium and compatibility will apply to both plane stress and plane strain. The theory of equilibrium and compatibility, appropriate to plane polar co-ordinates, will be derived later from transforming the Cartesian co-ordinate relationships.

2.3.2 Equilibrium

Let the stresses vary over the elemental dimensions δx and δy in the manner illustrated in Figs 2.6a and b. These act together, as shown in Fig. 2.5a,b, but, for clarity, the x and y forces are separated. Body forces, due to self-weight, are ignored.

(a) 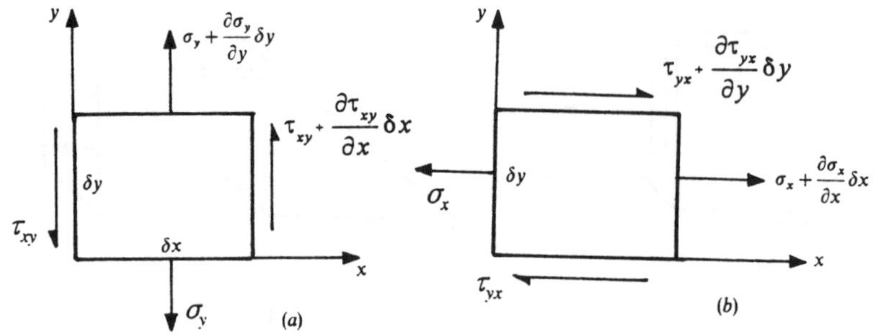 (b)

Figure 2.6 Stress variation in the x and y directions

Assuming unit thickness in Fig. 2.6a, vertical force equilibrium gives

$$[\sigma_y + (\partial \sigma_y /\partial y)\delta y - \sigma_y]\delta x + [\tau_{xy} + (\partial \tau_{xy} /\partial x)\delta x - \tau_{xy}]\delta y = 0$$
$$\partial \sigma_y /\partial y + \partial \tau_{xy} /\partial x = 0 \tag{2.16a}$$

and, from Fig. 2.6b, for horizontal force equilibrium

$$[\sigma_x + (\partial \sigma_x /\partial x)\delta x - \sigma_x]\delta y + [\tau_{yx} + (\partial \tau_{yx} /\partial y)\delta y - \tau_{yx}]\delta x = 0$$
$$\partial \sigma_x /\partial x + \partial \tau_{yx} /\partial y = 0 \tag{2.16b}$$

Equations(2.16a,b) are plane equilibrium equations. They apply to plane stress states in which there are x and y variations in the stress components from point to point.

2.3.3 Strain-Displacement Relations

In Chapter 1 it was shown that initially unstrained perpendicular line elements δx and δy stretch and rotate whilst translating into a new position under stress (see Fig. 1.12a). For plane deformation the u and v displacements defining the shift in the origin from O to O' are different functions of the x, y co-ordinates. Plane sections remain plane when all points in the body displace only within the x, y plane, i.e. u and v do not depend upon z. The displacement gradients in eqs(1.23a,b) and (1.24a), found from the partial differentiation of u and v with respect to x and y, define the normal and shear strain components. In confirmation of this, the normal strain components identify with the corresponding length changes in δx and δy

$$\varepsilon_x = \{[\delta x + (\partial u /\partial x)\delta x] - \delta x\}/\delta x = \partial u/\partial x \tag{2.17a}$$
$$\varepsilon_y = \{[\delta y + (\partial v /\partial y)\delta y] - \delta y\}/\delta y = \partial v/\partial y \tag{2.17b}$$

The small angular distortions are approximated by

$$e_{xy} = [(\partial u / \partial y)\delta y]/\delta y = \partial u/\partial y$$
$$e_{yx} = [(\partial v / \partial x)\delta x]/\delta x = \partial v/\partial x$$

Their sum gives the net change in the right angle, which, when expressed in radians, defines the engineering shear strain

$$\gamma_{xy} = \partial u/\partial y + \partial v/\partial x \tag{2.17c}$$

2.3.4 Compatibility

It is seen from eqs(2.17a,b and c) that the three strains ε_x, ε_y and γ_{xy} depend upon the two displacements u and v. The relationship between the strains expresses a condition of compatibility. This condition ensures that the material deforms as a continuum without a discontinuity arising in the strain distribution. A mathematical expression for strain compatibility is found by eliminating u and v between eqs(2.17 a-c).

$$\partial^2 \varepsilon_x /\partial y^2 + \partial^2 \varepsilon_y /\partial x^2 = \partial^2 \gamma_{xy} / \partial x\, \partial y \tag{2.18}$$

Equation (2.18) may further be written in terms of the stress components. With plane stress, for example, substituting from eqs(2.14a,b,d) leads to

$$\partial^2 (\sigma_x - \nu\sigma_y)/ \partial y^2 + \partial^2 (\sigma_y - \nu\, \sigma_x)/ \partial x^2 = 2(1 + \nu)\, \partial^2 \tau_{xy}/\partial x\, \partial y \tag{2.19a}$$

The corresponding plane strain relationship is, from eqs(2.15a,b,d)

$$\partial^2 [\sigma_x (1 - \nu^2) - \nu\, \sigma_y (1 + \nu) - \nu\, \varepsilon_o E]/ \partial y^2$$
$$+ \partial^2 [\sigma_y (1 - \nu^2) - \nu\, \sigma_x (1 + \nu) - \nu\, \varepsilon_o E]/ \partial x^2 = 2(1 + \nu)\, \partial^2 \tau_{xy} /\partial x\, \partial y \tag{2.19b}$$

in which the following substitution for σ_z has been made from eq(2.15c):

$$\sigma_z = E\varepsilon_o + \nu\, (\sigma_x + \sigma_y)$$

2.3.5 The Biharmonic Equation and Stress Components

The requirements that both equilibrium and compatibility are satisfied in plane stress and plane strain are met by combining eqs(2.16a,b) and (2.19a,b). For plane stress eqs(2.16a and 2.19a) lead to

$$\partial^2 \sigma_x / \partial y^2 - \nu\, \partial^2 \sigma_y / \partial y^2 + \partial^2 \sigma_y /\partial x^2 - \nu\, \partial^2 \sigma_x / \partial x^2 = - 2\, (1 + \nu)\, \partial^2 \sigma_y /\partial y^2 \tag{2.20a}$$

and, from eqs(2.16b) and (2.19a),

$$\partial^2 \sigma_x / \partial y^2 - \nu\, \partial^2 \sigma_y / \partial y^2 + \partial^2 \sigma_y /\partial x^2 - \nu\, \partial^2 \sigma_x / \partial x^2 = - 2(1 + \nu)\, \partial^2 \sigma_x / \partial x^2 \tag{2.20b}$$

Subtracting eq(2.20a) from (2.20b),

$$- 2\, (1 + \nu)\, \partial^2 \sigma_x / \partial x^2 + 2(1 + \nu)\, \partial^2 \sigma_y / \partial y^2 = 0$$
$$\partial^2 \sigma_y /\partial y^2 - \partial^2 \sigma_x / \partial x^2 = 0 \tag{2.21a}$$

Adding eqs(2.20a and b),

$$2[\partial^2\sigma_x/\partial y^2 - \nu\,\partial^2\sigma_y/\partial y^2 + \partial^2\sigma_y/\partial x^2 - \nu\,\partial^2\sigma_x/\partial x^2] = -2(1+\nu)(\partial^2\sigma_y/\partial y^2 + \partial^2\sigma_x/\partial x^2)$$
$$\partial^2\sigma_x/\partial x^2 + \partial^2\sigma_x/\partial y^2 + \partial^2\sigma_y/\partial y^2 + \partial^2\sigma_y/\partial x^2 = 0$$
$$(\partial^2/\partial x^2 + \partial^2/\partial y^2)(\sigma_x + \sigma_y) = 0 \qquad (2.21b)$$

It is seen that the equilibrium eq(2.21a) is satisfied when σ_x and σ_y are defined from a stress function $\phi = \phi(x, y)$ (G.B.Airy 1862) in the following manner

$$\sigma_x = \partial^2\phi/\partial y^2 \text{ and } \sigma_y = \partial^2\phi/\partial x^2 \qquad (2.22a,b)$$

and, from either eq(2.16a) or eq(2.16b), the shear stress is given by

$$\partial\tau_{xy}/\partial x = -\partial\sigma_y/\partial y = -\partial^3\phi/\partial x^2\partial y$$
$$\tau_{xy} = -\partial^2\phi/\partial x\,\partial y \qquad (2.22c)$$

The result of combining eq(2.21b) with eqs(2.22a and b) may be written as a biharmonic equation in any one of the following three forms:

$$(\partial^2/\partial x^2 + \partial^2/\partial y^2)(\partial^2\phi/\partial y^2 + \partial^2\phi/\partial x^2) = 0 \qquad (2.23a)$$
$$(\partial^2/\partial x^2 + \partial^2/\partial y^2)^2\phi = 0 \qquad (2.23b)$$
$$\nabla^2(\nabla^2\phi) = \nabla^4\phi = 0 \qquad (2.23c)$$

where 'del-squared' is defined as $\nabla^2 = \partial^2/\partial x^2 + \partial^2/\partial y^2$. The reader should check that identical eqs(2.22) and (2.23) apply to plane strain when eqs(2.16a,b) are combined with eq(2.19b).

2.3.6 Body Forces

The biharmonic equation $\nabla^4\phi = 0$ applies to plane stress and plane strain only in the absence of body forces. When the latter are present, e.g. in the form of components X and Y of self weight, the equilibrium equations (2.16a,b) are modified to

$$\partial\sigma_y/\partial y + \partial\tau_{xy}/\partial x + Y = 0 \qquad (2.24a)$$
$$\partial\sigma_x/\partial x + \partial\tau_{xy}/\partial y + X = 0 \qquad (2.24b)$$

where X and Y are the Cartesian body force components/unit volume. Individual solutions may be derived by combining eqs(2.24a,b) with the appropriate compatibility condition. A closed solution is possible in which the body forces are derivatives of a potential function Ω (x, y)

$$X = -\partial\Omega/\partial x,\ Y = -\partial\Omega/\partial y \qquad (2.25a,b)$$

Combining eqs(2.24a,b) and (2.25a,b)

$$\partial(\sigma_y - \Omega)/\partial y + \partial\tau_{xy}/\partial x = 0 \qquad (2.26a)$$
$$\partial(\sigma_x - \Omega)/\partial x + \partial\tau_{xy}/\partial y = 0 \qquad (2.26b)$$

A stress function $\phi(x, y)$ will satisfy eqs(2.26a,b) as follows:

$$\sigma_x = \Omega + \partial^2\phi/\partial y^2,\ \sigma_y = \Omega + \partial^2\phi/\partial x^2,\ \tau_{xy} = -\,\partial^2\phi/\partial x\,\partial y \qquad (2.27a,b)$$

In *plane strain* the compatibility condition (2.19b) remains unaltered. Combining this with eqs(2.26a,b) and noting that ε_o is either zero or a constant leads to

$$\partial^4\phi/\partial x^4 + 2\,\partial^4\phi/\partial x^2\partial y^2 + \partial^4\phi/\partial y^4 + [(1-2\nu)/(1+\nu)](\partial^2\Omega/\partial x^2 + \partial^2\Omega/\partial y^2) = 0 \quad (2.28a)$$

$$\nabla^4\phi + [(1-2\nu)/(1+\nu)]\nabla^2\Omega = 0 \qquad (2.28b)$$

Functions $\phi(x, y)$ and $\Omega(x, y)$ which satisfy eq(2.28b) will provide stress components (2.27a,b) which must also be made to match the boundary conditions.

Strictly, when body forces are present in *plane stress*, additional compatibility conditions accompany eq(2.19a). These arise from the dependence of the through-thickness strain ε_z upon the in-plane stresses

$$\partial^2\varepsilon_z/\partial x^2 = \partial^2\varepsilon_z/\partial y^2 = \partial^2\varepsilon_z/\partial x\,\partial y = 0 \qquad (2.29a)$$

which shows that ε_z should obey an equation of the form

$$\varepsilon_z = a_1 + a_2 x + a_3 y \qquad (2.29b)$$

where a_1, a_2 and a_3 are constants. Note, however, that eq(2.29b) will not be upheld when eqs(2.19a) and (2.26a,b) are combined to give a single equation to be satisfied by ϕ and Ω in plane stress

$$\partial^4\phi/\partial x^4 + 2\,\partial^4\phi/\partial x^2\partial y^2 + \partial^4\phi/\partial y^4 + (1-\nu)(\partial^2\Omega/\partial x^2 + \partial^2\Omega/\partial y^2) = 0 \qquad (2.30a)$$

$$\nabla^4\phi + (1-\nu)\nabla^2\Omega = 0 \qquad (2.30b)$$

This approximate solution is acceptable for thin plates. Otherwise, eq(2.29a) requires three additional equations which ϕ and Ω should satisfy:

$$2\,\partial^2\Omega/\partial x^2 + \partial^2(\nabla^2\phi)/\partial x^2 = 0$$
$$2\,\partial^2\Omega/\partial y^2 + \partial^2(\nabla^2\phi)/\partial y^2 = 0$$
$$2\,\partial^2\Omega/\partial x\,\partial y + \partial^2(\nabla^2\phi)/\partial x\,\partial y = 0$$

In the absence of body forces ($X = Y = 0$) we see again from eqs(2.28b) and (2.30b) that both plane stress and plane strain simplify to $\nabla^4\phi = 0$.

2.4 Cartesian Stress Functions

If a function $\phi(x, y)$ can be found that satisfies $\nabla^4\phi = 0$, it will satisfy both equilibrium and compatibility in plane stress and plane strain. The stresses are found from $\phi(x, y)$ using Airy's eqs(2.22a,b,c). Constants appearing in the function ϕ must meet a final requirement to satisfy the boundary conditions for a particular problem. That is, the internal stress distribution, which will in general vary throughout the volume of the body, must be made to contain the known forces, moments and torques applied to its boundary.

2.4.1 Polynomial Functions

The stress functions of most common use in plate and beam problems are taken from the polynomial function

$$\phi(x, y) = ax^2 + bxy + cy^2 + dx^3 + ex^2y + fxy^2 + gy^3 + hx^4 + ix^3y + jx^2y^2 + kxy^3 + ly^4 + ...$$

Linear and constant terms are excluded since these disappear with the second derivative. The following examples illustrate the usefulness of certain combinations of these terms.

Example 2.1 Show that the plane displacement system $u = ax^2y^2$, $v = byx^3$ is compatible.

The following strain components are found from eqs(2.17a-c):

$$\varepsilon_x = \partial u/\partial x = 2axy^2$$
$$\varepsilon_y = \partial v/\partial y = bx^3$$
$$\gamma_{xy} = \partial u/\partial y + \partial v/\partial x = 2ax^2y + 3byx^2$$

Applying the left- and right-hand sides of eq(2.18) separately, we see that

$$\partial^2\varepsilon_x/\partial y^2 + \partial^2\varepsilon_y/\partial x^2 = \partial(4\,axy)/\partial y + \partial(3bx^2)/\partial x = 4ax + 6bx$$
$$\partial^2\gamma_{xy}/\partial x\,\partial y = \partial(4axy + 6bxy)/\partial y = 4ax + 6bx$$

Example 2.2 Derive the displacements corresponding to the following stresses in plane strain $\sigma_x = c(y^2 + 2x)$, $\sigma_y = -cx^2$ and $\tau_{xy} = -2cy$ when $\varepsilon_o = 0$. Show that the corresponding strains are compatible.

Clearly, the given stress expressions satisfy the equilibrium equations(2.16a,b). Since $\varepsilon_o = 0$, we have, from eq(2.15c), $\sigma_z = \nu(\sigma_x + \sigma_y)$. Then, from eqs(2.15a,b,d),

$$\varepsilon_x = \partial u/\partial x = (1+\nu)[\sigma_x(1-\nu) - \nu\,\sigma_y]/E$$
$$= (1+\nu)[c(y^2 + 2x)(1-\nu) + \nu\,cx^2]/E$$
$$u = (1+\nu)[c(y^2x + x^2)(1-\nu) + \nu\,cx^3/3]/E + f(y) \quad \text{(i)}$$

$$\varepsilon_y = \partial v/\partial y = (1+\nu)[\sigma_y(1-\nu) - \nu\,\sigma_x]/E$$
$$= -(1+\nu)[cx^2(1-\nu) + \nu\,c(y^2 + 2x)]/E$$
$$v = -(1+\nu)[cx^2y(1-\nu) + \nu\,c(y^3/3 + 2xy)]/E + g(x) \quad \text{(ii)}$$

$$\gamma_{xy} = \partial u/\partial y + \partial v/\partial x = 2(1+\nu)\tau_{xy}/E$$
$$\therefore \partial u/\partial y + \partial v/\partial x = -4cy(1+\nu)/E \quad \text{(iii)}$$

Equations (i), (ii) and (iii) clearly satisfy the compatibility condition given in equation (2.18). Substituting eqs(i) and (ii) into (iii),

$$(1+\nu)2cyx(1-\nu)/E + f'(y) - (1+\nu)[2cyx(1-\nu) + 2\nu\,cy]/E + g'(x) = -4cy(1+\nu)/E$$
$$f'(y) - 2\nu\,cy(1+\nu)/E + g'(x) = -4cy(1+\nu)/E$$

from which it follows that $g'(x) = 0$ and

$$f'(y) = 2\nu cy(1+\nu)/E - 4cy(1+\nu)/E = 2cy(1+\nu)(\nu - 2)/E$$
$$f(y) = cy^2(1+\nu)(\nu - 2)/E$$

The displacement expressions are then, from eqs(i) and (ii)

$$u = (1+\nu)[c(y^2x + x^2)(1-\nu) + \nu cx^3/3]/E + cy^2(1+\nu)(\nu-2)/E$$
$$v = -(1+\nu)[cx^2y(1-\nu) + \nu c(y^3/3 + 2xy)]/E$$

Constants of integration have been omitted here as no boundary conditions are given. Note that u and v will differ for plane stress where $\sigma_z = 0$.

Example 2.3 The square plate $a \times a$ in Fig. 2.7 is subjected to the plane stresses shown. Determine, for the given boundary conditions, the general expressions for the u, v displacements at any point (x, y) in the plate. What are the displacements for a point originally at the origin?

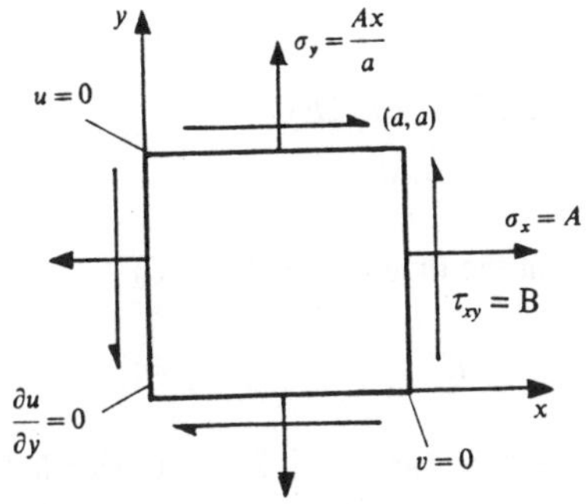

Figure 2.7 Plane stress plate

The given stresses obviously satisfy the equilibrium eqs(2.16a,b). For plane stress, eqs(2.14a,b,d) and (2.17a-c) give

$$\varepsilon_x = \partial u/\partial x = (\sigma_x - \nu\sigma_y)/E = A(1 - \nu x/a)/E$$
$$u = A[x - \nu x^2/(2a)]/E + f(y) + P \quad \text{(i)}$$

$$\varepsilon_y = \partial v/\partial y = (\sigma_y - \nu\sigma_x)/E = A(x/a - \nu)/E$$
$$v = A(xy/a - \nu y)/E + g(x) + Q \quad \text{(ii)}$$

$$\gamma_{xy} = \partial u/\partial y + \partial v/\partial x = 2(1+\nu)\tau_{xy}/E = 2B(1+\nu)/E \quad \text{(iii)}$$

Substituting (i) and (ii) in (iii) gives

$$f'(y) + Ay/(aE) + g'(x) = 2B(1+\nu)/E$$

Since the right-hand side is a constant, say $\alpha + \beta = 2B(1+\nu)/E$, it follows that

$$f'(y) + Ay/(aE) = \alpha, \quad \Rightarrow \quad f(y) = \alpha y - Ay^2/(2aE)$$

$g'(x) = \beta, \quad \Rightarrow \quad g(x) = \beta x$

Hence eqs(i) and (ii) become

$$u = A[x - \nu x^2/(2a)]/E + \alpha y - Ay^2/(2aE) + P \qquad \text{(iv)}$$
$$v = A(xy/a - \nu y)/E + \beta x + Q \qquad \text{(v)}$$

The following boundary conditions apply to eqs(iv) and (v)

$u = 0$ at $x = 0,\ y = a,$
$\therefore\ 0 = \alpha a - Aa^2/(2aE) + P, \qquad \Rightarrow \quad P = Aa/(2E) - \alpha a$

$v = 0$ at $x = a,\ y = 0$
$\therefore\ 0 = \beta a + Q, \qquad \Rightarrow \quad Q = -\beta a$

$\partial u/\partial y = 0$ at $x = 0,\ y = 0$
$\therefore\ \alpha = 0, \qquad \Rightarrow \quad \beta = 2B(1 + \nu)/E$

Finally, eqs(iv) and (v) become

$$u = Ax\,[1 - \nu x/(2a)]/E - Ay^2/(2aE) + Aa/(2E)$$
$$v = Ay(x/a - \nu)/E + 2B\,(1 + \nu)(x - a)/E$$

which give the displacements at the origin ($x = 0$, $y = 0$)

$u = Aa/(2E)$ and $v = -2Ba(1 + \nu)/E$

Example 2.4 The rectangular section bar in Fig. 2.8 is line loaded as shown. Assuming plane strain conditions, derive the stresses and displacements from the stress function $\phi = Ax^3$. If all the points along the z-axis are fixed, determine the displacements beneath the load.

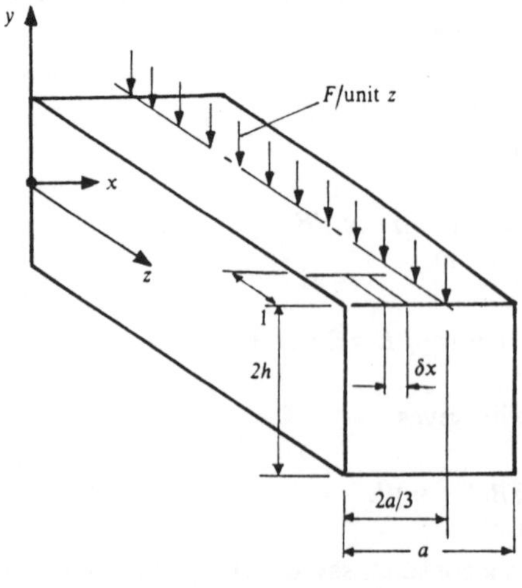

Figure 2.8 Plane strain bar

Since ϕ satisfies $\nabla^4\phi = 0$, the stress components are, from eqs(2.22a-c)

$$\sigma_y = \partial^2\phi/\partial x^2 = \partial^2(Ax^3)/\partial x^2 = \partial(3Ax^2)/\partial x = 6Ax$$
$$\sigma_x = \tau_{xy} = 0$$

Now, at the boundary, F is the resultant of the internal σ_y distribution. That is, for unit thickness in the z - direction,

$$F = \int_0^a \sigma_y\, \mathrm{d}x = 6A\int_0^a x\, \mathrm{d}x = 3Aa^2$$

$$\therefore A = F/(3a^2) \text{ and } \sigma_y = 2Fx/a^2$$

Alternatively, A may be found from moment balance

$$2aF/3 = \int_0^a \sigma_y x\, \mathrm{d}x = 6A\int_0^a x^2\, \mathrm{d}x, \Rightarrow A = F/(3a^2)$$

Since $\varepsilon_o = 0$ and $\sigma_x = 0$, eq(2.15c) gives $\sigma_z = \nu\sigma_y$. Then eqs(2.15a and b) become

$$\varepsilon_x = \partial u/\partial x = -\nu(1+\nu)\sigma_y/E = -2F\nu(1+\nu)x/(a^2E)$$
$$\varepsilon_y = \partial v/\partial x = (1-\nu^2)\sigma_y/E = 2F(1-\nu^2)x/(a^2E)$$

Integrating these leads to the u, v displacements

$$u = -F\nu(1+\nu)x^2/(a^2E) + f(y) + C_1 \qquad \text{(i)}$$
$$v = 2F(1-\nu^2)xy/(a^2E) + g(x) + C_2 \qquad \text{(ii)}$$

Since $\tau_{xy} = 0$, eqs(2.15d and 2.17c) become

$$\gamma_{xy} = \partial u/\partial y + \partial v/\partial x = 0 \qquad \text{(iii)}$$

Substituting eqs(i) and (ii) into (iii) gives

$$f'(y) + 2F(1-\nu^2)y/(a^2E) + g'(x) = 0$$
$$\therefore g'(x) = 0 \text{ and } f'(y) + 2F(1-\nu^2)y/(a^2E) = 0$$
$$f(y) = -F(1-\nu^2)y^2/(a^2E)$$

Hence eqs(i) and (ii) become

$$u = -F\nu(1+\nu)x^2/(a^2E) - F(1-\nu^2)y^2/(a^2E) + C_1 \qquad \text{(iv)}$$
$$v = 2F(1-\nu^2)xy/(a^2E) + C_2 \qquad \text{(v)}$$

where $C_1 = C_2 = 0$ since $u = v = 0$ for $x = y = 0$. At the load ($x = 2a/3$, $y = h$) eqs(iv) and (v) give

$$u = -F[h^2(1-\nu^2)/a^2 - 4\nu(1+\nu)/9]/E$$
$$v = 4Fh(1-\nu^2)/(3aE)$$

Example 2.5 Determine the stresses and displacements from the plane stress function: $\phi = Ax^2 + Bxy + Cy^2$. Apply these to the plate in Fig. 2.9.

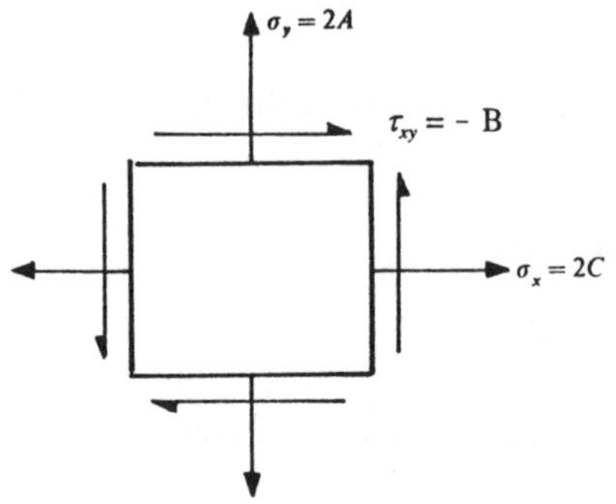

Figure 2.9 Thin plate under uniform stressing

Clearly ϕ satisfies $\nabla^4\phi = 0$. The stress components are, from eqs(2.22a-c)

$$\sigma_x = \partial^2\phi/\partial y^2 = \partial\,(Bx + 2C\,y)/\partial y = 2C$$
$$\sigma_y = \partial^2\phi/\partial x^2 = \partial(2Ax + B\,y)/\partial x = 2A$$
$$\tau_{xy} = -\,\partial^2\phi/\partial x\,\partial y = -\,\partial(2Ax + B\,y)/\partial y = -\,B$$

That is, the function here provides a solution to a thin plate under the action of uniform stresses along its sides (see Fig. 2.9). The displacements follow from eqs(2.14 and 2.17)

$$\varepsilon_x = \partial u/\partial x = (\sigma_x - \nu\,\sigma_y)\,/\,E = 2(C - \nu\,A)/E$$
$$u = 2\,(C - \nu\,A)x/E + f\,(y) + Q_1 \qquad \text{(i)}$$

$$\varepsilon_y = \partial v\,/\partial y = (\sigma_y - \nu\sigma_x)/\,E = 2\,(A - \nu\,C\,)/E$$
$$v = 2(\,A - \nu\,C\,)y/\,E + g(x) + Q_2 \qquad \text{(ii)}$$

$$\gamma_{xy} = \partial u/\partial y + \partial v/\partial x = 2\,(1 + \nu\,)\tau_{xy}\,/E = -\,2(1 + \nu\,)B/E \qquad \text{(iii)}$$

Substituting eqs(i) and (ii) into (iii) leads to

$$f'(y) + g'(x) = -\,2\,(1 + \nu\,)B/\,E = \alpha + \beta$$

Since the derivatives are the respective constants α and β, then $f\,(y) = \alpha y$ and $g(x) = \beta x$ and eqs(i) and (ii) become

$$u = 2(C - \nu\,A)x\,/E + \alpha\,y + Q_1$$
$$v = 2(A - \nu\,C\,)y/E + \beta\,x + Q_2$$

Example 2.6 Show that the stress function $\phi = Dxy^3 + Bxy$ provides the stress distribution for the cantilever of depth $2h$ and of unit thickness in Fig. 2.10, when it is loaded by a concentrated force P at its free end.

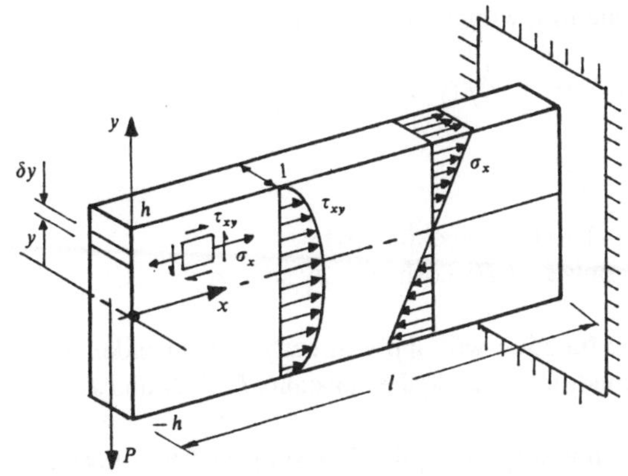

Figure 2.10 Plane cantilever with concentrated end load

Equations (2.27a-c) gives the stress components

$$\sigma_x = \partial^2\phi/\partial y^2 = \partial(3Dx\,y^2 + Bx)/\partial y = 6Dxy \qquad \text{(i)}$$

$$\sigma_y = \partial^2\phi/\partial x^2 = \partial(Dy^3 + By)/\partial x = 0$$

$$\tau_{xy} = -\,\partial^2\phi/\partial x\,\partial y = -\,\partial(3Dxy^2 + Bx)/\partial x = -\,(3Dy^2 + B\,) \qquad \text{(ii)}$$

The boundary conditions are

(a) $\tau_{xy} = 0$ for $y = \pm\, h$, since there is no shear stress along the free edges.
$-\,(B + 3Dh^2) = 0 \quad \Rightarrow \quad B = -\,3Dh^2$

(b) P is the resultant of the internal τ_{xy} distribution acting at the free end. That is

$$P = \int_{-h}^{h} \tau_{xy}\,\mathrm{d}y = 3D\int_{-h}^{h} (h^2 - y^2)\,\mathrm{d}y$$

$$= 3\mathrm{D}\;\Big|\; h^2y - y^3/3\;\Big|_{-h}^{h} = 4Dh^3, \quad \Rightarrow \quad D = P/(4h^3)$$

From eqs(i) and (ii)

$$\sigma_x = Pxy\,/(2h^3/3) = (Px)\,y\,/\,I = My/I$$

$$\tau_{xy} = 3P(h^2 - y^2)/(4h^3) = [P/(2I\,)](h^2 - y^2)$$

which are distributed in the manner shown in Fig. 2.10. Note that the stresses agree with the separate applications of bending and shear flow theory. In the stress function approach, x and not z is used to denote the length of a plane stress beam within the x, y plane. Then y bounds the depth and z bounds the thickness.

2.4.2 Sinusoidal Function

If a stress function of the form

$$\phi = f\,(y)\,\sin\,(n\pi x\,/L) = f\,(y)\,\sin\,(\alpha x) \qquad \text{(2.31a)}$$

is to satisfy $\nabla^4\phi = 0$, the following condition is found

$$d^4 f(y)/dy^4 - 2\alpha^2 d^2 f(y)/dy^2 + \alpha^4 f(y) = 0$$

The solution is

$$f(y) = (A + By)\exp(\alpha y) + (C + Dy)\exp(-\alpha y)$$
$$= E\cosh\alpha y + F\sinh\alpha y + Hy\cosh\alpha y + Jy\sinh\alpha y \qquad (2.31b)$$

where $\alpha = n\pi/L$. The function will apply directly to sinusoidal edge loadings where the particular boundary conditions enable the constants E, F, H and J to be found.

Example 2.7 A long thin strip of depth $2h$ is subjected to a normal stress distribution $p\sin(\alpha x/L)$ along its longer edges (see Fig. 2.11). Determine the constants in eq(2.31b) and the stress state along the x - axis.

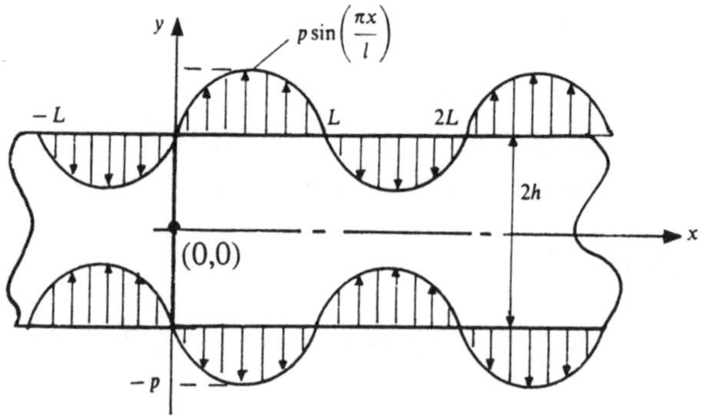

Figure 2.11 Sinusoidal wave edge loading

Here $\alpha = \pi/L$ since $n = 1$. The plane stress components are found from eqs(2.22a-c) and eqs(2.31a,b),

$$\sigma_y = \partial^2\phi/\partial x^2 = -\,[E\cosh\alpha y + F\sinh\alpha y + Hy\cosh\alpha y + Jy\sinh\alpha y]\alpha^2\sin\alpha x \qquad (i)$$

$$\tau_{xy} = -\,\partial^2\phi/\partial x\,\partial y = -\,[\alpha(E\sinh\alpha y + F\cosh\alpha y) + H(y\alpha\sinh\alpha y + \cosh\alpha y)$$
$$+ J(y\alpha\cosh\alpha y + \sinh\alpha y)]\alpha\cos\alpha x \qquad (ii)$$

$$\sigma_x = \partial^2\phi/\partial y^2 = [\alpha^2(E\cosh\alpha y + F\sinh\alpha y) + H\alpha(y\alpha\cosh\alpha y + 2\sinh\alpha y)$$
$$+ J\alpha(y\alpha\sinh\alpha y + 2\cosh\alpha y)]\sin\alpha x \qquad (iii)$$

In eq(i) the boundary condition is $\sigma_y = p\sin\alpha x$ (tensile) for $y = \pm h$ and all x. This gives

$$-\,\alpha^2[E\cosh\alpha h + F\sinh\alpha h + Hh\cosh\alpha h + Jh\sinh\alpha h] = p$$
$$-\,\alpha^2[E\cosh\alpha h - F\sinh\alpha h - Hh\cosh\alpha h + Jh\sinh\alpha h] = p$$

By addition and subtraction,

$$E = -\,(p + J\alpha^2 h\sinh\alpha h)/(\alpha^2\cosh\alpha h) \qquad (iv)$$
$$F = -\,(Hh)\cosh\alpha h/\sinh\alpha h \qquad (v)$$

In eq(ii) the boundary condition is $\tau_{xy} = 0$ for $y = \pm h$ and all x. Thus,

$$- [\alpha (E \sinh \alpha h + F \cosh \alpha h) + H (h\alpha \sinh \alpha h + \cosh \alpha h) + J (h\alpha \cosh \alpha h + \sinh \alpha h)] = 0$$
$$- [\alpha (- E \sinh \alpha h + F \cosh \alpha h) + H (h\alpha \sinh \alpha h + \cosh \alpha h) - J(h\alpha \cosh \alpha h + \sinh \alpha h)] = 0$$

By addition and subtraction,

$$H = - \alpha F (\cosh \alpha h) / (h\alpha \sinh \alpha h + \cosh \alpha h) \quad \text{(vi)}$$
$$J = - \alpha E (\sinh \alpha h) / (h\alpha \cosh \alpha h + \sinh \alpha h) \quad \text{(vii)}$$

Substituting eq(iv) into (vii) leads to

$$J = \alpha p (\sinh \alpha h) / [\alpha^2 (\sinh \alpha h \cosh \alpha h + \alpha h)]$$
$$E = - p (\sinh \alpha h + \alpha h \cosh \alpha h) / [\alpha^2 (\sinh \alpha h \cosh \alpha h + \alpha h)]$$

Substituting eq(v) into (vi) gives $F = H = 0$, which is a consequence of the common amplitude p for each edge distribution. When $y = 0$ the stress state along the x axis is, from eqs (i) - (iii)

$$\sigma_y = - E\alpha^2 \sin \alpha x$$
$$= p \sin \alpha x (\sinh \alpha h + \alpha h \cosh \alpha h) / (\sinh \alpha h \cosh \alpha h + \alpha h)$$

$$\sigma_x = (\alpha^2 E + 2J\alpha) \sin \alpha x$$
$$= p \sin \alpha x [- (\sinh \alpha h + \alpha h \cosh \alpha h) / (\sinh \alpha h \cosh \alpha h + \alpha h)$$
$$+ 2 \sinh \alpha h / (\sinh \alpha h \cosh \alpha h + \alpha h)]$$

$$\tau_{xy} = - (F\alpha + H)\alpha \cos \alpha x = 0$$

2.4.3 Fourier Series

(a) Full Range Series

When the edge loading is not sinusoidal it may still be represented by a sinusoidal series function $g(x)$ of period $2L$. The stress function in eqs(2.31a,b) then becomes

$$\phi (x, y) = [E \cosh \alpha y + F \sinh \alpha y + Hy \cosh \alpha y + Jy \sinh \alpha y]\, g(x) \quad (2.32)$$

where

$$g(x) = a_o/2 + a_1 \sin (\pi x/L) + a_2 \sin (2\pi x/L) + a_3 \sin (3\pi x/L)$$
$$+ b_1 \cos (\pi x/L) + b_2 \cos (2\pi x/L) + b_3 \cos (3\pi x/L).....$$
$$= a_o/2 + \sum_{n=1}^{\infty} [a_n \sin (n\pi x/L) + b_n \cos (n\pi x/L)] \quad (2.33a)$$

If $\psi(x)$ describes the actual edge loading, the coefficients a_o, a_n and b_n in eq(2.33a) are found by identifying $\psi(x)$ with each part of $g(x)$ and integrating over the range $\pm L$, as follows:

$$\int_{-L}^{L} [a_o/2 + a_n \sin (n\pi x/L) \sin (n\pi x/L) + b_n \cos (n\pi x/L) \cos (n\pi x/L)] \mathrm{d}x$$
$$= \int_{-L}^{L} \psi(x)\, \mathrm{d}x + \int_{-L}^{L} \psi(x) \sin (n\pi x/L)\, \mathrm{d}x + \int_{-L}^{L} \psi(x) \cos (n\pi x/L)\, \mathrm{d}x$$

This gives

$$\int_{-L}^{L} (a_o/2)\mathrm{d}x = \int_{-L}^{L} \psi(x)\,\mathrm{d}x$$

$$a_o = (1/L)\int_{-L}^{L} \psi(x)\,\mathrm{d}x \qquad (2.33b)$$

$$\int_{-L}^{L} a_n \sin(n\pi x/L)\sin(n\pi x/L)\mathrm{d}x = \int_{-L}^{L} \psi(x)\sin(n\pi x/L)\,\mathrm{d}x$$

$$a_n L = \int_{-L}^{L} \psi(x)\sin(n\pi x/L)\mathrm{d}x$$

$$a_n = (1/L)\int_{-L}^{L} \psi(x)\sin(n\pi x/L)\mathrm{d}x \qquad (2.33c)$$

$$\int_{-L}^{L} b_n \cos(n\pi x/L)\cos(n\pi x/L)\mathrm{d}x = \int_{-L}^{L} \psi(x)\cos(n\pi x/L)\mathrm{d}x$$

$$b_n L = \int_{-L}^{L} \psi(x)\cos(n\pi x/L)\mathrm{d}x$$

$$b_n = (1/L)\int_{-L}^{L} \psi(x)\cos(n\pi x/L)\mathrm{d}x \qquad (2.33d)$$

Note that when x lies outside the range $-L \le x \le L$, $g(x)$ in eqs(2.33a) will not, in general, represent $\psi(x)$ unless $\psi(x)$ is itself periodic.

Example 2.8 A rectangular wave loading of amplitude p and period $2L$ is applied normally to the longer edges of a thin plate of depth $2h$, as shown in Fig. 2.12. Determine the Fourier series loading function $g(x)$ in eq(2.33a). If $L = 4h$, determine σ_x and σ_y at $x = L/2$, $y = 0$ using the result from Example 2.7.

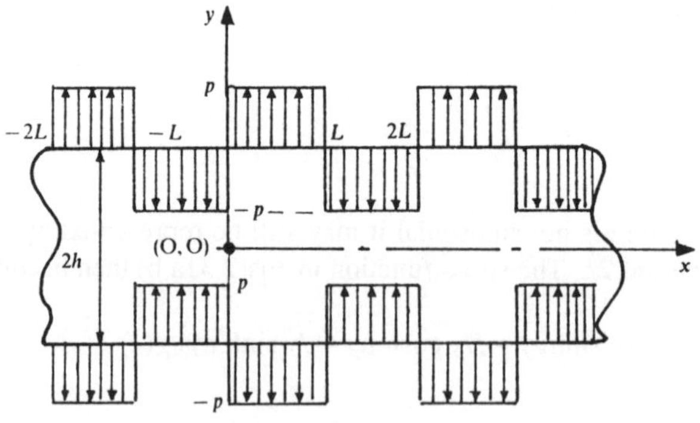

Figure 2.12 Rectangular wave

The loading function is

$$\psi(x) = \begin{cases} +p & \text{for } 0 \le x \le L \\ -p & \text{for } L \le x \le 2L \end{cases}$$

From eqs(2.33b - d),

$$a_o = (1/L)\left[\int_0^{L} (p)\mathrm{d}x + \int_L^{2L} (-p)\mathrm{d}x\right]$$

$= (1/L)[\,|\,px\,|_0^L - |\,px\,|_L^{2L}\,] = 0$

$a_n = (1/L)[\int_0^L p \sin(n\pi x/L)dx + \int_L^{2L}(-p)\sin(n\pi x/L)dx\,]$

$= (1/L)[Lp/(n\pi)][\,|-\cos(n\pi x/L)\,|_0^L + |\cos(n\pi x/L)\,|_L^{2L}\,]$

$= [p/(n\pi)][(1-\cos n\pi) + (\cos 2n\pi - \cos n\pi)]$

$a_n = [2p/(n\pi)](1-\cos n\pi)$ for $n = 1, 3, 5...$, $a_n = 0$ for $n = 2, 4, 6...$

$b_n = (1/L)[\int_0^L p\cos(n\pi x/L)dx + \int_L^{2L}(-p)\cos(n\pi x/L)dx]$

$= (1/L)[Lp/(n\pi)][\,|\sin(n\pi x/L)\,|_0^L - |\sin(n\pi x/L)\,|_L^{2L}\,]$

$= [p/(n\pi)][0 - 0] = 0$

Equation (2.33a) becomes

$$g(x) = \sum_{n=1}^{\infty} [2p/(n\pi)](1 - \cos n\pi)\sin(n\pi x/L) \text{ for } n = 1, 3, 5...$$

$$= (4\,p/\pi)[\sin(\pi x/L) + (1/3)\sin(3\pi x/L) + (1/5)\sin(5\pi x/L) + ...\,]$$

Now, since the functions x and y in eq(2.32) are separable, the hyperbolic function in y may be taken from the previous example where similar boundary conditions were applied. It is sufficient to take $n = 1$ within g(x), when it follows for $y = 0$, that the plane stresses are

$$\sigma_y = (4p/\pi)[\sin(\pi x/L) + (1/3)\sin(3\pi x/L) + (1/5)\sin(5\pi x/L) + ...\,]$$
$$\times [\sinh(\pi h/L) + (\pi h/L)\cosh(\pi h/L)] / [\sinh(\pi h/L)\cosh(\pi h/L) + \pi h/L]$$

$$\sigma_x = (4p/\pi)[\sin(\pi x/L) + (1/3)\sin(3\pi x/L) + (1/5)\sin(5\pi x/L) + ...\,]$$
$$\times [-\sinh(\pi h/L) - (\pi h/L)\cosh(\pi h/L) + 2\sinh(\pi h/L)] / [\sinh(\pi h/L)\cosh(\pi h/L) + \pi h/L]$$

Taking $x = L/2$, $h/L = 1/4$

$$\sigma_y = (4p/\pi)\,[1 - 1/3 + 1/5 - 1/7 + 1/9 - 1/11...]1.9801 / 1.835 = 1.036p$$
$$\sigma_x = (4p/\pi)[0.7543]\,[-1.9801 + 1.7373]/\,1.835 = -0.127p$$

(b) Odd and Even Functions

For an odd loading function $\psi(-x) = -\psi(x)$, $b_n = 0$ and the Fourier representation $g(x)$ contains sine terms only. For an even loading function $\psi(-x) = -\psi(x)$, which gives $a_n = 0$, resulting in a Fourier cosine series. In addition, when $\psi(x) = -\psi(x + L)$, as in this example, the Fourier series will contain odd harmonics only. When $\psi(x) = \psi(x + L)$, the series appears in even harmonics. Simplified Fourier sine or cosine series may be applied over the half-range $0 \le x \le L$. These are

$$g(x) = \sum_{n=1}^{\infty} a_n \sin \alpha x \qquad (2.34a)$$

where $\alpha = n\pi/L$, and

$$a_n = (2/L)\int_0^L \psi(x)\sin\alpha x\,dx$$

$$g(x) = a_o + \sum_{n=1}^{\infty} b_n \cos\alpha x \tag{2.34b}$$

$$a_o = (1/L)\int_0^L \psi(x)\,dx$$

$$b_n = (2/L)\int_0^L \psi(x)\cos\alpha x\,dx$$

Beyond the range of L, the sine series will extend $\psi(x)$ by alternating it in positive and negative y with period L about the x - axis. The cosine series duplicates $\psi(x)$ with period L along the x - axis. Half-range series' representations of odd and even functions are identical to the full series representation of each function.

Example 2.9 Find the expression $g(x)$ for the edge loading in Fig. 2.13.

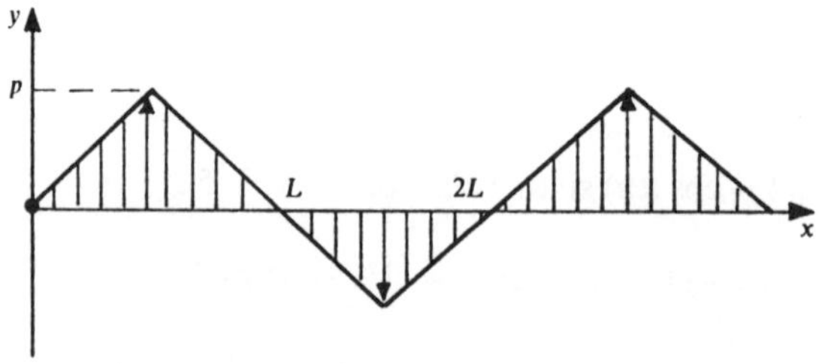

Figure 2.13 Triangular edge loading

Here $g(x)$ will follow from the half-range sine series, since $\psi(x)$ alternates about x with period L. The loading function becomes

$$\psi(x) = \begin{cases} 2px/L & \text{for } 0 \le x \le L/2 \\ 2p(1 - x/L) & \text{for } L/2 \le x \le L \end{cases}$$

Equations(2.34a,b) may be written as

$$a_n = (2/L)\int_0^{L/2} (2px/L)\sin\alpha x\,dx + (2/L)\int_{L/2}^{L} 2p\,(1 - x/L)\sin\alpha x\,dx \tag{i}$$

Using the following result from integrating by parts:

$$\int x(\sin\alpha x)\,dx = (1/\alpha^2)\sin\alpha x - (x/\alpha)\cos\alpha x$$

eq(i) becomes

$$a_n = (4p/L^2)\,|\,(1/\alpha^2)\sin\alpha x - (x/\alpha)\cos\alpha x\,\Big|_0^{L/2} - [4p/(L\alpha)]\,|\cos\alpha x\,\Big|_{L/2}^{L}$$

$$- (4p/L^2)\,|\,(1/\alpha^2)\sin\alpha x - (x/\alpha)\cos\alpha x\,\Big|_{L/2}^{L}$$

$$= (4p/L^2)[(1/\alpha^2)\sin(n\pi/2) - [L/(2\alpha)]\cos(n\pi/2)] - [4p/(L\alpha)][\cos(n\pi) - \cos(n\pi/2)]$$

$$- (4p/L^2)\{[(1/\alpha^2)\sin(n\pi) - (L/\alpha)\cos(n\pi)] - [(1/\alpha^2)\sin(n\pi/2) - [L/(2\alpha)]\cos(n\pi/2)]\}$$

$$= [8p/(L^2\alpha^2)]\sin(n\pi/2) = [8p/(\pi^2 n^2)]\sin(n\pi/2)$$

$$\therefore g(x) = \sum_{n=1}^{\infty} a_n \sin(n\pi x/L) = (8p/\pi^2)\sum_{n=1}^{\infty}(1/n^2)\sin(n\pi/2)\sin(n\pi x/L)$$

$$= (8p/\pi^2)[\sin(\pi x/L) - (1/3^2)\sin(3\pi x/L) + (1/5^2)\sin(5\pi x/L) - \ldots]$$

2.5 Cylindrical Plane Stress and Plane Strain

Many plane problems are better represented by cylindrical co-ordinates (r,θ). For this it is necessary to re-express the constitutive relations, the strain-displacement relations, equilibrium, compatibility and the stress functions in terms of r and θ. Fortunately, the number of derivations in the theory may be reduced by transforming the Cartesian co-ordinates' relations. Firstly, we should note that the constitutive relations are identical in form to the Cartesian eqs(2.14 and 2.15) where r and θ now replace x and y respectively (z is a common co-ordinate).

2.5.1 Strain-Displacement Relations

Strain components ε_r, ε_θ and $\gamma_{r\theta}$ are found from the distortion occurring to the cylindrical line elements $\delta r \times r\delta\theta$, as point O displaces to O' with radial and tangential components u and v respectively (see Fig. 2.14).

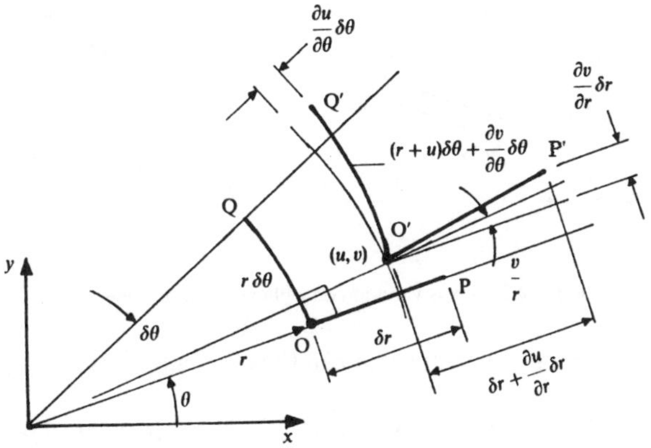

Figure 2.14 Plane deformation in cylindrical co-ordinates

The radial displacement of OP is $(\partial u/\partial r)\delta r$. Hence the radial strain is

$$\varepsilon_r = [\delta r + (\partial u/\partial r)\delta r - \delta r]/\delta r = \partial u/\partial r \qquad (2.35a)$$

For the tangential (hoop) strain in OQ, the increase in length of $r\delta\theta$ is due to both the rate of change of v with respect to θ and the change in radius from r to $r + u$ as O displaces to O'. This gives

$$\varepsilon_\theta = [(r + u)\delta\theta + (\partial v/\partial\theta)\delta\theta - r\delta\theta]/(r\delta\theta)$$
$$\varepsilon_\theta = u/r + (1/r)\,\partial v/\partial\theta \qquad (2.35b)$$

The true change in the angle ∠POQ defines the shear strain. This must omit the rigid rotation (v/r rad) which also contributes to the new position O'P'. That is,

$$\gamma_{r\theta} = (\partial v/\partial r)\delta r/\delta r - v/r + (\partial u/\partial\theta)[\delta\theta/(r\delta\theta)]$$
$$\gamma_{r\theta} = \partial v/\partial r - v/r + (1/r)(\partial u/\partial\theta) \qquad (2.35c)$$

2.5.2 *Equilibrium*

Equilibrium equations are found from radial and tangential force balance for the cylindrical element in Figs 2.15a,b. Note that the force components lying in each direction have been separated in (a) and (b) for clarity. The hoop σ_θ, radial σ_r and shear $\tau_{r\theta}$ stresses vary across the element in the manner shown when taking positive r and θ along the centre-line directions indicated. Since the inclinations of σ_θ and $\tau_{r\theta}$ vary with $\delta\theta$, these must be resolved parallel to r and θ. The following derivations apply in the absence of body forces, assuming unit thickness of the element.

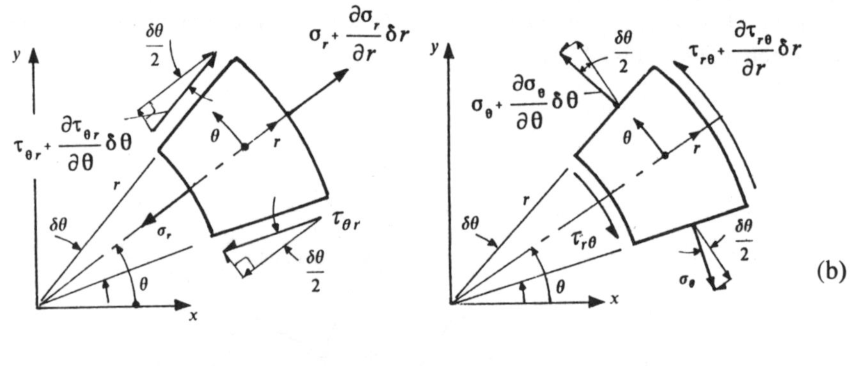

Figure 2.15 Stress variation on a plane cylindrical element

For radial equilibrium along the centre-line r in Figs 2.15a,b,

$$(\sigma_r + \delta r\,\partial\sigma_r/\partial r)(r + \delta r)\delta\theta - \sigma_r r\delta\theta - \tau_{\theta r}\delta r\cos(\delta\theta/2) + (\tau_{\theta r} + \delta\theta\,\partial\tau_{\theta r}/\partial\theta)\delta r\cos(\delta\theta/2)$$
$$- \sigma_\theta\delta r\sin(\delta\theta/2) - (\sigma_\theta + \delta\theta\,\partial\sigma_\theta/\partial\theta)\delta r\sin(\delta\theta/2) = 0$$

Now when $\delta\theta \to 0$, $\cos(\delta\theta/2) \to 1$ and $\sin(\delta\theta/2) \to \delta\theta/2$ rad. This gives

$$(\sigma_r + \delta r\,\partial\sigma_r/\partial r)(r + \delta r)\delta\theta - \sigma_r r\delta\theta - \tau_{r\theta}\delta r + (\tau_{r\theta} + \delta\theta\,\partial\tau_{r\theta}/\partial\theta)\delta r$$
$$- \sigma_\theta\delta r\delta\theta/2 - (\sigma_\theta + \delta\theta\,\partial\sigma_\theta/\partial\theta)\delta r\delta\theta/2 = 0$$

where $\tau_{r\theta} = \tau_{\theta r}$. Neglecting products of infinitesimals leads to

$$\sigma_r\delta r\delta\theta + r\delta r\delta\theta\,\partial\sigma_r/\partial r + \delta\theta\,\delta r\,\partial\tau_{r\theta}/\partial\theta - \sigma_\theta\delta r\delta\theta/2 - \sigma_\theta\delta r\delta\theta/2 = 0$$

Dividing through by $(r\delta\theta\,\delta r)$ gives the first equilibrium equation,

$$(\sigma_r - \sigma_\theta)/r + (1/r)\,\partial\tau_{r\theta}/\partial\theta + \partial\sigma_r/\partial r = 0 \tag{2.36a}$$

For equilibrium along the tangential centre-line θ in Figs 2.15a,b,

$$(\tau_{r\theta} + \delta r\,\partial\tau_{r\theta}/\partial r)(r + \delta r)\delta\theta - \tau_{r\theta} r\delta\theta + (\tau_{\theta r} + \delta\theta\,\partial\tau_{\theta r}/\partial\theta)\delta r\sin(\delta\theta/2)$$
$$+ \tau_{r\theta}\delta r\sin(\delta\theta/2) + (\sigma_\theta + \delta\theta\,\partial\sigma_\theta/\partial\theta)\delta r\cos(\delta\theta/2) - \sigma_\theta\delta r\cos(\delta\theta/2) = 0$$

and, with similar reductions, this leads to the second equilibrium equation

$$(2/r)\tau_{r\theta} + (1/r)\,\partial\sigma_\theta/\partial\theta + \partial\tau_{r\theta}/\partial r = 0 \tag{2.36b}$$

2.5.3 Bi-harmonic Equation

Figures 2.15a,b show that the Cartesian and cylindrical co-ordinates are related by $r^2 = x^2 + y^2$, $x = r\cos\theta$, and $y = r\sin\theta$. These relationships are used to derive the first and second derivatives

$$\partial/\partial x = (\partial r/\partial x)\,\partial/\partial r + (\partial\theta/\partial x)\,\partial/\partial\theta$$
$$= \cos\theta(\partial/\partial r) - (1/r)\sin\theta(\partial/\partial\theta)$$

$$\partial^2/\partial x^2 = [\cos\theta(\partial/\partial r) - (1/r)\sin\theta(\partial/\partial\theta)][\cos\theta(\partial/\partial r) - (1/r)\sin\theta(\partial/\partial\theta)]$$
$$= \cos\theta(\partial/\partial r)[\cos\theta(\partial/\partial r) - (1/r)\sin\theta(\partial/\partial\theta)]$$
$$- (1/r)\sin\theta(\partial/\partial\theta)[\cos\theta(\partial/\partial r) - (1/r)\sin\theta(\partial/\partial\theta)]$$
$$= \cos^2\theta(\partial^2/\partial r^2) + (1/r^2)\sin\theta\cos\theta(\partial/\partial\theta) + (1/r)\sin^2\theta(\partial/\partial r)$$
$$+ (1/r^2)\sin^2\theta(\partial^2/\partial\theta^2) + (1/r^2)\sin\theta\cos\theta(\partial/\partial\theta) \tag{2.37a}$$

$$\partial/\partial y = (\partial r/\partial y)\,\partial/\partial r + (\partial\theta/\partial y)\,\partial/\partial\theta$$
$$= \sin\theta(\partial/\partial r) + (1/r)\cos\theta(\partial/\partial\theta)$$

$$\partial^2/\partial y^2 = [\sin\theta(\partial/\partial r) + (1/r)\cos\theta(\partial/\partial\theta)][\sin\theta(\partial/\partial r) + (1/r)\cos\theta(\partial/\partial\theta)]$$
$$= \sin\theta(\partial/\partial r)[\sin\theta(\partial/\partial r) + (1/r)\cos\theta(\partial/\partial\theta)]$$
$$+ (1/r)\cos\theta(\partial/\partial\theta)[\sin\theta(\partial/\partial r) + (1/r)\cos\theta(\partial/\partial\theta)]$$
$$= \sin^2\theta(\partial^2/\partial r^2) - (1/r^2)\sin\theta\cos\theta(\partial/\partial\theta) + (1/r)\cos^2\theta(\partial/\partial r)$$
$$+ (1/r^2)\cos^2\theta(\partial^2/\partial\theta^2) - (1/r^2)\sin\theta\cos\theta(\partial/\partial\theta) \tag{2.37b}$$

Adding eqs(2.37a and b), the biharmonic eq(2.23b) is transformed to

$$[\,\partial^2/\partial r^2 + (1/r)(\partial/\partial r) + (1/r^2)(\partial^2/\partial\theta^2)]^2\phi = 0 \tag{2.38a}$$

If we let $\nabla^2 = [\ \ ]$, the LH side of eq(2.38a) becomes $[\nabla^2][\nabla^2]\phi$, which is abbreviated to

$$\nabla^4(r,\theta)\phi = 0 \tag{2.38b}$$

2.5.4 Stress Component Functions

When $\theta = 0$ in Figs 2.15a,b, we see that $\sigma_x = \sigma_r$ and $\sigma_y = \sigma_\theta$. Equations (2.37a,b) may then be used to transform σ_x and σ_y (eqs 2.22a and b) into σ_r and σ_θ. The shear stress $\tau_{r\theta}$ will follow from either of the equilibrium eqs(2.36a or b). This gives three plane polar stress functions:

$$\sigma_r = (1/r)\partial\phi/\partial r + (1/r^2)\partial^2\phi/\partial\theta^2 \quad (2.39a)$$
$$\sigma_\theta = \partial^2\phi/\partial r^2 \quad (2.39b)$$
$$\tau_{r\theta} = -\,\partial[(1/r)(\partial\phi/\partial\theta)]/\partial r$$
$$= (1/r^2)\partial\phi/\partial\theta - (1/r)(\partial^2\phi/\partial r\,\partial\theta) \quad (2.39c)$$

The reader should check that eqs(2.39a-c) satisfy the two equilibrium equations (2.36a,b).

2.5.5 Body Forces

Apart from self-weight, an example of a radial body force R is the centrifugal force produced in a rotating disc or cylinder. This force derives from a potential $\Omega(r)$ as $R = \partial\Omega/\partial r$, so that the equilibrium eqs(2.36a,b) are modified to

$$(\sigma_r - \sigma_\theta)/r + (1/r)\partial\tau_{r\theta}/\partial\theta + \partial\sigma_r/\partial r + \partial\Omega/\partial r = 0 \quad (2.40a)$$
$$(2/r)\tau_{r\theta} + (1/r)\,\partial\sigma_\theta/\partial\theta + \partial\tau_{r\theta}/\partial r = 0 \quad (2.40b)$$

Equations (2.40a,b) are satisfied by the stress functions

$$\sigma_r = (1/r)\partial\phi/\partial r + (1/r^2)\partial^2\phi/\partial\theta^2 \quad (2.41a)$$
$$\sigma_\theta = \partial^2\phi/\partial r^2 + r\,\partial\Omega/\partial r \quad (2.41b)$$
$$\tau_{r\theta} = -\,\partial[(1/r)(\partial\phi/\partial\theta)]/\partial r \quad (2.41c)$$

At this stage a distinction is made between plane stress and plane strain as their constitutive relations differ. For cylindrical plane stress (e.g. a thin disc), ϕ and Ω in eqs(2.39a-c) must satisfy

$$\nabla^4\phi(\theta, r) + (1 - \nu)\nabla^2\Omega(r) = 0 \quad (2.42a)$$

For cylindrical plane strain (a cylinder), ϕ and Ω in eqs(2.41a-c) must satisfy

$$\nabla^4\phi(\theta, r) + [(1 - 2\nu)/(1 + \nu)]\nabla^2\Omega(r) = 0 \quad (2.42b)$$

The expressions (2.42a,b) are identical to their Cartesian counterparts (2.28b) and (2.30b). However, the meaning of ∇^2 will now differ between the first and second terms of eqs(2.42a,b). They are respectively

$$\nabla^4 = (\nabla^2)^2 = [\partial^2/\partial r^2 + (1/r)(\partial/\partial r) + (1/r^2)(\partial^2/\partial\theta^2)]^2$$
$$\nabla^2 = \partial^2/\partial r^2 + (1/r)(\partial/\partial r)$$

These forms may be employed for the analysis of cylindrical bodies rotating about an axis of symmetry (see Chapter 3).

2.6 Polar Stress Functions

There is a number of useful stress functions ϕ satisfying eq(2.38b). These may depend upon θ and r alone or together, so that the following three classes of function are considered.

2.6.1 Stress Functions in θ

The function $\phi = \phi(\theta)$ is independent of r. Thus $\partial/\partial r = 0$, and the biharmonic eq(2.38a) becomes

$$[\partial^2/\partial r^2 + (1/r)(\partial/\partial r) + (1/r^2)(\partial^2/\partial\theta^2)][(1/r^2)\,\partial^2\phi/\partial\theta^2] = 0$$

which simplifies to

$$d^4\phi/d\theta^4 + 4\,d^2\phi/d\theta^2 = 0$$

The solution, which is obtained from a substitution of the form $\phi = N\exp(m\theta)$, where N and m are constants, has the final form

$$\phi = A\cos 2\theta + B\sin 2\theta + C\theta \tag{2.43}$$

where A, B and C are constants. The corresponding stress functions eqs(2.39a-c) gives $\sigma_\theta = 0$ with

$$\sigma_r = (1/r^2)\,d^2\phi/d\theta^2 \tag{2.44a}$$
$$\tau_{r\theta} = (1/r^2)\,d\phi/d\theta \tag{2.44b}$$

Any part of eq(2.43) is a valid stress function. The following functions are of interest.

(a) Torsion of a Disc
The simple function $\phi = C\theta$ applies to the torsion of a thin disc or rotating wheel. Here $\sigma_\theta = \sigma_r = 0$ and eq(2.44b) gives the shear stress as

$$\tau_{r\theta} = C/r^2$$

If the torque is applied to the axis of a solid disc of radius R, then C is found from the equilibrium condition

$$T = 2\pi\int_0^R \tau_{r\theta}\,r^2\,dr = 2\pi C\int_0^R dr = 2\pi CR$$

which gives $C = T/(2\pi R)$ and $\tau_{r\theta} = T/(2\pi R\,r^2)$. When a torque is applied to the axis of a thin annulus with inner radius R_i and outer radius R_o (see Fig. 2.16a) the equilibrium condition becomes

$$T = 2\pi\int_{R_i}^{R_o} \tau_{r\theta}\,r^2\,dr = 2\pi C\int_{R_i}^{R_o} dr = 2\pi C\,(R_o - R_i)$$

which gives $C = T/[2\pi(R_o - R_i)]$ and $\tau_{r\theta} = T/[2\pi r^2(R_o - R_i)]$. In each case $\tau_{r\theta}$ varies inversely with r^2. For the hollow tube $\tau_{r\theta}$ varies from a maximum value at R_i to a minimum at R_o, as shown in Fig. 2.16b.

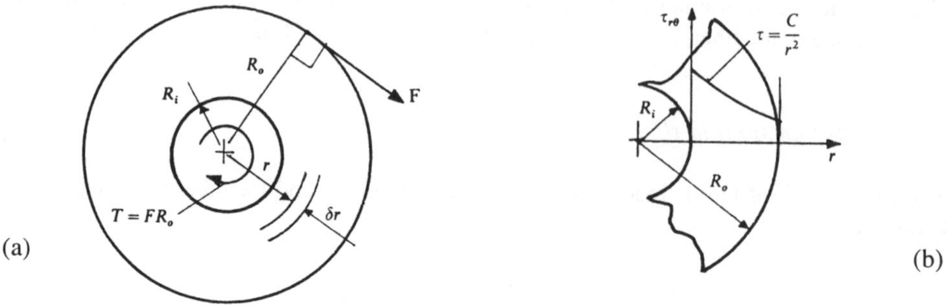

Figure 2.16 Hollow disc under axial torque

(b) Wedge with Tip Moment
The following example shows how the function $\phi = B\sin 2\theta + C\theta$ can be applied to a wedge with a moment applied to its tip.

Example 2.10 Determine the stress distribution in the body of a wedge of unit thickness and apex angle 2α when a moment M is applied at its tip in the direction of θ negative (see Fig. 2.17)

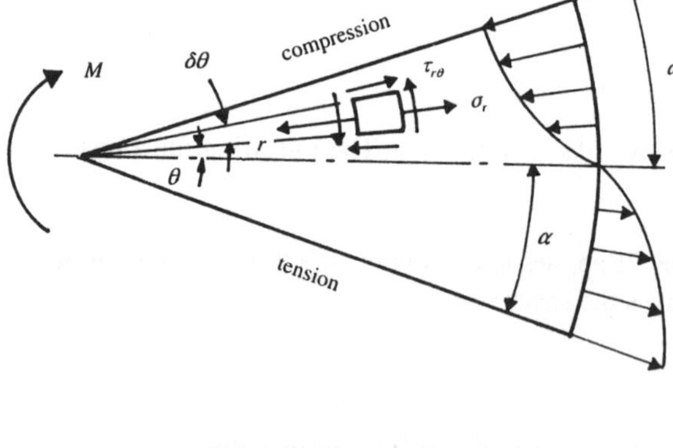

Figure 2.17 Wedge with tip moment

Applying eqs(2.44a,b) with $\phi = B\sin 2\alpha + C\alpha$ yields the radial and shear stress components,

$$\sigma_r = (-\,4B\,/\,r^2)\sin 2\theta \tag{i}$$
$$\tau_{r\theta} = (1/\,r^2)(2B\cos 2\theta + C\,) \tag{ii}$$

The constants B and C are found from the two conditions: (a) $\tau_{r\theta} = 0$ for $\theta = \pm\,\alpha$ in eq(ii) which gives

$$C = -\,2B\cos 2\alpha \tag{iii}$$

and (b) $M = \int_{-\alpha}^{\alpha} \tau_{r0}\, r^2\, d\theta$. Substituting from eq(ii) and (iii), we find the constants

$$B = M / [2(\sin 2\alpha - 2\alpha \cos 2\alpha)]$$
$$C = - M \cos 2\alpha / (\sin 2\alpha - 2\alpha \cos 2\alpha)$$

Substituting for B and C in eqs(i) and (ii) prescribes the stresses in the body of the wedge as

$$\sigma_r = - 2M \sin 2\theta / [r^2 (\sin 2\alpha - 2\alpha \cos 2\alpha)] \quad \text{(iv)}$$
$$\tau_{r0} = (M / r^2)[(\cos 2\theta - \cos 2\alpha) / [(\sin 2\alpha - 2\alpha \cos 2\alpha)] \quad \text{(v)}$$

Equations (iv) and (v) show (see Fig. 2.17) that σ_r is zero along the centre line of the wedge and reaches a maximum in tension and compression along the outer edges of magnitude

$$\sigma_r = \pm 2M / [r^2 (1 - 2\alpha \cot 2\alpha)]$$

In contrast, the shear stress is zero at the edges and with a maximum value on the centre-line

$$\tau_{r0} = (M / r^2)[(1 - \cos 2\alpha) / (\sin 2\alpha - 2\alpha \cos 2\alpha)]$$

Both stress components diminish rapidly with the square of radii away from the wedge tip.

2.6.2 Stress Functions in r

The stress function $\phi = \phi(r)$ refers to a body with symmetry about the z - axis. All points in the r, θ plane must displace radially. Thus $v = \partial v/\partial r = \partial u/\partial r = 0$ and eqs (2.35a-c) give $\tau_{r\theta} = 0$ with the strains

$$\varepsilon_r = du/dr, \quad \varepsilon_\theta = u/r \quad \text{(2.45a,b)}$$

These two strains depend upon a single displacement u. A compatibility condition, expresses this dependence as

$$\varepsilon_r = du/dr = d\,(r\varepsilon_\theta)/dr = r\, d\varepsilon_\theta/dr + \varepsilon_\theta$$
$$\therefore\ \varepsilon_r - \varepsilon_\theta = r\, d\varepsilon_\theta/dr$$

Since ϕ is independent of θ, then $\partial\phi/\partial\theta = 0$ and eq(2.38a) reduces to

$$[d^2/dr^2 + (1/r)\, d/dr][d^2\phi/dr^2 + (1/r)\, d\phi/dr] = 0$$

which gives

$$d^4\phi/dr^4 + (2/r)\, d^3\phi/dr^3 - (1/r^2) d^2\phi/dr^2 + (1/r^3)\, d\phi/dr = 0 \quad \text{(2.46a)}$$

Following a change of variable, with the substitution $r = \exp(t)$, the solution to eq(2.46a) is

$$\phi = A \ln r + Br^2 \ln r + C r^2 \quad \text{(2.46b)}$$

The stress function eqs(2.39a-c) yield $\tau_{r\theta} = 0$ and

$$\sigma_r = (1/r)\, \mathrm{d}\phi/\mathrm{d}r \tag{2.47a}$$

$$\sigma_\theta = \mathrm{d}^2\phi/\mathrm{d}r^2 \tag{2.47b}$$

where ϕ may be the sum of any combination of right-hand terms within eq(2.46b). The following three functions are of practical interest.

(a) Equi-biaxial Stressing

Applying eqs(2.47a,b) to the function $\phi = C\,r^2$ gives equal stresses $\sigma_r = \sigma_\theta = 2C$. These describe uniform plane stress states (i) within the wall of a thin-walled spherical vessel under pressure and (ii) in a thin plate subjected to co-planar, concurrent forces. Other functions that follow employ the $C\,r^2$ term to match boundary conditions for axi-symmetric bodies.

(b) Thick Cylinder Under Pressure

Taking the stress function $\phi = A\ \ln r + C\,r^2$, eqs(2.47a,b) supply radial and hoop stress components as

$$\sigma_r = 2C + A\,/\,r^2 \quad \text{and} \quad \sigma_\theta = 2C - A\,/\,r^2 \tag{2.48a,b}$$

These are the Lamé stresses (Gabriel Lamé 1795-1870). They exist in the wall of a long, thick cylinder subjected to internal (and external) pressure. The axial stress σ_z depends upon the end condition of the cylinder. The ends are said to be *open* when an internal pressure is contained by bore pistons. The cylinder wall in then unstressed axially ($\sigma_z = 0$). When *closed* ends contain the pressure, σ_z is found from a horizontal force equilibrium equation between the forces exerted by the axial stress in the cylinder wall and the internal pressure acting upon the end plates:

$$\sigma_z \pi (r_o^2 - r_i^2) = p\pi r_i^2$$

$$\therefore\ \sigma_z = p\,r_i^2 / (r_o^2 - r_i^2) \tag{2.49}$$

If the cylinder is built in to rigid end closures, its natural extension will be prevented. A *plane strain* condition may then be applied in which the axial strain ε_z is set either to zero (see Example 2.11) or to a constant value to find the axial wall stress.

Example 2.11 Determine the constants A and C in the Lamé equations, given that the cylinder is subjected to internal pressure only and that the axial strain is zero. Plot the distribution of σ_r, σ_θ and σ_z and derive a general expression for the radial displacement.

The boundary conditions are $\sigma_r = -\,p_i$ for $r = r_i$ and $\sigma_r = 0$ for $r = r_o$. Substituting into eq(2.48a) leads to the two simultaneous equations:

$$-\,p_i = 2C + A\,/\,r_i^2 \tag{i}$$

$$0 = 2C + A\,/\,r_o^2 \tag{ii}$$

Subtracting (ii) from (i) gives

$$A = -\,p r_i^2 r_o^2 / (\,r_o^2 - r_i^2), \quad 2C = p r_i^2 / (r_o^2 - r_i^2)$$

Substituting for A and $2C$ into eqs(2.48a and b)

$$\sigma_r = pr_i^2(1 - r_o^2/r^2)/(r_o^2 - r_i^2) \quad \text{(iii)}$$

$$\sigma_\theta = pr_i^2(1 + r_o^2/r^2)/(r_o^2 - r_i^2) \quad \text{(iv)}$$

When $\varepsilon_z = 0$

$$\sigma_z = \nu(\sigma_r + \sigma_\theta) = 2\nu pr_i^2/(r_o^2 - r_i^2) \quad \text{(v)}$$

Equation (v) differs slightly from the closed-end cylinder expression (2.49) but both give constant σ_z (independent of r). The stresses in eqs(iii)-(v) are distributed as shown in Fig. 2.18.

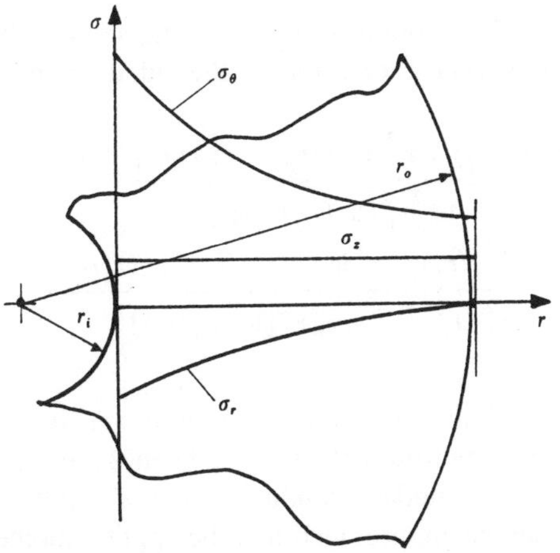

Figure 2.18 Stress distributions within the wall of a thick cylinder under internal pressure

At $r = r_i$ the radial stress σ_r is maximum in compession and the hoop stress σ_θ is a maximum in tension. The radial displacement follows from eq(iii)-(v) and either of eqs(2.45a or b):

$$u = r\varepsilon_\theta = r[\sigma_\theta - \nu(\sigma_r + \sigma_z)]/E$$

$$u = r[\sigma_\theta(1 - \nu^2) - \nu\sigma_r(1 + \nu)]/E$$

$$= prr_i^2[(1 - \nu^2)(1 + r_o^2/r^2) - \nu(1 + \nu)(1 - r_o^2/r^2)]/[E(r_o^2 - r_i^2)]$$

$$= pr_i^2[(1 - \nu - 2\nu^2)r + (1 + \nu)r_o^2/r]/[E(r_o^2 - r_i^2)]$$

(c) Bending of a Curved Beam

Taking the complete function ϕ from eq(2.46b), the following radial and tangential stresses are found from eq(2.47a,b):

$$\sigma_r = A/r^2 + B(1 + 2\ln r) + 2C \quad \text{(2.50a)}$$

$$\sigma_\theta = -A/r^2 + B(3 + 2\ln r) + 2C \quad \text{(2.50b)}$$

Golovin (1881) recognised that eqs(2.50a,b) are the exact solutions to the stresses in a curved beam under pure bending (see Fig. 2.19 and reference p.75). To find constants A, B and C, we let a hogging moment M be applied to a curved rectangular beam of unit thickness and with inner and outer radii r_i and r_o respectively. The first boundary condition is that of zero stress, normal to the outer surfaces, $\sigma_r = 0$ for $r = r_i$ and $r = r_o$. From eqs (2.47a) and (2.50a)

$$\sigma_r = [(1/r)\mathrm{d}\phi/\mathrm{d}r]_{r_i} = A/r_i^2 + B(1 + 2\ln r_i) + 2C = 0 \quad \text{(2.51a)}$$

$$\sigma_r = [(1/r)\mathrm{d}\phi/\mathrm{d}r]_{r_o} = A/r_o^2 + B(1 + 2\ln r_o) + 2C = 0 \quad \text{(2.51b)}$$

The second boundary condition refers to end moment equilibrium and uses eq(2.47b)

$$M = \int_{r_i}^{r_o} \sigma_\theta \, r \, dr = \int_{r_i}^{r_o} (d^2\phi/dr^2) r \, dr = | \, r \, d\phi/dr \, |_{r_i}^{r_o} - \int_{r_i}^{r_o} (d\phi/dr) \, dr \tag{2.52a}$$

It follows from eqs(2.51a and b), that the first term is zero. Integrating eq(2.52a) and substituting into eq(2.46b),

$$M = - |\phi|_{r_i}^{r_o} = - [\, A \ln (r_o/ r_i) + B\,(r_o^2 \ln r_o - r_i^2 \ln r_i) + C\,(\, r_o^2 - r_i^2)\,] \tag{2.52b}$$

The constants A, B and C are found from the simultaneous solution to eqs(2.51a,b) and (2.52b). This results in the final expression for the radial and hoop stresses

$$\sigma_r = \frac{4M\,[(r_i^2 r_o^2/r^2) \ln (r_o/r_i) + r_o^2 \ln (r/r_o) + r_i^2 \ln (r_i/r)]}{(r_o^2 - r_i^2)^2 - 4\, r_o^2 r_i^2 [\ln (r_o/r_i)]^2} \tag{2.53a}$$

$$\sigma_\theta = \frac{4M\,[-(r_i^2 r_o^2/r^2) \ln (r_o/r_i) + r_o^2 \ln (r/r_o) + r_i^2 \ln (r_i/r) + (r_o^2 - r_i^2)]}{(r_o^2 - r_i^2)^2 - 4\, r_o^2 r_i^2 [\ln (r_o/r_i)]^2} \tag{2.53b}$$

Equation (2.53b) is the bending stress with the distribution shown in Fig. 2.19. The position of the neutral axis can be found from setting $\sigma_\theta = 0$ in eq(2.53b). Equations (2.53a,b) satisfy both compatibility and the boundary conditions and will provide exact solutions to the stresses everywhere in the beam, provided that the applied moments correspond to the σ_θ end-distribution, as shown. Even if this end condition is not met exactly, *St Venant's principle* (Saint-Venant 1797-1886) assures us that the solution will be accurate at a distance equal to the beam depth beyond the ends. The principal cannot guarantee similar accuracy for approximate solutions to this problem where they do not satisfy the three fundamental requirements of elasticity theory. These are equilibrium of forces, compatibility of strain and a match to the boundary conditions.

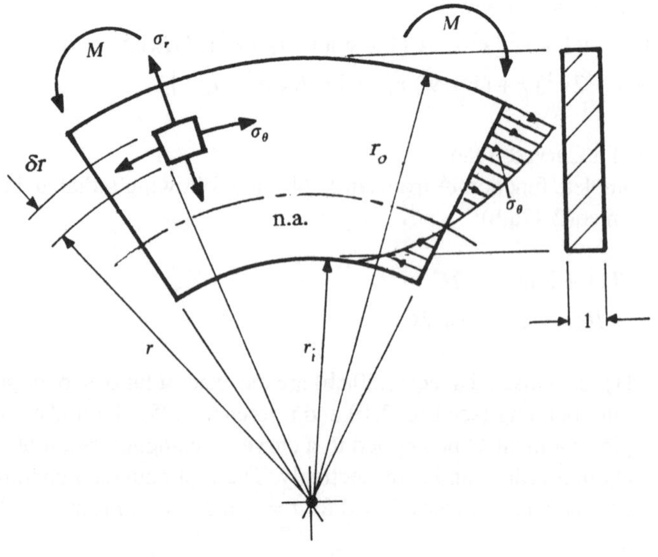

Figure 2.19 Bending of curved bar

2.6.3 General Function, $\phi = \phi(\theta, r)$

(a) Point Normal Force on Semi-infinite Plate
One of the most useful functions to satisfy eq(2.38b) is that attributed to Boussinesq (1885)[a] and Flambert (1892)[a]. Their function has the following form:

$$\phi = Cr\theta\sin\theta \tag{2.54}$$

Applying eqs(2.39a-c) to eq(2.54) gives the stresses as $\sigma_\theta = \tau_{r\theta} = 0$ and

$$\begin{aligned}\sigma_r &= (1/r)\,\partial\phi/\partial r + (1/r^2)\,\partial^2\phi/\partial\theta^2 \\ &= (C/r)\theta\sin\theta + (1/r^2)(\partial/\partial\theta)[Cr\sin\theta + Cr\theta\cos\theta] \\ &= (2C/r)\cos\theta \end{aligned} \tag{2.55a}$$

This radial stress arises from applying an outward point force normally to the straight edge of a large (semi-infinite) plate, as shown in Fig. 2.20a.

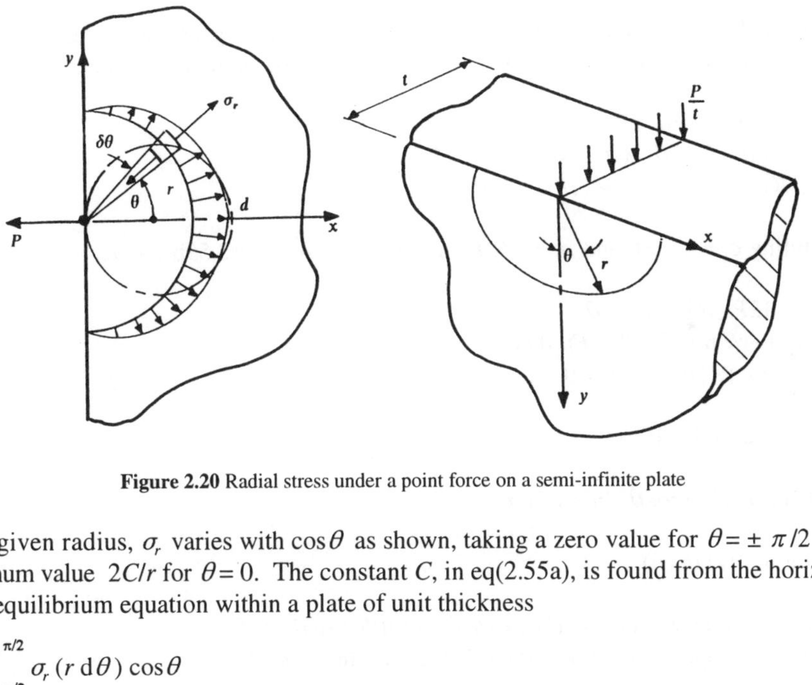

Figure 2.20 Radial stress under a point force on a semi-infinite plate

For a given radius, σ_r varies with $\cos\theta$ as shown, taking a zero value for $\theta = \pm\,\pi/2$ and a maximum value $2C/r$ for $\theta = 0$. The constant C, in eq(2.55a), is found from the horizontal force equilibrium equation within a plate of unit thickness

$$P = \int_{-\pi/2}^{\pi/2} \sigma_r\,(r\,\mathrm{d}\theta)\cos\theta$$

$$P = 2C\int_{-\pi/2}^{\pi/2} \cos^2\theta\,\mathrm{d}\theta = C\,|\theta + ½\sin 2\theta\,\Big|_{-\pi/2}^{\pi/2} = C\pi$$

In practice, it is more common for a compressive force to act in a line across a plate of thickness t (see Fig. 2.20b). Then P is replaced by $-P/t$ to give $-P/t = C\pi$, when eq(2.55a) supplies a compressive radial stress

$$\sigma_r = -\,[2P/(\pi tr)]\cos\theta \tag{2.55b}$$

Consider the circle of diameter d, lying tangential to y at P, in Fig.2.20a.

---

[a] see Timoshenko S. P. and Goodier J. N., *Theory of Elasticity*, 2nd edition, McGraw-Hill, 1984.

Since $r = d\cos\theta$, eq(2.55a) becomes

$$\sigma_r = 2C/d = 2P/(\pi d) = \text{constant} \tag{2.55c}$$

Note also that the maximum shear stress in the plate is given by $\tau_{max} = (\sigma_r - \sigma_\theta)/2$. Substituting from eq(2.55c), with $\sigma_\theta = 0$, gives $\tau_{max} = \sigma_r/2 = P/(\pi d)$. This means that both σ_r and τ_{max} are constant around the given tangent circle, each being inversely proportional to its diameter (τ_{max} is *not* the stress component $\tau_{r\theta}$). The theory is readily confirmed from point loading the straight edge of a birefringent transparent polymer and viewing the image under polarised light. The photoelastic method reveals the tangent circles as isochromatic fringes, each of a constant shear stress value. An example of an isochromatic fringe pattern is shown on the front cover of this book.

Example 2.12 Determine the displacement expressions in the body of a thin, semi-infinite plate when a compressive point force is applied normal to the straight edge.

Referring to Fig. 2.20b, the displacements follow from combining the plane stress constitutive relations eqs(2.14a,b,d) with the polar strain displacement relations (2.35a,b,c),

$$\varepsilon_r = \partial u/\partial r = (\sigma_r - \nu\sigma_\theta)/E$$
$$\varepsilon_\theta = u/r + (1/r)\,\partial v/\partial\theta = (\sigma_\theta - \nu\sigma_r)/E$$
$$\gamma_{r\theta} = \tau_{r\theta}/G = \partial v/\partial r - v/r + (1/r)\partial u/\partial\theta$$

Substituting $\sigma_\theta = \tau_{r\theta} = 0$ and $\sigma_r = -[2P/(\pi tr)]\cos\theta$, from eq(2.55b), gives

$$\partial u/\partial r = -[2P/(\pi Etr)]\cos\theta \tag{i}$$
$$u/r + (1/r)\,\partial v/\partial\theta = [2\nu P/(\pi Etr)]\cos\theta \tag{ii}$$
$$\partial v/\partial r - v/r + (1/r)\,\partial u/\partial\theta = 0 \tag{iii}$$

Integrating eq(i) leads to

$$u = -(2P/(\pi tE)]\,(\cos\theta)\ln r + f(\theta) \tag{iv}$$

Substituting (iv) in (ii)

$$-[2P/(\pi tE)]\,(\cos\theta)\ln r + f(\theta) + \partial v/\partial\theta = [2\nu P/(\pi tE)]\cos\theta$$
$$\therefore\ \partial v/\partial\theta = [2\nu P/(\pi tE)]\cos\theta + [2P/(\pi tE)]\,(\cos\theta)\ln r - f(\theta)$$
$$v = \int \{[2\nu P/(\pi tE)]\cos\theta + [2P/(\pi tE)]\,(\cos\theta)\ln r - f(\theta)\}\,\mathrm{d}\theta$$
$$= [2\nu P/(\pi tE)]\sin\theta + [2P/(\pi tE)]\,(\sin\theta)\ln r - \int f(\theta)\,\mathrm{d}\theta + g(r) \tag{v}$$

From eqs(iv) and (v), the partial derivatives of u and v are

$$\partial u/\partial\theta = [2P/(\pi tE)\}\,(\sin\theta)\ln r + f'(\theta) \tag{vi}$$
$$\partial v/\partial r = [2P/(\pi tEr)]\sin\theta + g'(r) \tag{vii}$$

Substituting eqs(v), (vi) and (vii) into (iii) gives

$$f'(\theta) + [2P/(\pi tE)](1 - \nu)\sin\theta + r\,g'(r) + \int f(\theta)\,\mathrm{d}\theta - g(r) = 0$$

The right-hand side is zero, so the sum of the separate functions in r and θ equals zero:

$$rg'(r) - g(r) = 0 \text{ or } \mathrm{d}g/g = \mathrm{d}r/r$$

$$\therefore \ln g = \ln r + \ln A_1 \Rightarrow g(r) = A_1 r \qquad \text{(viii)}$$

and

$$f'(\theta) + [2P/(\pi tE)](1 - \nu) \sin\theta + \int f(\theta)\, \mathrm{d}\theta = 0$$

$$\mathrm{d}^2 f / \mathrm{d}\theta^2 + f = -[2P/(\pi tE)](1 - \nu) \cos\theta \qquad \text{(ix)}$$

The solution to eq(ix) is

$$f(\theta) = A_2 \sin\theta + A_3 \cos\theta - (1 - \nu)[P/(\pi tE)]\theta \sin\theta \qquad \text{(x)}$$

The general u, v displacement expressions are found by substituting eqs(viii) and (x) into eqs(iv) and (v). This gives

$$u = A_2 \sin\theta + A_3 \cos\theta - [2P/(\pi tE)] (\cos\theta) \ln r - (1 - \nu)[P/(\pi tE)]\theta \sin\theta \qquad \text{(xi)}$$

$$v = (1 + \nu)[P/(\pi tE)] \sin\theta + [2P/(\pi tE)] (\sin\theta) \ln r - (1 - \nu)[P/(\pi tE)]\theta \cos\theta$$
$$+ A_1 r + A_2 \cos\theta - A_3 \sin\theta \qquad \text{(xii)}$$

Since $v = 0$ when $\theta = 0$ along the x - axis, then, from eq(xii), $A_2 = -A_1 r$. This condition can only be satisfied by $A_2 = A_1 = 0$ for all r. Because deformation is localised, it is assumed that $u = 0$ at some radius r_o from P along the x - axis (where $\theta = 0°$). Thus, from eq(xi),

$$u = 0 = A_3 - [2P/(\pi tE)] \ln r$$

$$\therefore A_3 = [2P/(\pi tE)] \ln r_o$$

Substituting for A_1, A_2 and A_3 into eqs(xi) and (xii), the plate displacements at a point θ and $r \le r_o$ are given by

$$u = [P/(\pi tE)][2 \ln (r_o / r) \cos\theta - (1 - \nu)\theta \sin\theta] \qquad \text{(xiii)}$$

$$v = [P/(\pi tE)][(1 + \nu) \sin\theta + 2 (\sin\theta) \ln (r/r_o) - (1 - \nu)\theta \cos\theta] \qquad \text{(xiv)}$$

When $\theta = 0°$, eqs(xiii) and (xiv) give $v = 0$ and $u = [2P/(\pi tE)]\ln(r_o/r)$. This shows that the displacement under P falls rapidly to zero as $r \to r_o$. The fact that both σ_r and u become infinite for $r = 0$ indicates that localised plasticity would occur if it were possible to achieve a true point force. Yielding is less likely to occur in practice where the force is distributed over a finite area. Any yielding that does occur at the load point will result only in a local disturbance to the predicted internal elastic stress distribution.

(b) Tangential Force on Semi-infinite Plate

The radial stress expression eq(2.55a) also applies when a tangential shear force is applied to the straight edge of a semi-infinite plate. This requires that θ be measured from an x - axis aligned with the force direction, as shown in Fig. 2.21a.

(c) Point Force Applied to Finite Bodies

Equation (2.55a) may also be applied to point loading of finite bodies when the constant C is re-defined. For example, this equation supplies the radial stress distribution within the body of a wedge when an inclined compressive force is applied to its tip. Figure 2.21b shows a wedge of unit thickness, apex angle 2α and inclination β of its compressive tip force P.

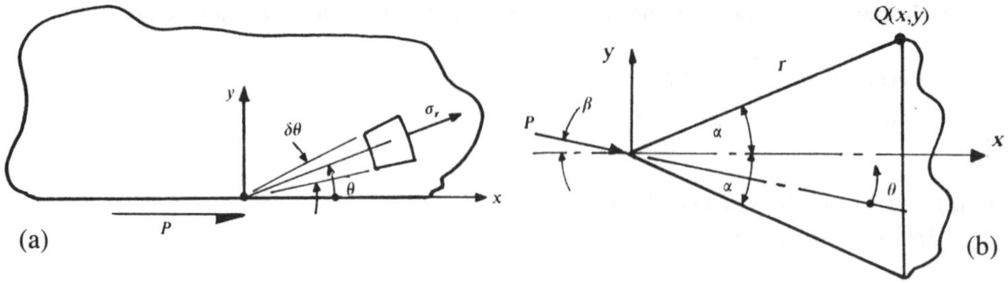

Figure 2.21 Further applications of the Boussinesq function

By taking θ anti-clockwise from the force line, the horizontal equilibrium equation becomes

$$P/t + \int_{-(\alpha-\beta)}^{(\alpha+\beta)} \sigma_r (r\,\mathrm{d}\theta)\cos\theta = 0$$

$$\therefore P/t = -2C\int_{-(\alpha-\beta)}^{(\alpha+\beta)} \cos^2\theta\,\mathrm{d}\theta$$

$$= -C\,|\theta + ½\sin 2\theta\,\Big|_{-(\alpha-\beta)}^{(\alpha+\beta)} = -C\,(2\alpha + \sin 2\alpha\cos 2\beta)$$

$$C = -P/[t\,(2\alpha + \sin 2\alpha\cos 2\beta)]$$

$$\therefore \sigma_r = -(2P\cos\theta)/[r\,t\,(2\alpha + \sin 2\alpha\cos 2\beta)] \qquad (2.56)$$

The wedge stresses arising from horizontal and vertical tip forces will be found by substituting $\beta = 0$ and $\beta = \pi/2$ respectively in eq(2.56). When these loads are combined, the σ_r distributions corresponding to each load applied separately may be superimposed to give a net stress distribution.

Example 2.13 The wedge in Fig. 2.21b has an apex angle of 30° and supports a vertical downward force of 10 kN. If the wedge thickness is 20 mm, find the maximum stress on a radius of 50 mm. Show that the simple theory of bending will provide 75% of the maximum radial stress when $2\alpha \le 10°$.

In this case we set $\beta = 90° = \pi/2$, in eq(2.56), to give

$$\sigma_r = -(P\cos\theta)/[r\,t\,(\alpha - \sin\alpha\cos\alpha)] \qquad (i)$$

Since θ is measured from the force line, the wedge interior is defined by $75° \le \theta \le 105°$. On the bottom edge $\theta = 75°$ and on the top edge $\theta = 105°$. In addition $P = 10 \times 10^3$ N, $t = 20$ mm, $r = 50$ mm and $\alpha = 15°$ in eq(i). The radial stress on the bottom edge becomes: $\sigma_r = -219.4$ MPa in compression and on the top edge $\sigma_r = 219.4$ MPa in tension. Also when $\theta = 90°$, $\sigma_r = 0$ along the wedge axis.

The radial stress distribution resembles that obtained from bending a cantilever beam of rectangular section (see Fig. 2.10). Indeed, we may apply bending theory to a slightly tapered cantilever beam to provide an approximation to these maxima. Let x define the wedge axis in Fig. 2.21b so that a point on the sloping side has co-ordinates $Q\,(x, y)$. When $\beta = 90°$, the bending stress at Q is

$$\sigma = My/I \qquad (ii)$$

where the bending moment $M = Px$ and the second moment of area for the transverse plane containing Q is $I = t\,(2\,y)^3 / 12$. Substituting into (ii) gives

$$\sigma = 3Px / (2\,t\,y^2) \tag{iii}$$

When $\alpha \le 5°$, we may write

$$y^2 = r^2 - x^2 = (r - x)(r + x) \approx 2x\,(r - x\,) \tag{iv}$$

Substituting eq(iv) into (iii) gives

$$\sigma = 3P / [4\,t\,(r - x)\,] \tag{v}$$

When α is small, $\sin\alpha = \alpha$, so that eq(i) approximates to a tensile bending stress,

$$\begin{aligned}\sigma_r &\simeq -\,P\cos(90 + \alpha) / [\,r\,t\,\alpha(1 - \cos\alpha\,)\,]\\ &= P / [r\,t\,(1 - \cos\alpha\,)\,]\end{aligned} \tag{vi}$$

and setting $\cos\alpha = x/r$ in eq(vi) gives

$$\sigma_r = P / [\,t\,(r - x\,)\,] \tag{vii}$$

Comparing eqs(v) and (vii) shows that bending theory approximates to 75% of the maximum tensile radial stress. The same approximation holds for the compressive side where negative signs accompany eqs(v) and (vii).

(d) Disc Under Diametral Compression

Consider a disc of diameter d and thickness t subjected to diametral compressive force P, as shown in Fig. 2.22a. A stress function of the following form applies:

$$\phi = Cr\theta\sin\theta + Dr^2 \tag{2.57}$$

Equations (2.39a-c) give the stress components as $\tau_{r\theta} = 0$

$$\sigma_r = (2C/r)\cos\theta + 2D \quad \text{and} \quad \sigma_\theta = 2D \tag{2.58a,b}$$

The term in D superimposes a uniform biaxial tension upon Boussinesq's radial stress field. This is necessary to account for the interruption caused to the stress field by the disc boundary. Neither a normal stress nor a tangential shear stress can exist around the boundary. This condition can be matched when eqs(2.58a,b) superimpose to give $\sigma_\theta = \sigma_r = 0$. Referring to Fig. 2.22a, the disc diameter is $d = r/\cos\theta = r'/\cos\theta'$.

The combined stress state at any point A on the boundary, resulting from each force P, is given by the sum of eqs(2.58a,b). This gives an equibiaxial tension as shown:

$$\sigma_r = \sigma_\theta = (2C/d) + 4D \tag{2.58c}$$

The resolution in σ_θ and σ_r normal and parallel to the boundary at A, will give two direct stresses of equal magnitude with no shear stress. It follows from eq(2.58c) that the normal stress will be zero when $D = -\,C/(2d)$. Hence, for any point (r, θ) in the disc, the stresses eqs(2.58a,b) due to a single force are

$$\sigma_r = (2C/r)\cos\theta - C/d, \quad \sigma_\theta = -C/d \tag{2.59a,b}$$

The constant C is found by applying the horizontal equilibrium condition across a section aligned with the y - axis

$$P + \int \sigma_x t \, dy = 0 \tag{2.60}$$

where x is aligned with horizontal force line.

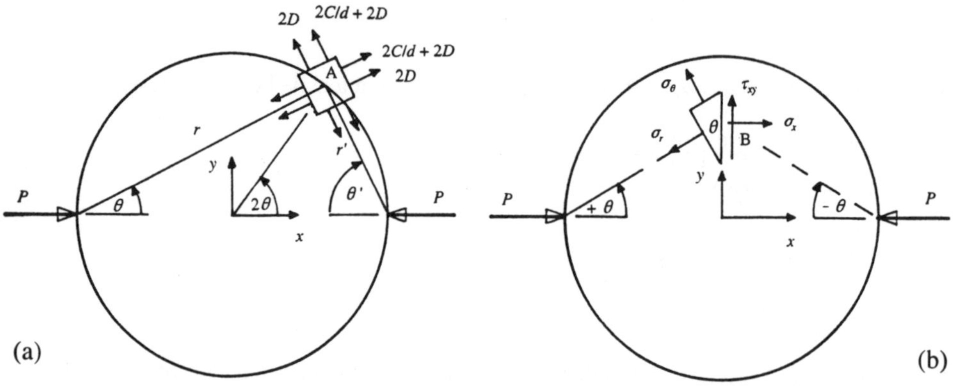

Figure 2.22 Stresses σ_r and σ_θ in a disc under compression

At point B in Fig. 2.22b, σ_x is the sum of the the radial and hoop stresses resulting from each force resolved into the x - direction. The stress transformations for the left half of the disc are

$$\sigma_x = \sigma_r \cos^2\theta + \sigma_\theta \sin^2\theta \tag{2.61a}$$
$$\sigma_y = \sigma_r \sin^2\theta + \sigma_\theta \cos^2\theta \tag{2.61b}$$
$$\tau_{xy} = \tfrac{1}{2}(\sigma_r - \sigma_\theta)\sin 2\theta \tag{2.61c}$$

For the right half of the disc, θ in eqs(2.61a-c) is replaced by $-\theta$, to give

$$\sigma_x = \sigma_r \cos^2\theta + \sigma_\theta \sin^2\theta \tag{2.62a}$$
$$\sigma_y = \sigma_r \sin^2\theta + \sigma_\theta \cos^2\theta \tag{2.62b}$$
$$\tau_{xy} = -\tfrac{1}{2}(\sigma_r - \sigma_\theta)\sin 2\theta \tag{2.62c}$$

The sum of eqs(2.61c) and (2.62c) confirms that $\sum \tau_{xy} = 0$ along an axis of symmetry. Substituting $\sum \sigma_x$ from eqs(2.61a) and (2.62a) into eq(2.60),

$$P + 2t \int (\sigma_r \cos^2\theta + \sigma_\theta \sin^2\theta)\, dy = 0 \tag{2.63a}$$

Since $y = (d/2)\tan\theta$, then $dy/d\theta = (d/2)\sec^2\theta$. Substituting for d$y$ and σ_r and σ_θ from eq(2.59a,b) into (2.63a) leads to

$$P + 2Ct \int_0^{\frac{\pi}{4}} (4\cos^2\theta - \sec^2\theta)\, d\theta = 0 \tag{2.63b}$$

Equation (2.63b) gives $C = -P/(\pi t)$. Thus, for the y - axis,

$$\sum \sigma_x = -\ [2P/(\pi dt)](4\cos^2\theta - 1)\cos^2\theta + (2P/\pi dt)\sin^2\theta \qquad (2.64a)$$
$$\sum \sigma_y = -\ [2P/(\pi dt)](4\cos^2\theta - 1)\sin^2\theta + (2P/\pi dt)\cos^2\theta \qquad (2.64b)$$

Finally, substituting $\cos\theta = d/\surd(d^2 + 4y^2)$, eqs(2.64a,b) lead to

$$\sigma_x = -\ 2P\,(3d^2 + 4y^2)(d^2 - 4y^2)\,/\,[(\pi dt)(d^2 + 4y^2)] \qquad (2.65a)$$
$$\sigma_y = 2P\,(d^2 - 4y^2)^2/\,[(\pi dt)(d^2 + 4y^2)] \qquad (2.65b)$$

Equations (2.65a,b) show that $\sigma_x = \sigma_y = 0$ at the boundary ($y = d/2$). At the disc centre ($y = 0$), these attain their maxima $\sigma_x = -\ 6P/(\pi dt)$ and $\sigma_y = 2P/(\pi dt)$, in compression and tension respectively.

(e) Stress Concentrations

A function of the following form

$$\phi(r,\theta) = f(r)\cos 2\theta \qquad (2.66a)$$

will satisfy eq(2.38b) when $f(r)$ is written as

$$f(r) = Ar^2 + Br^4 + C/r^2 + D \qquad (2.66b)$$

in which A, B, C and D are constants. Equation (2.66b) may be applied to the uniformly stressed plate with the small central hole, of radius a, shown in Fig. 2.23a.

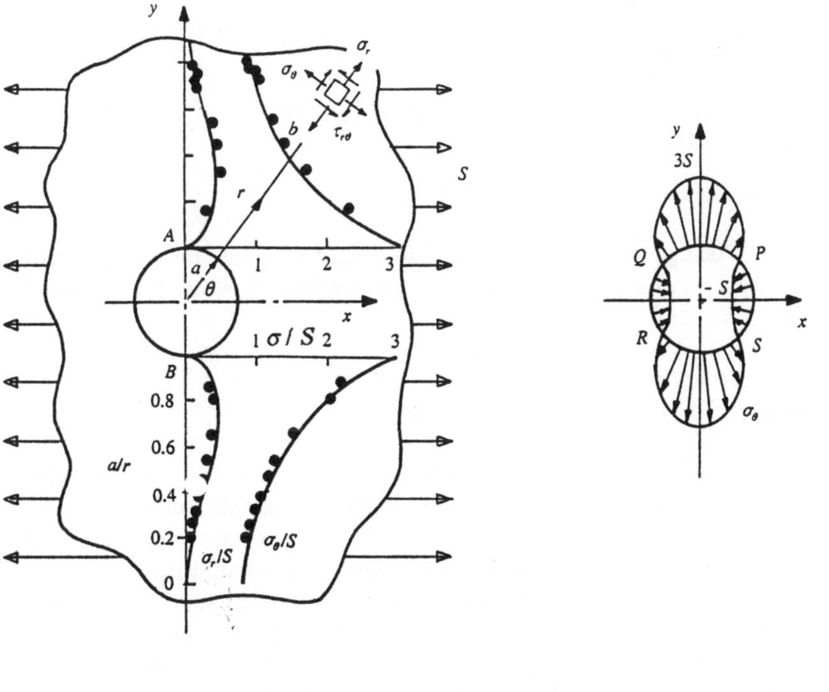

(a) (b)

Figure 2.23 Stress distributions around a hole in a uniformly stressed plate

The full effect that the hole has in concentrating stress is determined by superimposing the Lamé stress function (2.46b). This is written as

$$\phi = \phi(r) = E \ln r + F r^2 \tag{2.67}$$

where E and F are constants. Using St Venant's principle, it may be assumed that at any radius $b >> a$, the stress distribution is not affected by the hole. At radius b, a uniform radial Lamé pressure $S/2$ appears as a result of transforming the horizontally applied stress, S. Now at the position (b, θ) shown, the radial and shear stress components are

$$\sigma_r = S \cos^2\theta = (S/2)(1 + \cos 2\theta) \tag{2.68a}$$
$$\tau_{r\theta} = -(S/2) \sin 2\theta \tag{2.68b}$$

The accompanying tangential shear stress component (2.68b) and that part of eq(2.68a) which is $\sigma_r = (S/2)\cos 2\theta$ are derived from eq(2.66a) and the stress functions (2.39a,c). It follows that for $r \leq b$, the stress components σ_r, σ_θ and $\tau_{r\theta}$ are the result of summing the stress functions (2.66a) and (2.67). That is,

$$\phi(r,\theta) = (A r^2 + Br^4 + C/r^2 + D) \cos 2\theta + (E \ln r + F r^2) \tag{2.69a}$$

for which eqs(2.39a-c) supply the stress components. The boundary conditions for eqs(2.68a,b) are (i) $\sigma_r = \tau_{r\theta} = 0$ for $r = a$ and (ii) $\sigma_r = S/2$ and $\tau_{r\theta} = -(S/2)\sin 2\theta$ for $r = b$. Applying conditions (i) and (ii) to an infinitely large plate, where $a/b \sim 0$, leads to $A = -S/4$, $B = 0$, $C = -a^4 S/4$, $D = a^2 S/2$, $E = -a^2 S/2$ and $F = S/4$. Equation(2.69a) becomes

$$\phi(r,\theta) = (S/4)[r^2 - 2a^2 \ln r - (r - a^2/r)^2 \cos 2\theta] \tag{2.69b}$$

From eqs(2.39a,b,c),

$$\sigma_r = (S/2)[(1 - a^2/r^2) + (1 - 4a^2/r^2 + 3a^4/r^4) \cos 2\theta] \tag{2.70a}$$
$$\sigma_\theta = (S/2)[(1 + a^2/r^2) - (1 + 3a^4/r^4) \cos 2\theta] \tag{2.70b}$$
$$\tau_{r\theta} = -(S/2)(1 + 2a^2/r^2 - 3a^4/r^4) \sin 2\theta \tag{2.70c}$$

Equations (2.70a,b,c) are particularly useful for determining the stress concentration factor. On a transverse y - axis, passing through the centre of the hole, where $\theta = \pi/2$ and $\theta = 3\pi/2$, they give $\tau_{r\theta} = 0$, and

$$\sigma_\theta = (S/2)(2 + a^2/r^2 + 3a^4/r^4) \tag{2.71a}$$
$$\sigma_r = (S/2)(3a^2/r^2)(1 - a^2/r^2) \tag{2.71b}$$

Clearly when $r \approx a$, $\sigma_\theta = 3S$, $\sigma_r = 0$ and when $r \geq 10a$, $\sigma_\theta \approx S$, $\sigma_r \approx 0$. The manner in which stress is distributed according to eqs(2.71a,b), within intermediate radii $a < r < 10a$, is shown in Fig. 2.23a. The two distributions agree with experimental data given for a 10 s.w.g., 200 mm wide L72 duralumin plate with a 50.8 mm central hole. They confirm a threefold magnification (the stress concentration factor) in the axial stress for two transverse points A and B on the hole boundary. In Fig. 2.23b, the hoop stress distribution around the hole is found by substituting $r = a$ in eq(2.70b). This shows that the stress changes from tensile to compressive at the four singular points (P, Q, R and S) where $\sigma_\theta = 0$. Equation (2.70b) shows that these points occur at $\theta = ½ \cos^{-1}(0.5) = 30°, 150°, 210°$ and $330°$.

(a)

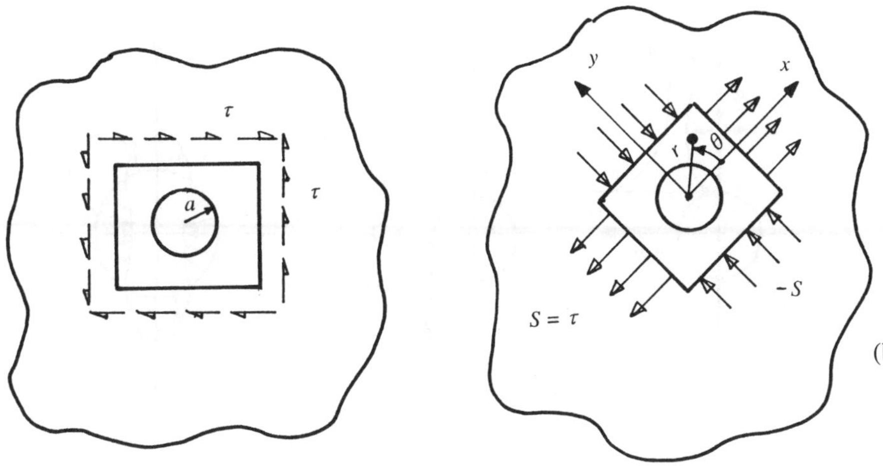

(b)

Figure 2.24 Hole in a plate under uniform shear

The solutions (2.70a-c) may be extended to pure shear loading of a plate with a hole, shown in Fig. 2.24a. For this, the equivalent stress system given in Fig. 2.24b is employed. It follows from eqs(2.70a-c) that the stress components under the uniform compressive stress ($-S$) for the y - direction are found by replacing θ with $\theta - \pi/2$. This gives upon factorisation,

$$\sigma_r = (-S/2)(1 - a^2/r^2)\{1 + (1 - 3a^2/r^2)\cos[2(\theta - \pi/2)]\} \quad (2.72a)$$

$$\sigma_\theta = (-S/2)\{(1 + a^2/r^2) - (1 + 3a^4/r^4)\cos[2(\theta - \pi/2)]\} \quad (2.72b)$$

$$\tau_{r\theta} = -(-S/2)(1 - a^2/r^2)(1 + 3a^2/r^2)\sin[2(\theta - \pi/2)] \quad (2.72c)$$

Adding eqs(2.70) and (2.72) gives the stress state under pure shear

$$\sigma_r = S(1 + 3a^4/r^4 - 4a^2/r^2)\cos 2\theta \quad (2.73a)$$

$$\sigma_\theta = -S(1 + 3a^4/r^4)\cos 2\theta \quad (2.73b)$$

$$\tau_{r\theta} = -S(1 - 3a^4/r^4 + 2a^2/r^2)\sin 2\theta \quad (2.73c)$$

Figures 2.25a,b show the manner in which eqs(2.73a and b) are distributed along the y ($\theta = \pi/2$) and x ($\theta = 0°$) axes respectively. Where these axes intersect the hole ($r = a$), σ_θ is magnified by a factor of 4 in both the x and y directions. Setting $r = a$ in eq(2.73b), the hoop stress around the hole is $\sigma_\theta = -4S\cos 2\theta$. This reveals unstressed points for $\theta = 45°$, 135°, 225° and 315° in Fig. 2.25c.

The principal of superposition can again be applied to the general case of in-plane biaxial stressing of a plate with a hole under under S_x and S_y (see Fig. 2.26a). Setting $\theta = \theta - 90°$ with $S = S_y$ in eqs(2.70a-c) gives σ_r, σ_θ and $\tau_{r\theta}$ for S_y acting alone. Adding these to eqs(2.70a-c), for when S_x acts alone, gives the stress components for the combined loading

$$\sigma_r = \tfrac{1}{2}(S_x + S_y)(1 - a^2/r^2) + \tfrac{1}{2}(S_x - S_y)(1 - 4a^2/r^2 + 3a^4/r^4)\cos 2\theta \quad (2.74a)$$

$$\sigma_\theta = \tfrac{1}{2}(S_x + S_y)(1 + a^2/r^2) - \tfrac{1}{2}(S_x - S_y)(1 + 3a^4/r^4)\cos 2\theta \quad (2.74b)$$

$$\tau_{r\theta} = -\tfrac{1}{2}(S_x - S_y)(1 + 2a^2/r^2 - 3a^4/r^4)\sin 2\theta \quad (2.74c)$$

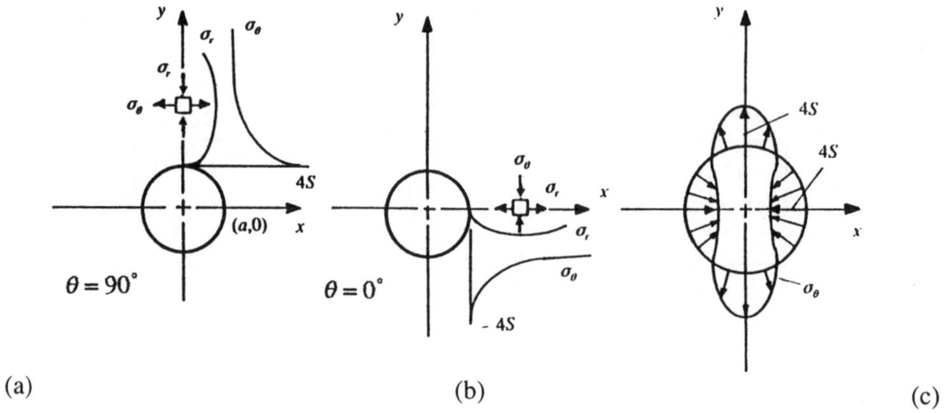

Figure 2.25 Stress distributions around a hole in a plate under pure shear

Equation (2.74b) allows the stress concentration at the hole ($r = a$) to be determined for given ratios $K = S_y/S_x$. For $\theta = 0°$ this gives $\sigma_\theta/S_x = 3K - 1$ and for $\theta = \pi/2$, $\sigma_\theta/S_x = 3 - K$. The linear dependence of a normalised stress concentration (σ_θ/S_x) upon K is shown in Fig. 2.26b. This includes the previous cases of uniaxial tension when $K = 0$ and also pure shear when $K = -1$. As K increases beyond 3, the stress concentration increases positively for $\theta = 0°$ under biaxial tension and decreases negatively for $\theta = \pi/2$.

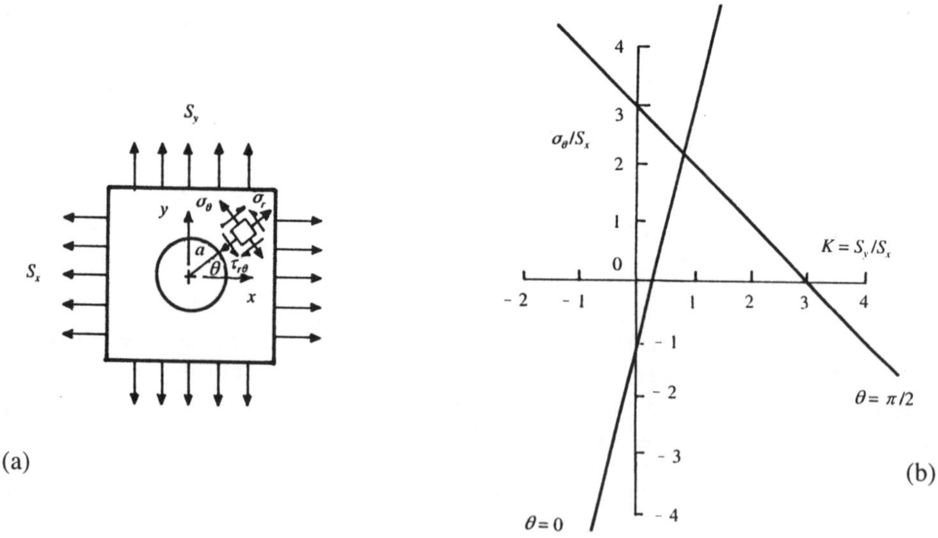

Fig. 2.26 Stress concentrations for a plate with hole under biaxial tension

Example 2.14 Determine the thickness of reinforcement necessary around a circular window in a pressurised fuselage in order to eliminate the stress concentration

In a thin-walled tubular fuselage, the pressure produces biaxial stressing in the ratio $S_y/S_x = 2$. In the absence of reinforcement, the greatest stress concentration would be 5 (see Fig. 2.26b). To eliminate this, the hole bead reinforcement must serve to reduce stresses to those values existing away from the hole, i.e for $r >> a$ in eqs(2.74a,b,c). That is,

$$\sigma_r = (S_x/2)(3 - \cos 2\theta) \quad \text{(i)}$$
$$\sigma_\theta = (S_x/2)(3 + \cos 2\theta) \quad \text{(ii)}$$
$$\tau_{r\theta} = (S_x/2)\sin 2\theta \quad \text{(iii)}$$

Let the reinforcement consist of a circular section bead bonded to the hole at radius a. Figure 2.27a shows that the radial and shear forces exerted by the fuselage plate (thickness t) on an adjacent element of this bead are $\sigma_r(a\delta\theta)t$ and $\tau_{r\theta}(a\delta\theta)t$ respectively. The inner surface is force free. Also shown are the variations in the tensile force T and transverse shear force Q carried by the bead. These are $(\mathrm{d}T/\mathrm{d}\theta)\delta\theta$ and $(\mathrm{d}Q/\mathrm{d}\theta)\delta\theta$ respectively in the direction θ-positive.

(a) 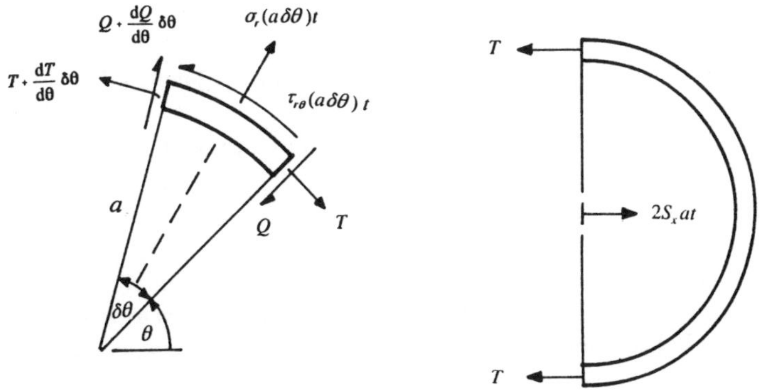 (b)

Figure 2.27 Forces on a reinforcing bead

Radial force equilibrium gives

$$\sigma_r t\,(a\delta\theta) + (\mathrm{d}Q/\mathrm{d}\theta)\delta\theta\cos(\delta\theta/2) - [T + (\mathrm{d}T/\mathrm{d}\theta)\delta\theta]\sin(\delta\theta/2) - T\sin(\delta\theta/2) = 0$$

As $\delta\theta \to 0$ we may set $\cos(\delta\theta/2) \to \delta\theta/2$ rad and $\sin(\delta\theta/2) \to 0$ to give

$$\sigma_r ta + \mathrm{d}Q/\mathrm{d}\theta - T = 0 \quad \text{(iv)}$$

where small products have been ignored. Tangential force equilibrium (Fig. 2.27a) leads to

$$\tau_{r\theta}\, ta + \mathrm{d}T/\mathrm{d}\theta + Q = 0 \quad \text{(v)}$$

Combining the two equilibrium equations (iv) and (v) leads to a differential equation

$$\mathrm{d}^2T/\mathrm{d}\theta^2 + T = ta\,(\sigma_r - \mathrm{d}\tau_{r\theta}/\mathrm{d}\theta) \quad \text{(vi)}$$

Substituting for σ_r and $\tau_{r\theta}$ from (i) and (iii) into (vi) gives

$$\mathrm{d}^2T/\mathrm{d}\theta^2 + T = (3\,S_x ta/2)(1 - \cos 2\theta) \quad \text{(vii)}$$

The solution to eq(vii) is

$$T = A\cos\theta + B\sin\theta + (3\,S_x t\,a/2)[1 + \tfrac{1}{3}\cos 2\theta] \quad \text{(viii)}$$

The constants A and B are found from a horizontal force equilibrium condition across the vertical section, i.e. $2T = 2aS_x t$ when $\theta = \pm\pi/2$ (see Fig. 2.27b). Substituting into eq(viii) gives $aS_x t = B + S_x ta$, from which $B = 0$. Similarly, for vertical force equilibrium across the horizontal section ($\theta = 0, \pi$), $2T = 2aS_y t = 2a(2S_x)\,t$. Equation (viii) gives $2aS_x t = A + 2aS_x t$, from which $A = 0$. Thus eq(viii) simplifies to

$$T = (3S_x ta/2)[1 + ⅓ \cos 2\theta] \qquad \text{(ix)}$$

A final compatibility condition is required to match the hoop strain in the sheet with that in the bead. The sheet is under biaxial tensile stresses σ_θ and σ_r and the bead is under a direct tensile force T. The corresponding strains are equalised when, for a bead of area A_b,

$$T/(A_b E) = (\sigma_\theta - \nu\sigma_r)/E \qquad \text{(x)}$$

Substituting from eqs(i) and (ii) and (viii) into (ix)

$$[3S_x t\,a/(2A_b E)][1 + ⅓ \cos 2\theta] = [S_x/(2E)][3(1 - \nu) + (1 + \nu)\cos 2\theta]$$

This shows that the cross-sectional area of the bead must vary with θ, according to

$$A_b = at\,(3 + \cos 2\theta)/\,[3(1 - \nu) + (1 + \nu)\cos 2\theta] \qquad \text{(xi)}$$

For $\theta = 0°$, $A_b = 2at/(2 - \nu)$ and for $\theta = \pi/2$, $A = at/(1 - 2\nu)$. This shows that by taking $\nu = ⅓$ for an aluminium alloy bead, its area increases from $6at/5$ to $3at$ as θ increases from 0° to 90°. Symmetry holds in the remaining quadrants. The volume of material required is

$$V = 4\int_0^{\pi/2} A_b\, a\, d\theta \qquad \text{(xii)}$$

Substituting eq(xi) into eq(xii), the bead volume becomes

$$V = 4\int_0^{\pi/2} \frac{[a^2 t(3 + \cos 2\theta)\, d\theta\,]}{[3(1 - \nu) + (1 + \nu)\cos 2\theta\,]} \qquad \text{(xiii)}$$

It is left as an exercise for the reader to evaluate eq(xiii) numerically, using Simpson's rule. This shows that $V \simeq 10\,a^2 t$. Since the volume of the disc removed is $\pi a^2 t$, then $10/\pi$ times the disc volume must be introduced as a bead to remove the stress concentration.

EXERCISES

Equilibrium and Compatibility in Cartesian Co-ordinates

2.1 Show that the following strains and displacements are all compatible in plane stress: (i) $\varepsilon_x = axy$, $\varepsilon_y = by^3$, $\gamma_{xy} = c - dy^2$, (ii) $\varepsilon_x = q\,(x^2 + y^2)$, $\varepsilon_y = qy^2$, $\gamma_{xy} = 2qxy$ and (iii) $u = ax^2y^2$, $v = byx^3$.

2.2 Analysis of plane strain ($\varepsilon_z = \gamma_{yz} = \gamma_{xz} = 0$) shows that non-zero strains can be represented by: $\varepsilon_x = a + b(x^2 + y^2) + x^4 + y^4$, $\varepsilon_y = c + d(x^2 + y^2) + x^4 + y^4$, $\gamma_{xy} = e + fxy\,(x^2 + y^2 - g^2)$ where a, b, c, d, e, f and g are constants. Show that these are only possible if $f = 4$ and $b + d + 2g^2 = 0$. Hence derive the displacements from the strains.
Answer: $u = Ax + b(x^3/3 + y^2x) + x^5/5 + y^4x + C_3 y + C_1$, $v = cy + d(x^2y + y^3/3) + x^4y + y^5/5 + C_4 x + C_2$

2.3 Find the conditions for which the shear strain $\gamma_{xy} = Ax^2y + Bxy + Cx^2 + Dy$ is compatible with the displacements $u = ax^2y^2 + bxy^2 + cx^2y$ and $v = ax^2y + bxy$.

Answer: $B = a(a + b), A = 2a, D = b, C = c$

2.4 A thin plate is subjected to the following in-plane stress field components: $\sigma_x = ay^3 + bx^2y - cx$, $\sigma_y = dy^3 - e$, $\tau_{xy} = fxy^2 + gx^2y - h$. What are the constraints on the constants a, b, c, d, e, f, g and h so that the stress field satisfies both equilibrium and compatibility?

Answer: $a = 2f/3, b = -f, d = -f/3, c = g = 0, e, f$ and h are unconstrained

2.5 The stress system in a plate consists of direct tensile stresses $\sigma_x = cyx^2$, $\sigma_y = cy^3/3$ together with a shear stress τ_{xy}. Determine τ_{xy}.

Answer: $\tau_{xy} = -cxy^2$

2.6 The general expressions for the two direct stress components of a plane stress field have the form $\sigma_x = ax^3 + bx^2y + cxy^2 + dy^3$, $\sigma_y = kx^3 + lx^2y + mxy^2 + ny^3$, where a, b, c, d, k, l, m, n are constants. Determine the necessary relationship for these constants if the stress distribution is to be compatible. For the special case where $a = c = d = k = m = 0$, determine expressions for the compatible stress components σ_x, σ_y and τ_{xy}. (CEI)

Answer: $(6a + 3k + c)(3a - b) - (m - 3a)(2b + l + 3d) = 0$; $\sigma_x = 3x^2y$; $\sigma_y = y^3 - 6x^2y$, $\tau_{xy} = 2x^3 - 3xy^2$

2.7 A square plate of side length a and with sides along the co-ordinate axes is fixed in position at the origin and with the side along the y - axis fixed in direction ($\partial u/\partial y = 0$). If the stresses in the plate are $\sigma_x = Ay/a$, $\sigma_y = A$ with a consistent value of shear stress, find expressions for this shear stress and the general displacements. What are the displacements existing at the corner (a,a)? (IC)

Answer: $\tau_{xy} = K$, $u = Axy/(aE) - \nu Ax/E$, $v = Ay/E - \nu Ay^2/(2aE) + 2Kx(1 + \nu)/E - Ax^2/(2aE)$ and at (a, a), $u = Aa(1 - \nu)/E$, $v = Aa(1 - \nu)/(2E) + 2Ka(1 + \nu)/E$

2.8 Derive general expressions for the displacements at any point (x, y) in the cantilever of Fig. 2.10. Find the constants and identify the terms in the final displacement expressions for each of the following boundary conditions: (i) $u = v = 0$ and $\partial u/\partial y = 0$ at $(l, 0)$, (ii) $u = v = 0$ and $\partial v/\partial x = 0$ at $(l, 0)$.

Answer: (i) $u = [P/(2EI)](x^2y - l^2y + \nu y^3/3) - Py^3/(6GI)$, $v = [P/(EI)](l^2x/2 - \nu xy^2/2 - x^3/6 - l^3/3) + [Ph^2/(2GI)](x - l)$; (ii) $u = [P/(2EI)](x^2y - l^2y + \nu y^3/3) - [Py^3/(6GI)](3h^2/y^2 - 1)$, $v = [P/(EI)](l^2x/2 - \nu xy^2/2 - x^3/6 - l^3/3) + [Ph^2/(2GI)](x - l)$. Terms with E are deflections due to bending. Terms with G are deflections due to shear

Cartesian Stress Functions

2.9 Derive the stress function supplying the stress components $\sigma_x = cy$, $\sigma_y = cx$, $\tau_{xy} = -k$, where c and k are constants.

Answer: $c(x^3 + y^3)/6 + kxy$

2.10 Find the condition for which the stress function $\phi = Axy^4 + Bx^3y^2$ is valid.

Answer: $A + B = 0$

2.11 The stress function $\phi = Ax^2 + Bxy + Cy^2$ provides the direct and shear stresses in a thin rectangular plate. One corner, which lies at the origin, is position- and direction-fixed. Show that the displacements for the diagonally opposite corner (a, b) are: $u = 2a(C - \nu A)/E$, $v = 2b(A - \nu C)/E$.

2.12 If ϕ in Exercise 2.11 applies to a thick plate in plane strain, where $\varepsilon_z = 0$, determine the displacements at the corner (a, b).

Answer: $u = 2x(1 + \nu)[C(1 - \nu) - A\nu]/E$, $v = 2y(1 + \nu)[A(1 - \nu) - C\nu]/E$

2.13 Show that the stress function $\phi = Dx^3$ matches in-plane bending of a thin plate of width $2a$ under end moments M (see Fig. 2.28). Taking the thicknesses to be unity, derive the displacements. Given that the plate centre is position fixed, show that the displacements at the point $(a,0)$ are given by $u = -M\nu a^2/(2EI)$ and $v = 0$.

Answer: $u = -M(\nu x^2 + y^2)/(2EI) + C_1 y + C_2$, $v = Mxy/(EI) + C_3 x + C_4$ where C_i are constants

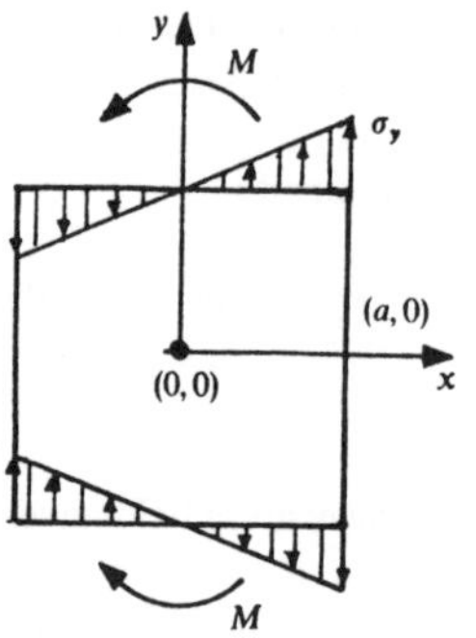

Figure 2.28

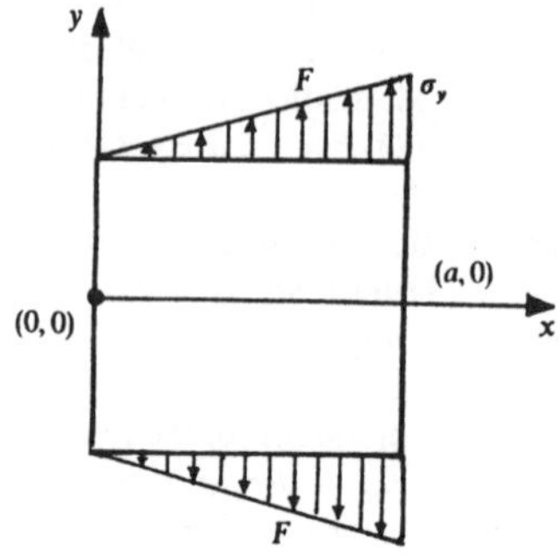

Figure 2.29

2.14. Show that the stress function $\phi = Dx^3$ applies to linearly varying edge loading of the plate shown in Fig. 2.29. F is the total force applied to the plate of width a. Taking the thicknesses to be unity, derive the displacements. Given that the plate origin (0,0) is position fixed show that the displacements at point $(a, 0)$ are $u = -\nu F/E$, $v = 0$.

Answer: $u = -F(\nu x^2 + y^2)/(Ea^2) + C_1 y + C_2$, $v = 2Fxy/(Ea^2) - C_3 x + C_4$ where C_i are constants

2.15 Derive the stress components from the stress function $\phi = Axy^3 + Bxy + Cy^3$, given that the stress resultants are a shear force F and a bending moment M at the section $x = 0$ of a beam. The latter has constant I and depth $\pm h$ relative to the neutral x-axis. How could these stress distributions be made to represent those for an encastre beam carrying a central concentrated load?

Answer: $\sigma_x = Fxy/I + My/I$, $\tau_{xy} = F(h^2 - x^2)/(2I)$

2.16 The stress function $\phi = Ax^2 + Bx^2y + Cx^2y^3$ provides the exact solution to the distribution of stress in a beam of depth $2h$ with unit thickness when it is simply supported over a length $2L$ and carries a uniformly distributed load q/unit length. With the origin at the beam centre, derive the stresses and compare these with those from simple bending theory. What is the vertical displacement at the centre?

Answer: $\sigma_x = [qy/(2I)](x^2 - l^2 + 2h^2/5 - 2y^2/3)$, $\sigma_y = [q/(2I)](y^3/3 - yh^2 - 2h^3/3)$, $\tau_{xy} = [qx/(2I)](h^2 - y^2)$, $v = [5ql/(24EI)][1 + [12h^2/(5l^2)](\nu/2 + 4/5)]$

2.17 Determine the state of stress represented by the stress function $\phi = Ax^2 - By^3$ when applied to a thin square plate $(a \times a)$ with origin at the centre of the left vertical side. Check that both equilibrium and compatibility are satisfied. Show the variation of stress on the plate surfaces. (CEI)

2.18 Show that the stress function $\phi = [q/(8c^3)][x^2(y^3 - 3c^2y + 2c^3) - (y^3/5)(y^2 - 2c^2)]$ is valid and determine the problem it solves when applied to the region: $y = \pm c$, $x = 0$ on the side x - positive.

2.19 A cantilever beam with rectangular cross-section occupies the region $-a \le z \le a$, $-h \le y \le h$ and $0 \le x \le l$. The end $x = l$ is built-in and a concentrated force P is applied at the free end $(x = 0)$ in the $+y$ direction. Given that the non-zero stress components are $\sigma_x = Cxy$ and $\tau_{xy} = A + By^2$ show that the constants become $2B + C = 0$ and that $C = 3P/(4ah^3)$ when the stresses satisfy force and moment equilibrium at the free end.

2.20 If, for the cantilever in Exercise 2.19, $a = 2$ mm, $h = 4$ mm, $l = 10$ mm and $P = 1$ kN determine (i) the Airy stress function from which the stresses derive, (ii) the principal stresses at the point $Q(5,2,0)$ and (iii) the normal and shear stresses at Q on an oblique plane whose edge is given by $y - x + 3 = 0$. (Hint: Use Table 1.1 in Chapter 1 for appropriate plane stress transformation equations.)

2.21 Establish the relationship between those constants in the quartic stress function $\phi = Ax^4 + Bx^3y + Cx^2y^2 + Dxy^3 + Ey^4$ if the function is to be valid. Determine further reltionships between the constants, from a consideration of known resultant normal forces X and Y and shear force S acting on the sides of a square plate of side length a and unit thickness, when the plate lies in the positive x, y quadrant with one corner at the origin. *Answer*: $3A + C + 3E = 0$, $X = (2C + 3D + 4E)d^3$, $Y = (4A + 3B + 2C)a^3$, $S = -(B + 2C + 3D)a^3 = (3B + 2C + D)a^3$, $B = D$

2.22 A rectangular strip of thickness t is simply supported at its ends ($x = \pm l$) and carries on its upper edge ($y = +d$) a uniformly distributed load, w per unit length, acting in the plane of the strip, as shown in Fig. 2.30. The expression $\phi = [3w/(4t)]\{[y^3/(30d^3)](5x^2 - y^2 - 5l^2 + 2d^2) - [x^2/(6d)](3y + 2d)]\}$ is proposed as an Airy stress function. Show that ϕ satisfies the compatibility condition and determine expressions for the direct and shear stresses in the strip. Investigate whether the direct stress expression satisfies the boundary equilibrium conditions. (CEI)

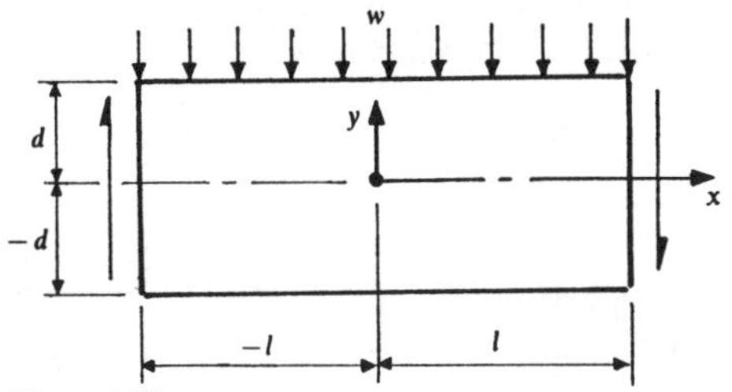

Figure 2.30

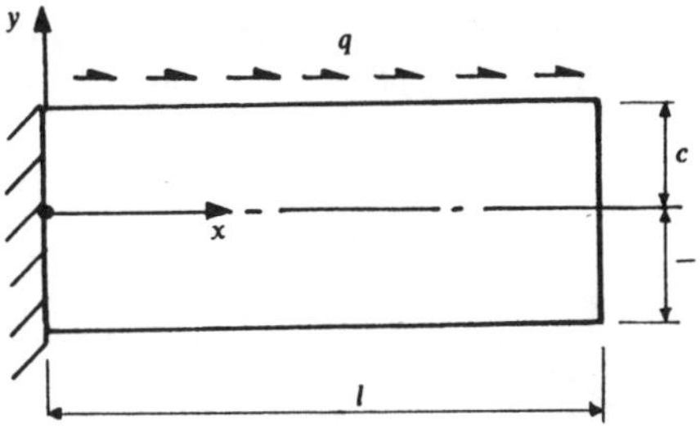

Figure 2.31

2.23 The rectangular plate shown in Fig. 2.31 is of uniform thickness t and is built in at the end $x = 0$. It is loaded along the upper edge $y = c$ by a tangential shear flow q. The lower edge of the plate is not loaded. The expression $\phi = [q/(4t)][xy - xy^2/c - xy^3/c^2 + l\,y^2/c + l\,y^3/c^2]$ is proposed as an appropriate stress function. To what extent does ϕ satisfy boundary conditions and compatibility? (CEI)

2.24 Show that the stress function $\phi = axy/2 - axy^3/(2c^2) + aby^3/(2c^2)$ is biharmonic and that it may be used to supply the stresses in the cantilever of Fig. 2.32. Is the condition that the cantilever is unloaded at its free end met by the function?

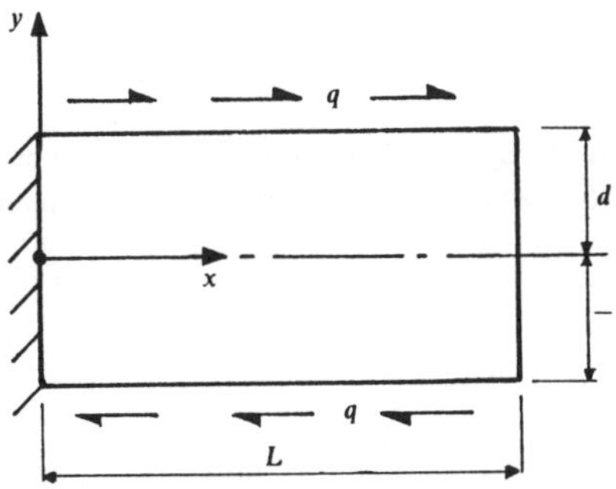

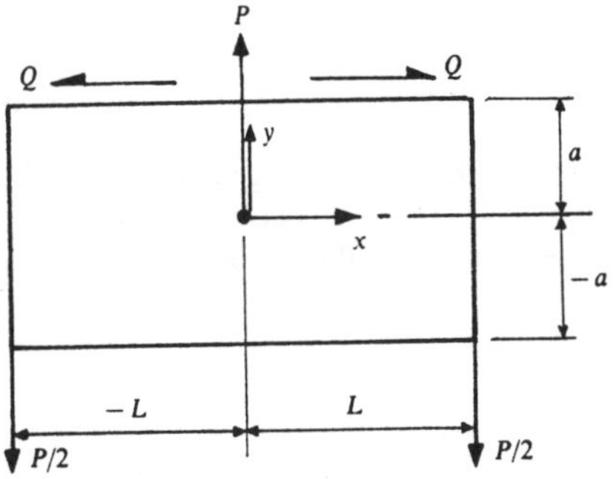

Figure 2.32

Figure 2.33

2.25 The function $\phi = -xy^3[P/(4a^3) + Q/(4a^2L)] + y^3[PL/(4a^3) + Q/(4a^2)] - xy^2Q/(4aL) + y^2Q/(4a) + xy[3P/(4a) + Q/(4L)]$ is proposed for the loaded beam in Fig. 2.33. Check that this satisfies $\nabla^4\phi = 0$ and determine the component stresses. To what extent are the boundary conditions satisfied in relation to the ratio L/a?

2.26 A cantilever beam of length l and depth $2h$ carries a downward acting uniformly distributed load w/unit area along its top and bottom faces under plane strain conditions. Using the stress function $\phi = Px^2y + Qx^2y^3 + Ry^3 + Sy^5$, find the constants from a consideration of the known conditions existing at the beam surface. Take the origin of co-ordinates to lie at the centre of the free end. (CEI)

2.27 The stress function $\phi = [3F/(2c)][xy - xy^3/(3c^2)] + Qy^3/(4c^3)$ is to be used to determine the stress field in a cantilever beam of length L, unit thickness and depth $2c$, subjected to a given loading. Determine the nature of that loading. (CEI)

Fourier Series Loading Functions

2.28 Identify full and half-range series for the edge loadings shown in Figs 2.34a-d and derive the appropriate Fourier functions $g(x)$ in eqs(2.33) and (2.34).

Answer: (a) $g(x) = (2p/\pi)\sin(\pi x/l) - [2p/(2\pi)]\sin(2\pi x/l) + [2p/(3\pi)]\sin(3\pi x/l) - \ldots$

(b) $g(x) = (2p/\pi)[\![[-\cos(\pi l/L) + (L/l\pi)\sin(\pi l/L)]\sin(\pi x/l)$
$+ \{½\cos(2\pi l/L) + [L/(4l\pi)]\sin(2\pi l/L)\}\sin(2\pi x/L) \ldots]\!]$

(c) $g(x) = p/2 - (4p/\pi^2)[\cos(\pi x/l) + (1/3^2)\cos(3\pi x/l) + (1/5^2)\cos(5\pi x/l) \ldots]$

(d) $g(x) = 3p/2 - (2p/\pi)[\sin(\pi x/l) + (1/3)\sin(3\pi x/l) + (1/5)\sin(5\pi x/l) \ldots]$

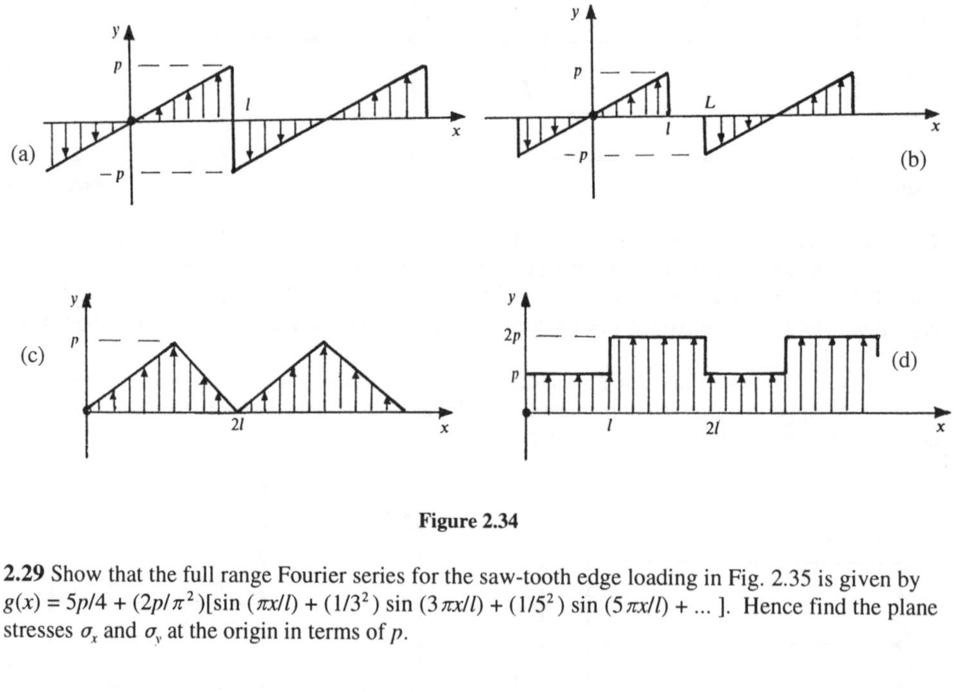

Figure 2.34

2.29 Show that the full range Fourier series for the saw-tooth edge loading in Fig. 2.35 is given by $g(x) = 5p/4 + (2p/\pi^2)[\sin(\pi x/l) + (1/3^2)\sin(3\pi x/l) + (1/5^2)\sin(5\pi x/l) + \ldots]$. Hence find the plane stresses σ_x and σ_y at the origin in terms of p.

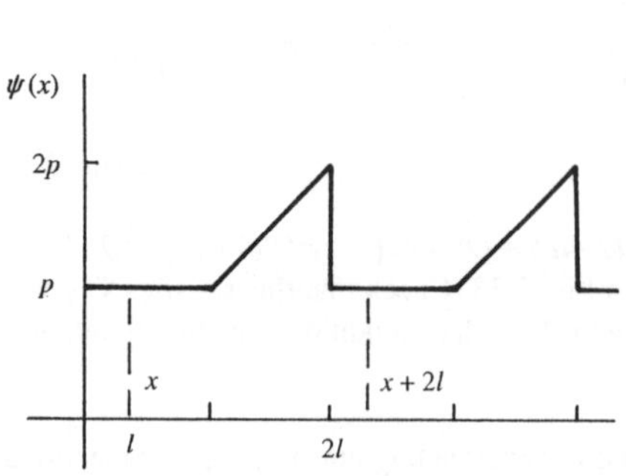

Figure 2.35

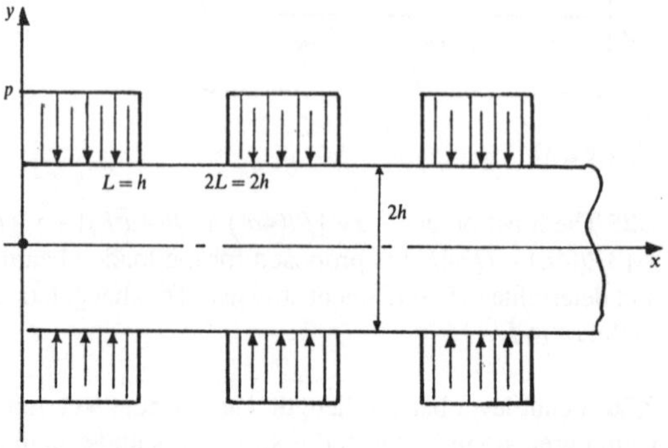

Figure 2.36

2.30 The stress function in eqs(2.31a,b) provides the solution to the problem of a long flat strip with its axis in direction x, under a sinusoidal edge loading of wavelength $2L$. Use this stress function to find the normal stress σ_y at the origin (see Fig. 2.36) in a strip of depth $2h$, loaded with a uniform pressure p under guides which are equally spaced along its edges, for the case when $h = L$. (IC)

Cylindrical Co-ordinates

2.31 For a particular plane strain problem the strain-displacement equations in cylindrical co-ordinates (r, θ, z) are $\varepsilon_r = du/dr$, $\varepsilon_\theta = u/r$, $\varepsilon_z = \gamma_{r\theta} = \gamma_{\theta z} = \gamma_{zr} = 0$. Show that the appropriate compatibility equation in terms of stress is $r\nu\, d\sigma_r/dr - r(1 - \nu)d\sigma_\theta/dr + \sigma_r - \sigma_\theta = 0$. State the nature of a problem represented by the above equations. (CEI)

2.32 A thick-walled cylinder with inner and outer radii r_i and r_o respectively is pressurised both internally (p_i) and externally (p_o). Find the expressions for the constants A and C in the corresponding stress function $\phi = A \ln r + C r^2$.

Answer: $A = -(p_i - p_o)\, r_o^2/[(r_o/r_i)^2 - 1]$, $2C = [p_i - p_o(r_o/r_i)^2]/[(r_o/r_i)^2 - 1]$

2.33 Determine the hoop stress at the inner and outer radii of a curved bar of 45 mm mean radius and 45 mm square cross-section when it is subjected to a pure moment of 300 Nm, inducing tension along the top edge.

Answer: 54.6 MPa, - 86.2 MPa

2.34 The stress function $\phi = C\theta$ prescribes the shear stress distribution in an annular disc of thickness t and inner radius r_i under torsion. If the torque is transmitted to the disc by a keyed shaft of radius r_i show that there is no discontinuity in shear stress at r_i when $t = r_i/4$.

2.35 Examine how to modify the stress distributions in the body of the wedge in Fig. 2.17 when the tip moment is produced by a force applied vertically to the wedge axis at a distance s from the wedge tip.

2.36 Find the maximum radial stress at a radius of 75 mm when the 30° wedge in Fig. 2.21b is loaded at its tip with 15 kN compression across its 5 mm thickness, with an inclination of 5° to the axis.

Answer: - 78.77 MPa on the load line

2.37. A very large plate of thickness t is subjected to a membrane tensile stress σ_o applied around its periphery. If a small circular hole is cut from the middle of the plate, show, from the Lamé equations, that the maximum stress at the edge of the hole is $2\sigma_o$.

2.38 The stress function $\phi = [qr^2/(2\pi)]\,(\tfrac{1}{2}\sin 2\theta - \theta)$ prescribes the stresses in a semi-infinite plate when a normal pressure q is applied over one half of the plate straight edge. Taking the origin at the load end with θ measured anti-clockwise from unloaded edge, show that the stresses at radius r in the plate are $\sigma_r = -(q/\pi)(\tfrac{1}{2}\sin 2\theta + \theta)$ and $\tau_{r\theta} = -[q/(2\pi)](\cos 2\theta - 1)$.

2.39 A hole of radius a in an infinite plate is loaded with an internal horizontal pressure p. Taking the origin of co-ordinates at the hole centre and assuming that the outer boundary is stress free, derive expressions for the polar stress components σ_r, σ_θ and $\tau_{r\theta}$ in the body of the plate and hence find the maximum stress concentration at the hole. (IC)

2.40 A window port in a large diameter, thin-walled pressurized cylinder is idealised as a large flat plate containing a hole with normal boundary stresses in the ratio $S_y/S_x = 3/2$ applied to the plate boundary. If the hole is to be free from stress concentration, what thickness of hole reinforcement is necessary?

2.41 Match the stress function $\phi = (Ar^3 + B/r + Cr + Dr \ln r) f(\theta)$ to each of the thin cantilever arcs in Figs 2.37a,b taking $f(\theta) = \sin\theta$ and $f(\theta) = \cos\theta$ respectively. Examine the local distributions of stress at the fixed and free ends and show their effects on the stress predictions from the stress function.

Answer: $\phi = F \sin\theta\,[r^3 - a^2b^2/r - 2(a^2 + b^2)\, r \ln r]/\ \{2[(a^2 - b^2) + (a^2 + b^2) \ln (b/a)]\}$
and $\phi = -\ \{p \cos\theta/[8(a + b)]\}[r^3 + a^3 (a + 2b)/r + 4abr \ln r]$

(a) 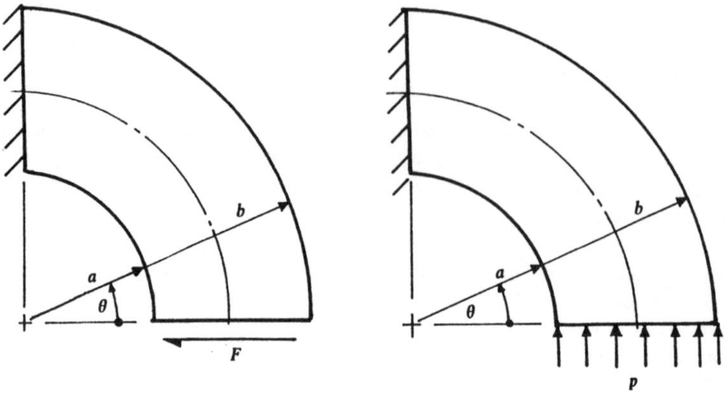 (b)

Figure 2.37

2.42 Employ the Boussinesq function $\phi = Cr\theta \sin\theta$ to obtain separate solutions for the stresses at point A in a semi-infinite plate of thickness t, due to two concentrated forces which constitute a couple Pa (see Fig. 2.38). Resolve forces at A to show that the stresses in the θ direction are given by $\sigma_r = [2P/(\pi t)][\,-(1/r)\cos\theta + (1/r_1)\cos\theta_1 \cos^2\alpha\,]$, $\sigma_\theta = [2P/(\pi t r_1)]\cos\theta_1 \sin^2\alpha$, $\tau_{r\theta} = [P/(\pi\, t r_1)]\cos\theta_1 \sin 2\alpha$. Show that when $a \to 0$ and the couple Pa is replaced by a moment M, the stresses are reduced to $\sigma_r = [2M/(\pi\, tr^2)]\sin 2\theta$, $\sigma_\theta = 0$ and $\tau_{r\theta} = [2M/(\pi\, tr^2)]\cos^2\theta$.

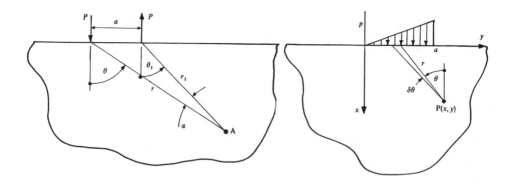

Figure 2.38 **Figure 2.39**

2.43 In Fig. 2.39 a linear distribution of loading is applied to the edge of a semi-infinite plate. Use the Boussinesq function and plane stress transformations for the given co-ordinates to derive integral expressions for the Cartesian stress components at a point $P(x, y)$ in the body of the plate.
(Hint: $\delta\sigma_x = -\ [2\delta p/(\pi\, r)]\cos^3\theta$, $\delta\sigma_y = -\ [2\delta p/(\pi\, r)]\cos\theta \sin^2\theta$ and $\delta\tau_{xy} = -\ [2\delta p/(\pi\, r)]\sin\theta \cos^2\theta$)

CHAPTER 3

STRUCTURES WITH SYMMETRY

There is a wide variety of load bearing structures possessing axes of symmetry. The membrane theory is concerned with pressurized thin-walled vessels having axes of revolution. It will be shown how this theory provides the membrane stresses in the wall of the more common vessels, including a cylinder, sphere, conic, ellipsoid and toroid. Thicker- walled vessels with symmetry axes, including cylinders, spheres, and discs, can sustain additional loadings. These loadings can be a combination of pressure, direct concentrated forces, distributed loading including self-weight coupled to a temperature-induced loading and centrifugal forces from rotation. We may solve for the internal stress and strain by employing first principles or by starting with an appropriate stress function. In the previous chapter a stress function was applied to plane stress and plane strain problems with axial symmetry. This chapter considers plane cylindrical elasticity with axial symmetry, body forces and temperature gradients.

3.1 Membrane Theory

Thin-walled pressure vessels, including the cylinder, sphere, ellipsoid, toroid and rotating rims, may be treated together within the membrane theory. Let an element of any membrane in Fig. 3.1, with thickness t, density ρ, wall radii r_θ and r_ϕ be subjected to the hoop and meridional stresses σ_θ and σ_ϕ due to combined pressure p, self-weight and centrifugal effects.

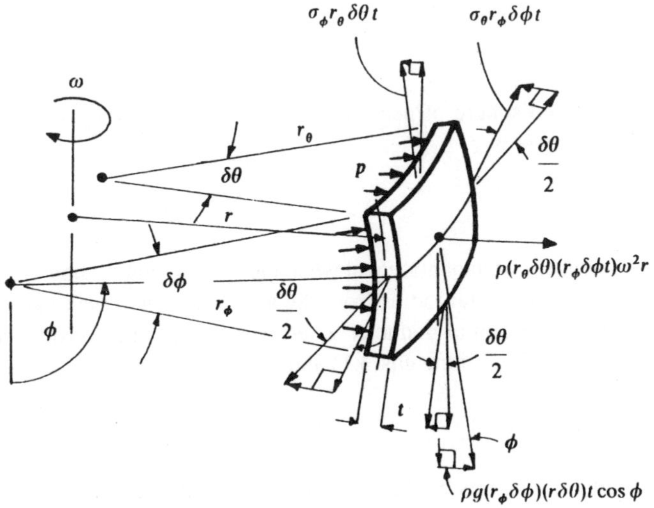

Figure 3.1 Membrane element

The sides subtend angles $\delta\theta$ and $\delta\phi$ at their centres of curvature. The solid may also rotate about its axis of symmetry with an angular velocity ω at radius r to induce body forces. Ignoring bending effects and assuming that the radial stress σ_r through the wall is negligibly small compared to σ_θ and σ_ϕ, the following radial equilibrium equation applies to the biaxial stress state in the plane of the wall:

$$(r_\theta \delta\theta)(r_\phi \delta\phi)(\rho t\omega^2 r + \rho g t \cos\phi + p) = 2\sigma_\theta r_\phi \delta\phi\, t \sin(\delta\theta/2) + 2\sigma_\phi r_\theta \delta\theta\, t \sin(\delta\phi/2)$$

In the limit, $\sin(\delta\theta/2) \rightarrow \delta\theta/2$ and $\sin(\delta\phi/2) \rightarrow \delta\phi/2$ and this equation simplifies to

$$r_\theta r_\phi(\rho\omega^2 r + \rho g \cos\phi + p/t) = \sigma_\theta r_\phi + \sigma_\phi r_\theta$$
$$\rho\omega^2 r + \rho g \cos\phi + p/t = \sigma_\theta/r_\theta + \sigma_\phi/r_\phi \tag{3.1}$$

By combining eq(3.1) with a second equilibrium equation isolating the action of one stress within a given vessel, both stresses σ_θ and σ_ϕ can be found. When the effect of bending is ignored, the deformation arising from membrane stresses at the mid-wall are found from

$$\varepsilon_\theta = (\sigma_\theta - \nu\sigma_\phi)/E \tag{3.2a}$$
$$\varepsilon_\phi = (\sigma_\phi - \nu\sigma_\theta)/E \tag{3.2b}$$
$$\varepsilon_r = -\nu(\sigma_\theta + \sigma_\phi)/E \tag{3.2c}$$

The dilatation, or volumetric strain $\delta V/V$, within a unit cube of unstrained material is

$$\delta V/V = (1 + \varepsilon_\theta)(1 + \varepsilon_\phi)(1 + \varepsilon_r) - 1 \approx \varepsilon_\theta + \varepsilon_\phi + \varepsilon_r \tag{3.3}$$

It follows that $\delta V/V$ may be expressed in terms of the two independent stresses using eqs(3.2a-c). The strains will also alter the internal volume of the vessel. A ratio between the change in internal volume to the original volume will depend upon the symmetry of the shell. So, too, will the dependence of the strains upon the displacements. The analysis of strain will be restricted to cylindrical and spherical shells where ε_θ and ε_ϕ take their simplest forms.

3.1.1 Sphere

For a stationary, pressurised sphere of negligible weight $\omega = 0$, $r_\theta = r_\phi = r$, when eq(3.1) reveals equal membrane stresses

$$\sigma_\theta = \sigma_\phi = pr/2t \tag{3.4a,b}$$

These stresses will be neither constant nor equal when the sphere rotates about a vertical axis (say). This is because the horizontal radius from the axis to the shell wall varies along the axis. For example, on the horizontal diameter where $r = r_\theta = r_\phi$, the vertical equilibrium equation is $p\pi r^2 = 2\pi r t\sigma_\phi$. Combining eqs(3.1) and (3.4) yields

$$\sigma_\phi = pr/2t \text{ and } \sigma_\theta = r\,(p/2t + \rho\omega^2 r)$$

When the internal radius r increases by the amount δr, the volumetric strain is

$$\delta V/V = (4\pi/3)[(r + \delta r)^3 - r^3] / (4\pi/3)r^3$$
$$\approx 3r^2(\delta r) / r^3 = 3\delta r/r$$

Now $\delta r/r$ is the hoop strain, defining the change in the circumferential length as

$$\varepsilon_\theta = [\pi(r + \delta r) - \pi r]/\pi r = \delta r/r$$

From eq(3.3),

$$\delta V/V = 3\varepsilon_\theta = (3/E)(\sigma_\theta - \nu\sigma_\phi) \quad (3.5a)$$

Clearly ε_θ depends upon the stress state existing at the mid-wall. In the simplest case of a stationary pressurised sphere, $\sigma_\theta = \sigma_\phi = pr/2t$. The radial stress can be neglected when ($r/t >$ 10) and eq(3.5a) becomes

$$\delta V/V = [3pr/(2tE)](1 - \nu) \quad (3.5b)$$

Equation (3.1) also applies when self-weight stresses are required. Take, for example, the element of a suspended spherical dish in Fig. 3.2 with radius r and solid angle 2α under no other external loading.

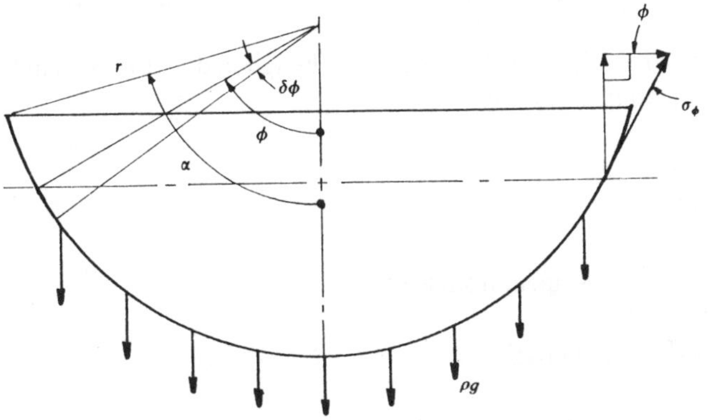

Figure 3.2 Self-weight loading

The self-weight term in the radial equilibrium equation (3.1) reveals that tensile stresses are induced. With $r = r_\theta = r_\phi$, this gives

$$\rho g \cos\phi = \sigma_\theta / r + \sigma_\phi / r$$

The stresses may be separated when the vertical equilibrium equation is referred to the horizontal section defined by ϕ. Noting that the elemental surface area is $2\pi(r\sin\phi)(r\delta\phi)$, then

$$2\pi(r\sin\phi)\, t\sigma_\phi \sin\phi = 2\pi r^2 t\rho g \int_0^\phi \sin\phi \, d\phi$$

$$\sigma_\phi = -\rho g r(1 - \cos\phi)/\sin^2\phi = \rho g r/(1 + \cos\phi) \quad (3.6a)$$

$$\sigma_\theta = \rho g r \cos\phi - \sigma_\phi = \rho g r[\cos\phi - 1/(1+\cos\phi)] \quad (3.6b)$$

It is seen from eqs(3.6a,b) that $\sigma_\phi = \sigma_\theta = \rho g r/2$ when $\phi = 0$ at the crown. The greatest stresses depend upon α, i.e. when $\alpha = 90°$, $\sigma_\phi = \rho g r$ and $\sigma_\theta = -\rho g r$.

3.1.2 Cylinder

For an internally pressurized cylinder of inner radius r and wall thickness t (Fig. 3.3a), σ_ϕ becomes the axial stress σ_z.

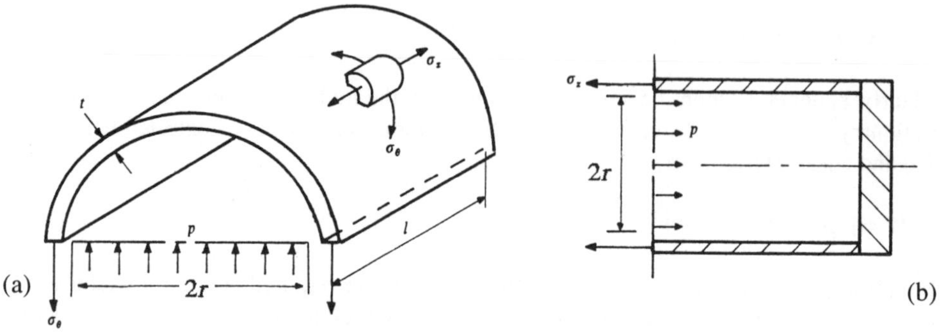

Figure 3.3 Membrane stresses for a thin-walled cylinder

Provided that $r/t \geq 10$, the radial stress may be ignored and σ_z is found from a horizontal equilibrium equation. From Fig. 3.3b,

$$p \times \pi r^2 = 2\pi r t \sigma_z$$
$$\sigma_\phi = \sigma_z = pr/2t \qquad (3.7a)$$

Since $r_\theta = r$ and $r_\phi = \infty$, then, from eq(3.1),

$$\sigma_\theta = pr/t + \rho\omega^2 r^2 + \rho g r\cos\phi \qquad (3.7b)$$

Equation (3.7b) gives $\sigma_\theta = pr\,/\,t$ for a stationary vessel with negligible self-weight. We can confirm this by taking a second equilibrium equation $2\sigma_\theta tl = 2prl$, in which the pressure acts on the plane projected area $2r \times l$ in Fig. 3.3a. In the absence of pressure and self-weight and with $\phi = 90°$, eqs(3.7a,b) give $\sigma_z = 0$ and $\sigma_\theta = \rho\omega^2 r^2$, which defines the hoop stress in a rotating rim or cylinder.

Let the initial length l and inner radius r of the cylinder increase by δl and δr respectively. The volumetric strain becomes

$$\delta V/V = \pi[(r + \delta r)^2(l + \delta l) - r^2 l]/(\pi r^2 l)$$
$$= 2\delta r\,/\,r + \delta l\,/\,l$$

The axial and circumferential strains are defined as $\varepsilon_z = \delta l/\,l$ and $\varepsilon_\theta = \delta r/r$.

$$\delta V/V = 2\varepsilon_\theta + \varepsilon_z \qquad (3.8a)$$

Substituting $\sigma_\theta = pr/t$ and $\sigma_\phi = pr/2t$ into eq(3.8a) leads to the volumetric strain for a stationary cylinder,

$$\delta V/V = [pr/(2tE)](5 - 4\nu) \qquad (3.8b)$$

Example 3.1 A cylindrical steel pressure vessel has two hemispherical caps welded to its ends. Given $\nu = ¼$, show that the thickness of the sphere must equal 3/7 that of the cylinder to ensure no mismatch in hoop strain at the weld. The cylinder portion is 800 mm inner diameter, 2 m long and 40 mm thick. What additional volume of oil could be pumped in to attain a working stress of 150 MPa within the vessel wall? Take E = 207 GPa for steel and K = 3500 MPa for oil.

Using subscripts s and c for sphere and cylinder respectively the equality in hoop strain at the interface follows from eq(3.2a)

$$\varepsilon_\theta = [(1/E)(\sigma_\theta - \nu\sigma_\phi)]_s = [(1/E)(\sigma_\theta - \nu\sigma_z)]_c$$

Substituting for $\sigma_\theta,$ σ_ϕ and σ_z from eqs(3.4a,b) and (3.7a,b), where p, d and E are common,

$$[pr/(2tE)]_s(1 - \nu)_s = [pr/(2tE)]_c(2 - \nu)_c$$
$$\therefore t_s / t_c = (1 - \nu)/(2 - \nu) = (3/4)/(7/4) = 3/7$$

For t_c = 40 mm and t_s = 17.14 mm, the limiting stress must correspond to the lesser of the two water pressures:

$$p_c = (t\sigma / r)_c = 40 \times 150 / 400 = 15 \text{ MPa (1.5 bar)}$$
$$p_s = (2t\sigma / r)_s = 2 \times 17.14 \times 150/400 = 12.86 \text{ MPa (1.286 bar)}$$

The total additional volume δV_t to be pumped in is given by

$$\delta V_t = \delta V_s + \delta V_c + \delta V_w$$

Substituting from eqs(3.5b, 3.8b, and 2.5a),

$$\delta V_t = [3prV_s/(2t_sE)](1 - \nu) + [prV_c/(2t_cE)](5 - 4\nu) + pV_t/K$$
$$= [pr/(2Et_c)][7(1 - \nu)V_s + (5 - 4\nu)V_c] + pV_t/K \qquad \text{(i)}$$

where

$$V_c = \pi \times 400^2 \times 2000 = 1005.3 \times 10^6 \text{ mm}^3$$
$$V_s = 4\pi r^3/3 = 4\pi \times 400^3/3 = 268.1 \times 10^6 \text{ mm}^3$$
$$V_t = V_c + V_s = 1273.4 \times 10^6 \text{ mm}^3$$

Substituting into eq(i)

$$\delta V_t = [(12.86 \times 400 \times 10^6)/(2 \times 40 \times 207 \times 10^3)][(7 \times 0.75 \times 268.1) + 1005.3]$$
$$+ (12.86 \times 1273.4 \times 10^6/3500)$$
$$= 310.63(1005.3 + 1407.53) + (4.68 \times 10^6)$$
$$= (0.75 + 4.68)10^6 \text{ mm}^3 = 5.43 \times 10^6 \text{ mm}^3$$

Oil compressibility has a far greater contribution tha the strained volume.

3.1.3 Conical Vessel

Consider a rotating, pressurised conical vessel as shown in Fig. 3.4.

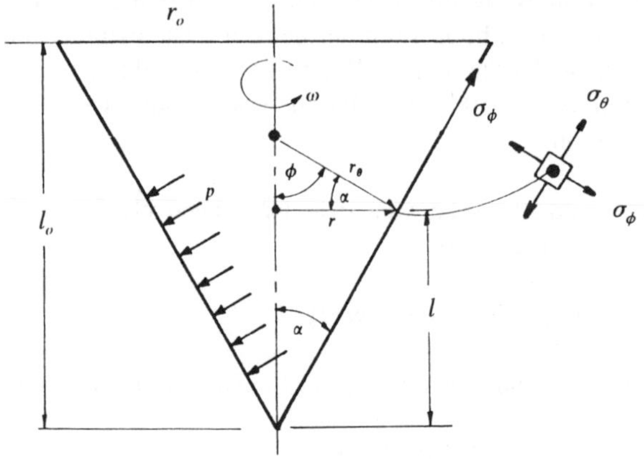

Figure 3.4 Rotating Cone

A second equilibrium equation is taken vertically above any horizontal plane of radius $r = r_0 \cos \alpha$. Since centrifugal forces act horizontally they will not appear. The balance between the force due to σ_ϕ and that due to pressure and self-weight of the cone wall become:

$$\sigma_\phi t\,(2\pi r)\cos\alpha = p\pi r^2 + \pi r^2 t\rho g / \tan\alpha$$

$$\sigma_\phi = pr/(2t\cos\alpha) + \rho g r/(2\sin\alpha) \qquad (3.9a)$$

Substituting eq(3.9a) into eq(3.1) with $r_\phi = \infty$ and $\phi = 90 - \alpha$ gives:

$$\sigma_\theta = (r/\cos\alpha)[p/t + \omega^2 \rho r + \rho g \sin\alpha] \qquad (3.9b)$$

Example 3.2 Determine the speed at which an upturned 30° cone, 2 m long and 5 mm thick, may rotate about a vertical axis when it contains a gas at pressure 3 bar, if the maximum allowable hoop stress is 150 MPa. Take a density $\rho = 7100$ kg/m³ for the cone material.

Equation (3.9b) shows that σ_θ is greatest at the base of the cone where $r = r_o = l_o \tan \alpha = 2 \tan 15° = 0.5358$ m. Substituting into eq(3.9b) reveals that the contribution from self-weight is negligible

$$\begin{aligned}
\omega^2 &= [(\sigma_\theta/r)\cos\alpha - p/t - \rho g \sin\alpha] / (\rho r) \\
&= [(150 \times 0.9659)/535.8 - (3 \times 10^5)/(10^6 \times 5) \\
&\quad - (7100 \times 9.81 \times 0.2588)/10^9)]10^{12}/(7100 \times 535.8) \\
&= [0.2704 - 0.06 - 0.000018](262.87 \times 10^3) = 55303.12
\end{aligned}$$

$\therefore\ \omega = 235.17$ rad/s or $N = 60\omega/(2\pi) = 2245.67$ rev/min

3.1.4 Ellipsoidal Vessel

For a stationary, ellipsoidal, pressurised vessel (Fig. 3.5) the meridional stress σ_ϕ may be found from a consideration of vertical equilibrium on horizontal plane. The ends of this plane have co-ordinates (x, y) as shown.

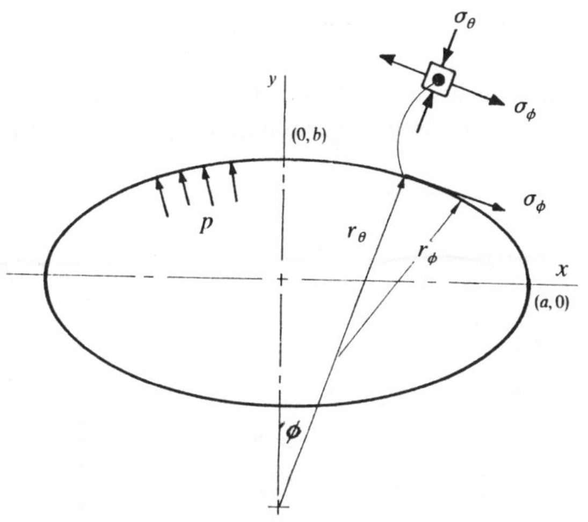

Figure 3.5 Ellipsoidal shell

Ignoring self-weight this gives

$$2\pi x\, t\, \sigma_\phi \sin\phi = \pi x^2 p$$

$$\sigma_\phi = px/(2t\sin\phi) = pr_\theta/(2t) \qquad (3.10a)$$

where radius $r_\theta = (1/b^2)\surd(a^4y^2 + b^4x^2)$ from the ellipse geometry. Substituting eq(3.10a) into eq(3.1) leads to the hoop stress

$$\sigma_\theta = (pr_\theta/t)[1 - r_\theta/(2r_\phi)] \qquad (3.10b)$$

in which the meridional radius $r_\phi = (b^2/a^4)r_\theta{}^3$. Particular stress values of interest, from eqs(3.10a,b), are (i) at the crown (0, b), where $r_\theta = r_\phi = a^2/b$, giving equal maximum stresses:

$$\sigma_\phi = \sigma_\theta = pa^2/(2bt)$$

and (ii) at the equator (a, 0), where $r_\theta = a$, $r_\phi = b^2/a$, giving different stresses:

$$\sigma_\phi = pa/(2t)$$
$$\sigma_\theta = (pa/t)[1 - a^2/(2b^2)]$$

The equatorial hoop stress σ_θ can produce buckling in the knuckle region when it becomes compressive for $a/b > \surd 2$.

Substituting eqs(3.10a,b) into eqs(3.2a,b,c) and (3.3) will give the strains and the dilations at these positions,

$$\varepsilon_\theta = [(pr_\theta/(Et))][1 - \tfrac{1}{2}(r_\theta/r_\phi + \nu)]$$
$$\varepsilon_\phi = [(pr_\theta/Et)[\tfrac{1}{2}(1 + r_\theta/r_\phi) - \nu]$$
$$\varepsilon_r = -[(\nu\, pr_\theta/(2Et)](3 - r_\theta/r_\phi)$$
$$\delta V/V = [pr_\theta/(2Et)]\{3 - \nu(6 - r_\theta/r_\phi)]\}$$

3.1.5 Toroid

In the case of a closed, stationary toroidal shell of thickness t and radii r and R (see Fig. 3.6), the hoop stress follows directly from the equilibrium equation referred to section X-X.

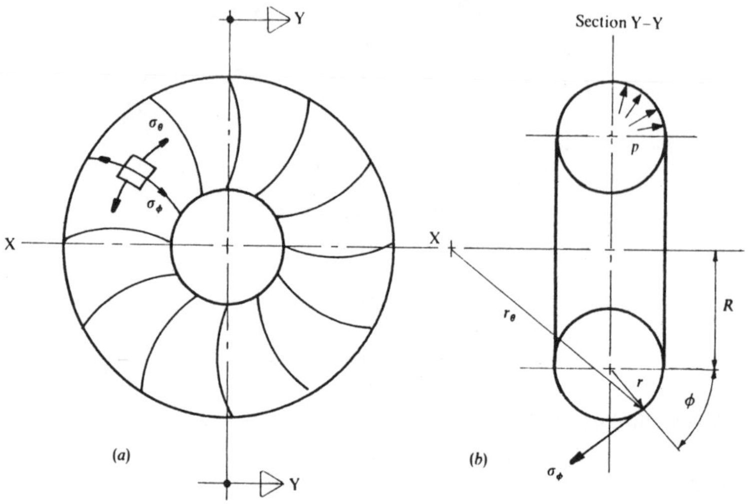

Figure 3.6 Toroidal shell under pressure

Ignoring self-weight, this gives $2 \times \sigma_\theta(2\pi rt) = 2\,p\pi r^2$

$$\therefore\ \sigma_\theta = pr/(2t\,) \tag{3.11a}$$

Substituting eq(3.11a) in eq(3.1) with $r_\phi = r$ and $r_\theta = r + R\,/\,\sin\phi$ leads to

$$\sigma_\phi\,/\,r_\phi = p\,/\,t - pr/(2tr_\theta) = (p/t\,)[1 - (r/2)(r + R\,/\,\sin\phi\,)]$$

$$\sigma_\phi = [pr\,/\,(2t\,)][(2R + r\,\sin\phi\,)\,/\,(R + r\,\sin\phi\,)] \tag{3.11b}$$

Example 3.3 Determine the maximum membrane stresses in a stationary toroid of thickness 10 mm, with cylinder and mean radii of 0.25 m and 1.5 m respectively, under a pressure of 10 bar. By how much do the inner and outer diameters change? Take $E = 207$ GPa, $\nu = 0.27$.

Equation (3.11b) is a maximum when $\phi = 3\pi/2$, giving

$$\sigma_\phi = [pr/(2t\,)][(2R - r)\,/\,(R - r)]$$
$$= [(1 \times 0.25)/(2 \times 0.01)][(3 - 0.25)\,/\,(1.5 - 0.25)] = 27.5 \text{ MPa}$$

and from eq(3.11a), the constant hoop stress is

$$\sigma_\theta = pr/(2t) = (1 \times 0.25)\,/\,(2 \times 0.01) = 12.5 \text{ MPa}$$

We can convert these stresses to ε_θ, from eq(3.2a),

$$\varepsilon_\theta = (1/E)(\sigma_\theta - \nu\,\sigma_\phi) = [12.5 - (0.27 \times 27.5)]\,/\,207000 = 24.5\mu$$

The change to the inner diameter is then

$$\Delta d_i = d_i\varepsilon_\theta = 2.5 \times 10^3 \times 24.5 \times 10^{-6} = 0.061 \text{ mm}$$

At the outer diameter $\phi = \pi/2$, when eq(3.11b) gives

$$\sigma_\phi = [pr/(2t)][2R + r)/(R + r)]$$
$$= [(1 \times 0.25)/(2 \times 0.01)][(3 + 0.25)/(1.5 + 0.25)] = 23.2 \text{ MPa}$$

The hoop strain and the outer diameter change are

$$\varepsilon_\theta = [12.5 - (0.27 \times 23.2)] / 207000 = 30.13\mu$$
$$\Delta d_o = \varepsilon_\theta d_o = 3.5 \times 10^3 \times 30.13 \times 10^{-6} = 0.105 \text{ mm}$$

3.2 Thick-Walled Cylinders

In thick-walled, cylinders it becomes necessary to consider the variation in radial stress σ_r through the wall. Unlike a thin-walled cylinder the radial stress is not negligible compared to the hoop (σ_θ) and axial (σ_z) stresses. The boundary conditions must ensure that σ_r is the pressure applied to the inner and outer diameters. With the axial and hoop stresses also present, a principal triaxial stress state exists within the wall. To convert this into the corresponding strain state requires the following 3D stress-strain equations:

$$\varepsilon_\theta = (1/E)[\sigma_\theta - \nu(\sigma_r + \sigma_z)] \quad (3.12a)$$
$$\varepsilon_r = (1/E)[\sigma_r - \nu(\sigma_\theta + \sigma_z)] \quad (3.12b)$$
$$\varepsilon_z = (1/E)[\sigma_z - \nu(\sigma_r + \sigma_\theta)] \quad (3.12c)$$

Consider, firstly, the solid cylinder under internal pressure p in Fig. 3.7a. Let cylindrical co-ordinates r, θ and z describe the triaxial stress state which we wish to find for any radius r in the wall.

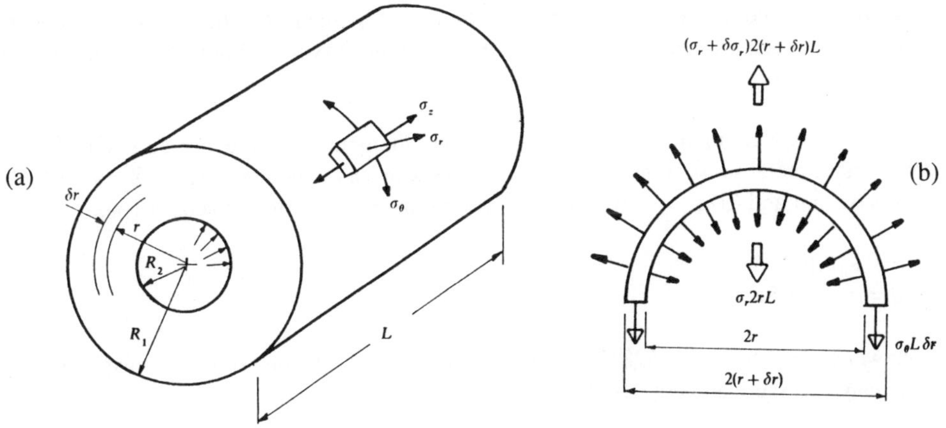

Fig. 3.7 Thick-walled pressurised cylinder

Vertical equilibrium about a horizontal diameter for a cylindrical element of radial thickness δr, over which σ_r varies (see Fig. 3.7b), is expressed by

$$2\sigma_\theta\, \delta r L + \sigma_r(2rL) = (\sigma_r + \delta\sigma_r) \times 2(r + \delta r)L$$

in which σ_r (assumed tensile) acts on the inner and outer projected areas. In the limit this leads to a radial equilibrium equation,

$$\sigma_\theta - \sigma_r = r\, \mathrm{d}\sigma_r / \mathrm{d}r \tag{3.13}$$

The stresses in eq(3.13) may be separated when it is assumed that σ_z is independent of r. This must be so if plane cross-sections are to remain plane, i.e. the longitudinal strain ε_z is constant in the length for all r. From eq(3.12c),

$$\sigma_\theta + \sigma_r = (\sigma_z - \varepsilon_z E)/\nu = 2a \tag{3.14}$$

where $2a$ is a constant. Eliminating σ_θ between eqs(3.13) and (3.14) and integrating leads to

$$2\int(\mathrm{d}r/r) = \int \mathrm{d}\sigma_r/(a - \sigma_r)$$
$$2\ln r = -\ln(a - \sigma_r) + \ln b \tag{3.15}$$

where b is a constant. Combining eqs(3.14) and (3.15) gives the radial and hoop stress expressions as

$$\sigma_r = a - b/r^2 \tag{3.16a}$$
$$\sigma_\theta = a + b/r^2 \tag{3.16b}$$

in which the constants a and b are found from the conditions existing at the boundary (where both internal and external pressures act). Equations (3.16a,b) are the Lamé (1852) equations which were previously derived from a stress function (see eqs 2.48a,b). Also discussed previously were the three end conditions used to determine σ_z and the distributions in σ_θ and σ_r through the wall (see Fig. 2.18).

Example 3.4 Find the necessary thickness of a hydraulic main, 130 mm internal diameter, required to contain a gauge pressure of 100 bar when the maximum hoop stress is limited to 15 MPa.

Since σ_θ is a maximum (tension) at the inner diameter and the pressure equals the compressive radial stress at the inner diameter, the boundary conditions are (i) $\sigma_\theta = +15$ MPa for $r = 65$ mm and (ii) $\sigma_r = -10$ MPa for $r = 65$ mm. Substituting into eqs(3.16a,b) gives

$$15 = a + b/(65)^2$$
$$-10 = a - b/(65)^2$$

from which $a = 2.5$ and $b = 52812.5$. At the outer diameter the pressure is atmospheric, i.e. zero gauge pressure and eq(3.16a) becomes

$$0 = 2.5 - 52812.5/r_o^2$$
$r_o = 145.4$ mm and $t = 80.4$ mm

Example 3.5 A pipe, 150 mm i.d. and 200 mm o.d., fails under hoop tension at an internal pressure of 500 bar. Determine, for a safety factor of 4, the safe internal pressure for a second pipe of the same material and internal diameter, but with walls 40 mm thick.

The boundary conditions for the first pipe are (i) $\sigma_r = -50$ MPa for $r = 75$ mm and (ii) $\sigma_r = 0$ for $r = 100$ mm. Applying these to eq(3.16a) gives

$$-50 = a - b\,(75)^2$$
$$0 = a - b/(100)^2$$

from which $a = 64.29$ and $b = 642.86 \times 10^3$. Then, from eq(3.16b) with $r = 75$ mm, the hoop stress that caused failure in the bore is

$$\sigma_\theta = a + b/r^2 = 64.29 + (642.86 \times 10^3)/75^2 = 178.6 \text{ MPa}.$$

Hence the safe working hoop stress for the second cylinder is $\sigma_\theta = 178.6/4 = 44.65$ MPa. Applying eqs(3.16a,b) with new constants a' and b',

$$44.65 = a' + b'/(75)^2$$
$$0 = a' - b'/(115)^2$$

from which $a' = 13.32$ and $b' = 176.16 \times 10^3$. The corresponding internal pressure $p = -\sigma_r$ for $r = 75$ mm is found from eq(3.16a)

$$-p = a' - b'/r^2 = 13.32 - (176.16 \times 10^3)/75^2$$
$$p = 18 \text{ MPa or } 180 \text{ bar}$$

3.3 Interference Fits

3.3.1 Shaft and Hub

When a solid cylindrical shaft or disc is forced into a hub, the external diameter of the shaft and the internal hub diameter are both subjected to compressive radial pressure p. The magnitude of this interface pressure is controlled by the initial difference in diameters. This interference (Δd) depends upon the difference between hoop strains at the common radius (r_c) which in turn is a function of each biaxial stress state (σ_θ, σ_r) in the adjacent components (see Fig. 3.8a).

Since the stress in the shaft cannot be infinite at $r = 0$, eqs(3.16a,b) become

$$\sigma_\theta = \sigma_r = -p = \text{constant}$$

Within the annular hub of outer radius r_o, the constants a and b in eqs(3.16a,b) may be found from the two conditions, (i) $\sigma_r = -p$ for $r = r_c$ and (ii) $\sigma_r = 0$ for $r = r_o$. These enable the hoop stress σ_θ at $r = r_c$ in the hub to be found. In the absence of axial stress σ_z in a thin disc/hub assembly, the relative hoop strain at r_c is given by

$$\Delta\varepsilon_\theta = [(\sigma_\theta - \nu\,\sigma_r)/E]_{\text{hub}} - [(\sigma_\theta - \nu\,\sigma_r)/E]_{\text{disc}} \quad (3.17a)$$
$$u/r_c = [(\sigma_\theta + \nu\,p)/E]_{\text{hub}} + p\,[(1 - \nu)/E]_{\text{disc}} \quad (3.17b)$$

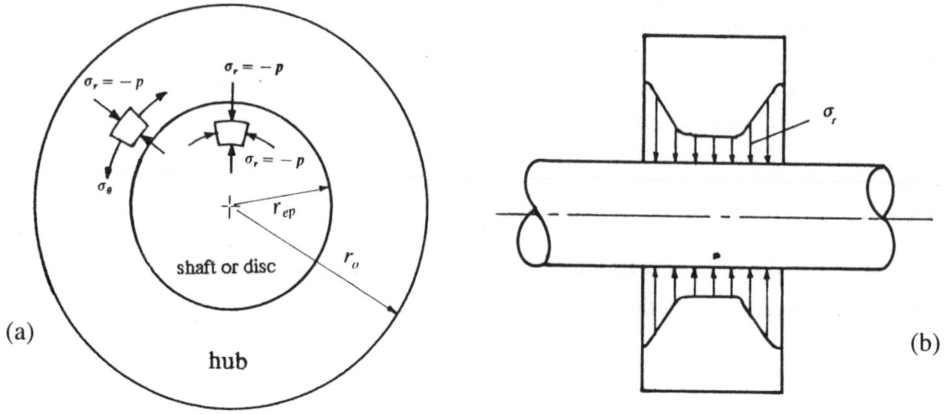

Figure 3.8 Interface stress states

Then $\Delta d_c = 2u$, in which u is the radial displacement of the common radius r_c. If the hub is pressed on a shaft, condition (i) ignores the concentration in σ_r occurring at the edges of the hub (see Fig. 3.8b). This problem may be complicated further in the presence of axial stress in the shaft, when eq(3.17a) is modified to

$$\Delta\varepsilon_\theta = [(\sigma_\theta - \nu\,\sigma_r)\,/E]_{\text{hub}} - \{[\sigma_\theta - \nu\,(\sigma_r + \sigma_z)]/E\}_{\text{shaft}} \tag{3.17c}$$

It will be seen that σ_z can usually be found from either a condition of zero axial force or a zero axial extension. Applications of eq(3.17b) appear in the following two examples.

Example 3.6 An oversized steel plug is pressed into a steel hub 12.5 mm i.d. and 38 mm o.d. If the maximum stress in the hub is not to exceed 465 MPa, calculate the necessary interference. Take E = 208 GPa, ν = 0.3.

Here σ_θ = 465 MPa for r_c = 6.25 mm. Also for the hub, $\sigma_r = 0$ for r_o = 19 mm. Equations (3.16a,b) supply the simultaneous equations

$$465 = a + b\,/\,(6.25)^2$$
$$0 = a - b\,/\,(19)^2$$

from which $a = 45.4$ and b = 16390.5. The radial stress σ_r at r_c = 6.25 mm is, from eq(3.16a),

$$\sigma_r = a - b/r^2 = 45.4 - 16390.5/\,(6.25)^2$$
$$= -\ 374.2 \text{ MPa}$$

Thus, p = 374.2 MPa and eq(3.17b) becomes

$$u\,/\,6.25 = [\{465 + (0.3 \times 374.2)\} + 374.2(1 - 0.3)]\,/\,208000$$
$$u = 6.25\,(465 + 374.2)\,/\,208000 = 0.025 \text{ mm}$$
$$\therefore\ \Delta d_c = 0.050 \text{ mm}$$

Example 3.7 A steel hub, 115 mm outer diameter and 75 mm inner diameter, is shrunk on to a solid aluminium shaft to give a circumferential strain at its outer diameter of 0.07%. Determine the interface pressure, the greatest tensile stress in the hub and the interference. For steel, take E = 207 GPa and ν = 0.27 and for aluminium, take E = 75 GPa and ν = 0.33.

The hub's boundary conditions are (i) $\sigma_r = 0$ for $r_o = 57.5$ mm and (ii) $\varepsilon_\theta = (\sigma_\theta - \nu\,\sigma_r)/E = 7 \times 10^{-4}$ for $r_o = 57.5$ mm. It follows from (i) and (ii) that the circumferential stress at the o.d. is $\sigma_\theta = 7 \times 10^{-4} \times 207 \times 10^3 = 144.9$ MPa. From eqs(3.16a,b),

$$0 = a - b/(57.5)^2$$
$$144.9 = a + b/(57.5)^2$$

giving $a = 72.45$ and $b = 239.54 \times 10^3$. At the interface radius, $r_c = 37.5$ mm, the hoop stress is a maximum in tension and the compressive radial stress equals the interface pressure:

$$\sigma_r = a - b\,/\,r_c^2 = 72.45 - 239.54 \times 10^3/(37.5)^2 = -97.89 \text{ MPa}$$
$$\sigma_\theta = a + b\,/\,r_c^2 = 72.45 + 239.54 \times 10^3/(37.5)^2 = 242.79 \text{ MPa}$$

It follows that the interface pressure is p = 97.89 MPa. The interference follows from eq(3.17b)

$$\Delta d = 2r_c\{[(\sigma_\theta + \nu\,p)/E]_{\text{hub}} + [p(1-\nu)/E]_{\text{shaft}}\}$$
$$= 2 \times 37.5 \times 10^{-3}\{[242.79 + (0.27 \times 97.89)]/207 + 97.89(1 - 0.33)/75\}$$
$$= 0.163 \text{ mm}$$

3.3.2 Thick-Walled Annular Discs

When one thick-walled disc is shrunk fit with interference on to another there results a compressive residual hoop stress distribution within the inner disc. This becomes effective in reducing the tensile hoop stress resulting from subsequent internal pressure. The effect is to produce a more even hoop stress distribution over the compound thickness. Where internal pressures reach high values, several discs may be compounded to enhance the benefit. For two thin annular discs in different materials, the relative hoop strain at the commom interface radius (r_c) is again found from the adjacent plane stress states (see Fig. 3.9).

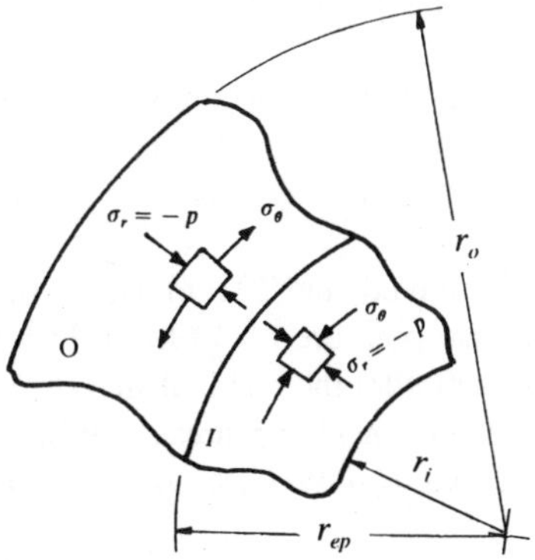

Figure 3.9 Stress state at interface

Taking respective subscripts I and O for the inner and outer discs, the relative hoop strain at the interface is

$$\Delta\varepsilon_\theta = [(1/E)(\sigma_\theta - \nu\,\sigma_r)]_O - [(1/E)(\sigma_\theta - \nu\,\sigma_r)]_I \quad (3.18a)$$
$$\Delta d/d_c = [(1/E)(\sigma_\theta + \nu\,p)]_O - [(1/E)[(\sigma_\theta + \nu\,p)]_I \quad (3.18b)$$

where Δd is the initial interference, p is the common interface pressure and σ_θ is supplied from eqs(3.16a,b). The constants are found from the boundary conditions. In I, $\sigma_r = 0$ for $r = r_i$ and $\sigma_r = -p$ for $r = r_c$. In O, $\sigma_r = 0$ for $r = r_o$ and $\sigma_r = -p$ for $r = r_c$.

3.3.3 Thin-Walled Rings

When two thin-walled rings are shrunk together with a common compressive pressure p at the interface diameter d_c, the membane theory gives

$(\sigma_\theta)_O = pd_c/(2t_O)$ (tensile)
$(\sigma_\theta)_I = pd_c/(2t_I)$ (compressive)

The interference is required to give an interface pressure $(\sigma_r)_I = (\sigma_r)_O = -p$. Neglecting the small lateral strain term $\nu p/E$, eq(3.18b) approximates to

$$\Delta d = (pd_c^2/2)[1/(E_I\, t_I) + 1/(E_O t_O)] \quad (3.19)$$

3.3.4 Thick-Walled Cylinders

The plane stress analyses in sub-sections 3.3.2 and 3.3.3 ignore residual axial stress σ_z at the interface following shrinkage. Plane strain conditions may be assumed when compounding long thick-walled cylinders, as with gun barrel manufacture. Taking $\varepsilon_z = 0$ gives an axial stress $\sigma_z = \nu(\sigma_\theta + \sigma_r)$. With a triaxial stress state $(\sigma_\theta, \sigma_r, \sigma_z)$ now present at the interface, eqs(3.18a,b) are modified to

$$\Delta\varepsilon_\theta = \{(1/E)[\sigma_\theta - \nu(\sigma_r + \sigma_z)]\}_O - \{(1/E)[\sigma_\theta - \nu(\sigma_r + \sigma_z)]\}_I$$
$$\Delta\varepsilon_\theta = \{(1/E)[\sigma_\theta(1 - \nu^2) - \nu\,\sigma_r(1 + \nu)]\}_O - \{(1/E)[\sigma_\theta(1 - \nu^2) - \nu\,\sigma_r(1 + \nu)]\}_I$$
$$\Delta d/d_c = \{(1/E)[\sigma_\theta(1 - \nu^2) + \nu\,p(1 + \nu)]\}_O - \{(1/E)[\sigma_\theta(1 - \nu^2) + \nu\,p(1 + \nu)]\}_I \quad (3.20)$$

The subsequent pressure stress distributions in compounded, thick-walled discs and cylinders are found from assuming a monobloc cylinder. The net stress under pressure is then the sum of the residual and pressure stresses. This is illustrated in the following example.

Example 3.8 A brass sleeve 55 mm i.d. and 70 mm o.d., is fitted within a steel hub 100 mm o.d. to give a common interface pressure of 30 MPa. Determine the initial difference in diameters between sleeve and hub. Plot the residual stresses resulting from the interference. Assuming plane stress, determine the net hoop stress distribution when a subsequent internal pressure of 1 kbar is applied. Take $E_b = 90$ GPa, $\nu_b = 0.33$, $E_s = 210$ GPa and $\nu_s = 0.28$.

Within the brass sleeve the boundary conditions are $\sigma_r = 0$ for $r = 27.5$ mm and $\sigma_r = -30$ MPa for $r = 35$ mm. Equation (3.16a) gives simultaneous equations

$0 = a - b/(27.5)^2$
$-30 = a - b/(35)^2$

from which the constants are $a = -78.5$ and $b = -59.5 \times 10^3$. The hoop stress at the brass interface (I in Fig. 3.9) is then

$\sigma_\theta = a + b/r_c^2 = -78.5 - (59.5 \times 10^3)/35^2 = -127.1$ MPa (compression)

Within the steel hub, the boundary conditions are $\sigma_r = 0$ for $r = 50$ mm and $\sigma_r = -30$ for $r = 35$ mm. These give

$0 = a' - b'/(50)^2$
$-30 = a' - b'/(35)^2$

These give $a' = 28.82$ and $b' = 72.1 \times 10^3$ from which the hoop stress at the steel interface is

$\sigma_\theta = a' + b'/r_c^2 = 28.82 + (72.1 \times 10^3)/35^2 = 87.62$ MPa (tension)

Now from eq(3.18b),

$$\begin{aligned}\Delta d / d_c &= [(1/E)(\sigma_\theta + \nu p)]_s - [(1/E)(\sigma_\theta + \nu p)]_b \\ &= [87.62 + (0.28 \times 30)]/(210 \times 10^3) - [-127.1 + (0.33 \times 30)]/(90 \times 10^3) \\ &= [0.4574 - (-1.302)]10^{-3} = 1.7595 \times 10^{-3}\end{aligned}$$
$$\Delta d = 70 \times 1.7595 \times 10^{-3} = 0.123 \text{ mm}$$

Note that if both these cylinders were long (plane strain), the resulting interference is, from eq(3.20), $\Delta d = 0.108$ mm. The stress state at $r = r_i$ in the brass is $\sigma_r = 0$ and

$$\begin{aligned}\sigma_\theta &= a + b/r_i^2 \\ &= -78.5 - (59.5 \times 10^3)/(27.5)^2 = -157.2 \text{ MPa}\end{aligned}$$

At $r = r_o$ in the steel, $\sigma_r = 0$ and

$$\begin{aligned}\sigma_\theta &= a' + b'/r_o^2 \\ &= 28.82 + (72.1 \times 10^3)/(50)^2 = 57.66 \text{ MPa}\end{aligned}$$

These residual stresses are distributed in the manner of Fig. 3.10.

For a subsequent internal pressure of 1 kbar the *monobloc* boundary conditions become $\sigma_r = -100$ MPa for $r = 27.5$ mm and $\sigma_r = 0$ for $r = 50$ mm. From eq(3.16a),

$-100 = a'' - b''/(27.5)^2$
$0 = a'' - b''/(50)^2$

Thus $a'' = 43.37$ and $b'' = 108.43 \times 10^3$. The hoop stresses due to the pressure at the three radii are ($\sigma_\theta = a'' + b''/r^2$)

$r = 27.5$ mm: $\sigma_\theta = 43.37 + (108.43 \times 10^3)/(27.5)^2 = 186.7$ MPa
$r = 35$ mm: $\sigma_\theta = 43.37 + (108.43 \times 10^3)/(35)^2 = 131.88$ MPa
$r = 50$ mm: $\sigma_\theta = 43.37 + (108.43 \times 10^3)/(50)^2 = 86.75$ MPa

Superimposing these upon the residuals in Fig. 3.10, the net hoop stresses at the inner and outer diameters become

$\sigma_\theta = -157.2 + 186.7 = 29.5$ MPa for $r_i = 27.5$ mm
$\sigma_\theta = 57.66 + 86.75 = 144.41$ MPa for $r_o = 50$ mm.

At the interface a discontinuity appears with two net hoop stress values

$\sigma_\theta = -127.1 + 131.8 = 4.7$ MPa
$\sigma_\theta = 87.62 + 131.8 = 219.4$ MPa

Figure 3.10 shows the benefit derived from compounding in reducing stress levels at the inner and outer radii. The design should ensure that the interface stress (219.4 MPa) remains elastic if the interference is to be effective under pressure.

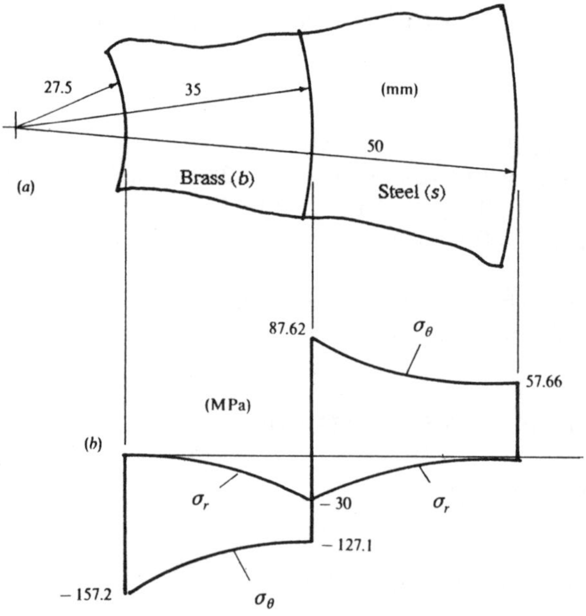

Figure 3.10 Residual and net stresses

3.4 Rotating Cylindrical Bodies

3.4.1 Annular Disc

(a) Solution from First Principles
When a thick-walled annular disc rotates with uniform angular speed ω rad/s about its axis, the centrifugal force will induce a radial (σ_r) and a hoop (σ_θ) stress distribution through the wall. Provided the disc has a small, constant thickness, a plane stress condition ($\sigma_z = 0$) applies. Consider unit thickness for the disc element shown in Fig. 3.11.

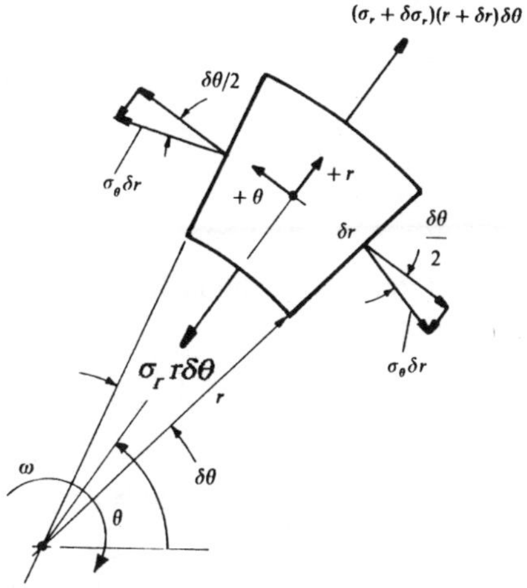

Figure 3.11 Rotating disc element

The radial force equilibrium equation is

$$(\sigma_r + \delta\sigma_r)(r + \delta r)\delta\theta - \sigma_r r\delta\theta - 2\sigma_\theta \delta r \sin(\delta\theta/2) + \rho\, \delta V\omega^2 r = 0$$

Putting $\delta V = r \times \delta\theta \times \delta r$ and $\sin(\delta\theta/2) \approx \delta\theta/2$ in the limit leads to

$$\sigma_r + r\, d\sigma_r / dr - \sigma_\theta + \rho\, \omega^2 r^2 = 0 \qquad (3.21)$$

The relations between stress, strain and displacement for axial symmetry are

$$\varepsilon_\theta = u/r = (1/E)(\sigma_\theta - \nu\, \sigma_r)$$
$$\varepsilon_r = du/dr = (1/E)(\sigma_r - \nu\, \sigma_\theta)$$

Combining these leads to the compatibility condition:

$$d\varepsilon_\theta /dr + (\varepsilon_\theta - \varepsilon_r) = 0 \qquad (3.22a)$$
$$r\,(d\sigma_\theta /dr - \nu\, d\sigma_r /dr) + (1 + \nu)(\sigma_\theta - \sigma_r) = 0 \qquad (3.22b)$$

Then, from eqs(3.21 and 3.22b)

$$r\,(d\sigma_\theta / dr + d\sigma_r / dr) + (1 + \nu)\rho\omega^2 r^2 = 0$$
$$d(\sigma_\theta + \sigma_r) / dr = - (1 + \nu)\rho\omega^2 r$$

Integration gives

$$\sigma_\theta + \sigma_r = 2a - (1 + \nu)\rho\omega^2 r^2 / 2 \qquad (3.23)$$

where $2a$ is a constant. Eliminating σ_θ between eqs(3.21) and (3.23) leads to

$$2\sigma_r + r\,\mathrm{d}\sigma_r / \mathrm{d}r = 2a - (3 + \nu)\rho\omega^2 r^2 / 2$$
$$(1/r)\,\mathrm{d}(r^2 \sigma_r) / \mathrm{d}r = 2a - (3 + \nu)\rho\omega^2 r^2 /2$$
$$\therefore r^2 \sigma_r = \int [2ar - (3 + \nu)\rho\omega^2 r^3 / 2]\mathrm{d}r$$
$$= ar^2 - (3 + \nu)\rho\omega^2 r^4 / 8 - b$$
$$\sigma_r = a - b/r^2 - (3 + \nu)\rho\omega^2 r^2 / 8 \qquad (3.24a)$$
$$\sigma_\theta = a + b/r^2 - (1 + 3\nu)\rho\omega^2 r^2 / 8 \qquad (3.24b)$$

The constants a and b are found from the condition that $\sigma_r = 0$ for $r = r_i$ and $r = r_o$. This gives

$$a = \rho\omega^2 (r_i^2 + r_o^2)(3 + \nu) / 8 \qquad (3.25a)$$
$$b = \rho\omega^2 r_i^2 r_o^2 (3 + \nu) / 8 \qquad (3.25b)$$

Hence the maximum hoop stress lies at the inner radius,

$$\sigma_\theta = [(3 + \nu)\rho\omega^2 r_o^2 /4]\{1 + [(1 - \nu) r_i^2] / [(3 + \nu) r_o^2]\} \qquad (3.25c)$$

The radial displacement u at any radius r follows as a result of setting $\sigma_z = 0$ in eq(3.12a). Substituting from eqs(3.24a,b) and (3.25a,b), we find

$$\varepsilon_\theta = u/r = (\sigma_\theta - \nu\sigma_r)/E$$
$$u = [\rho\omega^2 r (3 + \nu)(1 - \nu)/ (8E)][r_i^2 + r_o^2$$
$$+ (1 + \nu) r_i^2 r_o^2/(1 - \nu) r^2 - (1 + \nu) r^2/ (3 + \nu)] \qquad (3.26)$$

(b) Solution from a Stress Function

We can arrive at eqs(3.24a,b) by coupling a plane polar stress function Φ that accounts for the axial symmetry with a body force potential Ω. The terms within tor the stress function in eq(2.46b) provide the Lamé stresses. The centrifugal force derives from potential Ω, as follows:

$$\mathrm{d}\Omega /\mathrm{d}r = \rho\omega^2 r$$

so that:

$$\Omega = \rho\omega^2 r^2 / 2 \qquad (3.27a)$$

For axial symmetry, $\partial\Phi/ \partial\theta = 0$, and eq(2.42a) (where Φ now replaces ϕ) becomes

$$\mathrm{d}^4\Phi/\mathrm{d}r^4 + (2/r)\,\mathrm{d}^3\Phi/\mathrm{d}r^3 - (1/r^2)\,\mathrm{d}^2\Phi/\mathrm{d}r^2 + (1/r^3)\,\mathrm{d}\Phi/\mathrm{d}r + (1 - \nu)$$
$$[\mathrm{d}^2\Omega /\mathrm{d}r^2 + (1/r)\,\mathrm{d}\Omega /\mathrm{d}r] = 0 \qquad (3.27b)$$

Both Φ and Ω are to satisfy eq(3.27b). This gives Φ an additional rotational term:

$$\Phi = A \ln r + Cr^2 - (3 + \nu)\rho\omega^2 r^4 /32 \qquad (3.27c)$$

The radial and tangential stress components follow from eqs(2.47a,b) and (3.27a,c) as

$$\sigma_r = (1/r)\mathrm{d}\Phi/\mathrm{d}r = 2C + A/r^2 - (3 + \nu)\rho\omega^2 r^2/ 8$$
$$\sigma_\theta = \mathrm{d}^2\Phi/\mathrm{d}r^2 + r\,\mathrm{d}\Omega /\mathrm{d}r = 2C - A/r^2 - (1 + 3\nu)\rho\omega^2 r^2 / 8$$

Clearly these are identical to eqs(3.24a,b) when we set $a = 2C$ and $b = -A$. Note that both solutions are approximations as an exact plane stress analysis requires additional compatibility conditions to be satisfied (see eq 2.29a).

Example 3.9 Determine the hoop stress at the inner $r_i = 50$ mm and outer $r_o = 150$ mm radii for a disc when rotating uniformly at 5000 rev/min. Show the σ_θ and σ_r variation with r and the radius for which σ_r is a maximum. Take $\nu = 0.3$ and $\rho = 7480$ kg/m^3.

$$\omega = 2\pi N / 60 = 2\pi \times 5000 / 60 = 523.6 \text{ rad/s}$$

and, from eqs(3.25a and b),

$$a = 7480 \times (523.6)^2(50^2 + 150^2)10^{-6} \times 3.3 / 8 \quad \{\text{kg/m}^3 \times (\text{rad/s})^2 \times \text{m}^2 = \text{kgm/s}^2 \times \text{m}^{-2} \equiv \text{Pa}\}$$
$$= 21.15 \times 10^6 \text{ Pa} = 21.15 \text{ MPa}$$

$$b = 7480 \times (523.6)^2 \times 50^2 \times 150^2 \times 10^{-12} \times 3.3 / 8 \quad \{\text{kg/m}^3 \times (\text{rad/s})^2 \times \text{m}^4 = \text{kgm/s}^2 \equiv \text{N}\}$$
$$= 47.583 \times 10^3 \text{ N}$$

Then, from eq(3.24b) with $r = 50$ mm,

$$\sigma_\theta = 21.15 + (47.583 \times 10^3) / 50^2 - [1.9 \times 7480\,(523.6)^2 \times 50^2 \times 10^{-12}] / 8$$
$$= 21.15 + 19.033 - 1.218 = 38.97 \text{ MPa}$$

For $r = 150$ mm,

$$\sigma_\theta = 21.15 + (47.583 \times 10^3)/150^2 - [1.9 \times 7480\,(523.6)^2 \times 150^2 \times 10^{-12}] / 8$$
$$= 21.15 + 2.115 - 10.96 = 12.31 \text{ MPa}$$

The radial stress is a maximum when, from eq(3.24a),

$$d\sigma_r/dr = 2b/r^3 - \rho\omega^2 r\,(3 + \nu)/4 = 0$$
$$r^4 = 8b / [\rho\omega^2 (3 + \nu)]$$

Substituting for b from eq(3.25b) leads to

$$r = \surd(r_i r_o) \qquad \text{(i)}$$

The corresponding maximum radial stress is found from eqs(3.24a) and (3.25a,b)

$$(\sigma_r)_{\max} = a - b/(r_i r_o) - (3 + \nu)\rho\omega^2 r_i r_o / 8$$
$$= [(3 + \nu)\omega^2\rho / 8](r_i^2 + r_o^2 - r_i r_o - r_i r_o)$$
$$(\sigma_r)_{\max} = (3 + \nu)(r_o - r_i)^2\omega^2\rho / 8 \qquad \text{(ii)}$$

In this example eq(i) yields $r = \surd(50 \times 150) = 86.6$ mm, when from eq(ii),

$$(\sigma_r)_{\max} = (3.3 \times 100^2 \times 523.6^2 \times 7480 \times 10^{-12}) / 8 = 8.6 \text{ MPa}$$

The σ_r and σ_θ stress distributions are as shown in Fig. 3.12.

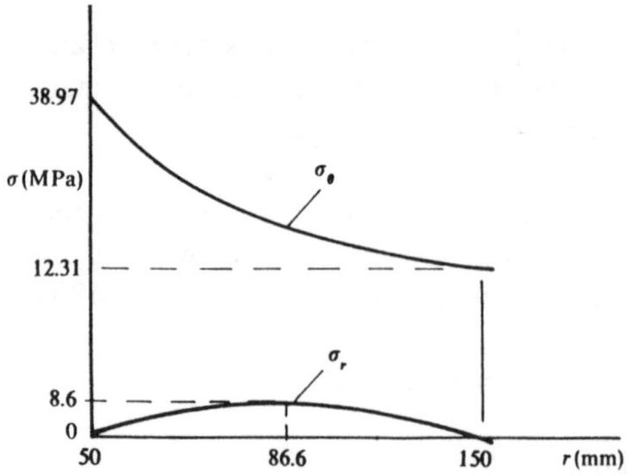

Figure 3.12 Stresses in rotating disc

3.4.2 Solid Disc

The stresses in a solid rotating disc are also given by eqs(3.24a,b) but with $b = 0$ as infinite stresses are not possible at the disc centre ($r = 0$). This gives

$$\sigma_r = a - (3 + \nu)\rho\omega^2 r^2/8 \qquad (3.28a)$$
$$\sigma_\theta = a - (1 + 3\nu)\rho\omega^2 r^2/8 \qquad (3.28b)$$

Since $\sigma_r = 0$ at the outer diameter, the constant a in eqs(3.28a,b) becomes

$$a = \rho\omega^2 r_o^2 (3 + \nu)/8 \qquad (3.28c)$$

Equation (3.28c) defines the stress state ($\sigma_r = \sigma_\theta = a$) at the disc centre. Compare this with the stress state at the inner boundary of a hollow disc, where the radial stress is absent but the hoop stress (first term in eq 3.25c) is doubled when the hole is small. For the radial displacement in a solid disc, set $r_i = 0$ in eq(3.26), to give

$$u = [\rho\omega^2 r(1 - \nu)/(8E)][(3 + \nu)r_o^2 - (1 + \nu)r^2] \qquad (3.29)$$

Equation (3.29) gives the change to the outer diameter $\Delta d = 2u$ for $r = r_o$, as $\Delta d = \rho\omega^2 r_o^3(1 - \nu)/(2E)$

3.4.3 Thick-Walled Cylinder

When a long cylinder rotates about its axis, a similar analysis to the annular disc applies but with an assumed plane strain condition, $\varepsilon_z = \varepsilon_o$ = constant. This leads to the triaxial stress distributions

$$\sigma_r = a - b/r^2 - (3 - 2\nu)\rho\omega^2 r^2/[8(1 - \nu)] \qquad (3.30a)$$
$$\sigma_\theta = a + b/r^2 - (1 + 2\nu)\rho\omega^2 r^2/[8(1 - \nu)] \qquad (3.30b)$$
$$\sigma_z = E\varepsilon_o + \nu(\sigma_\theta + \sigma_r) \qquad (3.30c)$$

Since $\sigma_r = 0$ for $r = r_i$ and $r = r_o$, the constants in eqs(3.30a,b) are

$$a = \rho\omega^2(r_i^{\,2} + r_o^{\,2})(3 - 2\nu)/[8(1 - \nu)] \tag{3.31a}$$

$$b = \rho\omega^2 r_i^2 r_o^2(3 - 2\nu)/[8(1 - \nu)] \tag{3.31b}$$

Substituting eqs(3.31a,b) into eq(3.30b) gives the maximum hoop stress in the bore

$$\sigma_\theta = \{[(3 - 2\nu)\rho\omega^2 r_o^{\,2}]/[4(1 - \nu)]\}\{1 + [(1 - 2\nu)r_i^{\,2}]/[(3 - 2\nu)r_o^{\,2}]\} \tag{3.31c}$$

Plane strain modifies only the compatibility condition (3.22). The reader should check that eqs(3.30a,b,c) satisfy the same equilibrium equation (3.21) appearing in the plane stress analysis. Alternatively, in seeking a stress function Φ for this problem, both Φ and the body force potential Ω in eq(3.27a) must now satisfy eq(2.42b). This gives

$$\mathrm{d}^4\Phi/\mathrm{d}r^4 + (2/r)\,\mathrm{d}^3\Phi/\mathrm{d}r^3 - (1/r^2)\,\mathrm{d}^2\Phi/\mathrm{d}r^2 + (1/r^3)\,\mathrm{d}\Phi/\mathrm{d}r$$
$$+ [(1 - \nu)/(1 + \nu)][\mathrm{d}^2\Omega/\mathrm{d}r^2 + (1/r)\,\mathrm{d}\Omega/\mathrm{d}r] = 0 \tag{3.32a}$$

from which

$$\Phi = A\ln r + Cr^2 - [(3 - 2\nu)\rho\omega^2 r^4]/[32(1 - \nu)] \tag{3.32b}$$

The radial and tangential stress components follow from eqs(3.32a,b) as

$$\sigma_r = (1/r)\,\mathrm{d}\Phi/\mathrm{d}r = 2C + A/r^2 - (3 - 2\nu)\rho\omega^2 r^2/[8(1 - \nu)] \tag{3.33a}$$

$$\sigma_\theta = \mathrm{d}^2\Phi/\mathrm{d}r^2 + r\,\mathrm{d}\Omega/\mathrm{d}r = 2C - A/r^2 - (1 + 2\nu)\rho\omega^2 r^2/[8(1 - \nu)] \tag{3.33b}$$

Clearly eqs(3.33a,b) are identical to eqs(3.30a,b) when we set $a = 2C$ and $b = -A$. Both solutions are exact since there are no further compatibility conditions to satisfy.

When $\varepsilon_o = 0$ in eq(3.30c), the axial extension is constrained, but if the cylinder is allowed to extend, an expression for ε_o follows from the absence of an axial force. That is,

$$\int_{r_i}^{r_o} 2\pi r\sigma_z\,\mathrm{d}r = 0 \tag{3.34a}$$

where, from eqs(3.30a-c),

$$\sigma_z = E\varepsilon_o + \nu\{2a - \rho\omega^2 r^2/[2(1 - \nu)]\} \tag{3.34b}$$

Substituting eq(3.34b) into eq(3.34a) leads to

$$\varepsilon_o = \nu\rho\omega^2(r_i^{\,2} + r_o^{\,2})/[4E(1 - \nu)] - 2a\,\nu/E \tag{3.34c}$$

Substituting eq(3.31a) into eq(3.34c) gives the axial strain

$$\varepsilon_o = -\nu\rho\omega^2(r_i^{\,2} + r_o^{\,2})/(2E) \tag{3.35a}$$

The axial stress follows from eqs(3.31a), (3.34b) and (3.35a) as

$$\sigma_z = \{\nu\rho\omega^2/[4(1 - \nu)]\}(r_i^{\,2} + r_o^{\,2} - 2r^2) \tag{3.35b}$$

Finally, eqs(3.12a) and (3.30a-c) provide the radial displacement

$$u = \{\rho\omega^2 r /[8E(1-\nu)]\}[(3-5\nu)(r_i^2 + r_o^2) + (3-2\nu)(1+\nu)r_i^2 r_o^2 / r^2 - (1+\nu)(1-2\nu)r^2] \quad (3.35c)$$

3.4.4 Solid Shaft

For a long, rotating solid shaft $b = 0$ in eqs(3.30a,b,c) as both σ_r and σ_θ must be finite at the shaft centre. This gives

$$\sigma_r = a - (3-2\nu)\rho\omega^2 r^2 / [8(1-\nu)] \quad (3.36a)$$
$$\sigma_\theta = a - (1+2\nu)\rho\omega^2 r^2 / [8(1-\nu)] \quad (3.36b)$$
$$\sigma_z = E\varepsilon_o + \nu(\sigma_\theta + \sigma_r) \quad (3.36c)$$

The constant a in eqs(3.36a,b) is found by setting $\sigma_r = 0$ in eq(3.36a) when $r = r_o$

$$a = (3-2\nu)\rho\omega^2 r_o^2 / [8(1-\nu)] \quad (3.36d)$$

It follows that a defines an equal stress state ($\sigma_\theta = \sigma_r = a$) at the shaft centre. Comparing eqs(3.31c) and (3.36d) shows that drilling a small hole along the shaft axis will again double the maximum hoop stress induced by the centrifugal force.

Setting $r_i = 0$ in eqs(3.35a,b,c) gives the axial strain, stress and radial displacement for a solid rotating shaft,

$$\varepsilon_o = -\nu\rho\omega^2 r_o^2/(2E)$$
$$\sigma_z = \{\nu\rho\omega^2 / [4(1-\nu)]\} \times (r_o^2 - 2r^2)$$
$$u = \{\rho\omega^2 r / [8E(1-\nu)]\} \times [(3-5\nu)r_o^2 - (1+\nu)(1-2\nu)r^2]$$

3.4.5 Disc and Hub with Interference Fit

When a thin disc-hub assembly is to rotate at high speed, the initial interference between their diameters may be chosen to prevent slipping and also to limit the maximum hoop stress. The initial interference and the shrinkage stresses have been found previously in section 3.3 (see example 3.8). When the assembly subsequently rotates, eqs(3.24a,b) supply the stresses in the annular hub and eqs(3.28a,b) supply those for the solid disc. The net stresses are the sum of the rotational (R) and shrinkage (S) values. For example, at a given radius r in the hub, the net hoop stress is

$$\begin{aligned}(\sigma_\theta)_N &= (\sigma_\theta)_S + (\sigma_\theta)_R \\ &= [a + b/r^2]_S + [a + b/r^2 - \rho\omega^2 r^2(1+3\nu)/8]_R \\ &= (a_S + a_R) + (b_S + b_R)/r^2 - \rho\omega^2 r^2(1+3\nu)/8 \\ &= a_N + b_N/r^2 - \rho\omega^2 r^2(1+3\nu)/8 \end{aligned} \quad (3.37a)$$

Similarly, the radial stress becomes

$$(\sigma_r)_N = a_N - b_N/r^2 - \rho\omega^2 r^2(3+\nu)/8 \quad (3.37b)$$

By putting $a_N = a_S + a_R$ and $b_N = b_S + b_R$, it follows that eqs(3.37a,b) supply net stresses directly. This approach can simplify the analysis considerably when net limiting stresses are specified at a given speed.

Example 3.10 A compound steel assembly comprises a hub with 510 mm i.d. and 710 mm o.d., shrunk on to a solid disc. Show that the initial diametral interference of 0.10 mm is maintained between the hub and disc at a speed of 10,000 rev/min (104.7 rad/s). Assume plane stress conditions and take $E = 207$ GPa, $\nu = 0.3$ and $\rho = 7835$ kg/m^3.

Following shrinkage, a residual compressive pressure $\sigma_r = -p$ exists at the interface radius (r_c) of the disc and hub (see Fig. 3.13a). Applying two boundary conditions, (i) $\sigma_r = 0$ for $r_o = 355$ mm and (ii) $\sigma_r = -p$ for $r_c = 255$ mm, to eqs(3.16a,b) give a residual hoop stress, $\sigma_\theta = 3.132p$, at the hub interface.

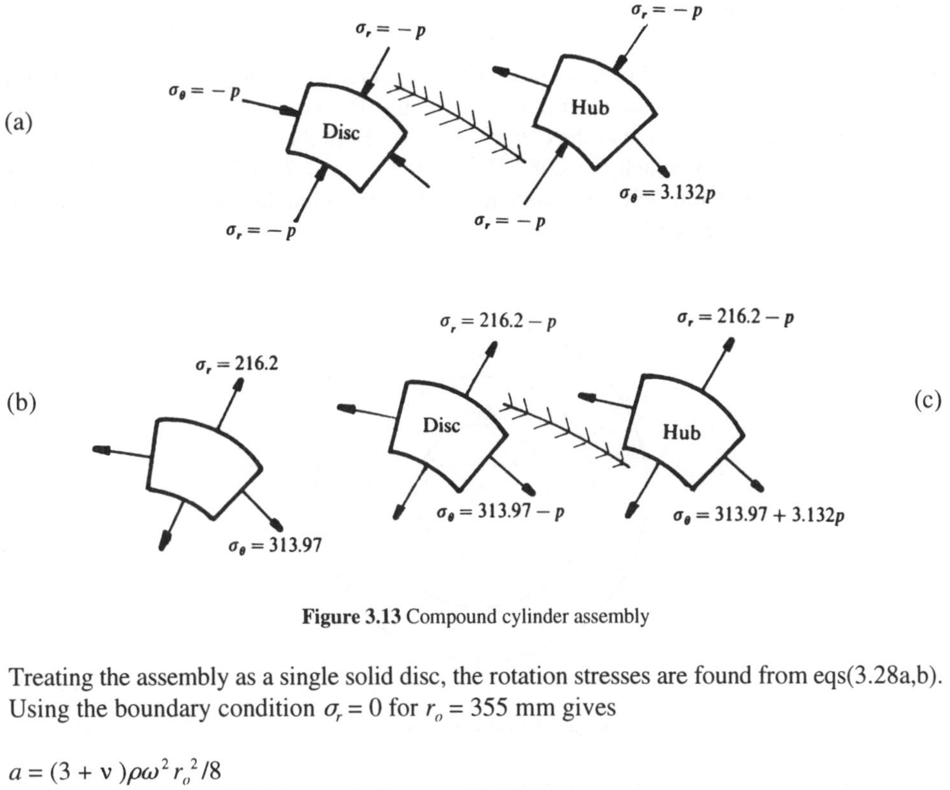

Figure 3.13 Compound cylinder assembly

Treating the assembly as a single solid disc, the rotation stresses are found from eqs(3.28a,b). Using the boundary condition $\sigma_r = 0$ for $r_o = 355$ mm gives

$$a = (3 + \nu)\rho\omega^2 r_o^2/8$$
$$= 3.3 \times 7835 \times (1047.2)^2 \times 355^2/(8 \times 10^{12}) = 446.66$$

The rotation stresses at $r_c = 255$ mm are

$$\sigma_r = 446.66 - (255/355)^2 \times 446.66$$
$$= 446.66 - 230.46 = 216.2 \text{ MPa}$$

$$\sigma_\theta = 446.66 - (1.9/3.3) \times 230.46$$
$$= 446.66 - 132.69 = 313.97 \text{ MPa}$$

The residual and rotational stress states at r_c are shown in Figs 3.13 (a) and (b) respectively. Figure 3.13(c) gives the net stresses at the interface between disc and hub. It follows from eq(3.18b) that the required interference is given by

$$0.10/510 = \{[(313.97 + 3.132p) - \nu\,(216.2 - p)] - [(313.97 - p) - \nu\,(216.2 - p)]\}/207000 \quad \text{(i)}$$

Hence $p = 9.823$ MPa. Applying eq(3.18b) to Fig. 3.13a, the original interference Δd is found from

$$\Delta d\,/510 = [(3.132p + \nu\,p) - (-p + \nu\,p)]/207000$$
$$\Delta d = (4.132 \times 9.823 \times 510)/207000 = 0.10 \text{ mm}$$

This condition will always apply to a disc-hub assembly in the same material because the rotation stresses cancel within eq(i).

Example 3.11 A steel disc of outer radius 250 mm is shrunk on to a hollow axle which has an outer radius of 40 mm and an inner radius of 20 mm. The radial interference between the disc and axle is 2.5×10^{-3} mm at the common radius. Both the disc and the axle are made of the same material, for which $E = 208$ GPa, $\nu = 0.3$ and $\rho = 7860$ kg/m^3. Find the radial stress at the common radius when the system is stationary. Find the angular speed at which the disc becomes loose. (CEI)

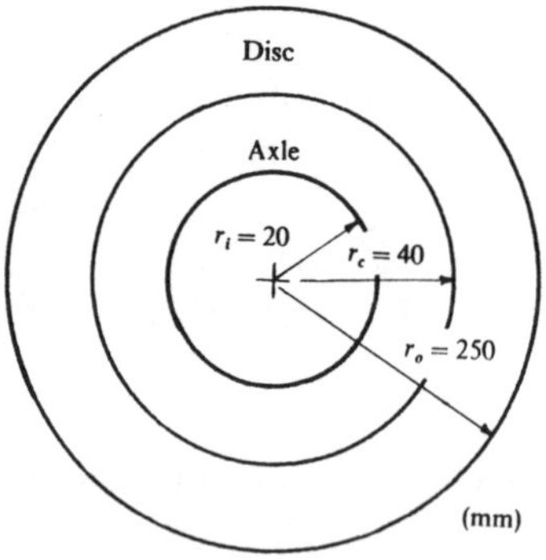

Figure 3.14 Disc-axle assembly

For the axle (Fig. 3.14) the boundary conditions within eq(3.16a) give

$$\sigma_r = 0 = a - b/20^2$$
$$\sigma_r = -p = a - b/40^2$$
$$\therefore a = -1.333p,\ b = -533.33p$$

and at $r_c = 40$ mm, assuming a plane strain condition

$$\sigma_\theta = a + b/r_c^{\,2}$$
$$\sigma_\theta = -1.333p - 533.33p/40^2 = -1.666p$$
$$\sigma_z = \nu\,(\sigma_\theta + \sigma_r) = 0.3(-1.666p - p) = -0.8p$$

For the disc's boundary conditions, eq(3.16a) gives

$\sigma_r = 0 = a' - b'/250^2$
$\sigma_r = -p = a' - b'/40^2$
$\therefore a' = 0.02627p, \; b' = 1642.04p$

and for $r_c = 40$ mm,

$\sigma_\theta = a' + b'/r_c^2 = 0.02627p + 1642.04\,p/40^2 = 1.0525p$

The radial interference u follows from a relation similar to eq(3.18b)

$E(u/r_c) = (1.0525p + 0.3\,p)_{\text{disc}} - [-1.666p - 0.3(-p - 0.8p)]_{\text{axle}} = 2.4785p$

from which $p = (208 \times 10^3 \times 2.5 \times 10^{-3})/(2.4785 \times 40) = 5.245$ MPa

$\therefore \sigma_r = -5.245$ MPa at $r_c = 40$ mm

It is assumed that when rotating with increasing speed, the disc will begin to slip on the axle when the interference is reduced to zero. This is a consequence of the increasing axial stress in the axle when its axial extension is prevented. Treating the rotating assembly as a solid annular disc the boundary conditions become $\sigma_r = 0$ for $r_i = 20$ mm and $r_o = 250$ mm. Equation(3.24a) gives

$\sigma_r = 0 = a'' - b''/20^2 - (7860 \times 3.3 \times 10^{-12} \times 20^2/8)\omega^2$
$0 = a'' - b''/20^2 - (1.297 \times 10^{-6})\omega^2$ (i)

$\sigma_r = 0 = a'' - b''/250^2 - (7860 \times 3.3 \times 10^{-12} \times 250^2/8)\omega^2$
$0 = a'' - b''/250^2 - (202.64 \times 10^{-6})\omega^2$ (ii)

Solving eqs(i) and (ii)

$a'' = (2.0394 \times 10^{-4})\omega^2, \quad b'' = 0.08106\omega^2$

At $r_c = 40$ mm, the radial and hoop stress in eq(3.24a,b) become

$\sigma_r = (2.0394 \times 10^{-4})\omega^2 - 0.08106\omega^2/40^2 - (7860 \times 3.3 \times 10^{-12} \times 40^2/8)\omega^2$
$\sigma_r = (148.09 \times 10^{-6})\omega^2$

$\sigma_\theta = (2.0394 \times 10^{-4})\omega^2 + 0.08106\omega^2/40^2 - (7860 \times 1.9 \times 10^{-12} \times 40^2/8)\omega^2$
$\sigma_\theta = (251.62 \times 10^{-6})\omega^2$

Hence, the axial stress at the interface becomes

$\sigma_z = \nu(\sigma_\theta + \sigma_r) = 0.3(251.62 + 148.09)10^{-6}\,\omega^2 = 119.91 \times 10^{-6}\,\omega^2$ (iii)

Compare this with $\sigma_z = 1.01 \times 10^{-6}\,\omega^2$ from a zero axial force condition in eq(3.35b). The relative hoop strain under the stresses at speed is simply:

$$\varepsilon_\theta = u/r_c = \nu\sigma_z/E \tag{iv}$$

Substituting σ_z from eq(iii) into eq(iv) and setting $u = 2.5 \times 10^{-3}$ mm, the slipping speed is

$$\omega^2 = (207 \times 10^3 \times 2.5 \times 10^{-3})/(40 \times 0.3 \times 119.91 \times 10^{-6})$$
$$\therefore\ \omega = 5430.5 \text{ rad/s} \Rightarrow N = 51{,}858 \text{ rev/min}$$

3.5 Thick-Walled Sphere under Pressure

Consider the equilibrium condition about a horizontal plane for the hemispherical element of radial thickness δr in Fig. 3.15.

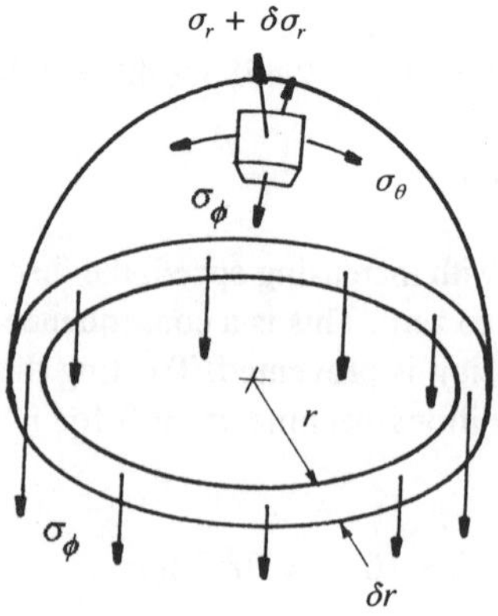

Figure 3.15 Element of sphere

Let the radial stress σ_r vary by the amount $\delta\sigma_r$ and assume the radial and hoop stresses are tensile. This gives

$$\sigma_\phi(2\pi r)\delta r + \sigma_r \pi r^2 = (\sigma_r + \delta\sigma_r)\pi(r + \delta r)^2$$

Neglecting products of small quantities, this reduces to

$$\sigma_\phi - \sigma_r = (r/2)\mathrm{d}\sigma_r/\mathrm{d}r \tag{3.38}$$

The constitutive relations for the inset triaxial stress state are

$$\varepsilon_r = \mathrm{d}u/\mathrm{d}r = (1/E)[\sigma_r - \nu(\sigma_\theta + \sigma_\phi)] \tag{3.39a}$$
$$\varepsilon_\phi = \varepsilon_\theta = u/r = (1/E)[\sigma_\phi - \nu(\sigma_r + \sigma_\theta)] \tag{3.39b}$$

where $\sigma_\theta = \sigma_\phi$. Since these two strains depend upon the single displacement u, the following compatibility condition applies

$$(\sigma_r - 2\nu\,\sigma_\theta) = \mathrm{d}[r\sigma_\phi(1 - \nu) - \nu r\sigma_r]/\mathrm{d}r$$
$$(1 + \nu)(\sigma_\phi - \sigma_r) = \nu r\,\mathrm{d}\sigma_r/\mathrm{d}r - (1 - \nu)\,r\,\mathrm{d}\sigma_\phi/\mathrm{d}r \tag{3.40}$$

Substituting eq(3.38) into eq(3.40) and integrating leads to

$$(r/2)(1 - \nu)\mathrm{d}\sigma_r/\mathrm{d}r + r(1 - \nu)\mathrm{d}\sigma_\phi/\mathrm{d}r = 0$$
$$\tfrac{1}{2}\,\mathrm{d}\sigma_r/\mathrm{d}r + \mathrm{d}\sigma_\phi/\mathrm{d}r = 0$$
$$\therefore \tfrac{1}{2}\sigma_r + \sigma_\phi = A \qquad (3.41)$$

where A is a constant. Combining eqs(3.38) and (3.41),

$$3\sigma_r + r\,\mathrm{d}\sigma_r/\mathrm{d}r = 2A$$
$$(1/r^2)\,\mathrm{d}(r^3\sigma_r)/\mathrm{d}r = 2A$$
$$\sigma_r = 2A/3 + B/r^3 \qquad (3.42a)$$
$$\sigma_\phi = A - \sigma_r/2 = 2A/3 - B/(2r^3) \qquad (3.42b)$$

Setting constants $a = 2A/3$ and $b = B$, eqs(3.42a,b) are written as

$$\sigma_r = a + b/r^3 \qquad (3.43a)$$
$$\sigma_\phi = \sigma_\theta = a - b/(2r^3) \qquad (3.43b)$$

where a and b depend upon whether the pressure is applied to one or both boundaries.

3.5.1 Internal and External Pressure

Applying the boundary conditions $\sigma_r = -p_i$ for $r = r_i$ and $\sigma_r = -p_o$ for $r = r_o$ to eq(3.43a) gives

$$a = (p_i r_i^3 - p_o r_o^3)/(r_o^3 - r_i^3) \qquad (3.44a)$$
$$b = (p_o - p_i)\, r_i^3 r_o^3/(r_o^3 - r_i^3) \qquad (3.44b)$$

Substituting eqs(3.44a,b) into eqs(3.43a,b)

$$\sigma_r = [-p_o + p_i(r_i/r_o)^3 + (p_o - p_i)(r_i/r)^3]\,/\,[1 - (r_i/r_o)^3] \qquad (3.45a)$$
$$\sigma_\phi = \sigma_\theta = [-p_o + p_i(r_i/r_o)^3 - \tfrac{1}{2}(p_o - p_i)(r_i/r)^3]\,/\,[1 - (r_i/r_o)^3] \qquad (3.45b)$$

If $r_i << r_o$, eqs(3.45a,b) approximate to

$$\sigma_r = -p_o + (p_o - p_i)(r_i/r)^3$$
$$\sigma_\phi = -p_o - \tfrac{1}{2}(p_o - p_i)(r_i/r)^3$$

3.5.2 External Pressure Only

Setting $p_i = 0$ in eqs(3.45a,b) gives

$$\sigma_r = [-1 + (r_i/r)^3]\,p_o/[1 - (r_i/r_o)^3]$$
$$\sigma_\phi = -[1 + \tfrac{1}{2}(r_i/r)^3]\,p_o/[1 - (r_i/r_o)^3]$$

The following approximate forms apply when $r_o >> r_i$:

$$\sigma_r = [-\ 1 + (r_i/r)^3]p_o \tag{3.46a}$$
$$\sigma_\phi = -[1 + ½\,(r_i/r)^3]p_o \tag{3.46b}$$

Equations (3.46a,b) provide the stresses around a spherical cavity when a uniform pressure is applied to the boundary. The shape of the boundary need not be spherical provided that it lies on all radii r_o for which $r_o >> r_i$. In particular, for $r = r_i$, eqs(3.46a,b) show that the hoop stress is concentrated to a magnitude $\sigma_\phi = -\ 3p_o/2$ around the hole.

3.5.3 Internal Pressure

Setting $p_o = 0$ in eqs(3.45a,b) gives

$$\sigma_r = [(r_i/r_o)^3 - (r_i/r)^3]\,p_i\,/\,[1 - (r_i/r_o)^3] \tag{3.47a}$$
$$\sigma_\phi = \sigma_\theta = [(r_i/r_o)^3 + ½\,(r_i/r)^3]p_i/\,[1 - (r_i/r_o)^3] \tag{3.47b}$$

Note that eqs(3.39a,b) apply when finding the strains and displacements in each case.

Example 3.12 Find expressions for the dilatation and the radial displacement at a point in the wall of a thick-walled sphere under internal fluid pressure p.

It follows from eq(2.6b) that the dilatation of the sphere material is

$$\delta V/V = \varepsilon_\phi + \varepsilon_\theta + \varepsilon_r = (2/E)[\sigma_\phi - \nu\,(\sigma_\theta + \sigma_r)] + (1/E)[\sigma_r - \nu\,(\sigma_\phi + \sigma_\theta)] \tag{i}$$

Setting $\sigma_\theta = \sigma_\phi$ and substituting eqs(3.47a,b) into eq(i), we find that $\delta V/V$ is independent of r,

$$\delta V/V = 3(1 - 2\nu)\,p_i\,r_i^{\,3}/[E(r_o^{\,3} - r_i^{\,3})] = p_i\,r_i^{\,3}/[K(r_o^{\,3} - r_i^{\,3})] \tag{ii}$$

Now as $V = 4\pi(r_o^{\,3} - r_i^{\,3})/3$ is the volume of the wall, eq(ii) shows that the wall material will change its volume elastically by $\delta V = 4\pi p r_i^{\,3}/3K$.

The radial displacement follows from eq(3.39b) as

$$u = (r/E)[\sigma_\phi - \nu\,(\sigma_r + \sigma_\theta)]$$

and substituting from eqs(3.47a,b),

$$u = \{p r_i^{\,3}/[E(r_o^{\,3} - r_i^{\,3})]\}[(1 - 2\nu)r + (1 + \nu)\,r_o^{\,3}/(2r^2)]$$

which may be checked from eq(3.39a).

3.6 Thermal Stresses in Cylindrical Bodies

Stresses can also arise in thin- and thick-walled cylindrical and spherical bodies due to two temperature effect, (i) a mismatch in expansion coefficients and (ii) radial temperature gradients. The following are of most practical interest.

3.6.1 Short, Thin-Walled Mating Rings

Consider an inner (I) and outer (O) ring that fit together initially without interference. The radial and hoop stresses induced at the common diameter d_c due to a temperature rise T are found from the compatibility condition:

$$\Delta\varepsilon_\theta = [(\sigma_\theta - \nu\,\sigma_r)/E]_O - [(\sigma_\theta - \nu\,\sigma_r)/E]_I = -(\alpha_O - \alpha_I)T \qquad (3.48)$$

when $\alpha_I > \alpha_O$. Now, at the interface, a common compressive radial stress $(\sigma_r)_O = (\sigma_r)_I = -p$ exists. The membrane hoop stresses are assumed constant within I and O as

$$(\sigma_\theta)_O = pd_c/(2t_O) \text{ and } (\sigma_\theta)_I = -pd_c/(2t_I) \qquad (3.49a,b)$$

Substituting eqs(3.49a,b) into eq(3.48) leads to

$$\{[pd_c/(2t) + \nu\,p]/E\}_O - \{[-pd_c/(2t) + \nu\,p]/E]\}_I = -(\alpha_O - \alpha_I)T$$

Neglecting the effect of the small lateral strain $\nu\,p/E$, this leads to

$$p = -(\alpha_O - \alpha_I)T\,/\,\{(d_c/2)[1/(E_O t_O) + 1/(E_I t_I)]\} \qquad (3.49c)$$

Equations (3.19) and (3.49a-c) may be used together to determine the net stresses following a temperature rise in discs with initial interference stresses at ambient temperature.

Example 3.13 Determine the membrane hoop stress for an inner copper / outer steel composite annular disc, with $d_c = 810$ mm, $t_c = 5$ mm, and $t_s = 7.5$ mm, when the temperature is raised by 25°C above the unstressed condition. Take $E_s = 210$ GPa, $\alpha_s = 12 \times 10^{-6}$ °C^{-1} and $E_c = 90$ GPa, $\alpha_c = 18 \times 10^{-6}$ °C^{-1}.

The interface pressure follows directly from eq(3.49c)

$$p = -[(12 - 18)10^{-6} \times 25 \times 10^3]\,/\{(810/2)[1/(210 \times 7.5) + 1/(90 \times 5)]\} = 0.13 \text{ MPa}$$

and the stresses from eq(3.49a,b)

$$(\sigma_\theta)_O = (0.13 \times 810)/\,(2 \times 7.5) = 7.02 \text{ MPa}$$
$$(\sigma_\theta)_I = -(0.13 \times 810)/\,(2 \times 5) = -10.53 \text{ MPa}$$

3.6.2 Thick-Walled Annular Disc

When the inner r_i and outer r_o radii of a thick-walled, stationary, unloaded disc are held at different, steady temperatures, radial and hoop stress variations will exist through the wall.

Assuming plane stress conditions, the radial equilibrium eq(3.13) again applies. Let the temperature vary with the radius r in the wall according to the function $T = T(r)$. The constitutive relations are

$$\varepsilon_r = \mathrm{d}u/\mathrm{d}r = (1/E)(\sigma_r - \nu\sigma_\theta) + \alpha T \tag{3.50a}$$
$$\varepsilon_\theta = u/r = (1/E)(\sigma_\theta - \nu\sigma_r) + \alpha T \tag{3.50b}$$

Eliminating u in eqs(3.50a,b) corresponds to the compatibility condition:

$$(1/E)(\sigma_r - \nu\sigma_\theta) + \alpha T = (1/E)\,\mathrm{d}[r(\sigma_\theta - \nu\sigma_r)]/\mathrm{d}r + \mathrm{d}(r\alpha T)/\mathrm{d}r$$
$$(1+\nu)(\sigma_r - \sigma_\theta) = r(\mathrm{d}\sigma_\theta/\mathrm{d}r - \nu\,\mathrm{d}\sigma_r/\mathrm{d}r) + E\alpha r(\mathrm{d}T/\mathrm{d}r) \tag{3.51a}$$

Substituting eq(3.13) into eq(3.51a) leads to

$$\mathrm{d}\sigma_\theta/\mathrm{d}r + \mathrm{d}\sigma_r/\mathrm{d}r = -E\alpha(\mathrm{d}T/\mathrm{d}r)$$
$$\therefore\ \sigma_\theta + \sigma_r = 2a - E\alpha T \tag{3.51b}$$

where $2a$ is a constant. Subtracting eq(3.13) from (3.51b) leads to

$$2\sigma_r + r\,\mathrm{d}\sigma_r/\mathrm{d}r = 2a - E\alpha T$$
$$(1/r)[\mathrm{d}(r^2\sigma_r)/\mathrm{d}r] = 2a - E\alpha T$$
$$r^2\sigma_r = \int (2ar - E\alpha T r)\,\mathrm{d}r - b$$
$$\sigma_r = a - b/r^2 - (E\alpha/r^2)\int Tr\,\mathrm{d}r \tag{3.52a}$$
$$\sigma_\theta = a + b/r^2 - E\alpha T + (E\alpha/r^2)\int Tr\,\mathrm{d}r \tag{3.52b}$$

Equations (3.52a,b) may be solved for a given function $T = T(r)$ with the boundary conditions $\sigma_r = 0$ at $r = r_i$ and $r = r_o$. This gives the constants a and b as

$$a = [E\alpha/(r_o^2 - r_i^2)]\int_{r_i}^{r_o} Tr\,\mathrm{d}r \tag{3.53a}$$

$$b = ar_i^2 = [E\alpha r_i^2/(r_o^2 - r_i^2)]\int_{r_i}^{r_o} Tr\,\mathrm{d}r \tag{3.53b}$$

For a solid disc $b = 0$ in eqs(3.52a,b) and $r_i = 0$ in eq(3.53a). Where both external and internal pressure loading exists with the temperature gradient, eqs(3.52a,b) remain valid, since they contain the Lamé eqs(3.16a,b) but with different constants due to the change in the boundary conditions. If the disc rotates in the presence of a temperature gradient, then the appropriate rotation terms from eqs(3.24a,b) may be superimposed upon eqs(3.52a,b). This gives

$$\sigma_r = a' - b'/r^2 - (E\alpha/r^2)\int Tr\,\mathrm{d}r - (3+\nu)\rho\omega^2 r^2/8$$
$$\sigma_\theta = a' + b'/r^2 - E\alpha T + (E\alpha/r^2)\int Tr\,\mathrm{d}r - (1+3\nu)\rho\omega^2 r^2/8$$

3.6.3 Thick-Walled Cylinder

Temperature variations arise within the long piping used in processing plant. Although the equilibrium eq(3.13) remains valid, a plane strain condition, with $\varepsilon_z = \varepsilon_o$ = constant, will apply here. Thus

$$\varepsilon_o = (1/E)[\sigma_z - \nu(\sigma_r + \sigma_\theta)] + \alpha T$$

from which the axial stress becomes

$$\sigma_z = E(\varepsilon_o - \alpha T) + \nu(\sigma_r + \sigma_\theta)$$

The radial and circumferential strains are now

$$\varepsilon_r = du/dr = (1/E)[\sigma_r - \nu(\sigma_\theta + \sigma_z)] + \alpha T$$
$$\varepsilon_\theta = u/r = (1/E)[\sigma_\theta - \nu(\sigma_r + \sigma_z)] + \alpha T$$

Eliminating u and σ_z the compatibility condition becomes

$$(1/E)[\sigma_r - \nu(\sigma_\theta + \sigma_z)] + \alpha T = (1/E)\, d\{r[\sigma_\theta - \nu(\sigma_r + \sigma_z)]\}/dr + d(\alpha r T)/dr$$

Substituting for σ_z and employing eq(3.13) leads to

$$(1 - \nu)(d\sigma_\theta/dr + d\sigma_r/dr) = -E\alpha\, dT/dr$$
$$\sigma_\theta + \sigma_r = 2a - E\alpha T/(1 - \nu)$$

Making the comparison with eq(3.51b) it can be deduced that the factor $(1 - \nu)$ will appear here in the final stress expressions:

$$\sigma_r = a - b/r^2 - \{E\alpha/[r^2(1 - \nu)]\}\int T r\, dr \tag{3.54a}$$
$$\sigma_\theta = a + b/r^2 - E\alpha T/(1 - \nu) + \{E\alpha/[r^2(1 - \nu)]\}\int T r\, dr \tag{3.54b}$$
$$\sigma_z = E(\varepsilon_o - \alpha T) + \nu[2a - \alpha E T/(1 - \nu)] \tag{3.54c}$$

It is known that $\sigma_r = 0$ when $r = r_i$ and $r = r_o$. Substituting these into eq(3.54a), the constants in eqs(3.54a,b) are found,

$$a = b/r_i^2 = \{E\alpha/[(1 - \nu)(r_o^2 - r_i^2)]\}\int_{r_i}^{r_o} T r\, dr \tag{3.55a}$$

$$b = \{E\alpha r_i^2/[(1 - \nu)(r_o^2 - r_i^2)]\}\int_{r_i}^{r_o} T r\, dr \tag{3.55b}$$

In eq(3.54c), when the cylinder is constrained, $\varepsilon_o = 0$ and the axial stress becomes

$$\sigma_z = -E\alpha T/(1 - \nu) + 2\nu a \tag{3.56a}$$

If the cylinder is allowed to extend freely ε_o follows from a zero axial force condition. Substituting eqs(3.55a) and (3.54c) into eq(3.34a) leads to

$$\varepsilon_o = [2\alpha/(r_o^2 - r_i^2)]\int T r\, dr \tag{3.56b}$$
$$\sigma_z = \{2E\alpha/[(1 - \nu)(r_o^2 - r_i^2)]\}\int T r\, dr - E\alpha T/(1 - \nu) \tag{3.56c}$$

Example 3.14 The temperatures at the inner 100 mm and outer 150 mm radii, in a cylinder are 200 and 100°C respectively. Assume that the absolute temperature T varies with radius r according to $T = A + B \ln r$. Determine the maximum radial and circumferential thermal stresses in the wall and examine the effect of free and fixed ends upon the axial stress. How are these stresses altered when the internal gauge pressure is raised to 2 kbar when all other conditions prevail? Take $E = 207$ GPa, $\nu = 0.29$, and $\alpha = 11 \times 10^{-6}/\,°C$.

With the given temperature function the integral in eqs(3.54a,b) is

$$\int T r\,\mathrm{d}r = \int r(A + B\ln r)\,\mathrm{d}r$$
$$= Ar^2/2 + (Br^2/4)(2\ln r - 1) \qquad \text{(i)}$$

The constants A and B in the temperature function $T = A + B\ln r$ follow from substituting the boundary values

$$473 = A + B\ln 100, \quad 373 = A + B\ln 150$$

$\therefore A = 1608.87$, $B = -246.65$. From eq(i) the constant a in eqs(3.55a) becomes (in units of mm and K)

$$a = \{E\alpha/[2(1-\nu)(r_o^2 - r_i^2)]\} \times | Ar^2 + (Br^2/2)(2\ln r - 1)\,\Big|_{100}^{150}$$
$$= 1.283[(3619.96 - 2503.23) - (1608.8 - 1012.54)]$$
$$= (1.283 \times 10^{-4})(520.4 \times 10^4) = 667.67$$
$$b = ar_i^2 = 667.67 \times 10^4$$

Equations (3.54a,b) become

$$\sigma_r = a - b/r^2 - \{E\alpha/[2r^2(1-\nu)]\} \times | Ar^2 + (Br^2/2)(2\ln r - 1)\,\Big|_{100}^{r}$$
$$= 667.67\,[1 - (100/r)^2] - (1.604/r^2)\,|\,1608.87r^2 - 123.33r^2(2\ln r - 1)\,\Big|_{100}^{r}$$
$$= -2110.8 + 395.5\ln r + (288.8 \times 10^4)/r^2 \qquad \text{(ii)}$$
$$\sigma_\theta = a + b/r^2 - E\alpha T/(1-\nu) + \{E\alpha/[2r^2(1-\nu)]\}\,|\,Ar^2 + (Br^2/2)(2\ln r - 1)\,\Big|_{100}^{r}$$
$$= 667.67[1 + (100/r)^2] - 3.21(1608.87 - 246.65\ln r) + (1.604/r^2)$$
$$\times \Big|\, 1608.87\,r^2 - 123.33\,r^2(2\ln r - 1)\,\Big|_{100}^{r}$$
$$= -1713.6 + 395.5\ln r - (288.8 \times 10^4)/r^2 \qquad \text{(iii)}$$

In the absence of axial strain, eq(3.56a) gives the axial stress as

$$\sigma_z = -E\alpha T/(1-\nu) + 2\nu a$$
$$= -3.207(1608.87 - 246.65\ln r) + (2 \times 667.67 \times 0.29)$$
$$= -4772.4 + 791\ln r \qquad \text{(iv)}$$

When the cylinder is allowed to extend freely, eq(3.56b) gives the axial strain

$$\varepsilon_o = \{2\alpha/[2(r_o^2 - r_i^2)]\} \times | Ar^2 + (Br^2/2)(2\ln r - 1)\,\Big|_{100}^{150}$$
$$= [(11 \times 10^{-6})/(150^2 - 100^2)]\,|\,1608.87\,r^2 - 123.33\,r^2(2\ln r - 1)\,\Big|_{100}^{150}$$
$$= 0.458 \times 10^{-2} = 0.48\%$$

and eq(3.56c) gives the axial stress

$$\sigma_z = \{2E\alpha/[2(1-\nu)(r_o^2 - r_i^2)]\} \times |\, Ar^2 + (Br^2/2)(2\ln r - 1)\,\Big|_{100}^{150} - [E\alpha/(1-\nu)](A + B\ln r)$$
$$= (2.566 \times 520.4) - 3.207\,(1608.87 - 246.65\ln r)$$
$$= -\,3824.5 + 791\ln r \qquad \text{(v)}$$

Figure 3.16a shows the σ_r and σ_θ distributions from eqs(ii) and (iii).

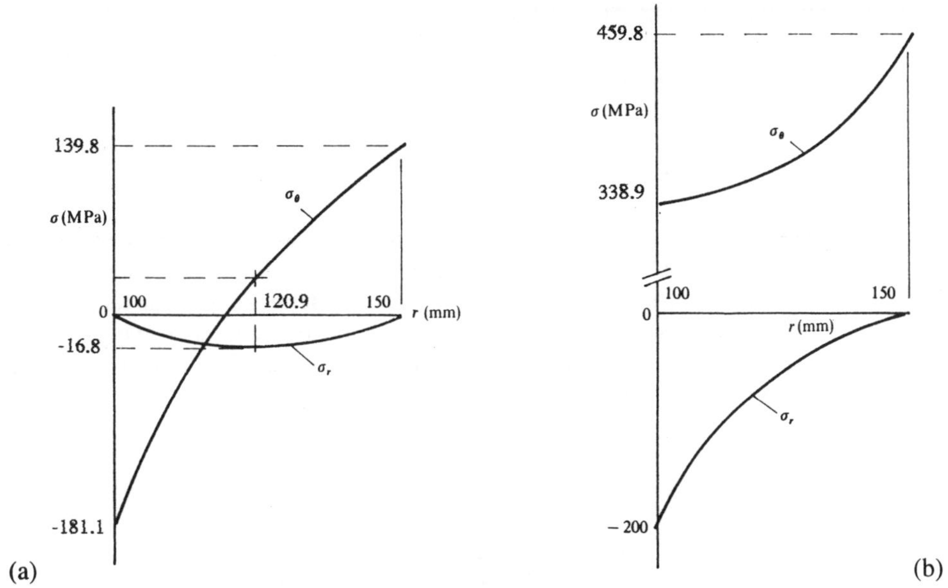

Figure 3.16 Radial and hoop stress distributions

Particular values of σ_θ are $-$ 181.1 MPa for r = 100 mm and 139.8 MPa for r = 150 mm. The maximum radial stress follows from eq(i)

$$\mathrm{d}\sigma_r/\mathrm{d}r = 395.5/r - 577.6 \times 10^4/r^3 = 0$$
$$\therefore\ r = 120.9\ \text{mm},\ \Rightarrow\ (\sigma_r)_{max} = -\,16.8\ \text{MPa}$$

Equation (iv) shows that the axial stresses become unacceptably high when the cylinder expansion is prevented. With free expansion, eq(v) shows that the axial stresses are reduced to equal the hoop stresses at the inner and outer radii.

A superimposed internal pressure of 2 kbar (i.e. $\sigma_r = -\,200$ MPa for r = 100 mm and σ_r = 0 for r = 150 mm) gives the Lamé pressure stresses (eqs(3.16a,b))

$$\sigma_r = 160 - (360 \times 10^4)/r^2 \qquad \text{(vi)}$$
$$\sigma_\theta = 160 + (360 \times 10^4)/r^2 \qquad \text{(vii)}$$

Adding eqs(vi) and (vii) to eqs(ii) and (iii) the net stresses become

$$\sigma_r = -\,1950.8 + 395.5\ln r - (71.2 \times 10^4)/r^2$$
$$\sigma_\theta = -\,1553.6 + 395.5\ln r + (71.2 \times 10^4)/r^2$$

These supply the stress distributions given in Fig. 3.16b. The boundary stress values are σ_θ

= 338.9 MPa, $\sigma_r \simeq$ - 200 MPa for r = 100 mm and σ_θ = 459.8 MPa, $\sigma_r \simeq 0$ for r = 150 mm. The axial stress (v) remains unchanged when the cylinder ends are free. When the axial extension is prevented we must add to eq(iv) $\sigma_z = \nu(\sigma_r + \sigma_\theta) = 92.8$ MPa from eq(vi) and (vii). This gives

$$\sigma_z = -4679.6 + 791 \ln r \tag{viii}$$

Equation (viii) shows that there is a slight reduction in the axial stress. These are - 1036.9 MPa at the i.d. and - 716.2 MPa at the o.d. Therefore, even in a high strength steel, the cylinder would yield in compression if it is prevented from extending.

3.6.4 Discs with Variable Breadth

Donath's numerical method (1929) provides thermal stresses in discs whose cross-section can be approximated with a number of stacked rectangles (see Fig. 3.17a). The elastic constants E and the thermal coefficient α for each rectangle must be referred to its mean temperature within the distribution over the disc (Fig.3.17b). Consider any two adjacent rectangles 1 and 2 shown in Fig. 3.17c. Their breadths b_1 and b_2 are assumed constant. Let r_i and r_o be the inner and outer radii of rectangle 1.

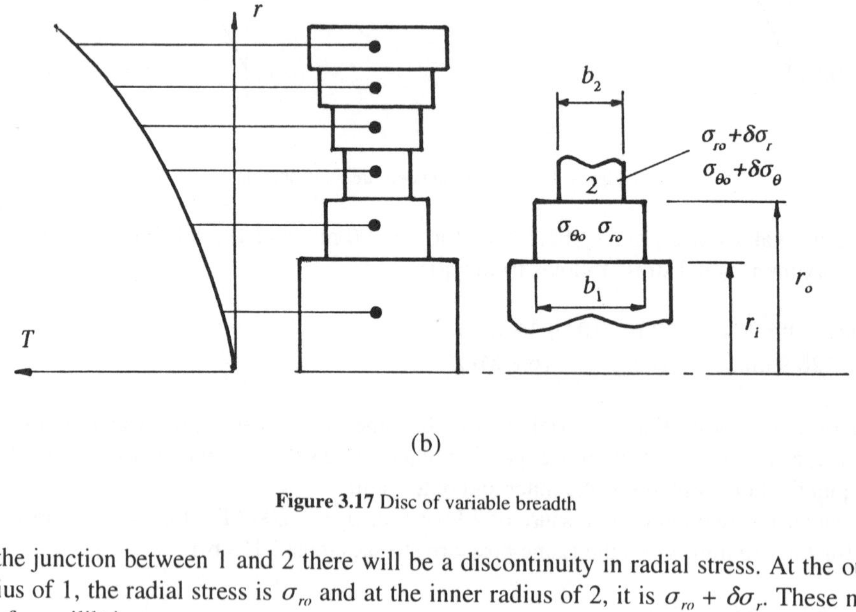

(a) (b) (c)

Figure 3.17 Disc of variable breadth

At the junction between 1 and 2 there will be a discontinuity in radial stress. At the outer radius of 1, the radial stress is σ_{ro} and at the inner radius of 2, it is $\sigma_{ro} + \delta\sigma_r$. These must satisfy equilibrium

$$2\pi r_o b_1 \sigma_{ro} = 2\pi r_o b_2(\sigma_{ro} + \delta\sigma_r)$$

Hence the change in radial stress across the junction is

$$\delta\sigma_r = \sigma_{ro}(b_1/b_2 - 1) \tag{3.57a}$$

The hoop stress is also discontinuous across this junction. At the outer radius of 1 the hoop stress is $\sigma_{\theta o}$ and at the inner radius of 2, it is $\sigma_{\theta o} + \delta\sigma_\theta$. Since the hoop strains must be compatible, this gives

$$(\sigma_{\theta o} - \nu_1 \sigma_{ro})/E_1 + \alpha_1 T_1 = [(\sigma_{\theta o} + \delta\sigma_\theta) - \nu_2(\sigma_{ro} + \delta\sigma_r)]/E_2 + \alpha_2 T_2$$

$$\delta\sigma_\theta = (E_2/E_1 - 1)(\sigma_{\theta o} - \nu\,\sigma_{ro}) + E_2(\alpha_1 T_1 - \alpha_2 T_2) + \nu\,\delta\sigma_r \qquad (3.57b)$$

in which it has been assumed that $\nu = \nu_1 = \nu_2$. The radial and hoop stresses at successive radii are given by the Lamé eqs(3.16a,b) as

$$\sigma_{\theta o} = a + b/r_o^2, \quad \sigma_{ro} = a - b/r_o^2$$

$$\sigma_{\theta i} = a + b/r_i^2, \quad \sigma_{ri} = a - b/r_i^2$$

Let S and D be sums and differences of the stresses, as follows:

$$S_o = S_i = 2a = (\sigma_{\theta o} + \sigma_{ro}) = (\sigma_{\theta i} + \sigma_{ri}) \qquad (3.57c)$$

$$D_i = (\sigma_{\theta i} - \sigma_{ri}) = 2b/r_i^2 \qquad (3.57d)$$

$$D_o = (\sigma_{\theta o} - \sigma_{ro}) = 2b/r_o^2 = D_i\,(r_i/r_o)^2 \qquad (3.57e)$$

Solving eqs(3.57c-e),

$$\sigma_{\theta o} = ½\,(S_o + D_o),\ \sigma_{ro} = ½\,(S_o - D_o) \qquad (3.57f,g)$$

The following example shows how to implement eqs(3.57a-g). Two calculations are required. Table 3.1 accounts for both thermal and mechanical effects assuming zero stress ($\sigma_r = \sigma_\theta = 0$) at the disc centre. Table 3.2 employs simplified calculations from eqs 3.57a-g) only for when the disc is at constant temperature and equal stresses ($\sigma_r = \sigma_\theta = 10$ MPa) are assumed for the disc centre. When the stresses in each calculation are then adjusted to give zero radial stress at the rim, the thermal stresses will remain. Note that the rows (a)-(g) in each table correspond to the calculations from eqs(3.51a-g) respectively. In Table 3.2, eq(3.37b) reduces to the single mechanical effect: $\delta\sigma_\theta = \nu\delta\sigma_r$.

Example 3.15 The smooth profile of a turbine disc is approximated in steps as shown in Fig. 3.18. The disc is heated asymmetrically so that the average temperature at each step is as given. Taking the corresponding values for E and α as given and assuming $\nu = 0.3$ is constant, calculate the stress distribution in the disc due to thermal effects. (IC)

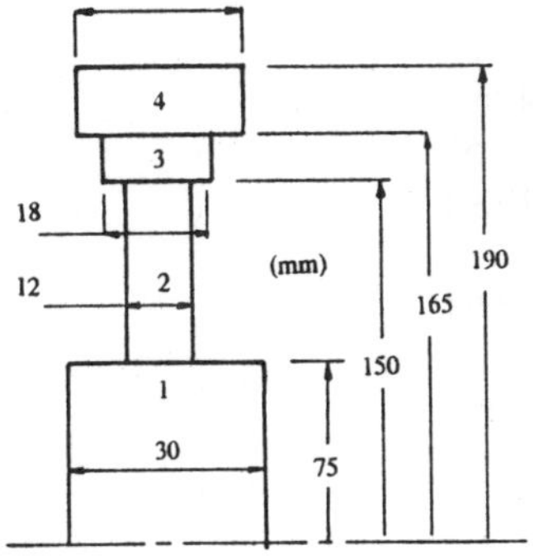

Step	1	2	3	4
T/°C	100	200	300	400
$\alpha/(10^{-6} \times °C^{-1})$	7.5	7.8	8.0	8.5
E/GPa	186.2	179.3	169	

Figure 3.18 Approximation to a turbine disc profile

Table 3.1 Thermal and mechanical effects

Quantity/Unit	1	2	3	4	RIM
r_i/mm	0	75	150	165	190
b/mm	30	12	18	25	25
T/°C	100	200	300	400	400
E/GPa	186.2	179.3	169.0	158.6	158.6
α/(°C × 10^{-6})	7.5	7.8	8.0	8.5	8.5
$\sigma_{\theta i}$/MPa	0	- 145.27	- 223.16	- 296.85	- 265.06
σ_{ri}/MPa	0	0	- 36.32	- 38.36	- 70.16
(c) $S_i = S_o$/MPa	0	- 145.27	- 259.48	- 335.21	- 335.21
(d) D_i/MPa	0	- 145.27	- 186.84	- 258.49	- 194.90
$(r_i/r_o)^2$	0	0.25	0.826	0.754	1
(e) D_o/MPa	0	- 36.32	- 154.33	- 194.90	- 194.90
(f) $\sigma_{\theta o}$/MPa	0	- 90.79	- 206.91	- 265.06	- 265.06
(g) σ_{ro}/MPa	0	- 54.48	- 52.58	- 70.16	- 70.16
$\nu\, \sigma_{ro}$/MPa	0	- 16.34	- 15.77	- 21.05	
$b_1/b_2 - 1$	1.5	- 0.333	- 0.28	0	
(a) $\delta\sigma_r$/MPa	0	18.16	14.22	0	
$\nu\delta\sigma_r$/MPa	0	5.45	4.42	0	
(h) $(\sigma_{\theta o} - \nu\sigma_{ro})$/MPa	0	- 74.45	- 191.14	- 244.01	
(i) $E_2/E_1 - 1$	- 0.037	- 0.0578	- 0.0612	0	
(h) × (i)/MPa	0	4.303	11.698	0	
$\alpha_1 T_1/10^{-6}$	750	1560	2400	3400	
$\alpha_2 T_2/10^{-6}$	1560	2400	3400	3400	
$\alpha_1 T_1 - \alpha_2 T_2$	- 810	- 840	- 1000	0	
$E_2(\alpha_1 T_1 - \alpha_2 T_2)$/MPa	- 145.27	- 141.94	- 158.64	0	
(b) $\delta\sigma_\theta$/MPa	- 145.27	- 132.19	- 142.52	0	

Table 3.2 Mechanical effects

Quantity/Unit	1	2	3	4	RIM
r_i/mm	0	75	150	165	190
b/mm	30	12	18	25	25
T/°C	100	200	300	400	400
E/GPa	186.2	179.3	169	158.6	158.6
α/(°C × 10^{-6})	7.5	7.8	8.0	8.5	8.5
$\sigma_{\theta i}$/MPa	10	14.5	- 20.55	- 18.47	- 15.22
σ_{ri}/MPa	10	25	14.04	7.94	4.69
(c) $S_i = S_o$/MPa	20	39.5	- 6.51	- 10.53	- 10.53
(d) D_i/MPa	0	- 10.5	- 34.59	- 26.41	- 19.91
$(r_i/r_o)^2$	0	0.25	0.826	0.754	1
(e) D_o/MPa	0	- 2.63	- 28.57	- 19.91	- 19.91
(f) $\sigma_{\theta o}$/MPa	10	- 18.44	- 17.54	- 15.22	- 15.22
(g) σ_{ro}/MPa	10	21.06	11.03	4.69	4.69
$b_1/b_2 - 1$	1.5	- 0.333	- 0.28	0	
(a) $\delta\sigma_r$/MPa	15	- 7.02	- 3.09	0	
(b) $\nu\delta\sigma_r = \delta\sigma_\theta$/MPa	4.5	- 2.11	- 0.93	0	

It follows that the radial stress at the rim is removed between Table 3.1 and Table 3.2 by writing $-70.16 + 4.69\,n = 0$, from which $n = 14.96$. Then, multiplying stresses in Table 3.2 by n and adding the result to the corresponding stresses in Table 3.1 leads to the thermal stresses given in Table 3.3. The correction applies to the outer radius stress components within rows (f) and (g). Across the step the stresses follow from adding rows (f) and (b) and (g) and (a) so that two stresses values appply to each radius.

Table 3.3 Thermal Stresses

r, mm	0	75		150		165		190	
σ_θ, MPa	149.4	149.4	71.3	- 365.6	- 530.9	- 468.3	- 573.2	- 492.1	- 492.1
σ_r, MPa	149.4	149.4	373.4	260.1	173.4	112.2	79.6	0	0

3.7 Thermal Stresses in Spheres

Referring to Fig. 3.15, the equilibrium eq(3.38) for a sphere is unaltered when there is a temperature gradient in its wall. A new compatibility condition is required to account for the thermal strains αT appearing in

$$\varepsilon_r = \mathrm{d}u/\mathrm{d}r = (1/E)[\sigma_r - \nu(\sigma_\theta + \sigma_\phi)] = (1/E)(\sigma_r - 2\nu\sigma_\phi) + \alpha T \tag{3.58a}$$

$$\varepsilon_\phi = \varepsilon_\theta = u/r = (1/E)[\sigma_\phi - \nu(\sigma_r + \sigma_\theta)] = (1/E)[\sigma_\phi(1 - \nu) - \nu\sigma_r] + \alpha T \tag{3.58b}$$

where $\sigma_\theta = \sigma_\phi$. The two strains in eqs(3.58a,b) depend upon the single displacement u:

$$(\sigma_r - 2\nu\sigma_\phi) + E\alpha T = \mathrm{d}\{r[\sigma_\phi(1 - \nu) - \nu\sigma_r] + rE\alpha T\}/\mathrm{d}r$$

$$(1 + \nu)(\sigma_r - \sigma_\phi) = E\alpha\, r\,(\mathrm{d}T/\mathrm{d}r) - \nu\, r\,(\mathrm{d}\sigma_r/\mathrm{d}r) + (1 - \nu)\, r\,(\mathrm{d}\sigma_\phi/\mathrm{d}r) \tag{3.59a}$$

Combining eq(3.38) and eq(3.59) and integrating leads to

$$\tfrac{1}{2}\sigma_r + \sigma_\phi = A - \mathrm{E}\alpha T/(1 - \nu) \tag{3.59b}$$

Subtracting eq(3.38) from (3.59b)

$$3\sigma_r + r\,\mathrm{d}\sigma_r/\mathrm{d}\,r = 2A - 2E\alpha T/(1 - \nu)$$

$$(1/r^2)\,\mathrm{d}\,(r^3\sigma_r)/\mathrm{d}r = 2A - 2E\alpha T/(1 - \nu)$$

$$\therefore\ \sigma_r = 2A/3 + B/r^3 - \{2E\alpha/[r^3(1 - \nu)]\}\int T r^2 \mathrm{d}\,r \tag{3.60a}$$

and, from eq(3.59b) and (3.60a)

$$\sigma_\phi = A - E\alpha T/(1 - \nu) - \sigma_r/2$$

$$= 2A/3 - B/2r^3 - E\alpha T/(1 - \nu) + \{E\alpha/[r^3(1 - \nu)]\}\int T r^2 \mathrm{d}\,r \tag{3.60b}$$

Setting $a = 2A/3$ and $b = B$, eqs(3.60a,b) are written as

$$\sigma_r = a + b/r^3 - \{2E\alpha/[r^3(1 - \nu)]\}\int T r^2 \mathrm{d}\,r \tag{3.61a}$$

$$\sigma_\phi = \sigma_\theta = a - b/(2r^3) - E\alpha T/(1 - \nu) + \{E\alpha/[r^3(1 - \nu)]\}\int T r^2 \mathrm{d}\,r \tag{3.61b}$$

The constants a and b in eqs(3.61a,b) will depend upon whether the sphere is solid or hollow.

3.7.1 Solid Sphere

The stresses in eqs(3.61a,b) cannot become infinite at the centre of a solid sphere. This gives $b = 0$. The remaining constant a is found from the boundary condition $\sigma_r = 0$ for $r = r_o$. Substituting into eq(3.61a) yields

$$a = \{2E\alpha / [r_o^3(1 - \nu)]\} \int_0^{r_o} T r^2 \mathrm{d}r$$

Equations (3.61a,b) then become

$$\sigma_r = [2E\alpha / (1 - \nu)][(1/r_o^3) \int_0^{r_o} T r^2 \mathrm{d}r - (1/r^3) \int_0^{r} T r^2 \mathrm{d}r]$$

$$\sigma_\phi = [E\alpha / (1 - \nu)][(2/r_o^3) \int_0^{r_o} T r^2 \mathrm{d}r + (1/r^3) \int_0^{r} T r^2 \mathrm{d}r - T]$$

3.7.2 Hollow Sphere

The boundary conditions are $\sigma_r = 0$ for $r = r_i$ and $r = r_o$. Substituting these into eq(3.61a) gives the constants a and b as

$$a = \{2E\alpha / [(1 - \nu)(r_o^3 - r_i^3)]\} \int_{r_i}^{r_o} T r^2 \mathrm{d}r \qquad (3.62a)$$

$$b = -a r_i^3 = -\{2E\alpha r_i^3 / [(1 - \nu)(r_o^3 - r_i^3)]\} \int_{r_i}^{r_o} T r^2 \mathrm{d}r \qquad (3.62b)$$

Substituting eqs(3.62a,b) into eqs(3.61a,b),

$$\sigma_r = [2E\alpha/(1 - \nu)] [\{(r^3 - r_i^3) / [r^3(r_o^3 - r_i^3)]\} \int_{r_i}^{r_o} T r^2 \mathrm{d}r - (1/r^3) \int_{r_i}^{r} T r^2 \mathrm{d}r]$$

$$\sigma_\phi = [E\alpha/(1 - \nu)][\{(2r^3 + r_i^3) / [r^3(r_o^3 - r_i^3)]\} \int_{r_i}^{r_o} T r^2 \mathrm{d}r + (1/r^3) \int_{r_i}^{r} T r^2 \mathrm{d}r - T]$$

in which $T = T(r)$ expresses a radial temperature gradient similar to that in the wall of a thick-cylinder (see Example 3.14).

References

[1] Lamé, G. *Lecons sur la theorie de l'elasticité*, Gauthier-Villars, Paris, 1852.
[2] Donath, M. *Die Berechnung Rotierender Scheiben und Ringe nach einem neuen Verfahren*, Berlin, 1929.

EXERCISES

Membrane Theory

3.1 Determine the thickness ratio required between the spherical ends and cylindrical body of a pressurised vessel when, across the weld, (a) the axial strain is to be the same and (b) the hoop stress is to be the same. Take $\nu = ¼$.
Answer: 3/2, 1/2

3.2 A thin-walled cylinder of internal diameter 75 mm and wall thickness 3 mm is 3 m long. Determine the changes in length, diameter and thickness for an applied internal pressure of 1.5 bar. Take the elastic constants $E = 210$ GPa, $\nu = 0.27$.
Answer: $\Delta l = 0.615$ mm, $\Delta d = 0.0579$ mm, $\Delta t = -1.083 \times 10^{-3}$ mm

3.3 A thin-walled steel sphere with 600 mm mean radius and 25 mm thickness contains oil at atmospheric pressure. To what depth may the sealed sphere be immersed in sea water of density 1040 kg/m^3 if the maximum compressive stress in the sphere wall is restricted to 92.5 MPa? Take constants $E = 207$ GPa, $\nu = 0.25$ for steel and $K = 3445$ MPa for oil. (Hint: $\delta V/V$ for oil and sphere are the same.)

3.4 A spherical vessel 1525 mm diameter with 75 mm wall thickness is made from a material whose ultimate compressive strength is 77 MPa. If the internal pressure of the sealed vessel remains constant as it is submerged in the sea ($\rho = 1040$ kg/m^3), determine the safe depth and the corresponding reduction to the internal volume. Take $E = 166$ GPa and $\nu = 0.25$.

3.5 What are the axial and hoop stresses induced in a long cylinder with 25 mm wall thickness and 1 m inside diameter, under an internal pressure of 10 bar, when the ends are (a) closed and (b) open, i.e. the end pressure is contained by moveable pistons?
Answer: 10, 20: 0, 20 MPa

3.6 A spherical balloon 0.2 mm thick and 250 mm diameter has an ultimate tensile strength of 9 MPa. Calculate the bursting pressure.

3.7 What is the bursting speed of a thin-rimmed cast iron flywheel of mean diameter 1 m if the u.t.s. of cast iron is 138 MPa and the specific weight is 7200 kg/m^3?

3.8 Find the diameter of the largest thin-walled rim that may revolve at 500 rev/min if it is to be made from steel of density 7756 kg/m^3 when the maximum tensile stress is limited to 46 MPa. Determine the increase in diameter. Take $E = 207$ GPa.

3.9 A thin steel cylinder, with walls 15 mm thick and diameter 2.1 m, is subjected to an internal pressure of 5.5 bar whilst being rotated about its axis at 300 rev/min. Find the maximum value of the hoop stress in the material and the factor of safety used in the design. The u.t.s. and density of the steel are 450 MPa and 7760 kg/m^3 respectively.
Answer: 45.8 MPa, 9.85

3.10 A thin steel rim, 1.5 m diameter and 150 mm wide, is constructed from 20 mm thick steel plate. Using a factor of safety of 8, determine the maximum permissible speed in rev/min, given that the u.t.s. for the rim material is 463 MPa. The rim is to be constructed in two semi-circular halves with each simple lap joint held together by a single bolt. Determine the bolt diameter when the permissible shear stress of the bolt material is limited to 69 MPa.
Answer: 1080 rev/min, 56 mm

3.11 A hemispherical concrete shell, of mean diameter 1.5 m and thickness 50 mm, rests on its annular rim. Determine the maximum stresses in the shell due to the combined action of self-weight and a normal pressure of 1 bar over its external surface. Take $\rho = 2400$ kg/m^3 for concrete.

3.12 Determine the maximum stresses in a hemispherical shell, 1.5 m diameter, 10 mm thickness, when it is filled with water of density 1000 kg/m^3 and simply suspended around its annular rim.
Answer: $\sigma_\theta = -184$ kN/m^2, $\sigma_\phi = 184$ kN/m^2

3.13 Derive the membrane stresses arising in the tori-spherical ends of a long thin cylindrical vessel containing pressure p, when the knuckle radius is a and the sphere radius is b.
Answer: sphere $\sigma_r = \sigma_\theta = pb/2$, knuckle $\sigma_r = -p(a + b\cos\theta)/(2\cos\theta)$,
$\sigma_\theta = -pr\,[2 - r/(b\cos\theta)]/(2\cos\theta)$ (with θ as for torus)

3.14 The vessel in Fig. 3.19 comprises part of a sphere of radius a, subjected to a ring of loading at radii r_1 and r_2, as shown. If the total effect of the load is W, derive the membrane stresses in the vessel.

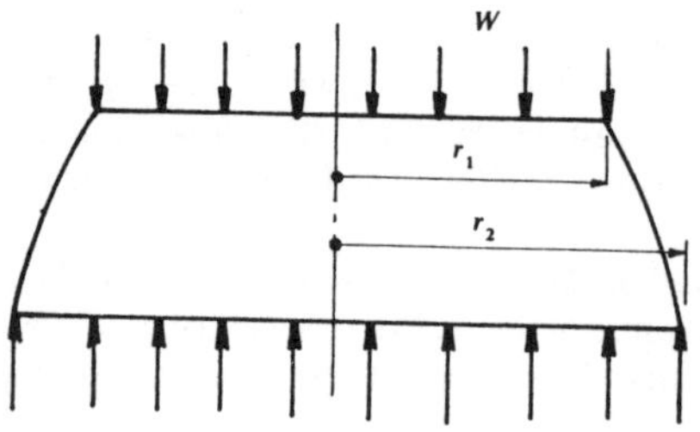

Figure 3.19

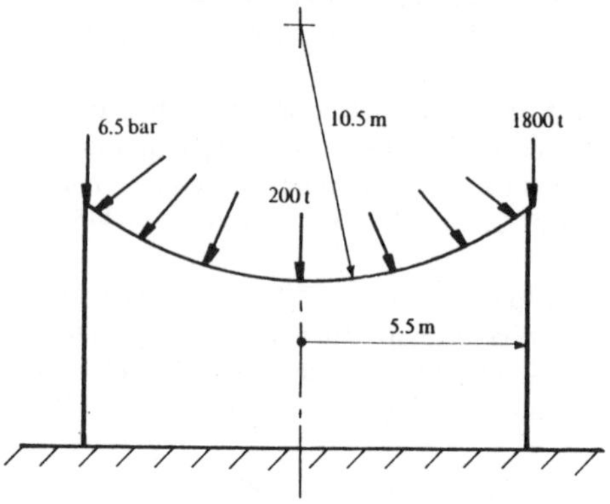

Figure 3.20

3.15 Part of a reactor vessel is simplified in Fig. 3.20. The spherical dish, 10.5 mm radius, rests on a cylinder, 5.5 m radius, both having a uniform thickness of 75 mm. The vessel supports a central concentrated core load of 200 t and a total rim load of 1800 t. Calculate the membrane stresses in the dish and cylinder when an additional pressure of 6.5 bar is applied over its top surface (a) when the dish self-weight is negligible and (b) when the dish self-weight is 78 kN/m^3.

Monobloc Thick-Walled Cylinders and Spheres

3.16 A steel cylinder 203 mm inside diameter is not to be stressed beyond 123.5 MPa under an internal pressure of 310 bar. Find a suitable external diameter and the magnitude of the strains at the outer surface. Take $E = 207$ GPa, $\nu = 0.28$.

3.17 A thick-walled cylinder of steel has an i.d. of 100 mm and an o.d. of 200 mm. The internal and external pressures are 550 and 70 bar respectively. Determine the maximum hoop and shear stresses in the cylinder and the change in the outer diameter. Take $E = 210$ GPa, and $\nu = 0.3$.

3.18 A cylindrical steel vessel 127 mm i.d. and 178 mm o.d., with closed ends, contains fluid at a pressure of 350 bar. Determine the axial, hoop and radial stresses and strains at the inner and outer diameters. Take $E = 200$ GPa and $\nu = 0.28$.

3.19 A cast iron closed cylinder, with i.d. of 114 mm and wall thickness 25 mm, can safely withstand an internal pressure of 170 bar. What is the safe internal pressure a second cast iron cylinder can withstand for an i.d. 152 mm and thickness 25 mm, when the greatest tensile stress remains the same?

3.20 An annular plate 305 mm o.d. and 12.5 mm thick is subjected to a radial compressive force of 2.35 kN/mm of inner circumference, whilst the outer edge remains unloaded. Compare the hoop stress distribution with that when a force is applied around the outer edge whilst the inner edge remains free.

3.21 A pressure of 400 bar exists within the 50 mm bore of a pipe-line. If the maximum stress is not to exceed 120 MPa, determine the outer diameter.

3.22 A closed cylinder, 150 mm i.d. with 75 mm wall thickness, is subjected to an internal gauge pressure of 700 bar. Calculate and plot, for each 25 mm of radius, the distribution of radial and hoop stress through the wall.
Answer: σ_θ (MPa) 116.67, 75.83, 56.93, 46.66; σ_r (MPa) − 70, − 29.17, − 10.27, 0: $\sigma_z = 23.33$ MPa

3.23 A steel bathysphere, 500 mm i.d., is to withstand an external fluid pressure of 1000 atmospheres. Find the necessary thickness if the maximum stress in the vessel is limited to 230 MPa. Find also the diameter and thickness changes under this pressure.

3.24 Derive expression for the stresses and displacements in a thick-walled sphere subjected to an external pressure when filled with (a) air and (b) an incompressible fluid which prevents a reduction to the internal volume.

Compound Cylinders with Interference

3.25 A steel sleeve, 51 mm i.d. and 102 mm o.d., is force fitted on to a solid shaft. If the external hoop strain is 0.06%, calculate the radial pressure at the common diameter and the maximum tensile stress in the sleeve. Take $E = 207$ GPa.

3.26 A copper bush, 20 mm i.d., 30 mm o.d. and 50 mm long, is pressed firmly into a cast iron hub, 50 mm o.d., with an interference of 0.03 mm. Determine the torque to cause slipping between hub and bush, assuming a coefficient of friction $\mu = 0.4$. $E = 140$ GPa, $\nu = 0.25$ for cast iron and $E = 100$ GPa, $\nu = 0.30$ for copper.

3.27 A brass sleeve, 50 mm i.d., 100 mm o.d. and 50 mm long, is to be force fitted onto a solid shaft of diameter 50 mm. The diametral interference of the assembly is 0.04 mm. Calculate the force required to fit the sleeve on to the shaft. Take $E = 200$ GN/m^2, $\nu = 0.3$ for steel and $E = 100$ GN/m^2, $\nu = 0.33$ for brass. The coefficient of friction for the pair is $\mu = 0.2$. (CEI)

3.28 A circular disc of outer diameter D, containing a central hole of diameter d, is fitted to a solid shaft of nominal diameter d using an interference fit. The disc and shaft are of the same material, with Young's modulus E and Poisson's ratio ν. The assembly is required to satisfy the condition that the maximum hoop stress in the disc shall not exceed a value of σ_{limit}. Derive the relationship for the diametral interference δ and show that if D/d is large, then δ is proportional to d. (CEI)

3.29 A steel wheel ring, 102 mm i.d., is to be shrunk on to a hollow steel wheel to give a radial pressure of 40 MPa. Calculate the necessary initial difference in the common diameter when the radial thicknesses are both 25 mm. Take $E = 207$ GPa.

3.30 A steel gear blank, 127 mm o.d., is pressed on to a 76 mm diameter hollow steel shaft whose internal diameter is 51 mm. If the hoop strain for the blank o.d. is 0.016%, determine the initial difference in the common diameter and the maximum hoop stress in the blank. What magnitude of torque would initiate slipping for a 50 mm length of contact between blank and shaft? Take $E = 207$ GPa and $\mu = 0.2$.

3.31 A hollow steel tube, 76 mm o.d. with 12.5 mm thick walls, has a bronze sleeve 102 mm o.d. shrunk on to it. If the common radial pressure is 27.5 MPa, calculate the initial interference. Plot the residual stress distributions. Take $E = 207$ GPa, $\nu = 0.3$ for steel and $E = 117$ GPa, $\nu = 0.33$ for bronze.

3.32 A compound tube is made from shrinking one steel tube onto another. The diameters of the assembly are 152 mm internal, 305 mm common and 355 mm external. If the shrinkage process results in 0.077% external hoop strain, determine the initial interference and the maximum hoop stress. Take $E = 207$ GPa.

3.33 A 25 mm thick steel cylinder, with an inside diameter of 200 mm, has shrunk on to it a second steel cylinder of 300 mm outside diameter. Calculate the interference fit required, prior to assembly, if both tubes are to suffer the same maximum hoop stress when the compound tube is pressurized to 60 N/mm^2. (CEI)

3.34 The two components of a compound cylinder have inner and outer radii of 175, 200 and 250 mm, and are made to give an interference on the common radius of 0.1 mm. Find the radial and tangential stresses at the common 200 mm radius due only to the shrinking process. Take E = 208 GPa. (CEI)

3.35 A hollow steel tube, 180 mm o.d. and 50 mm thick, has a bronze sleeve, 230 mm o.d., shrunk on to it. Find the required interference at the common diameter for a radial shrinkage pressure of 92.5 MPa. Plot the residual stress distributions following shrinkage and find the net hoop stress at the inner diameter of each tube when a pressure of 770 bar is then applied to the assembly. Take E = 210 GPa, ν = 0.3 for steel and E = 117 GPa, ν = 0.33 for bronze.

3.36 The nominal dimensions of two cylinders are 40 mm and 80 mm for the inner cylinder and 80 mm and 120 mm for the outer cylinder. If both cylinders are made from the same steel, determine the interference in order that the maximum bore stress for the compound cylinder does not exceed 60 MN/m^2 when it is pressurized to 60 MN/m^2. (CEI)

3.37 The nominal radii of a compound cylinder are 100, 120 and 140 mm. When an internal pressure of 15 MN/m^2 is applied to the assembly, it is required that the maximum circumferential stress shall be the same in both components. Determine the magnitude of the interaction pressure at this pressure and the initial interference. Take E = 200 GN/m^2 and ν = 0.3. (CEI)

Rotating Cylindrical Bodies

3.38 Find the maximum stress and the stress at the outer 228 mm o.d. of a solid disc when rotating at 12000 rev/min. Show graphically the radial and hoop stress distributions. Take ν = 0.3, and the density ρ = 7480 kg/m^3.

3.39 A steel disc 255 mm o.d., 127.5 mm i.d. and 12.5 mm thick rotates at 2000 rev/min. Ignoring the method of grip, determine the stresses due to centrifugal effects. Take ν = 0.3 and ρ = 7835 kg/m^3.

3.40 At what speed may an annular disc of 30 mm o.d. and 20 mm i.d. rotate if the maximum shear stress is limited to 35 MPa? Take ρ = 7600 kg/m^3 and ν = 0.28.

3.41 A thin uniform steel disc with o.d. 500 mm and i.d. 120 mm is subjected to a radial tensile stress of 110 MPa at its o.d. due to the blading when rotating at 7000 rev/min. Determine the maximum radial and circumferential stresses and the change in diameter, stating where these occur. Take material constants E = 200 GPa, ν = 0.3 and ρ = 7850 kg/m^3.

3.42 Determine the general expression for the circumferential stress in a solid disc of radius b, rotating at an angular frequency ω, when a radial stress component of magnitude σ_b (tensile) is applied to the periphery of the disc. Hence determine the position and magnitude of the maximum circumferential stress in the disc. (CEI)

3.43 An internal combustion engine has a cast iron flywheel that can be considered to be a uniform thickness disc of 230 mm o.d. and 50 mm i.d. Given that the u.t.s. and density of cast iron are 200 N/mm^2 and 7180 kg/m^3 respectively, calculate the speed at which the flywheel would burst. Take ν = 0.25. (CEI)

3.44 A thin turbine rotor has inner and outer nominal radii of 40 mm and 125 mm respectively. The effect of the rotating turbine blades is to set up a radial tensile stress of 25 MPa at the outer radius. The rotor is shrunk onto a solid shaft with a radial interference of δ. The fit is to be designed using the assumption that at a speed of 10000 rev/min the rotor would become loose (i.e. free to slide) on the shaft. Determine the necessary interference fit δ. Take E = 208 GPa, ν = 0.3 and ρ = 7500 kg/m^3 for both rotor and shaft. (CEI)

3.45 Figure 3.21 shows a disc with 36 integrally machined teeth of uniform cross-section arranged symmetrically around the disc. The toothed disc is cut from a single sheet of uniform thickness. Determine the maximum stress in the disc when it is rotating at 1500 rev/min, assuming that the combined centrifugal forces at the roots of the teeth due to rotation are uniformly distributed around the central 250 mm diameter. Take $E = 200$ GPa, $\nu = 0.3$ and $\rho = 7380$ kg/m^3. (CEI)

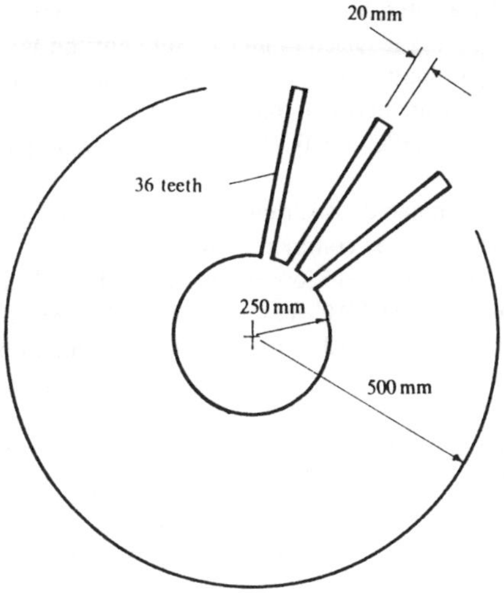

Figure 3.21

3.46 A thin solid steel disc, 455 mm diameter, has a steel ring of outer diameter 610 mm and of the same uniform thickness shrunk on to it. If the interference pressure is reduced to zero at a rotational speed of 3000 rev/min, calculate, for the common diameter, the diametral interference before assembly and the radial pressure due to shrinkage. Take $E = 207$ GPa, $\nu = 0.29$ and $\rho = 7756$ kg/m^3. (CEI)

3.47 A rotating disc and solid shaft steel assembly is designed for a pressure of 34.5 MPa and a hoop stress of 207 MPa at the common 508 mm diameter. If the disc outer diameter is 710 mm, at what maximum speed can the assembly be rotated? Find also the magnitude of the stresses at the shaft centre and at the disc outer diameter. Take $\nu = 0.28$ and $\rho = 7756$ kg/m^3.

3.48 A solid steel disc of thickness $t = 50$ mm and 760 mm diameter weighs 7844 kg/m^3. A central hole is to be machined in the disc, so that when it is pressed onto a solid steel shaft of diameter $d = 102$ mm, it will not slide on the shaft under an axial force of $W = 30$ kN. Determine the necessary hole size for the disc and the maximum allowable rotational speed if slippage does not occur. Constants are$E = 207$ GPa, $\nu = 0.3$ and $\mu = 0.3$. (Hint: The common interface pressure p is found from $\mu = W/(p\pi\, dt)$.)

Temperature Stresses

3.49 If, for the thin annular disc in example 3.13 the initial mismatch is 0.5 mm, calculate the mean hoop stress when the temperature of the assembly is raised by 25°C.

3.50 The temperature at the 60 mm i.d. of a turbine disc is 60°C whilst at the 200 mm o.d. it is 160°C, when the disc rotates at 6000 rev/min. If the outer diameter temperature is reduced to 140°C, determine the increase in rotational speed that is possible when the circumferential stress at the inside diameter remains constant. Take $E = 200$ GPa, $\nu = 0.25$, $\alpha = 11 \times 10^{-6}$/ °C and $\rho = 7500$ kg/m^3.

3.51 Use the principle of superposition to establish the net stress distribution in a thick walled cylinder, 100 mm i.d. and 200 mm o.d., when it carries oil at a pressure and temperature of 35 MPa and 800 K. The outer pressure is atmospheric and the temperature is 310 K. Find also the position and magnitude of the greatest hoop stress and the changes in the inner and outer diameters relative to a surrounding temperature of 295 K. Take $E = 200$ GPa, $\nu = 0.3$ and $\alpha = 12 \times 10^{-6}$/K.

3.52 A solid steel shaft, 0.2 m diameter, has a bronze bush of 0.3 m o.d. shrunk on to it. In order to remove the bush, the whole assembly is raised in temperature uniformly. After a rise of 100 °C, the bush can just be moved along the shaft. Neglecting any effect of temperature in the axial direction, calculate the original interface pressure between the bush and shaft. Take $E = 208$ GPa, $\nu = 0.29$, $\alpha = 12 \times 10^{-6}$/ °C for steel and $E = 112$ GPa, $\nu = 0.33$, $\alpha = 18 \times 10^{-6}$/ °C for bronze. (CEI)

3.53 An aluminium alloy ring, 400 mm o.d., is shrunk onto a solid steel shaft of 200 mm diameter. At 20°C, the diametral interference fit at the common diameter is 0.2 mm. Find (i) the temperature to which the assembly must be taken in order to loosen the ring and (ii) the angular speed at which the assembly just loosens when operating at 40°C. Take E = 207 GPa, $\nu = 0.3$, $\alpha = 11 \times 10^{-6}$/ °C, and $\rho = 7850$ kg/m^3 for steel and $E = 69$ GPa, $\nu = 0.3$, $\alpha = 23 \times 10^{-6}$/ °C and $\rho = 2720$ kg/m^3 for aluminium alloy. (CEI)

3.54 Derive the following expressions for σ_θ, σ_r and σ_z when the temperature gradient for a constrained thick-walled cylinder is given by $T(r) = (T_i - T_o)\ln(r_o/r) / \ln(r_o/r_i)$.

$$\sigma_r = -\{E\alpha T/[2(1-\nu)]\}\langle \ln(r_o/r) + \{r_i^2(r^2 - r_o^2)/[r^2(r_o^2 - r_i^2)]\} \times \ln(r_o/r_i)\rangle$$

$$\sigma_\theta = \{E\alpha T/[2(1-\nu)]\}\langle 1 - \ln(r_o/r) - \{r_i^2(r^2 + r_o^2)/[r^2(r_o^2 - r_i^2)]\} \times \ln(r_o/r_i)\rangle$$

$$\sigma_z = \{E\alpha T/[2(1-\nu)]\}\{\nu - 2\ln(r_o/r) - [2\nu r_i^2/(r_o^2 - r_i^2)] \times \ln(r_o/r_i)\}$$

CHAPTER 4

BENDING OF BEAMS AND PLATES

In Chapter 2 it was shown how a stress function provided the distribution of direct and shear stress in a cantilever beam (see Fig. 2.10). Flexural shear stresses in beams will be treated separately in Chapter 7. This chapter examines, from first principles, the direct stress due to bending of straight and curved beams and beams with asymmetric and composite sections. The engineer's theory of bending is also applied to the bending of circular and rectangular plates with various loadings and edge fixings.

4.1 Bending Of Straight Beams

Let an initially straight beam hog under an applied moment in the manner of Fig. 4.1a.

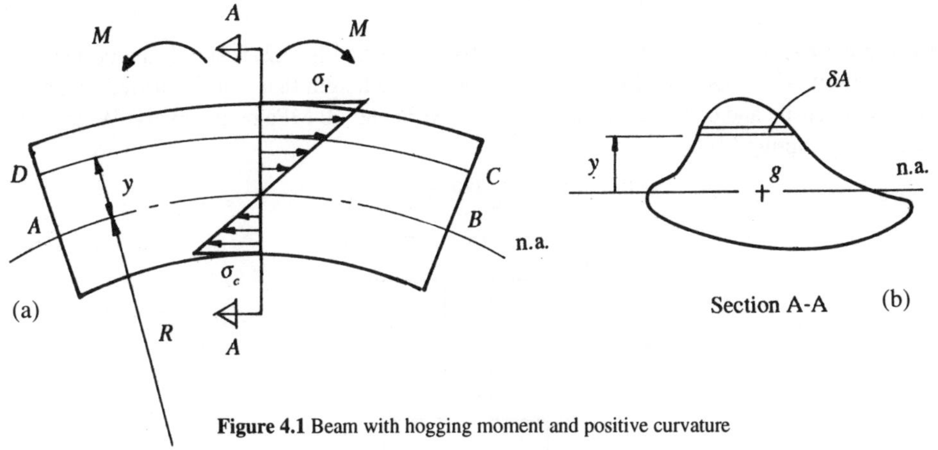

Figure 4.1 Beam with hogging moment and positive curvature

The longitudinal strain in the fibre CD is $\varepsilon = y/R$ where R is the radius of curvature of the unstrained (neutral) fibre AB. The longitudinal stress and strain are connected through the elasticity modulus $E = \sigma/\varepsilon$, to give

$$E/R = \sigma / y \tag{4.1}$$

The bending stress σ acts on an elemental area δA, distance y from the neutral axis in Fig. 4.1b. The moment balance (equilibrium) equation is

$$M = \int_A \sigma \, \mathrm{d}A \, y \tag{4.2a}$$

Substituting $\sigma = Ey/R$ from eq(4.1), into eq(4.2a):

$$M = (E/R) \int_A y^2 \, \mathrm{d}A = EI/R \tag{4.2b}$$

where $I = \int_A y^2\, dA$ is the second moment of area of the section. As the beam carries no axial force,

$$\int_A \sigma \, dA = (E/R) \int_A y\, dA = 0 \tag{4.2c}$$

The condition $\int_A y\, dA = 0$ from eq(4.2c), shows that the reference axis for y passes through the centroid, g, of the section. That is, the neutral axis (n.a.) passes through g in Fig. 4.1b. Combining eqs(4.1) and (4.2) leads to the engineer's theory of bending

$$M/I = E/R = \sigma / y \tag{4.3}$$

The longitudinal bending stress $\sigma = My/I$ applies to any section in any beam where the moment M is known (see Fig. 4.1a). The sign of M shown is positive, i.e. a hogging moment produces positive curvature. The sign of y is positive when locating fibres above the n.a. and negative for fibres beneath the n.a. Equation (4.1) will provide the tensile stresses $\sigma_t = (+M)(+y)/I$ in all fibres (including CD) lying on the positive side of the neutral axis. Compression $\sigma_c = (+M)(-y)/I$ exists in all fibres beneath the neutral axis, where y is negative. The signs for y must be independent of the direction of M. If we imagine a sagging (negative) moment instead of Fig. 4.1a, then we can expect, from eq(4.3), tension below the n.a., $\sigma_t = (-M)(-y)/I$ and compression above it, $\sigma_c = (-M)(+y)/I$. The following two examples will illustrate how eq(4.3) is applied in practice.

Example 4.1 A steel girder of unsymmetrical I-section (see Fig. 4.2) is simply supported over a 5 m length. Calculate the distributed load per metre length that can be carried given the maximum tensile and compressive stresses of 50 MPa and 40 MPa respectively. What is the radius of curvature?

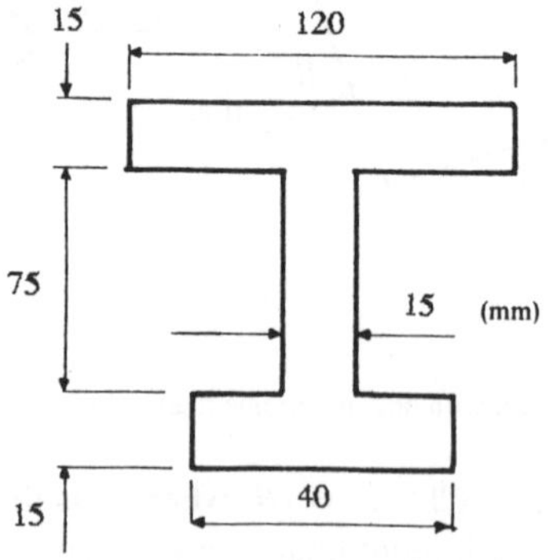

Figure 4.2 Unsymmetrical I-section

Taking first moments of area $\sum A_i y_i = A\bar{y}$ about the base, where the centroid position lies at a height $\bar{y}$,

$$(40 \times 15 \times 7.5) + (75 \times 15)(37.5 + 15) + (120 \times 15)(90 + 7.5)$$
$$= [(120 \times 15) + (75 \times 15) + (40 \times 15)]\,\bar{y}, \Rightarrow \bar{y} = 67.82 \text{ mm}$$

Second moment of area:

Method 1 - Divide the section into the sum and difference of rectangles with their bases lying on the centroidal (neutral) axis. Use the form $I = \sum(bd^3/3)$, as follows:

$I = [120(37.18)^3/3 - 105(22.18)^3/3] + [40(67.82)^3/3 - 25(52.82)^3/3]$
$= 4.605 \times 10^6\ \text{mm}^4$

Method 2 - Use the parallel axis theorem $I = I_g + Ah^2$ for each of the three rectangles which comprise the section. Here $I_g = bd^3/12$ and h is the distance between the centroidal axis and the neutral axis for each rectangle,

$I = [120(15)^3/12 + (120 \times 15)(29.68)^2] + [15(75)^3/12 + (75 \times 15)(15.32)^2]$
$+ [40(15)^3/12 + (15 \times 40)(60.32)^2] = 4.605 \times 10^6\ \text{mm}^4$

Two bending moments are calculated based upon the allowable stresses for the tensile (t) at the bottom and compressive (c) at the bottom edges of the section

$M_t = \sigma_t I / y_t = 50 \times 4.605 \times 10^6/(-67.82) = -3.4 \times 10^6$ Nmm
$M_c = \sigma_c I / y_c = (-40) \times 4.605 \times 10^6/37.18 = -4.95 \times 10^6$ Nmm

The lower value must be equated to the maximum sagging bending moment $M_t = -wl^2/8$ to find the uniformly distributed loading w. This gives

$w = -8M_t/l^2 = (8 \times 3.4 \times 10^6)/(25 \times 10^6) = 1.09$ N/mm (kN/m)

The radius of curvature is

$R = EI/M_t = (207 \times 10^3 \times 4.605 \times 10^6)/(-3.4 \times 10^6) = -280.36 \times 10^3$ mm

That is, a sagging curvature of 280.36 m. In this design the maximum tensile stress is reached, but the maximum compressive stress is not ($\sigma_c = M_t y_c/I = 27.45$ MPa). If the larger moment M_c were taken to attain the maximum compressive stress, then the maximum tensile stress exceeds its allowable value ($\sigma_t = M_c y_t/I = 72.9$ MPa).

Example 4.2 A 6 m long, simply supported beam carries a uniformly distributed load of 30 kN/m together with concentrated loads of 20 kN at 1.5 m from each end (Fig. 4.3a). The beam cross-section consists of a symmetrical I-section 0.3 m deep ($I_g = 198 \times 10^6$ mm^4) with top plates 0.3 m wide × 18 mm thick, riveted to the top and bottom flanges (see Fig. 4.3b). Calculate the maximum bending stress and find the percentage increase in strength owing to the addition of the plates.

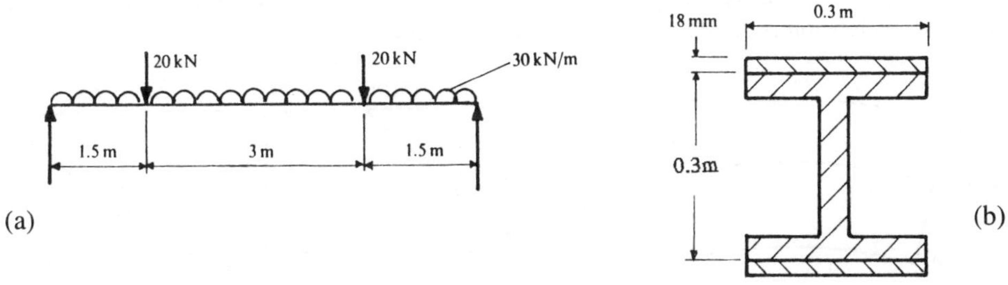

Figure 4.3 Fabricated I-section

Each support carries 110 kN. The maximum moment occurs at the centre, where

$M_{max} = (1.5 \times 20) + (3 \times 1.5 \times 30) - (3 \times 110) = -165$ kNm (sagging)

For the I-section alone, eq(4.3) gives the maximum tensile stress as

$\sigma = My/I = (-165 \times 10^6)(-150) / (198 \times 10^6) = 125$ N/mm^2 (MPa)

For the strengthened section, by parallel axes (see Appendix I)

$I = I_g = (198 \times 10^6) + 2[(300 \times 18^3)/12 + (300 \times 18 \times 159^2)] = 471.33 \times 10^6$ mm^4

The bending stress corresponding to $y = 150 + 18 = 168$ mm in eq(4.3) is

$\sigma = My/I = (-165 \times 10^6)(150 + 18) / (471.33 \times 10^6) = 58.81$ N/mm^2

The measure of strength is the maximum moment the beam can carry for a given allowable bending stress, but as $M/\sigma = I/y$, it is only necessary to employ a section modulus, $z = I/y$, for the strength comparison. This gives

Unstrengthened $z = I/y = (198 \times 10^6) / 150 = 1.32 \times 10^6$ mm^3
Strengthened $z = I/y = (471.33 \times 10^6) / 168 = 2.806 \times 10^6$ mm^3
$\therefore$ Strength increase = $(2.806 - 1.32) \times 100 / 1.32 = 112.54\%$

4.2 Combined Bending and Direct Stress

Combined loading occurs when a beam carries an axial force or when a column is eccentrically loaded. The direct stress arising from tensile and compressive forces are superimposed upon the bending stresses.

4.2.1 A Beam Carrying an Axial Force F and a Bending Moment M

The stresses produced by F and M both act along the beam in tension and compression. They are added to give the resultant stress,

$$\sigma = +F/A + My/I \tag{4.4}$$

We have seen how the sign of M and y can account for tension and compression on opposite sides of the n.a. A positive F refers to tension and a negative F to compression. For the beam in Fig. 4.4a, tension is combined with a hogging moment.

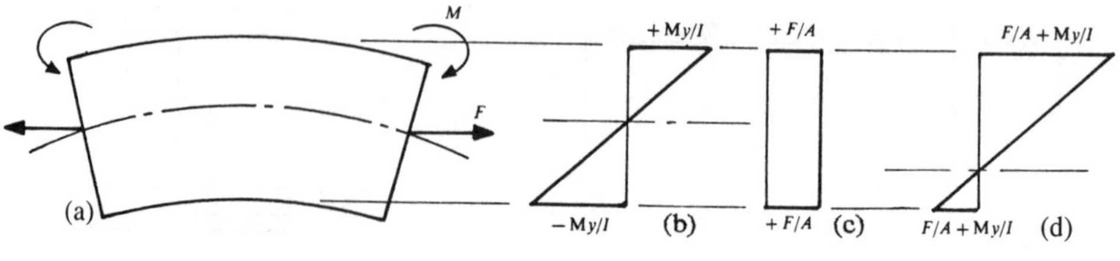

Figure 4.4 Beam under combined bending and direct stress

Thus both F and M are positive. The sign of y is positive above and negative below the n.a. The net stresses are respectively

$\sigma = + F/A + My/I$ (above n.a.)
$\sigma = + F/A - My/I$ (below n.a.)

Figure 4.4d shows that the net stress is the sum of the distributions due to bending in Fig. 4.4b and tension in Fig. 4.4c. Clearly, the addition of F modifies the position of zero stress which is found by equating (4.4) to zero.

Example 4.3 A simply supported beam has a rectangular section 1850 mm^2 in area, 75 mm deep with $I = 3 \times 10^6$ mm^4. The beam carries a uniformly distributed load of 45 kN/m over its 1.5 m length, together with an axial compressive force of 50 kN. Determine the distribution of stress at the mid-section and the depth in that section for which the stress is zero.

In this example, the directions of F and M in Fig. 4.3a are reversed and so they are negative in eq(4.4).

$M_{max} = - wl^2/8 = - (45 \times 1500^2)/8 = - (12.656 \times 10^6)$ Nmm (sagging)

when from eq(4.4), with $y = + 37.5$ mm for the top surface,

$\sigma = F/A + My/I = (- 50 \times 10^3/1850) + (- 12.656 \times 10^6)(+ 37.5)/(3 \times 10^6)$
$= - 27.03 - 158.2 = - 185.23$ MPa

and for the bottom surface, with $y = - 37.5$ mm,

$\sigma = F/A + My/I = (- 50 \times 10^3/1850) + (- 12.656 \times 10^6)(- 37.5) / (3 \times 10^6)$
$= - 27.03 + 158.2 = 131.17$ MPa

The zero stress position lies on the side of y (- ve)

$0 = F/A + My/I = - 27.03 + (- 12.656 \times 10^6)(- y) / (3 \times 10^6)$
$\therefore y = + 6.41$ mm beneath the centroid of the section

4.2.2 Eccentric Loading

Let a normal tensile force F lie in the positive x, y quadrant in Fig. 4.5. Within this quadrant, hogging moments, $M_x = Fk$ and $M_y = Fh$, are produced by the eccentricity (h, k) of the force to the centroidal axes x and y.

The resultant tensile stress at any point $P(x, y)$ is the sum of a direct stress due to F and bending stresses due to M_x and M_y. This gives

$$\sigma = F/A + M_x y / I_x + M_y x / I_y \qquad (4.5a)$$

The stresses in other quadrants are also determined from eq(4.5a) using the appropriate signs for the co-ordinates $P(x,y)$. For this, it is convenient to identify the first quadrant, i.e. positive x and y, with that in which the force is applied.

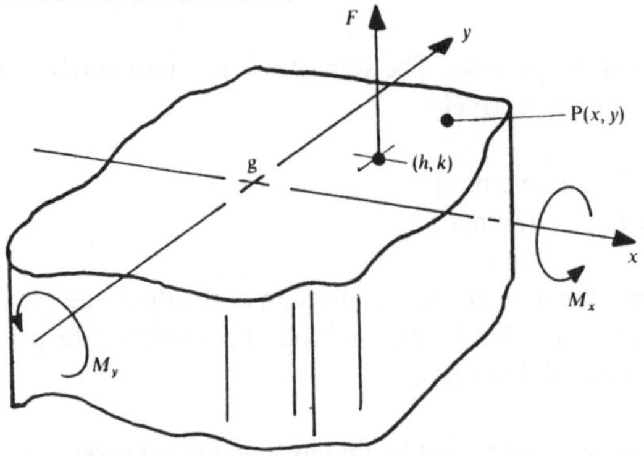

Figure 4.5 Eccentrically loaded column

When F is applied either along the x or the y-axis, only one bending stress term will appear. For example, when F lies on the x-axis, it is eccentric to y only. Then, $M_x = 0$, $M_y = Fh$ and eq(4.5a) becomes

$$\sigma = F/A + (Fh)x / I_y \tag{4.5b}$$

In a rectangular masonry compression column, (see Fig. 4.6), tensile stress is to be avoided. Hence the stress at point P $(-\ b/2,\ 0)$ should be zero when the force is applied at $F(h,\ 0)$. Equation (4.5b) gives the extreme position h for F along the x-axis as

$$0 = F/(bd) + (Fh)(-\ b/2) / (db^3/12)$$

from which $h = b/6$. Similarly, the extreme position for F applied along the y-axis is $k = d/6$. It can be shown further from eq(4.5a) that when F is eccentric to both x and y, it must be applied with the shaded area shown. In a circular column, diameter d, the corresponding safe area is a circle of radius d / 8.

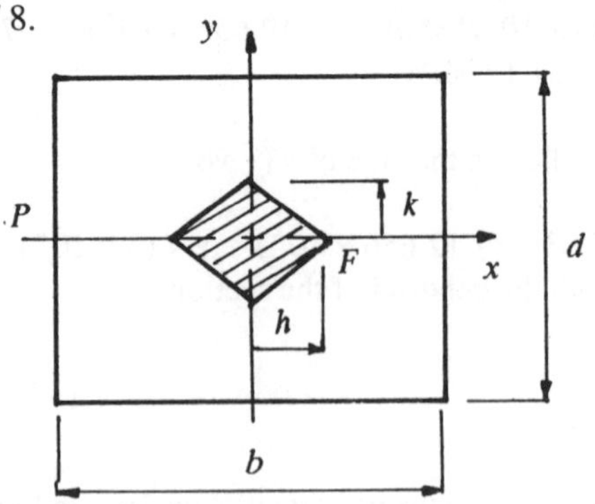

Figure 4.6 Rectangular section

Example 4.4 A compressive force F acts normally to the column section shown in Fig. 4.7. Determine the maximum eccentricity k if there is to be no tensile stress in the section. What is F if the maximum compressive stress is to be 92.5 MPa?

The $\bar{x}$ position of the centroid is found by taking first moments of area about the right vertical side. That is,

$$(150 \times 225 \times 112.5) - (100 \times 125 \times 137.5) = 21250\,\bar{x} \;\;\Rightarrow\;\; \bar{x} = 97.8 \text{ mm}$$

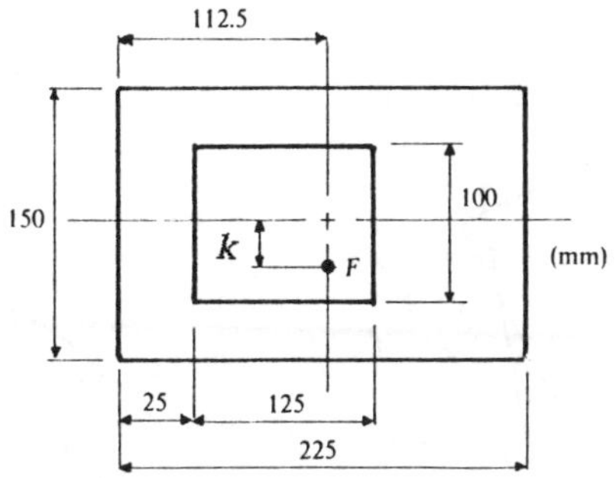

Figure 4.7 Column section

From this $h = 112.5 - 97.8 = 14.7$ mm, $M_x = Fk$ and $M_y = 14.7F$. Using $bd^3/12$, with parallel axes where appropriate, the second moments are:

$$I_x = 225 \times 150^3/12 - 125 \times 100^3/12 = 52.87 \times 10^6 \text{ mm}^4$$

$$I_y = [150 \times 225^3/12 + (150 \times 22.5 \times 14.7^2)] - [100 \times 125^3/12 + (125 \times 100 \times 39.7^2)]$$
$$= 113.7 \times 10^6 \text{ mm}^4$$

Taking x and y positive as shown, if stress at point $Q(-97.8, -75)$ is made zero, then tension cannot exist elsewhere in the section. From eq(4.5a),

$$0 = F/21250 + (Fk)(-75)/(52.87 \times 10^6) + (14.7F)(-97.8)/(113.7 \times 10^6) \qquad \text{(i)}$$

Note that with F as a negative compressive force, the first term on the RHS of eq(i) is negative, but the remaining two terms become positive. In fact F cancels in eq(i) leaving an equation from which $k = 24.26$ mm.

The maximum compressive stress occurs at point $P(127.2, 75)$. Setting this to 92.5 MPa, we have, from eq(4.5a)

$$-92.5 = F/21250 + (24.26F)(75)/(52.87 \times 10^6) + (14.7F)(127.2)/(113.7 \times 10^6)$$
$$= F(47.06 + 34.42 + 16.45)10^{-6}$$

from which $F = -944.6$ kN

4.3 Beams with Initial Curvature

The theory depends upon the relative magnitude of the radius of curvature R and the dimensions of the cross-section. Beams with small and large curvatures are treated separately.

4.3.1 Large Initial Radius (Small Curvature)

When R is much greater that the dimensions of the cross-section, as is the case with thin rings, the theory is similar to that for straight beams. The same analysis applies provided the initial curvature R_o is accounted for in the longitudinal strain expression. Referring to Fig. 4.8a, the centre of initial curvature C does not generally coincide with the centre C, induced by applied bending moments.

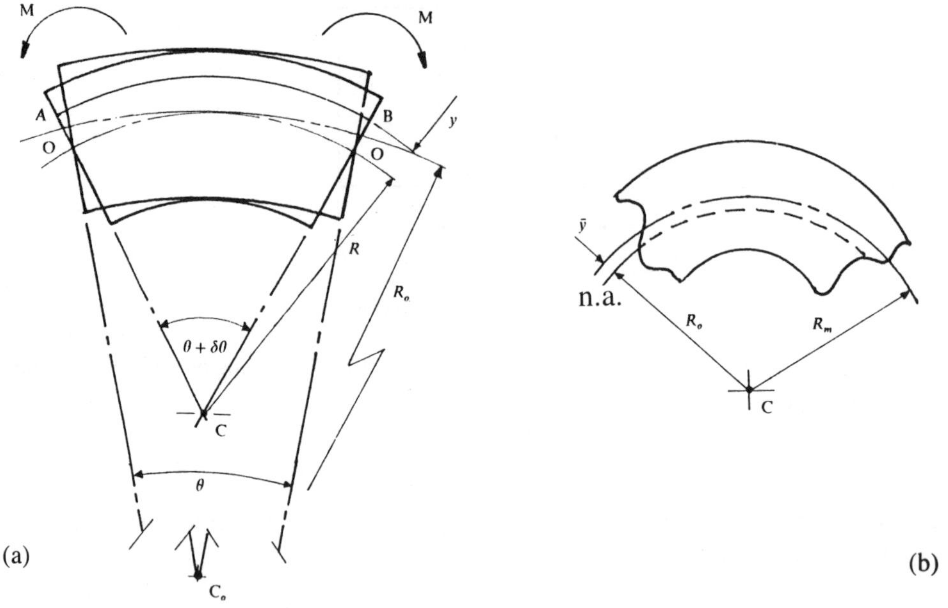

Figure 4.8 Bending of beams with (a) small curvature (b) large curvature

The tensile strain in fibre AB, distance y from the unstrained neutral axis OO, is

$$\varepsilon = [(R + y)(\theta + \delta\theta) - (R_o + y)\theta] / [(R_o + y)\theta] \tag{4.6}$$

where the denominator is the initial length of AB. Since the length of OO remains unchanged:

$$R(\theta + \delta\theta) = R_o\theta \tag{4.7}$$

Combining eqs(4.6) and (4.7) gives

$$\varepsilon = y(R_o - R)/[R(R_o + y)] \approx y[(1/R) - (1/R_o)] \tag{4.8}$$

provided $y << R_o$. The stress analysis is identical to that given for straight beams. Thus combining eqs(4,8) with eq(4.2) gives

$$M / I = E[(1/R) - (1/R_o)] = \sigma / y \tag{4.9}$$

It is seen from eq(4.9) that (i) the bending stress σ again varies linearly with dimension y and (ii) the n.a., from which y is measured, must pass though the centroid of the cross-section.

Example 4.5 The semi-circular steel arch in Fig. 4.9 is supported on rollers at each end and is 100 mm in width. Find the section depth by applying for a safety factor of 2.5 upon an allowable stress of ± 300 MPa. What is the radius of curvature under the maximum bending moment? Take E = 207 GPa.

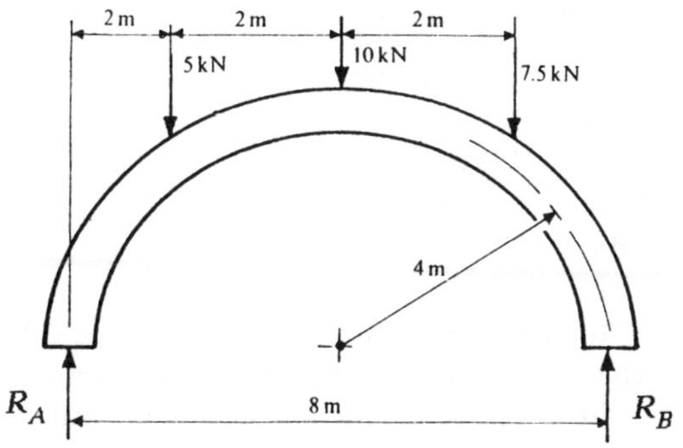

Figure 4.9 Steel arch

The support reactions are

$$8R_B = (5 \times 2) + (10 \times 4) + (7.5 \times 6), \quad \Rightarrow \; R_B = 11.875 \text{ kN}$$
$$R_A + R_B = 5 + 10 + 7.5, \quad \Rightarrow \; R_A = 10.625 \text{ kN}$$

The maximum moment (at centre) and the working stress are

$$M = -\,(4 \times 10.625) + (2 \times 5) = -\,32.5 \text{ kNm (sagging)}$$
$$\sigma = \sigma_a / S = 300/2.5 = 120 \text{ MPa}$$

Then, from eq(4.9), with M (– ve), σ(– ve) and $y = d/2$

$$M \,/\, (bd^3/12) = \sigma/(d/2)$$
$$\therefore\; d = \surd[6M/(b\sigma)] = \surd[(6 \times 32.5 \times 10^6)/(100 \times 120)] = 127.48 \text{ mm}$$

Now M_{max} is sagging (– ve), i.e. it increases the initial radius. From eq(4.9), with $I = bd^3/12$ $= 17.264 \times 10^6$ mm^4, the new radius R becomes

$$1/R = M/(EI) + 1/R_o = -\,(32.5 \times 10^6)/(207000 \times 17.264 \times 10^6) + 1/4000$$
$$R = 4.15 \text{ m}$$

4.3.2 Small Initial Radius (Large Curvature)

When R and the section dimensions are of the same order (see Fig. 4.8b), the approximation made in eq(4.8) becomes invalid. The bending stress must now be written as

$$\sigma = yE(R_o - R)/\,[R\,(R_o + y)] \qquad (4.10)$$

The total axial force is zero

$$F = \int \sigma \mathrm{d}A = [E(R_o - R)/R] \int [\, y\, \mathrm{d}A \,/\, (R_o + y)\,] = 0 \qquad (4.11)$$

where R and R_o are constants for a given cross-section. Furthermore, the resisting moment M for the section is given by

$$M = \int \sigma y \,\mathrm{d}A = [E(R_o - R)/R] \int [\, y^2 \mathrm{d}A/(R_o + y)\,]$$
$$= [E(R_o - R)/R] \{ \int y \,\mathrm{d}A - R_o \int [\, y \,\mathrm{d}A/(R_o + y)\,] \} \qquad (4.12)$$

Combining eqs(4.11 & 4.12), M becomes

$$M = E[(R_o - R)/R] \int y \,\mathrm{d}A \qquad (4.13)$$

However, although y is still measured from the n.a., eq(4.11) reveals that this axis will not now pass through the centroid g, of the section but shifts by an amount $\bar{y}$ towards the centre of curvature C in Fig. 4.8b. Let R_o be the initial radius of the n.a. and R_m the radius of the centroidal axis. Taking first moments of the cross-sectional area A about the n.a. in Fig. 4.10, $A\bar{y} = \int y \,\mathrm{d}A$ and eq(4.13) becomes

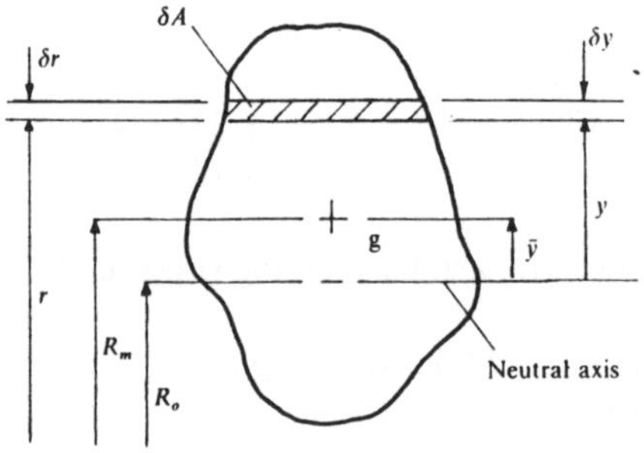

Figure 4.10 Neutral axis shift

$$M = E[(R_o - R)/R]\, A\bar{y} \qquad (4.14)$$

where R is the radius of the n.a. under M. The corresponding three-part formula is found from eqs(4.10 & 4.14) to complete the counterpart of eqs(4.3 and 4.9)

$$M / [A\bar{y}\,(R_o + y)] = E(R_o - R) / [R\,(R_o + y)] = \sigma / y \qquad (4.15)$$

Note that when $R < R_o$ in eq(4.15), a positive hogging moment will produce tension above the n.a. (y is positive) and compression below the n.a. (y is negative). Conversely, when $R > R_o$, a negative sagging moment produces compression above and tension below the n.a. Since eq(4.15) shows that σ is no longer proportional to y, the variation in σ across the section is non-linear. When applying eq(4.15), $\bar{y}$ must be found relative to the section's centroidal (or mean) radius, $R_m = R_o + \bar{y}$, as follows:

(a) Rectangular Section

For the rectangular section shown in Fig. 4.11, the integral in eq(4.11) gives

$$\int y \,\mathrm{d}A / (R_o + y) = \int y\,(b\,\mathrm{d}y) / [(R_m - \bar{y}) + y\,]$$

$$= b \int_{\bar{y} - d/2}^{\bar{y} + d/2} [1 - (R_m - \bar{y}) / (R_m - \bar{y} + y)\,]\,\mathrm{d}y = 0$$

$$\therefore\; b \left|\, y - (R_m - \bar{y}) \ln (R_m - \bar{y} + y) \right|_{\bar{y} - d/2}^{\bar{y} + d/2} = 0$$

$$\bar{y} = R_m - d / \{\ln [\,(R_m + d/2) / (R_m - d/2)\,]\} \qquad (4.16)$$

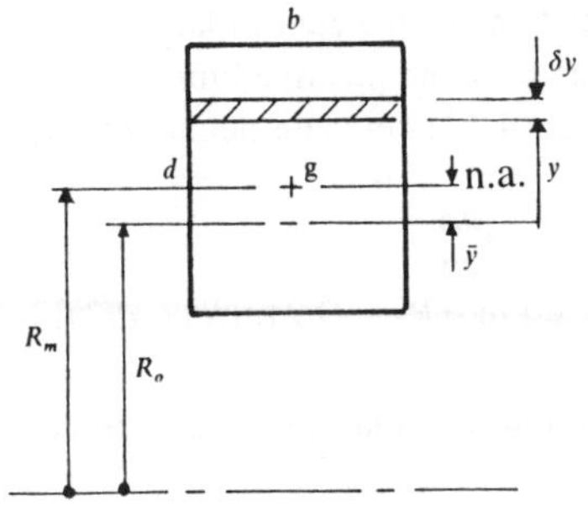

Figure 4.11 Rectangular section

(b) I - Section

In the I-section of Fig. 4.12 the web and two flanges are taken separately. It is convenient to introduce a parameter, $z = y - \bar{y}$, measured from the centroidal axis so that with mean radius $R_o = R_m - \bar{y}$, the integral in eq(4.11) becomes

$$i = \int (z + \bar{y})\,\mathrm{d}A/(R_m + z) = \int [1 - (R_m - \bar{y})/(R_m + z)]\mathrm{d}A \qquad (4.17)$$

Top flange ①, $\mathrm{d}A = B\,\mathrm{d}z$

$$i_{f1} = B\,|\,z - (R_m - \bar{y})\ln(R_m + z)\,\Big|_{d/2}^{D/2}$$
$$= B\{(D/2 - d/2) - (R_m - \bar{y})\ln[(R_m + D/2)/(R_m + d/2)]\} \qquad (4.18a)$$

Web ③, $\mathrm{d}A = b\,\mathrm{d}z$

$$i_w = b\,|\,z - (R_m - \bar{y})\ln(R_m + z)\,\Big|_{-d/2}^{d/2}$$
$$= b\{d - (R_m - \bar{y})\ln[(R_m + d/2)/(R_m - d/2)]\} \qquad (4.18b)$$

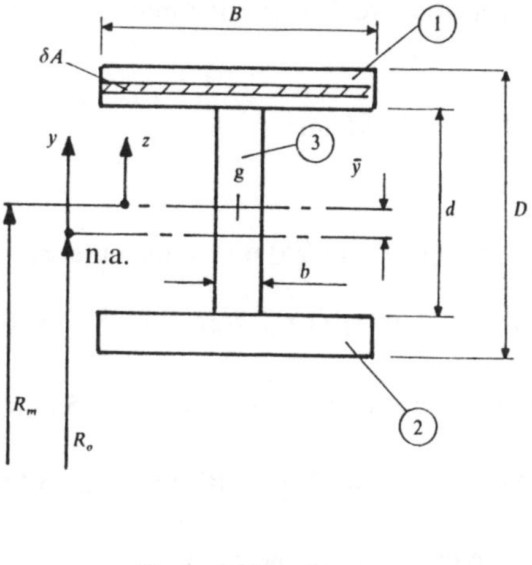

Figure 4.12 I-section

Bottom flange ②, $dA = B\,dz$,
Since the area now lies wholly beneath the centroidal axis it is simplest to change the sign of z in eq(4.17) and then integrate between positive limits for z of $D/2$ and $d/2$. Note that this is not equivalent to integrating eq(4.17) between negative limits

$$i_{f2} = B \,|\, z + (R_m - \bar{y}\,) \ln (R_m - z\,)\Big|_{d/2}^{D/2}$$
$$= B\left\{ (D/2 - d/2) + (R_m - \bar{y}\,) \ln [(R_m - D/2)/(R_m - d/2)] \right\} \qquad (4.18c)$$

The n.a. position $\bar{y}$ is then found by equating the sum of eqs(4.18a-c) to zero.

Example 4.6 Compare the maximum tensile and compressive stresses and the radii of curvature of the neutral axes for the initially curved beam in Fig. 4.13a for two sections given in Figs 4.13b and c under $M = +\,10$ kNm. Take $E = 207$ GPa.

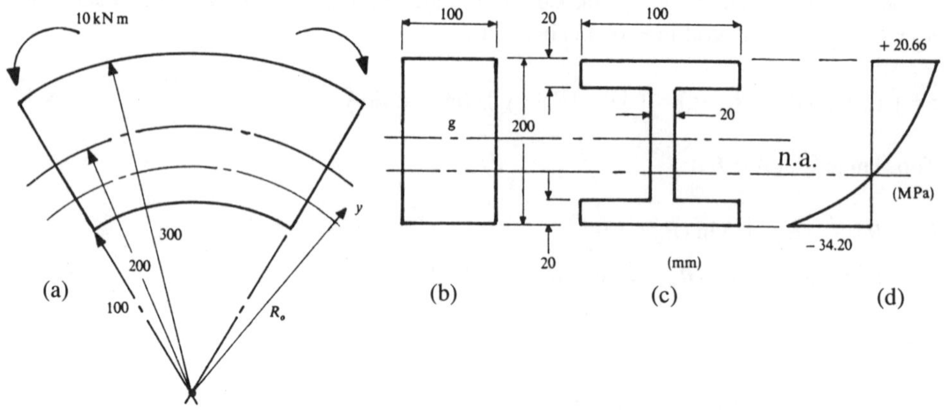

Figure 4.13 Beam with large initial curvature

The rectangular section properties are $A = 20 \times 10^3$ mm^2, $b = 100$ mm, $d = 200$ mm, $R_m = 200$ mm. From eqs(4.15) and (4.16),

$$\bar{y} = 200 - 200/ \ln (300/100) = 17.95 \text{ mm}$$
$$\sigma = My/[A\bar{y}\,(R_o + y)] \qquad \text{(i)}$$

For maximum tension, $y = +\,(100 + 17.95)$ mm at the outer radius. From eq(i)

$$\sigma = (10 \times 10^6 \times 117.95)/[(20 \times 10^3)17.95 \times 300] = +\,10.95 \text{ MPa}$$

For maximum compression, $y = -\,(100 - 17.95) = -\,82.05$ mm, when from eq(i)

$$\sigma = 10 \times 10^6 \times (-\,82.05)/(20 \times 10^3 \times 17.95 \times 100) = -\,22.85 \text{ MPa}$$

Re-arrange eq(4.15) to give the radius of the n.a. as

$R = R_o/[1 + M/(A E \bar{y})]$ (ii)

$= (200 - 17.95)/\{1 + [(10 \times 10^6)/(20 \times 10^3 \times 207 \times 10^3 \times 17.95)]\} = 182.03$ mm

For the I - section, $A = 7200$ mm^2, $B = 100$ mm, $D = 200$ mm, $b = 20$ mm, $d = 160$ mm and $R_m = 200$ mm. The respective eqs(4.18a-c) yield,

$i_{f1} = 100\,[(100 - 80) - (200 - \bar{y})\ln(300/280)] = 100\,[20 - 0.069\,(200 - \bar{y})]$ (iii)

$i_{f2} = 100\,[(100 - 80) + (200 - \bar{y})\ln(100/120)] = 100\,[20 - 0.1823\,(200 - \bar{y})]$ (iii)

$i_w = 20\,[160 - (200 - \bar{y})\ln(280/120)] = 20\,[160 - 0.8473\,(200 - \bar{y})]$ (iv)

Since $i_{f1} + i_{f2} + i_w = 0$, eqs (ii)-(iv) give $\bar{y} = 28.88$ mm. For maximum tension with $y = (100 + 28.88)$ mm, in eq(i),

$\sigma = (10 \times 10^6 \times 128.88)/(7200 \times 28.88 \times 300) = +\,20.66$ MPa

and for maximum compression with $y = -\,(100 - 28.88) = -\,71.12$ mm, in eq(i)

$\sigma = (10 \times 10^6) \times (-\,71.12)/(7200 \times 28.88 \times 100) = -\,34.20$ MPa

The typical variation in the stress between these limits across the section is shown in Fig. 6.10d. Equation (4.15) supplies the n.a. radius as

$R = (200 - 28.88)/[1 + (10 \times 10^6)/(7200 \times 207 \times 10^3 \times 28.88)] = 171.08$ mm

It is seen that for each section the radius of the strained n.a. differs little from its unstrained value R_o.

(c) Complex Sections

The shift in neutral axis $\bar{y}$, from the centroid g, for a large curvature beam with any given cross-section, is more conveniently found by the introduction of a new variable $r = R_o + y$ in Fig. 4.10. Since $R_m = R_o + \bar{y}$, the integral in eq(4.11) becomes

$$\int_A y\,\mathrm{d}A/(R_o + y) = \int_A (r - R_m + \bar{y})\,\mathrm{d}A/r = 0$$

$$\therefore A - (R_m - \bar{y})\int_A \mathrm{d}A/r = 0 \qquad (4.19)$$

from which $\bar{y}$ can be found once the integral $\int_A \mathrm{d}A/r$ is known. This integral may be evaluated for circular, triangular and trapezoidal sections. For example, taking the rectangular section in Fig. 4.11: $A = bd$, $\delta A = b \times \delta r$ and eq(4.19) becomes

$$bd - b(R_m - \bar{y})\int_{R_m - d/2}^{R_m + d/2} \mathrm{d}r/r = 0$$

from which $\bar{y}$ agrees with eq(4.16).

4.4 Elastic Bending Of Composite Beams

A composite beam section may be built up from vertical or horizontal layers of different materials. These arrangements can exploit the properties of each material to provide the strength required for a given weight. A beam can also be strengthened by reinforcement. For example, steel reinforcing rods embedded in concrete provide the bending strength required on the tensile side of a beam. Such designs must ensure that the strains between materials are compatible and that force and moment equilibrium is obeyed. Four common types of cross-section arise.

4.4.1 Vertical Layers

Let materials A and B comprise the balanced vertical layers of the beam section shown in Fig. 4.14.

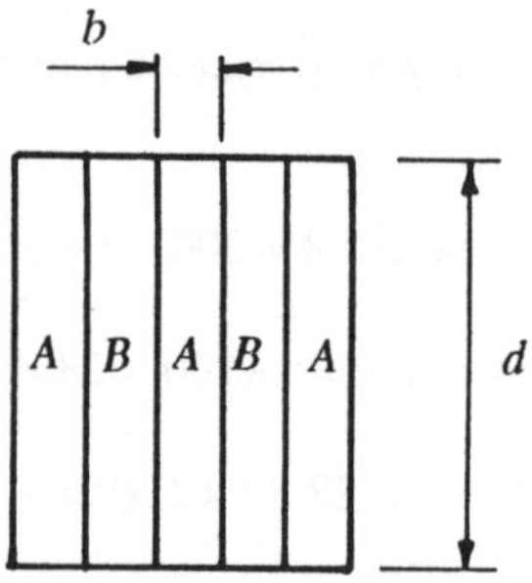

Figure 4.14 Composite beam

Provided A and B are fully bonded at their common interfaces, they will attain the same radius of curvature R in bending. When E, I and M in eq(4.3) are referred to each material, the curvature is

$$R = (EI/M)_A = (EI/M)_B$$

from which the moments carried by each fully stressed material are related in

$$M_A = M_B (EI)_A/(EI)_B \tag{4.20}$$

In eq(4.20), $I = \sum (bd^3/12)$, where b and d are the dimensions of an individual layer (Fig. 4.14). The bending stress in each material is

$$\sigma_A = (My/I)_A, \quad \sigma_B = (My/I)_B \qquad (4.21a,b)$$

The total moment carried by the section is its moment of resistance

$$M = M_A + M_B = M_A[1 + (EI)_B/(EI)_A] = M_B[1 + (EI)_A/(EI)_B] \tag{4.22}$$

Equations (4.21a,b) are not independent because of the relation (4.20) between the moments. Consequently, in this design, only one material can be fully stressed.

Example 4.7 The composite beam in Fig. 4.15 is fabricated from a central steel strip with outer brass strips. If the allowable stress for each material is 110 and 80 MPa respectively, determine the central load that can be applied when the beam is simply supported over 2 m. Which material is understressed and by how much? For steel take $E = 210$ GPa and for brass $E = 85$ GPa.

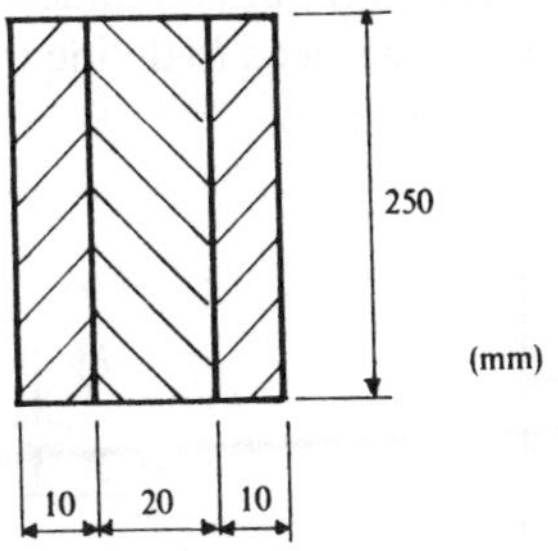

Figure 4.15 Composite section

$I_s = (bd^3/12)_S = 20 \times 250^3/12 = 26.04 \times 10^6$ mm^4
$I_b = 2(bd^3/12)_B = 2 \times 10 \times 250^3/12 = 26.04 \times 10^6$ mm^4

When the steel s and brass b are fully stressed, eqs(4.21a and b) give

$M_b = (\sigma I / y)_b = (80 \times 26.04 \times 10^6)/125 = 16.67$ kNm
$M_s = (\sigma I / y)_s = (110 \times 26.04 \times 10^6)/125 = 22.92$ kNm

From eq(4.20), when the brass is fully stressed, the steel moment is

$M_s = 16.67(210 \times 10^3 \times 26.04 \times 10^6) / (85 \times 10^3 \times 26.04 \times 10^6) = 41.19$ kNm

which shows that the steel would become overstressed. When the steel is fully stressed the brass moment is, from eq(4.20),

$M_b = 22.92[(85 \times 10^3)/(210 \times 10^3)] = 9.41$ kNm

For this condition, eq(4.21b) shows that the brass is safely understressed to a value

$\sigma_b = (My/I)_b = (9.41 \times 10^6 \times 125)/(26.04 \times 10^6) = 45.17$ MPa.

Equation (4.22) supplies the maximum moment for the beam as

$M = M_b + M_s = 9.41 + 22.92 = 32.33$ kNm

The corresponding central concentrated load is

$W = 4M/L = (4 \times 32.33)/2 = 64.66$ kN

4.4.2 Horizontal Layers

The approach given in Section 4.4.1 may be further applied to composite beam sections made up of horizontal unbonded layers. Provided the layers are thin and the interfaces between layers remains in contact under load, a common radius of curvature may be assumed (see Example 4.9).

With bonded horizontal layers calculations are based upon an equivalent section of the stiffest material. Consider a beam with three bonded layers in two materials (Fig. 4.16a). When A is stiffer than B ($E_A > E_B$), we need to define the web thickness for an equivalent I-section in material A (see Fig. 4.16b).

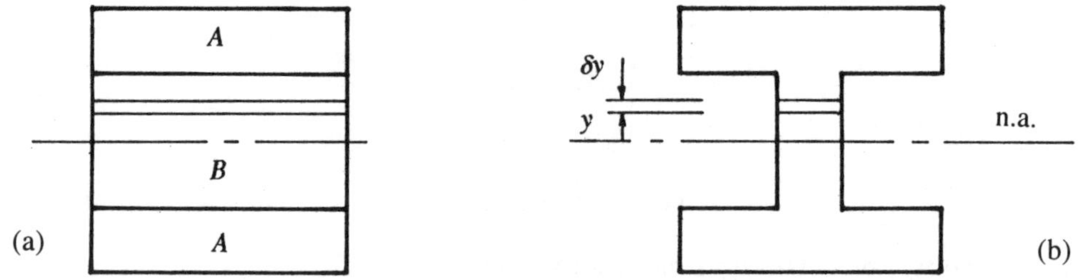

Figure 4.16 Composite beam in A and B with the equivalent section in A

At a given depth the fibres in the equivalent section are required to reach the strain in the original section. Thus, when σ_A and σ_B act on an elemental areas $\delta A = a\delta y$ and $\delta B = b\delta y$ shown, the strains are

$$\varepsilon = (\sigma/E)_A = (\sigma/E)_B \;\Rightarrow\; \sigma_B = \sigma_A E_B/E_A \tag{4.23}$$

Also the moments of the forces $\sigma_A\,\delta A$ and $\sigma_B\,\delta B$ must be the same across the original and the equivalent sections.

$$\sigma_A(a\delta y)y = \sigma_B(b\delta y)y = \sigma_A(E_B/E_A)(b\delta y)y$$
$$a\int \sigma_A\, y\, dy = b\,(E_B/E_A)\int \sigma_A\, y\, dy$$

With identical limits for y, the integrals cancel to give

$$a = b\,(E_B/E_A) \tag{4.24}$$

Equation (4.21a) supplies the bending stress in the equivalent section

$$\sigma_A = (My/I)_A \tag{4.25}$$

The true stress in material B follows from eqs(4.23) and (4.25) as

$$\sigma_B = (E_B/E_A)(My/I)_A \tag{4.26a}$$

Because of the dependence between σ_A and σ_B, the maximum allowable stresses for materials A and B cannot be reached simultaneously. The correct design ensures that one material is fully stressed while the other is understressed. For many layers of different materials we may generalise eq(4.26) to give the stress in the nth layer as

$$\sigma_n = (E_n/E_A)(My/I)_A \tag{4.26b}$$

where A defines the equivalent section in the stiffest material.

Example 4.8 The composite beam in Fig. 4.17a is composed of a top plate of duralumin (E = 70 GPa), a core of expanded plastic (E = 5 MPa) and a bottom plate of steel (E = 200 GPa). The beam length of 1 m is simply supported at its ends. Calculate the position of the neutral axis and determine the safe value of uniformly distributed loading when the stress in the core is restricted to 20 kPa and the maximum deflection is restricted to 10 mm. (CEI)

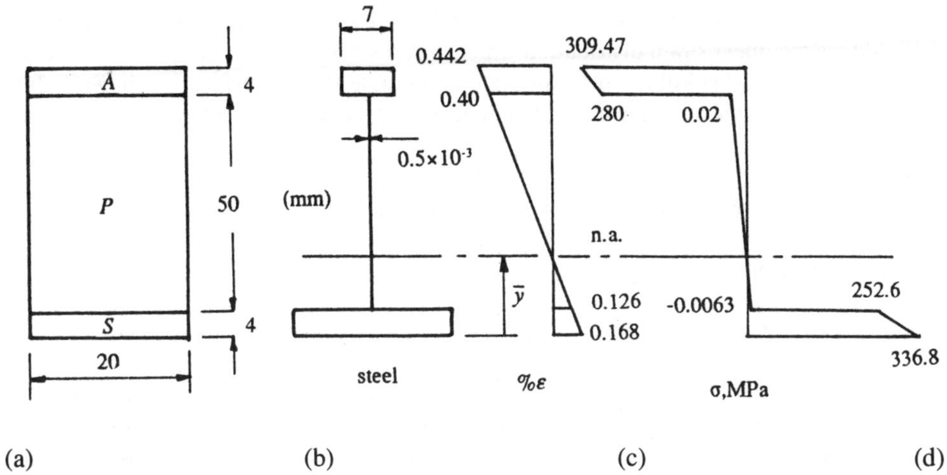

Figure 4.17 Three-layered beam

Using eq(4.24), replace the duralumin d and plastic p in Fig. 4.17a by the equivalent widths in steel s:

$s = p(E_p/E_s) = (20 \times 5)/(200 \times 10^3) = 0.5 \times 10^{-3}$ mm
$s = d(E_d/E_s) = (20 \times 70)/200 = 7$ mm

Figure 4.17b shows the equivalent section in steel. The position $\bar{y}$, of the neutral axis above the base is found from $\sum A_i y_i = A\bar{y}$

$(20 \times 4 \times 2) + (50 \times 0.5 \times 10^{-3} \times 29) + (7 \times 4 \times 56) = 108.025\,\bar{y} \;\Rightarrow\; \bar{y} = 16$ mm

The second moment of area follows from parallel axes:

$$I_g = [(7 \times 4^3/12) + (28 \times 40^2)] + [(0.5 \times 10^{-3} \times 50^3/12) + (0.5 \times 10^{-3} \times 50 \times 13^2)]$$
$$+ [(20 \times 4^3/12) + (20 \times 4 \times 14^2)] = 60633.43 \text{ mm}^4$$

The maximum bending stress in the plastic will be reached at the top interface where y = 38 mm. Using eq(4.26a) to find the allowable moment,

$\sigma_p = (E_p/E_s)(My/I)$
$M = (\sigma_p I E_s)/(yE_p) = (0.02 \times 60633.43 \times 200 \times 10^3)/(38 \times 5) = 1276.5$ Nm

in which unsubscripted terms refer to the equivalent section in steel. The corresponding distributed loading is:

$w = 8M/l^2 = 8 \times 1276.5 / 1^2 = 10212$ N/m = 10.21 N/mm

The central deflection is

$$\delta = 5wl^4 / (384EI) = (5 \times 10.21 \times 10^{12}) / (384 \times 200 \times 10^3 \times 60633.43) = 11.92 \text{ mm}$$

Since the deflection is restricted to 10 mm, w will need to be reduced to 8.5 N/mm. Figure 4.17c shows the linear strain distribution which ensures compatibility at the two interfaces. In order for this to be possible, stress discontinuities arise at the interfaces as shown in Fig. 4.17d. The stress values shown are found from the equivalent section in steel. From eq(4.26b),

$$\sigma_s = (My/I)_s,\ \sigma_p = (E_p/E_s)(My/I)_s,\ \sigma_d = (E_p/E_s)(My/I)_s$$

and the strains from

$$\varepsilon_s = \sigma_s/E_s,\ \varepsilon_p = \sigma_p/E_p,\ \varepsilon_d = \sigma_d/E_d$$

Example 4.9 A steel strip 50 × 12.5 mm in section is placed on top of a brass strip 50 × 25 mm in section to form a composite beam 50 mm wide × 37.5 mm deep. If the stresses in the steel and brass are not to exceed 140 and 70 MPa respectively, determine the maximum bending moment that the beam can carry and the maximum bending stresses in each material when the beams are (a) unbonded and (b) bonded at the interface. The elastic modulus of steel is 207 GPa and that for brass is 83 GPa.

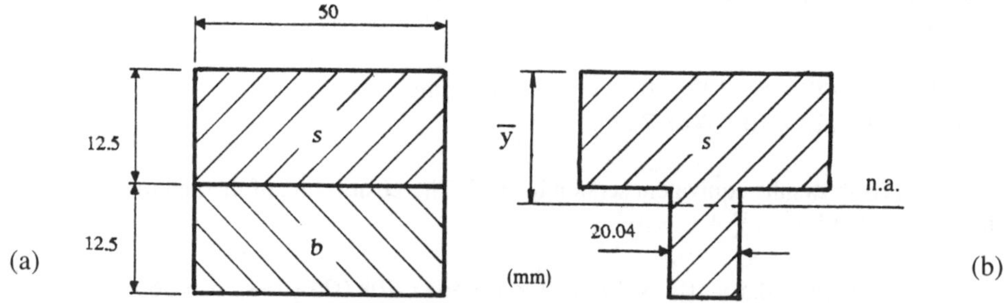

Figure 4.18 Rectangular composite with equivalent section

(a) In the unbonded case each beam can bend separately. The maximum moments for the given allowable stresses are, from eqs(4.21a,b),

$$M_s = (\sigma I / y)_s = 140 \times (50 \times 12.5^3/12)/6.25 = 65.1 \text{ Nm}$$
$$M_b = (\sigma I / y)_b = 70 \times (50 \times 25^3/12)/12.5 = 364.58 \text{ Nm}$$

For the same interface radius of curvature, eq(4.20) gives

$$M_s = M_b (EI)_s / (EI)_b$$
$$M_s = 364.58[(207 \times 12.5^3)/(83 \times 25^3)] = 113.66 \text{ Nm}$$

Thus, when the brass is fully stressed, the steel is overstressed. Therefore, the steel is fully stressed and the brass understressed. The moment for the brass is

$M_b = M_s(EI)_b/(EI_s) = 65.1[(83 \times 25^3)/(207 \times 12.5^3)] = 208.82$ Nm

Now from eq(4.22) the total moment carried by the section is

$M = 65.1 + 208.82 = 273.92$ Nm

The maximum stress reached in the brass is

$\sigma_b = (My/I)_b = (208.82 \times 10^3 \times 12.5 \times 12)/(50 \times 25^3) = 40.09$ MPa

(b) In the bonded case the equivalent width of steel is, from eq(4.24), $50 \times 83/207 = 20.04$ mm. For the equivalent steel section (see Fig. 4.18b), the centroidal position $\bar{y}$ from the top is found from

$(50 \times 12.5 \times 6.25) + (25 \times 20.04 \times 25) = (625 + 501)\bar{y} \Rightarrow \bar{y} = 14.59$ mm

and, using $I = bd^3/3$ for the second moment of area,

$I_g = (50 \times 14.59^3/3) - (29.96 \times 2.09^3/3) + (20.04 \times 22.91^3/3) = 132 \times 10^3$ mm^4

With the steel fully stressed along its top edge eq(4.25) gives the bending moment

$M = (\sigma I/y)_s = 140 \times 132 \times 10^3/14.59 = 1266.62$ Nm

Under this moment the maximum stress in the brass is found from eq(4.26a),

$$\sigma_b = (E_b/E_s)(My/I)_s \qquad \text{(i)}$$
$$= (83/207)[(1266.62 \times 10^3 \times 22.91)/(132 \times 10^3)] = 88.15 \text{ MPa}$$

As the brass is overstressed we must ensure that the stress in the brass is reduced to 70 MPa along the bottom edge. From eq(i) the corresponding bending moment is

$M = 70 \times 1266.62 / 88.15 = 1005.82$ Nm

The stress in the steel is correspondingly reduced to

$\sigma_s = My/I = (1005.82 \times 10^3 \times 14.59)/(132 \times 10^3) = 111.17$ MPa

This condition governs the design.

4.4.3 Bi-metallic Strip

Bi-metal beams can be used as a switch to control temperature. In a thermostat the beam is arranged as a cantilever that bends and deflects to operate a switch as the temperature changes. Consider two thin, straight strips of dissimilar metals 1 and 2 joined along their longer sides to form this beam. When the elastic moduli obey $E_1 > E_2$, the linear expansion coefficients will obey $\alpha_1 < \alpha_2$. The beam will bend as shown in Fig. 4.19a when it is subjected to a temperature change ΔT

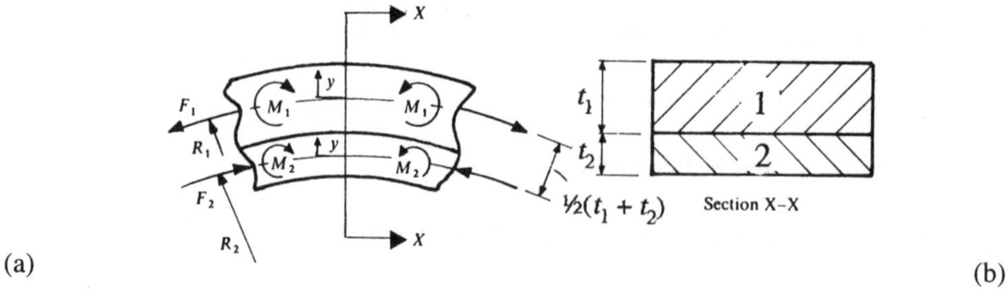

Figure 4.19 Bi-metallic strip (a) induced F and M, (b) section on X–X

Longitudinal tensile and compressive forces F_1 and F_2 respectively, arise due to restrained thermal expansion. This means that the beam fibres in 1 and 2 are stretched by greater and lesser amounts than their free expansions. Because there are no external forces or moments applied to the beam, two equilibrium conditions apply:

(i) Horizontal force equilibrium normal to a cross-section in Fig. 4 .19b requires that

$$F_1 = F_2 = F \tag{4.27a}$$

(ii) The couple $F(t_1 + t_2)/2$ produced from these forces must be resisted by internal moments M_1 and M_2 acting in the opposite sense. That is

$$M_1 + M_2 = F(t_1 + t_2)/2 \tag{4.27b}$$

In any fibre the total longitudinal strain is composed of a direct strain (F/AE) due to F, a bending strain (My/EI) due to M and a temperature strain ($\alpha l\Delta T / l = \alpha \Delta T$) due to ΔT. Since the total strain in 1 and 2 must be compatible at the interface then, from the F and M directions given,

$$F_1/(A_1E_1) + M_1(t_1/2)/(E_1I_1) + \alpha_1 \Delta T = -F_2/(A_2E_2) - M_2(t_2/2)/(E_2I_2) + \alpha_2 \Delta T \tag{4.28}$$

Assuming that the radii of curvature of the central neutral axes of each strip are the same, eq(4.3) supplies the relationships

$$R = R_1 = R_2 = (EI/M)_1 = (EI/M)_2 \tag{4.29a}$$
$$M_2 = M_1(E_2I_2)/(E_1I_1) \tag{4.29b}$$

Combining eqs(4.27b) and (4.29b),

$$M_1 = FE_1I_1(t_1 + t_2)/[2(E_1I_1 + E_2I_2) \tag{4.30}$$

The behaviour of the strip is thus reduced to the simultaneous solution to F from eqs(4.29b-4.30). The axial stress at distance y from the n.a. in either material is

$$\sigma = \pm F/A \pm My/I \tag{4.31}$$

Referring to Fig. 4.20, the corresponding end deflection δ_e of a cantilevered strip of length L may be found:

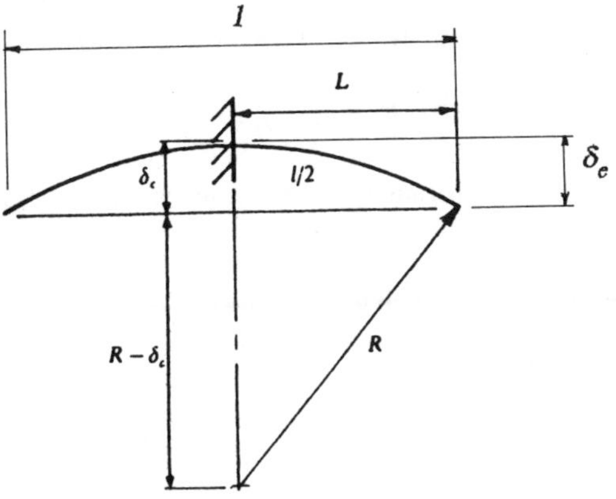

Figure 4.20 Strip deflection

$$R^2 = (R - \delta_e)^2 + (L)^2$$

Neglecting the small quantity δ_e^2

$$\delta_e \simeq L^2/(2R)$$

The deflection δ_c, at the centre of a simply supported bi-metallic beam of length l (see Fig. 4.20) follows from

$$R^2 = (R - \delta_c)^2 + (l/2)^2$$
$$\delta_c \simeq l^2/(8R)$$

Example 4.10 A bi-metallic strip in a temperature controller consists of brass bonded to steel, each material having the same rectangular cross-section and length. Calculate the stresses set up at the outer and interface surfaces in the longitudinal direction for an increase in temperature of 50°C in the strip. Neglect distortion and transverse stress. For steel, $\alpha = 11 \times 10^{-6}$ °C^{-1}, $E = 207$ GPa; for brass, $\alpha = 20 \times 10^{-6}$ °C^{-1}, $E = 103$ GPa. (CEI)

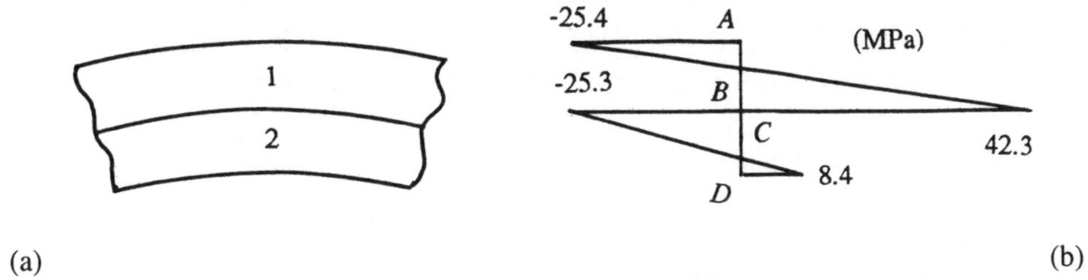

Figure 4.21 Strip showing stress distribution

In Fig. 4.21a we identify 1 with steel and 2 with copper so that $E_1 > E_2$ and $\alpha_1 < \alpha_2$ as in Fig. 4.19a. Hence eq(4.28) may be re-written as

$$[\alpha_1 \Delta T + F_1/(A_1 E_1)] - [\alpha_2 \Delta T - F_2/(A_2 E_2)] = -[M_1 t/(2EI)_1 + M_2 t/(2EI)_2]$$

Substituting $M_1 = E_1 I_1/R_1$ and $M_2 = E_2 I_2/R_2$ where $I_1 = I_2$ and $R_1 = R_2$

$$[\alpha_1 \Delta T + F_1/(A_1 E_1)] - [\alpha_2 \Delta T - F_2/(A_2 E_2)] = -t/R \tag{i}$$

where the areas A_1 and A_2 for steel and brass are the same. Equations (4.27b) and (4.29a) yield

$$(EI/R)_1 + (EI/R)_2 = Ft$$
$$F = bt^2 (E_1+E_2)/(12R) \tag{ii}$$

in which $R = R_1 = R_2$ and $I = I_1 = I_2 = bt^3/12$. Substituting eq(ii) into eq(i) leads to

$$(\alpha_1 - \alpha_2)\Delta T + I(E_1 + E_2)^2/(AtRE_1E_2) = -t/R \tag{iii}$$

Substituting $I/A = (bt^3/12)/(bt) = t^2/12$ into (iii) provides the ratio R/t

$$R/t = [12E_1E_2 + (E_1 + E_2)^2]/[12E_1E_2(\alpha_2 - \alpha_1)\Delta T] \tag{iv}$$

Equation(4.31) provides the stresses at points A, B, C and D as

$$\sigma_A = F_1/A_1 - M_1 t/(2I_1) = F(E_2 - 5E_1)/[bt(E_1 + E_2)] \tag{v}$$
$$\sigma_B = F_1/A_1 + M_1 t/(2I_1) = F(7E_1 + E_2)/[bt(E_1 + E_2)] \tag{vi}$$
$$\sigma_C = -F_2/A_2 - M_2 t/(2I_2) = -F(7E_2 + E_1)/[bt(E_1 + E_2)] \tag{vii}$$
$$\sigma_D = -F_2/A_2 + M_2 t/(2I_2) = -F(E_1 - 5E_2)/[bt(E_1 + E_2)] \tag{viii}$$

where $M_1 = FtE_1/(E_1 + E_2)$ and $M_2 = FtE_2/(E_1 + E_2)$. Now from eq(ii) and (iv),

$$F/(bt) = [E_1E_2(E_1 + E_2)(\alpha_2 - \alpha_1)\Delta T]/[12E_1E_2 + (E_1 + E_2)^2] \tag{ix}$$

Substituting eq(ix) into eqs(v)-(viii) provides the stresses from the information supplied

$$\sigma_A = [(E_2 - 5E_1)E_1E_2(\alpha_2 - \alpha_1)\Delta T]/[12E_1E_2 + (E_1 + E_2)^2] = -25.4 \text{ MPa}$$
$$\sigma_B = [(7E_1 + E_2)E_1E_2(\alpha_2 - \alpha_1)\Delta T]/[12E_1E_2 + (E_1 + E_2)^2] = 42.3 \text{ MPa}$$
$$\sigma_C = -[(7E_2 + E_1)E_1E_2(\alpha_2 - \alpha_1)\Delta T]/[12E_1E_2 + (E_1 + E_2)^2] = -25.3 \text{ MPa}$$
$$\sigma_D = -[(E_1 - 5E_2)E_1E_2(\alpha_2 - \alpha_1)\Delta T]/[12E_1E_2 + (E_1 + E_2)^2] = 8.4 \text{ MPa}$$

and these are distributed in the manner of Fig. 4.21b.

4.5 Reinforced Sections (Steel in Concrete)

The following theory is concerned specifically with reinforced concrete. Concrete is weak in tension so when used for a beam section it requires steel reinforcement on the tensile side of the neutral axis. A similar theory will apply wherever a horizontal line of reinforcing rods is inserted into a section of material which is inherently weak in bending. Both the tensile and compressive sides sides may be reinforced where necessary.

4.5.1 Rectangular Section With Single Steel Line

The presence of steel, of total area A_s, on the tensile side of the section of breadth B with steel line depth D, will modify the n.a. to a non-central position h (see Fig. 4.22a).

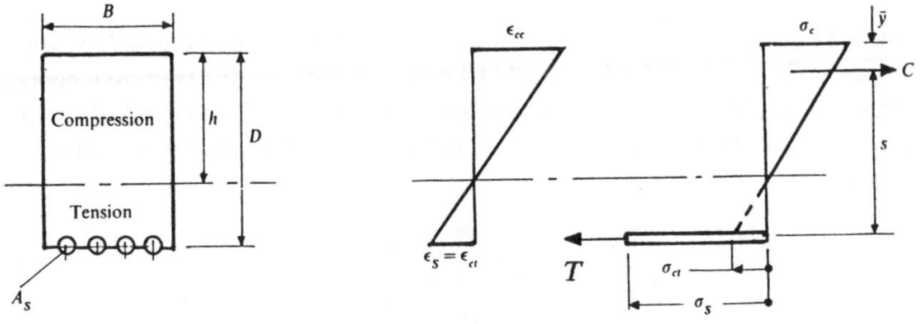

Figure 4.22 Single line reinforcement

Since the steel strain ε_S and the tensile strain in the concrete ε_{ct} are compatible at the interface (Fig. 4.22b), the corresponding stresses σ_S and σ_{ct} (Fig. 4.22c) are found from

$$\varepsilon_S = \varepsilon_{ct}\,, \Rightarrow \quad \sigma_S = \ (E_S/E_c)\sigma_{ct} = m\sigma_{ct} \qquad (4.32a,b)$$

where $m = E_S/E_c$ is the modular ratio. Now from Fig. 4.22c

$$\sigma_{ct}/(D - h) = \sigma_c/h \qquad (4.33)$$

when, from eq(4.32b and 4.33), h is found from

$$\sigma_S = m(D - h)\sigma_c/\,h \qquad (4.34)$$

where σ_S and σ_c are the maximum allowable stresses for the steel and concrete respectively. If the section is to be designed economically, then A_S is found from the balance between the average compressive force, $C = \sigma_c\,Bh/2$, acting at the centroid ($\bar{y} = h/3$) of the stress distribution for the concrete area above the n.a. and the total tensile force $T = \sigma_S A_S$ in the steel. This neglects any tensile stress in the concrete (broken line in Fig. 4.22c), so that

$$\sigma_c Bh/2 = \sigma_S\,A_S\,, \quad \Rightarrow \quad A_S = (Bh/2)(\sigma_c/\sigma_S) \qquad (4.35)$$

If $s = D - \bar{y}$, is the distance between C and T, then, because $C = T$, the moment of resistance M is given by either one of the expressions

$$M_c \ = Cs = \sigma_c(Bh/2)(D - h/3) \qquad (4.36a)$$

$$M_S = Ts = A_S\,\sigma_S(D - h/3) \qquad (4.36b)$$

With an uneconomic section A_S will not obey eq(4.35). When A_S is specified, h is again found from the zero net axial force condition $C = T$ where eq(4.34) applies with the concrete understressed. That is,

$\sigma_c Bh/2 = A_S[m(D - h)\sigma_c/h]$

from which a quadratic in h results:

$$Bh^2 + 2A_S mh - 2A_S mD = 0 \tag{4.37}$$

Example 4.11 Determine the moment of resistance for the section in Fig. 4.23 when the steel reinforcement (i) consists of 4 bars each 10 mm diameter and (ii) is chosen economically based upon allowable stresses in the steel and concrete of 125 and 4.5 MPa respectively. Find the maximum distributed loading an 8 m length s.s. beam in each section could carry. Take $m = 15$.

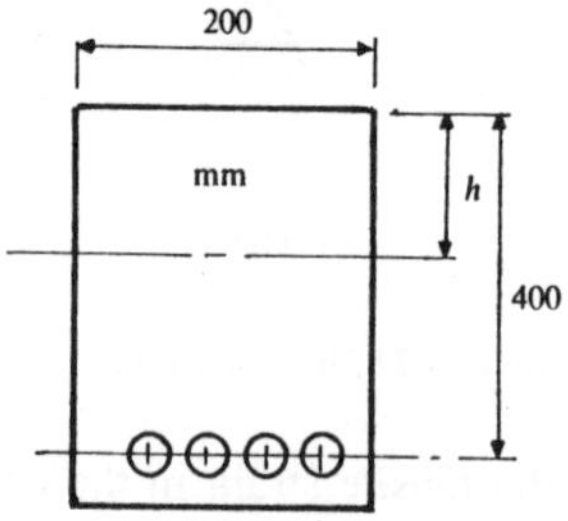

Figure 4.23 Reinforced section

(i) Uneconomic Design

$A_S = 4 \times \pi(10)^2/4 = 314.14\ \text{mm}^2$

From eq(4.37) the n.a. position h is

$$200h^2 + (2 \times 314.14 \times 15)h - (2 \times 314.14 \times 15 \times 400) = 0$$
$$h^2 + 47.12h - 18848.4 = 0, \quad \Rightarrow h = 115.74\ \text{mm}$$

M is then the lesser of

$$M_c = Cs = \sigma_c(Bh/2)(D - h/3)$$
$$= 4.5(200 \times 115.74 / 2)(400 - 115.74/3)10^{-6} = 18.82\ \text{kNm}$$

$$M_t = Ts = A_S\,\sigma_S(D - h/3)$$
$$= 314.14 \times 125\,(400 - 115.74 / 3)10^{-6} = 14.19\ \text{kNm}$$

Taking $M_c = M_t = 14.19$ kNm ensures that the steel is fully stressed while the concrete remains understressed. Finally, the concrete stress is

$$\sigma_c = M_t / [(Bh/2)(D - h/3)]$$
$$= (2 \times 14.19 \times 10^6)/[200 \times 115.74(400 - 115.74 / 3)] = 3.39\ \text{MPa}$$

$$w = 8M / l^2 = (8 \times 14.19 \times 10^6)/(8000)^2 = 1.77\ \text{N/mm (kN/m)}$$

(ii) Economic Design
With both steel and concrete fully stressed, eq(4.34) defines the n.a. position h as

$125 = 15(200 - h)4.5/h, \Rightarrow h = 140.26$ mm

Then the steel area is supplied by eq(4.35),

$A_S = (4.5 \times 200 \times 140.26) / (2 \times 125) = 504.94$ mm^2

and the moment of resistance by eq(4.36b),

$M = 504.94 \times 125(400 - 140.26 / 3)10^{-6} = 22.3$ kNm

$w = (8 \times 22.3 \times 10^6)/(8000)^2 = 2.79$ kN/m

4.5.2 T-section With Single Steel Line

With a single line of reinforcement at the web bottom (Fig. 4.24a) it is seen that the same strain and stress distributions (Fig. 4.24c and d) apply as with the rectangular section.

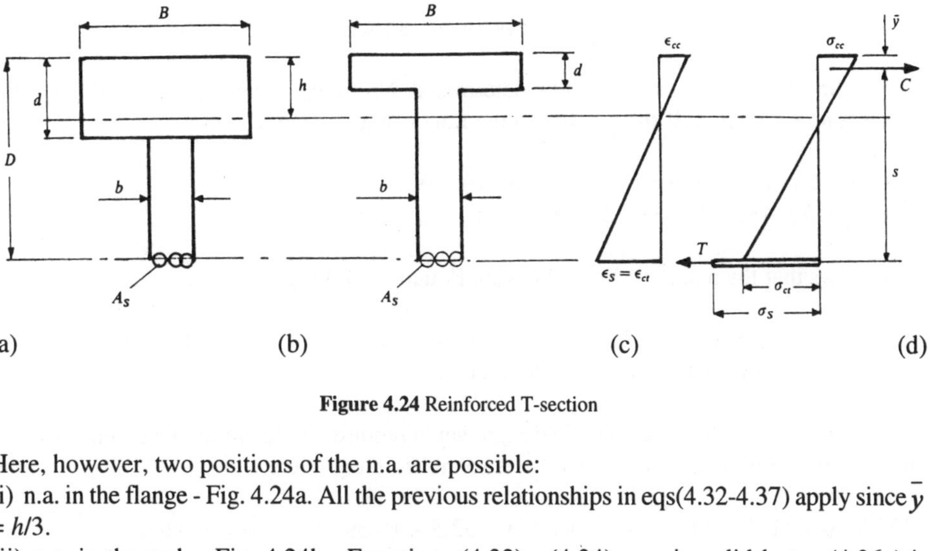

(a) (b) (c) (d)

Figure 4.24 Reinforced T-section

Here, however, two positions of the n.a. are possible:
(i) n.a. in the flange - Fig. 4.24a. All the previous relationships in eqs(4.32-4.37) apply since $\bar{y} = h/3$.
(ii) n.a. in the web - Fig. 4.24b. Equations (4.32) - (4.34) remain valid but eq(4.36a) is invalid since C now acts upon the area above the n.a. That is, with $C = T$,

$$(\sigma_c/2)[Bd + b(h - d)] = \sigma_S A_s = mA_S\sigma_c(D - h)/h:$$
$$bh^2 + h(Bd - bd + 2mA_S) - 2mA_S = 0 \qquad (4.38)$$

The moments of resistance expressions (4.36a,b) are also modified to

$$M_c = Cs = (\sigma_c/2)[Bd + b(h - d)](D - \bar{y}) \qquad (4.39a)$$
$$M_t = Ts = \sigma_S A_S(D - \bar{y}) \qquad (4.39b)$$

where $\bar{y}$ locates the centroid of the compressive area from the top edge

$$[\,Bd + (h - d)\,b]\,\bar{y} = Bd(d/2) + b(h - d)[d + (h - d)/2] \qquad (4.40)$$

Again, eqs(4.39a and b) may be equated for economical design. Otherwise, selecting the lesser value of M avoids overstressing the concrete.

Example 4.12 Find the moment of resistance for the T-section in Fig. 4.25 and establish the magnitude of the stresses in the steel and concrete, given respective allowable values of 125 and 7 MPa. Take $m = 15$.

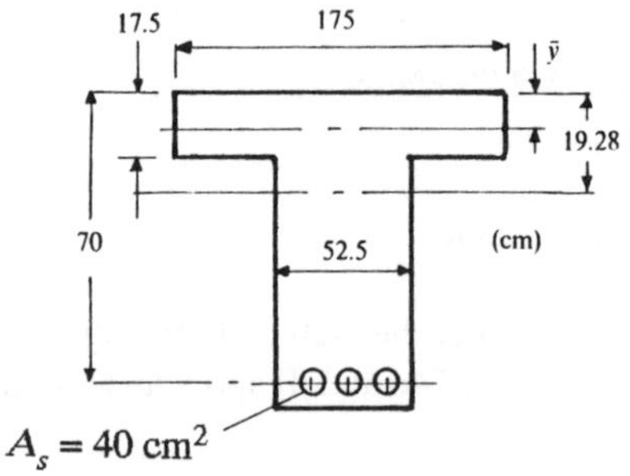

Figure 4.25 Reinforced T-section

Firstly, the position of the n.a. must be determined. Normally, when A_S is given this implies an uneconomic design. If the n.a. lies in the flange, eq(4.37) gives

$$175h^2 + (2 \times 40 \times 15)h - (2 \times 40 \times 15 \times 70) = 0$$
$$h^2 + 6.857h - 480 = 0, \quad \Rightarrow \quad h = 18.75 \text{ cm}$$

This shows that the n.a. must lie in the web. Equation (4.38) give:

$$52.5h^2 + h\,[(175 \times 17.5) - (52.5 \times 17.5) + (2 \times 15 \times 40)] - (2 \times 15 \times 40 \times 70) = 0$$
$$h^2 + 63.69h - 1600 = 0, \quad \Rightarrow \quad h = 19.28 \text{ cm}$$

Note from eq(4.34) that an economic design would require $h \approx 32$ cm for the given allowable stresses. Working in cm, the centroid $\bar{y}$ is found from eq(4.40),

$$3155.95\,\bar{y} = (175 \times 17.5 \times 8.75) + (1.78 \times 52.5 \times 18.39), \quad \Rightarrow \quad \bar{y} = 9.026 \text{ cm}$$

The moment of resistance is the lesser of eqs(4.39 a and b)

$$M_c = (7/2)[(175 \times 17.5) + 52.5(19.28 - 17.5)](70 - 9.026) \times 10^3/10^6 = 673.51 \text{ kNm}$$
$$M_t = 125 \times 40(70 - 9.026)10^{-3} = 304.87 \text{ kNm}$$

Selecting M_t ensures that only the steel is fully stressed. The concrete stress is found either from eq(4.39a) with $M_c = M_t$

$$\sigma_c = 2M_t/\{[Bd + b(h - d)](D - \bar{y})\} = (2 \times 304.87 \times 10^6)/(192431 \times 10^3) = 3.17 \text{ MPa}$$

or, by the proportion of the allowable concrete stress σ_{ca},

$$\sigma_c = (M_t/M_c)\sigma_{ca} = (304.87 / 673.51)7 = 3.17 \text{ MPa}$$

4.5.3 Rectangular Section With Double Reinforcement

The stress and strain distributions in the steel (S) and concrete (C) with a line of reinforcement on the tensile (t) and compressive (c) sides are shown in Figs 4.26a-c. Let σ_{St} and σ_{Sc} be the respective tensile and compressive stresses in the steel lines and σ_c be the maximum concrete stress at the compressive surface. The strains in the steel and concrete at the tensile interface are denoted by ε_{St} and ε_{Ct} and at the compressive interface by ε_{Sc} and ε_{Cc} respectively.

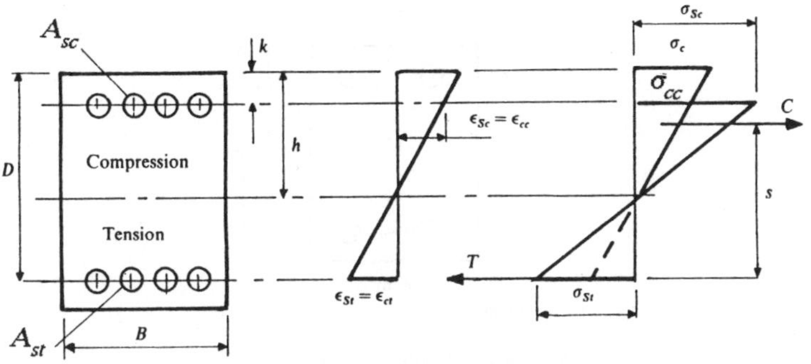

(a) (b) (c)

Figure 4.26 Double reinforcement in rectangular section

Strain compatibility at each interface provides the relationships:

$$\varepsilon_{St} = \varepsilon_{Ct}, \;\Rightarrow\; \sigma_{St} = m\sigma_c(D - h)/h \tag{4.41a}$$
$$\varepsilon_{Sc} = \varepsilon_{Cc}, \;\Rightarrow\; \sigma_{Sc} = m\sigma_c(h - k)/h \tag{4.41b}$$

For an economic section with the same maximum allowable tensile and compressive stresses ($\sigma_{St} = \sigma_{Sc} = \sigma_S$), it follows from eqs(4.41a and b) that $D - h = h - k$. That is, the n.a. passes through the centroid. Assuming the same steel area A_S on the tensile and compressive sides, this area is found by equating horizontal forces C and T, where these are

$$C = \sigma_c Bh/2 + m(h - k)\sigma_c A_S/h - (h - k)\sigma_c A_S/h = \sigma_c[Bh/2 + (m - 1)A_S(h - k)/h] \tag{4.42}$$
$$T = A_S\sigma_S = A_S m\sigma_c(D - h)/h \tag{4.43}$$

Since $M_t = M_c$ for an economic section, the moment of resistance is found from either the concrete or the steel. That is,

$$M_t = Ts = A_S\sigma_S s \tag{4.44a}$$
$$\begin{aligned} M_c = Cs &= (D - k)A_S\sigma_S - (D - k)A_S(h - k)\sigma_c/h + (\sigma_c Bh/2)(D - h + 2h/3) \\ &= (D - k)A_S m(h - k)\sigma_c/h - (D - k)A_S(h - k)\sigma_c/h + (D - h/3)Bh\sigma_c/2 \\ &= \sigma_c(D - k)[(Bh/2)(D - h/3)/(D - k) + (m - 1)A_S(h - k)/h] \end{aligned} \tag{4.44b}$$

The moment arm s is found by combining eqs(4.42) and eq(4.44b). If the steel areas on each side are different, it becomes necessary to solve eqs(4.42-4.44) simultaneously for h and each area, using $C = T$ and $M_t = M_c$ if the section is to remain economic. When the same steel areas are employed with an uneconomic section, the position h of the n.a. follows from equating

(4.42 and 4.43). This gives the following quadratic in h, from which the moment of resistance is the lesser of eqs(4.44a and b):

$$Bh^2 + 2(h - k)(m - 1)A_S - 2mA_S(D - h) = 0 \tag{4.45}$$

Example 4.13 Determine the moment of resistance for the uneconomic reinforced concrete section in Fig. 4.27, given $\sigma_c = 6$ MPa, $\sigma_S = 125$ MPa and $m = 15$. Find the maximum stress in the understressed material.

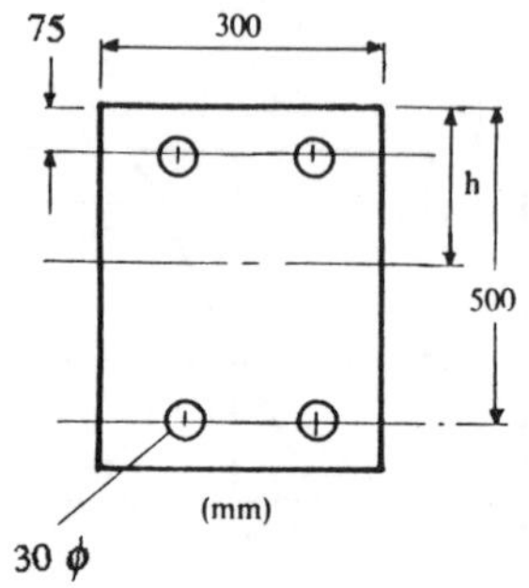

Figure 4.27 Double reinforced section

With $B = 300$ mm, $D = 500$ mm, $k = 75$ mm and $A_S = 2(\pi \times 30^2)/4 = 1413.72$ mm², the n.a. position is found from eq(4.45)

$$h^2 + 273.32h - 80582.1 = 0 \Rightarrow h = 178.39 \text{ mm}$$

For the lever arm s, eq(4.42) gives

$$C = 6[(300 \times 178.39/2) + (15 - 1)1413.72(178.39 - 75) / 178.39] = 229.38 \text{ kN}$$

and from eq(4.44b) the concrete moment is

$$M_c = Cs = 6(500 - 75)[(300 \times 178.39 / 2)(500 - 178.39 / 3)/(300 - 75)$$
$$+ (14 \times 1413.73)(178.39 - 75) / 178.39] = 99.98 \text{ kNm}$$
$$\therefore s = M_c/C = 99.98 \times 10^3 / 229.38 = 435.88 \text{ mm}$$

Note that as σ_c cancels, s is independent of the concrete stress actually achieved. The lesser M is, from eq(4.44a),

$$M_t = 1413.72 \times 125 \times 435.88 = 77.03 \text{ kNm}$$

This is the allowable moment which indicates that the steel is fully stressed in tension while the concrete is understressed. In compression the steel stress is $\sigma_{Sc} = (h - k)\sigma_S/(D - h) =$ 40.19 MPa. The maximum compressive concrete stress is $6(77.03 / 99.98) = 4.62$ MPa. The latter may also be checked from eq(4.34)

$$\sigma_{Cc} = (h - k)\sigma_C/k = \sigma_S h/[m(D - h)] = 125/[15(500 - 178.39)] = 4.62 \text{ MPa}$$

The reader should confirm that the removal of the compressive steel line results in $M = 76.3$

kNm (h = 204.4 mm), which only slightly impairs the bending strength of this section. With an economic design, however, the area of an additional compressive steel line may be chosen to stress the steel fully on both sides and so improve the resistive moment.

4.6 Asymmetric Bending

Bending theory has so far been applied where the moment axes for a given section have coincided with its principal axes. In beam bending, for example, this arises when the n.a. is a horizontal axis of symmetry about which the bending moment has been applied. In the more general case (see Fig. 4.28), the moment axis x in beam bending, or axes x and y in eccentric loading, do not align with the principal axes u and v for the section. It then becomes necessary to resolve the applied moments M_x and M_y into M_u and M_v.

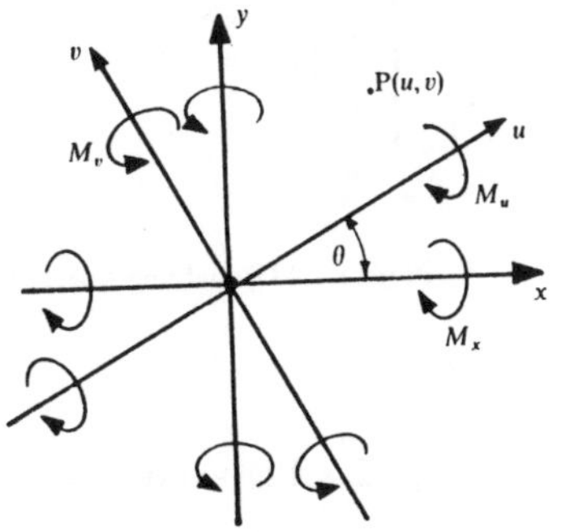

Figure 4.28 Axes of bending

Let M_x and M_y be hogging within the positive x, y quadrant. In the positive u, v quadrant the right hand screw rule gives the net hogging bending moments as a sum of their components

$$M_u = M_x \cos\theta - M_y \sin\theta \tag{4.46a}$$
$$M_v = M_x \sin\theta + M_y \cos\theta \tag{4.46b}$$

where θ is the inclination between positive u and x. Appendix I shows that θ is found from the second moments of area for axes x and y as

$$\tan 2\theta = 2I_{xy}/(I_y - I_x) \tag{4.46c}$$

The bending stress σ at any point P(u,v) is then

$$\sigma = M_u v / I_u + M_v u / I_v \tag{4.47a}$$

where the co-ordinates are

$$u = x\cos\theta + y\sin\theta \tag{4.47b}$$
$$v = y\cos\theta - x\sin\theta \tag{4.47c}$$

and the principal second moments of area are

$$I_{u,v} = ½(I_x + I_y) \pm ½\surd[I_x - I_y)^2 + 4\,I_{xy}{}^2] \tag{4.47d}$$

The determination of the principal second moments of area I_u and I_v from I_x, I_y and I_{xy} has been outlined in Appendix 1. The signs all follow from those positive directions given in Fig. 4.28. Note that M_x and M_y in eq(4.46) are negative when sagging relate to positive x and y.

Example 4.14 Calculate the greatest stresses induced at points A, B and C when the equal-angle section in Fig. 4.29 is mounted as a cantilever 1 m long carrying a downward vertical end load of 1 kN through its centroid g.

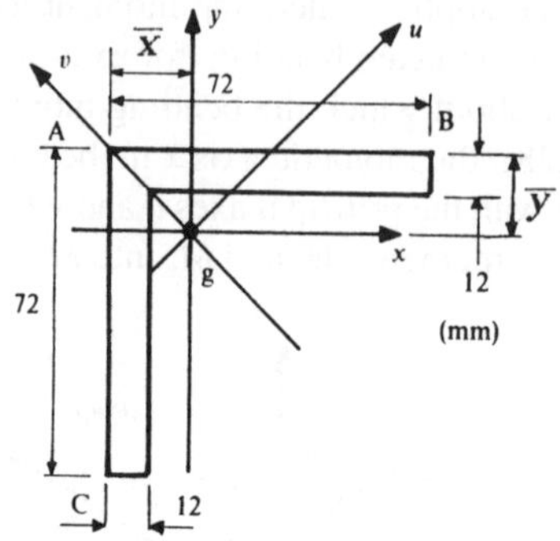

Figure 4.29 Equal-angle section

First moments of area about AB give

$$(72 \times 12 \times 6) + (60 \times 12 \times 42) = [(72 \times 12) + (60 \times 12)]\bar{y}$$
$$\therefore \bar{y} = 22.4 \text{ mm} = \bar{x}$$

The second moments of area are, using $I = \sum bd^3/3$

$$I_x = I_y = 72(22.4)^3/3 - 60(10.4)^3/3 + 12(49.6)^3/3 = 73.46 \times 10^4 \text{ mm}^4$$
$$I_{xy} = (72 \times 12)(+13.6)(+16.4) + (60 \times 12)(-19.6)(-16.4) = 42.41 \times 10^4 \text{ mm}^4$$

Then from eqs(4.4b,c) and (4.47d), $\tan 2\theta = \infty$, $\theta = 45°$ and

$$I_{u,v} = ½(2 \times 73.46 \times 10^4) \pm ½\surd[0 + (4 \times 42.41 \times 10^4)]$$
$$I_u = 31.05 \times 10^4 \text{ mm}^4, \quad I_v = 115.87 \times 10^4 \text{ mm}^4$$

At the fixed end $M_x = WL = (1 \times 1) = +1$ kNm, $M_y = 0$. Equations(4.46a,b) and (4.47a) give

$$M_u = 1 \times \cos 45° - 0 = 0.7071 \text{ kNm}, \quad M_v = 1 \times \sin 45° + 0 = 0.7071 \text{ kNm}$$

$$\sigma = (0.7071 \times 10^6)\, v\,/\,(31.05 \times 10^4) + (0.7071 \times 10^6)\, u\,/\,(115.87 \times 10^4)$$
$$= 2.277\, v + 0.6103\, u \qquad \text{(i)}$$

The stresses follow from eq(i) and the u, v co-ordinates of points A, B and C. They are, from eq(4.47b),

A: $x = -22.4$ mm, $y = 22.4$ mm,
$u = -22.4 \cos 45° + 22.4 \sin 45° = 0$, $v = 22.4 \cos 45° + 22.4 \sin 45° = 31.68$ mm,
$\sigma_A = 2.277\,(31.68) + 0.6103\,(0) = 72.13$ MPa

B: $x = 49.6$ mm, $y = 22.4$ mm, $\Rightarrow u = 50.92$ mm, $v = -19.24$ mm
$\sigma_B = 2.277(-19.24) + 0.6103(50.92) = -12.73$ MPa

C: $x = -22.4$ mm, $y = -49.6$ mm, $\Rightarrow u = -50.92$ mm, $v = -19.24$ mm
$\sigma_C = 2.277(-19.24) + 0.6103(-50.92) = -74.87$ MPa

Example 4.15 The cross-section in Fig. 4.29 is mounted as a vertical column to support a compressive force of 50 kN at the centre of its vertical web. Determine the stresses at points A, B and C.

The x, y co-ordinates of the force point are (-16.4, -13.6). Then, the corresponding e_u, e_v eccentricities follow from eqs(4.47b,c)

$$e_u = -16.4/\surd 2 - 13.6/\surd 2 = -21.22 \text{ mm}$$
$$e_v = -13.6/\surd 2 + 16.4/\surd 2 = 1.98 \text{ mm}$$

The bending moments about u and v are simply the product of the force and its respective eccentricity. For the positive u, v quadrant in Fig. 4.29,

$$M_u = Fe_v = -50 \times 1.98 = -99 \text{ Nm (sagging)},$$
$$M_v = \text{F}e_u = -50 \times (-21.22) = 1061 \text{ Nm (hogging)}$$

The stresses are, from eq(4.47a),

$$\sigma = (-99 \times 10^3)v / (31.05 \times 10^4) + (1061 \times 10^3)\, u / (115.87 \times 10^4)$$
$$= -0.319\, v + 0.916\, u \text{ (MPa)}$$

At A (0, 31.68), $\sigma_A = -10.11$ MPa, at B (50.92,$-$ 19.24), $\sigma_B = 52.78$ MPa and at C ($-$50.92,$-$ 19.24), $\sigma_C = -40.51$ MPa

4.6.1 Equivalent Moments

In the above examples it was necessary to evaluate I_u, I_v, θ, u and v separately. This may be avoided, however, if θ is first eliminated between eqs(4.46a-c) and the resulting expressions for u and v are substituted, together with eq(4.47d), into eq(4.47a). This results in a bending stress expression in terms of the x, y quantities

$$\sigma = \bar{M}_x\, y / I_x + \bar{M}_y\, x / I_y \tag{4.48a}$$

where the equivalent moments are

$$\bar{M}_x = [M_x - M_y(I_{xy}/I_y)]/[1 - I_{xy}^2/(I_x I_y)] \tag{4.48b}$$
$$\bar{M}_y = [M_y - M_x(I_{xy}/I_x)]/[1 - I_{xy}^2/(I_x I_y)] \tag{4.48c}$$

Example 4.16 A beam with the asymmetric channel section in Fig. 4.30 has bending moments of $M_x = 600$ Nm (hogging) $M_y = 400$ Nm (sagging) applied along positive x and y as shown. Determine the maximum tensile and compressive stresses and the inclination of the true n.a. of bending

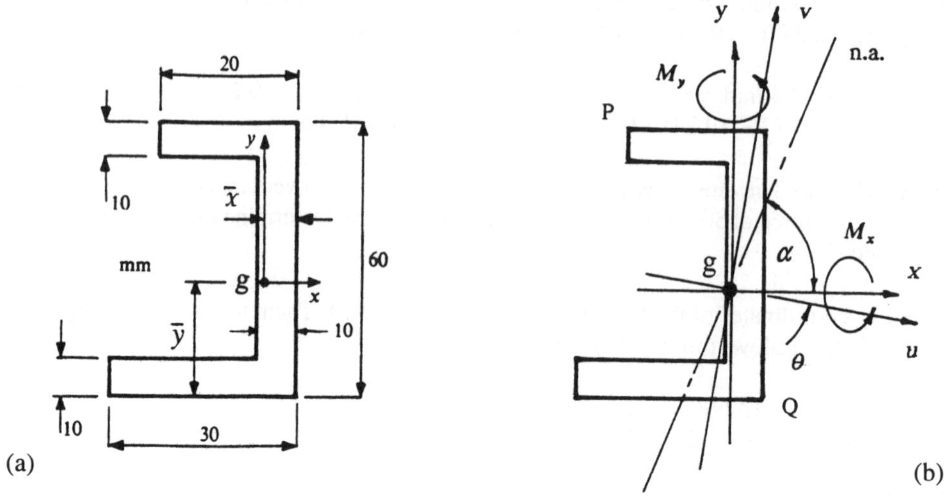

Figure 4.30 Asymmetric channel

Taking first moments about axes x, y to find the centroid position in Fig. 4.30:

$$(30 \times 10 \times 5) + (40 \times 10 \times 30) + (20 \times 10 \times 55) = 900\bar{y}$$
$$\therefore \bar{y} = 27.22 \text{ mm}$$

$$(60 \times 10 \times 5) + (10 \times 10 \times 15) + (20 \times 10 \times 20) = 900\bar{x}$$
$$\therefore \bar{x} = 9.45 \text{ mm}$$

$$I_x = 20(32.78)^3/3 - 10(22.78)^3/3 + 30(27.22)^3/3 - 20(17.22)^3/3$$
$$= 36.31 \times 10^4 \text{ mm}^4$$

$$I_y = 60(9.45)^3/3 + 10(20.55)^3/3 + 10\,(10.55)^3/3 + 40(0.55)^3/3$$
$$= 4.97 \times 10^4 \text{mm}^4$$

$$I_{xy} = 600(+\,2.78)(+\,4.45) + 100(+\,27.78)(-\,5.55) + 200(-\,22.22)(-\,10.55)$$
$$= 3.89 \times 10^4 \text{ mm}$$

Analytically, from eqs(4.46c) and (4.47d)

$$\tan 2\theta = 2(+\,3.89)/\,(4.97 - 36.31)$$
$$\therefore\ \theta = -\,6.97°$$

$$I_u, I_v = ½\,(36.31 + 4.97) \pm ½\,\sqrt{[(36.31 - 4.97)^2 + 4\,(3.89)^2]} \times 10^4$$
$$\therefore I_v = 4.49 \times 10^4 \text{ mm}^4,\ I_u = 36.79 \times 10^4 \text{ mm}^4$$

Method 1 Referring to x, y co-ordinates, the equivalent moments are, from eq(4.48b,c),

$$\bar{M}_x = [600 - (-\,400)\,(3.89 / 4.97)] / [1 - 3.89^2/(36.31 \times 4.97)] = 996.65 \text{ Nm}$$
$$\bar{M}_y = [-\,400 - 600\,(3.89 / 36.31)] / [1 - 3.89^2/(36.31 \times 4.97)] = -\,506.77 \text{ Nm}$$

The maximum tensile stress will occur at the corner P(− 10.55, 32.78) for the second x, y quadrant, where the actions of M_x and M_y are both hogging (Fig. 4.30b). From eq(4.48a)

$$\sigma_P = (996.65 \times 10^3 \times 32.78) / (36.31 \times 10^4) + (-506.77 \times 10^3)(-10.55) / (4.97 \times 10^4)$$
$$= 89.98 + 107.57 = 197.55 \text{ MPa}$$

The maximum compressive stress occurs at the corner Q (9.45, − 27.22) for the fourth x, y quadrant, where the actions of M_x and M_y in Fig. 4.30b are both sagging.

$$\sigma_Q = (996.65 \times 10^3)(-27.22)/(36.31 \times 10^4) + (-506.77 \times 10^3)(9.45)/(4.97 \times 10^4)$$
$$= -74.71 - 96.35 = -171.07 \text{ MPa}$$

Along the n.a. it follows from eq(4.48a) that

$$\bar{M}_x\, y / I_x + \bar{M}_y x / I_y = 0$$
$$(996.65 \times 10^3) y / (36.31 \times 10^4) + (-506.77 \times 10^3)\, x / (4.97 \times 10^4) = 0$$
$$2.745\, y - 10.196\, x = 0$$
$$\therefore\ \alpha = \tan^{-1}(y/x) = \tan^{-1}(10.196 / 2.745) = 74.93° \text{ (anti-clockwise, see Fig. 4.30b)}$$

Method 2 Referring the moments u, v co-ordinates, eqs (4.46a,b) give

$$M_u = 600 \cos(-6.97°) - (-400) \sin(-6.97°) = 547.03 \text{ Nm}$$
$$M_v = 600 \sin(-6.97°) - 400 \cos(-6.97°) = -469.85 \text{ Nm}$$

The co-ordinates P(u, v) follow from eqs(4.47b,c) as

$$u = -10.55 \cos(-6.97°) + 32.78 \sin(-6.97°) = -14.45 \text{ mm}$$
$$v = 32.78 \cos(-6.97°) + 10.55 \sin(-6.97°) = 31.26 \text{ mm}$$

Applying eq(4.47a)

$$\sigma_P = (547.03 \times 10^3 \times 31.26) / (36.79 \times 10^4) + (-469.85 \times 10^3)(-14.47) / (4.49 \times 10^4)$$
$$= 46.48 + 151.21 = 197.69 \text{ MPa}$$

Similarly, for point Q eqs (4.47a,b) yield its co-ordinates as

$$u = 9.45 \cos(-6.97°) - 27.22 \sin(-6.97°) = 12.68 \text{ mm}$$
$$v = -27.22 \cos(-6.97°) - 9.45 \sin(-6.97°) = -25.87 \text{ mm}$$
$$\sigma_Q = (547.03 \times 10^3)(-25.87) / (36.79 \times 10^4) + (-469.85 \times 10^3)(12.68) / (4.49 \times 10^4)$$
$$= -38.47 - 132.69 = -171.15 \text{ MPa}$$

4.6.2 Idealised Sections

Thin-walled beam cross-sections may be idealised into concentrated areas, known as booms, inter-connected with thin webs. Booms carry the bending stresses and webs carry shear stress only (see chapter 7). The longitudinal bending stresses in the booms are found from simple bending theory where, in the case of an asymmetric section, eqs(4.48a-c) again applies.

Example 4.17 The idealised section in Fig. 4.31 is subjected to bending moments $M_x = 10$ kNm (sagging), $M_y = 5$ kNm (hogging) w.r.t. the first x, y quadrant. If the areas of booms A and C are 500 mm^2 and the areas of booms B and D are 200 mm^2, find the bending stress in each boom. Assume that the webs do not contribute to the total x, y moments of area.

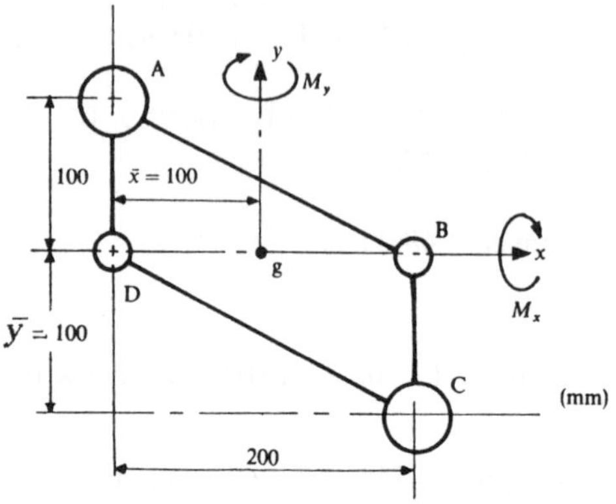

Figure 4.31 Section idealised with boom areas

To locate the centroid g, take first moments of area about a horizontal through C:

$$2(200 \times 100) + (200 \times 500) = 2(200 + 500)\,\bar{y}, \quad \Rightarrow \quad \bar{y} = 100\text{mm}$$

and about the vertical passing through A and D:

$$200(500 + 200) = 2(200 + 500)\,\bar{x}, \quad \Rightarrow \quad \bar{x} = 100\text{mm}$$

To calculate the second moments at g use the simplified form $I = Ah^2$,

$$I_x = 2 \times 500 \times 100^2 = 10 \times 10^6 \text{ mm}^4,$$
$$I_y = 2[(500 \times 100^2) + (200 \times 100^2)] = 14 \times 10^6 \text{ mm}^4$$
$$I_{xy} = 500(-100)(100) + 500(100)(-100) = -10 \times 10^6 \text{ mm}^4$$

The equivalent moments are, from eq(4.48b,c),

$$\bar{M}_x = [-10 - 5(-10/14)]/[1 - (-10)^2/(10 \times 14)] = -22.5 \text{ kNm}$$
$$\bar{M}_y = [5 - (-10)(-10/10)]/[1 - (-10)^2/(10 \times 14)] = -17.5 \text{ kNm}$$

The stresses are then given by eq(4.48a), i.e. $\sigma = -2.25\,y - 1.25\,x$. Then,

Boom A (− 100, 100): $\sigma_A = -225 + 125 = -100$ MPa; Boom B (100, 0): $\sigma_B = -125$ MPa;
Boom C (100,− 100): $\sigma_C = 225 - 125 = 100$ MPa; Boom D (− 100, 0): $\sigma_D = 125$ MPa.

4.7 Bending of Circular Plates

Consider a thin, initially flat, circular plate of outer radius r_o and uniform thickness t (Fig. 4.32a), subjected to a given axi-symmetric loading applied normally to its faces. For a given radius r, the radial σ_r and hoop σ_θ stresses will be constant in the same plane, i.e. at a perpendicular distance z from the neutral middle plane. The neutral plane lies mid-way between top and bottom surfaces of the plate and is unstressed.

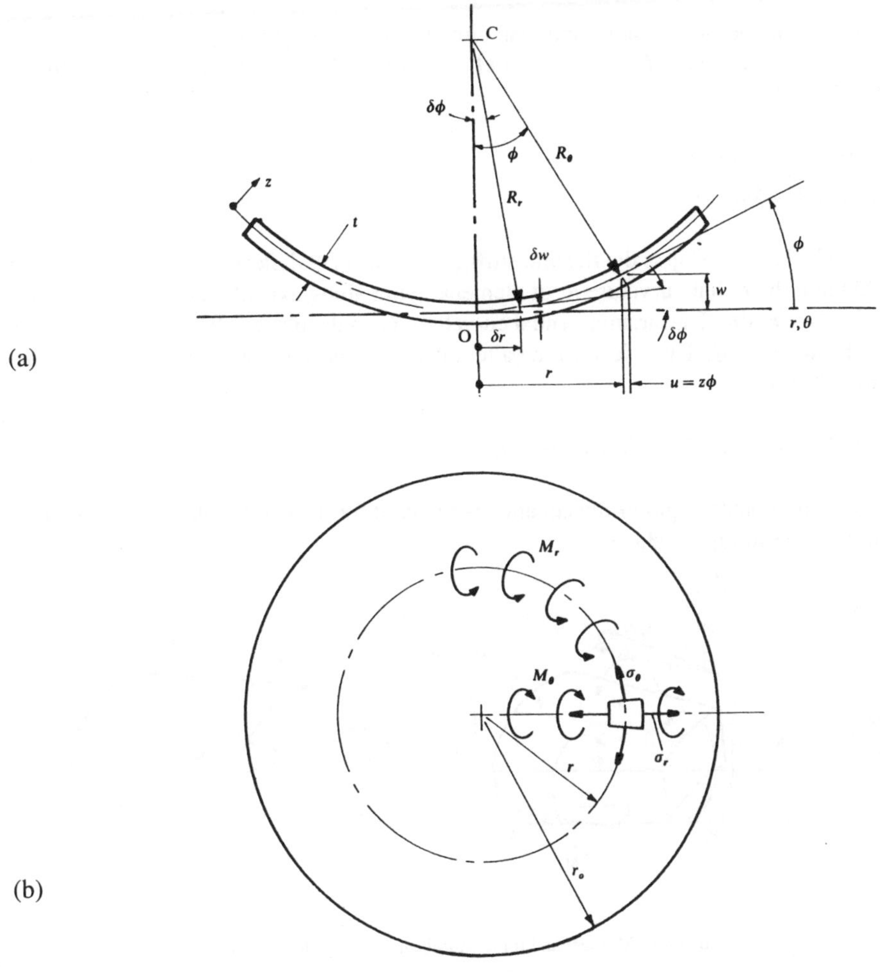

Figure 4.32 Plate Bending

When the vertical deflection w of the middle plane is small compared to the plate thickness, the following simplified theory enables w, σ_r and σ_θ to be found. If u is the radial displacement, distance z from the neutral plane, it follows from Fig. 4.32a that $u = z\phi$ in which $\phi = \mathrm{d}w/\mathrm{d}r$, relative to the plate centre where $\phi = 0$ and $w = 0$. Conventionally, the deflection w is *positive upwards* from the plate centre. For any point (r, z) in the plane of the plate (see Fig. 4.32b), the biaxial constitutive relations are

$$\varepsilon_r = \mathrm{d}u/\mathrm{d}r = (\sigma_r - \nu\sigma_\theta)/E$$
$$\varepsilon_\theta = u/r = (\sigma_\theta - \nu\sigma_r)/E$$

Inverting these, the stresses become

$$\sigma_r = E(\varepsilon_r + \nu\,\varepsilon_\theta)/(1-\nu^2) = E(\mathrm{d}u/\mathrm{d}r + \nu\, u/r)/(1-\nu^2)$$
$$= Ez\,(\mathrm{d}\phi/\mathrm{d}\,r + \nu\,\phi/r)/(1-\nu^2) \quad (4.49a)$$
$$\sigma_\theta = E(\varepsilon_\theta + \nu\,\varepsilon_r)/(1-\nu^2) = E(u/r + \nu\,\mathrm{d}u/\mathrm{d}r)/(1-\nu^2)$$
$$= Ez\,(\phi/r + \nu\,\mathrm{d}\phi/\mathrm{d}\,r)/(1-\nu^2) \quad (4.49b)$$

The respective moments/unit radial and /unit circumferential lengths (see Fig. 4.32b), which produce these stresses are $M_r = \sigma_r I / z$ and $M_\theta = \sigma_\theta I / z$, where $I = 1 \times t^3/12$. Substituting from eqs(4.49a,b),

$$M_r = D\,(\mathrm{d}\phi/\mathrm{d}\,r + \nu\phi/r) \quad (4.50a)$$
$$M_\theta = D\,(\phi/r + \nu\,\mathrm{d}\phi/\mathrm{d}\,r) \quad (4.50b)$$

where $D = Et^3/[12(1-\nu^2)]$, is the flexural stiffness. For a small deflections w, it follows from Fig. 4.32a that the radial curvature R_r of elements lying along axis Or and in the plane zOr is given by $\mathrm{d}r \simeq R_r\mathrm{d}\phi$. The circumferential curvature R_θ in the plane $zO\theta$ for these elements is found from $r \simeq R_\theta\,\phi$. These lead to the alternative expressions for the bending moments from eqs(4.50a,b)

$$M_r = D(1/R_r + \nu/R_\theta), \quad M_\theta = D(1/R_\theta + \nu/R_r)$$

Finally, a radial moment equilibrium equation is required for the central plane of the annular element $r\delta\theta \times \delta r$ in Fig. 4.33a.

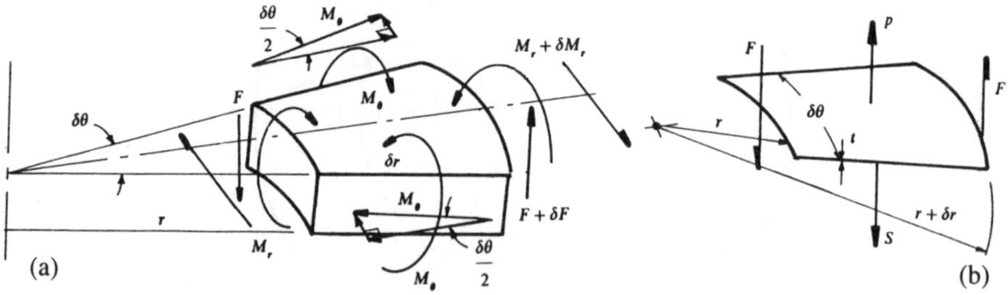

Figure 4.33 Moment and force equilibrium for a plate element

For the axi-symmetric loading shown, the vertical shear force F varies with r and not θ. Thus, when $F = F(r)$ is a downward shear force/unit circumference at the inner annular radius r, moment equilibrium gives

$$(M_r + \delta M_r)(r + \delta r)\delta\theta - M_r r\delta\theta - 2M_\theta\,\delta r \sin(\delta\theta/2) + (F + \delta F)(r + \delta r)^2\delta\theta - Fr^2\delta\,\theta = 0$$

in which the right hand screw rule determines the indicated moment vector directions and the effect of the normal loading over δr is ignored. In the limit this becomes

$$M_r + r\,\mathrm{d}M_r/\mathrm{d}\,r - M_\theta + Fr = 0 \quad (4.51)$$

Substituting eqs(4.50a,b) into eq(4.51) leads to

$$(1/r)(\mathrm{d}\phi/\mathrm{d}r + v\phi/r + r\,\mathrm{d}^2\phi/\mathrm{d}r^2 - v\phi/r + v\mathrm{d}\phi/\mathrm{d}r - \phi/r - v\mathrm{d}\phi/\mathrm{d}r) = -F/D$$

$$\therefore \quad (\mathrm{d}/\mathrm{d}r)[(1/r)\times \mathrm{d}(r\phi)/\mathrm{d}r] = -F/D \tag{4.52a}$$

$$\text{or,} \quad (\mathrm{d}/\mathrm{d}r)[(1/r)\times \mathrm{d}(r\,\mathrm{d}w/\mathrm{d}r)/\mathrm{d}r] = -F/D \tag{4.52b}$$

The vertical deflection w is found from successive integration of eq(4.52a) once the function $F(r)$ has been established from vertical force equilibrium. That is, if $p = p(r)$ is a net upward pressure applied normally to the surfaces of the elemental ring in Fig. 4.33b, for which S is the self-weight/unit volume. Then, in general,

$$\delta\theta\,(r+\delta r)(F+\delta F) + (p\times r\,\delta\theta\,\delta r) = r\delta\theta\,F + r\delta\theta\,\delta r\,t\,S$$

$$\mathrm{d}(Fr)/\mathrm{d}r + pr - Str = 0 \tag{4.53}$$

where F is found from integration. Equations (4.52b) and (4.53) will now be employed to derive expressions for the maximum deflection and the maximum bending stresses in common cases.

4.7.1 Central Concentrated Force - Edges Simply Supported

In the absence of self-weight and normal pressure, eq(4.53) gives Fr = constant. F now acts upwards around the sides of the plate opposing the central force P. Applying vertical equilibrium for all radii r in Fig. 4.34 gives $2\pi rF = P$, from which the shear force is $F = P/(2\pi r)$.

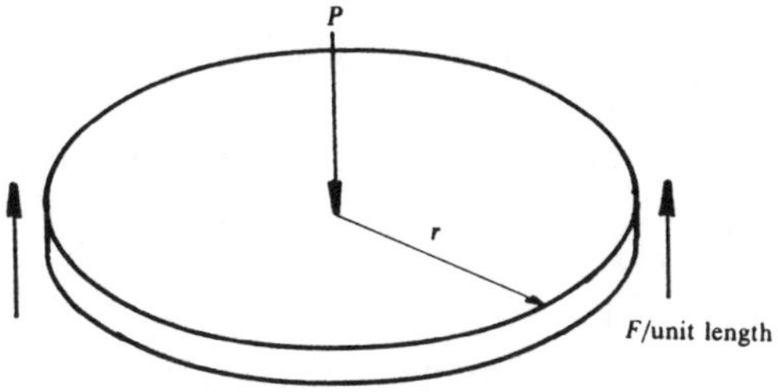

Figure 4.34 Concentrated Loading

From the successive integration of eq(4.52b),

$$(\mathrm{d}/\mathrm{d}r)[(1/r)\,\mathrm{d}(r\,\mathrm{d}w/\mathrm{d}r)/\mathrm{d}r] = -P/(2\pi rD)$$

$$[(1/r)\,\mathrm{d}(r\,\mathrm{d}w/\mathrm{d}r)/\mathrm{d}r] = -[P/(2\pi D)]\ln r + C_1$$

$$\mathrm{d}(r\,\mathrm{d}w/\mathrm{d}r)/\mathrm{d}r = -[P/(2\pi D)]r\ln r + C_1 r$$

$$(r\,\mathrm{d}w/\mathrm{d}r) = -[Pr^2/(8\pi D)](2\ln r - 1) + C_1 r^2/2 + C_2$$

$$\mathrm{d}w/\mathrm{d}r = -[Pr/(8\pi D)](2\ln r - 1) + C_1 r/2 + C_2/r$$

$$w = -[Pr^2/(8\pi D)](\ln r - 1) + C_1 r^2/4 + C_2\ln r + C_3 \tag{4.54}$$

The following conditions determine constants C_i (i = 1, 2, 3): $\mathrm{d}w/\mathrm{d}r = 0$ for $r = 0$, $\Rightarrow C_2 = 0$, $w = 0$ for $r = 0$, $\Rightarrow C_3 = 0$. Also $M_r = 0$ for $r = r_o$, when, from eq(4.50a) and eq(4.54),

$\mathrm{d}\phi/\mathrm{d}r + \nu\,\phi/r = \mathrm{d}^2 w/\mathrm{d}r^2 + (\nu/r)\,\mathrm{d}w/\mathrm{d}r = 0$

$-\,[P/(8\pi D)](1 + 2\ln r) + C_1/2 + (\nu/r)\{-\,[Pr/(8\pi D)](2\ln r - 1) + C_1 r/2\} = 0$

$C_1 = [P/(4\pi D)][2\ln r_o + (1-\nu)/(1+\nu)]$

Hence eq(4.54) becomes

$$w = -\,[Pr^2/(8\pi D)](\ln r - 1) + [Pr^2/(16\pi D)][2\ln r_o + (1-\nu)/(1+\nu)] \qquad (4.55)$$

Equation(4.55) supplies the maximum central deflection for $r = r_o$

$$w_{\max} = [Pr_o^2/(16\pi D)][2 + (1-\nu)/(1+\nu)] = [3Pr_o^2/(4\pi Et^3)](1-\nu)(3+\nu)$$

From eqs(4.49a,b),

$$\sigma_r = Ez\,[\mathrm{d}^2w/\mathrm{d}r^2 + (\nu/r)\,\mathrm{d}w/\mathrm{d}r]/(1-\nu^2)$$
$$\sigma_\theta = Ez\,[(1/r)\,\mathrm{d}w/\mathrm{d}r + \nu\mathrm{d}^2w/\mathrm{d}r^2]/(1-\nu^2)$$

Substituting from eq(4.55), with $z = t/2$ for the compressive side,

$$\sigma_r = -\,[3P(1+\nu)/(2\pi t^2)]\ln(r_o/r)$$
$$\sigma_\theta = -\,[3P(1+\nu)/(2\pi t^2)][\ln(r_o/r) + (1-\nu)/(1+\nu)]$$

These apply only to finite r values and so they avoid the infinite stress values at the plate centre. In practice, the stresses will be greatest within the small surface area on which P acts.

4.7.2 Central Concentrated Load - Edges Clamped

Equation (4.54) remains valid for the following boundary conditions:

$w = 0$ for $r = 0$, $\Rightarrow C_3 = 0$

$\mathrm{d}w/\mathrm{d}r = 0$ for $r = 0$, $\Rightarrow C_2 = 0$

$\mathrm{d}w/\mathrm{d}r = 0$ for $r = r_o$, $\Rightarrow C_1 = [P/(4\pi D)](2\ln r_o - 1)$

These give

$$w = [Pr^2/(16\pi D)][1 - 2\ln(r/r_o)] \qquad (4.56)$$

Equation (4.56) is a maximum at the plate centre for $r = r_o$

$$w_{\max} = Pr_o^2/(16\pi D) = [3Pr_o^2/(4\pi Et^3)](1-\nu^2)$$

Differentiating eq(4.56) and substituting into eqs(4.49a,b) gives the maximum tensile and compressive stresses for the top surface at the fixed edge, i.e. $r = r_o$ and $z = \pm t/2$,

$$\sigma_r = PEt/[(8\pi D)(1-\nu^2)] = 3P/(2\pi t^2)$$
$$\sigma_\theta = \nu\,PEt/[(8\pi D)(1-\nu^2)] = 3\nu P/(2\pi t^2)$$

4.7.3 Uniformly Distributed Loading - Edges Simply Supported

Let the total distributed loading, W / unit area, act downward to include self-weight. Substituting $p = -W$ = constant and $S = 0$ in eq(4.53) gives $F = Wr/2$. Alternatively, from Fig. 4.35, $2\pi Fr = W\pi r^2$, $\Rightarrow$ $F = Wr/2$.

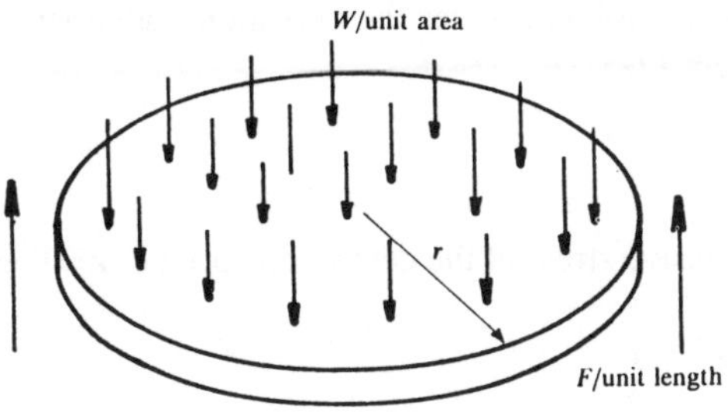

Figure 4.35 Distributed Loading

Substituting F into eq(4.52b) with successive integration leads to

$$w = -Wr^4/(64D) + C_1 r^2/4 + C_2 \ln r + C_3 \tag{4.57}$$

The following conditions apply: $w = 0$ for $r = 0$, $\Rightarrow C_3 = 0$, $dw/dr = 0$ for $r = 0$, $\Rightarrow C_2 = 0$. Also $M_r = 0$ for $r = r_o$, when, from eqs(4.50a) and (4.57),

$$W r_o^2 (3 + \nu)/(16D) + C_1 (1 + \nu)/2 = 0$$
$$\therefore C_1 = W r_o^2 (3 + \nu)/[8D(1 + \nu)]$$

Substituting for C_1 in eq(4.57), the general displacement expression is found

$$w = [W r^2/(64D)][2r_o^2 (3 + \nu)/(1 + \nu) - r^2] \tag{4.58}$$

and for $r = r_o$,

$$w_{max} = [W r_o^4/(64D)](5 + \nu)/(1 + \nu) = [3W r_o^4/(16E t^3)](5 + \nu)(1 - \nu)$$

Substituting eq(4.58) into eqs(4.49a,b) and using $\phi = dw/dr$ gives the maximum stresses at the centre, i.e. $r = 0$ and $z = \pm t/2$

$$\sigma_r = \sigma_\theta = 3W r_o^2 (3 + \nu)/(8t^2)$$

4.7.4 Uniformly Distributed Loading - Edges Fixed

Equation (4.57) remains valid and again $C_2 = C_3 = 0$. The third condition is

$$dw/dr = 0 \text{ for } r = r_o, \quad \Rightarrow \quad C_1 = W r_o^2/(8D)$$

The general deflection expression is

$$w = [W r^2/(64D)](2r_o^2 - r^2) \tag{4.59}$$

and for $r = r_o$,

$$w_{max} = W r_o^4/(64D) = [3W r_o^4/(16E t^3)](1 - \nu^2)$$

Substituting eq(4.59) into eq(4.49a) gives the maximum radial stress for the top surface of the fixed edge ($r = r_o$, $z = t/2$),

$$\sigma_r = 3W r_o^2 / (4 t^2)$$

and a maximum compressive stress at the centre top surface ($r = 0$, $z = t/2$)

$$\sigma_\theta = - 3W r_o^2(1 + \nu) / (8 t^2)$$

The expressions derived in sections 4.8.1-4.8.4 may be superimposed when a plate is subjected to combined loading. For example, the sum of the stresses and displacements for cases 1 and 3 apply to a simply supported plate with concentrated and distributed loading. With a fixed plate edge under similar combined loading, cases 2 and 4 may be superimposed.

4.7.5 Non-Standard Cases

Many plate problems require separate treatment for their particular geometry, loading and boundary conditions. The following examples illustrate how eqs(4.49-4.53) are applied.

Example 4.18 An annular plate, inner radius r_i = 38 mm and outer radius r_o =165 mm is loaded by a uniform pressure p (N/mm^2) on its top surface whilst being clamped around its outer edge and free around its inner edge (see Fig. 4.36a). Determine the distribution of M_r and M_θ throughout the plate. Take ν = 0.27.

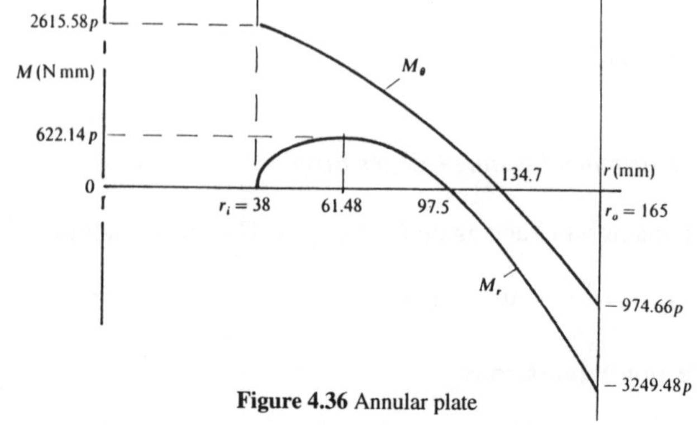

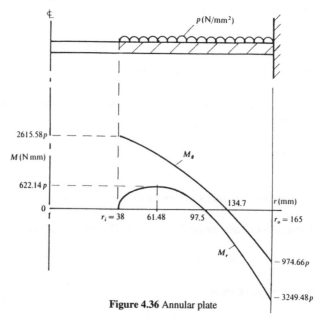

Figure 4.36 Annular plate

For any radius $38 \le r \le 165$ (mm), vertical equilibrium gives

$$2\pi r F = \pi p(r^2 - r_i^2)$$
$$F = (p/2)(r - r_i^2/ r)$$

Substituting into eq(4.52b) and integrating leads to

$$(\mathrm{d}/\mathrm{d}r)[(1/r)\,\mathrm{d}\,(r\phi)/\,\mathrm{d}r] = -\,[\,p/(2D)](r - r_i^2/\,r)$$
$$\phi = [\,p/(2D)][(r_i^2 r/4)\,(\,\ln r^2 - 1) - r^3/8\,] + C_1 r/2 + C_2/\,r \qquad \text{(i)}$$
$$\mathrm{d}\phi/\mathrm{d}r = (p/2D)[(r_i^2/4)\,(\ln r^2 - 1) - 3r^2/8 + r_i^2/2\,] + C_1/2 - C_2/r^2 \qquad \text{(ii)}$$

Substituting eqs(i) and (ii) into eq(4.50a),

$$M_r = \mathrm{D}\Big\{[\,p/(2D)][(r_i^2/4)\,(\ln r^2 - 1) - 3r^2/8 + r_i^2/2\,] + C_1/2 - C_2/\,r^2$$
$$+ (\nu/r)\{[\,p/(2D\,)]\,[(r_i^2 r/4)\,(\ln r^2 - 1) - r^3/8\,] + C_1 r/2 + C_2/r\,\}\Big\} \qquad \text{(iii)}$$

The following conditions apply to eqs(i) and (iii):

$\phi = \mathrm{d}w/\mathrm{d}r = 0$ for $r_o = 165$ mm,

$$[\,p/(2D)]\,\{(38^2 \times 165/4)\,[\ln (165)^2 - 1] - (165)^3/8\} + 165C_1/2 + C_2/165 = 0$$
$$-\,6404.67\,p/D + 82.5C_1 + C_2/165 = 0 \qquad \text{(iv)}$$

$M_r = 0$ for $r_i = 38$ mm,

$$1504.37p/D + 0.635C_1 - (5.0554 \times 10^{-4})C_2 = 0 \qquad \text{(v)}$$

Solving eqs(iv) and (v) gives $C_1 = -\,129.08\,p/D$ and $C_2 = (2813.64 \times 10^3)p/D$. Substituting C_1 and C_2 into eq(iii) leads to the radial moment expression for all r,

$$M_r = p\,[229.24\,(\ln r^2 - 1) - 0.2044\,r^2 - (2.054 \times 10^6)/r^2 + 279.03] \qquad \text{(vi)}$$

This gives the distribution in Fig. 4.36b where $M_r = 0$ for $r_i = 38$ mm (as a check), $M_r = 0$ for $r = 97.5$ mm (by trial) and $M_r = -\,3249.48\,p$ for $r_o = 165$ mm. A stationary value of M_r in eq(vi) occurs for

$$\mathrm{d}M_r/\mathrm{d}r = 0 = (229.24 \times 2\,/\,r) - 0.4088\,r + 2\,(2.054 \times 10^6)\,/\,r^3$$
$$0.4088\,r^4 - 458.48\,r^2 - (4.108 \times 10^6) = 0$$

from which $r = 61.48$ mm where, from eq(vi), $M_r = 622.14\,p$.

The circumferential moment expression is found by substituting eqs(i) and (ii) into eq(4.50b),

$$M_\theta = (r_i^2 p/8)(1 + \nu)(\ln r^2 - 1) - (\,pr^2/16)(1 + 3\nu) + D(1 + \nu)C_1/2 + D(1 - \nu)C_2/r^2$$
$$= p\,[229.24\,(\ln r^2 - 1) - 0.1131r^2 + (2.054 \times 10^6)\,/\,r^2 - 81.97\,]$$

This gives the distribution in Fig. 4.36b where $M_\theta = 2615.58p$ for $r_i = 38$ mm, $M_\theta = -\,974.66p$ for $r_o = 165$ mm and $M_\theta = 0$ for $r = 134.7$ mm.

Example 4.19 A solid circular steel plate, 300 mm diameter and 12 mm thick, is clamped around its outer edge and loaded by a force ring P = 20 kN at a 50 mm radius (Fig. 4. 37). Determine the central plate deflection. Take E = 207 GPa and ν = 0.27 for steel.

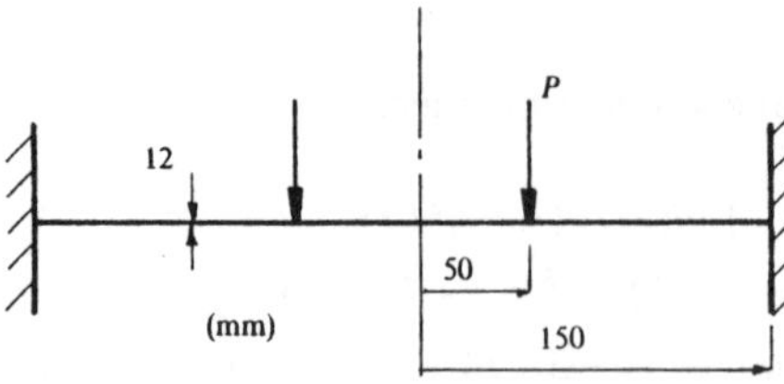

Figure 4.37 Clamped ring

Working in N and mm the plate is considered in two sections:

(I) $0 \le r \le 50$ mm, for which $F = 0$. Integrating eq(4.52a) gives

$$\phi = A_1 r/2 + A_2/r \qquad \text{(i)}$$

Since $\phi = \mathrm{d}w/\mathrm{d}r = 0$ for $r = 0$, this gives $A_2 = 0$ in eq(i).

(II) $50 \le r \le 150$ mm, for which the vertical equilibrium equation is

$$2\pi r F = P, \;\Rightarrow\; F = P/(2\pi r)$$

Equation (4.52b) gives

$$\phi = -\,[Pr/(8\pi D)]\,(2\ln r - 1) + B_1 r/2 + B_2/r \qquad \text{(ii)}$$

Now $\phi = 0$ for $r = 150$ mm in eq(ii), which gives

$$0 = -\,(1.0768\times 10^{6})/D + 75B_1 + (6.667\times 10^{-3})\,B_2 \qquad \text{(iii)}$$

Moreover, eqs(i) and (ii) may be equated for $r = 50$ mm

$$25A_1 = -\,(0.2715\times 10^{6})/D + 25B_1 + (20\times 10^{-3})\,B_2 \qquad \text{(iv)}$$

Also M_r in eq(4.50a) must be the same for $r = 50$ mm. This leads to

$$A_1(1+\nu)/2 = -\,[P/(8\pi D)][2(1+\nu)\ln r + (1-\nu)] + B_1(1+\nu)/2 - (B_2/r^2)(1-\nu)$$

$$0.635A_1 = -\,(8.488\times 10^{3})/D + 0.635B_1 - (0.292\times 10^{-3})\,B_2 \qquad \text{(v)}$$

Solving eqs(iii, iv and v) provides the constants

$$A_1 = (2.082\times 10^{3})/D,\;\; B_1 = (14.534\times 10^{3})/D \text{ and } B_2 = -\,(1990.7\times 10^{3})/D$$

The central deflection is found from matching the deflections at radius $r = 50$ mm. Firstly, integrating eq(i)

$$w = A_1 r^2/4 + A_3$$

but as $w = 0$ for $r = 0$ then $A_3 = 0$. For $r = 50$ mm

$$w = (2.082 \times 10^3)(50)^2/(4D) = (1.3013 \times 10^6)/D \qquad \text{(vi)}$$

and integrating eq(ii)

$$w = -[Pr^2/(8\pi D)](\ln r - 1) + B_1 r^2/4 + B_2 \ln r + B_3 \qquad \text{(vii)}$$

which becomes, for $r = 50$ mm,

$$w = -(4.4972 \times 10^6)/D + B_3 \qquad \text{(viii)}$$

Equating (vi) and (viii) gives $B_3 = (5.7985 \times 10^6)/D$. Finally, putting $r = 150$ mm in eq(vii) gives the central deflection

$$w = (5.767 \times 10^6)/D$$
$$= (5.767 \times 10^6) \times 12(1 - 0.27^2)/(207 \times 10^3 \times 12^3) = 0.179 \text{ mm}$$

4.8 Rectangular Plates

The foregoing theory of circular plates and the theory of elliptical plates are particular cases of the general theory of rectangular plates. The new convention used is summarised from Fig. 4.38.

(a)

(b)

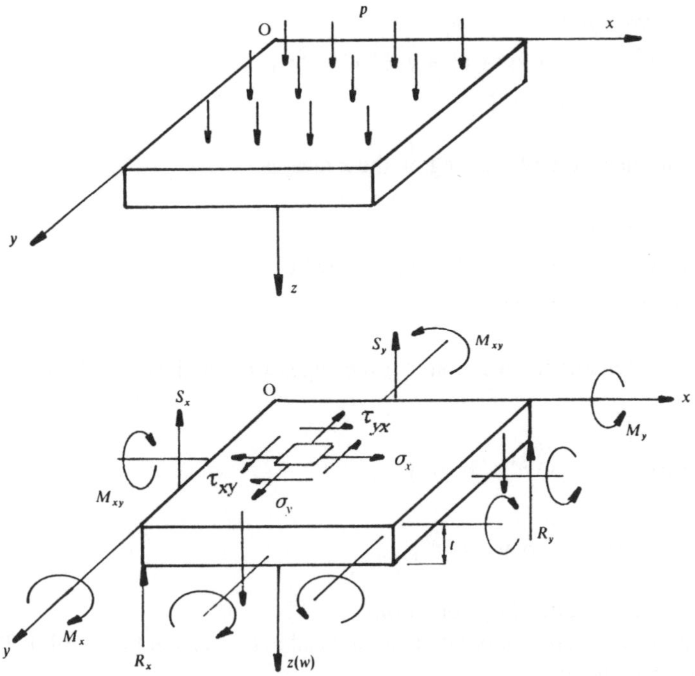

Figure 4.38 Bending and deflection of a rectangular plate

When a uniform rectangular plate, lying in the x, y, z co-ordinate frame in Fig. 4.38a, is loaded with a normal pressure p over its top surface, the *positive downward* displacement $w = w(x, y)$ must satisfy the governing equation in any one of the following three forms:

$$D\,(\partial^4 w/\partial x^4 + 2\,\partial^4 w/\partial x^2 \partial y^2 + \partial^4 w/\partial y^4) = p \tag{4.60a}$$
$$D\,(\partial^2 w/\partial x^2 + \partial^2 w/\partial y^2)^2 w = p \tag{4.60b}$$
$$D\,(\nabla^2)^2 w = D\,\nabla^4 w = p \tag{4.60c}$$

where $\nabla^2 = \partial^2 w/\partial x^2 + \partial^2 w/\partial y^2$ and $D = E\,t^3/\,[12(1-\nu^2)]$. The Cartesian stress components in the plane of the plate (Fig. 4.38b) are given by

$$\sigma_x = -\,Ez\,(\partial^2 w/\partial x^2 + \nu\,\partial^2 w/\partial y^2)/(1-\nu^2) \tag{4.61a}$$
$$\sigma_y = -\,Ez\,(\partial^2 w/\partial y^2 + \nu\,\partial^2 w/\partial x^2)/(1-\nu^2) \tag{4.61b}$$
$$\tau_{xy} = -\,Ez\,(\partial^2 w/\partial x \partial y)/(1+\nu) \tag{4.61c}$$

The corresponding moment components/length of side are

$$M_x = -\,D\,(\partial^2 w/\partial x^2 + \nu\,\partial^2 w/\partial y^2) \tag{4.62a}$$
$$M_y = -\,D\,(\partial^2 w/\partial y^2 + \nu\,\partial^2 w/\partial x^2) \tag{4.62b}$$
$$M_{xy} = D\,(1-\nu)\,\partial^2 w/\partial x\,\partial y \tag{4.62c}$$

In general, the shear force S_x, acting on the y face in Fig. 4.38b, is found from moment equilibrium

$$\begin{aligned} S_x &= [\,\partial M_x/\partial x - \partial M_{xy}/\partial y] \\ &= -\,D\,[\,\partial^3 w/\partial x^3 + \partial^3 w/\partial x\,\partial y^2 + (1-\nu)\,\partial^3 w/\partial x\,\partial y^2\,] \\ &= -\,D\,[\,\partial^3 w/\partial x^3 + \partial^3 w/\partial x\,\partial y^2\,] \end{aligned} \tag{4.63a}$$

and similarly, the shear force S_y, acting on the x face, is

$$\begin{aligned} S_y &= [\,\partial M_y/\partial y - \partial M_{xy}/\partial x\,] \\ &= -\,D\,[\,\partial^3 w/\partial y^3 + \partial^3 w/\partial y\,\partial x^2 + (1-\nu)\,\partial^3 w/\partial y\,\partial x^2\,] \\ &= -\,D\,[\,\partial^3 w/\partial y^3 + \partial^3 w/\partial y\,\partial x^2\,] \end{aligned} \tag{4.63b}$$

Equations (4.63a,b) enable the reactions for the edges $x = 0$ and $y = 0$ to be found,

$$\begin{aligned} R_x &= [S_x - \partial M_{xy}/\partial y] \\ &= -\,D\,[\,\partial^3 w/\partial x^3 + (2-\nu)\,\partial^3 w/\partial x\,\partial y^2\,]_{x=0} \end{aligned} \tag{4.64a}$$
$$\begin{aligned} R_y &= [S_y - \partial M_{xy}/\partial x] \\ &= -\,D\,[\partial^3 w/\partial y^3 + (2-\nu)\,\partial^3 w/\partial y\,\partial x^2\,]_{y=0} \end{aligned} \tag{4.64b}$$

There are three common boundary conditions:
(i) where the plate is simply supported along its edge (y - axis in Fig. 4.39a). Then $w = 0$ and $M_x = 0$ for $x = 0$ and all y
(ii) where the edge is fixed along its edge (y - axis in Fig. 4.39b). Then $w = 0$ and $\partial w/\partial x = 0$ for $x = 0$ and all y
(iii) where one edge, say $x = 0$, is free then M_x and R_x are both zero in the respective eqs(4.62a) and (4.64a).

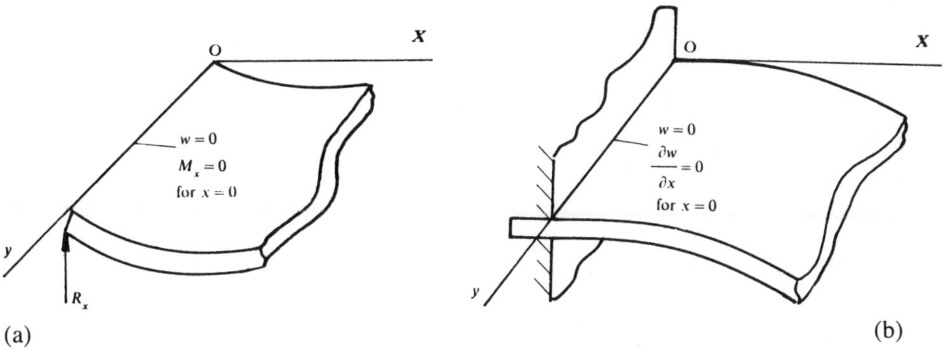

Figure 4.39 Typical plate boundary conditions

Example 4.20 Show that the deflection function $w = Axy(x - a)(y - b)$ is valid for a rectangular plate $a \times b$ under normal pressure p. Determine the loading and moment distributions along the edges and the stresses at the plate bottom centre.

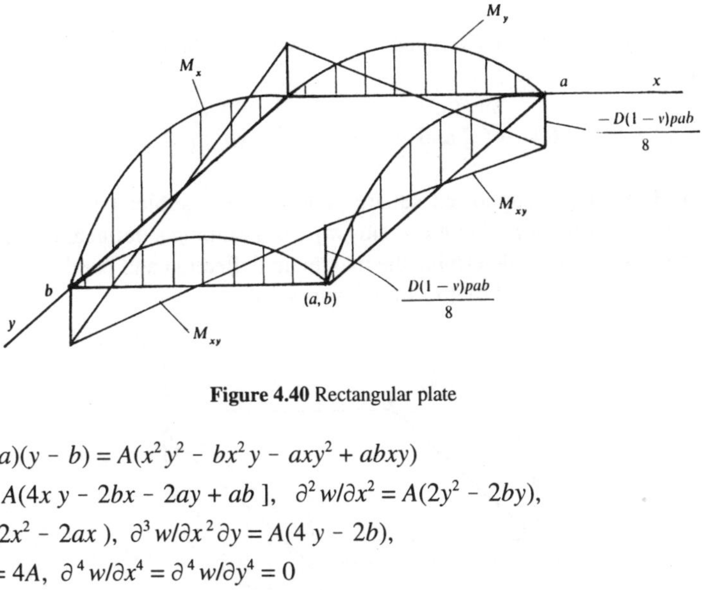

Figure 4.40 Rectangular plate

$$w = Axy(x - a)(y - b) = A(x^2y^2 - bx^2y - axy^2 + abxy)$$

$$\partial^2 w/\partial x\,\partial y = A(4xy - 2bx - 2ay + ab],\quad \partial^2 w/\partial x^2 = A(2y^2 - 2by),$$

$$\partial^2 w/\partial y^2 = A(2x^2 - 2ax),\quad \partial^3 w/\partial x^2\partial y = A(4y - 2b),$$

$$\partial^4 w/\partial x^2\partial y^2 = 4A,\quad \partial^4 w/\partial x^4 = \partial^4 w/\partial y^4 = 0$$

Substituting these derivatives into eq(4.60a) gives $p = 8A$ = constant, i.e. a uniform pressure. From eqs(4.62a,b,c),

$$M_x = -2AD[(y^2 - by) + \nu(x^2 - ax)]$$

$$= -pEt^3[(y^2 - by) + \nu(x^2 - ax)]/[48(1 - \nu^2)]$$

$$M_y = -2AD[(x^2 - ax) + \nu(y^2 - by)]$$

$$= -pEt^3[(x^2 - ax) + \nu(y^2 - by)]/[48(1 - \nu^2)]$$

$$M_{xy} = D(1 - \nu)A(4xy - 2bx - 2ay + ab)$$

$$= pEt^3(4xy - 2bx - 2ay + ab)/[96(1 + \nu)]$$

For edges $x = 0$ and $x = a$,

$$M_x = -\,pEt^3 y(y - b)/[48(1 - \nu^2)]$$
$$M_y = -\,pEt^3 y(y - b)/[48(1 - \nu^2)]$$
$$M_{xy} = \pm\, pEt^3 a\,(b - 2y)/[96(1 + \nu)]$$

For edges $y = 0$ and $y = b$:

$$M_x = -\,pEt^3 x\,(x - a)/[48(1 - \nu^2)]$$
$$M_y = -\,pEt^3 x\,(x - a)/[48(1 - \nu^2)]$$
$$M_{xy} = \pm\, pEt^3 b\,(a - 2x)/[96(1 + \nu)]$$

These moment distributions are shown in Fig. 4.40. The following stress components are found from eqs(4.61a,b,c):

$$\sigma_x = -\,2AEz\,[\nu\, x(x - a) + y\,(y - b)]\,/\,(1 - \nu^2)$$
$$\sigma_y = -\,2AEz\,[\nu\, y\,(y - b) + x\,(x - a)]\,/\,(1 - \nu^2)$$
$$\tau_{xy} = -\,AEz\,[4\,x\,y - 2bx - 2ay + ab]\,/\,(1 + \nu)$$

Substituting $A = p/8$, $x = a/2$, $y = b/2$ and $z = t/2$ gives $\tau_{xy} = 0$ and the tensile stress state at the bottom centre

$$\sigma_x = pE\,t\,(a^2\nu + b^2)/[32(1 - \nu^2)] \text{ and } \sigma_y = pE\,t\,(a^2 + \nu b^2)/[32(1 - \nu^2)]$$

Example 4.21 A simply supported square plate $a \times a$ is subjected to a normal pressure distribution of the form $p = p_o \sin(\pi x/a)\sin(\pi y/a)$. Assuming a deflected shape of the form $w = c\sin(\pi x/a)\sin(\pi y/a)$, determine the maximum deflection and bending moment. What are the edge reactions exerted by the simple supports?

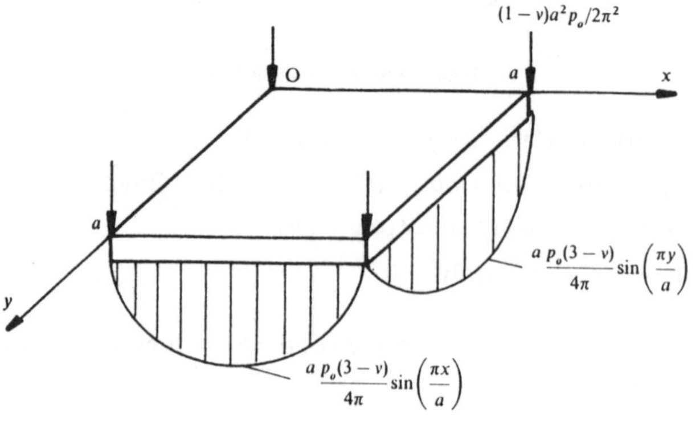

Figure 4.41 Plate reactions

The derivatives of the deflection function are

$$\partial w/\partial x = (\pi c/a)\cos(\pi x/a)\sin(\pi y/a),\quad \partial^2 w/\partial x^2 = -\,(2\pi c/a^2)\sin(\pi x/a)\sin(\pi y/a)$$
$$\partial^3 w/\partial x^3 = -\,(\pi^3 c/a^3)\cos(\pi x/a)\sin(\pi y/a),\quad \partial^2 w/\partial x\,\partial y = (\pi^2 c/a^2)\cos(\pi x/a)\cos(\pi y/a)$$
$$\partial^3 w/\partial x\partial y^2 = -\,(\pi^3 c/a^3)\cos(\pi x/a)\sin(\pi y/a),\quad \partial^3 w/\partial y\partial x^2 = -\,(\pi^3 c/a^3)\sin(\pi x/a)\cos(\pi y/a)$$

$\partial^4 w/\partial x^4 = (\pi^4 c/a^4)\sin(\pi x/a)\sin(\pi y/a) = \partial^4 w/\partial y^4$ (i)

$\partial^4 w/\partial x^2 \partial y^2 = (\pi^4 c/a^4)\sin(\pi x/a)\sin(\pi y/a)$ (ii)

Substituting eqs(i) and (ii) into eq(4.60a) in order to validate the assumed deflection function,

$4[(\pi^4 c/a^4)\sin(\pi x/a)\sin(\pi y/a)] = p/D$

$4[(\pi^4 c/a^4)\sin(\pi x/a)\sin(\pi y/a)] = (p_o/D)\sin(\pi x/a)\sin(\pi y/a)$

$4\pi^4 c/a^4 = p_o/D \;\Rightarrow\; c = a^4 p_o/(4\pi^4 D)$

$w = [a^4 p_o/(4\pi^4 D)]\sin(\pi x/a)\sin(\pi y/a)$ (iii)

The maximum deflection occurs at the plate centre, where $x = y = a/2$. Substituting into eq(iii), this gives

$w_{max} = p_o a^4/(4D\pi^4)$

From eqs(4.62a,b),

$M_x = M_y = (D\pi^2 c/a^2)(1 + \nu)\sin(\pi x/a)\sin(\pi y/a)$

$= [a^2 p_o/(4\pi^2)](1 + \nu)\sin(\pi x/a)\sin(\pi y/a)$

This is a maximum for $x = a/2$ and $y = a/2$, giving

$(M_x)_{max} = (M_y)_{max} = p_o a^2(1 + \nu)/(4\pi^2)$

Equations (iii) and eqs(4.64a,b) supply the following edge reactions for this plate

$R_x = [D\pi^3 c(3 - \nu)/a^3]\sin(\pi y/a) = [ap_o(3 - \nu)/(4\pi)]\sin(\pi y/a)$

$R_y = [D\pi^3 c(3 - \nu)/a^3]\sin(\pi x/a) = [ap_o(3 - \nu)/(4\pi)]\sin(\pi x/a)$

which are shown graphically in Fig. 4.41. Note, from moment equilibrium, that forces of magnitude $2M_{xy} = (1 - \nu)a^2 p_o/(2\pi^2)$ will prevent the plate from rising at the corners (a moment/unit length has force units). These forces are found either from the substitution $x = 0$ and $y = 0$ or $x = a$ and $y = a$ in eq(4.62c)

EXERCISES

Bending of Straight Beams

4.1 Derive expressions for the position and magnitude of the maximum bending stress in a tapered cantilever with solid circular section carrying a concentrated load at its end.

4.2 A solid circular stepped shaft 1.5 m long is supported in bearings at each end. It carries concentrated forces 25 and 30 kN at respective distances of 0.5 and 1 m one end. Determine suitable shaft diameters if the maximum stress in the material is everywhere limited to 65 MPa.

4.3 A 5 m square floor is simply supported on parallel timber joists 150 mm deep × 50 mm wide spaced at 500 mm intervals. Calculate the greatest floor pressure that can be carried when the bending stress in the joist is limited to 6 MPa.

4.4 A beam with a T-section is 5 m long and rests upon simple supports with its flange at the top. The flange is 160 mm wide and 30 mm deep. The web depth is 150 mm and its thickness is 30 mm. Calculate the greatest distributed loading the beam can support when the maximum tensile and compressive stresses are limited to 75 and 40 MPa respectively.

4.5 A steel pipe-line, mean diameter $d = 1$ m, wall thickness $t = 20$ mm and density 5600 kg/m^3, is rigidly supported at 10 m intervals. Calculate the maximum stress in the pipe from bending and pressure effects when it carries liquid at a pressure and density 7 bar and 800 kg/m^3 respectively. Take the maximum central moment to be $wl^2/24$ and the hoop stress due to pressure as $pd/2t$.

4.6 The bored rectangular section in Fig. 4.42 is to act as a 1 m long simply supported beam. Find the maximum bending moment and the greatest stress in the material when a concentrated force of 20 kN is applied at centre span.

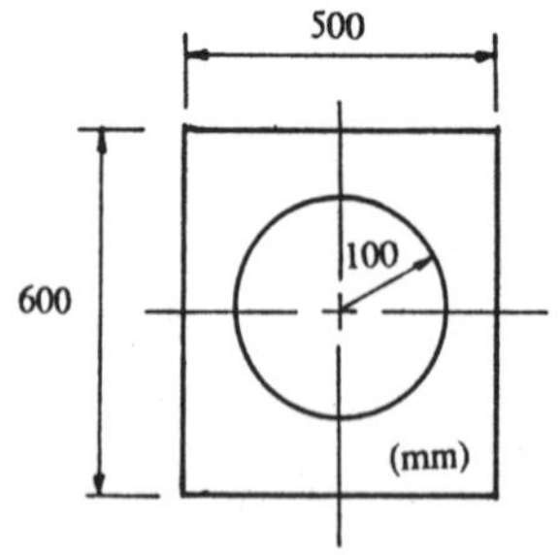

Figure 4.42

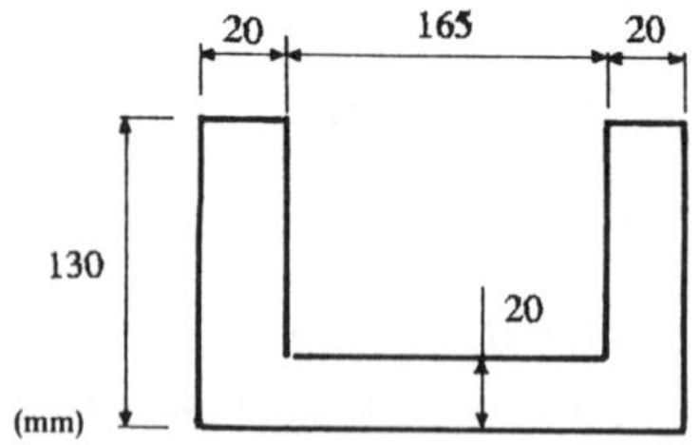

Figure 4.43

4.7 The section shown in Fig. 4.43 is to carry soil within its channel over a length of 3 m. If the ends are simply supported, calculate the greatest distributed loading in N/m the beam can carry if the tensile and compressive stresses are limited to 40 and 60 MPa respectively.

4.8 A cantilever beam, 2 m long, has the channel section shown in Fig. 4.44. It is to support a uniformly distributed loading of 300 N/m together with a concentrated load of 250 N at a point 0.5 m from the free end. What are the maximum tensile and compressive stresses due to bending?

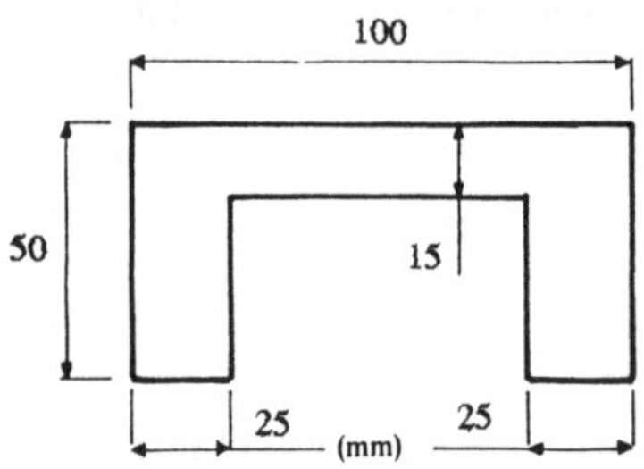

Figure 4.44

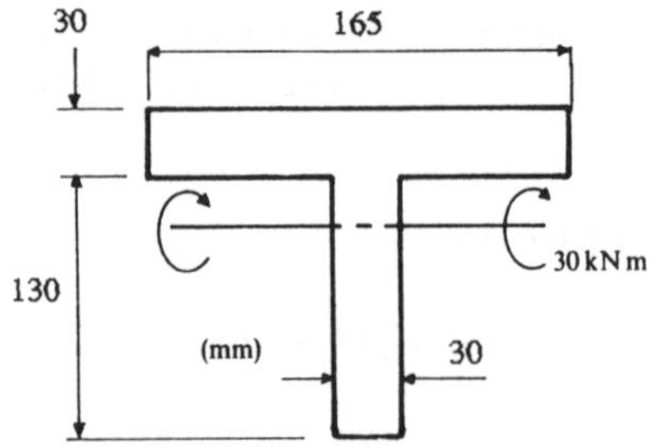

Figure 4.45

4.9 An applied moment of 30 kNm causes compression in the flange of T-section shown in Fig. 4.45. Find the maximum bending stress and the radius of curvature under this moment. Take $E = 210$ GPa.

4.10 Calculate the maximum tensile and compressive stresses in a cast iron beam due to its own weight. The beam is 10 m long, simply supported, with the section given in Fig. 4.46. Take the density of cast iron as 6900 kg/m^3.

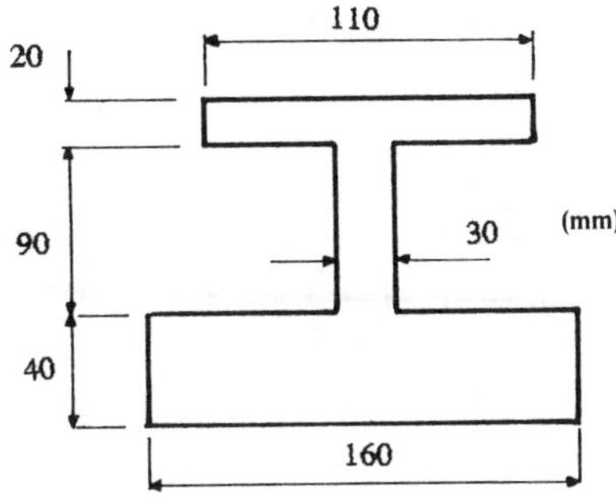

Figure 4.46

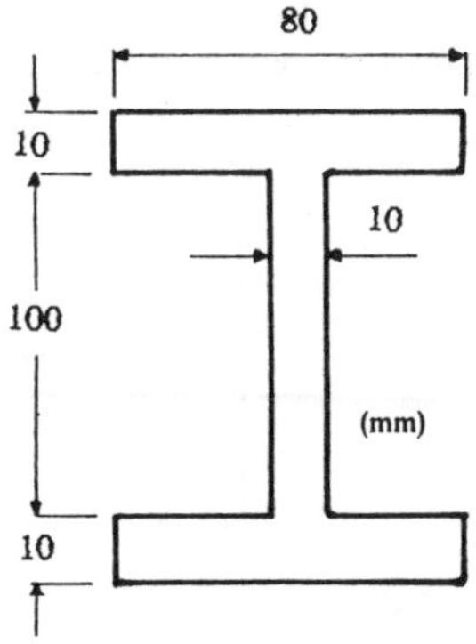

Figure 4.47

4.11 The working stress for the I-section in Fig. 4.47 is limited to 100 MPa. Find the maximum moment and the radius of curvature under this moment. Take E = 210 GPa.

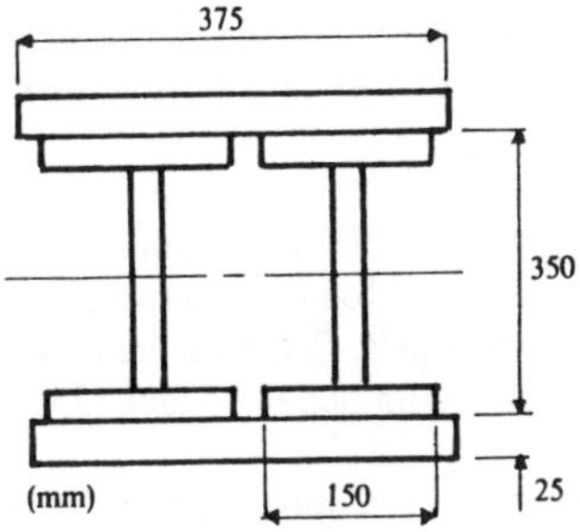

Figure 4.48

4.12 The section in Fig. 4.48 is fabricated from 350 × 150 mm joists with $I = 2 \times 10^{4}\ \text{m}^4$ and plates 370 × 150 mm as shown. The section is used as a 8 m long, simply supported beam, carrying a total load of 200 kN distributed uniformly over the length. Determine the maximum stress in the material.

Combined Bending and Direct Stress

4.13 A cantilever beam is subjected to an end force of 15 kN inclined at 15° to the horizontal axis of the beam. The beam is 150 mm long with a rectangular section 50 mm wide × 25 mm deep. Determine the net stress distribution across the fixed end section.

4.14 A simply supported I-beam, 2 m long, carries a central concentrated force of W = 15 kN in addition to a longitudinal tensile force of F = 60 kN. The depth of the section is 100 mm, its cross-sectional area is 1800 mm^2 and its second moment of area is 3.25×10^6 mm^4. Determine the maximum tensile and compressive stresses and the position of the neutral axis.

4.15 A vertical load of 25 tonne is applied to a tubular cast iron column, 150 mm o.d. and 100 mm i.d. What is the permissible eccentricity of the load if the maximum tensile stress is not to exceed 40 MPa? Determine the maximum stress in compression.

4.16 In a tensile test on a round bar of 30 mm diameter, it was found that points, originally 150 mm apart, separated by 0.75 mm along one side, but decreased by 0.25 mm on the opposite side. Find the axial stress distribution, the axial elongation and the eccentricity of the load. Take E = 207 GPa.

4.17 The central section of a riveter is shown in Fig. 4.49. What is the maximum tensile and compressive stress in the section when a compressive force of 300 kN acts at the anvil position C.

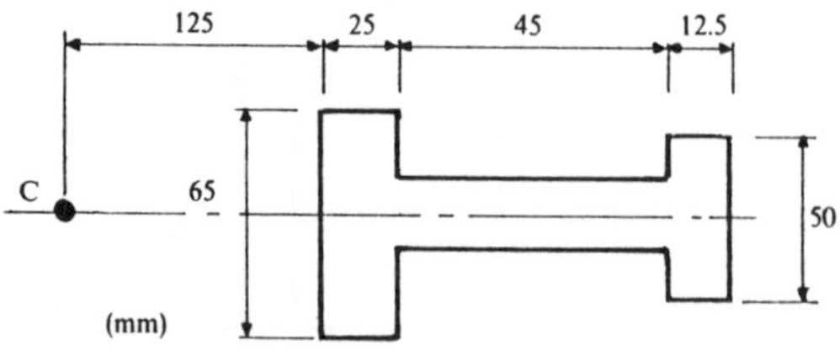

Figure 4.49

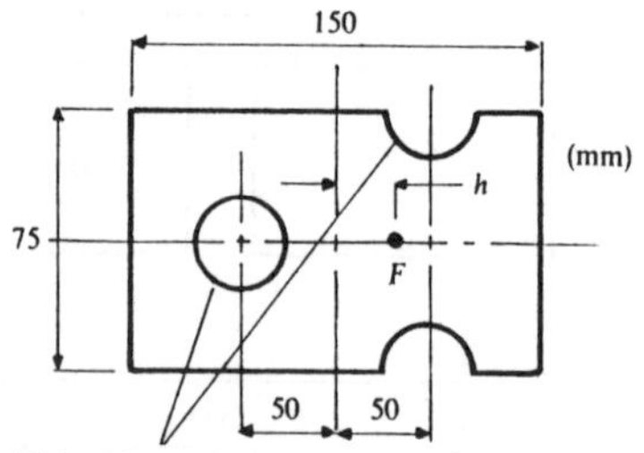

Figure 4.50

4.18 Figure 4.50 shows the cross-section of a short column. When a compressive force F is applied at position h shown, the left edge of the section is to remain unstressed. Find a value for h and the maximum compressive stress in the section when $F = 250$ kN.

4.19 Derive an expression for the maximum eccentricity h from the centroid for a rectangular section $b \times d$, so that when a compressive load is applied along a centroidal axis parallel to d, the stress nowhere becomes tensile. Repeat for the case of solid and tubular circular sections.
Answer: $d/6$, $d/8$, $(D^2 + d^2)/(2D)$

4.20 A compressive force P is applied at 300 mm from the axis of a tubular cast iron column with dimensions, 200 mm external diameter and 165 mm internal diameter. In addition, an axial compressive load of 750 kN is applied along the tube axis. Calculate the maximum value of P when the tensile and compressive stresses in the material are limited to 35 and 125 MPa respectively.

4.21 A short tube, o.d. = 200 mm, i.d. = 165 mm, is subjected to an eccentric compressive load of 60 kN applied at a distance of 330 mm from the cylinder axis. If the maximum allowable compressive and tensile stresses for the material are 120 and 30 MPa respectively, determine the maximum additional axial compressive load that may be applied. Draw the net stress distribution for the tube cross-section.

4.22 The short column of rectangular cross-section in Fig. 4.51 has a vertical compressive force of 200 kN, applied at F. Calculate the resultant stress at each corner of the section.
Answer: $\sigma_A = -49.62$ MPa, $\sigma_B = -11.22$ MPa, $\sigma_C = 17.64$ MPa, $\sigma_D = -20.78$ MPa

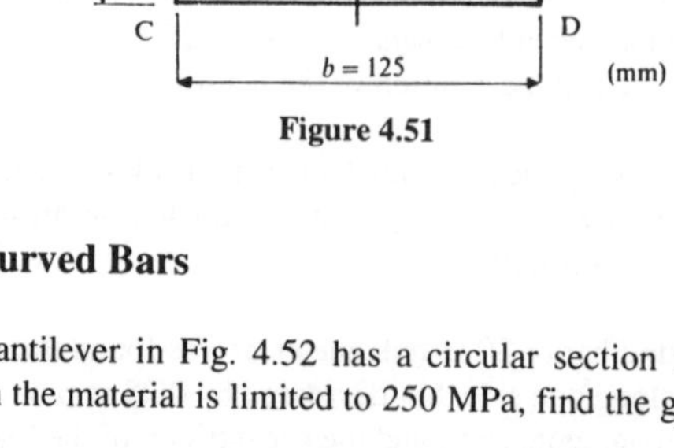

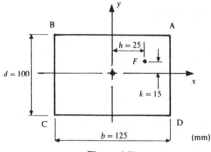

Figure 4.51

Bending of Initially Curved Bars

4.23 The semi-circular cantilever in Fig. 4.52 has a circular section 100 mm in diameter. If the maximum tensile stress in the material is limited to 250 MPa, find the greatest end force that can be carried.

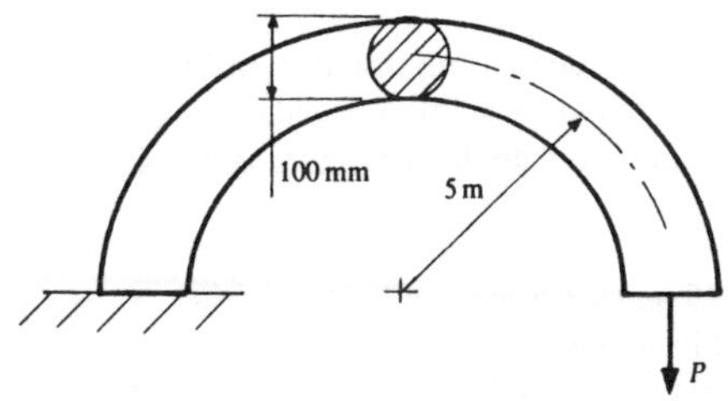

Figure 4.52

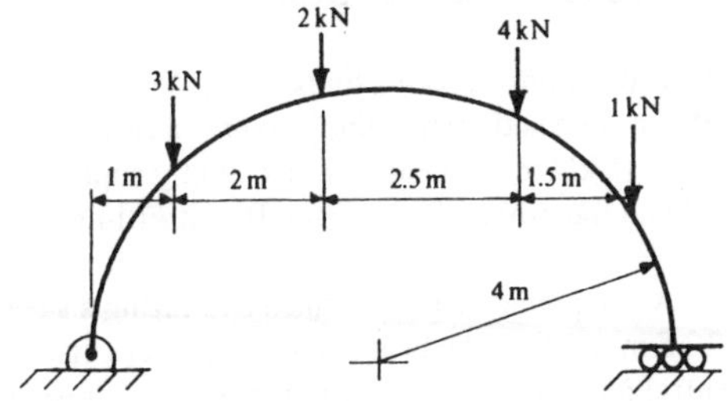

Figure 4.53

4.24 The arched beam in Fig. 4.53 carries the concentrated vertical forces shown. Determine, for the point of greatest bending moment, the maximum stress in the material and the radius of curvature for a uniform cross-section 100 mm wide × 20 mm deep. Take $E = 200$ GPa.

4.25 Compare the maximum stresses from the large and small initial curvature theories in the case of a split-ring of trapezoidal section when a 5 kN force is applied, as shown in Fig. 4.54. Do not neglect the additive effect of the direct stress acting on the section *X-X* where the moment is a maximum.

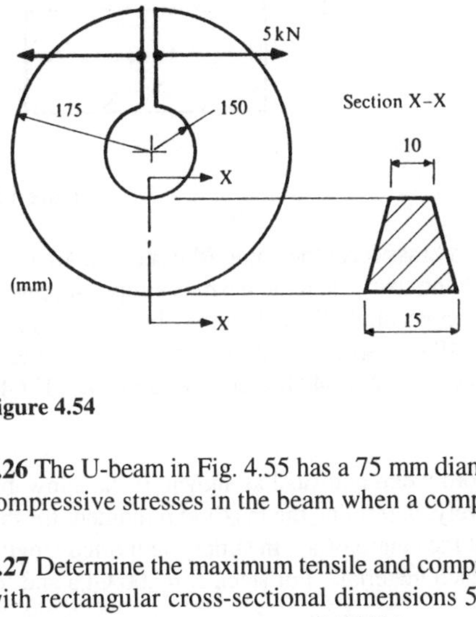

Figure 4.54

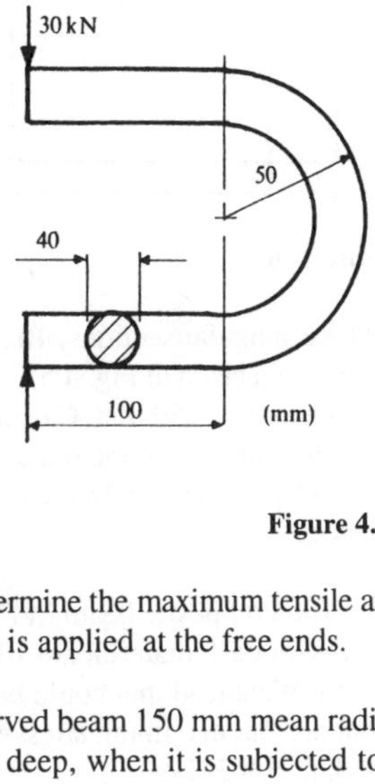

Figure 4.55

4.26 The U-beam in Fig. 4.55 has a 75 mm diameter cross-section. Determine the maximum tensile and compressive stresses in the beam when a compressive force of 30 kN is applied at the free ends.

4.27 Determine the maximum tensile and compressive stresses for a curved beam 150 mm mean radius with rectangular cross-sectional dimensions 50 mm wide × 150 mm deep, when it is subjected to a decreasing curvature moment of 15 kNm.

4.28 Plot the variation in stress across the depth of a curved bar with rectangular section dimensions 75 mm wide × 50 mm deep, with mean radius 125 mm, when it is subjected to a sagging moment of 15 kNm tending to straighten it.

4.29 Show, for a solid circular section of radius r, in a beam with large initial curvature of mean radius R_m, that the neutral axis shifts by $y = R_m - \frac{1}{2}r^2[R_m - \sqrt{(R_m^2 - r^2)}]$.

4.30 Find the necessary wall thickness for a tube with mean section radius 100 mm when it is to be subjected to a pure bending moment of 10 kNm which decreases the initial 200 mm mean radius, without the maximum compressive stress exceeding 65 MPa. What would be the greatest compressive stress in a straight beam with the same cross-sectional dimensions?

Bending of Composite Beams

4.31 The bottom edge of a timber joist, 150 mm wide and 250 mm deep, is reinforced with steel plate, 150 mm wide and 10 mm thick, to give a composite beam, 260 mm deep × 150 mm wide. A distributed load of 15 kN/m is spread over its simply supported 5 m span. Calculate the maximum stresses in the steel and the timber. Take E(steel) = 20E(timber).

4.32 A compound beam is formed by brazing a steel strip, 25 mm × 6.5 mm, on to a brass strip, 25 mm × 12.5 mm. When the beam is bent into a circular arc about its longer side, find: (a) the position of the neutral axis, (b) the ratio of the maximum stress in the steel to that in the brass and (c) the radius of curvature of the neutral axis when the steel stress reaches its maximum allowable value of 70 MPa. Take E = 207 GPa for steel, E = 82.8 GPa for brass.

4.33 Find the maximum allowable bending moment for the composite section in Fig. 4.56, given maximum permissible stresses of 150 MPa for steel and 8 MPa for wood. Their elastic moduli are 210 and 8.5 GPa respectively.

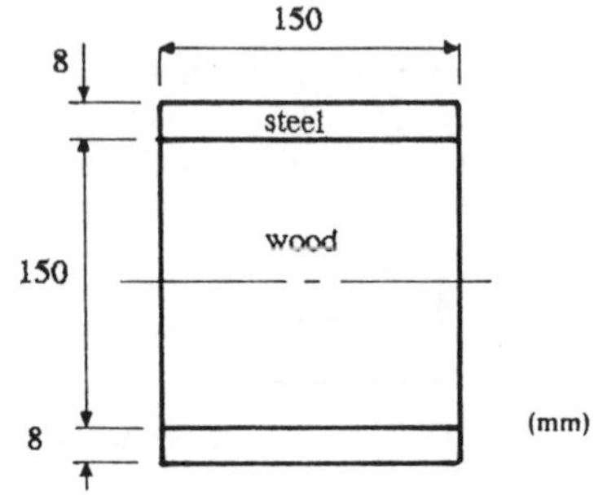

Figure 4.56

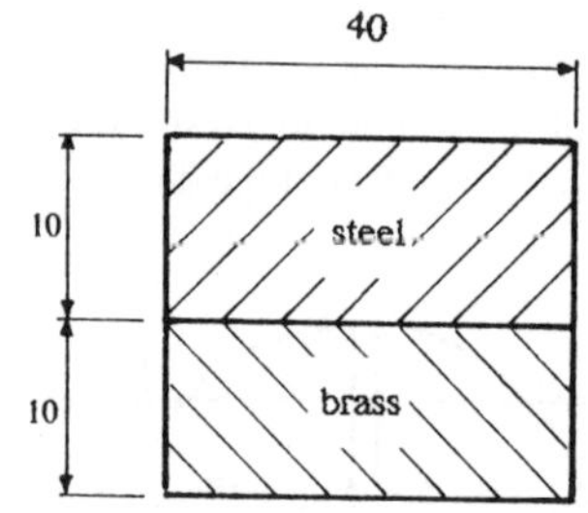

Figure 4.57

4.34 Rectangular sections, 40 mm × 10 mm, of steel and brass form the composite section, 40 mm × 20 mm, of a beam in Fig. 4.57. The brass is to lie beneath the steel when the beam is simply supported over a length of 750 mm. Calculate the maximum central concentrated load that may be applied, given allowable stresses for the brass and steel of 70 and 105 MPa respectively, when the materials (i) each bend independently and (ii) are bonded along the interface. Take E = 84 GPa for brass and E = 207 GPa for steel.

4.35 The composite beam in Fig. 4.58 is fabricated from wood and steel as shown. If the allowable stresses for each material are 10 and 150 MPa respectively, determine the maximum moment for this section. What load that could be applied to this beam at the centre of a 2 m simply supported length? What are the maximum stresses actually achieved in each material? For steel, E = 200 GPa and for wood, E = 8.5 GPa.

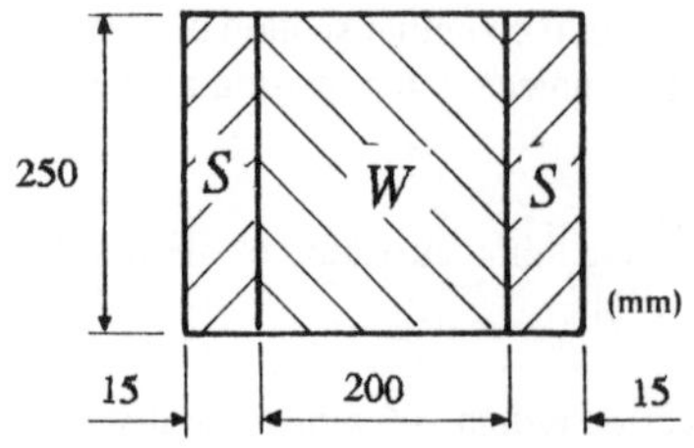

Figure 4.58

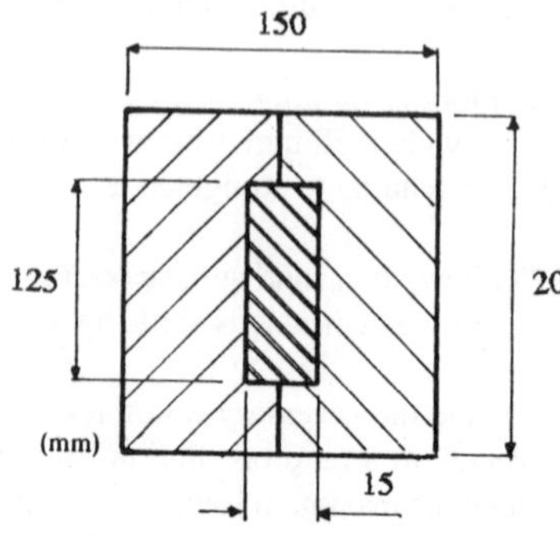

Figure 4.59

4.36 A composite beam is constructed from a steel strip surrounded by timber joists as shown in Fig. 4.59. Calculate the moment of resistance for the section when the maximum bending stress in the timber is 8.5 MPa. What then is the maximum bending stress in the steel? Take $E = 210$ GPa for steel, $E = 12.5$ for timber.

4.37 Aluminium sheet is attached all along the top and bottom surfaces of a timber beam, 80 mm wide × 160 mm deep. The moment of resistance of the composite beam is to be four times that of the timber alone whilst maintaining the same value of maximum bending stress in the timber. Determine the thickness of the aluminium sheet and the ratio of the maximum bending stresses in each material. Take $E(\mathrm{Al}) = 7 \times E(\mathrm{timber})$.

4.38 A 500 mm long glass-fibre beam is reinforced along its outer edges with carbon-fibre to the same 15 mm width and occupying 25% of the total beam volume. The beam length is 20 times its total depth. Determine the maximum central concentrated load a simply supported beam can carry when the allowable bending stresses for each material are 210 MPa for glass-fibre and 1050 MPa for carbon-fibre. The respective moduli are 15 GPa and 125 GPa.

4.39 The copper-steel bi-metallic strip in Fig. 4.60 is subjected to a temperature rise of 65°C. Calculate the net stresses induced at the free and interface surfaces. For Cu: $E = 105$ GPa and $\alpha = 17.5 \times 10^{-6}$/ °C. For steel: $E = 205$ GPa and $\alpha = 11 \times 10^{-6}$/ °C.

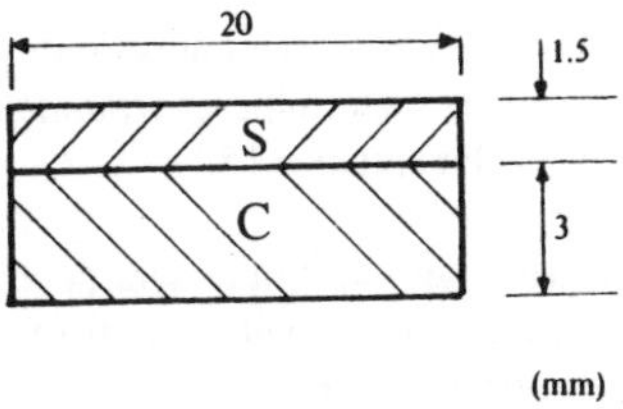

Figure 4.60

4.40 A bi-metallic cantilevered strip is to operate a switch at its free end. The steel and bronze strips are each 100 mm long and 15 mm × 1 mm in cross-section. Calculate the common radius of curvature and deflection at the free end for a temperature rise of 50°C. Take $E = 200$ GPa, $\alpha = 10 \times 10^{-6}$/ °C for steel and $E = 120$ GPa, $\alpha = 20 \times 10^{-6}$/ °C for bronze.

Reinforced Concrete

4.41 A reinforced concrete beam is 300 mm wide and 500 mm deep to the centre of the single steel line. If stresses in the concrete and steel of 5 and 125 MPa are to be reached under economic design conditions, find the steel area and the distributed weight that can be simply supported over a span of 10 m. Ignore self weight of the beam and take $m = 15$. Explain why there should be several steel rods each of small cross-sectional area rather than one of large area.

4.42 A reinforced concrete beam section is 250 mm wide, 500 mm deep and 50 mm deep to the centre of the steel line where there are 4 rods each 12.5 mm diameter. If the maximum permissible stresses for steel and concrete are 125 and 5 MPa respectively and the modular ratio is 15, calculate the moment of resistance for the section and the stress in the concrete. Where in the cross-section does the neutral axis lie?

4.43 A concrete beam with steel reinforcement is 150 mm wide and 375 mm deep to the steel line, where there are 3 rods each 12.5 mm diameter. If the maximum permissible stresses for the steel and concrete are 125 and 4 MPa respectively, and the modular ratio is 15, calculate the moment of resistance for the section. If the section were to be economically designed, what would the steel area and the moment of resistance then be?

4.44 A reinforced concrete beam covers a span of 12 m and carries a uniformly distributed load of 1.5 kN/m in addition to the beam's self-weight of 1.2 kN/m. The maximum allowable stresses for the steel and concrete are 125 and 5 MPa respectively. Assuming economic design, calculate the section dimensions and the steel area when the breadth is 0.6 × the depth to the steel line. Take $m = 15$.

4.45 A rectangular section concrete beam is 300 mm wide and 200 mm deep to the steel line. The limiting stresses are 7 MPa for the concrete and 125 MPa for the steel. If both materials are to be fully stressed, calculate the steel area and the moment of resistance for the section when the modular ratio is 15.

4.46 A reinforced concrete beam with steel line depth d and breadth $b = d/2$, carries a uniformly distributed load of 1.5 kN/m over a span of 6 m in addition to its self-weight of 1.2 kN/m. Find d and the steel area, given that the allowable stresses for steel and concrete are 124 and 4 MPa respectively and $m = 15$.

4.47 A loading bay 18 m × 6 m is to be supported by 12 evenly spaced tension-reinforced concrete cantilevers. Each beam is 6 m long, 300 mm wide with limiting stresses in the steel and concrete of 138 and 10 MPa respectively. If the maximum uniformly distributed load plus self weight is 9.6 kN/m^2, find the remaining dimensions of the rectangular cross-section that will support the bay. Take $m = 15$. (CEI)

4.48 A 500 mm × 250 mm rectangular concrete section is doubly reinforced with two pairs of 28 mm diameter steel rods set 400 mm apart. Determine the moment of resistance for the section with permissible stresses of 125 and 5.5 MPa and $m = 15$.

4.49 A concrete beam, 300 mm wide with 450 mm steel depth, spans 9 m and carries a uniformly distributed load of 8 kN/m. The steel reinforcement consists of four 25 mm bars. Find the maximum stresses in the steel and concrete given that $m = 15$.

4.50 A reinforced T-beam has flange and web widths of 750 and 150 mm respectively. The web steel line is 300 mm from the flange top. If the neutral axis is to lie within the flange width, find its position when the maximum stresses in the steel and concrete are limited to 110 and 4 MPa respectively. What then is the area of steel reinforcement and the moment of resistance for the section?

4.51 A reinforced concrete T-beam has a flange 1.5 m wide × 150 mm deep. The web width is 450 mm and the depth to the steel line is 600 mm from the flange top. If the allowable stresses for the steel and concrete are 125 and 7 MPa respectively and the modular ratio is 15, find the stresses actually sustained by each material, given that the area of tensile reinforcement is 3750 mm^2.

4.52 The flange of a reinforced concrete T-beam is 1.5 m wide and 100 mm deep and the steel line in the web lies 375 mm from the flange top. If the limiting stresses for steel and concrete are 110 and 4 MPa respectively and the neutral axis is to coincide with the lower edge of the flange, calculate the moment of resistance for the beam cross-section and the area of steel reinforcement. What are the stresses attained in each material? Take $m = 15$.

4.53 A doubly reinforced concrete beam is to sustain a bending moment of 135 kNm. The width of the beam is half the effective depth and the compressive steel line lies at 1/10 this depth. The tensile reinforcement is to be 2% of the effective section area. If the permissible stresses in the steel and concrete are 125 and 7 MPa respectively, find the concrete section dimensions and the steel areas. Take $m = 15$.

4.54 The short column of reinforced concrete in Fig. 4.61 supports a compressive load F, applied at the centroid O. What is the safe value of F if the maximum allowable stresses for steel and concrete are 125 and 4 MPa respectively? If F is moved distance h away from O as shown, find the value of h and the limiting F if tension in the concrete is to be avoided. Take $m = E_s/E_c = 15$.

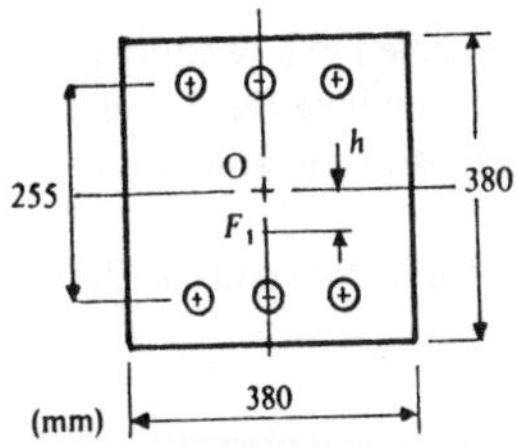

Figure 4.61

Asymmetric Bending

4.55 A 100 × 75 × 12.5 mm angle section rests on two supports 4.5 m apart with its longer side vertical and the shorter side at the top. Calculate the maximum stress in the angle when a concentrated force of 10 kN is applied at mid-span.

4.56 A 1.5 m long steel bar of rectangular section 100 × 37.5 mm is supported by bearings and carries a load of 9 kN at mid-span. If the beam is rotated slowly, find the slope of the longer side with respect to x when the bending stress reaches a maximum.

4.57 The stress at point P for the beam cross-section in Fig. 4.62 is restricted to 30 MPa. Find the maximum moment which may be applied about the x-axis.

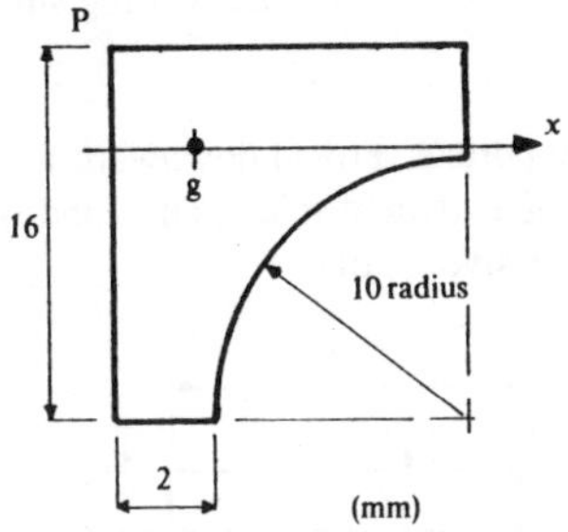

Figure 4.62

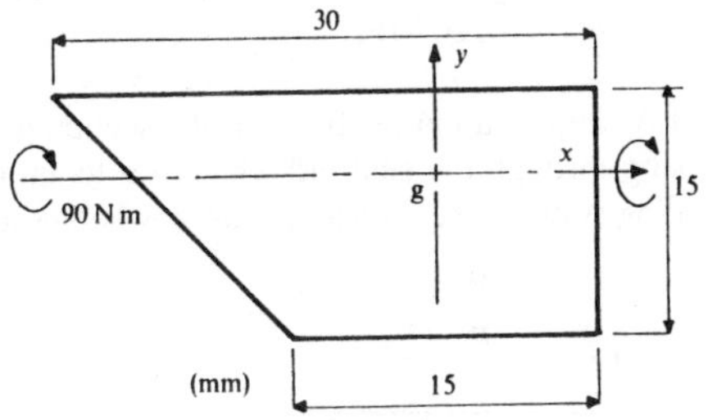

Figure 4.63

4.58 The section in Fig. 4.63 is subjected to a bending moment of 90 Nm about the x-axis. Find the greatest bending stress and the inclination of the n.a. to x.

4.59 The thin-walled triangular tube in Fig. 4.64 is subjected to a bending moment M_y, as shown. Find the inclination of the neutral axis with respect to the x-axis.

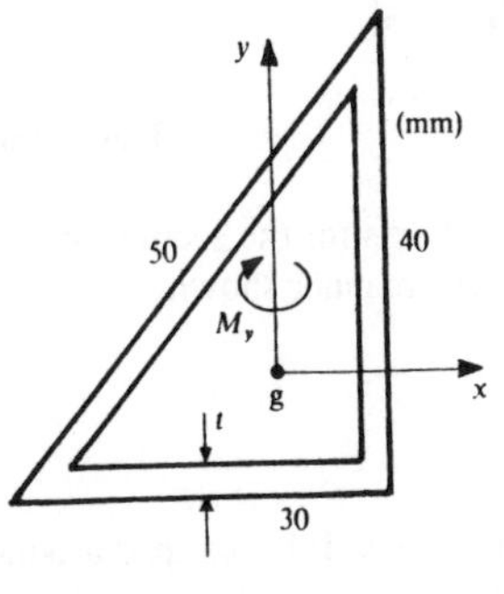

Figure 4.64

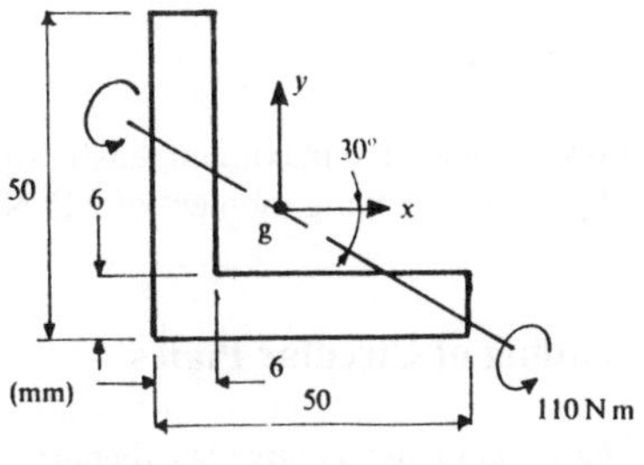

Figure 4.65

4.60 The angle section in Fig. 4.65 is subjected to a bending moment of 110 Nm about a centroidal axis inclined at 30° to the x-axis, as shown. Calculate the principal second moments of area, the position of the neutral plane and the maximum tensile and compressive stresses. (CEI)

4.61 A cantilever 500 mm long is loaded at its free end, as shown in Fig. 4.66. Calculate the maximum tensile and compressive bending stresses.

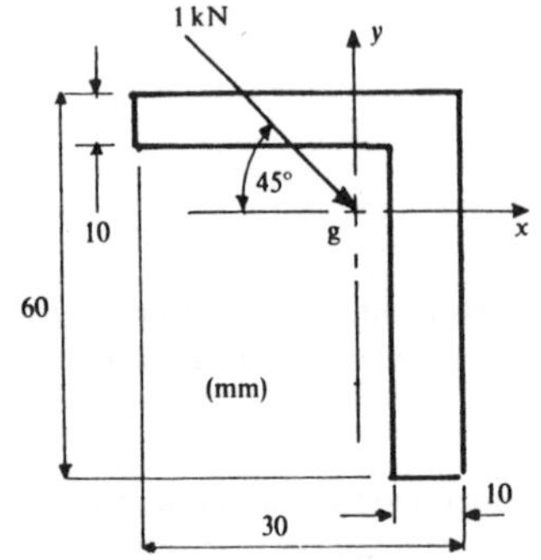

Figure 4.66

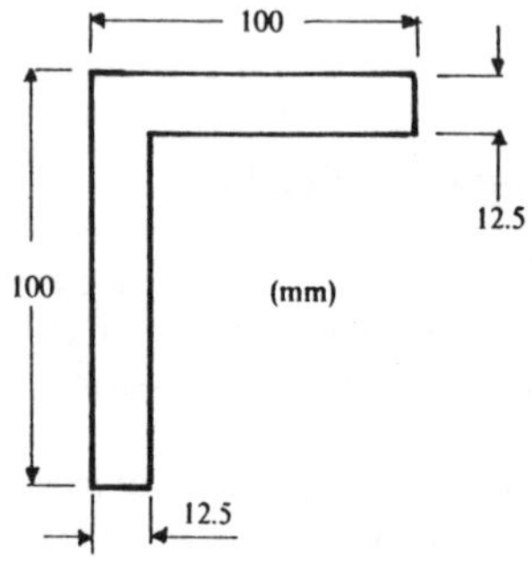

Figure 4.67

4.62 The equal-angle section in Fig. 4.67 is to be used as a 1 m long cantilever for carrying a vertical end load passing through the centroid. Determine the maximum value for this load if the bending stress is not to exceed 120 MPa.

4.63 A simply supported beam of the section given in Fig. 4.68 supports a vertical downward force passing through the centroid which results in a bending moment of 1 kNm at mid-span. Find the bending stress at the corner A, stating whether it is tensile or compressive. (CEI)

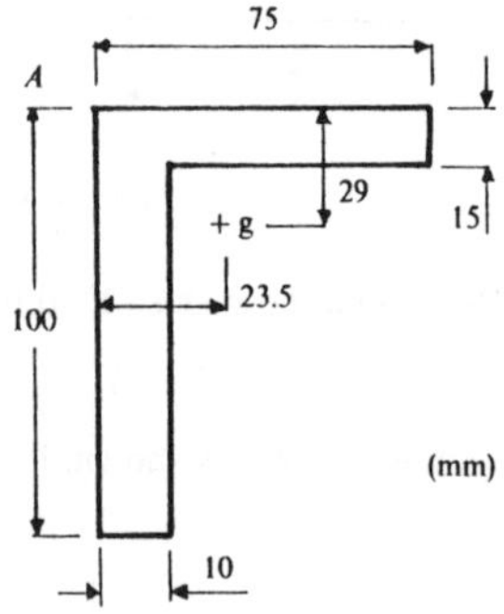

Figure 4.68

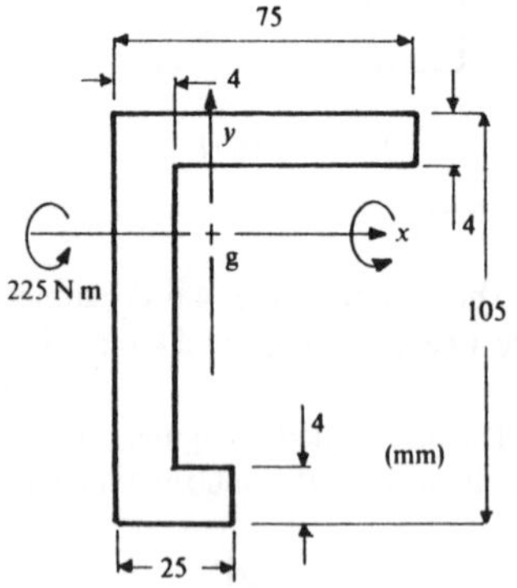

Figure 4.69

4.64 Calculate the maximum tensile stress and the position of the neutral axis for the section in Fig. 4.69 when a bending moment of 225 Nm is applied about the x-axis in the manner shown.

Bending of Circular Plates

4.65 An aluminium alloy plate diaphragm is 510 mm diameter and 6.5 mm thick. It is clamped around its periphery and subjected to a uniform pressure of 0.7 bar. Calculate the values of the maximum bending stress and deflection. Take, for aluminium, $E = 70$ GPa and $\nu = 0.3$. (CEI)

4.66 A circular plate, 330 mm o.d. and 76 mm i.d., is loaded all around its inner edge by a total vertical force P (kN) whilst being simply supported around its outer edge. Determine the distribution of radial bending moment and find the position and magnitude of the maximum value. Take $\nu = 0.27$.

4.67 The loading for the plate in Exercise 4.66 is replaced by a ring of edge bending moments M_{r1} per unit of outer circumference, whilst the inner edge remains free. Plot the radial moment distribution and repeat for the similar problem when a ring of inner edge bending moments $M_{r\,2}$ per unit of circumference is applied in the absence of outer edge loading. (Ans, gradual falls in $M_{r\,1}$ and M_{r2} with $1/r^2$ to zero)

4.68 A simply supported circular steel plate of diameter 1 m and thickness 50 mm is subjected to an increase in pressure of 0.5 bar above atmospheric. Given that density of steel is 7750 kg/m^3, calculate the total deflection at the plate centre due to the pressure and self weight. What is the maximum tensile stress in the plate and the safe pressure when the deflection is restricted to 0.1 mm? Take, for steel, $E = 207$ GPa, $\nu = 0.27$.

4.69 A diaphragm of diameter 220 mm is clamped around its periphery and subjected to a normal gas pressure of 2 bar on one surface. If the central deflection is not to exceed 0.65 mm determine the thickness and maximum bending stress. Take $E = 210$ GPa and $\nu = 0.3$.

4.70 A uniform circular flat plate is rigidly built in around the boundary at radius a, and is subjected to a pressure p acting normally on one surface. It is required to double the central deflection by adding a central concentrated normal force W. Find an expression for W. (CEI)

4.71 The flat end of a 2 m diameter container can be regarded as clamped around its edge. Under operating conditions the plate will be subjected to a uniformly distributed pressure of 0.02 N/mm^2. Calculate from first principles the maximum deflection and the required thickness of the end plate if the bending stress is not to exceed 150 N/mm^2. $E = 200$ GPa and $\nu = 0.3$. (CEI)

4.72 A thin, horizontal circular flat plate is rigidly built in around its circumference at radius R. It supports a vertical central concentrated load W. In order to reduce the central deflection under W to one quarter, a uniform pressure p is applied to the opposite side of the plate. Find an expression for the required value of p in terms of W and the relevant constants. (CEI)

4.73 Figure 4.70 shows the arrangement for an overload pressure warning device. A contact block X is centrally positioned 0.05 mm from the surface of a clamped circular plate which is then subjected to a uniform pressure on one side. Electrical contact is made between the block and the plate when the pressure exceeds a set value. The plate is 30 mm diameter, thickness 0.2 mm with $E = 250$ GPa, $\nu = 0.3$ and a tensile yield stress of 200 MN/m^2. Determine the pressure required to activate the warning device and check whether deformation is elastic at this limit. (CEI)

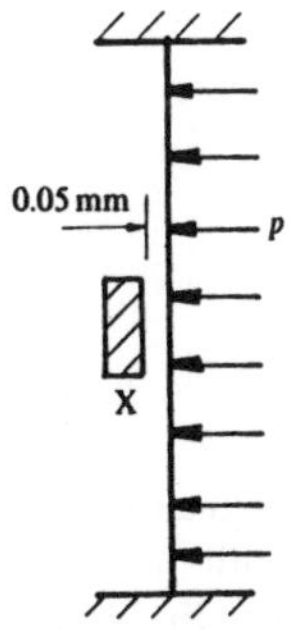

Figure 4.70

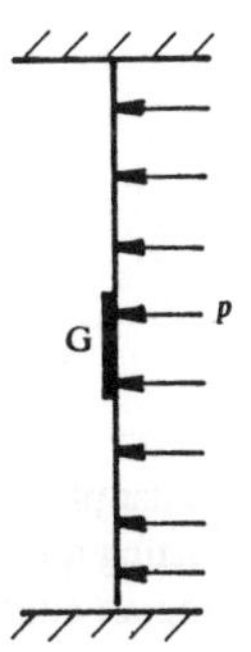

Figure 4.71

4.74 Derive an expression relating the radial strain measured by the gauge G in Fig.4.71 to the applied pressure p for a plate rigidly clamped at a 50 mm radius with thickness 2 mm. Take $E = 200$ GPa and $\nu = 0.3$. (CEI)

4.75 Identify clearly the loading condition for each plate in Fig. 4.72a-c by deriving the shear force per unit length of circumference at radius r. State the particular boundary conditions that should be invoked to obtain the bending stresses.

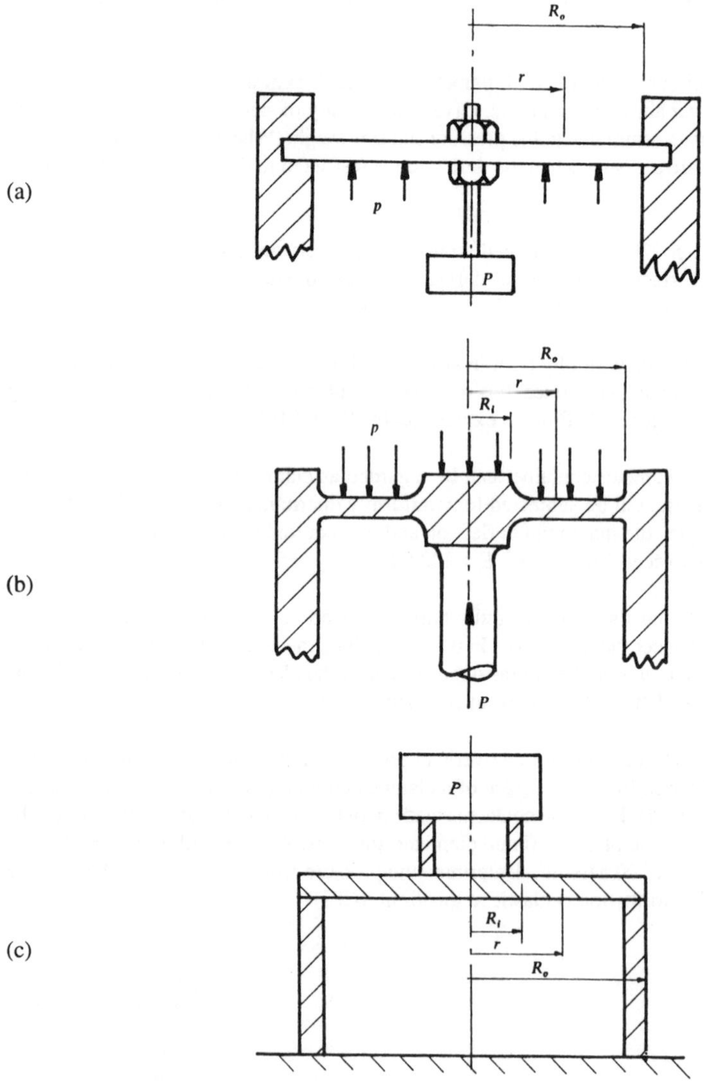

Figure 4.72

4.76 A flat circular plate, with radius 200 mm and thickness 10 mm, is subjected to a uniform pressure $p = 1$ MN/m^2 acting normally on the surface of one side only. It is semi-rigidly built in around its outer edge, producing an edge moment $m_e = 2$ kNm per metre of circumferential length. Find the largest radial stress in the plate and the central deflection. Take $E = 208$ GPa and $\nu = 0.3$ (CEI)

4.77 A cylinder head valve (Fig. 4.73) of diameter 38 mm is subjected to a gas pressure of 1.4 MN/m². Assuming simple supports and that the stem applies a concentrated force P at the centre of the plate, calculate the movement of the stem necessary to lift the plate from its supports. Take $D = 260$ Nm and $\nu = 0.3$. (CEI)

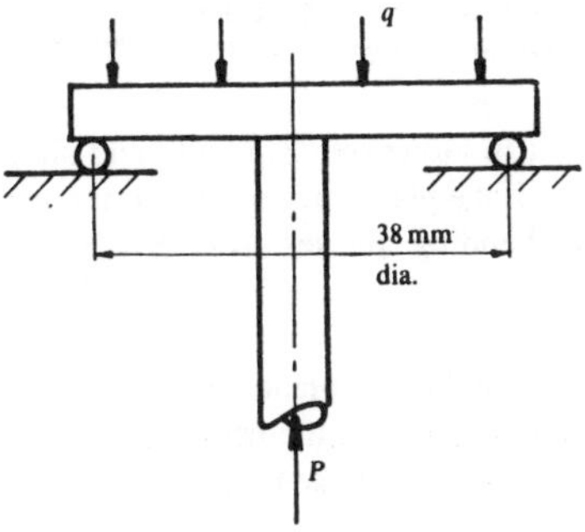

Figure 4.73

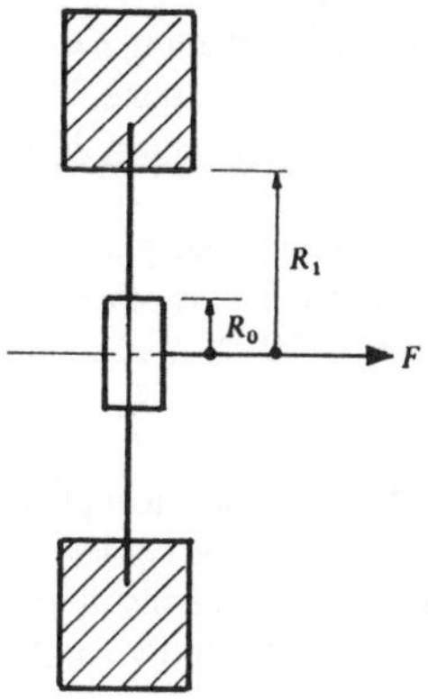

Figure 4.74

4.78 A thin circular plate is rigidly clamped at radius R_1. Two circular metal blocks of radii $R_o << R_1$ are firmly clamped concentrically on either side, as shown in Fig. 4.74. Derive an expression for the block deflection when a normal force F is applied at the centre. (CEI)

Bending of Rectangular Plates

4.79 Show, for a simply supported circular plate subjected to a normal pressure p, that $D\,\nabla^4 w = p$ reduces to eq(4.52b) where $F = pr/2$. (Hint: Use the same conversions from Cartesian to polar co-ordinates as with $\nabla^4\phi = 0$ given in Chapter 2.)

4.80 Given that the equilibrium condition $\partial\sigma_x/\partial x + \partial\tau_{xy}/\partial y + \partial\tau_{xz}/\partial z = 0$ applies to a thin rectangular plate, show that the distribution of shear stress τ_{xz} over the thickness t is given by the expression $\tau_{xz} = (2S_{xz}/2\,t)(1 - 4z^2/t^2)$, where S_{xz} is the shear force/unit length of y.

4.81 The deflection w of a square plate $a \times a$ lying in the x-y plane, with centre at the origin, is given by $w = w_o \cos(\pi x/a)\cos(3\pi y/a)$, where w_o is the central deflection. Find the surface loading, the edge support reactions and the bending moments at the plate centre. (CEI)

4.82 A thin elliptic plate, with thickness t and semi-axes a and b ($a > b$), is rigidly clamped around its edges and subjected to a uniform normal pressure p. Show that the deflection is given by the expression $w = w_o(1 - x^2/a^2 - y^2/b^2)^2$ and find the value for w_o. Determine the top surface stresses, σ_x and σ_y, at the plate centre and at the ends of the major axis.

Answer: $w_o = (pa^4b^4)/[8D(3b^4 + a^2b^2 + 3a^4)]$, $\sigma_x = -\,12pa^2b^2(b^2 + \nu a^2)/[t^2(3b^4 + a^2b^2 + 3a^4)]$, $\sigma_y = -\,12\,pa^2b^2(a^2 + \nu b^2)/[t^2(3b^4 + a^2b^2 + 3a^4)]$, $\sigma_x = 6pa^2b^4/[t^2(3a^4 + a^2b^2 + 3b^4)]$, $\sigma_y = 6pa^4b^2/[t^2(3a^4 + a^2b^2 + 3b^4)]$

4.83 A simply supported rectangular plate $a \times 2a$ lies in the x-y plane so that the x-axis bisects the longer sides and the y-axis is coincident with one longer side. If the surface loading of the plate is given by $q = q_o \sin(\pi x/a)$ and the resulting deflection by $w = [qa^4/(D\pi^4)][1 + A\cosh(\pi y/a) + B(\pi y/a)\sinh(\pi y/a)]$, find q_o, A and B and calculate the central deflection and the bending moments M_x and M_y. Take $\nu = 0.30$. (CEI)

4.84 A simply supported rectangular steel plate, with plane dimensions 3 m × 4 m and 25 mm thick, carries a uniform normal pressure of 15 kN/m^2. Find the maximum deflection and the greatest bending stresses. Take E = 210 GPa, ν = 0.28.

4.85 A simply supported thin plate of edge dimensions 1 × 1.5 m and thickness 0.025 m is subjected to a sinusoidal loading $p = p_o \sin(\pi x/a)\sin(\pi y/b)$ with maximum amplitude 2 kN/m^2 for the origin at one corner of the plate. Find the maximum values of the bending moment and deflection. Take the elastic constants as E = 200 GPa, ν = 0.35.

4.86 A rectangular plate, 3 m × 2 m and 30 mm thick, is built in along one 3 m side and simply supported along the other 3 m side. The remaining edges are free. If the plate carries a uniform normal pressure of 20 kN/m^2, determine the maximum deflection and the bending stresses. Take E = 207 GPa and ν = 0.27.

4.87 A rectangular steel plate, 1.25 × 1.75 m and 30 mm thick, lies in the x-y plane with the origin of co-ordinates at one corner. Find the maximum values of deflection and bending moment when the plate carries the pressure distribution $q = 2\sin(\pi x/1.25)\sin(\pi y/1.75)$ (kN/m^2). Take the elastic constants as E = 210 GPa and ν = 0.28.

CHAPTER 5

THEORIES OF TORSION

5.1 Torsion of Circular Sections

Figure 5.1a shows the cross-section of a solid circular shaft of radius R subjected to opposing axial torques T about the z-axis, over a length L. Consider the action of the torque upon a cylindrical core of this shaft at radius r (Fig. 5.1b). The core will experience a proportion of the applied torque. Consequently, a line AB shears through an angle ϕ into position A' B' so that the angle of twist of B' relative to A' is θ.

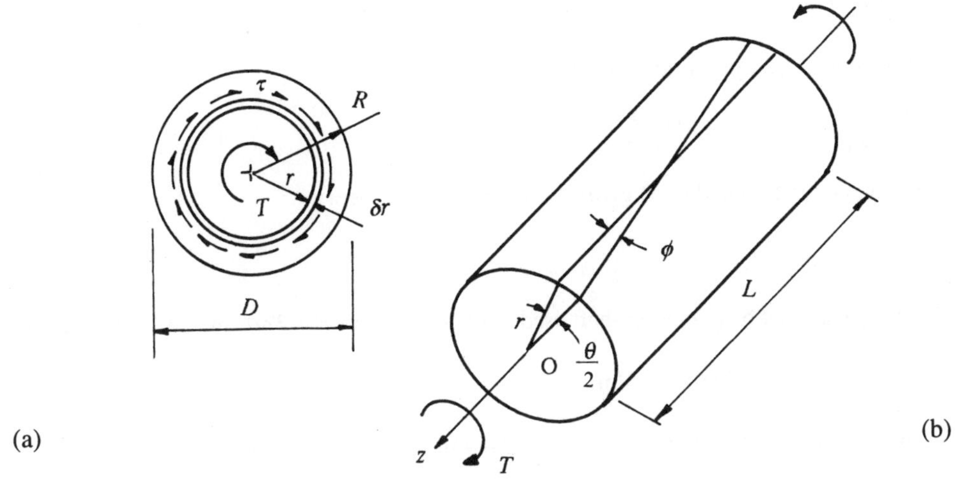

Figure 5.1 Core element and section of a shaft under torsion

The length of the arc AA' in Fig. 5.1b connects ϕ to θ when they are given in radians. That is AA' = $r\theta/2 = L\phi/2$ and from this, the shear angle is

$$\phi = r\theta/L \tag{5.1}$$

The strain γ is the tangent of the shear angle. For elastic deformation γ can also be expressed as the ratio between the shear stress τ in the shaft at radius r and the shear modulus G. This gives

$$\gamma = \tau/G = \tan\phi \tag{5.2}$$

Because ϕ is small, $\tan\phi \approx \phi$ and eqs(5.1) and (5.2) combine to give

$$\gamma = \tau/G = r\theta/L \tag{5.3}$$

The core material resists the torque by becoming stressed in shear. At radius r, the shear stress τ is induced within an annulus of radial thickness δr (Fig. 5.1a). The torque carried by the annular area is $\delta T = 2\pi r\, \delta r \times \tau \times r$. Substituting from eq(5.3)

$$\delta T = 2\pi (G\theta / L)\, r^3\, \delta r \tag{5.4}$$

Extending eq(5.4) to the solid section of outer radius R, gives the torque equilibrium condition:

$$T = 2\pi (G\theta / L) \int r^3\, \mathrm{d}r$$

which is written as

$$T / J = G\theta / L \tag{5.5}$$

where J is the polar second moment of area. For the solid shaft,

$$J = 2\pi \int_0^R r^3\, \mathrm{d}r = 2\pi \left| r^4 / 4 \right|_0^R = \pi R^4 / 2 = \pi D^4 / 32$$

With a tubular section of inner and outer radii R_i and R_o

$$J = 2\pi \int_{R_i}^{R_o} r^3\, \mathrm{d}r = 2\pi \left| r^4 / 4 \right|_{R_i}^{R_o} = (\pi / 2)(R_o^4 - R_i^4)$$

Combining eqs(5.3) and (5.5) gives the simple torsion theory

$$T / J = G\theta / L = \tau / r \tag{5.6}$$

The torque carried by a rotating shaft used to transmit power $P = T\omega$ (unit - W) is found from

$$T = P / \omega = 60P / (2\pi N) \text{ (unit - Nm)} \tag{5.7}$$

where the speed of the shaft is ω rad/s or N rev/min.

Example 5.1 In a generating plant a tubular drive shaft with a diameter ratio 3/5 is to transmit 450 kW at 120 rev/min. If the maximum allowable shear stress is restricted to 100 MPa and the angle of twist is not to exceed 2° over a 3 m length, calculate the diameters required. What diameter of solid shaft can meet both conditions? Take G = 83 GPa.

Equation (5.7) gives the torque as

$$T = (60 \times 450 \times 10^3) / (2\pi \times 120) = 35.81 \times 10^3 \text{ Nm}$$

The maximum shear stress occurs at the outside radius. Equation (5.6) gives

$$2T / [\pi (R_o^4 - R_i^4)] = \tau_{max} / R_o \tag{i}$$

Rearranging eq(i), the outer radius is

$$R_o^3 = 2T / \{\pi \tau_{max} [1 - (R_i / R_o)^4]\} = 2 \times 35.81 \times 10^6 / \{\pi \times 100 [1 - (3/5)^4]\}$$

$\therefore R_o = 63.98$ mm, $R_i = 38.39$ mm

Also, the given angle of twist appears within eq(5.6) as

$$2T/[\pi(R_o^4 - R_i^4)] = G\theta / L \qquad \text{(ii)}$$

Rearranging eq(ii) and converting θ to radians, the outer radius is

$$R_o^4 = 2TL/\{\pi G\theta[1 - (R_i/R_o)^4]\}$$
$$= (2 \times 35.81 \times 10^6 \times 3 \times 10^3 \times 180)/\{\pi^2 \times 83 \times 10^3 \times 2[1 - (3/5)^4]\}$$

$\therefore R_o = 72.16$ mm, $R_i = 43.29$ mm

Setting $R_i = 0$ in eqs(i) and (ii) gives outer radii of a solid shaft to meet each condition as

$$R = [2T/(\pi\tau)]^{1/3} = 61.06 \text{ mm}$$
$$R = [2TL/(\pi G\theta)]^{1/4} = 69.7 \text{ mm}$$

The greater radii of each shaft must be selected in order to keep within the twist allowed. The radii of the hollow shaft are 72.16 mm and 43.29 mm. This design saves a considerable amount of material when compared to a solid shaft of radius 69.7 mm. It is only the outer fibres that reach the maximum shear stress and therefore the understressed core material may be removed without altering the angular twist and torque-carrying capacity significantly.

5.1.1 Stepped Shaft

When a stepped circular shaft is subjected to an axial torque T, the total twist θ_t is the sum of the twists in each each bar. Let subscripts 1, 2, 3 etc refer to lengths over which the diameter is constant. The total angular twist is, from eq(5.6)

$$\theta_t = [TL/(JG)]_1 + [TL/(JG)]_2 + [TL/(JG)]_3 + \ldots \text{ (radians)}$$
$$\theta_t = (T/G)[(L/J)_1 + (L/J)_2 + (L/J)_3 + \ldots](180/\pi) \text{ (degrees)} \qquad (5.8)$$

The maximum shear stresses in each section occur at the outer radii r_1, r_2, r_3 etc as

$$\tau_1 = T r_1 / J_1,\ \tau_2 = T r_2 / J_2 \text{ and } \tau_3 = T r_3 / J_3 \text{ etc.} \qquad (5.9)$$

Example 5.2 An aluminium alloy tube, inner diameter 20 mm and outer diameter 25 mm, is welded to a nimonic alloy tube, 10 mm inner diameter and 20 mm outer diameter (see Fig. 5.2). If the maximum allowable shear stresses for aluminium and nimonic are 35 and 75 MPa respectively and their rigidity moduli are 30 and 80 GPa respectively, determine the maximum torque that the assembly can carry. Given that the length of the nimonic tube is to be twice that of the aluminium tube, find these lengths when the total twist is to be 5°.

Using subscripts A and N for aluminium and nimonic,

$$J_A = \pi(12.5^4 - 10^4)/2 = 14.726 \times 10^3 \text{ mm}^4$$
$$J_N = \pi(10^4 - 5^4)/2 = 22.641 \times 10^3 \text{ mm}^4$$

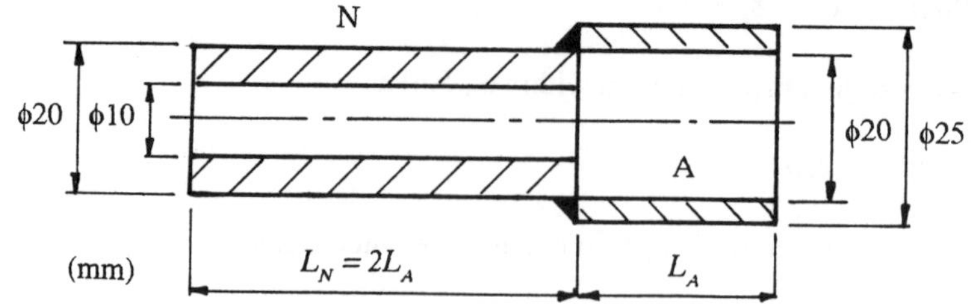

Figure 5.2 Composite alloy bar

Applying eq(5.9), the maximum torque in each bar becomes

$$T_N = (\tau J / r)_N = (75 \times 22.64 \times 10^3)/10 = 110.45 \text{ Nm}$$
$$T_A = (\tau J / r)_A = (35 \times 14.73 \times 10^3)/12.5 = 63.4 \text{ Nm}$$

The lower torque is the maximum permitted for the assembly. Equation(5.8) gives the total twist as

$$\theta_t = T\{[L/(GJ)]_N + [L/(GJ)]_A\} = TL\{[2/(GJ)]_N + [1/(GJ)]_A\}$$

and therefore the length of aluminium is

$$L_A = \theta_t / (T\{[2/(GJ)]_N + [1/(GJ)]_A\}) \quad \text{(i)}$$

where $\theta_t = 5 \times \pi/180 = 0.0873$ rad. Equation (i) gives

$$L_A = 0.0873 / \{(63.4 \times 10^3)[2/(80 \times 10^3 \times 22.641 \times 10^3) + 1/(30 \times 10^3 \times 14.726 \times 10^3)]\}$$
$$= 408.9 \text{ mm} \quad \therefore L_N = 817.8 \text{ mm}$$

5.1.2 Composite Shaft

Co-axial shafts, either fixed together at their ends or bonded along a common diameter (see Figs 5.3a,b), may be treated in a similar manner.

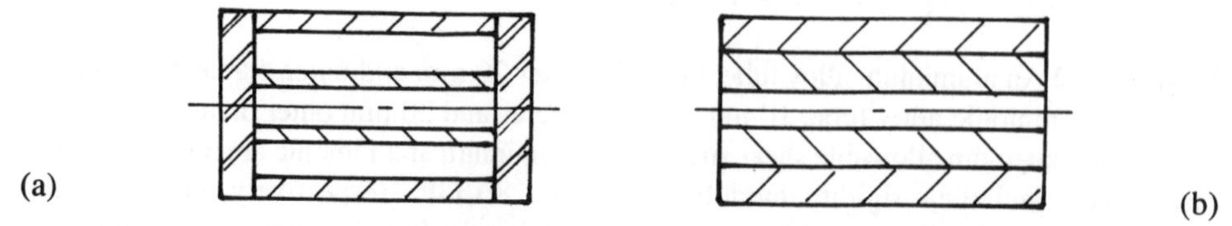

Figure 5.3 Composite shaft section

There are two common features to the torsion of each composite shaft: (i) the angular twist in each shaft will be the same, provided no slipping occurs, and (ii) the torque applied will equal the sum of the torques taken by each shaft. If we let subscripts 1 and 2 refer to the two different materials comprising the composite shafts in Figs 5.3a,b and apply (i) and (ii), then

$$\theta_1 = [TL/(GJ)]_1, \quad \theta_2 = [TL/GJ)]_2 \tag{5.10a,b}$$
$$T = T_1 + T_2 = T_1(1 + T_2/T_1) \tag{5.10c}$$

where, from eq(5.5), at the outer diameter of each bar,

$$T_1 = \tau_1 J_1 / r_1, \; T_2 = \tau_2 J_2 / r_2 \text{ and } T_1/T_2 = G_1 J_1/(G_2 J_2) \tag{5.11a,b,c}$$

Substituting eqs(5.11a-c) into (5.10c),

$$T = (2\tau_1/d_1)[J_1 + (G_2/G_1) J_2] = (2\tau_2/d_2)[J_2 + (G_1/G_2) J_1] \tag{5.12a,b}$$

Example 5.3 The materials and allowable shear stresses given in Example 5.2 form a composite shaft comprising the two bonded concentric cylinders shown in Fig. 5.4a. The outer cylinder is aluminium with an outer diameter of 25 mm and the inner cylinder is nimonic with an inner diameter of 10 mm. The cylinders are bonded along a common diameter of 20 mm for a 500 mm length. Find the torque carrying capacity, the angular twist and the distribution of shear stress through the wall.

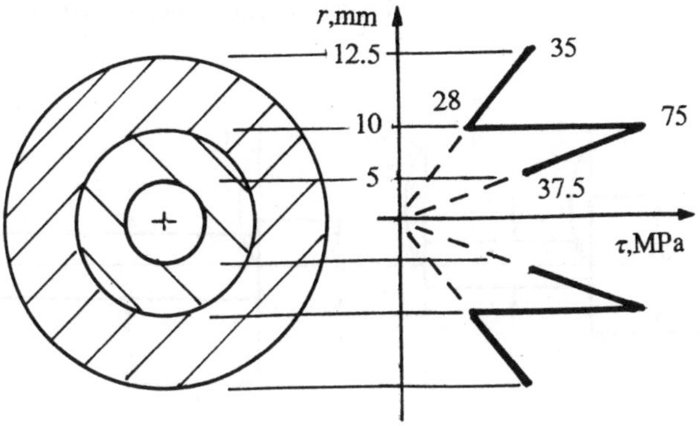

Figure 5.4 Composite shaft showing shear stress distribution

Replacing 1 with A (aluminium) and 2 with N (nimonic), eq(5.10c) and (5.11a,b) give

$$T = (\tau J / r)_A + (\tau J / r)_N$$
$$= (35 \times 14.726 \times 10^3)/12.5 + (75 \times 22.64 \times 10^3)/10$$
$$= (41.233 + 169.808) \times 10^3 \text{ Nmm} = 211.04 \text{ Nm}$$

The twist in the aluminium, is from eq(5.10a),

$$\theta_A = [TL/GJ)]_A = (41.233 \times 10^3 \times 500) / (30 \times 10^3 \times 14.726 \times 10^3)$$
$$= 0.0466 \text{ rad} = 2.674°$$

Check, from eq(5.10b), for the twist in the nimonic

$$\theta_N = [TL/(GJ)]_N = (169.808 \times 10^3 \times 500) / (80 \times 10^3 \times 22.641 \times 10^3)$$
$$= 0.0466 \text{ rad} = 2.674°$$

Equation (5.11a) provides the shear stress at the inner and outer diameters of the aluminium

$(\tau_i)_A = (T\,r_i/J\,)_A = (41.233 \times 10^3 \times 10) \,/\, (14.726 \times 10^3) = 28$ MPa
$(\tau_o)_A = (T\,r_o/J\,)_A = (41.233 \times 10^3 \times 12.5) \,/\, (14.726 \times 10^3) = 35$ MPa

The shear stresses at the inner and outer diameters of the nimonic are

$(\tau_i)_N = (Tr_i\,/J\,)_N = (169.808 \times 10^3 \times 5) \,/\, (22.641 \times 10^3) = 37.5$ MPa
$(\tau_o)_N = (Tr_o/J\,)_N = (169.808 \times 10^3 \times 10) \,/\, (22.641 \times 10^3) = 75$ MPa

These show that as the allowable shear stresses are imposed at each outer diameter, there must exist a drop in the shear stress from 75 MPa to 28 MPa, across the interface (see Fig. 5.4b). The stress in each material varies linearly with r and therefore each distribution passes through the shaft centre.

5.1.3 Varying Torque

Figure 5.5 shows a stepped shaft with fixed ends carrying concentrated torques T_1, T_2, T_3 and T_4 at the positions shown.

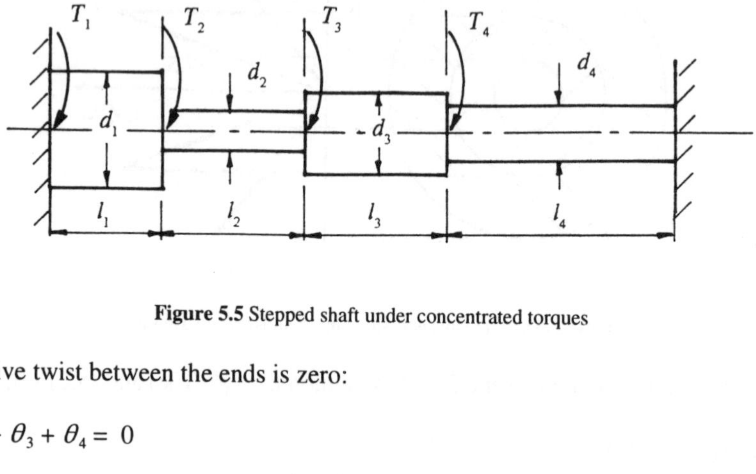

Figure 5.5 Stepped shaft under concentrated torques

The relative twist between the ends is zero:

$$\theta_1 + \theta_2 + \theta_3 + \theta_4 = 0 \qquad (5.13a)$$

The twist in each parallel portion, relative to the r.h. end, is found from eq(5.5),

$$\theta_1 = T_1\,l_1/(G_1J_1),\;\; \theta_2 = (T_1 + T_2)\,l_2/(G_2J_2) \qquad (5.13b,c)$$
$$\theta_3 = (T_1 + T_2 + T_3)\,l_3/(G_3J_3),\;\; \theta_4 = (T_1 + T_2 + T_3 + T_4)\,l_4/(G_4J_4) \qquad (5.13d,e)$$

Substituting eqs(5.13b-e) into (5.13a) and taking all torques to be clockwise,

$$T_1\,l_1/(G_1J_1) + (T_1 + T_2)\,l_2/(G_2J_2) + (T_1 + T_2 + T_3)l_3/(G_3J_3)$$
$$+ (T_1 + T_2 + T_3 + T_4)\,l_4/(G_4J_4) = 0 \qquad (5.14)$$

The maximum shear stress in each portion is found from eq(5.6),

$\tau_1 = T_1 d_1/(2 J_1),\ \ \tau_2 = (T_1 + T_2) d_2/(2 J_2),$ (5.15a,b)

$\tau_3 = (T_1 + T_2 + T_3) d_3/(2 J_3),\ \ \tau_4 = (T_1 + T_2 + T_3 + T_4) d_4/(2 J_4)$ (5.15c,d)

The denominators in eq(5.14) will cancel in the case of a uniform shaft in a single material.

Example 5.4 Construct the torque and twist diagrams for the stepped shaft in Fig. 5.6a and find the maximum shear stress and angular twist in the shaft. Take G = 80 GPa.

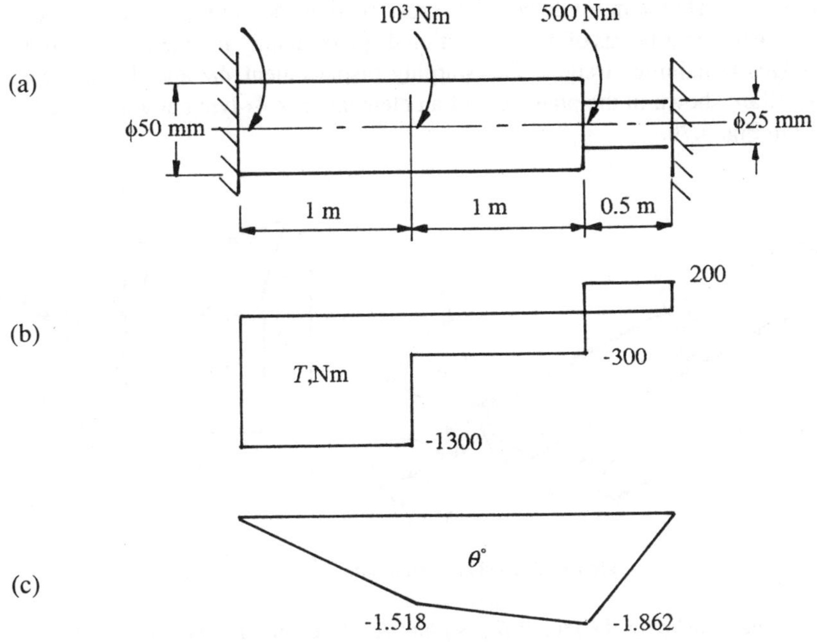

Figure 5.6 Stepped shaft under concentrated torques

When $d_1 = d_2 = 2d_3$ and $l_1 = l_2 = 2l_3$, eq(5.14) reduces to

$$l_1/(G_1 J_1)[T_1 + (T_1 + T_2) + 8(T_1 + T_2 + T_3)] = 0$$
$$10\,T_1 + 9T_2 + 8T_3 = 0$$
$$T_1 = -\ (8T_3 + 9T_2)/10 = -\ 1300 \text{ Nm}$$

indicating that the direction of T_1 is reversed. Figure 5.6b shows the variation in T with shaft length, from which it is evident that the maximum torque is 1300 Nm. Using $J = \pi d^4/32$, eqs(5.13b-d) give the relative twists as

$$\theta_1 = -\ (1300 \times 10^3 \times 10^3) / (80 \times 10^3 \times 61.36 \times 10^4) = -\ 0.026 \text{ rad} = -\ 1.517°$$
$$\theta_2 = -\ (300 \times 10^3 \times 10^3) / (80 \times 10^3 \times 61.36 \times 10^4) = -\ 0.006 = -\ 0.350°$$
$$\theta_3 = (200 \times 10^3 \times 500) / (80 \times 10^3 \times 38.35 \times 10^3) = 0.0326 = 1.867°$$

Their sums are distributed in the manner of Fig. 5.6c. The maximum shear stress in each section follows from eqs(5.15a-c). The sign change implies stress reversal

$\tau_1 = -\ (1300 \times 10^3 \times 25) / (61.36 \times 10^4) = -\ 52.97$ MPa
$\tau_2 = -\ (300 \times 10^3 \times 25) / (61.36 \times 10^4) = -\ 12.22$ MPa
$\tau_3 = (200 \times 10^3 \times 12.5) / (38.35 \times 10^3) = 65.19$ MPa

5.2 Torsion of Thin Strips

Consider the twisting of a thin strip of width b and thickness t as a torque T is applied about its centroidal axis z, as shown in Fig.5.7a. Plane cross-sections do not remain plane due to warping, i.e. adjacent corner points of a cross-section displace axially in opposite directions but with the edges remaining straight. The warping displacement δw and the spanwise distortion $y \times \delta\theta$ are shown in the plan view of an element $\delta z \times \delta x$ lying at a distance $+\ y$ from the x-axis in Fig. 5.7b.

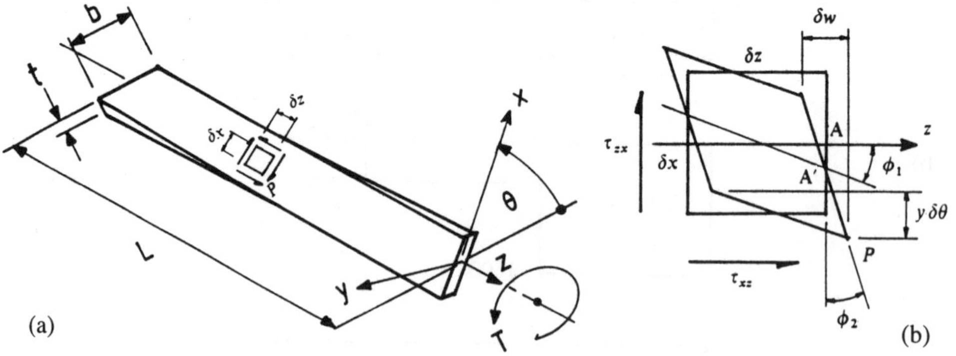

Figure 5.7 Torsion of a thin strip

Let $\delta\theta$ be the angular twist over the length δz as point A rotates into A'. The total change in the angle between originally perpendicular sides in Fig. 5.7b defines the shear strain γ as

$$\gamma = \tan\phi_1 + \tan\phi_2 \tag{5.16}$$

where ϕ_1 is the angular rotation in the chord and ϕ_2 is an equal rotation in the span due to the complementary nature of the shear stress ($\tau_{xz} = \tau_{zx}$) acting on the sides. Elastic distortions are small, so that $\tan\phi_1 \simeq \phi_1 = \phi_2$ (rad) and eq(5.16) becomes

$$\gamma = y\delta\theta/\delta z + \delta w/\delta x = 2y\ \delta\theta/\delta z \tag{5.17}$$

where $\delta\theta/\delta z$ is the rate of twist. Note that the centroidal plane $y = 0$ retains its original rectangular shape, since $\gamma = 0$. From eq(5.17) the distortions to an element beneath this plane, i.e. with y negative, become an inversion of Fig. 5.7b. The horizontal shear stress $\tau = \tau_{xz} = \tau_{xz} = 0$, lying in the plane at height $+\ y$ is, from eq(5.17),

$$\tau = G\gamma = 2Gy\ \delta\theta/\delta z \tag{5.18a}$$

Equation(5.18a) shows that τ varies linearly with y and thus τ_{max} is reached at the longer edges (see Fig. 5.8). This figure also shows the linear distribution in vertical shear stress τ' in which the maximum value τ'_{max} is assumed to be related to τ_{max} by $\tau'_{max} = t\tau_{max}/b$.

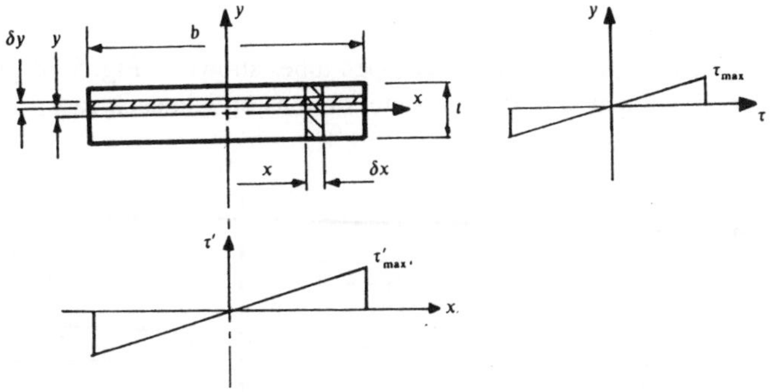

Figure 5.8 Shear stress distributions from torsion of a thin strip

For any point (x, y) in the cross-section, τ and τ' are similarly related, $\tau' = t\tau / b$. Substituting from eq(5.18a) gives an expression for the vertical shear stress distribution,

$$\tau' = 2Gx\,(t/b)^2\delta\theta/\delta z \tag{5.18b}$$

in which $y = (t/b)x$. Since $\delta w/\delta x = y\ \delta\theta/\delta z$, the warping displacements w can be found from

$$w = \int y\,(\delta\theta/\delta z)\,\mathrm{d}x \tag{5.19}$$

An integration constant is unnecessary when the origin for x, y coincides with a point that does not warp. This point is known as the centre of twist and will lie at the intersection between axes of symmetry.

The applied torque is the resultant of the two shear stress distributions. Let torque T_1 be due to τ in eq(5.18a). Since τ will act on an elemental strip of thickness δy, distance y from x (see Fig. 5.8), then

$$T_1 = \int (\tau b\,\mathrm{d}\,y)\,y = 2Gb\,(\delta\theta/\delta z)\int_{-t/2}^{t/2} y^2\,\mathrm{d}\,y = 2Gb\,(\delta\theta/\delta z)(t^3/12)$$

The vertical shear stress τ' in eq(5.18b) acts on an elemental strip of thickness δx and distance x from y (see Fig. 5.8). The additional torque contribution T_2 is therefore

$$T_2 = \int (\tau'\,t\,\mathrm{d}x)\,x = 2G\,t\,(t/b)^2(\delta\theta/\delta z)\int_{-b/2}^{b/2} x^2\,\mathrm{d}x = 2Gb(\delta\theta/\delta z)(t^3/12) = T_1$$

$$\therefore\ T = T_1 + T_2 = 4Gb\,(\delta\theta/\delta z)(t^3/12) \tag{5.20}$$

Combining eqs(5.18a and 5.20) and introducing the *St Venant torsion constant* $J = bt^3/3$ gives a three-part formula:

$$G\ \delta\theta/\delta z = \tau/(2y) = T/J \tag{5.21}$$

Equation (5.21) is similar to eq(5.6), for circular shafts. Setting $y = \pm t/2$ in eq(5.21) gives the angular twist and maximum shear stress for a uniformly thin rectangular sections as

$$\theta = TL/(JG) \text{ and } \tau_{\max} = 3T/(bt^2) \tag{5.22a,b}$$

5.2.1 Thin-Walled Open Tubes

Equation (5.21) will also apply to thin-walled open tubes shown in Fig. 5.9a, provided that the thickness t is small compared to the perimeter dimension b in each case.

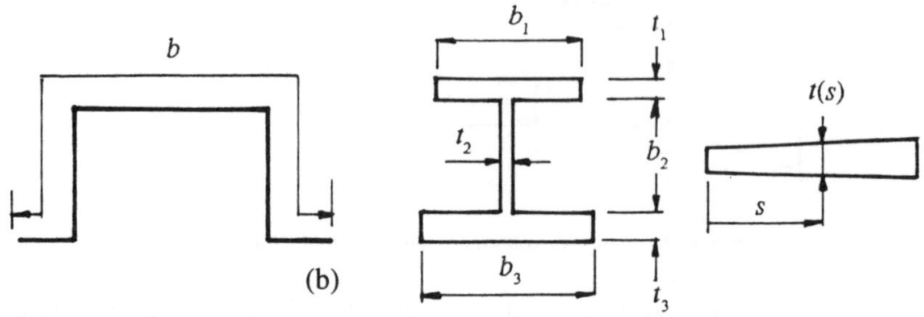

Figure 5.9 Thin-walled open sections under torsion

In the case of Fig. 5.9b, where the thickness of the rectangular components varies, the torsion constant in eq(5.22a) is written as

$$J = \tfrac{1}{3}\sum(b\,t^3) = b_1 t_1^3/3 + b_2 t_2^3/3 + b_3 t_3^3/3 \tag{5.23a}$$

When the thickness tapers gradually with the perimeter dimension s in Fig. 5.9c, then $t = t\,(s)$ and the torsion constant is found from the integral,

$$J = \tfrac{1}{3}\int t^3 \mathrm{d}s \tag{5.23b}$$

The primary warping displacement that occurring in these tubes and the effect of constraining this warping is examined in Section 5.5.

5.2.2 Thicker-Walled Open Tubes

For a uniform rectangular section, in which t is not small compared to b, eqs(5.22a and b) are written as

$$\theta = TL/(\beta\, bt^3 G) \quad \text{and} \quad \tau_{max} = T/(\alpha\, b\, t^2) \tag{5.24a,b}$$

where Table 5.1 shows that the coefficients α and β depend upon $b\,/\,t$. As $b\,/\,t$ increases both α and β approach the ⅓ values for a thin strip.

Table 5.1 Coefficients α and β in eqs(5.24a,b)

b/t	1	2	4	6	8	10	∞
α	0.208	0.246	0.282	0.299	0.307	0.313	0.333
β	0.141	0.299	0.281	0.299	0.307	0.313	0.333

For thicker sections composed of n rectangles, e.g. I and U extrusions and others fabricated from welded plates, in which t is not small compared to b, eq(5.24a) becomes

$$\theta = TL/[\sum_{i=1}^{n} (\beta bt^3)_i G] \tag{5.25}$$

The twist experienced by each component rectangle will be the same

$$\theta = T_1 L/[(\beta bt^3)_1 G] = T_2 L/[(\beta bt^3)_2 G] = T_3 L/[(\beta bt^3)_3 G] \tag{5.26a,b,c}$$

where $T = T_1 + T_2 + T_3$ and τ_{max} is the greatest of

$$\tau_1 = T_1/(\alpha b t^2)_1,\ \tau_2 = T_2/(\alpha b t^2)_2,\ \tau_3 = T_3/(\alpha b t^2)_3 \tag{5.27a,b,c}$$

Example 5.5 Figure 5.10 gives the dimensions of an extruded cross-section in light alloy that is to withstand a torque of 75 Nm applied about its longitudinal centroidal axis. If G = 29 GPa calculate the position and magnitude of the maximum shear stress in each component rectangular area and the rate of twist.

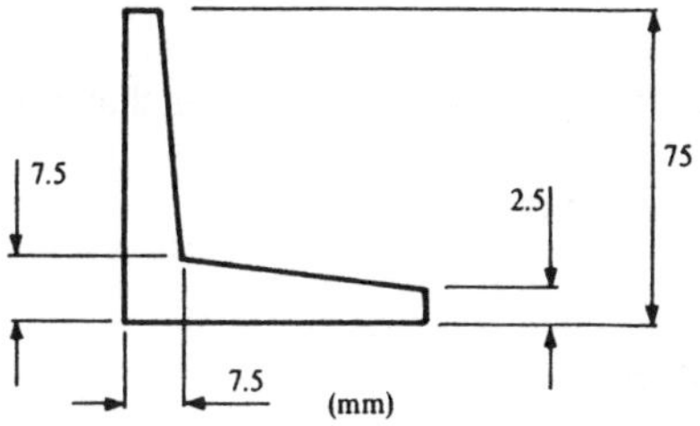

Figure 5.10 Angle section with tapered edges

The section may be split into two trapezia and one square. To find J for the section we must evaluate

$$J = (2/3) \int [t(s)]^3 ds + \beta bt^3 \tag{i}$$

Let s originate from the bottom right and top left corners so that the thickness of the sloping sides of the trepezia varies as $t = s/13.5 + 2.5$ for $0 \le s \le 67.7$ mm. For the square $b/t = 1$ and $\beta = 0.141$ from Table 5.1. Equation (i) becomes

$$J = (2/3) \int_0^{67.5} (s/13.5 + 2.5)^3 ds + 0.141(7.5)^4$$
$$= (1/3686.7)\,|\, s^4/4 + 33.74\, s^3 + 1707.59\, s^2 + 38409.3 \,\Big|_0^{67.5} + 446.13$$
$$= 7036.2 + 446.13 = 7482.23 \text{ mm}^4$$

The rate of twist is

$$\delta\theta/\delta z = T/(GJ)$$
$$= (75 \times 10^3) / (29 \times 10^3 \times 7482.23)$$
$$= 0.3457 \times 10^{-3} \text{ rad/mm} = 19.8°/\text{m}$$

The torque carried by each trapezium is

$T_t = (\delta\theta / \delta z) J_t G = 0.3457 \times 10^{-3} \times 3518.10 \times 29 \times 10^3 = 35.27$ Nm

Equation (5.21) gives a maximum shear stress at the thicker end, where $t = 7.5$ mm,

$\tau = (2\,y) T_t / J_t = T_t\, t / J_t = (35.27 \times 10^3) 7.5 / 3518.03 = 75.19$ MPa

The torque carried by the square is

$T_s = (\delta\theta / \delta z) J_s G = 0.3457 \times 10^{-3} \times 446.13 \times 29 \times 10^3 = 4.47$ Nm

Taking $\alpha = 0.208$ in eq(5.24b), the maximum shear stress in the square sides is

$\tau = (4.47 \times 10^3) / (0.208 \times 7.5^3) = 51$ MPa

These maximum shear stresses occur at a point along the longer edges nearest the centroid. The effect that the fillet radius has on raising τ_{max} may be assessed from Tefftz's equation:

$$\tau = (7/4)\, \tau_{max} (t / r)^{1/3} \qquad \text{(ii)}$$

For example, if the fillet radius was 1 mm, eq(ii) shows that τ is raised to

$\tau = (7/4)\; 75.19\; (7.5 / 1)^{1/3} = 255$ MPa

5.3 Torsion of Prismatic Bars

Figure 5.11a shows a solid shaft of any arbitrary cross-section carrying a torque T, applied about a longitudinal axis z, passing throught the centre of twist O.

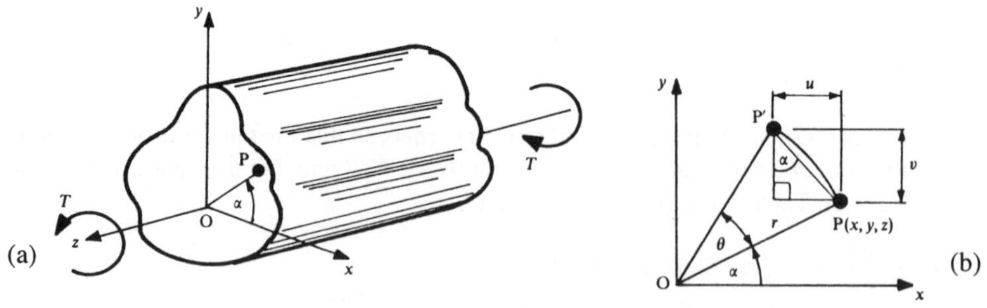

Figure 5.11 Prismatic bar under torsion

Figure 5.11b shows that a point P(x, y, z), not co-incident with O, will displace to P' by the amounts u and v in the x and y - directions respectively. These displacements take the signs of x and y, so that for a unit length of bar

$$u = -\,r\theta \sin\alpha = -\,r\theta (y / r) = -\,y\theta \qquad (5.28a)$$
$$v = r\theta \cos\alpha = r\theta (x / r) = x\theta \qquad (5.28b)$$

in which θ is the angle of twist of the bar defining the anti-clockwise rotation of P to P' in the direction of the torque. Point P will also displace in the z - direction as the cross-section warps out of the x - y plane. The warping displacement w, is assumed to be proportional to the rate of twist $\delta\theta/\delta z$ and a function $\psi(x, y)$ describing the variation in w over the cross-section. This gives the warping function

$$w = \psi(x, y)\delta\theta/\delta z \tag{5.28c}$$

where ψ is to be determined. Next, examine the strains associated with a displacement vector **PP'**. The distortion of the cross-section out of its own plane is defined by the following two shear strain components

$$\begin{aligned}\gamma_{zx} &= \partial u/\partial z + \partial w/\partial x \\ &= -\partial(y\theta)/\partial z + (\partial\psi/\partial x)(\delta\theta/\delta z) = (\partial\psi/\partial x - y)(\delta\theta/\delta z)\end{aligned} \tag{5.29a}$$

$$\begin{aligned}\gamma_{zy} &= \partial v/\partial z + \partial w/\partial y \\ &= \partial(x\theta)/\partial z + (\partial\psi/\partial y)(\delta\theta/\delta z) = (x + \partial\psi/\partial y)(\delta\theta/\delta z)\end{aligned} \tag{5.29b}$$

It follows from eqs(5.28a-c) and eqs(2.17a-c) that direct strains $\varepsilon_x = \partial u/\partial x$, $\varepsilon_y = \partial v/\partial y$ and $\varepsilon_z = \partial w/\partial z$ are absent and that, since $\gamma_{xy} = \partial u/\partial y + \partial v/\partial x = -\theta + \theta = 0$, the cross-section does not shear in the x, y plane. The components of shear stress present are those associated with the non-zero shear strains. These are, from eqs(5.29a and b),

$$\tau_{zx} = G\gamma_{zx} = G(\partial\psi/\partial x - y)\delta\theta/\delta z \tag{5.30a}$$

$$\tau_{zy} = G\gamma_{zy} = G(\partial\psi/\partial y + x)\delta\theta/\delta z \tag{5.30b}$$

which act together with their complements ($\tau_{xz} = \tau_{zx}$, $\tau_{yz} = \tau_{zy}$) in the directions shown in Fig. 5.12a. From eqs(5.30a,b),

$$\partial\tau_{zx}/\partial y = G(\partial^2\psi/\partial x\,\partial y - 1)\delta\theta/\delta z \tag{5.31a}$$

$$\partial\tau_{zy}/\partial x = G(\partial^2\psi/\partial x\,\partial y + 1)\delta\theta/\delta z \tag{5.31b}$$

Subtracting eq(5.31a) from eq(5.31b) gives

$$\partial\tau_{zy}/\partial x - \partial\tau_{zx}/\partial y = 2G(\delta\theta/\delta z) \tag{5.32}$$

Introducing the Prandtl stress function ϕ,

$$\tau_{zx} = \partial\phi/\partial y,\quad \tau_{zy} = -\partial\phi/\partial x \tag{5.33a,b}$$

eq(5.32) becomes

$$\partial^2\phi/\partial x^2 + \partial^2\phi/\partial y^2 = -2G(\delta\theta/\delta z) \tag{5.34}$$

The determination of the shear stress distribution consists of finding a function ϕ to satisfy eq(5.34). Once ϕ has been found, the warping function ψ follows from eq(5.28c). Lines of constant ϕ in the section (Fig. 5.12b) are trajectories of constant resultant shear stress $\tau_{max} = \sqrt{(\tau_{zx}^2 + \tau_{zy}^2)}$. The boundary is also a trajectory, usually associated with $\phi = 0$, since τ_{zy} will be zero for the exterior surface.

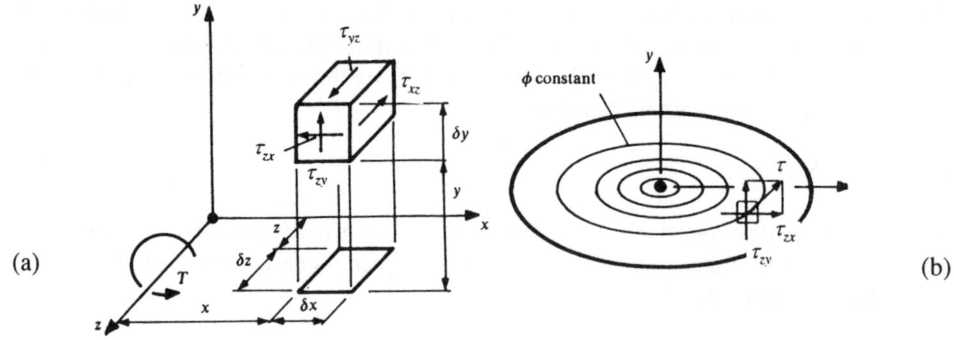

Figure 5.12 Shear stresses and their resultant trajectories

Figure 5.12a shows how the resistive torque is the resultant of the two shear stress distributions. Integrating over x and y, T is found from the double integral

$$T = \int_x \int_y (x\tau_{zy} - y\tau_{zx})\, \mathrm{d}x\, \mathrm{d}y$$

Substituting the stress function ϕ from eqs(5.33a,b),

$$\begin{aligned} T &= -\int\int (x\, \partial\phi/\partial x + y\, \partial\phi/\partial y)\, \mathrm{d}x\, \mathrm{d}y \\ &= -\int \mathrm{d}y \int (\partial\phi/\partial x)\, x\, \mathrm{d}x - \int \mathrm{d}x \int (\partial\phi/\partial y)\, y\, \mathrm{d}y \\ &= -\int \mathrm{d}y\, [x\phi - \int \phi\, \mathrm{d}x] - \int \mathrm{d}x\, [\, y\phi - \int \phi\, \mathrm{d}y] \end{aligned}$$

and, with $\phi = 0$ around the boundary, $[x\phi] = [y\phi] = 0$. Hence, the applied torque becomes

$$T = 2 \int\int \phi\, \mathrm{d}x\, \mathrm{d}y = GJ\, (\delta\theta / \delta z) \tag{5.35}$$

The final term in eq(5.35) applies to all torsion problems including thin strips and open and closed tubes. Equations(5.33) - (5.35) apply to all prismatic bars but they are unable to provide closed solutions to all but a few bar sections.

Example 5.6 Show that the function $\phi = C\,(1 - x^2/A^2 - y^2/B^2)$ provides a solution to the torsion of an elliptical shaft whose section, given in Fig. 5.13a, is $x^2/A^2 + y^2/B^2 = 1$. Determine the stress distribution and the manner of warping and show that the equations reduce to simple torsion theory in the case of a circular shaft.

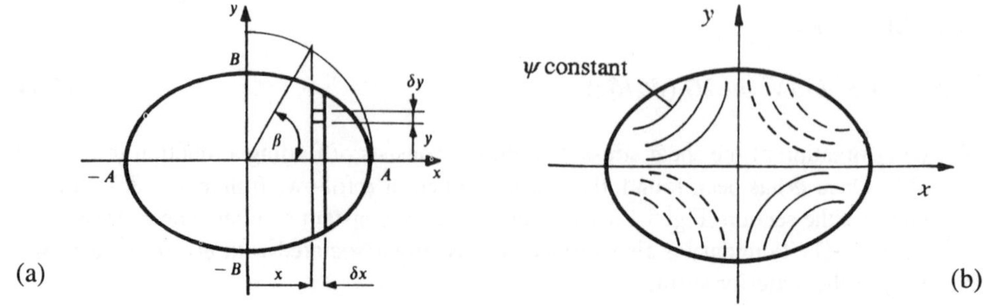

Figure 5.13 Elliptical section

The constant C is found from eq(5.34) when

$$- 2C / A^2 - 2C / B^2 = - 2G\,(\delta\theta / \delta z)$$
$$\therefore\ C = A^2 B^2 G\,(\delta\theta / \delta z)/(A^2 + B^2) \qquad \text{(i)}$$

The check on the function is made through eq(5.35) where

$$T = 2 \int\int \phi\ \mathrm{d}x\,\mathrm{d}y = 2C \int\int (1 - x^2/A^2 - y^2/B^2)\,\mathrm{d}x\,\mathrm{d}y$$

Substituting $\mathrm{d}\,y = -\,(x/y)(B^2/A^2)\,\mathrm{d}x$ where $y = B\,\sqrt{(1 - x^2/A^2)}$

$$T = (8CB/3) \int_{-A}^{A} (1 - x^2/A^2)^{3/2}\,\mathrm{d}x$$

With a further trigonometric substitution $x = A \sin \beta$, the integral is evaluated using the result in eq(i), to give

$$T = \pi ABC = [\pi A^3 B^3/(A^2 + B^2)]G\,(\delta\theta / \delta z) \qquad \text{(ii)}$$

The fact that the term in square brackets is the polar second moment of area for an elliptical section confirms the validity of the function ϕ. The stresses are then found from eqs(5.33a,b),

$$\tau_{zx} = \partial\phi / \partial y = C\,\partial[1 - x^2/A^2 - y^2/B^2]/\,\partial y$$
$$= -\,2y\,A^2 G\,(\delta\theta / \delta z)/(A^2 + B^2) \qquad \text{(iii)}$$

$$\tau_{zy} = -\,\partial\phi / \partial x = -\,C\,\partial[1 - x^2/A^2 - y^2/B^2]/\,\partial x$$
$$= 2x\,B^2 G\,(\delta\theta / \delta z)/(A^2 + B^2) \qquad \text{(iv)}$$

The warping function is found from equating (iii) to 5.30a

$$\tau_{xz} = G(\partial\psi / \partial x - y)(\delta\theta / \delta z) = -\,2y\,A^2 G\,(\delta\theta / \delta z)/(A^2 + B^2)$$
$$\therefore\ \partial\psi / \partial x - y = -\,2y\,A^2/(A^2 + B^2)$$
$$\partial\psi / \partial x = y[1 - 2A^2/(A^2 + B^2)] = y\,(B^2 - A^2)/(B^2 + A^2)$$
$$\therefore\ \psi(x, y) = y\,x\,(B^2 - A^2)/(B^2 + A^2)$$

from which the warping displacements are given by

$$w = \psi(x, y)\delta\theta / \delta z = x\,y\,(\delta\theta / \delta z)(B^2 - A^2) / (B^2 + A^2) \qquad \text{(v)}$$

Equation (v) may be checked from eqs (iv) and (5.30b). Thus, for a given twist rate $\delta\theta / \delta z$, eq(v) describes the manner in which every cross-section warps. There is no warping at points along the x and y axes, i.e. when x or y is zero. Equation(v) describes hyperbolas of constant w. For example, Fig. 5.13b shows these to be upward in quadrant 2, downward in quadrant 3. Putting $A = B = R$ in eq(ii) facilitates a circular section of radius R,

$$T = (\pi / 2)R^4 G\,(\delta\theta / \delta z) = JG\,(\delta\theta / \delta z) \qquad \text{(vi)}$$

and from eqs(iii and iv), the resultant shear stress is

$$\tau_{max} = \sqrt{(\tau_{zy}^2 + \tau_{zx}^2)} = G\,(\delta\theta/\delta z)\,\sqrt{(x^2 + y^2)} = GR\,(\delta\theta/\delta z) \qquad \text{(vii)}$$

Combining eqs(vi) and (vii) recovers eq (5.6) in the form

$$T/J = G\,(\delta\theta/\delta z) = \tau/R$$

where it also becomes apparent from eq(v) that when $w = 0$ for circular sections, plane cross-sections remain plane.

Example 5.7 Show that when a prismatic bar has a rectangular section for which breadth $b \gg$ thickness t (Fig. 5.14a), the corresponding stress function confirms the thin strip theory given in Section 5.2. Find the warping function for the strip when the dimensions are $b = 100$ mm and $t = 5$ mm. Find the twist rate and the manner in which the section warps under a torque of 10 Nm. Take $G = 80$ GPa.

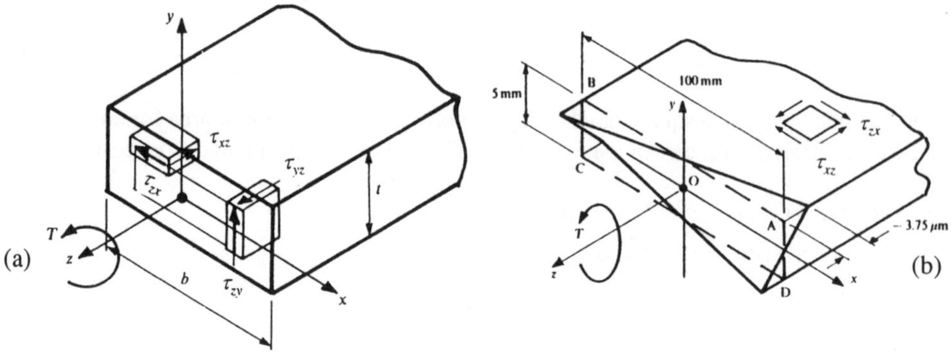

Figure 5.14 Rectangular section bar

The curvature of a thin strip allows a simplification $\partial^2\phi/\partial x^2 = 0$ for the x-direction. Equation(5.34) becomes

$$\partial^2\phi/\partial y^2 = -2G\,(\delta\theta/\delta z)$$

which may be integrated directly, to give

$$\phi = -G\,(\delta\theta/\delta z)\,y^2 + A\,y + B$$

Taking the boundary $y = \pm\,t/2$ to be the contour $\phi = 0$ gives

$$0 = -G\,(\delta\theta/\delta z)\,t^2/4 + At/2 + B$$
$$0 = -G\,(\delta\theta/\delta z)\,t^2/4 - At/2 + B$$

Adding and subtracting gives $A = 0$ and $B = G(\delta\theta/\delta z)(t^2/4)$

$$\therefore\ \phi = G(\delta\theta/\delta z)(t^2/4 - y^2)$$

It follows from eq(5.32b) that the vertical shear stress component is

$$\tau_{zy} = -\,\partial\phi/\partial x = 0$$

That is, the vertical shear stress is disregarded when $b \gg t$ in Fig. 5.14a. The non-zero horizontal shear stress is, from eq(5.33a),

$$\tau_{zx} = \partial \phi / \partial y = -\ 2G\, y\, (\delta\theta / \delta z) \qquad \text{(i)}$$

which reveals a linear variation across the thickness, having a maximum value along the longer sides for $y = \pm\, t/2$. The torque-twist relationship is supplied from eq(5.35a)

$$T = -\ 2 \int\int \phi\, \mathrm{d}x\, \mathrm{d}y = 2G(\delta\theta / \delta z) \int_{-t/2}^{t/2} (t^2/4 - y^2)\, \mathrm{d}y \int_{-b/2}^{b/2} \mathrm{d}x$$

$$= -\ 2bG\, (\delta\theta / \delta z) \int_{-t/2}^{t/2} (t^2/4 - y^2)\, \mathrm{d}y = 2bG\, (\delta\theta / \delta z)\ \Big|\ t^2 y/4 - y^3/3 \Big|_{-t/2}^{t/2}$$

$$T = G\, (bt^3/3)(\delta\theta / \delta z) = GJ\, (\delta\theta / \delta z) \qquad \text{(ii)}$$

Combining eqs(i) and (ii) confirms the previous derivation in eq(5.21). Equating (i) to (5.30a) supplies the warping function ψ

$$\tau_{xz} = G\, (\partial \psi / \partial x - y)(\delta\theta / \delta z) = -\ 2G\, y\, (\delta\theta / \delta z)$$
$$\therefore\ \partial \psi / \partial x = -\ y \ \Rightarrow\ \psi = -\ yx$$

This is readily confirmed from eq(5.30b). The integration applies to x, y axes of symmetry with an origin at the centre of twist where $w = 0$. Note, from the comparison with eq(5.19), that the minus sign arises because this derivation employs a positive anti-clockwise torque. The twist rate is;

$$\delta\theta / \delta z = T/(GJ) = (10 \times 10^3)/[(100 \times 5^3/3)(80 \times 10^3)] = 0.03 \times 10^{-3}\ \text{rad/mm}$$

and the warping displacements, shown in Fig. 5.14b, are found from eqs (iii) and (5.28c):

$w = 0.03 \times 10^{-3} (-\ x\, y)$
@ A (+ 50, + 2.5) $w_A = -\ 3.75 \times 10^{-3}$ mm
@ B (+ 50, - 2.5) $w_B = +\ 3.75 \times 10^{-3}$ mm
@ C (- 50, - 2.5) $w_C = -\ 3.75 \times 10^{-3}$ mm
@ D (- 50, + 2.5) $w_D = +\ 3.75 \times 10^{-3}$ mm

Along each axis of symmetry, $x = y = 0$ and $w = 0$ (see Fig. 5.14b).

5.3.1 Membrane Analogy

Prandtl showed that eq(5.34) is similar in form to the differential equation governing the deflection of a membrane pressurized from one side. The plan and sectional elevations of the membrane, (Figs 5.15a,b), reveal the forces acting on an element of the deflected surface.

The pressure p deflects the membrane in the z-direction and stretches it with a uniform surface tension S (N/mm). The surface element $\delta x \times \delta y$ has forces $S\delta x$ and $S\delta y$ acting parallel to x and y (Fig. 5.15a) and tangential to the surface in Fig. 5.15b. The gradients of the left edge of the element in Fig. 5.15b is $\partial z/\partial x$. This gradient increases by the amount $\partial/\partial x(\partial z/\partial x)\delta x = (\partial^2 z/\partial x^2)\delta x$ at the right edge as shown. Resolving the tangential forces leads to a vertical equilibrium equation for the element.

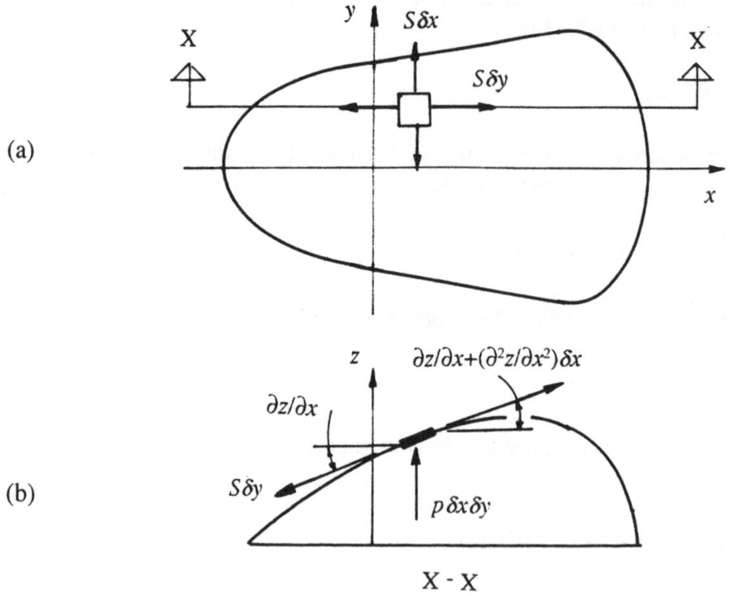

Figure 5.15 Deflection of a membrane under lateral pressure

$$- S\,\delta y(\partial z/\partial x) + S\,\delta y[\partial z/\partial x + (\partial^2 z/\partial x^2)\delta x] - S\,\delta x\,(\partial z/\partial y)$$
$$+ S\,\delta x\,[\partial z/\partial y + (\partial^2 z/\partial y^2)\delta y] + p\delta x\delta y = 0$$

which simplifies to

$$\partial^2 z/\partial x^2 + \partial^2 z/\partial y^2 = - p/S \tag{5.36}$$

The volume beneath the deflected membrane is given by

$$V = \int\int z\,\mathrm{d}x\,\mathrm{d}y \tag{5.37}$$

Now, when we compare eqs(5.36) with (5.34) and eq(5.37) with (5.35), the membrane analogy becomes clear. For convenience if we set $p/S = 2$, the stress function becomes

$$\phi = z \times G\,(\delta\theta/\delta z) = \text{Deflection of membrane} \times G\,(\delta\theta/\delta z) \tag{5.38}$$

and the torque becomes

$$T = \text{Twice the volume beneath the membrane} \times G\,(\delta\theta/\delta z) \tag{5.39}$$

The St Venant torsion constant follows from eqs(5.35) and (5.39) as

$$J = T/[G\,(\delta\theta/\delta z)] = \text{Twice the volume under the membrane} \tag{5.40}$$

The shear stress components, which are partial derivatives of ϕ (see eqs 5.33a,b), become

$\tau_{zx} = \partial\phi/\partial y = G\,(\delta\theta/\delta z) \times \partial z/\partial y$

$= G\,(\delta\theta/\delta z) \times$ slope of membrane in the y-direction (5.41a)

$\tau_{zy} = -\,\partial\phi/\partial x = -\,G\,(\delta\theta/\delta z) \times \partial z/\partial x$

$= -\,G\,(\delta\theta/\delta z) \times$ slope of membrane in the x-direction (5.41b)

Equations (5.38 - 5.41) apply to the deflection of a membrane covering a hole of the same shape as the bar cross-section. If a bubble is blown over the hole then the stress function ϕ can be identified with contour lines of constant deflection z. Taking measurements of the gradient of tangents lying on a contour enables the components of shear stress to be obtained experimentally for sections of complex shape. At a given point on the contour, the maximum shear stress is found from the gradient of the normal to the contour. To show this, refer to the plan of a contour in Fig. 5.16a.

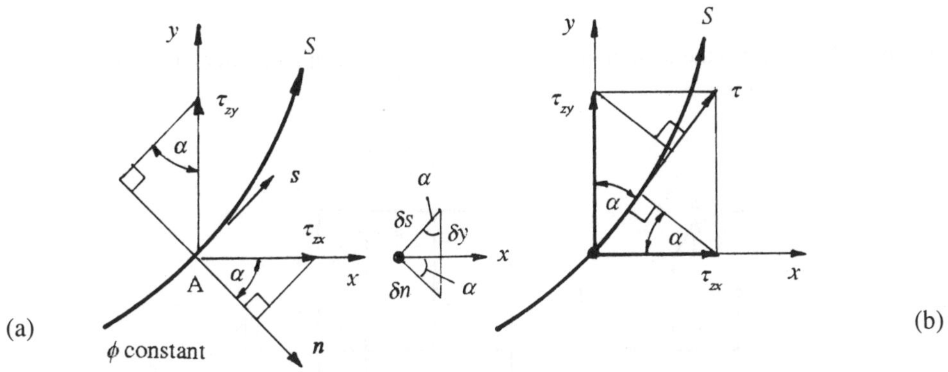

Figure 5.16 Membrane analogy

At point A, the normal and tangential vectors are **n** and **s** with magnitudes n and s respectively. Now along the contour line S, the gradient is

$$\mathrm{d}\phi/\mathrm{d}s = (\partial\phi/\partial y)(\mathrm{d}y/\mathrm{d}s) + (\partial\phi/\partial x)(\mathrm{d}x/\mathrm{d}s)$$

Substituting from eqs(5.41a,b) and noting (inset Fig. 5.16b) that $\sin\alpha = \mathrm{d}x/\mathrm{d}s$, $\cos\alpha = \mathrm{d}y/\mathrm{d}s$

$$\mathrm{d}\phi/\mathrm{d}s = \tau_{zx}\cos\alpha - \tau_{zy}\sin\alpha \tag{5.42a}$$

The left-hand side of eq(5.42a) is zero, since $\mathrm{d}\phi/\mathrm{d}s$ is zero when ϕ is constant for a contour line (by analogy z is constant in the deflected membrane). The right-hand side of eq(5.42a) defines the projection of the two shear stress components upon the **n**-direction. That is, from Fig. 5.16a,

$$\tau_n = \tau_{zx}\cos\alpha - \tau_{zy}\sin\alpha \tag{5.42b}$$

The shear stress is zero normal to the contour. It follows from eqs(5.42a,b) that $\tau_n = 0$. Hence the shear stress vector must lie tangential to the contour line. Projecting the two shear stresses at point A into the **s** - direction (see Fig. 5.16b) gives

$$\tau = \tau_{zx} \sin \alpha + \tau_{zy} \cos \alpha \qquad (5.43a)$$

Substituting from eqs(5.41a,b) and noting from Fig. 5.16b that $\sin \alpha = -\, dy/dn$, $\cos \alpha = dx/dn$

$$\tau = (\partial \phi / \partial y)(-\, dy/dn) + (-\, \partial \phi / \partial x)(dx/dn) = -\, d\phi / dn \qquad (5.43b)$$

Equation (5.43b) shows that the magnitude of the shear stress is the slope of the normal to the membrane. Also, we see from the geometry in Fig. 5.16b that the resultant shear stress is

$$\tau_{max} = \surd(\tau_{zx}^{\,2} + \tau_{zy}^{\,2}) \qquad (5.43c)$$

We shall first apply the membrane analogy to confirm the known solution to torsion of a thin strip and then show how the result can be employed to describe the torsion of an aerofoil section divided into thin strips.

Example 5.8 Confirm the torsion theory of a thin strip breadth b and thickness t ($b \gg t$) from the membrane analogy.

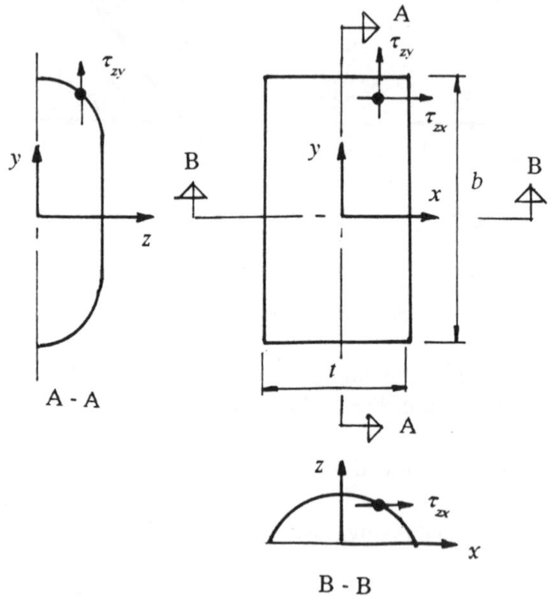

Figure 5.17 Torsion of a thin rectangular strip

Imagine a membrane stretched over a thin rectangular aperture $b \times t$. Figure 5.17 shows two further views of the membrane deflection under lateral pressure. It is clear from this that contour lines are flat in the central region of section A-A but not of section B-B. Thus $\partial^2 z / \partial y^2 = 0$ and eq(5.36) reduces to $\partial^2 z / \partial x^2 = -\, 2$. Integrating these for z

$$z = -\, x^2 + C_1 x + C_2 y + C_3$$

where both C_1 and C_2 are zero, since for $x = 0$, $\partial z/\partial x = 0$ and for $y = 0$, $\partial z/\partial y = 0$. Also for $x = \pm t/2$, $z = 0$ which gives $C_3 = t^2/4$. Hence the deflection equation becomes

$$z = -x^2 + t^2/4$$

The encosed volume eq(5.37) becomes

$$V = \int\int z\,dx\,dy = \int_{-b/2}^{b/2}[\int_{-t/2}^{t/2}(t^2/4 - x^2)\,dx\,]\,dy$$

$$= (1/6)\int_{-b/2}^{b/2} t^3\,dy = bt^3/6 \qquad \text{(i)}$$

By the analogy, it follows from eqs(5.39) - (5.41) that

$$J = bt^3/3$$
$$T = 2G\,(\delta\theta/\delta z)(bt^3/6) = GJ\,(\delta\theta/\delta x)$$
$$\tau_{zy} = 2Gx\,(\delta\theta/\delta x),\ \tau_{zx} = 0$$

This shows that τ_{zy} is the the resultant shear stress for which a maximum occurs at the edges where $x = t/2$

$$\tau_{max} = Gt\,(\delta\theta/\delta z) = 3T/(bt^2)$$

5.3.2 Aerofoil Section

There is no stress function appropriate to a prismatic bar with an aerofoil section, but two approximate solutions are available from eqs(5.21) and (5.39). The section is divided into thin strips, as shown in Figs 5.18a,b. Here it is necessary to align the long side of the strip with the y-direction and call this the thickness as shown.

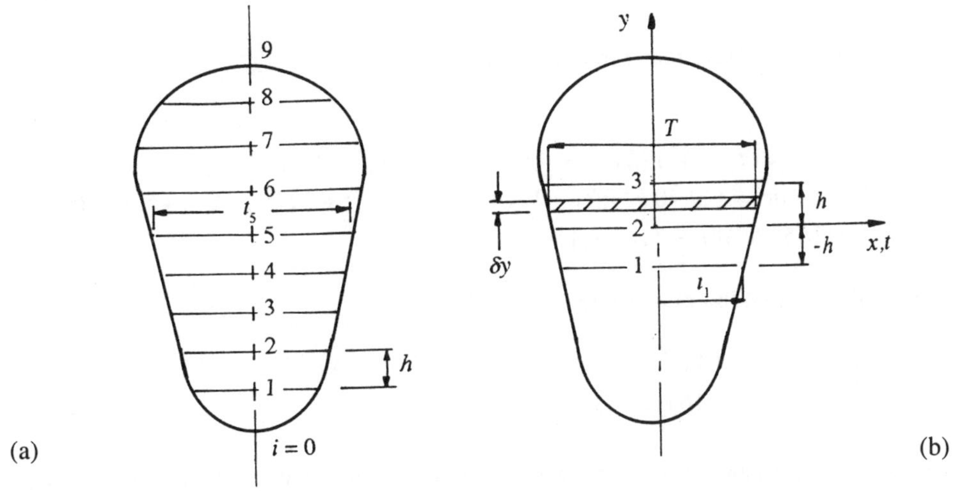

Figure 5.18 Aerofoil section

In the first solution, the torsion constant eq(5.23a) is applied to strips of equal spacing, h in Fig. 5.18a as

$$J = (h/3) \sum_{i=1}^{N} t_i^{\,3} \tag{5.44}$$

where $i = 1, 2 \ldots N$, where N is the number of strips and t_i is the "thickness" at the i th position. The second solution avoids the error associated with a varying thickness within each strip. This applies the membrane analogy across two strips adjacent to x-axis (see Fig. 5.18b). Let the profile within the region $-h \le y \le h$, with t positive, be given by the parabola:

$$t = a + by + cy^2 \tag{5.45a}$$

where the constants a, b and c are found from the following conditions: (i) $y = -h$, $t = t_1$, (ii) $y = 0$, $t = t_2$ and (iii) $y = h$, $t = t_3$. Substituting (i) - (iii) into (5.45a), gives

$$\begin{aligned} a &= t_2, \\ b &= (t_3 - t_1)/(2h) \\ c &= (t_1 - 2t_2 + t_3)/(2h^2) \end{aligned} \tag{5.45b,c,d}$$

Using the result in Example 5.8, the volume beneath a membrane covering the elemental strip (shaded) $T \times \delta y$, is given by eq(i)

$$\delta V = \Big(\int_x z\,\mathrm{d}x\Big)\delta y = \Big[\int_{-T/2}^{T/2} (T^2/4 - x^2)\,\mathrm{d}x\Big]\,\delta y = (T^3/6)\,\delta y \tag{5.46}$$

Substituting eq(5.45a) into eq(5.46) with $T = 2t$, the volume beneath the two strips becomes

$$V = (4/3) \int_{-h}^{h} (a + by + cy^2)^3\,\mathrm{d}y$$

The integration leads to

$$V = (8/3)(A\,h + B\,h^3 + C\,h^5 + D\,h^7) \tag{5.47a}$$

where, A, B, C and D are found from eqs(5.45b - d) as

$$A = a^3 = t_2^{\,3} \tag{5.47b}$$
$$B = a\,(ac + b^2) = [\,t_2/(4h^2)](2\,t_1 t_2 - 4\,t_2^{\,2} + 2\,t_2 t_3 + t_3^{\,2} - 2\,t_1 t_3 + t_1^{\,2}) \tag{5.47c}$$
$$C = (3c/5)(b^2 + ac) = [\,3/(40\,h^4)](t_1 - 2\,t_2 + t_3)[(t_3 - t_1)^2 + 2\,t_2(t_1 - 2\,t_2 + t_3)] \tag{5.47d}$$
$$D = c^3/7 = [1/(56\,h^6)](t_1 - 2\,t_2 + t_3)^3 \tag{5.47e}$$

Substituting eqs(5.47b - e) into eq(5.47a) leads to

$$\begin{aligned} V = (2h/105)[13\,t_1^{\,3} + 64\,t_2^{\,3} + 13\,t_3^{\,3}) + t_1^{\,2}(20\,t_2 - 3\,t_3) + 4\,t_2^{\,2}(4\,t_3 - 11\,t_1) \\ + t_3^{\,2}(20\,t_2 - 3\,t_1) - 16\,t_1 t_2 t_3] \end{aligned} \tag{5.48}$$

By taking further pairs of strips, the whole volume beneath an aerofoil membrane may be found. Recall from eq(5.40) that J is twice this total volume. The maximum shear stress in the aerofoil is, from eq(5.21), $\tau = 2tT/J$ when t is its maximum semi-thickness (i.e. the semi-width).

5.4 Circular Shaft With Variable Diameter

Polar co-ordinates r, θ and z are used to describe the torsion of a shaft whose diameter varies with length (see Fig. 5.19a).

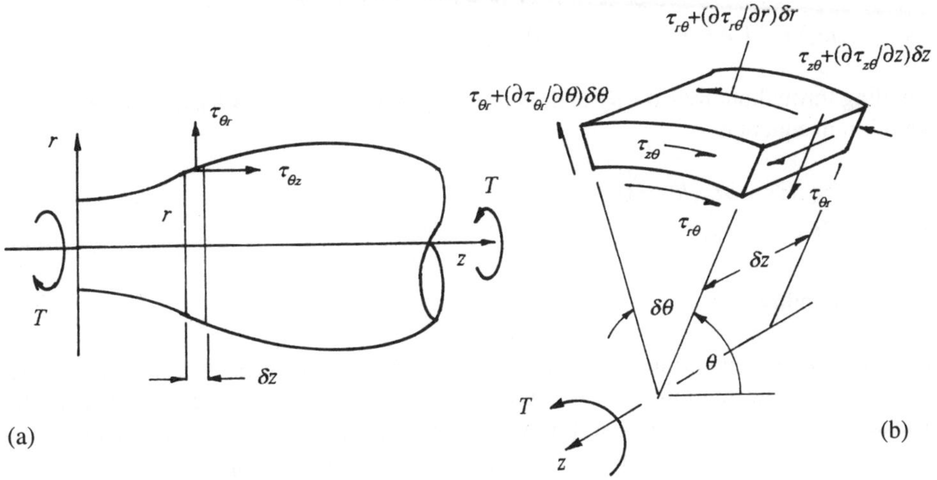

Figure 5.19 Torsion of a shaft with varying diameter

(a) Twist Rate

Since all cross-sections are circular, eq(5.6) may be applied to find the twist rate in this shaft.

$$\delta\phi/\delta z = T/(JG) \tag{5.49a}$$

in which ϕ now replaces θ as the symbol for twist (θ is a polar co-ordinate). To find ϕ over a finite length, eq (5.49a) may be expressed in terms of the single variable z by substituting the relationship $r = r(z)$ describings the sides. This gives $J = ½\pi\,[r(z)]^4$ from which the angular twist ϕ for the shaft is

$$\phi = [2T/(\pi G)]\int_z \mathrm{d}z/[r(z)]^4 \tag{5.49b}$$

in which T and G are assumed constant.

(b) Equilibrium

The only non-zero stresses are the shear components $\tau_{\theta z}$ and $\tau_{\theta r}$ acting with their complements $\tau_{\theta z} = \tau_{z\theta}$ and $\tau_{\theta r} = \tau_{r\theta}$, as shown in Fig. 5.19b. These stresses must be allowed to vary with r, θ and z, as shown, so that the tangential equilibrium equation for the element becomes

$$\tau_{\theta r}\,\delta r\,\delta z\sin(\delta\theta/2) + [\tau_{\theta r} + \delta\theta(\partial\tau_{\theta r}/\partial\theta)]\,\delta r\,\delta z\sin(\delta\theta/2) + [\tau_{r\theta} + \delta r(\partial\tau_{r\theta}/\partial r)](r+\delta r)\,\delta\theta\,\delta z$$
$$+ [\tau_{z\theta} + \delta z\,(\partial\tau_{z\theta}/\partial z)](r + \delta r/2)\,\delta\theta\,\delta r - \tau_{r\theta}\,r\,\delta\theta\,\delta z = 0$$

When $\delta\theta \to 0$, $\sin(\delta\theta/2) \to \delta\theta/2$ and we can cancel $\delta\theta$ throughout and also ignore terms in $(\delta r)^2$, to give

$$\tau_{\theta r}\,\delta r\,\delta z + \tfrac{1}{2}(\partial\tau_{\theta r}/\partial\theta)\,\delta\theta\,\delta r\,\delta z + \tau_{r\theta}\,r\,\delta z + (\partial\tau_{r\theta}/\partial r)\,r\,\delta r\,\delta z + \tau_{r\theta}\,\delta r\,\delta z$$
$$+ (\partial\tau_{z\theta}/\partial z)\,r\,\delta\theta\,\delta r\,\delta z - \tau_{\theta r}\,r\,\delta z = 0$$

Dividing throughout by $r\,\delta r\,\delta z$ and ignore the product $(\partial\tau_{z\theta}/\partial z)\delta\theta$ leads to the required equilibrium equation

$$2\tau_{\theta r}/r + \partial\tau_{\theta r}/\partial r + \partial\tau_{\theta z}/\partial z = 0$$
$$\partial(r^2\tau_{\theta r})/\partial r + r^2(\partial\tau_{\theta z}/\partial z) = 0 \tag{5.50}$$

(c) Compatibility

The shear strain occurs within two planes in Fig. 5.19b, r - θ and and θ - z. This means that the front face and the top faces will both distort in shear. Let v describe the tangential displacement accompanying twist. It is assumed that this point does not displace radially or axially, i.e. $u = w = 0$. Thus, the shear strain in the θ - z plane is simply

$$\gamma_{\theta z} = \partial v/\partial z \tag{5.51a}$$

The distortion in the r - θ plane is found by setting $\partial u/\partial\theta = 0$ in eq(2.35c). This gives

$$\gamma_{r\theta} = \partial v/\partial r - v/r \tag{5.51b}$$

Since the two shear strains depend upon a single displacement the compatibility condition follows from eliminating v from eqs(5.51a,b). Partial differentiation of eq(5.51b) with respect to z leads to

$$\partial\gamma_{r\theta}/\partial z = \partial^2 v/\partial r\partial z - (1/r)(\partial v/\partial z)$$

and substituting from eq(5.51a)

$$\partial\gamma_{r\theta}/\partial z = \partial\gamma_{\theta z}/\partial r - \gamma_{\theta z}/r \tag{5.52a}$$

When compatibility is written in terms of stress, we set $\tau_{r\theta} = G\gamma_{r\theta}$ and $\tau_{\theta z} = G\tau_{\theta z}$ in eq(5.52a), to give

$$\partial\tau_{r\theta}/\partial z = \partial\tau_{\theta z}/\partial r - \tau_{\theta z}/r \tag{5.52b}$$

(d) Stress Function

The exact solution must satisy both eqs(5.50) and (5.52b). Clearly eq(5.50) will be satisfied by a stress function $\chi(r, z)$ in which the two shear stress components are given by

$$\tau_{\theta z} = (1/r^2)(\partial\chi/\partial r) \tag{5.53a}$$

$$\tau_{r\theta} = -(1/r^2)(\partial\chi/\partial z) \tag{5.53b}$$

Substituting eqs(5.53a,b) into eq(5.52b) leads to

$$\partial^2\chi/\partial r^2 - (3/r)\,\partial\chi/\partial r + \partial^2\chi/\partial z^2 = 0 \tag{5.54}$$

Looking at the side face of Fig. 5.19b we see that both $\tau_{\theta r}$ and $\tau_{\theta z}$ act on the same area $\delta z \times \delta r$. Thus, for a given position (r, z), the maximum shear stress is the resultant of eqs(5.53a,b).

$$\tau_{\max} = \sqrt{(\tau_{r\theta}^{\ 2} + \tau_{\theta z}^{\ 2})} \tag{5.55}$$

and is independent of θ.

(e) Boundary Conditions

The complete solution to the problem is reduced to finding a stress function $\chi = \chi(r, z)$ that satisfies eq(5.54) and the following two boundary conditions imposed upon χ. In the first, it is known that there can be no force acting normal to the boundary. Hence the projection of the forces due to $\tau_{\theta z}$ and $\tau_{\theta r}$ in a normal direction **n** to the boundary (see Fig. 5.20a) must sum to zero.

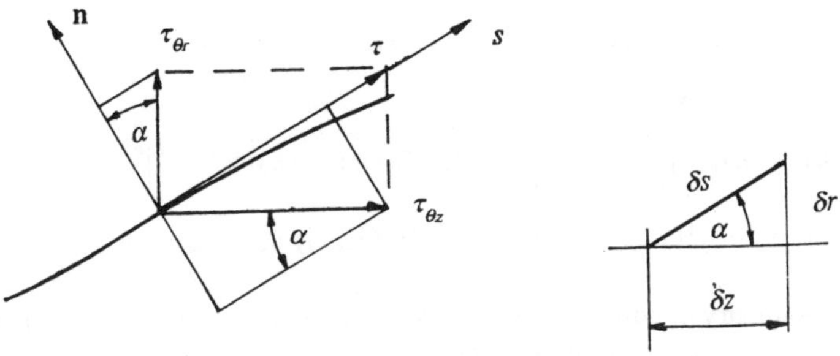

Figure 5.20 Boundary conditions

This gives

$$(2\pi r\,\delta s \times \tau_{\theta r})\cos\alpha - (2\pi r\,\delta s \times \tau_{\theta z})\sin\alpha = 0$$

$$\tau_{\theta r}\cos\alpha - \tau_{\theta z}\sin\alpha = 0 \tag{5.56a}$$

where $\cos\alpha = \delta z/\delta s$ and $\sin\alpha = \delta r/\delta s$ (see Fig. 5.20b). Figure 5.20a also shows that $\tau_{\max}$ from eq(5.55) lies tangential to the boundary. Substituting eqs(5.53a,b) into (5.56a) gives

$$(2/r^2)(\mathrm{d}\chi/\mathrm{d}s) = 0 \tag{5.56b}$$

Equation (5.56b) shows that $\chi(r, z)$ must remain constant along the boundary. The second condition is that eq(5.53a) must obey torque equilibrium. Referring to Fig. 5.19b, for any section of outer radius a

$$T = 2\pi r^2 \tau_{0z}\,\mathrm{d}r = 2\pi \int_0^a (\partial\chi/\partial r)\,\mathrm{d}r$$

Thus, on every cross-section defined by z, the function χ must satisfy

$$T = 2\pi[\chi(a, z) - \chi(0, z)] \qquad (5.57)$$

5.4.1 Conical Shaft

The torsion of a conical shaft (Fig. 5.21) is a special case of the foregoing theory. The twist θ, over length L, follows from substituting $r(z) = a + (b - a)z/L$ and $\mathrm{d}r/\mathrm{d}z = (b - a)/L$ in eq(5.49b). The latter substitution gives the twist as

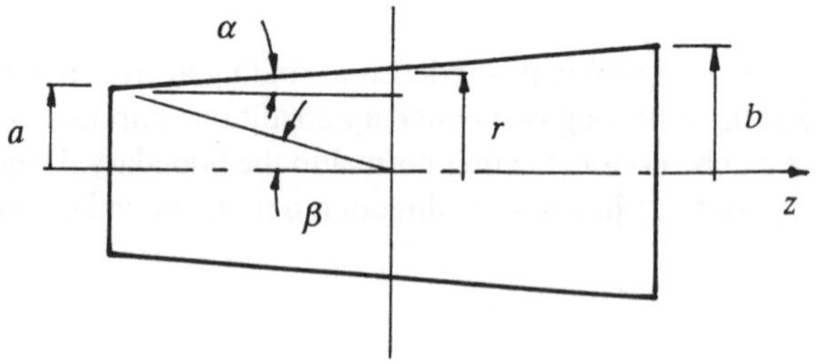

Figure 5.21 Conical shaft under axial torque

$$\phi = \{2TL/[G\pi(b - a)]\} \int_a^b r^{-4}\,\mathrm{d}r = -\{2TL/[3G\pi(b - a)]\}\ \Big| r^{-3} \Big|_a^b$$

$$\phi = \{2TL/[3G\pi(b - a)]\}(1/a^3 - 1/b^3) = [\,2TL/(3\pi G)\,](a^2 + ab + b^2)/(a^3 b^3) \qquad (5.58)$$

From the first boundary condition (5.56a,b), we can derive the stress function $\chi(r, z)$ from the constant gradient to the sloping boundary as

$$\cos\alpha = z/\{\sqrt{[(z^2 + (r - a)^2]}\} = \text{constant} \qquad (5.59a)$$

The correct function $\chi(r, z) = \chi(\cos\alpha)$, which satisfies eq(5.54), can be shown to be

$$\chi(\cos\alpha) = C(\cos\alpha - ⅓\cos^3\alpha) \qquad (5.59b)$$

where the constant C is found from the second boundary condition (5.57) to be

$$T = 2\pi[C(\cos\alpha - ⅓\cos^3\alpha) - C(\cos\beta - ⅓\cos^3\beta)] \qquad (5.60a)$$

This gives

$$C = T/\{2\pi[(\cos\alpha - ⅓\cos^3\alpha) - (\cos\beta - ⅓\cos^3\beta)]\} \qquad (5.60b)$$

where $\cos\beta = z/\sqrt{(z^2 + a^2)}$ defines the angle shown. Finally, the shear stresses are, from eqs(5.53) and (5.55),

$\tau_{z\theta} = -\ C\,r\,z\,/\,(r^2 + z^2)^{5/2},$

$\tau_{r\theta} = -\ C\,r^2/\,(r^2 + z^2)^{5/2}$ (5.61a,b)

$\tau_{\max} = C\,r\,/\,(r^2 + z^2)^2$ (5.61c)

Example 5.9 A steel shaft tapers for a length of 1.5 m between end diameters one of which is twice the other. If the maximum shear stress is to attain 60 MPa as the angle of twist reaches 4°, find (i) the end diameters and (ii) the maximum permissible axial torque. Take $G = 80$ GPa.

When $b = 2a$ (see Fig. 5.21), eq(5.58) gives the angle of twist as

$$\phi = 7TL\,/\,(12G\pi a^4) \qquad \text{(i)}$$

The maximum shear stress occurs in the outer fibres of the smaller diameter. At this position $z = 0$ and $r = a$, when eq(5.60b) gives $C = 3T/(4\pi)$. From eq(5.61c),

$$\tau_{\max} = C/a^3 = 3T/(4\pi a^3) \qquad \text{(ii)}$$

Equation (ii) gives the maximum torque as

$$T = 4\pi\tau_{\max}\,a^3/3 \qquad \text{(iii)}$$

Substituting eq(iii) into (i) and re-arranging for a,

$$\begin{aligned} a &= (7\,\tau_{\max} L)/(9G\phi) \\ &= (7 \times 60 \times 1.5 \times 10^3 \times 180)\,/\,(9 \times 80 \times 10^3 \times \pi \times 4) \\ &= 12.53 \text{ mm} \end{aligned}$$

and from eq(iii) the permissible torque is

$$\begin{aligned} T &= 4\pi \times 60 \times (12.53)^3/3 \\ &= 0.495 \text{ kNm} \end{aligned}$$

Equations (5.61a,b) show that at the smaller end, where $z = 0$, the components of the maximum shear stress are $\tau_{z\theta} = 0$ and $\tau_{r\theta} = \tau_{\max}$.

5.5 Torsion of Thin-Walled Closed Sections

Closed, thin-walled, single and multi-cell tubular sections are able to withstand torsion by means of a constant shear flow around the wall. In the Bredt-Batho torsion theory the torque is assumed to act about a longitudinal axis passing through a point in the cross-section known as the centre of twist. The latter point does not rotate or displace axially and will lie at the intersection between the axes of symmetry within the cross-section. Where the torque axis is displaced from the centre of twist we may take a statically equivalent system of a torque plus a transverse shear force at the centre of twist. The shear force introduces a flexural shear dealt with in Chapter 7.

5.5.1 Single Cell Tube

Let the closed tube in Fig. 5.22a enclose an area A within its mean wall centre line. Firstly, it is assumed that the thickness t varies with the perimeter length s in a similar manner for every cross-section at position z in the length. When the tube is subjected to a torque T, this produces shear stress τ in the wall which varies over an element ABCD of dimension $\delta s \times \delta z$ in the manner shown.

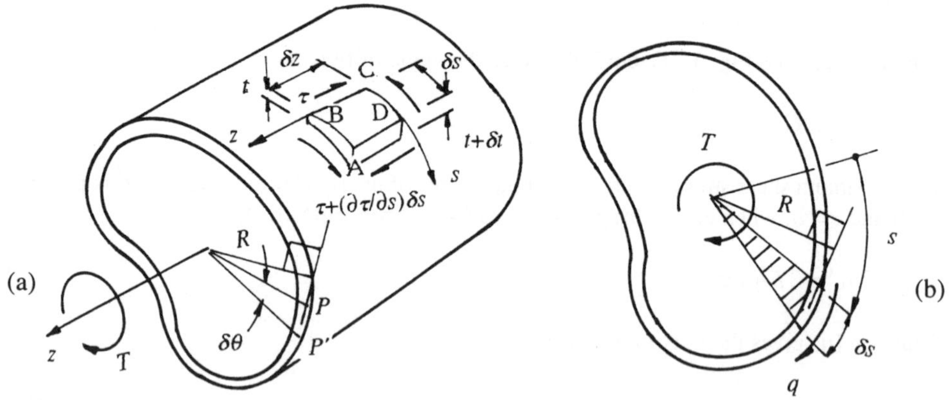

Figure 5.22 Torsion of a closed single-cell tube

(a) Shear Flow

The following equation expresses the force equilibrium of this element in the z - direction:

$$\tau \delta z\, t = [\tau + (\mathrm{d}\tau/\mathrm{d}s)\delta s\,](t + \delta t)\delta z \tag{5.62a}$$

Expanding (5.62a) and neglecting the product $(\mathrm{d}\tau/\mathrm{d}s)\delta t$ leads to

$$\tau(\delta t/\delta s) + t\,(\mathrm{d}\tau/\mathrm{d}s) = 0 \tag{5.62b}$$

In the limit, as δt and δs approach zero, eq(5.62b) becomes a differential product

$$\mathrm{d}(\tau t)/\mathrm{d}s = 0$$

from which it follows that

$$q = \tau t = \text{constant} \tag{5.63}$$

where q is known as the shear flow and remains constant around a thin-closed section regardless of its shape. Note that $\tau = q/t$ is constant only when t is constant.

(b) Torque

It is apparent from Fig. 5.22b that the torque T is resisted by the shear flow, as follows:

$$\delta T = (q\ \mathrm{d}s)R$$

$$T = q \oint R\ \mathrm{d}s \tag{5.64a}$$

Now R ds is twice the shaded area shown, and therefore the path integral $\oint R\,\mathrm{d}s$ describes twice the area enclosed by the section. The torque is simply

$$T = 2Aq = 2A\tau t \tag{5.64b}$$

(c) Twist Rate

The rate of twist $\delta\theta/\delta z$ is found from a similar analysis to that previously employed for thin strips.

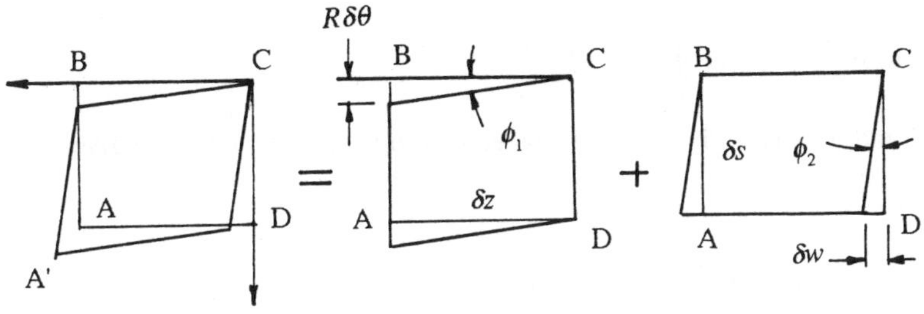

(a) (b) (c)

Figure 5.23 Shear distortion within a wall element

Figures 5.23a,b show that the distorted shape of the surface element ABCD is the sum of two distortions (Fig. 5.23a) ϕ_1 due to twist and (Fig.5.23b) ϕ_2 due to warping. For example, in Fig.5.23c point A displaces to A' by rotating an amount $R\delta\theta$ in the s direction and by warping an amount δw in the z - direction. The shear strain becomes

$$\begin{aligned}\gamma &= \tan\phi_1 + \tan\phi_2 \\ &= R\,(\delta\theta/\delta z) + \delta w/\delta s \end{aligned} \tag{5.65}$$

Equation (5.65) gives the warping displacement, for the element,

$$\begin{aligned}\delta w &= \gamma\,\delta s - (R\delta s)(\delta\theta/\delta z) \\ &= (\tau/G)\delta s - (R\delta s)(\delta\theta/\delta z)\end{aligned} \tag{5.66a}$$

When eq(5.66a) is integrated around the section, the warping displacement will be zero. Substituting from eq(5.64b),

$$0 = [T/(2AG)]\oint \delta s/t - (\delta\theta/\delta z)\oint R\delta s \tag{5.66b}$$

Equations (5.64a) and (5.66b) give the twist rate as

$$\mathrm{d}\theta/\mathrm{d}z = [T/(4A^2G)]\oint(\mathrm{d}s/t) = [q/(2AG)]\oint(\mathrm{d}s/t) \tag{5.67}$$

Introducing the St Venant torsion constant J,

$$J = 4A^2/[\oint(\mathrm{d}s/t)] \tag{5.68}$$

we may combine eqs(5.64) and (5.67a) to give a convenient formula:

$$T = GJ\,(\mathrm{d}\theta/\mathrm{d}z) = 2\,A\,q \tag{5.69}$$

For a uniform section in the same material, eq(5.67) may be integrated directly, to give

$$\theta = [TL\,/\,(4\,A^2 G\,)]\oint(\mathrm{d}s/t\,) = [qL/(2AG)]\oint(\mathrm{d}s/t\,) \tag{5.70a}$$

The torque T will vary with z in uniform tubes of length L under wind loading and under multiple torques. The integration in eq(5.67) is modified to

$$\theta = [1\,/\,(4A^2 G\,)]\int_0^L T\,(z)\,\mathrm{d}z\oint(\mathrm{d}s\,/\,t\,) \tag{5.70b}$$

When both T and A vary with z (say) in a tapered composite tube, eq(5.67) gives the angle of twist as

$$\theta = \tfrac{1}{4}\,\{\int_0^L T\,(z)\,\mathrm{d}z\,/\,[A(z)]^2\,\}\oint[\mathrm{d}s/(Gt\,)] \tag{5.70c}$$

(d) Warping

Assume that at the datum for s in Fig. 5.22a the warping displacement is w_o. Equation (5.66a) supplies the warping displacements w at distance s from this datum. Substituting from eqs(5.64b) and (5.67),

$$\mathrm{d}w = (\,\tau/G\,)\,\mathrm{d}s - (R\,\mathrm{d}s)(\mathrm{d}\theta/\mathrm{d}z)$$

$$= [T\,/(2AG\,)]\,\mathrm{d}s/\,t - (R\,\mathrm{d}s)\,[\,T\,/\,(4\,A^2 G\,)]\oint(\mathrm{d}s/t\,)$$

$$w - w_o = [T\,/(2AG)]\int_0^s(\mathrm{d}s/\,t) - [T\,/\,(4A^2 G\,)]\oint(\mathrm{d}s/\,t\,)\int_0^s R\,\mathrm{d}s$$

$$w = [T\,/(2AG\,)][\,\int_0^s \mathrm{d}s\,/\,t - [1/\,(2A)]\oint\mathrm{d}s/\,t\int_0^s R\,\mathrm{d}s\,] + w_o \tag{5.71a}$$

When the datum is chosen to coincide with an axis of symmetry, $w_o = 0$ and eq(5.71a) may be written as

$$w = [T\,i\,/\,(2AG\,)]\,[\,i_{os}/\,i - A_{os}/A\,] \tag{5.71b}$$

where $i = \oint \mathrm{d}s/t$, $i_{os} = \int_o^s \mathrm{d}s/t$, A is the total area enclosed by the section and A_{os} is the segment of area enclosed between δs and the centre of twist (see Fig. 5.22b). Where a warping displacement w_o occurs at the origin for s, then w_o is added to the right side of eq(5.71b). Otherwise, it follows from eq(5.71b) that a section will not warp when

$$i_{os}/\,i = A_{os}\,/\,A \tag{5.72}$$

Consider a circular tube with mean radius R and of uniform thickness. The area enclosed is $A = \pi R^2$ and the path integral is $i = 2\pi R/t$. Within a segment of area $A_{os} = Rs/2$, and $i_{os} = s/t$, we see that eq(5.72) is satisfied

$$(s/t)\,/\,(2\pi R/t\,) = (Rs/2)/(\pi R^2)$$

Equation (5.72) is also satisfied by a uniform thickness triangular tube and a rectangular tube in which the ratio of the side lengths equals the ratio of their thicknesses. Tubes that do not warp are known as *Neuber tubes*.

Example 5.10 Figure 5.24 shows a composite aluminium section with vertical webs in steel. A tube of similar section is held at one end when subjected to a uniformly distributed anti-clockwise torque of 3 kNm/m over a length of 5 m. Assuming the section is free to warp, find the greatest values of shear stress attained in each material, the angular twist and the twist rate. Take $G = 30$ GPa for aluminium and $G = 80$ GPa for steel.

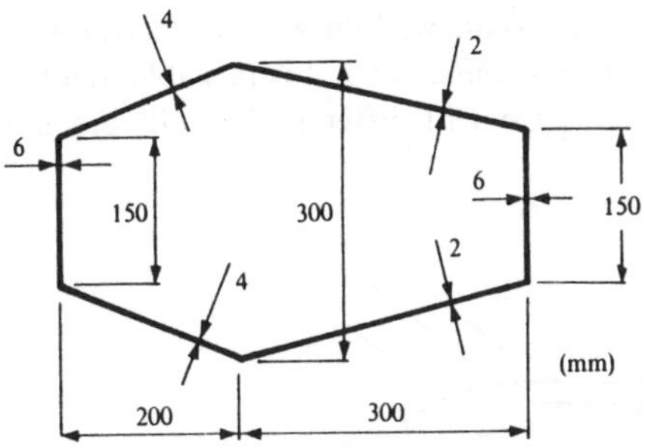

Figure 5.24 Single-cell tube

$$A = ½\,(150 + 300)200 + ½(150 + 300)300 = 112.5 \times 10^3\ \text{mm}^2$$

$$\begin{aligned}\oint \mathrm{d}s/(Gt) &= 2\,[150/(80 \times 10^3 \times 6) + (75^2 + 300^2)^{½}/(2 \times 30 \times 10^3) \\ &\quad + (75^2 + 200^2)^{½}/(4 \times 30 \times 10^3)\,] \\ &= 2 \times 10^{-3}\,(0.3125 + 5.154 + 1.780) \\ &= 14.493 \times 10^{-3}\ \text{mm}^2/\text{N}\end{aligned}$$

With the origin for z at the fixed end, the torque varies as

$$T(z) = 3(5000 - z)\ \text{(Nm)}$$

Setting $z = 0$ for the free end gives a maximum torque $T_{max} = 15$ kNm. Using eq(5.64b), the maximum shear stresses within the walls at the fixed end can be found

$$\tau = T_{max}/(2A\,t) = (15 \times 10^6)/(2 \times 112.5 \times 10^3\,t) = 66.67/\,t$$

steel webs: $\tau = 66.67/6 = 10.11$ MPa
aluminium spars: $\tau = 66.67/2 = 33.33$ MPa, $\tau = 66.67/4 = 16.67$ MPa

In this example the twist rate follows from eq(5.67). Working in units of N and mm

$$\begin{aligned}\mathrm{d}\theta/\mathrm{d}z &= [T(z)/(4A^2)\,]\oint[\mathrm{d}s/(Gt)] \\ &= [3 \times 10^3 \times (5000 - z) \times 14.493 \times 10^{-3}\,]/[\,4 \times (112.5 \times 10^3)^2\,]\end{aligned}$$

$$\mathrm{d}\theta/\mathrm{d}z = 8.59(5000 - z) \times 10^{-10}\ \text{rad/mm} \qquad \text{(i)}$$

Integrating eq(i) over the length gives the free end twist,

$$\theta = 8.59 \times 10^{-10} \int_0^{5000} (5000 - z)\, dz$$

$$= 8.59 \times 10^{-10} \left| 5000\, z - z^2/2 \right|_0^{5000}$$

$$= 8.59 \times 10^{-10} \times 5000^2/2 = 0.0107 \text{ rad } (0.615°)$$

5.5.2 Multi-Cell Tube

In a multi-cell tube (see Fig. 5.25) the wall thicknesses t_1, t_2, t_3 and t_4 and the web thicknesses t_{12}, t_{23} and t_{34} can vary between cells. Also, it is possible that the shear moduli for the wall and the webs G_1 , G_2 , G_{12} etc will differ if the cells are manufactured from different materials.

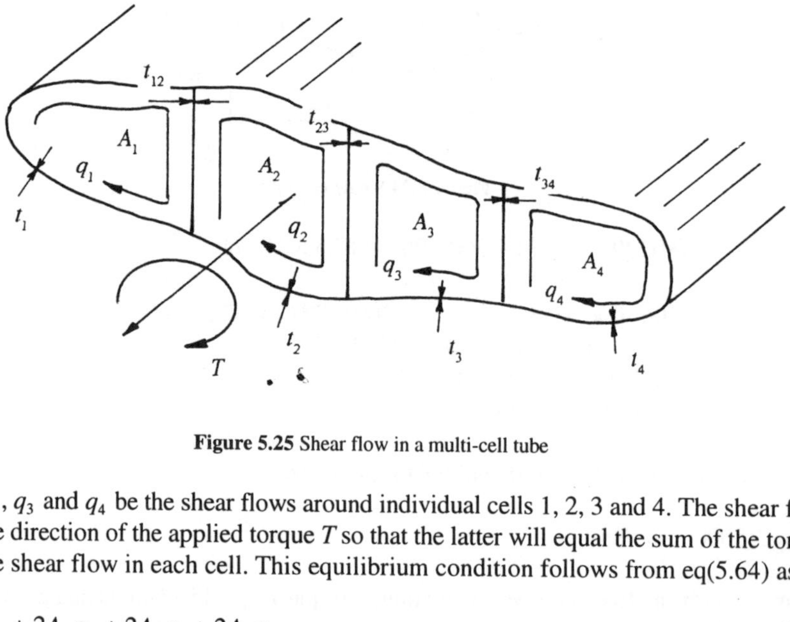

Figure 5.25 Shear flow in a multi-cell tube

Let q_1, q_2, q_3 and q_4 be the shear flows around individual cells 1, 2, 3 and 4. The shear flows follow the direction of the applied torque T so that the latter will equal the sum of the torques due to the shear flow in each cell. This equilibrium condition follows from eq(5.64) as

$$T = 2A_1 q_1 + 2A_2 q_2 + 2A_3 q_3 + 2A_4 q_4 \tag{5.73}$$

Because the q's in eq(5.73) cannot be found from equilibrium alone, the structure is *statically indeterminate*. The solution to the shear flows employs an additional compatibility condition - that the twist rate $\delta\theta/\delta z$ within each cell is constant. To find $\delta\theta/\delta z$ we must modify eq(5.67) to account for the opposing shear flows on each side of each web. The twist rate for the end cells 1 and 4 is modified by the shear flow in a single adjacent web. The twist rate for cells 2 and 3 is modified by the shear flow in two adjacent webs. These rates are

$$(\mathrm{d}\theta/\mathrm{d}z)_1 = [q_1/(2A_1)] \oint_1 [\mathrm{d}s/(Gt)] - [q_2/(2A_1)] \int_{12} [\mathrm{d}s/(Gt)] \tag{5.74a}$$

$$(\mathrm{d}\theta/\mathrm{d}z)_2 = [q_2/(2A_2)] \oint_2 [\mathrm{d}s/(Gt)] - [q_1/(2A_2)] \int_{21} [\mathrm{d}s/(Gt)] - [q_3/(2A_2)] \int_{23} [\mathrm{d}s/(Gt)] \tag{5.74b}$$

$$(\mathrm{d}\theta/\mathrm{d}z)_3 = [q_3/(2A_3)] \oint_3 [\mathrm{d}s/(Gt)] - [q_2/(2A_3)] \int_{32} [\mathrm{d}s/(Gt)] - [q_4/(2A_3)] \int_{34} [\mathrm{d}s/Gt)] \tag{5.74c}$$

$$(\mathrm{d}\theta/\mathrm{d}z)_4 = [q_4/(2A_4)] \oint_4 [\mathrm{d}s/(Gt)] - [q_3/(2A_4)] \int_{43} [\mathrm{d}s/(Gt)] \tag{5.74d}$$

The compatibility condition is

$$(\mathrm{d}\theta/\mathrm{d}z)_1 = (\mathrm{d}\theta/\mathrm{d}z)_2 = (\mathrm{d}\theta/\mathrm{d}z)_3 = (\mathrm{d}\theta/\mathrm{d}z)_4 \tag{5.74e}$$

The six eqs(5.73) and (5.74a-e) are sufficient to solve q_1, q_2, q_3 and q_4. Once the q's are solved the shear stresses in the walls are given by

$$\tau_1 = q_1/t_1,\ \tau_2 = q_2/t_2,\ \tau_3 = q_3/t_3,\ \tau_4 = q_4/t_4 \quad (5.75\text{a,b,c})$$

and in the webs,

$$\tau_{12} = (q_1 - q_2)/t_{12},\ \tau_{23} = (q_2 - q_3)/t_{23},\ \tau_{34} = (q_3 - q_4)/t_{34} \quad (5.75\text{d–f})$$

There are many simplifications that can be made to these equations. For example, with n identical cells, each with area A and in a single material, $q_1 = q_2 = q_3 = q_4 = q$ and $T = 2nAq$.

Example 5.11 The cross-section of a thin-walled, single-cell tube is shown in Fig. 5.26a Determine, when the tube supports a torque of 1.85 kNm over a length of 10 m (a) the shear flow, (b) the shear stresses in the walls, (c) the twist rate, (d) the angle of twist and (e) the warping displacements at points A, B, C and D when the torque is applied at the position shown. Examine the influence on the twist and the shear stress when a vertical web is placed at the position shown in Fig. 5.26b to make a two-cell tube. Take G = 79 GPa.

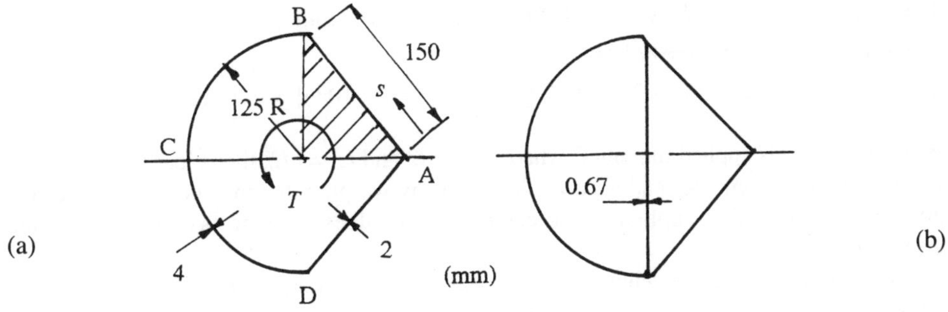

Figure 5.26 Thin-walled closed tubes section under torsion

The enclosed area within Fig. 5.26a is

$$A = \pi(125)^2/2 + (250/2)[(150)^2 - (125)^2]^{1/2}$$
$$= (24.54 + 10.36) \times 10^3 = 34.9 \times 10^3\ \mathrm{mm}^2$$

The line integral appearing in eq(5.67) is

$$\oint (\mathrm{d}s/t) = (\pi \times 125/4) + 2(150/2) = 248.2$$

(a) $q = T/(2A) = (1.85 \times 10^6)/(2 \times 34.9 \times 10^3) = 26.5$ N/mm

(b) $\tau_1 = q/t_1 = 26.5/4 = 6.63$ MPa

$\tau_2 = q/t_2 = 26.5/2 = 13.25$ MPa

(c) $\mathrm{d}\theta / \mathrm{d}z = [T/(4A^2G)]\oint (\mathrm{d}s/t)$

$= (1.85 \times 10^6 \times 248.2) / [4 \times (34.9 \times 10^3)^2 \times 79 \times 10^3]$

$= 1.193 \times 10^{-6}$ rad/mm $= 0.6835 \times 10^{-4}$ °/mm

(d) For a 10 m length under a uniform torque the angular twist becomes
$\theta = 0.6835 \times 10^{-4} \times 10 \times 10^3 = 0.684°$

(e) Warping is absent at points A and C along the axis of symmetry. Taking A as the datum for s measured anti-clockwise (i.e. the same direction as T) eq(5.71b) is applied to find the warping displacement at B. The coefficient in this equation is

$Ti/(2AG) = (1.85 \times 10^6 \times 248.2) / (2 \times 34.9 \times 10^3 \times 79 \times 10^3) = 0.08327$ mm

The area OAB is $A_{os} = (125/2)\,(150^2 - 125^2)^{1/2} = 5.182 \times 10^3$ mm^2. The line integral from A to B is $i_{os} = 150/2 = 75$. From eq(5.71b),

$w_B = 0.08327[75/248.2 - (5.182 \times 10^3)/(34.9 \times 10^3)] = 0.0128$ mm

The positive sign for w_B means it follows positive z, i.e. out from the page in Fig. 5.26a. Equation (5.71b) shows further that warping at D is of similar magnitude but of opposite direction. Here we have A_{os} = Areas ($OAB + OBC + OCD$) and i_{os} for the path ABCD

$A_{os} = (125/2)\,(150^2 - 125^2)/2 + \pi(125)^2/2 = 29.73 \times 10^3$ mm^2

$i_{os} = 150/2 + (\pi \times 125/4) = 173.18$

$w_D = 0.08327\,[173.175/248.2 - (29.726 \times 10^3)/(34.9 \times 10^3)] = -\,0.0128$ mm

When eq(5.71b) is applied to point C where $A_{os} = A/2$ and $i_{os} = i/2$, then clearly $w_C = 0$. The warping displacements w_B and w_D are relative to any warping that may occur at the torque point. Only when T is applied at the centre of twist does warping become an absolute displacement.

For the two-cell tube, the cell areas are $A_1 = 24.54 \times 10^3$, $A_2 = 10.36 \times 10^3$ and their path integrals are

$(\oint \mathrm{d}s/t)_1 = (\pi \times 125/4) + 250/0.67 = 471.3$

$(\oint \mathrm{d}s/t)_2 = 2(150/2) + 250/0.67 = 523.1$

and for the vertical web,

$(\int \mathrm{d}s/t)_{12} = 250/0.67 = 373.1$

Equation(5.73) becomes (units of N and mm)

$T = 2A_1q_1 + 2A_2q_2$

$1.85 \times 10^6 = (2 \times 24.54 \times 10^3)\,q_1 + (2 \times 10.36 \times 10^3)\,q_2$

$1850 = 49.08\,q_1 + 20.72\,q_2$ (i)

The twist rates in each cell (see eqs 5.74a,b) become

$$(\mathrm{d}\theta/\mathrm{d}z)_1 = [1/(2A_1 G)][q_1 \oint_1 (\mathrm{d}s/t) - q_2 \int_{12} (\mathrm{d}s/t)]$$
$$= [1/(2 \times 24.54 \times 10^3 \times 79 \times 10^3)][(q_1 \times 471.3) - (q_2 \times 373.1)] \qquad \text{(ii)}$$

$$(\mathrm{d}\theta/\mathrm{d}z)_2 = [1/(2A_2 G)][q_2 \oint_2 (\mathrm{d}s/t) - q_1 \int_{12} (\mathrm{d}s/t)]$$
$$= [1/(2 \times 10.36 \times 10^3 \times 79 \times 10^3)][(q_2 \times 523.1) - (q_1 \times 373.1)] \qquad \text{(iii)}$$

Now, as $(\mathrm{d}\theta/\mathrm{d}z)_1 = (\mathrm{d}\theta/\mathrm{d}z)_2$, we may equate (ii) to (iii), to give

$$[(q_2 \times 523.1) - (q_1 \times 373.1)] = (10.36/24.54)[(q_1 \times 471.3) - (q_2 \times 373.1)]$$
$$\therefore q_1 = 1.1897 q_2 \qquad \text{(iv)}$$

Substituting eq(iv) into eq(i) gives $q_2 = 23.39$ N/mm and $q_1 = 27.82$ N/mm. The shear stresses in the walls and web are, from eqs(5.75a,b,d),

$$\tau_1 = q_1 / t_1 = 27.82/4 = 6.95 \text{ MPa}$$
$$\tau_2 = q_2 / t_2 = 23.39/2 = 11.69 \text{ MPa}$$
$$\tau_{12} = (q_1 - q_2)/ t_{12} = (27.82 - 23.39)/ 0.67 = 6.62 \text{ MPa}$$

The twist rate follows from eq(ii) as

$$\delta\theta/\delta z = [1/(2 \times 24.54 \times 10^3 \times 79 \times 10^3)][(q_1 \times 471.3) - (q_2 \times 373.1)]$$
$$= [1/(3877.32 \times 10^6)][27.82 \times 471.3) - (23.39 \times 373.1)]$$
$$= 1.131 \times 10^{-6} \text{ rad/mm} = 0.648 \times 10^{-4}\ ^\circ/\text{mm}$$

The twist in a 10 m length of this tube becomes

$$\theta = 0.648 \times 10^{-4} \times 10 \times 10^3 = 0.648^\circ$$

The web has only a small influence in reducing τ_2 and $\delta\theta/\delta z$, i.e. it stiffens the tube.

5.6 Wagner-Kappus Torsion of Open Restrained Tubes

Consider a thin-walled open tube subjected to an axial torque about a longitudinal z-axis passing through the centre of twist. The latter refers to a point in the cross-section where there are to be no resultant moments about axes, other than z in Fig. 5.27a. The centre of twist will not coincide with the centroid of the section in general. The analysis of an open tube is simplified by knowing that the centre of twist coincides with the shear centre E of the section (see Chapter 7). The shear centre refers to that point in the section through which a transverse shear force must pass if it is to bend and not twist the section. Similarly, a torque must be applied at the centre of twist if it is not to bend the section. This reciprocal relationship between the two centres was used to show their coincidence (Hoff 1943). Let the co-ordinate axis z pass through E, length s lie along the mid-wall perimeter and n be the direction of the normal to the mid-wall (see Fig. 5.27a).

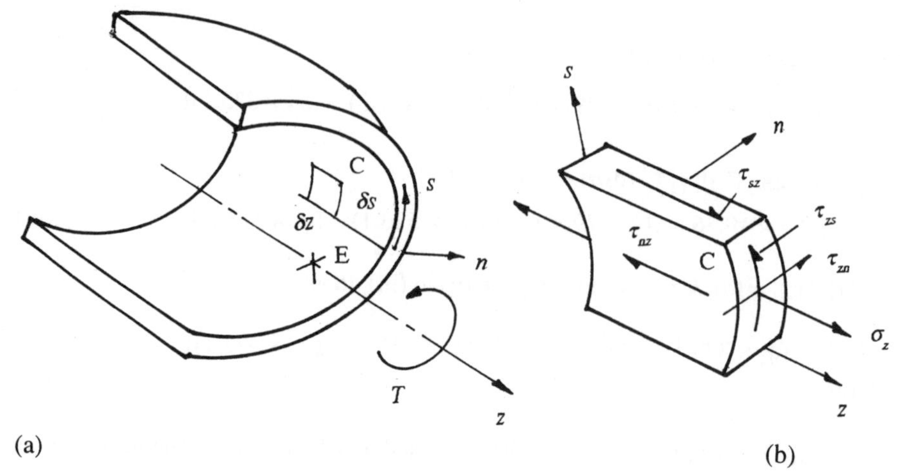

Figure 5.27 Element of a thin walled-open tube under torsion

In Fig. 5.27b the shear stresses τ_{nz} and τ_{sz} within an element $\delta s \times \delta z$ of the wall are shown with their complementary actions. Restraining warping will introduce an axial stress σ_z as shown. Consider a point C lying on the median line s. If we draw the tangent to s at this point then normal and tangential radii, R_n and R_t respectively, are centred for C at E, as shown in Fig. 5.28a.

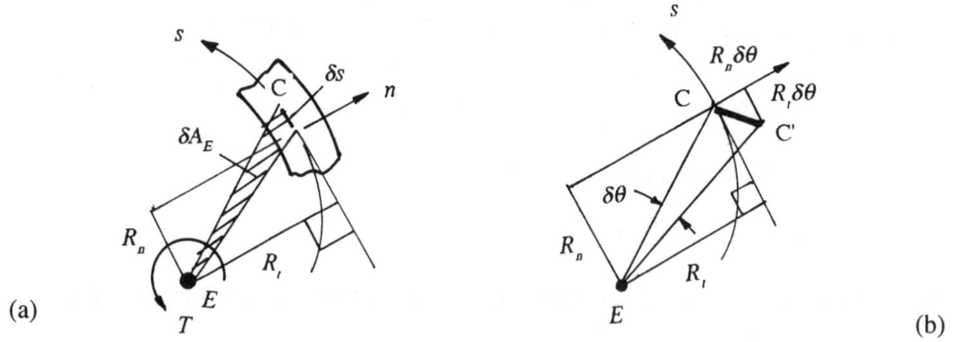

Figure 5.28 Normal and tangential radii to median line

Note that the elemental length δs subtends an area at E:

$$\delta A_E = \tfrac{1}{2} R_t\, \delta s \tag{5.76}$$

5.6.1 Unconstrained Warping

When there are no constraints to twist, the tube will warp freely in its length. Let the tube twist by the small amount $\delta\theta$ at E so that point C moves to C' in Fig. 5.28b. The tangential and normal components of this displacement are $R_t\,\delta\theta$ and $R_n\,\delta\theta$ respectively. Two warping displacements of C: δw_s and δw_n, are aligned with the length. They are found from the shear distortion of the element in the s - z and n - z planes as shown in Figs 5.29a,b.

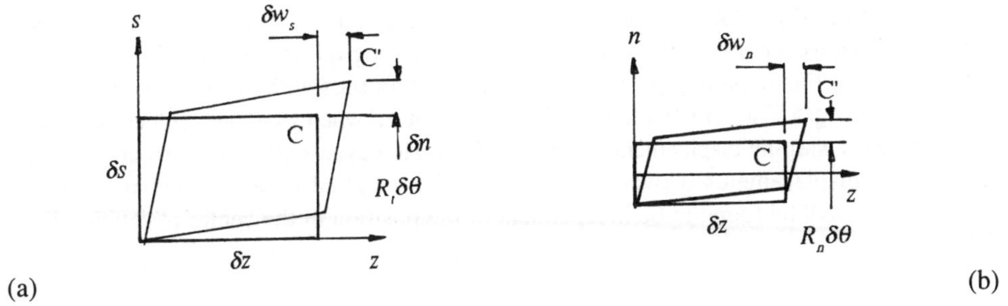

Figure 5.29 Distortion to the wall of an open tube under torsion

Primary warping (Fig. 5.29a) occurs in the s - z plane and is constant across the wall. The corner C will warp by an amount δw_s in the direction of z. C also displaces tangentially by the amount $R_t\,\delta\theta$. The shear strain γ_{sz} in the s - z plane is the sum of the two shear angles

$$\gamma_{sz} = \delta w_s / \delta s + R_t\, \delta\theta / \delta z$$

Secondary warping occurs in the n - z plane and varies through the wall in the manner of Fig. 5.29b. The shear strain is given as

$$\gamma_{nz} = \tfrac{1}{2}\,(\delta w_n / \delta n) + R_n\,(\delta\theta / \delta z)$$

If the shear stresses $\tau_{sz} = G\gamma_{sz}$ and $\tau_{nz} = G\gamma_{nz}$ are to equal zero along the mid-line it follows that the warping increments are

$$\delta w_s = -\ (\delta\theta / \delta z)\, R_t\, \delta s, \qquad \delta w_n = -\ 2R_n \delta n\, (\delta\theta / \delta z)$$

Now from eq(5.76) $R_t\, \delta s = 2\ \delta A_E$. Taking R_n to be independent of n, the *primary* and *secondary* warping displacements are respectively,

$$w_s = -\ (\delta\theta / \delta z) \int_0^s R_t\, \mathrm{d}s = -\ (2\, A_E)(\delta\theta / \delta z) \tag{5.77a}$$

$$w_n = -\ 2R_n n\, (\delta\theta / \delta z) \tag{5.77b}$$

where A_E is the area swept between E, the datum $s = 0$ (where $w = 0$) and the perimeter length s (see Fig. 5.30a).

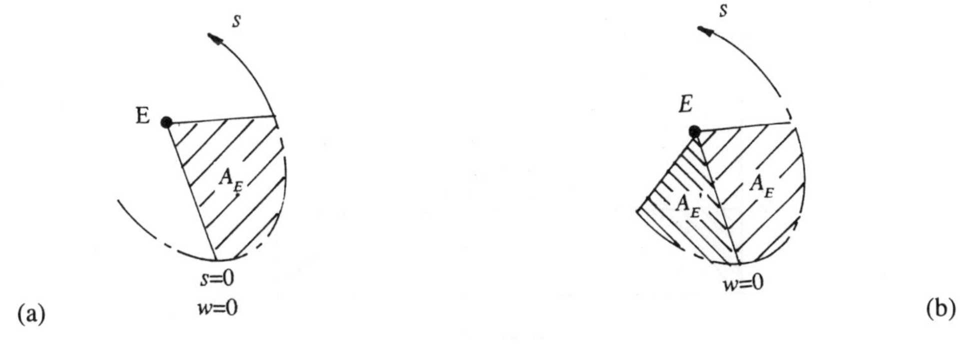

Figure 5.30 Areas enclosed by mean wall perimeter

If we wish to find the total warping displacements at point (s, n) then $w = w_s + w_n$ from eq(5.77a,b). However, secondary warping w_n is usually negligible although we shall look at the influence of this in paragraph 5.6.3. The analysis proceeds with primary warping w_s but note that A_E in eq(5.77a) will be incorrect by an amount A_E' when $s = 0$ does not coincide with a point of zero warping displacment (see Fig. 5.30b). However, eq(5.77a) correctly supplies the relative displacements between points, and when applied between free surfaces, this displacement will be found to be independent of the position of the centre of rotation E.

5.6.2 Constrained Primary Warping

It may not be obvious where a point of zero warping exists and so by taking an arbitrary origin for s, the swept volume required is overestimated by an amount A_E' (see Fig. 5.30b). To correct for this we must use the fact that axial stresses induced from constraining warping have no resultant force. That is,

$$\int_0^s \sigma_z \, t \, ds = 0 \tag{5.78a}$$

where σ_z follows from

$$\sigma_z = E\varepsilon_z = E\,(\,\partial w/\partial z) \tag{5.78b}$$

Substituting eq(5.78b) into eq(5.78b),

$$\sigma_z = -\,2\,A_E E\,(\mathrm{d}^2\theta/\mathrm{d}z^2) \tag{5.78c}$$

Now θ is independent of s, varying only with z and, therefore, from eqs(5.78a,c),

$$\int_0^s 2\,A_E\, t\, ds = 0 \tag{5.79a}$$

Let $A_{os} = A_E + A_E'$ be the total area swept from $s = 0$ in Fig. 5.30b. Equation(5.79a) becomes

$$\int_0^s 2\,A_{os}\, t\, ds \; - \; 2A_E' \int_0^s t\, ds = 0 \tag{5.79b}$$

Writing $\bar{y} = 2A_E'$ and $y = 2A_{os}$ and taking t as a constant, eq(5.79b) gives $\bar{y}$ as

$$\bar{y} = \int y\, ds \;/\; \int ds \tag{5.79c}$$

We can then interpret eq(5.79c) in the manner of Fig. 5.31.

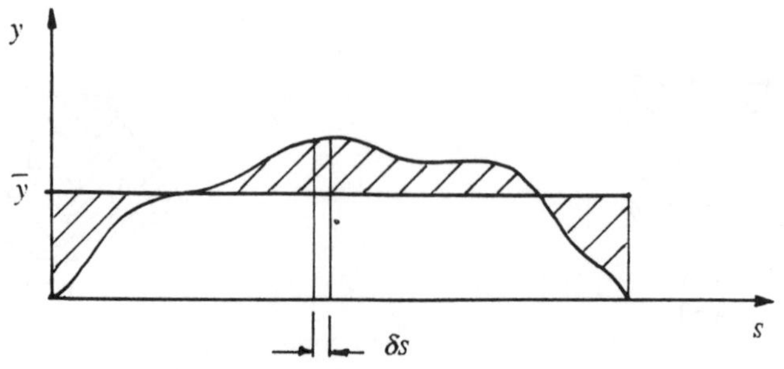

Figure 5.31 Graph of y versus s

This shows that $\bar{y}$ is the ordinate of a rectangle with the same area as that enclosed between y and the perimeter length s. Hence in eq(5.77a),

$$2A_E = 2A_{os} - 2A_E' = y - \bar{y} \tag{5.80}$$

Figures 5.32a,b show the two planes s - z and n - z within Fig. 5.27b, in which the axial and shear stresses are allowed to vary with δs in the manner shown.

(a) 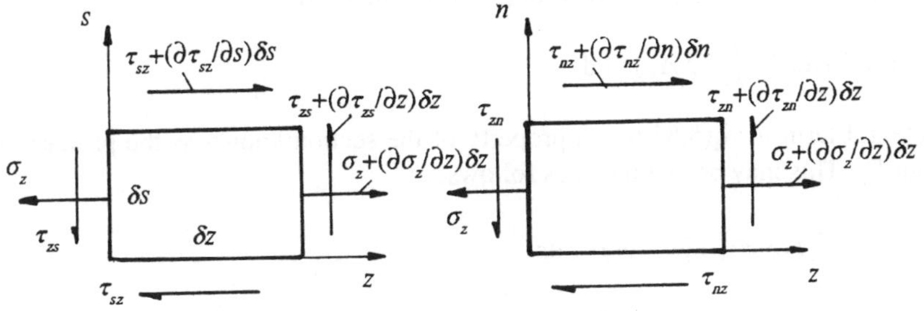 (b)

Figure 5.32 Stress variations across element

The horizontal equilibrium equation for the element s - z in Fig. 5.32a becomes

$$[\sigma_z + (\partial\sigma_z/\partial z)\delta z]\delta s\, t + [\tau_{sz} + (\partial\tau_{sz}/\partial s)\delta s]\delta z\, t = \sigma_z\,\delta s\, t + \tau_{sz}\,\delta z\, t$$

which gives

$$\partial\sigma_z/\partial z + \partial\tau_{sz}/\partial s = 0 \tag{5.81a}$$

The horizontal equilibrium equation for the n - z element in Fig. 5.32b becomes

$$[\sigma_z + (\partial\sigma_z/\partial z)\delta z]\delta n\, t + [\tau_{nz} + (\partial\tau_{nz}/\partial n)\delta n]\delta z\, t = \sigma_z\,\delta n\, t + \tau_{nz}\,\delta z\, t$$

which gives

$$\partial\sigma_z/\partial z + \partial\tau_{nz}/\partial n = 0 \tag{5.81b}$$

Equation (5.81b) is used later (see section 5.6.3) for a secondary warping analysis. In primary warping, eq(5.81a) supplies the shear flow as $q = \tau_{sz} \times t$ as

$$q = -\int (\partial\sigma_z/\partial z)\, t\, \mathrm{d}s \tag{5.82a}$$

Substituting eq(5.78c) into eq(5.82a),

$$q = E\,(\mathrm{d}^3\theta/\mathrm{d}z^3)\int (2A_E)\, t\, \mathrm{d}s \tag{5.82b}$$

The Wagner-Kappus torque then follows from the shear flow as

$$T_w = \int_c q\, R_t\, \mathrm{d}s$$

$$= E\,(\mathrm{d}^3\theta/\mathrm{d}z^3)\int_c R_t\left[\int_0^s (2A_E)\, t\, \mathrm{d}s\right]\mathrm{d}s \tag{5.83a}$$

where c refers to the whole length of the mid-wall perimeter. Integrating eq(5.83a) by parts and substituting from eq(5.76),

$$T_w = E\,(\mathrm{d}^3\theta/\mathrm{d}z^3)[(2A_E)\int_c (2\,A_E)\,t\,\mathrm{d}s - \int_c (2\,A_E)^2\,t\,\mathrm{d}s\,] \tag{5.83b}$$

For a tube of constant thickness eq(5.80) shows that $2A_E$ can be identified with the ordinate $y - \bar{y}$ in Fig. 5.31. Hence the first integral in eq(5.83b) becomes zero, i.e. the sum of the areas lying above and below $\bar{y}$, as shown. Equation(5.83b) reduces to

$$T_w = -\,E\,(\mathrm{d}^3\theta/\mathrm{d}z^3)\int_c (2\,A_E)^2\,t\,\mathrm{d}s \tag{5.83c}$$

The integral term in eq(5.83c) is a property of the section known as the primary warping constant Γ_1. This may be evaluated as follows:

$$\Gamma_1 = \int (2\,A_E)^2\,t\,\mathrm{d}s = \int (y - \bar{y}\,)^2\,t\,\mathrm{d}s \tag{5.84a}$$

$$= \int y^2 t\,\mathrm{d}s - 2\bar{y}\int y\,t\,\mathrm{d}s + \bar{y}^{\,2}\int t\,\mathrm{d}s$$

Substituting from eq(5.79b),

$$\Gamma_1 = \int y^2\,t\,\mathrm{d}s - \bar{y}^2\int t\,\mathrm{d}s \tag{5.84b}$$

If t is constant in eq(5.84b) we may interpret the first integrand as $t \times$ the square of the ordinate in Fig. 5.31 and the second integrand as $t \times$ the perimeter length. Finally, the total torque is the sum of the Wagner and St Venant torques. The latter follows from eq(5.21) as $T_v = GJ\,(\delta\theta/\delta z)$. This gives $T = T_v + T_w$, where from eq(5.83c),

$$T = GJ\,(\mathrm{d}\theta/\mathrm{d}z) - E\,\Gamma_1\,\mathrm{d}^3\theta/\mathrm{d}z^3 \tag{5.85}$$

where the form of J has been defined previously in eqs(5.23a,b). Equation(5.85) supplies a solution to the rate of twist as

$$\mathrm{d}\theta/\mathrm{d}z = [T/(GJ\,)]\{1 - \cosh[\mu\,(L - z)]\,/\cosh(\mu L)\} \tag{5.86a}$$

where $\mu = \sqrt{[GJ\,/\,(E\,\Gamma_1\,)]}$. Thus, in a tube constrained at one end, the twist rate varies with length z. Integrating eq(5.86a) and substituting $z = L$ gives the angular twist at the free end,

$$\theta = [TL/(GJ\,)]\{[1 - [1/(\mu L)]\tanh(\mu L\}\} \tag{5.86b}$$

The second term is the amount by which the free twist in the tube is reduced by the constraint. Writing, from eqs(5.83c) and (5.86a), the Wagner torque,

$$T_w = -\,E\,\Gamma_1\,(\mathrm{d}^3\theta/\mathrm{d}\,z^3) = T\cosh[\mu\,(L - z)]\,/\cosh(\mu L) \tag{5.87}$$

Equation(5.87) shows that at the fixed end, where $z = 0$, $T_w = T$, all the torque is due to bending. At the free end ($z = L$) T_w diminishes to $T/\cosh\mu L$, i.e. it does not disappear. The axial stress is given from eqs(5.78c) and (5.86a) as

$$\sigma_z = -\,[2\,A_E T\,/\,(\mu\,\Gamma_1\,)]\{\sinh[\mu\,(L - z)\,]\,/\cosh(\mu L)\} \tag{5.88a}$$

Hence σ_z is zero at the free end and attains a maximum at the fixed end. Using eq(5.77a) we may write the coefficient in eq(5.88a) in terms of the free warping displacements w_o

$$(2A_E)T/(\mu\Gamma_1) = -\ w_o GJ\ T/(\mu\Gamma_1) = -\ \mu\ E\ w_o \tag{5.88b}$$

Combining eqs(5.88a,b), we see that

$$\sigma_z = \mu\ E\ w_o\ \{\sinh\ [\mu\ (L - z)]/\cosh\ (\mu\ L)\} \tag{5.88c}$$

which shows that σ_z is proportional to w_o. Equation(5.78b) gives the constrained warping displacements at position z,

$$w = (1/E) \int_0^z \sigma_z\ \mathrm{d}z \tag{5.89a}$$

Substituting eq(5.88c) into eq(5.89a) and integrating,

$$w = [\mu\ w_o / \cosh\ (\mu L)] \int_0^z \sinh\ [\mu\ (L - z)]\ \mathrm{d}z$$

$$w = w_o\ \{1 - \cosh\ [\mu\ (L - z)]\ /\ \cosh\ (\mu L)\} \tag{5.89b}$$

Equation(5.89b) shows that free warping displacement w_o exists only at the free end and is eliminated at the fixed end. The problem of constrained warping in closed tubes is more complex than in open tubes. However, for doubly symmetric tubes, eqs(5.88c) and (5.89b) still apply. The following examples will show how to apply the Wagner-Kappus theory to open tubes.

Example 5.12 Determine the warping constant and the unconstrained warping displacements for the I - section in Fig. 5.33a. Show that the flanges carry all the bending torque.

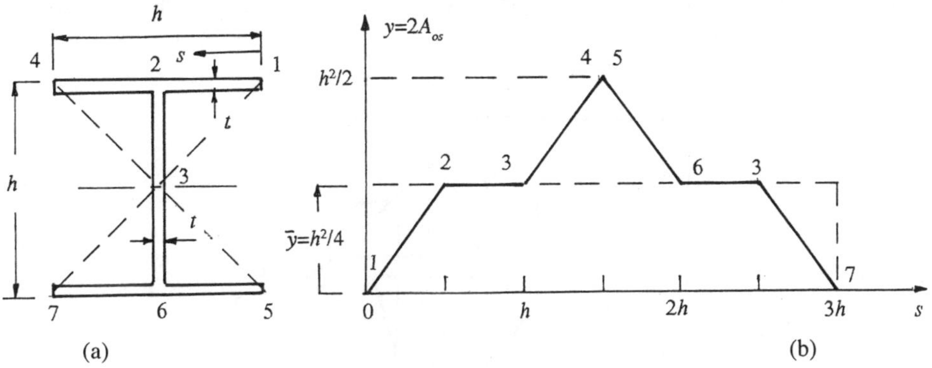

Figure 5.33 I - section showing swept areas

With the origin for s at point 1, Fig. 5.33b shows twice the area swept out from a centre of twist at point 3. Applying eq(5.79c) where $\int y\ \mathrm{d}s$ are the enclosed areas and $\int \mathrm{d}s$ is the perimeter length,

$$\bar{y} \int \mathrm{d}s = \int y\ \mathrm{d}s$$

$$3h\,\bar{y} = 2(h/4)(h^2/4) + 4\,(h^2/4)(h/2) + (h/2)(h^2/4)$$

from which $\bar{y} = h^2/4$. To find Γ_1 the integral $\int y^2\,ds$ in eq(5.84b) requires the equations of the straight lines in Fig. 5.33b

1-2: $y = hs\,/\,2$

$$\int_s y^2\,ds = (h^2/4)\int_0^{h/2} s^2\,ds = h^5/96$$

2-3: $y = h^2/4$

$$\int_s y^2\,ds = (h^4/16)\int_{h/2}^{h} ds = h^5/32$$

3-4: $y = (h/2)(s - h/2)$

$$\int_s y^2\,ds = (h^2/4)\int_h^{3h/2} (s - h/2)^2\,d\,s = 7h^5/96$$

Substituting into eq(5.84b),

$$\Gamma_1 = 2t\,(h^5/96 + h^5/32 + 7h^5/96) - t\,(h^2/4)^2 \times 3h = t\,h^5/24$$

Employing eqs(5.21) and (5.80), the warping displacements follow from eq(5.77a) as

$$w = -\,[T\,/\,(GJ\,)]\,(y - \bar{y})$$

where $y - \bar{y}$ at points 1, 2 ... 7 follow from Fig. 5.33b, so tha,

$$(GJ\,/\,T\,)w_1 = (GJ\,/T\,)w_7 = -\,(0 - h^2/4) = h^2/4$$
$$(GJ\,/\,T\,)w_2 = (GJ\,/T\,)w_3 = (GJ\,/T\,)w_6 = -\,(h^2/4 - h^2/4\,) = 0$$
$$(GJ\,/\,T\,)w_4 = (GJ\,/T\,)w_5 = -\,(h^2/2 - h^2/4\,) = -\,h^2/4$$

Thus a plane cross-section does not remain plane but warps, as shown for the free end of the cantilever in Fig. 5.34.

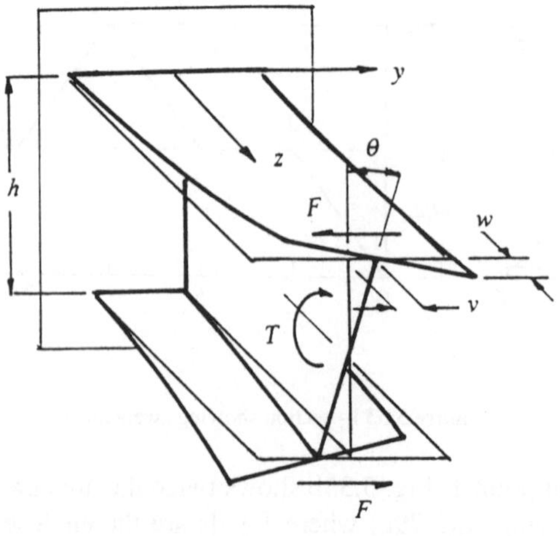

Figure 5.34 Warping and twist at the free end of a I-section under torsion

When the free warping displacements are constrained, say through building in one end, axial stresses are induced along the length. These stresses are proportional to w, so they are distributed in a similar manner to w within the flange of an I - beam (see Fig. 5.34). At the section shown the bending torque is found from the shear force F in each flange

$$T_b = - Fh = - (\delta M / \delta z)\, h \qquad \text{(i)}$$

T_b is negative since it opposes the applied torque T. The flexure equation will apply to the lateral displacement v of the z - axis

$$(E I_f)\, \mathrm{d}^2 v/\mathrm{d}z^2 = M \qquad \text{(ii)}$$

where $I_f = th^3/12$ is the second moment of area of one flange. Noting that $v = (h/2)\theta$ and substituting eq(ii) into eq(i)

$$T_b = - (h^2/2) E I_f\, \mathrm{d}^3\theta/\mathrm{d}z^3 \qquad \text{(iii)}$$

The contribution to torque T from a St Venant torque has the usual unconstrained form $T_v = GJ\,(\delta\theta/\delta z)$. Adding this to eq(iii) gives the total torque,

$$T = GJ\,(\delta\theta/\delta z) - (h^2/2) E I_f\, \mathrm{d}^3\theta/\mathrm{d}z^3 \qquad \text{(iv)}$$

Comparing eqs(iv) and (5.85) shows that we can again write $\Gamma_1 = (h^2/2) I_f = th^5/24$ thus confirming that the flanges carry the Wagner torque.

Example 5.13 The thin-walled S-section shown in Fig. 5.35a is rigidly fixed at one end and subjected to an axial torque of 2 Nm about a longitudinal axis passing through the shear centre at the free end. Show that the primary warping constant for the section is $13\,t\,h^5/12$. If $h = 50$ mm, $t = 1$ mm, $L = 2$ m. Determine, for a length of 2 m, (i) the angular twist and the warping displacements at the free end and (ii) the axial stress distribution and the shear flow at the fixed end. Take $E = 80 \times 10^3$ MPa and $G = 30 \times 10^3$ MPa.

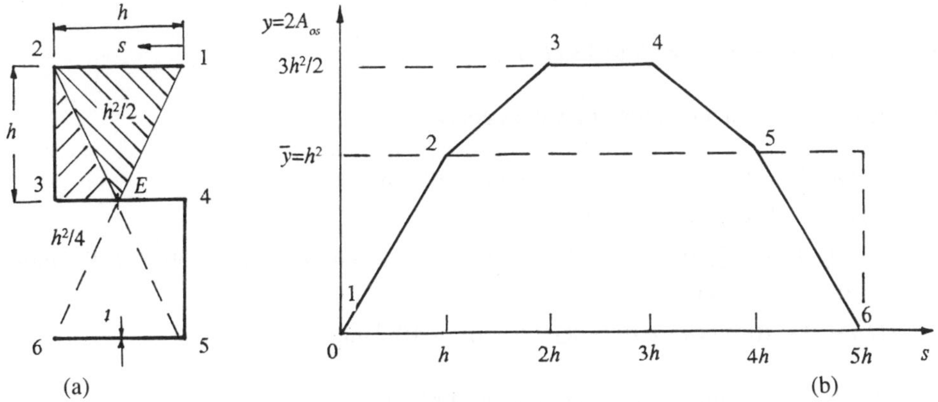

Figure 5.35 S-section cantilever

Let the origin for s lie at point 1. We first need to find the area ($\bar{y} = 2A_E'$) enclosed by point 1, the shear centre E and a point of zero warping. Construct Fig. 5.35b for twice the swept areas between points E, 1 and points 2, 3 ... 6, as shown. Applying eq(5.79c),

$\bar{y} \int ds = \int y\, ds$

$\bar{y}(5h) = 2\,(h^2 \times h/2) + (3h^2/2 \times h) + 2(h^2 + 3h^2/2)h/2$

$$\bar{y} = 2A_E' = h^2 \tag{i}$$

The torsion bending constant employs both ordinates $\bar{y} = 2A_E'$ and $y = 2A_{os}$. With t constant, eq(5.84) becomes

$$\Gamma_1 = t\,[\int y^2 ds - \bar{y}^2 \int ds\,] \tag{ii}$$

The first integral requires equations of the lines 1-2, 2-3 etc within the co-ordinates y and s. From Fig. 5.35b, these are

1-2: $y = h\,s$ for $0 \le s \le h$

2-3: $y = (h/2)(s + h)$ for $h \le s \le 2h$

3-4: $y = 3h^2/2$ for $2h \le s \le 3h$

4-5: $y = -\,h(s/2 - 3h)$ for $3h \le s \le 4h$

5-6: $y = -\,h(s - 5h)$ for $4h \le s \le 5h$

The integral in eq(ii) is evaluated between the appropriate limits

$$\int_0^s y^2\, ds = h^2 \int_0^h s^2 ds + (h/2)^2 \int_h^{2h} (s + h)^2\, ds + (9h^4/4) \int_{2h}^{3h} ds$$
$$+\, h^2 \int_{3h}^{4h} (s/2 - 3h)^2 ds + h^2 \int_{4h}^{5h} (s - 5h)^2 ds$$
$$= h^5/3 + (h^2/4)(19h^3/3) + 9h^5/4 + h^2(19h^3/12) + h^2(h^3/3)$$
$$= 73h^5/12 \tag{iii}$$

Substituting eqs(i) and (iii) into eq(ii),

$\Gamma_1 = t\,[73\,h^5/12 - (h^2)^2 \times 5h\,] = 13\,t\,h^5/12$

For the section given, the following values apply

$\Gamma_1 = 13\,t\,h^5/12 = 13 \times 1 \times 50^5/12 = 338.54 \times 10^6\ \text{mm}^6$

$J = \sum(h\,t^3/3) = 5 \times 50 \times 1^3/3 = 83.33\ \text{mm}^4$

$\mu = \sqrt{[GJ/(E\,\Gamma_1)]} = \sqrt{[(30 \times 10^3 \times 83.33)/(80 \times 10^3 \times 338.54 \times 10^6)]} = 0.3038 \times 10^{-3}\ \text{mm}^{-1}$

$\mu L = 0.3038 \times 2 = 0.6076$

$T/(GJ) = (2 \times 10^3)/(80 \times 10^3 \times 83.33) = 0.3 \times 10^{-3}\ \text{mm}^{-1}$

The *angular twist* at the free end is found by substituting $z = L$ in eq(5.86b),

$$\theta = [TL/(GJ)]\{1 - [1/(\mu L)]\tanh(\mu L)\}$$
$$= [T/(\mu GJ)][(\mu L - 1) + (\mu L + 1)e^{-2\mu L}]/(1 + e^{-2\mu L}) \tag{iv}$$

Substituting the values given above into eq(iv) gives $\theta = 0.1716$ rad (9.83°). Compare this with an unconstrained twist of

$\theta = TL/(GJ) = (2 \times 10^3 \times 2 \times 10^3)/(30 \times 10^3 \times 83.33) = 1.6$ rad (91.68°)

Substituting $z = 0$ in eq(5.88a) gives the axial stress at the fixed end

$$\sigma_z = -\,[2A_E T/(\mu\,\Gamma_1)]\tanh(\mu L)$$
$$= -\,[2A_E T/(\mu\,\Gamma_1)](1 - e^{-2\mu L})/(1 + e^{-2\mu L}) \qquad \text{(v)}$$

Substituting values into eq(v) shows that

$$\sigma_z = -\,10.5 \times 10^{-3}\,(2A_E)$$

where $2A_E = 2A_{0S} - 2A_E' = y - \bar{y}$. We can recognise $y - \bar{y}$ as the ordinate in Fig. 5.35b when $\bar{y} = h^2$ is taken as the abscissa (dotted line). This gives

$$\sigma_z = -\,10.5 \times 10^{-3}(y - \bar{y}) \qquad \text{(vi)}$$

where, for points 1, 2 ... 6,

$$(\sigma_z)_1 = (\sigma_z)_6 = -\,10.5 \times 10^{-3}(-h^2) = 26.25 \text{ MPa}$$
$$(\sigma_z)_2 = (\sigma_z)_5 = 0$$
$$(\sigma_z)_3 = (\sigma_z)_4 = -\,10.5 \times 10^{-3}(h^2/2) = -\,13.25 \text{ MPa}$$

At the fixed end, the stress is distributed in the manner of Fig. 5.36a. Equation(5.87) shows that a similar distribution will apply elsewhere along the beam but that the magnitudes of σ_z will diminish to zero at the free end.

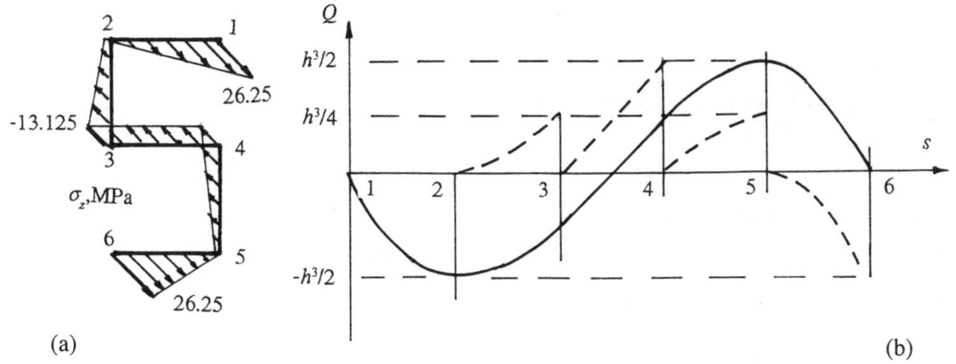

Figure 5.36 Axial stress and shear flow distributions

Equation (5.77a) provides the uncontrained *primary warping* displacement. Writing this as

$$w_o = -\,(\delta\theta/\delta z)(2A_E) = -\,[T/(GJ)](y - \bar{y})$$
$$w_o = -\,0.3 \times 10^{-3}(y - \bar{y}) \qquad \text{(vii)}$$

Using Fig. 5.35b, the displacements at points 1, 2 ... 6 are,

$$(w_o)_1 = (w_o)_6 = -0.3 \times 10^{-3}(-h^2) = 0.75 \text{ mm}$$
$$(w_o)_2 = (w_o)_5 = 0$$
$$(w_o)_3 = (w_o)_4 = -0.3 \times 10^{-3}(h^2/2) = -0.375 \text{ mm}$$

Equations (vi) and (vii) confirm that $\sigma_z \propto w_o$ and therefore the w_o distribution is similar to that for σ_z shown in Fig. 5.36a.

At any section within the length of a constrained open tube the *shear flow* follows from writing eq(5.82b) as

$$q = E\,(\mathrm{d}^3\theta/\mathrm{d}z^3)\int_0^s (2A_E)\,t\,\mathrm{d}s$$
$$= Et\,(\mathrm{d}^3\theta/\mathrm{d}z^3)\int_0^s (y - \bar{y})\,\mathrm{d}s \qquad \text{(viii)}$$

Differentiating eq(5.86a) it is seen that $\mathrm{d}^3\theta/\mathrm{d}z^3$ varies with length. The integral (say Q) in eq(viii) is independent of length and may be evaluated separately. For this it is necessary to express $y - \bar{y}$ from Fig. 5.35b in terms of s and add an integration constant from limb to limb, as follows:

1-2: $y - \bar{y} = h\,(s - h)$
$$Q_{1\text{-}2} = h\int_0^s (s - h)\,\mathrm{d}s = h\,(s^2/2 - hs)$$
$Q_1 = 0$ and $Q_2 = -h^3/2$

2-3: $y - \bar{y} = (h/2)(s - h)$
$$Q_{2\text{-}3} = (h/2)\int_h^s (s - h)\,\mathrm{d}s + Q_2 = (h/2)(s^2/2 - hs + h^2/2) - h^3/2$$
$Q_3 = -h^3/4$

3-4: $y - \bar{y} = h^2/2$
$$Q_{3\text{-}4} = (h^2/2)\int_{2h}^s \mathrm{d}s + Q_3 = (h^2/2)(s - 2h) - h^3/4$$
$Q_4 = h^3/4$

4-5: $y - \bar{y} = -h\,(s/2 - 2h)$
$$Q_{4\text{-}5} = -h\int_{3h}^s (s/2 - 2h)\,\mathrm{d}s + Q_4 = h\,(2hs - s^2/4 - 15h^2/4) + h^3/4$$
$Q_5 = h^3/2$

5-6: $y - \bar{y} = -h\,(s - 4h)$
$$Q_{5\text{-}6} = -h\int_{4h}^s (s - 4h)\,\mathrm{d}s + Q_5 = h\,(4hs - s^2/2 - 8h^2) + h^3/2$$
$Q_6 = 0$

The distribution in shear flow (see Fig. 5.36b) is found from $q = Et\,(\mathrm{d}^3\theta/\mathrm{d}z^3) \times Q$ where Q are the expressions within each limb given above. We see from eqs(viii) and (5.85) that $q = (t/\Gamma_1)T_w Q$. This shows that q varies in proportion with T_w over the length, this being a measure of the degree of axial constraint.

5.6.3 Secondary Warping

Up to now we have ignored the small contribution to axial warping occurring across the wall thickness. This is acceptable for most open sections but not for L and T sections. The reader should confirm that $\Gamma_1 = 0$ where the shear centre lies at the intersection between the limbs. These sections employ a secondary warping constant Γ_2 in the Wagner-Kappus theory. To derive Γ_2 generally, we must admit contributions to the torque from both τ_{sz} and τ_{nz} in Fig. 5.27b. That is, for components τ_{zs} and τ_{zn} within the shaded area $\delta s \times \delta n$ shown

$$\delta T = \tau_{zs}(\delta s \times \delta n)R_t - \tau_{zn}(\delta s \times \delta n)\,R_n \qquad (5.90a)$$

Recall that eq(5.81a) was accompanied by the second equilibrium equation (5.81b) for the s - z plane. Substituting $\tau_{zs} = -\,(\partial\sigma_z/\partial z)\,\mathrm{d}s$ and $\tau_{zn} = -\,(\partial\sigma_z/\partial z)\,\mathrm{d}n$ into eq(5.90a) gives

$$\mathrm{d}T = -\int_0^s (\partial\sigma_z/\partial z)\,\mathrm{d}s \times (\mathrm{d}s \times \mathrm{d}n)\,R_t + \int_0^n (\partial\sigma_z/\partial z)\,\mathrm{d}n \times (\mathrm{d}s \times \mathrm{d}n)\,R_n \qquad (5.90b)$$

For the axial stress we must write the total warping displacement from eqs(5.77a,b) as

$$w = -\,2\,(\delta\theta/\delta z)(A_E + R_n n) \qquad (5.91)$$

from which the axial stress and its derivatives are

$$\sigma_z = E\,(\partial w/\partial z) = -\,2E\,(\partial^2\theta/\partial z^2)(A_E + R_n n)$$
$$\partial\sigma_z/\partial z = -\,2E\,(\partial^3\theta/\partial z^3)(A_E + R_n n) \qquad (5.92)$$

Substituting eq(5.92) into eq(5.90b),

$$\mathrm{d}T = E\,(\delta^3\theta/\delta z^3)\{2\int_0^s [A_E + R_n n]\mathrm{d}s \times (\mathrm{d}s \times \mathrm{d}n)\,R_t\}$$
$$-\,E\,(\delta^3\theta/\delta z^3)\{2\int_0^n [A_E + R_n n\,]\mathrm{d}n \times (\mathrm{d}s \times \mathrm{d}n)\,R_n\}$$

Integrating for T,

$$T = E\,(\delta^3\theta/\delta z^3)\int_0^n \{2\int_0^s\int_0^s [A_E + R_n n\,]\mathrm{d}s \times (R_t \times \mathrm{d}s)\}\mathrm{d}n$$
$$-\,E\,(\delta^3\theta/\delta z^3)\int_0^s \{2\int_0^n\int_0^n [A_E + R_n n\,]\mathrm{d}n \times (R_n \times \mathrm{d}n)\}\mathrm{d}s$$

Writing this in the Wagner torque form,

$$T = -\,E\,(\Gamma_1 + \Gamma_2)(\delta^3\theta/\delta z^3) = -\,E\,\Gamma_E\,(\delta^3\theta/\delta z^3)$$

where the total warping constant Γ_E is the sum of the primary and secondary warping constants appearing as the respective integrals,

$$\Gamma_1 = \int_0^n \{2\int_0^s\int_0^s [A_E + R_n n\,]\mathrm{d}s \times (R_t \times \mathrm{d}s)\}\mathrm{d}n \qquad (5.93a)$$

$$\Gamma_2 = \int_0^s \{2\int_0^n\int_0^n [A_E + R_n n\,]\mathrm{d}n \times (R_n \times \mathrm{d}n)\}\mathrm{d}s \qquad (5.93b)$$

Integrating eq(5.93a) by parts between limits 0 to s and n from $-t/2$ to $t/2$ again leads to the eq(5.84a). We integrate eq(5.93b) as a sum

$$\Gamma_2 = \int_0^s [\int_{-t/2}^{t/2} (\int_0^n 2A_E \mathrm{d}n) \times R_n \mathrm{d}n] \,\mathrm{d}s + \int_0^s [\int_{-t/2}^{t/2} (\int_0^n 2R_n n \,\mathrm{d}n) \times (R_n \times \mathrm{d}n)] \,\mathrm{d}s$$

$$= \int_0^s (\int_{-t/2}^{t/2} 2A_E R_n n \,\mathrm{d}n) \,\mathrm{d}s + \int_0^s (\int_{-t/2}^{t/2} R_n^2 n^2 \,\mathrm{d}n) \,\mathrm{d}s$$

where the first integral is zero between the given limits of t. This leaves the the secondary warping constant defined by the second integral as

$$\Gamma_2 = \frac{1}{12} \int_0^s R_n^2 t^3 \mathrm{d}s \tag{5.94}$$

Hence the total warping constant for any open section becomes $\Gamma_E = \Gamma_1 + \Gamma_2$

$$\Gamma_E = \int_0^s (2A_E)^2 t \,\mathrm{d}s + \frac{1}{12} \int_0^s R_n^2 t^3 \mathrm{d}s \tag{5.95}$$

We have shown earlier how to evaluate the first integral Γ_1. The secondary warping constant Γ_2 is usually small enough to be ignored except for sections where $\Gamma_1 = 0$.

Example 5.14 Show that $\Gamma_1 = 0$ for the angle section given in Fig. 5.37 and evaluate the secondary warping constant.

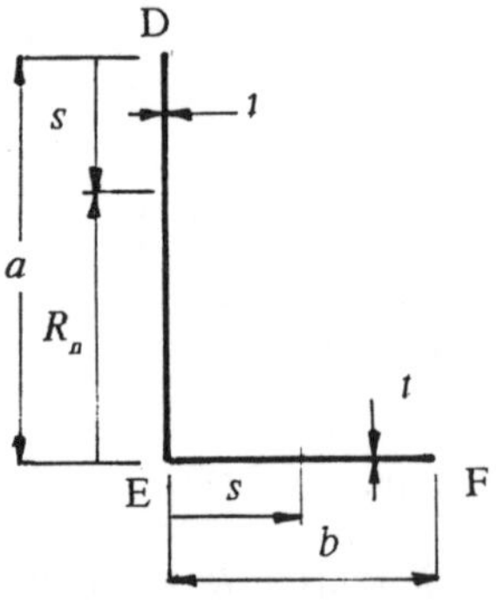

Figure 5.37 Thin angle section with unequal limbs

Clearly, in moving around the mid-wall line from the edge D to E no area is subtended when the corner co-incides with the shear centre E. The same applies as we move with s from E to F. Hence $\Gamma_1 = 0$ and there is no primary warping in this section.

We determine Γ_2 from eq(5.94) where R_n is measured within each limb from E as shown. This gives $R_n = a - s$ for limb DE and $R_n = s$ for limb EF.

$$\Gamma_2 = (t^3/12) \int_0^a (a - s)^2 \mathrm{d}s + (t^3/12) \int_0^b s^2 \mathrm{d}s$$

$$= (t^3/12) \,[\,|\, a^2 s - as^2 + s^3/3 \,|_0^a + |\, s^3/3 \,|_0^b \,]$$

$$= (t^3/36)(a^3 + b^3)$$

Bibliography

Hoff N. J., *Jl Roy. Ae Soc*, **47**, (1943), 35-83.
Megson T. H. G., *Aircraft Structures for Engineering Students*, E. Arnold, 1972.
Ross C. T. F., *Advanced Applied Stress Analysis*, Ellis Horwiood, 1987.
Saada A. S., *Elasticity, Theory and Applications*, Pergamon, 1974.
Williams D., *Theory of Aircraft Structures*, E. Arnold, 1960.

EXERCISES

Torsion of Circular Sections

5.1 At its limiting shear stress a solid steel shaft, 40 mm diameter, can transmit a torque of 3 kNm. What diameter shaft of a similar steel can transmit 50 kW when rotating at 250 rev/min?
Answer: 34.4 mm

5.2 Find the maximum power that can be transmitted by a 150 mm diameter shaft running at 240 rev/min, given that the allowable shear stress is 55 MPa. The shaft is coupled to a motor with a flanged joint containing 6 bolts on a 265 mm pitch circle diameter. Find the required bolt diameter when the maximum shear stress for the bolt material is restricted to 100 MPa.

5.3 Compare the ratio of the weights for solid and hollow shafts in the same material when transmitting the same torque, if, for the hollow shaft, the outer diameter is twice the inner. What power can a hollow shaft of 300 mm o.d. and 200 mm i.d. transmit at a speed of 200 rev/min, when the maximum shear stress is restricted to 65 MPa?

5.4 A solid drive shaft of diameter d is replaced by a hollow tube of mean diameter D, which can sustain the same torque without exceeding the maximum shear stress. Show that these conditions are achieved when $D = \sqrt{[d^3/(8\,t)]}$ and that the ratio between the solid and hollow twist rates is equal to d/D under a given torque.

5.5 The drive shaft of a motor car is required to transmit 20 kW at 250 rev/min. If the shaft is hollow, 75 mm o.d and 62.5 mm i.d., calculate the maximum shear stress in the shaft when the maximum torque exceeds the mean torque by 20%. Find the greatest angle of twist for a 2 m length of shaft. Note that T in eq(5.7) is the mean torque. Take $G = 82$ GPa.

5.6 A solid circular shaft is connected to the drive shaft of an electric motor with a flanged coupling. The drive is taken by eight bolts each 12.5 mm diameter on a pitch circle diameter of 230 mm. Calculate the shaft diameter if the maximum shear stress in the shaft is to equal that in the bolts.

5.7 A solid shaft 200 mm diameter is subjected to a torque of 50 kNm. This shaft is to be replaced in the same material by a hollow shaft with diameter ratio 2 for the same torque and maximum shear stress. Compare the ratio of their weights and angles of twist over a length of 3 m. Take $G = 80$ GPa for both shafts.

5.8 A hollow steel shaft is required to transmit 6 MW at 110 rev/min. If the allowable shear stress is 60 MPa and the i.d. = (3/5)(o.d.), calculate the dimensions of the shaft and the angle of twist on a 3 m length. Take $G = 83$ GPa.

5.9 A tubular turbine shaft is to transmit power at 240 rev/min. If the shaft is 1 m outer diameter and 25 mm thick, find the power transmitted at a maximum shear stress of 70 MPa. What is the diameter of the equivalent solid shaft and the percentage saving in weight of the hollow shaft?

5.10 A 30 mm diameter solid steel spindle is fitted with a 2 m long tubular extension, 40 mm o.d. and 30 mm i.d. If, when transmitting torque, the maximum shear stress in the spindle is 35 MPa, calculate the maximum shear stress and the angle of twist for the tube. Take $G = 82$ GPa.

5.11 A hollow steel shaft, 100 mm o.d. and 10 mm thick, is welded to a solid 110 mm diameter steel shaft. If the length of the hollow shaft is 1 m, find the length of solid shaft which limits the total twist to 2° under a torque of 30 kNm. Take $G = 79$ GPa.

5.12 A steel shaft, 1 m long and 80 mm diameter, is bored to 50 mm diameter over one quarter of its length. Find the greatest torque the shaft can withstand when the maximum shear stress is limited to 90 MPa. What is the angle of twist between the ends? Take $G = 80$ GPa.

5.13 A solid 25 mm steel bar is surrounded by an aluminium alloy tube so that both twist together without slip. If their respective allowable shear stresses are not to exceed 120 and 60 MPa under an axial torque of 2 kNm, determine the aluminium outer diameter and the angle of twist over a 2 m length. Take $G = 28$ GPa for aluminium and $G = 80$ GPa for steel.

5.14 A composite shaft consists of two concentric cylinders: nimonic within aluminium firmly bonded at the interface. The dimensions are, nimonic: 10 mm i.d., 20 mm o.d. and aluminium: 20 mm i.d., 25 mm o.d. Given the respective allowable shear stresses are 35 MPa and 75 MPa, what is the maximum torque that the shaft can carry? Take $G = 30$ GPa for aluminium and and $E = 80$ GPa for nimonic.

5.15 Draw the torque and twist diagrams for the stepped shaft in Fig. 5.38, showing maximum values.

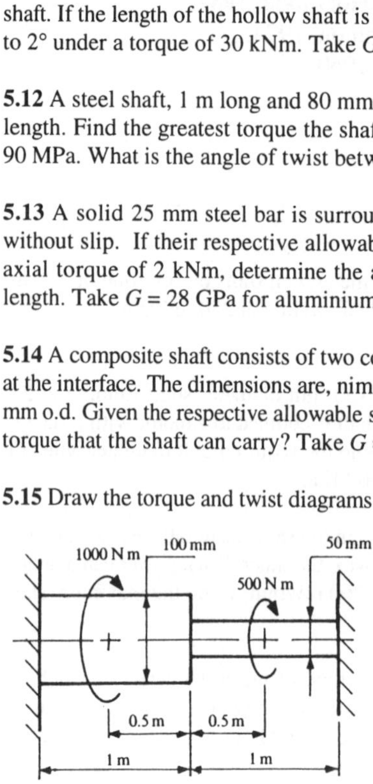

Figure 5.38

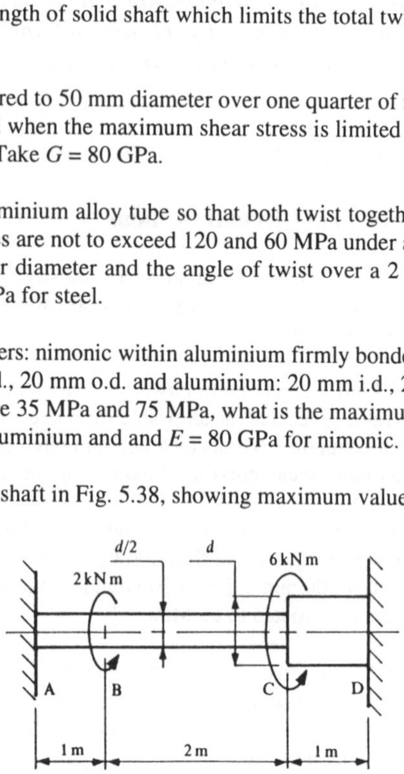

Figure 5.39

5.16 Figure 5.39 shows a stepped shaft ABCD which is rigidly fixed at both ends. The shaft is subjected to applied torques of 2 kNm and 6 kNm at B and C respectively. The diameter of the portion AC is one half that of portion CD. Calculate the torque distribution in the shaft. Determine the diameters of the shaft if it is required that neither the maximum shear stress nor the angle of twist should exceed 50 N/mm^2 and 1 degree respectively. Assume $G = 80 \times 10^3$ N/mm^2. (CEI)

Torsion of Thin Strips and Open Sections

5.17 A torque of 7.5 Nm is applied along the axis of a thin rectangular strip, 45 mm wide × 1.5 mm thick. Determine the maximum shear stress and the angle of twist over a 300 mm length.

5.18 A 300° circular steel arch is bent from 3 mm sheet metal. It is to support a torque of 40 Nm without the maximum shear stress exceeding 55 MPa whilst restricting the angular twist to 3°. Find the necessary radius and length of the sheet and the warping displacements between its free ends.

5.19 A strip of metal of rectangular section, 10 mm × 60 mm, length 0.3 m and shear modulus 80 GPa, is clamped in a torsion testing machine and subjected to a steadily increasing torque. If the material yields when the shear stress exceeds 100 MPa, determine the torque required to just cause yielding in the material, and also determine the corresponding relative angular rotation of the two ends of the strip.

5.20 Determine the torsional stiffness for the steel drive shaft in Fig. 5.40, when 0.75 m of its length is a square section whose diagonal is equal to the remaining 100 mm circular shaft diameter. Assume that the square section is free to warp for the coefficients in Table 5.1. Take $G = 80$ GPa.

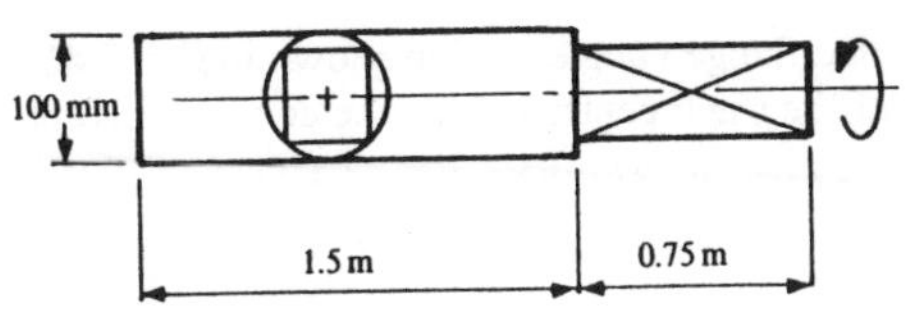

Figure 5.40

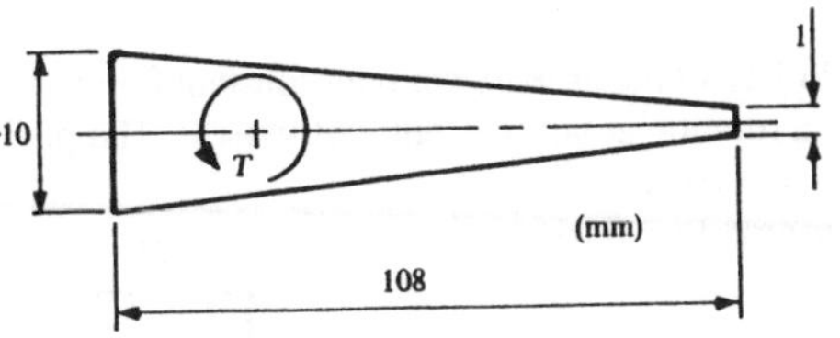

Figure 5.41

5.21 Determine the St Venant torsion constant for the thin tapered section in Fig. 5.41. Hence derive an expression for the shear stress acting along the longer edges in terms of the torque T when the thickness $10 \le t \le 1$. If $T = 15$ Nm estimate the position and magnitude of the maximum shear stress from a consideration of contours of constant shear stress.

5.22 Figure 5.42 shows the free end of an equal-angle cantilever beam section. The dimensions increase uniformly over its 500 mm length so that at the fixed end all dimensions become 50% greater. If the maximum shear stress is limited to 100 MPa, determine the allowable free end torque and the angle of twist between the ends.

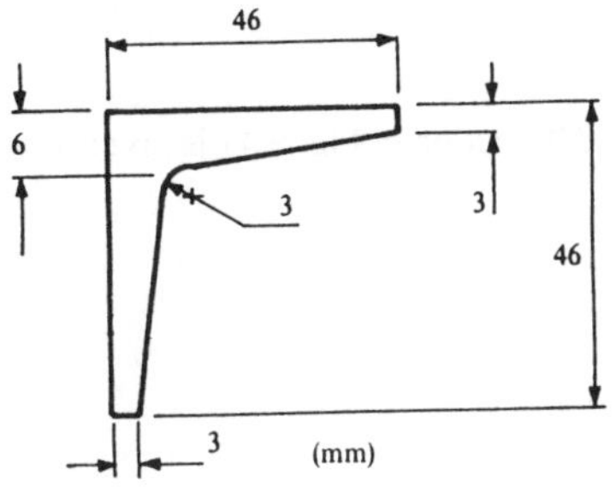

Figure 5.42

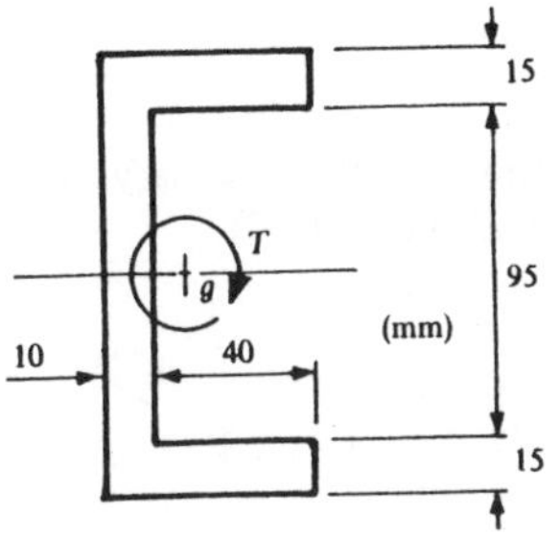

Figure 5.43

5.23 Find the torque T that can be withstood by the steel channel section in Fig. 5.43 when the greatest shear stress is limited to 120 MPa. What angle would this section twist through in a length of 3 m? Take $G = 83$ GPa and use the coefficients given in Table 5.1.

5.24 Torque is applied about an axis passing through the centre of the vertical web for the aluminium channel section shown in Fig. 5.44. Show that the Wagner-Kappus torsion-bending constant is given as $\Gamma_1 = tH^5/3$ and determine the free warping displacements when $H = 50$ mm, $t = 1$ mm and the torque is 2 Nm. Take $G = 30$ GPa.

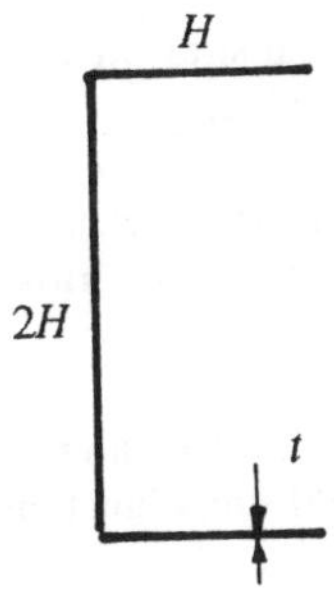

Figure 5.44

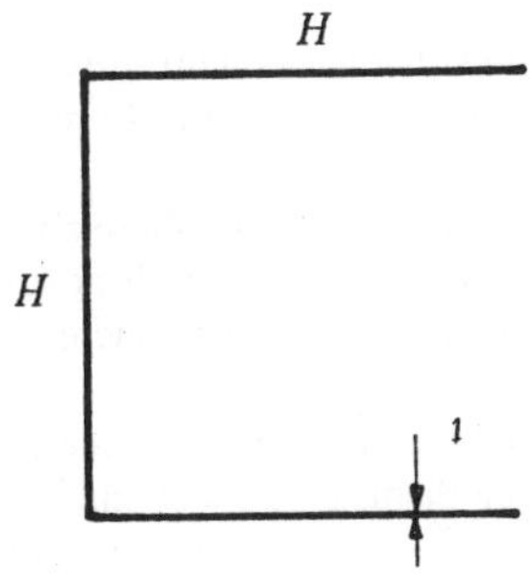

Figure 5.45

5.25 Show that the torsion-bending constant for the steel channel in Fig. 5.45 is $\Gamma_1 = tH^5/15$. A similar section, with $t = 1.5$ mm and $H = 60$ mm, is mounted as a 1 m long cantilever with one end fixed. Determine (a) the angular twist at the free end, (b) the axial stress distribution at the fixed end and (c) the shear flow variation around the perimeter at mid-span. Take $G = 30$ GPa and $E = 80$ GPa.

5.26 Derive the primary warping constant for the thin-walled flanged angle section shown in Fig. 5.46 for a rotation about the centre of twist E. The position of the shear centre E from the corner is given as $e = 0.108a$.

Answer: $\Gamma_1 = 0.0184\, ta^5$

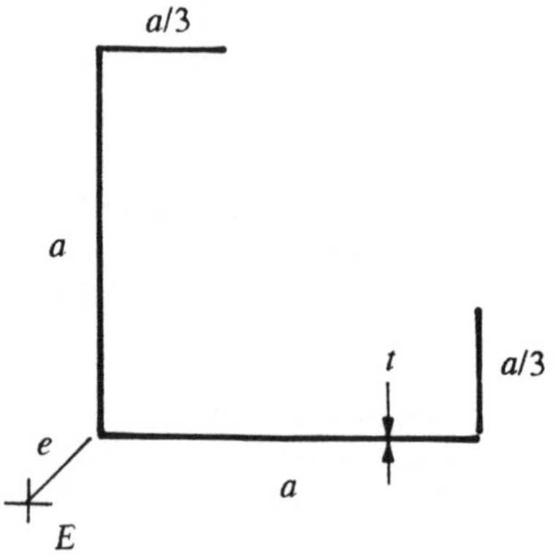

Figure 5.46

5.27 Show that the secondary warping constant for the inverted T-section in Fig. 5.47 is given as: $\Gamma_2 = (t^3/36)(a^3 + b^3/4)$

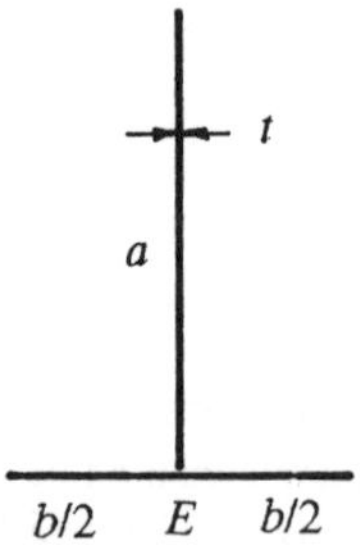

Figure 5.47

Torsion of Prismatic Bars

5.28 Show that the Prandtl stress function $\phi = K\,(x^2/R^2 + y^2/R^2)$ defines the shear stress in a solid circular section of radius R. Confirm simple torsion theory and the absence of warping.

5.29 The function $\phi = G\,(\delta\theta/\delta z)(t^2/4 - y^2)$ represents torsion of a thin rectangular strip where $b \gg t$. Determine the maximum shear stress under a torque of 2 Nm for $b = 50$ mm and $t = 2$ mm. How is the theory altered when t is increased to 10 mm? Take $G = 80$ GPa. (Ans 30 MPa)

5.30 Find the rate of twist under a torque of 175 Nm for a solid shaft of elliptical section with major and minor diameters $2a = 50$ mm and $2b = 35$ mm respectively. Assume the Prandtl stress function $\phi = D\,(x^2/a^2 - y^2/b^2 - 1)$ where D is a constant and take $G = 80$ GPa.

5.31 Show that the Prandtl stress function $\phi = A[(x^2 + y^2)/2 - (x^3 - 3xy^2)/2a - 2a^2/27]$ provides a solution to the torsion of a solid equilateral triangular section of height a. Derive from the constant A, the torsion constant J and plot the distribution of shear stress τ_{yz} along the vertical to locate the maximum value. If the side length is 50 mm, determine the angle of twist on a 2 m length under a pure axial torque of 100 Nm. Take G = 80 GPa.

5.32 Derive the warping function for the solid triangular section in Exercise 5.31 and hence establish the manner in which this cross-section warps. [Ans $\psi = (\delta\theta/\delta z)(y^3 - 3x^2y)/2a$]

5.33 A solid steel shaft tapers for a length of 1.5 m between end diameters for which one is twice the other. If the maximum shear stress and the angle of twist are limited to 60 MPa and 4° respectively under a torque of 10 kNm, determine the diameters. Take G = 80 GPa.

5.34 The diameter of a 0.5 m length steel shaft increases uniformly from 50 mm at one end to 75 mm at the other end. If the maximum shear stress in the shaft is limited to 95 MPa, find the torque which may be transmitted and the angle of twist between the ends. Take G = 80 GPa.

5.35 Figure 5.48 shows the dimensions (mm) of one half of an aerofoil section in aluminium alloy. Determine the St Venant torsion constant and find the maximum shear stress under an axial torque of 500 Nm. Take G = 80 GPa.

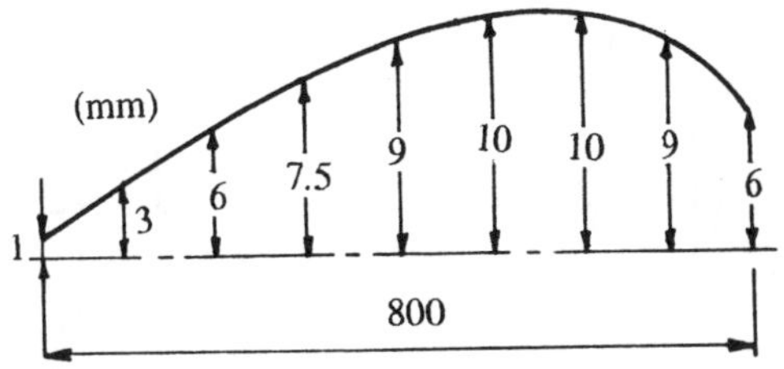

Figure 5.48

Torsion of Thin-Walled Closed Sections

5.36 Calculate the angular twist and thickness t for a thin-walled circular steel tube of mean radius R = 0.2 m and length L = 1.5 m that is required to withstand a torque of 75 kNm, when the shear stress is limited to 35 MPa. Examine the possibility of torsional buckling given the critical stress, $\tau_{cr} = 0.272E(t/R)^{3/2}/(1 - \nu^2)^{3/4}$. Take E = 208 GPa and ν = 0.27.

5.37 A rectangular box-section cantilever has walls 5 mm thick. The 200 mm depth is constant but the breadth varies over its 3 m length from 100 mm at the free end to 200 mm at the fixed end. Find the maximum shear stress and derive an expression for the twist as a function of length, when a torque of 100 Nm is applied at the free end. Take G = 28 GPa.

5.38 A thin-walled equilateral triangular section is to withstand an axial torque of 500 Nm without the shear stress exceeding 40 MPa. If the material cost C in pence per mm^2 of tube section per metre length is given by $C = 15 + s/(10t)$, determine the tube side length s and thickness t that will minimise cost whilst sustaining this torque.

5.39 The material cost of a thin-walled square tube is $[0.4 + s/(100t)]$ pence per mm^2 of the section area per metre length. Determine the side length s and the thickness t to minimise cost when the applied torque is 660 Nm for a maximum shear stress of 75 MPa.

5.40 Compare the theoretical torsional stiffnesses according to the Batho and simple torsion theories for a circular tube whose wall thickness is 1/20 of the mean wall diameter.

5.41 An I-section, 120 mm wide × 300 mm deep, with web and flange thicknesses 10 mm and 4 mm respectively, sustains a torque of 300 Nm over a length of 3 m. Find the angle of twist in degrees and the maximum shear stress. Compare these with the corresponding values for a welded box section formed by splitting the web into two 5 mm plates. Take $G = 80$ GPa.

5.42 Develop a relationship between torque and angle of twist for a closed uniform tube of thin-walled, non-circular section and use this to derive the twist per unit length for a strip of thin rectangular cross-section. Use the above relationship to show that for the same torque the ratio of angular twist per unit length for a closed square section tube to that for the same section but opened by a longitudinal slit and free to warp is approximately $4\,t^2/(3b^2)$, where t, the material thickness, is much less than the mean width b of the cross-section. (CEI) (Hint: In the derivation apply Batho to an elementary tube of thickness δy to find δT and $\delta\theta/\delta z$ noting that $y \ll b$.)

5.43 Calculate the twist in a 2 m length of section shown in Fig. 5.49 when it is subjected to a torque of 100 kNm. Find the position and magnitude of τ_{max}. Take $G = 30$ GPa.

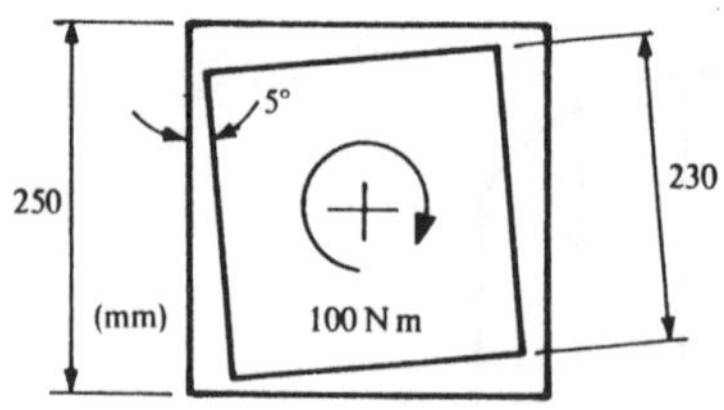

Figure 5.49

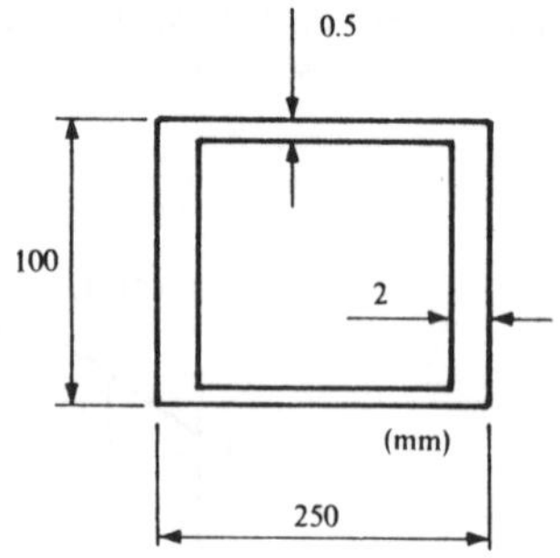

Figure 5.50

5.44 Determine the warping displacements at the four corners of the box section in Fig. 5.50 when it is subjected to a pure torque of 100 Nm. Take $G = 27$ GPa.

5.45 The two-bay alloy wing structure in Fig. 5.51 is subjected to concentrated torques at the two positions shown. Calculate the position and magnitude of the maximum shear stress in the skin when the given free-end thicknesses are uniform throughout. Take $G = 28$ GPa

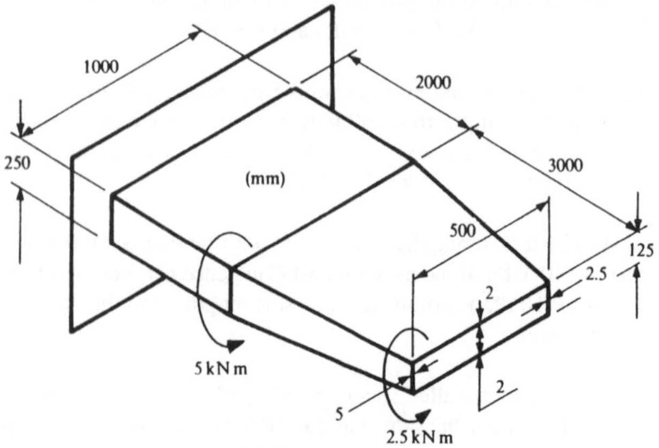

Figure 5.51

5.46 Figure 5.52 shows the uniform cross-section of a 5 m long tailplane. If the tube is subjected to a torque of 50 kNm, determine the shear stress in each of the webs, the angle of twist and the warping displacements at the corners, assuming that the centre of twist lies at the centroid. Take $G = 30$ GPa.
Answer: 55.56 MPa, 111.11 MPa, 37.04 MPa, 4.39°

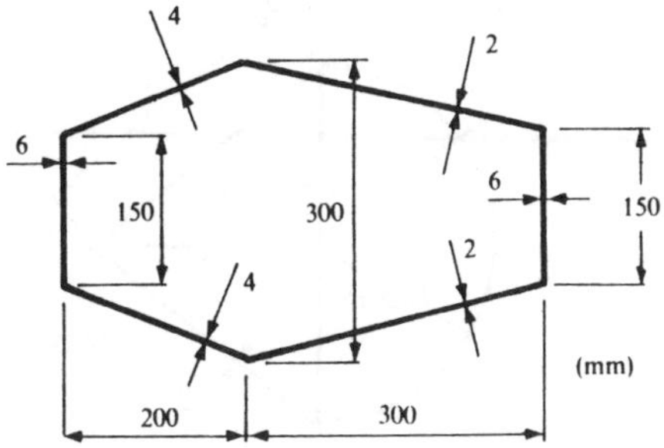

Figure 5.52

5.47 Figure 5.53 represents the cross-section of two, thin-walled uniformly thick tubes differing only in that one has a closed section while the other is open, being slit axially at the centre of web 2-3. Each tube is subjected to a constant torque and is constrained to twist about an axis passing through the middle of web 1-4. Calculate the warping of each cross-section in terms of the rate of twist, illustrating the distribution with a sketch on which principal values are indicated. Assume that the shear stress in the closed tube is given by Batho and in the open tube by St. Venant. (CEI)

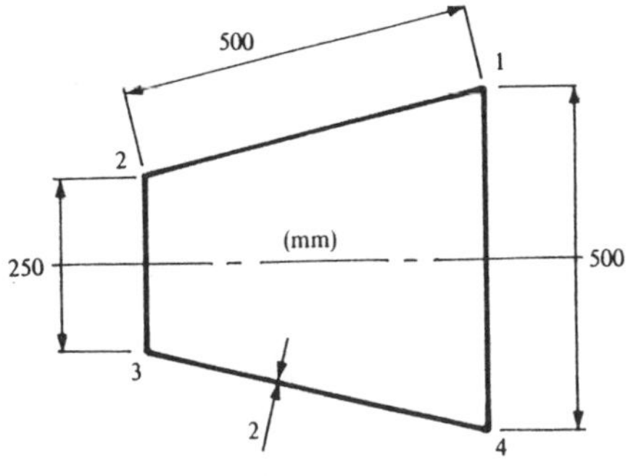

Figure 5.53

5.48 A 5m length aluminium tube, with a two-cell section shown in Fig. 5.54, is subjected to a torque of 200 Nm. Assuming that no buckling occurs, determine (a) the angular twist and (b) and the maximum shear stress in each wall of the tube. Take $G = 30$ GPa. (CEI)
Answer: 0.76 MPa, 1.01 MPa, 0.195 MPa, 0.106°

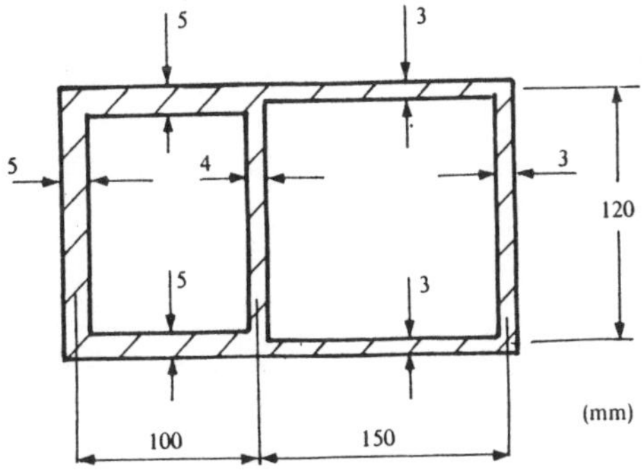

Figure 5.54

5.49 The three-cell tube in Fig. 5.55 supports a torque of 2 kNm. Calculate the position and magnitude of the maximum shear stress, the rate of twist and the torsional stiffnesses/m length. Compare these values with those for a similar single cell found from removing the webs. Take $G = 79$ GPa.

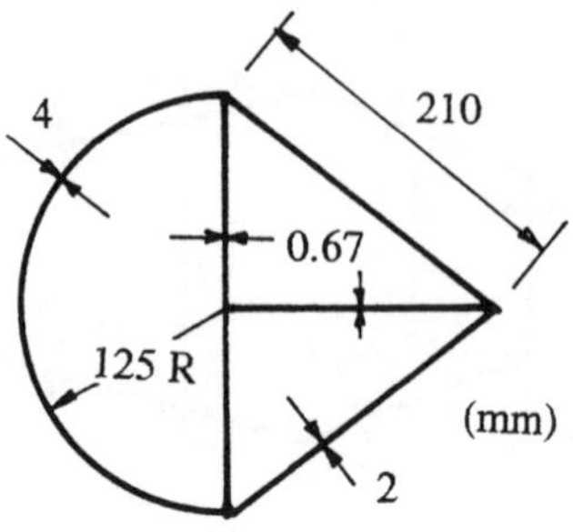

Figure 5.55

CHAPTER 6

MOMENT DISTRIBUTION

In the analysis of the stresses in loaded beams and structures it is necessary to know the manner in which moments and forces vary throughout their length. A single bay beam resting upon two simple supports is *statically determinate*. It will be shown that the application of equilibrium principles alone is sufficient to determine their force and moment distributions. Beams with encastre fixings, two or more bays and many types of structure are *statically indeterminate*. Their analysis becomes more complex requiring both equilibrium and compatibility conditions to be satisfied. Two methods are outlined: (i) the theorem of three moments, which employs known relationships between moments, slopes and deflections and (ii) moment distribution in which moments at supports are balanced by trial. Method (ii) is more versatile and will be applied to both continuous beams and structures.

6.1 Single Span Beams

6.1.1 Relationships Between F and M

In any beam the shear force F and bending moment M obey a differential relationship. Consider an elemental length δz of beam, over which the force and moments vary by δF and δM in the manner shown in Fig. 6.1. Taking moments about O and applying vertical equilibrium,

$$(M + \delta M) + \delta z\,(w\delta z)/2 = M + (F + \delta F)\delta z$$

$$F + w\,\delta z = F + \delta F$$

$$F = \delta M/\delta z \text{ and } w = \delta F/\delta z \qquad (6.1\text{a,b})$$

Equations (6.1a,b) show that w is the derivative of F and F is the derivative of M, or that M is the area beneath the F- diagram. Note also from eq(6.1b) that the maximum M will occur where F is zero.

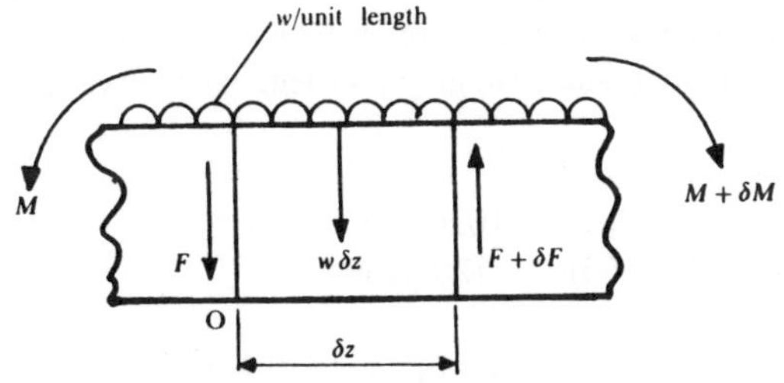

Figure 6.1 Beam element

The following examples show how the distributions in shear force F and bending moment M may be presented graphically.

Example 6.1 Draw the F and M diagrams for the simply supported beam in Fig. 6.2.

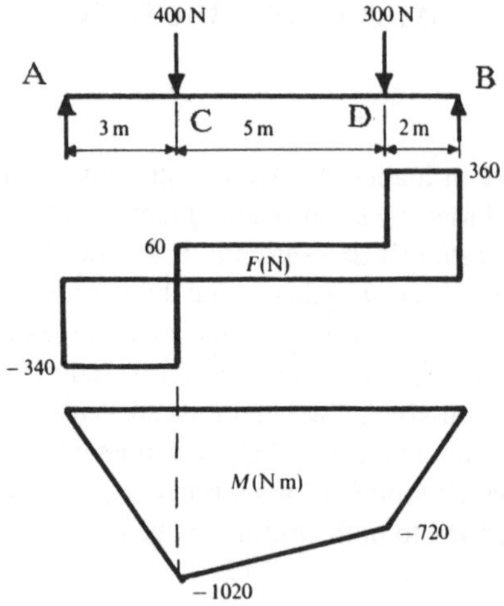

Figure 6.2 F and M distributions

The support reactions R_A and R_B are found from force and moment equilibrium. Take moments about the left support to find the right support reaction,

$$(3 \times 400) + (8 \times 300) = 10\, R_B \;\Rightarrow\; R_B = 360 \text{ N}$$

Apply vertical force equilibrium to find the left support reaction,

$$400 + 300 = R_A + R_B \;\Rightarrow\; R_A = 340 \text{ N}$$

F - diagram: Plot F (↑ + ve) from right to left along the length. The F - diagram must close on the datum line at the l.h. end. The ordinate in the diagram is the shear force at a given section. It may also be calculated independently as the net force arising from all forces to the right (or left) of that section.

M - diagram: An ordinate in the M - diagram is the sum of the moments exerted by all forces lying to one side of a beam at a given position. This net moment may be calculated from working to the left or right depending upon which is easier. A convention applies that a hogging moment is positive and a sagging moment is negative, as follows:

@ r.h. & l.h. ends $M = 0$
@ C to left, $M_C = -\ (3 \times 340) = -\ 1020$ Nm
@ D to right, $M_D = -\ (2 \times 360) = -\ 720$ Nm

Both M_C and M_D will sag the beam. $M_{max} = 1020$ Nm where the F - diagram crosses the horizontal datum. In this example, the bending moment diagram is all sagging under concentrated forces and simple end-supports.

6.1.2 Point of Contraflexure

With other forms of loading, the bending moment diagrams may show both hogging and sagging of the beam. A point of contraflexure (or inflection) lies at positions of zero bending moment. This is the point in the length of a beam where its curvature changes from hogging to sagging.

Example 6.2 Sketch the F and M diagrams for the beam in Fig. 6.3 and determine the position and magnitude of the maximum bending moment and the position of the point of contraflexure.

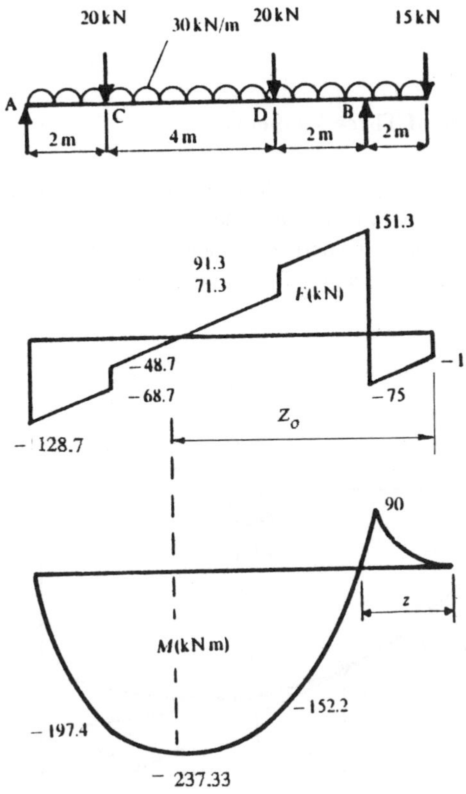

Figure 6.3 Inflection in M - diagram

Reactions R_A and R_B

$$(2 \times 20) + (6 \times 20) + (10 \times 15) + (30 \times 10 \times 5) = 8R_B \quad \Rightarrow \quad R_B = 226.3 \text{ kN}$$
$$R_A = [20 + 20 + 15 + (30 \times 10)] - 226.3 = 128.7 \text{ kN}$$

F - diagram: Use the construction method where the distributed loading defines the gradient of the F - diagram (from eq 6.1a). Discontinuities arise at points of concentrated forces. The diagram reveals the position of zero F as

$$F = 0 = -15 + 226.3 - 30\,z_o - 20 = 0\,, \quad \Rightarrow \quad z_o = 6.376 \text{ m}$$

M - diagram: Working to left or right, the ordinates are

$M_B = + (2 \times 15) + (2 \times 30 \times 1) = 90$ kNm,
$M_C = - (128.7 \times 2) + (30 \times 2 \times 1) = - 197.4$ kNm,
$M_D = - (128.7 \times 6) + (20 \times 4) + (6 \times 30 \times 3) = - 152.2$ kNm
$M_{max} = (15 \times 6.376) + (2.376 \times 20) - (4.376 \times 226.3) + 20(6.376)^2/2$
$= - 237.33$ kNm, where $F = 0$

Point of Contraflexure: let this lie distance z from RH end

$M = 15z - 226.3(z - 2) + 30z^2/2 = 0$
$15z^2 - 211.3z + 452.6 = 0$
$\therefore z = 2.64$ m.

6.2 Clapeyron's Theorem of Three Moments

Clapeyron's theorem applies to any continuous beam but is derived by taking spans in pairs. The term *three moments* refers to the unknown moments at the central support and at the two ends of any pair of adjacent spans. To derive the theorem it is first necessary to revise Mohr's two theorems for finding the slope and deflection of beams.

(a)

(b)

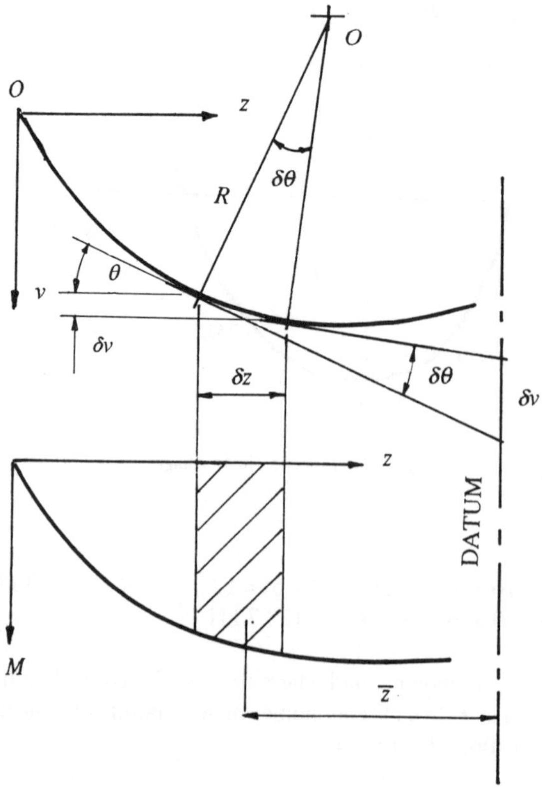

Figure 6.4 Mohr's theorems

Referring to Fig. 6.4a, the theorems provide the change in slope θ and deflection v over length δz of the beam as

$$\delta\theta = \delta z/R = (M\,\delta z)/(EI) \tag{6.2a}$$

$$\delta\theta = (1/EI)(\text{Area of the } M \text{ diagram over } \delta z) \tag{6.2b}$$

$$\delta v = \delta\theta\,\bar{z} = (\delta z/R)\,\bar{z} = (1/EI)\,(M\delta z)\bar{z} \tag{6.3a}$$

$$\delta v = (1/EI)(\text{Moment of area of the } M \text{ diagram over } \delta z) \tag{6.3b}$$

The geometric interpretations (6.2b) and (6.3b) require that I is constant, so that $M \times \delta z$ becomes the shaded area of the M diagram in the region δz (see Fig. 6.4b). In taking the moment of that area, its centroidal distance $\bar{z}$ is measured from a datum where v is required.

Clapeyron applied Mohr's theorems to a beam resting on more than two supports. Consider any two successive spans L_1 and L_2 in Fig. 6.5a under arbitrary loading where the moments existing at their ends A, B and C are M_A, M_B and M_C respectively.

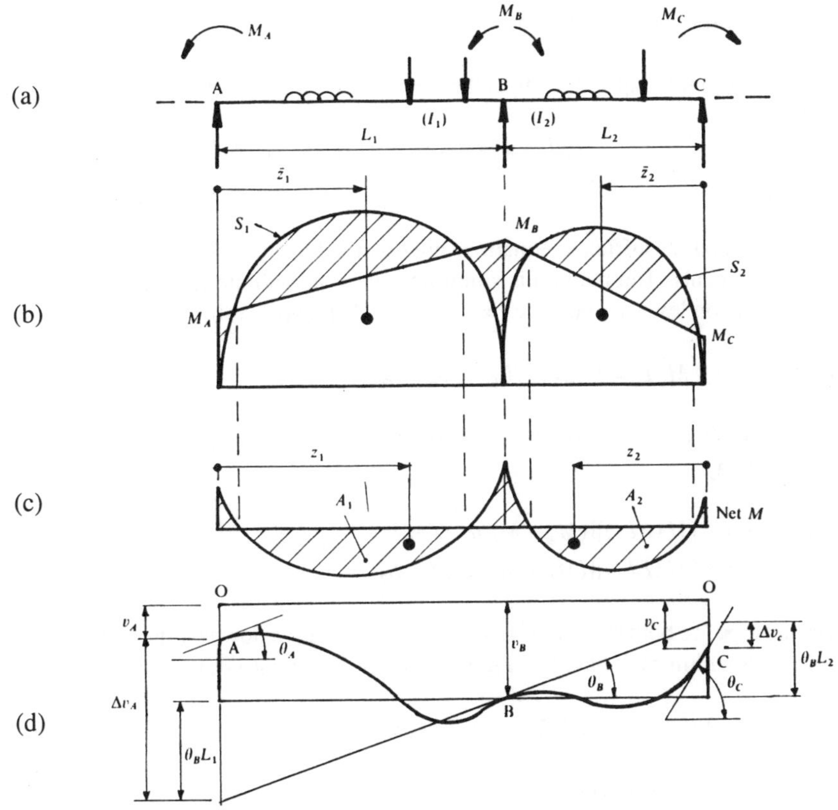

Figure 6.5 Continuous beam

The corresponding free and fixing moment diagrams are superimposed in Fig. 6.5b and take opposing signs. Thus, shaded areas A_1 and A_2 in each bay, that are not common to both diagrams, form the net moment diagram, given in Fig. 6.5c. In the general case let the beam rest on non-level supports at positive vertical distances v_A, v_B and v_C from any horizontal datum OO, as shown in Fig. 6.5d. Furthermore, let the second moments of area I_1 and I_2 for each bay differ.

The centroidal positions $\bar{z}_{1,2}$ and $z_{1,2}$ refer to the free and net moment areas S and A in each bay of Figs 6.5b,c respectively. The slope of the tangent at B is θ_B and the intercepts made by this tangent relative to the ends A and C are Δv_A and Δv_C. Taking the origin at B and with the datum in turn at A and C, eq(6.3a) yields

$$\Delta v_A = A_1 z_1/(EI_1) = \theta_B L_1 + (v_B - v_A) \tag{6.4a}$$
$$\Delta v_C = A_2 z_2/(EI_2) = -[\theta_B L_2 - (v_B - v_C)] \tag{6.4b}$$

in which Δv_C is negative upwards. Eliminating θ from eqs(6.4a,b) leads to

$$A_1 z_1/(EI_1 L_1) + A_2 z_2/(EI_2 L_2) = (v_B - v_A)/L_1 + (v_B - v_C)/L_2 \tag{6.5}$$

where, from Figs 6.5b,c, with hogging positive,

$$A_1 z_1 = \{[M_A L_1^2/2 + (M_B - M_A)L_1^2/3] - S_1 \bar{z}_1\} \tag{6.6a}$$
$$A_2 z_2 = \{[M_C L_2^2/2 + (M_B - M_C)L_2^2/3] - S_2 \bar{z}_2\} \tag{6.6b}$$

Substituting eqs(6.6a,b) into eq(6.5) leads to the theorem of three moments,

$$M_A L_1/I_1 + 2M_B(L_1/I_1 + L_2/I_2) + M_C L_2/I_2 = 6[S_1 \bar{z}_1/(I_1 L_1) + S_2 \bar{z}_2/(I_2 L_2)]$$
$$+ 6E[(v_B - v_A)/L_1 + (v_B - v_C)/L_2] \tag{6.7a}$$

By taking the spans in pairs, sufficient equations are obtained to solve them simultaneously for all the fixing moments. The method is simplified when a continuous beam of uniform section ($I_1 = I_2$) rests on level supports ($v_A = v_B = v_C$), when eq(6.7a) becomes

$$M_A L_1 + 2M_B(L_1 + L_2) + M_C L_2 = 6(S_1 \bar{z}_1/L_1 + S_2 \bar{z}_2/L_2) \tag{6.7b}$$

Consider now the application of Mohr's first theorem to Fig. 6.5. From eq(6.2b), with the origin at B for each bay,

$$\theta_A - \theta_B = A_1/(EI_1) = -[1/(EI_1)][(M_A + M_B)L_1/2 - S_1] \tag{6.8a}$$
$$\theta_C - \theta_B = A_2/(EI_2) = -[1/(EI_2)][(M_B + M_C)L_2/2 - S_2] \tag{6.8b}$$

These changes in slope are both shown to be negative for v positive downwards with positive z originating from B within each bay. Within the first bay, θ_B in eq(6.8a) is

$$\theta_B = \Delta v_A/L_1 - (v_B - v_A)/L_1 = A_1 z_1/(EI_1 L_1) - (v_B - v_A)/L_1 \tag{6.8c}$$

and within the second bay, θ_B in eq(6.8b) is

$$\theta_B = \Delta v_C/L_2 + (v_B - v_C)/L_2 = A_2 z_2/(EI_2 L_2) + (v_B - v_C)/L_2 \tag{6.8d}$$

Substituting eqs(6.6a,b) and (6.8c,d) into eqs (6.8a,b), the slopes become

$$\theta_A = [S_1/(EI_1)](1 - \bar{z}_1/L_1) - [L_1/(6EI_1)](M_B + 2M_A) - (v_B - v_A)/L_1 \tag{6.9a}$$
$$\theta_B = -S_1 \bar{z}_1/(EI_1 L_1) + [L_1/(6EI_1)](M_A + 2M_B) - (v_B - v_A)/L_1 \tag{6.9b}$$
$$= -S_2 \bar{z}_2/(EI_2 L_2) + [L_2/(6EI_2)](M_C + 2M_B) + (v_B - v_C)/L_2 \tag{6.9c}$$

$$\theta_C = [S_2/(EI_2)](1 - \bar{z}_2/L_2) - [L_2/(6EI_2)](M_B + 2M_C) + (v_B - v_C)/L_2 \tag{6.9d}$$

Equations (6.9a-d) are again simplified for a uniform beam resting on level supports, when they become

$$\theta_A = [S_1/(EI)](1 - \bar{z}_1/L_1) - [L_1/(6EI)](M_B + 2M_A) \tag{6.10a}$$

$$\theta_B = - S_1 \bar{z}_1/(EIL_1) + [L_1/(6EI)](M_A + 2M_B)$$

$$= - S_2 \bar{z}_2/(EIL_2) + [L_2/(6EI)](M_C + 2M_B) \tag{6.10b,c}$$

$$\theta_C = [S_2/(EI)](1 - \bar{z}_2/L_2) - [L_2/(6EI)](M_B + 2M_C) \tag{6.10d}$$

When the moments M_A, M_B and M_C have been found from Clapeyron's theorems (6.7a,b), the slopes θ_A, θ_B and θ_C at the support points follow from eqs(6.9) or (6.10). The deflection and slope at any other points in a continuous beam may be found by integrating Mohr's theorems with the origin O located at a support point of known slope. Let the datum lie at a point within the first bay, distance z from O co-incident with A. Equations (6.2b) and (6.3b) become

$$\theta = \theta_A + [1/(EI)][\text{Area of net } M \text{ diagram from O to } z] \tag{6.11a}$$

$$v = \theta_A z + [1/(EI)][\text{Moment of net } M \text{ diagram from O to } z \text{ about the datum}] \tag{6.11b}$$

The areas and moment of areas of net M diagrams refer to the difference between the areas and moments of areas for the free and fixing moment diagrams. Normally these are sagging (negative) and hogging (positive) diagrams, so that, according to this convention, the shaded areas in Figs 6.5b,c compose the net bending moment diagram.

Example 6.3 Determine the bending moment diagram for the two-bay beam in Fig. 6.6a using the three-moment theorem. Find the support reactions, the slopes and deflections beneath the 10 kN force in the first bay and at the centre of the second bay. Take $EI = 100$ MNm^2.

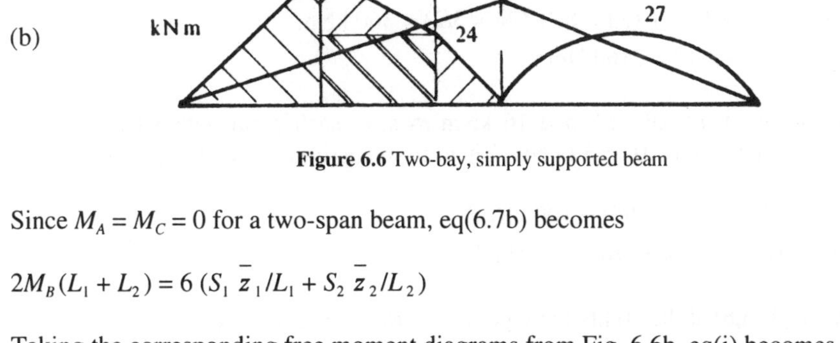

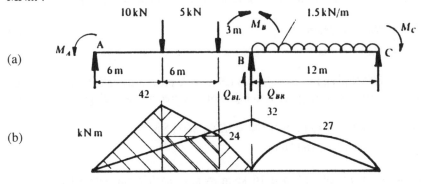

Figure 6.6 Two-bay, simply supported beam

Since $M_A = M_C = 0$ for a two-span beam, eq(6.7b) becomes

$$2M_B(L_1 + L_2) = 6\,(S_1\,\bar{z}_1/L_1 + S_2\,\bar{z}_2/L_2) \tag{i}$$

Taking the corresponding free moment diagrams from Fig. 6.6b, eq(i) becomes

$2M_B(15 + 12) = 6[(1/15)$(Moment of area of free LH M diagram about A)
$+ (1/12)$(Moment of area of free RH M - diagram about C)] (ii)

Taking LH triangular and rectangular constituent areas in kNm^2 and noting that the r.h. area of a parabola is $2Lh/3$ (centroid $5L/8$ from tip), eq(ii) yields

$$(54/6)\, M_B = (1/15)[(42 \times 6 \times 2 \times 6) / (2 \times 3) + (18 \times 6 \times 8/2) + (24 \times 6 \times 9) + (24 \times 3 \times 13/2)\} + (1/12)\{2 \times 12 \times 27 \times 6/3]$$

$$9M_B = 2700/15 + 1296/12, \Rightarrow M_B = 32 \text{ kNm} \quad \text{(iii)}$$

This value is confirmed later from the moment distribution solution (see Table 6.2 in Example 6.5). Let the support reactions and fixing moment M_B act as shown in Fig. 6.6a. The central reaction R_B is composed of its left- and right-hand bay components Q_{BL} and Q_{BR}. Although M_A and M_C are both zero here, they are included to extend the generality to any continuous beam.

Moments about B in bay AB:
$(5 \times 3) + (10 \times 9) - 15R_A + (M_A - M_B) = 0, \quad \Rightarrow R_A = 4.867$ kN
Moments about A in bay AB:
$(6 \times 10) + (12 \times 5) + 32 - 15Q_{BL} = 0, \quad \Rightarrow Q_{BL} = 10.13$ kN
Moments about B in bay BC:
$(1.5 \times 12 \times 6) + (M_C - M_B) - 12R_C = 0, \quad \Rightarrow R_C = 6.333$ kN
Moments about C in bay BC:
$(1.5 \times 12 \times 6) + (M_B - M_C) - 12Q_{BR} = 0, \quad \Rightarrow Q_{BR} = 11.67$ kN
$\therefore R_B = Q_{BL} + Q_{BR} = 10.13 + 11.67 = 21.80$ kN.
Check: $\sum F = [10 + 5 + (12 \times 1.5)] - [4.867 + 21.80 + 6.333] = 0$

The slopes at A and C follow from eqs(6.10a) and (6.10d) with $M_A = M_C = 0$

$$\theta_A EI = S_1(1 - \bar{z}_1/L_1) - L_1 M_B/6 \quad \text{(iv)}$$
$$\theta_C EI = S_2(1 - \bar{z}_2/L_2) - L_2 M_B/6 \quad \text{(v)}$$

where $S_1\, \bar{z}_1/L_1 = 180$ kNm^2 and $S_2\, \bar{z}_2/L_2 = 108$ kNm^2 were previously calculated in the three moment expression (iii). Referring again to Fig. 6.6b, the free moment diagram areas are

$$S_1 = (42 \times 6/2) + (18 \times 6/2) + (24 \times 6) + (24 \times 3/2) = 360 \text{ kNm}^2$$
$$S_2 = 2Lh/3 = 2 \times 12 \times 27/3 = 216 \text{ kNm}^2$$

Note that fixing moments of 12.8 and 16 kNm exist at each point within bay 1 and 2 respectively in proportion to $M_B = 32$ kNm. Then, from eqs(iv) and (v), the slopes are

$$\theta_A EI = (360 - 180) - (15 \times 32/6) = 100 \text{ kNm}^2$$
$$\theta_C EI = (216 - 108) - (12 \times 32/6) = 44 \text{ kNm}^2$$

Now from eqs(6.11a,b) at the 10 kN force point, where $z = 6$ m from A,

$\theta EI = 100 + [(12.8 \times 6/2) - (42 \times 6/2)] = 12.4 \text{ kNm}^2$

$\theta = 12.4 \times 10^{-5}$ rad

$vEI = (100 \times 6) + [(12.8 \times 6 \times 6) / (2 \times 3) - (42 \times 6 \times 6) / (2 \times 3)] = 424.8 \text{ kNm}^3$

$v = 4.25$ mm

and at the centre of the second bay, where $z = 6$ m from C,

$\theta EI = 44 + [(16 \times 6/2) - (2 \times 6 \times 27/3)] = -16 \text{ kNm}^2$

$\theta = -16 \times 10^{-5}$ rad

$vEI = (44 \times 6) + [(16 \times 6 \times 6) / (2 \times 3) - (2 \times 6 \times 27 \times 3 \times 6) / (3 \times 8)] = 117 \text{ kNm}^3$

$v = 1.17$ mm

Example 6.4 Determine the fixing moments for the beam in Fig. 6.7a using Clapeyron's theorem.

(a)

(b)

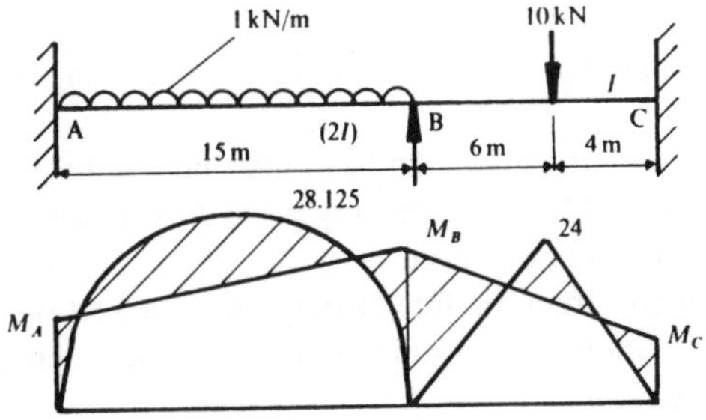

Figure 6.7 Two-bay encastre beam

The free moment diagrams are given in Fig. 6.7b. The fixing moments M_A and M_C, together with the central support moment M_B, are unknowns for this continuous beam. M_C is assumed to be the greatest of these to correspond with Fig. 6.5b so allowing the application of the three-moment theorem. Since $S_1 \bar{z}_1$ and $S_2 \bar{z}_2$ are the moments of area of the free M diagrams about the left- and right-hand ends respectively, eq(6.7a) becomes

$$15M_A/(2I) + 2M_B[15/(2I) + 10/I] + 10M_C/I = 6\{(2 \times 15 \times 28.125 \times 7.5) / (3 \times 2I \times 15) + [(24 \times 6 \times 6/2) + (24 \times 4 \times 2 \times 4)/(2 \times 3)] / (I \times 10)\}$$

$$M_A + 4.67M_B + 1.33M_C = 101.05 \qquad \text{(i)}$$

Two further relationships are found from the application of eq(6.3) to each bay. The intercept between the tangents at A and B is zero. Taking the origin at A and the datum at B,

$$\Delta v_B = [1/(2EI)](\text{Moment of } M \text{ diagram from A to B about B}) = 0$$

$$(15M_A \times 7.5) + (M_B - M_A)(15 \times 15)/(2 \times 3) - (2 \times 15 \times 28.125 \times 7.5/3) = 0$$

$$2M_A + M_B = 56.25 \qquad \text{(ii)}$$

Furthermore, the intercept between the tangents at C and B is zero with the origin at C and the datum at B. Thus,

$\Delta v_B = [1/(EI)]$(Moment of M diagram from C to B about B) = 0

$(10M_C \times 5) + (M_B - M_C)(10{\times}10)/(2 \times 3) - [(24 \times 4 \times 7.333/2) + (24{\times}6{\times}2{\times}6)/(2 \times 3)] = 0$

$M_B + 2M_C = 38.4$ (iii)

Solving eqs(i) - (iii) gives $M_A = 21.365$, $M_B = 13.52$ and $M_C = 12.44$ kNm (least). The values are confirmed in Example 6.6 (Table 6.3) using the moment distribution method.

6.3 The Moment Distribution Method

Where a beam has two or more spans, it is necessary to ensure that the moments to the left and right sides of all its inner supports are equal. In a structure the sum of the moments at all joints should be zero. Because an initial estimate of these moments does not balance, it is necessary to distribute the imbalance until the equality is achieved. The method is as follows:

(i) Fix the beam at all supports and calculate the LH and RH fixing moments for each bay from Table 6.1
(ii) Where a beam has simply supported ends, release the fixing moment and carry over (eqs 6.14a,b)
(iii) Use the distribution factors (eqs 6.18a,b) to distribute unbalanced moments at inner supports and carry over. Moments are only carried over to ends that are encastré
(iv) Construct the fixing moment diagram
(v) Isolate each bay and construct the free M-diagram (see Section 6.1)
(vi) The net M-diagram is then found from the difference between ordinates in the fixing and free moment diagrams
(vii) Calculate the supporting reactions from the balanced fixing moments and construct the F-diagram

6.3.1 Fixing Moments (F.M.)

Standard expressions given in Table 6.1 provide fixing moments for the single-span encastre beams shown. They act in the directions shown to hog the beam. The convention adopted is that clockwise F.M's are positive. Therefore the signs of the F.M. alternate from negative to positive within a bay and on either side of each support.

6.3.2 Carry-Over Moments (C.O.M.)

In Fig. 6.8, it is assumed that the adjacent F.M's, as calculated from fixing support B, do not balance. The out-of-balance moment M, at B, is distributed into bays BA and BC as

$$M = M_{BA} + M_{BC} \tag{6.12}$$

where the directions of M_{BA} and M_{BC} must oppose the directions of M to maintain moment equilibrium at B. A further compatibility requirement is that one half of each distributed moment will be carried over in the same direction to the opposite end. That is, M_{AB} is carried over from M_{BA}, and M_{CB} is carried over from M_{BC}.

Table 6.1 Fixed-End Moment Expressions (F.M.)

BEAM	R.H.F.M.	L.H.F.M.
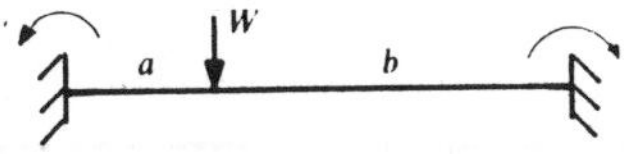	Wa^2b/L^2	Wab^2/L^2
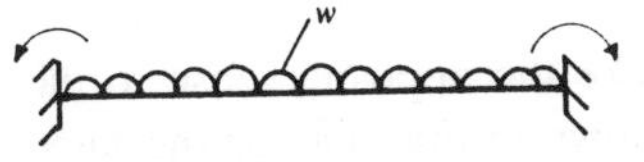	$wL^2/12$	$wL^2/12$
	$wa^2(6L^2 - 8aL + 3a^2)/(12L^2)$	$wa^2(4aL - 3a^2)/(12L^2)$
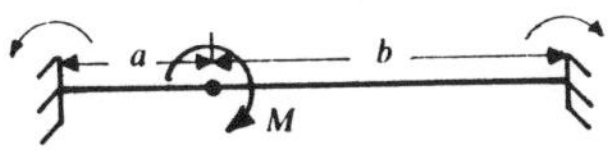	$(Mb/L)(1 - 3a/L)$	$(Ma/L)(1 - 3b/L)$
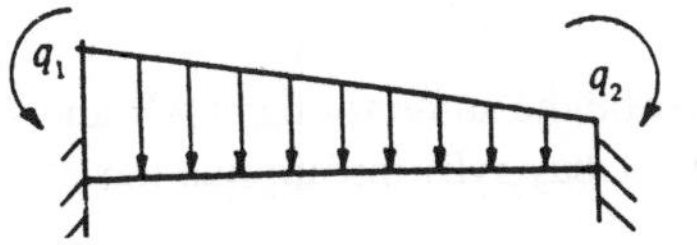	$(2q_1 + 3q_2)L^2/60$	$(3q_1 + 2q_2)L^2/60$
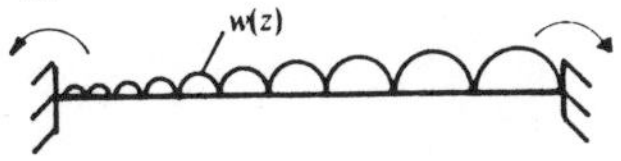	$(1/L^2)\int wx(L-x)^2\,dx$	$(1/L^2)\int wx^2(L-x)\,dx$
	0	$3EI\delta/L^2$
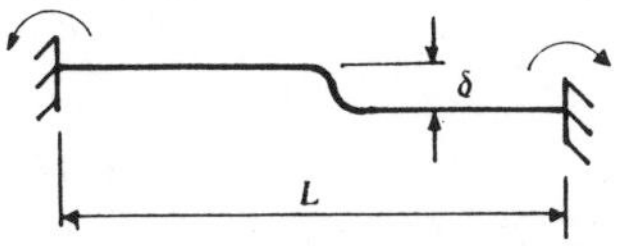	$6EI\delta/L^2$	$6EI\delta/L^2$

To show this, both M_{AB} and M_{CB} must ensure that the deflection v, at B, is zero. Applying Mohr's deflection theorem (6.3b) to bay AB where the M-diagram is composed of two parts: (i) the hogging moment M_{BA}, carried over, and (ii) a sagging moment, due to the reaction R_B at B,

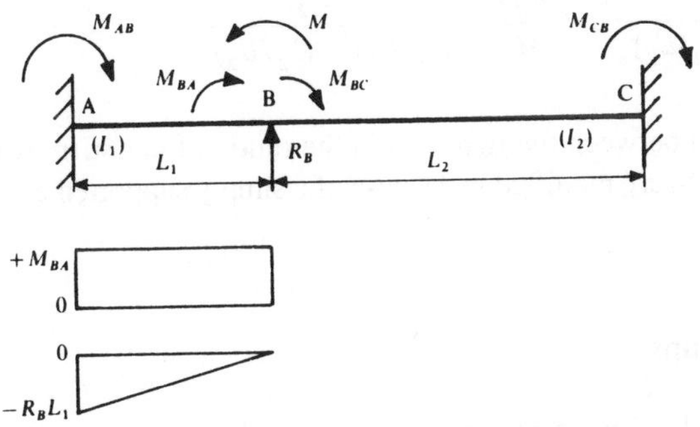

Figure 6.8 Distribution and carry-over of unbalanced moment M

$v_B = [1/(EI)]$(Moment of area of M diagram from A to B about B) = 0

$$M_{BA} L_1 (L_1/2) - (R_B L_1^2/2)(2L_1/3) = 0$$

$$R_B = 3M_{BA}/(2L_1) \qquad (6.13)$$

The net moment applied at A is then

$$M_{BA} - [3M_{BA}/(2L_1)]L_1 = -M_{BA}/2$$

The reacting C.O.M. at A and similarly the C.O.M. at C take the opposite sense to the moments applied at A and C. It follows that each C.O.M. shown in Fig. 6.8 has the same sense with a magnitude,

$$M_{AB} = M_{BA}/2$$

$$M_{CB} = M_{BC}/2 \qquad (6.14,b)$$

6.3.3 Distribution Factors (D.F.)

A further compatibility condition is that the slope θ, at B, must be the same for bays AB and BC. If L_1 and I_1 are the length and second moment of area for bay AB, the application of Mohr's slope theorem (6.3a) to Fig. 6.8 gives

$\theta_B = [1/(EI)]$(Area of M diagram from A to B]

$$= [1/(EI_1)](M_{BA} L_1 - R_B L_1^2/2) = -M_{BA} L_1/(2EI_1) \qquad (6.15)$$

in which R_B is given by eq(6.13). Similarly, if L_2 and I_2 are the respective length and second moment of area for bay BC, then

$$\theta_B = -M_{BC} L_2/(2EI_2) \qquad (6.16)$$

Equating (6.15 and 6.16) leads to

$$M_{BA}/M_{BC} = (I_1 L_2)/(I_2 L_1) \qquad (6.17)$$

Combining eqs(6.12) and (6.17) provides the *distribution factors*:

$$\frac{M_{BA}}{M} = \frac{(I_1/L_1)}{(I_1/L_1) + (I_2/L_2)}, \quad \frac{M_{BC}}{M} = \frac{(I_2/L_2)}{(I_1/L_1) + (I_2/L_2)} \qquad (6.18a,b)$$

That is, M is divided between the two bays in the ratio of their stiffnesses, I/L. It should be noted that eqs(6.18a,b) are modified in the case of a simply supported end bay whose stiffness becomes ¾(I/L).

6.4 Continuous Beams

The following examples illustrate the application of Table 6.1 and eqs(6.14a,b) and (6.18a,b) to continuous beams.

Example 6.5 Construct the net bending moment diagram for the two-bay beam in Fig. 6.9a showing the maximum values. Find the support reactions.

Fixed end-moments from Table 6.1:

Bay AB, L.H.F.M.=$(10 \times 6^2 \times 9/15^2) + (5 \times 12^2 \times 3/15^2) = 24$ kNm
Bay AB, L.H.F.M. = $(10 \times 9^2 \times 6/15^2) + (5 \times 3^2 \times 12/15^2) = 24$ kNm
Bay BC, R.H.F.M. = L.H.F.M. = $(1.5 \times 12^2/12) = 18$ kNm

D.F's at joint B, from eqs(6.18a,b). Note that the ¾ factor cancels for a beam with simple supports throughout.

$M_{BA}/M = (3/4)(I/15)/[(3/4)(I/12 + I/15)] = 4/9$
$M_{BC}/M = (3/4)(I/12)/[(3/4)(I/12 + I/15)] = 5/9$

Fixing Moments: The distribution and carry-over of moments given in Table 6.2 ensure that the moments at B become equal and that there can be no moments at the ends A and C.

Table 6.2 Distributions for Example 6.5

Joint	A	B		C
Member	AB	BA	BC	CA
D.F.		4/9	5/9	
Initial F.E.M.	-24	+24	-18	+18
Release @ A & C & C.O.M.	+24	+12	-9	-18
Modified F.E.M.	0	+36	-27	0
Unbalanced moment		+9		
1st Distribution		-4	-5	
Final fixing moments	0	+32	-32	0

The fixing moment diagram is superimposed upon the inverted free moment diagrams in Fig. 6.9b. The shaded areas in Fig. 6.6b represent the net moment diagram which has been re-based in Fig. 6.9c.

(a)

(b)

(c)

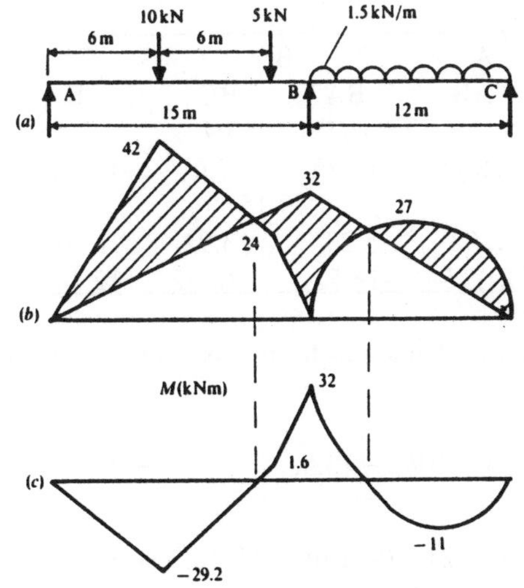

Figure 6.9 Free, fixing and net moment diagrams

The support reactions follow from applying moment equilibrium to the left and right of B. This must include the fixing moments:

$\sum M_B = 0 = +32 + 15R_A - (9 \times 10) - (3 \times 5), \quad \Rightarrow R_A = 4.87$ kN
$\sum M_B = 0 = -32 - 12R_C + (1.5 \times 12 \times 6), \quad \Rightarrow R_C = 6.33$ kN

Apply vertical force equilibrium,

$\sum F = 0 = 4.87 + R_B + 6.33 - 10 - 5 - (12 \times 1.5), \Rightarrow R_B = 21.8$ kN

Example 6.6 Find the fixing moments and support reactions for the beam in Fig. 6.10.

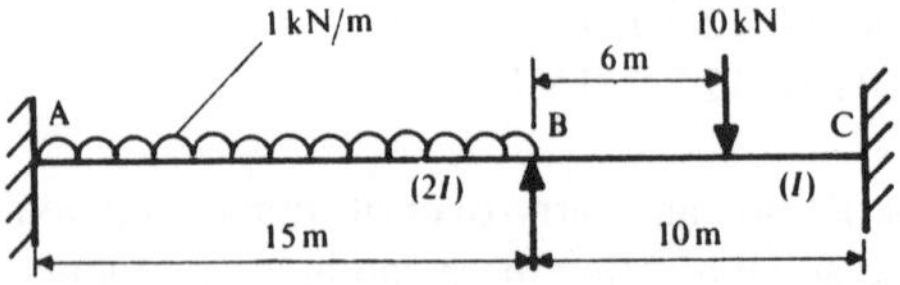

Figure 6.10 Two-bay encastre beam

Fixing moments from Table 6.1:
Bay AB: L.H.F.M.= R.H.F.M. = $1 \times 15^2/12 = 18.75$ kNm
Bay BC: R.H.F.M. = $10 \times 6^2 \times 4/10^2 = 14.4$ kNm
L.H.F.M. = $10 \times 4^2 \times 6/10^2 = 9.6$ kNm

D.F.'s at B from eqs(6.18a,b):
$M_{BA}/M = (2I/15)/[(2I/15) + (I/10)] = 4/7$
$M_{BC}/M = (I/10)/[(2I/15) + (I/10)] = 3/7$

Moment Distribution: Table 6.3 shows how to balance the moments at B and carry over to establish the moments at A and C.

Table 6.3 Distributions for Example 6.6

Joint	A	B		C
Member	AB	BA	BC	CB
D.F.		4/7	3/7	
Initial F.E.M.	- 18.75	+18.75	- 9.6	+14.40
Unbalanced moment		+9.15		
Distribution & C.O.M.	- 2.62	-5.23	- 3.92	- 1.96
Final fixing moments	- 21.37	+13.52	- 13.52	+12.44

These F.M.'s are employed in calculations for the reactions. Take moments about B to right (positive clockwise),

$\sum M_B = 0 = (10 \times 6) + 12.44 - 10R_C - 13.5, \quad \Rightarrow R_C = 5.894$ kN
Moments about B to left,
$\sum M_B = 0 = -(15 \times 1 \times 7.5) - 21.37 + 15R_A + 15.32, \quad \Rightarrow R_A = 8.03$ kN

Vertical force equilibrium,

$\sum F = 0 = 8.023 + R_B + 5.894 - 25,\ R_B = 11.083$ kN

Example 6.7 Construct the moment diagram for the beam in Fig. 6.11a. Find the four support reactions.

F.M. from Table 6.1:

Bay AB, L.H.F.M. = $5 \times 3^2 \times 2/5^2 = 3.6$ kNm, R.H.F.M. = $5 \times 2^2 \times 3/5^2 = 2.4$ kNm

Bay BC, L.H.F.M. = R.H.F.M. = $1 \times 5^2/12 = 2.08$ kNm

Bay CD, L.H.F.M. = $(3 \times 3^2 \times 2/5^2) + (3 \times 1^2 \times 4/5^2) = 2.64$ kNm

R.H.F.M. = $(3 \times 2^? \times 3/5^2) + (3 \times 4^2 \times 1/5^2) = 3.36$ kNm

D.F. at B and C. Introduce ¾ factor for end bays within eqs(6.18a,b),

$M_{BA}/M = (3/4)(I/5)/[(3/4)(I/5)+I/5] = 3/7$

$M_{BC}/M = (I/5)/[(3/4)(I/5) + I/5] = 4/7$

$M_{CB}/M = (I/5)/[(3/4)(I/5) + I/5] = 4/7$

$M_{CD}/M = (3/4)(I/5)/[(3/4)(I/5) + I/5] = 3/7$

Moment distribution: In Table 6.4 the moments are distributed until they balance at B and C and equal zero at A and D.

Table 6.4 Distributions for Example 6.7

Joint	A	B		C		D
Member	AB	BA	BC	CB	CD	DC
D.F.		3/7	4/7	4/7	3/7	
Initial F.E.M.	- 3.6	+2.4	- 2.08	+2.08	- 2.64	+3.36
Release A & D & C.O.	+3.6	+1.8			- 1.68	- 3.36
Net fixing moments	0	+4.2	-2.08	+2.08	- 4.32	0
Unbalanced moment		+2.12		-2.24		
1st Distribution		- 0.91	-1.21	+1.28	+0.96	
C.O.M.		0	+0.64	-0.61	0	
Unbalanced moment		+0.64		- 0.61		
2nd Distribution		- 0.27	- 0.37	+0.34	+0.26	
C.O.M.		0	+0.17	-0.19	0	
Unbalanced moment		+0.17		-0.19		
3rd Distribution		- 0.07	- 0.10	+0.11	+0.08	
C.O.M.		0	+0.06	-0.05	0	
Unbalanced moment		+0.06		-0.05		
4th Distribution		- 0.025	- 0.035	+0.03	+0.02	
C.O.M.		0	+0.015	-0.017		
Final fixing moments	0	+2.925	-2.91	+2.98	- 3.0	0

The free and fixing moments are superimposed in Fig. 6.11b using average F.M. values. The net moment diagram is given by the shaded regions.

(a)

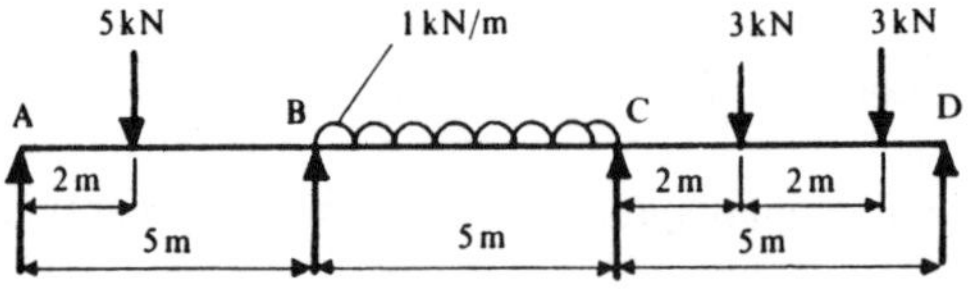

(b)

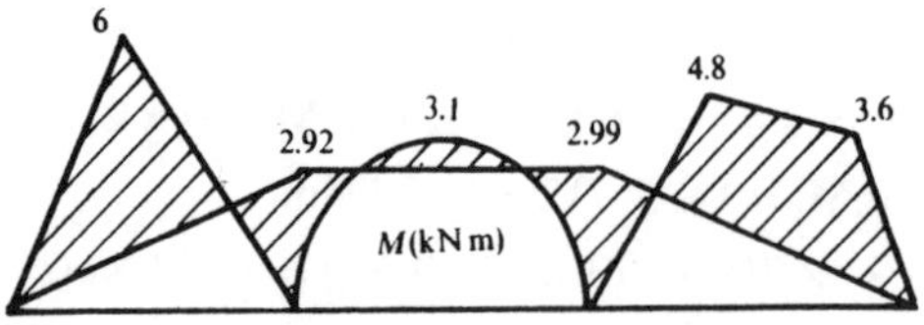

Figure 6.11 Free, fixing and net moment diagrams for a three-bay s.s. beam

Reactions: Take moments to left and right of B and C including the final F.M.'s

$M_C = 0 = (3 \times 2) + (3 \times 4) - 2.99 - 5R_D, \quad \Rightarrow \quad R_D = 3$ kN
$M_B = 0 = 5R_A + 2.92 - (3 \times 5), \quad \Rightarrow \quad R_A = 2.42$ kN
$M_C = 0 = 2.99 + 5R_B + (10 \times 2.42) - (8 \times 5) - (5 \times 1 \times 2.5), \quad \Rightarrow \quad R_B = 5.06$ kN
$M_B = 0 = -2.92 + (5\times1\times2.5) - 5R_C + (7\times3) + (9\times3) - (10\times3) \quad \Rightarrow \quad R_C = 5.52$ kN

Example 6.8 Draw the bending moment diagram for the continuous beam in Fig. 6.12a. Calculate the support reactions and draw the F - diagram.

Fixed-end moments from Table 6.1:
Bay AB: F.M's $= 0$
Bay BC: L.H.F.M. = R.H.F.M. = $6 \times 1^3/2^2 = 1.5$ kNm
Bay CD: " " $= 3 \times 2^2/12 = 1.0$ kNm
Bay DE: $M_D = 3 \times 1^2/2 = 1.5$ kNm (hogging), $M_E = 0$

D.F.'s at B and C, from eqs(6.18a,b):
$M_{BA}/M = (2I/2)/[(2I/2) + (2I/2)] = 0.5 = M_{AB}/M$
$M_{CB}/M = (2I/2)/[(2I/2) + (3/4)(I/2)] = 0.73$
$M_{CD}/M = (3/4)(I/2)[(2I/2) + (3/4)(I/2)] = 0.27$

Free Sagging Moments:
Bay BC, $M_{\max} = Wl/4 = 6 \times 2/4 = 3$ kNm,
Bay CD, $M_{\max} = wl^2/8 = 3 \times 2^2/8 = 1.5$ kNm

In Table 6.5 the moments are balanced at B and C and D. Further, it must be ensured that $M_D = 1.5$ kNm remains the net moment at D. These calculations reveal a net bending moment diagram within the shaded regions of Figs 6.12b. The re-based net moments in Fig. 6.12c are consistent with the convention that hogging moments are positive. The shear force diagram (see Fig. 6.12d) construction starts from the RH end, taking downward forces to be negative.

Table 6.5 Distributions fro Example 6.8

Joint	A	B		C		D		E
Member	AB	BA	BC	CB	CD	DC	DE	ED
D.F.	0	0.5	0.5	0.73	0.27			
Initial F.E.M.	0	0	-1.5	+1.50	-1.00	+1.0	-1.5	0
Balance @ D						+0.5		
C.O.M.					+0.25			
Net fixing moments	0	0	-1.5	+1.5	-0.75	+1.5	-1.5	0
Unbalanced moment		-1.5		+0.75				
1st Distribution		+0.75	+0.75	-0.55	-0.20			
C.O.M.	+0.37		-0.28	+0.37				
2nd Distribution		+0.14	+0.14	-0.27	-0.10			
C.O.M.	+0.07		-0.14	+0.07				
3rd Distribution		+0.07	+0.07	-0.05	-0.02			
C.O.M.	+0.035		-0.025	+0.035				
4th Distribution		+0.012	+0.012	-0.026	-0.01			
C.O.M.	+0.006		-0.013	+0.006				
Final F.M's	+0.48	+0.972	-0.986	+1.085	-1.080	+1.5	-1.5	0

$\sum M_C = 0 = -\ 1.082 + (3 \times 3 \times 1.5) - 2R_D$, from which reaction is $\Rightarrow R_D = 6.2$ kN
$\sum M_B = 0 = -\ 0.98 + (6 \times 1) + (3 \times 3 \times 3.5) - 2R_C - (4 \times 6.2)$, $\Rightarrow R_C = 5.86$ kN
$\sum M_A = 0 = +\ 0.48 + (6 \times 3) + (3\times3\times5.5) - 2R_B - (4\times5.86) - (6 \times 6.2)$, $\Rightarrow R_B = 3.67$ kN
$\sum F = 0 = 6 + (3 \times 3) - 3.67 - 5.86 - 6.2 - R_A$, $\Rightarrow R_A = -\ 0.73$ kN

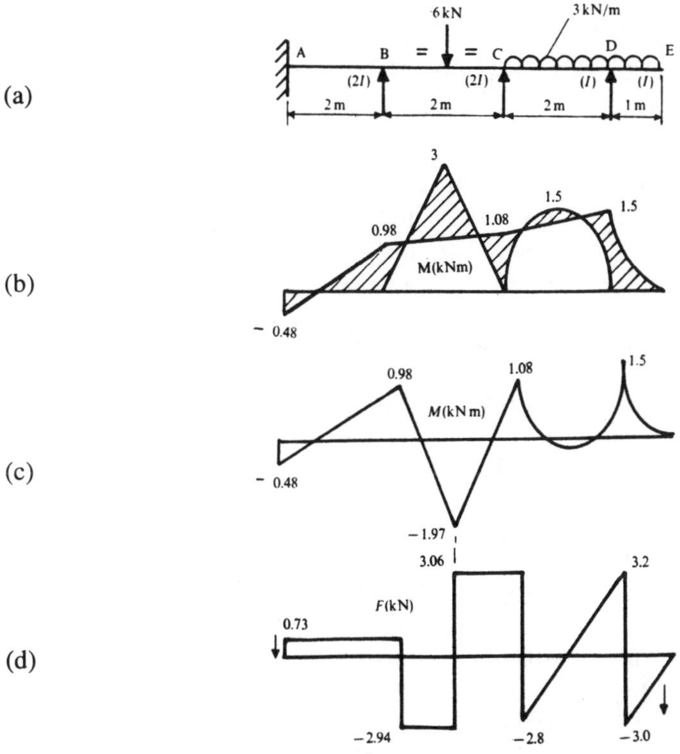

Figure 6.12 Continuous beam with overhang

6.5 Beams With Misaligned Supports

If the supports for a continuous beam are not level, moments are induced at these points. This applies to all misaligned interior supports and to the ends if these are encastré.

Example 6.9 Find the moments at B and C for the beam in Fig. 6.13 when the support at B lies 40 mm below the level of A, C and D. Take $E = 200$ GPa, $I = 4 \times 10^6$ mm^4.

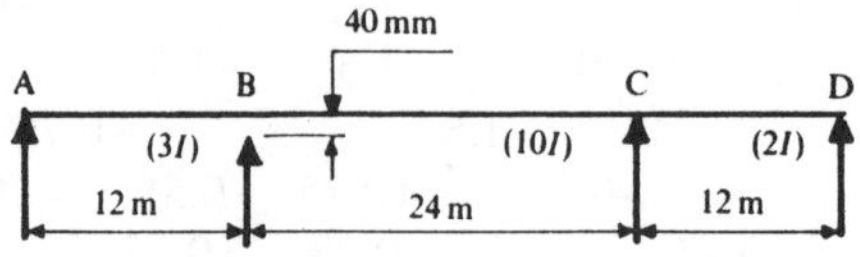

Figure 6.13 Continuous beam with misaligned supports

F.M.'s are found from $M = 6EI\delta/L^2$ (see Table 6.1)

Bay AB, L.H.F.M. = $(6 \times 200 \times 10^3 \times 3 \times 4 \times 10^6 \times 40)/(12^2 \times 10^6)$
$= 4 \times 10^6$ Nmm = 4 kNm = R.H.F.E.M.

Bay BC, L.H.F.M. = $(6 \times 200 \times 10^3 \times 10 \times 4 \times 10^6 \times 40)/(24^2 \times 10^6)$
$= (3.333 \times 10^6)$ Nmm = 3.333 kNm = R.H.F.M.

Bay CD, F.M's = 0

D.F.'s at B and C, from eqs(6.9):

$M_{BA}/M = (3/4)(3I/12)/[(3/4)(3I/12) + 10I/24] = 0.31$

$M_{BC}/M = (10I/24)/[[(3/4)(3I/12) + 10I/24] = 0.69$

$M_{CB}/M = (10I/24)/[(3/4)(2I/12) + 10I/24] = 0.77$

$M_{CD}/M = (3/4)(2I/12)/[(3/4)(2I/12) + 10I/24] = 0.23$

Table 6.6 Distributions for Example 6.9

Joint	A	B		C		D
Member	AB	BA	BC	CB	CD	DC
D.F		0.31	0.69	0.77	0.23	
Initial F.E.M.	-4	-4	+3.333	+3.333	0	0
Rel. at A & D & C.O.	+4	+2				
Net F.E.M.	0	-2	+3.333	+3.333	0	0
Unbalanced Moment		+1.333		+3.333		
1st Distribution	0	-0.413	-0.920	-2.566	-0.767	
C.O.M.			-1.283	-0.460		
2nd Distribution	0	+0.398	+0.885	+0.354	+0.106	
C.O.M.			+0.177	+0.443		
3rd Distribution	0	-0.055	-0.122	-0.341	-0.102	
C.O.M.			-0.171	-0.061		
4th Distribution	0	+0.053	+0.118	+0.047	+0.014	
C.O.M.			+0.024	+0.059		
5th Distribution	0	-0.007	-0.017	-0.045	-0.014	
C.O.M.			-0.023	-0.009		
6th Distribution	0	+0.007	+0.016	+0.007	+0.002	
C.O.M.			+0.004	+0.008		
Final F.E.M.	0	-2.017	+2.021	+0.769	-0.761	0

Because the end A is simply supported, it is first necessary to release the moment at this point and carry over. Table 6.6 shows that, without external loading, the initial F.M's will have the same sense within each bay but alternating from bay to bay.

6.6 Moment Distribution for Structures

The moment distribution method can be extended to structures. The F.M.'s are found for all bays from Table 6.1. The joints in a structure may connect two or more bars with various inclinations. The D.F.'s are found from extending eqs(6.18a,b) to all bars at each joint. For $n = 1, 2 \ldots N$ bars the fraction of the joint imbalance moment M, carried by each bar, defines the D.F. as

$$\text{D.F.} = M_n / M = (I/L)_n \Big/ \sum_{n=1}^{N} (I/L)_n \tag{6.19}$$

The following examples illustrate the principles involved in attaining the required moment balance $\sum_{n=1}^{N} M_n = 0$ at joints for structures that are rigid and for those that sway.

Example 6.10 Find the vertical reactions at A and B for the structure in Fig. 6.14a. The ends A and B are encastre, C is pinned and O is a rigid joint.

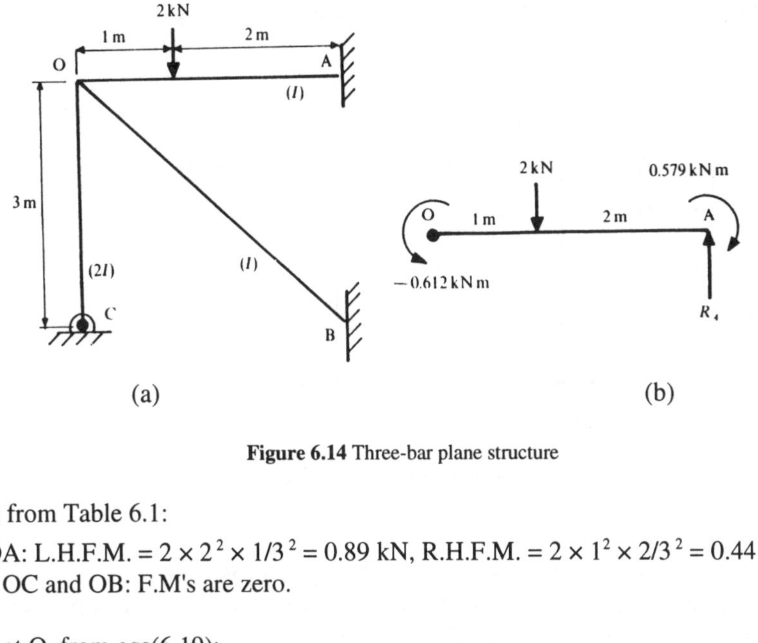

Figure 6.14 Three-bar plane structure

F.M.'s from Table 6.1:

Bay OA: L.H.F.M. = $2 \times 2^2 \times 1/3^2 = 0.89$ kN, R.H.F.M. = $2 \times 1^2 \times 2/3^2 = 0.44$ kN.
Struts OC and OB: F.M's are zero.

D.F.'s at O, from eqs(6.19):

$M_{OA} / M = (I/3) / [(I/3) + I/(3\surd 2) + (3/4)(2I/3)] = 0.313$

$M_{OB} / M = [I/(3\surd 2)] / [(I/3) + I/(3\surd 2) + (3/4)(2I/3)] = 0.220$

$M_{OC} / M = (3/4)(2I/3) / [(I/3) + I/(3\surd 2) + (3/4)(2I/3)] = 0.467$

The L.H.F.M. for bay OA is the unbalanced moment for the joint O, which is distributed within Table 6.7:

Table 6.7 Distributions fro Example 6.10

Joint	C		O		A	B
Member	CO	OC	OB	OA	AO	BO
D.F		0.467	0.220	0.313		
Initial F.E.M.	+0	-0	+0	-0.89	+0.44	-0
Unbalanced moment			-0.89			
1st Distribution		+0.416	+0.196	+0.278		
C.O.M.	+0				+0.139	+0.098
Final F.M's	+0	+0.416	+0.196	-0.612	+0.579	+0.098

At the joint O, Table 6.7 shows that the final F.M's in the three bars OC, OB and OA sum to zero. The reaction at A is found by taking moments about O, with OA as a free body, see Fig. 6.14b,

$$\sum M_O = 0 = -0.612 + (2 \times 1) + 0.579 - 3R_A, \quad \Rightarrow R_A = 0.657 \text{ kN}$$

Similarly for OB as a free body,

$$\sum M_O = 0 = +0.196 + 0.098 - 3R_B, \quad \Rightarrow R_B = 0.098 \text{ kN}$$

Example 6.11 The bridge structure in Fig. 6.15 is built in at A, D and F and is hinged at E. Find the bending moments at A, F and D when I is uniform throughout.

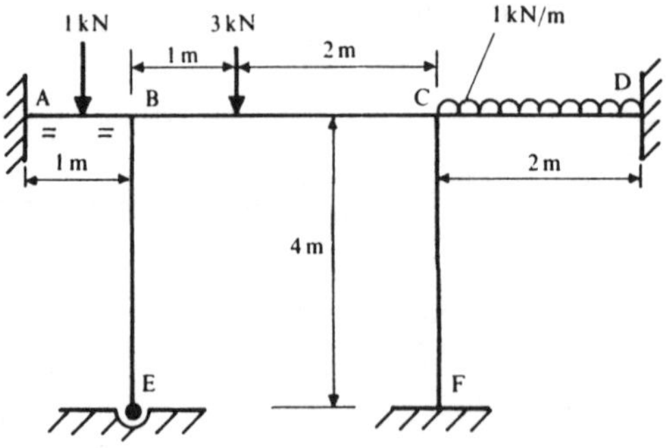

Figure 6.15 Bridge structure

F.M.'s from Table 6.1

Bay AB: L.H.F.M. = R.H.F.M. = $1 \times (1/2) \times (1/2)^2/12 = 0.125$ kNm
Bay BC: L.H.F.M. = $3 \times 2^2 \times 1/3^2 = 1.33$ kNm, R.H.F.M = $3 \times 1^2 \times 2/3^2 = 0.67$ kNm
Bay CD: L.H.F.M. = R.H.F.M. = $1 \times 2^2/12 = 0.33$ kNm
Bays BE and CF: F.M's are zero.

D.F.'s at joint B, from eqs(6.19):

$M_{BA}/M = (I/1)/[(I/1) + (3/4)(I/4) + (I/3)] = 0.658$

$M_{BE}/M = (3/4)(I/4)/[(I/1) + (3/4)(I/4) + (I/3)] = 0.123$

$M_{BC}/M = (I/3)/[(I/1) + (3/4)(I/4) + (I/3)] = 0.219$

D.F.'s at joints C:

$M_{CB}/M = (I/3)/[(I/3) + (I/4) + (I/2)] = 0.308$
$M_{CF}/M = (I/4)/[(I/3) + (I/4) + (I/2)] = 0.231$
$M_{CD}/M = (I/2)/[(I/3) + (I/4) + (I/2)] = 0.461$

The pinned end E cannot fix a moment and therefore need not enter into Table 6.8. If, however, a horizontal force were applied along BE then the F.M. calculated for E would need to be released and carried over to B.

Table 6.8 Distributions for Example 6.11

Joint	A		B			C		F	D
Member	AB	BA	BE	BC	CB	CF	CD	FC	DC
D.F's		.658	.123	.219	.308	.231	.461		
Init.F.E.M.	-.125	+.125	0	-1.33	+.67	0	-.33	0	+.33
Imbalance			-1.205			+.340			
1st Dibn		+.793	+.148	+.264	-.105	-.079	-.157		
C.O.M.	+.397			-.053	+.132			-.040	-.0785
2nd Dibn		+.033	+.006	+.011	-.040	-.030	-.060		
C.O.M.	+.017			-.020	+.006			-.015	-.030
3rd Dibn		+.013	+.003	+.004	-.0019	-.0014	-.0028		
C.O.M.	+.007			-.001	+.002			-.0007	-.0014
4th Dibn		+.0007	+.00012	+.00022	-.00062	-.00046	-.00092		
C.O.M.	+.0003			-.0003	+.00011			-.0002	-.0005
Final F.M.	+.296	+.965	+.157	-1.127	+.663	-.111	-.551	-.056	+.220

The distribution is continued until the unbalanced carry-over moments at B and C become negligibly small. Final F.M's at A, F and D lie in the first and final two columns respectively.

Example 6.12 Construct the bending moment diagram for the portal frame in Fig. 6.16a. Joints A and B are pinned and joints B and C are rigid. Assume *EI* is constant.

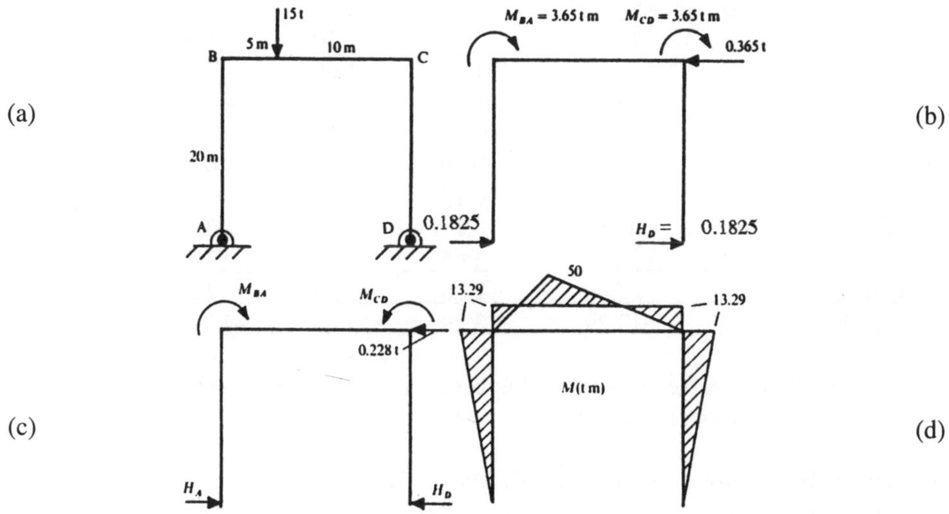

Figure 6.16 Moment analysis of a portal frame

F.M.'s for bay BC

R.H.F.M. = $15 \times 5^2 \times 10/15^2 = 16.67$ tm
L.H.F.M. = $5 \times 10^2 \times 15/15^2 = 33.33$ tm
F.M's = 0 for struts AB and CD.

D.F.'s for joints B and C

$M_{BA}/M = (3/4)(I/20)/[(3/4)(I/20) + (I/15)] = 0.36 = M_{CD}/M$
$M_{BC}/M = (I/15)/[(3/4)(I/20) + (I/15)] = 0.64 = M_{CB}/M$

Table 6.9 shows the F.M.'s are distributed in the usual way, but they must then be corrected to account for the effect of horizontal side sway.

Table 6.9 Distribution for Example 6.12

Joint	A	B		C		D
Member	AB	BA	BC	CB	CD	DC
D.F.		0.36	0.64	0.64	0.36	
Initial F.M.	0	0	-33.33	+16.67	0	0
1st Dibn		+12	+21.3	-10.7	-6	
C.O.M.			-5.4	+10.7		
2nd Dibn		+1.95	+3.45	-6.85	-3.85	
C.O.M.			-3.43	+1.73		
3rd Dibn		+1.24	+2.19	-1.10	-0.63	
C.O.M.			-0.55	+1.10		
4th Dibn		+0.20	+0.35	-0.70	-0.40	
C.O.M.			-0.40	+0.20		
5th Dibn		+0.15	+0.25	-0.13	-0.07	
C.O.M.			-0.07	+0.13		
6th Dibn		+0.03	+0.04	-0.08	-0.05	
Non sway F.M.		+15.57	-15.57	+11.0	-11.0	0
Init sway *M*	0	+5		+5		0
1st Dibn		-1.8	-3.20	-3.20	-1.80	
C.O.M.			-1.60	-1.60		
2nd Dibn		+0.60	+1.00	+1.00	+0.60	
C.O.M.			+0.50	+0.50		
3rd Dibn		-0.20	-0.30	-0.30	-0.20	
C.O.M.			-0.15	-0.15		
4th Dibn		+0.05	+0.10	+0.10	+0.05	
Sway *M*	0	+3.65	-3.65	-3.65	+3.65	0
Corr sway *M*	0	-2.28	+2.28	+2.28	-2.28	0
Final F.M.	0	+13.29	-13.29	+13.28	+13.28	0

The non-sway F.M's enable the horizontal forces at A and D to be found by treating AB and DC as free bodies. From Fig. 6.16b,

$\sum M_B = 0 = M_{BA} - 20H_A, \quad \Rightarrow H_A = 15.57/20 = 0.778$ t
$\sum M_C = 0 = M_{CD} - 20H_D, \quad \Rightarrow H_D = 11.0/20 = 0.55$ t

Horizontal force equilibrium of the frame yields

$\sum H = 0 = H_A - H_D - H_C, \quad \Rightarrow \quad H_C = 0.778 - 0.55 = 0.228$ t

The non-sway F.M's in Table 6.9 require a horizontal force of 0.228 t to be applied at B. Without this force, the frame sways sideways by an amount δ at C. If δ is known, moments M_{BA} and M_{CD} at B and C are given by $M = 3EI\delta / L^2$ (see Table 6.1), each with the same sense. Since δ is unknown, we assume a value for M_{BA} and M_{CD} (say 5 tm) for a second moment distribution within Table 6.9 in order to find the sway moments of 3.65 tm at the rigid joints at B and C. These sway moments produce equal horizontal reactions $H_A = H_D =$ 0.1825 t, balanced by a force of 0.365 t applied to C (see Fig. 6.16c). However, there can be no horizontal force at C since the horizontal reactions at A and D must balance. That is, between Figs 5.17 b and c, $0.228 + 0.365C = 0$, where $C = -0.625$ is the multiplication factor used to correct the sway moments in Table 6.8. The sign of C depends upon the sense assumed for M_{BA} and M_{CD}. The final F.M.'s become the sum of the corrected sway and non-sway moments in Table 6.9. The B.M. diagram is constructed in Fig. 6.16d. For bay BC, the net moments lie within the shaded areas, these being the difference between the free and fixing moment diagrams.

EXERCISES

Clapeyron's Theorem

6.1 Verify the solutions previously found from moment distribution for the continuous beams in Figs 6.11 and 6.12, using the three-moment theorem.

6.2 Draw the shear force and bending moment diagrams, showing maximum values for the continuous beam in Fig. 6.17. Determine also the slope and deflection beneath the 2 kN force. Take $E = 200$ GPa, $I = 80 \times 10^6$ mm^4.

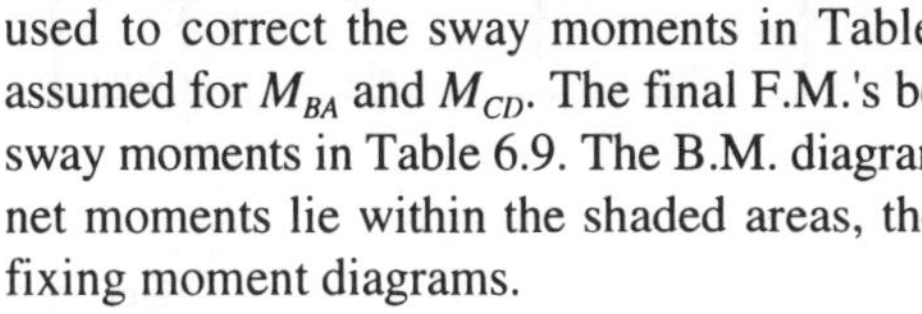

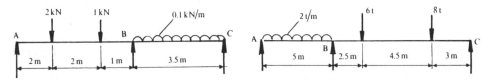

Figure 6.17

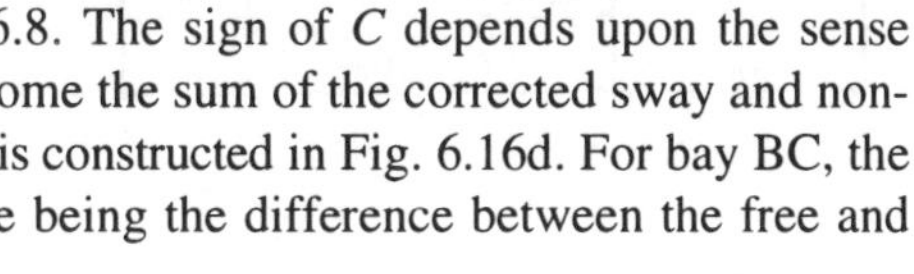

Figure 6.18

6.3 Find the central fixing moment and the position and magnitude of (i) the maximum bending moment and (ii) the maximum deflection for the continuous beam in Fig. 6.18. Take $EI = 15$ MNm^2.

6.4 A beam of flexural stiffness 30 kNm^2 is fixed horizontally at the left end A and is simply supported at the same level at distances of 4 m and 6 m from that end, as shown in Fig. 6.19. A uniformly distributed load of 2500 N/m is carried between supports B and C. Determine the deflection at a distance of 3 m from A.

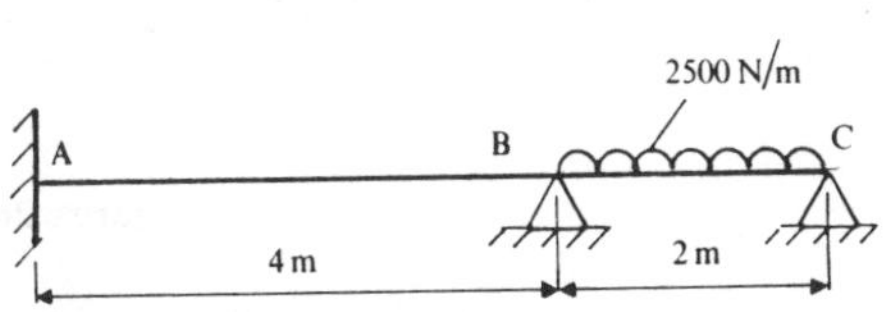

Figure 6.19

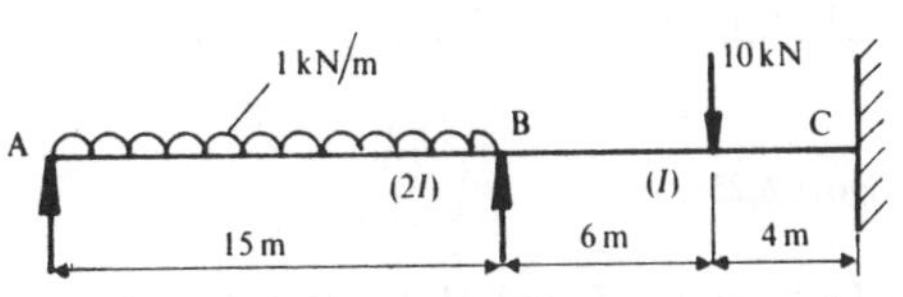

Figure 6.20

6.5 Use the three-moment theorem to construct the bending moment diagram for the beam in Fig. 6.20. Derive from the shear force diagram from the moment diagram.

Moment Distribution - Continuous Beams

6.6 A simply supported beam of constant cross-section is loaded as shown in Fig. 6.21. Draw the *F* and *M* diagrams, showing the maximum values.

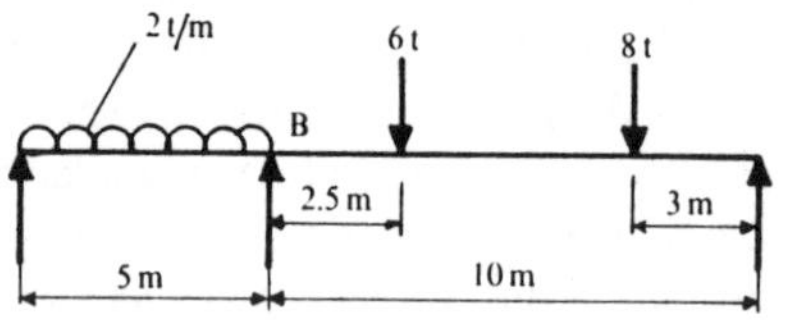

Figure 6.21

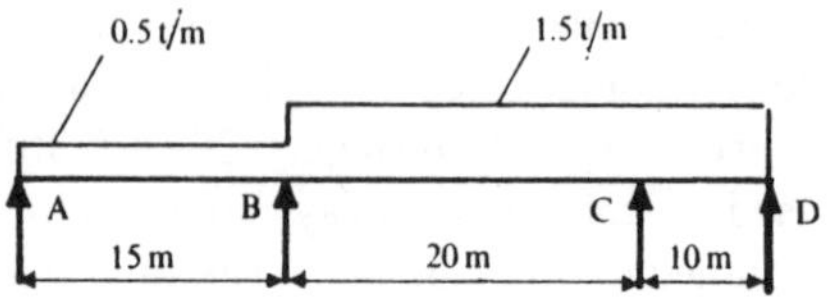

Figure 6.22

6.7 The continuous beam in Fig. 6.22 rests on four level supports. Determine the bending moments and the reaction at each support.

6.8 The beam in Fig. 6.23 is of uniform section. Draw the shear force and bending moment diagrams, indicating principal values.

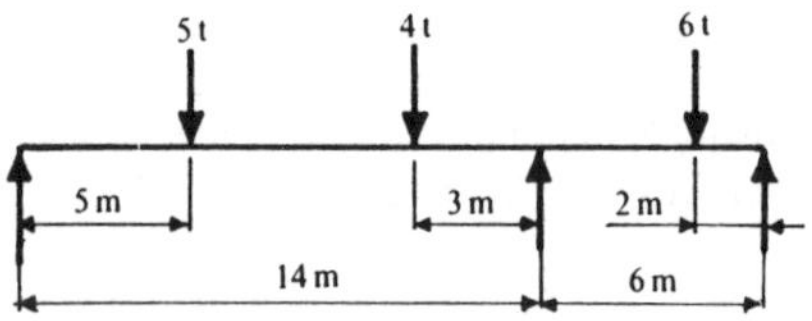

Figure 6.23

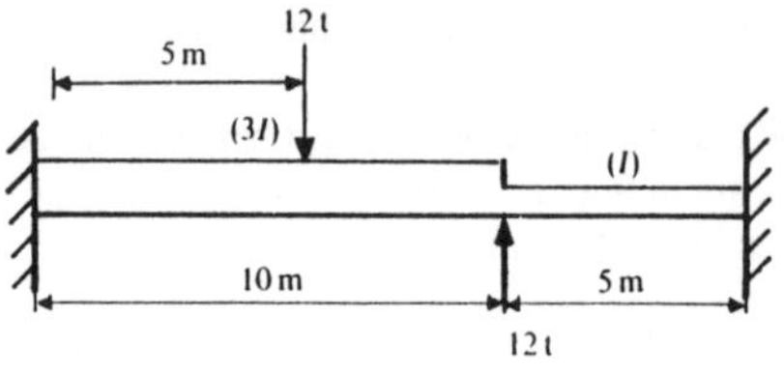

Figure 6.24

6.9 Calculate the fixed-end moments at each end of the continuous beam given in Fig. 6.24.

6.10 The continuous beam in Fig. 6.25 is built in at one end and propped at the three positions shown. If the section varies as shown, draw the *M* and *F* diagrams, indicating salient values.

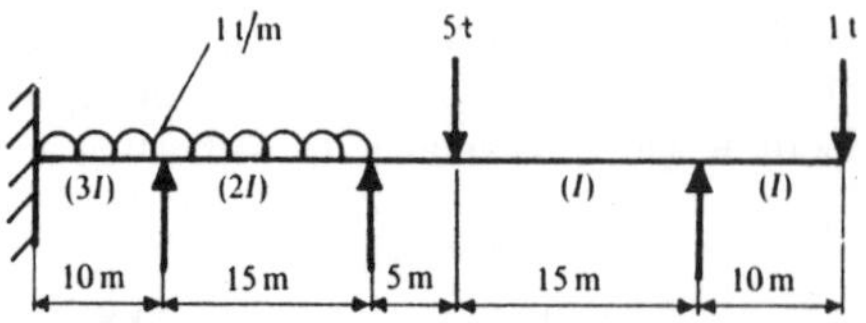

Figure 6.25

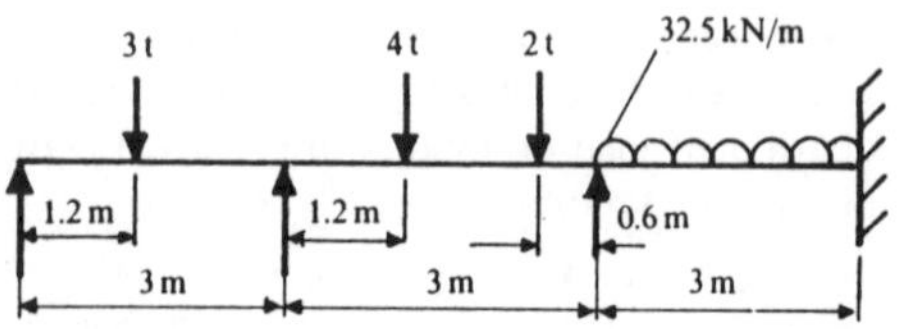

Figure 6.26

6.11 Draw the *F* and *M* diagrams for the propped cantilever in Fig. 6.26. What is the maximum moment?

6.12 The beam in Fig. 6.27 has a uniform section throughout. Draw the F and M diagrams, showing the maximum values.

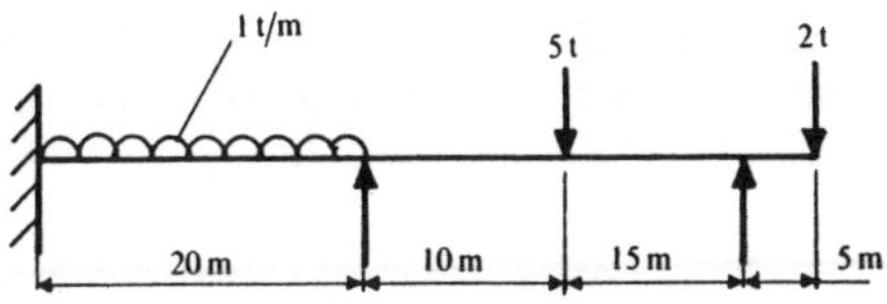

Figure 6.27

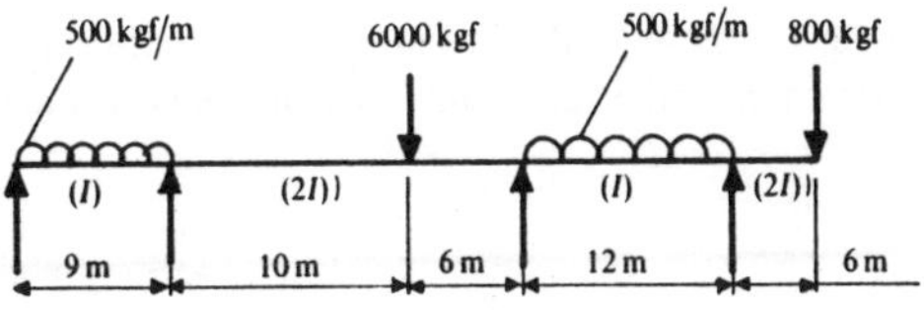

Figure 6.28

6.13 Determine the magnitude and position of the maximum bending moment for the beam in Fig. 6.28.

6.14 Calculate the bending moments at points B and C for the beam in Fig. 6.29 and hence construct the M-diagram, showing maximum values.

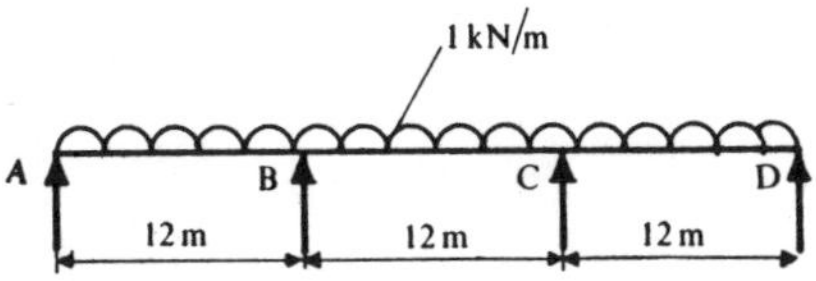

Figure 6.29

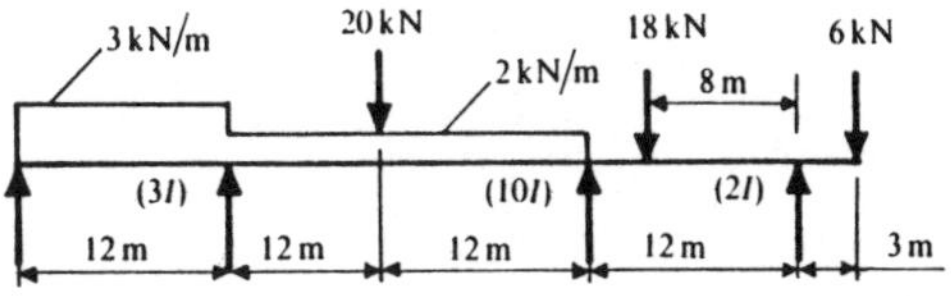

Figure 6.30

6.15 Determine all the fixing moments for the continuous beam in Fig. 6.30.

6.16 A continuous beam rests on three simple supports A, B and C. If the level of B is 20 mm below that of A and C, determine the fixing moment at B, given that lengths AB = 5 m and BC = 10 m. Take $I_{AB} = 12 \times 10^8\ \text{mm}^4$, $I_{BC} = 40 \times 10^8\ \text{mm}^4$ and $E = 210$ GPa.

Moment Distribution for Structures

6.17 Construct the bending moment diagrams for the equal-legged portal frame in Fig. 6.31 where EI is constant and the legs are pinned to the foundations.

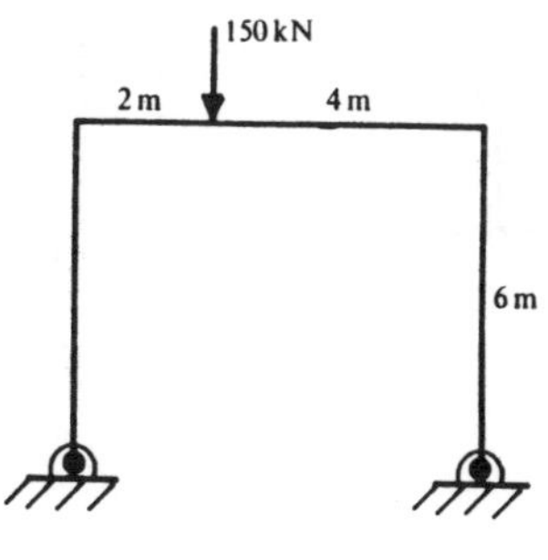

Figure 6.31

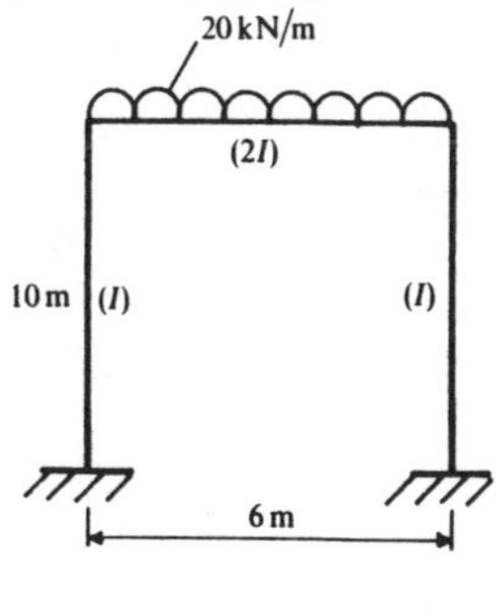

Figure 6.32

6.18 The symmetrical portal frame in Fig. 6.32 is rigidly fixed into its foundations. Establish the moment distribution diagram for the frame.

6.19 The frame ABC in Fig. 6.33 is simply supported at A and pinned at C. Find the horizontal and vertical reactions at A and C when a concentrated moment of 10 kNm is applied at the rigid joint B.

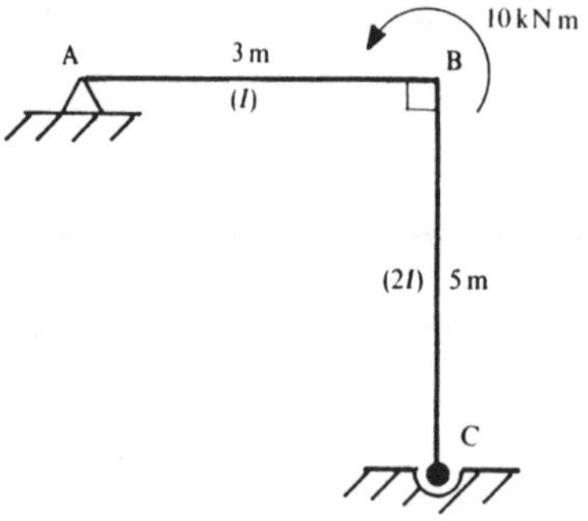

Figure 6.33

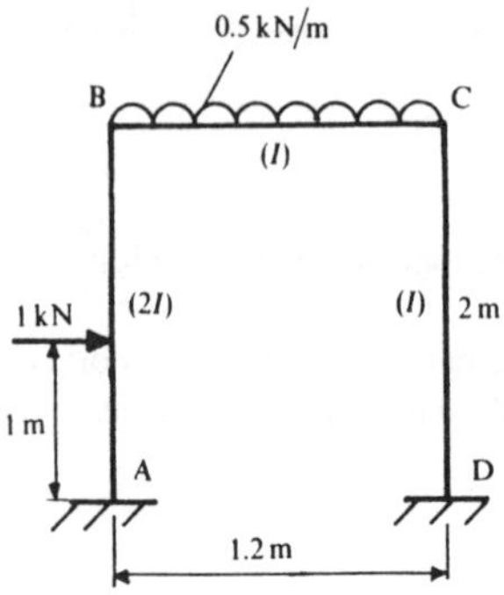

Figure 6.34

6.20 The frame in Fig. 6.34 carries a concentrated horizontal force of 1 kN in addition to uniformly distributed loading of 0.5 kN/m as shown. Determine the reactions at the rigid foundations and the horizontal displacement at C. Take $E = 75$ GPa, $I = 2.5 \times 10^3$ mm^4.

6.21 The two-storey frame in Fig. 6.35 is manufactured with a central horizontal member 4 mm too short. Determine the resulting moment distribution for the frame when this bar is elastically stretched into position. The cross-section of each bar is rectangular 100 mm × 25 mm and $E = 200$ GPa.

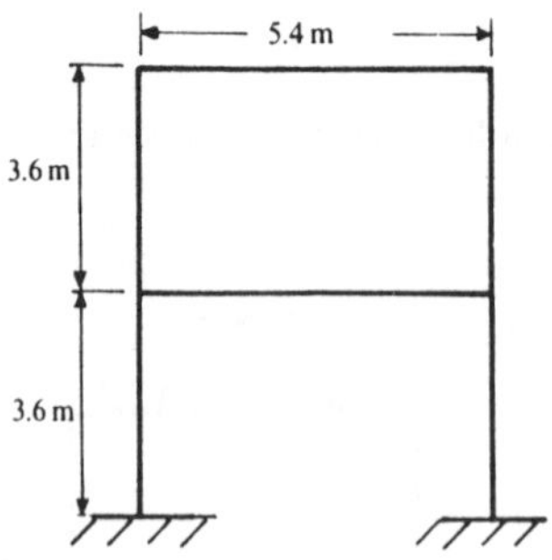

Figure 6.35

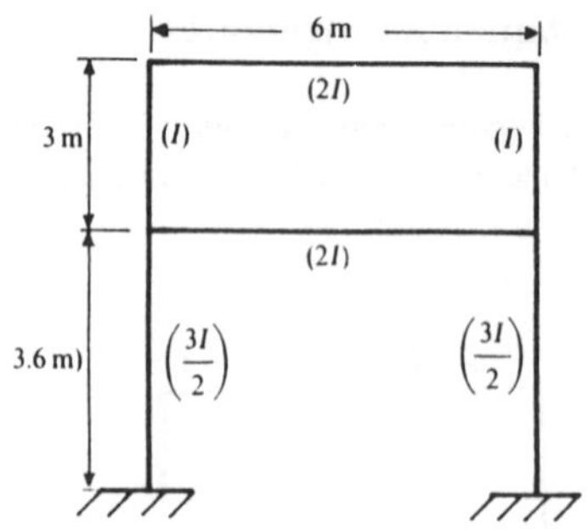

Figure 6.36

6.22 The two-storey frame shown in Fig. 6.36 is subjected to a temperature increase of 25°C. Determine the resulting induced moment distribution given that $\alpha = 15 \times 10^{-6}$/°C, $E = 210$ GPa and $I = 20 \times 10^6$ mm^4.

6.23 Construct the bending moment diagram for the unsymmetrical portal frame in Fig. 6.37, where *EI* is constant, taking account of side sway. Show the position and magnitude of the maximum bending moment.

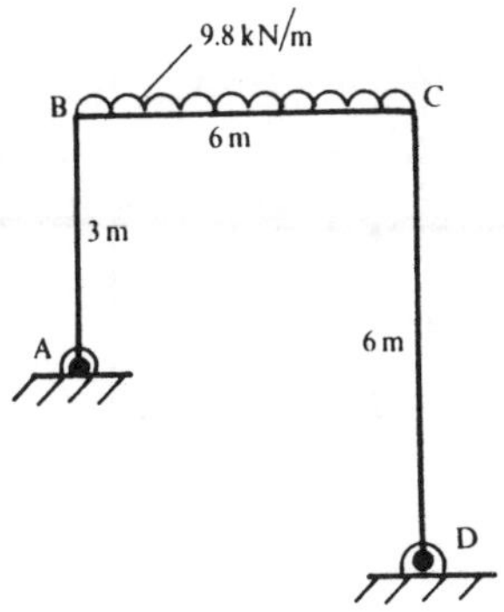

Figure 6.37

6.24 Find the position and magnitude of the maximum bending moment for the structure in Fig. 6.38 when it is fixed at A, rigid at B and hinged at C.

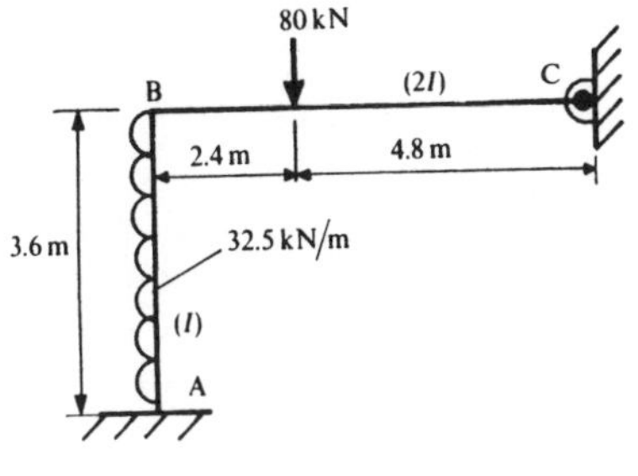

Figure 6.38

6.25 Determine the fixing moments and construct the bending moment diagram for the structure in Fig. 6.39, given that *I* is constant.

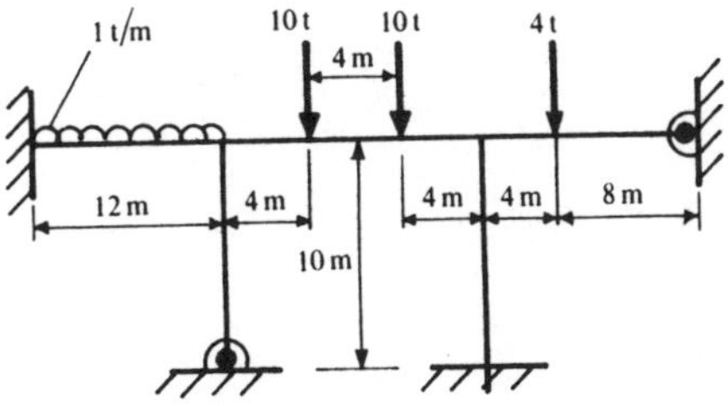

Figure 6.39

6.26 Determine the maximum of the true fixing moments for the structure in Fig. 6.40 when the effect of side sway is accounted for.

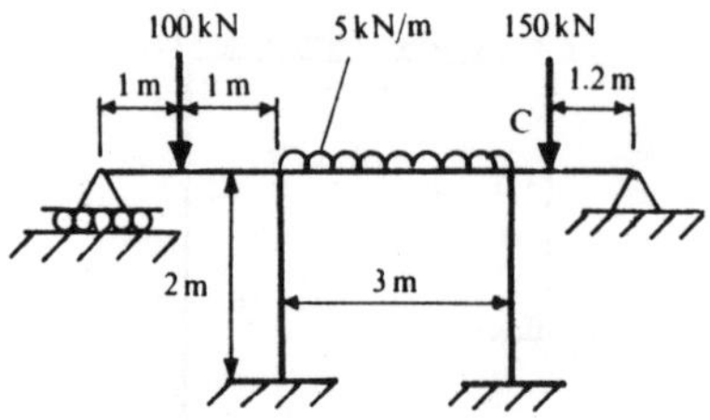

Figure 6.40

CHAPTER 7

FLEXURAL SHEAR FLOW

7.1 Shear Stress Due to Shear Force in Beams

In Fig. 7.1a an element δz in the length of a beam is shown, over which the shear force and bending moment vary by δF and δM respectively. These produce corresponding variations in the shear and bending stresses of $\delta\tau$ and $\delta\sigma$ over the length δz. Because both σ and τ will vary with y, it is necessary to consider the forces they produce when acting over any given slice, $\delta z \times \delta y \times x$, lying at a distance y from the neutral axis, as shown in Fig. 7.1b.

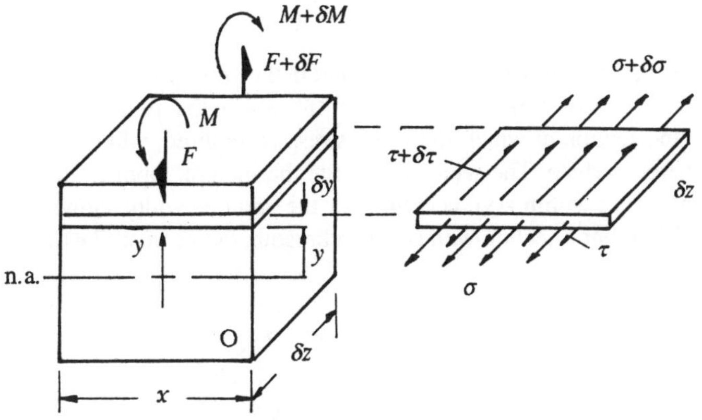

(a) (b)

Figure 7.1 Flexural shear in a beam element

Neither σ nor τ varies with x and therefore, to find an expression for the vertical shear stress τ across X-X, it is necessary to express force equilibrium between the horizontal complementary shear actions and the bending stress σ, shown in Fig. 7.1b. This gives:

$$(\sigma + \delta\sigma)\, x\, \delta y + (\tau + \delta\tau)\, x\, \delta z = \sigma x\, \delta y + \tau x\, \delta z$$
$$\delta\sigma\, x\, \delta y + \delta\tau\, x\, \delta z = 0$$
$$\therefore\ \delta\tau = -\ \delta\sigma\, x\, \delta y/(x\, \delta z) \qquad (7.1)$$

Taking moments about point O, in Fig.7.1a,

$$-\ M + (M + \delta M) - (F + \delta F)\delta z = 0$$

which gives $F = \delta M/\delta z$. With the hogging action shown, eq(4.3) gives $\delta\sigma = \delta M\, y/I = (F\delta z)y/I$.

Substituting into eq(7.1) gives

$$\delta\tau = -\ [F/(I\,x\,)](y\,x\,\delta\,y) \tag{7.2}$$

When eq(7.2) is integrated between the neutral axis ($y = 0$) and the top surface ($y = y_1$) we should expect a zero value of shear stress for the free surface. The shear stress is not zero at the neutral axis. To achieve this we integrate eq(7.2), as follows:

$$\tau = -\ [F/(Ix)]\int_0^y yx\,\mathrm{d}y + C \tag{7.3a}$$

Now, $\tau = 0$ when $y = y_1$ and therefore C becomes

$$C = [F/(Ix)]\int_0^{y_1} yx\,\mathrm{d}y \tag{7.3b}$$

Combining eqs(7.3a,b),

$$\tau = -\ [F/(Ix)]\int_0^y yx\,\mathrm{d}y + [F/(Ix)]\int_0^{y_1} yx\,\mathrm{d}y$$

$$\tau = [F/(Ix)]\int_y^{y_1} y\,x(y)\,\mathrm{d}y \tag{7.4}$$

Equation(7.4) provides the vertical shear stress at a point in the cross-section distance y from the n.a. It is seen that the limits of y apply to the shaded area above that point. Thus, the lower limit is the height in the section at which the shear stress is required, while upper limit y_1 defines the height of free surface. The width x, outside the integral, applies to the chosen section, while that appearing within $x(y)$l accounts for the variable width shown within the shaded area (see Fig. 7.2a). For a section with constant breadth, the x's in eq(7.4) will cancel.

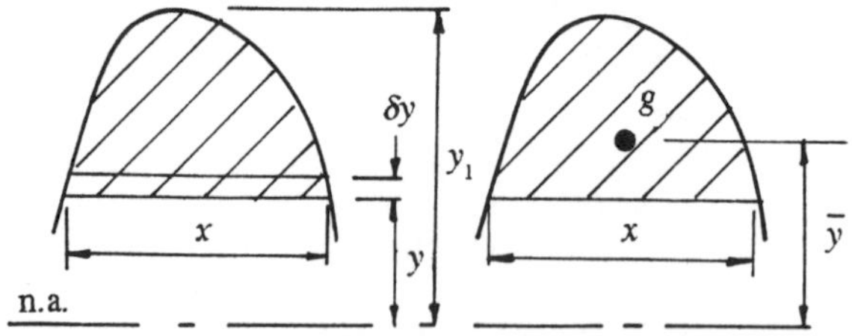

(a) (b)

Figure 7.2 Effective section area in shear

Recognising that the integral in eq(7.4) is the first moment of the shaded area A about the centroidal axes in Fig. 7.2b provides the equivalent alternative expression for τ:

$$\tau = F(A\bar{y})/(I\,b) \tag{7.5}$$

where $\bar{y}$ is the distance between the centroid of area A and the centroid of the whole cross-section as shown in Fig. 7.2b. Note that I in eq(7.5) is the second moment of area for the whole cross-section and F is positive when acting vertically downwards.

Example 7.1 Determine an expression for the vertical shear stress distribution in a beam of solid circular section of radius R (Fig. 7.3a).

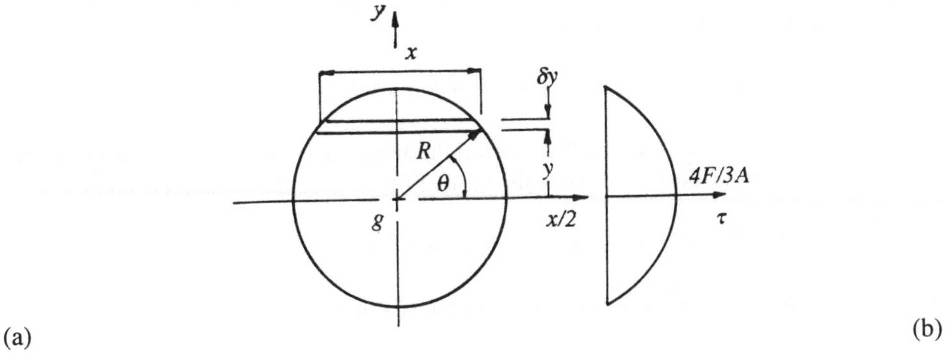

Figure 7.3 Circular section

Taking an elemental strip distance y from the n.a. and of width $2x$, eq(7.4) becomes

$$\tau = [F/I\,(2x)]\int_y^R 2y\,(R^2 - y^2)^{1/2}\,dy \qquad \text{(i)}$$

Converting eq(i) to polar co-ordinates (r,θ) from the following: $y = R\sin\theta$, $dy = R\cos\theta\, d\theta$ and $(R^2 - y^2)^{1/2} = R\cos\theta$,

$$\tau = [FR^3/(IR\cos\theta)]\int_\theta^{\pi/2}\cos^2\theta\sin\theta\, d\theta$$

$$= -\,[FR^3/(3IR\cos\theta)]\;\left|\cos^3\theta\right|_\theta^{\pi/2} = [FR^2/(3I)]\cos^2\theta$$

$$= [FR^2/(3I)](1 - y^2/R^2) = [4F/(3A)](1 - y^2/R^2) \qquad \text{(ii)}$$

in which $I = \pi R^2/4$. Equation (ii) is zero at the edges and reaches a maximum of $4F/3A$ at the n.a. The distribution across the depth is parabolic (see Fig. 7.2b)

Example 7.2 Determine the distribution of shear stress for the trapezoidal section in Fig. 7.4a when it is subjected to a vertical shear force of 20 kN.

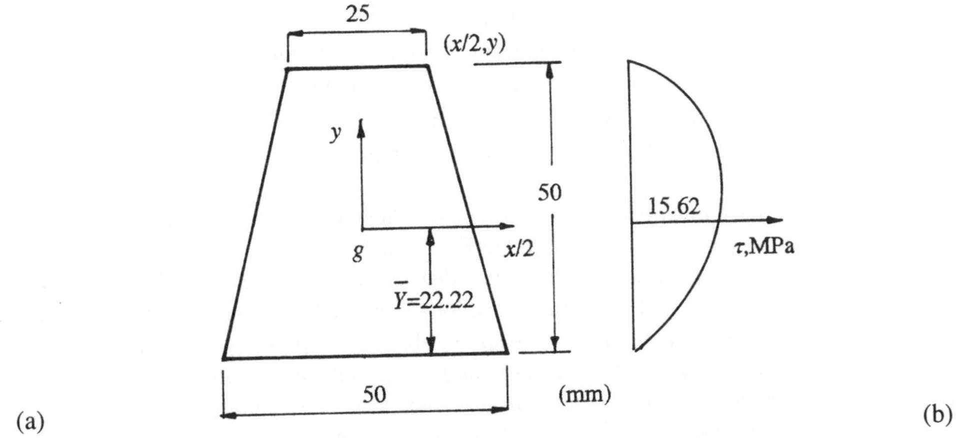

Figure 7.4 Trapezoidal section

The position of the centroidal axis is found by taking first moments of area about the base. Let g lie at distance $\bar{y}$ above the base,

$$(50 \times 25 \times 25) + 2\,(12.5 \times 50/2)(50/3) = [(25 \times 50) + (12.5 \times 50)]\bar{y}$$

from which $\bar{y} = 22.22$ mm. Noting that, for a right-angled triangle base b and height h, $I = bh^3/12$ about the base, the total I about the base of the trapezoidal section becomes

$$I = (25 \times 50^3/3) + (2 \times 12.5 \times 50^3/12) = 1.3021 \times 10^6\ \text{mm}^4$$

Transferring this to the centroid using the parallel axis theorem,

$$I_x = I - A\bar{y}^2 = 1.3021 \times 10^6 - 1875\,(22.22)^2 = 37.636 \times 10^4\ \text{mm}^4$$

When applying eq(7.4) it is necessary to find the function $x(y)$. For the co-ordinates $(x/2, y)$ shown, the equation of the right sloping side passing through the top corner $(x_1/2, y_1)$ is

$$y - y_1 = m(x/2 - x_1/2)$$
$$y - 27.78 = -\,(50/12.5)(x/2 - 12.5)$$
$$x = 38.89 - y/2 \qquad \text{(i)}$$

Substituting eq(i) into eq(7.4) with $y_1 = 27.78$ mm,

$$\tau = [F/(Ix)] \int_y^{27.78} (38.89 - y/2)\, y\, \mathrm{d}y$$
$$= [F/(Ix)] \left| \, 38.89\, y^2/2 - y^3/6 \right|_y^{27.78}$$
$$= [F/(Ix)]\, [11433.16 - (38.89\, y^2/2 - y^3/6)] \qquad \text{(ii)}$$

where x is given by eq(i) and $F/I = (20 \times 10^3)/(37.636 \times 10^4) = 53.14 \times 10^{-3}$ N/mm^4. Equations (i) and (ii) apply to positive and negative y for the co-ordinates shown. They give the following values:

$\tau = 0$ at the top free surface where $y = 27.78$ mm
$\tau = 15.14$ MPa an intermediate value where $y = 10$ mm, $x = 33.89$ mm
$\tau = 15.62$ MPa at the centroidal axis where $y = 0$, $x = 38.89$ mm
$\tau \approx 0$ at the bottom free surface where $y = -\,22.22$ mm
$\tau = 3.681$ MPa an intermediate value where $y = -\,10$ mm

Equation(ii) supplies a maximum value $\tau = 16.04$ MPa at 4 mm above the centroidal axis. The variation across the depth is illustrated in Fig. 7.4b.

Figure 7.5a,b shows the vertical shear stress distribution (τ_v) when eq(7.5) is applied to the flanges and webs of T and U-sections. Thus, for a point in the web, the τ_v ordinate shown is proportional to the first moment of the area above that point (widely shaded). The distribution is parabolic in y with a discontinuity at the web-flange intersection. Equation (7.5) may further be applied to find the horizontal shear stress distribution within the flanges of I, T and U-sections under a vertical shear force. Here we need to work back from the free edge of the flange so that $A\bar{y}$ becomes the moment of the cross shaded areas in Figs 7.5a,b.

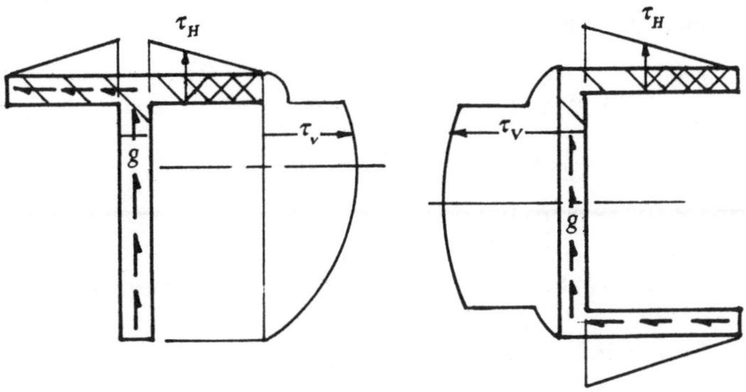

(a) (b)

Figure 7.5 Transverse shear stresses in T and U beams

Clearly $\bar{y}$ is now constant and therefore the horizontal shear stress τ_h increases linearly from zero to a maximum at the wedge intersections.

Example 7.3 A symmetrical I-section beam, 300 mm deep, 200 mm wide with 25 mm thick flanges and 10 mm thick web (upper half in Fig. 7.6a), sustains at a certain section a shear force F (N) and a bending moment M (Nm). Derive expressions for the distribution of shear stress in the web and flange and show them diagrammatically when F = 30 kN. What percentage of F is carried by the web and what percentage of M is carried by the flanges?

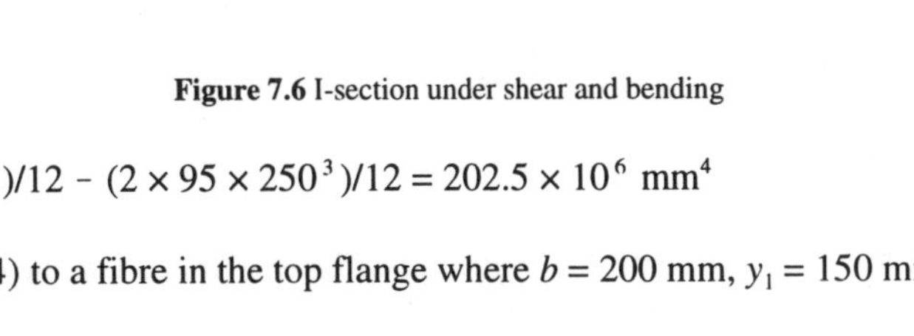

(a) (b)

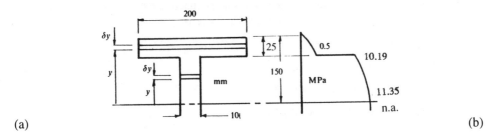

Figure 7.6 I-section under shear and bending

$I = (200 \times 300^3)/12 - (2 \times 95 \times 250^3)/12 = 202.5 \times 10^6 \text{ mm}^4$

Applying eq(7.4) to a fibre in the top flange where b = 200 mm, y_1 = 150 mm, x = 200 mm gives

$$\tau_f = [F/(I \times 200)] \int_y^{150} 200\, y\, dy = (F/I) \int_y^{150} y\, dy$$

$$= (F/I)\, \left| y^2/2 \right|_0^{150} = (F/2I)\, [150^2 - y^2] \qquad \text{(i)}$$

The distribution is parabolic in y. When $F = 30 \times 10^3$ N in eq(i),

$\tau_f = (0.7407 \times 10^{-4})\, [(2.25 \times 10^4) - y^2]$

$\tau_f = 0$ at the flange top where y = 150 mm

$\tau_f = 0.51$ MPa at the flange bottom where y = 125 mm

In applying eq(7.4) to a fibre in the web above the NA, both the web and flange areas above this section contribute to τ. That is,

$$\tau_w = [F/(10I)]\{ \int_{125}^{150} 200\, y\, dy + \int_{y}^{125} 10\, y\, dy\}$$

$$= [F/(10I)][200\,(150^2 - 125^2)/2 + 10\,(125^2 - y^2)/2]$$
$$= (F/I)[(7.6563 \times 10^4) - y^2/2\,] \qquad \text{(ii)}$$

This distribution is again parabolic in y. When $F = 30 \times 10^3$ N in eq(ii),

$$\tau_w = (1.482 \times 10^{-4})\,[(7.6563 \times 10^4) - y^2/2\,]$$
$\tau_w = 10.19$ MPa with $y = 125$ mm,
$\tau_w = 11.35$ MPa with $y = 0$.

Equation (7.5) gives the horizontal shear stress in the flange as

$$\tau_h = [F/(Ix)]\,A_f\,\bar{y}$$
$$= [(30 \times 10^3)/(202.5 \times 10^6 \times 25)]\,A_f \times 137.5 = 0.815 \times 10^{-3} A_f$$

where A_f varies from a zero value at the flange tip to (100×25) mm^2 at the flange centre. Thus τ_h varies linearly with flange length between these positions from zero to 2.04 MPa (e.g. see Fig. 7.5a).

These τ values (see Fig. 7.6b) are mirrored in the bottom half of a symmetrical section. Figure7.6b shows the discontinuity in shear stress at the junction between the flange and web. For the elemental web strip in Fig. 7.6a, the shear force δF_w is

$$\delta F_w = 10\tau_w\,\delta y$$

Substituting from eq(ii) gives

$$\delta F_w = 10(F/I)\,[(7.6563 \times 10^4) - y^2/2\,]\delta y$$

Integrating for the whole length of web gives

$$F_w = (10F/I)\Big|\,(7.6563 \times 10^4)\,y - y^3/6\Big|_{-125}^{125}$$
$$F_w = (184.9 \times 10^6)(F/I)$$
$$F_w/F = 184.9/202.5 = 0.913$$

This shows that the web carries 91.3% of the applied shear force. For the elemental flange strip shown in Fig. 7.6a, the bending moment δM_f is

$$\delta M_f = (200\delta y)\sigma y = (200\delta y)(My/I)\,y$$

Integrating this across each flange depth gives

$$M_f = 2 \times 200(M/I) \int_{125}^{150} y^2\, d\,y$$

$$= 2 \times 200[M/(3I)]\,[150^3 - 125^3] = 2\,(94.792 \times 10^6)M/I$$
$$M_f/M = 2 \times 94.792/202.5 = 0.936$$

That is, the flanges carry 93.6% of the applied bending moment. This example illustrates that I-sections exhibit the desirable combination of high strength and low weight. Hence their common use as beams in structural work. It is shown in Section 7.5, from the fact that the web carries most of the shear force and the flange most of the bending moment, that a structural idealisation can simplify the analysis of shear stress distributions.

7.2 St Venant Shear in Prismatic Bars

The vertical shear stress distributions found in Examples 7.1 and 7.2 are not strictly accurate as they ignore horizontal shear. At the edges of these sections the resultant shear stress is directed along the curved or sloping sides and therefore a greater horizontal component exists in these regions. This behaviour will apply to all non-rectangular sections under shear and is accentuated where shear stresses follow the boundaries of circular and triangular sections of beams. St Venant's theory admits the horizontal component of shear. Figure 7.7a shows the point in the arbitrary cross-section of a prismatic beam under flexural shear.

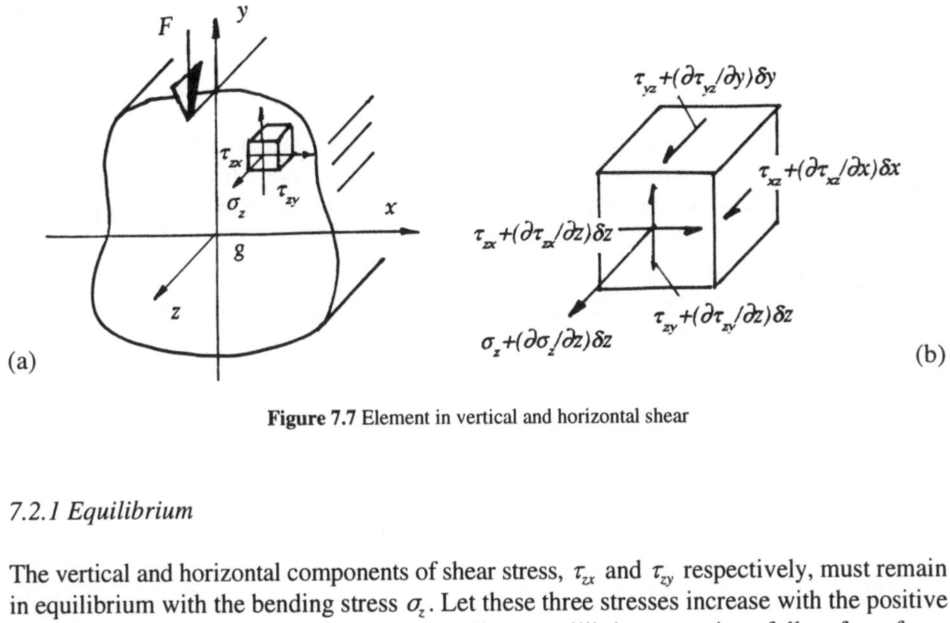

Figure 7.7 Element in vertical and horizontal shear

7.2.1 Equilibrium

The vertical and horizontal components of shear stress, τ_{zx} and τ_{zy} respectively, must remain in equilibrium with the bending stress σ_z. Let these three stresses increase with the positive directions *x*, *y* and *z*, as shown in Fig. 7.7b. Three equilibrium equations follow from force balance in the *x*, *y* and *z* directions. From Fig. 7.7b,

$$\sum F_z = 0: (\sigma_z + \delta z \times \partial\sigma_z/\partial z)\delta x\delta y - \sigma_z\,\delta x\delta y + (\tau_{zy} + \delta y \times \partial\tau_{zy}/\partial y)\delta x\delta z - \tau_{zy}\,\delta x\delta z$$
$$+ (\tau_{zx} + \delta x \times \partial\tau_{zx}/\partial x)\delta y\delta z - \tau_{zx}\delta y\delta z = 0$$
$$\sum F_y = 0: (\tau_{zy} + \delta z \times \partial\tau_{zy}/\partial z)\delta x\delta y - \tau_{zy}\delta x\delta y = 0$$
$$\sum F_x = 0: (\tau_{zx} + \delta z \times \partial\tau_{zx}/\partial z)\delta x\delta y - \tau_{zx}\delta x\delta y = 0$$

which yield

$$\partial\sigma_z/\partial z + \partial\tau_{zx}/\partial x + \partial\tau_{zy}/\partial y = 0 \quad (7.6a)$$
$$\partial\tau_{zx}/\partial z = 0, \;\; \partial\tau_{zy}/\partial z = 0 \quad (7.6b,c)$$

Equations(7.6b,c) require that τ_{zx} and τ_{zy} do not vary with z and they can therefore be applied to regions of a beam between concentrated forces where the shear force is constant. They do not apply when body forces are present, i.e. due to distributed loading and self-weight. Writing $F = \mathrm{d}M/\mathrm{d}z$, it follows that $\partial \sigma_z/\partial z = \partial(My/I)/\partial z = (y/I)F$. Equation (7.6a) becomes

$$\partial \tau_{zx}/\partial x + \partial \tau_{zy}/\partial y = -(y/I)F \tag{7.7}$$

7.2.2 Compatibility

Six components of strain, ε_x, ε_y, ε_z, γ_{xy}, γ_{yz} and γ_{xz}, exist and these depend upon the three displacements u, v and w. Hence a relationship between the strains expresses a compatibility condition. This condition may be converted to stress from Hooke's law. For example, we see that $\varepsilon_z = \partial w/\partial z$, $\gamma_{yz} = \partial v/\partial z + \partial w/\partial y$ and $\gamma_{xy} = \partial u/\partial y + \partial v/\partial x$ are satisfied by

$$2\partial^2 \varepsilon_z/\partial x\,\partial y = (\partial/\partial z)(\partial \gamma_{yz}/\partial x + \partial \gamma_{xz}/\partial y - \partial \gamma_{xy}/\partial z) \tag{7.8}$$

Similarly $\varepsilon_x = \partial u/\partial x$, $\varepsilon_y = \partial v/\partial y$ and $\gamma_{xy} = \partial u/\partial y + \partial v/\partial x$ are satisfied by

$$\partial^2 \gamma_{xy}/\partial x\,\partial y = \partial^2 \varepsilon_x/\partial y^2 + \partial^2 \varepsilon_y/\partial x^2 \tag{7.9}$$

There are two further compatibility equations of the the type (7.8) and (7.9), making 6 in all. When we convert these to our three, non-zero components of stress from eqs(2.15a-d) and combine with eqs(7.6a-c), two compatibility conditions become appropriate to this problem:

$$(\partial^2/\partial x^2 + \partial^2/\partial y^2)\tau_{zx} = 0, \tag{7.10a}$$

$$(1 + \nu)(\partial^2/\partial x^2 + \partial^2/\partial y^2)\tau_{zy} + \partial^2 \sigma_z/\partial y \partial z \tag{7.10b}$$

where $\partial^2 \sigma_z/\partial y\partial z = F/I$

7.2.3 Stress Function

Introduce a stress function $\phi(x,y)$ to satisfy eqs(7.10a,b) as

$$\tau_{zy} = \partial \phi/\partial x - Fy^2/(2I) + g(x), \quad \tau_{zx} = -\partial \phi/\partial y \tag{7.11a,b}$$

Substitute eqs(7.11a,b) into eqs(7.10a,b) leads to

$$(\partial/\partial y)(\partial^2/\partial x^2 + \partial^2/\partial y^2)\phi = 0 \tag{7.12a}$$

$$(\partial/\partial x)(\partial^2/\partial x^2 + \partial^2/\partial y^2)\phi = F\nu/[I(1 + \nu)] - \partial^2 g/\partial x^2 \tag{7.12b}$$

Integrating eq(7.12b),

$$(\partial^2/\partial x^2 + \partial^2/\partial y^2)\phi = F\nu x/[I(1 + \nu)] - \partial g/\partial x + f(y) + C \tag{7.13}$$

It follows from eqs(7.12a) and (7.13) that $f(y) = 0$. The constant $C = 0$ when F is aligned with y as an axis of symmetry. Otherwise, C becomes the constant rate of twist about the z - axis when F is offset to y.

7.2.4 Boundary Conditions

Figure7.8a shows that when the element $\delta x \times \delta y \times \delta z$ intersects the boundary, the latter makes an inclined plane with x, y and z.

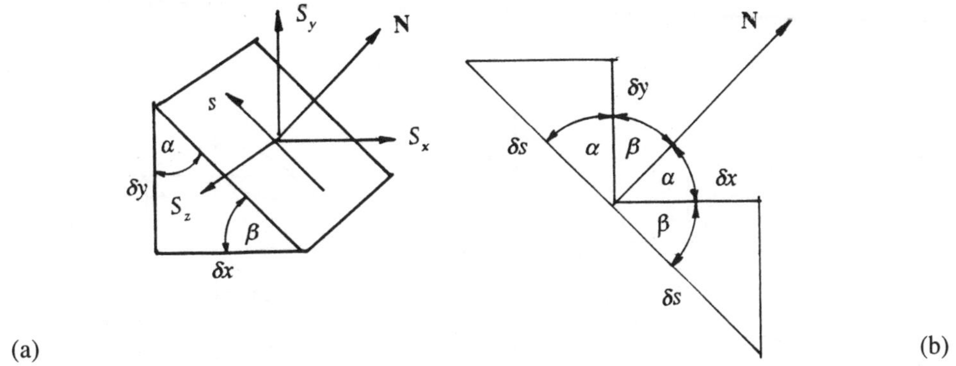

Figure 7.8 Intersection of element with boundary

Equation (1.4c) gives the forces on this plane as

$$S_z = \tau_{zx}\, l + \tau_{zy}\, m + \sigma_z\, n = 0 \tag{7.14a}$$

The bar is free from axial forces, so we can set $S_z = 0$. Let s define the perimeter so that the direction cosines are $l = \cos\alpha = \mathrm{d}y/\mathrm{d}s$, $m = \cos\beta = -\,\mathrm{d}x/\mathrm{d}s$ and $n = \cos\gamma = \cos 90° = 0$ (see Fig. 7.8b). Equation (7.14a) reduces to

$$\tau_{zx}(\mathrm{d}y/\mathrm{d}s) - \tau_{zy}(\mathrm{d}x/\mathrm{d}s) = 0 \tag{7.14b}$$

The stress function (7.11a,b) must also satisfy the condition at the boundary. We can therefore combine eqs(7.11a) and (7.14b), to give

$$(\partial\phi/\partial y)(\mathrm{d}y/\mathrm{d}s) + (\partial\phi/\partial x)(\mathrm{d}x/\mathrm{d}s) = [Fy^2/(2I) - g(x)](\mathrm{d}x/\mathrm{d}s)$$
$$\partial\phi/\partial s = [Fy^2/(2I) - g(x)](\mathrm{d}x/\mathrm{d}s)$$

It follows that ϕ along the boundary is given by

$$\phi = \int [Fy^2/(2I) - g(x)]\mathrm{d}x \tag{7.15a}$$

If we make $\phi = 0$ for a boundary curve $f(x, y) =$ constant in which a function $y = y(x)$ applies, it follows from eq(7.15a) that

$$g(x) = F[\,y(x)]^2/(2I) \tag{7.15b}$$

Equation(7.15b) will supply $g(x)$ within the stress function (7.11a) pertaining to all points (x, y) in the section. Between eqs(7.11a) and (7.15b) equilibrium, compatibility and the boundary conditions have been satisfied. Thus, the two shear stress resultants (7.11a,b) will match an external vertical force F in the absence of an axial force. From Fig. 7.7b

$$F = \int_x \int_y \tau_{zy}\, \mathrm{d}x\, \mathrm{d}y \quad \text{and} \quad \int_x \int_y \tau_{zx}\, \mathrm{d}x\, \mathrm{d}y = 0$$

Example 7.4 Examine the vertical and horizontal shear stress distributions in a solid circular section of outer radius R. Compare with the distribution found in example 7.1 from the elementary theory.

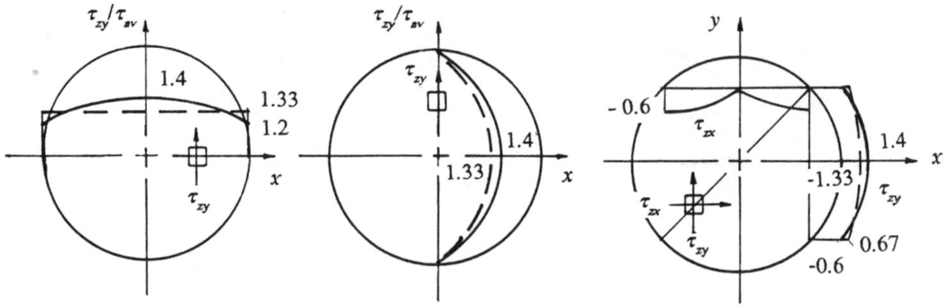

Figure 7.9 Vertical and horizontal shear stress distributions in a circular section

The equation of a circle with centre at the origin is $x^2 + y^2 = R^2$. Hence from eq(7.15b) we can make $\phi = 0$ at the boundary, when

$$g(x) = [(F/2I)](R^2 - x^2) \tag{i}$$

For interior points (x,y), the stress function follows from substituting eq(i) into eq(7.13),

$$\partial^2\phi / \partial x^2 + \partial^2\phi / \partial y^2 = (1 + 2\nu)Fx / [I\,(1 + \nu)] \tag{ii}$$

The solution to eq (ii) is

$$\phi = \{(1 + 2\nu)F / [8I\,(1 + \nu)]\}(x^2 + y^2 - R^2)x \tag{iii}$$

From eqs(7.11a,b), the shear stresses are found from eq(iii) as

$$\tau_{zx} = -\,(1 + 2\nu)Fxy / [4I\,(1 + \nu)] \tag{iv}$$
$$\tau_{zy} = [(3 + 2\nu)(R^2 - y^2) - (1 - 2\nu)x^2]F / [8I\,(1 + \nu)] \tag{v}$$

Let us apply eqs(iv) and (v) to a material with $\nu = ¼$ and normalise with an average shear stress $\tau_{av} = F/A$. Noting that $I = Ar^2/4$, these give

$$\tau_{zx}/\tau_{av} = -\,1.2\,xy/R^2 \tag{vi}$$
$$\tau_{zy}/\tau_{av} = (1.4/R^2)(R^2 - y^2 - x^2/7) \tag{vii}$$

Equations (vi) and (vii) have been used to plot the distributions of shear stresses on 0, 90 and 45° diameters (see Figs 7.9 a-c). The horizontal shear stress τ_{zx} is absent in 7.9a and b. The vertical component τ_{zy} varies parabolically with x and y, reaching a maximum at the centre. Predictions from the elementary theory are shown to underestimate the maximum shear stress and do not account for a variation across the horizontal diameter. Figure7.9c shows that both components exist along a 45° diameter. The horizontal component τ_{zx} is zero at the centre and is equal in magnitude to τ_{yz} where the 45° diameter intersects the boundary. This must be so if their resultant is to lie tangential to the boundary at this point.

7.3 The Shear Centre and Flexural Axis

In Examples 7.1 - 7.3 the sections were symmetrical about the y - axis. Thus when the line of action of the vertical shear force passes through the centroid of these sections, there will be no twisting effect. If longitudinal twisting is to be avoided, for a section that is not symmetrical about a vertical axis, then the shear force must be displaced to pass through a point called the *shear centre*. This ensures that the shear stress distribution, which is statically equivalent to the applied shear force, has zero moment about any point in the line of that force. Twisting can be avoided in a non-uniform beam section when the forces lie on a flexural axis which is the locus of the shear centres for all cross-sections. The shear centre is a property of the section not generally coincident with the centroid. It is seen from the above examples that the shear centre of a doubly or a multiply symmetric section does lie at the centroid. The flexural axis is then coincident with the centroidal axis. The shear centre of a singly symmetrical section lies on its axis of symmetry and the flexural axis will lie in the plane of symmetry. It is often obvious from inspection where the shear centre is located. It will lie at the intersection of the limbs in T, angle and crucifix sections as this is the point of intersection between the force resultants of the shear stress distributions for those limbs.

In practice, the shear forces are not concurrent with the shear centre E (see Fig. 7.10a). Shear stresses due to torsion as well as flexural shear are therefore induced. The principle of superposition enables the net shear stress to be found by adding the separate effects of pure flexural shear and pure torsion (see Figs 10.7b and c).

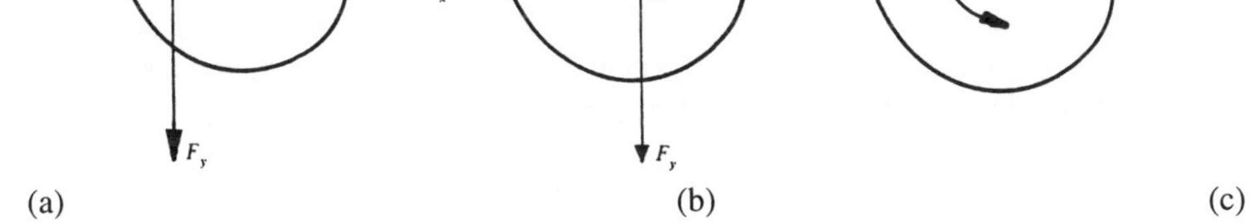

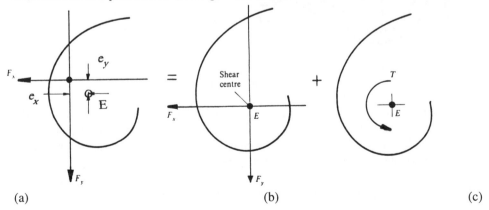

Figure 7.10 Superposition of pure flexural shear and torsion

The shear stress due to torsion for the thin-walled, open section shown has been dealt with previously in Chapter 5. That is, we can apply eq(5.21) to find the shear stress within Fig. 7.10c when torque is $T = F_x\, e_y + F_y\, e_x$.

The following discussion illustrates how the shear flow distribution and the position of the shear centre may be found in the case of open and closed thin-walled sections.

7.4 Shear Flow in Thin-Walled Open Sections

Consider a beam with an open section of arbitrary shape. The wall thickness may vary but must remain thin compared to the other dimensions (Fig. 7.11a). Let axes x and y pass through the centroid g and lie parallel to the directions of the transverse shear forces F_x and F_y. If the section is not to twist, both shear forces must act through the shear centre E. We shall let them act in the negative x and y directions as shown, so that positive hogging moments M_x and M_y appear in the first quadrant of x, y. In a beam these forces and their accompanying bending moments will generally vary with length. Within the plane of the cross-section there will be a component of shear stress τ_{zs} parallel to the mid-section centre line, where the material is assumed to be concentrated, and one normal to this mid-line, τ_{zn}. Because τ_{zn} is zero at the free edges, its variation with t can be ignored. Also τ_{sz} is assumed to be uniform between the edges of a thin section but will vary with the perimeter length s. (In contrast, eq(5.21) shows that the shear stress due to torsion of an open tube varies linearly through the thickness). A shear flow $q = t\tau_{sz}$ will account for variations in τ_{sz} with t around the section in the positive s direction shown.

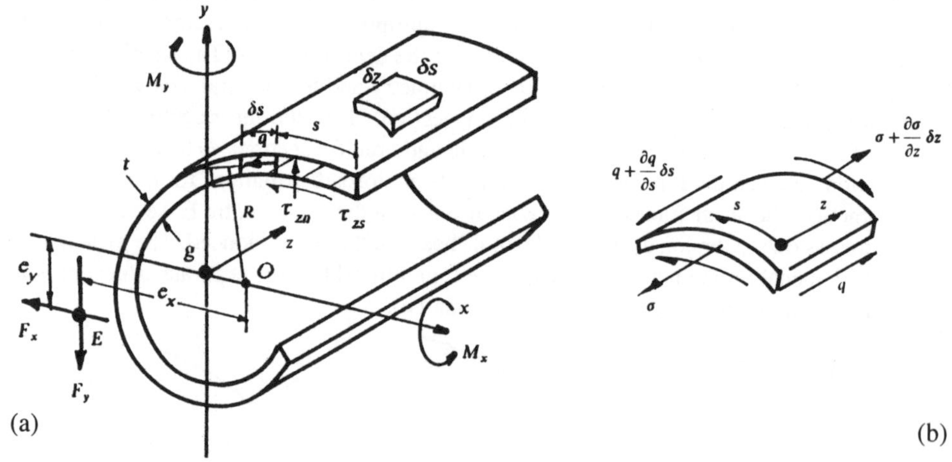

Figure 7.11 Open section in shear

Acting in the z - direction there is a mid-line complementary shear flow, $q = t\tau_{zs}$, and a bending stress σ due to the bending moment. The variation in these actions across an element $\delta s \times \delta z$ of wall is shown in Fig. 7.11b. When the slight variation in t with δs is ignored, the equilibrium equation for the z - direction becomes

$$[\sigma + (\partial\sigma/\partial z)\delta z](\delta s \times t) - \sigma(\delta s \times t) = [q + (\partial q/\partial s)\delta s]\delta z - q\delta z$$

This leads to the following equilibrium equation where q and σ increase with positive s and z, as shown:

$$\partial q/\partial s = t\,\partial\sigma/\partial z \qquad (7.16)$$

Since both M_x and M_y are positive (hogging) within first x, y quadrant in Fig. 7.11a, the longitudinal bending stress is

$$\sigma = M_x\, y/I_x + M_y\, x/I_y \tag{7.17}$$

Substituting eq(7.17) into eq(7.16) and integrating for the shear flow,

$$q = (1/I_x)\int (\mathrm{d}M_x/\mathrm{d}z)(y\,t\,\mathrm{d}s) + (1/I_y)\int (\mathrm{d}M_y/\mathrm{d}z)(x\,t\,\mathrm{d}s) \tag{7.18}$$

The hogging moments for any section at a distance z from the origin in Fig.7.11(a) are $M_x = F_y\, z$ and $M_y = F_x\, z$.

$$\therefore\ \mathrm{d}M_x/\mathrm{d}z = F_y \text{ and } \mathrm{d}M_y/\mathrm{d}z = F_x \tag{7.19a,b}$$

That is, F_x and F_y are positive shear forces associated with the hogging moments. Equations (7.19a,b) will connect the forces to the moments for any section distance z along the beam. Substituting eqs(7.19a,b) into eq(7.18) gives,

$$q = (F_y/I_x)D_x + (F_x/I_y)D_y \tag{7.20}$$

where $D_x = \int y\,t\,\mathrm{d}s$ and $D_y = \int x\,t\,\mathrm{d}s$ are the respective first moments of the shaded area about the x and y axes in Fig. 7.11a. Equation (7.20) will give a positive q for the positive F_x and F_y directions shown and with s measured counter-clockwise from the free surface. Equation (7.20) is not restricted to positive forces. The sign of q will indicate its true direction relative to the chosen direction for s.

7.4.1 Principal Axes

So far we have specified that x and y are centroidal axes lying parallel to the shear forces. Equation (7.20) also applies when x and y coincide with the principal axes for the section, i.e. I_x and I_y are principal second moments of area of the section, shown in Fig. 7.12a.

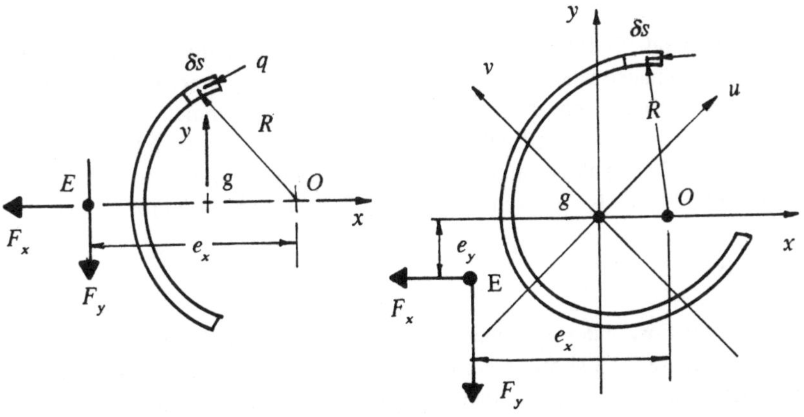

(a) (b)

Figure 7.12 Principal axes coincident and not coincident with shear force directions

Due to the asymmetric nature of the cross-section in Fig. 7.12b, it is unlikely that shear forces will be aligned with the principal directions u and v. Take the common case where a single vertical shear force F_y is applied though the shear centre E but is not aligned with either direction u or v in Fig. 7.12b. Equation (7.20) becomes

$$q = (F_y'/I_x)D_x + (F_x'/I_y)D_y \tag{7.21a}$$

where equivalent shear force components F_x' and F_y' may be derived from the equivalent moments employed with the asymmetric bending of beams (see Section 4.6.1). Applying eqs(7.19a,b) to eqs(4.48b,c) gives:

$$F_x' = -\,F_y/(I_x/I_{xy} - I_{xy}/I_y),\;\; F_y' = F_y/(1 - I_{xy}^{\,2}/I_xI_y) \tag{7.21b,c}$$

In the more general case both vertical F_y and horizontal F_x shear forces are applied at the shear centre E for an asymmetric section, in the negative x and y directions (see Fig. 7.12b). The principal of superposition gives equivalent shear force components F_x' and F_y' in eqs(7.21b,c) as

$$F_x' = \frac{-F_y}{(I_x/I_{xy} - I_{xy}/I_y)} + \frac{F_x}{(1 - I_{xy}^2/I_xI_y)} = \frac{F_x - (I_{xy}/I_x)F_y}{(1 - I_{xy}^2/I_xI_y)} \tag{7.22a}$$

$$F_y' = \frac{-F_x}{(I_y/I_{xy} - I_{xy}/I_x)} + \frac{F_y}{(1 - I_{xy}^2/I_xI_y)} = \frac{F_y - (I_{xy}/I_y)F_x}{(1 - I_{xy}^2/I_xI_y)} \tag{7.22b}$$

7.4.2 Shear Centre

To find the position of the shear centre E, the following principle applies to any point O in the section: the moment due to q will be equivalent to the resultant moment produced by shear forces acting at the shear centre. For example, taking moments about a point O on the x-axis in Fig. 7.12b, the resultant moment is

$$F_y\, e_x - F_x\, e_y = \int q\, R\, \mathrm{d}s \tag{7.23a}$$

where R is the perpendicular distance of q from O. The moment due to one of the forces in eq(7.23a) can be eliminated when O is chosen to lie along a force line. We shall see that this method is adopted to find the shear centre for a section that is symmetrical about the x-axis. For example, in Fig.7.12a we need only apply a single vertical force to find the position of E from O. Equation (7.23a) alone cannot locate the position of E for an asymmetric section. If one of the applied loads is removed (say F_x), eq(7.23a) is modified to

$$F_y\, e_x = \int q'\, R\, \mathrm{d}s \tag{7.23b}$$

where q' is the shear flow under F_y. Equations (7.23a,b) may then be solved for e_x and e_y (see Example 7.14). Figure 7.13 shows a convenient method of finding the shear centre for an asymmetric section when no applied loading is given. Two equations are required to define the co-ordinates of E(a,b) from point O and a further equation must describe the orientation of one of the assumed forces F_x* and F_y*.

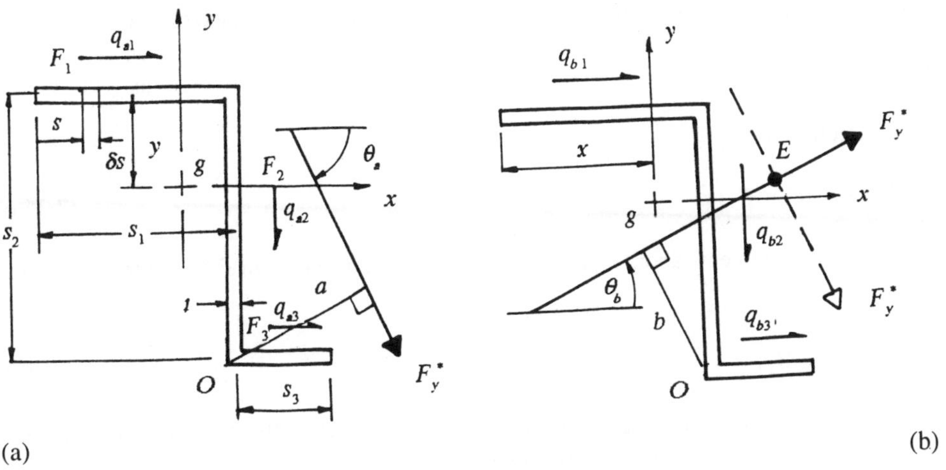

Figure 7.13 Shear Centre

In Fig. 7.13a we must find q_a due to F_y*, and in Fig.7.13b q_b due to F_x*. These are

$$q_a = F_y^* D_x/I_x \quad q_b = F_x^* D_y/I_y \tag{7.24a,b}$$

where F_x* is inclined so that it will bend the section only about the y-axis. F_y* is inclined to bend the section only about the x-axis. F_x* is perpendicular to F_y*. Referring to Fig. 7.13a, eq(7.24a) yields

$$q_{a1} = (F_y^* / I_x)\int \bar{y}\, t\, ds = (F_y^* \bar{y}\, t /I_x)s$$

$$q_{a2} = (F_y^* / I_x) \int (\bar{y} - s)\, t\, ds + (q_{a1})_{s=s_1}$$

$$= (F_y^* t / I_x)\, [s\,(\bar{y} - s/2) + \bar{y}\, s_1]$$

$$q_{a3} = (F_y^* / I_x) \int (s_2 - \bar{y})\, t\, ds + (q_{a2})_{s=s_2}$$

$$= (F_y^* t / I_x)\, [s\,(s_2 - \bar{y}) + s_2(\bar{y} - s_2/2) + \bar{y}\, s_1]$$

The forces F_1 and F_3 in the flanges and the force F_3 in the web are resultants of their shear flows.

$$F_1 = \int_0^s q_{a1}\, ds = (F_y^* \bar{y}\, t /I_x)\int_0^{s_1} s\, ds = [F_y^* \bar{y}\, t /(2I_x)]s_1^{\,2}$$

$$F_2 = \int_0^s q_{a2}\, ds = (F_y^* t /I_x) \int_0^{s_2}[\, s\,(\bar{y} - s/2) + \bar{y}\, s_1\,]\, ds$$

$$= (F_y^* t\, s_2 /I_x)[\,\bar{y}\, s_2/2 - s_2^{\,2}/6 + \bar{y}\, s_1\,]$$

$$F_3 = \int_0^s q_{a3}\, ds = (F_y^* t /I_x)(s_2 - \bar{y})\int_0^{s_3} s\, ds$$

$$+ (F_y^* t /I_x)[\, s_2(\bar{y} - s_2/2) + \bar{y}\, s_1\,]\, s_3$$

$$= (F_y^* t /I_x)\{(s_2 - \bar{y})\, s_3^{\,2}/2 + s_3[\, s_2(\bar{y} - s_2/2) + \bar{y}\, s_1\,]\}$$

Moreover, since $F_y{*}$ is the resultant of F_1, F_2 and F_3, it has a magnitude of

$$F_y{*} = [(F_1 + F_3)^2 + F_2^2]^{½}$$

and a direction of

$$\tan\theta_a = F_2 / (F_1 + F_3) \tag{7.25a}$$

Finally, taking moments about O, the perpendicular distance a from $F_y{*}$ to O follows from

$$F_y{*}\, a = F_1 s_2$$
$$a = s_2 t \bar{y} s_1^2/(2I_x) \tag{7.25b}$$

It is left as an exercise for the reader to confirm, from eq(7.24b), that the perpendicular distance b from $F_x{*}$ to O (Fig. 7.13b) is given by

$$b = (t s_2 s_1^2/(2I_y))(\bar{x} - s_1/3) \tag{7.25c}$$

Equations (7.25a,b,c) are sufficient to locate the shear centre, i.e. the intersection at E between $F_x{*}$ and $F_y{*}$. Note that eqs(7.25a-c) do not require the magnitudes $F_x{*}$ and $F_y{*}$ to be known. The essence of this solution to E(a,b) lies in the use of single forces whose inclinations bend the section only about x or y. An alternative method is to locate the centre from forces lying perpendicular to the principal axes, but the method outlined above is more convenient for sections with straight sides. When O is chosen at the intersection between two sides in Fig.7.13b, their shear flows do not contribute to the moments. All that is required is the position ($\bar{x}$, $\bar{y}$) of the centroid g and I_x and I_y.

Example 7.5 Derive an expression for the shear flow distribution along the mid-line of the the equal-angle section in Fig. 7.14a. Locate the position of the shear centre.

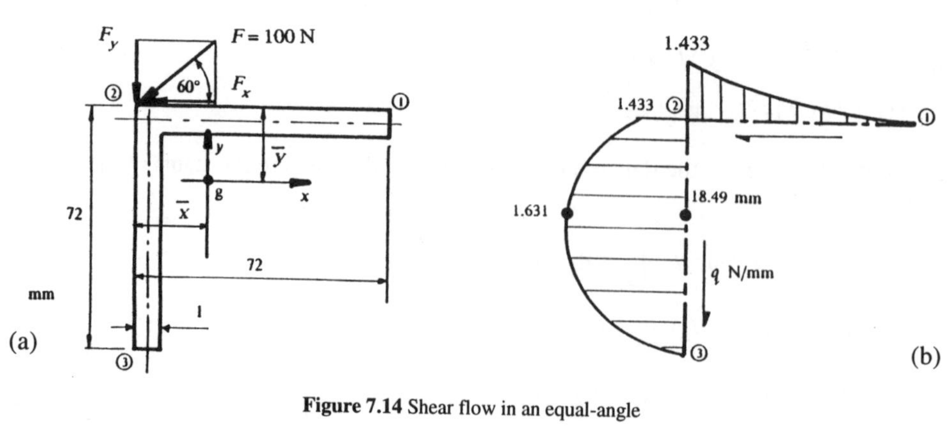

Figure 7.14 Shear flow in an equal-angle

The centroid position g and the inertias are found from

$$(71 + 72)\bar{y} = (72 \times 1 \times 1/2) + (71 \times 1 \times 36.5), \Rightarrow \bar{y} = \bar{x} = 18.37 \text{ mm}$$
$$I_x = I_y = (72 \times 17.87^2) + (71 \times 18.13^2) + (1 \times 71^3/12) = 76.16 \times 10^3 \text{ mm}^4$$
$$I_{xy} = (72 \times 17.63 \times 17.87) + [71(-18.13)(-17.87)] = 45.69 \times 10^3 \text{ mm}^4$$

The 10 N force is applied at the shear centre, giving

$F_x = 100 \cos 60° = +50$ N (in the x - direction)
$F_y = 100 \sin 60° = +86.6$ N (in the y - direction)

when, from eqs(7.22a,b),

$$F_x' = 50 - 86.6(45.69 / 76.16)/ [1 - (45.69 / 76.16)^2] = -3.056 \text{ N}$$
$$F_y' = 86.6 - 50(45.69 / 76.16)/ [1 - (45.69 / 76.16)^2] = +88.433 \text{ N}$$
$$F_x'/I_y = -0.0401 \times 10^{-3} \text{ N/mm}^4, \; F_y'/I_x = 1.161 \times 10^{-3} \text{ N/mm}^4$$

For the flange 1 → 2, $y = 17.87$ mm, $x = 53.56 - s$, $t = 1$ mm

$$q = (F_y/I_x)\int_0^s 17.87 \, ds + (F_x/I_y)\int_0^s (53.63 - s)\, ds$$
$$= (1.161 \times 10^{-3} \times 17.87)\, s - (0.401 \times 10^{-3})(53.63s - s^2/2)$$
$$q = (18.6s + 0.0201s^2)10^{-3}$$

The distribution is therefore parabolic, varying from zero to a maximum value $q_2 = +1.433$ N/mm for $s = 71.5$ mm .

For the web 2 → 3, $y = -(s - 17.87)$, $x = -17.87$ mm, $t = 1$ mm

$$q = (F_y/I_x)\int_0^s -(s - 17.87)\, ds + (F_x/I_y)\int_0^s -17.87 ds + q_2$$
$$= -(1.161 \times 10^{-3})(s^2/2 - 17.87s) + (0.0401 \times 10^{-3} \times 17.87)\, s + 1.433$$
$$q = (1433 + 21.464s - 0.5805s^2)10^{-3}$$

The web q distribution is again parabolic, varying from + 1.433 N/mm at 2 to a maximum of + 1.631 N/mm at $s = 18.49$ mm. At the free surface 3, where $s = 71.5$ mm, $q_3 \approx 0$. Figure 7.14b shows these shear flow variations. Obviously, point 2 is the shear centre since the moment due to F and to the two shear flows is equivalent to zero at this point.

7.4.3 Singly Symmetrical Sections

The analysis is simplified when a section is symmetric about either of the axes x or y. This means that x and y are principal axes and that both the centroid G and the shear centre E will lie along the axis of symmetry (x in Fig. 7.12a). In the case of a symmetrical section under a single shear force F_x (or F_y), one or other term is omitted from eq(7.20). For example, in Fig. 7.12a, if F_x is absent, the shear flow becomes

$$q = (F_y/I_x)D_x \tag{7.26a}$$

With x an axis of symmetry, the shear centre is found from

$$F_y\, e_x = \int q\, R\, ds \tag{7.26b}$$

and this applies irrespective of whether F_x is present or not. The sections considered in the following examples all possess x-axis symmetry.

Example 7.6 Determine the shear flow and the maximum shear stress for the section in Fig. 7.15a, when a downward shear force of 1 kN is applied parallel to the vertical side. Where must this force be applied if the section is not to be subjected to a simultaneous torque?

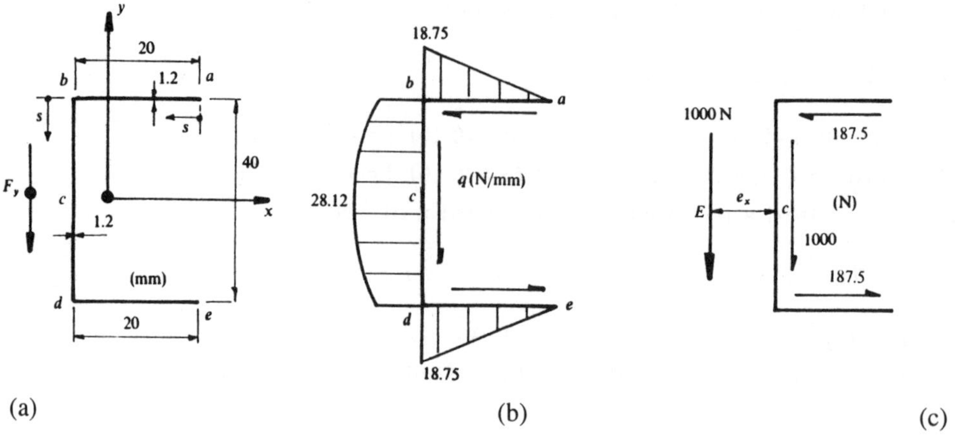

Figure 7.15 Shear flow in channel section

This is an example of a singly symmetric section with a single force $F_y = +\ 1000$ N applied parallel to a principal axis. Equation (7.26a) is applied for

$I_x = (1.2 \times 40^3/12) + 2(20 \times 1.2 \times 20^2) = 25600$ mm^4 and so $F_y/I_x = 3.906 \times 10^{-2}$ N/mm^4. For any point, distance s from a in the flange ab,

$$
\begin{aligned}
q_{ab} &= (F_y/I_x)D_x \\
&= 3.906 \times 10^{-2} \int y\,t\,ds \\
&= (3.906 \times 10^{-2})(20 \times 1.2) \int ds = 0.9374\, s \text{ N/mm}
\end{aligned}
$$

It follows that the variation in q_{ab} is linear, being zero at a and reaching a maximum value at b of $q_b = 0.9374 \times 20 = 18.75$ N/mm. The maximum shear stress in the top flange becomes $\tau_b = q_b/t = 18.75/1.2 = 15.63$ MPa.

Working from b to point c ensures that y remains positive upwards. Now, for any point, distance s from b in the web, $y = 20 - s$ and the shear flow is

$$
\begin{aligned}
q_{bc} &= (F_y/I_x)D_x + q_b \\
&= (3.906 \times 10^{-2}) \times 1.2 \int (20 - s)\,ds + 18.75 \\
&= 0.04687\,(20\,s - s^2/2) + 18.75
\end{aligned}
$$

The variation in q_{bc} is now parabolic with extreme values of $q_b = 18.75$ N/mm for $s = 0$ at b and $q_c = 9.374 + 18.75 = 28.14$ N/mm for $s = 20$ mm at c.

Beneath c, the shear flow mirrors that found above. This may be confirmed, as with the previous example, either by continuing from c in an a.c.w. direction or by working c.w. from the free surface at e. The resulting shear flow is that shown in Fig. 7.15b, from which it follows that the greatest shear stress in the section occurs at c with magnitude $\tau_c = q_c/t = 28.12 / 1.2 = 23.43$ MPa. The resultant applied shear forces along each side are then

$$F_{ab} = \int q_{ab}\,\mathrm{d}s = (F_y/I_x)\int_0^{20} 24s\,\mathrm{d}s$$

$$= 3.906\times 10^{-2}\times 24\,(20^2/2) = 187.5\ \mathrm{N} = F_{de}$$

$$F_{bc} = \int q_{bc}\,\mathrm{d}s = \int_0^{20}[0.04687\,(20s - s^2/2) + 18.75]\,\mathrm{d}s = 500\ \mathrm{N}$$

$\therefore\ F_y = 2F_{bc} = 1000$ N. The resultant forces are exerted by the cross-section in the same direction as q, shown in Fig. 7.15c. If the section is not to be subjected to torque, F_y must act at the shear centre, E, distance e_x from the web. This ensures that the applied torque at any point in the plane of the section is balanced by the moment effects of the resultant forces due to q at that point. For convenience, take moments about point c so that eq(7.235b) again applies,

$$1000\,e_x = (187.5\times 20) + (187.5\times 20)$$

$$\therefore\ e_x = 7.5\ \mathrm{mm}.$$

Example 7.7 Determine the shear flow distribution and the position of the shear centre for the thin-walled semi-circular section in Fig. 7.16a, when it is loaded through the shear centre by shear forces F_x and F_y lying parallel to the principal axes x and y as shown.

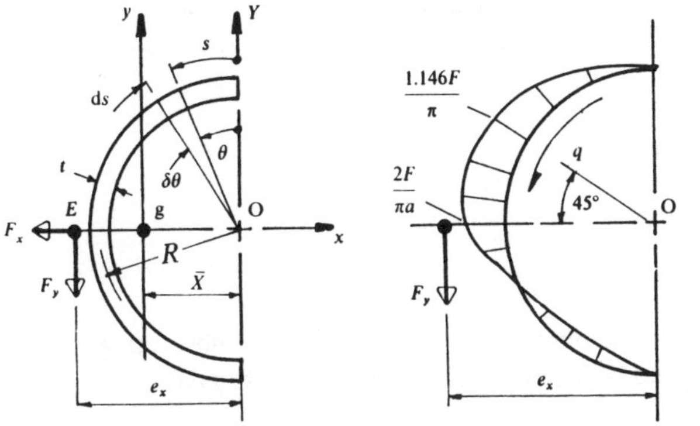

(a) (b)

Figure 7.16 Semi-circular section under shear

We must first find the position of the centroid g by taking the first moments of area about the Y- axis shown. Let s and δs subtend angles θ and $\delta\theta$ at O as shown in Fig. 7.16a. Using $A\overline{X} = \int x\,\mathrm{d}A$ and $\delta s = R\delta\theta$

$$(\pi R\,t)\overline{X} = \int_0^{\pi}(R\sin\theta)(R\,\mathrm{d}\theta\,t)$$

$$\overline{X} = (R/\pi)\int_0^{\pi}\sin\theta\,\mathrm{d}\theta = 2R/\pi$$

Both I_x and I_y are required in this example:

$$I_x = \int y^2\,\mathrm{d}A = 2\int_0^{\pi/2}(R\cos\theta)^2\,(R\,\mathrm{d}\theta\,t) = \pi R^3 t/2$$

$$I_Y = \int X^2\,\mathrm{d}A = \int_0^{\pi}(R\sin\theta)^2\,(R\,\mathrm{d}\theta\,t) = \pi R^3 t/2$$

and, from parallel axes

$$I_y = I_Y - A\overline{X}^2$$
$$= \pi R^3 t/2 - (\pi R\, t)(2R/\pi)^2 = R^3 t\,(\pi/2 - 4/\pi)$$

The first moments in eq(7.20) are

$$D_x = \int t\, y\, \mathrm{d}s = \int_0^\theta t\,(R\cos\theta)\,(R\,\mathrm{d}\theta) = \int_0^\theta R^2 t\cos\theta\,\mathrm{d}\theta = R^2\, t\sin\theta$$

$$D_y = \int tx\, \mathrm{d}s = \int_0^\theta t\,(2R/\pi - R\sin\theta)\,(R\,\mathrm{d}\theta)$$

$$= R^2 t\ \left| 2\theta/\pi + \cos\theta \right|_0^\theta = R^2 t\,(2\theta/\pi + \cos\theta - 1)$$

Both F_x and F_y are conventionally positive, acting in the negative x and y directions. Substituting into eq(7.20) results in the following expression for q in terms of θ:

$$q = [2F_y/(\pi R)]\sin\theta + 2\pi F_x\,(2\theta/\pi + \cos\theta - 1)/[R\,(\pi^2 - 8)]$$

which is positive for the direction of s shown. When $F_x = F_y = F$, the shear flow varies around the perimeter as shown in Fig. 7.16b. The shear flow distribution is statically equivalent to the two applied shear forces, provided that they act through the shear centre E. In this case E will lie along the x-axis, say distance e_x from O. By taking moments about O, it is only necessary to consider moment equilibrium between F_y and the component of shear flow due to F_y. This is

$$q = [2F_y/(\pi R)]\sin\theta$$

From eq(7.25b),

$$F_y\, e_x = \int q\, R\, \mathrm{d}s$$

where the integral is applied around the perimeter. Substituting $\delta s = R\delta\theta$, the integral is conveniently applied between limits of 0 and π rad, to give

$$F_y\, e_x = R^2 \int_0^\theta q\,\mathrm{d}\theta$$
$$= (2RF_y/\pi)\int_0^\pi \sin\theta\,\mathrm{d}\theta = -\,(2RF_y/\pi)\,|\cos\theta|_0^\theta = 4RF_y/\pi$$

$$\therefore\ e_x = 4R/\pi$$

Example 7.8 Determine the shear flow distribution for the 1.5 mm thick extrusion in Fig. 7.17 when a downward vertical force of 1 kN passes through its shear centre. At what distance from the vertical web does the shear centre lie?

Since x is a principal axis of symmetry and F_x is absent, the solution applies eq(7.26a) to each leg of the section, identified as 1 - 5 in Fig. 7.17a. Now as F_y = + 1000 N (acting in the negative y - direction) and

$$I_x = 2\,[(20\times 1.5)35^2 + (15\times 1.5)25^2] + (1.5\times 70^3/12) = 144.5\times 10^3\ \mathrm{mm}^4$$

Firstly establish the $D_x = \int ytds$ distribution, which is later converted to q with the multiplying factor $F_y/I_x = 6.92\times 10^{-3}$ N/mm^4. Then,

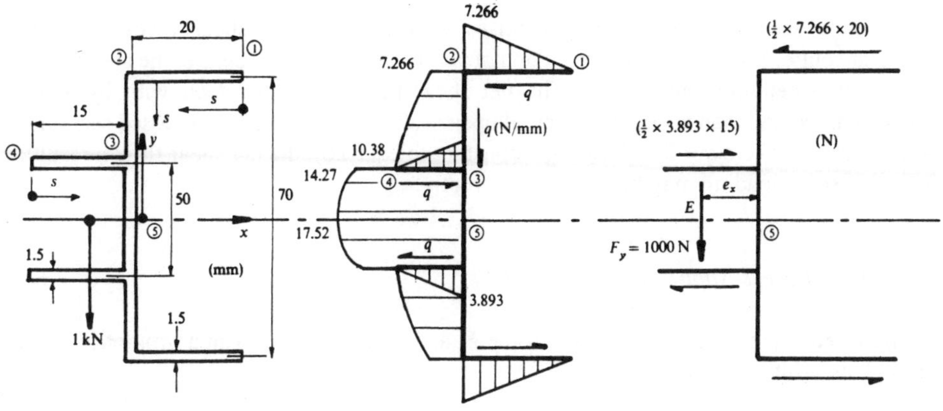

(a) (b) (c)

Figure 7.17 Shear flow distribution in an extruded section

For flange 1→2, $y = 35$ mm, $t = 1.5$ mm

$$D_x = (35 \times 1.5) \int ds = 52.5\, s$$

$$@\ 1,\ s = 0\ \therefore\ D_{x1} = 0$$

$$@\ 2,\ s = 20 \text{ mm}\ \therefore\ D_{x2} = (52.5 \times 20) = 1050 \text{ mm}^3$$

For web 2→3, $y = 35 - s$, $t = 1.5$ mm

$$D_x = 1.5 \int (35 - s)\, ds + D_{x2} = 1.5\,(35s - s^2/2) + 1050$$

$$@\ 3,\ s = 10 \text{ mm}\ \therefore\ D_{x3} = 1500 \text{ mm}^3$$

For flange 4→3, $y = 25$ mm, $t = 1.5$ mm

$$D_x = (25 \times 1.5) \int ds = 37.5s$$

$$@\ 4,\ s = 0\ \therefore\ D_{x4} = 0$$

$$@\ 3,\ s = 15 \text{ mm}\ \therefore\ D_{x3} = (37.5 \times 15) = 562.5 \text{ mm}^3$$

$$\sum D_{x3} = 1500 + 562.5 = 2062.5 \text{ mm}^3$$

For web 3→5, $y = 25 - s$, $t = 1.5$ mm

$$D_x = 1.5 \int (25 - s)\, ds + \sum D_{x3} = 1.5(25s - s^2/2) + 2062.5$$

$$@\ 5,\ s = 25 \text{ mm}\ \therefore\ D_{x5} = 2531.25 \text{ mm}^3$$

The corresponding q distribution is that shown in Fig. 7.17b. To find the shear centre, let F_y act at distance e_x from the web and take moments about point 5, in order to eliminate the torque produced by q in the web. The resultant coupling forces due to q in the horizontal limbs are given in Fig. 7.17c. From these eq(7.26b) becomes

$$1000\, e_x = (½ \times 7.266 \times 20 \times 70) - (½ \times 3.893 \times 15 \times 50) \ \Rightarrow\ e_x = 3.626 \text{ mm}.$$

7.5 Shear Flow in Thin-Walled Closed Sections

Pure flexural shear flow in a closed tube arises when the transverse shear forces act through the shear centre. When the shear forces do not act through the shear centre, the shear flow is due to both bending and torsion. The net shear flow is usually dealt with by a more convenient method than is employed for an open section (see Fig. 7.10). A closed section has no free surface where $q = 0$. By adding a constant to eq(7.20), the net shear flow q resulting from any shear loading may be calculated directly.

7.5.1 Shear Forces Applied at any Point P

In Fig. 7.18a two components of shear force F_x and F_y are applied to an asymmetric section through an arbitrary point P.

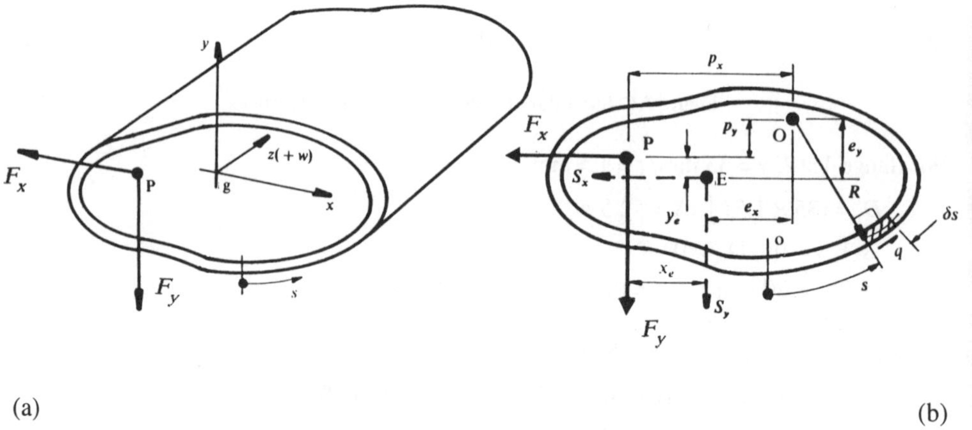

Figure 7.18 Closed tube under shear

Equation (7.20) becomes

$$q = q_b + q_o \tag{7.27a}$$

where

$$q_b = (F_x'/I_y)\, D_{yb} + (F_y'/I_x)\, D_{xb} \tag{7.27b}$$

in which F_x' and F_y' are equivalent forces referred to axes x and y when the latter are not principal axes. They can be defined in terms of the applied forces F_x and F_y and the second moments of area I_x, I_y and I_{xy} (see eqs(7.22a,b)). We shall work with the applied forces for the following analysis. The reduction to a section with x - symmetry follows from putting $F_x = F_x'$ and $F_y = F_y'$ in eq(7.27b). Now q_o is equal to a constant value q at the origin for s. Both the position of this origin and point P will govern the magnitude of q_o. With a fixed origin for s, the torsional effects are accounted for solely in the effect that the position of P has on q_o. Now q_b is actually the pure flexural shear flow for the equivalent tube when the wall is split at $s = 0$. This must correspond to a shift in the position of the applied forces in order to become concurrent with the shear centre for the opened section. In order to restore the original loading, a torque must also act when forces are translated to the shear centre of the opened section. It follows that q_o is effectively the constant shear flow within a closed tube produced by this torque. The reasoning is similar to that in Fig. 7.10. Different opening positions are associated with particular q_o values because of the resulting change in position of the shear centre. The method amounts to converting a structure with a single degree of redundancy into a statically determinate one. A similar approach may be employed to determine the shear flow distribution in a closed, multi-cell tube (section 5.5.2). A yield criterion provides the necessary check that the shear stresses in the wall remain elastic (see Chapter 11).

Initially, it is convenient to assume that $q_o = 0$ for $s = 0$, when finding q_b from eq(7.27a). The value of q_o, for any point O in the plane of the cross-section, can be found later from an equilibrium condition that moments produced by the net q are statically equivalent to the applied shear forces. Thus, for the directions given in Fig. 7.18b,

$$F_y\, p_x - F_x\, p_y = \oint q\, R\, \mathrm{d}s$$

Substituting from eq(7.27a,b),

$$F_y\, p_x - F_x\, p_y = \oint q_b R\, \mathrm{d}s + q_o \oint R\, \mathrm{d}s \qquad (7.28)$$

where $\oint R\, \mathrm{d}s$ is twice the area enclosed by the mid-line of the wall section. Note that when point O coincides with P, the LH side of eq(7.28) becomes zero.

7.5.2 Shear Forces Applied at the Shear Centre

Equation (7.28) cannot supply q_o directly when the shear forces are applied at the shear centre E in Fig. 7.18b when the position of E is unknown. This requires a compatibility condition to be satisfied, i.e. that the rate of twist will be zero. A derivation similar to eq(5.70b) applies but where q now varies with s. In general, the twist rate $\delta\theta / \delta z$ due to the net shear flow q is

$$\mathrm{d}\theta / \mathrm{d}z = [1/(2AG)] \oint q\, \mathrm{d}s/t \qquad (7.29)$$

For pure flexural shear, q_e becomes the particular q_o value in eq(7.27a). Substituting into eq(7.29) and equating to zero gives

$$\oint q_b\, \mathrm{d}s/t + q_e \oint \mathrm{d}s/t = 0 \qquad (7.30)$$

Once q_e is found from eq(7.30), the position of the shear centre E, follows from taking moments about any convenient point O. That is, with the corresponding co-ordinates e_x and e_y in Fig. 7.18b,

$$F_y\, e_x - F_x\, e_y = \oint q_b\, R\, \mathrm{d}s + q_e \oint R\, \mathrm{d}s \qquad (7.31)$$

When E is known to lie on a horizontal axis of symmetry, and O is also chosen to lie along this axis, $e_y = 0$ and e_x may be determined from F_y.

7.5.3 Warping

Unrestrained warping accompanies twisting of the section, when shear forces are applied away from the shear centre. The derivation of these displacements is similar to that leading to eq(5.67). To determine the contribution to warping from shear flow we replace the Bredt-Batho shear flow $q = T/(2A)$ in this equation, with the net shear flow distribution $q = q(s)$. This gives the relative warping displacement:

$$\Delta w = \int_0^s q\, \mathrm{d}s/(Gt) - (A_{os}/A) \oint q\, \mathrm{d}s/(Gt) \tag{7.32}$$

where A_{os} is the area enclosed between the centre of twist (coincident with the shear centre), a peripheral origin o, that may itself warp and a point on the periphery, distance s from o (see Fig. 7.18b). If, however, there is no warping for $s = 0$, usually at a point of symmetry, then eq(7.32) will supply absolute w values. Example 7.13 shows how to apply eq(7.32) to a closed idealised tube consisting of webs and booms. This shows that the calculation of warping in an idealised tube is simplified as the shear flow is constant within webs connecting booms.

When warping displacements are prevented at an encastre support, a system of self-equilibrating forces is induced which modifies the shear flow distribution. We saw in the Wagner-Kappus theory (Section 5.6) how axial stress arose from the supression of warping in an open tube under torsion. Axial restraint in asymmetric tubes under shear requires a more complex theory. Although simplifications can be made for doubly symmetric idealised tubes, these will not be covered here.

Example 7.9 The thin-walled tubular section $r = 50$ mm, $t = 1$ mm in Fig. 7.19a is subjected to a vertical shear force of $F_y = 500$ N as shown. Determine the shear flow distribution, the maximum shear stress, the rate of twist and the free warping distribution. Take $G = 80$ GPa.

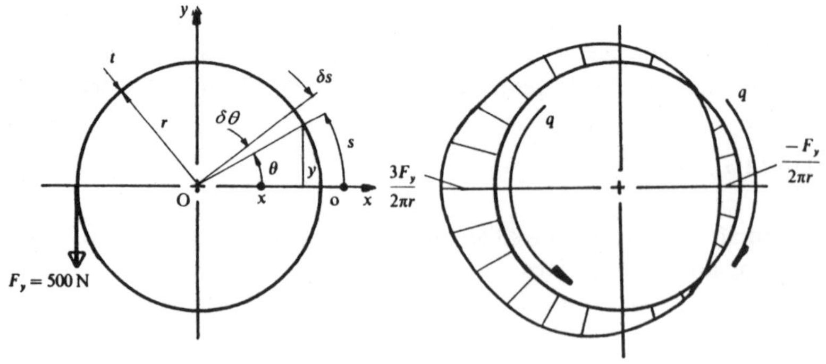

(a) (b)

Figure 7.19 Thin-walled tubular section

The section is symmetric with F_y positive and $F_x = 0$. From eq(7.20), the net shear flow is

$$q = (F_y/I_x)D_{xb} + q_o$$

Now, as $y = r\sin\theta$ and $\mathrm{d}s = r\,\mathrm{d}\theta$ in Fig. 7.19a, then

$$I_x = \int y^2\, t\, ds = 2\, r^3 t \int_0^{\pi} \sin^2\theta\, d\theta = \pi r^3 t$$

$$D_{xb} = \int y\, t\, ds = r^2 t \int_0^{\theta} \sin\theta\, d\theta = r^2\, t\,(1 - \cos\theta)$$

$$\therefore q_b = (F_y/I_x)\, D_{xb} = [F_y/(\pi r)](1 - \cos\theta)$$

From eq(7.28), with the moment centre O at the centre of the circle,

$$F_y\, r = (F_y\, r/\pi) \oint (1 - \cos\theta) d\theta + 2(\pi r^2)\, q_o = 2F_y\, r + 2\pi r^2 q_o$$

$$\therefore q_o = -\, F_y/(2\pi r)$$

Hence the net shear flow, $q = q_o + q_b$, becomes

$$q = [F_y/(\pi r](1 - \cos\theta) - F_y/(2\pi r) = [F_y/(2\pi r)](1 - 2\cos\theta)$$

This has the distribution shown in Fig. 7.19b for the given r, t and F_y values. The greatest q occurs at $\theta = 3\pi/2$ rad, giving

$$\tau_{\max} = q\,/\,t = 3F_y/(2\pi r\, t) = (3 \times 500)/(2\pi \times 500 \times 1) = 4.775 \text{ MPa}$$

Here the centroid O is clearly the shear centre. The section will therefore twist under F_y at the rate supplied by eq(7.29),

$$d\theta/dz = [1/(2\pi r^2 G)][F_y/(2\pi r\, t)] \oint (1 - 2\cos\theta)\, r\, d\theta = F_y/(2\pi r^2 G\, t)$$

$$= 500/(2\pi \times 50^2 \times 80000 \times 1) = 0.3978 \times 10^{-6} \text{ rad/mm}$$

Applying eq(7.32) to find the warping distribution with the given origin for s in Fig.7.19a,

$$\int q\, ds = [F_y/(2\pi r)] \int_0^{\theta} (1 - 2\cos\theta)\, r\, d\theta = [F_y/(2\pi)](\theta - 2\sin\theta)$$

$$\oint q\, ds = [F_y/(2\pi)]\, |\theta - 2\sin\theta|_0^{2\pi} = F_y$$

$$A_{os} = ½\int r\, ds = (r^2/2) \int_0^{\theta} d\theta = r^2\theta/2$$

$$\therefore Gt\Delta w = (F_y/(2\pi)(\theta - 2\sin\theta) - [\, r^2\theta/(2\pi r^2)]F_y$$

$$\Delta w = -\, [F_y/(\pi Gt)]\, \sin\theta$$

which shows that all points on the periphery, except those on the horizontal diameter, displace axially, relative to + z (i.e. into the page within Fig. 7.18a) .

Example 7.10 Find the pure flexural shear flow distribution and the shear centre for the thin-walled, semi-circular, closed tube of uniform thickness t and radius a, in Fig. 7.20a.

I_x is found from the sum of the I's for a rectangle and a semi-circle:

$$I_x = t\,(2a)^3/12 + \int y^2\, dA$$

and substituting $y = a\cos\theta$, $\delta A = t\delta s = t\,(a\delta\theta)$

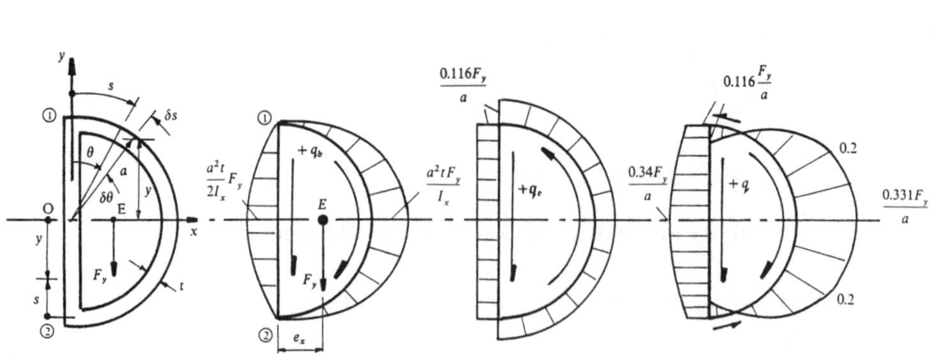

Figure 7.20 Thin-walled closed tube

$$I_x = t\,(2a)^3/12 + 2a^3 t\int_0^{\pi/2}\cos^2\theta\,\mathrm{d}\theta$$

$$I_x = \pi a^3 t/2 + t\,(2a)^3/12 = a^3 t\,(\pi/2 + 2/3)$$

With F_y applied through the shear centre, E, eq(7.27a) is written as

$$q = q_b + q_e \qquad \text{(i)}$$

Since, for this singly symmetric section $F_x = 0$ and x is a principal axis,

$$q_b = (F_y/I_x)\,D_{xb} \qquad \text{(ii)}$$

With the origin at 1 and s measured clockwise from 1 → 2 in the semi-circular portion,

$$D_{xb} = \int_0^s y\,t\,\mathrm{d}s = \int_0^\theta (a\cos\theta)\,t\,(a\,\mathrm{d}\theta) = a^2 t\int_0^\theta \cos\theta\,\mathrm{d}\theta = a^2 t\sin\theta$$

@ 1 $\theta = 0$, $D_{xb1} = 0$
@ 2 $\theta = \pi$ rad, $D_{xb2} = 0$
@ $\theta = \pi/2$ rad, $D_{xb} = a^2 t$, a maximum

and for 2 → 1 in the vertical web,

$$D_{xb} = \int_0^s yt\,\mathrm{d}s + D_{xb2} = \int_0^s (s - a)\,t\,\mathrm{d}s + 0 = t\,|\,s^2/2 - as\,\Big|_0^s = t\,s\,(s/2 - a)$$

@ 2 $s = 0$, $D_{xb2} = 0$
@ 1 $s = 2a$, $D_{xb1} = 0$
@ $s = a$, $D_{xb} = -\,a^2 t/2$

Correspondingly, the q_b shear flow distribution in Fig. 7.20b is found from eq(ii),

$q_b = F_y \sin\theta/[a(\pi/2 + 2/3)]$ for 1 → 2
$q_b = [F_y s\,(s/2 - a)]/[a^3(\pi/2 + 2/3)]$ for 2 → 1

Because the latter gives negative q_b for $0 < s < 2a$, this q_b direction is reversed, as shown in Fig. 7.20b. Substituting for q_b in eq(7.30), to find q_e

$$\{F_y / [at(\pi/2 + 2/3)]\} \int_0^{\pi} \sin\theta (a\,\mathrm{d}\theta) + \{F_y / [a^3 t(\pi/2 + 2/3)]\} \int_0^{2a} (s^2/2 - as)\,\mathrm{d}s$$

$$+ q_e [\int_0^{\pi} (a\,\mathrm{d}\theta)/t + \int_0^{2a} \mathrm{d}s/t] = 0$$

$$2F_y / [t(\pi/2 + 2/3)] - 2F_y / [3t(\pi/2 + 2/3)] + q_e a(\pi + 2)/t = 0$$

$$\therefore q_e = - F_y / [a(1 + 3\pi/4)(1 + \pi/2)]$$

which represents the constant shear flow around the section shown in Fig. 7.20c. The net shear flow for the semi-circular portion 1 → 2 in Fig. 7.20d is found from eq(i),

$$q = F_y \sin\theta / [a(\pi/2 + 2/3)] - F_y / [a(1 + 3\pi/4)(1 + \pi/2)]$$

$$= (F_y/a)(0.447 \sin\theta - 0.116)$$

For the web,

$$q = F_y\, s(s/2 - a) / [a^3(\pi/2 + 2/3)] - F_y / [a(1 + 3\pi/4)(1 + \pi/2)]$$

$$= (F_y/a)[0.447(s/a^2)(s/2 - a) - 0.116]$$

Provided the direction chosen for s remains the same for the determination of q_b and q_e then q is given by their sum. Now apply eq(7.31) to find the shear centre. Taking moments about the web centre O in Fig. 7.20d eliminates F_x (when present) and q in the web, to give

$$F_y\, e_x = \int_0^{s} qa\,\mathrm{d}s = a^2 \int_0^{\pi} q\,\mathrm{d}\theta$$

$$= a^2 \{F_y / [a(\pi/2 + 2/3)]\} \int_0^{\pi} \sin\theta\,\mathrm{d}\theta - a^2 \{F_y / [a(1 + 3\pi/4)(1 + \pi/2)]\} \int_0^{\pi} \mathrm{d}\theta$$

$$F_y\, e_x = 2aF_y/(\pi/2 + 2/3) - \pi a F_y/(1 + \pi/2)(1 + 3\pi/4)$$

$$\therefore e_x = a(3 + \pi/2)/[(1 + \pi/2)(1 + 3\pi/4)] = 0.53a$$

Note that (i) it is not necessary to calculate the x-position of the centroid, (ii) the magnitudes of shear forces are not required to find the position of the shear centre and (iii) only the net shear flow for the web is needed to establish e_x.

Example 7.11 The wing box structure in Fig. 7.21a is loaded in shear along its left vertical web as shown. Calculate the distribution of shear flow, the rate of twist and the position of the shear centre from this web. Take $G = 27$ GPa.

Here x is a principal axis of symmetry, so that

$$q_b = (F_y / I_x) D_{xb}$$

where $F_y = -20000$ N. Refer to Example I.6 in Appendix I for the calculation of I_x. With $\sin\alpha = 110/610$, the contributions to I_x from the web and sloping sides appear in the summation

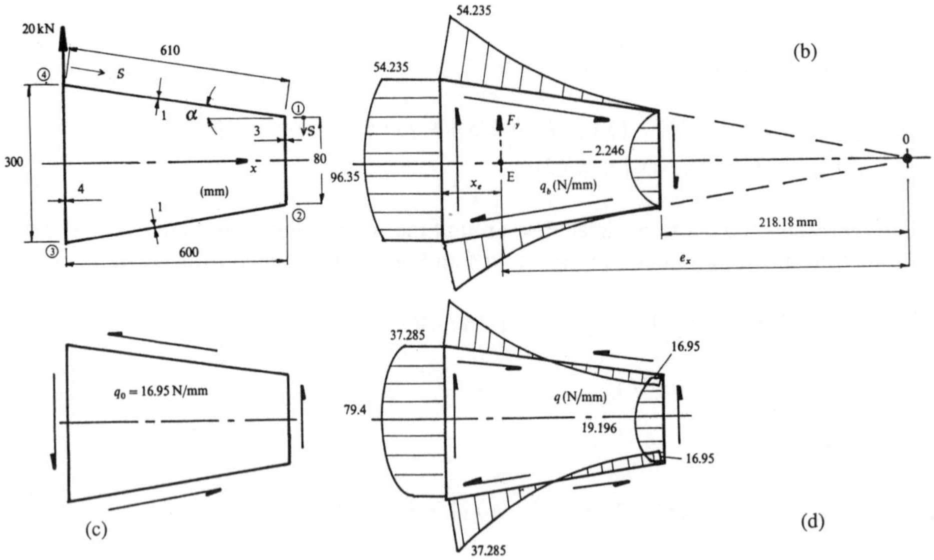

Figure 7.21 Shear flow in a wing box structure

$I_x = 4\,(300)^3/12 + 3\,(80)^3/12$ (for web)

$\quad + (1 \times 610)[80^2/2 + (80 \times 610)(110/610) + (2 \times 610^2/3)(110/610)^2]$ (for sides)

$\quad = (9 + 0.128 + 12.24)10^6 = 21.37 \times 10^6\ \text{mm}^4$

$1 \to 2,\ y = 40 - s,\ t = 3$ mm

$D_{xb} = \int y\,t\,\mathrm{d}s = 3\int_0^s (40 - s)\,\mathrm{d}s = 3(40s - s^2/2)$

@1 $s = 0,\ D_{xb1} = 0$

@2 $s = 80$ mm, $D_{xb2} = 0$

@ $s = 40$ mm, $D_{xb} = 2400\ \text{mm}^3$ (a maximum)

$2 \to 3,\ y = -\,(40 + 110s/610),\ t = 1$ mm

$D_{xb} = -\int_0^s (40 + 110s/610)\,\mathrm{d}s + D_{xb2} = -\,(40s + 11s^2/122) + 0$

@2 $s = 0,\ D_{xb2} = 0$

@3 $s = 610$ mm, $D_{xb3} = -\,57950\ \text{mm}^3$ (a maximum)

$3 \to 4,\ y = s - 150,\ t = 4$ mm

$D_{xb} = 4\int_0^s (s - 150)\,\mathrm{d}s + D_{xb3} = 2s^2 - 600s - 57950$

@3 $s = 0,\ D_{xb3} = -\,57950\ \text{mm}^3$

@4 $s = 300$ mm, $D_{xb4} = -\,57950\ \text{mm}^3$

@ $s = 150$ mm, $D_{xb} = -\,102950\ \text{mm}^3$ (a maximum)

$4 \to 1$, $y = 150 - 110s/610$, $t = 1$ mm

$D_{xb} = \int_0^s (150 - 110s/610)\, ds + D_{xb4} = 150s - 11s^2/122 - 57950$

@4, $s = 0$, $D_{xb4} = -57950$ (a maximum)

@1, $s = 610$ mm, $D_{xb} = 0$

The q_b distribution in Fig. 7.21b is found by multiplying D_{xb} by the constant $F_y/I_x = -9.359 \times 10^{-4}$ N/mm^4. The given directions correspond to positive q_b. In the application of eq(7.28) to find q_o, it is convenient to take the moment centre O where the cover lines intersect so that no moments are exerted by them. For the webs, eq(7.28) becomes

$$F_y\, p_x = \oint q_b R\, ds + 2Aq_o \qquad \text{(i)}$$

$$20\times10^3(600+218.18) = -9.359 \times 10^{-4} \int_0^{300} (2s^2 - 600s - 57950)(818.18)ds \text{ (c.w. for } 3\to4)$$
$$- [-9.359 \times 10^{-4} \int_0^{80} 3(40s - s^2/2)(218.18)\, ds] \text{ (a.c.w. for } q_b\ 1\to 2)$$
$$+ (300 + 80)600q_o \ (q_o \text{ assumed c.w.}).$$

$$1636.36\times10^4 = -0.7657\, |2s^3/3 - 300s^2 - 57950s\,|_0^{300} + 0.6126|20s^2 - s^3/6|_0^{80} + (22.8 \times 10^4)q_o$$
$$= (2020.3 \times 10^4) + (2.6137 \times 10^4) + (22.8 \times 10^4)\, q_o$$

$\therefore q_o = -16.95$ N/mm (i.e. a.c.w. in Fig. 7.21c).

Note that particular care must be taken with signs. Remember that the q_b expressions were derived for a given s direction. This direction controls the sense of the moments $\oint q_b R$ ds in eq(i) before numerical values are substituted. These may be equated to $F_y p_x$ in this way, provided it is recognised that this product must be the the resultant of the moments due to the net $q = q_b + q_o$ distribution. Alternatively, with the given q_b directions in Fig. 7.21b, the moments about O may be found by multiplying the net web forces by the perpendicular distance. Since

Force in 12 is $2 \times 80 \times 2.246/3 = 119.79$ N

Force in 34 is $[54.235 + 2(96.35 - 54.235)/3]300 = 24693.93$ N

$\therefore (20 \times 10^3 \times 818.18) = (24693.93 \times 818.18) + (119.79 \times 218.18) + (22.8 \times 10^4)q_o$

Hence, $q_o = -16.95$ N/mm and the net shear flow $q = q_b + q_o$ is given in Fig. 7.21d. Applying eq(7.29) for the rate of twist,

$$d\theta/dz = [1/(2AG)] \oint q\, ds/t = [1/(2AG)][\oint q_b\, ds/t + q_o \oint ds/t\,] \qquad \text{(ii)}$$

where

$A = (300 + 80)600/2 = 114 \times 10^3$ mm^2

$q_o \oint ds/t = -16.95[80/3 + 610/1 + 300/4 + 610/1] = -16.95 \times 1321.67 = -22.403 \times 10^3$ N/mm

$$\oint q_b\, ds/t = (F_y/I_x)\, [\int_0^{80} 3(40s - s^2/2)(ds/3) + \int_0^{610} -(40s + 11s^2/122)(ds/1)$$
$$+ \int_0^{300} (2s^2 - 600s - 57950)(ds/4) + \int_0^{610} (150s - 11s^2/122 - 57950)(ds/1)]$$
$$= -(9.359 \times 10^{-4})[\,|20s^2 - s^3/6|_0^{80} - |20s^2 + 11s^3/366|_0^{610}$$

$$+ \tfrac{1}{4} \mid 2s^3/3 - 300s^2 - 57950 \Big|_0^{300} + \mid 150s^2/2 - 11s^3/366 - 57950s \Big|_0^{610} \;]$$

$$= - (9.359 \times 10^{-4})[4.2666 - 1426.3833 - 659.6248 - 1426.3833\,]10^4$$
$$= - (9.359 \times 10^{-4})[- 3508.1248 \times 10^4] = 32.833 \times 10^3 \text{ N/mm}$$

Substituting into eq(ii),

$$\mathrm{d}\theta/\mathrm{d}z = (- 22.403 + 32.833)10^3/(2 \times 114 \times 10^3 \times 27 \times 10^3)$$
$$= 1.6943 \times 10^{-6} \text{ rad/mm} = 0.0971 \text{ °/m}$$

To find the shear centre E, let F_y acting through this point. Equation (7.30) is applied in which q_e accounts for the transference of F_y to E but q_b and the line integrals remain the same. Thus, from the previous calculations eq(7.30) gives

$$\oint q_b \, \mathrm{d}s/t + q_e \oint \mathrm{d}s/t = 0$$
$$(32.833 \times 10^3) + 1321.67 q_e = 0$$
$$\therefore q_e = - 24.842 \text{ N/mm}$$

Taking moments about the same cover intersection point O in Fig. 7.21b, distance e_x from E,

$$(20\, e_x)10^3 = (119.79 \times 218.18) + (24693.93 \times 818.18) + [2 \times 114(- 24.842)10^3]$$

from which $e_x = 728.09$ mm. That is, the shear centre lies on the x - axis 90 mm to the right of the longer vertical web. Alternatively, by Maxwell's reciprocal law, the position of the shear centre may be found from the pure torque T required to produce the same rate of twist as under F_y. From eq(5.70a),

$$[T / (4A^2 G)] \oint \mathrm{d}s/t = \delta\theta/\delta z = 1.6973 \times 10^{-6}$$
$$T = [1.6943 \times 10^{-6} \times 4 \times (114 \times 10^3)^2 \times 27 \times 10^3]/1321.67 = 1.80 \times 10^6 \text{ Nmm}$$

This is counterbalanced to give zero twist rate when F_y is translated distance x_e to the shear centre. That is,

$$F_y \, x_e = 1.80 \times 10^6$$
$$\therefore x_e = (1.80 \times 10^6)/(20 \times 10^3) = 90 \text{ mm, as before.}$$

7.6 Web-Boom Idealisation for Symmetrical Sections

It was seen in Example 7.3 how little the flanges contributed to the vertical shear force in I-section beams. Also, the vertical shear stress distribution (see Fig. 7.6b) does not vary greatly over the depth of the web. This leads to the idealisations given in Figs 7.22a-d for thin web sections.

The areas above and below the neutral axis in Fig. 7.22a, which resist the bending moment, may be considered to be concentrated as "booms" separated by the depth d of the shear web, as shown in Fig. 7.22b. This is a particularly useful approximation to make for the analysis of the effects of longitudinal stiffeners and stringers on the shear flow in thin-walled aircraft structures.

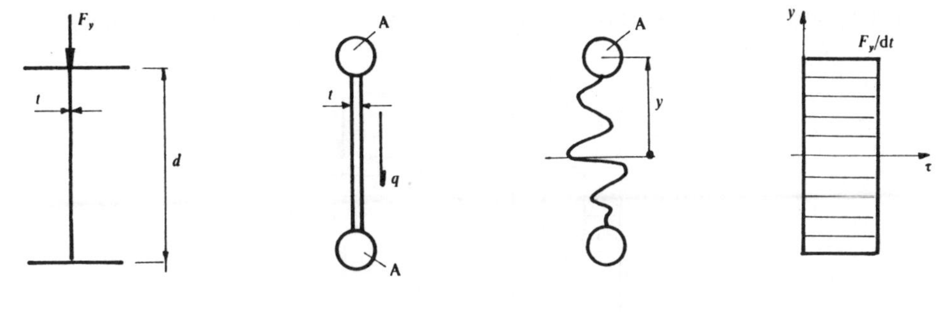

Figure 7.22 Web-boom idealisation for an I-beam

7.6.1 Open Section

Consider an idealised, open structure in which a single shear force F_y is applied parallel to the principal y - axis, where x is an axis of symmetry. Equation (7.20) gives the web or skin shear flow as

$$q = (F_y/I_x)D_x \tag{7.33}$$

where, with the contribution to D_x from both the web and boom areas A_r (r = 1, 2, 3 ... n), the first moment of area becomes

$$D_x = \int_0^s y\,t\,\mathrm{d}s + \sum_{r=1}^{n} A_r y_r$$

The total second moment of area is given by

$$I_x = I_x \text{ (for web or skin)} + \sum_{r=1}^{n} A_r y_r^2 \text{ (for booms)}$$

The greater change in shear flow will be caused by the booms when

$$\sum_{r=1}^{n} A_r y_r > \int_0^s y\,t\ \mathrm{d}s$$

which is normally the case for thin-walled sections under any flexural shear loading. In the simplest of idealisations, $\int y\,t\,\mathrm{d}s$ in eq(7.33) is removed by lumping the web or skin area into that of the booms. Equation (7.33) becomes

$$q = (F_y/I_x)\sum_{r=1}^{n}(A_r y_r) \tag{7.34}$$

This gives a constant shear flow over any shape of web connecting two adjacent boom areas. In Fig. 7.22c, for example, with $Ay = Ad/2$, $I = 2A(d/2)^2$ for an open section, the web shear flow becomes

$$q = F_y(Ad/2)/[2A\,(d/2)^2] = F_y/d$$

acting in a direction parallel to its mid-line. Here $\tau = q/t = F_y/(dt)$ is the average shear stress for the web area (see Fig. 7.22d).

In the application of eq(7.34) to Fig. 7.23a, $\sum(A_r\, y_r)$ is the sum of the first moments about the n.a. of only those boom areas lying above the connecting web being considered.

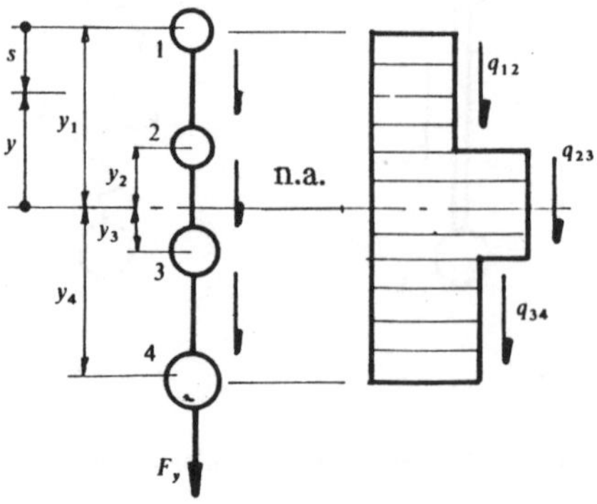

(a) (b)

Figure 7.23 Web-boom idealisation showing constant q between booms

Thus, with $I = A_1 y_1^2 + A_2 y_2^2 + A_3 y_3^2 + A_4 y_4^2$ in Fig. 7.23a

$$q_{12} = (F/I)A_1 y_1$$
$$q_{23} = (F/I)(A_1 y_1 + A_2 y_2) = (F/I)(-A_4 y_4 - A_3 y_3)$$
$$q_{34} = (F/I)(-A_4 y_4) = (F/I)(A_1 y_1 + A_2 y_2 - A_3 y_3)$$

If the section supports a second shear force F_x at right angles to F_y, where x and y are principal axes, the effect on the web shear flow is additive. Thus with F_x and F_y acting in the negative x and y directions, the summation of equations, similar to eq(7.34), gives

$$q = (F_y/I_x)\sum_{r=1}^{n}(A_r\, y_r) + (F_x/I_y)\sum_{r=1}^{n}(A_r x_r) \tag{7.35}$$

where $\sum(A\, y)$ and $\sum(A\, x)$ are sums of the first moments of those boom areas lying "above or below" the web being considered, about the principal centroidal axes. In Fig. 7.24, the boom areas are balanced above and below the x-axis. Equation(7.35) will supply the following mean line web shear flows when there is no twisting of the section:

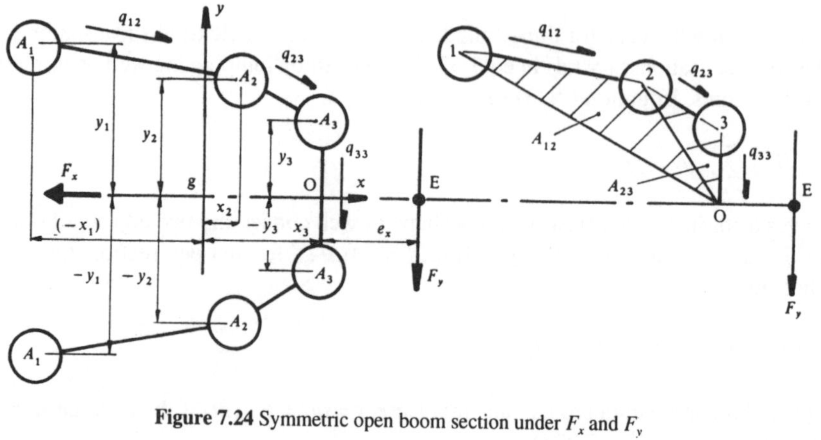

(a) (b)

Figure 7.24 Symmetric open boom section under F_x and F_y

Now from eq(7.35),

$$q_{12} = (F_y/I_x)(A_1y_1) + (F_x/I_y)(-A_1x_1)$$
$$= (F_y/I_x)(-A_1y_1) + (F_x/I_y)[2(A_2x_2) + 2(A_3x_3) - (A_1x_1)]$$
$$q_{23} = (F_y/I_x)[(A_1y_1) + (A_2y_2)] + (F_x/I_y)[-(A_1x_1) + (A_2x_2)]$$
$$= (F_y/I_x)[-(A_1y_1) - (A_2y_2)] + (F_x/I_y)[2(A_3x_3) + (A_2x_2) - (A_1x_1)]$$
$$q_{33} = (F_y/I_x)[(A_1y_1) + (A_2y_2) + (A_3y_3)] + (F_x/I_y)[-(A_1x_1) + (A_2x_2) + (A_3x_3)]$$

where $I_x = 2(A_1y_1^2 + A_2y_2^2 + A_3y_3^2)$, $I_y = 2(A_1x_1^2 + A_2x_2^2 + A_3x_3^2)$

In this case let the shear centre E lie along the x-axis distance e_x from the vertical web as shown in Fig. 7.24b. With the moment centre O, chosen to lie at the centre of the vertical web, neither F_x nor the shear flow in the web will exert a moment. Equation (7.23b) becomes

$$F_y\, e_x = \int qR\, ds = 2\sum (qA) \tag{7.36a}$$

The summation follows from q being constant along each web for which $\int R ds$ is twice the area enclosed by the web and O, as shown for q_{12} and q_{23} in Fig. 7.24b. Thus, for the relevant areas above and below the neutral axis in Fig. 7.24b,

$$\sum (qA) = 2\,(q_{12}A_{12} + q_{23}A_{23}) \tag{7.36b}$$

7.6.2 Closed Tube

When a singly-symmetric closed tube is subjected to shear forces F_x and F_y, the shear flow follows from eqs 7.27a,b.

$$q = (F_y/I_x)\sum_{r=1}^{n}(A_ry_r) + (F_x/I_y)\sum_{r=1}^{n}(A_rx_r) + q_o \tag{7.37a}$$

where the determination of q_o has been discussed in section 7.5.1. Figure 7.25 shows a particular loading where a single force F_y is applied at the shear centre, E, and x is an axis of symmetry (principal axis). Equation (7.37a) then reduces to

$$q = (F_x/I_y)\sum_{r=1}^{n}(A_rx_r) + q_e \tag{7.37b}$$

where the summation term accounts for the boom areas passed by the peripheral co-ordinate s, when taken anti-clockwise (say) from its origin at boom 1.

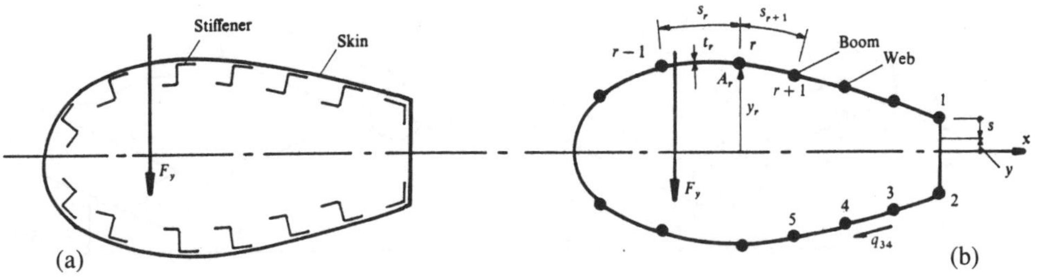

Figure 7.25 Idealisation of a single-cell closed tube

For example, in the determination of q_{34}

$$\sum_{r=1}^{3} A_r\, y_r = A_1\, y_1 - A_2\, y_2 - A_3\, y_3 \tag{7.38a}$$

The position of the shear centre for this loading is supplied by eq(7.31), where similar summations replace the path integrals,

$$F_y\, e_x = 2\sum\, (q_b A) + 2Aq_e \tag{7.38b}$$

Example 8.22 considers shear flow in a symmetrical multi-cell tube within the general consideration of the application of the principal of virtual work to redundant structures.

7.6.3 Calculation of Boom Areas

A structure under flexural shear will also carry direct stress due to bending. In the idealised structure the boom areas carry this direct stress. Their areas reflect the combined effects that the skin or web, the flanges and longitudinal stiffeners (see Figs 7.25 and 7.29) have on resisting bending. The following simplifications are made in idealising a structure

(i) The boom centroids lie in the plane of the web or the skin mid-line.
(ii) The skin or web carries only shear stress while the booms carry only bending stress. The boom areas abruptly alter the first moment of area, thereby interrupting the shear flow.
(iii) The shear stress is uniform across the web or skin thickness and the direct stress is constant across the boom area.

In addition, the net value of I about the neutral axis of the section must remain unchanged. Thus in Fig. 7.25, for example, when the booms are to represent the finite stiffner areas at their given positions in the original structure, at the rth boom,

$$(A_r\, y_r^{\,2})_{\text{boom}} = (I_{xr})_{\text{stiffner}}$$

$$\therefore\ A_r = I_{xr} / \, y_r^2 \tag{7.39}$$

from which it follows that A_r will approximate to the actual stiffener area only when y_r is large. The web or skin makes an additional contribution to I_x by increasing the net area of the boom. This is assessed from the following consideration of bending of the section about its neutral axis.

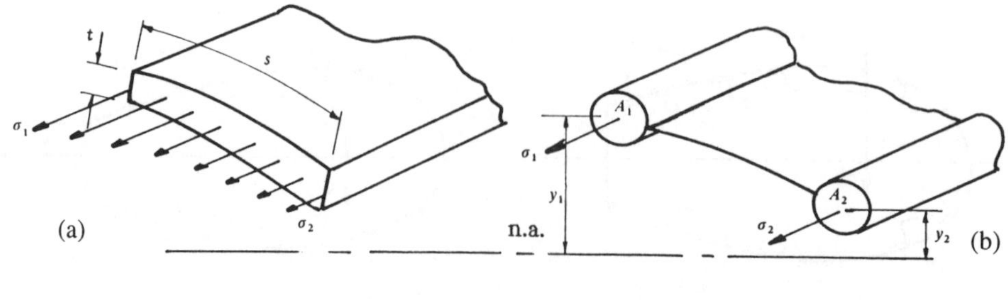

Figure 7.26 Web-boom idealisation of a single panel

The panel element in Fig. 7.26a is idealised into two booms connected by a shear web of zero bending stiffness in Fig. 7.26b. The following conditions are to apply:

(i) the net axial force must be the same. With a linear variation in bending stress assumed between the two booms then,

$$st\,(\sigma_1 + \sigma_2)/2 = A_1\sigma_1 + A_2\,\sigma_2 \tag{7.40}$$

where, from bending theory, the stress is proportional to the distance from the neutral axis

$$\sigma_1/\sigma_2 = y_1/y_2 \tag{7.41}$$

(ii) the moments about the neutral axis must be the same

$$\sigma_2\,s\,t\,(y_1 + y_2)/2 + [(\sigma_1 - \sigma_2)\,s\,t/2][\,y_2 + 2\,(\,y_1 - y_2)/3] = A_1\sigma_1 y_1 + A_2\,\sigma_2\,y_2 \tag{7.42}$$

Combining eqs(7.40-7.42) leads to the boom area contributions,

$$A_1 = (s\,t/6)(2 + \sigma_2/\sigma_1) \tag{7.43a}$$
$$A_2 = (s\,t/6)(2 + \sigma_1/\sigma_2) \tag{7.43b}$$

The reader should verify that implicit in these conditions is the fact that I_x for the panel will remain unchanged. In extending eqs(7.43a,b) to the multi-boom idealisation in Fig. 7.25, the area of the rth boom is influenced by both the adjacent webs r and $r + 1$. Then,

$$A_r = (s_r t_r/6)(2 + \sigma_{r-1}/\sigma_r) + (s_{r+1}\,t_{r+1}/6)(2 + \sigma_{r+1}/\sigma_r) \tag{7.44}$$

where:

$$\sigma_{r-1}/\,\sigma_r = y_{r-1}/\,y_r \text{ and } \sigma_{r+1}/\,\sigma_r = y_{r+1}/\,y_r$$

The net boom areas are then found by adding eqs(7.39) and (7.44). In practice, the skin or web thickness construction may not be fully effective in resisting direct stress where there is a risk of buckling, for example. This leads to the use of an effective skin thickness in eqs(7.43) and (7.44), which differs from the actual thickness, for the assessment of its contribution to the boom area. However, the following examples illustrate that the actual thickness is normally retained in the line integral $\oint$ (ds/t) when determining warping displacements and the position of the shear centre.

Example 7.12 Determine the web shear flow for Fig. 7.17a using a four-boom idealisation.

Firstly the flanges have been assumed to be concentrations of area at the web centre line, as shown in Fig. 7.27a. The contributions to these areas from the web are found from eqs(7.43) and (7.44). The former deals with the end booms,

$$A_1 = (s_1\,t_1/6)(2 + y_2/y_1) = (10 \times 1.5/6)(2 + 25/35) = 6.786 \text{ mm}^2 = A_4$$

and the latter deals with the intermediate boom areas in the table:

Position	1	2	3	4
s, mm	-	10	50	10
t, mm	-	1.5	1.5	1.5
y, mm	35	25	- 25	- 35

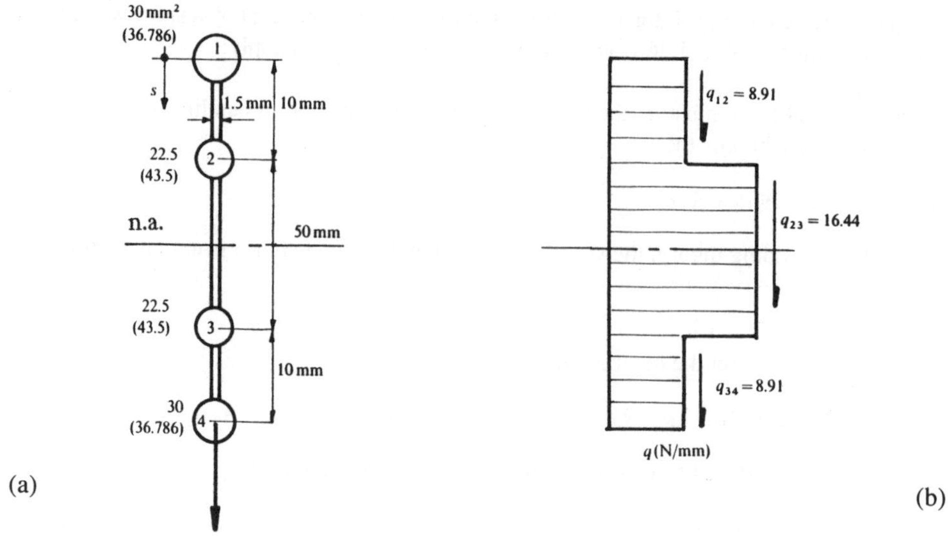

Figure 7.27 Shear flow in four-boom idealisation

$A_2 = (s_2 t_2/6)(2 + y_1/y_2) + (s_3 t_3/6)(2 + y_3/y_2)$
$= (10 \times 1.5/6)(2 + 35/25) + (50 \times 1.5/6)(2 - 25/25) = 21\ \text{mm}^2.$

$A_3 = (s_3 t_3/6)(2 + y_2/y_3) + (s_4 t_4/6)(2 + y_4/y_3)$
$= (50 \times 1.5/6)[2 + 25/(-25)] + (10 \times 1.5/6)[2 + (-35)/(-25)] = 21\ \text{mm}^2$

The net boom areas are those shown in Fig. 7.27a, from which

$I_x = 2[(36.786 \times 35^2) + (43.5 \times 25^2)] = 144.5 \times 10^3\ \text{mm}^4$ (as before in Example 7.8)

and $F_y / I_x = (1 \times 10^3)/(144.5 \times 10^3) = 144.5^{-1}\ \text{N/mm}^4$. Now from eq(7.34),

$$q = (F_y/I_x)\sum_{r=1}^{n} A_r y_r$$

where $\sum A_r y_r$ and q are

$1 \rightarrow 2\quad A_1 y_1 = 36.786 \times 35 = 1287.5\ \text{mm}^3, \quad \Rightarrow q_{12} = 8.91\ \text{N/mm}$
$2 \rightarrow 3\quad A_1 y_1 + A_2 y_2 = 1287.5 + (43.5 \times 25) = 2375\ \text{mm}^3, \quad \Rightarrow q_{23} = 16.44\ \text{N/mm}$
$3 \rightarrow 4\quad A_1 y_1 + A_2 y_2 + A_3 y_3 = 2375 + 43.5(-25) = 1287.5\ \text{mm}^3, \quad \Rightarrow q_{34} = 8.91\ \text{N/mm}$

The web shear flow in Fig. 7.27b is a reasonable approximation to that shown in Fig. 7.17b.

Example 7.13 Using a four-boom idealisation for the tube in Fig. 7.21, determine the shear flow distribution, the position of the shear centre and the warping displacements at the four corners. Take $G = 30$ GPa.

With reference to the chosen boom positions in Fig. 7.28a, the following table applies:

Boom	1	2	3	4
y, mm	+40	− 40	− 150	+150
s, mm	610	80	610	300
t, mm	1	3	1	4

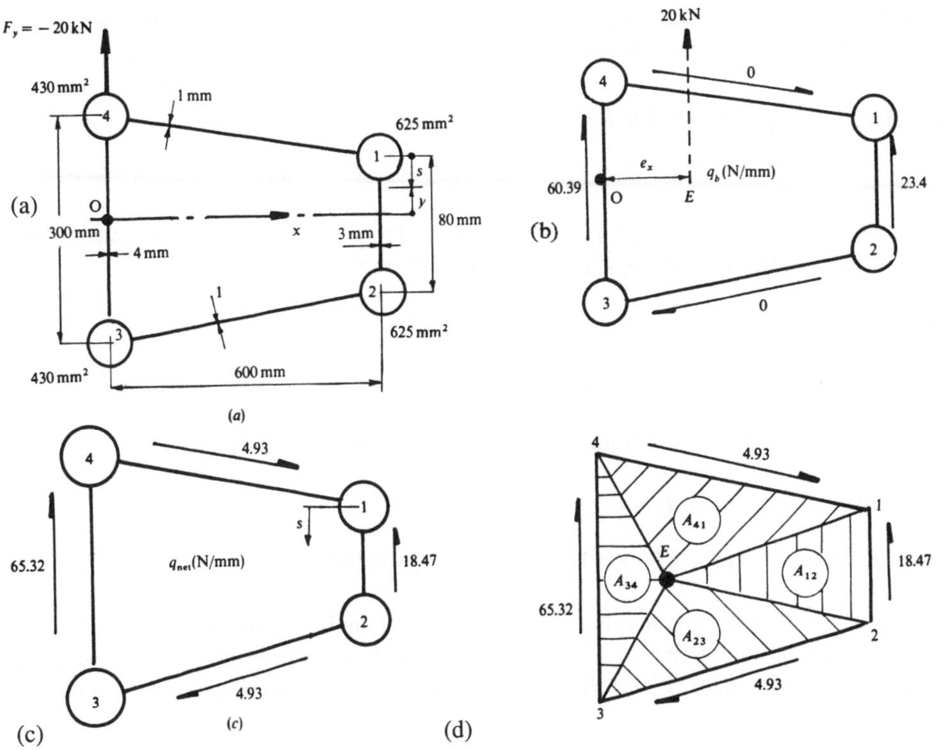

Figure 7.28 Shear centre and warping displacements

Since there are no stiffeners to deal with in this case, the boom areas are directly supplied by eq(7.44),

$$A_1 = (s_1 t_1/6)(2 + y_4/y_1) + (s_2 t_2/6)(2 + y_2/y_1)$$
$$= (610 \times 1/6)(2 + 150/40) + (80 \times 3/6)[2 + (-40)/40] = 625 \text{ mm}^2 = A_2$$
$$A_3 = (s_3 t_3/6)(2 + y_2/y_3) + (s_4 t_4/6)(2 + y_4/y_3)$$
$$= (610 \times 1/6)[2 + (-40)/(-150)] + (300 \times 4/6)[2 + 150/(-150)] = 430 \text{ mm}^2 = A_4$$
$$I_x = 2[(625 \times 40^2) + (430 \times 150^2)] = 21.37 \times 10^6 \text{ mm}^4$$

so agreeing with the previous value for Fig. 7.21 in Example 7.11. From eq(7.37a),

$$q_b = (F_y/I_x)\Big(\sum_{r=1}^{n} A_r y_r\Big)_b + q_0$$

where $F_y/I_x = -20000/(21.37 \times 10^6) = -9.363 \times 10^{-4}$ N/mm⁴. D_{xb} and q become

$1\rightarrow2$ $D_{xb} = 625 \times 40 = 2.5 \times 10^4$ mm³, $\therefore q_b = -9.363 \times 2.5 = -23.4$ N/mm
$2\rightarrow3$ $D_{xb} = (2.5 \times 10^4) - (625 \times 40) = 0$ $\therefore q_b = 0$
$3\rightarrow4$ $D_{xb} = 0 - (430 \times 150) = -6.45 \times 10^4$ mm³ $\therefore q_b = (-9.363)(-6.45) = 60.39$ N/mm
$4\rightarrow1$ $D_{xb} = -6.45 \times 10^4 + (430 \times 150) = 0,$ $\therefore q_b = 0$

This establishes the q_b distribution shown in Fig. 7.28b. For the determination of q_o it is convenient to take moments about the centre O, of the left hand web. Thus, $F_y p_x = 0$ and eq (7.28) becomes, (for the positive s - direction),

$$0 = \oint q_b R \, ds + q_o \oint R \, ds = \oint q_b R \, ds + 2A \, q_o$$

$$0 = -(23.4 \times 80 \times 600) + (300 + 80)600 q_o \Rightarrow q_o = +4.93 \text{ N/mm}$$

Thus the net shear flow, $q = q_b + q_o$, is that shown in Fig. 7.28c. For the shear centre E, with F_y translated distance e_x to E in Fig. 7.28b, eq(7.30) becomes

$$\oint q_b \, ds/t + q_e \oint ds/t = 0$$

$$[-(23.4 \times 80/3) + 0 + (60.39 \times 300/4) + 0] + q_e(80/3 + 610/1 + 300/4 + 610/1) = 0$$

Hence $q_e = -2.955$ N/mm. Taking moments about O, eq(7.31) becomes

$$F_y e_x = \oint q_b R \, ds + q_e \oint R \, ds = \oint q_b R \, ds + 2A \, q_e$$

$$-(20 \times 10^3)e_x = -(23.4 \times 80 \times 600) - 600(300 + 80)2.955$$

and again $e_x = 90$ mm. The warping displacements are found by applying eq(7.32) to the net q shear flow distribution in Fig. 7.28c. Starting at position 1, where there will be a warping displacement w_1, eq(7.32) is applied by measuring A_{os} w.r.t. the shear centre E (see Fig. 7.28d. Now,

$$\oint q \, ds/t = (-18.47 \times 80/3) + (4.93 \times 610/1) + (65.32 \times 300/4) + (4.93 \times 610/1) = 10421.1$$

$$A = (300 + 80)600/2 = 114 \times 10^3 \text{ mm}^2$$

$$(A_{os}/A)\oint q \, ds/t = (A_{os} \times 10421.1)/(114 \times 10^3)$$

$$\therefore G\Delta w = \int_0^s q \, ds/t - (91.4 \times 10^{-3}) A_{os}$$

$1 \rightarrow 2$ $A_{os} = A_{12} = 80 \times 510/2 = 20.4 \times 10^3$ mm^2

$$\int q \, ds/t = (-18.47/3) \int_0^s ds = -6.156 \, s$$

$$@ \, 2 \; s = 80 \text{ mm}, \; q\int_0^s ds/t = (-6.158 \times 80) = -492.5$$

$$G(w_2 - w_1) = -492.5 - (91.4 \times 20.4) = -2.357 \times 10^3 \text{ N/mm}$$

$2 \rightarrow 3$ $A_{23} = ½[114 - 20.4 - (300 \times 90/2)]10^3 = 40.05 \times 10^3$ mm^2

$$A_{os} = A_{12} + A_{23} = (20.4 + 40.05)10^3 = 60.45 \times 10^3 \text{ mm}^2$$

$$\int q \, ds/t = -492.5 + (4.93/1) \int_0^s ds = -492.5 + 4.93s$$

$$@ \, 3 \; s = 610 \text{ mm}, \; q\int_0^s ds/t = 2514.8$$

$$G(w_3 - w_1) = 2514.8 - (91.4 \times 60.45) = -3.01 \times 10^3 \text{ N/mm}$$

$3 \rightarrow 4$ $A_{34} = (300 \times 90)/2 = 13.5 \times 10^3$ mm^2

$$A_{os} = A_{12} + A_{23} + A_{34} = (20.4 + 40.05 + 13.5)10^3 = 73.95 \times 10^3 \text{ mm}^2$$

$\int q\, ds/t = 2514.8 + (65.32/4)\int_0^s ds = 2514.8 + 16.33s$

@ 4 $s = 300$ mm, $q\int_0^s ds/t = 7413.8$

$G(w_4 - w_1) = 7413.8 - (91.4 \times 73.95) = +0.655 \times 10^3$ N/mm

$4 \rightarrow 1 \quad A_{os} = 114 \times 10^3\ \text{mm}^2$

$\int q\, ds/t = 7413.8 + (610/1)\int_0^s ds = 7413.8 + 610s$

@1 $s = 610$ mm, $\oint q\, ds/t = 10421.1$

$G(w_1 - w_1) = 10421.1 - (91.4 \times 114) \approx 0$ (provides a check).

Assuming that the x-axis of symmetry does not warp and noting that Δw is linear in s throughout, then

$Gw_1 = +2.357 \times 10^3/2 = +1.179 \times 10^3$, $\quad Gw_2 = (-2.357 + 1.179)10^3 = -1.179 \times 10^3$
$Gw_3 = (-3.01 + 1.179)10^3 = -1.832 \times 10^3$, $\quad Gw_4 = (+0.655 + 1.179)10^3 = +1.833 \times 10^3$

and finally, setting the shear modulus $G = 30 \times 10^3$ MPa, the warping displacements are $w_1 = +0.039$, $w_2 = -0.039$, $w_3 = -0.061$ and $w_4 = +0.061$ mm. Points above the x - axis displace axially in the positive z - direction (see Fig. 7.18a) while those points beneath the x - axis displace axially in the opposite direction.

7.7 Web-Boom Idealisation for Asymmetric Sections

The basic form of eq(7.35) applies when shear forces F_x and F_y are applied to an idealised section in negative x and y directions. However, because x and y are not principal axes it becomes necessary to employ equivalent shear forces F_x' and F_y' if we wish to refer the first and second moments of area to these axes. Equation (7.35) is modified to

$$q = (F_y'/I_x)\sum_{r=1}^{n}(A_r y_r) + (F_x'/I_y)\sum_{r=1}^{n}(A_r x_r) + q_o \tag{7.45}$$

where F_x' and F_y' have been previously defined in eqs(7.22a,b). Equation (7.45) will apply to both open and closed sections provided we set $q_o = 0$ at a free surface. The determination of q_o when forces are applied to a closed tube either at, or away from, its shear centre has been previously outlined in section 7.5.1.

Example 7.14 Determine the shear flow for the flanges and the web in the unsymmetrical channel section of Fig. 7.29a, when shear forces act at the shear centre E as shown. Locate the position of E.

The centroidal position $\bar{x}$ of the y - axis is found from

$(500 \times 100) + (500 \times 200) = (4 \times 500)\,\bar{x}$, $\Rightarrow \bar{x} = 75$ mm

and, as the x-axis passes through the centre of the depth,

$I_x = 4 \times 500 \times 200 = 80 \times 10^6\ \text{mm}^4$
$I_y = (2 \times 75^2 \times 500) + (25^2 \times 500) + (125^2 \times 500) = 13.75 \times 10^6\ \text{mm}^4$
$I_{xy} = (500{\times}200{\times}25) - (500{\times}75{\times}200) + (500{\times}200{\times}75) - (500{\times}125 \times 200) = -10 \times 10^6\ \text{mm}^4$

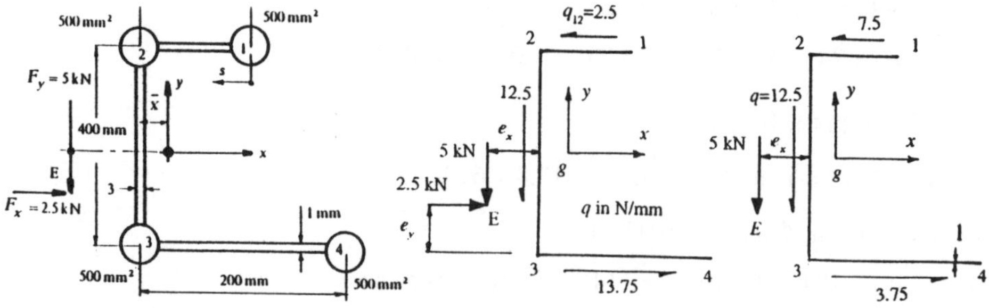

(a) (b) (c)

Figure 7.29 Unsymmetrical channel section

Now from eqs(7.22), with $F_y = +5$ kN and $F_x = -2.5$ kN,

$$F_x' = [-2.5 - 5(-10/80)]/[1 - 10^2/(80 \times 13.75)] = -2.0625 \text{ kN}$$
$$F_{y'} = [5 + 2.5(-10/13.75)]/[1 - 10^2/(80 \times 13.75)] = 3.5 \text{ kN}$$
$$F_x'/I_y = -(2.0625 \times 10^3)/(13.75 \times 10^6) = -150 \times 10^{-6} \text{ N/mm}^4$$
$$F_y'/I_x = (3.5 \times 10^3)/(80 \times 10^6) = 43.75 \times 10^{-6} \text{ N/mm}^4$$

In the application of eq(7.45) with $q_o = 0$, the shear flow between booms 1 and 2, i.e. from 1 to 2, is found by applying the summation terms to boom 1 only. That is,

$$q_{12} = (F_y'/I_x)(500 \times 200) + (F_x'/I_y)(500 \times 25) \tag{i}$$

That is, $q_{12} = 2.50$ N/mm, acting in the sense shown in Fig. 7.29b. The shear flow between booms 2 and 3 is found by applying the summations to booms 1 and 2. That is, working anti-clockwise from 1,

$$q_{23} = (F_y'/I_x)[(500 \times 200) + (500 \times 200)] + (F_x'/I_y)[(500 \times 25) - (500 \times 75)] \tag{ii}$$

This gives q_{23}= 12.5 N/mm positive in the sense shown in Fig. 7.29b. Alternatively, working clockwise from 4,

$$q_{23} = (F_y'/I_x)[-(500{\times}200)-(500{\times}200)]+(F_x'/I_y)[(500 \times 125) - (500 \times 75)] = -12.5 \text{ N/mm}$$

confirming the positive direction previously found. Working anti-clockwise from 1, the shear flow between 3 and 4 is given by

$$q_{34} = (F_y'/I_x)[(500 \times 200) + (500 \times 200) - (500 \times 200)]$$
$$+ (F_x'/I_y)[+(500 \times 25) - (500 \times 75) - (500 \times 75)] \tag{iii}$$

from which q_{34}= 13.75 N/mm. Alternatively, working clockwise from 4,

$q_{34} = (F_y'/I_x)[-(500 \times 200)] + (F_x'/I_y)(500 \times 125) = -13.75$ N/mm

The magnitudes of q and the directions shown in Fig. 7.29b must be statically equivalent to the applied forces. It is only necessary to multiply q by the respective web lengths in order to check this because $q = \tau t$. Then,

$F_x = (2.5 \times 100) - (13.75 \times 200) = -2500$ N, $F_y = (12.5 \times 400) = 5000$ N.

We shall use the applied loading to find the co-ordinates (e_x, e_y) of the shear centre. Taking moments about O in Fig. 7.29b gives

$$(2.5 \times 100) \times 400 = (5 \times 10^3)\, e_x - (2.5 \times 10^3)\, e_y$$
$$10 = 0.5e_x - 0.25e_y \qquad \text{(iv)}$$

A second equation is required to separate e_x from e_y. This is derived from a re-evaluation of the shear flow in the absence of one force. Thus in Fig.7.29c, with F_x absent, we have from eqs(7.22) the equivalent forces $F_x' = 0.6875$ kN and $F_y' = 5.5$ kN. Hence $F_x'/I_y = 50 \times 10^{-6}$ N/mm^4 and $F_y'/I_x = 68.75 \times 10^{-6}$ N/mm^4. Substituting into eqs(i), (ii) and (iii), the shear flows become $q_{12} = 7.5$ N/mm, $q_{23} = 12.5$ N/mm and $q_{34} = 3.75$ N/mm, as shown. The horizontal forces are self-equilibrating and the applied force is the resultant of the vertical flow. Taking moments about O,

$$(7.5 \times 100) \times 400 = (5 \times 10^3)\, e_x$$

from which $e_x = 60$ mm. Substituting into eq(iv) gives $e_y = 80$ mm (independent of F_y).

Bibliography

Boresi,A.P.,Schmidt,R.J. and Sidebottom,O.M., *Advanced Mechanics of Materials*, 5th edition, Wiley, 1993.

Kuhn,P., *Stresses in Aircraft and Shell Structures*, McGraw-Hill, 1956.

Megson,T.H,G., *Aircraft Structures for Engineering Students*, Arnold, 1972.

Perry, D.J., *Aircraft Structures*, McGraw-Hill 1950.

Timoshenko,S.P and Goodier,J.N., *Theory of Elasticity*, McGraw-Hill, 1951.

Ugural,A.C. and Fenster,S.K., *Advanced Strength and Applied Elasticity*, Arnold, 1981.

Williams,D., *Theory of Aircraft Structures*, 1960.

EXERCISES

Shear Stress in Beams

7.1 Show that the maximum flexural shear stress in a rectangular beam is 50% greater than the mean value for the section.

7.2 A rectangular channel section, 500 mm wide, 300 mm deep and 25 mm thick, is used as a beam in the form of a trough. Derive equations in which the shear stress at any layer in the vertical sides may be found under a vertical shear force of 9 kN. Plot the distribution showing major values. Find the fraction of the total shear force carried by the vertical sides.

7.3 An hexagonal bar section with 25 mm sides is used as a cantilever with one diagonal lying horizontally. Plot the distribution of shear stress through the depth, when a concentrated vertical force of 20 kN is applied at the free end. Where in the depth is the shear stress a maximum and what is its value at the neutral axis?

7.4 The section of an I-beam is subjected to a vertical shear force of 40 kN. Derive an equation for which the shear stress at any layer in the web may be determined. Calculate the shear stress acting at the top of the web and the maximum shear stress value. What percentage of the shear force is carried by the web? Section details are flanges 175 mm wide × 25 mm deep, total depth 225 mm, web thickness 10 mm.

7.5 Plot the distributions of vertical and horizontal shear stress for the channel sections in Figs 7.30a and b, when they are each subjected to a vertical shear force of 50 kN. Determine the shear centres for each channel.

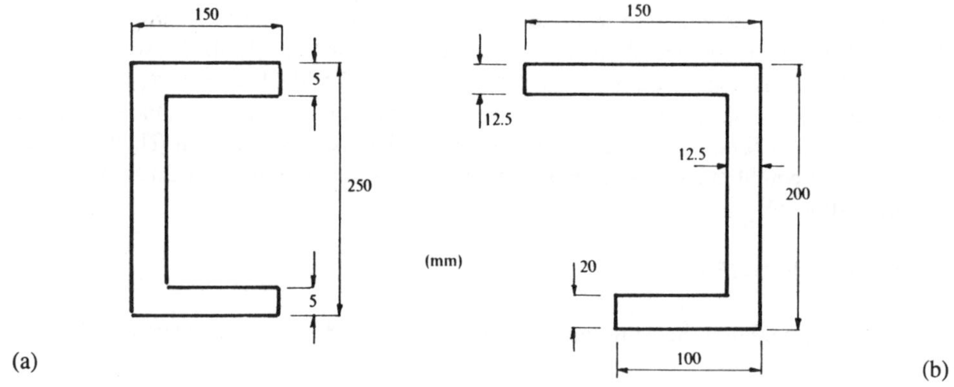

Figure 7.30

7.6 Find an expression for the shear stress at any point in the flange and the web for the I-section in Fig. 7.31. Plot the distribution of shear stress through the depth, inserting major values corresponding to a vertical shear force of 400 kN.

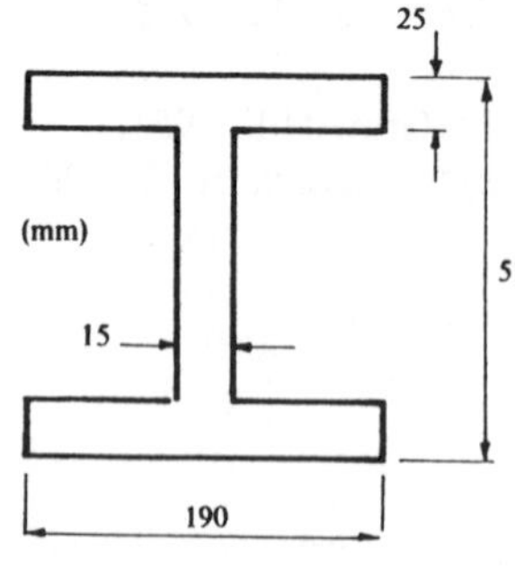

Figure 7.31

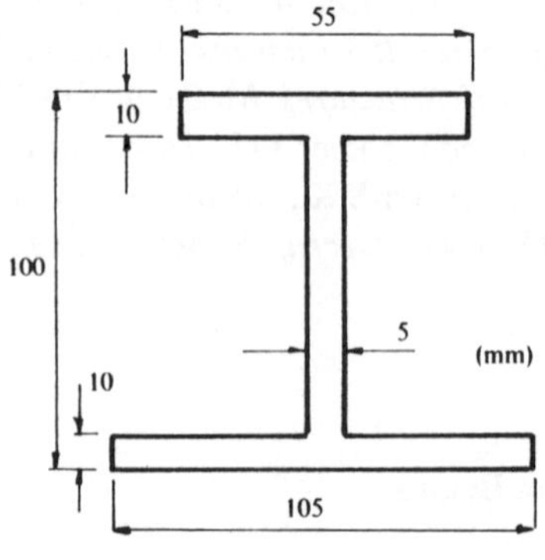

Figure 7.32

7.7 Draw the distribution of vertical shear stress for the cross-section shown in Fig. 7.32, corresponding to a shear force of 300 kN.

7.8 Find the maximum shear stress when the I-section in Fig. 7.33 is subjected to a vertical shear force of 80 kN. What is the percentage error when this stress is taken to be the mean value calculated from force/area?

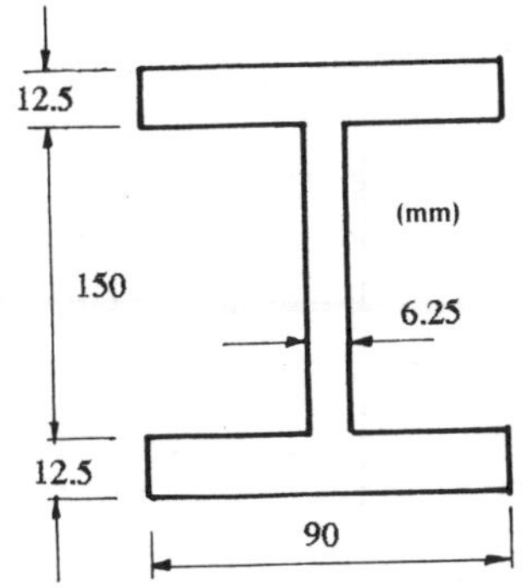

Figure 7.33

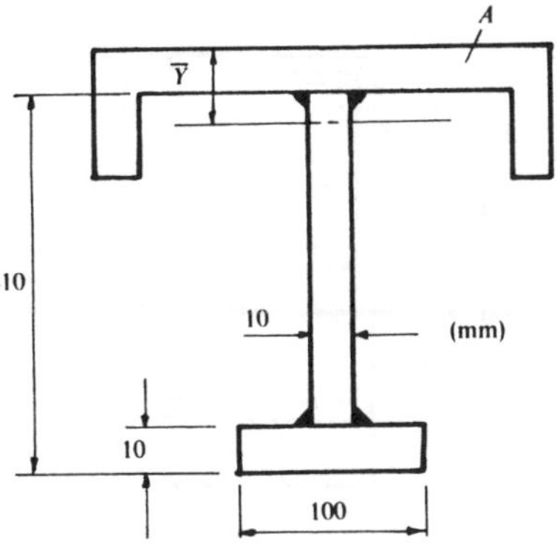

Figure 7.34

7.9 The welded composite beam in Fig. 7.34 is subjected to a vertical force of 250 kN. Find the magnitude of the shear flow at the two welded flange joints if for the channel section $A = 3000$ mm^2, $I = 150 \times 10^6$ mm^4 and $\bar{y} = 12.5$ mm, as shown.

7.10 Calculate for the cross-section shown in Fig. 7.35a and b, the maximum shear stress and the shear flow carried by the rivets.

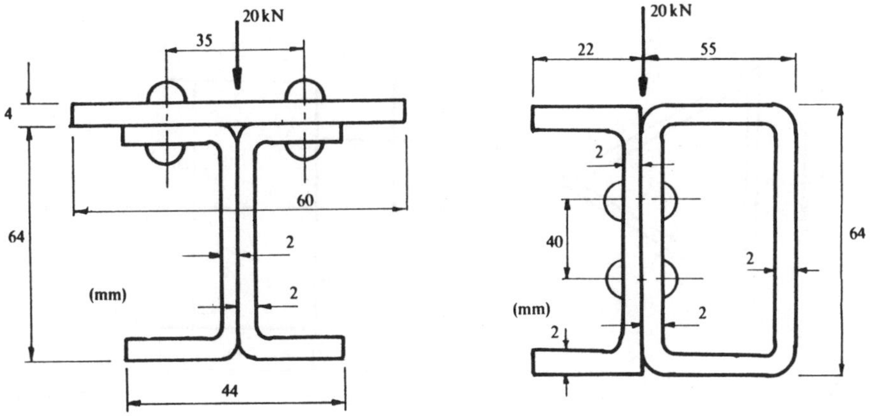

(a) (b)

Figure 7.35

7.11 A solid elliptical section with lengths of semi-major and semi-minor axes a and b respectively, is mounted as a beam, with the major axis aligned vertically. Determine the distribution of vertical and horizontal shear stresses within the section, when it carries a vertical shear force F. Examine the variation in vertical shear stress along the horizontal (minor) axis. Show that $(\tau_{yz})_{max} = 4F/(3A) =$ constant when $b \ll a$, $(\tau_{yz})_{max} = 2F/[(1 + \nu)A]$ at the centre and $(\tau_{yz})_{max} = 4F/[(1 + \nu)A]$ at the ends when $b \simeq a$. Compare with the elementary theory in each case.

Shear Flow in Thin-Walled Open Sections

7.12 A 0.5 m length cantilever with the channel section in Fig. 7.36 carries an inclined force of 25 N through the shear centre E, as shown. Determine the position for E and the magnitude and position of the maximum shear stress.

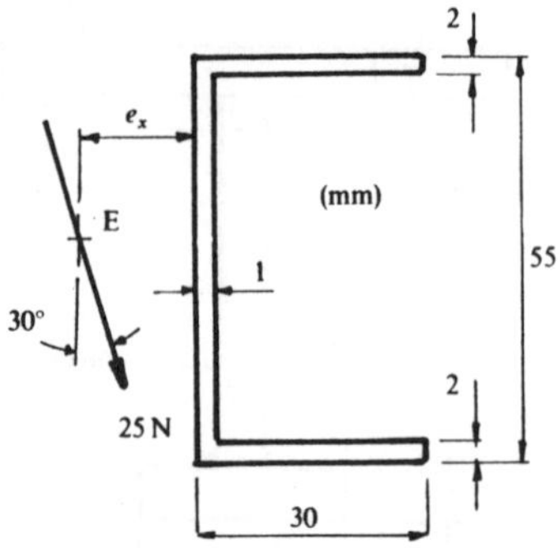

Figure 7.36

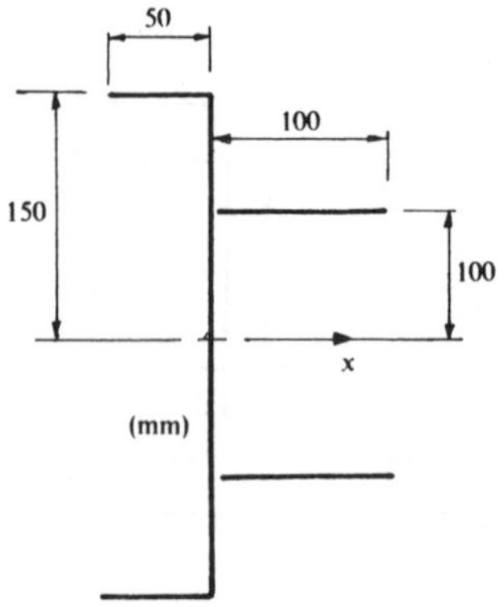

Figure 7.37

7.13 The extruded section in Fig. 7.37 is symmetrical about the x - axis and has a constant thickness of 1 mm. Calculate and plot the shear flow distribution when a vertical force of 1 kN is applied downwards, through the shear centre. Where does the shear centre lie?

7.14 A thin tube of equilateral triangular section with side length a and thickness t is opened by a longitudinal slit, as shown in Fig. 7.38. Show that the shear flow distribution along the sloping sides is $q = F_y s^2/a^3$ and that the position of the shear centre is $e = 0.289a$ when a vertical shear force F_y is applied at the centre.

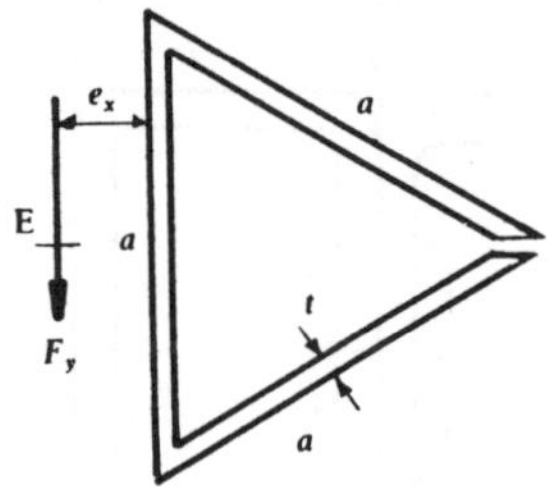

Figure 7.38

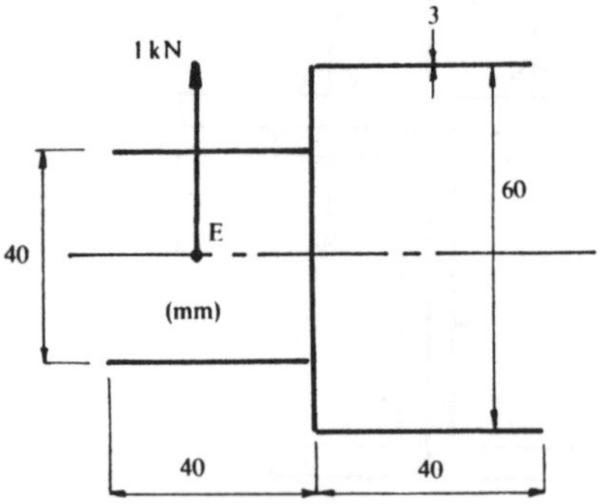

Figure 7.39

7.15 The extruded beam section in Fig. 7.39 is subjected to an upward vertical shear force of 1 kN through the shear centre. Determine the maximum shear flow and the horizontal distance the shear centre lies from the vertical web.

7.16 The open semi-circular tube in Fig. 7.40 is loaded through the shear centre as shown. Draw the shear flow distribution and find the horizontal position of the shear centre.

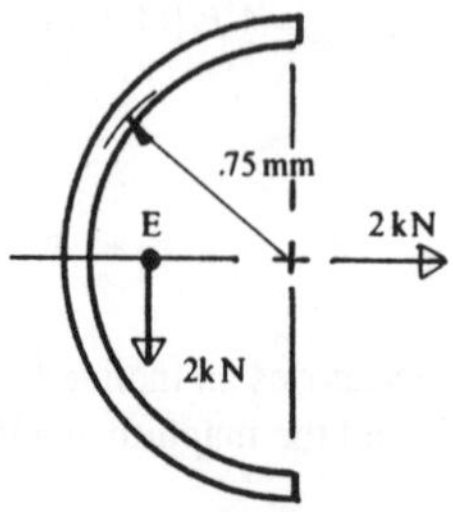

Figure 7.40

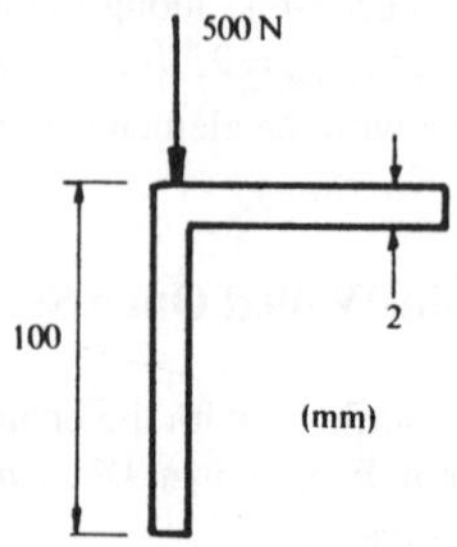

Figure 7.41

7.17 Establish the shear flow distribution for the equal-angle section in Fig. 7.41, when a vertical shear force of 500 N is applied downwards along the vertical web.

7.18 Find the position of the shear centre for each of the open tubes in Figs 7.42a-c.

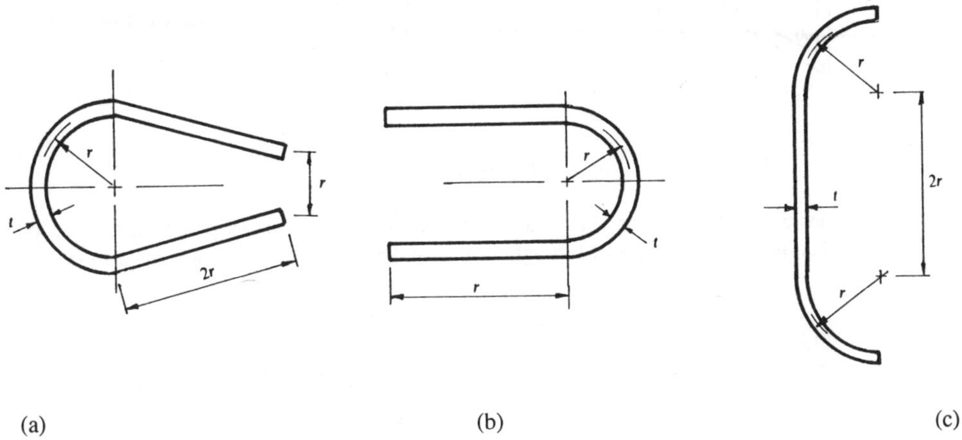

Figure 7.42

7.19 A uniform cantilever has a thin tubular cross-section of mean radius r and thickness t with a narrow slit cut through the wall at the right-hand side of the horizontal diameter, as shown in Fig. 7.43. Show that the twisting moment set up under a vertical shear force Q is $2Qr^4\pi\, t/I$ and hence that, in order to prevent twisting taking place, the force should be placed at $e = 2r$ from the centre, as shown. (CEI)

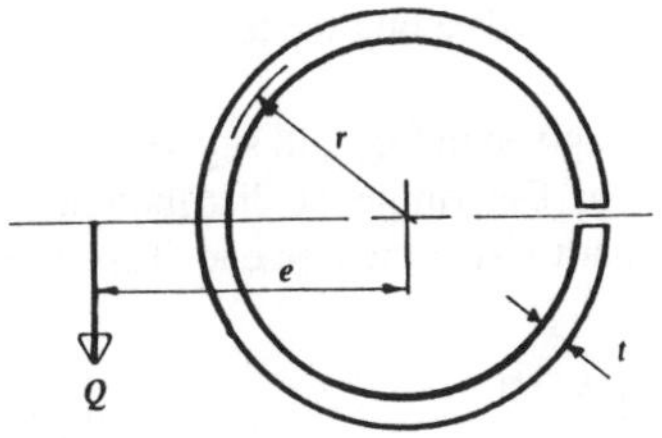

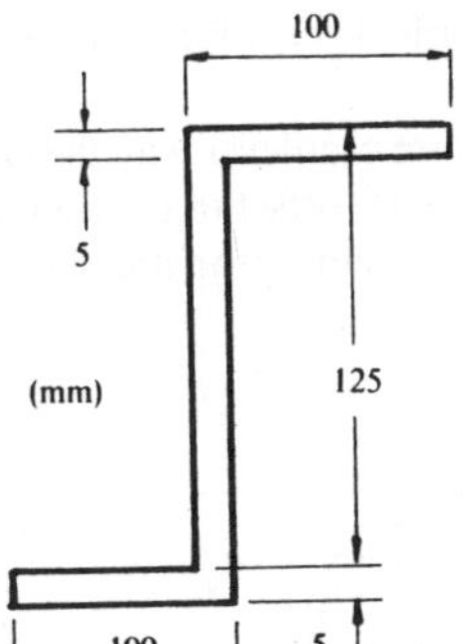

Figure 7.43 **Figure 7.44**

7.20 Determine the distribution of shear stress in the flanges and web and their maxima for the Z - section in Fig. 7.44, when a downward shear force of 1 kN is applied along the web.

Shear Flow in Thin-Walled Closed Sections

7.21 The thin-walled tubes in Fig. 7.45a and b each have a downward vertical shear force of 500 N applied along their left vertical sides. Determine the shear flow distribution around the section, the rate of twist and the position of the shear centre.

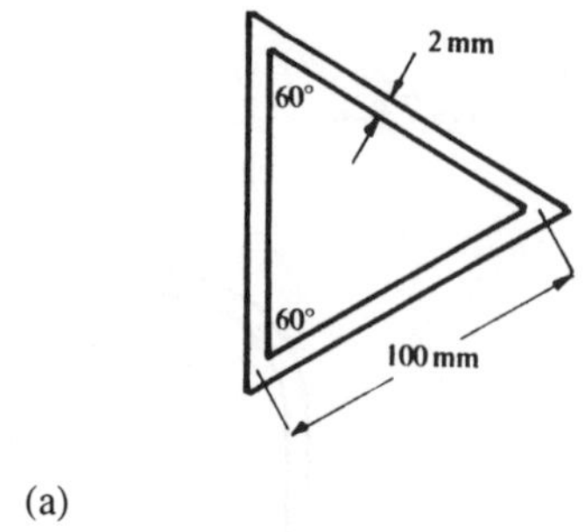

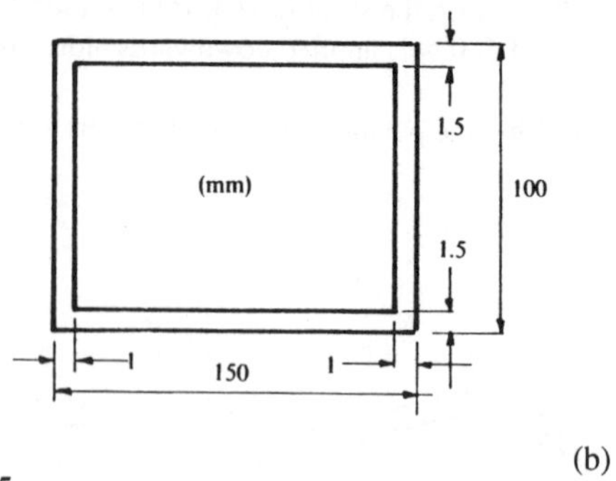

Figure 7.45

7.22 Find the position and magnitude of the maximum shear stress in the tubular parallelogram section in Fig. 7.46 when it is loaded vertically through the centroid with a 1 kN downward shear force.

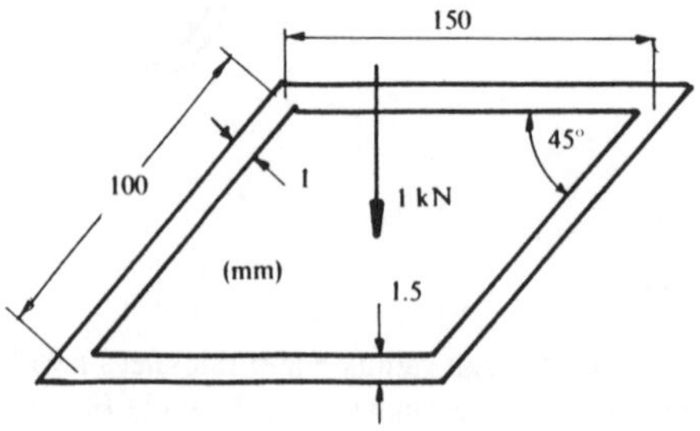

Figure 7.46

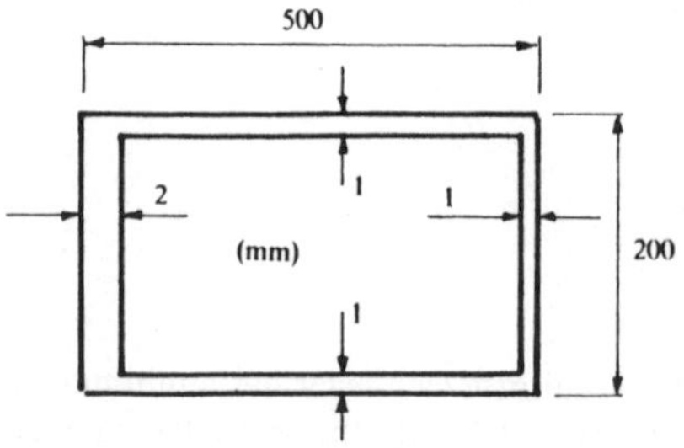

Figure 7.47

7.23 Plot the distribution of shear flow, inserting principal values, for the tubular rectangular section in Fig. 7.47, when a vertical shear force of 6 kN acts downwards along the left web.

7.24 A 2.5 m long uniform cantilever with the cross-section given in Fig.7.48 supports a 60 kN vertical upward force at its free end, offset to the right by 62.5 mm. Determine the distribution of shear flow over the section, the maximum shear stress and the angular twist at the free end. Take $G = 30$ GPa.

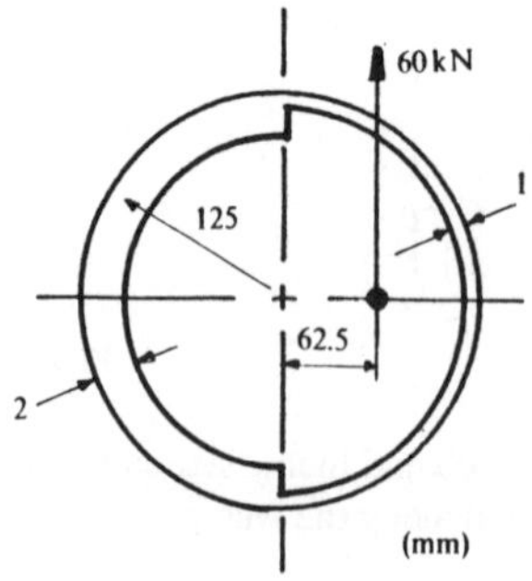

Figure 7.48

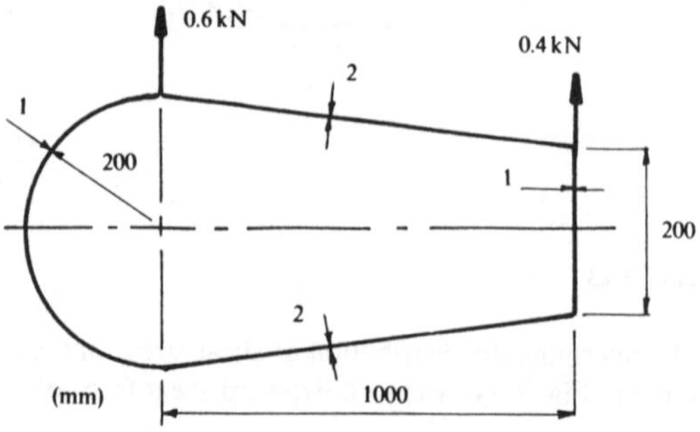

Figure 7.49

7.25 Determine the shear flow distribution and the position of the shear centre for the tube section in Fig. 7.49 when two vertical shear forces are applied as shown.

7.26 Determine the position of the shear centre for the trapezoidal tubular section given in Fig. 7.50.

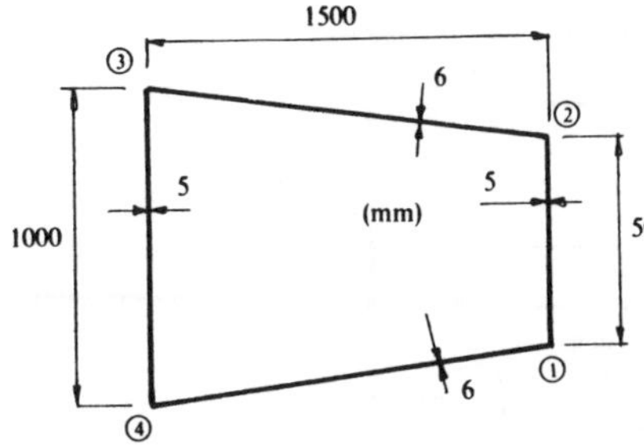

Figure 7.50

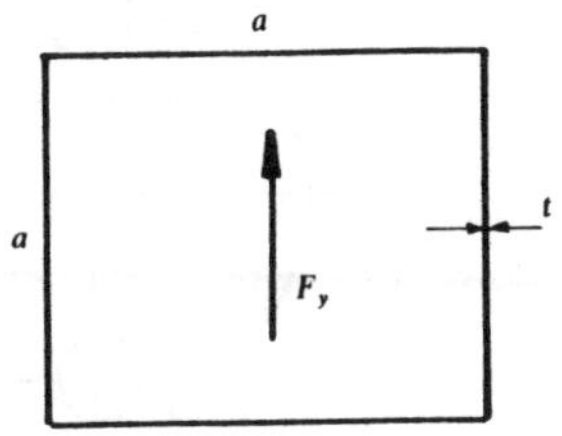

Figure 7.51

7.27 The square section tube in Fig.7.51 is loaded through the centroid by an upward vertical shear force F_y. Determine the warping displacements at the four corners.

Web-Boom Idealisation For Thin-Walled Structures

7.28 Determine the web shear flow for the open section in Fig. 7.37, using a simple four boom idealisation, when a downward force of 1 kN acts through the shear centre . Compare with the shear flow found from Exercise 7.13.

7.29 Confirm the position of the shear centre found in Exercise 7.15 using a suitable idealisation of the section in Fig. 7.39.

7.30 A 50 kN vertical shear force acts downwards through the shear centre of the section in Fig. 7.52. Determine the flexural shear flow in each web and the position of the shear centre.

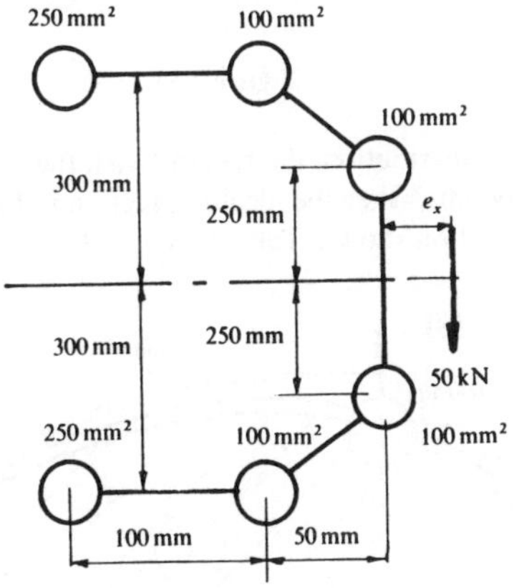

Figure 7.52

7.31 Idealise the box section in Fig. 7.50 when the given thicknesses are all effective in shear. Webs 1-2 and 2-3 carry no direct stress but covers 1-4 and 2-3 carry direct stress equivalent to a thickness of 5 mm. Hence determine the position of the shear centre.

7.32 The cross-section of a beam is simplified into the web-boom idealisation given in Fig. 7.53. Calculate the flexural shear flow distribution corresponding to a vertical downward shear force of 200 kN applied through the shear centre. What is the horizontal distance of the centre from the open ends?

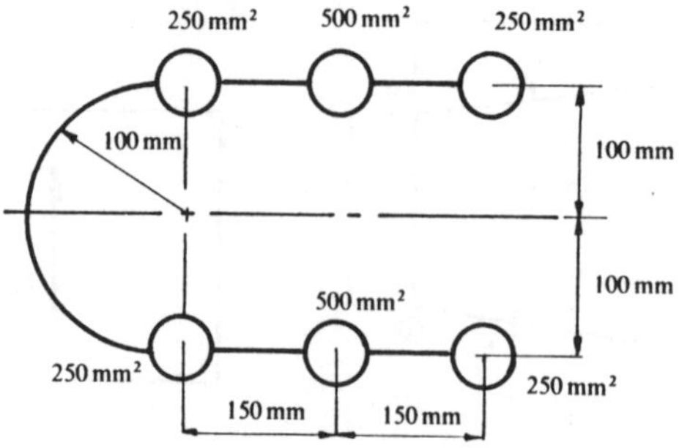

Figure 7.53

7.33 Determine the position e_x of the shear centre for the section in Fig. 7.54, assuming that the booms carry direct stress and the walls carry shear stress. Take $A_1 = A_6 = 500$ mm², $A_2 = A_5 = 200$ mm², $A_3 = A_4 = 800$ mm², $t_{12} = t_{23} = t_{45} = t_{56} = 1.5$ mm, $t_{16} = 2$ mm and $t_{34} = 4$ mm.

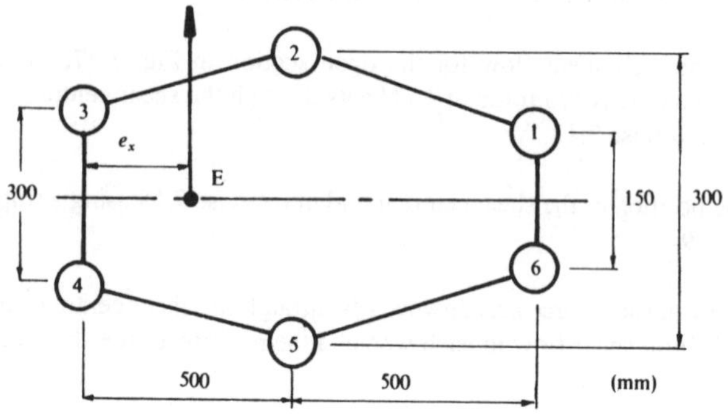

Figure 7.54

7.34 Determine the shear flow distribution, the rate of twist, the position of the shear centre and the warping displacements at the booms, when the idealised section in Fig. 7.55 is subjected to an upward shear force of 2.5 kN in the position shown. Take $A_1 = 3 \times 10^3$ mm², $A_2 = 5 \times 10^3$ mm².

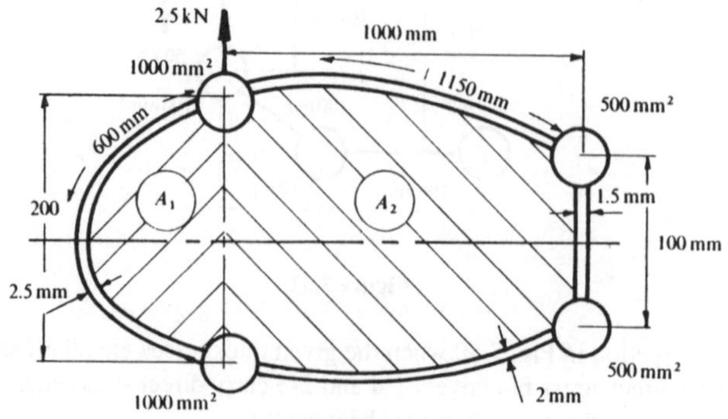

Figure 7.55

7.35 Determine the shear flow distribution and the warping displacements when the octagonal box section in Fig. 7.56 supports a 5 kN downward force at its centroid. All areas are 250 mm².

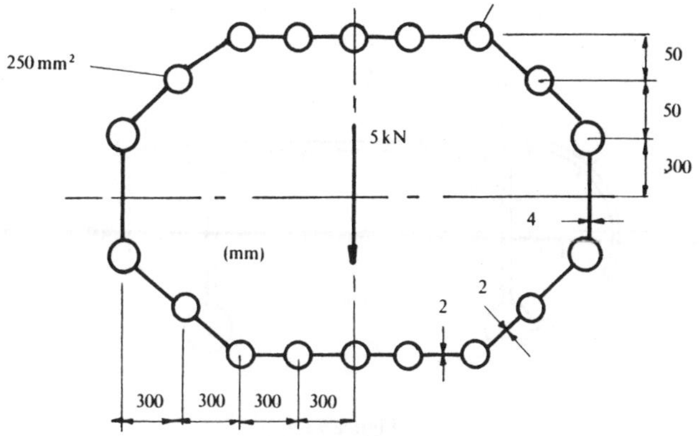

Figure 7.56

7.36 Determine the position of the shear centre for the idealised wing section in Fig.7.57. Each 200 mm^2 boom area carries direct stress and the 2 mm thick webs carry shear stress.

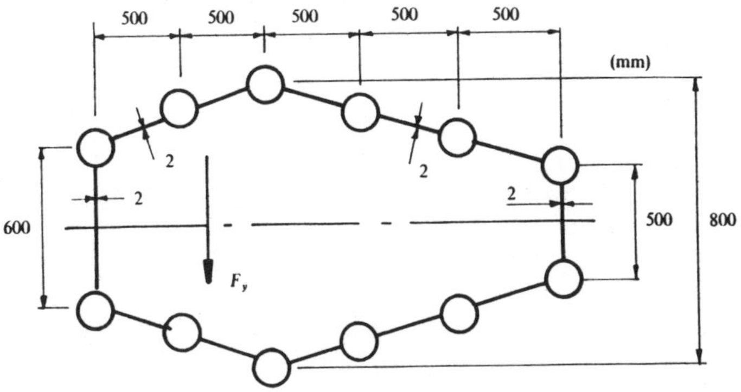

Figure 7.57

7.37 The idealised, singly symmetric, closed tube in Fig. 7.58 carries an 5 kN upward vertical force at the position shown. Determine the shear flow distribution and the unrestrained warping displacements resulting from this force. All boom areas are 500 mm^2.

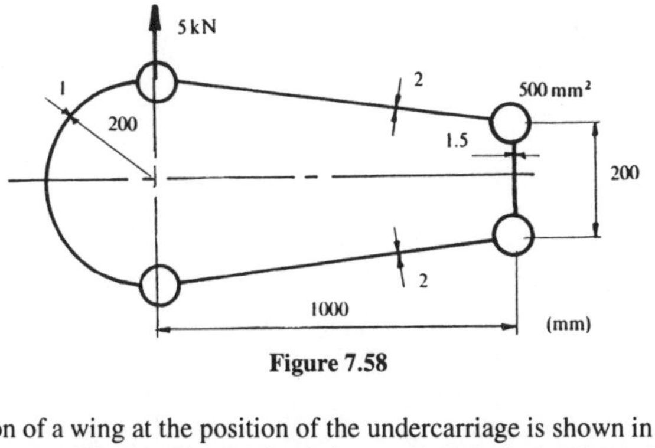

Figure 7.58

7.38 The cross-section of a wing at the position of the undercarriage is shown in Fig. 7.59. The front cell is closed and the rear cell open, the section being symmetrical apart from the missing skin 1 - 4. Walls 1-2, 2-3, 3-4 are straight, wall 3-4 is curved. All walls have the same shear modulus and are assumed to be effective in carrying only shear stress. Direct stress is carried by four equal booms, 1 to 4, each of area 1000 mm^2. The enclosed area of the front cell is 120×10^3 mm^2. Calculate the position of the shear centre of the tube cross-section. (CEI)

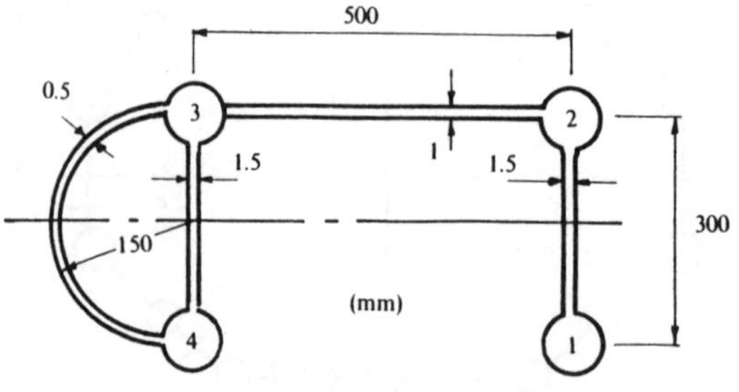

Figure 7.59

7.39 The given thicknesses for the closed tube section in Fig. 7.58 are those effective in shear. The 500 mm^2 boom areas, which carry direct stress, are further increased by a contribution from the upper and lower panels whose effective thicknesses are both 4 mm. Establish the net boom areas and the position of the shear centre.

CHAPTER 8

ENERGY METHODS

8.1 Strain Energy and External Work

The first law of thermodynamics states that energy is conserved within a closed system in the following manner:

$$\delta Q - \delta W = \delta U \tag{8.1}$$

where δQ and δW are respectively heat and work transfers to and/or from a system which change to its internal energy U by an amount δU. Here the "system" is a body in equilibrium under applied forces. With an increase in these forces, external work is done on the body ($-\delta W$) that will result in a positive change to its internal energy δU. When this process is adiabatic ($\delta Q = 0$) and it follows from eq(8.1) that

$$\int \delta W = \int \delta U \tag{8.2}$$

i.e. the total work done by these forces is equal to the increase in internal energy. When the forces are slowly applied, such that kinetic energy due to the rate of deformation is negligible, δU becomes the increase in the elastic strain energy stored within the body. It then becomes possible to derive an expression for δU. In the general case consider a volume element, $\delta V = \delta x \times \delta y \times \delta z$, for any solid or structure in Fig. 8.1 subjected to a generalised stress system σ_{ij}. The equivalence between the stress components in the tensor notation ($i, j = 1, 2, 3$) and the engineering notation ($i, j = x, y, z$) is expressed in the three independent normal components: $\sigma_{11} = \sigma_x$, $\sigma_{22} = \sigma_y$ and $\sigma_{33} = \sigma_z$ and in the three independent shear stresses: $\sigma_{12} = \tau_{xy}$, $\sigma_{23} = \tau_{yz}$ and $\sigma_{31} = \tau_{zx}$.

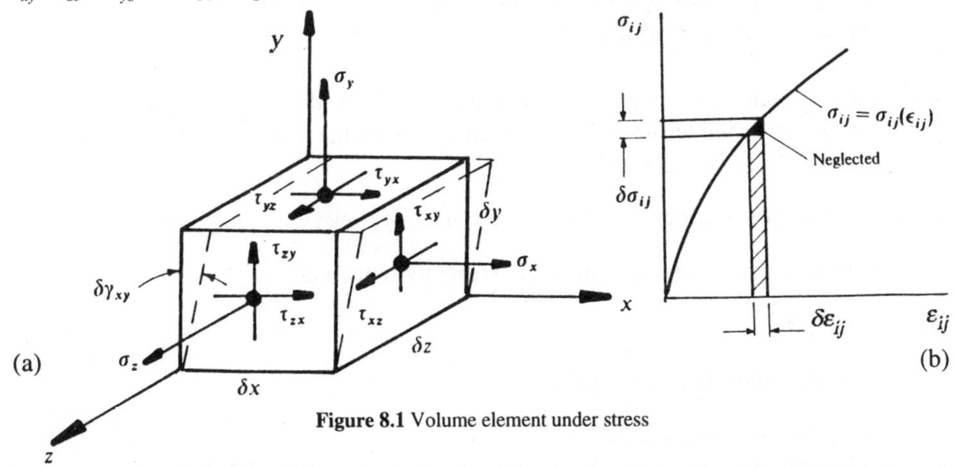

Figure 8.1 Volume element under stress

Dependent complements of shear stress $\sigma_{21} = \tau_{yx}$, $\sigma_{32} = \tau_{zy}$ and $\sigma_{31} = \tau_{zx}$ also exist to give a 3 × 3 stress matrix in either notation:

$$\sigma_{ij} = \begin{bmatrix} \sigma_{11} & \sigma_{12} & \sigma_{13} \\ \sigma_{21} & \sigma_{22} & \sigma_{23} \\ \sigma_{31} & \sigma_{32} & \sigma_{33} \end{bmatrix} \equiv \begin{bmatrix} \sigma_x & \tau_{xy} & \tau_{xz} \\ \tau_{yx} & \sigma_y & \tau_{yz} \\ \tau_{zx} & \tau_{zy} & \sigma_z \end{bmatrix} \tag{8.3a}$$

Let the corresponding strain increments $\delta\varepsilon_{ij}$ $(i, j = 1, 2, 3)$ be produced by an elastic stress change $\delta\sigma_{ij}$, where

$$\delta\varepsilon_{ij} = \begin{bmatrix} \delta\varepsilon_{11} & \delta\varepsilon_{12} & \delta\varepsilon_{13} \\ \delta\varepsilon_{21} & \delta\varepsilon_{22} & \delta\varepsilon_{23} \\ \delta\varepsilon_{31} & \delta\varepsilon_{32} & \delta\varepsilon_{33} \end{bmatrix} \equiv \begin{bmatrix} \delta\varepsilon_x & \frac{1}{2}\delta\gamma_{xy} & \frac{1}{2}\delta\gamma_{xz} \\ \frac{1}{2}\delta\gamma_{yx} & \delta\varepsilon_y & \frac{1}{2}\delta\gamma_{yz} \\ \frac{1}{2}\delta\gamma_{zx} & \frac{1}{2}\delta\gamma_{zy} & \delta\varepsilon_z \end{bmatrix} \tag{8.3b}$$

from which it is seen that each component of engineering shear strain $(\gamma_{xy}, \gamma_{yz}, \gamma_{xz})$ is twice the corresponding tensor component $(\varepsilon_{12}, \varepsilon_{23}, \varepsilon_{13})$. Work is done by the normal stresses when forces $(\sigma_x + \delta\sigma_x)(\delta y \times \delta z)$, $(\sigma_y + \delta\sigma_y)(\delta x \times \delta z)$ and $(\sigma_z + \delta\sigma_z)(\delta x \times \delta y)$ move their points of application through their respective extensions $(\delta\varepsilon_x \times \delta x)$, $(\delta\varepsilon_y \times \delta y)$ and $(\delta\varepsilon_z \times \delta z)$. The work done by shear forces becomes $(\tau_{yx} + \delta\tau_{yx})(\delta x \times \delta z)$, $(\tau_{zy} + \delta\tau_{zy})(\delta y \times \delta x)$ and $(\tau_{xz} + \delta\tau_{xz})(\delta z \times \delta y)$ as they move their points of application through the respective shear displacements $(½\delta\gamma_{yx})\delta y$, $(½\delta\gamma_{zy})\delta z$ and $(½\delta\gamma_{xz})\delta x$. In addition, the action of complementary shear forces $(\tau_{xy} + \delta\tau_{xy})(\delta y \times \delta z)$, $(\tau_{yz} + \delta\tau_{yz})(\delta x \times \delta z)$ and $(\tau_{zx} + \delta\tau_{zx})(\delta y \times \delta x)$ do work against their respective displacements $(½\delta\gamma_{xy})\delta x$, $(½\delta\gamma_{yz})\delta y$ and $(½\delta\gamma_{zx})\delta z$. The net increase in strain energy is, by superposition, the sum of the products of these forces and the displacements along their lines of action. That is,

$$\begin{aligned} \delta U = {} & (\sigma_x + \delta\sigma_x)(\delta y \times \delta z)(\delta\varepsilon_x \times \delta x) + (\sigma_y + \delta\sigma_y)(\delta x \times \delta z)(\delta\varepsilon_y \times \delta y) \\ & + (\sigma_z + \delta\sigma_z)(\delta x \times \delta y)(\delta\varepsilon_z \times \delta z) + (\tau_{yx} + \delta\tau_{yx})(\delta x \times \delta z)(½\delta\gamma_{yx})\delta y \\ & + (\tau_{zy} + \delta\tau_{zy})(\delta y \times \delta z)(½\delta\gamma_{zy})\delta z + (\tau_{xz} + \delta\tau_{xz})(\delta z \times \delta y)(½\delta\gamma_{xz})\delta x \\ & + (\tau_{xy} + \delta\tau_{xy})(\delta y \times \delta z)(½\delta\gamma_{xy})\delta x + (\tau_{yz} + \delta\tau_{yz})(\delta x \times \delta z)(½\delta\gamma_{yz})\delta y \\ & + (\tau_{zx} + \delta\tau_{zx})(\delta y \times \delta x)(½\delta\gamma_{zx})\delta z \end{aligned} \tag{8.4a}$$

Since elastic strains are small, the infinitesimal products $\delta\sigma \times \delta\varepsilon$ in eq(8.4a), may be neglected to give a simpler form of δU, which embodies the properties of complementary shear, $\tau_{xy} = \tau_{yx}$, $\delta\gamma_{xy} = ½\delta\gamma_{xy} + ½\delta\gamma_{yx}$ etc.

$$\delta U = (\sigma_x\delta\varepsilon_x + \sigma_y\delta\varepsilon_y + \sigma_z\delta\varepsilon_z + \tau_{xy}\,\delta\gamma_{xy} + \tau_{yz}\,\delta\gamma_{yz} + \tau_{zx}\,\delta\gamma_{zx})\delta V \tag{8.4b}$$

Note, in the tensor notation, eq(8.4b) is conveniently written as

$$\delta U = (\sigma_{ij}\,\delta\varepsilon_{ij})\delta V \tag{8.5a}$$

because, by the summation convention

$$\begin{aligned} \sigma_{ij}\,\delta\varepsilon_{ij} &= \sigma_{1j}\,\delta\varepsilon_{1j} + \sigma_{2j}\,\delta\varepsilon_{2j} + \sigma_{3j}\,\delta\varepsilon_{3j} \\ &= [\sigma_{11}\,\delta\varepsilon_{11} + \sigma_{12}\,\delta\varepsilon_{12} + \sigma_{13}\,\delta\varepsilon_{13}] + [\sigma_{21}\,\delta\varepsilon_{21} + \sigma_{22}\,\delta\varepsilon_{22} + \sigma_{23}\,\delta\varepsilon_{23}] \\ &\quad + [\sigma_{31}\,\delta\varepsilon_{31} + \sigma_{32}\,\delta\varepsilon_{32} + \sigma_{33}\,\delta\varepsilon_{33}] \end{aligned}$$

There are alternative ways of writing the scalar measure of work in eq(8.4a). If we define column vectors of stress and strain increment as follows: $\boldsymbol{\sigma} = [\sigma_x\ \sigma_y\ \sigma_z\ \tau_{xy}\ \tau_{xz}\ \tau_{yz}]^{\mathrm{T}}$ and $\boldsymbol{\varepsilon} = [\varepsilon_x\ \varepsilon_y\ \varepsilon_z\ \gamma_{xy}\ \gamma_{xz}\ \gamma_{yz}]^{\mathrm{T}}$ respectively, then

$$\delta U = \boldsymbol{\sigma}^{\mathrm{T}}\,\delta\boldsymbol{\varepsilon}\,\delta V = \boldsymbol{\sigma}\,(\delta\boldsymbol{\varepsilon})^{\mathrm{T}}\,\delta V \tag{8.5b}$$

Equation (8.5a) is identical to the the sum of the diagonal components when the matrices of stress $\mathbf{S} = \sigma_{ij}$ and strain increment $\delta\mathbf{E} = \delta\varepsilon_{ij}$, eqs(8.3a,b), are multiplied. This is written as the trace of the matrix product:

$$\delta U = \mathrm{tr}\,(\mathbf{S}\,\delta\mathbf{E})\delta V = \mathrm{tr}\,(\delta\mathbf{E}\,\mathbf{S})\delta V \tag{8.5c}$$

Figure 8.1b shows the change in all the stress and strain components when an elastic relationship exists between them. Clearly the shaded area represents the change in strain energy per unit volume, or the *strain energy density* $\delta U/\delta V = \sigma_{ij}\,\delta\varepsilon_{ij}$. Elasticity implies that strains and the strain energy are wholly recoverable. When the same elastic stress-strain path is followed between loading and unloading, the formulation $\sigma_{ij} = \sigma_{ij}\,(\varepsilon_{ij})$ admits materials that are elastic and non-linear. Only in the particular case where Hooke's law is obeyed will the stress be linearly related to strain by elastic constants. The total strain energy stored in a body may be found by integrating any of eqs(8.5a,b,c) along the strain path and over the whole volume

$$U = \int_{\varepsilon_{ij}}\int_V (\sigma_{ij}\,\mathrm{d}\varepsilon_{ij})\,\mathrm{d}V = \int_{\boldsymbol{\varepsilon}}\int_V \boldsymbol{\sigma}^{\mathrm{T}}\delta\boldsymbol{\varepsilon}\ \mathrm{d}V = \int_{\mathbf{E}}\int_V \mathrm{tr}\,(\mathbf{S}\,\delta\mathbf{E})\,\mathrm{d}V \tag{8.6}$$

For Hookean materials the component of $\delta\varepsilon_{ij}$ will depend upon each acting stress, by the constitutive laws given in eqs(2.15a-d). Elementary texts derive the integral (8.6) for simple elasic loadings in tension, bending, shear and torsion. These are summarised in column two of Table 8.1 for each action: a direct force W, a bending moment M, a shear force F and a torque T. Generally, the area properties vary with the length z by writing these as $A(z)$, $I\,(z)$ and $J(z)$. When these variations are absent, A, I and J become constants and the integrals yield simple expressions all of similar form appearing in the final column. They will be applied in the following examples.

Table 8.1 Strain Energy Expressions

Loading	U variable loading/section	U constant loading/section
Tension/Compression	$\dfrac{1}{2E}\displaystyle\int_0^L \dfrac{[W(z)]^2\mathrm{d}z}{A(z)}$	$\dfrac{W^2L}{2EA}$
Bending	$\dfrac{1}{2E}\displaystyle\int_0^L \dfrac{[M(z)]^2\mathrm{d}z}{I(z)}$	$\dfrac{M^2L}{2EI}$
Shear	$\dfrac{1}{2G}\displaystyle\int_0^L \dfrac{[F(z)]^2\mathrm{d}z}{A(z)}$	$\dfrac{F^2L}{2GA}$
Torsion	$\dfrac{1}{2G}\displaystyle\int_0^L \dfrac{[T(z)]^2\mathrm{d}z}{J(z)}$	$\dfrac{T^2L}{2GJ}$

Example 8.1 The 20 mm diameter, cranked steel cantilever in Fig. 8.2 carries a vertical end force of $F = 50$ N. Determine the displacement v, under F, from a consideration of the strain energy stored in each limb under bending, shear and torsional effects. Take $E = 200$ GPa and $G = 80$ GPa.

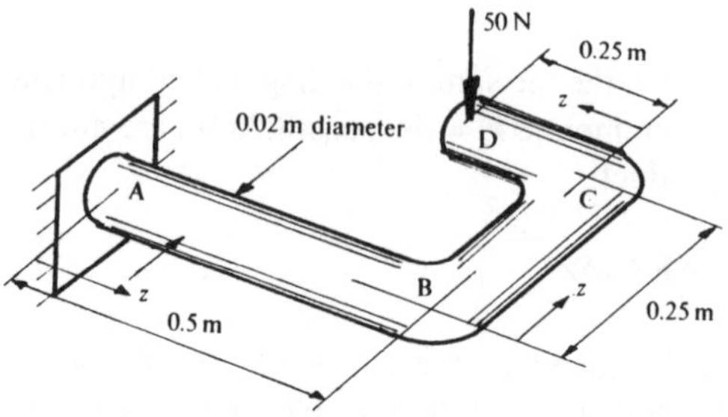

Figure 8.2 Cranked cantilever beam

$A = (\pi/4)(0.02)^2 = 0.3142 \times 10^{-3}\ \mathrm{m}^2$
$I = (\pi/64)(0.02)^4 = 0.7854 \times 10^{-8}\ \mathrm{m}^4$
$J = 2I = 1.571 \times 10^{-8}\ \mathrm{m}^4$

With the origin for z at A, limb AB is subjected to the three effects

$M = 50\,(0.50 - z)$ Nm for $0 \le z \le 0.5$ m
$T = 50 \times 0.25 = 12.5$ Nm for $0 \le z \le 0.5$ m
$F = 50$ N for $0 \le z \le 0.5$ m

Substituting into Table 8.1 expressions and adding, the net strain energy stored in AB is

$$U_{AB} = \int_0^{0.50} (50)^2 (0.5 - z)^2 \mathrm{d}z/(2EI) + \int_0^{0.50} (12.5)^2 \mathrm{d}z/(2GJ) + \int_0^{0.50} (50)^2 \mathrm{d}z/(2GA) \quad \text{(i)}$$

Limb BC is also subjected to all three effects. With the origin for z at B, they are

$M = 50\,(0.25 - z)$ Nm for $0 \le z \le 0.25$
$T = 50 \times 0.25 = 12.5$ Nm for $0 \le z \le 0.25$ m
$F = 50$ N for $0 \le z \le 0.25$ m

$$U_{BC} = \int_0^{0.25} (50)^2 (0.25 - z)^2 \mathrm{d}z/(2EI) + \int_0^{0.25} (12.5)^2 \mathrm{d}z/(2GJ) + \int_0^{0.25} (50)^2 \mathrm{d}z/(2GA) \quad \text{(ii)}$$

Limb CD is subjected to bending and shear. With the origin for z at C,

$M = 50(0.25 - z)$ Nm for $0 \le z \le 0.25$
$F = 50$ N for $0 \le z \le 0.25$ m

$$\therefore U_{CD} = \int_0^{0.25} (50)^2 (0.25 - z)^2 \mathrm{d}z/(2EI) + \int_0^{0.25} (50)^2 \mathrm{d}z/(2GA) \quad \text{(iii)}$$

Adding eqs(i),(ii) and (iii) to give the total strain energy stored,

$$U_T = \int_0^{0.50} (50)^2 (0.5 - z)^2 \mathrm{d}z /(2EI) + 2\int_0^{0.25} (50)^2 (0.25 - z)^2 \mathrm{d}z /(2EI)$$

$$+ 3\int_0^{0.25} (12.5)^2 \mathrm{d}z /(2GJ) + 4\int_0^{0.25} (50)^2 \mathrm{d}z /(2GA)$$

$$= [33.20 + 8.29 + 46.62 + 0.05]10^{-3}\ \mathrm{Nm} = 88.16 \times 10^{-3}\ \mathrm{J}$$

in which the contribution from shear is negligible. The external work done is $W = \int F\,\mathrm{d}\Delta$, and when a linear law $F = K\Delta$ (K is a constant stiffness) is followed, the integration leads to $W = F\Delta/2$, i.e. the triangular area beneath the F versus Δ diagram for loading. Then from eq(8.2), $F\Delta/2 = U_T$ and the deflection is

$$\Delta = 2U_T/F = 2 \times 88.16/50 = 3.53\ \mathrm{mm}$$

Example 8.2 Find an expression for the shear deflection of a cantilever of length L with rectangular section $b \times d$ carrying (a) a vertical concentrated force F at its free end and (b) a uniformly distributed load w/unit length (see Fig. 8.3a,b). Compare with the deflection due to bending.

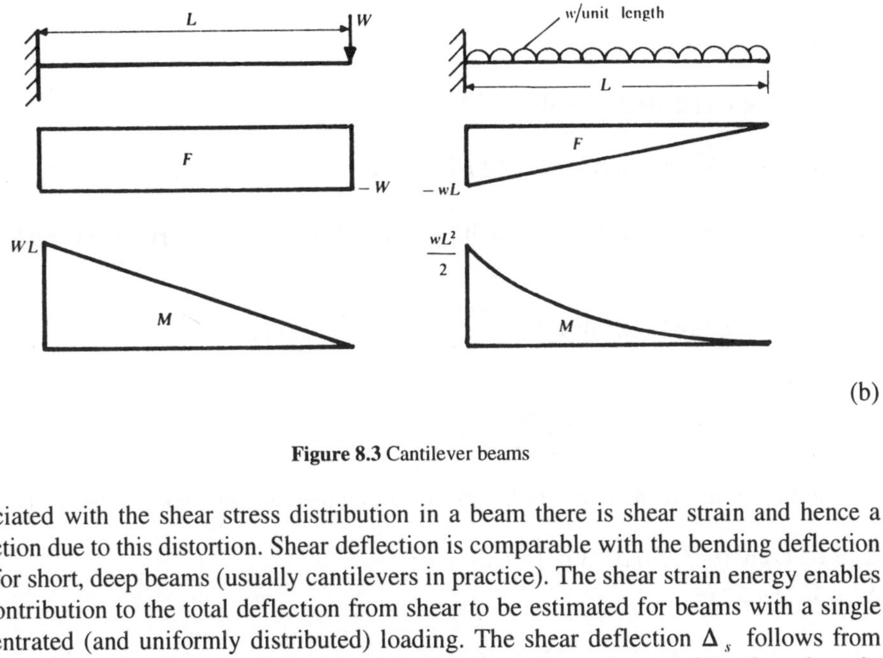

Figure 8.3 Cantilever beams

Associated with the shear stress distribution in a beam there is shear strain and hence a deflection due to this distortion. Shear deflection is comparable with the bending deflection only for short, deep beams (usually cantilevers in practice). The shear strain energy enables the contribution to the total deflection from shear to be estimated for beams with a single concentrated (and uniformly distributed) loading. The shear deflection Δ_s follows from eq(8.2) with $W = F\Delta_s/2$ and U defined in Table 8.1. For volume element $\delta V = \delta x \times \delta y \times \delta z$,

$$F\Delta_s/2 = [1/(2G)] \int_x \int_y \int_z \tau^2\, \mathrm{d}x\, \mathrm{d}y\, \mathrm{d}z \qquad \text{(i)}$$

where τ is the expression for the shear stress due to shear force in eq(7.4). In the general case, τ is a function of x, y and z because the shear force F varies with the length z, and the first moment integral in eq(7.4) varies with x and y. However, in a rectangular section $b \times d$, $I = bd^3/12$, $x = b$ and $y_1 = d/2$, so eq(7.4) reduces to

$$\tau = [12F/(bd^3)] \int_y^{d/2} y\, \mathrm{d}y$$

$$\tau = [6F/(bd^3)][d^2/4 - y^2] \qquad \text{(ii)}$$

(a) In Fig. 8.3a the shear force F remains constant for the length of the cantilever. Thus τ varies only with y. Substituting eq(ii) and $\delta V = bL\delta y$ into (i) gives

$$F\Delta_s/2 = [1/(2G)][6F/(bd^3)]^2 bL \int_{-d/2}^{d/2} [d^2/4 - y^2]^2 \mathrm{d}\,y \qquad \text{(iii)}$$
$$= [18F^2L/(Gbd^6)] \mid d^4 y/16 - d^2y^3/6 + y^5/5 \Big|_{-d/2}^{d/2}$$
$$= 3F^2L/(5bdG)$$
$$\therefore \Delta_s = 6FL/(5bdG)$$

The free-end bending deflection Δ_b is also found from eq(8.2) and Table 8.1 as follows:

$$F\Delta_b/2 = [1/(2EI)] \int_0^L M^2 \mathrm{d}z \qquad \text{(iv)}$$

where, with the origin for z at the fixed end (see Fig. 8.3a) $M = F(L - z)$. Hence

$$F\Delta_b/2 = [F^2/(2EI)] \int_0^L (L^2 - 2Lz + z^2)\, \mathrm{d}z$$
$$= F^2L^3/(6EI)$$
$$\therefore \Delta_b = FL^3/(3EI)$$

The total free-end deflection is then

$$\Delta_t = \Delta_s + \Delta_b = 6FL/(5bdG) + FL^3/(3EI)$$

(b) For the distributed loading in Fig. 8.3b, F varies with both y and z. Hence $\delta V = b\delta y\delta z$ and eq(iii) becomes

$$F\Delta_s/2 = [1/(2G)][6F/(bd^3)]^2 b \int_0^L [\int_{-d/2}^{d/2} (d^2/4 - y^2)^2 \mathrm{d}\,y]\mathrm{d}z$$

from which

$$\Delta_s = (1/G)[6/(bd^3)]^2 b \int_0^L F[\int_{-d/2}^{d/2} (d^2/4 - y^2)^2 \mathrm{d}\,y]\mathrm{d}z$$

Substituting from Fig. 8.3b, $F = w(L - z)$ and noting that the inner integral term [ ] = $d^5/30$

$$\Delta_s = (w/G)[6/(bd^3)]^2 (bd^5/30) \int_0^L (L - z)\, \mathrm{d}z$$
$$\Delta_s = 3wL^2/(5Gbd)$$

It follows from the M-diagram in Fig. 8.3b that the strain energy for bending is found from substituting $M = w(L - z)^2/2$ into eq(iv)

$$U = [w^2/(8EI)] \int_0^L (L - z)^4\, \mathrm{d}z$$

However, the external work cannot be found when the load is distributed and we need to use the principal of virtual work to show that the end deflection is $\Delta_b = wL^4/(8EI)$ (see Section 8.3 for this derivation).

8.2 Castigliano's Theorems

Equating the strain energy to external work in the manner of the previous two examples will allow the displacement at the load point to be found. This method cannot be used to determine deflections at the load points of a multiply loaded structure. For this Castigliano's theorems (1879) may be employed. The first theorem follows from the application of eq(8.2) to an equilibrium structure carrying n external forces F_i (i = 1, 2, 3 ... n), for which the associated in-line displacements are Δ_i. Let $\delta\Delta_i$ correspond to a change $\mathrm{d}F_i$ in these forces. From eq(8.2)

$$U = \int_0^{\Delta_i} F_i \, \mathrm{d}\Delta_i$$

in which the following summation is implied

$$U = \int_0^{\Delta_1} F_1 \mathrm{d}\Delta_1 + \int_0^{\Delta_2} F_2 \mathrm{d}\Delta_2 + \int_0^{\Delta_3} F_3 \mathrm{d}\Delta_3 + \dots \int_0^{\Delta_n} F_n \mathrm{d}\Delta_n$$

It can be seen that if all but one displacement be held fixed, then

$$\partial U / \partial \Delta_i = \partial (\int_0^{\Delta_i} F_i \, \mathrm{d}\Delta_i) / \partial \Delta_i, \quad \text{or}$$

$$F_i = \partial U / \partial \Delta_i \tag{8.7}$$

Equation (8.7) states that the force F_i at point i may be obtained from differentiating U partially with respect to the displacement Δ_i in-line with that force.

The second part of the theorem, which leads to a more useful result than eq(8.7), employs the concepts of complementary energy and work. The external work integral $W = \int F \, \mathrm{d}\Delta$ has previously been identified with the area beneath the F versus Δ diagram, as shown in Fig. 8.4.

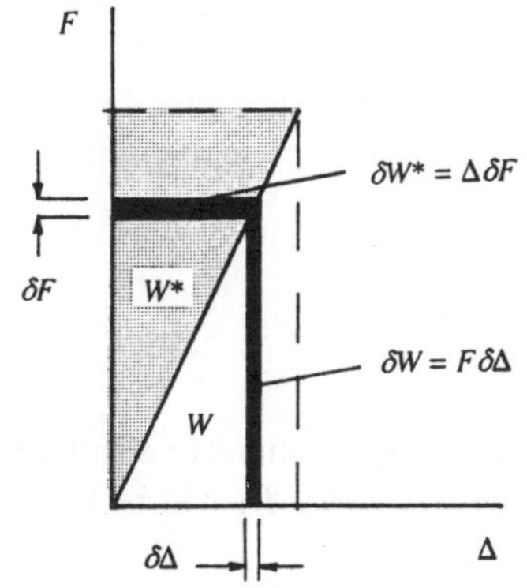

Figure 8.4 External work W and complementary work W^*

The complementary work increment $\delta W^* = \Delta \delta F$ is that area W^* in Fig. 8.4 defined as

$$W^* = \int_0^F \Delta \, \mathrm{d}F \tag{8.8}$$

In physical terms W^* may be identified with the work done when the point of application of a variable force F undergoes a displacement Δ. This is stored as complementary energy U^* such that

$$U^* = W^* = \int_0^F \Delta \, \mathrm{d}F$$

In the case of a number of externally applied forces F_i with their corresponding displacements Δ_i, eq(8.8) becomes

$$U^* = \int_0^{F_i} \Delta_i \, dF_i$$

and, if all but one force remains fixed, then

$$\partial U^* / \partial F_i = \partial \left(\int_0^{F_i} \Delta_i \, dF_i \right) / \partial F_i$$

$$\Delta_i = \partial U^* / \partial F_i \tag{8.9}$$

That is, the deflection Δ_i, at a given load point i, may be obtained by differentiating U^* with respect to that force F_i. Equation(8.9) applies to any non-linear elastic body. In the particular case of a linear Hookean material, where $\sigma = E\varepsilon$, the strain and complementary energies may be written from eq(8.6) as

$$U = \int_\varepsilon \int_V \sigma \, d\varepsilon \, dV = E \int_\varepsilon \int_V \varepsilon \, d\varepsilon \, dV = E\varepsilon^2 V/2 = \sigma \, \varepsilon \, V/2$$

$$U^* = \int_\sigma \int_V \varepsilon \, d\sigma \, dV = (1/E) \int_\sigma \int_V \sigma \, d\sigma \, dV = \sigma^2 V/(2E) = \sigma \, \varepsilon \, V/2$$

Thus, $U = U^*$ and eq(8.9) appears in a more useful form:

$$\Delta_i = \partial U / \partial F_i \tag{8.10a}$$

The theorem may be extended to include the rotations ϕ_i under externally applied moments M_i. For a Hookean material, the rotation equivalent to eq(8.10a) follows a similar proof:

$$\phi_i = \partial U / \partial M_i \tag{8.10b}$$

Castigliano's theorems are subject to the limitation that displacements cannot be found very easily at points where no loads act. For this we use the virtual work principle, as outlined in section 8.3.

8.2.1 Statically Determinate Structures

The following examples illustrate two applications of eqs(8.10a,b) where the *action* within the energy expression W, M, F and T in Table 8.1 will be known from static equilibrium.

Example 8.3 The thin strip in Fig. 8.5 is fixed at C and free to slide without friction at A. Calculate the horizontal reaction F_A at A, when the strip supports a vertical force F_B at point B, as shown. EI is constant.

Assume that all the strain energy stored is due to bending. Within the arcs AB and BC shown θ is measured anti-clockwise from A. Taking clockwise moments as positive, the moment expressions at points p and q apply to AB and BC as follows:

$$M_{AB} = F_A R\,(1 - \cos\theta) \text{ for } 0 \le \theta \le \pi/2 \tag{i}$$

$$M_{BC} = F_A R\,[1 + \sin(\theta - \pi/2)] - F_B R\,[1 - \cos(\theta - \pi/2)] \tag{ii}$$

$$= F_A R\,(1 - \cos\theta) - F_B R\,(1 - \sin\theta) \text{ for } \pi/2 \le \theta \le 5\pi/4$$

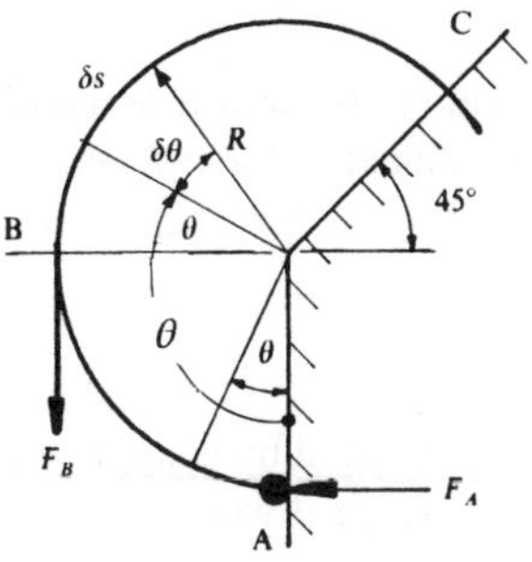

Figure 8.5 Thin strip

The total strain energy of bending is, from Table 8.1

$U = U_{AB} + U_{BC}$
$= [1/(2EI)] \int M_{AB}{}^2 \mathrm{d}s + [1/(2EI)] \int M_{BC}{}^2 \mathrm{d}s$

The deflection in-line with F_A is prevented when from eq(8.10a),

$\Delta_A = 0 = \partial U / \partial F_A$
$= [1/(EI)] \int M_{AB} (\partial M_{AB} / \partial F_A) \mathrm{d}s + [1/(EI)] \int M_{BC} (\partial M_{BC} / \partial F_A) \mathrm{d}s$

Substituting from eqs(i) and (ii) with $\mathrm{d}s = R\,\mathrm{d}\theta$,

$$[R^3/(EI)] \int_0^{\pi/2} F_A(1-\cos\theta)^2 \mathrm{d}\theta + [R^3/(EI)] \int_{\pi/2}^{5\pi/4} [F_A(1-\cos\theta)^2 - F_B(1-\sin\theta)(1-\cos\theta)]\mathrm{d}\theta = 0$$

$$F_A | 3\theta/2 - 2\sin\theta + ¼ \sin 2\theta \Big|_0^{\pi/2} + F_A | 3\theta/2 - 2\sin\theta + ¼ \sin 2\theta \Big|_{\pi/2}^{5\pi/4}$$
$$- F_B | \theta + \cos\theta - \sin\theta - ¼ \cos 2\theta \Big|_{\pi/2}^{5\pi/4} = 0$$

$F_A(3\pi/4 - 2) + F_A[(15\pi/8 + \sqrt{2} + 1/4) - (3\pi/4 - 2)]$
$\quad - F_B[(5\pi/4 - 1/\sqrt{2} + 1/\sqrt{2}) - (\pi/2 - 1 + 1/4)] = 0$

$0.3562\,F_A + 7.1985 F_A - 3.1062 F_B = 0$

$F_A = 3.1062 F_B / 7.5547 = 0.411 F_B$

Example 8.4 Determine the vertical deflection at point Q for the Warren bay girder in Fig. 8.6 when it is hinged at S and supported on rollers at P. Take E = 208 GPa for each bar with identical 10 m length and area 1280 mm^2.

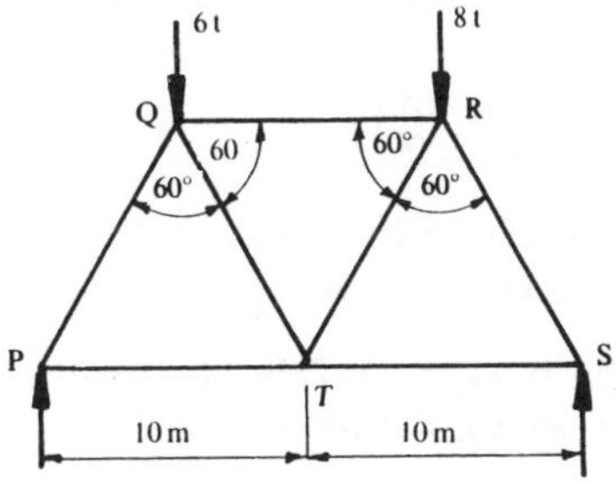

Figure 8.6 Warren bay girder

The total strain energy stored is the sum of the tensile and compressive strain energies stored in the bars. From Table 8.1, with a constant force P induced in each bar of length L and section area A, the total energy stored by the frame is

$$U = \Sigma\, [P^2 L/(2AE)] \tag{i}$$

In the case where the bar forces P are induced by a single external force F, the deflection Δ beneath F is found from equating the external work to U. That is, from eq(8.2),

$$F\Delta/2 = \Sigma\, [P^2 L/(2AE)] \tag{ii}$$

If a unit force replaces F to produce bar forces k, it follows that $P = kF$ and eq(ii) becomes

$$F\Delta/2 = (F^2/2)\Sigma\, [\, k^2 L/(AE)]$$
$$\Delta = F\, \Sigma\, [\, k^2 L/(AE)] \tag{iii}$$

In this problem two external forces F are applied to the frame. The deflection Δ_Q is required at a node point Q where the force is F_Q. It is convenient to take the induced forces P as the sum of the two components, $P = P' + P''$ where P' are the bar forces in the absence of F_Q but in the presence of the other external force(s) and P'' are the bar forces when F_Q acts in isolation. When a unit force acts in isolation at Q, to induce bar forces k, then $P'' = kF_Q$. From eq(i), the total strain energy in the frame is

$$U = \Sigma\, [(P' + P'')^2 L/(2AE)] = \Sigma\, [(P' + kF_Q)^2 L/(2AE)] \tag{iv}$$

Applying eq(8.10a) to eq(iv), the deflection in the direction of F_Q is

$$\Delta_Q = \partial U / \partial F_Q = \Sigma\, 2\, (P' + kF_Q)[kL/(2AE)]$$
$$\Delta_Q = \Sigma\, P' kL/(AE) + F_Q \Sigma\, k^2 L/(AE) \tag{v}$$

Note that k is a dimensionless force coefficient and if $F_Q = 0$, it is still possible to obtain the deflection at Q from the first term in eq(v). The symmetry of the given frame allows a further simplification to eq(v):

$$\Delta_Q = [L/(AE)][F_R \Sigma(k\, k') + F_Q \Sigma\, k^2] \tag{vi}$$

where k' are the bar forces when a unit force acts at R. Bar forces k are found for a unit force at Q from the following joint equilibrium equations by assuming tension at each joint.

At P: $\uparrow$, $0.75 + k_{PQ} \sin 60° = 0, \Rightarrow k_{PQ} = -\ 0.866$
$\rightarrow$ $k_{PT} + k_{PQ} \cos 60° = 0, \quad \Rightarrow k_{PT} = 0.433$

At Q: $\uparrow$, $-\ 1 - k_{PQ} \sin 60° - k_{QT} \sin 60° = 0, \Rightarrow k_{QT} = -\ 0.289$
$\rightarrow$ $k_{QR} + k_{QT} \cos 60° - k_{PQ} \cos 60° = 0, \Rightarrow k_{QR} = -\ 0.289$

At S: $\uparrow$, $0.25 + k_{RS} \sin 60° = 0, \Rightarrow k_{RS} = -\ 0.289$
$\rightarrow$, $-\ k_{RS} \cos 60° - k_{TS} = 0, \Rightarrow k_{TS} = 0.144$

At R: $\uparrow$, $-\ k_{RT} \sin 60° - k_{RS} \sin 60° = 0, \Rightarrow k_{RT} = 0.289$

With a unit force applied at R bar forces k' are obvious from inspection within Table 8.2.

Table 8.2 Bar force coefficients

Bar	k	k'	k^2	kk'
PQ	-0.866	-0.289	0.750	0.250
QT	-0.289	0.289	0.0835	-0.0835
QR	-0.289	-0.289	0.0835	0.0835
PT	0.433	0.144	0.188	0.0622
RT	0.289	-0.289	0.0835	-0.0835
RS	-0.289	-0.866	0.0835	0.250
TS	0.144	0.433	0.02074	0.0622
			1.2927	0.5409

Applying eq(vi) to the sum of the final two columns in Table 8.2,

$$\Delta_Q = [(10 \times 10^3 \times 9.81 \times 10^3)/(1280 \times 208 \times 10^3)][(8 \times 0.5409) + (6 \times 1.2927)] = 4.45 \text{ mm}$$

8.2.2 Statically Indeterminate Structures

When the actions *W*, *M*, *F* and *T* in Table 8.1 cannot be found from equilibrium equations it becomes necessary to employ the compatibility conditions. The latter relates to what is known about the displacement or slope in the deformed structure so that we achieve the same number of equations as there are unknown actions. There are many so-called redundant or hyperstatic structures each requiring individual treatment by Castigliano, as the following examples show. Later in this chapter it will be seen that virtual work provides an alternative solution.

Example 8.5 During the elastic calibration of the proving ring in Fig. 8.7a, the in-line deflection is measured under a vertical diametral force *F*. Establish the theoretical stiffness factor.

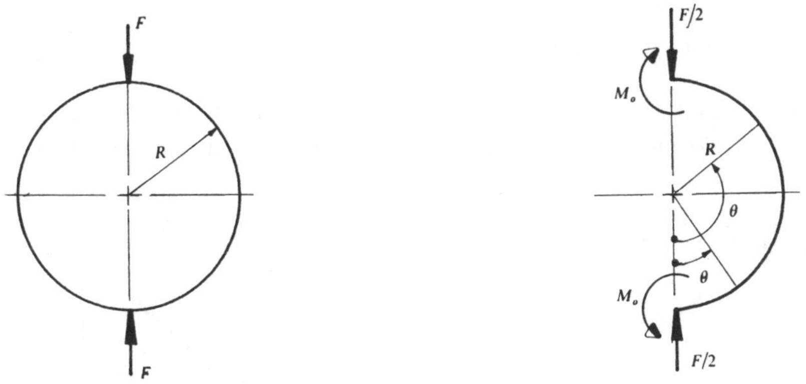

(a) (b)

Figure 8.7 Proving ring calibration

To find the strain energy it is only necessary to consider one symmetrical half of the ring in Fig. 8.7b. However, the induced moment M_o results in static indeterminancy. A compatibility condition requires that there must be zero slope ϕ under M_o. The strain energy for the whole ring is given by Table 8.1 as

$$U = 2[1/(2EI)] \int M^2 \mathrm{d}s$$

Under F the deflection and slope are, from eqs(8.10a,b),

$$\Delta = [2/(EI)] \int M\,(\partial M/\partial F)\,\mathrm{d}s \quad \text{(i)}$$

$$\phi = [2/(EI)] \int M(\partial M/\partial M_o)\,\mathrm{d}s = 0 \quad \text{(ii)}$$

where, from Fig. 8.6b, with the given origin for θ,

$$M = (FR/2)\sin\theta - M_o \quad \text{for } 0 \le \theta \le \pi/2$$

$$M = (FR/2)\cos(\theta - \pi/2) - M_o = (FR/2)\sin\theta - M_o \quad \text{for } \pi/2 \le \theta \le \pi$$

i.e. a single expression for M applies. Substituting this into eq(ii),

$$\int_0^{\pi} [(FR/2)\sin\theta - M_o](-1)R\,\mathrm{d}\theta = 0$$

$$| M_o\theta + (FR/2)\cos\theta \Big|_0^{\pi} = M_o\pi - FR = 0$$

$$\therefore M_o = FR/\pi$$

Substituting for M and M_o in eq(i),

$$\Delta = [2/(EI)] \int_0^{\pi} [(FR/2)\sin\theta - FR/\pi][(R/2)\sin\theta - R/\pi](R\,\mathrm{d}\theta)$$

$$= [2FR^3/(EI)] \int_0^{\pi} [\tfrac{1}{2}\sin\theta - (1/\pi)]^2\,\mathrm{d}\theta$$

$$= (2FR^3/(EI)\,|(\tfrac{1}{8}(\theta - \tfrac{1}{2}\sin 2\theta) + (1/\pi)\cos\theta + \theta/\pi^2 \Big|_0^{\pi}$$

$$= [2FR^3/(EI)](\pi/8 - 1/\pi)$$

$$K = F/\Delta = EI/[R^3(\pi/4 - 2/\pi)]$$

Example 8.6 If for the structure in Fig. 8.8a: $F = 2$ t, AB = DC = 10 m, AD = BC = 8 m and for all bars $A = 500$ mm^2, $E = 210$ GPa, determine the bar forces and the vertical deflection at C.

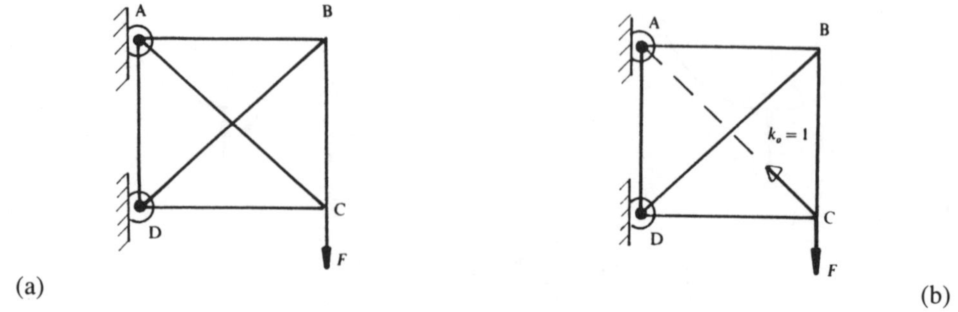

Figure 8.8 Redundant structure

The structure in Fig. 8.8a is redundant because either of the bars AC or BD could be removed without lessening the ability of the structure to support external forces. Castigliano postulated that the forces in the members of a redundant structure adjust themselves to minimise the strain energy. First remove the redundant bar AC (or BD) to make a statically determinate structure in Fig. 8.8b. Determine the bar forces P' under the externally applied loading in Fig. 8.8b. Then remove all external forces and apply a unit tensile force k_o at either joint A or C in the direction of the missing bar. Establish the bar forces k under the unit force. If now the redundant bar force R replaces k_o, it follows that the bar forces become $P'' = kR$. The net bar forces are $P = P' + P''$, so that the total strain energy of the frame in Fig. 8.8b becomes

$$U = \sum_1^{n-1} (P' + kR)^2 L / (2AE) \qquad \text{(i)}$$

where n is the number of bars in Fig. 8.8a. Applying eq(8.10a) to eq(i), the deflection of C in the line of R is

$$\Delta_C = \partial U / \partial R = \sum_1^{n-1} (P' + kR) kL / (AE)$$

$$\Delta_C = (1/E) \left[\sum_1^{n-1} (LkP'/A) + R \sum_1^{n-1} (k^2 L/A) \right] \qquad \text{(ii)}$$

The free extension of the redundant bar under R is

$$x = RL/(AE) = R\,L\,k_o^2/(AE) \qquad \text{(iii)}$$

When the redundant bar is replaced, it must fit exactly between A and C. This condition is ensured from eqs(ii) and (iii) when $\Delta_c + x = 0$,

$$(1/E)\left[\sum_1^{n-1} (LkP'/A) + R \sum_1^{n-1} (k^2 L/A) \right] + RLk_o^2/(AE) = 0$$

$$\sum_1^{n-1} (LkP'/A) + R\left[\sum_1^{n-1} (Lk^2/A) + Lk_o^2/A \right] = 0$$

$$R = \frac{-\sum_1^{n-1} (LkP'/A)}{\sum_1^{n-1} (Lk^2/A) + Lk_o^2/A} = \frac{-\sum_1^{n} (LkP'/A)}{\sum_1^{n} (Lk^2/A)} \qquad \text{(iv)}$$

The numerator may contain all bars when we set $P' = 0$ to include the redundant bar. The P' and k values are given in Table 8.3.

Table 8.3 Forces coefficients and lengths used within eq(iv)

Member	L(m)	P' (t)	k	LkP'	Lk^2	$P'' = kR$	$P = P' + P''$
AB	10	2.5	- 0.78	-19.50	6.084	- 1.428	1.072
BC	8	2	- 0.624	- 9.984	3.115	- 1.142	0.858
CD	10	0	- 0.780	0	6.084	- 1.428	- 1.428
BD	12.81	- 3.21	1.0	- 41.152	12.820	1.830	- 1.380
AD	8	2	- 0.624	- 9.984	3.115	- 1.142	0.858
AC	12.81	0	1.0	0	12.820	1.830	1.830
				$\sum = -80.62$	$\sum = 44.038$		

Following the procedure outlined above, eq(iv) is applied to the sum of the fifth and sixth columns noting that the bar areas are constant,

$$R = -\sum_1^n (LkP')/\sum_1^n (Lk^2) = 80.62 / 44.038 = 1.83 \text{ t}$$

This enables the determination of the bar forces P in the final column of Table 8.3. Finally the vertical deflection at point C under a single external force is (see Example 8.4)

$$\Delta_C = F_C \sum_1^n k^2 L/(AE) \qquad \text{(v)}$$

where k are the bar forces when a unit load is applied at C. They are found from dividing P in Table 8.3 by $F = 2$ t. That is,

Table 8.4 Final bar force coefficients

Member	AB	BC	CD	BD	AD	AC
k	0.536	0.429	- 0.714	- 0.690	0.429	0.915
L(m)	10	8	10	12.82	8	12.82
k^2L	2.873	1.472	5.098	6.104	1.472	10.733

$\therefore \sum_1^n k^2 L = 27.75$ (m) and, for constant bar areas in the same material, eq(v) gives

$$\Delta_C = (2 \times 10^3 \times 9.81 \times 27.75 \times 10^3)/(500 \times 210 \times 10^3) = 5.19 \text{ mm}$$

The deflection at any point Q in a redundant structure under a number of externally applied forces is found from writing eq(iv) in example 8.3 as

$$\Delta_Q = \sum_1^n [P' kL/(AE)] + F_Q \sum_1^n [k^2 L/(AE)] \qquad \text{(vi)}$$

First, the force in the redundant bar must be calculated twice from eq(iv) under each of the following loading conditions:

(i) for all external forces present except F_Q to give P'.

(ii) with only F_Q present to give k.

Associated with each redundant force calculation there will be separate P' and k bar force calculations. In section 8.7 it will be shown how the unit load method may be employed to find deflections more easily in structures with more than one redundancy.

8.3 The Principle of Virtual Work

When a system of co-planar forces is in equilibrium then the resultant force is zero. Assume that a similar system of forces F_k ($k = 1, 2, 3 \ldots n$) act on the body in Fig. 8.9 which does not itself deform. Let both the body and the forces be displaced by the virtual in-line displacements Δ^v_k, under the action of some external agency that does not change the magnitude of F_k.

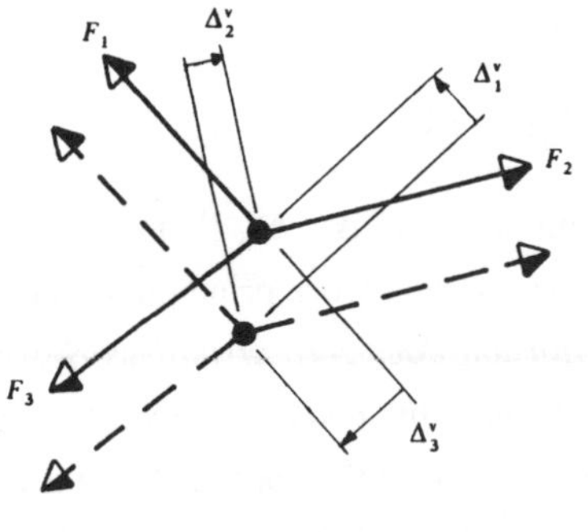

Figure 8.9 Virtual work principle

Clearly, since no external work is done by the forces, F_k, the principle of virtual work states that the virtual work is zero under a virtual displacement as a consequence of the equilibrium state. That is,

$$F_1\Delta^v{}_1 + F_2\Delta^v{}_2 + F_3\Delta^v{}_3 + \ldots F_n\Delta^v{}_n = 0 \tag{8.11a}$$
$$F_k\Delta^v{}_k = 0 \tag{8.11b}$$

Equations (8.11a,b) express the virtual work principle its simplest mathematical form for a rigid body. Now consider the deformable (non-rigid) volume element in Fig. 8.1a. The above principle will further apply where each element in a body is subjected to an equilibrium system of internal stresses and external forces which each contribute to the net work done. The expression (8.6) previously derived for the strain energy stored, U, represents the work done *on* the element. It follows that the internal work done, W_I, *by* the stresses is identical to U but with a change in sign. That is,

$$W_I = -\int_V \int_{\varepsilon_{ij}} (\sigma_{ij}\,\mathrm{d}\varepsilon_{ij})\,\mathrm{d}V \tag{8.12}$$

Let the stress system σ_{ij} be produced by a system of external forces F_k and the strain increment system $\delta\varepsilon_{ij}$ be produced by the virtual incremental displacements $\delta\Delta_k$. When the agency responsible for the latter is temperature, for example, it follows in the general case, that σ_{ij} and $\delta\varepsilon_{ij}$ are independent. Now, since the external work done is

$$W_E = \int_0^{\Delta_k} F_k\,\mathrm{d}\Delta_k \tag{8.13}$$

and the virtual work is zero, as a consequence of equilibrium, it follows from eqs(8.12) and (8.13)

$$W_E + W_I = 0$$

$$\int_0^{\Delta_k} F_k\,\mathrm{d}\Delta_k - \int_V \int_{\varepsilon_{ij}} (\sigma_{ij}\,\mathrm{d}\varepsilon_{ij})\,\mathrm{d}V = 0 \tag{8.14a}$$

Because the real and virtual quantities F_k and $\delta\Delta_k$ are independent so too are σ_{ij} and $\delta\varepsilon_{ij}$. Equation (8.14a) integrates directly to

$$F_k\Delta_k - \int_V (\sigma_{ij}\varepsilon_{ij})\,\mathrm{d}V = 0 \tag{8.14b}$$

in which the summation convention is implied over like subscripts. Alternatively, the corresponding matrix form of eq(8.14b) is

$$\mathbf{F}^{\mathrm{T}}\boldsymbol{\Delta} - \int_V \boldsymbol{\sigma}^{\mathrm{T}} \boldsymbol{\varepsilon}\, \mathrm{d}V = \boldsymbol{\Delta}^{\mathrm{T}}\mathbf{F} - \int_V \boldsymbol{\varepsilon}^{\mathrm{T}} \boldsymbol{\sigma}\, \mathrm{d}V = 0 \tag{8.14c}$$

where $\mathbf{F} = [F_1\ F_2\ F_3\,]^{\mathrm{T}}$ and $\boldsymbol{\Delta} = [\Delta_1\ \Delta_2\ \Delta_3\,]^{\mathrm{T}}$ are column matrices of the force and displacements components. The virtual work principle is stated thus:

A system of forces in equilibrium does zero virtual work when moved through a virtual displacement due to deformation of the body on which those forces act.

Note that the principle was derived from the application of a virtual displacement (strain) in the presence of a real or actual force (stress) system. Applications of this approach (i.e. the principle of virtual displacements) are considered in section 8.7. Equation (8.14b) could also have been derived from any one of three alternative force-displacements systems:

(i) A virtual, equilibrium, force-stress system with a real but independent, compatible, displacement-strain system. This is the principal of virtual forces which is particularly useful for the determination of displacements and forces in redundant structures.

(ii) Both systems real as with Castigliano's theorems.

(iii) Both systems virtual - which is of limited use.

Applications of the particular principles of virtual forces and displacements are now considered.

8.4 The Principle of Virtual Forces (PVF)

The real strain (ε_{ij}) and displacement (Δ_k) system is produced by the actual loading. Since the virtual force-stress system is independent, it will apply when the actual loading is removed and a single virtual or dummy force F^v is applied at the point where the actual displacement Δ is required. Equations(8.14b and c) become

$$F^v\Delta - \int_V (\sigma_{ij}{}^v \varepsilon_{ij})\, \mathrm{d}V = 0, \quad (i, j = 1, 2, 3) \tag{8.15a}$$

$$F^v\Delta - \int_V (\boldsymbol{\sigma}^v)^{\mathrm{T}} \boldsymbol{\varepsilon}\, \mathrm{d}V = 0 \tag{8.15b}$$

The second term in eqs(8.15a,b) has been re-expressed as $-\ W_I$ (see eq(8.14a)) in Table 8.5 for common states of virtual stress and real strain. The expressions in the final column apply to the unit load method (ULM) and will be discussed in the following section.

Table 8.5 Common forms of $W_I = -\int_V (\sigma_{ij}{}^v \varepsilon_{ij})\, \mathrm{d}V = -\int_V (\boldsymbol{\sigma}^v)^{\mathrm{T}} \boldsymbol{\varepsilon}\, \mathrm{d}V$

Loading	$\sigma_{ij}{}^v$	σ_{ij}	$\mathrm{d}V$	$-\ W_I$ (PVW)	$-\ W_I$(ULM)
Direct stress	$\sigma^v = P^v/A$	$\varepsilon = P/(AE)$	$\mathrm{d}A\ \mathrm{d}L$	$P^v PL/(AE)$	$pPL/(AE)$
Bending	$\sigma^v = M^v y/I$	$\varepsilon = My/(IE)$	$\mathrm{d}A\ \mathrm{d}z$	$\int M^v M \mathrm{d}z/(IE)$	$\int mM\mathrm{d}z/(IE)$
Shear	$\tau^v = F^v/A$	$\gamma = F/(AG)$	$\mathrm{d}A\ \mathrm{d}z$	$\int F^v F \mathrm{d}z/(AG)$	$\int f F\mathrm{d}z/(AG)$
Torsion	$\tau^v = T^v r/J$	$\gamma = Tr/(JG)$	$2\pi r\ \mathrm{d}r\ \mathrm{d}z$	$\int T^v T\mathrm{d}z/(JG)$	$\int tT\,\mathrm{d}z/(JG)$
Shear flow	$\tau^v = q^v/t$	$\gamma = q/(tG)$	$t\ \mathrm{d}s\ \mathrm{d}z$	$\oint q^v q\ \mathrm{d}s/(tG)$	$\oint q_1 q\ \mathrm{d}s/(tG)$

Example 8.7 Find the deflection at the free end of a cantilever in Fig. 8.10a when the temperature varies linearly in both the y and z directions, as shown.

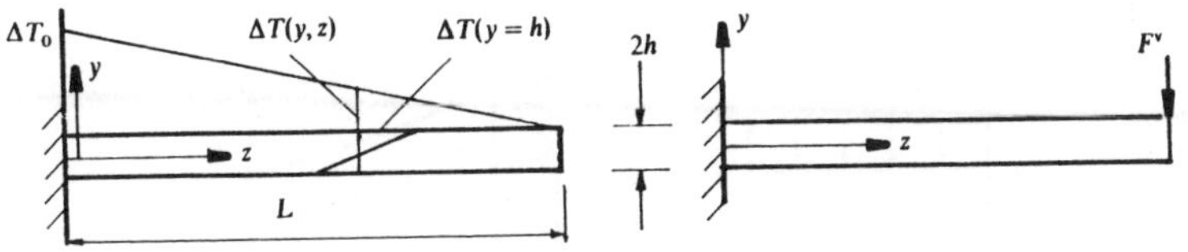

(a) (b)

Figure 8.10 Cantilever beam

A virtual force F^v applied at the free end produces the virtual uniaxial bending stress,

$$\sigma_{ij}^{\,v} \equiv \sigma = M^v y/I = F(L - z)\, y/I \qquad \text{(i)}$$

The actual (real) uniaxial strain produced by a temperature variation in both y and z is

$$\varepsilon_{ij} \equiv \alpha\,\Delta T(y, z) = \alpha\,[\Delta T_o y\,(L - z)/(hL)] \qquad \text{(ii)}$$

The volume element of beam is here: $\mathrm{d}V = \mathrm{d}A \times \mathrm{d}z$. Substituting eqs(i) and (ii) into eq(8.15a)

$$F^v\Delta - \int_A \int_z [F(L - z)\, y/I\,][\alpha\,\Delta T_o\, y\,(L - z)\,/\,(hL)](\mathrm{d}A \times \mathrm{d}z) = 0$$

$$F^v\Delta - [F^v\alpha\,\Delta T_o/(I\,hL)] \int_0^L (L - z)^2\,\mathrm{d}z \int_A y^2\,\mathrm{d}A = 0$$

but $I = \int_A y^2\,\mathrm{d}A$

$$\therefore\ F^v\Delta - [F^v\alpha\,\Delta T_o/(hL)] \int_0^L (L - z)^2\,\mathrm{d}z = 0$$

$$F^v\Delta - F^v\alpha\,\Delta T_o L^2/(3h) = 0$$

$$\Delta = \alpha\Delta T_o L^2\,(/3h)$$

Example 8.8 Determine the slope and deflections at the free-end of the cantilever structure in Fig. 8.11 from a consideration of bending only. What is the largest source of error from neglecting other effects?

Table 8.5 supplies the appropriate form of eq(8.15a) to give both the vertical Δ_y and horizontal Δ_x deflection at A. That is,

$$F^v\Delta - \int M^v M\,\mathrm{d}z/(EI) = 0 \qquad \text{(i)}$$

For Δ_y use is made of the actual M - diagram under the distributed loading in Fig. 8.11b and the M_y^v diagram (Fig. 8.11c) under an isolated virtual vertical force F_y^v at A. Applying the integral in eq(i) to limbs AB and BC separately,

$$F_y^v\Delta_y - [\int_0^L M_y^v M\,\mathrm{d}z/(EI\,) + \int_0^{3L/2} M_y^v M\,\mathrm{d}z/(EI\,)] = 0$$

$$F_y^{\,v}\Delta_y - [\int_0^L (F_y^{\,v}z)(wz^2/2)\mathrm{d}z/(EI) + \int_0^{3L/2} (F_y^{\,v}L)(wL^2/2)\mathrm{d}z/(2EI)] = 0$$

$$F_y^{\,v}\Delta_y - [wF_y^{\,v}/(2EI)\mid z^4/4\big|_0^L + [wF_y^{\,v}L^3/(4EI)]\mid z\big|_0^{3L/2}] = 0$$

$$\Delta_y = wL^4/(2EI)$$

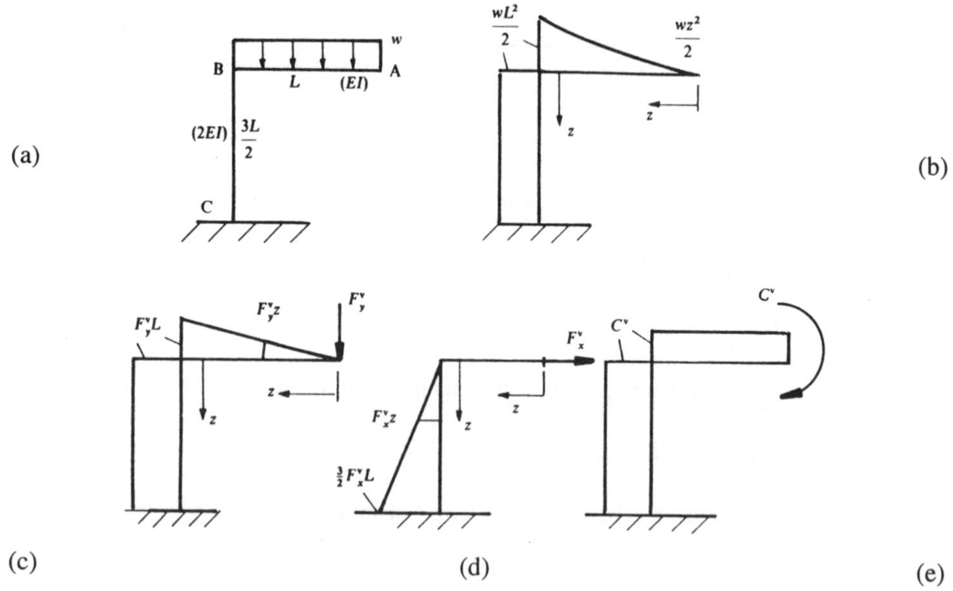

Figure 8.11 Cantilever structure

For Δ_x, the real moments (Fig. 8.11b) are used in conjunction with the virtual $M_x^{\,v}$ diagram (Fig. 8.11d) with a virtual horizontal force $F_x^{\,v}$ applied at A. Equation(i) becomes

$$F_x^{\,v}\Delta_x - [\int_0^L M_x^{\,v}M\,\mathrm{d}z/(EI) + \int_0^{3L/2} M_x^{\,v}M\,\mathrm{d}z/(EI)] = 0$$

$$F_x^{\,v}\Delta_x - [\int_0^L (0)(wz^2/2)\mathrm{d}z/(EI) + \int_0^{3L/2} (F_x^{\,v}z)(wL^2/2)\mathrm{d}z/(2EI)] = 0$$

$$F_x^{\,v}\Delta_x - [wF_x^{\,v}L^2/(4EI)]\mid z^2/2\,\big|_0^{3L/2} = 0$$

$$\Delta_x = 9wL^4/(32EI)$$

The following form of eq(8.15a) determines the free-end slope ϕ when a virtual couple C^v is applied at that point. This gives

$$C^v\phi - \int M^v M\,\mathrm{d}z/(EI) = 0 \qquad \text{(ii)}$$

Applying the integral in eq(ii) separately to each limb,

$$C^v\phi - [\int_0^L M^v M\,\mathrm{d}z/(EI) + \int_0^{3L/2} M^v M\,\mathrm{d}z\,/(EI)] = 0$$

in which M^v are the virtual moments produced in each limb under C^v (Fig. 8.11e) and M are the real moments in Fig. 8.11b. Then,

$$C^v\phi - [\int_0^L C^v(wz^2/2)\mathrm{d}z/(EI) + \int_0^{3L/2} C^v(wL^2/2)\mathrm{d}z/(2EI)] = 0$$

$$C^v\phi - \{[wC^v/(2EI)] \mid z^3/3\Big|_0^L + [wC^vL^2/(4EI)] \mid z\Big|_0^{3L/2}\} = 0$$

$$C^v\phi - \{[wC^vL^3/(6EI)] + [3wC^vL^3/(8EI)]\} = 0$$

$$\phi = 13wL^3/(24EI)$$

Effect of shear force - if we draw the shear force diagrams under the real and virtual loading, it will be seen that shear increases only the vertical displacement by an amount:

$$F_y^v\Delta_y - \int_0^L F^vF\,\mathrm{d}z/(GA) = 0$$

$$F_y^v\Delta_y - [1/(GA_{AB})]\int_0^L(-F_y^v)(-wz)\,\mathrm{d}z = 0$$

$$\Delta_y = wL^2/(2GA_{AB})$$

Effect of axial loading - again Δ_x is unaltered but a further contribution to Δ_y arises from compression in BC

$$F_y^v\,\Delta_y - P^vPL/(AE) = 0:$$
$$F_y^v\,\Delta_y - (-F_y^v)(-wL)(3L/2)/(A_{BC}E) = 0$$
$$\Delta_y = 3wL^2/(2A_{BC}E)$$

Both these contributions to Δ_y are small compared to that from bending.

Example 8.9 Derive an expression for the angle of twist and the vertical deflection under the end force P for the cantilever arc in Fig. 8.12a.

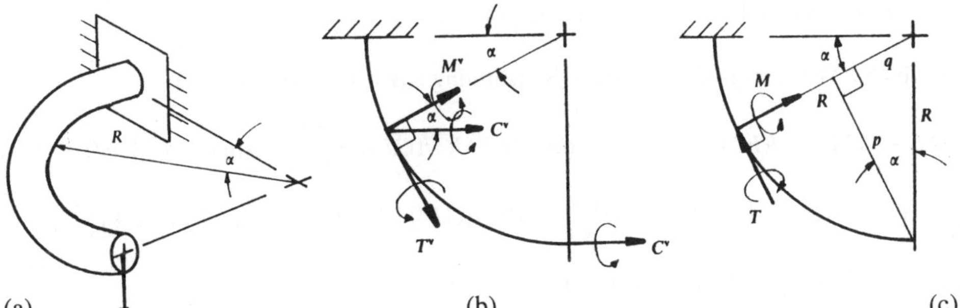

Figure 8.12 Cantilever arc

Ignoring shear force effects, the deformation arises from bending and torsion. The particular form of equation (8.15a), required for the angular twist, follows from Table 8.5,

$$C^v\theta - [\int M^vM\,\mathrm{d}s/(EI) + \int T^vT\,\mathrm{d}s/(GJ)] = 0 \qquad \text{(i)}$$

where s, the perimeter distance, replaces z and C^v is a virtual couple applied at the free end.

The virtual moment M^v and virtual torque T^v produced by C^v are found from resolution in Fig. 8.12b,

$$M^v = C^v \cos\alpha,\quad T^v = C^v \sin\alpha \qquad \text{(ii)}$$

The actual moment M and the torque T produced by the applied force P at any section defined by R and α are, from Fig. 8.12c,

$$M = Pp = PR\cos\alpha \text{ and } T = -P(R-q) = -PR(1-\sin\alpha) \qquad \text{(iii)}$$

The moment vectors show that M and M^v act in the same sense but the torque vectors T and T^v oppose. Hence the need for the minus sign in the T expression. Substituting (ii) and (iii) in (i) with $\mathrm{d}s = R\,\mathrm{d}\alpha$,

$$C^v\theta - [\textstyle\int (C^v\cos\alpha)(PR\cos\alpha)(R\,\mathrm{d}\alpha)/(EI) - \int (C^v\sin\alpha)PR(1-\sin\alpha)(R\,\mathrm{d}\alpha)/(GJ)] = 0$$

$$C^v\theta - \{[C^vPR^2/(EI)]\int_0^{\pi/2}\cos^2\alpha\,\mathrm{d}\alpha - [C^vPR^2/(GJ)]\int_0^{\pi/2}(\sin\alpha - \sin^2\alpha)\,\mathrm{d}\alpha\} = 0$$

$$C^v\theta - \{[C^vPR^2/(2EI)]\,|\alpha + \tfrac{1}{2}\sin 2\alpha|_0^{\pi/2} - [C^vPR^2/(GJ)]\,|\tfrac{1}{4}\sin 2\alpha - \cos\alpha - \alpha/2|_0^{\pi/2}\} = 0$$

$$\theta - \{[PR^2/(2EI)](\pi/2) - [PR^2/(GJ)](1-\pi/4)\} = 0$$
$$\theta = PR^2[\pi/(4EI) - (1-\pi/4)/(GJ)]$$

The vertical deflection Δ is supplied from eq(8.15a) and Table 8.5,

$$F^v\Delta - [\textstyle\int M^vM\,\mathrm{d}s/(EI) + \int T^vT\,\mathrm{d}s/(GJ)] = 0 \qquad \text{(iv)}$$

where eq(iii) again gives M and T under the actual loading. The virtual force F^v applied at the free end produces the virtual moment and torque

$$M^v = F^vR\cos\alpha \text{ and } T^v = -F^vR(1-\sin\alpha) \qquad \text{(v)}$$

in which F^v has replaced P in eq(iii). Substituting eqs(iii) and (v) into (iv),

$$F^v\Delta - \{[PF^vR^3/(EI)]\int_0^{\pi/2}\cos^2\alpha\,\mathrm{d}\alpha + [PF^vR^3/(GJ)]\int_0^{\pi/2}(1-\sin\alpha)^2\,\mathrm{d}\alpha\} = 0$$

$$\Delta - \{[PR^3/(2EI)]\,|\alpha + \tfrac{1}{2}\sin 2\alpha|_0^{\pi/2} + [PR^3/(GJ)]\,|3\alpha/2 + 2\cos\alpha - \tfrac{1}{4}\sin 2\alpha|_0^{\pi/2}\} = 0$$

$$\Delta - [\pi PR^3/(4EI) + (3\pi/4 - 2)PR^3/(GJ)] = 0$$
$$\Delta = PR^3[\pi/(4EI) + (3\pi/4 - 2)/(GJ)]$$

Example 8.10 Confirm the vertical deflection at point Q for the Warren bay girder in Fig. 8.6 (see Example 8.4), using the principle of virtual work.

With n - bars in the frame, each under direct stress, the required combination of eq(8.15) and Table 8.5 is

$$F^v\Delta - \sum_{j=1}^{n}[P^vPL/(AE)]_j = 0 \qquad \text{(i)}$$

where P^v are the bar forces when a virtual vertical force F^v is applied at Q, and P are the real bar forces under the actual loading. When calculating P, the force at Q is not removed as in Castigliano's approach. The reader should check the P and P^v bar forces given in Table 8.6 by any suitable method.

Table 8.6 Real and virtual bar forces

Bar	P^v	P	P^vP
PQ	$-0.866F^v$	-7.506	$6.500F^v$
QT	$-0.289F^v$	0.577	$-0.167F^v$
QR	$-0.289F^v$	-4.042	$1.168F^v$
PT	$0.433F^v$	3.753	$1.625F^v$
RT	$0.289F^v$	-0.577	$-0.167F^v$
RS	$-0.289F^v$	-8.660	$2.503F^v$
TS	$0.144F^v$	4.330	$0.624F^v$

Then $\sum P^vP = 12.086F^v$ and eq(i) gives

$$F^v\Delta_Q = \frac{(12.086\,F^v \times 10^3 \times 9.81 \times 10 \times 10^3)}{(1280 \times 208 \times 10^3)}$$

$$\Delta_Q = 4.45 \text{ mm}$$

8.5 The Unit Load Method (ULM)

It will be seen that in each of the above examples the virtual force F^v cancels within the virtual work expression. It follows that F^v could be replaced by a unit virtual load (i.e. a force, moment or torque) applied at the point where the displacement Δ, rotation ϕ or twist θ is required. Equation (8.15a) becomes

$$\Delta\ (\text{or } \phi \text{ or } \theta) - \int_V (\sigma_{ij}{}^v \varepsilon_{ij})\, dV = 0 \tag{8.16}$$

In order to distinguish between the virtual forces, moments and torques etc produced by the unit load, the corresponding lower case symbols appear with $-W_I$ (ULM) in Table 8.5. The following examples illustrate this widely used simplified form of the PVF.

Example 8.11 Determine the vertical deflection at the free end of the cantilever in Fig. 8.13a when, for the cross-section, $I = 10^9$ mm^4 and $E = 207$ GPa.

(a)

(b)

(c)

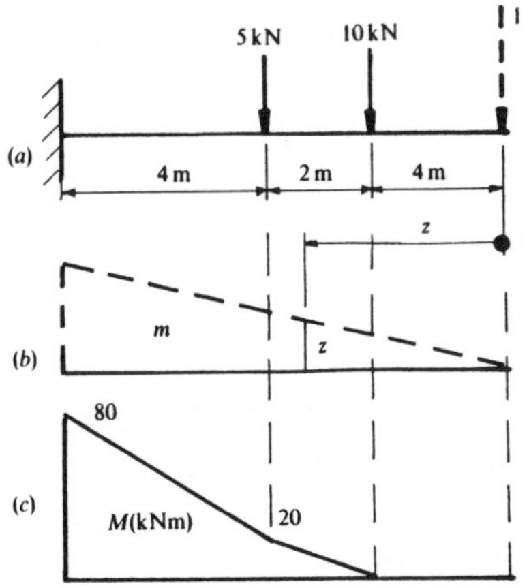

Figure 8.13 Cantilever deflection

Assuming that deformation is solely due to bending, eq(8.16) becomes

$$\Delta - \int mM\,dz/(EI) = 0 \qquad \text{(i)}$$

where, from Fig. 8.13b, under a unit force at the free end:
$m = z$ for $0 \le z \le 10$
and from Fig. 8.13c under the applied loading,
$M = 0$ for $0 \le z \le 4$ m
$M = 10(z - 4)$ (kNm) for $4 \le z \le 6$ m
$M = 10(z - 4) + 5(z - 6)$ (kNm) for $6 \le z \le 10$ m

Substituting into eq(i),

$$EI\Delta = \int_4^6 10(z - 4)\,z\,dz + \int_6^{10} [10(z - 4) + 5(z - 6)]\,z\,dz$$

$$= 10\,|\,z^3/3 - 2z^2\,|_4^6 + 10\,|\,z^3/3 - 2z^2\,|_6^{10} + 5|\,z^3/3 - 3z^2\,|_6^{10}$$

$$= 10(10.667) + 10(133.33) + 5(33.33 + 36)$$

$$= 1786.62 \text{ kNm}^3$$

$$\Delta = (1786.62 \times 10^{12})/(207 \times 10^3 \times 10^9) = 8.66 \text{ mm}$$

Example 8.12 Determine the horizontal displacement at the free end of the cantilevered arc in Fig. 8.14. Take $EI = 2 \times 10^{12}$ Nmm^2.

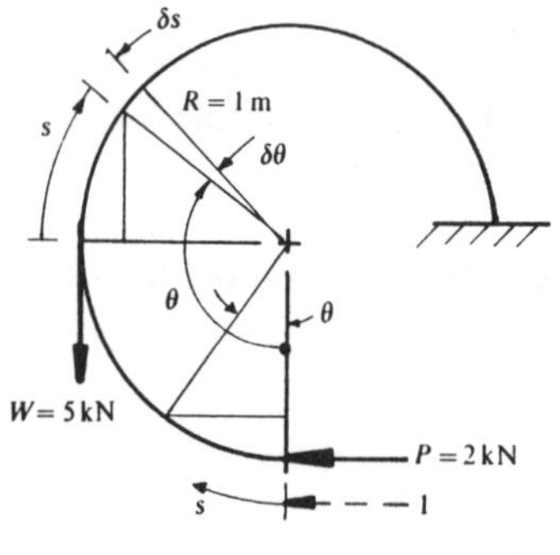

Figure 8.14 Cantilevered arc

Equation (8.16) is written as

$$\Delta - \int mM\,ds/(EI) = 0 \qquad \text{(i)}$$

where, for a unit horizontal force at the free end,

$m = R\,(1 - \cos\theta)$ (c.w. + ve) for $0 \le \theta \le 3\pi/2$

Under the applied loading in Fig. 8.14,

$M = PR(1 - \cos\theta)$ for $0 \le \theta \le \pi/2$

$M = PR[1 + \sin(\theta - \pi/2) - WR[1 - \cos(\theta - \pi/2)]$
$\quad = PR(1 - \cos\theta) - WR(1 - \sin\theta)$ for $\pi/2 \le \theta \le 3\pi/2$

Substituting into eq(i) with $ds = R\,d\theta$,

$$\Delta = [PR^3/(EI)]\int_0^{\pi/2}(1-\cos\theta)^2 d\theta + [R^3/(EI)]\int_{\pi/2}^{3\pi/2} P[(1-\cos\theta)^2 - W(1-\cos\theta)(1-\sin\theta)]d\theta$$

$$\Delta = [PR^3/(EI)] \mid 3\theta/2 - 2\sin\theta + \tfrac{1}{4}\sin 2\theta \Big|_0^{\pi/2}$$
$$+ [R^3/(EI)]\{P \mid 3\theta/2 - 2\sin\theta + \tfrac{1}{4}\cos 2\theta \Big|_{\pi/2}^{3\pi/2} - W \mid \theta - \sin\theta + \cos\theta - \tfrac{1}{4}\cos 2\theta \Big|_{\pi/2}^{3\pi/2}\}$$
$$= [PR^3/(EI)](3\pi/4 - 2) + [R^3/(EI)][P(3\pi/2 + 4) - W(\pi + 2)]$$
$$= [(9\pi/4 + 2)P - (\pi + 2)W]R^3/(EI)$$

Substituting $P = 2$ kN, $W = 5$ kN and $R = 1$ m,

$$\Delta = [(9.0685 \times 2) - (5.1417 \times 5)](10^9 \times 10^3) / (2 \times 10^{12})$$
$$= -\ 3.79 \text{ mm (opposing the direction of } P)$$

Example 8.13 Derive expressions for both the vertical and horizontal displacements at the load point C for the davit in Fig. 8.15.

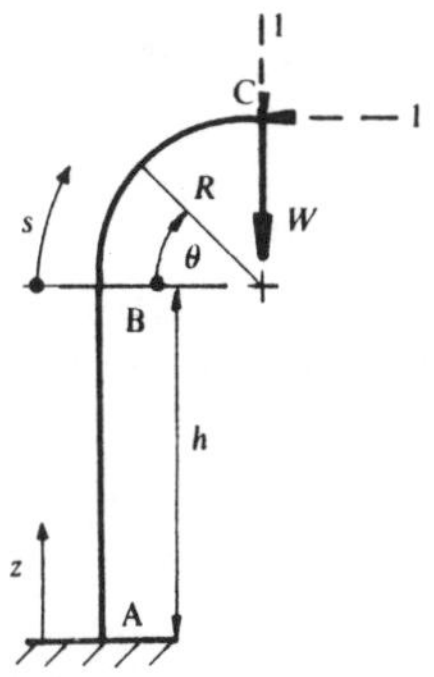

Figure 8.15 Davit deflection

Ignoring the effects of shear and compression, the vertical deflection is found from:

$$\Delta_y - \int m_y M \, ds/(EI) = 0 \qquad \text{(i)}$$

where, under a unit vertical load ↓ at point C,

$m_y = 1 \times R$ (CW for $0 \le z \le h$ in the straight portion AB)
$m_y = 1 \times R\cos\theta$ (CW for $0 \le \theta \le \pi/2$ in the curved portion BC)

Correspondingly, under the actual load W, the clockwise moments are

$M = WR$ (clockwise in AB) and $M = WR\cos\theta$ (clockwise in BC).

Substituting into eq(i) with $ds = dz$ for AB and $ds = R\,d\theta$ for BC,

$$\Delta_y = \int_0^h WR^2 dz/(EI) + \int_0^{\pi/2} WR^3 \cos^2\theta \, d\theta/(EI)$$
$$= WR^2 h/(EI) + WR^3/(2EI) \mid \theta + \tfrac{1}{2}\sin 2\theta \Big|_0^{\pi/2}$$
$$= WR^2 h/(EI) + \pi WR^3/(4EI)$$
$$= [WR^2/(EI)](h + \pi R/4) \text{ downwards} \downarrow.$$

The horizontal deflection is found from

$$\Delta_x - \int m_x M \, ds/(EI) = 0 \qquad \text{(ii)}$$

where, under a unit horizontal load ← at C,

$m_x = -\ (R + h - z)$ anticlockwise for $0 \le z \le h$ in AB
$m_x = -\ R(1 - \sin\theta)$ anticlockwise for $0 \le \theta \le \pi/2$ in BC.

Substituting, together with M in eq(ii),

$$\Delta_x = -\int_0^h (R+h-z)(WR)\,dz/(EI) - \int_0^{\pi/2} WR^3(1-\sin\theta)\cos\theta\,d\theta/(EI)$$

$$= -[WR/(EI)]\,|\,Rz + hz - z^2/2\big|_0^h - [WR^3/(EI)]\,|\,\sin\theta + \tfrac{1}{4}\cos 2\theta\big|_0^{\pi/2}$$

$$= -[WR/(EI)](Rh + h^2 - h^2/2) - WR^3/(2EI)$$

$$= -WR(R+h)^2/(2EI) \text{ forwards} \rightarrow$$

Example 8.14 Determine the vertical deflection at B for the simply supported beam in Fig. 8.16, given $EI = 40 \times 10^3$ kNm^2.

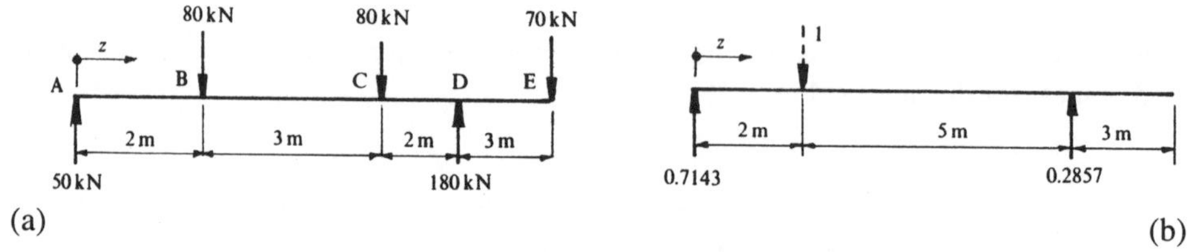

Figure 8.16 Simply supported beam

The ULM provides an alternative to the Macaulay method for the determination of beam deflections where

$$\Delta = \int mM\,dz/(EI) \qquad \text{(i)}$$

With the reactions under the actual and virtual loading given in Figs 8.16a and b and the origin for z at the left hand end, the moments are

AB ($0 \le z \le 2$): $M = -50z$, $m = -0.7143z$
BC ($2 \le z \le 5$): $M = -50z + 80(z-2)$, $m = -0.7143z + 1(z-2)$
CD ($5 \le z \le 7$): $M = -50z + 80(z-2) + 80(z-5)$, $m = -0.7143z + 1(z-2)$
DE ($7 \le z \le 10$): $M = -50z + 80(z-2) + 80(z-5) - 180(z-7)$,
$m = -0.7143z + 1(z-2) - 0.2857(z-7) = 0$

Substituting in eq(i)

$$EI\Delta = 35.715\int_0^2 z^2 dz + \int_2^5 (0.2857z - 2)(30z - 160)\,dz + \int_5^7 (0.2857z - 2)(110z - 560)\,dz$$

$$= 35.715\,|\,z^3/3\big|_0^2 + 8.571\,|\,z^3/3\big|_2^5 - 105.712\,|\,z^2/2\big|_2^5 + 320\,|\,z\big|_2^5 + 31.43\,|\,z^3/3\big|_5^7$$

$$- 78\,|\,z^2/2\big|_5^7 + 1120\,|\,z\big|_5^7$$

$EI\Delta = 243.45$ kNm^3
$\therefore \Delta = (243.45 \times 10^{12})/(40 \times 10^{12}) = 6.09$ mm

Example 8.15 The cranked cantilever in Fig. 8.17 carries an inclined force F at its tip. Taking bending and direct stress into account, determine the horizontal and vertical components of the free-end deflection. Ignoring the effect of direct stress, what is the inclination θ for which the tip deflection becomes aligned with P ?

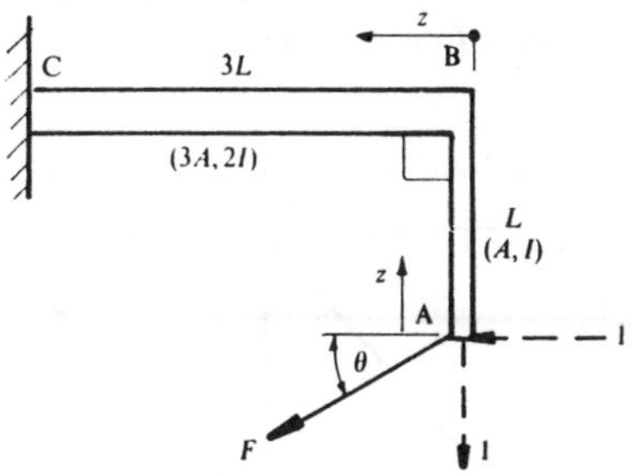

Figure 8.17 Cranked cantilever

The horizontal deflection is found from

$$\Delta_x = \int M\, m_x\, \mathrm{d}z/(EI) + P p_x L/(AE) \qquad \text{(i)}$$

where M and P are found from the actual loading, m_x and p_x for a unit force ← at A. That is, for AB,

$M = Fz\cos\theta$ and $m_x = 1 \times z$ (both clockwise)
$P = F\sin\theta$ (tensile) and $p_x = 0$

for BC, $M = FL\cos\theta + Fz\sin\theta$ and $m_x = 1 \times L$ (both clockwise)
$P = -F\cos\theta$ and $p_x = -1$ (both compressive)

Substituting into eq(i),

$$\Delta_x = \int_0^L Fz^2\cos\theta\,\mathrm{d}z/(EI) + \int_0^{3L} (FL\cos\theta + Fz\sin\theta)L\mathrm{d}z/(2EI) + (F\cos\theta)(3L)/(3AE)$$
$$= 11FL^3\cos\theta/(6EI) + 9FL^3\sin\theta/(4EI) + FL\cos\theta/(AE) \qquad \text{(ii)}$$

The vertical deflection is found from

$$\Delta_y = \int M\, m_y\, \mathrm{d}z/(EI) + P\, p_y L/(AE) \qquad \text{(iii)}$$

where, with a unit force ↓ at A, $m_y = 0$, $p_y = +1$ for AB and $m_y = z$, $p_y = 0$ for BC. Substituting with the previous M and P expressions into eq(iii),

$$\Delta_y = \int_0^{3L} (FL\cos\theta + Fz\sin\theta)\, z\, \mathrm{d}z/(2EI) + F(3L)\sin\theta/(3AE)$$
$$= 9FL^3\cos\theta/(4EI) + 9FL^3\sin\theta/(2EI) + FL\sin\theta/(AE) \qquad \text{(iv)}$$

Dividing eq(iv) by (ii) and omitting the final direct stress term,

$\tan\theta = \Delta_y/\Delta_x = [(9/4) + (9/2)\tan\theta] \,/\, [(11/6) + (9/4)\tan\theta]$

$27\tan^2\theta - 32\tan\theta - 27 = 0$

$\tan\theta = 1.755$ or -0.569

$\theta = 60.33°$ or $-29.67°$

Only the first value corresponds with the $\Delta_x(\leftarrow)$ and $\Delta_y(\downarrow)$ directions.

Example 8.16 Find the vertical and horizontal displacements at the force point for the cantilevered frame in Fig. 8.18. EI is constant.

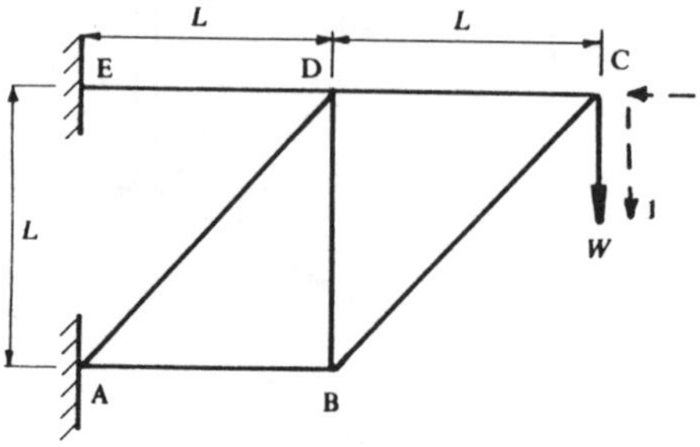

Figure 8.18 Cantilever frame

Since bar deformation is caused by direct stress, the ULM gives the displacements

$$\Delta_x = \sum p_x PL/(AE) \text{ and } \Delta_y = \sum p_y PL/(AE) \qquad \text{(i)}$$

where, in the Table 8.7, p_x and p_y are the bar forces under unit loads applied at C in the horizontal (←) and vertical (↓) directions respectively and P are the bar forces under the actual loading.

Table 8.7 Unit load method

Bar	p_x	p_y	P	L	p_xPL	p_yPL
AB	0	- 1	- W	L	0	WL
BC	0	- √2	- √2W	√2L	0	2 √2WL
CD	- 1	1	W	L	- WL	WL
BD	0	1	W	L	0	WL
AD	0	- √2	- √2W	√2L	0	2 √2WL
DE	- 1	2	2W	L	-2WL	4WL
					$\sum = -\ 3WL$	$\sum = 12.657WL$

Then, from eqs(i),

$$\Delta_x = -\ 3WL/(AE) \text{ and } \Delta_y = 12.657WL/(AE)$$

The minus indicates that the direction of Δ_x is →.

Example 8.17 Use ULM to determine the rate of twist in a thin-walled tube of any cross-section (see Fig. 5.23) subjected to an axial torque.

This alternative derivation of eq(5.70) employs ULM corresponding to deformation under torsion. Equation (8.15a) is written as

$$T^v\theta - \int \tau^v \gamma \, dV = 0 \qquad \text{(i)}$$

where, from the Bredt-Batho theory (eq 5.64)

$\tau^v = T^v/(2A\,t)$ $(T^v = 1$ in the ULM)
$\gamma = \tau/G = T/(2AG\,t)$
$\mathrm{d}V = t \times \mathrm{d}s \times \mathrm{d}z$

Putting $\theta = \int (\mathrm{d}\theta/\mathrm{d}z)\,\mathrm{d}z$ and substituting into eq(i) gives,

$$\int_0^z (\mathrm{d}\theta/\mathrm{d}z)\,\mathrm{d}z = \int_0^z \oint_s [1/(2A\,t)][(T/(2AG\,t)](t\,\mathrm{d}s\,\mathrm{d}z)$$

Differentiating with respect to z,

$$\mathrm{d}\theta/\,\mathrm{d}z = \oint_s (T/(4A^2Gt)\,\mathrm{d}s = [T/(4A^2)] \oint_s \mathrm{d}s/(Gt)$$

8.6 Redundant Structures

The PVF provides an alternative to Castigliano's theorem for the solution to redundant pin-jointed structures. Moreover, the unit load method may be adapted to many other forms of redundant structure, i.e. where the stresses cannot be determined from the equilibrium equations. The method ensures that additional compatibility conditions between internal strain and displacements are also satisfied. The ULM is applied as follows:

(i) Render the structure statically determinate by removing the redundancies, e.g. by cutting the redundant bars.
(ii) Determine the virtual stresses σ_{ij}^v corresponding to isolated applications of unit loads to each cut in turn. These stresses produce relative displacements and rotations at each cut.
(iii) With the introduction of each unknown redundant load (force, torque or moment) determine the strain ε_{ij} expression under the applied loading.
(iv) Apply the compatibility conditions necessary to eliminate the relative displacements. These correspond to zero virtual work at each cut. That is, from eq(8.16a),

$$\int_V \sigma_{ij}{}^v \varepsilon_{ij}\,\mathrm{d}V = 0 \tag{8.17}$$

Equation (8.17) enables the redundant loading to be found. The particular form of eq(8.17) depends upon the applied loading (see Table 8.5). The following examples illustrate this common approach for a various redundant structures.

8.6.1 Propped Cantilever

A propped cantilever (see Fig. 8.19) has three unknowns M_A, R_A and R_B that cannot be determined from two equations expressing force and moment equilibrium. Prior knowledge of the prop displacement is used as the additional compatibility condition in the ULM.

Example 8.18 Determine the deflection at the centre of the propped cantilever in Fig. 8.19a when the end displacement is zero and *EI* is constant.

Remove the redundant prop (reaction R_B). Then, for the actual distributed loading at distance z from B, the moment is

$M_o = wz^2/2$ for $0 \le z \le L$ (i)

and, with a unit upward force applied at B in the absence of w (see Fig. 8.19b),

$m = -1 \times z$ for $0 \le z \le L$ (ii)

(a)

(b)

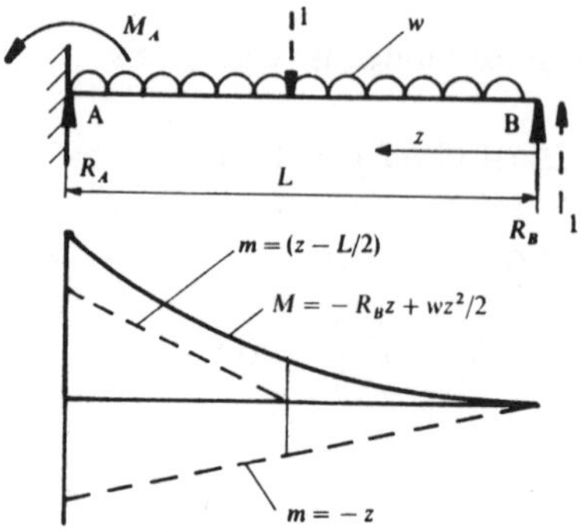

Figure 8.19 Propped cantilever

Thus, for the redundant structure it follows from eqs (i) and (ii) that the moment at z is

$M = R_B m + M_o = -R_B z + wz^2/2$ for $0 \le z \le L$

Now, from eq(8.17) and Table 8.5, zero displacement at B is ensured when

$\int mM \, dz/(EI) = 0$ (iii)

Substituting for m and M into eq(iii),

$$\int_0^L (-R_B z + wz^2/2)(-z)\, dz/(EI) = 0$$

$$| wz^4/8 - R_B z^3/3 |_0^L = 0$$

$R_B = 3wL/8$

Displacements are found from

$\Delta = \int mM \, dz/(EI)$ (iv)

Now, with a unit downward force applied at the centre,

$m = 0$ for $0 \le z \le L/2$

$m = 1 \times (z - L/2)$ for $L/2 \le z \le L$

$M = -3wLz/8 + wz^2/2$ for $0 \le z \le L$

Substituting into eq(iv),

$$\Delta = \int_{L/2}^{L} (z - L/2)(-3wLz/8 + wz^2/2)\,dz/(EI)$$

$$= [1/(2EI)] \int_{L/2}^{L} (wz^2 - 3wLz/4)(z - L/2)\, dz$$

$$= [1/(2EI)] \int_{L/2}^{L} (wz^3 - 5wLz^2/4 + 3wL^2 z/8)\, dz$$

$= [1/(2EI)] \mid wz^4/4 - 5wLz^3/12 + 3wL^2z^2/16 \Big|_{L/2}^{L}$
$= [1/(2EI)] [wL^4/64 - 5wL^4/96 + 3wL^4/64]$
$\Delta = wL^4/(192EI)$

Note that if the origin for z was at the left hand end, compatibility must ensure zero slope ϕ at A. Equation(iii) becomes $\phi = \int mM\, dz/(EI) = 0$, where m is the moment under a unit couple applied at A.

8.6.2 Pin-Jointed Structures

With a single degree of redundancy, let p be the bar forces when a unit load replaces the redundant bar. If the bar forces under the applied loading in the absence of the redundant bar are P_o and the force in the redundant bar is R, the net bar forces are $P = P_o + Rp$. It follows that

$\sigma_{ij}^{\ v} = p/A$ and $\varepsilon_{ij} = P/(AE) = (P_o + Rp)/(AE)$

Substituting into eq(8.17),

$\int \sigma_{ij}^{\ v} \varepsilon_{ij} dV = \sum[(p/A)(P_o + Rp)/(AE)](AL) = 0$

$\sum p\,(P_o + Rp)L/(AE) = 0$

$\sum [pP_oL/(AE)] + R\sum p^2L/(AE) = 0$

$$R = -\frac{\sum_1^n (pP_oL/AE)}{\sum_1^n (p^2L/AE)} \tag{8.18}$$

Clearly, with E constant throughout the structure, eq(8.18) is identical to Castigliano's theorem (iv) employed in Example 8.10.

With two or more redundancies we let P_o be the statically determinate bar forces and also p_1, p_2, p_3 ... be the bar forces when separately applied unit loads replace the redundant bars. R_1, R_2, R_3 ... denote the redundant bar forces to be found. The net bar forces and the strains are then

$P = P_o + R_1p_1 + R_2p_2 + R_3p_3 \ldots$
$\varepsilon_{ij} = P/(AE)$

and the virtual stresses are

$(\sigma_{ij}^{\ v})_1 = p_1/A$
$(\sigma_{ij}^{\ v})_2 = p_2/A$
$(\sigma_{ij}^{\ v})_3 = p_3/A \ldots.$

Substitutung into eq(8.17) gives expressions that eliminate displacements due to bar removal

$$\Sigma[P_o p_1 L/(AE)] + R_1\Sigma[p_1^2 L/(AE)] + R_2\Sigma[p_1 p_2 L/(AE)] + R_3\Sigma[p_1 p_3 L/(AE)] + .. = 0$$
$$\Sigma[P_o p_2 L/(AE)] + R_1\Sigma[p_1 p_2 L/(AE)] + R_2\Sigma(p_2^2 L/(AE)] + R_3\Sigma[p_2 p_3 L/(AE)] + .. = 0$$
$$\Sigma[P_o p_3 L/(AE)] + R_1\Sigma[p_1 p_3 L/(AE)] + R_2\Sigma[p_2 p_3 L/(AE)] + R_3\Sigma[p_3^2 L/(AE)] + .. = 0 \quad (8.19)$$
$$\cdot \qquad \cdot \qquad \cdot \qquad \cdot \qquad .. = 0$$
$$\cdot \qquad \cdot \qquad \cdot \qquad \cdot \qquad .. = 0$$

which are solved simultaneously for the redundant bar forces R_1, R_2, R_3 ...

Example 8.19 Determine the forces in the bars of the aluminium frame in Fig. 8.20 when the lengths and areas are those given in the Table 8.8. Find also the in-line vertical deflection beneath the 2.4 kN force? Take E = 70 GPa.

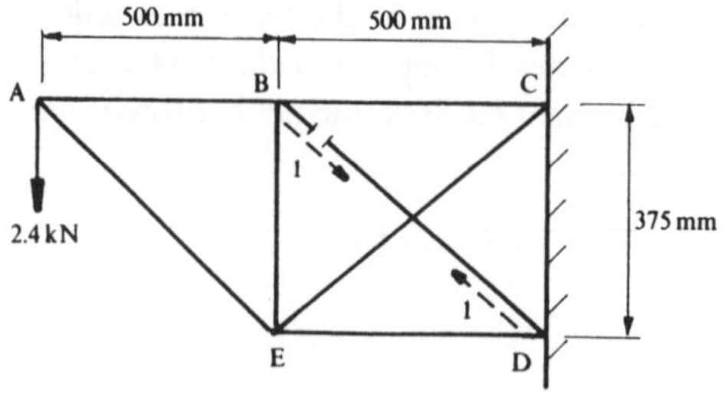

Figure 8.20 Frame forces

Cut BD and let this redundant bar force be R. The forces under the resulting statically determinate structure are those in the P_o column. With unit loads acting in tension at B and D in the line of the missing bar BD, the remaining bar forces are those in the p column.

Table 8.8 Calculations for redundant frame

Bar	L (mm)	A (mm^2)	P_o (kN)	p	$P_o pL/A$	p^2L/A	$P = P_o + Rp$ (kN)
AB	500	62.5	+ 3.2	0	0	0	3.2
BC	500	125	+ 3.2	- 0.8	- 10.24	2.56	5.48
AE	625	62.5	- 4.0	0	0	0	- 4.0
BE	375	62.5	0	- 0.6	0	2.16	1.71
CE	625	62.5	+ 4.0	1	40	10	1.15
DE	500	62.5	- 6.4	- 0.8	40.96	5.12	- 4.12
BD	625	125	0	1	0	5	- 2.85
					Σ = 70.72	Σ = 24.84	

Then, from eq(8.18), R = - 70.72/24.84 = - 2.85 kN, which enables the final column to be completed. The vertical deflection Δ_y follows from

$$F^v\Delta_y - \Sigma[pPL/(AE)] = 0 \qquad \text{(i)}$$

The bar forces P follow from the above table and p are now the bar forces under a unit force at A. Since these are $p = P/2.4$, eq(i) gives

$$\Delta_y = [1/(2.4E)]\,\Sigma\,P^2L/A \qquad \text{(ii)}$$

where $P^2 L/A$ for each bar is compiled in Table 8.9

Table 8.9 Bar forces

Bar	AB	BC	AE	BE	CE	DE	BD
P(kN)	3.2	5.48	- 4.0	1.71	1.15	- 4.12	- 2.85
P^2	10.24	30.03	16	2.924	1.323	16.974	8.123
$P^2 L/A$	81.92	120.12	160	17.54	13.23	135.79	40.62

This gives $\sum P^2 L/A = 569.22$ kN^2/mm, when, from eq(ii),

$\Delta_y = (569.22 \times 10^3)/(2.4 \times 70 \times 10^3) = 3.34$ mm

which is less than if bar BD were absent, i.e. the structure is stiffened with its redundant bar.

Example 8.20 Determine the forces in each bar of the steel space frame in Fig. 8.21 and the vertical deflection at A. Bars AC, AD and AE are each 200 mm^2 in area while AB is 1200 mm^2. Take E = 200 GPa.

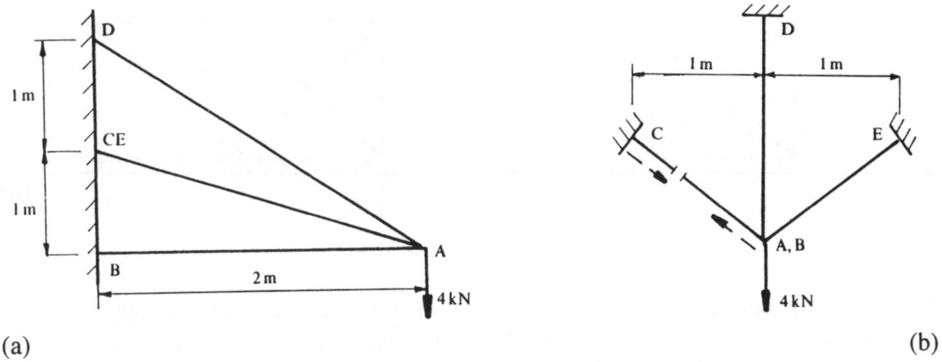

Figure 8.21 Space frame

Let AC be the redundant bar. The bar forces P_o and p in Table 8.10 may be found by any method, e.g. tension coefficients.

Table 8.10 Bar geometry and forces

Bar	L (mm)	A (mm^2)	P_o(kN)	p	pP_oL/A	p^2L/A	$P=P_o+Rp$	P^2L/A
AB	2000	1200	- 4	- 0.8165	5.443	1.111	- 5.596	52.19
AC	2449.5	200	0	1	0	12.248	1.955	46.81
AD	2828.4	200	5.657	- 1.155	- 92.38	18.866	3.400	163.48
AE	2449.5	200	0	1	0	12.248	1.955	46.81
					$\sum = -86.937$	$\sum = 44.473$		$\sum = 309.29$

From eq(8.18), $R = 86.937/44.473 = 1.955$ kN, which enables the final column of bar forces to be completed. As with the previous example the vertical deflection at A is given by

$$\Delta = [\,1/(4E)]\sum P^2 L/A$$
$$= (309.29 \times 10^3)/(4 \times 200 \times 10^3) = 0.386 \text{ mm}$$

Alternatively, Δ may be calculated from the above P forces in the simplest statically determinate frame, say AB and AD. With a downward unit force at A, $p_{AB} = -1$, $p_{AD} = \sqrt{2}$ and

$$\Delta = [\,pPL/(AE)]_{AB} + [\,pPL/(AE)]_{AD}$$
$$= (-1)(-5.596)2000\,/\,(1200E) + (\sqrt{2} \times 3.4 \times 2828.4)\,/\,(200E) = 0.386 \text{ mm}$$

Example 8.21 Determine the bar forces for the pin-jointed structure in Fig. 8.22a. *EI* is constant.

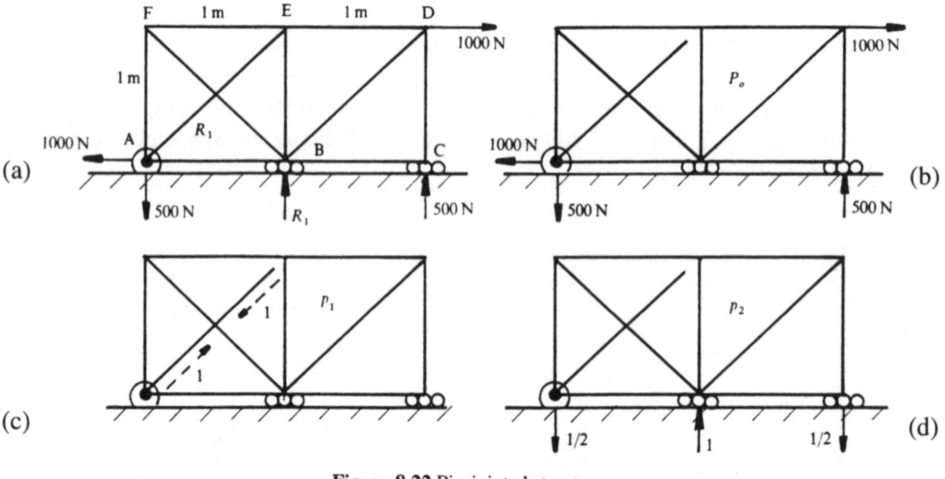

Figure 8.22 Pin-jointed structure

The two redundancies present in this structure may be deduced from the application of moment and force equilibrium equations. Taking these to be the force in bar AE (R_1) and the reaction at support B (R_2), eq(8.19) becomes

$$\sum [\,P_o p_1 L/(AE)] + R_1 \sum [\,p_1^2 L/(AE)] + R_2 \sum [(p_1 p_2 L/(AE)] = 0 \qquad \text{(i)}$$
$$\sum [\,P_o p_2 L/(AE)] + R_1 \sum [\,p_1 p_2 L/(AE)] + R_2 \sum [\,p_2^2 L/(AE)] = 0 \qquad \text{(ii)}$$

where bar forces; P_o, p_1 and p_2 in Table 8.11, apply to Figs 8.22b,c and d respectively. Substituting into eqs(i) and (ii),

$$-2414 + 4.828\,R_1 - 1.7080\,R_2 = 0$$
$$500 - 1.708\,R_1 + 2.4142\,R_2 = 0$$

from which $R_1 = 569.2$ N and $R_2 = 195.6$ N. Hence the final column of net bar forces is found from $P = P_o + 569.2\,p_1 + 195.6\,p_2$.

In order to find the deflection at any joint it is only necessary to take a statically determinate part of the frame and apply

$$\Delta = \sum pPL/(AE)$$

where P are the actual bar forces in the table and p are the bar forces under a unit load applied at the joint in the required direction (see Example 8.20). Note that, where necessary, the reactions to the unit load must be included in the calculation of p.

Table 8.11 Bar forces for doubly redundant frame

Bar	L(m)	P_o	p_1	p_2	$P_o p_1 L$	$P_o p_2 L$	$p_1 p_2 L$	$p_1^2 L$	$p_2^2 L$	P (N)
AB	1	1000	- 0.707	0	- 707	0	0	0.5	0	597.5
AE	√2	0	1	0	0	0	0	1.414	0	569.2
AF	1	500	- 0.707	0.5	- 353.6	250	- 0.354	0.5	0.25	195.3
BC	1	0	0	0	0	0	0	0	0	0
BD	√2	707	0	- 0.707	0	- 707	0	0	0.707	568.7
BE	1	0	- 0.707	0	0	0	0	0.5	0	- 402.5
BF	√2	- 707	1	- 0.707	- 1000	707	- 1	1.414	0.707	- 276
CD	1	- 500	0	0.5	0	- 250	0	0	0.25	- 402.2
DE	1	500	0	0.5	0	250	0	0	0.25	597.8
EF	1	500	- 0.707	0.5	- 353.6	250	- 0.354	0.5	0.25	195.3
				$\sum =$	- 2414.2	500	- 1.708	4.828	2.4142	

8.6.3 Torsion of a Thin-Walled, Multi-Cell Tube

The two-cell problem has been solved previously in Chapter 5 (section 5.3) from the separate application of equilibrium and compatibility. It is instructive to confirm this theory from our general approach to redundant structures. Cut cell 2 as shown in Fig. 8.23a to give the statically determinate Batho shear flow in cell 1 as

$$q_o = T / (2A_1) \tag{8.20}$$

This will cause a relative displacement between the cut faces. By applying isolated, opposing unit shear forces at the cut a a complementary unit shear flow will be induced in cell 2 (see Fig. 8.23b).

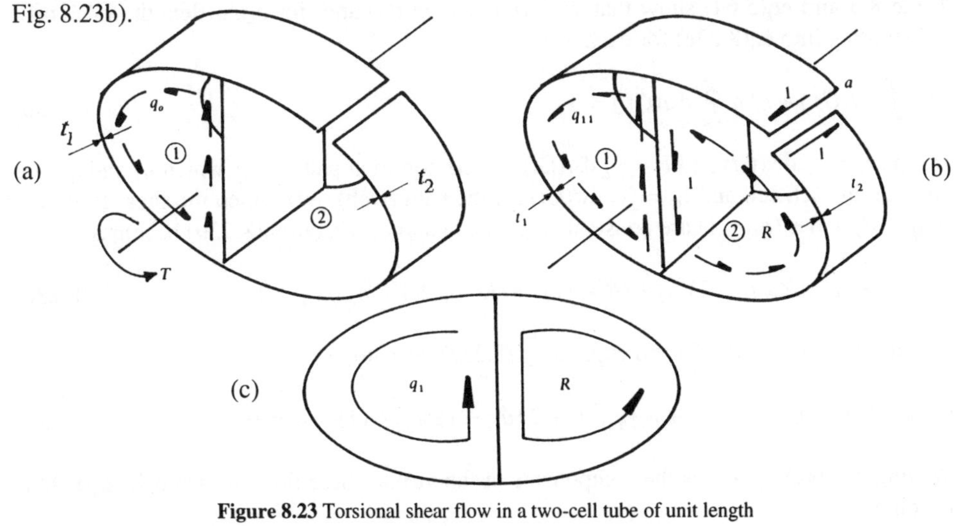

Figure 8.23 Torsional shear flow in a two-cell tube of unit length

The unit shear flow will be reacted by a shear flow q_{11} in cell 1 according to the equilibrium condition $T = 0$ in eq(5.73). That is

$$0 = 2A_1 q_{11} + 2A_2, \Rightarrow q_{11} = -A_2/A_1 \tag{8.21}$$

If q_1 is the net shear flow in cell 1 and R ($= q_2$) the shear flow in the redundant cell 2 (see Fig. 8.22c), then

$$q_1 = q_o + q_{11} R \tag{8.22}$$

The actual and virtual shear flows for each cell are shown in Fig. 8.24a and b.

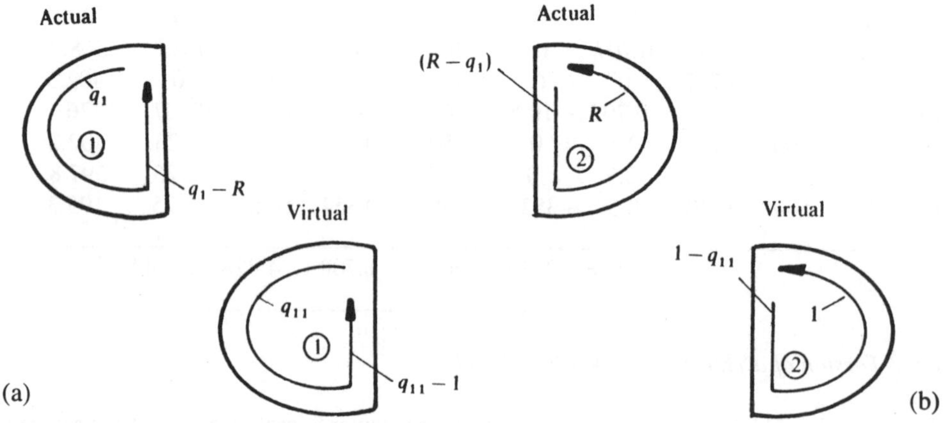

Figure 8.24 Actual and virtual shear flow in each cell

Compatibility is restored by imposing zero relative displacement at the cut. That is, for zero virtual work, eq(8.17) becomes

$$\int_V \sigma_{ij}{}^v \varepsilon_{ij}\, dV = \int_V \tau^v \gamma\, dV = 0 \tag{8.23a}$$

Table 8.5 and eq(5.64) show that $\tau^v = q^v/t$, $\gamma = q/(tG)$ and, for a unit length, $dV = t\, ds$. Substituting into eq(8.23a) for each tube,

$$q^v q \oint_1 ds/(Gt) + q^v q \oint_2 ds/(Gt) = 0 \tag{8.23b}$$

Define $i_1 = \int_1 ds/(Gt)$ and $i_2 = \int_2 ds/(Gt)$ for the common paths over which q_1 and q_{11} are constant in each cell and $i_{12} = \int_{12} ds/(Gt)$ for the path (web) over which the shear flows are $\pm(q_1 - R)$ and $\pm(q_{11} - 1)$ are constant (Figs 8.24a and b). Equation(8.23b) becomes

$$q_1 q_{11} i_1 + (q_1 - R)(q_{11} - 1)\, i_{12} + (R \times 1)\, i_2 + (R - q_1)(1 - q_{11})\, i_{12} = 0 \tag{8.23c}$$

Substituting from eq(8.22) and expanding (8.23c) leads to

$$q_o[\, q_{11}\; i_1 + 2\,(q_{11} - 1)\, i_{12}] + R\,[q_{11}{}^2\, i_1 + 2\,(q_{11} - 1)^2 i_{12} + (1)^2\, i_2] = 0 \tag{8.23d}$$

Writing, for brevity, q_{1a} as the components of the virtual shear flow in each cell, eq(8.23d) becomes

$$\sum q_o\, q_{1a}\, i + R \sum q_{1a}^{\,2}\, i = 0 \tag{8.23e}$$

from which ($R = q_2$) can be found and the similarity with eq(8.18) is evident. Although this approach may be extended to multi-cell tubes, it is more convenient to solve these from the fact that each cell will suffer the same rate of twist. The simultaneous solution to cell shear flows follows more readily from eqs(5.75a - c) than from virtual work.

8.6.4 Flexural Shear Flow in Multicell Tubes

The ULM is particularly useful for determining shear flow in multi-cell, idealised tubes when a transverse shear force F acts through the shear centre (see Chapter 7). The single cell tube which has one degree of shear flow redundancy R is again solved by making a single cut in one web. If q_o is the resulting shear flow under F and $q^v = 1$ is the virtual shear flow due to opposing forces, acting in isolation at the cut, then the net shear flow is

$$q = q_o + R\, q^v$$

where R is found from eliminating the relative displacement at the cut with zero virtual work. From eq(8.17),

$$\int \sigma_{ij}^{\ v}\, \varepsilon_{ij}\, \mathrm{d}V = \sum \oint\, q^v q\, \mathrm{d}s/(Gt) = 0$$

$$\sum q^v\, (q_o + R\, q^v) \oint\, \mathrm{d}s/(Gt) = 0$$

$$\sum q^v q_o \oint \mathrm{d}s/(Gt) + R \sum (q^v)^2 \oint \mathrm{d}s/(Gt) = 0 \tag{8.24a}$$

which is clearly similar to eq(8.23e) for a singly redundant, two-cell tube under torsion. The summations account for variations in q_o and in the path integral $\oint \mathrm{d}s/(Gt)$, within each cell. In the case of the two-cell tube under flexural shear in Fig. 8.25a, two cuts a and b are necessary to achieve static determinancy.

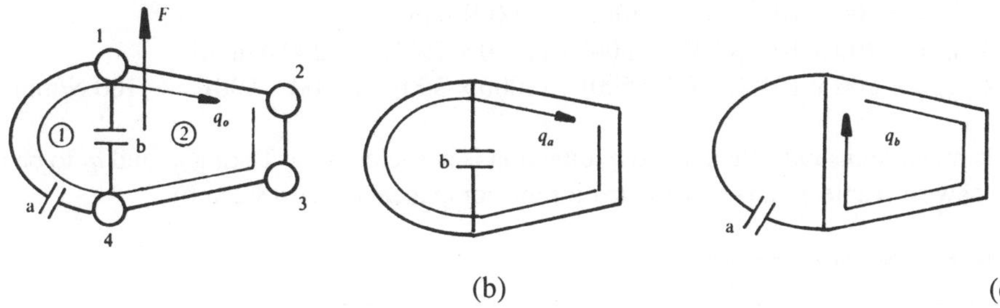

Figure 8.25 Two-cell tube under flexural shear

The two degrees of shear flow redundancy are now denoted by R_a and R_b respectively, giving the net shear flow in any cell as

$$q = q_o + R_a\, q_a + R_b\, q_b \tag{8.24b}$$

where $q_a = 1$ and $q_b = 1$ are the unit virtual shear flows arising from the application of unit forces to each cut in isolation (Figs 8.25 b and c). Compatibility is ensured when the cut displacements are eliminated through eq(8.24a):

$$\sum q_o q_a \oint \mathrm{d}s/(Gt) + R_a \sum q_a^2 \oint \mathrm{d}s/(Gt) + R_b \sum q_a q_b \oint \mathrm{d}s/(Gt) = 0 \tag{8.25a}$$

$$\sum q_o q_b \oint \mathrm{d}s/(Gt) + R_a \sum q_a q_b \oint \mathrm{d}s/(Gt) + R_b \sum q_b^2 \oint \mathrm{d}s/(Gt) = 0 \tag{8.25b}$$

which are solved simultaneously for R_a and R_b.

Example 8.22 Determine the shear flows for the idealised wing box structure in Fig. 8.26a when a shear force of 200 kN acts vertically upwards through the shear centre. Take G and t as constants throughout.

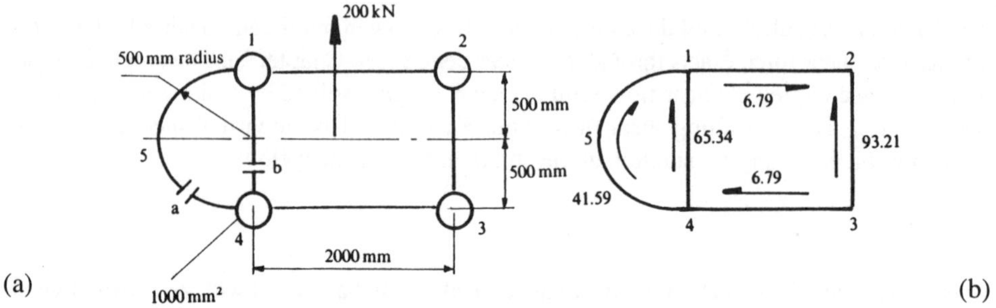

Figure 8.26 Wing box structure

The cuts correspond to points a and b in Fig. 8.25a. The following q_o values are found from

$$q_o = (F_y / I_x) \sum (A_r y_r)$$

where $F_y/I_x = -(200 \times 10^3)/(4 \times 1000 \times 500^2) = -200 \times 10^{-6}$ N/mm^4

Working clockwise from the "free surface" at a

1-4 and 1-5-4, $q_o = 0$
1-2, $q_o = -200 \times 10^{-6}(1000 \times 500) = -100$ N/mm
2-3, $q_o = -200 \times 10^{-6}[(1000 \times 500) + (1000 \times 500)] = -200$ N/mm
3-4, $q_o = -200 \times 10^{-6}[(1000 \times 500) + (1000 \times 500) - (1000 \times 500)] = -100$ N/mm

The minus signs indicate that the q_o direction is anti-clockwise Taking q_a and q_b to be anti-clockwise, Table 8.12 is constructed for the components in eqs(8.25a,b).

Table 8.12 Shear flow components

Web	q_o	q_a	q_b	s (mm)	$q_o q_a s$	$q_o q_b s$	$q_a^2 s$	$q_b^2 s$	$q_a q_b s$
1-2	100	1	1	2000	2×10^5	2×10^5	2000	2000	2000
2-3	200	1	1	1000	2×10^5	2×10^5	1000	1000	1000
3-4	100	1	1	2000	2×10^5	2×10^5	2000	2000	2000
1-4	0	0	1	1000	0	0	0	1000	0
1-5-4	0	1	0	1571	0	0	1571	0	0
				$\sum =$	(6×10^5)	(6×10^5)	6571	6000	5000

Substituting into eqs(8.25a and b),

$$(6 \times 10^5) + 6571\, R_a + 5000\, R_b = 0$$
$$(6 \times 10^5) + 5000\, R_a + 6000\, R_b = 0$$

from which $R_a = -\ 41.59$, $R_b = -\ 65.34$ N/mm. Equation(8.24b) supplies the final shear flows, illustrated with anti-clockwise positive in Fig. 8.25b:

1-2, $q = 100 - 41.45 - 65.34 = -\ 6.79$ N/mm
2-3, $q = 200 - 41.45 - 65.34 = 93.21$ N/mm
3-4, $q = 100 - 41.45 - 65.34 = -\ 6.79$ N/mm
1-4, $q = 0 - 0 - 65.34 = -\ 65.34$ N/mm
1-5-4, $q = 0 - 41.59 - 0 = -\ 41.59$ N/mm

8.7 The Principle of Virtual Displacements (PVD)

Recall that the virtual work principle in eq(8.14) was derived using a virtual displacement system with real loads. That means that for externally applied forces F and a virtual displacement Δ^v eq(18.14b) is written as,

$$\sum F\Delta^v - \int_V (\sigma_{ij}\,\varepsilon_{ij}{}^v)\, \mathrm{d}V = 0 \tag{8.26a}$$

Alternatively, when external moments M_E comprise the loading, the principle employs a virtual rotation θ^v,

$$\sum M_E\,\theta^v - \int_V (\sigma_{ij}\,\varepsilon_{ij}{}^v)\, \mathrm{d}V = 0 \tag{8.26b}$$

The PVD or the unit displacement method could be employed for finding forces analytically in structures including those that are redundant, but the PVF or the unit load method is more often used for this purpose. The PVD is however particularly useful for providing the internal forces and displacements numerically from an assumed displacement mode, this being the basis for the "displacement" or "stiffness" methods of finite element analysis (see chapter 10). The following examples illustrate these applications.

Example 8.23 Determine the force in the bar CD for the frame in Fig. 8.27a using the PVD given that EI is constant.

(a)

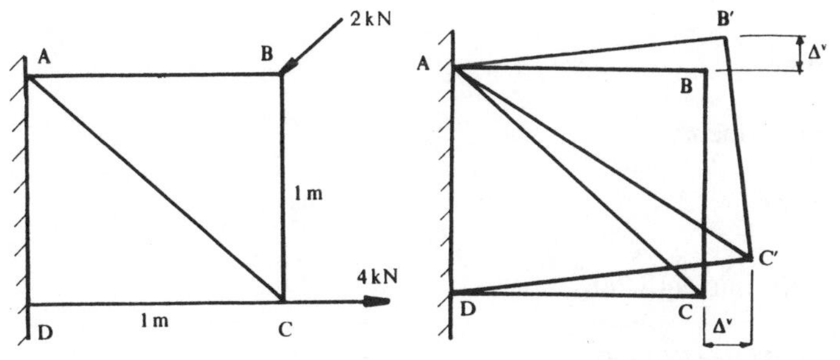

(b)

Figure 8.27 Displacements of a plane frame

Impose a virtual displacement Δ^v at C on CD so that the lengths of all other bars remain unchanged. It follows that C is displaced to C ' along an arc of radius AC. The geometry is such that B is displaced by Δ^v to B' as shown in Fig. 8.27b. Equation (8.26a) becomes

$$\sum F\Delta^v - \sum (P/A)(\delta^v/L)(AL) = 0$$

$$\therefore \sum F\Delta^v = \sum P\delta^v \qquad \text{(i)}$$

where δ^v are the displacements of the bars necessary to accommodate Δ^v. The external work (left-hand side of eq(i)) is found from the product of the applied forces F and their corresponding in-line Δ^v. With a minus sign to denote opposing in-line directions,

$$\sum F\,\Delta^v = (4 \times \Delta^v) + (-\ 2 \sin 45° \times \Delta^v)$$

The internal work (right-hand side of eq(i)) is $P_{CD}\,\delta^v = P_{CD}\Delta^v$ since no other bar is strained.

$$P_{CD}\Delta^v = 4\Delta^v - \surd 2\Delta^v$$

$$\therefore P_{CD} = 2.586 \text{ kN}$$

Note that since Δ^v cancels, a unit displacement could have replaced it.

Example 8.24 Find from PVD the central deflection v_c of a simply supported beam of length L, when a concentrated moment M_o is applied at one end (Fig. 8.28a). Assume a deflected shape of the form $v = v_c \sin(\pi z/L)$. Compare with the true value from the PVF. Take EI as constant.

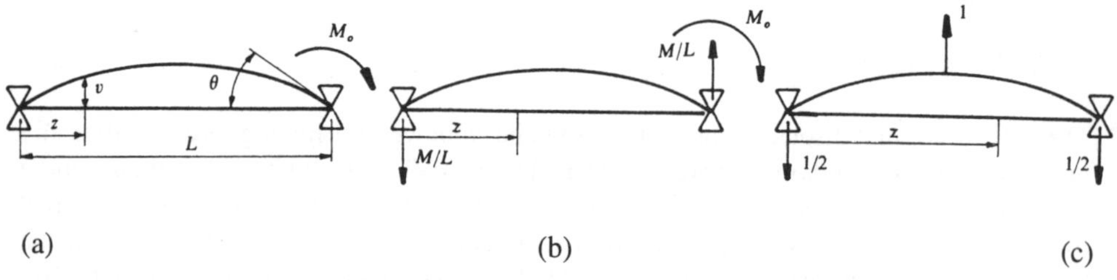

Figure 8.28 Deflection of a simply supported beam

The form assumed for the real displacement v clearly satisfies the boundary conditions that $v = 0$ for $z = 0$ and $z = L$. The virtual displacement function v^v may be any shape that matches the boundary conditions. Here it is taken to have the same form as v

$$v^v = v_c^v \sin(\pi z/L)$$

Now from flexure theory,

$$(EI)\,\mathrm{d}^2v/\mathrm{d}z^2 = M \qquad \text{(i)}$$

Substituting eq(i) into eq(8.26b),

$$M_o\,\theta^v - \int_0^L M^v M\,\mathrm{d}z/(EI) = 0$$

$$M_o\,\theta^v - EI\int_0^L (\mathrm{d}^2v^v/\mathrm{d}z^2)(\mathrm{d}^2v/\mathrm{d}z^2)\,\mathrm{d}z = 0 \qquad \text{(ii)}$$

where here,

$$d^2 v^v/dz^2 = -\ v_c^v (\pi/L)^2 \sin(\pi z/L) \qquad \text{(iii)}$$
$$d^2 v / dz^2 = -\ v_c (\pi/L)^2 \sin(\pi z/L) \qquad \text{(iv)}$$

and since

$$\theta^v = dv^v/dz = v_c^v (\pi/L) \cos(\pi z/L)$$

the virtual rotation to be applied at the end point $z = L$, where the external moment is M_o, is

$$\theta^v = -\ v_c^v \pi/L \qquad \text{(v)}$$

Substituting eqs(iii),(iv) and (v) into (ii),

$$M_o(-\ v_c^v \pi/L) = EI\, v_c v_c^v (\pi/L)^4 \int_0^L \sin^2(\pi z/L)\, dz$$

$$M_o = -\ v_c \pi^3 EIL/(2L^3)$$

$$v_c = -\ M_o L^2/[(\pi^3/2)(EI)] \text{ (upwards)} \qquad \text{(vi)}$$

The true value may be found from the ULM:

$$\Delta = \int_0^L mM\, dz/(EI) \qquad \text{(vii)}$$

where, from the real and virtual "forces" in Figs 8.28b and c respectively,

$$M = M_o z/L$$
$$m = z/2 \text{ for } 0 \le z \le L/2$$
$$m = z/2 - (z - L/2) \times 1 \text{ for } L/2 \le z \le L$$

Substituting into eq(vii),

$$\Delta = \int_0^{L/2} (z/2)(M_o z/L)\, dz/(EI) + \int_{L/2}^{L} [z/2 - (z - L/2)](M_o z/L)\, dz/(EI)$$

$$= [M_o/(2EIL)] \int_0^{L/2} z^2\, dz + [M_o/(EIL)] \int_{L/2}^{L} z[z/2 - (z - L/2)]\, dz$$

$$= [M_o/(2EIL)] \left| z^3/3 \right|_0^{L/2} + [M_o/(EIL)\left[\left| z^3/6 \right|_{L/2}^{L} - \left| z^3/3 - z^2 L/4 \right|_{L/2}^{L} \right]$$

$$= M_o L^2/(48EI) + [M_o/(EIL)]\{(L^3/6 - L^3/48) - [(L^3/3 - L^3/4) - (L^3/24 - L^3/16)]\}$$

$$= M_o L^2/(16EI)$$

Since $\pi^3/2 = 15.5$ in the denominator of the approximation expression (vi), it is seen that the PVD approach yields an acceptable solution.

Example 8.25 Derive an expression for the force in bar BD for the redundant pin-jointed plane frame in Fig. 8.29a when a vertical load F is applied at point D and the link lengths are in the ratio BD/BC/CD = 3/4/5, given that the bar material obeys (a) the non-linear elastic law $\sigma = a\varepsilon(1 - b\varepsilon^2)$ and (b) Hooke's law. All bar areas A are constant.

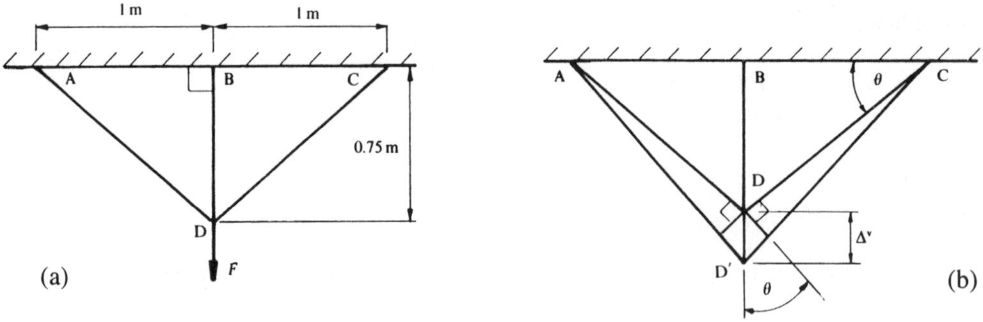

Figure 8.29 Redundant frame

Employing the PVD eq(8.26a) for part (a),

$$F\Delta^{v} = \sum P\delta^{v} \qquad \text{(i)}$$

Applying Δ^{v} at D (Fig. 8.29b) produces the virtual bar displacements δ^{v}

$$\delta_{CD}{}^{v} = \delta_{AD}{}^{v} = \Delta^{v}\sin\theta = (3/5)\Delta^{v}$$
$$\delta_{BD}{}^{v} = \Delta^{v}$$

Substituting into eq(i),

$$F\Delta^{v} = (P\delta^{v})_{AD} + (P\delta^{v})_{BD} + (P\delta^{v})_{CD} = (3/5)P_{AD}\Delta^{v} + P_{BD}\Delta^{v} + (3/5)P_{CD}\Delta^{v}$$
$$\therefore F = (3/5)P_{AD} + P_{BD} + (3/5)P_{CD} \qquad \text{(ii)}$$

where the bar forces P, may be expressed in terms of the actual bar displacements $\delta = \varepsilon L$, from the particular σ, ε law. Here,

$$P = \sigma A = a\varepsilon A\,(1 - b\varepsilon^{2}) = aA\,(\delta/L)[1 - b(\delta/L)^{2}]$$
$$\therefore P_{AD} = aA\,(\delta/L)_{AD}\,[1 - b\,(\delta/L)_{AD}{}^{2}] = P_{CD}$$
$$P_{BD} = aA\,(\delta/L)_{BD}\,[1 - b\,(\delta/L)_{BD}{}^{2}]$$

Now if Δ is the actual displacement under F,

$$\delta_{CD} = \delta_{AD} = (3/5)\Delta,\ \delta_{BD} = \Delta$$
$$\therefore P_{AD} = aA(3/5)(\Delta/L_{AD})[1 - b(9/25)(\Delta/L_{AD})^{2}] = P_{CD}$$
$$P_{BD} = aA\,(\Delta/L_{BD})[1 - b\,(\Delta/L_{BD})^{2}]$$

Substituting into eq(ii),

$$F = (18/25)aA\,(\Delta/L_{AD})[1 - b(9/25)(\Delta/L_{AD})^{2}] + aA\,(\Delta/L_{BD})[1 - b(\Delta/L_{BD})^{2}] \qquad \text{(iii)}$$
$$\therefore P_{BD}/F = \{1 + (54/125)[1 - b(9/25)(\Delta/L_{AD})^{2}]/\,[1 - b\,(\Delta/L_{BD})^{2}]\}^{-1} \qquad \text{(iv)}$$

Thus for given values of A, F, a, b and L, eqs(iii) and (iv) may be solved to provide a numerical value for the ratio P_{BD}/F. This is unecessary for a Hookean (linear elastic) material, when for $b = 0$, eq(iv) becomes

$$P_{BD}/F = [1 + (54/125)]^{-1} = 0.698$$

A similar PVD approach may be employed for any redundant structure. In general, the number of simultaneous equations to be solved with this method equates to the total number of degrees of freedom for the frame. This is the sum of the degrees of freedom for each joint. In Fig. 8.29 the number is only one since joint B moves in the y - direction. The choice between PVD and PVF for the determination of bar forces depends upon whether the frame has a smaller number of degrees of freedom in PVD than the number of redundancies in PVF. For example, PVF is the better choice for the frame in Fig. 8.20 since the two degrees of freedom at each of joints A, B and E total six in this singly redundant frame.

Example 8.26 A thin pipe is simply supported at length intervals L when carrying fluid of density ρ at pressure p and velocity V (Fig. 8.30a). Assuming a deflected shape of the form $v = a\,(1 - \cos 2\pi z/L)$ between any pair of supports, derive an expression that will enable the prediction of lateral instability. The weight of the pipe is negligible compared to the fluid.

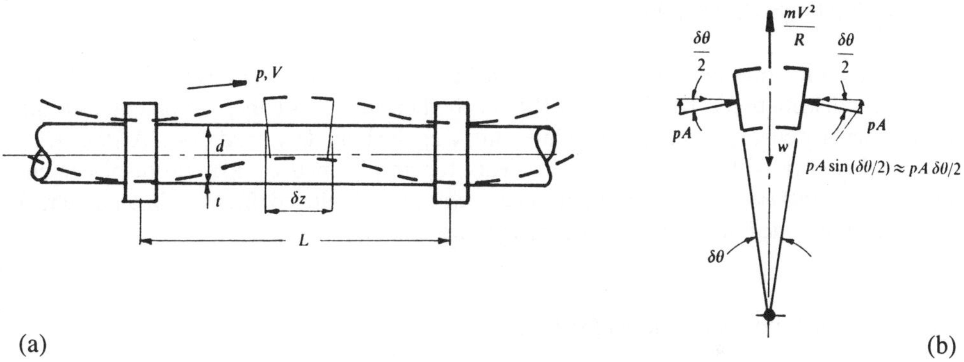

Figure 8.30 Simply supported pipe

When a variable external vertical loading w is distributed over a given length L, eq(8.26a) becomes

$$\int_0^L (\mathrm{d}w/\mathrm{d}z)\, v^{\mathrm{v}}\, \mathrm{d}z = EI \int_0^L (\mathrm{d}^2 v^{\mathrm{v}}/\mathrm{d}z^2)\,(\mathrm{d}^2 v/\mathrm{d}z^2)\, \mathrm{d}z \qquad \text{(i)}$$

in which the derivation of the right-hand side has been given in Example 8.24. Now, for a length δz of the deflected pipe shown in Fig. 8.30b, w is composed of inertia and pressure effects. That is

$$\delta w = m\,(V^2/R) + pA\,\delta\theta = (\rho A\,\delta z)(V^2/R) + pA\,\delta\theta$$

The curvature is given by $1/R = \mathrm{d}^2v/\mathrm{d}z^2 = \mathrm{d}\theta/\mathrm{d}z$ so that

$$\therefore\ \delta w/\delta z = (\rho A V^2 + pA)\, \mathrm{d}^2 v/\mathrm{d}z^2$$

Substituting into eq(i) gives

$$-\int_0^L (\rho A V^2 + pA)(\mathrm{d}^2 v/\mathrm{d}z^2)\, v^{\mathrm{v}}\, \mathrm{d}z = EI \int_0^L (\mathrm{d}^2 v^{\mathrm{v}}/\mathrm{d}z^2)(\mathrm{d}^2 v/\mathrm{d}z^2)\, \mathrm{d}z \qquad \text{(ii)}$$

in which the directions of v^{v} and w oppose eachother. Assuming that the virtual deflection has the same form as v,

$v = a\,[1 - \cos(2\pi z/L)],\quad d^2v/dz^2 = (4\pi^2/L^2)\,a\cos(2\pi z/L)$
$v^{v} = a^{v}[1 - \cos(2\pi z/L)],\ d^2v^{v}/dz^2 = (4\pi^2/L^2)\,a^{v}\cos(2\pi z/L)$

Substituting into eq(ii) and integrating leads to the condition

$$- A(\rho V^2 + p)\,aa^{v}(4\pi^2/L^2)\int_0^L \cos(2\pi z/L)[1 - \cos(2\pi z/L)]dz$$
$$= EI\,(4\pi^2/L^2)^2\,aa^{v}\int_0^L \cos(2\pi z/L)$$
$$V^2\rho + p = 4\pi^2 EI/(AL^2)$$

and thus, fluid properties v, ρ and p are related to the properties E, A and I for the beam section at the lowest point of instability.

8.8 The Rayleigh-Ritz Method

Variational principles may be used to find the best approximate solutions to displacements and stress distributions. The Rayleigh-Ritz method ensures that either the potential or the complementary energy is minimised while satisfying the boundary conditions. A *stationary potential energy* will provide the coefficients for an assumed compatible displacement function. A *stationary complementary energy* will provide the coefficients in an assumed stress field provided the equilibrium equations are satisfied.

8.8.1 Stationary Potential Energy (SPE)

Since $\delta W_I = -\,\delta U$ from eq(8.12), the principle of virtual work (eq 8.14a) may be rewritten in the following form:

$$\delta W_E - \delta U = 0 \tag{8.27}$$

Now, as potential energy (V) is the energy stored in the body for its displaced position under the external forces, it follows that

$$\delta V = -\,\delta W_E \tag{8.28}$$

Combining eqs(8.27) and (8.28),

$$\delta U + \delta V = 0$$

That is, for a conservative force system where a complete cycle of loading-unloading results in zero net work, the principle of virtual work becomes

$$\delta(U + V) = 0 \tag{8.29}$$

Taking U to mean the PE of strain, eq(8.29) shows that a stationary value of the total potential energy ($U + V$) exists when the system is in equilibrium. The principle states:
Of all the possible compatible displacement systems which satisfy the boundary conditions, that which also satisfies the equilibrium conditions gives a stationary value to the PE.

The SPE equation (8.29) is normally combined with eq(8.28) in the following form:

$$\delta[U - \sum(F_k \Delta_k)] = 0 \qquad (8.30)$$

Equation (8.30) is particularly useful for finding good approximations to displacement functions, where exact solutions may not be available.

Example 8.27 Find the deflection v, at any point in the length z, of a simply supported beam of length L which carries a vertical concentrated force P at a distance c from the end (Fig. 8.31). Assume the deflected shape function $v = \sum a_n \sin(n\pi z/L)$ for $n = 1, 2 \ldots \infty$. Compare with the accepted expression when $z = c = L/2$.

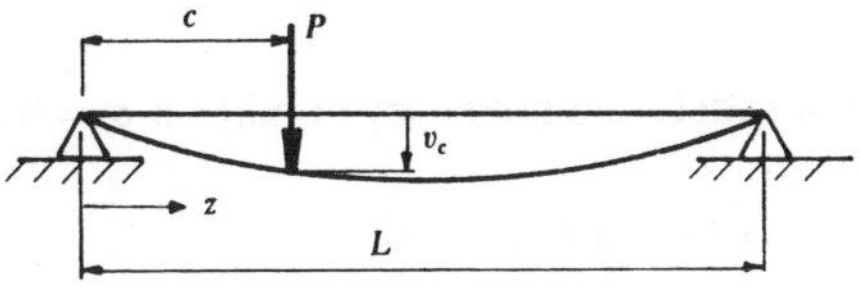

Figure 8.31 Simply supported beam

The boundary conditions $v = 0$ for $z = 0$ and $z = L$ are satisfied. Now,

$$\sum F_k \Delta_k = Pv$$

and from Table 8.1 and the flexure equation,

$$U = \int_0^L M^2 \mathrm{d}z/(2EI) = (EI/2)\int_0^L (\mathrm{d}^2v/\mathrm{d}z^2)^2 \mathrm{d}z$$

$$U + V = -Pv + (EI/2)\int_0^L (\mathrm{d}^2v/\mathrm{d}z^2)^2\,\mathrm{d}z$$

$$= -P\sum a_n \sin(n\pi c/L) + (EI/2)\int_0^L [\sum a_n (n\pi/L)^2 \sin(n\pi z/L)]^2\,\mathrm{d}z$$

$$= -P\sum a_n \sin(n\pi c/L) + (EI/2)\sum a_n^2 (n\pi/L)^4 (L/2)$$

As a_n controls the magnitude of $U + V$, for equilibrium, eq(8.30) becomes

$$\delta(U + V)/\delta a_n = 0 = -P\sum \sin(n\pi c/L) + (EI)\sum a_n (n\pi/L)^4 (L/2)$$

$$\therefore a_n = P\sum \sin(n\pi c/L)/\,[(EI)\sum(n\pi/L)^4(L/2)] = [2PL^3/(EI\pi^4)]\sum n^{-4} \sin(n\pi c/L)$$

$$\therefore v = [2PL^3/(EI\pi^4)]\sum n^{-4} \sin(n\pi c/L)\sin(n\pi z/L) \qquad \text{(i)}$$

Equation (i) supplies the deflection at any point. Putting $z = c = L/2$ gives the deflection beneath a central concentrated load. That is,

$$v = [2PL^3/(EI\pi^4)][1 + 3^{-4} + 5^{-4} + \ldots]$$
$$= PL^3/(48.035EI)$$

which compares with the exact value of 48 in the denominator.

Example 8.28 Determine the constants A and B within an expression $u = A(r + B/r)$ assumed for the radial displacements at radius r in the wall of a thick cylinder under internal pressure p, with zero outer pressure.

The radial and hoop strains are

$$\varepsilon_r = \partial u/\partial r \text{ and } \varepsilon_\theta = u/r \tag{i}$$

which give:

$$\varepsilon_r = A\,(1 - B/r^2),\ \varepsilon_\theta = A\,(1 + B/r^2) \tag{ii}$$

The constitutive relations are:

$$\sigma_\theta = E(\varepsilon_\theta + \nu\,\varepsilon_r)\,/\,(1 - \nu^2) \tag{iii}$$
$$\sigma_r = E(\varepsilon_r + \nu\,\varepsilon_\theta)\,/\,(1 - \nu^2) \tag{iv}$$

Equations (ii) and (iv) must satisfy the boundary condition $\sigma_r = 0$ for $r = r_2$,

$$\varepsilon_r + \nu\,\varepsilon_\theta = A(1 - B\,/\,r_2^{\,2}) + A\nu\,(1 + B\,/\,r_2^{\,2}) = 0$$
$$\therefore\ B = r_2^{\,2}\,(1 + \nu\,)/\,(1 - \nu\,) \tag{v}$$

Though B could be found from a second condition: $\sigma_r = -\,p$ for $r = r_1$, SPE will now be used for this purpose. U for a principal biaxial stress state (σ_r, σ_θ) follows from eq(8.4b), (iii) and (iv) as

$$\begin{aligned}\delta U\,/\,\delta V' &= \sigma_\theta\delta\varepsilon_\theta + \sigma_r\,\delta\varepsilon_r\\ &= [E/\,(1 - \nu^2)][(\varepsilon_\theta + \nu\varepsilon_r)\delta\varepsilon_\theta + (\varepsilon_r + \nu\varepsilon_\theta)\delta\varepsilon_r]\\ &= [E/(1 - \nu^2)][(\varepsilon_\theta\,\delta\varepsilon_\theta + \nu\,\mathrm{d}(\varepsilon_r\varepsilon_\theta) + \varepsilon_r\,\delta\varepsilon_r\,]\\ &= [E/(1 - \nu^2)][(\varepsilon_\theta^{\,2}/2 + \nu\,\mathrm{d}(\varepsilon_r\,\varepsilon_\theta) + \varepsilon_r^{\,2}/2\,]\end{aligned}$$

where V' is the volume.

$$U = \{E/[2(1 - \nu^2)]\}\int_{V'}[\varepsilon_r^{\,2} + 2\nu\,\varepsilon_r\,\varepsilon_\theta + \varepsilon_\theta^{\,2}\,]\mathrm{d}V'$$

Since $\sum F_k\Delta_k = 2\pi r_1 p u_1$ and $\delta V' = r\times\delta\theta\times\delta r$ per unit length, eq(8.30) gives

$$U + V = -\,2\pi r_1\,p\,u_1 + \{E/[2(1 - \nu^2)]\}\int_0^{2\pi}\int_{r_1}^{r_2}[\varepsilon_r^{\,2} + 2\nu\,\varepsilon_r\,\varepsilon_\theta + \varepsilon_\theta^{\,2}]\,r\,\mathrm{d}\theta\mathrm{d}r$$

Substituting from eqs (i) and (ii),

$$U + V = -\,2\pi r_1 pA\,(r_1 + B/r_1) + [2\pi EA^{\,2}/(1 - \nu^2)]\int_{r_1}^{r_2}[(1 + B^2/r^4) + \nu\,(1 - B^2/\,r^4)]\,r\,\mathrm{d}r$$

$$= -\,2\pi r_1 pA\,(r_1 + B/r_1) + [\pi EA^{\,2}/(1 - \nu^2)]\,|\,r^2 - B/r^2 + \nu\,(r^2 + B/r^2)\,\Big|_{r_1}^{r_2} \tag{vi}$$

Now from eq(8.29), since B is a constant, the stationary value will be found from eq(vi) corresponding to

$$\mathrm{d}\,(U + V\,)/\mathrm{d}A = -\,2\pi r_1 p\,(r_1 + B/r_1\,) + [2\pi EA\,/(1 - \nu^{\,2})]\,|\,r^2 - B/r^2 + \nu\,(r^2 + B/r^2)\,\Big|_{r_1}^{r_2} = 0$$

$$A = p\,r_1\,(r_1 + B/r_1\,)(1 - \nu^2)\,/\,[\,E\,|\,r^2 - B/r^2 + \nu\,(\,r^2 + B/r^2)\,\Big|_{r_1}^{r_2}\,]$$

$$= p\,r_1\,(r_1 + B/\,r_1\,)(1 - \nu^2)\,/\,\{E\,[(1 + \nu\,)\,(r_2^2 - r_1^2) - B\,(1 - \nu\,)\,(1/\,r_2^2 - 1/\,r_1^2)]\}$$

Substituting for B from eq(v) finally leads to

$$A = pr_1^2(1 - \nu) / [E(r_2^2 - r_1^2)]$$

8.8.2 Stationary Complementary Energy (SCE)

It is also possible to formulate a stationary energy principle based upon the corresponding complements of δU and δV in eq(8.29). Thus if δU^* is the complementary strain energy and $\delta V^* = -\delta W_E^*$ where δW_E^* is the complementary work of the external forces, then the total complementary energy for a strained body is $U^* + V^*$. The stationary value of $U^* + V^*$ becomes equivalent to the principle of virtual complementary work:

$$\delta(U^* + V^*) = 0 \qquad (8.31)$$

The SCE principle is then stated as

Of all the equilibrium force (stress) systems satisfying boundary conditions, that which also satisfies the compatibility condition gives a stationary value to the complementary energy.

For Hookean materials where $U = U^*$ and $\delta W_E^* = \Delta\, \delta F$, eq(8.31) becomes

$$\delta(U - \textstyle\sum \Delta_k F_k) = 0 \qquad (8.32)$$

SCE can be used as an alternative to the ULM for solving redundant structures. A further simplification is often made that, in the presence of small boundary displacements, the summation term in eq(8.32) is negligibly small. However, this will only strictly apply to a rigid boundary, when eq(8.32) becomes

$$\delta U = \delta U^* = 0 \qquad (8.33)$$

which is particularly useful for obtaining approximate stress distributions where exact solutions are not available. The reduced form of SCE in eq(8.33) is often referred to as the *principle of least work.*

Example 8.29 The cantilever in Fig. 8.32 is propped at mid-position when carrying an end load P. Determine from SCE, the prop reaction R_A when the prop deflects the beam by the amount: (i) Δ_A upwards and (ii) zero. EI is constant.

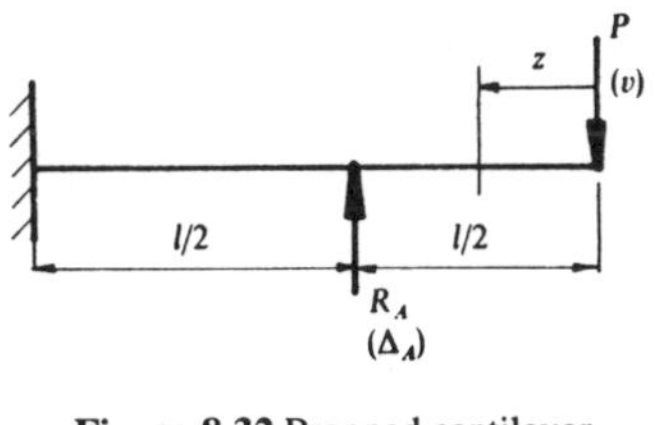

Figure 8.32 Propped cantilever

Taking case (i) generally,

$$U^* = U = \int M^2 \mathrm{d}z/(2EI)$$

$$= \int_0^{l/2} (Pz)^2 \mathrm{d}z/(2EI) + \int_{l/2}^{l} [Pz - R_A(z - l/2)]^2 \mathrm{d}z/(2EI)$$

$$V^* = -\sum\Delta_k F_k = -R_A\Delta_A - P\,v$$

The total complementary energy is then

$$U^* + V^* = \int_0^{l/2}(Pz)^2\,\mathrm{d}z/(2EI) + \int_{l/2}^{l}[Pz - R_A(z - l/2)]^2\,\mathrm{d}z/(2EI) - R_A\Delta_A - P\,v$$

This has a stationary value in respect of the single redundancy R_A. That is, from eq(8.31),

$$\mathrm{d}(U^* + V^*)/\mathrm{d}R_A = \int_{l/2}^{l}[Pz - R_A(z - l/2)]\,(z - l/2)\,\mathrm{d}z/(EI) + \Delta_A = 0$$

$$\int_{l/2}^{l} P(z^2 - lz/2) - R_A(z^2 - lz + l^2/4)\,\mathrm{d}z + EI\,\Delta_A = 0$$

$$P\,|\,z^3/3 - lz^2/4\,\Big|_{l/2}^{l} - R_A\,|\,z^3/3 - lz^2/2 + l^2z/4\,\Big|_{l/2}^{l} + EI\,\Delta_A = 0$$

$$P\,[(l^3/3 - l^3/4) - (l^3/24 - l^3/16)]$$
$$- R_A[(l^3/3 - l^3/2 + l^3/4) - (l^3/24 - l^3/8 + l^3/8)] + EI\,\Delta_A = 0$$
$$5Pl^3/48 - R_A\,l^3/24 + EI\,\Delta_A = 0$$
$$R_A = 5P/2 + 24EI\,\Delta_A/l^3$$

When in part (ii) $\Delta_A = 0$, $R_A = 5P/2$. For a structure with m redundancies, eq(8.32) may be applied to each redundancy in turn to yield m simultaneous equations.

Example 8.30 A linear increase in hoop stress $\sigma_\theta = A + B\,r$ is assumed from the inner (r_1) to outer (r_2) radii in a thick-walled annular disc under internal pressure p. Determine from SCE the constants A and B when the effect of radial stress is ignored.

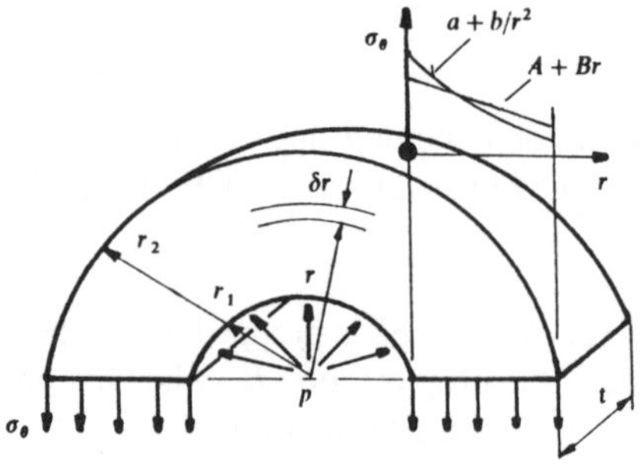

Figure 8.33 Thick cylinder

The boundary condition is satisfied by equating forces along a horizontal diameter in Fig. 8.33 to give

$$p\,(2r_1)\,t = 2\int_{r_1}^{r_2}\sigma_\theta\,t\,\mathrm{d}r = 2\int_{r_1}^{r_2}(A + Br)\,t\,\mathrm{d}\,r$$

where p acts on the projected area $2r_1\,t$. This gives

$$A = pr_1/(r_2 - r_1) - B(r_2 + r_1)/2 \qquad \text{(i)}$$
$$\sigma_\theta = pr_1/(r_2 - r_1) + B[\,r - (r_1 + r_2)/2] \qquad \text{(ii)}$$

Assuming a rigid boundary the total complementary energy is simply $U^* = U$. Since the stress system is assumed to be uniaxial, Table 8.1 gives $U = U^*$ as

$$U^* = [1/(2E)] \int_V \sigma_\theta^2 \, dV$$

The stationary value applies in respect of B,

$$dU^*/dB = (1/E) \int_V \sigma_\theta (d\sigma_\theta / dB) \, dV = 0$$

Substituting from eq(ii) with $dV = (2\pi r \delta r)\, t$,

$$\int_{r_1}^{r_2} \{ p r_1 / (r_2 - r_1) + B [r - (r_1 + r_2)/2] \} [r - (r_1 + r_2)/2] r \, dr = 0$$

Integrating and introducing the radius ratio $K = r_2 / r_1$ leads to

$$B = \frac{6p(K+1)^2 - 8p(K^2 + K + 1)}{r_1[6(K^4 - 1) - 8(K^3 - 1)(K+1) + 3(K^2 - 1)(K+1)^2]}$$

and, from eq(i), A is written as

$$A = p/(K-1) - B r_1 (K+1)/2$$

Satisfying equilibrium and compatibility in this way provides the most acceptable linear hoop stress distribution. It is compared in Fig. 8.33 with Lame's exact solution, $\sigma_\theta = 2C - A/r^2$ (see Eq 2.48b).

EXERCISES

Castigliano's Theorems

8.1 Two views of a cantilever arc are shown in Fig. 8.34a,b. It is made from a solid 10 mm diameter rod by bending into the quadrant of a circle of mean radius 200 mm. Find the slope and deflection in the plane of an oblique 10 kN force, when applied at its free end as shown in Fig. 8.34b. Take $E = 210$ GPa and $G = 84$ GPa.

Figure 8.34

8.2 Find the force H necessary to eliminate horizontal deflection at A and B for each of the parabolic steel arches shown in Fig. 8.35a and b. Draw the bending moment diagram in M,θ co-ordinates inserting major values in each case. What is the increase in H, if the additional horizontal displacement due to a temperature rise of 50°C is to be prevented? Take $I = 300 \times 10^6$ mm^4, $\alpha = 11 \times 10^{-6}$ °C^{-1}.

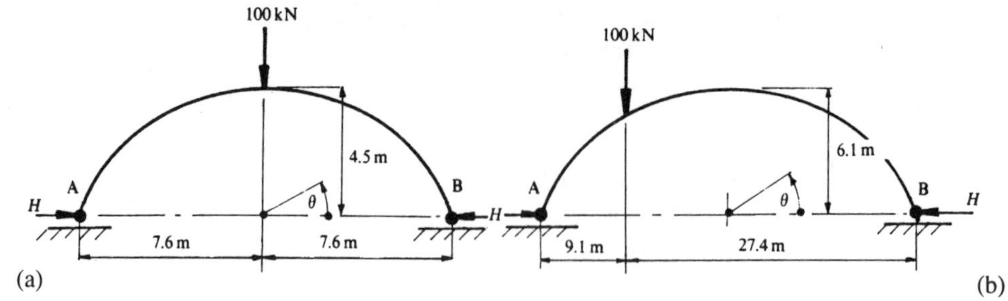

Figure 8.35

8.3 Determine the vertical displacement beneath each load for the frames in Figs 8.36a and b when all bar areas are 1280 mm^2 and $E = 207$ GPa.

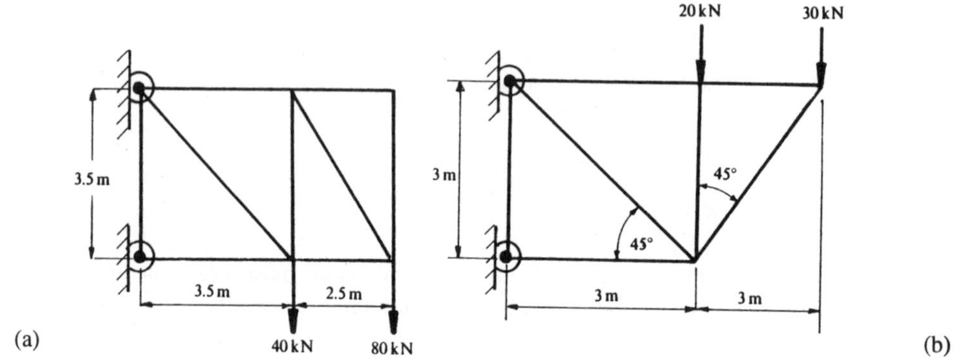

Figure 8.36

8.4 Show, for the frame in Fig. 8.37, that the vertical deflection at point P is given by $\delta_A = 6.22FL/(EA)$ when EA is constant throughout the frame.

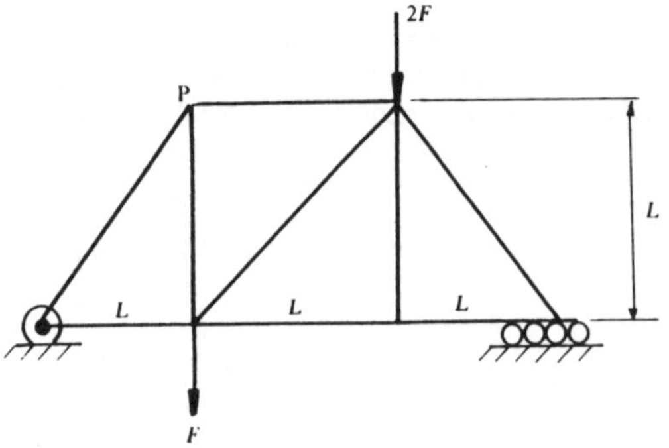

Figure 8.37

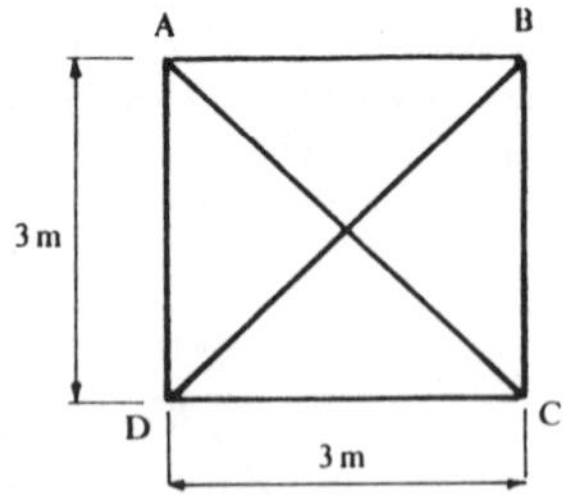

Figure 8.38

8.5 Calculate the forces for the frame in Fig. 8.38 given that bar BD, which was initially 3 mm short, was elastically strained into position. Take the area $A = 970$ mm^2 and Young's modulus $E = 204$ GPa as constants throughout.

8.6 Find the forces in each bar of the redundant structure in Fig. 8.39, when the walls react to both vertical and horizontal forces at points A and B. Take $A = 500$ mm^2 and $E = 200$ GPa.

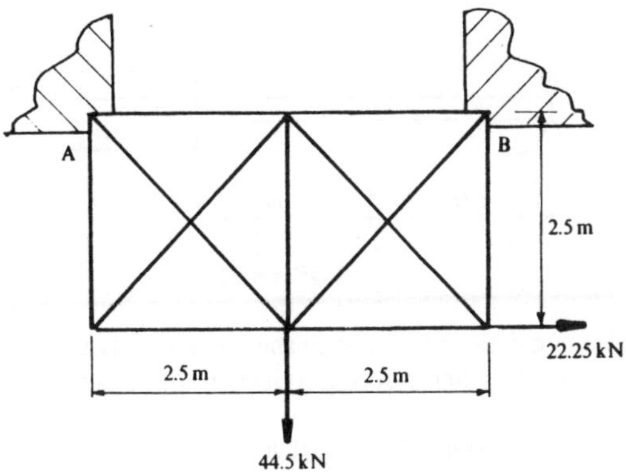

Figure 8.39

The Principle of Virtual Forces (Including ULM)

8.7 Apply the PVF or ULM to confirm Castigliano's answers to the loaded arcs given in Figs 8.34 and 8.35.

8.8 A semi-circular arch, 3.5 m radius, supports a central vertical load of 10 kN. If the arch is pinned at one end and rests on rollers at the other end, find the horizontal deflection at the rollers. If both ends are pinned, what are the horizontal reactions? Sketch the polar shear force and bending moment diagrams, inserting the major values. Take $EI = 4725$ kNm^2.

8.9 Examine the effect of replacing the semi-circular arch in Exercise 8.8 with an arch of 2 m rise over the same base length and under the same loading.

8.10 If the cantilever in Fig. 8.12 carries, instead of P, a uniformly distributed load w throughout its length show that the vertical deflection at the free end due to bending and torsional effects is given by $(wR^4/2)[1/(EI) + (1 - \pi - \pi^2/4)/(GJ)]$

8.11 Find the vertical deflection at the free end of the cantilevered bracket in Fig. 8.40. Take the moment of area as $I = 42 \times 10^3$ mm^4 and the modulus as $E = 207$ GPa.

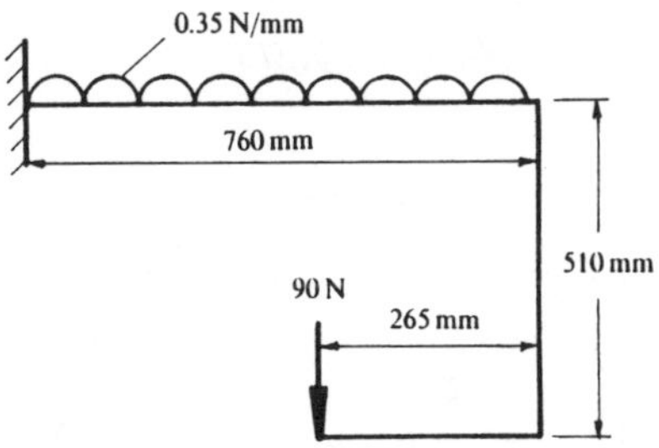

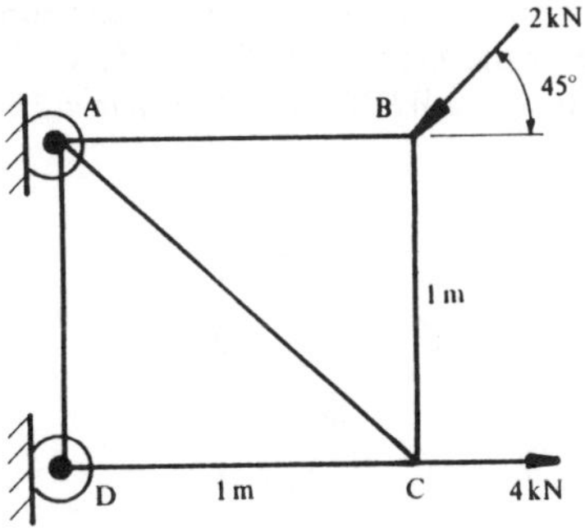

Figure 8.40

Figure 8.41

8.12 Determine the vertical deflection at C for the cantilever frame in Fig. 8.41.

8.13 Find the vertical deflection at point B for the stepped section beam in Fig. 8.42. Take $E = 207$ GPa, $I = 50 \times 10^3$ mm^4.

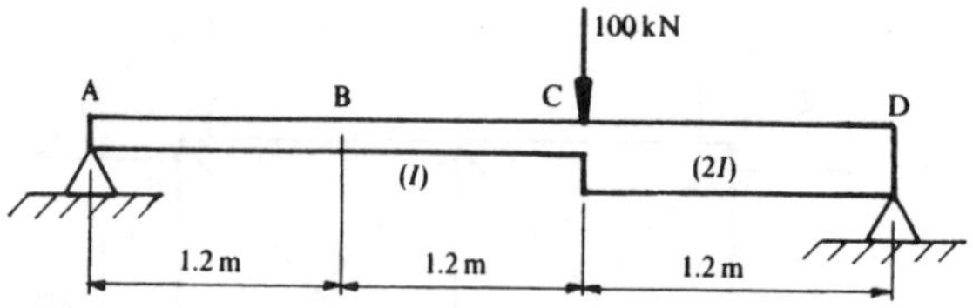

Figure 8.42

8.14 Determine the maximum deflection Δ due to shear in a simply supported beam length L, breadth b, depth d and shear modulus G, when carrying (i) a central concentrated load W and (ii) a uniformly distributed load w/unit length.
Answer: (i) $\Delta = 3FL/(10bdG)$, (ii) $\Delta = 3wL^2/(20bdG)$

8.15 The cantilever quadrant in Fig. 8.43 is loaded at its free end through two cables under equal tensions T. The tip is not to deflect vertically, so that the horizontal tension T remains in position. At what angle β must the second tension be applied to ensure this condition? EI is constant.

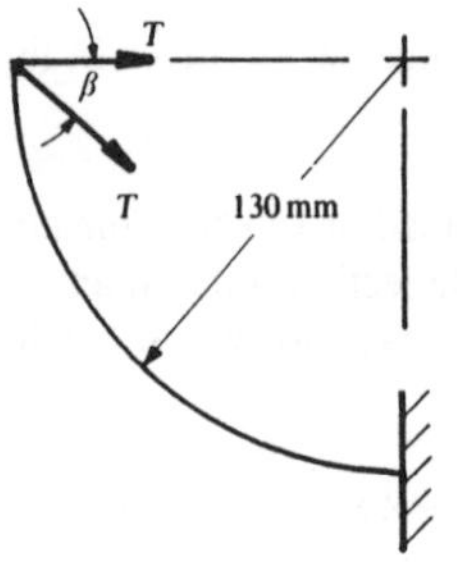

Figure 8.43

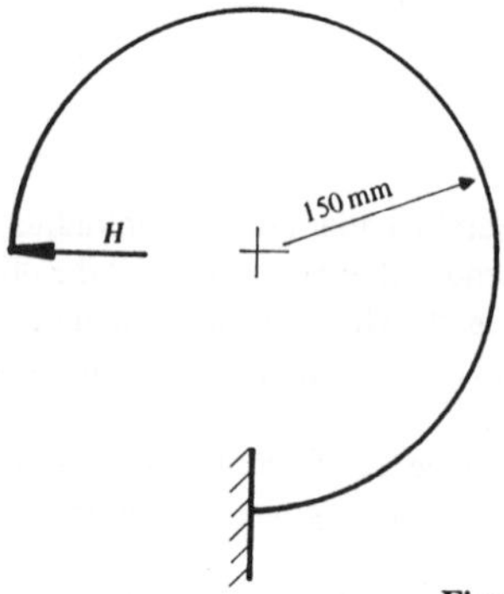

Figure 8.44

8.16 The 3/4 circular arc in Fig. 8.44 with mean radius 150 mm, deflects 10 mm horizontally at its tip under the action of a horizontal force H applied as shown. If the maximum bending stress in the material is limited to 90 MPa find (i) the cross-section dimensions when the breadth is five times the thickness, (ii) the force H and (iii) the vertical deflection at the tip. Take $E = 210$ GPa.

8.17 Derive expressions for the vertical and horizontal components of deflection at the load point in each of the cantilevered arcs in Figs 8.45a and b.
Answer: (a) $\delta_V = \pi WR^3/(4EI)$, $\delta_H = WR^3/(2EI)$ (b) $\delta_V = WR^3(\pi - 2)/(EI)$

(a)

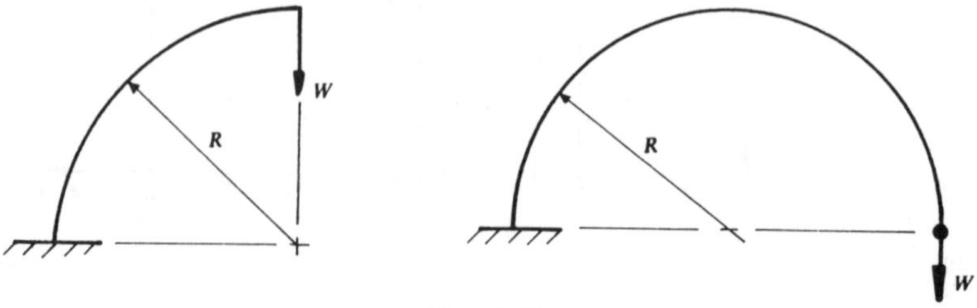

(b)

Figue 8.45

8.18 Calculate the horizontal displacement at the rollers for the portal frame in Fig. 8.46 if, for the vertical cross-sections, $A = 9000$ mm^2, $I = 1 \times 10^8$ mm^4, and, for the horizontal section with area properties, $A = 8500$ mm^2, $I = 3 \times 10^8$ mm^4. Take $E = 200$ GPa.

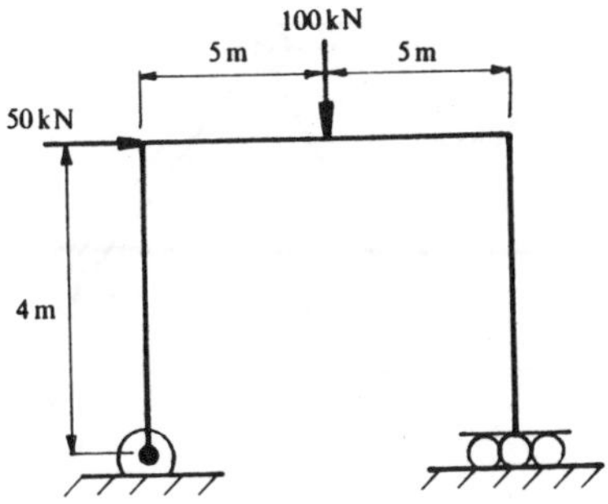

Figure 8.46

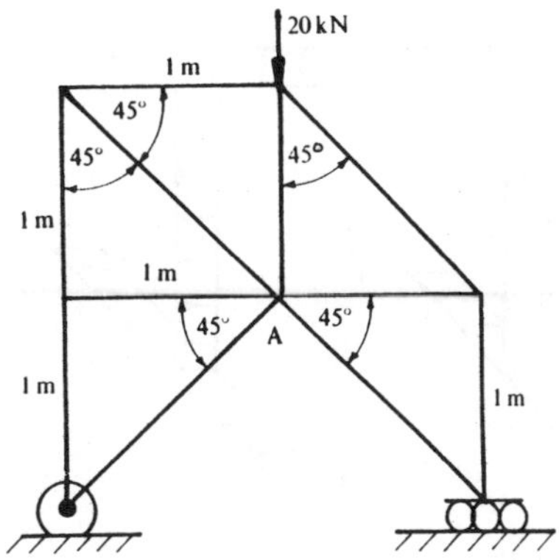

Figure 8.47

8.19 Find the horizontal deflection at A for the pin-jointed frame in Fig. 8.47 when for all bars, $A = 200$ mm^2 and $E = 200$ GPa.

8.20 The bar lengths for the inclined pin-jointed frame in Fig. 8.48 are DC = CE = AB = BE = √2 m and DE = CB = EA = 2 m. If all bar areas are 100 mm^2, with $E = 100$ GPa, determine the vertical and horizontal deflections at point E when a vertical force of 2 kN is applied at B.

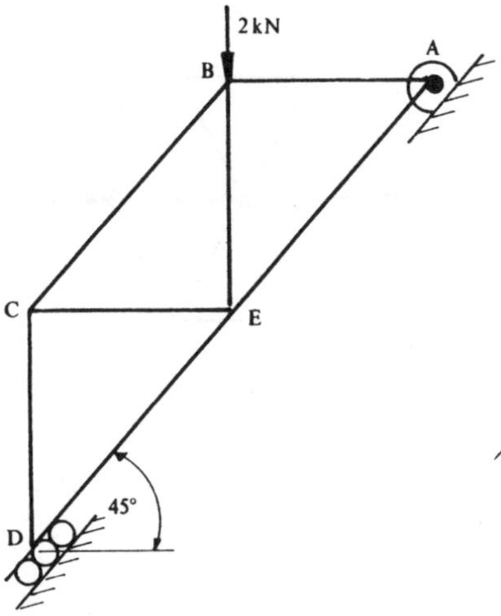

Figure 8.48

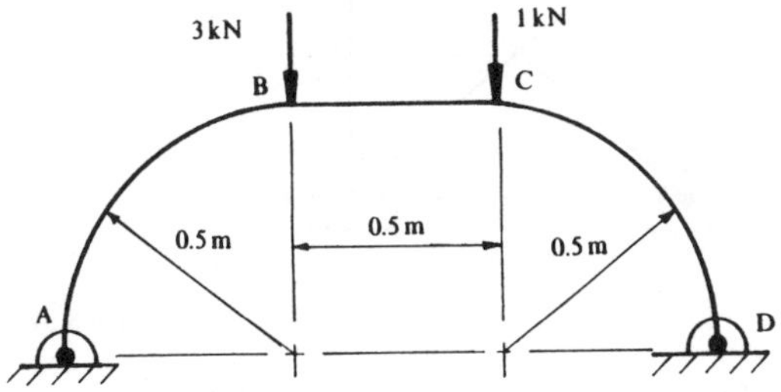

Figure 8.49

8.21 The arch in Fig. 8.49 is loaded by two vertical forces at B and C and pinned at A and D as shown. If the section is a tube 40 mm o.d. with $I = 3.37 \times 10^4$ mm^4 and $E = 207$ GPa, determine, from the polar bending moment diagram, the position and magnitude of the maximum bending stress when A and D are free to move apart. What horizontal force would be required at A and D to prevent this movement?

8.22 The structure in Fig. 8.50 consists of a vertical portal frame resting on two·parallel, horizontal simply supported beams. A uniformly distributed load w acts outwards throughout, as shown. Determine the horizontal reaction at the feet of the frame based upon bending effects only. EI is constant.

Answer: $p\,(8a^4 - 5L^4)/(20a^3 + 4L^3)$

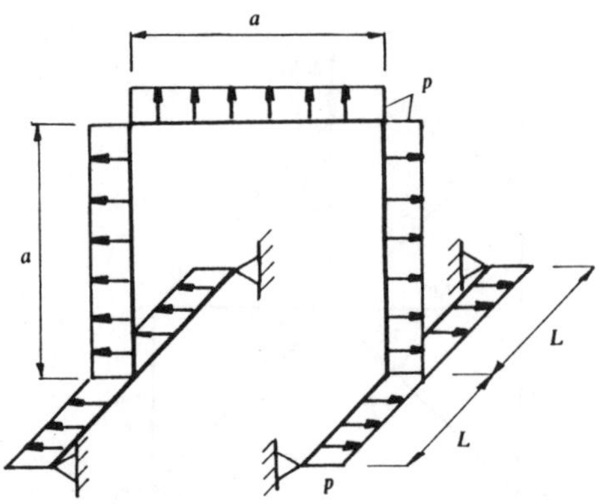

Figure 8.50

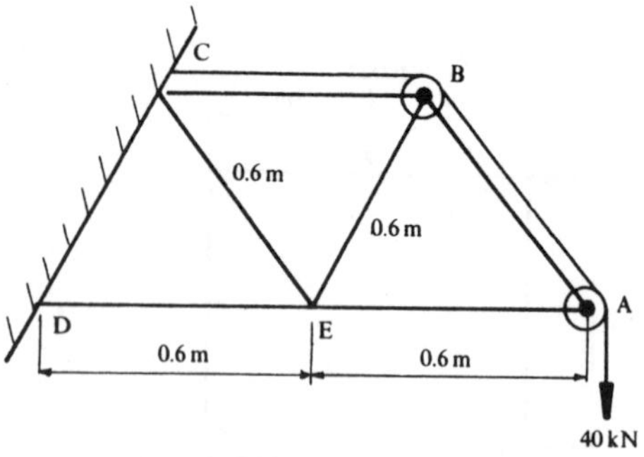

Figure 8.51

8.23 The Warren girder in Fig. 8.51 is built in to a wall at C and D. The structure is used to lift a weight of 40 kN with a wire passing over pulleys at A, B and C. Find the vertical deflection at A and the force acting in the direction AB that will eliminate this deflection. Bar areas are 180 mm^2 and $E = 70$ GPa.

Redundant Structures

8.24 Find the forces in all the bars of the pin-jointed structure in Fig. 8.52a when *EA* is constant throughout.

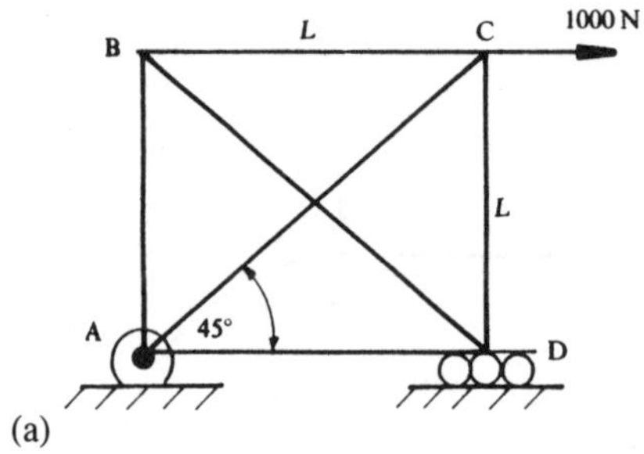

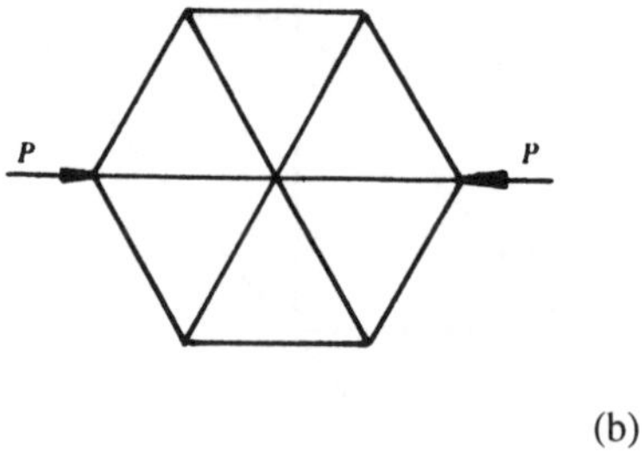

Figure 8.52

8.25 Find the forces in all the members of the plane, pin-jointed, hexagonal frame in Fig. 8.52b when *A*, *E* and *L* for each member are constant.
Answer: outers - P/6, inner horizontals - 5P/6, others + *P*/6)

8.26 The plane frame in Fig. 8.53 is loaded through a cable passing over frictionless pulleys at A, B and C. If *E* and *A* are constant, determine the force in each bar.

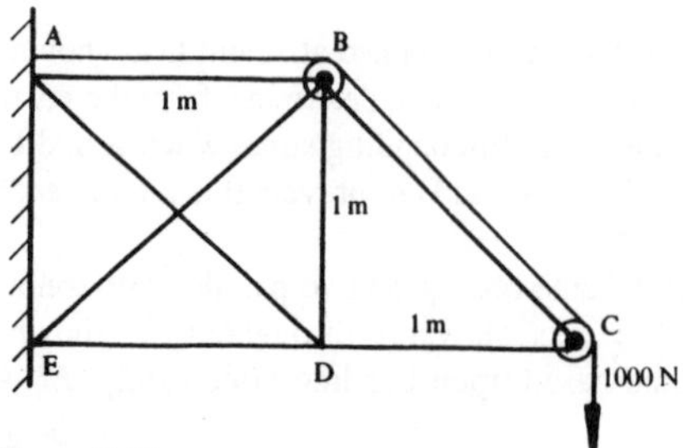

Figure 8.53

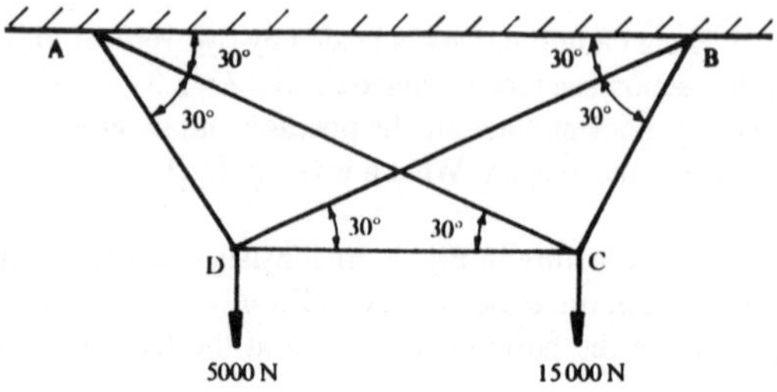

Figure 8.54

8.27 Steel tubes of length 2 m and 100 mm^2 in area comprise the outer bars of the frame in Fig. 8.54. Determine the minimum area for steel wires to remain taut when connecting points A to C and B to D as shown. E is constant.

8.28 Find the forces in the bars of the pin-jointed frame in Fig. 8.55 when the expansion of AD is prevented by built-in hinges and the area of bar AC is twice that of the remaining bars. (Hint - there are two redundancies, say the force in BD and a horizontal reaction.)

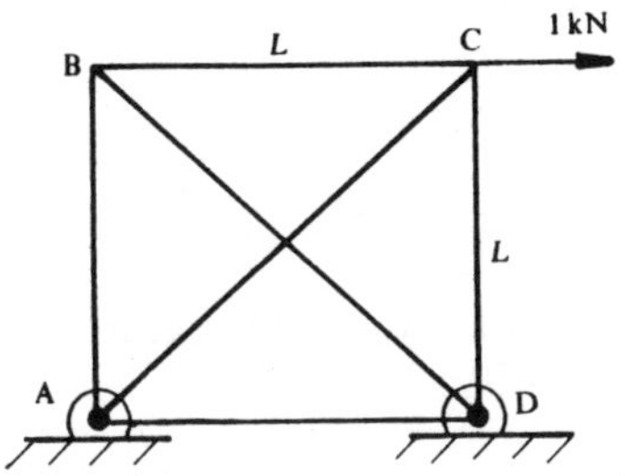

Figure 8.55

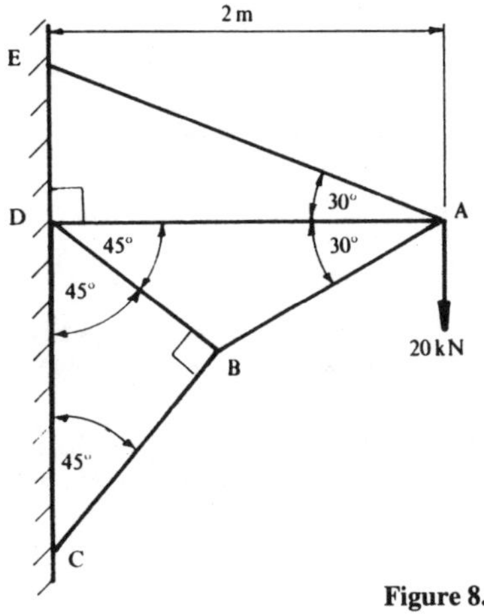

Figure 8.56

8.29 Calculate the force in each member of the redundant structure in Fig. 8.56 and examine the possibility of unstable Euler strut failure when each bar is made from 20 mm square section with a constant $E = 200$ GPa. What then is the vertical deflection at A?

8.30 Calculate the horizontal deflection at the load point for the portal frame in Fig. 8.57a when A is built in and D is (a) hinged and (b) built in. Take EI as constant.
Answer: (a) $1.836PL^3/(EI)$; (b) $1.315PL^3/(EI)$

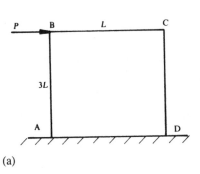

(a) (b)

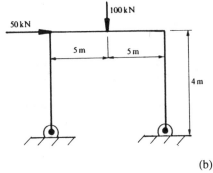

Figure 8.57

8.31 Find the horizontal base reactions for the portal frame in Fig. 8.57b when, for the verticals, A = 9000 mm^2, $I = 10^8$ mm^4, and, for the horizontal bars, A = 8500 mm^2, $I = 3 \times 10^8$ mm^4.
Answer: 42.4 kN

8.32 The aircraft fin in Fig. 8.58a is built in along its base with all other connections free to transmit horizontal shear forces only. Determine the bending moments at the built in connections when a horizontal force F is applied at the central connection normal to the plane of the fin. EI is constant throughout.
Answer: 0.32PL outer, 0.364PL inner

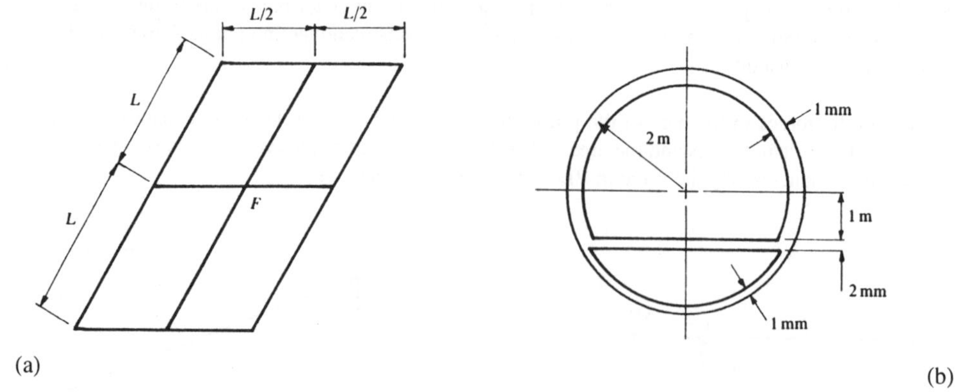

Figure 8.58

8.33 Find the shear stress in the walls and the torsional stiffness when the two-cell fuselage in Fig. 8.58b is subjected to a torque of 2 MNm over a 25 m length. Take $G = 30$ GPa.

8.34 Find the shear stress in the walls and the torsional stiffness for the two-cell tube in Fig. 8.59, when it is subjected to a torque of 100 Nmm. Take $G = 27$ GPa as constant.

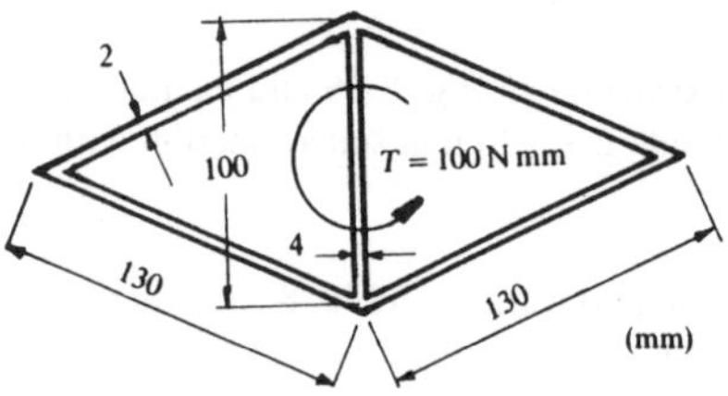

Figure 8.59

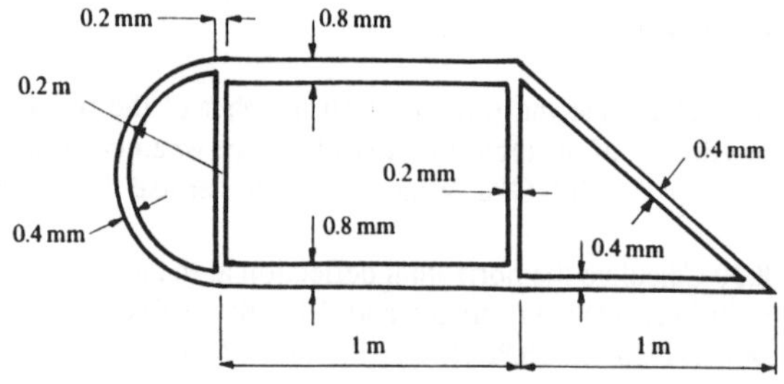

Figure 8.60

8.35 Determine the shear flows for the multi-cell tube in Fig. 8.60 when it carries a pure torque of 10 kNm. What is the relative angle of twist in a 10 m length? Take $G = 30$ GPa.

8.36 Using ULM, determine the shear flow in each web of the idealised structure in Fig. 8.61, when a vertically upward shear force of 20 kN is applied through the shear centre.

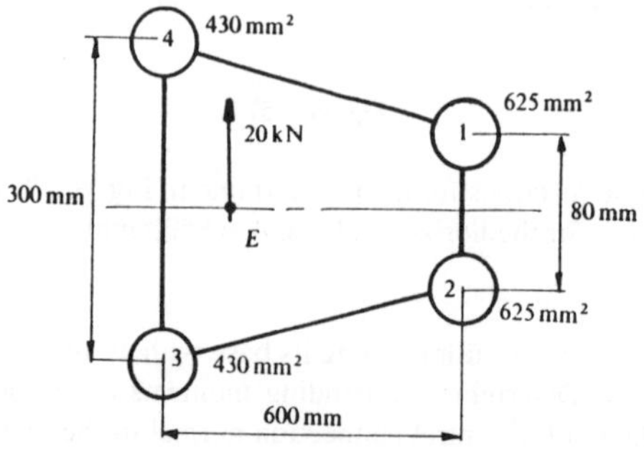

Figure 8.61

Principle of Virtual Displacements

8.37 Find the free-end deflection v_f for a cantilever of length L, when it supports a concentrated end-load P. Assume a deflected shape of the form $v = v_f[1 - \cos(\pi z/L)]$ and take EI to be constant. Compare with the exact value and that value found from assuming a virtual displacement for a cantilever that deflects rigidly about a stiff hinge at its centre.
Answer: $v = 32PL^3/(\pi^4 EI)$, $v = 2\sqrt{2}PL^3/(\pi^2 EI)$

8.38 Decide on suitable displacement functions in each of Figs 8.62a - d and hence derive the maximum displacement in each case. (EI is constant.)

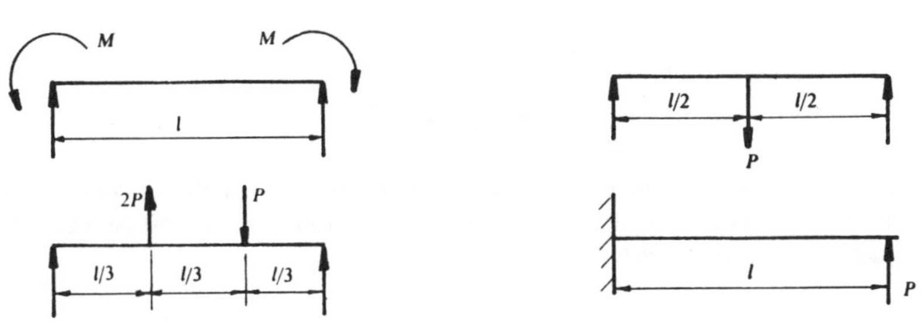

Figure 8.62

8.39 Determine the deflection at 1/4 and 1/2 length positions for a simply supported beam of length L with a central concentrated force W, assuming a displacement function of the form $v = a\sin(\pi z/L)$ and a virtual function of the same form. Compare with the exact values.
Answer: $WL^3/(68.9EI)$, $WL^3/(48.7EI)$

8.40 Use PVD to find the deflection at the $z = L/2$ and $z = 3L/2$ positions for a beam of length $2L$ when a clockwise moment M is applied at its centre. Assume that a sine wave describes the deflected shape. Compare with the corresponding exact values. EI is constant.
Answer: $ML^2/(32EI)$, $-ML^2/(32EI)$

8.41 A strut of length L is built in at one end and carries an axial compressive force P at its free end. If the section varies from I for $0 \le z \le L/2$ to $2I$ for $L/2 \le z \le L$ where z is measured from the free end, determine the critical buckling load. Use the PVD and assume a deflected shape of the following form $v = \Delta\{1 - \cos[\pi z/(2L)]\}$ where Δ is the lateral deflection at the free end.
Answer: $P_{cr} = 2\pi^2 EI/[(3\pi - 2)L^2]$

8.42 A beam is simply supported at both its centre and ends. The beam is partially constrained against rotation θ at its centre by a torque spring with a torsional stiffness $K = T/\theta$. Determine the rotation at the centre when an external moment M acts against this spring. Take EI to be constant.
Answer: $\theta = M/(3EI/L + K)$

8.43 A beam material obeys the non-linear law $\sigma = E(\varepsilon - \beta\varepsilon^3)$. Determine the deflection beneath a central concentrated load of 12.5 kN when the beam is simply supported over a length of 5 m, given that $E = 70.3$ GPa, $\beta = 10^4$ and $I = 1.83 \times 10^6$ mm^4. Assume a deflected shape of the following form $v = a\sin(\pi z/L)$. If the linear law $\sigma = E\varepsilon$ applies during gradual unloading, determine the deflection remaining in the beam for its unloaded state.

8.44 The tapered beam in Fig. 8.63a is propped and loaded by a concentrated moment M_o at its free end. Assuming a deflected shape of the form $v = a[(z/L)^3 - (z/L)^2]$, determine the rotation θ_o under M_o when the stiffness varies as $EI = K(2 - z/L)^3$. Take the virtual deflection to have (i) the same form as v and (ii) that shown in Fig. 8.63b, where C is a point of contraflexure.
Answer: (i) $M_oL/(9K)$, (ii) $M_oL/(9.48K)$

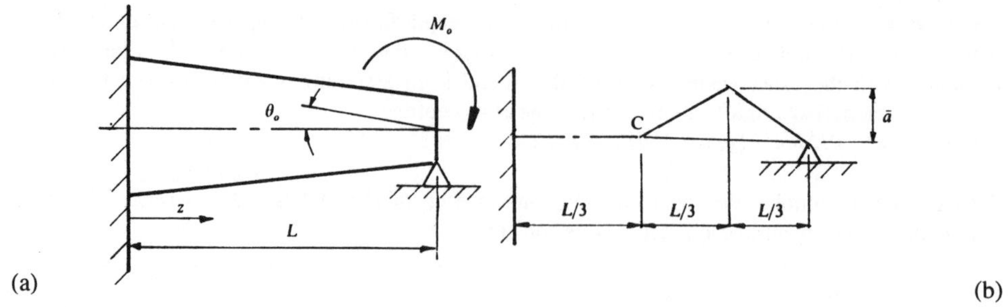

Figure 8.63

8.45 From a consideration of the number of redundancies and the total number of degrees of freedom for each frame in Figs 8.64a and b, determine whether the bar forces would best be found from PVF or PVD. Outline the solution in each case.

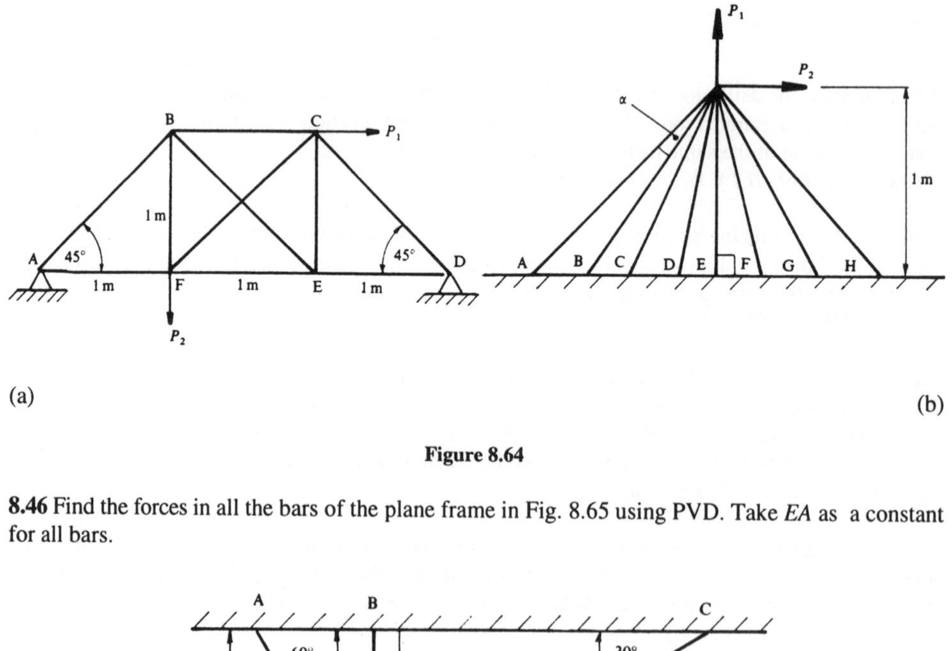

Figure 8.64

8.46 Find the forces in all the bars of the plane frame in Fig. 8.65 using PVD. Take EA as a constant for all bars.

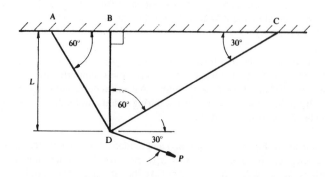

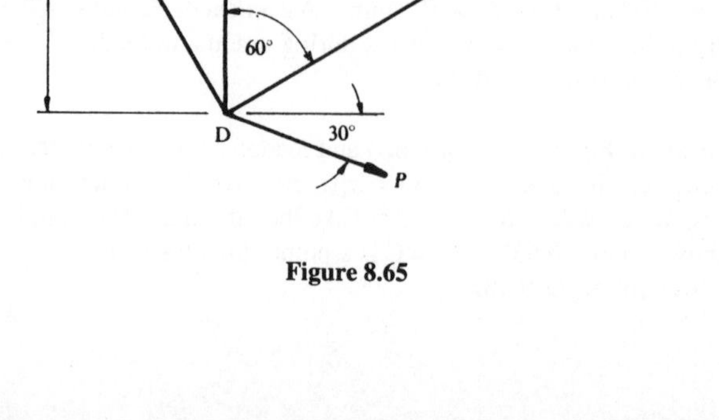

Figure 8.65

Rayleigh-Ritz Methods

8.47 In SPE the series $v = \sum b_n \cos(n\pi z/L)$ with n odd is to represent the deflected shape of a simply supported beam carrying a u.d.l.(w) over its length L. Taking the origin for z to be at mid-length, derive the expression for the displacement at any point. Compare with the accepted value at the mid-length.
Answer: $[4wL^4/(\pi^5 EI)](1 - 1/3^5 - 1/5^5 - 1/7^5 \ldots)$

8.48 An encastre beam of length L rests on an elastic support of stiffness K (N/m) at its centre, where it carries a central concentrated load P. Assume a deflection of the form $v = \sum a_n[\cos(2\pi nz/L) - 1]$ where $n = 1, 2 \ldots N$, to show that the Rayleigh-Ritz method supplies the coefficients a_n in the form of the linear simultaneous equation $(n^4 a_n)\, 4\pi^4 EI/L^3 + 2K \sum a_n + p$ where n is odd. What value for N would be acceptable?

8.49 The expression $u = K(1 - r_2)$ is assumed to represent the radial displacement at any radius r when a thin annular disc is held rigidly around its outer radius r_2 and subjected to a pressure p around its inner radius r_1. Apply the principle of SPE to determine A in terms of E, r_1, r_2 and p. (IC)
Answer: $A = pr_1(1 - \nu^2)/[E(1 - r_2)\ln(r_2/r_1)]$

8.50 Use SCE to find the forces in the frame of Fig. 8.65.

8.51 Confirm that if the expression $\sigma = R(y + Dy^3)$ is used to represent the stress distribution for a beam under a bending moment M, then the application of the principle of SCE leads to the engineer's theory of bending.

8.52 Use SCE to determine the shear stress due to shear force F in a beam of rectangular section breadth b, depth d, assuming that this is approximated by $\tau = A + By + Cy^2$. Compare graphically with the exact solution: $\tau = (6F/bd^3)(d^2/4 - y^2)$.

8.53 If the expression $\tau = Mr^2 + Nr^3$ approximates to the shear stress distribution for a solid cylindrical bar of radius R under a pure torque, find M and N from the principle of SCE. Compare with the exact solution from torsion theory and show that a 6.6% error arises in τ for $r = R$.
Answer: $M = 45T/(8\pi R^5)$, $N = -90T/(24\pi R^6)$

8.54 The distribution of hoop stress $\sigma_\theta = C_0 + C_1 y + C_2 y^2$ is assumed for a sharply curved beam of mean radius $R = (R_1 + R_2)/2$ (from which y is measured) with depth h and unit thickness. It is subjected to a pure bending moment M, tending to decrease its curvature (Fig. 8.66). Ignoring the effect of radial stress, determine C_0, C_1 and C_2 from the principle of SCE and compare graphically with that found from eq(4.15) when $R = h$ and $R_2/R_1 = 3$. (IC)
Answer: $C_0 = M/(hR)$, $C_1 = 12M/h^3$, $C_2 = -12M/(h^3 R)$

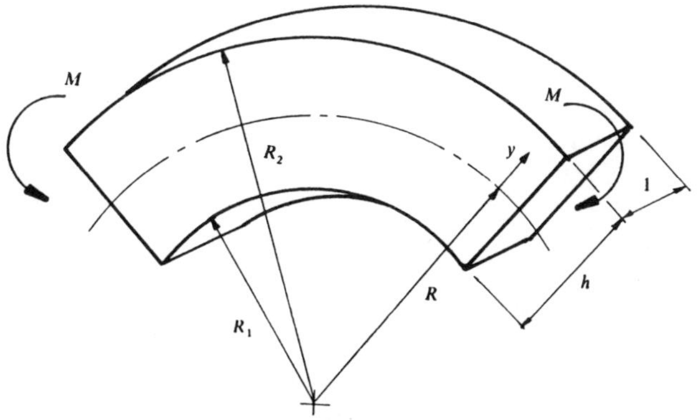

Figure 8.66

8.55 Determine the σ_x and τ_{xy} stress distributions for a point P (x, y) within the plate subjected to sinusoidal external loading shown in Fig. 8.67. Use horizontal equilibrium to find B, and the SCE principle to find A when the σ_x distribution is assumed to be $\sigma_x = A \sin(\pi x/a) + B \sin(\pi y/a)$. Ignore σ_y and note that σ_x and τ_{xy} are related through the equilibrium condition $\partial\sigma_x/\partial x + \partial\tau_{xy}/\partial y = 0$. (IC)
Answer: $A = 4\sigma(\pi - 4) / [\pi(3\pi - 8)]$

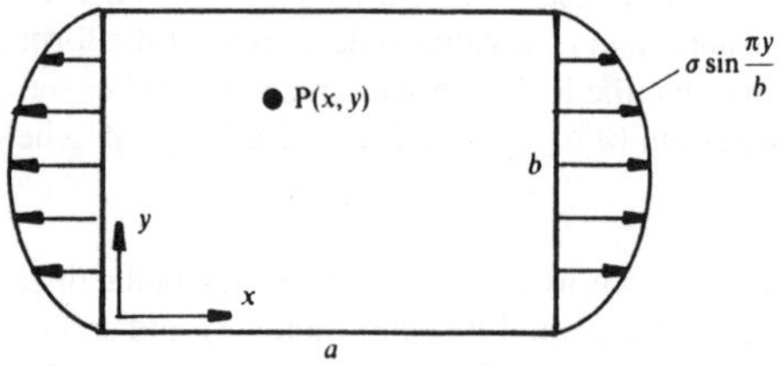

Figure 8.67

CHAPTER 9

INSTABILITY OF COLUMNS AND PLATES

In the consideration of the buckling behaviour of long thin members, i.e. struts under compression, both analytical and empirical approaches are presented for a variety of different end fixings. The additional examination of the buckling behaviour of long thin plates under direct compression is restricted here to relatively simple edge fixings. The common objective is that of finding the critical buckling load. Under purely elastic buckling conditions, where theoretical solutions are available, this load appears as a function of the length, elastic constants and some property of the resisting area, i.e. the second moment of area of a strut and an aspect ratio for a plate. Semi-empirical solutions to the buckling load are available to account for the influence of plasticity when, with shorter members, the net section stress exceeds the yield stress.

9.1 Perfect Euler Strut

Ideally, a perfectly straight strut would compress but not buckle under a purely axial compressive load. Lateral displacement is required to make a strut unstable and this inevitably arises from any slight eccentricity of loading and lack of initial straightness. The effect of such imperfections on the buckling load are consistent and predictable for long struts operating under elastic conditions. A differential equation governs the instability behaviour and the solution to this equation results in an Euler critical buckling load. This applies to struts with various combinations of pinned, fixed and free ends.

9.1.1 Pinned or Hinged Ends

The flexure equation is derived for a point (z, v) on the deflected strut in Fig. 9.1 with v positive in the direction shown and hogging moments positive.

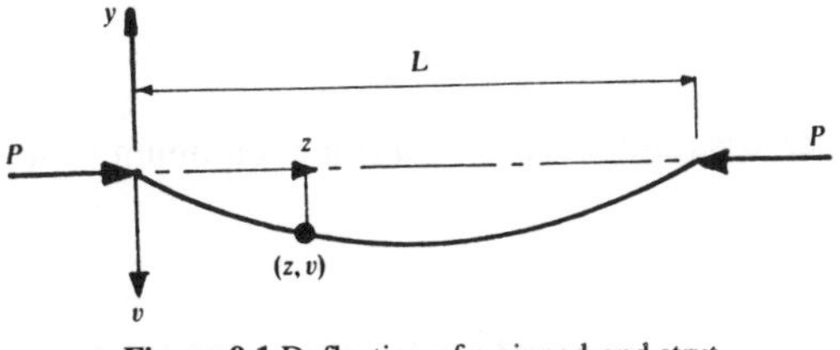

Figure 9.1 Deflection of a pinned-end strut

Taking moments to the left of that point and setting $\theta = \delta v/dz$ in eq(6.2a) gives,

$EI\ \mathrm{d}^2v/\mathrm{d}z^2 = -\ Pv$ (sagging)

which is written as

$$d^2v / dz^2 + \alpha^2 v = 0 \tag{9.1}$$

where $\alpha^2 = P/(EI)$. The solution to eq(9.1) is

$$v = A \sin \alpha z + B \cos \alpha z \tag{9.2}$$

which must satisfy the following boundary conditions:

(i) $v = 0$ at $z = 0$, $\therefore\ B = 0$
(ii) $v = 0$ at $z = L$, $\therefore\ A \sin \alpha L = 0$

In order to satisfy condition (ii), either $A = 0$ when, unrealistically, $v = 0$ for all z in eq(9.2), or $\sin \alpha L = 0$ when $\alpha L = \pi, 2\pi, 3\pi.....n\pi$. Buckling of the strut will commence when the least of these values for αL is achieved. Thus

$$\alpha L = \pi, \quad \Rightarrow \quad \surd[(P/(EI))]L = \pi$$

from which the critical buckling load is

$$P_E = \pi^2 EI / L^2 \tag{9.3}$$

where I is the second moment of area about the axis in the cross-section offering the least resistance to buckling, i.e. the minimum principal axis u. For example, in a rectangular section I in eq(9.3) applies to an axis passing through the centroid parallel to the longer side. The following alternative form of eq(9.3) employs the definition that $I = Ak^2$ where k is the radius of gyration of the given section area A:

$$P_E = \pi^2 EA / (L/k)^2 \tag{9.4}$$

where the slenderness ratio L/k becomes the determining factor in checking the validity of elastic buckling analysis. Thus, from eq(9.4) the net section stress for an Euler strut is:

$$\sigma_E = P_E/A = \pi^2 E / (L/k)^2 \tag{9.5}$$

It follows that the elastic Euler theory remains valid provided σ_E does not exceed the compressive yield stress σ_c. That is,

$$\pi^2 E / (L/k)^2 \le \sigma_c$$

and therefore the slenderness ratio must exceed some minimum value defined as

$$L/k \ge \surd(\pi^2 E/\sigma_c) \tag{9.6}$$

which is a material constant, e.g. for mild steel with $\sigma_c = 325$ MPa and $E = 207$ GPa:

$$L/k \ge \surd(\pi^2 \times 207000/325) \approx 80$$

The semi-empirical forms discussed later (see section 9.3) become more appropriate when L/k is less than the minimum Euler value supplied by eq(9.6).

9.1.2 Other End Fixings with Axial Compressive Loads

It is possible to deduce the effect that different end fixings have on the buckling load without detailed derivation. Thus, in respect of eq(9.3), L is simply replaced by an effective length L_e, over which the previous pinned-end analysis would apply. Table 9.1 gives the effective buckling length and load for each of the five struts shown.

Table 9.1 Predictions from Euler for other end fixings

ENDS	STRUT	L_e	P_E
Fixed-fixed	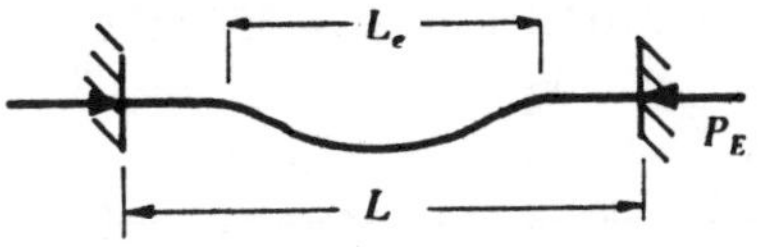	$L/2$	$4\pi^2 EI/L^2$
Pinned-fixed	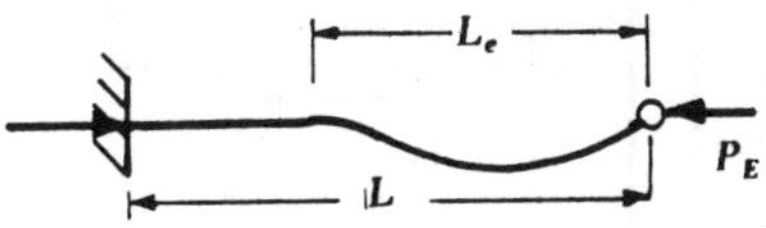	$L/\sqrt{2}$	$2\pi^2 EI/L^2$
Fixed-free	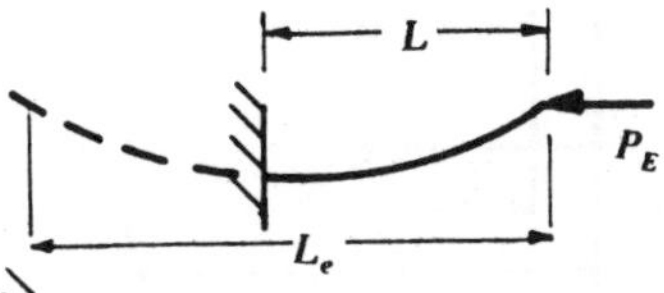	$2L$	$\pi^2 EI/4L^2$
Fixed-fixed with mis-alignment	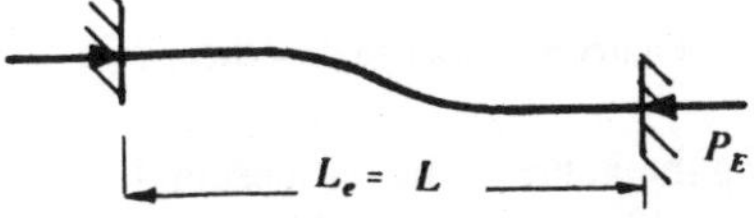	L	$\pi^2 EI/L^2$
Fixed-pinned with mis-alignment	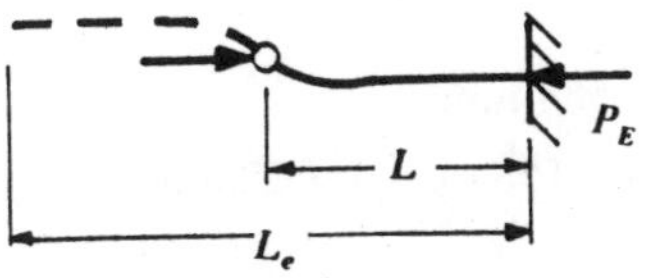	$2L$	$\pi^2 EI/4L^2$

With imperfect fixed ends an effective length of 0.6 - 0.8 is used to account for the degree of rotational restraint. When L is replaced by the appropriate L_e, eq(9.6) again applies to the smallest slenderness ratio for which the Euler theory remains valid in each of these struts. This ensures that the resistance offered by direct compression is small in comparison to the flexural strength.

Example 9.1 A strut 3 m long has a tubular cross-section of 50 mm outside diameter and 44 mm inside diameter. Determine the Euler critical load when the ends are fixed. If the tube material has a compressive yield stress of 310 MPa, find the shortest length of this tube for which the theory applies. Take E = 207 GPa.

$A = \pi(d_o^2 - d_i^2)/4 = (\pi/4)(50^2 - 44^2) = 442.97 \text{ mm}^2$

$I = \pi(d_o^4 - d_i^4)/64 = \pi[(50)^4 - (44)^4]/64 = 122.7 \times 10^3 \text{ mm}^4$

From Table 9.1, the fixed-end buckling load is

$P_E = 4\pi^2 EI/L^2 = (4\pi^2 \times 207 \times 10^3 \times 122.7 \times 10^3)/(3000)^2 = 111.4 \times 10^3\ \text{N} = 111.4\ \text{kN}$

and from eq(9.6) the shortest permissible length is

$L = \surd(4\pi^2 Ek^2/\sigma_c) = \surd[4\pi^2 EI/(A\sigma_c)]$
$= \surd[(4\pi^2 \times 207 \times 10^3 \times 122.7 \times 10^3)/(442.97 \times 310)] = 2702\ \text{mm} = 2.702\ \text{m}$

It follows that the Euler theory is valid for a 3 m length strut.

Example 9.2 If the cross-section given in Fig. 9.2 is to be used as a strut with one end pinned and the other end fixed, calculate the buckling load based upon the minimum permissible length of an Euler strut. Take $E = 210$ GPa and $\sigma_c = 350$ MPa for the strut material.

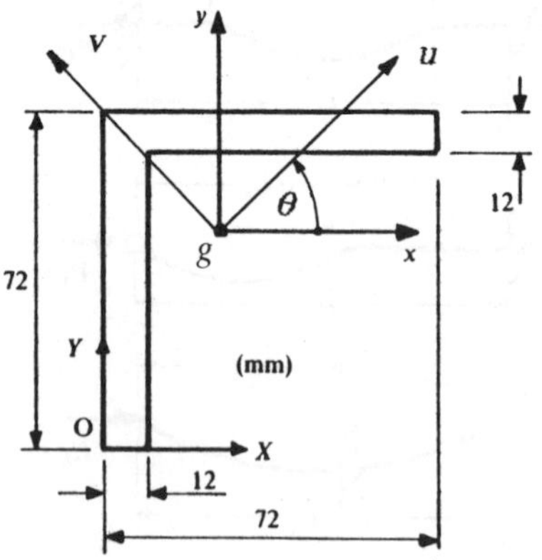

Figure 9.2 Equal-angle section

The problem is solved by finding the value of the least second moment of area for the section, i.e. I_u or I_v, by the method oulined in Appendix I. The centroidal co-ordinates are:

$A = [(72 \times 12) + (60 \times 12)] = 1584\ \text{mm}^2$
$(72 \times 12 \times 6) + (60 \times 12 \times 42) = 1584\,\bar{y},\ \bar{y} = 22.4\ \text{mm}\ (= \bar{x})$

$I_x = 72 \times (22.4)^3/3 - 60 \times (10.4)^3/3 + 12 \times (49.6)^3/3 = 73.5 \times 10^4\ \text{mm}^4\ (= I_y)$
$I_{xy} = [(72 \times 12)(+13.6)(+16.4)] + [(60 \times 12)(-19.6)(-16.4)] = 42.4 \times 10^4\ \text{mm}^4$
$I_{u,v} = \tfrac{1}{2}(I_x + I_y) \pm \tfrac{1}{2}\surd[(I_x - I_y)^2 + 4I_{xy}^2] = \{73.5 \pm \tfrac{1}{2}\surd[4(42.4)^2]\} \times 10^4$
$I_u = 31.1 \times 10^4\ \text{mm}^4,\ I_v = 115.9 \times 10^4\ \text{mm}^4$

The lesser value I_u is the lowest value for the section. Axes u and v are perpendicular with

$\tan 2\theta = 2I_{xy}/(I_y - I_x) = \infty$

so that u is inclined at 45° to x. The minimum permissible length (see Table 9.1) is

$L = \surd 2L_e = \surd[2\pi^2 EI_u/(A\sigma_c)]$
$= \surd[(2\pi^2 \times 210 \times 10^3 \times 31.1 \times 10^4)/(1584 \times 350)] = 655.8\ \text{mm}$

and the corresponding buckling load is

$P_E = 2\pi^2 EI/L^2 = (2\pi^2 \times 210 \times 10^3 \times 31.1 \times 10^4)/(655.8)^2 = 2.998\ \text{MN}$

9.2 Imperfect Euler Struts

Practical struts may differ from perfect Euler struts in the ways outlined in the following cases. The onset of buckling is normally associated with the attainment of the compressive yield stress at the section subjected to the greatest bending moment.

9.2.1 Pinned-End Strut with Eccentric Loading

When the line of action of the compressive force P is applied with a measurable amount of eccentricity e to the strut axis, as shown in Fig. 9.3, then the bending moment at point (z, v) will be modified.

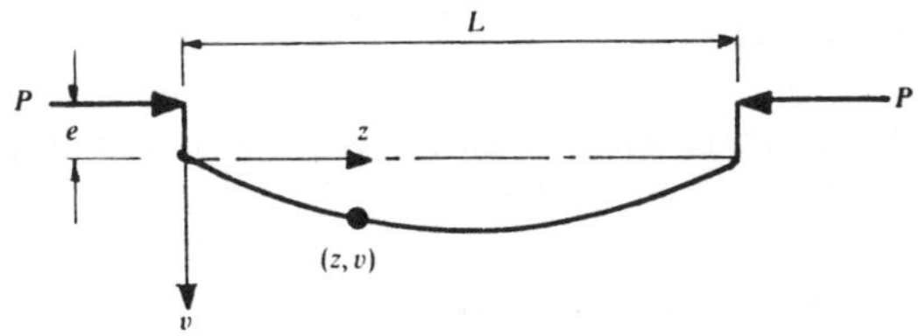

Figure 9.3 Deflection of an eccentrically loaded pinned-end strut

The governing flexure equation becomes

$$EI\, d^2v/dz^2 = -P(e + v)$$

This now takes the form

$$d^2v/dz^2 + \alpha^2 v = -\alpha^2 e$$

and with the corresponding solution

$$v = A \cos \alpha z + B \sin \alpha z - e \quad (9.7)$$

This solution may be employed to find the safe axial load for a given allowable stress σ. In general, for any point in the strut the stress is composed of direct compressive stress and a bending stress which changes sign across the section. The net stress is expressed as

$$\sigma = -P/A \pm My/I \quad (9.8)$$

This stress is greatest on the compressive side of the section where the bending moment is a maximum. From eq(9.8),

$$\sigma = -P/A \pm M_{max}\, y/I \quad (9.9)$$

where y is the perpendicular distance from the buckling axis to the furthest compressive edge. Now, from Fig. 9.3, at the central section

$$M_{max} = -P(e + v_{max}) \quad (9.10)$$

The maximum deflection at $z = L/2$ follows from eq(9.7) where the following boundary conditions apply:

(i) $v = 0$ at $z = 0$, $\therefore A = e$
(ii) $dv/dz = 0$ at $z = L/2$. This gives

$$- A\alpha \sin\alpha L/2 + B\alpha \cos \alpha L/2 = 0, \quad \therefore \quad B = e \tan(\alpha L/2)$$
$$\therefore v_{max} = e \cos(\alpha L/2) + e \tan(\alpha L/2) \sin(\alpha L/2) - e$$
$$v_{max} = e[\sec(\alpha L/2) - 1] \tag{9.11}$$

Then, from eqs(9.9) and (9.10) the maximum compressive stress is

$$M_{max} = -(Pe) \sec(\alpha L/2) \text{ (sagging)},$$
$$\sigma = -P/A - (Pey/I) \sec(\alpha L/2) \tag{9.12}$$

Since $\alpha = \sqrt{(P/EI)}$ this form is not convenient when P is required. For this Webb's approximation to the secant function may be used:

$$\sec\theta \simeq [(\pi/2)^2 + (\pi^2/8 - 1)\theta^2] / [(\pi/2)^2 - \theta^2] \tag{9.13}$$

Putting $\theta = \alpha L/2$ and introducing from eq(9.3), the Euler Buckling load $P_E = \pi^2 EI / L^2$, eq(9.13) gives

$$\sec(\alpha L/2) \simeq [P_E + P(\pi^2/8 - 1)] / (P_E - P) \tag{9.14}$$

Substituting eq(9.14) into eq(9.12) supplies the quadratic

$$aP^2 + bP + c = 0 \tag{9.15a}$$

in which P is the positive root with σ taken to be negative within the coefficients:

$$a = eyA(\pi^2/8 - 1) - I, \quad b = eyAP_E + I P_E - I\sigma A, \quad c = I\sigma AP_E \tag{9.15b,c,d}$$

Example 9.3 The 75 mm × 75 mm equal-angle section shown in Fig. 9.4 has an area of 1775 mm^2 and a least radius of gyration of 14.5 mm about a 45° v - axis passing through the centroid G. When used as pinned-end strut 1.5 m long, the section supports a compressive load through point O, 6.5 mm from G as shown. If at the mid-length cross-section the net compressive stress is not to exceed 123.5 MPa, find the maximum compressive load. Take $E = 207$ GPa.

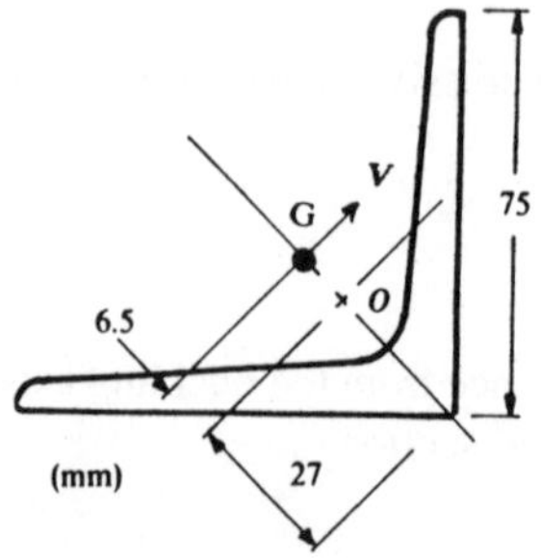

Figure 9.4 Equal-angle section

$I_v = Ak^2 = 1775 \times 14.5^2 = 37.32 \times 10^4$ mm^4
$P_E = \pi^2 EI_v / L^2 = (\pi^2 \times 207 \times 10^3 \times 37.32 \times 10^4)/(1500)^2 = 338.876$ kN

and from eq(9.15b,c,d) the coefficients are

$a = 6.5 \times 27 \times 1775(\pi^2/8 - 1) - 37.32 \times 10^4 = -30.04 \times 10^4$ mm^4
$b = (6.5 \times 27 \times 1775 \times 338.876 \times 10^3) + (37.32 \times 10^4)(338.876 \times 10^3)$
$\quad - (37.32 \times 10^4)(-123.5)(1775) = 31.38 \times 10^{10}$ Nmm^4
$c = (37.32 \times 10^4)(-123.5)(1775)(338.867 \times 10^3) = -27.723 \times 10^{15}$ N^2 mm^4

The quadratic (9.15a) becomes

$$P^2 - (1.045 \times 10^6)P + (9.229 \times 10^{10}) = 0$$

from which the lowest root is $P = 97.39$ kN.

9.2.2 Strut, One End Fixed, the Other End Free, Carrying an Eccentric Compressive Load

Let v_L be the free-end deflection with the origin of z, v co-ordinates lying at the fixed end as shown in Fig. 9.5.

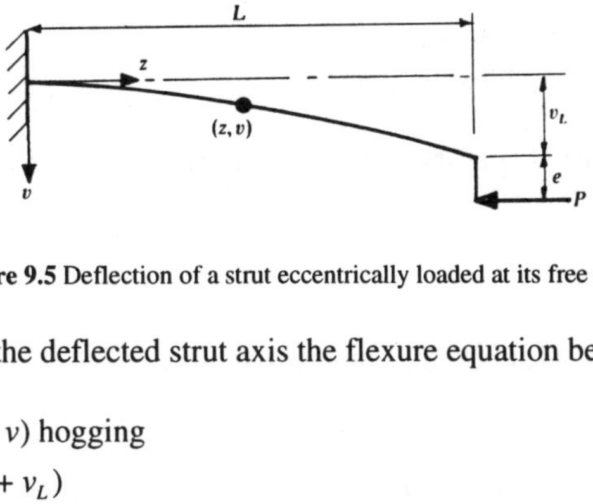

Figure 9.5 Deflection of a strut eccentrically loaded at its free end

For any point (z, v) on the deflected strut axis the flexure equation becomes

$EI\, d^2v/dz^2 = P(e + v_L - v)$ hogging
$d^2v/dz^2 + \alpha^2 v = \alpha^2(e + v_L)$

and the solution is

$$v = A \cos \alpha z + B \sin \alpha z + (v_L + e)$$

Applying the boundary conditions,

(i) $dv/dz = 0$ at $z = 0$, $\therefore B = 0$
(ii) $v = 0$ at $z = 0$, $\therefore A = -(v_L + e)$

$$\therefore v = (v_L + e)(1 - \cos \alpha z)$$

but $v = v_L$ at $z = L$

$$v_L = e(\sec \alpha L - 1) \qquad (9.16)$$

The bending moment is a maximum at the fixed end when

$$M_{max} = P(e + v_L) = Pe \sec(\alpha L) \qquad (9.17)$$

where from Webb's approximation (9.13), with $\theta = \alpha L$

$$\sec \alpha L \approx [P_E + 4P(\pi^2/8 - 1)]/(P_E - 4P) \qquad (9.18)$$

Substituting eqs(9.17 & 9.18) into eq(9.9), P is the positive root of the quadratic for a given maximum compressive (negative) stress,

$$aP^2 + bP + c = 0 \qquad (9.19a)$$

where $a = 4[eyA(\pi^2/8 - 1) - I]$, $b = eyAP_E + IP_E - 4I\sigma A$, $c = I\sigma AP_E$ (9.19b,c,d)

Example 9.4 A tubular strut 75 mm o.d. and 65 mm i.d. is 3 m long. It is loaded eccentrically at its free end with a compressive force at a radial distance of 6.5 mm from the axis. Determine the instability load that gives zero resultant tensile stress at the fixed end. What is the strut end deflection and the maximum compressive stress? Take $E = 210$ GPa.

The solution requires the application of eqs(9.16 and 9.19), where

$$A = (\pi/4)(75^2 - 65^2) = 1099.56 \text{ mm}^2$$
$$I = (\pi/64)(75^4 - 65^4) = 67.69 \times 10^4 \text{ mm}^4$$

Zero stress on the tensile side of the fixed end, where $M = M_{max}$, is expressed from eq(9.8) and (9.17) as

$$\sigma = -P/A + [Pe \sec(\alpha L)]\, y/I = 0$$

Substituting $\alpha^2 = P/(EI)$ leads to the compressive load:

$$P = (EI/L^2)\{\sec^{-1}[I/(Aye)]\}^2$$
$$= (210000 \times 67.69 \times 10^4/3000^2)\{\sec^{-1}[(67.69 \times 10^4)/(1099.56 \times 37.5 \times 6.5)]\}^2$$
$$= 15.794 \times 10^3 [\sec^{-1}(2.5256)]^2$$
$$= 21.39 \text{ kN}$$

$$\alpha = \sqrt{[P/(EI)]} = \sqrt{[(21.39 \times 10^3)/(210000 \times 67.69 \times 10^4)]} = 3.8789 \times 10^{-4} \text{ mm}^{-1}$$
$$\sec(\alpha L) = \sec(3.8789 \times 10^{-4} \times 3000 \times 180/\pi) = 2.5254$$

and from eq(9.16) the maximum deflection is $v_L = 6.5\,(2.5254 - 1) = 9.915$ mm. The compressive stress under this load is found from

$$\sigma = -P/A - [Pe \sec(\alpha L)]y/I$$
$$= -(21.39 \times 10^3)[1/1099.56 + (6.5 \times 2.5254 \times 37.5)/(67.69 \times 10^4)]$$
$$= -38.91 \text{ MPa}$$

9.2.3 Pin-Ended Strut, Lateral Load at Mid-Span in Addition to an Axial Compressive Load

For a point (z, v) on the deflected strut in Fig. 9.6, the left hand moment is composed of two sagging components Pv and $Wz/2$ due to the support reaction.

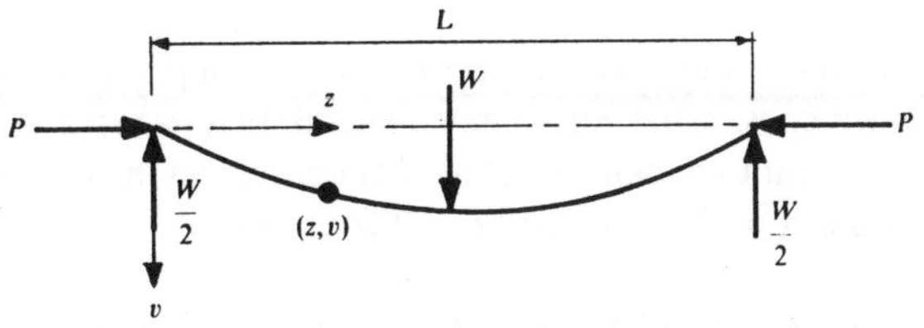

Figure 9.6 Pin-ended strut with axial and lateral loading

The flexure equation becomes

$$EI\, d^2v/dz^2 = -Pv - Wz/2 \text{ (sagging)}$$
$$d^2v/dz^2 + \alpha^2 v = -Wz/(2EI)$$

The corresponding solution is

$$v = A\cos\alpha z + B\sin\alpha z - Wz/(2P)$$

Applying the boundary conditions

(i) $v = 0$ at $z = 0$, $\quad\therefore A = 0$
(ii) $dv/dz = 0$ at $z = L/2$, $\therefore B = W/[(2P\alpha)\cos(\alpha L/2)]$

This leads to a maximum displacement at $z = L/2$ of

$$v_{max} = [W/(2P)]\,[(1/\alpha)\tan(\alpha L/2) - L/2] \tag{9.20}$$

and a maximum sagging moment at $z = L/2$ of

$$M_{max} = -P\,v_{max} - WL/4 = -(W/2\alpha)\tan(\alpha L/2) \tag{9.21}$$

Equation(9.9) gives the maximum compressive stress in the central section

$$\sigma = -P/A - [Wy/(2I\alpha)]\tan(\alpha L/2) \tag{9.22}$$

This equation may be more conveniently solved for P using Webb's tangent approximation:

$$\tan\theta = [(\pi/2)^2\theta - (\pi^2/12 - 1)\theta^3]/[(\pi/2)^2 - \theta^2]$$

With $\theta = \alpha L/2$ and $P_E = \pi^2 EI/L^2$ this provides

$$\tan(\alpha L/2) = (\alpha L/2)[P_E - P(\pi^2/12 - 1)]/(P_E - P)$$

when from eq(9.22), the following quadratic in positive P is found:

$aP^2 + bP + c = 0$

where $a = 1$, $b = [WLyA/(4I)](\pi^2/12 - 1) + \sigma A - P_E$, $c = -P_E A[\sigma + Wl\,y/(4I)]$

in which σ is the maximum compressive (negative) stress.

Example 9.5 A steel strut 25 mm diameter and 1.5 m long supports an axial compressive load of 8 kN through pinned ends. What concentrated lateral force, when applied normally at mid-span, would cause the strut to buckle? Find the deflection and the maximum moment at the point of buckling. Take $\sigma_c = 278$ MPa and $E = 207$ GPa.

$I = \pi d^4/64 = \pi(25)^4/64 = 19174.8\ \text{mm}^4$, $A = \pi d^2/4 = \pi(25)^2/4 = 490.87\ \text{mm}^2$
$\alpha = \sqrt{[P/(EI)]} = \sqrt{[(8\times10^3)/(207\times10^3\times19174.8)]} = 1.42\times10^{-3}\ \text{mm}^{-1}$
$\alpha L = 2.13$

Substituting into eq(9.22),

$-278 = -(8\times10^3)/490.87 - [12.5W/(2\times19174.8\times1.42\times10^{-3})]\tan(1.065\times180/\pi)$
$-278 = -16.298 - 0.4145W$

From which $W = 631.5$ N. From eq(9.20) the maximum deflection is

$v_{max} = [631.5/(2\times8000)][(10^3/1.42)\tan(1.065\times180/\pi) - (1.5\times10^3)/2] = 20.58$ mm

and from eq(9.21) the maximum central moment is

$M_{max} = -[631.5/(2\times1.42\times10^{-3})]\tan(1.065\times180/\pi) = -401.48$ Nm

9.2.4 Pin-Ended Strut Carrying a Uniformly Distributed Lateral Load in Addition to an Axial Compressive Load

Figure 9.7 shows that the bending moment to the left of a point (z, v). This is composed of two sagging components Pv and $wLz/2$, and a hogging component $wz^2/2$.

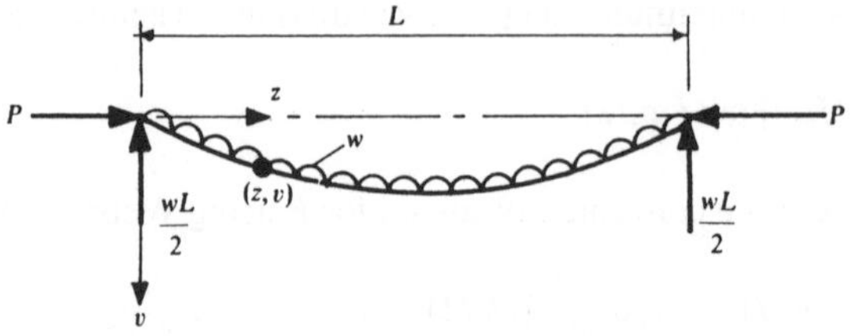

Figure 9.7 Pin-ended strut with laterally distributed and axial loading

The flexure equation becomes

$EI\,d^2v/dz^2 = -Pv - wL\,z/2 + wz^2/2$
$d^2v/dz^2 + \alpha^2 v = -[w/(2EI)](Lz - z^2)$

for which the solution is

$$v = A\cos\alpha z + B\sin\alpha z + [w/(2P)]\,(z^2 - Lz - 2EI/P)$$

Applying the boundary conditions,

(i) $v = 0$ at $z = 0$, $\Rightarrow A = EI\,w/P^2$
(ii) $v = 0$ at $z = L$, $\Rightarrow B = (EI\,w/P^2)\tan(\alpha L/2)$

The maximum displacement at $z = L/2$ is then

$$v_{max} = (EI\,w/P^2)\cos(\alpha L/2) + (EI\,w/P^2)\tan(\alpha L/2)\sin(\alpha L/2) - (w/2P)(L^2/4 + 2EI/P)$$
$$= (EI\,w/P^2)[\sec(\alpha L/2) - 1] - wL^2/(8P)$$

The maximum, central sagging bending moment is, correspondingly,

$$M_{max} = -Pv_{max} - wL^2/4 + wL^2/8$$
$$= -(EI\,w/P)[\sec(\alpha L/2) - 1] + wL^2/8 - wL^2/4 + wL^2/8$$
$$M_{max} = -(EI\,w/P)[\sec(\alpha L/2) - 1] \tag{9.23}$$

Equation(9.9) gives the maximum compressive stress at the central section:

$$\sigma = -P/A - (Ewy/P)[\sec(\alpha L/2) - 1] \tag{9.24}$$

If it is required to solve for P then Webb's approximation to $\sec(\alpha L/2)$ in eq(9.14) results in the quadratic

$$aP^2 + bP + c = 0$$

where $a = 1$, $b = -(P_E - A\sigma)$, $c = -A(\sigma P_E + Ewy\pi^2/8)$

Note that if P in Fig. 9.7 is tensile the flexure equation becomes

$$EI\,d^2v/dz^2 = Pv - wLz/2 + wz^2/2$$

and the solution gives the maximum central moment:

$$M_{max} = -(EI\,w/P)[1 - \text{sech}(\alpha L/2)]$$

9.2.5 Laterally Loaded Struts with Fixed Ends

Figures 9.8a and b shows similar loading the former two cases but when each end is fixed. This results in a fixed-end moment M_o that modifies the net moment expression. Thus, in Fig. 9.8a at a point (z, v) on the deflected strut,

$$EI\,d^2v/dz^2 = M = -Pv - Wz/2 + M_o$$

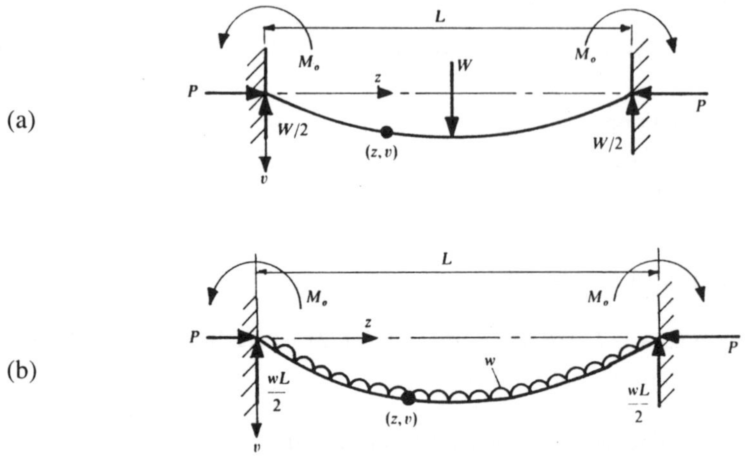

Figure 9.8 Encastré struts with lateral loading

Integration and the determination of the constants from the boundary conditions leads to central sagging and hogging end moments of similar magnitude:

$$M_c = -M_o = -[W/(2\alpha)][\text{cosec}(\alpha L/2) - \cot(\alpha L/2)]$$

Similarly in Fig. 9.8b,

$$EI\, d^2v/dz^2 = M = -Pv - wLz/2 + wz^2/2 + M_o$$

from which the corresponding central and end moments are

$$M_c = -[wL/(2\alpha)][\text{cosec}(\alpha L/2) - 2/(\alpha L)] \text{ sagging} \qquad (9.25a)$$

$$M_o = [wL/(2\alpha)][\cot(\alpha L/2) - 2/(\alpha L)] \text{ hogging} \qquad (9.25b)$$

and the maximum stresses are again found from eq(9.9).

Example 9.6 Find the maximum bending moment and the maximum compressive and tensile stresses for the strut in Fig. 9.8b when $P = 200$ kN, $w = 5$ kN/m, and $L = 3$ m. The I-section has an area of 2900 mm^2, depth 100 mm and a least I value of 3.55×10^6 mm^4. Take $E =$ 210 GPa.

$$\alpha = \sqrt{[P/(EI)]} = \sqrt{[(200 \times 10^3)/(210 \times 10^3 \times 3.55 \times 10^6)]} = 5.18 \times 10^{-4}\ \text{mm}^{-1}$$

$$wL/(2\alpha) = (5 \times 3 \times 10^3)/(2 \times 5.18 \times 10^{-4}) = 14.48 \times 10^6\ \text{Nmm}$$

$$\alpha L/2 = 5.18 \times 10^{-4} \times 3 \times 10^3/2 = 0.777$$

and from eqs(9.25a,b) the central and end moments become

$$M_c = -14.48\,[\text{cosec}(0.777 \times 180/\pi) - 1/0.777] = -2.016\ \text{kNm}$$

$$M_o = 14.48\,[\cot(0.777 \times 180/\pi) - 1/0.777] = -3.911\ \text{kNm}$$

The compressive stresses at these two positions are found from eq(9.8),

$\sigma_o = -\ P/A \pm M_o\ y\ /\ I$
$= -\ (200 \times 10^3)/2900 \pm (3.911 \times 10^6 \times 50)\ /\ (3.55 \times 10^6)$
$= -\ 68.97 \pm 55.08$
$= -\ 13.89$ MPa and $-\ 124.05$ MPa

$\sigma_c = -\ P/A \pm M_c\ y\ /\ I$
$= -\ (200 \times 10^3)\ /\ 2900 \pm (2.016 \times 10^6 \times 50)/(3.55 \times 10^6)$
$= -\ 68.97 \pm 28.39$
$= -\ 40.58$ MPa and $-\ 97.36$ MPa

9.2.6 Pinned-End Strut with Initial Curvature

Let v_o be the deviation for any position $0 \le z \le L$. It is assumed that the initial curvature for the strut in Fig. 9.9 is described by the sinusoidal function $v_o = h \sin(\pi z/L)$ where h is the value of the maximum central deviation.

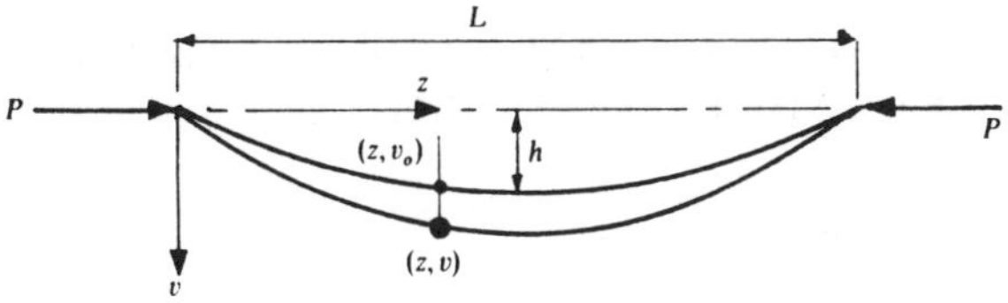

Figure 9.9 Deflection of an initially curved pinned strut

With further deflection v, from the application of a compressive axial force P, the flexure equation becomes

$$EI\,\mathrm{d}^2(v - v_o)/\mathrm{d}z^2 = -\ Pv$$
$$EI\,\mathrm{d}^2 v/\mathrm{d}z^2 + Pv = EI\,\mathrm{d}^2 v_o/\mathrm{d}z^2 = (EI\,h)\,\mathrm{d}^2[\sin(\pi z\,/L)]\ /\ \mathrm{d}z^2$$
$$\mathrm{d}^2 v\,/\mathrm{d}z^2 + \alpha^2\,v = -\ (h\pi^2/L^2)\sin(\pi z\,/L)$$

The solution is

$$v = A\cos\alpha z + B\sin\alpha z + [h\sin(\pi z/L)]\ /\ [1 - (\alpha L/\pi)^2]$$

Applying the boundary conditions $v = 0$ at both $z = 0$ and $z = L$ leads to $A = B = 0$ and

$$v = [\ h\sin(\pi z/L)]\ /\ [1 - (\alpha L/\pi)^2]$$

which has its maximum at $z = L/2$

$$v_{\max} = h\ /[1 - (\alpha L/\pi)^2] \qquad (9.26)$$

Correspondingly, the maximum bending moment at this position is

$$M_{\max} = -\ Pv_{\max} = -Ph\ /\ [1 - (\alpha L/\pi)^2]$$

Since $(\alpha L/\pi)^2 = P/P_E$, the maximum compressive stress is, from eq(9.9),

$$\sigma = -P/A - Ph\,y/\{I[1-(\alpha L/\pi)^2]\} = -P/A - Ph\,y/[I(1-P/P_E)]$$

This equation may be directly solved for P in the resulting quadratic:

$$aP^2 + bP + c = 0 \tag{9.27}$$

where $a = 1$, $b = -(h\,y\,A\,P_E/I + P_E - \sigma A)$, $c = -\sigma A P_E$.

Example 9.7 Find P for a 2.5 m long strut in Fig. 9.9 that would just produce yielding in the mid-span 20 mm square section given that the maximum amplitude of initial curvature is 10 mm and the compressive yield stress is 320 MPa. What is the additional central deflection under this load? Take $E = 207$ GPa.

$A = 20^2 = 400\ \text{mm}^2$, $I = 20^4/12 = 13.33\times10^3\ \text{mm}^4$
$P_E = \pi^2 EI/L^2 = \pi^2 \times 207000 \times 13.33\times10^3/2500^2 = 4358.4$ N

Substituting into eq(9.27) with $\sigma = -320$ MPa, $h = 10$ mm and $y = 10$ mm,

$b = -[(10\times10\times400\times4358.4)/(13.33\times10^3) + 4358.4 - (-320\times400)] = -145.43\times10^3$ N
$c = -\sigma A P_E = 320\times400\times4358.4 = 557.88\times10^6\ \text{N}^2/\text{mm}^2$

Then, in units of N and mm,

$$P^2 - (145.43\times10^3)P + (557.88\times10^6) = 0$$

from which $P = 3.943$ kN. Now, from eq(9.26), the maximum deflection is

$$v_{max} = h/[1-(\alpha L/\pi)^2] = h/(1-P/P_E)$$
$$= 10/(1-3942.86/4358.4) = 104.89\ \text{mm}$$

9.2.7 Composite Strut

In Capey [1] and Walter's [2] theory, two lengths 1 and 2 of a stepped strut have constant flexural rigidity EI where $E_2 I_2 > E_1 I_1$ with a rigid attatchment at their join. When the centroids of the two sections are coaxial and the axes of minimum inertia are co-planar, the following parameters are employed:

$$\alpha^2 = P L_1^2/(E_1 I_1),\ \beta^2 = P L_2^2/(E_2 I_2),\ \lambda = L_2/L_1$$

The Euler theory shows that buckling occurs at the lowest value of $P(L_1 + L_2)^2/(E_1 I_1)$ which satisfies the appropriate condition.

(a) Both ends simply supported

$$\lambda\alpha\cos\alpha\sin\beta + \beta\cos\beta\sin\alpha = 0$$

(b) End 1 simply supported, end 2 fixed

$$\cos\alpha\sin\beta + (\alpha\lambda/\beta)\cos\alpha\sin\beta + \beta(1 + 1/\lambda)\sin\alpha\sin\beta) - \alpha(1+\lambda)\cos\alpha\cos\beta = 0$$

(c) End 2 simply supported, end 1 fixed

$$\sin\alpha\cos\beta + [\beta/(\alpha\lambda)]\sin\alpha\cos\beta + \alpha(1+\lambda)\sin\alpha\sin\beta - \beta(1+1/\lambda)\cos\alpha\cos\beta = 0$$

(d) Both ends clamped

$$2 + [\alpha\lambda/\beta + \beta/(\alpha\lambda)]\sin\alpha\sin\beta - 2\cos\alpha\cos\beta - \alpha(1+\lambda)\sin\alpha\cos\beta - \beta(1+1/\lambda)\cos\beta\sin\beta = 0$$

(e) End 1 clamped, end 2 free

$$\beta\cos\alpha\cos\beta - \alpha\lambda\sin\alpha\sin\beta = 0$$

(f) End 2 clamped, end 1 free

$$\alpha\lambda\cos\alpha\cos\beta - \beta\sin\alpha\sin\beta = 0$$

Graphical solutions to eqs(a) - (f) employ plots $P(L_1 + L_2)^2/(E_1I_1)$ versus $L_1/(L_1 + L_2)$ for various values of the ratio $(E_2I_2)/(E_1I_1)$ for each support [3]. Figure 9.10 applies, for example, to a simply supported stepped strut so that if, for $(E_2I_2)/(E_1I_1) = 2$, $E_1I_1 = 750$ Nm^2 and $E_2I_2 = 1500$ Nmm^2, $L_1 = 0.5$ m, $L_2 = 1$ m we, have from Fig. 9.10, $P(L_1 + L_2)^2/(E_1I_1) = 16$ when $P = 5333$ N.

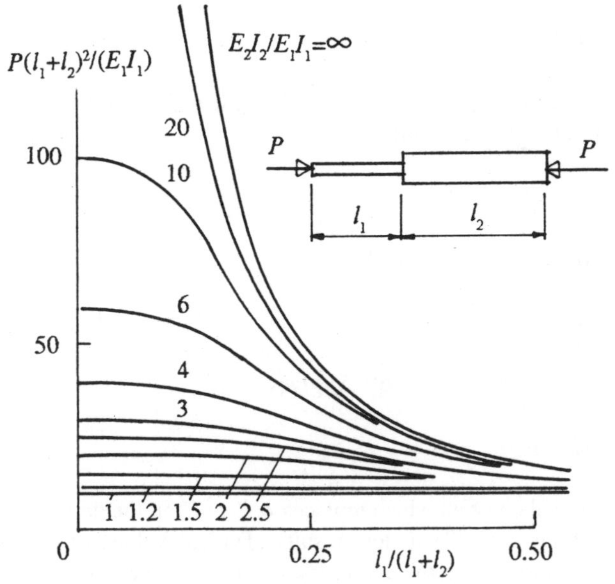

Figure 9.10 Stepped strut with simply supported ends

9.3 Semi-Empirical Approaches

As the slendeness ratio decreases so the Euler theory diverges from the observed behaviour. Figure 9.11 shows the region below and beyond the limiting L/k value for the Euler theory in eq(9.6). The following empirical approaches, represented graphically in Fig. 9.11, account for the increasing effect that compressive yielding has on the buckling behaviour with decreasing L/k in the invalid Euler region (broken line).

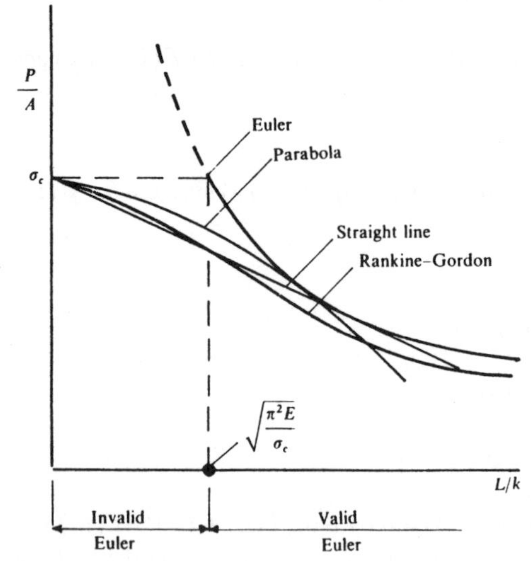

Figure 9.11 Comparison between Euler theory and various empirical approaches

9.3.1 Rankine-Gordon

In the case of a pinned-end strut, the Euler buckling load P_E and the compressive yield load P_c are combined in reciprocal form:

$$1/P_R = 1/P_E + 1/P_c$$

from which the Rankine buckling load is

$$P_R = P_c / (1 + P_c / P_E)$$

Substituting $P_c = A\sigma_c$ and P_E from eq(9.5) gives

$$P_R = A\sigma_c / [1 + a\,(L/k)^2] \tag{9.28}$$

where a is an empirical constant which replaces $\sigma_c/(\pi^2 E)$ in the derivation. This constant is found from matching eq(9.28) to test results. Typical values of a, for the pinned-end condition, are given along with the compressive yield stress σ_c in Table 9.2.

Table 9.2 Rankine-Gordon constants in Eq(9.28)

Material	σ_c / MPa	a^{-1}
Mild steel	325	7500
Wrought iron	247	9000
Cast iron	557	1600
Timber	35	750

For other end fixings the effective length concept may again be employed. Thus L in eq(9.28) is replaced with the L_e values given in Table 9.1. Note, from the graphical representation of eq(9.28) in Fig. 9.11, that as the Rankine stress P_R/A does not exceed σ_c, it applies over the whole L/k range and lies on the safe side of the Euler curve.

Example 9.8 Find the axis about which buckling takes place for the section given in Fig. 9.12. What is the safe axial load that may be applied to an 11 m length of this strut when the ends are fixed? Decide between the Euler and Rankine predictions given the area properties for a single channel $I_x = 34.5 \times 10^6$ mm^4, $I_y = 1.66 \times 10^6$ mm^4, $A = 3660$ mm^2, $\bar{x} = 18.8$ mm. Take $E = 207$ GPa, $a = 1/7500$ and $\sigma_c = 324$ MPa.

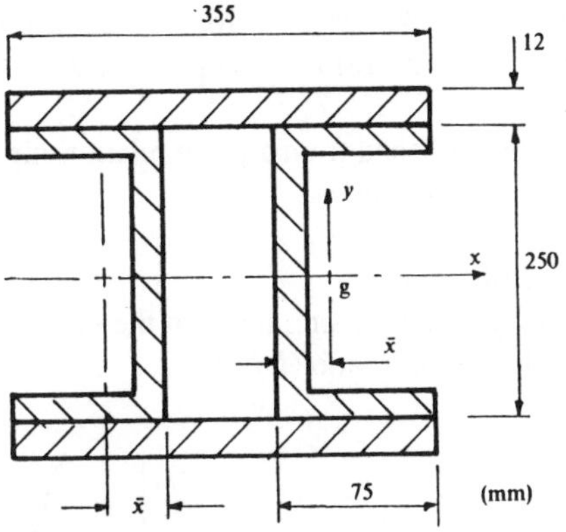

Figure 9.12 Fabricated section

$L_e = L/2 = 5.5$ m
$A = 2[(355 \times 12) + 3660] = 15.84 \times 10^3$ mm^2

Buckling occurs about the axis with the lesser I value, where

$$I_x = (2 \times 34.5 \times 10^6) + 2(355 \times 12^3/12 + (355 \times 12)131^2)] = 215.31 \times 10^6 \text{ mm}^4$$
$$I_y = 2 \times 12 \times 355^3/12 + 2[(1.66 \times 10^6) + (3660 \times 121.3^2)] = 200.5 \times 10^6 \text{ mm}^4$$
$$k^2 = I_y / A = (200.5 \times 10^6) / (15.84 \times 10^3) = 12.66 \times 10^3 \text{ mm}^2$$
$$\therefore L_e / k = 5500 / \sqrt{(12.66 \times 10^3)} = 5500 / 112.5 = 48.88$$

Then from eq(9.4),

$$P_E = \pi^2 EA / (L_e / k)^2 = \pi^2 \times 207000 \times 15.84 \times 10^3/(48.88)^2 = 13.54 \text{ MN}$$

and from eq(9.28),

$$P_R = (15.84 \times 10^3 \times 324) / [1 + (48.88)^2 / 7500] = 3.89 \text{ MN}$$

Only the Rankine prediction is valid since, from eq(9.6), Euler ceases to apply for lengths below

$$L = \sqrt{(4\pi^2 Ek^2/\sigma_c)} = \sqrt{[(4\pi^2 \times 207000 \times 12.66 \times 10^3)/ 324]} = 17.87 \text{ m}$$

9.3.2 Straight Line Formulae

Commonly used in the U.S.A. for a particular L/k range, the linear prediction shown in Fig. 9.11 has the form

$$P = A\sigma_c[1 - n(L/k)] \tag{9.29}$$

where n and σ_c are constants with typical values for a pin-ended strut:

Mild steel:	$n = 0.004$,	$\sigma_c = 258$ MPa	for	$L/k \le 150$
Cast iron:	$n = 0.004$,	$\sigma_c = 234$ MPa	for	$L/k \le 100$
Duralumin:	$n = 0.009$,	$\sigma_c = 302$ MPa	for	$L/k \le 85$
Oak:	$n = 0.005$,	$\sigma_c = 37$ MPa	for	$L/k \le 65$

These constants have been established from a given relationship between the straight line and the Euler curve for a pinned-end strut.

(i) Intersections with the Euler curve
Figure 9.13a shows a straight line passing through the compressive yield stress σ_c and the Euler curve at a chosen slenderness ratio $(L/k)_c$.

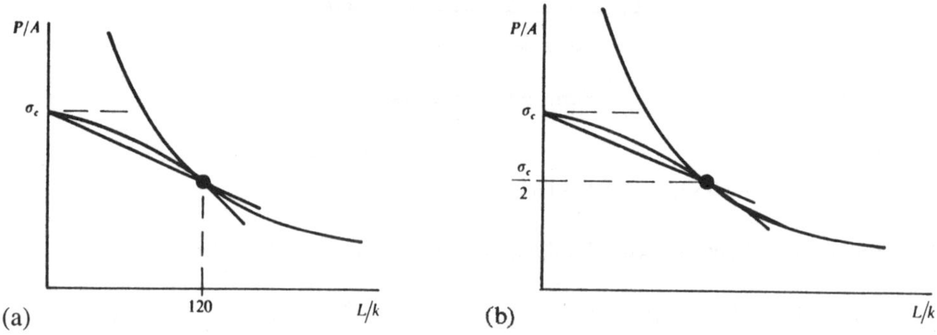

Figure 9.13 Intersections with Euler curve

Since the stress is common at the intersection, eqs(9.5) and (9.29) give

$$P/A = \sigma_c[1 - n(L/k)_c] = \pi^2 E/(L/k)_c^2$$
$$n = \{1 - \pi^2 E/[\sigma_c (L/k)_c^2]\}/(L/k)_c$$

Alternatively, Fig. 9.13b shows an intersection at $\sigma_c/2$. At the intersection point the Euler curve gives the slenderness ratio

$$P/A = \sigma_c/2 = \pi^2 E/(L/k)_c^2$$
$$(L/k)_c = \surd(2\pi^2 E/\sigma_c)$$

and eq(9.29) gives the n - value

$$n = 1/[2(L/k)_c] = \surd[\sigma_c/(8\pi^2 E)]$$

(ii) Tangent to the Euler curve
A further interpretation to the constant n arises when the straight line becomes a tangent to the Euler curve (see Fig. 9.14).

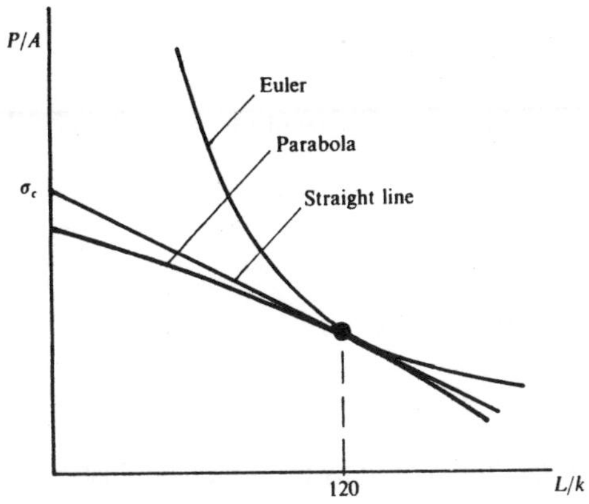

Figure 9.14 Tangents to the Euler curve

If we pre-select the tangent point to occur at $(L/k)_c$, the intersection with the stress axis will not, in general, equal σ_c. A better method is to find $(L/k)_c$ and n for which the ordinate is σ_c as shown. The gradients to each curve are

Euler: $\mathrm{d}\sigma/\mathrm{d}(L/k) = -2\pi^2 E/(L/k)^3$
Straight line: $\mathrm{d}\sigma/\mathrm{d}(L/k) = -n\sigma_c$

Equating these gradients and the stress ordinate for each curve at $L/k = (L/k)_c$ gives

$$2\pi^2 E/(L/k)_c^3 = n\sigma_c$$
$$\pi^2 E/(L/k)_c^2 = \sigma_c[1 - n(L/k)_c]$$

from which,

$$(L/k)_c = \sqrt{(3\pi^2 E/\sigma_c)}, \quad n = (2/9)\sqrt{(3\sigma_c/\pi^2 E)} \tag{9.30a,b}$$

9.3.3 Johnson's Parabolic Formula

The following parabolic equation is employed with struts of smaller slenderness ratios within the invalid Euler region:

$$P = A\sigma_c[1 - b(L/k)^2] \tag{9.31}$$

where b and σ_c are constants, with typical values $b = (23 - 30) \times 10^{-6}$ and $\sigma_c = 275$ MPa for $L/k \le 150$ in mild steel. This form often takes σ_c to be less than the actual compressive yield stress of the strut material to offset the need for a safety factor. We can again make intersections and tangents between eq(9.31) and the Euler curve as appropriate to a chosen design of a pinned-end strut (see Figs 9.13 and 9.14).

(i) Intersections with Euler curve

Equating (9.5) and Johnson's parabola (eq 9.31) for an intersection at a given slenderness ratio,

$$P/A = \sigma_c\,[1 - b\,(L/k)_c^{\,2}\,] = \pi^2 E/(L/k)_c^{\,2}$$
$$b = (L/k)_c^{\,-2}\,\{1 - \pi^2 E/[\sigma_c (L/k)_c^{\,2}]\}$$

Alternatively, Fig. 9.13b shows an intersection at $\sigma_c/2$. The Euler curve gives the corresponding slenderness ratio

$$(L/k)_c = \sqrt{(2\pi^2 E/\sigma_c)}$$

and eq(9.31) gives the b - value

$$b = \sigma_c/(4\pi^2 E) \tag{9.32}$$

The particular form of eq(9.31) employed by the American Institute of Steel Construction is

$$P = A\,\sigma_c\{1 - [1/(2C_c^{\,2})](L/k)^2\} \tag{9.33}$$

Comparing eqs(9.31) and (9.33) we see from eq(9.32) that $C_c = \sqrt{[1/(2b)]} = \sqrt{(2\pi^2 E/\sigma_c)}$.

(ii) Tangent to the Euler Curve

In Fig. 9.14 the tangent point is pre-selected to occur at $(L/k)_c$ (typically 120 or less) so that the intersection with the stress axis is less than σ_c (say σ_o). The gradients to each curve are

Euler: $\quad \mathrm{d}\sigma/\mathrm{d}\,(L/k) = -\,2\pi^2 E/\,(L/k)^3$

Parabola: $\mathrm{d}\sigma/\mathrm{d}\,(L/k) = -\,2\sigma_o b\,(L/k)$

Equating these gradients and the stress ordinate for each curve at $L/k = (L/k)_c$ gives

$$2\pi^2 E/\,(L/k)_c^{\,3} = 2b\sigma_o\,(L/k)_c$$
$$\pi^2 E/\,(L/k)_c^{\,2} = \sigma_o[1 - b\,(L/k)_c^{\,2}]$$

from which,

$$\sigma_o = 2\pi^2 E/\,(L/k)_c^{\,2}, \quad b = 1/[\,2(L/k)_c^{\,2}] \tag{9.34a,b}$$

This shows that the intercept σ_o is dependent upon the chosen slenderness ratio. The point of tangency is thus chosen so that $\sigma_o \le \sigma_c$.

9.3.4 Perry-Robertson

The Perry-Robertson equation is is recommended in a British Standard [4] for the determination of allowable compressive loads in structural steel columns. In common with the Rankine-Gordon method, the following expression represents a transition curve between pin-ended strut buckling at high L/k and short column yielding for low L/k in the range $80 \le L/k \le 350$.

$$P = [A/(2K)]\left([\sigma_c + (\mu + 1)\sigma_E] - \sqrt{\{[\sigma_c + (\mu + 1)\sigma_E]^2 - 4\sigma_E\sigma_c\}}\right) \tag{9.35a}$$

where σ_c is the minimum yield strength, and the load (i.e. safety) factor K is typically taken as 1.7 - 2.0. An account of initial sinusoidal curvature and eccentricity of loading appear indirectly in a deformation factor adopted by the British Standard $\mu = a\,(L/k)$ where a = 0.001 - 0.003 based upon Robertson's experiments. Alternatively, the Dutheil factor

$$\mu = (0.3/\pi^2)(\sigma_c/E)(L/k)^2 \tag{9.35b}$$

agreed with the lowest failure loads observed in his buckling experiments.

9.3.5 Fidler

A particular simplified form of eq(9.35a) was developed by Fidler for the design of columns used in bridge construction. This is,

$$P = (A/q)\left\{(\sigma_c + \sigma_E) - \sqrt{[(\sigma_c + \sigma_E)^2 - 2q\,\sigma_E\sigma_c]}\right\} \tag{9.36}$$

where q is a load factor with an average value of 1.2.

9.3.6 Engesser

Engesser's modification to the Euler theory accounts for inelastic buckling by simply replacing the elastic modulus with the tangent modulus. Equation (9.3) becomes:

$$\sigma = P/A = \pi^2 E_t / (cL/k)^2 \tag{9.37}$$

where c accounts for the effect of end fixing. In the range $50 \le L/k \le 100$ it is known that the effect of end constraint on the plastic buckling load P is much less than the equivalent elastic constraint given in Table 9 .1. The tangent modulus $E_t = d\sigma/d\varepsilon$ is the gradient of the tangent to the uniaxial tensile or compressive stress-strain curve (see Fig. 9.15).

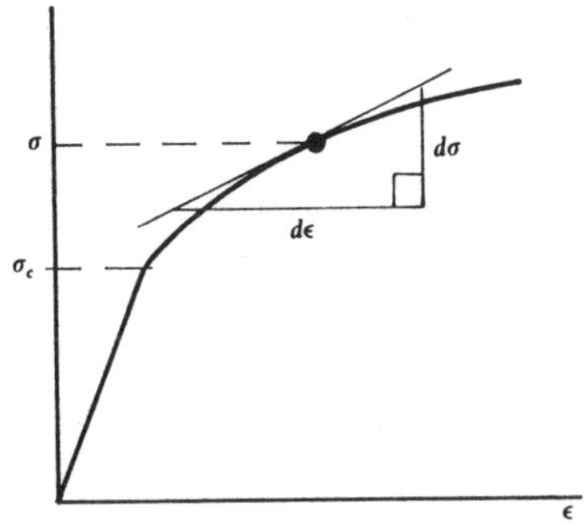

Figure 9.15 Tangent modulus

It follows from this definition of E_t that the buckling stress in eq(9.37) must satisfy the following condition:

$$\sigma = (\pi / c)^2 \, (L / k)^{-2} \, (\mathrm{d}\sigma / \mathrm{d}\varepsilon) \tag{9.38}$$

which may be established either graphically or from a suitable empirical representation to the σ versus ε curve, the simplest being the Hollomon law [5]:

$$\sigma = B\varepsilon^n \tag{9.39}$$

where B and n are the respective hardening coefficient and exponent, which are constants for a given material. Now, from eq(9.39),

$$\mathrm{d}\sigma / \mathrm{d}\varepsilon = (n\,B)\,\varepsilon^{n-1} = (n\,B)\,(\sigma / B)^{(n-1)/n} \tag{9.40}$$

Substituting eq(9.40) into eq(9.38), with $L_e = cL$, results in an equation that is soluble in σ.

$$\sigma = B\,[\,n\pi^2 / (L_e / k)^2\,]^n \tag{9.41}$$

Example 9.9 A 7.5 m length steel column with fixed ends is constructed by joining two channels sections together as shown in Fig. 9.16. If for one section the area is 6950 mm^2, $I_x = 91.1 \times 10^6$ and $I_y = 5.74 \times 10^6$ mm^4, compare the axial load that the column may carry according to the straight line, parabola, Perry-Robertson, Fidler and Engesser predictions. Take $E = 207$ GPa, $K = 1.7$, $q = 1.2$, $\sigma_c = 310$ MPa. The straight line and the parabola make tangents with the Euler curve in the manner outlined above. Take the straight line intercept as σ_c and the parabola tangent at $L / k = 120$. The stress-strain curve for the material is given by $\sigma = 650\,\varepsilon^{0.2}$.

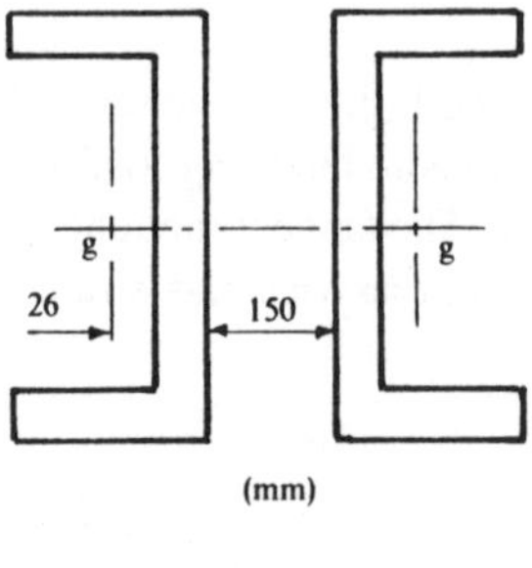

Figure 9.16 Fabricated section

$A = 2 \times 6950 = 13900$ mm^2,

$I_X = 2 \times 91.1 \times 10^6 = 182.2 \times 10^6$ mm^4

$I_Y = 2[(5.74 \times 10^6) + (6950 \times 101^2)] = 153.27 \times 10^6$ mm^4

$k = \sqrt{(I_Y/A)} = \sqrt{[(153.27 \times 10^6) / 13900\,]} = 148.5$ mm,

$L_e / k = 3750 / 148.5 = 25.25$

$\sigma_E = \pi^2 E / (L_e / k)^2 = \pi^2 \times 207000 / 25.25^2 = 3203.76$ MPa

Straight Line and Parabola - using the constants n and σ_c, found previously in eq(9.30a,b), the respective buckling load predictions are

$n = (2/9)\surd[3\sigma_c/(\pi^2E)] = (2/9)\surd[(3 \times 310)/(\pi^2 \times 207000)] = 4.74 \times 10^{-3}$

$P_{sl} = (13900 \times 310)\,[1 - 4.74 \times 10^{-3})\,25.25] = 3.794$ MN

$\sigma_o = 2\pi^2E/(L/k)_c^2 = (2\pi^2 \times 207000)/120^2 = 283.75$ MPa

$b = 1/[2\,(L/k)_c^2] = 1/(2 \times 120^2) = 3.472 \times 10^{-5}$

$P_p = (13900 \times 283.75)[1 - (3.472 \times 10^{-5})\,(25.25)^2] = 3.857$ MN

Perry-Robertson - the Dutheil deformation factor, eq(9.35b), becomes

$$\mu = (0.3/\pi^2)(\sigma_c/E)(L_e/k)^2$$
$$= (0.3/\pi^2)(310/207000)(25.25)^2 = 0.029$$

Then from eq(9.35a), with $K = 1.7$ the buckling load is

$$P = [13900/(2 \times 1.7)]\big([\,310 + (1.029 \times 3203.76)\,] - \surd\{[\,310 + (1.029 \times 3203.76)]^2 - (4 \times 310 \times 3203.76)\}\big)$$
$$= 4088.24\,(3606.67 - 3005.89) = 2.456 \text{ MN}$$

Fidler's - from eq(9.36), with $q = 1.2$,

$$P = (13900/1.2)\big\{(310 + 3203.76) - \surd[(310 + 3203.76)^2 - (2.4 \times 310 \times 3203.76)\,]\big\}$$
$$= 11583\,(3513.76 - 3156.41) = 4.138 \text{ MN}$$

Engesser - from eq(9.41),

$$\sigma = B\,[\,n\pi^2/(L_e/k)^2]^{1/5} = 650\,[0.2 \times \pi^2/(25.25)^2\,]^{1/5} = 204.68$$
$$P_E = 162.8 \times 13900 = 2.845 \text{ MN}$$

These calculations show that Perry-Robertson becomes the most conservative throughout the introduction of load factor K. Engesser becomes more realistic as the stress levels approach the yield stress of the material.

9.4 Buckling Theory of Plates

This introduction to plate buckling omits the full theoretical analysis that may be found in more specialist texts (see bibliography). The edge fixings considered are limited to simple and clamped supports and varying degrees of rotational restraint. Use is made of design data sheets that account for combinations of these fixings when an allowance for plasticity effects is necessary.

9.4.1 Plate as a Wide Strut with Long Edges Unsupported

When in a thin plate the width dimension b is large, i.e. of the same order as the length a, then biaxial in-plane stresses σ_x and σ_y will exist in the body of the plate (see Fig. 9.17a). These ensure that the cross-section remains rectangular during bending. Now from eq(2.14a,b), with no dimensional change in the y - direction,

$$\varepsilon_y = 0 = (1/E)(\sigma_y - \nu\sigma_x)$$
$$\varepsilon_x = (1/E)(\sigma_x - \nu\sigma_y) = (1 - \nu^2)\sigma_x/E \qquad (9.42)$$

The absence of σ_y in thin rectangular section struts with a smaller width results in *anticlastic curvature*. Here the section does not remain rectangular under bending but distorts due to the opposite sense of the induced y - direction strains between the tensile and compressive surfaces (the Poisson effect). Here the strains in the length and width direction are $\varepsilon_x = \sigma_x/E$ and $\varepsilon_y = -\nu\sigma_x/E$.

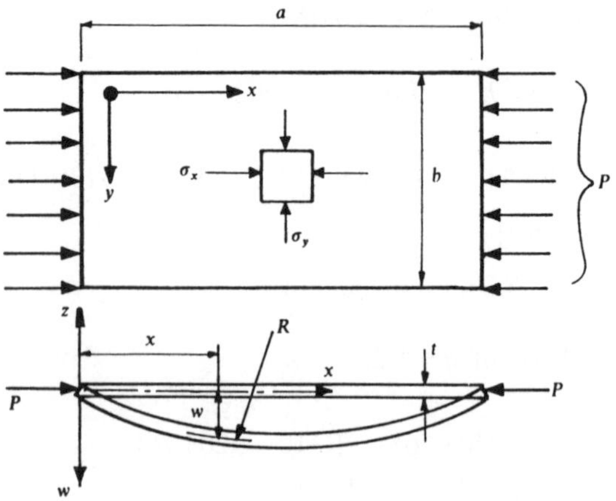

Figure 9.17 Buckling of a wide thin plate

Comparing ε_x and ε_y with the strains in eq(9.42) it becomes necessary to modify the beam flexure equation $d^2w/dx^2 = M/EI$ by the factor $(1 - \nu^2)$ in the case of the wide strut. Combining this with bending theory $M/(EI) = 1/R$ we have for the co-ordinates given in Fig. 9.17

$$d^2w/dx^2 = 1/R = \varepsilon_x / z \qquad (9.43)$$

Substituting eq(9.42) into (9.43) with $M/I = \sigma_x/z$, it follows that

$$d^2w/dx^2 = (1 - \nu^2)\sigma_x/(Ez) = (1 - \nu^2)M/(EI)$$

Then, with $M = -Pw$ (see Fig. 9.17b), the critical buckling load for a plate with pinned-ends is deduced from the solution to eq(9.1). This is

$$P_{cr} = \pi^2 EI / [(1 - \nu^2)a^2]$$

Substituting $P_{cr} = \sigma_{cr}bt$ and $I = bt^3/12$ leads to the more common form of expression in terms of aspect ratios t/b and $r = a/b$:

$$\sigma_{cr} = (1/12)(\pi/r)^2 [E/(1 - \nu^2)](t/b)^2 \qquad (9.44)$$

Other edge fixings may be accounted for by replacing the 1/12 factor for pinned ends in eq(9.44) with a buckling coefficient K.

9.4.2 Bi-directional Compressive Loading with Edges Supported

Consider a thin rectangular plate $a \times b$ with thickness t subjected to uniform compressive stresses σ_x acting upon area $b \times t$ and σ_y acting upon area $a \times t$. as shown in Fig. 9.18.

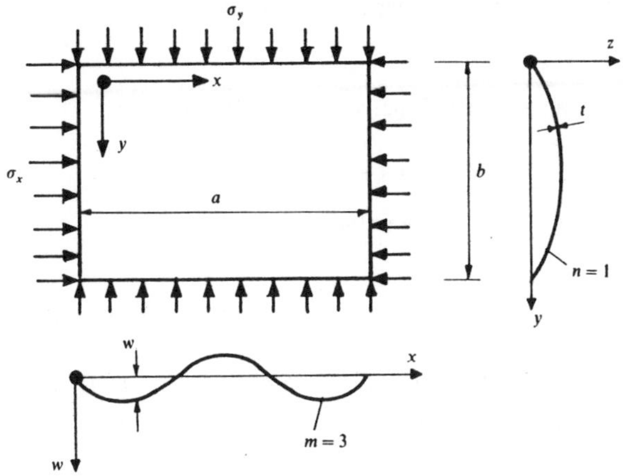

Figure 9.18 Buckling of a thin plate under biaxial stress

In Chapter 4 (see eq 4.60) it was shown that when the edges are simply supported the equation governing the dependence of the lateral deflection w upon x and y is

$$D\,(\partial^2 w/\partial x^2 + \partial^2 w/\partial y^2)^2 = (\sigma_x\, t)\,\partial^2 w/\partial x^2 + (\sigma_y\, t)\,\partial^2 w/\partial y^2 \tag{9.45a}$$

where the flexural stiffness $D = E\,t^3/[12(1-\nu^2)]$. The solution to eq(9.45a) is

$$w(x, y) = \sum_{m=1}^{\infty} \sum_{n=1}^{\infty} A_{mn} \sin(m\pi x/a) \sin(n\pi y/b) \tag{9.45b}$$

in which the respective number of half-waves m and n for buckling in the x and y plate directions determine the number of terms and associated coefficients A_{mn}. In Fig. 9.18, for example, $m = 3$ and $n = 1$ is shown. When stresses increase in a constant ratio $\alpha = \sigma_y/\sigma_x$, the actual number of half-waves of buckling is that which minimises the potential energy for a given edge fixing. Employing the Raleigh-Ritz energy method (see Chapter 8) we differentiate the potential energy with respect to the unknown coefficient A_{mn} in eq(9.45b). This leads to a solution to the critical bucking stress:

$$(\sigma_x)_{cr} = \frac{D\pi^2[(m/a)^2 + (n/b)^2]^2}{t[(m/a)^2 + \alpha\,(n/b)^2]} \tag{9.46}$$

Example 9.10 Find the critical buckling stress for a square plate of side length a with simply supported edges subjected to equal in-plane loading.

With simply supported edges, i.e. no rotational restraint, a square plate buckles with one half-wave $m = n = 1$ in each direction. Substituting $r = a/b = 1$ and $\alpha = \sigma_y/\sigma_x = 1$ in eq(9.46) leads to the buckling stress:

$$(\sigma_x)_{cr} = [D\pi^2/(t\,a^2)](1 + r^2)^2/(1 + r^2\alpha)$$

$(\sigma_x)_{cr} = 2D\pi^2/(t\,a^2)$
$= (\pi^2/6)[E/(1-\nu^2)](t/a)^2$

Example 9.11 If the plate in Fig. 9.18 has simply supported edges with $b << a$, find $(\sigma_x)_{cr}$ and the half wavelength of the buckled shape in the direction of x, when (i) $\alpha = \sigma_y/\sigma_x < ½$ and (ii) $\alpha \to ½$.

Taking $n = 1$ for the much smaller b dimension, eq(9.46) becomes

$$(\sigma_x)_{cr} = \frac{D\pi^2(m^2 + r^2)^2}{a^2 t(m^2 + \alpha r^2)} = \frac{D\pi^2(m/r + r/m)^2}{b^2 t[1 + \alpha(r/m)^2]} \qquad \text{(i)}$$

Then m is found from the condition that $(\sigma_x)_{cr}$ is a minimum. From eq(i),

$d(\sigma_x)_{cr}/dm = [1 + \alpha(r/m)^2]2(m/r + r/m)(1/r - r/m^2) - (m/r + r/m)^2 2\alpha(r/m)(-r/m^2) = 0$
$[1 - (r/m)^2][1 + \alpha(r/m)^2] + \alpha(r/m)^2[1 + (r/m)^2] = 0$
$\therefore (r/m)^2(1 - 2\alpha) = 1$

This gives

$$m = r(1 - 2\alpha)^{1/2} \qquad \text{(ii)}$$

with the corresponding half wavelength $a/m \simeq r/m = (1 - 2\alpha)^{-1/2}$. Substituting eq(ii) in (i),

$(\sigma_x)_{cr} = \{D\pi^2[(1 - 2\alpha) + 2 + 1/(1 - 2\alpha)]\}/\{bt^2[1 + \alpha/(1 - 2\alpha)]\}$
$= \{D\pi^2[(1 - 2\alpha)(3 - 2\alpha) + 1]\}/[b^2 t(1 - \alpha)]$

It is apparent from this solution that as $\alpha \to ½$, the half wavelength in the x-direction approaches infinity ($a/m \to \infty$) and the buckling stress becomes

$(\sigma_x)_{cr} \to 2D\pi^2/(b^2 t)$

9.4.3 Uni-Directional Compressive Loading

(a) Simple Supports
With only a uniform compressive stress σ_x acting, the critical buckling stress is found by putting $\sigma_y = 0$ ($\alpha = 0$) in eq(9.46). This gives

$(\sigma_x)_{cr} = (D\pi^2/t)[(m/a)^2 + (n/b)^2]^2(a/m)^2$

$$(\sigma_x)_{cr} = [D\pi^2/(t\,b^2)][mb/a + n^2 a/(mb)]^2 \qquad (9.47)$$

When the plate is simply supported along its unstressed sides it buckles with one half-wave ($n = 1$) in the y - direction as shown in Fig. 9.19. From eq(9.47) the buckling stress becomes

$$(\sigma_x)_{cr} = [D\pi^2/(t\,b^2)][m/r + r/m]^2 \qquad (9.48)$$

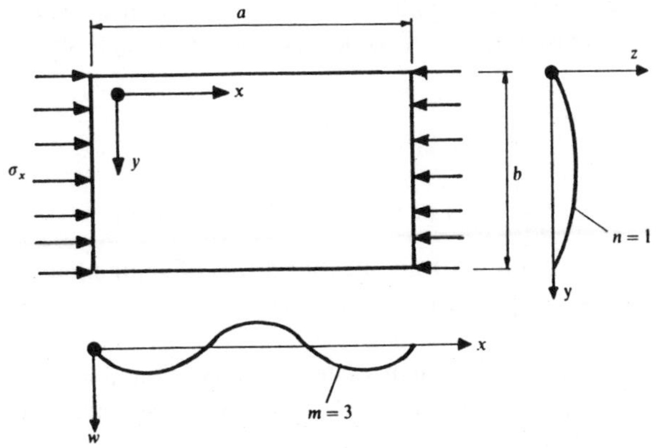

Figure 9.19 Buckling of a thin plate under uniaxial compression

The question then arises as to what values of m minimise eq(9.48) for integral values of $r = a/b$. This condition is expressed in

$$d(\sigma_x)_{cr}/dm = 2(m/r + r/m)(1/r - r/m^2) = 0$$
$$m/r^2 + 1/m - 1/m - r^2/m^3 = 0$$
$$m^4 - r^4 = 0$$
$$(m - r)(m + r)(m^2 + r^2) = 0$$

Thus $m = r$ implies that the plate will buckle into an integral number of square cells $a \times a$ under the same stress. That is, from eq(9.48),

$$(\sigma_x)_{cr} = 4D\pi^2/(t\,b^2) = \{\pi^2 E/[3(1 - \nu^2)]\}(t/b)^2 \tag{9.49}$$

Furthermore, when, for non-integral values of r, eq(9.48) is applied to particular values of m, the graph in Fig. 9.20 represents the buckling stress.

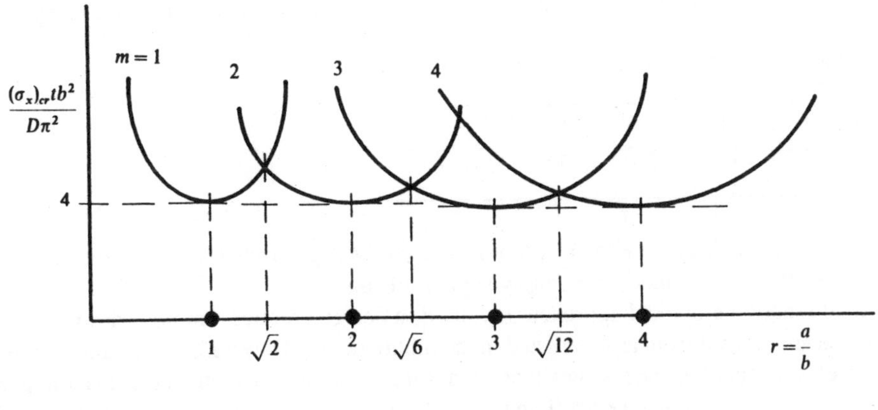

Figure 9.20 Effect of r and m on uniaxial buckling stress

The trough in each curve corresponds to eq(9.49) and the points of intersection determine the r - value for which m is increased by one. This occurs when, from eq(9.48),

$$[m/r + r/m] = [(m+1)/r + r/(m + 1)]$$
$$\therefore\ r = \sqrt{[m(m + 1)]}$$

from which, $r = \sqrt{2}$ for $m = 2$, $r = \sqrt{6}$ for $m = 3$, $r = \sqrt{12}$ for $m = 4$ etc.

Equation (9.49) should not be confused with buckling stress of a thin plate acting as a strut with its longer parallel sides unsupported (see eq.9.44).

Example 9.12 Find the thickness of aluminium plate with dimensions $a = 1.5$ m, $b = 1$ m and simply supported along its edges that will buckle elastically under a compressive stress of 210 MPa if $E = 70$ MPa and $\nu = 0.3$.

Figure 9.20 shows that with $r = a/b = 1.5$ the plate will buckle into $m = 2$ half-waves of length 0.75 m in the x - direction. From eq(9.48) with $D = Et^3/[12(1 - \nu^2)]$,

$$t = \sqrt{\{[12\,(1 - \nu^2)b^2(\sigma_x)_{cr}]/\,[\pi^2 E\,(m/r + r/m)^2]\}}$$
$$= \sqrt{[(\,12 \times 0.91 \times 1^2 \times 210)]/\,[\pi^2 \times 70 \times 10^3(2/1.5 + 1.5/2)^2]} = 0.0277\text{m} = 27.7\text{ mm}$$

(b) Other Edge Fixings

A number of approaches have been proposed. The simplest of these employs the analytical expression:

$$(\sigma_x)_{cr} = K_r\,\pi^2 D/(\,t\,b^2)$$

where, for all possible edge fixings, the restraint coefficient is

$$K_r = (m/r)^2 + p + q\,(r/m)^2$$

which contains eq(9.48) as a special case when the restraint factors are $p = 2$ and $q = 1$ for simple supports. The dependence of K_r upon three factors: the rotational edge restraint (p and q), the plate dimensions ($r = a/b$) and the buckling mode m, has been established experimentally in certain cases. Table 9.3 applies to the case of fixed sides for example.

Table 9.3 Restraint coefficients for a plate with fixed sides

$r = a/b$	0.75	1.0	1.5	2.0	2.5	3.0
K_r	11.69	10.07	8.33	7.88	7.57	7.37

As r increases, the effect of edge restraint lessens and K_r approaches a minimum value of 4 found in eq(9.49) for a plate with simply supported edges.

An alternative graphical approach, favoured in design practice, employs design curves [6]. These are derived from articles listed in the bibliography. They supply the ratio between critical elastic buckling stress and that of a simply supported plate (eq.9.49) for plate geometry $r = a/b$ with a given edge fixing. For example, Fig. 9.21 shows how this stress ratio varies with r for clamped and various mixed edge fixings. The sheet extends this type of presentation to biaxially loaded plates with various edge fixings.

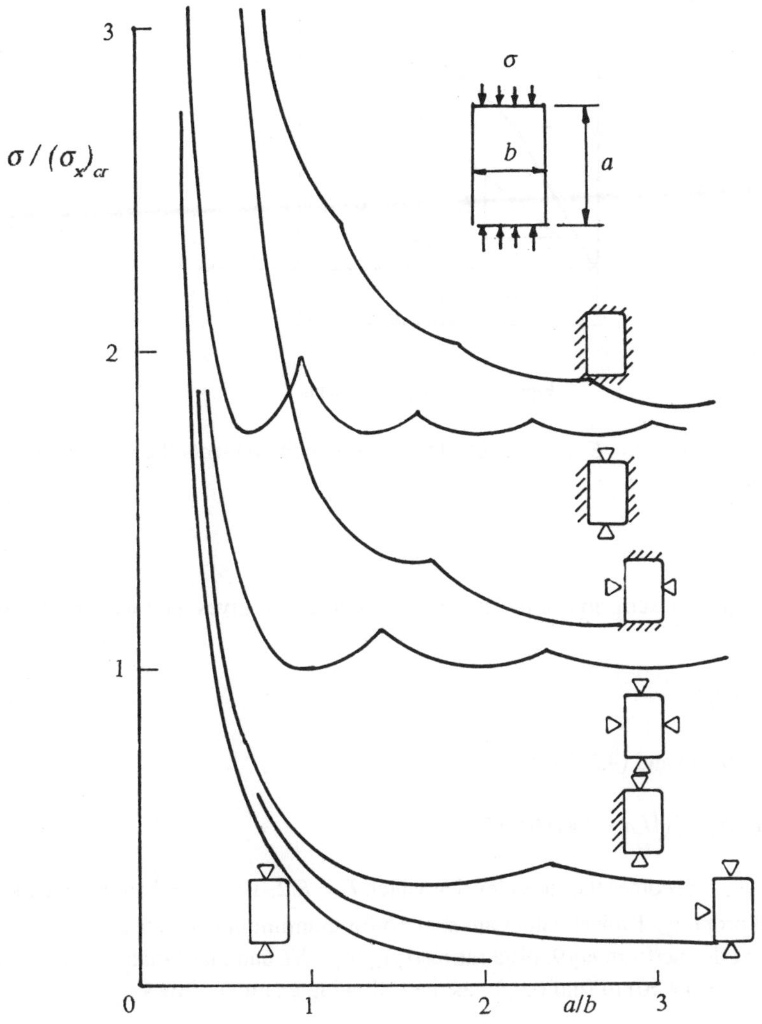

Figure 9.21 Plate buckling under compression

(c) Inelastic Buckling.
In thicker plates the critical elastic stress, $(\sigma_{cr})_e$ calculated from eq(9.49), can exceed the yield stress of the plate material. The solution will be invalid because with the use of E and ν in eq(9.49) linear elasticity will have been assumed in reaching this stress level. Fig. 9.22 shows that some plasticity will have occurred and this reduces the buckling stress to a lower level $(\sigma_{cr})_p$. To account for this a plasticity reduction factor is defined as follows:

$$\mu = (\sigma_{cr})_p / (\sigma_{cr})_e \tag{9.50}$$

where $\mu < 1$. Figure 9.22 shows further how the tangent modulus is used to account for a stress level in excess of the yield stress

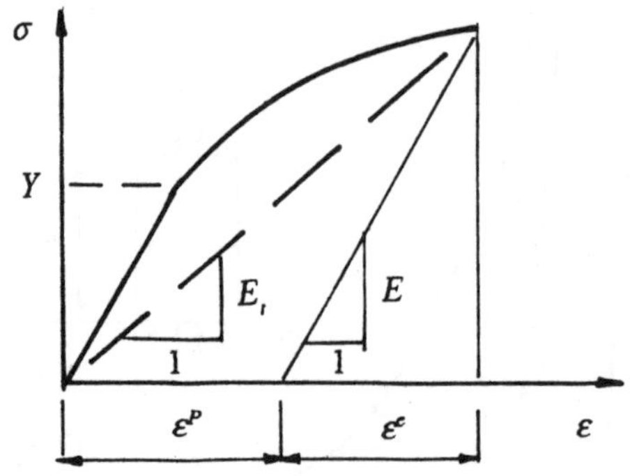

Figure 9.22 Tangent modulus

This gives $\sigma = E_t\,\varepsilon$. Let $n = E_t/E$ be a selected value which connects the plastic stress and strain levels as

$$\sigma_n = nE\varepsilon_n \tag{9.51}$$

A Ramberg-Osgood description of the stress-strain curve also gives the total strain under a plastic stress as

$$\varepsilon = \sigma/E + \alpha(\sigma/E)^m \tag{9.52}$$

Combining eqs(9.51) and (9.52) leads to

$$\varepsilon = (\sigma_n/E)[(\sigma/\sigma_n) + (1/n - 1)(\sigma/\sigma_n)^m] \tag{9.53}$$

Set $n = ½$ so that σ_n becomes the stress level at which $E_t = E/2$. The work-hardening exponent m is a material property. Typical values are $m \approx 16$ for aluminium alloy sheet and $m \approx 10$ for steel sheet. Next define from eq(9.49) a ratio $(\sigma_{cr})_e/\sigma_n$. We can interpolate a μ value from Fig. 9.23 and use eq(9.50) to find $(\sigma_{cr})_p$ as the following example shows.

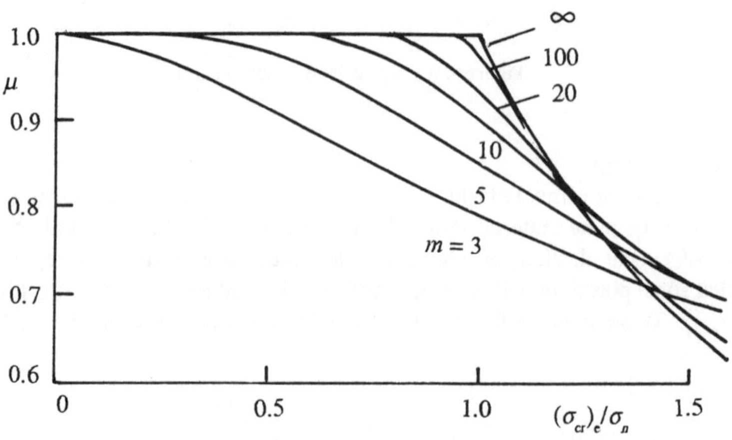

Figure 9.23 Plasticity reduction factor

Example 9.13 A plate 400 × 200 × 7.5 mm is simply suppored along its longer sides and carries a compressive stress along it shorter sides. Determine the critical buckling stressses when the shorter sides are (i) simply supported and (ii) clamped, given the material properties $E = 73$ GPa, $Y = 250$ MPa, $\sigma_n = 400$ MPa, ν = 0.34 and $m = 20$.

(i) The plate aspect ratio is an integral number $r = 2$. Hence, from eq(9.49),

$$(\sigma_{cr})_e = \{(\pi^2 \times 73 \times 10^3)/[3(1 - 0.34^2)]\}(7.5/200)^2 = 381.87 \text{ MPa}$$

Since this exceeds the yield stress Y, a correction for plasticity is necessary. Thus

$$(\sigma_{cr})_e/\sigma_n = 381.87 / 400 = 0.955$$

when Fig. 9.23 gives $\mu = 0.96$. Hence we find from eq(9.50)

$$(\sigma_{cr})_p = 0.96 \times 381.87 = 364.6 \text{ MPa}$$

(ii) With the shorter sides clamped, Fig. 9.21 gives

$$(\sigma_{cr})_e = 1.214 \times 381.87 = 463.59 \text{ MPa}$$

Correcting for plasticity as before,

$$(\sigma_{cr})_e/\sigma_n = 463.59/400 = 1.159$$

Figure 9.23 gives $\mu = 0.86$. Hence we find from eq(9.50)

$$(\sigma_{cr})_p = 0.86 \times 463.59 = 398.7 \text{ MPa.}$$

9.4.4 Local Buckling of Plate Sections

The straight thin walls of an open section may distort without translation or rotation. Localised stresses concentrated at corners may exceed the yield stress and become sufficiently high either to cause a local crippling failure or to reduce the stiffness to resistance against buckling by other modes. Strut sections made up with straight sides such as I, Z, U, as well as thin-walled closed tubes, are prone to local buckling at their sharp corners. When the strut length is at least four times greater than the cross-section web depth h, the flanges and weds of these sections may be treated as plates with a simple support along one or both sides from their neighbours. A supported side will restrain an unsupported side. The solution to the local compressive elastic buckling stress takes the common form:

$$\sigma_{be} = KE\,(t/h)^2 \qquad (9.54)$$

where the thickness t of the sides $t < d/5$ where d is the flange length (see Fig. 9.24). The buckling coefficients K depend upon the d/h ratio and the neighbouring restraints within the four sections, as shown in Fig. 9.24. K must again be corrected with a multiplication factor $0.91/(1 - \nu^2)$ when Poisson's ratio is different from 0.3. If σ_{be} from eq(9.54) is found to exceed the yield stress of the strut material it becomes necessary to reduce this using a plasticity reduction factor ($\sigma_b = \mu\, \sigma_{be}$) and a suitable description to the stress-strain curve as in the previous example.

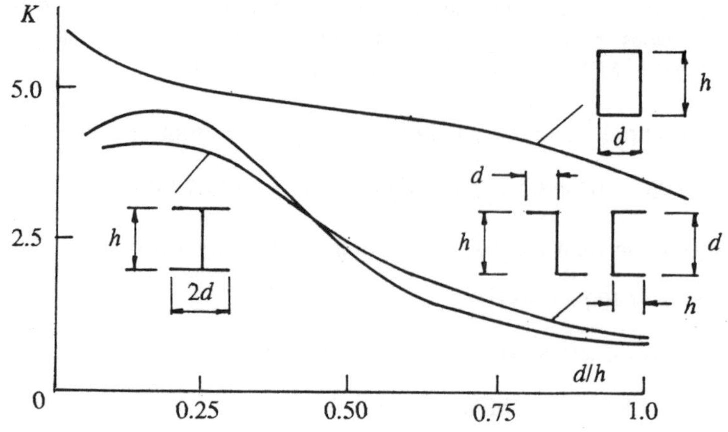

Figure 9.24 Local buckling coefficient for uniform thin sections

In medium length struts local instability results in a loss of stiffness but without complete failure. The influence upon any global buckling mode through flexure and torsion becomes important. However, if the strut is very long the global buckling stresses are attained well before the onset of local buckling. On the other hand, a short strut may carry more compression beyond local buckling. Final failure is estimated to occur under a crippling stress σ_c found from

$$\sigma_c = (\sigma_b Y)^{1/2} \tag{9.55}$$

where Y is the 0.2% compressive proof stress of strut material. Test data shows that eq(9.55) is accurate to within 10%.

Example 9.14 Find the buckling and crippling stresses of a short strut with an I-section given $h = 50$ mm, $d = 25$ mm and $t = 2$ mm. Take the material constants as follows: $E = 73.8$ GPa, $Y = 310$ MPa, $m = 10$ and $\sigma_n = 264$ MPa.

From Fig. 9.24 we find $K = 2.38$ for $d/h = 0.5$. Equation (9.54) gives the local elastic buckling stress as

$$\sigma_{be} = 2.38 \times 73.8 \times 10^3 (2/50)^2 = 281 \text{ MPa}$$

This exceeds the yield stress because $\sigma_n = 264$ MPa corresponds to $E_t = 2E$ (see Fig. 9.22). The ratio $\sigma_{be}/\sigma_n = 281/264 = 1.064$ is used with Fig. 9.23 to give a plasticity reduction factor $\mu = 0.885$. Equation (9.50) gives the local buckling stress as

$$\sigma_b = \mu\, \sigma_{be} = 0.885 \times 281 = 248.7 \text{ MPa}$$

Using eq(9.55), the crippling stress is estimated as

$$\sigma_c = (248.7 \times 310)^{1/2} = 277.7 \text{ MPa}$$

9.4.5 Post-Buckling of Flat Plates

When a plate buckles, the load may be increased further as the stress increases in the material near the side supports. Only a slight increase in axial stress occurs in the central buckled material (see Fig. 9.25).

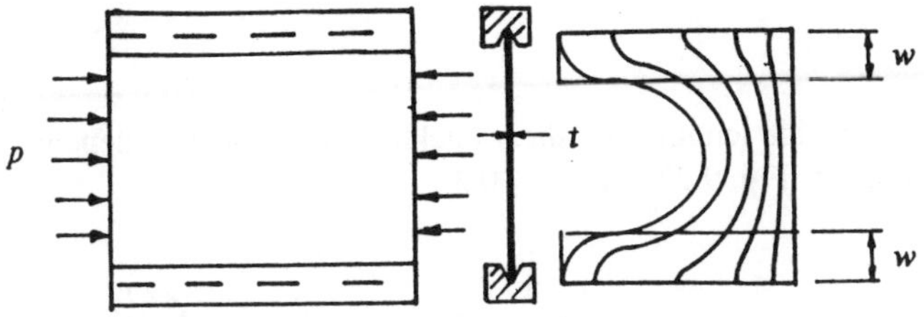

Figure 9.25 Stress distributon in buckled plate

If we assume that the whole load P is carried by two edge strips of effective width $2w$ over which σ is assumed constant, then the load is

$$P = 2wt\sigma$$

With the edges of our equivalent elastic plate all simply supported, eq(9.49) gives

$$\sigma_{cr} = \{ \pi^2 E /[3(1 - \nu^2)]\}[t / (2 w)]^2 \tag{9.56a}$$

from which w may be found as σ_{cr} attains the yield stress Y.

$$w = (\pi t /2)\sqrt{\{E / [3Y (1 - \nu^2)]\}} \tag{9.56b}$$

Taking $\nu = 0.3$ in eq(9.56b) gives $w = 0.95t\sqrt{(E/Y)}$ but experiment shows that the coefficient is nearer 0.85. Other edge fixings are dealt with in a similar manner using Fig. 9.21 to determine a critical elastic buckling stress ratio for use with eq(9.56a).

9.5 Buckling in Shear

When the sides of a thin plate $a \times b \times t$ are subjected to shear forces, the principal stress state within the plate is diagonal tension and compression of equal magnitude $\sigma_1 = \tau$, $\sigma_2 = -\tau$ (see Fig. 9.26).

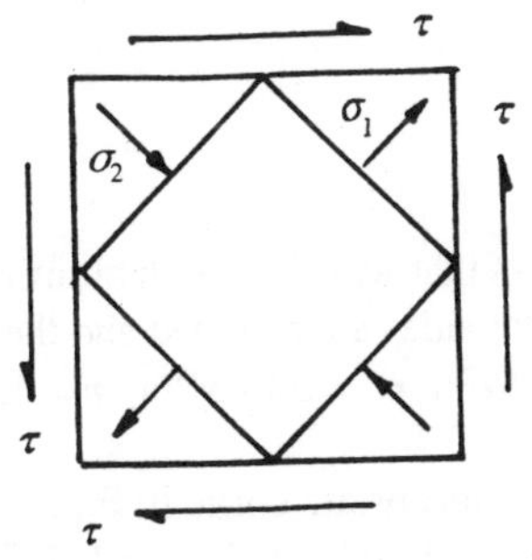

Figure 9.26 Plate in shear

For the biaxial stress state of ratio $\alpha = -1$, shear buckling occurs when wrinkles run parallel at intervals across the plate lying perpendicular to the compressive stress. Under this condition the plate cannot sustain a further increase in diagonal compression although an increase in diagonal tension is possible. In the buckling of flat plates under shear loading it can be shown [7] that the critical elastic shear stress is

$$(\tau_{cr})_e = K_s\, E\,(t/b)^2 \tag{9.57}$$

where b is the lesser side length. The shear buckling coefficient K_s depends upon the edge fixing in the manner of Fig. 9.27 when $\nu = 0.3$

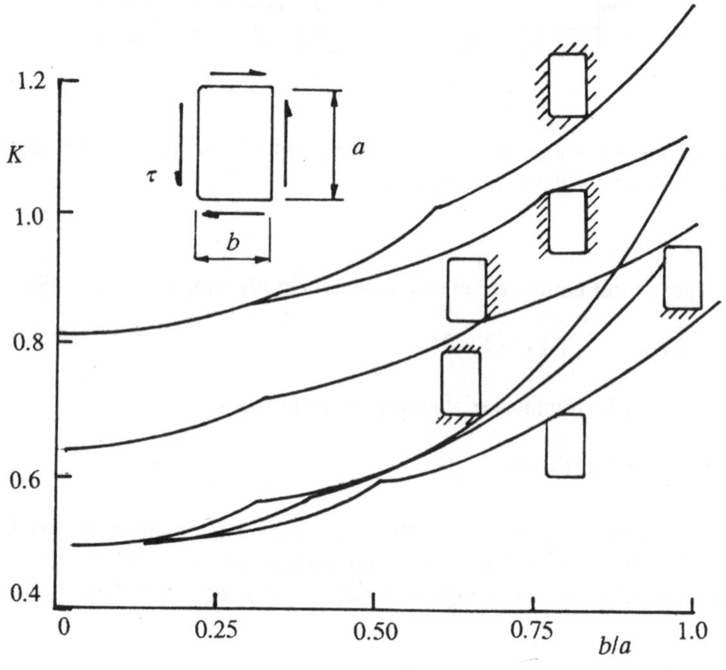

Figure 9.27 Shear buckling coefficient

If $\nu \neq 0.3$ the K_s ordinate value is multiplied by $0.91/(1-\nu^2)$. Equation (9.57) will then supply $(\tau_{cr})_e$ which will need further correction if it appears to exceed the shear yield stress k for the plate material. The plasticity reduction factor μ is obtained from Fig. 9.28 knowing σ_n and m for the material. This gives

$$(\tau_{cr})_p = \mu\,(\tau_{cr})_e \tag{9.58}$$

Example 9.15 Find the shear stress that will buckle an aluminium plate 200 mm wide × 260 mm long × 5 mm thick. The shorter sides are clamped and the longer sides simply supported. The plate properties are k = 150 MPa, σ_n = 340 MPa, m = 15, E = 73 GPa, ν = 0.34.

For $b/a = 200/260 = 0.769$, the appropriate curve in Fig. 9.27 gives $K_s = 8.08$. Correcting for Poisson's ratio gives $K_s = 8.08 \times 0.91/(1 - 0.34^2) = 8.313$. The elastic buckling stress is therefore, from eq(9.57)

$(\tau_{cr})_e = 8.313 \times 73 \times 10^3 (5/200)^2 = 379.3$ MPa

Since this exceeds the shear yield stress $k = 150$ MPa, a plasticity reduction factor is required. The abcissa in Fig. 9.28 now has a value:

$(\tau_{cr})_e / \sigma_n = 379.3 / 340 = 1.116$

The graph for $m = 16$ gives an ordinate of $\mu = 0.545$. Therefore, from eq(9.58),

$(\tau_{cr})_p = 0.545 \times 379.3 = 206.7$ MPa

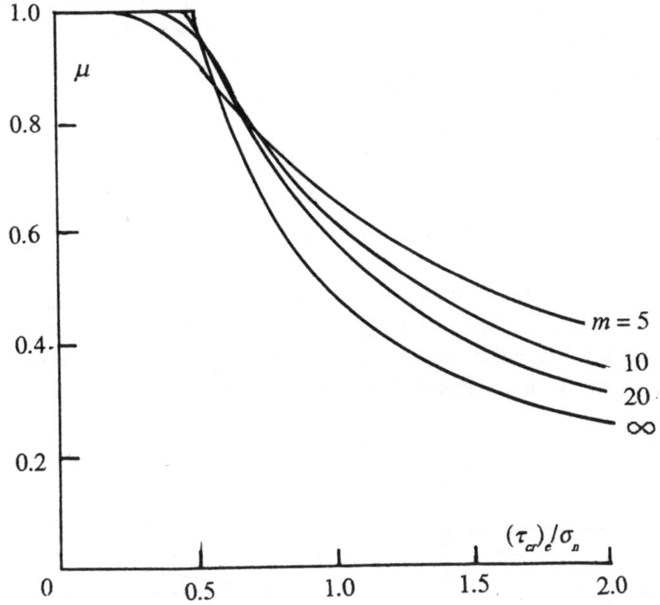

Figure 9.28 Plasticity reduction factor in shear

9.6 Flexural and Torsional Instability of Thin Open Sections

A critical value of the Euler buckling stress, eq(9.5), for a compressed strut may not be reached when a thin-walled open section is torsionally weak. A cross-section away from the ends can both translate and rotate and therefore buckling may occur by a combination of torsion and flexure. The analysis requires a consideration of the axial stress arising from torsion. We saw in Chapter 5 that an axial stress arises in torsion when the ends of a strut are not free to warp. Recall that the torque is given by the sum of St Venant and Wagner torques along the bar (see eq(5.85)). Wagner [8] further derived the critical end thrust for torsional buckling of sections with symmetry. Goodier [9] extended the analysis to any arbitrary thin-walled open section under compression by the analysis which follows. Let the unsymmetrical channel section in Fig. 9.29a support a compressive force P at its centroid g. Locate the directions of the principal axes in the plane of the section by the usual method (see Appendix I). Here x and y denote principal axes and u and v refer to displacements. When considering torsional instability recall that the centre of twist will lie at the shear centre E. Therefore, let axes X and Y be parallel to x and y with their origin at E along with that of the strut axis Z. A point in the cross-section is defined by co-ordinates (X, Y). Displacements of that point (u, v) may be referred to either axes x, y or X,Y while the accompanying rotation θ must be referred to E.

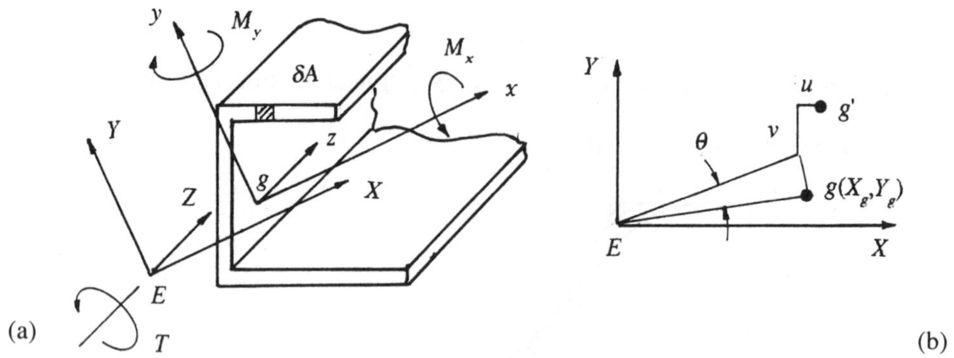

Figure 9.29 Open section strut under compression

The strut both twists and bends under a load applied at its centroid g. Given its initial co-ordinates $g(X_g, Y_g)$, they become $u - Y_g\theta$ and $v + X_g\theta$, when g displaces to g' as shown in Fig. 9.29b. Figures 9.30a and b show these displacements in the two planes X - Z (or x - z) and Y - Z (or y - z) for which g' lies at distance z from one end.

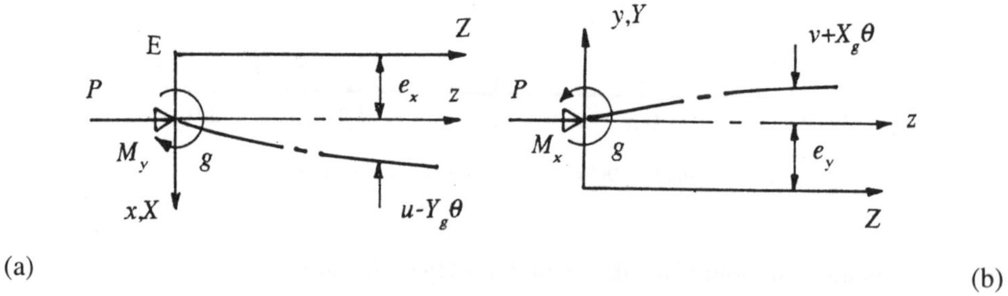

Figure 9.30 Flexure of the strut axis in the two principal planes

The bending moments M_y and M_x associated with the displacements are given by the flexure equations

$$EI_y\,\mathrm{d}^2u/\mathrm{d}z^2 = M_y = -P(u - Y_g\,\theta) \tag{9.59a}$$
$$EI_x\,\mathrm{d}^2v/\mathrm{d}z^2 = M_x = -P(v + X_g\,\theta) \tag{9.59b}$$

where E is the elastic modulus and I_x, I_y are the principal second moments of area referred to the centroid g of the section. Figure 9.29a shows an element δA of the wall area upon which a uniform compressive stress σ arises from the force P. The normal force $\sigma\,\delta A$ is inclined to the z - axis when the strut is in its deflected position. This force may be resolved into components parallel and perpendicular to the Z - axis. The latter component produces an elemental torque about the Z - axis of

$$\delta T = \sigma\,\delta A\,[Y\,(du/dz - Y\,d\theta/dz) - X\,(dv/dz + X\,d\theta/dz)\,] \tag{9.60a}$$

Integrating eq(9.60a) over the section area leads to

$$T = PY_g(du/dz) - PX_g(dv/dz) - Pk_p^{\,2}(d\theta/dz) \tag{9.60b}$$

where $P = \sigma A$ and k_p is the polar radius of gyration of the section about the shear centre. A shear stress also acts across δA due to the torque in eq(5.85) but since no torque is applied $\sum T = 0$ and it follows from eq(9.60b) that

$$PY_g\,(du/dz) - PX_g\,(dv/dz) - Pk_p^{\,2}\,(d\theta/dz) + (GJ\,d\theta/dz - E\,\Gamma_E\,d^3\theta/dz^3) = 0$$

which leads to the differential equation:

$$E\Gamma_E\,d^3\theta/dz^3 + (Pk_p^{\,2} - GJ\,)\,d\theta/dz + PX_o\,dv/dz - PY_o\,du/dz = 0 \tag{9.61}$$

where E and G are elastic moduli and J is St Venant's torsion constant (see eq(5.68). The torsion-bending constant Γ_E has been defined in Chapter 5 (see eq(5.95)). It is given as the sum of the primary (Γ_1) and secondary (Γ_2) warping constants. Usually $\Gamma_E \simeq \Gamma_1$ since Γ_2 is very small (see Example 9.16).

The solutions to eqs(9.59a,b) and (9.61) show that the precise mode of instability will depend upon the proximity of the centroid G to the shear centre E (see Figs 9.31a-c).

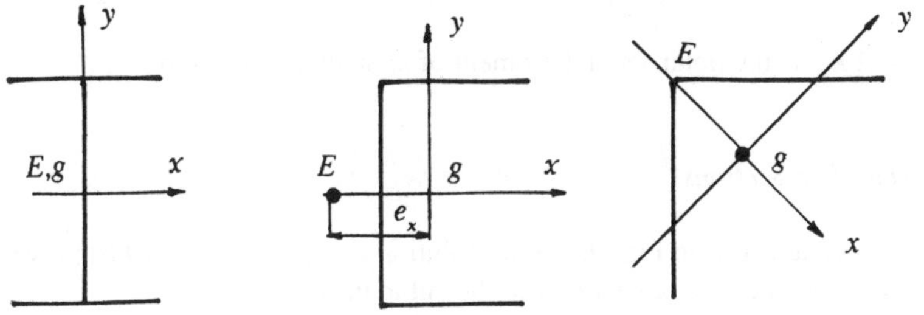

(a) (b) (c)

Figure 9.31 Centroid and shear centre for thin-walled open sections

Their buckling modes can be purely torsional, purely flexural or a combination of these. Thus, the critical compressive stress will be one of, or a combination of, the Euler buckling stresses from eq(9.7) and the Wagner axial stress arising from eq(9.61). Goodier [10] showed that the buckling mode of a pinned-end strut will depend upon the presence of symmetry within the cross-section.

9.6.1 Sections with Symmetry

The shear centre will coincide with the centroid for sections with two axes of symmetry such as an I - section (Fig. 9.31a). When a compressive force is applied at the centroid of a pinned-end strut of section, $X_g = Y_g = 0$ and eqs(9.59a,b) supply the usual Euler stresses for flexural buckling, These are, respectively,

$$\sigma_x = \pi^2 E/(L / k_x)^2 \tag{9.62a}$$
$$\sigma_y = \pi^2 E/(L / k_y)^2 \tag{9.62b}$$

where the radii of gyration k_x and k_y refer to the principal axes $k_x = \surd(I_x/A)$ and $k_y = \surd(I_y/A)$. In addition, the solution to eq(9.61) gives Wagner's critical thrust for torsional buckling as

$$\sigma_p = GJ/I_p + \pi^2 E\Gamma_E/(I_p L^2) \tag{9.63}$$

With a strut under torsion, the first term on the right-hand side of eq(9.63) applies to ends that are free to warp. The second term is the increase in axial stress arising when the ends are constrained. I_p is the polar second moment of area about the centre of twist, i.e. the shear centre. With a strut under compression, eq(9.63) applies to its full length when the pinned ends are free to warp. A half length applies to torsional buckling when fixed ends cannot warp or rotate. For the section concerned, the critical stress is the least of eqs(9.62a,b) and (9.63). For a section with point symmetry, e.g. an I-section with equal flange lengths (Fig. 9.31a), one of the two Euler stresses (9.62a,b) will always yield the smaller critical stress value. For sections with a single axis of symmetry, say the x - axis (see Fig. 9.31b), flexural instability can occur about the y - axis where eq(9.62b) applies. Alternatively, instability may arise from flexure about the x - axis combined with torsion about the shear centre. The lower of the two buckling stresses is the true solution, i.e. the critical buckling stress σ is the lesser root to the quadratic:

$$(1 - \sigma_x/\sigma)(1 - \sigma_p/\sigma) = (X_g / k_p)^2 \tag{9.64}$$

where $I_p = A\,k_p^2$ is the polar second moment of area about the Z - axis.

9.6.2 Asymmetric Sections

For the thin-walled strut in Fig. 9.29a the solution to eqs(9.59) and (9.61) reveals that the critical stress becomes the largest root to the cubic in σ:

$$(1- \sigma_x/\sigma)(1- \sigma_y/\sigma)(1- \sigma_p/\sigma) - (1- \sigma_x/\sigma)(Y_g/k_p)^2 - (1- \sigma_y/\sigma)(X_g/k_p)^2 = 0 \tag{9.65}$$

In this case σ will be smaller than σ_x, σ_y and σ_p, but if the latter three stresses vary widely σ will be close to the smallest of these which dictates the buckling mode. Equation (9.65) reduces to (9.64) when $Y_g = 0$.

A special case of eq(9.65) arises for sections where the shear centre lies at the intersection between limbs. The simplest of these are thin-walled right angles and tees in which the shear centre E is coincident with the single point of intersection between their straight limbs (Fig. 9.31c). Note that for each section $\Gamma_1 = 0$, so that $\Gamma_E = \Gamma_2$ in eq(9.63). Thus σ_p is low and buckling occurs by a purely torsional mode.

Example 9.16 Determine the torsion-bending constant for the lipped channel section shown in Fig. 9.32a given that the shear centre E is 24 mm to the left of centre-line of the vertical web. If a 2 m length of this section in titanium alloy carries an axial compressive force, determine the likely mode of buckling. Take $E = 117$ GPa, $G = 43.3$ GPa and all thicknesses to be 5mm.

In Fig. 9.32b the areas enclosed between E and perimeter points 1, 2 ... 6 are doubled and plotted against the perimeter length. We then determine the centroid $\bar{y}$ of Fig. 9.32b from eq(5.79c) as

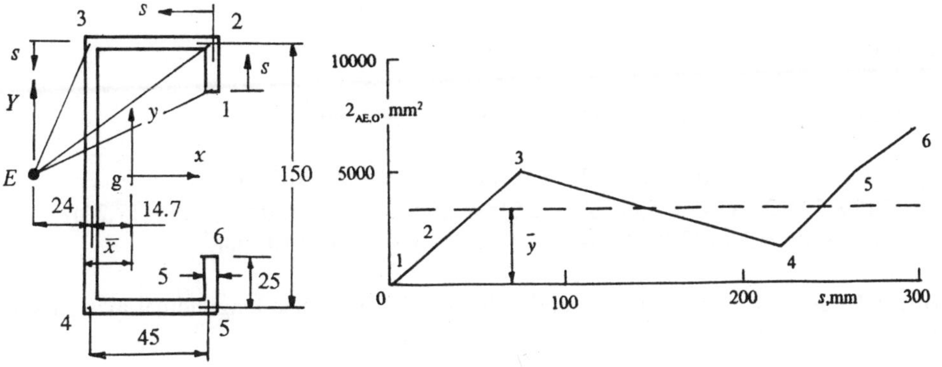

(a) (b)

Figure 9.32 Lipped Section

$$290\bar{y} = (1725 \times 25/2) + (50 \times 1725) + (50 \times 3375 / 2) + (195 \times 1500) + (3600 \times 150/2) + (45 \times 3375 / 2) + (25 \times 4875) + (1727 \times 25/2)$$

$$\bar{y} = 3358.84 \text{ mm}^2$$

In eq(5.84b) the first integral is evaluated from the line segments in Fig. 9.32b as:

$$\int_0^s y^2 \, ds = \int_0^{25} (69\, s)^2 \, ds + \int_{25}^{70} (75\, s - 150)^2 \, ds + \int_{70}^{220} (-24\, s + 6780)^2 \, ds$$
$$+ \int_{220}^{265} (75\, s - 15000)^2 \, ds + \int_{265}^{290} (69\, s - 13410)^2 \, ds$$
$$= 37.17 \times 10^8 \text{ mm}^5$$

Substituting into eq(5.84b) gives the primary warping constant

$$\Gamma_1 = 5[37.173 - 290(0.3358)^2] \times 10^8 = 22.28 \times 10^8 \text{ mm}^6$$

The secondary warping constant for this section follows from eq(5.94) as

$$\Gamma_2 = (1/12) \int R_n^{\,2} t^3 \, ds$$

where R_n is the perpendicular distance from the shear centre, E to the median line 1234 at s. This gives

$$\Gamma_2 = (5^3/12)[\int_0^{25} (s + 50)^2 \, ds + \int_0^{45} (69 - s)^2 \, ds + \int_0^{150} (75 - s)^2 \, ds + \int_0^{45} (24 + s)^2 \, ds\,]$$
$$= 32552.1 \text{ mm}^5$$

Comparing Γ_1 and Γ_2 shows that Γ_2 may be neglected and therefore $\Gamma_E = \Gamma_1$.

Further properties of the area required for the buckling calculations are (i) the position of the centroid ($\bar{x}$ from the vertical edge):

$$\bar{x} = \frac{(145 \times 5 \times 2.5) + 2\,(50 \times 5) + 2\,(22.5 \times 5)\,47.5}{(145 \times 5) + 2\,(50 \times 5) + 2\,(22.5 \times 5)} = 17.24 \text{ mm}$$

$$X_g = 17.24 + 24 - 2.5 = 38.74 \text{ mm}$$

(ii) the principal second moments of area about the centroid

$$I_x = 50 \times 155^3/12 - 40 \times 145^3/12 - 5 \times 100^3/12 = 4.937 \times 10^6 \text{ mm}^4$$
$$I_y = 145 \times 5^3/12 + (145 \times 5)(17.24 - 2.5)^2 + 2[5 \times 50^3/12 + (50 \times 5)\,(25 - 17.24)^2]$$
$$+ 2[22.5 \times 5^3/12 + (22.5 \times 5)\,(47.5 - 17.24)^2] = 0.49798 \times 10^6 \text{ mm}^4$$

(iii) polar second moment of area and radius of gyration for a longitudinal axis passing through shear centre

Transfer I_y to a parallel axis Y through E, distance X_g apart,

$$I_Y = I_y + AX_g^{\,2} = (0.4998 \times 10^6) + (1450 \times 38.74^2) = 2.658 \times 10^6 \text{ mm}^4$$

and from perpendicular axes,

$$I_p = I_x + I_Y = (4.937 + 2.658) \times 10^6 = 7.595 \times 10^6 \text{ mm}^4$$
$$k_p = \sqrt{(I_p/A)} = \sqrt{(7.595 \times 10^6/\,1450)} = 72.4 \text{ mm}$$

(iv) the St Venant torsion constant:

$$J = (1/3)\textstyle\sum b\,t^3 = 290 \times 5^3/3 = 12083.3 \text{ mm}^4$$

Assuming purely flexural buckling about the x - axis, eq(9.62b) gives

$$\sigma_y = (\pi^2 \times 0.497 \times 10^6 \times 117 \times 10^3)\,/\,(2000^2 \times 1450) = 98.95 \text{ MPa}$$

For a torsion-flexure failure eqs(9.62a) and (9.63) give the respective stress components

$$\sigma_x = (\pi^2 \times 4.93 \times 10^6 \times 117 \times 10^3)/(2000^2 \times 1450) = 981.5 \text{ MPa}$$
$$\sigma_p = (43.3{\times}10^3 \times 2083.3)/(7.6{\times}10^6) + (\pi^2 \times 22.28{\times}\,10^8 \times 117{\times}\,10^3)/(7.6 \times 10^6 \times 2000)^2$$
$$= 68.84 + 84.62 = 153.46 \text{ MPa}$$

Strictly, a correction is required for the plasticity that occurs at the higher stress level. This will also apply where all stress components in shorter struts are found to exceed the yield stress. The usual method of correction is to replace E in eqs(9.62a,b) with the tangent modulus. This procedure requires a description suitable to the stress-strain curve, e.g. eq(9.39), as with the Engesser method. Applying eq(9.64), the combined buckling stress becomes the lesser root to the quadratic:

$$(1 - 981.5/\sigma)(1 - 153.46/\sigma) = (38.7\,/\,72.4)^2$$
$$\sigma^2 - 1589.58\sigma + (210.95 \times 10^3) = 0 \quad \Rightarrow \quad \sigma = 146.15 \text{ MPa}$$

The lesser of the two stresses within each mode indicates that this strut would buckle by flexure about its x - axis.

9.6.3 Enforced Axis of Rotation

If the strut is constrained to rotate about a longitudinal axis displaced from the shear centre we must refer Γ_1 and I_p in eq(9.63) to this axis of twist [9]. The secondary warping constant is unaffected. Let O be the axis of enforced rotation with co-ordinates (X_o, Y_o) relative to E. The theorem for parallel axes passing through E and O leads to

$$(\Gamma_1)_o = \Gamma_1 + X_o^2 I_x + Y_o^2 I_y - 2 X_o Y_o I_{xy} \tag{9.66a}$$

$$(I_p)_o = I_p + A\,(X_o^2 + Y_o^2 - 2 X_o X_g - 2 Y_o Y_g) \tag{9.66b}$$

Equations (9.66a,b) are referred to any pair of perpendicular axes, x and y, passing through the centroid G. When these become aligned with the principal axes then $I_{xy} = 0$. This restraint imposes a purely torsional buckling mode in which the critical stress becomes

$$\sigma_p = GJ / (I_p)_o + \pi^2 E\, \Gamma_o / (I_p)_o L^2$$

where $\Gamma_o = (\Gamma_1)_o + \Gamma_2$

References

[1] Capey, E.C., unpublished *RAe Note* (ESDU01.01.23, June 1960).
[2] Waters, H., *J Roy. Aero Soc.*, **66**, (1962), 724-726.
[3] *The buckling of a composite strut*, ESDU Note 01.01.23, March 1969.
[4] BS 449, 1969, Pt 2. *Allowable compressive loads in structural steel columns.*
[5] Hollomon, J., *Trans AIME*, **162**, (1945), 268.
[6] *Buckling of flat isotropic plates under uniaxial and biaxial loading*, ESDU Note72019, August 1972.
[7] *Buckling of flat plates in shear*, ESDU Note 71005, February 1971.
[8] Wagner, H., *NACA Tech Memo* 807, 1936.
[9] Goodier, J.N., *Jl Appl Mech.* **9** (1942), A103-A107.
[10] *Flexural and torsional instability of thin-walled open section struts*, ESDU Note 89007, March 1989.

Bibliography

Budiansky, B. and Connor, R.W., *NACA Tech Note*, 1559, 1947.
Bulson, P.S., *The stability of flat plates*, Chatto and Windus, 1970
Cook, I.T. and Rockey, K.C., *Aero Q*, **15**(4) (1963), 349-356.
Gerard, G., *J.Appl Mech.* **15**(1) (1948), 7-12.
Stein, M. and Neff, J., *NACA Tech Note* 1222, 1946.
Stowell, E.Z., *NACA Report* 898, 1948.
Timoshenko, S., *Theory of Elastic Stability*, McGraw-Hill, 1961.
Vinson, J.R., *Structural Mechanics*, Wiley, 1974.

EXERCISES

Euler Theory

9.1 A strut 2 m long with inner and outer diameters 44 and 50 mm respectively has pinned ends. Determine the Euler critical load and the shortest length for which this theory is valid. Take $E = 207$ GPa and $\sigma_c = 310$ MPa.

9.2 Determine the Euler critical load for the strut in exercise 9.1 when the ends are fixed. If the tube material has a compressive yield stress of 310 MPa, find the shortest length of this tube for which the theory applies. Take $E = 207$ GPa.

9.3 If the cross-section given in Fig. 9.33 is to be used as a strut with one end pinned and the other end fixed, calculate the buckling load based upon the minimum permissible length of an Euler strut. Take $E = 207$ GPa and $\sigma_c = 310$ MPa for the strut material.
Answer: 809.1 mm, 279 kN

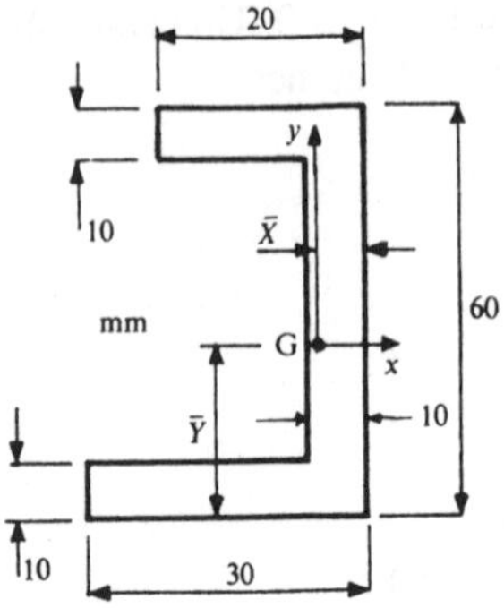

Figure 9.33

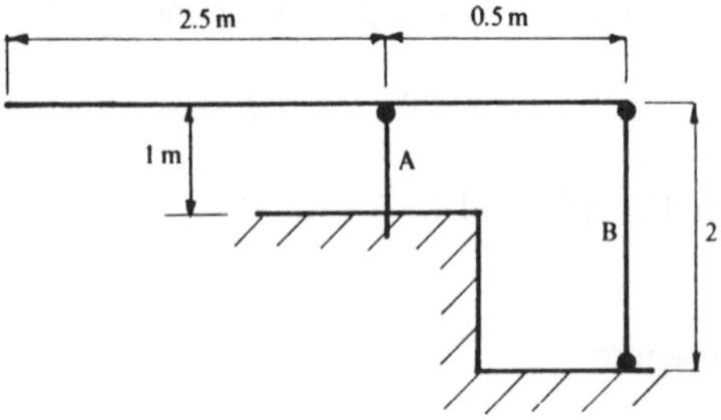

Figure 9. 34

9.4 Figure 9.34 shows a horizontal beam of mass 15 kg/m supported by two vertical members, A and B. Member A is pin-jointed to the beam and built in at its lower end. Member B is pin-jointed to both the beam and the lower anchor point. The least second moment of area of A is 420×10^3 mm^4 and that of B is 200×10^3 mm^4. Determine the maximum value of a moveable vertical point force carried by the beam if collapse of the structure is to be avoided. Take $E = 200$ GPa. (CEI)

9.5 A tubular strut 75 mm o.d. and 65 mm i.d. is 3m long. Calculate the maximum stress and deflection when the strut carries a compressive force of 50 kN offset by 3mm from its longitudinal centroidal axis.

9.6 A 3 m long strut with tubular section 50 mm o.d. and 25 mm i.d. is loaded eccentrically at its free end with a compressive force at a radial distance of 75 mm from the centroidal axis. Find the maximum deflection and the safe axial load when (i) the maximum compressive stress is limited to 35 MPa and (ii) the net stress on the tensile side is to be zero at the fixed end. Take $E = 210$ GPa.

9.7 The ends of a thin-walled, elliptical, light alloy tube are fixed in the plane of its major 60 mm diameter and pinned in the plane of its minor 20 mm diameter. Given that the slenderness ratio is 100, determine the wall thickness needed to support an axial compressive load of 50 kN. Take $E = 70$ GPa.

9.8 The vertical column of length L in Fig. 9.35 is pinned at the top and bottom. It consists of a lower part of length nL which may be regarded as rigid and this is fixed to an upper part which is slender and of stiffness EI. Show that at the instability point $\tan(1 - n)kL = -nkL$, and determine the buckling load for $n = ½$. (CEI)

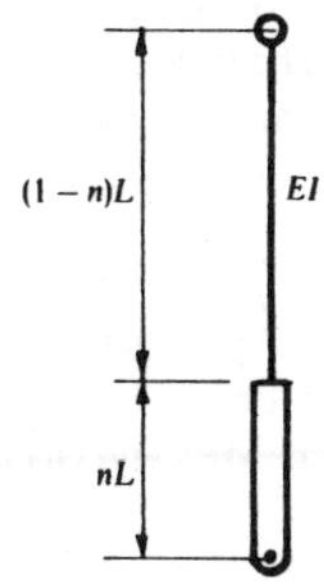

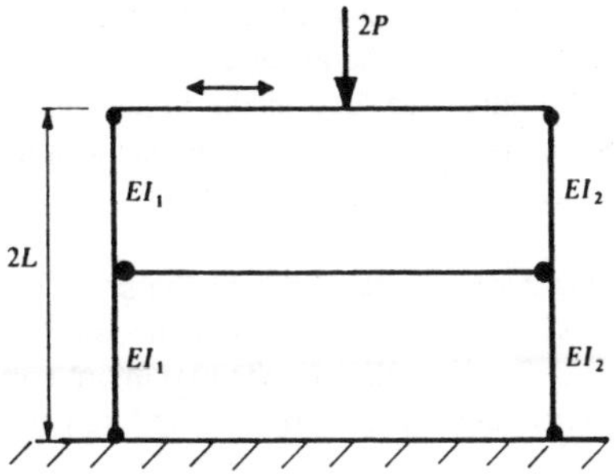

Figure 9.35

Figure 9.36

9.9 Figure 9.36 shows two vertical struts, each of length $2L$, joined at their mid-point by a rigid horizontal connection. The struts have flexural rigidity of EI_1 and EI_2 about axes normal to the plane of the figure and are constructed from the same material. The struts are pin- jointed at one end to a rigid horizontal beam which slides in a frictionless guide, while the opposite ends are pin-jointed to the ground. Show that the condition for simultaneous buckling of both struts in the plane of the figure, when a central compressive force $2P$ is applied at mid-spad of the beam, is given by the expression $\alpha_1 \tan \alpha_2 L + \alpha_2 \tan \alpha_1 L = 2\alpha_1 \alpha_2 L$, where $\alpha_1{}^2 = P/(EI_1)$ and $\alpha_2{}^2 = P(/EI_2)$. (CEI)

9.10 A straight strut of length L is fixed at one end and pinned at the other. If, in calculating the buckling load, it is assumed that the effective length is $2L/3$, what is the % error incurred?
Answer: 12.5%

9.11 A straight strut of length L, rigidly built in at one end and free at the other, is subjected to an axial compressive force P, and a side force S, perpendicular to P, both applied at the free end. Show that the free end deflection is given by $v_o = (S/P)\,[(1/\alpha)\tan \alpha L - L]$ where $\alpha^2 = P/(EI)$. Hence show that buckling failure occurs when the load $P \rightarrow \pi^2 EI/(4L^2)$. (CEI)

9.12 A strut of length L has each end fixed in an elastic material which can exert a restraining moment μ/rad. Prove that the critical load P is given by the following expression: $P + \mu\sqrt{[P/(EI)]} \times \tan\{(L/2)\sqrt{[P/(EI)]}\} = 0$. The designed buckling load of a 1 m long strut, assuming the ends to be rigidly fixed, was 2.5 kN. If during service, the ends were found to rotate, with each mounting exerting a restraining moment of 10 kNm/rad, show that the buckling load decreases by 20%. (CEI)

Imperfect Euler Struts

9.13 A tubular steel strut with 65 mm and 50 mm outer and inner diameters respectively is 1 m long and fixed at one end. Find the maximum permissible eccentricity for a 130 kN compressive force when applied at the free end if the allowable stress for the strut material is 325 MPa. If the eccentricity is to be eliminated find (i) the length of strut that could support this force and (ii) the force that could be applied to a length of 1 m.

9.14 A tubular section strut with 75 mm outer and 65 mm inner diameters is 3m long. Calculate the maximum stress and deflection when the strut carries a compressive force of 50 kN offset by 3 m from its longitudinal axis. Take $E = 200$ GPa.

9.15 The axis of a 3 m upright column, built in at one end and free at the other deviates slightly from the vertical. The cross-section is equal angle: 100 mm side length × 10 mm thickness. When the strut supports a vertical compressive load of 20 kN at its free end, a maximum eccentricity of 20 mm from the centroid is allowable. What is the greatest angular deviation of the strut to the vertical axis if the maximum stress may not exceed 80 MPa? Take $E = 207$ GPa.

9.16 A vertical, 3 m long steel tube whose external and internal diameters are 76.2 mm and 63.5 mm respectively is fixed at the lower end and completely unrestrained at the top. The tube is subjected to a vertical compressive load at the upper end and eccentric by 6.35 mm from the centroidal axis. Determine the value of the instability load which gives a point of zero resultant stress at the base of the tube. Take $E = 207$ GPa. (CEI)

9.17 A tubular strut 50 mm o.d. and 25 mm i.d. is 3 m long. It is loaded eccentrically at its free end with a compressive force at a radial distance of 75 mm from the axis. Determine the safe load and the maximum deflection when (i) the maximum compressive stress is limited to 35 MPa and (ii) there is to be zero resultant tensile stress at the fixed end. Take $E = 210$ GPa.

9.18 A long vertical slender strut of uniform solid circular section is of length L and diameter d. The strut is rigidly built in at the base and unrestrained at the top. Derive an expression for the free-end deflection of the strut when a vertical compressive force P is applied there at a point on the circumference. Hence determine the vertical force required to cause a horizontal end deflection of magnitude d in terms of the Euler buckling load P_E for the same strut. (CEI)

9.19 A long thin uniform bar of length L is lifted by 45° wires attached to its ends and leading to a crane hook. If the mass of the bar is w/unit length and the ends are assumed to be pin-jointed, show that the mid-span suspended deflection is $[4EI/(wL^2)][\sec(\alpha L/2) - 1] - L/4$ where $\alpha = \sqrt{[wL/(2EI)]}$. Hence derive the mid-span bending moment. (CEI)

9.20 A pinned-end strut 2 m long and 30 mm diameter supports an axial compressive force of 20 kN. If the maximum compressive stress is 200 MPa, calculate the additional load that may be applied laterally at mid-span and the maximum deflection and moment. Take $E = 207$ GPa.

9.21 A tie bar supports a total load of 8 kN uniformly distributed along its 4 m length together with an axial tensile load of 40 kN. The cross-section is symmetrical with $I = 10 \times 10^6$ mm^4 and area $A = 10 \times 10^3$ mm^2 for an overall depth of 200 mm. Determine the maximum bending moment and the stress in the cross-section. Take $E = 100$ GPa.

9.22 A 3 m long strut of section depth of 100 mm with $I = 10 \times 10^6$ mm^4 and $A = 3000$ mm^2 carries a distributed loading varying linearly from zero at one end to a maximum of 1.5 kN at the other end. What magnitude of additional axial compressive thrust would cause the strut to buckle between pinned ends under a compressive yield stress of 280 MPa? (Take $E = 210$ GPa)

9.23 Find the maximum bending moment and the maximum compressive and tensile stresses for the strut in Fig. 9.7 when $P = 125$ kN, $w = 5$ kN/m and $L = 2.5$ m. The I-section has flange dimensions 50 × 20 mm and the web measures 60 × 15 mm. Take $E = 210$ GPa.

9.24 A uniform horizontal strut with fixed ends carries a load that varies linearly from zero at one end to a maximum at the other. Show that the governing equation is $\mathrm{d}^2M/\mathrm{d}z^2 + [P/(EI)]M = z/2$ and hence solve to find the fixing moments, knowing that $F = \mathrm{d}M/\mathrm{d}z$ at the fixed ends where the shear forces F are known.

9.25 Find P for a 2 m long strut in Fig. 9.9 that would just produce yielding in the mid-span 25 mm diameter circular section given a 5 mm maximum amplitude of initial curvature and a compressive yield stress of 300 MPa. What additional central deflection occurs under this load? $E = 210$ GPa.

Semi-Empirical Methods

9.26 A tubular steel strut 50 mm outer diameter and 3.2 mm thick is pinned at its ends. Find the length below which the Euler theory ceases to apply given a yield stress of 324 MPa. Compare the Euler and Rankine-Gordon buckling loads for a 1.83 m length of this strut. Take $E = 207$ GPa, $a = 7500^{-1}$.

9.27 A steel column of solid circular section is 1.83 m long. If this supports a compressive load of 50 kN with pinned ends, find the necessary strut diameter from the Euler theory using a safety factor of 6. What then is the Rankine buckling load with $\sigma_c = 324$ MPa and $a = 7500^{-1}$. Take $E = 201$ GPa.

9.28 Compare predicted values of the Euler and Rankine-Gordon buckling loads for a 5 m length mild steel strut for the cross-section in Fig. 9.37. Show that the Euler value is invalid. Take $E = 210$ GPa and the Rankine constants from Table 8.2.

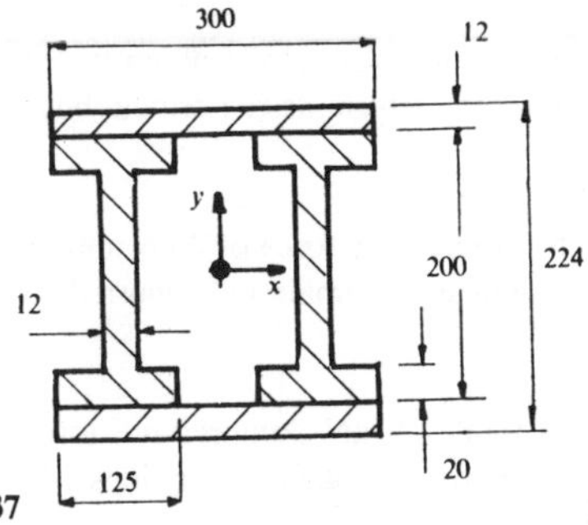

Figure 9.37

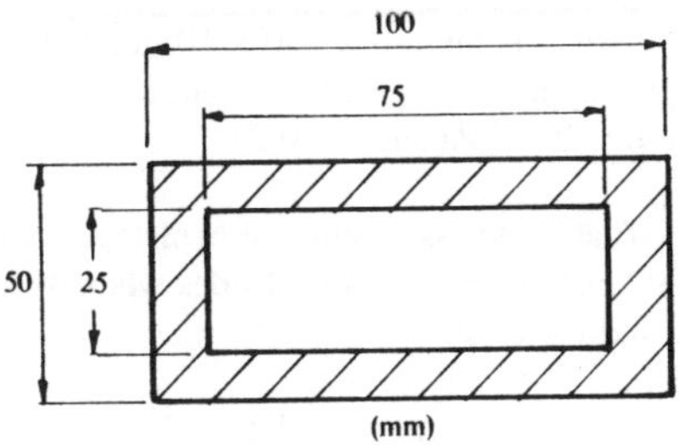

Figure 9.38

9.29 A pinned-end strut 1.5 m long has the cross-section shown in Fig. 9.38. Find the inelastic Engesser buckling load given that the law $\sigma / \sigma_c = 5\varepsilon^{1/3}$ describes the compressive stress-strain curve in the region beyond the yield stress $\sigma_c = 300$ MPa. Compare this with the corresponding Perry-Robertson and Fidler predictions with $E = 207$ GPa.

9.30 A steel pipeline connects with two flanged plates each with a lateral stiffness of 833.3×10^6 N/m in the direction of the strut axis. If, when fluid flows through the pipe under room temperature conditions, the pipe material is unstressed, calculate the maximum allowable temperature rise of the fluid if buckling is to be avoided. Material constants are $E = 200$ GPa, $\alpha = 10 \times 10^{-6}$/ °C, $\sigma_c = 300$ MPa and $a = 7500^{-1}$ (pinned ends). Show that the Euler theory is unrealistic in this case.

9.31 Compare the permissible Euler and Rankine-Gordon compressive buckling loads that can be applied to a tubular steel strut 50 mm outer diameter and 3.2 mm wall thickness when it is pinned at its ends over a length of 3 m. For what length of this strut does the Euler theory cease to apply? Take $\sigma_c = 324$ MPa, $a = 7500^{-1}$ and $E = 207$ GPa.

9.32 The equal-angle section in Fig. 9.2 is used in a mild steel strut 2 m in length with one end fixed and the other end pinned. Calculate the safe load from the Rankine-Gordon theory using a safety factor of 4 and the constants in Table 9.2.

9.33 A short piece of steel tube 38 mm o.d and 32 mm i.d. failed in compression under an axial load of 115 kN. When a 1.15 m length of the same tubing was tested as a fixed-end strut it buckled under a load of 90 kN. Determine the constants employed with the Rankine-Gordon theory and hence determine the buckling load for a 1.85 m length of this tubing when used as a strut with pinned ends. Compare with the Euler prediction where $E = 208$ GPa.

9.34 Given that rectangular sections of steel $b = 2d$ are available with $\sigma_c = 400$ MPa, $a = 9000^{-1}$ to replace fixed ended tubular struts 50 mm outer and 40 mm inner diameters with $\sigma_c = 300$ MPa, $a = 7500^{-1}$ and $L/k = 96$, determine the size of an appropriate rectangular section to the nearest mm.

9.35 A 150 × 200 mm tubular box-section with 10 mm wall thickness is to act as a steel strut of length 6 m with fixed ends. Find the safe axial compressive load using a load factor of 4 based upon the Rankine-Gordon, straight line, parabola and Perry-Robertson formulae. Where appropriate take $E = 207$ GPa, $\sigma_c = 325$ MPa, $a = 7500^{-1}$ and $\mu = 0.003L / k$. Let the straight line and parabola pass through σ_c and $L / k = 120$.

9.36 A brass strut with elliptical thin-walled section is 800 mm long. The major and minor axes are 75 mm and 25 mm respectively and the wall thickness is 2.5 mm. Estimate the plastic buckling load for the minor axis where the ends may be assumed pinned. The stress-strain behaviour of brass is given by $\sigma/\sigma_c = 8\varepsilon^{0.3}$ where $\sigma_c = 150$ MPa is the yield stress.

Buckling of Plates

9.37 A steel plate 400 × 250 × 10 mm is subjected to a compressive stress along its 200 × 10 mm simply supported edges. If the longer edges are unsupported, determine the critical buckling stress. Take $E = 207$ GPa and $\nu = 0.27$.

9.38 What thickness of aluminium plate can withstand a compressive force of 2 kN applied normal to its 100 mm simply supported sides when its longer 300 mm sides remain unsupported? Take $E = 71$ GPa and $\nu = 0.32$.

9.39 A vertical standing plate 1.25 m high carries a total 1.5 kN compressive force uniformly distributed along its 0.5 m breadth. If the plate is simply supported along all four sides what thickness would result in plate buckling? Take $E = 74$ GPa, $\nu = 0.32$.

9.40 Determine the buckling stress for a plate 300 × 200 × 5 mm with all sides fixed when it is loaded in compression normal to its shorter sides. Take $E = 70$ GPa, $\nu = 0.33$.

9.41 If the plate in Exercise 9.40 is simply supported along its shorter sides and free to translate but not to rotate along its longer sides, find the elastic compressive buckling stress given that the rotational restraint is 14×10^3 Nm/rad per metre of the longer sides.

9.42 An aluminium alloy plate 1.5 × 0.5 m with clamped edges supports a compressive load of 650 kN on its shorter sides. Find the necessary thickness that will prevent buckling from occurring. If the allowable compressive stress is 210 MPa, does the calculated thickness ensure an elastic stress state? Take $E = 70$ GPa and $\nu = 0.3$.

9.43 Determine the stress for which a 500 × 250 × 5 mm steel plate buckles under the action of uniform compression along all four sides. Take $E = 207$ GPa and $\nu = 0.27$. How would you employ simple bending theory to determine the critical buckling stress for this plate?

9.44 Determine the elastic compressive stress which, when applied to the shorter sides of a thin plate 390 × 200 × 7.5 mm, will cause it to buckle in the presence of a constant stress of 80 MPa acting (i) in compression and (ii) in tension along its longer sides. Take $E = 72.4$ GPa and $\nu = 0.33$ with all sides simply supported.

9.45 Correct the stresses in exercise 9.44 for plasticity effects using Fig. 9.23. Take $\sigma_n = 405$ MPa and $m = 16$.

9.46 A plate $a \times b \times t$ is subjected to a normal tensile stress $\sigma_y = \sigma$ on its $a \times t$ sides and a compressive stress $\sigma_x = -\sigma$ on its $b \times t$ sides. If all edges are simply supported show that the minimum value of critical buckling stress is given by $(\sigma_x)_{cr} = 8\pi^2 D/(t\,b^2)$ for one half-wave in the y-direction. Show that the values of $r = a/b$ for which this minimum applies are: $r = 1/\surd 3,\ 2/\surd 3$ etc.

9.47 Find the shear stress at which a rectangular plate 200 mm wide × 260 mm long × 4.5 mm thick buckles when the short sides are clamped and the long sides are simply supported. Correct for plasticity given $\sigma_n = 340$ MPa, $m = 16$, $E = 73$ GPa and $\nu = 0.3$.

CHAPTER 10

FINITE ELEMENTS

10.1 The Stiffness Method

This introduction to finite elements (FE) will show how the theory of elasticity, as outlined in Chapter 2, is employed for the sub-division of a body into smaller elements inter-connected throughout at nodal points. In this technique of discretization the initial unstrained shape of the body is described by an assemblage of elements. When the assembly is elastically strained, the behaviour of the entire body may be computed from the known elastic behaviour of its elements. FE provides for equilibrium and compatibility of its internal stress and strain distributions and matches external boundary conditions of known force and displacement.

In the *stiffness or displacement method* of finite elements the displacements at the nodal points are the unknowns. These displacements are solved numerically from their relation to the nodal forces. This relation is established by assembling a stiffnesss matrix which connects vectors of nodal point forces and displacements for the element in question. The strains follow from the strain-displacement relations and, finally, the stresses are found from the constitutive relations. Thus, from an assumed displacement function the stiffness method provides numerical solutions to the stress and strain distributions within a loaded body. Alternatively, the *force or flexibility method* of finite elements enables the displacements to be found from an assumed stress distribution. This method may offer certain advantages in its application to statically indeterminate structures but, is less often used. The reader is referred to a specialist text [1], since the force method will not be considered further here.

The purpose of this chapter is to illustrate how to develop the stiffness matrix for different types of plane elements using a routine procedure. Triangular and rectangular elements are employed with plates and bars in plane stress and strain, torsion of prismatic bars, bodies with axial symmetry and plate flexure. The finite element method is used when solutions are not readily available from the classical theory of elasticity. It should be emphasised, however, that FE methods cannot improve on known stress function solutions. One should always check, therefore, on the availability of a stress function before employing finite elements. The approach adopted here is firstly to derive an element stiffness matrix $\mathbf{K}^e$ connecting nodal point force and displacement vectors $\mathbf{f}^e$ and $\boldsymbol{\delta}^e$ respectively, as follows:

$$\mathbf{f}^e = \mathbf{K}^e\,\boldsymbol{\delta}^e \tag{10.1a}$$

Then, in selected examples, it is shown how the overall stiffness matrix $\mathbf{K}$ is assembled for a given mesh.

$$\mathbf{f} = \mathbf{K}\,\boldsymbol{\delta} \tag{10.1b}$$

Bold upper-case Roman letters **A, B, C, K**, denote matrices and bold lower case Roman and Greek symbols $\mathbf{f}$, $\boldsymbol{\alpha}$, $\boldsymbol{\varepsilon}$, $\boldsymbol{\delta}$, $\boldsymbol{\sigma}$, denote column matrices or column vectors. The vectors take a physical meaning as with force $\mathbf{f}$ and displacement $\boldsymbol{\delta}$. We begin with a summary of simple bar elements where it is possible to write down the matrix $\mathbf{K}^e$ in eq(10.1a) by inspection.

10.2 Bar Elements

10.2.1 Uniaxial Stress

This simplest loading of a uniform bar element is either an axial tensile or compressive force. The element is particularly useful for the analysis of the forces and displacements in the ties and struts of framed structures. Consider a bar element of length, L, with uniform cross-sectional area, A, connecting nodal points 1 and 2, in Fig. 10.1.

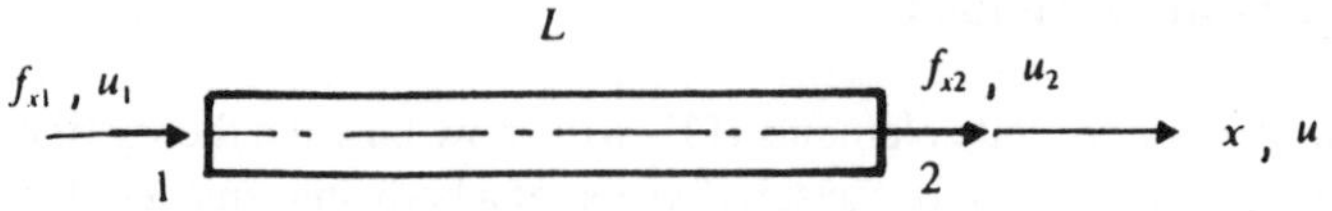

Figure 10.1 Bar element

Let the axis of the bar be aligned with the x-direction. Nodal forces f_{x1} at node 1 and f_{x2} at node 2 act in the direction x positive. These produce positive displacements u_1 and u_2. In this case $\mathbf{K}^e$ is obvious from inspection since $f = (AE/L)u$. This reveals that the form of eq(10.1a) is simply

$$\begin{bmatrix} f_{x_1} \\ f_{x_2} \end{bmatrix} = \frac{AE}{L}\begin{bmatrix} 1 & -1 \\ -1 & 1 \end{bmatrix}\begin{bmatrix} u_1 \\ u_2 \end{bmatrix} \tag{10.2}$$

where $\mathbf{f}^e = [\,f_{x1}\ f_{x2}\,]^T$ and $\delta^e = [\,u_1\ u_2\,]^T$. The element stiffness matrix, $\mathbf{K}^e$, is symmetrical. It is 2×2 because each of the two nodes has a single degree of freedom. The stresses and strains follow from $\sigma = Eu/L$ and $\varepsilon = \sigma/E$.

10.2.2 Torsion

Consider a uniform circular shaft element of length L subjected to nodal torques t_1 and t_2 as shown in Fig. 10.6. The element nodes 1 and 2 twist by amounts θ_1 and θ_2 relative to the undeformed bar.

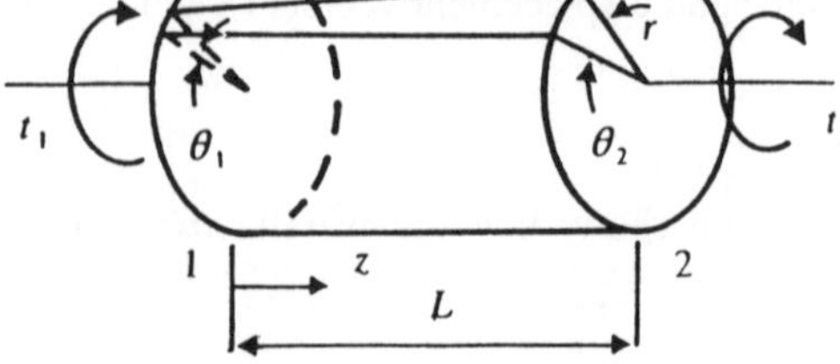

Figure 10.2 Shaft element under torsion

Torsion theory (eq(5.5)) supplies directly the components of the torsional stiffness matrix as

$$\delta t / \delta\theta = JG/L \tag{10.3a}$$

where $J = \pi d^4/32$ is the polar second moment of area and G is the rigidity modulus. We may deduce that the element stiffness matrix is 2 × 2 since there are two nodes with a single degree of freedom at each node, similar to that derived previously for the bar element under tension. For a single element, $\mathbf{K}^e$ follows directly from eq(10.3a) as

$$\begin{bmatrix} t_1 \\ t_2 \end{bmatrix} = \frac{JG}{L}\begin{bmatrix} 1 & -1 \\ -1 & 1 \end{bmatrix}\begin{bmatrix} \theta_1 \\ \theta_2 \end{bmatrix} \qquad (10.3b)$$

where in eq(10.1a) we identify $\mathbf{f}^e = [\, t_1 \;\; t_2\,]^T$ and $\boldsymbol{\delta}^e = [\theta_1 \;\; \theta_2]^T$. Once the nodal twists have been found, the shear strain $\boldsymbol{\gamma}^e = [\,\gamma_1 \;\; \gamma_2]^T$ and stress $\boldsymbol{\tau}^e = [\tau_1 \;\; \tau_2]^T$ will follow from $\gamma = \tau/G = r\theta/L$ as

$\boldsymbol{\gamma}^e = (r/L)\, \boldsymbol{\delta}^e \quad \boldsymbol{\tau}^e = (Gr/L)\, \boldsymbol{\delta}^e$

10.2.3 Beam Bending

A beam element (see Fig. 10.3a) is used to determine the displacements, strains and stresses under bending. At nodes 1 and 2 a shear force, q, and a moment, m, act to produce a deflection, v, and a rotation, θ, for the positive directions shown.

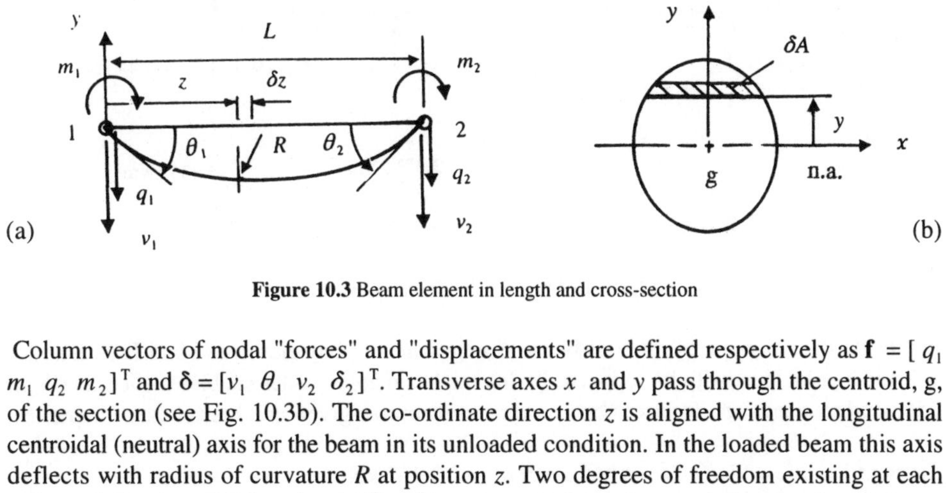

Figure 10.3 Beam element in length and cross-section

Column vectors of nodal "forces" and "displacements" are defined respectively as $\mathbf{f} = [\, q_1 \;\; m_1 \;\; q_2 \;\; m_2\,]^T$ and $\boldsymbol{\delta} = [v_1 \;\; \theta_1 \;\; v_2 \;\; \delta_2\,]^T$. Transverse axes x and y pass through the centroid, g, of the section (see Fig. 10.3b). The co-ordinate direction z is aligned with the longitudinal centroidal (neutral) axis for the beam in its unloaded condition. In the loaded beam this axis deflects with radius of curvature R at position z. Two degrees of freedom existing at each node and therefore $\mathbf{K}^e$ is a 4 × 4 ($2^2 \times 2^2$) symmetrical matrix connecting $\mathbf{f}$ to $\boldsymbol{\delta}$ as follows:

$$\begin{bmatrix} q_1 \\ m_1 \\ q_2 \\ m_2 \end{bmatrix} = \begin{bmatrix} K_{11} & K_{12} & K_{13} & K_{14} \\ K_{21} & K_{22} & K_{23} & K_{24} \\ K_{31} & K_{32} & K_{33} & K_{34} \\ K_{41} & K_{42} & K_{43} & K_{44} \end{bmatrix}\begin{bmatrix} v_1 \\ \theta_1 \\ v_2 \\ \theta_2 \end{bmatrix} \qquad (10.4a)$$

where the elements of the stiffness matrix K_{ij} can be shown to be [2]

$$\mathbf{K}^e = \frac{EI}{L^3}\begin{bmatrix} 12 & 6L & -12 & 6L \\ 6L & 4L^2 & -6L & 2L^2 \\ -12 & -6L & 12 & -6L \\ 6L & 2L^2 & -6L & 4L^2 \end{bmatrix} \qquad (10.4b)$$

Once the nodal displacements have been found, the bending stress and strain at each node follow from bending theory (see eq(4.3)).

Example 10.1 Derive the moment and shear force diagrams for the beam in Fig. 10.4a using a single element for each bay. Check from moment distribution.

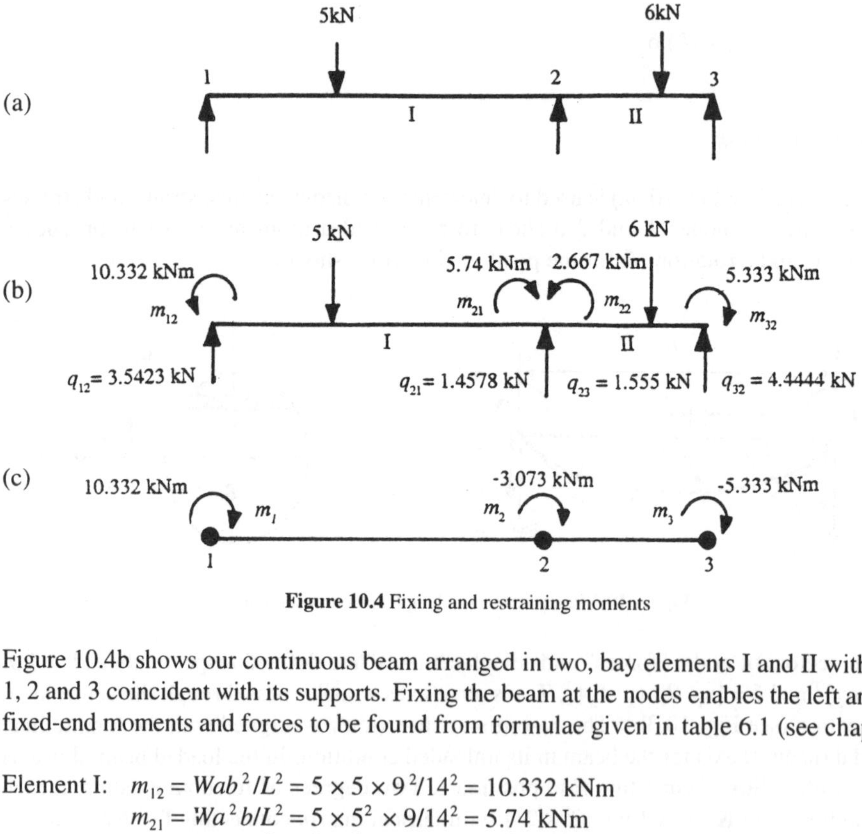

Figure 10.4 Fixing and restraining moments

Figure 10.4b shows our continuous beam arranged in two, bay elements I and II with nodes 1, 2 and 3 coincident with its supports. Fixing the beam at the nodes enables the left and right fixed-end moments and forces to be found from formulae given in table 6.1 (see chapter 6):

Element I: $m_{12} = Wab^2/L^2 = 5 \times 5 \times 9^2/14^2 = 10.332$ kNm
$m_{21} = Wa^2b/L^2 = 5 \times 5^2 \times 9/14^2 = 5.74$ kNm
$q_{12} = W(3ab^2 + b^3)/L^3 = 5(3 \times 5 \times 9^2 + 9^3/14^3 = 3.542$ kN
$q_{21} = W(3a^2b + a^3)/L^3 = 5(3 \times 5^2 \times 9 + 5^3/14^3 = 1.458$ kN

Element II: $m_{23} = 6 \times 4 \times 2^2/6^2 = 2.667$ kNm
$m_{32} = 6 \times 4^2 \times 2/6^2 = 5.333$ kNm
$q_{23} = 6(3 \times 4 \times 2^2 + 2^3)/6^3 = 1.556$ kN
$q_{32} = 6(3 \times 4^2 \times 2 + 4^3)/6^3 = 4.444$ kN

These moments act in the sense shown in Fig. 10.4b to produce hogging curvature. When a finite element analysis is made of the nodal moments m_1, m_2 and m_3, they must act in opposition to release the beam at these nodes (see Fig.10.4c). That is, from eq(10.4b), for element I,

$K_{11}{}^{\mathrm{I}} = 12E(2I)/14^3 = 0.00875EI$
$K_{12}{}^{\mathrm{I}} = 6E(2I)/14^2 = 0.0612EI$
$K_{22}{}^{\mathrm{I}} = 4E(2I)/14 = 0.5714EI$
$K_{24}{}^{\mathrm{I}} = 2E(2I)/14 = 0.286EI$

and, for element II,

$K_{11}{}^{\mathrm{II}} = 12EI/6^3 = 0.0556EI$
$K_{12}{}^{\mathrm{II}} = 6EI/6^2 = 0.1667EI$
$K_{22}{}^{\mathrm{II}} = 4EI/6 = 0.667EI$
$K_{24}{}^{\mathrm{II}} = 2EI/6 = 0.333EI$

Thus the stiffness matrices for these two elements are

$$\mathbf{K}^{\mathrm{I}} = \frac{EI}{10^{-3}}\begin{bmatrix} 8.75 & 61.2 & -8.75 & 61.2 \\ 61.2 & 571.4 & -61.2 & 286 \\ -8.75 & -61.2 & 8.75 & -61.2 \\ 61.2 & 286 & -61.2 & 571.4 \end{bmatrix}, \quad \mathbf{K}^{\mathrm{II}} = \frac{EI}{10^{-3}}\begin{bmatrix} 55.6 & 167 & -55.6 & 167 \\ 167 & 667 & -167 & 333 \\ -55.6 & -167 & 55.6 & -167 \\ 167 & 333 & -167 & 667 \end{bmatrix} \qquad \text{(i,ii)}$$

Assembling eqs(i,ii) within an overall stiffness matrix, **K**, leads to

$$\begin{bmatrix} q_1 \\ m_1 \\ q_2 \\ m_2 \\ q_3 \\ m_3 \end{bmatrix} = \frac{EI}{10^{-3}}\begin{bmatrix} 8.75 & 61.2 & -8.75 & 61.2 & 0 & 0 \\ 61.2 & 571.4 & -61.2 & 286 & 0 & 0 \\ -8.75 & -61.2 & 64.35 & 105.8 & -55.6 & 167 \\ 61.2 & 286 & 105.8 & 1238.4 & -167 & 333 \\ 0 & 0 & -55.6 & -167 & 55.6 & -167 \\ 0 & 0 & 167 & 333 & -167 & 667 \end{bmatrix}\begin{bmatrix} v_1 \\ \theta_1 \\ v_2 \\ \theta_2 \\ v_3 \\ \theta_3 \end{bmatrix} \qquad \text{(iii)}$$

Since $v_1 = v_2 = v_3 = 0$ in eq(iii), the three rows in m lead to three equations necessary for the solution to θ_1, θ_2 and θ_3. They are

$10.332/(EI) = 0.5714\theta_1 + 0.286\theta_2$
$-3.073/(EI) = 0.286\theta_1 + 1.2384\theta_2 + 0.333\theta_3$
$-5.333/(EI) = 0.333\theta_2 + 0.667\theta_3$

where, conventionally, clockwise moments are positive. The solutions to these equations are $\theta_1 = 21.089/(EI)$, $\theta_2 = -6.0082/(EI)$ and $\theta_3 = -4.996/(EI)$. Recovering the nodal forces and moments from the individual element stiffness matrices in eqs(i) and (ii) gives

$q_1{}^{\mathrm{I}} = 0.923$ kN, $m_1{}^{\mathrm{I}} = 10.332$ kNm, $q_2{}^{\mathrm{I}} = -0.923$ kN, $m_2{}^{\mathrm{I}} = 2.598$ kNm, (iv)
$q_2{}^{\mathrm{II}} = -1.8377$ kN, $m_2{}^{\mathrm{II}} = -5.671$ kNm, $q_3{}^{\mathrm{II}} = 1.8377$ kN, $m_3{}^{\mathrm{II}} = -5.333$ kNm,

The moments in eq(iv) are employed in Fig. 10.5a - c for the construction of the net fixing moment diagram (c).

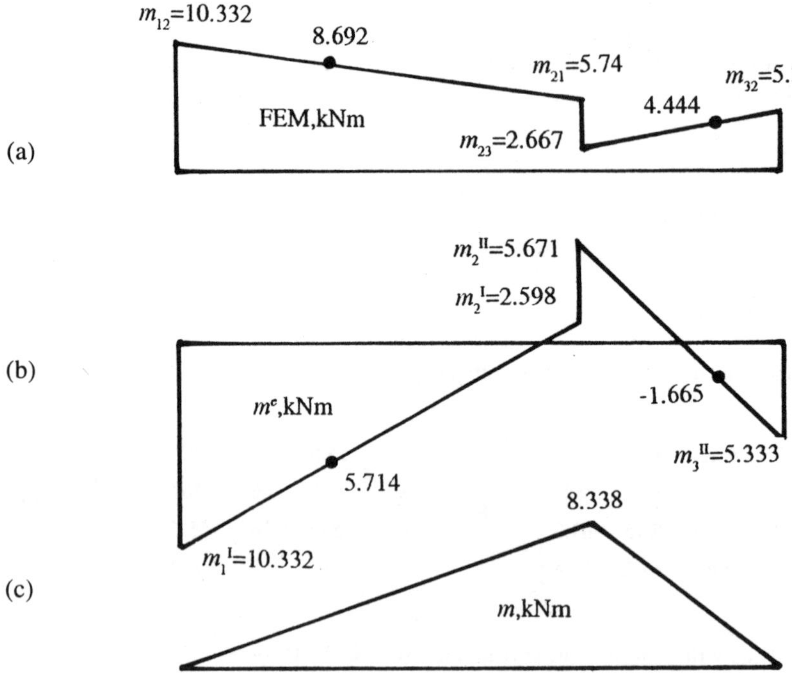

Figure 10.5 Construction of the fixing moment diagram

Figures 10.5a and b show positive hogging moments and negative sagging moments. Adding the fixed-end moment diagram in (a) to the restraining moment diagram in (b) gives the net fixing moment diagram in (c). The latter shows zero moments at each end and a moment of 8.338 kNm (5.74 + 2.598 = 2.667 + 5.671) at the inner support. Next, superimpose upon Fig. 10.5c the free-moment diagrams that apply to each beam element when taken in isolation. The ordinates within the shaded areas in Fig. 10.6a give the net moments. These ordinates are shown separately to a common datum within the m_{net} diagram in Fig. 10.6b. The net forces are found from eqs(iv) and Fig. 10.4b as follows:

$$q_1 = q_{12} + q_1^{\,I} = -\ 3.5423 + 0.923 = -\ 2.6193 \text{ kN} \uparrow$$

$$q_2 = q_{21} + q_2^{\,I} + q_{23} + q_2^{\,II} = -\ 1.4578 = 0.923 - 1.5555 - 1.8377 = -\ 5.774 \text{ kN} \uparrow$$

$$q_3 = q_{32} + q_3^{\,II} = -\ 4.4444 + 1.8377 = -\ 2.6067 \text{ kN} \uparrow$$

These may be checked from the gradients to the m_{net} diagram (Fig. 10.6b) since the shear force diagram (Fig. 10.6c) is the derivative $q = \mathrm{d}m_{net}/\mathrm{d}z$. In the case of uniformly distributed loading w/unit length, the fixed-end moments and reactions are $wl^2/12$ and $wl/2$ respectively. If a concentrated force also acts in the span then superposition may be used to give the net fixing values. This will always apply where the elements are the bays of a continuous beam. Where the nodes do not coincide with the supports it is necessary to lump distributed loading into equal concentrated forces at each node (see Example 10.4).

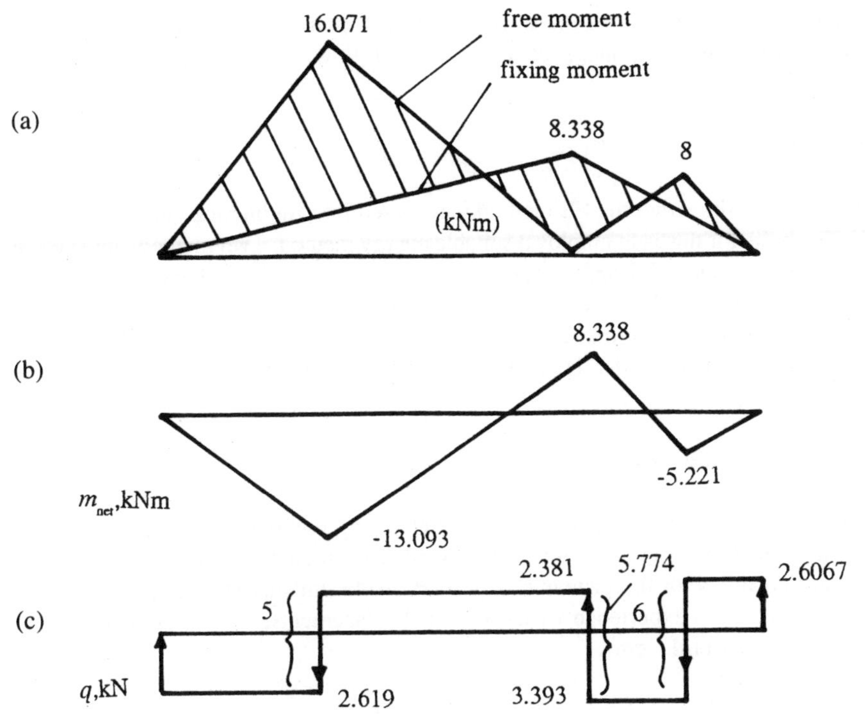

Figure 10.6 Bending moment and shear force diagrams

The moment distribution method provides a rapid check on the FE solution for our continuous beam. This method employs the fixed-end moments previously calculated together with the following distribution factors $m = (I/L)/\sum(I/L)$ for each bay. They are

$$m_{21}/m = (2I/14)/[(2I/14) + (I/6)] = 0.461$$
$$m_{23}/m = (I/6)/[(2I/14) + (I/6)] = 0.539$$

Moment distributions are applied to this beam, as shown in Table 10.1, where clockwise fixed-end moments are taken to be positive. The moments carried over to node 2 retain the sense of, and equal, half the magnitude of the release moments. The latter are injected at nodes 1 and 3 to ensure zero net moments at these points. The resulting net moment at node 2 confirms the corresponding value shown in Fig. 10.5c.

Table 10.1 Moment distribution for beam in Fig. 10.4a

Node	1	2		3
Distribution Factors		0.461	0.539	
Fixed-end Moments	- 10.332	5.740	- 2.667	5.333
Release at A and C	10.332			- 5.333
Carry over		5.166	- 2.666	
Modified Moments	0	10.906	- 5.333	0
Unbalanced Moment		5.573		
Distribute		- 2.569	- 3.004	
Net Moments	0	8.337	- 8.337	0

Up to now we have merely stated the form of stiffness matrices in eqs(10.2) - (10.4). A detailed derivation of the components of $\mathbf{K}^e$ will now follow for more complex elements.

10.3 Energy Methods

Equations (10.1a,b) show that the FE method is resolved into finding the element stiffness matrix $\mathbf{K}^e$. The basis for this is to employ a suitable energy method. Two methods most often employed are the principal of virtual work (PVW) and stationary potential energy (SPE) [2]. It will be seen that either method reveals that $\mathbf{K}^e$ can be expressed in terms of the product of matrices in the nodal point co-ordinates and the elastic constants for the material. The choice of method will depend upon which is the more convenient to apply.

10.3.1 Principle of Virtual Work

The stiffness FE method employs a virtual displacement with real loads applied to the nodes of an element. Virtual forces are used with the flexibility FE method. Recall from Chapter 8 that we identified this with the principle of virtual displacements. Here a system of real forces, f_k (k = 1, 2, 3 ...) experience virtual in-line displacements, $\delta_k{}^v$. Thus we can write eq(8.26a) in its most useful forms:

$$\mathbf{f}^{eT}\,\boldsymbol{\delta}^{ev} - \int \boldsymbol{\sigma}^{T}\,\boldsymbol{\varepsilon}^{v}\,\mathrm{d}V = 0 \tag{10.5a}$$

$$\boldsymbol{\delta}^{evT}\,\mathbf{f}^{e} - \int \boldsymbol{\varepsilon}^{vT}\,\boldsymbol{\sigma}\,\mathrm{d}V = 0 \tag{10.5b}$$

In eqs(10.5a,b) **f**, $\boldsymbol{\delta}$, $\boldsymbol{\sigma}$ and $\boldsymbol{\varepsilon}$ are all column matrices. The reversal in the order of their multiplication governs the matrix to transpose. This does not alter the magnitude of the resulting scalar products since $\mathbf{f}^T\,\boldsymbol{\delta} = \boldsymbol{\delta}^T\,\mathbf{f}$ and $\boldsymbol{\varepsilon}^T\,\boldsymbol{\sigma} = \boldsymbol{\sigma}^T\,\boldsymbol{\varepsilon}$. The superscript *e* refers to the element's nodal forces and displacements. Superscript *v* denotes that displacements and strains are virtual. The nodal force $\mathbf{f}^e$ and internal stress $\boldsymbol{\sigma}$ are real and in equilibrium. The force vector will contain the applied forces acting at nodal points usually around the boundary. The virtual internal strains $\boldsymbol{\varepsilon}^v$ and nodal displacements $\boldsymbol{\delta}^{ev}$ are compatible but are independent of the real force-stress system. We can therefore integrate the strain independently of stress. The integral is applied over volume, *V*, through which the stress and strain will generally vary. In the FE analyses that follow, eq(10.5b) is the most convenient form for deriving $\mathbf{K}^e$. In this equation $\boldsymbol{\delta}^{evT}$ is to be read as the transpose of the virtual nodal displacement vector. The nodal "displacement" vector $\boldsymbol{\delta}^{ev}$ contains all the deflections, rotations and twists at the nodal points. The "force" vector $\mathbf{f}^e$ will contain the real nodal forces, moments and torques. Where forces are distributed, we shall see (section 10.5.6) that $\mathbf{f}^e$ it is compiled from an equivalent system of concentrated nodal forces using the virtual work principle.

10.3.2 Stationary Potential Energy

A stationary value of the total potential energy *P* exists under equilibrium conditions. For a deformable element we write *P* as the sum of the strain energy stored *U* and the work of external forces *V*. SPE can be expressed from eq(8.29) in the alternative forms:

$$\mathrm{d}\left(\int\!\!\int \boldsymbol{\sigma}^{T}\,\mathrm{d}\boldsymbol{\varepsilon}\,\mathrm{d}V - \mathbf{f}^{eT}\,\boldsymbol{\delta}^{e}\right) = 0 \tag{10.6a}$$

For a Hookean material there is proportionality between stress and strain (both are real). For the integration over strain eq(10.6a) becomes

$$\mathrm{d}\,(\tfrac{1}{2}\int \sigma^{\mathrm{T}}\,\varepsilon\,\mathrm{d}V - \mathbf{f}^{e\mathrm{T}}\,\delta^{e}) = 0 \qquad (10.6b)$$

Stationary PE can then be applied using a suitable partial derivative. For example, when applied with respect to the element displacement vector δ^{v}, eq(10.6b) gives

$$\partial P/\partial\delta^{e} = \tfrac{1}{2}\int_{V}[\,\sigma^{\mathrm{T}}(\partial\varepsilon/\partial\delta^{e}) + \varepsilon^{\mathrm{T}}(\partial\sigma/\partial\delta^{e})]\,\mathrm{d}V - \mathbf{f}^{e\mathrm{T}} = 0 \qquad (10.6c)$$

where the relationship $\sigma^{\mathrm{T}}\varepsilon = \varepsilon^{\mathrm{T}}\sigma$ has been applied.

10.4 Prismatic Torsion

St Venant's theory of prismatic bars deals with a uniform torque applied to a solid section of arbitrary cross-section (see Fig. 10.7).

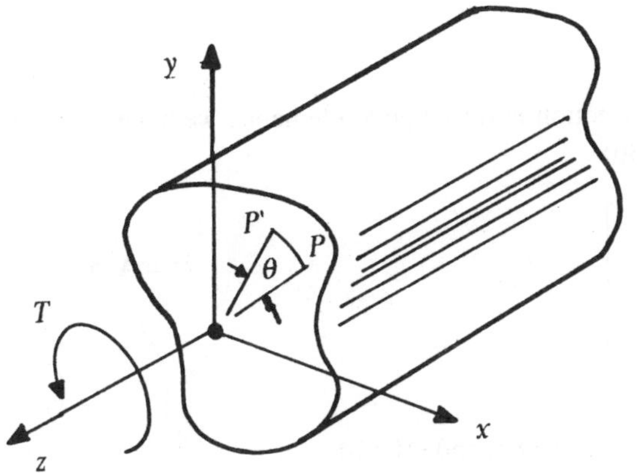

Figure 10.7 Prismatic bar

In a prismatic bar, where the section remains constant over length z, the twist rate, $\alpha = \delta\theta/\delta z$, is a constant. The in-plane displacements of point P to P ' are u and v as shown in Fig. 10.8a.

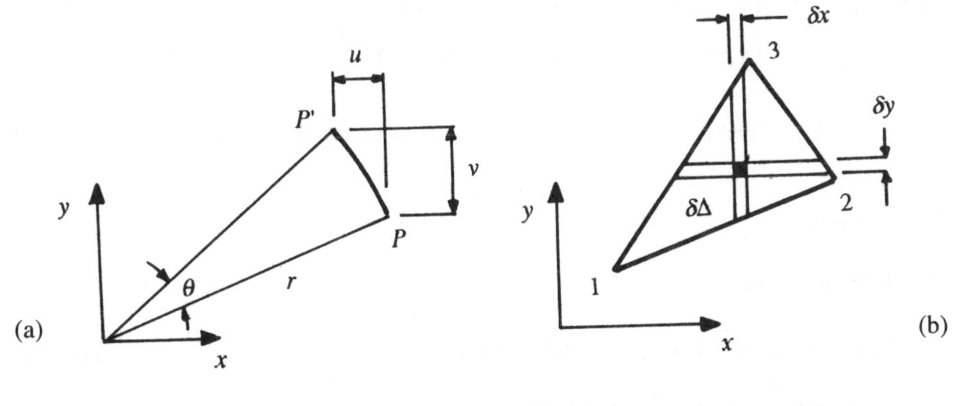

Figure 10.8 Bar cross-section

10.4.1 Displacements

In addition to u and v there is an axial warping displacement w of P' relative to P. In eqs(5.28a - c) the three displacements are given as [3]

$$u = -\,æ\, z\, y,\; v = æ\, z\, x \quad \text{and} \quad w = \psi(x, y)\, æ \tag{10.7a,b,c}$$

The warping displacement function $\psi(x, y)$ in eq(10.7c) is taken as

$$\psi(x, y) = \alpha_1 + \alpha_2 x + \alpha_3 y \tag{10.8a}$$

which we write as

$$\psi = \begin{bmatrix} 1 & x & y \end{bmatrix} \begin{bmatrix} \alpha_1 \\ \alpha_2 \\ \alpha_3 \end{bmatrix} \qquad \boldsymbol{\delta} = \mathbf{A}\, \boldsymbol{\alpha} \tag{10.8b,c}$$

Subdividing the cross-section into triangular elements, we have for the three nodes of any element (see Fig. 10.8b)

$$\begin{bmatrix} \psi_1 \\ \psi_2 \\ \psi_3 \end{bmatrix} = \begin{bmatrix} 1 & x_1 & y_1 \\ 1 & x_2 & y_2 \\ 1 & x_3 & y_3 \end{bmatrix} \begin{bmatrix} \alpha_1 \\ \alpha_2 \\ \alpha_3 \end{bmatrix} \qquad \boldsymbol{\delta}^e = \mathbf{A}^e\, \boldsymbol{\alpha} \tag{10.9a,b}$$

Eliminating $\boldsymbol{\alpha}$ between eqs(10.8c) and (10.9b),

$$\boldsymbol{\delta} = \mathrm{A}\, (\mathrm{A}^e)^{-1}\, \boldsymbol{\delta}^e \tag{10.10a}$$

where

$$(\mathbf{A}^e)^{-1} = \frac{1}{2\Delta} \begin{bmatrix} (x_2 y_3 - x_3 y_2) & (x_3 y_1 - x_1 y_3) & (x_1 y_2 - x_2 y_1) \\ (y_2 - y_3) & (y_3 - y_1) & (y_1 - y_2) \\ (x_3 - x_2) & (x_1 - x_3) & (x_2 - x_1) \end{bmatrix} \tag{10.10b}$$

where $2\Delta = (y_2 - y_3)(x_1 - x_2) - (y_1 - y_2)(x_2 - x_3)$ is twice the element area (Δ).

10.4.2 Strains

The shear strains are

$$\gamma_{xy} = 0,\; \gamma_{xz} = (\partial\psi/\partial x - y)\, æ,\; \gamma_{yz} = (\partial\psi/\partial y + x)\, æ \tag{10.11a}$$

which are written from eq(10.8a) as follows:

$$\begin{bmatrix} \gamma_{xy} \\ \gamma_{xz} \\ \gamma_{yz} \end{bmatrix} = \begin{bmatrix} 0 & 0 & 0 \\ 0 & 1 & 0 \\ 0 & 0 & 1 \end{bmatrix} \begin{bmatrix} \alpha_1 \\ \alpha_2 \\ \alpha_3 \end{bmatrix} æ + \begin{bmatrix} 0 & 0 & 0 \\ 0 & 1 & 0 \\ 0 & 0 & 1 \end{bmatrix} \begin{bmatrix} z \\ -y \\ x \end{bmatrix} æ \qquad \boldsymbol{\varepsilon} = \mathbf{C}(\boldsymbol{\alpha} + \boldsymbol{\beta})æ \qquad (10.11b,c)$$

Writing, for brevity, $A^{e-1} = (A^e)^{-1}$ we substitute $\boldsymbol{\alpha} = \mathbf{A}^{e-1}\boldsymbol{\delta}^e$ from eq(10.9b) into eq(10.11c)

$$\boldsymbol{\varepsilon} = [(\mathbf{CA}^{e-1})\,\boldsymbol{\delta}^e + \mathbf{C}\boldsymbol{\beta}\,]æ = (\mathbf{B}\boldsymbol{\delta}^e + \mathbf{C}\boldsymbol{\beta}\,)æ \qquad (10.12a)$$

where $\mathbf{B} = \mathbf{CA}^{e-1}$ takes the explicit form:

$$\mathbf{B} = \frac{1}{2\Delta}\begin{bmatrix} 0 & 0 & 0 \\ (y_2 - y_3) & (y_3 - y_1) & (y_1 - y_2) \\ (x_3 - x_2) & (x_1 - x_3) & (x_2 - x_1) \end{bmatrix} \qquad (10.12b)$$

10.4.3 Stresses

The shear stresses follow from

$$\tau_{xy} = 0,\ \tau_{xz} = G\gamma_{xz} \text{ and } \tau_{yz} = G\gamma_{yz} \qquad (10.13a)$$

which take the corresponding matrix form

$$\begin{bmatrix} \tau_{xy} \\ \tau_{xz} \\ \tau_{yz} \end{bmatrix} = G \begin{bmatrix} 0 & 0 & 0 \\ 0 & 1 & 0 \\ 0 & 0 & 1 \end{bmatrix} \begin{bmatrix} \gamma_{xy} \\ \gamma_{xz} \\ \gamma_{yz} \end{bmatrix} \qquad \boldsymbol{\sigma} = G\,\mathbf{C}\,\boldsymbol{\varepsilon} \qquad (10.13b,c)$$

Substituting eq(10.12a) into eq(10.13c) gives the stress vector as

$$\boldsymbol{\sigma} = G\,\mathbf{C}\,[(\mathbf{CA}^{e-1})\,\boldsymbol{\delta}^e + \mathbf{C}\,\boldsymbol{\beta}\,]æ = G\,\mathbf{C}\,(\mathbf{B}\boldsymbol{\delta}^e + \mathbf{C}\boldsymbol{\beta}\,)æ \qquad (10.14)$$

10.4.4 Stiffness Matrix

There are no nodal forces in this case but we can apply the stationary energy principle to find a pseudo-stiffness equation for warping. In eq(8.29) we have a contribution to potential energy P from the work of external forces

$$V = -\int T\,(\mathrm{d}\theta/\mathrm{d}z)\,\mathrm{d}z = -\,T\,æ\,L \qquad (10.15)$$

and from the strain energy stored

$$U = (1/2)\int \boldsymbol{\varepsilon}^{\mathrm{T}}\,\boldsymbol{\sigma}\,\mathrm{d}V = (L/2)\int \boldsymbol{\varepsilon}^{\mathrm{T}}\,\boldsymbol{\sigma}\,\mathrm{d}\Delta \qquad (10.16a)$$

where $d\Delta = dx \times dy$ in Fig. 10.8b. Substitute eqs(10.12a) and (10.14) into eq(10.16a),

$$U = (GL/2)\alpha^2 \int (\mathbf{C}\mathbf{A}^{e-1}\,\delta^e + \mathbf{C}\,\beta)^T \mathbf{C}\,(\mathbf{C}\mathbf{A}^{e-1}\delta^e + \mathbf{C}\,\beta)\, d\Delta$$

$$= (GL/2)\alpha^2 \int [\,\mathbf{C}\,(\mathbf{A}^{e-1}\,\delta^e + \beta)]^T \mathbf{C}^2\,(\mathbf{A}^{e-1}\,\delta^e + \beta)\, d\Delta$$

$$= (GL/2)\alpha^2 \int [\delta^{eT}\,(\mathbf{A}^{e-1})^T + \beta^T\,]\,\mathbf{C}^T\mathbf{C}^2\,(\mathbf{A}^{e-1}\delta^e + \beta)\, d\Delta \tag{10.16b}$$

Now as $\mathbf{C}^T\mathbf{C}^2 = \mathbf{C}^3 = \mathbf{C}$, the expansion of the integrand in (10.16b) becomes

$$U = (GL/2)\alpha^2 \int [\delta^{eT}(\mathbf{A}^{e-1})^T\,\mathbf{C}\mathbf{A}^{e-1}\,\delta^e + \delta^{eT}(\mathbf{A}^{e-1})^T\,\mathbf{C}\beta + \beta^T\,\mathbf{C}\mathbf{A}^{e-1}\delta^e + \beta^T\,\mathbf{C}\beta]d\Delta \tag{10.16c}$$

Employing the symmetrical product relation $(\mathbf{A}^{e-1}\,\delta^e)^T\,\mathbf{C}\beta = [(\mathbf{A}^{e-1}\delta^e)^T\,(\mathbf{C}\beta\,)]^T$ in the second and third terms of eq(10.16c) leads to

$$U = (GL/2)\alpha^2 \int [\delta^{eT}(\mathbf{A}^{e-1})^T\,\mathbf{C}\mathbf{A}^{e-1}\,\delta^e + 2(\mathbf{A}^{e-1}\delta^e\,)^T\,\mathbf{C}\beta + \beta^T\,\mathbf{C}\beta\,]\,d\Delta \tag{10.16d}$$

Thus, from eqs(10.15) and (10.16d), the total PE is

$$P = U + V = (GL/2)\alpha^2 \int [\delta^{eT}(\mathbf{A}^{e-1})^T\mathbf{C}\mathbf{A}^{e-1}\delta^e + 2(\mathbf{A}^{e-1}\,\delta^e)^T\,\mathbf{C}\beta + \beta^T\,\mathbf{C}\beta\,]d\Delta - T\alpha L \tag{10.17}$$

Equation (10.17) is stationary for (i) $\partial P/\partial\delta^e = 0$ and (ii) $\partial P/\partial\alpha = 0$. Condition (i) enables the warping displacements to be found as follows:

$$\frac{\partial}{\partial\delta^e}\int_\Delta \delta^{eT}(\mathbf{A}^{e-1})^T\,\mathbf{C}\mathbf{A}^{e-1}\,\delta^e\, d\Delta + 2\,\frac{\partial}{\partial\delta^e}\int_\Delta \delta^{eT}(\mathbf{A}^{e-1})^T\,\mathbf{C}\,\beta\, d\Delta + \frac{\partial}{\partial\delta^e}\int_\Delta \beta^T\,\mathbf{C}\beta\, d\Delta = 0$$

Since $\mathbf{A}^e$ and δ^e are independent of the area Δ, this becomes

$$\frac{\partial}{\partial\delta^e}[\delta^{eT}(\mathbf{A}^{e-1})^T\,\mathbf{C}(\int_\Delta d\Delta)\,\mathbf{A}^{e-1}\,\delta^e\,] + 2\frac{\partial}{\partial\delta^e}[\delta^{eT}(\mathbf{A}^{e-1})^T\int_\Delta(\mathbf{C}\beta)\,d\Delta\,] = 0$$

Now as $\partial\delta^{eT}/\,\partial\delta^e = \mathbf{I}$ and $\mathbf{C}\beta = \beta$, the differentiation gives

$$(\mathbf{A}^{e-1})^T\,\mathbf{C}(\int_\Delta d\Delta)\,\mathbf{A}^{e-1}\,\delta^e + 2(\mathbf{A}^{e-1}\,)^T\int_\Delta(\beta\, d\Delta) = 0$$

The integration yields

$$(\mathbf{A}^{e-1})^T\,\mathbf{C}\Delta\,\mathbf{A}^{e-1}\,\delta^e + 2(\mathbf{A}^{e-1})^T\,\bar{\beta}\Delta\,] = 0 \tag{10.18a}$$

where $\bar{\beta} = [\,0\;\; -y\;\; x\,]^T$ is referred to the co-ordinates of the centroid: $\bar{x} = (x_1 + x_2 + x_3)/3$ and $\bar{y} = (y_1 + y_2 + y_3)/3$. Equation (10.18a) is written as a warping stiffness equation:

$$\mathbf{K}^e\,\delta^e - \mathbf{F}^e = 0 \tag{10.18b}$$

where, from eqs(10.18a,b), the stiffness matrix (units of force/length) is

$$\mathbf{K}^e = (GL/2)(\mathbf{A}^{e-1})^T\,\mathbf{C}\Delta\,\mathbf{A}^{e-1} \tag{10.19a}$$

$$= \frac{GL}{8\Delta}\begin{bmatrix} x_2y_3 - x_3y_2 & y_2 - y_3 & x_3 - x_2 \\ x_3y_1 - x_1y_3 & y_3 - y_1 & x_1 - x_3 \\ x_1y_2 - x_2y_1 & y_1 - y_2 & x_2 - x_1 \end{bmatrix}\begin{bmatrix} 0 & 0 & 0 \\ 0 & 1 & 0 \\ 0 & 0 & 1 \end{bmatrix}\begin{bmatrix} x_2y_3 - x_3y_2 & x_3y_1 - x_1y_3 & x_1y_2 - x_2y_1 \\ y_2 - y_3 & y_3 - y_1 & y_1 - y_2 \\ x_3 - x_2 & x_1 - x_3 & x_2 - x_1 \end{bmatrix} \tag{10.19b}$$

$$= \frac{GL}{8\Delta}\begin{bmatrix} (y_2-y_3)^2+(x_3-x_2)^2 & (y_2-y_3)(y_3-y_1)+(x_3-x_2)(x_1-x_3) & (y_2-y_3)(y_1-y_2)+(x_3-x_2)(x_2-x_1) \\ (y_3-y_1)(y_2-y_3)+(x_1-x_3)(x_3-x_2) & (y_3-y_1)^2+(x_1-x_3)^2 & (y_3-y_1)(y_1-y_2)+(x_1-x_3)(x_2-x_1) \\ (y_1-y_2)(y_2-y_3)+(x_2-x_1)(x_3-x_2) & (y_1-y_2)(y_3-y_1)+(x_2-x_1)(x_1-x_3) & (y_1-y_2)^2+(x_2-x_1)^2 \end{bmatrix}$$

Comparing eqs(10.18a,b), the pseudo "force" matrix (torque units: force × length) is

$$\mathbf{F}^e = -\,GL\,(\mathbf{A}^{e-1})^T\,\bar{\beta}\Delta \tag{10.20a}$$

$$= -\frac{GL}{2}\begin{bmatrix} x_2y_3-x_3y_2 & y_2-y_3 & x_3-x_2 \\ x_3y_1-x_1y_3 & y_3-y_1 & x_1-x_3 \\ x_1y_2-x_2y_1 & y_1-y_2 & x_2-x_1 \end{bmatrix}\begin{bmatrix} 0 \\ -\bar{y} \\ \bar{x} \end{bmatrix} = -\frac{GL}{2}\begin{bmatrix} -\bar{y}(y_2-y_3)+\bar{x}(x_3-x_2) \\ -\bar{y}(y_3-y_1)+\bar{x}(x_1-x_3) \\ -\bar{y}(y_1-y_2)+\bar{x}(x_2-x_1) \end{bmatrix} \tag{10.20b}$$

It follows from eqs(10.19b) and (10.20b) that the nodal warping displacements $\delta^e = \mathbf{K}^{e-1}\mathbf{F}^e$, depend only upon the co-ordinates of the element nodes and its centroid (G and L will cancel). No boundary conditions are required but the solution must match an absence of warping at some point in the cross-section, e.g. it is known that no warping occurs at the intersection between axes of symmetry.

Applying condition (ii) $\partial P/\partial æ = 0$, to eq(10.17) gives the angle of twist as follows:

$$GLæ\int_\Delta [\delta^{eT}(\mathbf{A}^{e-1})^T\mathbf{C}\mathbf{A}^{e-1}\delta^e + 2(\mathbf{A}^{e-1}\delta^e)^T\mathbf{C}\beta + \beta^T\mathbf{C}\beta\,]\mathrm{d}\Delta - TL = 0$$

$$GLæ\,[\delta^{eT}(\mathbf{A}^{e-1})^T\mathbf{C}\mathbf{A}^{e-1}\delta^e\Delta + 2(\mathbf{A}^{e-1}\delta^e)^T\,\bar{\beta}\Delta + \int_\Delta \beta^T\mathbf{C}\beta\,\mathrm{d}\Delta\,] - TL = 0$$

$$æ\,(2\,\delta^{eT}\,\mathbf{K}^e\,\delta^e - 2\,\delta^{eT}\,\mathbf{F}^e + GL\int_\Delta \beta^T\,\mathbf{C}\,\beta\,\mathrm{d}\Delta\,) - TL = 0 \tag{10.21a}$$

The integral term may be approximated with centroidal co-ordinates as

$$\int_\Delta \beta^T\mathbf{C}\,\beta\,\mathrm{d}\Delta = \int_\Delta [\,z\;\; -y\;\; x\,]\begin{bmatrix} 0 & 0 & 0 \\ 0 & 1 & 0 \\ 0 & 0 & 1 \end{bmatrix}\begin{bmatrix} z \\ -y \\ x \end{bmatrix}\mathrm{d}\Delta \simeq (\bar{y}^2 + \bar{x}^2)\Delta \tag{10.21b}$$

Equations (10.21a,b) supply the angular twist rate as

$$æ = \theta/L = TL/\,[2\,\delta^{eT}\,\mathbf{K}^e\,\delta^e - 2\,\delta^{eT}\,\mathbf{F}^e + GL\,(\bar{y}^2 + \bar{x}^2)\Delta\,] \tag{10.22}$$

This FE solution is particularly useful for determining stress concentrations in drive shafts machined with splines and keyways [3,4]. The key region may require a finer subdivision with a more refined triangular element having three additional nodes at the centre of each side. This particular element requires a warping function with quadratic terms in x and y.

10.5 Plane Triangular Element

To find the stiffness matrix for the plane triangular element (e) in Fig. 10.9a, we firstly number the nodes 1, 2 and 3 in an anti-clockwise direction. Within an assembly of elements e = I, II, III, IV etc, the resulting stress, strain and displacement relations are applied in the same sense as the element node numbering.

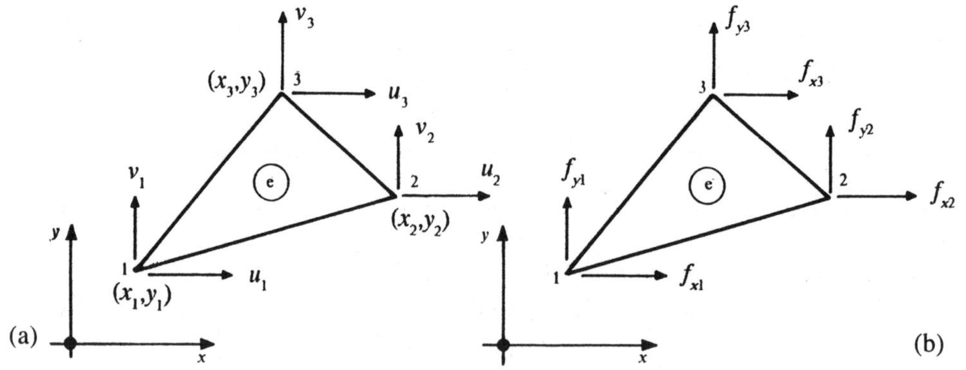

Figure 10.9 Nodal displacements and forces for a triangular element

10.5.1 Displacements

The co-ordinates of nodes 1, 2 and 3 are (x_1, y_1), (x_2, y_2) and (x_3, y_3) respectively. The displacements of each node may be expressed in an assumed function $u(x, y)$ and $v(x, y)$ of their co-ordinates. For any point (x, y), a simple linear displacement function is assumed [5,6]

$$u = \alpha_1 + \alpha_2 x + \alpha_3 y \qquad (10.23a)$$
$$v = \alpha_4 + \alpha_5 x + \alpha_6 y \qquad (10.23b)$$

where the six coefficients $\alpha_1, \alpha_2 \ldots \alpha_6$, match the total number of degrees of freedom for the element, i.e. two per node. The linear polynomial will also ensure displacement continuity along the sides of adjacent elements. The two constants, α_1 and α_4, in eqs(10.23a,b) account for any rigid body translation. Equations (10.23a,b) are combined in the matrix form to give the displacement vector $\boldsymbol{\delta} = [u \;\; v]^T$ for any point (x, y) in the element as

$$\boldsymbol{\delta} = \begin{bmatrix} u \\ v \end{bmatrix} = \begin{bmatrix} 1 & x & y & 0 & 0 & 0 \\ 0 & 0 & 0 & 1 & x & y \end{bmatrix} \begin{bmatrix} \alpha_1 \\ \alpha_2 \\ \alpha_3 \\ \alpha_4 \\ \alpha_5 \\ \alpha_6 \end{bmatrix} \qquad \boldsymbol{\delta} = \mathbf{A}\,\boldsymbol{\alpha} \qquad (10.24a,b)$$

In eq(10.24b), $\boldsymbol{\alpha} = [\alpha_1 \;\; \alpha_2 \;\; \alpha_3 \;\; \alpha_4 \;\; \alpha_5 \;\; \alpha_6]^T$ is a column matrix and $\mathbf{A}$ is a 6×2 matrix in x and y. Substituting x and y for each node in eq(10.24a) gives the nodal point displacement vectors: $\boldsymbol{\delta}_1 = [\,u_1 \;\; v_1\,]^T$, $\boldsymbol{\delta}_2 = [\,u_2 \;\; v_2\,]^T$ and $\boldsymbol{\delta}_3 = [\,u_3 \;\; v_3\,]^T$. These may be combined to give the element's nodal point displacement vector:

$$\delta^e = \begin{bmatrix} \delta_1 \\ \delta_2 \\ \delta_3 \end{bmatrix} = \begin{bmatrix} u_1 \\ v_1 \\ u_2 \\ v_2 \\ u_3 \\ v_3 \end{bmatrix} = \begin{bmatrix} 1 & x_1 & y_1 & 0 & 0 & 0 \\ 0 & 0 & 0 & 1 & x_1 & y_1 \\ 1 & x_2 & y_2 & 0 & 0 & 0 \\ 0 & 0 & 0 & 1 & x_2 & y_2 \\ 1 & x_3 & y_3 & 0 & 0 & 0 \\ 0 & 0 & 0 & 1 & x_3 & y_3 \end{bmatrix} \begin{bmatrix} \alpha_1 \\ \alpha_2 \\ \alpha_3 \\ \alpha_4 \\ \alpha_5 \\ \alpha_6 \end{bmatrix} \qquad \delta^e = \mathbf{A}^e \alpha \qquad (10.25a,b)$$

where $\delta^e = [\ \delta_1\ \ \delta_2\ \ \delta_3\ ]^T$ and $\mathbf{A}^e$ is a 6 × 6 matrix, derived from **A**. The coefficients α_1, α_2 ... α_6, are expressed in terms of the displacements (u, v) from solving the six equations in (10.25a). That is, by inverting matrix **A** in eq(10.25b)

$$\alpha = (\mathbf{A}^e)^{-1}\, \delta^e \qquad (10.26a)$$

where

$$(\mathbf{A}^e)^{-1} = \frac{1}{2\Delta} \begin{bmatrix} x_2y_3 - x_3y_2 & 0 & x_3y_1 - x_1y_3 & 0 & x_1y_2 - x_2y_1 & 0 \\ y_2 - y_3 & 0 & y_3 - y_1 & 0 & y_1 - y_2 & 0 \\ x_3 - x_2 & 0 & x_1 - x_3 & 0 & x_2 - x_1 & 0 \\ 0 & x_2y_3 - x_3y_2 & 0 & x_3y_1 - x_1y_3 & 0 & x_1y_2 - x_2y_1 \\ 0 & y_2 - y_3 & 0 & y_3 - y_1 & 0 & y_1 - y_2 \\ 0 & x_3 - x_2 & 0 & x_1 - x_3 & 0 & x_2 - x_1 \end{bmatrix} \qquad (10.26b)$$

in which

$$\Delta = \tfrac{1}{2}\,[(x_1 - x_2)\,(y_2 - y_3) - (x_2 - x_3)\,(y_1 - y_2)\,] \qquad (10.26c)$$

is the area of the element. A check on eq(10.26b) is provided by the square matrix relation $\mathbf{A}^e\,\mathbf{A}^{e-1} = \mathbf{I}$. Combining eqs(10.24b) and (10.26a) expresses displacements $\delta = [u\ \ v\ ]^T$ within the element in terms of the displacements $\delta^e = [\delta_1\ \ \delta_2\ \ \delta_3\,]^T$ at its nodes. This gives

$$\delta = \mathbf{A}\,\mathbf{A}^{e-1}\,\delta^e \qquad (10.27a)$$

Equation (10.27a) is similar to eq(10.10a) and will appear again for plane elements. In fact, the matrix product

$$\mathbf{N} = \mathbf{A}\,\mathbf{A}^{e-1} \qquad (10.27b)$$

defines the element shape function as the following example shows.

Example 10.2 Show that the triangular element displacements u, v may be expressed as $u = N_i u_i$ and $v = N_i v_i$ (i = 1, 2, 3) where $N_i = a_i + b_i x + c_i y$. Determine the coefficients a_i, b_i and c_i in terms of the nodal point co-ordinates.

Substituting eqs(10.24a) and (10.26b) into eq(10.27a), the matrix multiplication leads to

$$u = [1/(2\Delta)]\{[(x_2 y_3 - x_3 y_2) + x(y_2 - y_3) + y(x_3 - x_2)]u_1 + [(x_3 y_1 - x_1 y_3) + x(y_3 - y_1) + y(x_1 - x_3)]u_2 + [(x_1 y_2 - x_2 y_1) + x(y_1 - y_2) + y(x_2 - x_1)]u_3\}$$

$$v = [1/(2\Delta)]\{[(x_2 y_3 - x_3 y_2) + x(y_2 - y_3) + y(x_3 - x_2)]v_1 + [(x_3 y_1 - x_1 y_3) + x(y_3 - y_1) + y(x_1 - x_3)]v_2 + [(x_1 y_2 - x_2 y_1) + x(y_1 - y_2) + y(x_2 - x_1)]v_3\}$$

These show that the coefficients are the same so we can write these in an abbreviated form:

$$u = N_i u_i = N_1 u_1 + N_2 u_2 + N_3 u_3 \text{ and } v = N_i v_i = N_1 v_1 + N_2 v_2 + N_3 u_3 \tag{i,ii}$$

where

$$N_1 = a_1 + b_1 x + c_1 y,\ N_2 = a_2 + b_2 x + c_2 y \text{ and } N_3 = a_3 + b_3 x + c_3 y \tag{iii-iv}$$

and

$$a_1 = (x_2 y_3 - x_3 y_2)/(2\Delta),\ b_1 = (y_2 - y_3)/(2\Delta),\ c_1 = (x_3 - x_2)/(2\Delta)$$
$$a_2 = (x_3 y_1 - x_1 y_3)/(2\Delta),\ b_2 = (y_3 - y_1)/(2\Delta),\ c_2 = (x_1 - x_3)/(2\Delta)$$
$$a_3 = (x_1 y_2 - x_2 y_1)/(2\Delta),\ b_3 = (y_1 - y_2)/(2\Delta),\ c_3 = (x_2 - x_1)/(2\Delta)$$

Equations (i) and (ii) must of course give the nodal displacements $u = u_1$, $v = v_1$ for $x = x_1$ and $y = y_1$ etc. This requires the substitution of Δ from eq(10.26c). The coefficients a_i, b_i and c_i in eqs(iii)-(iv), which appear in terms of nodal co-ordinates, depend upon the displacement function assumed for the element.

10.5.2 Strains

The displacement derivatives, given in eqs(2.17a-c), provide compatible, Cartesian strain components. From eqs(10.23a,b):

$$\varepsilon_x = \partial u/\partial x = \alpha_2 \tag{10.28a}$$
$$\varepsilon_y = \partial v/\partial y = \alpha_6 \tag{10.28b}$$
$$\gamma_{xy} = \partial u/\partial y + \partial v/\partial x = \alpha_3 + \alpha_5 \tag{10.28c}$$

In matrix form, the strain components (10.28a-c) are written as

$$\begin{bmatrix} \varepsilon_x \\ \varepsilon_y \\ \gamma_{xy} \end{bmatrix} = \begin{bmatrix} 0 & 1 & 0 & 0 & 0 & 0 \\ 0 & 0 & 0 & 0 & 0 & 1 \\ 0 & 0 & 1 & 0 & 1 & 0 \end{bmatrix} \begin{bmatrix} \alpha_1 \\ \alpha_2 \\ \alpha_3 \\ \alpha_4 \\ \alpha_5 \\ \alpha_6 \end{bmatrix} \qquad \boldsymbol{\varepsilon} = \mathbf{C}\boldsymbol{\alpha} \tag{10.29a,b}$$

In eq(10.29b), $\boldsymbol{\varepsilon} = [\varepsilon_x\ \varepsilon_y\ \gamma_{xy}]^{\mathrm{T}}$ and $\boldsymbol{\alpha} = [\alpha_1\ \alpha_2\ \alpha_3\ \alpha_4\ \alpha_5\ \alpha_6]^{\mathrm{T}}$ are column matrices and **C** is a 6×3 matrix. Substituting eq(10.26a) into eq(10.29b) expresses strains $\boldsymbol{\varepsilon} = [\varepsilon_x\ \varepsilon_y\ \gamma_{xy}]^{\mathrm{T}}$ at any point (x, y) within the element in terms of the nodal point displacements $\boldsymbol{\delta}^e$. That is

$$\boldsymbol{\varepsilon} = \mathbf{C}(\mathbf{A}^e)^{-1}\boldsymbol{\delta}^e = \mathbf{B}\,\boldsymbol{\delta}^e \tag{10.30a}$$

where $\mathbf{B} = \mathbf{C}(\mathbf{A}^e)^{-1}$ is found from matrix multiplication of eqs(10.26b) and (10.29a). This gives

$$\mathbf{B} = \frac{1}{2\Delta}\begin{bmatrix} y_2 - y_3 & 0 & y_3 - y_1 & 0 & y_1 - y_2 & 0 \\ 0 & x_3 - x_2 & 0 & x_1 - x_3 & 0 & x_2 - x_1 \\ x_3 - x_2 & y_2 - y_3 & x_1 - x_3 & y_3 - y_1 & x_2 - x_1 & y_1 - y_2 \end{bmatrix} \qquad (10.30b)$$

Note, that **B** in eq(10.30b) appears only in terms of the nodal point co-ordinates. The three strain components are therefore constant throughout the element. This is a consequence of the linear displacement functions in eqs(10.23a,b) where matrix **C** in eq(10.30a) does not contain x or y.

10.5.3 Constitutive Relations

Firstly, recall the constitutive relations (2.14a,b) for plane stress. The "through thickness" strain ε_z in eq(2.14c) does not appear in the matrix since it depends upon σ_x and σ_y. For a plane strain condition, with $\varepsilon_z = 0$, the corresponding relations appear in eq(2.15a,b,c). We have seen in eq(2.26) that these relations may be expressed in the symbolic form, $\boldsymbol{\varepsilon} = \mathbf{P}\,\boldsymbol{\sigma}$. In each case the square matrix **P** may be inverted to give the internal stresses in terms of known strain components as $\boldsymbol{\sigma} = \mathbf{P}^{-1}\,\boldsymbol{\varepsilon}$ where $\boldsymbol{\sigma} = [\,\sigma_x \;\; \sigma_y \;\; \tau_{xy}]^T$. Writing $\mathbf{D} = \mathbf{P}^{-1}$ and substituting from eq(10.30a),

$$\boldsymbol{\sigma} = \mathbf{D}\,\boldsymbol{\varepsilon} = \mathbf{D}\,\mathbf{B}\,\boldsymbol{\delta}^e \qquad (10.31a)$$

The components of the inverse 3 × 3 matrix **D** are found from expressing the stress components σ_x, σ_y and τ_{xy} in terms of the strain components ε_x, ε_y and γ_{xy}. For plane stress, eqs(2.14a,b,d) show that

$$\mathbf{D} = \frac{E}{(1-\nu^2)}\begin{bmatrix} 1 & \nu & 0 \\ \nu & 1 & 0 \\ 0 & 0 & \tfrac{1}{2}(1-\nu) \end{bmatrix} \qquad (10.31b)$$

For plane strain, eqs(2.15a,b,d) give

$$\mathbf{D} = \frac{E(1-\nu)}{(1+\nu)(1-2\nu)}\begin{bmatrix} 1 & \nu/(1-\nu) & 0 \\ \nu/(1-\nu) & 1 & 0 \\ 0 & 0 & (1-2\nu)/[2(1-\nu)] \end{bmatrix} \qquad (10.31c)$$

The analysis may proceed with the single matrix **D** provided it is remembered that its components, D_{ij} in eqs(10.31b and c), will depend upon the nature of the plane problem. Because **D** and **B** do not contain the generalised co-ordinates x, y, it follows from eq(10.31a) that the three stress components σ_x, σ_y, τ_{xy} are also constant throughout a triangular element for given nodal displacement $\boldsymbol{\delta}^e$. Normally, these are taken to apply to the centroid or mid-side length of the element when plotting stress distributions.

10.5.4 Element Stiffness Matrix

An element stiffness matrix $\mathbf{K}^e$, of dimensions 6 × 6, relates the nodal point force vectors $\mathbf{f}_1 = [f_{x1}\, f_{y1}]^T$, $\mathbf{f}_2 = [f_{x2}\; f_{y2}]^T$ and $\mathbf{f}_3 = [f_{x3}\; f_{y3}]^T$ (see Fig. 10.9b) to the nodal point displacements $\boldsymbol{\delta}_1 = [u_1\;\; v_1]^T$, $\boldsymbol{\delta}_2 = [u_2\;\; v_2]^T$ and $\boldsymbol{\delta}_3 = [u_3\;\; v_3]^T$ (see Fig. 10.9a) through the relation

$$\mathbf{f}^e = \mathbf{K}^e\, \boldsymbol{\delta}^e \tag{10.32a}$$

The element force vector $\mathbf{f}^e$ is

$$\mathbf{f}^e = [\mathbf{f}_1\;\; \mathbf{f}_2\;\; \mathbf{f}_3]^T = [f_{x1}\;\; f_{y1}\;\; f_{x2}\;\; f_{y2}\;\; f_{x3}\;\; f_{y3}]^T$$

and the element displacement vector $\boldsymbol{\delta}^e$ is

$$\boldsymbol{\delta}^e = [\boldsymbol{\delta}_1\;\; \boldsymbol{\delta}_2\;\; \boldsymbol{\delta}_3]^T = [u_1\;\; v_1\;\; u_2\;\; v_2\;\; u_3\;\; v_3]^T$$

To obtain $\boldsymbol{\delta}^e$ from eq(10.32a), $\mathbf{K}^e$ will need to be inverted:

$$\boldsymbol{\delta}^e = (\mathbf{K}^e)^{-1}\, \mathbf{f}^e \tag{10.32b}$$

The problem is again resolved into defining $\mathbf{K}^e$ in eq(10.32a) from known nodal point co-ordinates (x_i, y_i, for i = 1, 2, 3) and the elastic constants E and ν for the material. In this case we shall use both the principal of virtual work and stationary potential energy to derive $\mathbf{K}^e$.

(a) Virtual Work
Substitute eqs(10.30a) and (10.31a) into eq(10.5b) to give

$$\boldsymbol{\delta}^{evT}\, \mathbf{f}^e = \int (\mathbf{B}\, \boldsymbol{\delta}^{ev})^T (\mathbf{D}\, \mathbf{B}\, \boldsymbol{\delta}^e)\, \mathrm{d}V \tag{10.33a}$$

Because neither $\mathbf{B}$ nor $\mathbf{D}$ depends upon x and y and $(\mathbf{B}\, \boldsymbol{\delta}^{ev})^T = \boldsymbol{\delta}^{evT}\, \mathbf{B}^T$, eq(10.33a) becomes

$$\boldsymbol{\delta}^{evT}\, \mathbf{f}^e = \boldsymbol{\delta}^{evT}\, (\mathbf{B}^T\, \mathbf{D}\, \mathbf{B}\, \boldsymbol{\delta}^e) \int \mathrm{d}V \tag{10.33b}$$

Cancelling the virtual displacements and putting $\int \mathrm{d}V = V = \Delta t$, where Δ and t are respectively the area and thickness of the triangle, eq(10.33b) leads to

$$\mathbf{f}^e = \mathbf{B}^T\, \mathbf{D}\, \mathbf{B}\, \boldsymbol{\delta}^e\, V \tag{10.33c}$$

Comparing eqs(10.32a) and (10.33c) defines the element stiffness matrix as

$$\mathbf{K}^e = \mathbf{B}^T\, \mathbf{D}\, \mathbf{B}\, V \tag{10.34a}$$

(b) Stationary Potential Energy
Substituting eqs(10.30a) and (10.31a) into eq(10.6c) gives

$$\tfrac{1}{2} \int_V \Big[(\mathbf{D}\, \boldsymbol{\varepsilon})^T \frac{\partial}{\partial \boldsymbol{\delta}^e} (\mathbf{B}\, \boldsymbol{\delta}^e) + \boldsymbol{\varepsilon}^T \frac{\partial}{\partial \boldsymbol{\delta}^e} (\mathbf{D}\, \mathbf{B}\, \boldsymbol{\delta}^e) \Big]\, \mathrm{d}V - \mathbf{f}^{eT} = 0$$

$$\tfrac{1}{2} \int_V [(\mathbf{D}\, \boldsymbol{\varepsilon})^T\, \mathbf{B} + \boldsymbol{\varepsilon}^T\, \mathbf{D}\, \mathbf{B}\,]\mathrm{d}V - \mathbf{f}^{eT} = 0$$

$$\mathbf{f}^{eT} = \int_V \boldsymbol{\varepsilon}^T\, \mathbf{D}\, \mathbf{B}\, \mathrm{dV} = \int_V (\mathbf{B}\, \boldsymbol{\delta}^e)^T\, \mathbf{D}\, \mathbf{B}\, \mathrm{d}V$$

in which $\mathbf{D} = \mathbf{D}^T$. Since we can write this as

$$\mathbf{f}^{eT} = \delta^{eT}(\mathbf{B}^T\,\mathbf{D}\,\mathbf{B})\,V$$

the transpose of both sides gives

$$\mathbf{f}^e = [\delta^{eT}(\mathbf{B}^T\,\mathbf{D}\,\mathbf{B})]^T\,V = (\mathbf{B}^T\mathbf{D}\,\mathbf{B})^T\,\delta^e\,V = (\mathbf{D}\,\mathbf{B})^T\mathbf{B}\,\delta^e\,V = (\mathbf{B}^T\mathbf{D}^T\mathbf{B}\,V\,)\,\delta^e$$

and it again follows that $\mathbf{K}^e = \mathbf{B}^T\mathbf{D}\,\mathbf{B}\,V$, as in eq(10.34a).

Thus, the components of $\mathbf{K}^e$ may be found explicitly from multiplying the matrices $\mathbf{B}^T$, $\mathbf{D}$ and $\mathbf{B}$ strictly in the order of the right side of eq(10.34a). This results in the following 6 × 6 symmetrical matrix:

$$\mathbf{K}^e = \begin{bmatrix} K^e_{11} & K^e_{12} & K^e_{13} & K^e_{14} & K^e_{15} & K^e_{16} \\ K^e_{21} & K^e_{22} & K^e_{23} & K^e_{24} & K^e_{25} & K^e_{26} \\ K^e_{31} & K^e_{32} & K^e_{33} & K^e_{34} & K^e_{35} & K^e_{36} \\ K^e_{41} & K^e_{42} & K^e_{43} & K^e_{44} & K^e_{45} & K^e_{46} \\ K^e_{51} & K^e_{52} & K^e_{53} & K^e_{54} & K^e_{55} & K^e_{56} \\ K^e_{61} & K^e_{62} & K^e_{63} & K^e_{64} & K^e_{65} & K^e_{66} \end{bmatrix} \tag{10.34b}$$

In eq(10.34b) the components of the leading diagonal are

$$\begin{aligned}
K_{11}^e &= [t/(4\Delta)][D_{11}(y_2 - y_3)^2 + D_{33}(x_3 - x_2)^2] \\
K_{22}^e &= [t/(4\Delta)][D_{22}(x_3 - x_2)^2 + D_{33}(y_2 - y_3)^2] \\
K_{33}^e &= [t/(4\Delta)][D_{11}(y_3 - y_1)^2 + D_{33}(x_1 - x_3)^2] \\
K_{44}^e &= [t/(4\Delta)][D_{22}(x_1 - x_3)^2 + D_{33}(y_3 - y_1)^2] \\
K_{55}^e &= [t/(4\Delta)][D_{11}(y_1 - y_2)^2 + D_{33}(x_2 - x_1)^2] \\
K_{66}^e &= [t/(4\Delta)][D_{22}(x_2 - x_1)^2 + D_{33}(y_1 - y_2)^2]
\end{aligned} \tag{10.34c}$$

and the off-diagonal components are

$$\begin{aligned}
K_{12}^e = K_{21}^e &= [t/(4\Delta)][D_{12}(x_3 - x_2)(y_2 - y_3) + D_{33}(x_3 - x_2)(y_2 - y_3)] \\
K_{13}^e = K_{31}^e &= [t/(4\Delta)][D_{11}(y_3 - y_1)(y_2 - y_3) + D_{33}(x_3 - x_2)(x_1 - x_3)] \\
K_{14}^e = K_{41}^e &= [t/(4\Delta)][D_{12}(x_1 - x_3)(y_2 - y_3) + D_{33}(x_3 - x_2)(y_3 - y_1)] \\
K_{15}^e = K_{51}^e &= [t/(4\Delta)][D_{11}(y_1 - y_2)(y_2 - y_3) + D_{33}(x_3 - x_2)(x_2 - x_1)] \\
K_{16}^e = K_{61}^e &= [t/(4\Delta)][D_{12}(x_2 - x_1)(y_2 - y_3) + D_{33}(x_3 - x_2)(y_1 - y_2)] \\
K_{23}^e = K_{32}^e &= [t/(4\Delta)][D_{12}(x_3 - x_2)(y_3 - y_1) + D_{33}(x_1 - x_3)(y_2 - y_3)] \\
K_{24}^e = K_{42}^e &= [t/(4\Delta)][D_{22}(x_3 - x_2)(x_1 - x_3) + D_{33}(y_3 - y_1)(y_2 - y_3)] \\
K_{25}^e = K_{52}^e &= [t/(4\Delta)][D_{12}(x_3 - x_2)(y_1 - y_2) + D_{33}(x_2 - x_1)(y_2 - y_3)] \\
K_{26}^e = K_{62}^e &= [t/(4\Delta)][D_{22}(x_2 - x_1)(x_3 - x_2) + D_{33}(y_1 - y_2)(y_2 - y_3)] \\
K_{34}^e = K_{43}^e &= [t/(4\Delta)][D_{12}(x_1 - x_3)(y_3 - y_1) + D_{33}(x_1 - x_3)(y_3 - y_1)] \\
K_{35}^e = K_{53}^e &= [t/(4\Delta)][D_{11}(y_1 - y_2)(y_3 - y_1) + D_{33}(x_1 - x_3)(x_2 - x_1)] \\
K_{36}^e = K_{63}^e &= [t/(4\Delta)][D_{12}(x_2 - x_1)(y_3 - y_1) + D_{33}(x_1 - x_3)(y_1 - y_2)] \\
K_{45}^e = K_{54}^e &= [t/(4\Delta)][D_{12}(x_1 - x_3)(y_1 - y_2) + D_{33}(x_2 - x_1)(y_3 - y_1)] \\
K_{46}^e = K_{64}^e &= [t/(4\Delta)[]D_{22}(x_1 - x_3)(x_2 - x_1) + D_{33}(y_1 - y_2)(y_3 - y_1)] \\
K_{56}^e = K_{65}^e &= [t/(4\Delta)][D_{12}(x_2 - x_1)(y_1 - y_2) + D_{33}(x_2 - x_1)(y_1 - y_2)]
\end{aligned} \tag{10.34d}$$

where $D_{ij} = D_{ji}$ $(i, j = 1, 2, 3)$ have been previously defined in eqs(10.31b and c). These show, for example, (i) $D_{11} = E/(1 - \nu^2) = D_{22}$ in plane stress, (ii) $D_{11} = (1 - \nu)E/[(1+\nu)(1-2\nu)] = D_{22}$ in plane strain and (iii) $D_{13} = D_{23} = 0$ for both conditions. Replacing ε with ε^e in eqs(10.30a) refers the constant nodal strain components to each of the three nodes as

$$\boldsymbol{\varepsilon}^e = \mathbf{B}^e \boldsymbol{\delta}^e \tag{10.35a}$$

where $\mathbf{B}^e = \mathbf{B}$ in eq(10.30b). Similarly, from eq(10.31a) the constant nodal stresses are

$$\boldsymbol{\sigma}^e = \mathbf{D}\,\mathbf{B}^e \boldsymbol{\delta}^e = \mathbf{H}^e \boldsymbol{\delta}^e \tag{10.35b}$$

The product $\mathbf{H}^e = \mathbf{D}\,\mathbf{B}^e$ in eq(10.35b) is again defined, explicitly, in terms of the nodal point co-ordinates and the elastic constants as

$$\mathbf{H}^e = \frac{1}{2\Delta}\begin{bmatrix} D_{11}(y_2-y_3) & D_{12}(x_3-x_2) & D_{11}(y_3-y_1) & D_{12}(x_1-x_3) & D_{11}(y_1-y_2) & D_{12}(x_2-x_1) \\ D_{21}(y_2-y_3) & D_{22}(x_3-x_2) & D_{21}(y_3-y_1) & D_{22}(x_1-x_3) & D_{21}(y_1-y_2) & D_{22}(x_2-x_1) \\ D_{33}(x_3-x_2) & D_{33}(y_2-y_3) & D_{33}(x_1-x_3) & D_{33}(y_3-y_1) & D_{33}(x_2-x_1) & D_{33}(y_1-y_2) \end{bmatrix} \tag{10.35c}$$

Equation (10.35c) applies to both plane stress and strain in which eqs(10.31b,c) again supply their respective D_{ij} components.

10.5.5 Overall Stiffness Matrix

Finally, the overall stiffness matrix $\mathbf{K}$ is assembled from the individual element stifness matrices $\mathbf{K}^e$ (e = I, II, III etc). The nodal displacement vector can then be found from the following inversion process:

$$\mathbf{f} = \mathbf{K}\,\boldsymbol{\delta}, \;\Rightarrow\; \boldsymbol{\delta} = \mathbf{K}^{-1}\mathbf{f} \tag{10.36a,b}$$

The assembly of an overall stiffness matrix is now illustrated for the 4-element plane stress cantilever shown in Fig. 10.10.

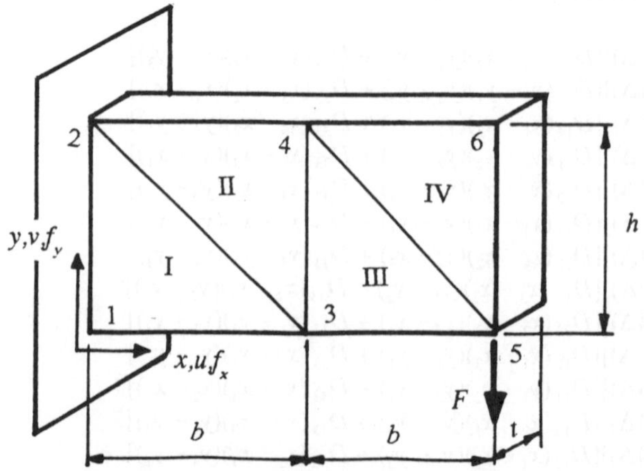

Figure 10.10 Cantilever with four triangular elements

First note that $\mathbf{K}^e$ for each element is 6 × 6. The dimensions of $\mathbf{K}^e$ follow from the fact that there are two degrees of freedom in the three nodes of an element. Since there are two degrees of freedom at each of 6 nodes for the beam, the total of 12 degrees of freedom implies that the overall stiffness matrix **K** will have dimensions of 12 × 12. Consider element e = I, eq(10.32a) becomes $\mathbf{f}^{\mathrm{I}} = \mathbf{K}^{\mathrm{I}}\,\delta^{\mathrm{I}}$. Since the numbering is clockwise then 3 replaces 2 and 2 replaces 3 in the foregoing analyses. This gives

$$\begin{bmatrix} f_{x1} \\ f_{y1} \\ f_{x3} \\ f_{y3} \\ f_{x2} \\ f_{y2} \end{bmatrix} = \begin{bmatrix} K_{11}^{\mathrm{I}} & K_{12}^{\mathrm{I}} & K_{13}^{\mathrm{I}} & K_{14}^{\mathrm{I}} & K_{15}^{\mathrm{I}} & K_{16}^{\mathrm{I}} \\ K_{21}^{\mathrm{I}} & K_{22}^{\mathrm{I}} & K_{23}^{\mathrm{I}} & K_{24}^{\mathrm{I}} & K_{25}^{\mathrm{I}} & K_{26}^{\mathrm{I}} \\ K_{31}^{\mathrm{I}} & K_{32}^{\mathrm{I}} & K_{33}^{\mathrm{I}} & K_{34}^{\mathrm{I}} & K_{35}^{\mathrm{I}} & K_{36}^{\mathrm{I}} \\ K_{41}^{\mathrm{I}} & K_{42}^{\mathrm{I}} & K_{43}^{\mathrm{I}} & K_{44}^{\mathrm{I}} & K_{45}^{\mathrm{I}} & K_{46}^{\mathrm{I}} \\ K_{51}^{\mathrm{I}} & K_{52}^{\mathrm{I}} & K_{53}^{\mathrm{I}} & K_{54}^{\mathrm{I}} & K_{55}^{\mathrm{I}} & K_{56}^{\mathrm{I}} \\ K_{61}^{\mathrm{I}} & K_{62}^{\mathrm{I}} & K_{63}^{\mathrm{I}} & K_{64}^{\mathrm{I}} & K_{65}^{\mathrm{I}} & K_{66}^{\mathrm{I}} \end{bmatrix} \begin{bmatrix} u_1 \\ v_1 \\ u_3 \\ v_3 \\ u_2 \\ v_2 \end{bmatrix} \tag{10.37a}$$

The matrix components K_{ij}^{I} are again defined by eqs(10.34c and d) when subscripts 2 and 3 on x and y are interchanged. These single element stiffnesses are re-assembled within the overall stiffness matrix **K** in eq(10.36a) as follows:

$$\begin{bmatrix} f_{x1} \\ f_{y1} \\ f_{x2} \\ f_{y2} \\ f_{x3} \\ f_{y3} \\ f_{x4} \\ f_{y4} \\ f_{x5} \\ f_{y5} \\ f_{x6} \\ f_{y6} \end{bmatrix} = \begin{bmatrix} K_{11}^{\mathrm{I}} & K_{12}^{\mathrm{I}} & K_{15}^{\mathrm{I}} & K_{16}^{\mathrm{I}} & K_{13}^{\mathrm{I}} & K_{14}^{\mathrm{I}} & \cdots \\ K_{21}^{\mathrm{I}} & K_{22}^{\mathrm{I}} & K_{25}^{\mathrm{I}} & K_{26}^{\mathrm{I}} & K_{23}^{\mathrm{I}} & K_{24}^{\mathrm{I}} & \cdots \\ K_{51}^{\mathrm{I}} & K_{52}^{\mathrm{I}} & K_{55}^{\mathrm{I}} & K_{56}^{\mathrm{I}} & K_{53}^{\mathrm{I}} & K_{54}^{\mathrm{I}} & \cdots \\ K_{61}^{\mathrm{I}} & K_{62}^{\mathrm{I}} & K_{65}^{\mathrm{I}} & K_{66}^{\mathrm{I}} & K_{63}^{\mathrm{I}} & K_{64}^{\mathrm{I}} & \cdots \\ K_{31}^{\mathrm{I}} & K_{32}^{\mathrm{I}} & K_{35}^{\mathrm{I}} & K_{36}^{\mathrm{I}} & K_{33}^{\mathrm{I}} & K_{34}^{\mathrm{I}} & \cdots \\ K_{41}^{\mathrm{I}} & K_{42}^{\mathrm{I}} & K_{45}^{\mathrm{I}} & K_{46}^{\mathrm{I}} & K_{43}^{\mathrm{I}} & K_{44}^{\mathrm{I}} & \cdots \\ \cdot & \cdot & \cdot & \cdot & \cdot & \cdot & \cdots \\ \cdot & \cdot & \cdot & \cdot & \cdot & \cdot & \cdots \\ \cdot & \cdot & \cdot & \cdot & \cdot & \cdot & \cdots \\ \cdot & \cdot & \cdot & \cdot & \cdot & \cdot & \cdots \\ \cdot & \cdot & \cdot & \cdot & \cdot & \cdot & \cdots \\ \cdot & \cdot & \cdot & \cdot & \cdot & \cdot & \cdots \end{bmatrix} \begin{bmatrix} u_1 \\ v_1 \\ u_2 \\ v_2 \\ u_3 \\ v_3 \\ u_4 \\ v_4 \\ u_5 \\ v_5 \\ u_6 \\ v_6 \end{bmatrix} \tag{10.37b}$$

Applying eq(10.32a) to the second element e = II, gives $\mathbf{f}^{\mathrm{II}} = \mathbf{K}^{\mathrm{II}}\ \delta^{\mathrm{II}}$. Since the node numbering 234 here is anti-clockwise, this equation becomes

$$\begin{bmatrix} f_{x2} \\ f_{y2} \\ f_{x3} \\ f_{y3} \\ f_{x4} \\ f_{y4} \end{bmatrix} = \begin{bmatrix} K_{11}^{II} & K_{12}^{II} & K_{13}^{II} & K_{14}^{II} & K_{15}^{II} & K_{16}^{II} \\ K_{21}^{II} & K_{22}^{II} & K_{23}^{II} & K_{24}^{II} & K_{25}^{II} & K_{26}^{II} \\ K_{31}^{II} & K_{32}^{II} & K_{33}^{II} & K_{34}^{II} & K_{35}^{II} & K_{36}^{II} \\ K_{41}^{II} & K_{42}^{II} & K_{43}^{II} & K_{44}^{II} & K_{45}^{II} & K_{46}^{II} \\ K_{51}^{II} & K_{52}^{II} & K_{53}^{II} & K_{54}^{II} & K_{55}^{II} & K_{56}^{II} \\ K_{61}^{II} & K_{62}^{II} & K_{63}^{II} & K_{64}^{II} & K_{65}^{II} & K_{66}^{II} \end{bmatrix} \begin{bmatrix} u_2 \\ v_2 \\ u_3 \\ v_3 \\ u_4 \\ v_4 \end{bmatrix} \qquad (10.37c)$$

in which K_{ij}^{II} are defined from eqs(10.34c and d) where subscripts 1, 2 and 3 on x and y are replaced by 2, 3 and 4 respectively. These are assembled within eq(10.36a) as follows:

$$\begin{bmatrix} f_{x1} \\ f_{y1} \\ f_{x2} \\ f_{y2} \\ f_{x3} \\ f_{y3} \\ f_{x4} \\ f_{y4} \\ f_{x5} \\ f_{y5} \\ f_{x6} \\ f_{y6} \end{bmatrix} = \begin{bmatrix} \cdot & \cdot & \cdot & \cdot & \cdot & \cdot & \cdot & \cdot & \cdot & \cdot & \cdot & \cdot \\ \cdot & \cdot & \cdot & \cdot & \cdot & \cdot & \cdot & \cdot & \cdot & \cdot & \cdot & \cdot \\ \cdot & \cdot & K_{11}^{II} & K_{12}^{II} & K_{13}^{II} & K_{14}^{II} & K_{15}^{II} & K_{16}^{II} & \cdot & \cdot & \cdot & \cdot \\ \cdot & \cdot & K_{21}^{II} & K_{22}^{II} & K_{23}^{II} & K_{24}^{II} & K_{25}^{II} & K_{26}^{II} & \cdot & \cdot & \cdot & \cdot \\ \cdot & \cdot & K_{31}^{II} & K_{32}^{II} & K_{33}^{II} & K_{34}^{II} & K_{35}^{II} & K_{36}^{II} & \cdot & \cdot & \cdot & \cdot \\ \cdot & \cdot & K_{41}^{II} & K_{42}^{II} & K_{43}^{II} & K_{44}^{II} & K_{45}^{II} & K_{46}^{II} & \cdot & \cdot & \cdot & \cdot \\ \cdot & \cdot & K_{51}^{II} & K_{52}^{II} & K_{53}^{II} & K_{54}^{II} & K_{55}^{II} & K_{56}^{II} & \cdot & \cdot & \cdot & \cdot \\ \cdot & \cdot & K_{61}^{II} & K_{62}^{II} & K_{63}^{II} & K_{64}^{II} & K_{65}^{II} & K_{66}^{II} & \cdot & \cdot & \cdot & \cdot \\ \cdot & \cdot & \cdot & \cdot & \cdot & \cdot & \cdot & \cdot & \cdot & \cdot & \cdot & \cdot \\ \cdot & \cdot & \cdot & \cdot & \cdot & \cdot & \cdot & \cdot & \cdot & \cdot & \cdot & \cdot \\ \cdot & \cdot & \cdot & \cdot & \cdot & \cdot & \cdot & \cdot & \cdot & \cdot & \cdot & \cdot \\ \cdot & \cdot & \cdot & \cdot & \cdot & \cdot & \cdot & \cdot & \cdot & \cdot & \cdot & \cdot \end{bmatrix} \begin{bmatrix} u_1 \\ v_1 \\ u_2 \\ v_2 \\ u_3 \\ v_3 \\ u_4 \\ v_4 \\ u_5 \\ v_5 \\ u_6 \\ v_6 \end{bmatrix} \qquad (10.37d)$$

In applying $\mathbf{f}^{III} = \mathbf{K}^{III}\,\delta^{III}$ to element e = III, eq(10.32a) becomes

$$\begin{bmatrix} f_{x3} \\ f_{y3} \\ f_{x5} \\ f_{y5} \\ f_{x4} \\ f_{y4} \end{bmatrix} = \begin{bmatrix} K_{11}^{III} & K_{12}^{III} & K_{13}^{III} & K_{14}^{III} & K_{15}^{III} & K_{16}^{III} \\ K_{21}^{III} & K_{22}^{III} & K_{23}^{III} & K_{24}^{III} & K_{25}^{III} & K_{26}^{III} \\ K_{31}^{III} & K_{32}^{III} & K_{33}^{III} & K_{34}^{III} & K_{35}^{III} & K_{36}^{III} \\ K_{41}^{III} & K_{42}^{III} & K_{43}^{III} & K_{44}^{III} & K_{45}^{III} & K_{46}^{III} \\ K_{51}^{III} & K_{52}^{III} & K_{53}^{III} & K_{54}^{III} & K_{55}^{III} & K_{56}^{III} \\ K_{61}^{III} & K_{62}^{III} & K_{63}^{III} & K_{64}^{III} & K_{65}^{III} & K_{66}^{III} \end{bmatrix} \begin{bmatrix} u_3 \\ v_3 \\ u_5 \\ v_5 \\ u_4 \\ v_4 \end{bmatrix} \qquad (10.37e)$$

Finally, for element e = IV, $\mathbf{f}^{IV} = \mathbf{K}^{IV}\,\boldsymbol{\delta}^{IV}$ becomes

$$\begin{bmatrix} f_{x4} \\ f_{y4} \\ f_{x5} \\ f_{y5} \\ f_{x6} \\ f_{y6} \end{bmatrix} = \begin{bmatrix} K_{11}^{IV} & K_{12}^{IV} & K_{13}^{IV} & K_{14}^{IV} & K_{15}^{IV} & K_{16}^{IV} \\ K_{21}^{IV} & K_{22}^{IV} & K_{23}^{IV} & K_{24}^{IV} & K_{25}^{IV} & K_{26}^{IV} \\ K_{31}^{IV} & K_{32}^{IV} & K_{33}^{IV} & K_{34}^{IV} & K_{35}^{IV} & K_{36}^{IV} \\ K_{41}^{IV} & K_{42}^{IV} & K_{43}^{IV} & K_{44}^{IV} & K_{45}^{IV} & K_{46}^{IV} \\ K_{51}^{IV} & K_{52}^{IV} & K_{53}^{IV} & K_{54}^{IV} & K_{55}^{IV} & K_{56}^{IV} \\ K_{61}^{IV} & K_{62}^{IV} & K_{63}^{IV} & K_{64}^{IV} & K_{65}^{IV} & K_{66}^{IV} \end{bmatrix} \begin{bmatrix} u_4 \\ v_4 \\ u_5 \\ v_5 \\ u_6 \\ v_6 \end{bmatrix} \qquad (10.37f)$$

When the components of $\mathbf{K}^{III}$ and $\mathbf{K}^{IV}$ are further added to the appropriate locations and combined with eqs(10.37b) and (10.37d), the overall stiffness eq(10.36a) for the cantilever becomes

$$\begin{bmatrix} f_{x1} \\ f_{y1} \\ f_{x2} \\ f_{y2} \\ f_{x3} \\ f_{y3} \\ f_{x4} \\ f_{y4} \\ f_{x5} \\ f_{y5} \\ f_{x6} \\ f_{y6} \end{bmatrix} = \begin{bmatrix}
K_{11}^{I} & K_{12}^{I} & K_{15}^{I} & K_{16}^{I} & K_{13}^{I} & K_{14}^{I} & \cdot & \cdot & \cdot & \cdot & \cdot & \\
K_{21}^{I} & K_{22}^{I} & K_{25}^{I} & K_{26}^{I} & K_{23}^{I} & K_{24}^{I} & \cdot & \cdot & \cdot & \cdot & \cdot & \cdot \\
K_{51}^{I} & K_{52}^{I} & K_{55}^{I}+K_{11}^{II} & K_{56}^{I}+K_{12}^{II} & K_{53}^{I}+K_{13}^{II} & K_{54}^{I}+K_{14}^{II} & K_{15}^{II} & K_{16}^{II} & \cdot & \cdot & \cdot & \cdot \\
K_{61}^{I} & K_{62}^{I} & K_{65}^{I}+K_{21}^{II} & K_{66}^{I}+K_{22}^{II} & K_{63}^{I}+K_{23}^{II} & K_{64}^{I}+K_{24}^{II} & K_{25}^{II} & K_{26}^{II} & \cdot & \cdot & \cdot & \cdot \\
K_{31}^{I} & K_{32}^{I} & K_{35}^{I}+K_{31}^{II} & K_{36}^{I}+K_{32}^{II} & K_{33}^{I}+K_{33}^{II}+K_{11}^{III} & K_{34}^{I}+K_{34}^{II}+K_{12}^{III} & K_{35}^{II}+K_{15}^{III} & K_{36}^{II}+K_{16}^{III} & K_{13}^{III} & K_{14}^{III} & \cdot & \cdot \\
K_{41}^{I} & K_{42}^{I} & K_{45}^{I}+K_{41}^{II} & K_{46}^{I}+K_{42}^{II} & K_{43}^{I}+K_{43}^{II}+K_{21}^{III} & K_{44}^{I}+K_{44}^{II}+K_{22}^{III} & K_{45}^{II}+K_{25}^{III} & K_{46}^{II}+K_{26}^{III} & K_{23}^{III} & K_{24}^{III} & \cdot & \cdot \\
\cdot & \cdot & K_{51}^{II} & K_{52}^{II} & K_{53}^{II}+K_{51}^{III} & K_{54}^{II}+K_{52}^{III} & K_{55}^{II}+K_{55}^{III}+K_{11}^{IV} & K_{56}^{II}+K_{56}^{III}+K_{12}^{IV} & K_{53}^{III}+K_{13}^{IV} & K_{54}^{III}+K_{14}^{IV} & K_{15}^{IV} & K_{16}^{IV} \\
\cdot & \cdot & K_{61}^{II} & K_{62}^{II} & K_{63}^{II}+K_{61}^{III} & K_{64}^{II}+K_{62}^{III} & K_{65}^{II}+K_{65}^{III}+K_{21}^{IV} & K_{66}^{II}+K_{66}^{III}+K_{22}^{IV} & K_{63}^{III}+K_{23}^{IV} & K_{64}^{III}+K_{24}^{IV} & K_{25}^{IV} & K_{26}^{IV} \\
\cdot & \cdot & \cdot & \cdot & K_{31}^{III} & K_{32}^{III} & K_{35}^{III}+K_{31}^{IV} & K_{36}^{III}+K_{32}^{IV} & K_{33}^{III}+K_{33}^{IV} & K_{34}^{III}+K_{34}^{IV} & K_{35}^{IV} & K_{36}^{IV} \\
\cdot & \cdot & \cdot & \cdot & K_{41}^{III} & K_{42}^{III} & K_{45}^{III}+K_{41}^{IV} & K_{46}^{III}+K_{42}^{IV} & K_{43}^{III}+K_{43}^{IV} & K_{44}^{III}+K_{44}^{IV} & K_{45}^{IV} & K_{46}^{IV} \\
\cdot & \cdot & \cdot & \cdot & \cdot & \cdot & K_{51}^{IV} & K_{52}^{IV} & K_{53}^{IV} & K_{54}^{IV} & K_{55}^{IV} & K_{56}^{IV} \\
\cdot & \cdot & \cdot & \cdot & \cdot & \cdot & K_{61}^{IV} & K_{62}^{IV} & K_{63}^{IV} & K_{64}^{IV} & K_{65}^{IV} & K_{66}^{IV}
\end{bmatrix} \begin{bmatrix} u_1 \\ v_1 \\ u_2 \\ v_2 \\ u_3 \\ v_3 \\ u_4 \\ v_4 \\ u_5 \\ v_5 \\ u_6 \\ v_6 \end{bmatrix}$$

$$(10.38a)$$

The large number of stiffness calculations that arises here with just four elements shows that a computer is an essential requirement in FE practice involving several hundred elements. Note that **K,** like $\mathbf{K}^e$, is again symmetrical about the leading diagonal. With careful node-numbering, a narrow banding may be achieved thus reducing the amount of computer memory required. Cost may further be reduced by increasing the speed when solving for $\boldsymbol{\delta}$ between the many simultaneous equations that eq(10.38a) contains. For this, sub-matrices are used in the assembly of **K**. For example, with the partitioning of eqs(10.37a,c,e and f) submatrices $\mathbf{S}_{ij}^{\,e}$, are identified within eq(10.37g) as follows:

$$
\begin{bmatrix} \mathbf{f}_1 \\ \mathbf{f}_2 \\ \mathbf{f}_3 \\ \mathbf{f}_4 \\ \mathbf{f}_5 \\ \mathbf{f}_6 \end{bmatrix} = \begin{bmatrix} \mathbf{S}_{11}^{\mathrm{I}} & \mathbf{S}_{12}^{\mathrm{I}} & \mathbf{S}_{13}^{\mathrm{I}} & \cdot & \cdot & \cdot \\ \mathbf{S}_{21}^{\mathrm{I}} & \mathbf{S}_{22}^{\mathrm{I}}+\mathbf{S}_{22}^{\mathrm{II}} & \mathbf{S}_{23}^{\mathrm{I}}+\mathbf{S}_{23}^{\mathrm{II}} & \mathbf{S}_{24}^{\mathrm{II}} & \cdot & \cdot \\ \mathbf{S}_{31}^{\mathrm{I}} & \mathbf{S}_{32}^{\mathrm{I}}+\mathbf{S}_{32}^{\mathrm{II}} & \mathbf{S}_{33}^{\mathrm{I}}+\mathbf{S}_{33}^{\mathrm{II}}+\mathbf{S}_{33}^{\mathrm{III}} & \mathbf{S}_{34}^{\mathrm{II}}+\mathbf{S}_{34}^{\mathrm{III}} & \mathbf{S}_{35}^{\mathrm{III}} & \cdot \\ \cdot & \mathbf{S}_{42}^{\mathrm{II}} & \mathbf{S}_{43}^{\mathrm{II}}+\mathbf{S}_{43}^{\mathrm{III}} & \mathbf{S}_{44}^{\mathrm{II}}+\mathbf{S}_{44}^{\mathrm{III}}+\mathbf{S}_{44}^{\mathrm{IV}} & \mathbf{S}_{45}^{\mathrm{III}}+\mathbf{S}_{45}^{\mathrm{IV}} & \mathbf{S}_{46}^{\mathrm{IV}} \\ \cdot & \cdot & \mathbf{S}_{53}^{\mathrm{III}} & \mathbf{S}_{54}^{\mathrm{III}}+\mathbf{S}_{54}^{\mathrm{IV}} & \mathbf{S}_{55}^{\mathrm{III}}+\mathbf{S}_{55}^{\mathrm{IV}} & \mathbf{S}_{56}^{\mathrm{IV}} \\ \cdot & \cdot & \cdot & \mathbf{S}_{64}^{\mathrm{IV}} & \mathbf{S}_{65}^{\mathrm{IV}} & \mathbf{S}_{66}^{\mathrm{IV}} \end{bmatrix} \begin{bmatrix} \delta_1 \\ \delta_2 \\ \delta_3 \\ \delta_4 \\ \delta_5 \\ \delta_6 \end{bmatrix} \tag{10.38b}
$$

where $\mathbf{f} = [f_x \; f_y]^{\mathrm{T}}$ and $\delta = [u \;\; v]^{\mathrm{T}}$. Individual stiffnesses in eq(10.38a) will reappear with the addition of the sub-matrices. For example, in eq(10.38b), $\mathbf{S}_{33}{}^{\mathrm{I}} + \mathbf{S}_{33}{}^{\mathrm{II}} + \mathbf{S}_{33}{}^{\mathrm{III}}$, represents

$$
\begin{bmatrix} K_{33}^{\mathrm{I}} & K_{34}^{\mathrm{I}} \\ K_{43}^{\mathrm{I}} & K_{44}^{\mathrm{I}} \end{bmatrix} + \begin{bmatrix} K_{33}^{\mathrm{II}} & K_{34}^{\mathrm{II}} \\ K_{43}^{\mathrm{II}} & K_{44}^{\mathrm{II}} \end{bmatrix} + \begin{bmatrix} K_{11}^{\mathrm{III}} & K_{12}^{\mathrm{III}} \\ K_{21}^{\mathrm{III}} & K_{22}^{\mathrm{III}} \end{bmatrix} = \begin{bmatrix} (K_{33}^{\mathrm{I}}+K_{33}^{\mathrm{II}}+K_{11}^{\mathrm{III}}) & (K_{34}^{\mathrm{I}}+K_{34}^{\mathrm{II}}+K_{12}^{\mathrm{III}}) \\ (K_{43}^{\mathrm{I}}+K_{43}^{\mathrm{II}}+K_{21}^{\mathrm{III}}) & (K_{44}^{\mathrm{I}}+K_{44}^{\mathrm{II}}+K_{22}^{\mathrm{III}}) \end{bmatrix}
$$

The final step is to ensure that the boundary conditions are met. There are various procedures to maintain zero displacement at support points. In a manual assembly of **K** we can account for $u_1 = v_1 = u_2 = v_2 = 0$ by eliminating the row and column corresponding to the fixed displacements. Alternatively, negligible displacements can be achieved numerically by multiplying the appropriate diagonal stiffnesses, K_{ii}, by a large number, say 10^{10}. This is equivalent to directly setting $K_{ii} = 1$ and the remaining $K_{ij} = 0$ in the row and column, as the following example illustrates.

Example 10.3 Given $b = 120$ mm, $h = 80$ mm and $t = 10$ mm for the cantilever in Fig. 10.10, derive the overall stiffness matrix and find the nodal displacements when $F = 10$ kN. Take $E = 200$ GPa and $\nu = 0.3$.

Table 10.2 gives specific data for the stiffness calculations. The origin of x, y is co-incident with node 1. The applied force is reacted by vertical concentrated forces at nodes 1 and 2.

Table 10.2 Nodal force and co-ordinate input data (mm, kN)

Node	x	y	f_x	f_y
1	0	0	0	5
2	0	80	0	5
3	120	0	0	0
4	120	80	0	0
5	240	0	0	−10
6	240	80	0	0

It is shown later in Example 10.5 that the fixed-end reaction follows a symmetrical parabolic distribution which can be replaced by an equal lumping of forces.

The components K_{ij} of the individual element stiffness matrix (10.34b) are expressed in eqs(10.34c,d). The element area $\Delta = bh/2 = 120 \times 80/2 = 4800\ \text{mm}^2$ is constant and, under plane stress, D_{ij} follows from eq(10.31b) as

$$\mathbf{D} = \frac{200 \times 10^3}{(1 - 0.3^2)} \begin{bmatrix} 1 & 0.3 & 0 \\ 0.3 & 1 & 0 \\ 0 & 0 & 0.35 \end{bmatrix}$$

A constant coefficient is used for stiffness calcuations with elements of similar geometry and a single beam material

$$[E/(1 - \nu^2)] \times t/(4\Delta) = [200 \times 10^3/(1 - 0.3^2)] \times [10/(4 \times 4800)] = 114.47\ \text{N/mm}^3$$

Equations (10.34c,d) were derived for counter-clockwise numbering of nodes. Thus for element e = I in Fig. 10.10, it is necessary to interchange the subscripts 2 and 3. Equation (10.34c) gives, for example, the diagonal stiffnesses:

$$\begin{aligned} K_{11}{}^{\text{I}} &= 114.47[D_{11}(y_3 - y_2)^2 + D_{33}(x_2 - x_3)^2] \\ &= 114.47[1(0 - 80)^2 + 0.35(0 - 120)^2] = 1.3095 \times 10^6\ \text{N/mm} \end{aligned}$$

$$\begin{aligned} K_{22}{}^{\text{I}} &= 114.47[D_{22}(x_2 - x_3)^2 + D_{33}(y_3 - y_2)^2] \\ &= 114.47[1(0 - 120)^2 + 0.35\,(0 - 80)^2] = 1.9048 \times 10^6\ \text{N/mm} \end{aligned}$$

and from eq(10.34d) we have a sample of the off-axis stiffnesses:

$$\begin{aligned} K_{12}{}^{\text{I}} &= 114.47\,[D_{12}(x_2 - x_3)(y_3 - y_2) + D_{33}(x_2 - x_3)(y_3 - y_2)] \\ &= 114.47[0.3(0 - 120)(0 - 80) + 0.35(0 - 120)(0 - 80)] = 0.7143 \times 10^6\ \text{N/mm} \end{aligned}$$

$$\begin{aligned} K_{13}{}^{\text{I}} &= 114.47[D_{11}(y_2 - y_1)(y_3 - y_2) + D_{33}(x_2 - x_3)(x_1 - x_2)] \\ &= 114.47[1(80 - 0)(0 - 80) + 0.35(0 - 120)(0 - 0)] = -\,0.7326 \times 10^6\ \text{N/mm} \end{aligned}$$

For element e = II (Fig. 10.10), counter-clockwise node numbers 2, 3 and 4 replace subscripts 1, 2 and 3 respectively in eqs(10.34c,d). These give, for example,

$$\begin{aligned} K_{11}{}^{\text{II}} &= 114.47[D_{11}(y_3 - y_4)^2 + D_{33}(x_4 - x_3)^2] \\ &= 114.47\,[1\,(0 - 80)^2 + 0.35(120 - 120)^2] = 0.7326 \times 10^6\ \text{N/mm} \end{aligned}$$

$$\begin{aligned} K_{12}{}^{\text{II}} &= 114.47[D_{12}(x_4 - x_3)(y_3 - y_4) + D_{33}(x_4 - x_3)(y_3 - y_4)] \\ &= 114.47\,[0.3(120 - 120)(0 - 80) + 0.35(120 - 120)(0 - 80)] = 0 \end{aligned}$$

The stiffness components $K_{ij}{}^{\text{III}}$ and $K_{ij}{}^{\text{IV}}$ are found from eqs(10.34c,d) in a similar manner. Assembling the 6×6 matrices $\mathbf{K}^{\text{I}}$, $\mathbf{K}^{\text{II}}$, $\mathbf{K}^{\text{III}}$ and $\mathbf{K}^{\text{IV}}$ within eq(10.38a) gives the 12×12 symmetrically banded, overall stiffness matrix $\mathbf{K}$ for the cantilever beam (units $\times\ 10^6$ N/mm):

$$\begin{bmatrix}
1.3095 & 0.7143 & -0.5769 & -0.3297 & -0.7326 & -0.3846 & \cdot & \cdot & \cdot & \cdot & \cdot & \cdot \\
0.7143 & 1.9048 & -0.3846 & -1.6484 & -0.3297 & -0.2564 & \cdot & \cdot & \cdot & \cdot & \cdot & \cdot \\
-0.5769 & -0.3846 & 1.3095 & 0 & 0 & 0.7143 & -0.7326 & -0.3297 & \cdot & \cdot & \cdot & \cdot \\
-0.3297 & -1.6484 & 0 & 1.9048 & 0.7143 & 0 & -0.3846 & -0.2564 & \cdot & \cdot & \cdot & \cdot \\
-0.7326 & -0.3297 & 0 & 0.7143 & 2.6190 & 0.7143 & -1.1538 & -0.7143 & -0.7326 & -0.3846 & \cdot & \cdot \\
-0.3846 & -0.2564 & 0.7143 & 0 & 0.7143 & 3.8096 & -0.7143 & -3.2968 & -0.3297 & -0.2564 & \cdot & \cdot \\
\cdot & \cdot & -0.7326 & -0.3846 & -1.1538 & -0.7143 & 2.6190 & 0.7143 & 0 & 0.7143 & -0.7326 & -0.3297 \\
\cdot & \cdot & -0.3297 & -0.2564 & -0.7143 & -3.2968 & 0.7143 & 3.8096 & 0.7143 & 0 & -0.3846 & -0.2564 \\
\cdot & \cdot & \cdot & \cdot & -0.7326 & -0.3297 & 0 & 0.7143 & 1.3122 & 0 & -0.5769 & -0.3846 \\
\cdot & \cdot & \cdot & \cdot & -0.3846 & -0.2564 & 0.7143 & 0 & 0 & 1.9048 & -0.3297 & -1.6484 \\
\cdot & \cdot & \cdot & \cdot & \cdot & \cdot & -0.7326 & -0.3846 & -0.5769 & -0.3297 & 1.3095 & 0.7143 \\
\cdot & \cdot & \cdot & \cdot & \cdot & \cdot & -0.3297 & -0.2564 & -0.3846 & -1.6484 & 0.7143 & 1.9048
\end{bmatrix} \tag{i}$$

Before the displacements $\boldsymbol{\delta}$ can be found from the solution to $\mathbf{f} = \mathbf{K}\boldsymbol{\delta}$, it is necessary to impose the fixed displacements $u_1 = v_1 = 0$ and $u_2 = v_2 = 0$ within the matrix (i). In the direct method, matrix $\mathbf{K}$ and the force vector $\mathbf{f}$ are modified numerically to become (in units of N and mm)

$$\begin{bmatrix} 0 \\ 0 \\ 0 \\ 0 \\ 0 \\ 0 \\ 0 \\ 0 \\ 0 \\ -0.01 \\ 0 \\ 0 \end{bmatrix} = \begin{bmatrix}
1 & 0 & 0 & 0 & 0 & 0 & \cdot & \cdot & \cdot & \cdot & \cdot & \cdot \\
0 & 1 & 0 & 0 & 0 & 0 & \cdot & \cdot & \cdot & \cdot & \cdot & \cdot \\
0 & 0 & 1 & 0 & 0 & 0 & 0 & 0 & \cdot & \cdot & \cdot & \cdot \\
0 & 0 & 0 & 1 & 0 & 0 & 0 & 0 & \cdot & \cdot & \cdot & \cdot \\
0 & 0 & 0 & 0 & 2.6190 & 0.7143 & -1.1538 & -0.7143 & -0.7326 & -0.3846 & \cdot & \cdot \\
0 & 0 & 0 & 0 & 0.7143 & 3.8096 & -0.7143 & -3.2968 & -0.3297 & -0.2564 & \cdot & \cdot \\
\cdot & \cdot & 0 & 0 & -1.1538 & -0.7143 & 2.6190 & 0.7143 & 0 & 0.7143 & -0.7326 & -0.3297 \\
\cdot & \cdot & 0 & 0 & -0.7143 & -3.2968 & 0.7143 & 3.8096 & 0.7143 & 0 & -0.3846 & -0.2564 \\
\cdot & \cdot & \cdot & \cdot & -0.7326 & -0.3297 & 0 & 0.7143 & 1.3122 & 0 & -0.5769 & -0.3846 \\
\cdot & \cdot & \cdot & \cdot & -0.3846 & -0.2564 & 0.7143 & 0 & 0 & 1.9048 & -0.3297 & -1.6484 \\
\cdot & \cdot & \cdot & \cdot & \cdot & \cdot & -0.7326 & -0.3846 & -0.5769 & -0.3297 & 1.3095 & 0.7143 \\
\cdot & \cdot & \cdot & \cdot & \cdot & \cdot & -0.3297 & -0.2564 & -0.3846 & -1.6484 & 0.7143 & 1.9048
\end{bmatrix} \begin{bmatrix} u_1 \\ v_1 \\ u_2 \\ v_2 \\ u_3 \\ v_3 \\ u_4 \\ v_4 \\ u_5 \\ v_5 \\ u_6 \\ v_6 \end{bmatrix} \tag{ii}$$

Equation (ii) then contains eight equations in the eight unknown displacements. Alternatively, in the assembly of $\mathbf{K}$ we could eliminate the row and column corresponding to the fixed displacements:

$$\begin{bmatrix} 0 \\ 0 \\ 0 \\ 0 \\ 0 \\ -0.01 \\ 0 \\ 0 \end{bmatrix} = \begin{bmatrix}
2.6190 & 0.7143 & -1.1538 & -0.7143 & -0.7326 & -0.3846 & - & - \\
0.7143 & 3.8096 & -0.7143 & -3.2968 & -0.3297 & -0.2564 & - & - \\
-1.1538 & -0.7143 & 2.6190 & 0.7143 & - & 0.7143 & -0.7326 & -0.3297 \\
-0.7143 & -3.2968 & 0.7143 & 3.8096 & 0.7143 & - & -0.3846 & -0.2564 \\
-0.7326 & -0.3297 & - & 0.7143 & 1.3122 & - & -0.5769 & -0.3846 \\
-0.3846 & -0.2564 & 0.7143 & - & - & 1.9048 & -0.3297 & -1.6484 \\
- & - & -0.7326 & -0.3846 & -0.5769 & -0.3297 & 1.3095 & 0.7143 \\
- & - & -0.3297 & -0.2564 & -0.3846 & -1.6484 & 0.7143 & 1.9048
\end{bmatrix} \begin{bmatrix} u_3 \\ v_3 \\ u_4 \\ v_4 \\ u_5 \\ v_5 \\ u_6 \\ v_6 \end{bmatrix} \tag{iii}$$

The solution to eq(iii) provides the following nodal displacements (in mm): $u_3 = -0.0173$, $v_3 = -0.045$, $u_4 = 0.0186$, $v_4 = -0.0477$, $u_5 = -0.0215$, $v_5 = -0.127$, $u_6 = 0.0239$ and $v_6 = -0.127$. Negative v's are downward, i.e. opposing $+y$ in Fig. 10.10. Similarly negative u's are inward, opposing the $+x$ direction.

10.5.6 Force Vector

In order to invert **K** within eq(10.36b) nodal points must co-incide with concentrated force points. Where nodal points can be chosen to coincide with concentrated forces applied around the boundary, the assembly of the force vector $\mathbf{f}^e = [f_{x1}\ f_{y1}\ f_{x2}\ f_{y2} \ldots]^T$ is straightforward. This requires only that each nodal force be resolved into its x and y components. However, forces may be distributed over the boundary edge of an element. Distributed loading, which appears as a pressure p/unit edge area, must be replaced with statically equivalent forces acting at the element nodes. Figure 10.11 shows how a pressure may be resolved into x and y components within the vector $\mathbf{p} = [\,p_x\ p_y\,]^T$.

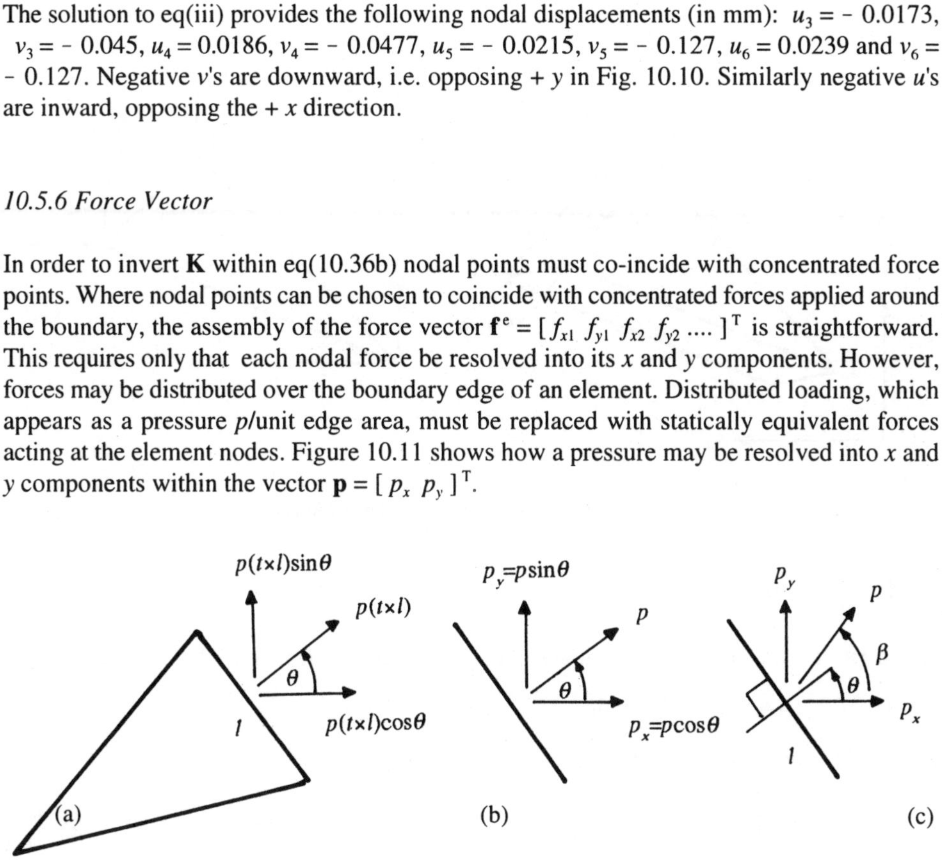

Figure 10.11 Force and pressure components

When p acts normally to the edge, the force resolution given in Fig. 10.11a allows these components to be expressed as $p_x = p\cos\theta$ and $p_y = p\sin\theta$. Figure 10.11b shows that p_x and p_y provide the correct x and y component forces when multiplied by the side area $l \times t$. When the pressure p is inclined to the side, as shown in Fig. 10.11c, then

$$p_x = p\,[\cos(\beta - \theta)\cos\beta], \quad p_y = p\,[\cos(\beta - \theta)\sin\beta]$$

In the case of a varying pressure distribution, p_x and p_y denote functions of x and y respectively. In addition, body forces b/unit volume can arise in every element from self-weight and centrifugal forces. These body forces may be resolved into a vector $\mathbf{b} = [b_x\ b_y]^T$. We shall now consider separately how to replace **p** and **b** with an equivalent system of nodal forces $\mathbf{f}^e$ using the virtual work principle.

(a) Pressure loading

Let a uniformly distributed loading $\mathbf{p} = [\,p_x\ p_y]^T$ be applied to the side 23 of the triangular element as shown in Fig. 10.12a. We wish to replace **p** with equivalent concentrated nodal forces $\mathbf{f}^e = [f_{x1}\ f_{y1}\ f_{x2}\ f_{y2}\ f_{x3}\ f_{y3}]^T$ aligned with an x, y co-ordinate system shown in Fig. 10.12b. Employing the principal of virtual displacements, we equate the work done by the actual loading to the virtual work of the equivalent forces [7]:

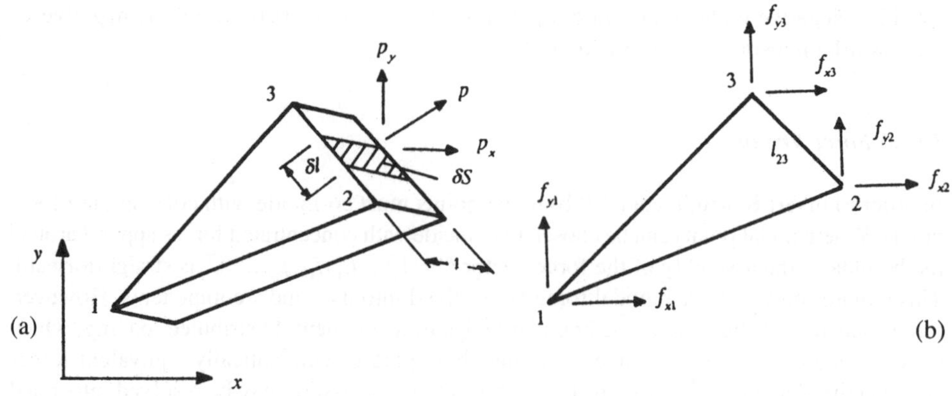

Figure 10.12 Equivalent system of nodal forces

$$\delta^{evT}\,\mathbf{f}^e = \int \delta^{vT}\,\mathbf{p}\,(t\,\mathrm{d}l\,) \tag{10.39a}$$

where $t \times \mathrm{d}l$ is an elemental area along side 23. Substituting from eq(10.27a) gives

$$\delta^{evT}\,\mathbf{f}^e = t\int [(\mathbf{AA}^{e-1})\,\delta^{e}\,]^{vT}\,\mathbf{p}\,\mathrm{d}l$$

$$\delta^{evT}\,\mathbf{f}^e = t\int \delta^{evT}\,(\mathbf{AA}^{e-1})^{T}\,\mathbf{p}\,\mathrm{d}l$$

$$= t\int \delta^{evT}\,(\mathbf{A}^{e-1})^{T}\,\mathbf{A}^{T}\,\mathbf{p}\,\mathrm{d}l \tag{10.39b}$$

where $(\mathbf{A}^{e-1})^T$ means $[(\mathbf{A}^e)^{-1}]^T$. Equations (10.25a) and (10.26b) show that the elements of both δ^{ev} and $(\mathbf{A}^{e-1})^T$ are determined by the nodal values. Hence, they may be removed from the integration in (10.39b). However, from eq(10.24a), $\mathbf{A}^T$ depends upon both x and y. Cancelling δ^{evT} in eq(10.39b) gives the equivalent boundary force vector:

$$\mathbf{f}^e = t\int (\,\mathbf{A}^{e-1})^T\,\mathbf{A}^T\,\mathbf{p}\,\mathrm{d}l = (\mathbf{A}^{e-1})^T\,\mathbf{p}^e \tag{10.40a}$$

where $\mathbf{p}^e = t\int \mathbf{A}^T\,\mathbf{p}\,\mathrm{d}l$ is a 1×6 colummn matrix. Thus $\mathbf{p} = [\,p_x\ p_y\,]^T$ should be pre-multiplied by $\mathbf{A}^T$ and integrated over x and y term by term. This gives

$$\mathbf{p}^e = t\int \begin{bmatrix} 1 & 0 \\ x & 0 \\ y & 0 \\ 0 & 1 \\ 0 & x \\ 0 & y \end{bmatrix} \begin{bmatrix} p_x \\ p_y \end{bmatrix} \mathrm{d}l = t\int \begin{bmatrix} p_x \\ xp_x \\ yp_y \\ p_y \\ xp_y \\ yp_y \end{bmatrix} \mathrm{d}l = t\,l_{23} \begin{bmatrix} p_x \\ \tfrac{1}{2}p_x(x_2+x_3) \\ \tfrac{1}{2}p_x(y_2+y_3) \\ p_y \\ \tfrac{1}{2}p_y(x_2+x_3) \\ \tfrac{1}{2}p_y(y_2+y_3) \end{bmatrix} \tag{10.40b}$$

In converting eq(10.40b) to the single variable to x, with limits of x_2 and x_3, use has been made of the equation of the side 23 in Fig. 10.12b:

$y = (y_3 - y_2)(x - x_2)/\ (x_3 - x_2) + y_2$

and of the relationships:

$dl = l_{23}\, dx\ /\ (x_3 - x_2)$, where $l_{23} = \surd[(x_2 - x_3)^2 + (y_3 - y_2)^2]$

where l_{23} is the length of side 23. We next substitute eqs(10.26b) and (10.40b) into eq(10.40a). Since $(\mathbf{A}^{e-1})^T$ is 6×6, it follows $\mathbf{f}^e$ is 1×6 giving $\mathbf{f}^e = [\, f_{x1}\ f_{y1}\ f_{x2}\ f_{y2}\ f_{x3}\ f_{y3}\,]^T$ in which

$$
\begin{aligned}
f_{x1} &= [t\, l_{23}/(2\Delta)][p_x(x_2y_3 - x_3y_2) + \tfrac{1}{2}p_x(x_2 + x_3)(y_2 - y_3) + \tfrac{1}{2}p_x(y_2 + y_3)(x_3 - x_2)]\\
f_{y1} &= [t\, l_{23}/(2\Delta)][p_y(x_2y_3 - x_3y_2) + \tfrac{1}{2}p_y(x_2 + x_3)(y_2 - y_3) + \tfrac{1}{2}p_y(y_2 + y_3)(x_3 - x_2)]\\
f_{x2} &= [t\, l_{23}/(2\Delta)][p_x(x_3y_1 - x_1y_3) + \tfrac{1}{2}p_x(x_2 + x_3)(y_3 - y_1) + \tfrac{1}{2}p_x(y_2 + y_3)(x_1 - x_3)]\\
f_{y2} &= [t\, l_{23}/(2\Delta)][p_y(x_3y_1 - x_1y_3) + \tfrac{1}{2}p_y(x_2 + x_3)(y_3 - y_1) + \tfrac{1}{2}p_y(y_2 + y_3)(x_1 - x_3)] \quad (10.41a\text{-}f)\\
f_{x3} &= [t\, l_{23}/(2\Delta)][p_x(x_1y_2 - x_2y_1) + \tfrac{1}{2}p_x(x_2 + x_3)(y_1 - y_2) + \tfrac{1}{2}p_x(y_2 + y_3)(x_2 - x_1)]\\
f_{y3} &= [t\, l_{23}/(2\Delta)][p_y(x_1y_2 - x_2y_1) + \tfrac{1}{2}p_y(x_2 + x_3)(y_1 - y_2) + \tfrac{1}{2}p_y(y_2 + y_3)(x_2 - x_1)]
\end{aligned}
$$

Equations (10.41a-f) satisfy force equilibrium

$$
\begin{aligned}
f_{x1} + f_{x2} + f_{x3} &= p_x\, l_{23}\, t\\
f_{y1} + f_{y2} + f_{y3} &= p_y\, l_{23}\, t
\end{aligned}
$$

Boundary loading of the adjacent elements e + 1 and e − 1 will increase the equivalent nodal forces for element e. Equations (10.41a - f) will apply to the elements e + 1 and e − 1 with an appropriate change of subscripts for their node numbering. A summation is required to give the net equivalent nodal forces.

Example 10.4. Show, when side 23 in Fig. 10.12a is (i) horizontal and (ii) vertical, that a normal uniformly distributed loading is equivalent to lumping equal concentrated forces at nodes 2 and 3. Hence determine the nodal forces for the distributed beam loading in Fig. 10.13a.

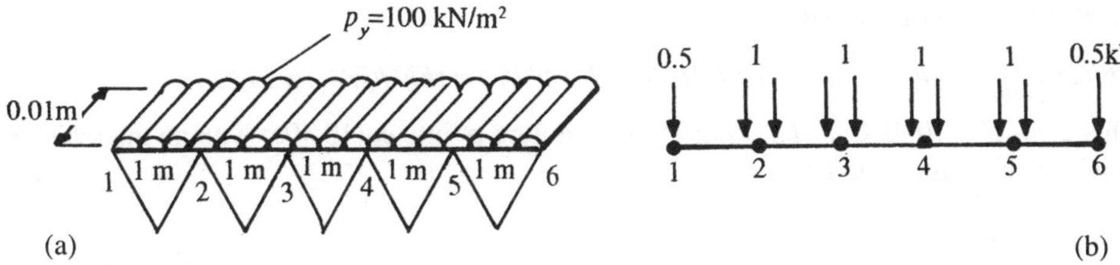

Figure 10.13 Equivalent nodal forces for a beam with uniformly distributed loading

Case (i): Using the previous analysis let the side 23 in Fig. 10.12a lie horizontally. We then set $p_x = 0$, $y_2 = y_3 =$ constant and $l_{23} = x_2 - x_3$ in eqs(10.41a - f). These show that: $f_{x1} = f_{x2} = f_{x3} = 0$ and $f_{y1} = 0$ and $f_{y2} = f_{y3} = \tfrac{1}{2} p_y t\, (x_2 - x_3)$ in which $p_y\, t\, (x_2 - x_1)$ is the total force acting on side 23. The equal lumping provides the equivalent system of nodal forces for the beam shown in Fig. 10.13b. Each equivalent nodal force is $\tfrac{1}{2} \times 100 \times 0.01 \times 1 = 0.5$ kN.

Case (ii): Next, let the side 23 lie vertically. We then set $p_y = 0$, $x_2 = x_3$ = constant and $l_{23} = y_3 - y_2$ in eqs(10.41a - f). These show that $f_{y1} = f_{y2} = f_{y3} = 0$ and $f_{x1} = 0$, leaving only $f_{x2} = f_{x3} = \tfrac{1}{2} p_x t (y_3 - y_2)$. Thus, again, we can lump one half the total horizontal force carried by the sides at each node.

Example 10.5 In the beam of Fig. 10.10 the vertical side 12 of element I reacts the vertical force F applied at node 5. Show that this reaction is equivalent to equal, vertically upward, forces $F/2$ at nodes 1 and 2.

The reaction is distributed parabolically in the manner of the shear stress. This is zero at the edges and a maximum at the centre (see eq 10.46b). Let t be the thickness, h be the length of the vertical side 12 and b be the length of the horizontal side 13. With the origin for y at node 1, eq(10.46b) expresses the pressure, p_y (load distribution/unit edge area) as

$$p_y = [F/(2I)](yh - y^2) \qquad \text{(i)}$$

where $d = h/2$, $y' = y - h/2$ and $I = t\,h^3/12$ is the second moment of the section area. Substituting eq(i) with $p_x = 0$ into eq(10.40b) gives

$$\mathbf{p}^e = t\int \begin{bmatrix} 1 & 0 \\ x & 0 \\ y & 0 \\ 0 & 1 \\ 0 & x \\ 0 & y \end{bmatrix} \begin{bmatrix} 0 \\ \dfrac{F}{2I}(yh - y^2) \end{bmatrix} \mathrm{d}y = t\int \begin{bmatrix} 0 \\ 0 \\ 0 \\ \dfrac{F}{2I}(yh - y^2) \\ \dfrac{Fx}{2I}(yh - y^2) \\ \dfrac{Fy}{2I}(yh - y^2) \end{bmatrix} \mathrm{d}y \qquad \text{(ii)}$$

The non-zero integrands in eq(ii) are

$[tF/(2I)]\int_0^h (yh - y^2)\,\mathrm{d}y = [tF/(2I)](h^3/2 - h^3/3) = tFh^3/(12I) = F$

$[tF/(2I)]\,x\int_0^h (yh - y^2)\,\mathrm{d}y = Fx$

$[tF/(2I)\int_0^h y\,(yh - y^2)\,\mathrm{d}y = [tF/(2I)](h^4/3 - h^4/4) = tFh^4/(24I) = Fh/2$

This gives $\mathbf{p}^e = [0\ \ 0\ \ 0\ \ F\ \ Fx\ \ Fh/2\,]^T$. Substituting $\mathbf{p}^e$ into eq(10.40a) with an interchange between subscripts 1 and 3

$$\begin{bmatrix} f_{x3} \\ f_{y3} \\ f_{x2} \\ f_{y2} \\ f_{x1} \\ f_{y1} \end{bmatrix} = \frac{1}{2\Delta} \begin{bmatrix} (x_2y_1 - x_1y_2) & (y_2 - y_1) & (x_1 - x_2) & 0 & 0 & 0 \\ 0 & 0 & 0 & (x_2y_1 - x_1y_2) & (y_2 - y_1) & (x_1 - x_2) \\ (x_1y_3 - x_3y_1) & (y_1 - y_3) & (x_3 - x_1) & 0 & 0 & 0 \\ 0 & 0 & 0 & (x_1y_3 - x_3y_1) & (y_1 - y_3) & (x_3 - x_1) \\ (x_3y_2 - x_2y_3) & (y_3 - y_2) & (x_2 - x_3) & 0 & 0 & 0 \\ 0 & 0 & 0 & (x_3y_2 - x_2y_3) & (y_3 - y_2) & (x_2 - x_3) \end{bmatrix} \begin{bmatrix} 0 \\ 0 \\ 0 \\ F \\ Fx \\ Fh/2 \end{bmatrix} \qquad \text{(iii)}$$

where $\Delta = bh/2$. Multiplying out eq(iii) and substituting the nodal co-ordinates (x, y) gives the equivalent nodal forces $f_{x1} = f_{x2} = f_{x3} = f_{y3} = 0$ and $f_{y1} = f_{y2} = F\,h\,b/[2(2\Delta)] = F/2$.

(b) Body Forces

In Fig. 10.14a body weight acts vertically downwards and hence is identified as a b_y component. In the case of an element rotating about the z-axis, the radial centrifugal force at any instant is resolved into b_x and b_y components. Thus b_x and b_y in Fig. 10.14a are the net components of all such body forces in an individual element. Figure 10.14a,b show the six equivalent nodal forces $\mathbf{f}^e = [f_{x1}\ f_{y1}\ f_{x2}\ f_{y2}\ f_{x3}\ f_{y3}]^T$ arising from body forces $\mathbf{b} = [b_x\ \ b_y]^T$ per unit volume.

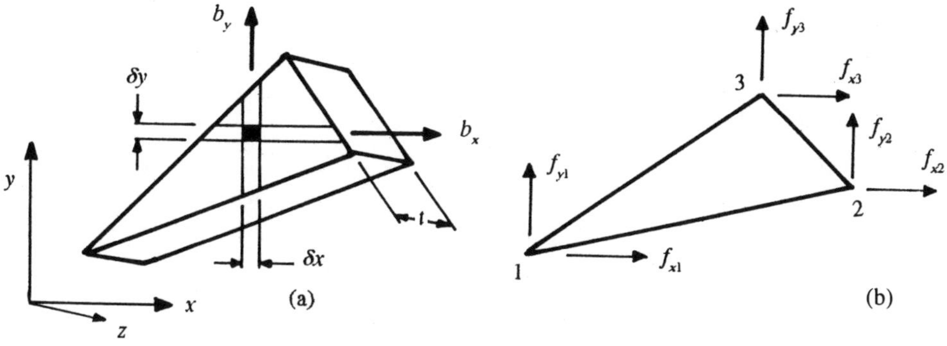

Figure 10.14 Equivalent nodal body forces

The equivalent nodal forces $\mathbf{f}^e$ (see Fig. 10.14b) are found from the PVW. Since each force system is in equilibrium it does no work under a virtual displacement. This gives

$$\delta^{evT}\mathbf{f}^e = \int_V \delta^{vT}\,\mathbf{b}\,\mathrm{d}V \tag{10.42a}$$

Substituting $\mathrm{d}V = t\,\mathrm{d}A$ and $\delta = \mathbf{A}\mathbf{A}^{e-1}\,\delta^e$ into the right side of eq(10.42a) gives

$$\begin{aligned}\delta^{evT}\mathbf{f}^e &= t\int_A \mathbf{A}\mathbf{A}^{e-1}\delta^e)^{vT}\,\mathbf{b}\,\mathrm{d}A \\ &= t\,\delta^{evT}(\mathbf{A}^{e-1})^T\int_A \mathbf{A}^T\,\mathbf{b}\,\mathrm{d}A\end{aligned} \tag{10.42b}$$

from which

$$\mathbf{f}^e = t\,(\mathbf{A}^{e-1})^T\int_A \mathbf{A}^T\,\mathbf{b}\,\mathrm{d}A = (\mathbf{A}^{e-1})^T\,\mathbf{b}^e \tag{10.43a}$$

where $\mathbf{b}^e = t\int_A \mathbf{A}^T\mathbf{b}\,\mathrm{d}A$ becomes

$$\mathbf{b}^e = t\int\begin{bmatrix}1 & 0\\ x & 0\\ y & 0\\ 0 & 1\\ 0 & x\\ 0 & y\end{bmatrix}\begin{bmatrix}b_x\\ b_y\end{bmatrix}\mathrm{d}A = t\int\begin{bmatrix}b_x\\ x\,b_x\\ y\,b_x\\ b_y\\ x\,b_y\\ y\,b_y\end{bmatrix}\mathrm{d}A \tag{10.43b}$$

In the case of a uniformly thin element, b_x and b_y are assumed constant. Writing dA = dx × dy, the integration employs standard double integrals for a triangle of area Δ as follows:

$$\int_A t\, b_x\, \mathrm{d}A = t\, b_x \int_x \int_y \mathrm{d}x\, \mathrm{d}y = (\, t\, b_x)\Delta$$

$$t \int_A x\, b_x\, \mathrm{d}A = t\, b_x \int_x \int_y x\, \mathrm{d}y\, \mathrm{d}x = (t\, b_x)(x_1 + x_2 + x_3)\Delta\, /3$$

$$t \int_A y\, b_x\, \mathrm{d}A = t\, b_x \int_x \int_y y\, \mathrm{d}y\, \mathrm{d}x = (t\, b_x)(y_1 + y_2 + y_3)\Delta\, /3$$

$$t \int_A b_y\, \mathrm{d}A = t\, b_y \int_x \int_y \mathrm{d}x\, \mathrm{d}y = (t\, b_y)\Delta$$

$$t \int_A x\, b_y\, \mathrm{d}A = t\, b_y \int_x \int_y x\, \mathrm{d}y\, \mathrm{d}x = (t\, b_y)(x_1 + x_2 + x_3)\Delta\, /3$$

$$t \int_A y\, b_y\, \mathrm{d}A = t\, b_y \int_x \int_y y\, \mathrm{d}y\, \mathrm{d}x = (t\, b_y)(y_1 + y_2 + y_3)\Delta\, /3$$

Thus $\mathbf{b}^e$ in eq(10.43b) becomes

$$\mathbf{b}^e = t\Delta \begin{bmatrix} b_x \\ \tfrac{1}{3}\, b_x(x_1 + x_2 + x_3) \\ \tfrac{1}{3}\, b_x(y_1 + y_2 + y_3) \\ b_y \\ \tfrac{1}{3}\, b_y(x_1 + x_2 + x_3) \\ \tfrac{1}{3}\, b_y(y_1 + y_2 + y_3) \end{bmatrix} \qquad (10.43c)$$

Substituting eqs(10.26b) and (10.43c) into eq(10.43a) leads to

$$f_{x1} = (tb_x/2)[(x_2 y_3 - x_3 y_2) + (x_1 + x_2 + x_3)(y_2 - y_3)/3 + (y_1 + y_2 + y_3)(x_3 - x_2)/3]$$
$$f_{y1} = (tb_y/2)[(x_2 y_3 - x_3 y_2) + (x_1 + x_2 + x_3)(y_2 - y_3)/3 + (y_1 + y_2 + y_3)(x_3 - x_2)/3] = (b_y/b_x)\, f_{x1}$$
$$f_{x2} = (tb_x/2)[(x_3 y_1 - x_1 y_3) + (x_1 + x_2 + x_3)(y_3 - y_1)/3 + (y_1 + y_2 + y_3)(x_1 - x_3)/3] \qquad (10.44\text{a–f})$$
$$f_{y2} = (tb_y/2)[(x_3 y_1 - x_1 y_3) + (x_1 + x_2 + x_3)(y_3 - y_1)/3 + (y_1 + y_2 + y_3)(x_1 - x_3)/3] = (b_y/b_x)\, f_{x2}$$
$$f_{x3} = (tb_x/2)[(x_1 y_2 - x_2 y_1) + (x_1 + x_2 + x_3)(y_1 - y_2)/3 + (y_1 + y_2 + y_3)(x_2 - x_1)/3]$$
$$f_{y3} = (tb_y/2)[(x_1 y_2 - x_2 y_1) + (x_1 + x_2 + x_3)(y_1 - y_2)/3 + (y_1 + y_2 + y_3)(x_2 - x_1)/3] = (b_y/b_x)\, f_{x3}$$

Equations(10.44a - f) satisfy equilibrium:

$$f_{x1} + f_{x2} + f_{x3} = t\Delta\, b_x \quad \text{and} \quad f_{y1} + f_{y2} + f_{y3} = t\Delta\, b_y$$

where Δ is the triangle area. For multiply connected elements, the nodal forces of element e are influenced by common sides of three adjacent elements. Summation of net nodal forces for element e ($\sum f_{x1}$, $\sum f_{y1}$ .....) amounts to adding to the RH sides of eqs(10.44a - f) two similar expressions derived from the cyclic permutation 1→2, 2→3 and 3→1. The final force vector appearing in the element stiffness equation $\mathbf{f}^e = \mathbf{K}^e\, \boldsymbol{\delta}^e$ will be the sum of the net boundary and body forces in eqs(10.41) and (10.44).

10.5.7 Mesh Size and Curved Boundaries

The simple, plane triangular element does not allow stress and strain to vary within its area. Generally, stress and strain values are referred to the centroid or to a mid-side position. It follows that many elements will be needed to discretise an area when plotting stress distributions from one triangle to the next. Also, errors will arise when approximating a curved boundary with the element's straight side unless the element size is reduced. To examine these effects a plane stress/strain FE Fortran program was executed in order to:

1. Determine $\mathbf{K}^e$ for each element.
2. Assemble the overall stiffness matrix $\mathbf{K}$ from each $\mathbf{K}^e$.
3. Apply boundary conditions and solve for the nodal displacements $\boldsymbol{\delta} = \mathbf{K}^{-1}\mathbf{f}$.
4. Determine element strain and stress from $\sigma = \mathbf{D}\,\mathbf{B}\,\delta$ and $\varepsilon = \mathbf{B}\,\delta$ respectively.

The computer provides a rapid solution to the many simultaneous equations involved within step 3. The various simplifying procedures used nowadays to reduce memory store and hence cost from either direct or iterative solutions have been discussed elsewhere [8-10]. The particular method employed here is Gauss-Siedel iteration with successive over-relaxation (SOR). The latter ensures a convergent solution for a symmetric matrix $\mathbf{K}$. That is, for the r th nodal point and the i th iteration,

$$\delta_r^{(i+1)} = \delta_r^{(i)} + \omega\,\Delta\,\delta_r^{(i+1)} \tag{10.45a}$$

where $\delta_r = [\, u_r \;\; v_r \,]^T$ is the displacement vector and ω is the SOR factor, lying in the following range $0 \le \omega \le 2$. Now, $\Delta\,\delta_r^{(i+1)}$ in eq(10.45a) is expressed from the equilibrium of the node forces $\mathbf{f}_r = [\, f_{xr} \;\; f_{yr} \,]^T$ as

$$\Delta\,\delta_r^{(i+1)} = \mathbf{K}_{rr}^{-1}\,[\,\mathbf{f}_r - \sum_{s=1}^{r-1} \mathbf{K}_{rs}\,\delta_s^{(i+1)} - \sum_{s=r}^{N} \mathbf{K}_{rs}\,\delta_s^{(i)}\,] \tag{10.45b}$$

where the terms within [ ] are out-of-balance forces due to an assumed displacement. The convergence requires that [ ] → 0. In practice, eq(10.45b) employs the lower triangular part of $\mathbf{K}$ applicable to node r. The SOR method therefore involves less memory and is faster than a direct solution by Gaussian elimination. An optimum ω value, one that minimises the number of iterations in eq(10.45a), may only be found by trial. Normally, for any given ω value, eq(10.45a) would be used with a tolerance limit to specify when iteration should cease.

Example 10.6 Compare with the engineer's theory of bending, the mid-span FE stress distributions for a cantilever subdivided into 8 × 24 and 16 × 48 right-angled, isosceles meshes shown in Figs 10.15a and b.

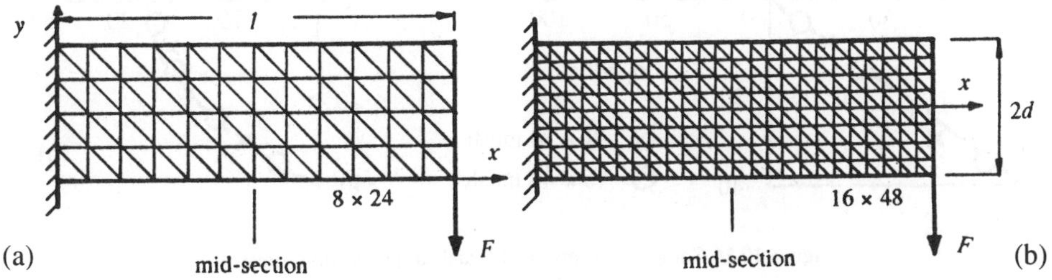

Figure 10.15 Finite element meshes for a cantilever

The elastic constants and beam dimensions are similar to those given in Example 10.3 but with a corresponding increase in the number of elements.
The bending stress σ_x and shear stress τ_{xy} from the plane stress programme (setting $\sigma_z = 0$) are given in Table 10.3. These apply to the mid-sides of triangles lying along the vertical mid-sections shown in Fig. 10.15a,b.

Table 10.3 Bending and shear stresses across depth at mid-span

8 × 24 MESH			16 × 48 MESH		
y (mm)	σ_x (MPa)	τ_{xy} (MPa)	y (mm)	σ_x (MPa)	τ_{xy} (MPa)
			5	- 94.75	4.53
10	- 73.7	8.06			
			15	- 67.85	11.25
20					
			25	- 40.95	15.73
30	- 26.0	16.05			
			35	- 14.06	17.98
40					
			45	12.84	17.99
50	21.7	16.16			
			55	39.37	15.76
60					
			65	66.62	11.27
70	69.34	7.05			
			75	93.51	4.54

A comparison is made in Fig. 10.16 between the FE stress values from Table 10.3 and the classical solution. The latter is supplied by either the engineer's theory of bending (e.t.b.) or the appropriate stress function [2].

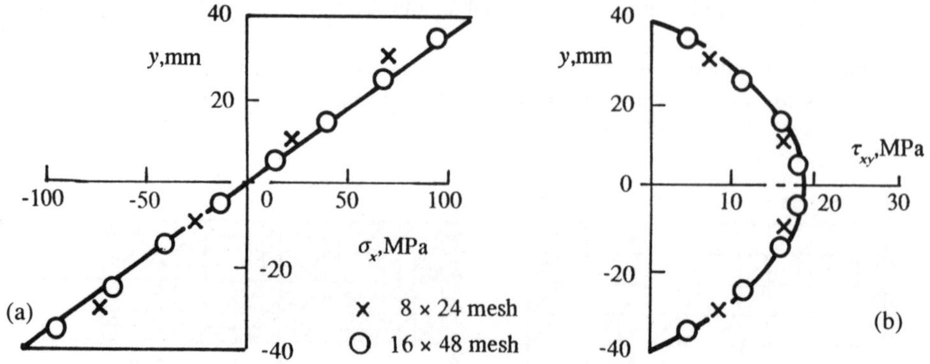

Figure 10.16 Comparison between FE and simple bending stresses

The bending theory gives the respective bending and shear stresses as

$$\sigma_x = My'/I \tag{10.46a}$$
$$\tau_{xy} = [F/(2I)](d^2 - y'^2) \tag{10.46b}$$

where $y' = y - 40$ mm is measured from the neutral axis in Figs 10.16a,b. The stresses found apply to $F = 10 \times 10^3$ N, $2d = 80$ mm, $t = 10$ mm and $l = 240$ mm with

$$I = t\,(2d)^3/12 = 10 \times 80^3/12 = 426.67 \times 10^3 \text{ mm}^4$$
$$M = F(l - x) = 10 \times 10^3(240 - x) = 1200 \text{ Nmm}$$

Figure 10.16a shows that the 16 × 48 FE mesh provides closer agreement with the linear bending stress distribution σ_x from eq(10.46a). A similar observation is made in Fig. 10.16b between the finer mesh and the parabolic distribution in shear stress τ_{xy} from eq(10.46b). Provided the sub-divided element nodes contain those of the coarser mesh, then the FE stresses will converge to indicate where it is unnecessary to refine the mesh further.

Example 10.7 A disc of diameter d = 35mm and thickness, t = 3.8 mm is subjected to a vertical, diametral compressive force P = 890 N. Two plane triangular finite element meshes are shown for one quadrant of the disc in Fig. 10.17a and b. Establish the stress values σ_y at the mid-points of nodes lying along the x-axis. Compare these with solutions from the stress function (see eqs (2.65a,b)).

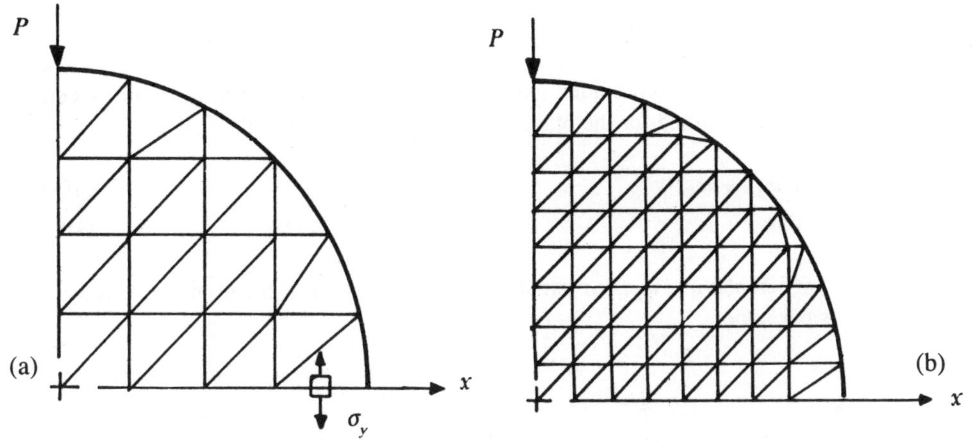

Figure 10.17 FE mesh for a disc under diametral compression

Because our theory applies to straight-sided triangles, it is best to avoid having sides of sharp curvature by selecting smaller elements adjacent to the curved boundary. Fig. 10.18 shows the four FE σ_y values from mesh (a) and eight values from mesh (b). These are compared with classical elasticity solution (see eqs(2.65a,b)). The common normalising factor is the greatest theoretical compressive stress at the disc centre $-\ 6P/(\pi\, d\, t\,)$ (found from substituting $y = 0$ in eq(2.65a)). Figure 10.18 shows that the stresses from the three solutions remain in good agreement as their magnitudes increase towards the centre. At the centre, the relatively coarse 28 FE mesh and the fine 112 mesh are equally representative of the theoretical stress values. It has been shown [3] that a second order triangular element, with three additional nodes at its mid-side positions, would permit greater accuracy to be achieved with a similar number of elements in the region of boundaries. Alternatively, Fig. 10.18 reveals that improved accuracy is achieved with the finer three-noded triangular mesh (Fig. 10.17b.)

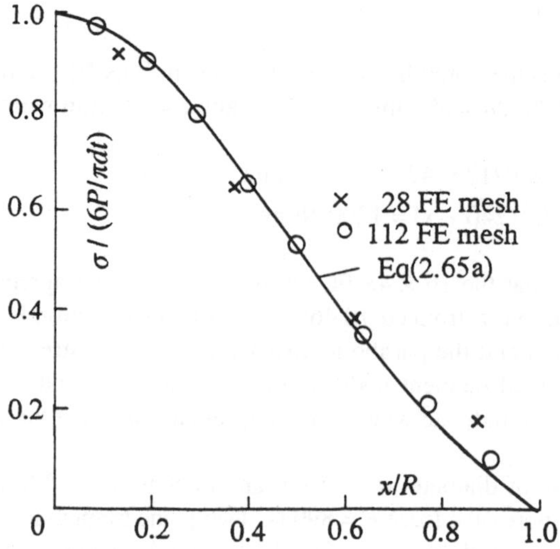

Figure 10.18 Comparison between FE and classical disc compression solution

10.6 Plane Rectangular Element

The x, y plane may also be subdivided into a rectangular mesh. The following displacement functions [5,6] are assumed for the plane rectangular element in Fig. 10.19a:

$$u = \alpha_1 + \alpha_2 x + \alpha_3 y + \alpha_4 x\,y \tag{10.47a}$$

$$v = \alpha_5 + \alpha_6 x + \alpha_7 y + \alpha_8 x\,y \tag{10.47b}$$

The eight coefficients; α_1, α_2.... α_8 match the eight degrees of freedom, i.e. 4 nodes with two degrees per node.

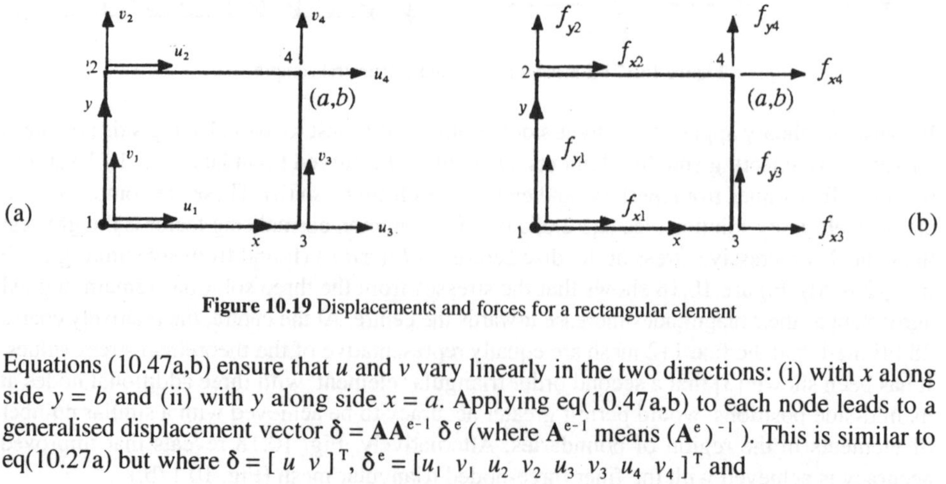

Figure 10.19 Displacements and forces for a rectangular element

Equations (10.47a,b) ensure that u and v vary linearly in the two directions: (i) with x along side $y = b$ and (ii) with y along side $x = a$. Applying eq(10.47a,b) to each node leads to a generalised displacement vector $\boldsymbol{\delta} = \mathbf{A}\mathbf{A}^{e-1}\,\boldsymbol{\delta}^e$ (where $\mathbf{A}^{e-1}$ means $(\mathbf{A}^e)^{-1}$). This is similar to eq(10.27a) but where $\boldsymbol{\delta} = [\,u\ \ v\,]^{\mathrm{T}}$, $\boldsymbol{\delta}^e = [u_1\ \ v_1\ \ u_2\ \ v_2\ \ u_3\ \ v_3\ \ u_4\ \ v_4\,]^{\mathrm{T}}$ and

$$\mathbf{A} = \begin{bmatrix} 1 & x & y & xy & 0 & 0 & 0 & 0 \\ 0 & 0 & 0 & 0 & 1 & x & y & xy \end{bmatrix} \tag{10.48a}$$

Substituting the x and y co-ordinates of each node in eq(10.48a), $\mathbf{A}^e$ is derived from $\mathbf{A}$ as

$$\mathbf{A}^e = \begin{bmatrix} 1 & 0 & 0 & 0 & 0 & 0 & 0 & 0 \\ 0 & 0 & 0 & 0 & 1 & 0 & 0 & 0 \\ 1 & 0 & b & 0 & 0 & 0 & 0 & 0 \\ 0 & 0 & 0 & 0 & 1 & 0 & b & 0 \\ 1 & a & 0 & 0 & 0 & 0 & 0 & 0 \\ 0 & 0 & 0 & 0 & 1 & a & 0 & 0 \\ 1 & a & b & ab & 0 & 0 & 0 & 0 \\ 0 & 0 & 0 & 0 & 1 & a & b & ab \end{bmatrix} \tag{10.48b}$$

The strain components (2.17a-c) follow from eqs(10.47a,b) as

$$\begin{aligned} \varepsilon_x &= \partial u/\partial x = \alpha_2 + \alpha_4 y \\ \varepsilon_y &= \partial v/\partial y = \alpha_7 + \alpha_8 x \\ \gamma_{xy} &= \partial u/\partial y + \partial v/\partial x = \alpha_3 + \alpha_4 x + \alpha_6 + \alpha_8 y \end{aligned} \tag{10.49a}$$

It is seen that as the linear variation in these strains satisfies the compatibility condition (2.18), they will ensure strain continuity at the interface between adjacent elements when the nodal displacements are matched. It follows that the matrix $\mathbf{C}$, appearing in $\boldsymbol{\varepsilon} = \mathbf{C}\boldsymbol{\alpha}$, where $\boldsymbol{\varepsilon} = [\varepsilon_x\ \ \varepsilon_y\ \ \gamma_{xy}]^{\mathrm{T}}$ and $\boldsymbol{\alpha} = [\alpha_1\ \ \alpha_2\ \ \alpha_3\ \ \alpha_4\ \ \alpha_5\ \ \alpha_6\ \ \alpha_7\ \ \alpha_8]^{\mathrm{T}}$, becomes

$$\mathbf{C} = \begin{bmatrix} 0 & 1 & 0 & y & 0 & 0 & 0 & 0 \\ 0 & 0 & 0 & 0 & 0 & 0 & 1 & x \\ 0 & 0 & 1 & x & 0 & 1 & 0 & y \end{bmatrix} \tag{10.49b}$$

The dependence of $\mathbf{C}$ upon x and y shows that both stress and strain will vary throughout a rectangular element. This contrasts with the constant-strain triangle where $\mathbf{C}$, in eq(10.29a), does not contain x or y. The general strain-nodal displacement relation is $\boldsymbol{\varepsilon} = \mathbf{B}\,\boldsymbol{\delta}^e$ where $\mathbf{B} = \mathbf{C}\,\mathbf{A}^{e-1}$. The constitutive relation (10.31a) remain unaltered since they depend upon the plane condition and not upon the shape of the element. Thus, the general stress-strain relation is $\boldsymbol{\varepsilon} = \mathbf{D}\,\mathbf{B}\,\boldsymbol{\delta}^e$. Recall that general stress and strain matrices are required for the derivation of the element stiffness matrix $\mathbf{K}^e$. Since $\mathbf{B}$ now depends upon x and y, the matrix product $\mathbf{B}^{\mathrm{T}}\mathbf{DB}$ must be integrated over the volume V. To do this, first write $\delta V = t \times \delta x \times \delta y$ for an elemental volume of uniform thickness t. Applying PVW, eq(10.5b) becomes

$$\boldsymbol{\delta}^{ev}\,\mathbf{f}^e = t\int_x\int_y (\mathbf{B}\boldsymbol{\delta}^{ev})^{\mathrm{T}}(\mathbf{DB})\,\boldsymbol{\delta}^e\ \mathrm{d}x\,\mathrm{d}y$$

$$\mathbf{f}^e = \{t\int_x\int_y \mathbf{B}^{\mathrm{T}}\mathbf{DB}\ \mathrm{d}x\,\mathrm{d}y\} \times \boldsymbol{\delta}^e$$

from which

$$\mathbf{K}^{\mathrm{e}} = t \int_x \int_y \mathbf{B}^{\mathrm{T}} \mathbf{D} \, \mathbf{B} \, \mathrm{d}x \, \mathrm{d}y \tag{10.50a}$$

Setting $r = a/b$ for node 4, the integration yields the following symmetrical $\mathbf{K}^{\mathrm{e}}$ matrix:

$$\begin{bmatrix}
4(D_{11}r^{-1}+D_{33}r) & \cdot & \cdot & \cdot & \cdot & \cdot & \cdot & \cdot \\
3(D_{12}+D_{33}) & 4(D_{22}r+D_{33}r^{-1}) & \cdot & \cdot & \cdot & \cdot & \cdot & \cdot \\
2(D_{11}r^{-1}-2D_{33}r) & 3(D_{21}-D_{33}) & 4(D_{11}r^{-1}+D_{33}r^{-1}) & \cdot & \cdot & \cdot & \cdot & \cdot \\
3(D_{33}-D_{12}) & 2(D_{33}r^{-1}-2D_{22}r) & 3(D_{33}-D_{21}) & 4(D_{33}r^{-1}+D_{22}r) & \cdot & \cdot & \cdot & \cdot \\
2(D_{33}r-2D_{11}r^{-1}) & 3(D_{33}-D_{21}) & 2(D_{33}r-D_{11}r^{-1}) & 3(D_{33}+D_{12}) & 4(D_{11}r^{-1}+D_{33}r & \cdot & \cdot & \cdot \\
3(D_{12}-D_{33}) & 2(D_{22}r-2D_{33}r^{-1}) & 3(D_{12}+D_{33}) & 2(D_{33}r^{-1}-D_{22}r) & 3(D_{33}-D_{12}) & 4(D_{22}r+D_{33}r^{-1}) & \cdot & \cdot \\
2(D_{33}r-D_{11}r^{-1}) & 3(D_{33}-D_{12}) & 2(D_{33}r-2D_{11}r^{-1}) & 3(D_{33}-D_{12}) & 2(D_{11}r^{-1}-2D_{33}r) & 3(D_{33}-D_{12}) & 4(D_{11}r^{-1}+D_{33}r) & \cdot \\
3(D_{33}-D_{12}) & 2(D_{33}r^{-1}-D_{22}r) & 3(D_{33}-D_{12}) & 2(D_{22}r-2D_{33}r^{-1}) & 3(D_{12}-D_{33}) & 2(D_{33}r^{-1}-2D_{22}r) & 3(D_{12}+D_{33}) & 4(D_{22}r+D_{33}r^{-1})
\end{bmatrix} \tag{10.50b}$$

In arriving at eq(10.50b), the integration must follow matrix multiplication $\mathbf{B}^{\mathrm{T}} \mathbf{D} \, \mathbf{B}$. A similar procedure was followed when deriving $\mathbf{K}^{\mathrm{e}}$ for the beam element. When $\mathbf{K}^{\mathrm{e}}$ from eq(10.50b) is substituted into eq(10.32a), it relates the following nodal point displacements and forces: $\delta^{\mathrm{e}} = [\, u_1 \;\; v_1 \;\; u_2 \;\; v_2 \;\; u_3 \;\; v_3 \;\; u_4 \;\; v_4 \,]^{\mathrm{T}}$ and $\mathbf{f}^{\mathrm{e}} = [\, f_{x1} \; f_{y1} \; f_{x2} \; f_{y2} \; f_{x3} \; f_{y3} \; f_{x4} \; f_{y4} \,]^{\mathrm{T}}$, shown in Fig. 10.18a and b. The overall stiffness matrix $\mathbf{K}$ for a rectangular mesh may be assembled as before and eq(10.36a) solved for the nodal displacements δ^{e}. The u, v displacements at any point x, y in the element are supplied by $\delta = \mathbf{A}\mathbf{A}^{\mathrm{e}-1}\,\delta^{\mathrm{e}}$. Finally, eq(10.31a) supplies the general element stresses $\sigma = [\sigma_x \;\; \sigma_y \;\; \tau_{xy}\,]^{\mathrm{T}}$, as follows:

$$\sigma = \mathbf{D}\mathbf{B}\delta^{\,\mathrm{e}} = \mathbf{D}\,(\,\mathbf{C}\mathbf{A}^{\,\mathrm{e}-1}\,)\,\delta^{\mathrm{e}} = \mathbf{H}\,\delta^{\mathrm{e}} \tag{10.51a}$$

The explicit form of the 3×8 matrix $\mathbf{H}$ is found from performing the matrix multiplication in eq(10.51a). This gives

$$\mathbf{H} = \frac{1}{ab}\begin{bmatrix}
-D_{11}(b-y) & -D_{12}(a-x) & -D_{11}y & D_{12}(a-x) & D_{11}(b-y) & -D_{12}x & D_{11}y & D_{12}x \\
-D_{12}(b-y) & -D_{22}(a-x) & -D_{12}y & D_{22}(a-x) & D_{12}(b-y) & -D_{22}x & D_{12}y & D_{22}x \\
-D_{33}(a-x) & -D_{23}(b-y) & D_{33}(a-x) & -D_{33}y & -D_{33}x & D_{33}(b-y) & D_{33}x & D_{33}y
\end{bmatrix} \tag{10.51b}$$

where x and y are the co-ordinates of a node for which the three stress components are required. It follows that the full nodal stress vector is $\sigma^{\mathrm{e}} = [\sigma_1 \;\; \sigma_2 \;\; \sigma_3 \;\; \sigma_4\,]^{\mathrm{T}}$, where $\sigma_1 = [\sigma_{x1} \;\; \sigma_{y1} \;\; \tau_{xy1}\,]^{\mathrm{T}}$ etc. is found from $\sigma^{\mathrm{e}} = \mathbf{H}^{\mathrm{e}}\,\delta^{\mathrm{e}}$ where σ^{e} is 12×1 and δ^{e} is 8×1. The connecting 12×8 matrix $\mathbf{H}^{\mathrm{e}}$ is derived from $\mathbf{H}$ in eq(10.51b) following substitution of the x, y nodal co-ordinates for each of the four nodes. Neither the stress nor the strain is constant within a rectangular element. Thus, large rectangular meshes are undesirable since they lead to discontinuities in the stresses at common nodal points.

10.7 Triangular Elements with Axial Symmetry

A triangular toroidal element is used to describe axially symmetric bodies in co-ordinates of r, θ and z. In Fig. 10.20, the respective force and displacement vectors $\mathbf{f}^e = [f_{r1}\ f_{z1}\ f_{r2}\ f_{z2}\ f_{r3}\ f_{z3}]^T$ and $\delta^e = [\,u_1\ w_1\ u_2\ w_2\ u_3\ w_3\,]^T$ are referred to the nodes of an element in section.

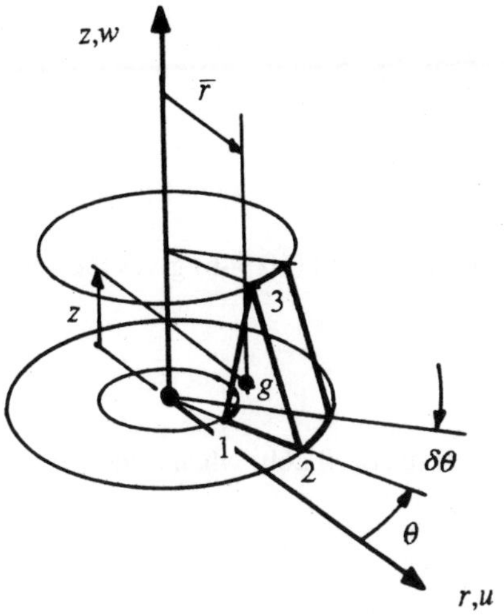

Figure 10.20 Toroidal element with axial symmetry about z-axis

It is seen that as the element rotates about the z - axis it describes a toroid of triangular cross-section.

10.7.1 Displacements

The nodes 1, 2 and 3 lie in the z, r plane so that δ^e describes the displacements u and w for each node in the directions of r and z. The general displacements are [10]

$$u = \alpha_1 + \alpha_2 r + \alpha_3 z \tag{10.52a}$$

$$w = \alpha_4 + \alpha_5 r + \alpha_6 z \tag{10.52b}$$

The matrix form of eqs(10.52a,b) is

$$\begin{bmatrix} u \\ w \end{bmatrix} = \begin{bmatrix} 1 & r & z & 0 & 0 & 0 \\ 0 & 0 & 0 & 1 & r & z \end{bmatrix} \begin{bmatrix} \alpha_1 \\ \alpha_2 \\ \alpha_3 \\ \alpha_4 \\ \alpha_5 \\ \alpha_6 \end{bmatrix} \qquad \delta = \mathbf{A}\,\alpha \tag{10.53a,b}$$

where $\delta = [u \quad w]^T$. Substituting the r, z co-ordinates for each of the three nodes into eq(10.53a) gives

$$\begin{bmatrix} u_1 \\ w_1 \\ u_2 \\ w_2 \\ u_3 \\ w_3 \end{bmatrix} = \begin{bmatrix} 1 & r_1 & z_1 & 0 & 0 & 0 \\ 0 & 0 & 0 & 1 & r_1 & z_1 \\ 1 & r_2 & z_2 & 0 & 0 & 0 \\ 0 & 0 & 0 & 1 & r_2 & z_2 \\ 1 & r_3 & z_3 & 0 & 0 & 0 \\ 0 & 0 & 0 & 1 & r_3 & z_3 \end{bmatrix} \begin{bmatrix} \alpha_1 \\ \alpha_2 \\ \alpha_3 \\ \alpha_4 \\ \alpha_5 \\ \alpha_6 \end{bmatrix} \qquad \boldsymbol{\delta}^e = \mathbf{A}^e \boldsymbol{\alpha} \qquad (10.54a,b)$$

Eliminating $\boldsymbol{\alpha}$ between eqs(10.53b) and (10.54b) gives the general displacement vector $\boldsymbol{\delta}$ in terms of $\boldsymbol{\delta}^e$:

$$\boldsymbol{\delta} = \mathbf{A}\mathbf{A}^{e-1}\,\boldsymbol{\delta}^e \qquad (10.55)$$

In eq(10.55), $\mathbf{A}^{e-1}$ is identical to eq(10.26b) when r and z replace x and y.

10.7.2 Stress and Strain

The strain components are found from eqs(10.52a,b) and the following polar strain-displacement relations [11]:

$$\varepsilon_r = \partial u/\partial r = \alpha_2\,,\ \varepsilon_z = \partial w/\partial z = \alpha_6$$
$$\varepsilon_\theta = u/r = \alpha_1/r + \alpha_2 + \alpha_3 z/r$$
$$\gamma_{rz} = \partial u/\partial z + \partial w/\partial r = \alpha_3 + \alpha_5$$

These show that nodal points 1, 2 and 3 may displace radially and axially but not tangentially. The corresponding matrix forms for these general strains are

$$\begin{bmatrix} \varepsilon_z \\ \varepsilon_r \\ \varepsilon_\theta \\ \gamma_{rz} \end{bmatrix} = \begin{bmatrix} 0 & 0 & 0 & 0 & 0 & 1 \\ 0 & 1 & 0 & 0 & 0 & 0 \\ 1/r & 1 & z/r & 0 & 0 & 0 \\ 0 & 0 & 1 & 0 & 1 & 0 \end{bmatrix} \begin{bmatrix} \alpha_1 \\ \alpha_2 \\ \alpha_3 \\ \alpha_4 \\ \alpha_5 \\ \alpha_6 \end{bmatrix} \qquad \boldsymbol{\varepsilon} = \mathbf{C}\,\boldsymbol{\alpha} \qquad (10.56a,b)$$

Combining eqs(10.54b) and (10.56b) gives the general strain $\boldsymbol{\varepsilon}$ for any point (r, z) in terms of the nodal displacements $\boldsymbol{\delta}^e$

$$\boldsymbol{\varepsilon} = \mathbf{C}\,\mathbf{A}^{e-1}\boldsymbol{\delta}^e = \mathbf{B}\,\boldsymbol{\delta}^e \qquad (10.57)$$

where $\mathbf{B} = \mathbf{C}\,\mathbf{A}^{e-1}$. The elastic constitutive relations are

$$
\begin{aligned}
\varepsilon_r &= (1/E)[\sigma_r - \nu(\sigma_\theta + \sigma_z)] \\
\varepsilon_\theta &= (1/E)[\sigma_\theta - \nu(\sigma_r + \sigma_z)] \\
\varepsilon_z &= (1/E)[\sigma_z - \nu(\sigma_\theta + \sigma_r)] \\
\gamma_{rz} &= \tau_{rz}/G = 2(1+\nu)\tau_{rz}/E
\end{aligned}
\tag{10.58}
$$

Inverting eq(10.58) gives the stress components in matrix form

$$
\begin{bmatrix} \sigma_z \\ \sigma_r \\ \sigma_\theta \\ \tau_{rz} \end{bmatrix} = \frac{E(1-\nu)}{(1+\nu)(1-2\nu)} \begin{bmatrix} 1 & \nu/(1-\nu) & \nu/(1-\nu) & 0 \\ \nu/(1-\nu) & 1 & \nu/(1-\nu) & 0 \\ \nu/(1-\nu) & \nu/(1-\nu) & 1 & 0 \\ 0 & 0 & 0 & (1-2\nu)/[2(1-\nu)] \end{bmatrix} \begin{bmatrix} \varepsilon_z \\ \varepsilon_r \\ \varepsilon_\theta \\ \gamma_{rz} \end{bmatrix} \quad \sigma = \mathbf{D}\,\varepsilon
\tag{10.59a,b}
$$

Combining eqs(10.57) and (10.59b), general stresses follow from the nodal displacements:

$$
\sigma = \mathbf{D}\,\mathbf{C}\,\mathbf{A}^{e\,-1}\,\delta^e = \mathbf{D}\,\mathbf{B}\,\delta^e
\tag{10.60}
$$

Since $\mathbf{C}$ contains r and z, it follows that the strain and stress for any node may be found by substituting the co-ordinates (r, z) of that node into the respective eqs(10.57) and (10.60).

10.7.3 Element Stiffness Matrix

The element stiffness matrix $\mathbf{K}^e$ follows from substituting eqs(10.57) and (10.60) into (10.5b). Setting $\delta V = 2\pi r \times \delta r \times \delta z$ in eq(10.5b) leads to

$$
\mathbf{K}^e = 2\pi \int_r \int_z \mathbf{B}^T \mathbf{D}\,\mathbf{B}\, r\,\mathrm{d}r\,\mathrm{d}z
\tag{10.61a}
$$

It is seen that $\mathbf{K}^e$ is similar to eqs(10.34a) and (10.50a) for plane triangular and rectangular elements respectively. Integration of eq(10.61a) must be performed term by term following matrix multiplication $\mathbf{B}^T\mathbf{D}\mathbf{B}$. Inverting $\mathbf{A}^e$ in eq(10.54a), the matrix $\mathbf{B} = \mathbf{C}\,\mathbf{A}^{e\,-1}$ is found from

$$
\mathbf{B} = \frac{1}{2\Delta} \begin{bmatrix} 0 & 0 & 0 & 0 & 0 & 1 \\ 0 & 1 & 0 & 0 & 0 & 0 \\ 1/r & 1 & z/r & 0 & 0 & 0 \\ 0 & 0 & 1 & 0 & 1 & 0 \end{bmatrix} \begin{bmatrix} (z_3r_2 - z_2r_3) & 0 & (z_1r_3 - z_3r_1) & 0 & (z_2r_1 - z_1r_2) & 0 \\ (z_2 - z_3) & 0 & (z_3 - z_1) & 0 & (z_1 - z_2) & 0 \\ (r_3 - r_2) & 0 & (r_1 - r_3) & 0 & (r_2 - r_1) & 0 \\ 0 & (z_3r_2 - z_2r_3) & 0 & (z_1r_3 - z_3r_1) & 0 & (z_2r_1 - z_1r_2) \\ 0 & (z_2 - z_3) & 0 & (z_3 - z_1) & 0 & (z_1 - z_2) \\ 0 & (r_3 - r_2) & 0 & (r_1 - r_3) & 0 & (r_2 - r_1) \end{bmatrix}
$$

$$
= \frac{1}{2\Delta} \begin{bmatrix} 0 & (r_3 - r_2) & 0 & (r_1 - r_3) & 0 & (r_2 - r_1) \\ (z_2 - z_3) & 0 & (z_3 - z_1) & 0 & (z_1 - z_2) & 0 \\ \begin{matrix}(z_3r_2 - z_2r_3)/r + (z_2 - z_3) \\ + z(r_3 - r_2)/r\end{matrix} & 0 & \begin{matrix}(z_1r_3 - z_3r_1)/r + (z_3 - z_1) \\ + z(r_1 - r_3)/r\end{matrix} & 0 & \begin{matrix}(z_2r_1 - z_1r_2)/r + (z_1 - z_2) \\ + z(r_2 - r_1)/r\end{matrix} & 0 \\ (r_3 - r_2) & (z_2 - z_3) & (r_1 - r_3) & (z_3 - z_1) & (r_2 - r_1) & (z_1 - z_2) \end{bmatrix}
\tag{10.61b}
$$

where 2Δ is twice the area of the triangular element. This area is given by

$$2\Delta = \det \begin{vmatrix} 1 & r_1 & z_1 \\ 1 & r_2 & z_2 \\ 1 & r_3 & z_3 \end{vmatrix}$$

Writing the elasticity matrix **D**, in eq(10.59b), in the following form

$$\mathbf{D} = \begin{bmatrix} D_{11} & D_{12} & D_{13} & 0 \\ D_{21} & D_{22} & D_{23} & 0 \\ D_{31} & D_{32} & D_{33} & 0 \\ 0 & 0 & 0 & D_{44} \end{bmatrix} \tag{10.61c}$$

the components D_{ij} are identified with eq(10.59a) as

$$D_{11} = D_{22} = D_{33} = E(1 - \nu) / [(1 + \nu)(1 - 2\nu)]$$
$$D_{12} = D_{13} = D_{21} = D_{31} = D_{23} = D_{32} = E\nu / [(1 + \nu)(1 - 2\nu)]$$
$$D_{44} = E(1 - \nu)(1 - 2\nu) / [2(1 - \nu^2)(1 - 2\nu)]$$

Integrating the matrix product $\mathbf{B}^{\mathrm{T}}\mathbf{D}\,\mathbf{B}$ in eq(10.61a) will supply the elements of the stiffness matrix. For example, substituting from eqs(10.61b and c), the component $K_{11}{}^{e}$ of $\mathbf{K}^{e}$ is found from

$$\begin{aligned} K_{11}{}^{e} &= [2\pi/(4\Delta^2)]\int_r\int_z [\,(z_2 - z_3)\{D_{22}(z_2 - z_3) + D_{23}[(z_3 r_2 - z_2 r_3)/r + (z_2 - z_3) + z\,(r_3 - r_2)/r]\} \\ &\quad + [(z_3 r_2 - z_2 r_3)/r + (z_2 - z_3) + z\,(r_3 - r_2)/r][D_{32}(z_2 - z_3) + D_{33}[(z_3 r_2 - z_2 r_3)/r \\ &\quad + (z_2 - z_3) + z\,(r_3 - r_2)/r] + (r_3 - r_2)^2 D_{44}\,]\; r\,\mathrm{d}r\,\mathrm{d}z \\ &= (\pi/\Delta^2)[\,m_1\int_r\int_z r\,\mathrm{d}r\,\mathrm{d}z + m_2\int_r\int_z \mathrm{d}r\,\mathrm{d}z + m_3\int_r\int_z z\,\mathrm{d}r\,\mathrm{d}z + m_4\int_r\int_z (z/r)\,\mathrm{d}r\,\mathrm{d}z \\ &\quad + m_5\int_r\int_z (1/r)\,\mathrm{d}r\,\mathrm{d}z + m_6\int_r\int_z (z^2/r)\,\mathrm{d}r\,\mathrm{d}z\,] \end{aligned} \tag{10.62a}$$

Equation (10.62a) is abbreviated to

$$K_{11}{}^{e} = (\pi/\Delta^2)[\,m_1 I_1 + m_2 I_2 + m_3 I_3 \;....\; + m_6 I_6\,] \tag{10.62b}$$

where the constants m_1 .... m_6 are

$$m_1 = D_{22}(z_2 - z_3)^2 + D_{23}(z_2 - z_3)^2 + D_{32}(z_2 - z_3)^2 + D_{33}(z_2 - z_3)^2 + D_{44}(r_3 - r_2)^2$$
$$\begin{aligned} m_2 &= D_{23}(z_2 - z_3)(z_3 r_2 - z_2 r_3) + D_{32}(z_3 r_2 - z_2 r_3)(z_2 - z_3) + D_{33}(z_3 r_2 - z_2 r_3)(z_2 - z_3) \\ &\quad + D_{33}(z_2 - z_3)(z_3 r_2 - z_2 r_3) \end{aligned}$$
$$m_3 = D_{23}(z_2 - z_3)(r_3 - r_2) + D_{33}(r_3 - r_2)(z_2 - z_3) + D_{32}(r_3 - r_2)(z_2 - z_3) + D_{33}(r_3 - r_2)(z_2 - z_3)$$
$$m_4 = D_{33}(r_3 - r_2)(z_3 r_2 - z_2 r_3) + D_{33}(r_3 - r_2)(z_3 r_2 - z_2 r_3)$$
$$m_5 = D_{33}(z_3 r_2 - z_2 r_3)^2$$
$$m_6 = D_{33}(r_3 - r_2)^2$$

The double integrals $I_1 \dots I_6$ in eq(10.62b) are applied over the area of the triangular element as shown in Fig. 10.21.

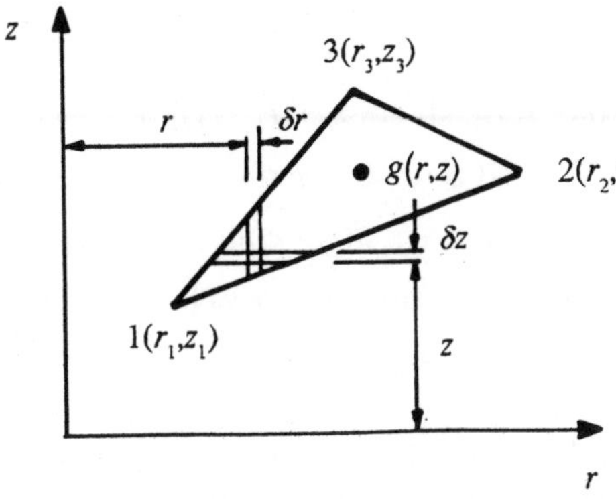

Figure 10.21 Variables in double integration

For example, writing the first integral in eq(10.62b) as

$$I_1 = \int_r \int_z r\,\mathrm{d}r\,\mathrm{d}z = \int_{z_1}^{z_2} \{ \int_{r_{1,3}}^{r_{1,2}} r\,\mathrm{d}r\}\mathrm{d}z + \int_{z_2}^{z_3} \{ \int_{r_{1,3}}^{r_{2,3}} r\,\mathrm{d}r\}\,\mathrm{d}z$$

$$= ½ \{ \int_{z_1}^{z_2} | r^2 |_{r_{1,3}}^{r_{1,2}} \mathrm{d}z + \int_{z_2}^{z_3} | r^2 |_{r_{1,3}}^{r_{2,3}} \mathrm{d}z \}$$

$$= ½ \int_{z_1}^{z_2} \{[(r_2 - r_1)(z - z_1)/(z_2 - z_1) + r_1]^2 - [(r_3 - r_1)(z - z_1) / (z_3 - z_1) + r_1]^2 \}\, \mathrm{d}z$$

$$+ ½ \int_{z_2}^{z_3} \{[(r_3 - r_2)(z - z_2) / (z_3 - z_2) + r_2]^2 - [(r_3 - r_1)(z - z_1) / (z_3 - z_1) + r_1]^2 \}\, \mathrm{d}z$$

in which the equation of a straight line $z - z_1 = (z_2 - z_1)(r - r_1) / (r_2 - r_1)$ has been used to define the variation in r with z over the side 12. This gives a standard form:

$$I_1 = (\Delta /3)(r_1 + r_2 + r_3)$$

The integrals remaining in eq(10.62b) can be found in a similar manner:

$$I_2 = \int_r \int_z \mathrm{d}r\,\mathrm{d}z = \Delta$$

$$I_3 = \int_r \int_z z\,\mathrm{d}r\,\mathrm{d}z = (\Delta /3)(z_1 + z_2 + z_3)$$

$$I_4 = \int_r \int_z (z/r)\,\mathrm{d}r\,\mathrm{d}z$$

$$= \sum (r_1/2)^2 \{[(z_2 - z_1) / (r_2 - r_1)]^2 - [(z_1 - z_3) / (r_1 - r_3)]^2 \}$$

$$+ r_1 \{[(z_2 - z_1) / (r_2 - r_1)][(z_1 r_2 - z_2 r_1) / (r_2 - r_1)] - [(z_1 - z_3) / (r_1 - r_3)]$$

$$\times [(z_1 r_3 - z_3 r_1) / (r_1 - r_3)]\} + ½ \ln r_1 \{[(z_1 r_2 - z_2 r_1)/(r_2 - r_1)]^2 - [(z_1 r_3 - z_3 r_1)/(r_1 - r_3)]^2 \}$$

$$+ \sum_{231} + \sum_{312}$$

$$I_5 = \int_r \int_z (1/r)\,\mathrm{d}r\,\mathrm{d}z$$
$$= \sum r_1 [(z_2 - z_1)/(r_2 - r_1) - (z_1 - z_3)/(r_1 - r_3)]$$
$$+ \ln r_1 [(z_1 r_2 - z_2 r_1)/(r_2 - r_1) - (z_1 r_3 - z_3 r_1)/(r_1 - r_3)] + \sum_{231} + \sum_{312}$$

$$I_6 = \int_r \int_z (z^2/r)\,\mathrm{d}r\,\mathrm{d}z$$
$$= \sum (r_1^3/9)\{[(z_2 - z_1)/(r_2 - r_1)]^3 - [(z_1 - z_3)/(r_1 - r_3)]^3\}$$
$$+ (r_1^2/2)\{[(z_2 - z_1)/(r_2 - r_1)]^2[(z_1 r_2 - z_2 r_1)/(r_2 - r_1)] - [(z_1 - z_3)/(r_1 - r_3)]^2$$
$$\times (z_1 r_3 - z_3 r_1)/(r_1 - r_3)\} + r_1\{[(z_2 - z_1)/(r_2 - r_1)][(z_1 r_2 - z_2 r_1)/(r_2 - r_1)]^2$$
$$- [(z_1 - z_3)/(r_1 - r_3)][(z_1 r_3 - z_3 r_1)/(r_1 - r_3)]^2\} + ⅓ \ln r_1 \{[(z_1 r_2 - z_2 r_1)/(r_2 - r_1)]^3$$
$$- [(z_1 r_3 - z_3 r_1)/(r_1 - r_3)]^3\} + \sum_{231} + \sum_{312}$$

The additional summations $\sum_{231} + \sum_{312}$ in I_4, I_5 and I_6 imply a similar number of further terms found from rotating subscripts in the foregoing terms, i.e. 1 to 2, 2 to 3 and 3 to 1. Note that the expressions for I_4, I_5 and I_6 apply where r_1, r_2 and r_3 are non-zero and different. Where the orientation of the element is such that radii for two of their nodes are equal or are both zero when lying on the axis of symmetry, it becomes necessary to re-define the integrals [3]. Alternatively, numerical integration can be be employed. The simplest method is to evaluate the integrand in eq(10.61a) for the centroid of each element. Matrix **B** in eq(10.61b) is replaced with a mean matrix, $\bar{\mathbf{B}}$, whose components are based upon the centroidal co-ordinates $(\bar{r}, \bar{z})$ of the element. In Fig. 10.20 these are

$$\bar{r} = (r_1 + r_2 + r_3)/3 \text{ and } \bar{z} = (z_1 + z_2 + z_3)/3$$

Using PVW we may write eq(10.5b) as

$$\delta^{evT}\,\mathbf{f}^e = \int_V \boldsymbol{\varepsilon}^{vT}\sigma\,\mathrm{d}V$$
$$= \int_V (\bar{\mathbf{B}}\delta^{ev})^T(\mathbf{D}\bar{\mathbf{B}}\delta^e)\,\mathrm{d}V$$

from which

$$\mathbf{f}^e = \{\bar{\mathbf{B}}^T\mathbf{D}\bar{\mathbf{B}}\int_V \mathrm{d}V\} \times \delta^e$$

and hence

$$\mathbf{K}^e = \{\bar{\mathbf{B}}^T\,\mathbf{D}\,\bar{\mathbf{B}}\int_V \mathrm{d}V\}$$

Putting $\int_V \mathrm{d}V = 2\pi\bar{r}\Delta$, the element stiffness matrix becomes

$$\mathbf{K}^e = 2\pi(\bar{\mathbf{B}}^T\,\mathbf{D}\bar{\mathbf{B}})(\bar{r}\Delta) \tag{10.63}$$

The appproximate method is sufficiently accurate for fine meshes. Moreover, eq(10.63) will provide a superior numerical solution for larger radii when the logarithmic terms I_4, I_5 and I_6 in eq(10.61a) are found to be unreliable.

10.7.4 Force Vector

The components of the force vector $\mathbf{f}^e$ are the nodal point forces equivalent to pressure, body weight and centrifugal effects. These loadings can readily arise from, say, a large spinning toroid supporting distributed loading on its top surface. It becomes important to establish their separate contributions to $\mathbf{f}^e$.

(a) Pressure Loading
Let a pressure p act normally to the side 23 as shown in Fig. 10.22a. The band of area δS is formed from one rotation of an elemental length δl at radius r about the z - axis.

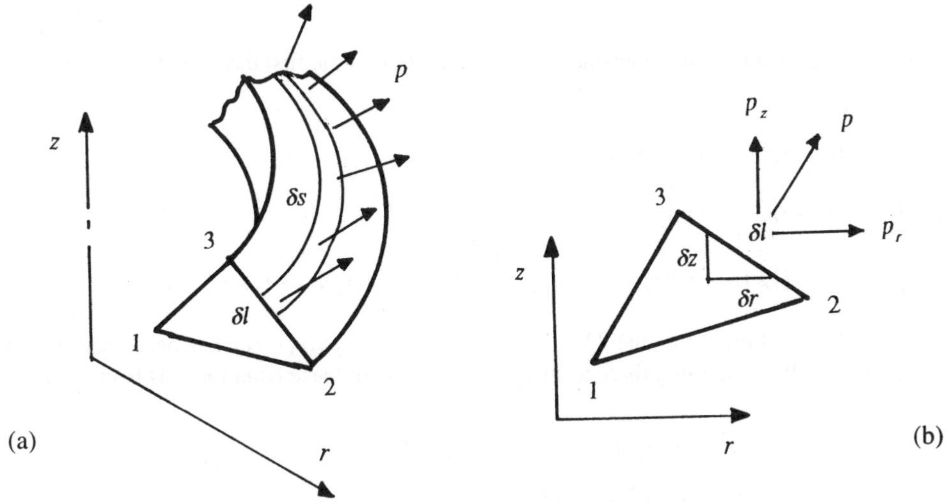

Figure 10.22 Surface pressure on an element with axial symmetry

When δS replaces $\delta l \times t$ in Fig. 10.12a, a similar analysis allows for the resolution of p into the components p_r and p_z to give a vector $\mathbf{p} = [\,p_r \;\; p_z\,]^T$. Referring to eq(10.39a), the PVW gives

$$\delta^{evT}\,\mathbf{f}^e = \int_S \delta^{vT}\,\mathbf{p}\;\mathrm{d}S$$
$$= \int_S [(\mathbf{A}\,\mathbf{A}^{e-1})\,\delta^e\,]^{vT}\,\mathbf{p}\;\mathrm{d}S$$
$$= \delta^{evT}(\mathbf{A}^{e-1})^T \int_S \mathbf{A}^T\,\mathbf{p}\;\mathrm{d}S$$

Cancelling δ^{evT}, the equivalent force vector is written as

$$\mathbf{f}^e = (\mathbf{A}^{e-1})^T\,\mathbf{p}^e \tag{10.64}$$

where $\mathbf{p}^e = \int_S \mathbf{A}^T\,\mathbf{p}\;\mathrm{d}S$. Setting $\mathrm{d}S = 2\pi\,r\,\mathrm{d}l$ gives

$$\mathbf{p}^e = 2\pi \int_l \mathbf{A}^T\mathbf{p}\;r\;\mathrm{d}l \tag{10.65a}$$

Converting eq(10.65a) to a single variable r we have, from Fig. 10.21b, $\mathrm{d}l = l_{23}\,\mathrm{d}r/(r_2 - r_3)$, which gives

$$\mathbf{p}^e = [2\pi\,l_{23}/\,(r_2 - r_3)] \int_r \mathbf{A}^T\,\mathbf{p}\;r\;\mathrm{d}r \tag{10.65b}$$

Taking the transpose of **A** in eq(10.53a), the integrand in (10.65a) becomes

$$\mathbf{A}^{\mathrm{T}}\mathbf{p} = \begin{bmatrix} 1 & 0 \\ r & 0 \\ z & 0 \\ 0 & 1 \\ 0 & r \\ 0 & z \end{bmatrix} \begin{bmatrix} p_r \\ p_z \end{bmatrix} = \begin{bmatrix} p_r \\ rp_r \\ zp_r \\ p_z \\ rp_z \\ zp_z \end{bmatrix} \qquad (10.65c)$$

Assuming that p_r and p_z are independent of r, integration of the first three terms in eq(10.65b) gives

$$p_r \int_{r_2}^{r_3} r\,\mathrm{d}r = (r_3^2 - r_2^2)\,p_r/2$$

$$p_r \int_{r_2}^{r_3} r^2\,\mathrm{d}r = (r_3^3 - r_2^3)\,p_r/3$$

$$p_r \int_{r_2}^{r_3} z\,r\,\mathrm{d}r = [⅓\,(z_3 - z_2)(r_3^2 + r_2 r_3 + r_2^{\,2}) + ½\,(z_2 r_3 - z_3 r_2)(r_3 + r_2)$$

in which the equation of the line 23, $z - z_3 = (r - r_3)(z_3 - z_2)/(r_3 - r_2)$, has been used. In the integration of the remaining three terms, p_z replaces p_r in these equations. This gives p_e as

$$\mathbf{p}^e = \pi\, l_{23} \begin{bmatrix} p_r(r_2 + r_3) \\ (2p_r/3)(r_3^2 + r_2 r_3 + r_2^2) \\ (p_r/3)[z_3(2r_3 + r_2) + z_2(2r_2 + r_3)] \\ p_z(r_2 + r_3) \\ (2p_z/3)(r_3^2 + r_2 r_3 + r_2^2) \\ (p_z/3)[z_3(2r_3 + r_2) + z_2(2r_2 + r_3)] \end{bmatrix} \qquad (10.65d)$$

Substituting $(\mathbf{A}^{e-1})^{\mathrm{T}}$ from eq(10.26b) and eq(10.65d) into eq(10.64) leads to the equivalent nodal forces

$$f_{r1} = [\pi\, l_{23} p_r/(2\Delta)][(r_2 z_3 - r_3 z_2)(r_2 + r_3) + (2/3)(z_2 - z_3)(r_3^2 + r_2 r_3 + r_2^2) + ⅓(r_3 - r_2)[z_3(2r_3 + r_2) + z_2(2r_2 + r_3)] \qquad (10.66a)$$

$$f_{z1} = [\pi\, l_{23} p_z/(2\Delta)][(r_2 z_3 - r_3 z_2)(r_2 + r_3) + (2/3)(z_2 - z_3)(r_3^2 + r_2 r_3 + r_2^2) + ⅓\,(r_3 - r_2)[z_3(2r_3 + r_2) + z_2(2r_2 + r_3)] = f_{r1} \times p_z/p_r \qquad (10.66b)$$

$$f_{r2} = [\pi\, l_{23} p_r/(2\Delta)][(r_3 z_1 - r_1 z_3)(r_2 + r_3) + (2/3)(z_3 - z_1)(r_3^2 + r_2 r_3 + r_2^2) + ⅓\,(r_1 - r_3)[z_3(2r_3 + r_2) + z_2(2r_2 + r_3)] \qquad (10.66c)$$

$$f_{z2} = [\pi\, l_{23} p_z/(2\Delta)][(r_3 z_1 - r_1 z_3)(r_2 + r_3) + (2/3)(z_3 - z_1)(r_3^2 + r_2 r_3 + r_2^2) + ⅓\,(r_1 - r_3)[z_3(2r_3 + r_2) + z_2(2r_2 + r_3)] = f_{r2} \times p_z/p_r \qquad (10.66d)$$

$$f_{r3} = [\pi\, l_{23} p_r/(2\Delta)][(r_1 z_2 - r_2 z_1)(r_2 + r_3) + (2/3)(z_1 - z_2)(r_3^2 + r_2 r_3 + r_2^2) + ⅓\,(r_2 - r_1)[z_3(2r_3 + r_2) + z_2(2r_2 + r_3)] \qquad (10.66e)$$

$$f_{z3} = [\pi\, l_{23} p_z/(2\Delta)][(r_1 z_2 - r_2 z_1)(r_2 + r_3) + (2/3)(z_1 - z_2)(r_3^2 + r_2 r_3 + r_2^2) + ⅓\,(r_2 - r_1)[z_3(2r_3 + r_2) + z_2(2r_2 + r_3)] = f_{r3} \times p_z/p_r \qquad (10.66f)$$

These six forces satisfy the two equilibrium conditions:

$$f_{r1} + f_{r2} + f_{r3} = 2\pi \bar{r}\, l_{23}\, p_r$$
$$f_{z1} + f_{z2} + f_{z3} = 2\pi \bar{r}\, l_{23}\, p_z$$

where $\bar{r} = ½\,(r_2 + r_3)$. The manner in which the nodal forces replace the distributed loading will depend upon the orientation of the side 23. When side 23 is vertical, with normal pressure loading p_r, we set $p_z = 0$ and $r_2 = r_3 = c$ (constant). Equations (10.66a,b,d,f) show that $f_{z1} = f_{z2} = f_{z3} = 0$ and $f_{r1} = 0$. This leaves eqs(10.66c,e) as

$$f_{r2} = f_{r3} = ½\,[2\pi c\,(z_3 - z_2)\, p_r]$$

This gives an equal weighting to nodes 2 and 3. However in the case of 23 lying horizontally we set $p_r = 0$ and $z_2 = z_3 = c$ (constant). Equations (10.66a,b,c,e) show that $f_{r1} = f_{r2} = f_{r3} = 0$ and $f_{z1} = 0$. Equations (10.66d,f) leave an unequal weighting:

$$f_{z2} = (\pi/3)(r_2 - r_3)(2r_2 + r_3)\, p_z \qquad (10.67a)$$
$$f_{z3} = (\pi/3)(r_2 - r_3)(2r_3 + r_2)\, p_z \qquad (10.67b)$$

The following example compares this with equally weighted nodal forces:

$$f_{z2} = f_{z3} = (\pi/2)(r_2 - r_3)(r_2 + r_3) \qquad (10.67c)$$

Example 10.8. An axi-symmetric ring 10 m diameter is loaded with a uniform pressure of $p_z = 1$ kN/m² on its top surface. Using the five elements shown in Fig. 10.23a, compare the exact equivalent nodal forces with approximate values found from equal weighting.

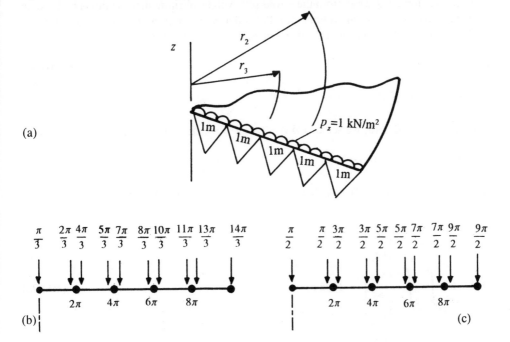

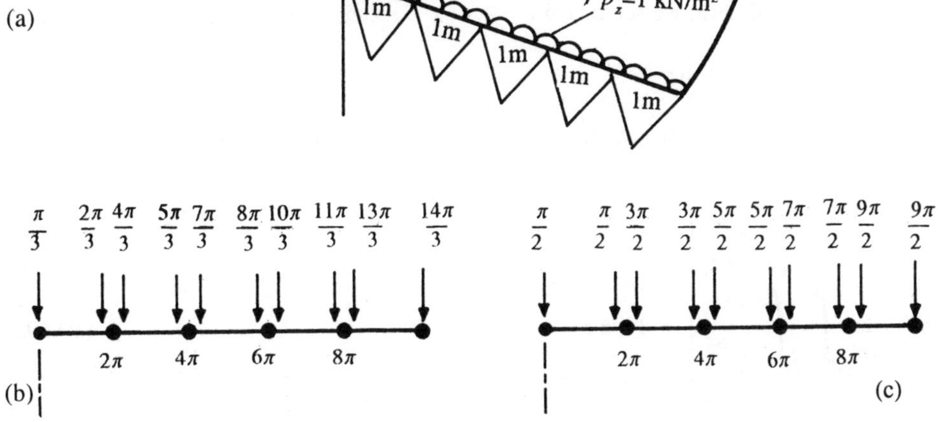

Figure 10.23 Ring under pressure

Equations (10.67a,b,c) are applied to each element in turn (e.g. $r_2 = 3$ m and $r_3 = 2$ m as shown in Fig. 10.23a)). This gives an exact and approximate system of nodal forces shown in Figs 10.23b and c. The net nodal forces given in brackets reveal only a small error for the end nodes by the approximation (10.67c).

(b) Body Forces

Consider the toroid in Fig. 10.24a of mass m with a triangular cross-section (nodes 123). Let the toroid rotate at ω rad/s about the z - axis.

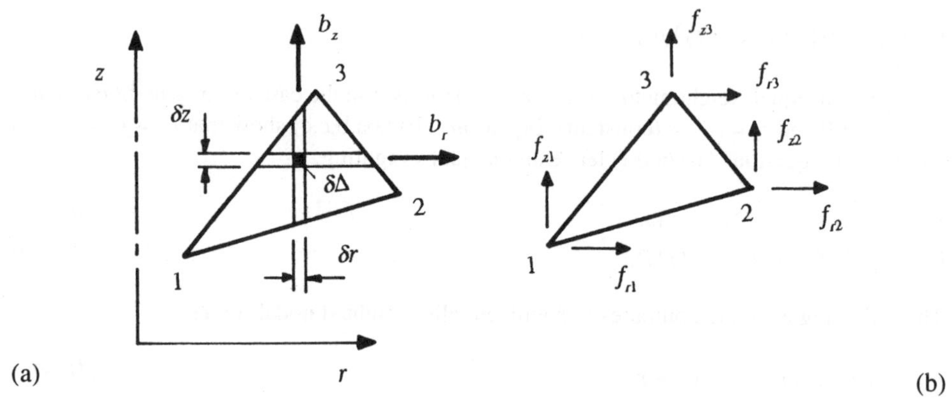

Figure 10.24 Body forces in a rotating toroid

The horizontal centrifugal force is $m\omega^2 \bar{r}$, where $\bar{r} = (r_1 + r_2 + r_3)/3$ locates the centroid. A vertical body force mg arises from the toroid self-weight. Writing these within a body force vector $\mathbf{b} = [b_r \;\; b_z]^{\mathrm{T}}$ per unit volume, it follows that $b_r = \rho\omega^2 r$ and $b_z = -\rho g$. Substituting eq(10.55) into PVW, eq(10.42a) leads to an equivalent nodal force vector:

$$\mathbf{f}^e = (\mathbf{A}^{e-1})^{\mathrm{T}} \mathbf{b}^e \tag{10.68}$$

$$\text{where } \mathbf{b}^e = \int_V \mathbf{A}^{\mathrm{T}} \mathbf{b} \, \mathrm{d}V \tag{10.69a}$$

Now, from Fig. 10.23(a) we see that $\delta V = 2\pi r \times \delta r \times \delta z$, so that eq(10.69a) becomes

$$\mathbf{b}^e = 2\pi \int_r \int_z \mathbf{A}^{\mathrm{T}} \mathbf{b} \, r \, \mathrm{d}r \, \mathrm{d}z \tag{10.69b}$$

where, guided by eq(10.65c): $\mathbf{A}^{\mathrm{T}} \mathbf{b} = [b_r \;\; rb_r \;\; zb_r \;\; b_z \;\; rb_z \;\; zb_z]^{\mathrm{T}}$. Thus $\mathbf{b}^e$ is

$$\mathbf{b}^e = 2\pi \int_r \int_z \begin{bmatrix} b_r \\ rb_r \\ zb_r \\ b_z \\ rb_z \\ zb_z \end{bmatrix} r \, \mathrm{d}r \, \mathrm{d}z \tag{10.69c}$$

Taking b_r and b_z as constants, the first three integrals in eq(10.69c) are

$$2\pi b_r \int_r \int_z r\,\mathrm{d}r\,\mathrm{d}z = (2\pi/3)\, b_r \Delta(r_1 + r_2 + r_3)$$
$$2\pi b_r \int_r \int_z r^2\,\mathrm{d}r\,\mathrm{d}z = (\pi/6)\, b_r \Delta[(r_1 + r_2 + r_3)^2 + (r_1^2 + r_2^2 + r_3^2)]$$
$$2\pi b_r \int_r \int_z z\, r\,\mathrm{d}r\,\mathrm{d}z = (\pi/6)\, b_r \Delta[(r_1 + r_2 + r_3)(z_1 + z_2 + z_3) + (r_1 z_1 + r_2 z_2 + r_3 z_3)]$$

and in the remaining three integrals b_z replaces b_r. Substitute $(\mathbf{A}^{e-1})^{T}$ (see eq(10.26b)) and eq(10.69c) into eq(10.68). Matrix multiplication leads to the six nodal forces shown in Fig. 10.23b:

$$f_{r1} = (\pi b_r/3)\{(\ 1\ )(r_2 z_3 - r_3 z_2) + [\ 2\ ](z_2 - z_3)/4 + [\ 3\ ](r_3 - r_2)/4\}$$
$$f_{z1} = (\pi b_z/3)\{(\ 1\ )(r_2 z_3 - r_3 z_2) + [\ 2\ ](z_2 - z_3)/4 + [\ 3\ ](r_3 - r_2)/4\} = f_{r1} \times b_z / b_r$$
$$f_{r2} = (\pi b_r/3)\{(\ 1\ )(r_3 z_1 - r_1 z_3) + [\ 2\ ](z_3 - z_1)/4 + [\ 3\ ](r_1 - r_3)/4\} \qquad (10.70\text{a-f})$$
$$f_{z2} = (\pi b_z/3)\{(\ 1\ )(r_3 z_1 - r_1 z_3) + [\ 2\ ](z_3 - z_1)/4 + [\ 3\ ](r_1 - r_3)/4\} = f_{r2} \times b_z / b_r$$
$$f_{r3} = (\pi b_r/3)\{(\ 1\ )(r_1 z_2 - r_2 z_1) + [\ 2\ ](z_1 - z_2)/4 + [\ 3\ ](r_2 - r_1)/4\}$$
$$f_{z3} = (\pi b_z/3)\{(\ 1\ )(r_1 z_2 - r_2 z_1) + [\ 2\ ](z_1 - z_2)/4 + [\ 3\ ](r_2 - r_1)/4\} = f_{r3} \times b_z / b_r$$

where the numbered brackets are

$$(\ 1\ ) = (r_1 + r_2 + r_3)$$
$$[\ 2\ ] = [(r_1 + r_2 + r_3)^2 + (r_1^2 + r_2^2 + r_3^2)]$$
$$[\ 3\ ] = [(r_1 + r_2 + r_3)(z_1 + z_2 + z_3) + (r_1 z_1 + r_2 z_2 + r_3 z_3)]$$

The six forces in eqs(10.70a-f) satisfy the three equilibrium equations

$$f_{r1} + f_{r2} + f_{r3} = 2\pi \bar{r} \Delta\, b_r$$
$$f_{z1} + f_{z2} + f_{z3} = 2\pi \bar{r} \Delta\, b_z$$

in which $\bar{r} = (r_1 + r_2 + r_3)/3$ and $\bar{z} = (z_1 + z_2 + z_3)/3$ are the centroidal co-ordinates. The contributions to nodal forces from adjoining elements is made with a cyclic permutation of the subscripts as for the plane triangle (see Section 10.5.5).

10.8 Rectangular Element for Plate Flexure

Figure 10.25 shows the system of co-ordinates for a single rectagular plate element. The x, y plane lies at the mid-thickness co-incident with the neutral plane.

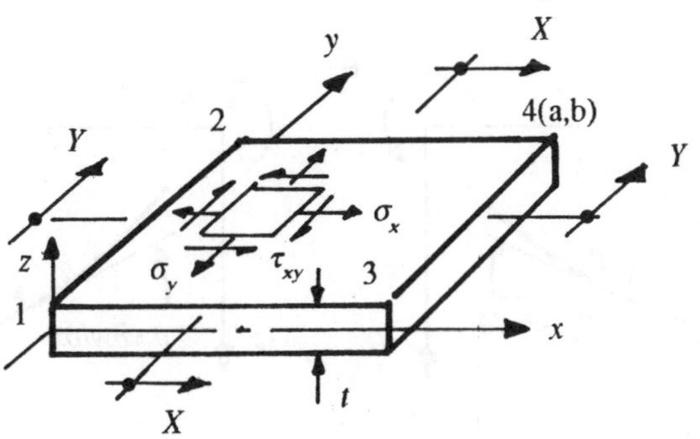

Figure 10.25 Plate element in x, y plane

In the analysis that follows, corner 1 will be taken at the origin of x, y where z is is positive upwards. The forces and deflections at each node are shown in Figs 10.26a and b.

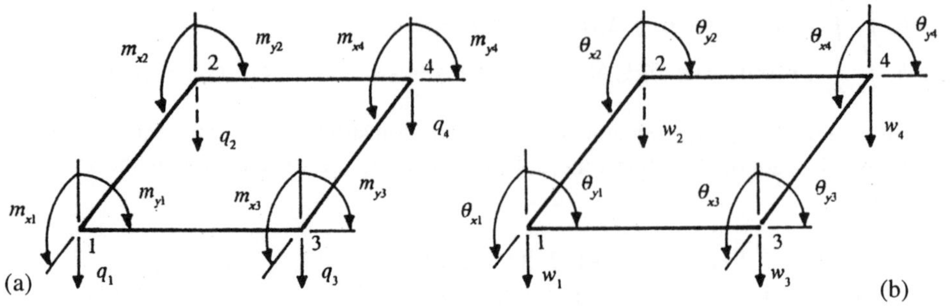

Figure 10.26 Plate nodal point forces and deflections

The forces in Fig. 10.26a comprise moments m and shear forces f. The positive directions of m shown are found from applying the right-hand screw rule in the directions of increasing x and y. The subscripts x and y refer to the axis about which m acts. Shear forces are positive when acting downwards. In Fig. 10.26b it is seen that a similar convention has been used for the rotations θ and deflections w at each node. The force and displacement vectors for the element are

$$\mathbf{f}^e = [\,\mathbf{f}_1 \;\; \mathbf{f}_2 \;\; \mathbf{f}_3 \;\; \mathbf{f}_4\,]^T \;\text{ and }\; \boldsymbol{\delta}^e = [\,\boldsymbol{\delta}_1 \;\; \boldsymbol{\delta}_2 \;\; \boldsymbol{\delta}_3 \;\; \boldsymbol{\delta}_4\,]^T$$

where $\mathbf{f}_1 = [\,m_{x1} \;\; m_{y1} \;\; q_1\,]^T$ etc and $\boldsymbol{\delta}_1 = [\,\theta_{x1} \;\; \theta_{y1} \;\; w_1\,]^T$ etc.

10.8.1 Displacements

The following fourth-order displacement function is assumed [12]:

$$w = \alpha_1 + \alpha_2 x + \alpha_3 y + \alpha_4 x^2 + \alpha_5 xy + \alpha_6 y^2 + \alpha_7 x^3 + \alpha_8 x^2 y + \alpha_9 xy^2 + \alpha_{10} y^3 + \alpha_{11} x^3 y + \alpha_{12} xy^3 \tag{10.71a}$$

Small rotations θ (rad) are the gradients of the tangents to the deflected plate. Figures 10.27a,b show the sign convention used with respective sections X-X and Y-Y taken from Fig. 10.25.

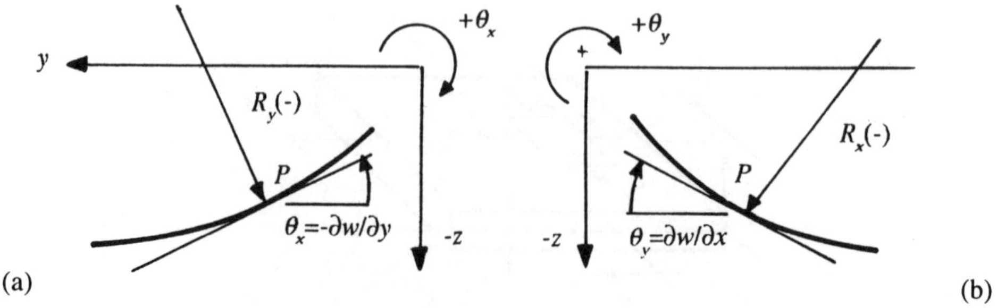

Figure 10.27 Convention for slopes in the y - z and x - z planes

For the sagging plate shown, it is necessary to compare the gradients at any point P with the positive directions assumed for slopes θ_x and θ_y. From this it follows that $\theta_x = -\,\partial w/\partial y$ and $\theta_y = \partial w/\partial x$. Thus, from eq(10.71a), we have

$$\theta_y = \alpha_2 + 2\alpha_4 x + \alpha_5 y + 3\alpha_7 x^2 + 2\alpha_8 xy + \alpha_9 y^2 + 3\alpha_{11} x^2 y + \alpha_{12} y^3 \qquad (10.71b)$$

$$\theta_x = -\,(\alpha_3 + \alpha_5 x + 2\alpha_6 y + \alpha_8 x^2 + 2\alpha_9 x\,y + 3\alpha_{10} y^2 + \alpha_{11} x^3 + 3\alpha_{12} x\,y^2) \qquad (10.71c)$$

Equations (10.71a-c) can be expressed in matrix form as

$$\begin{bmatrix} \theta_x \\ \theta_y \\ w \end{bmatrix} = \begin{bmatrix} 0 & 0 & -1 & 0 & -x & -2y & 0 & -x^2 & -2xy & -3y^2 & -x^3 & -3xy^2 \\ 0 & 1 & 0 & 2x & y & 0 & 3x^2 & 2xy & y^2 & 0 & 3x^2y & y^3 \\ 1 & x & y & x^2 & xy & y^2 & x^3 & x^2y & xy^2 & y^3 & x^3y & xy^3 \end{bmatrix} \begin{bmatrix} \alpha_1 \\ \alpha_2 \\ \alpha_3 \\ \alpha_4 \\ \alpha_5 \\ \alpha_6 \\ \alpha_7 \\ \alpha_8 \\ \alpha_9 \\ \alpha_{10} \\ \alpha_{11} \\ \alpha_{12} \end{bmatrix} \qquad \boldsymbol{\delta} = \mathbf{A}\,\boldsymbol{\alpha} \qquad (10.72a,b)$$

Setting x and y in eq(10.72a) to the co-ordinates of the nodes (see Fig. 10.25a) gives the element nodal displacements $\boldsymbol{\delta}^e$ as

$$\begin{bmatrix} \theta_{x1} \\ \theta_{y1} \\ w_1 \\ \theta_{x2} \\ \theta_{y2} \\ w_2 \\ \theta_{x3} \\ \theta_{y3} \\ w_3 \\ \theta_{x4} \\ \theta_{y4} \\ w_4 \end{bmatrix} = \begin{bmatrix} 0 & 0 & -1 & 0 & 0 & 0 & 0 & 0 & 0 & 0 & 0 & 0 \\ 0 & 1 & 0 & 0 & 0 & 0 & 0 & 0 & 0 & 0 & 0 & 0 \\ 1 & 0 & 0 & 0 & 0 & 0 & 0 & 0 & 0 & 0 & 0 & 0 \\ 0 & 0 & -1 & 0 & 0 & -2b & 0 & 0 & 0 & -3b^2 & 0 & 0 \\ 0 & 1 & 0 & 0 & b & 0 & 0 & 0 & b^2 & 0 & 0 & b^3 \\ 1 & 0 & b & 0 & 0 & b^2 & 0 & 0 & 0 & b^3 & 0 & 0 \\ 0 & 0 & -1 & 0 & -a & 0 & 0 & -a^2 & 0 & 0 & -a^3 & 0 \\ 0 & 1 & 0 & 2a & 0 & 0 & 3a^2 & 0 & 0 & 0 & 0 & 0 \\ 1 & a & 0 & a^2 & 0 & 0 & a^3 & 0 & 0 & 0 & 0 & 0 \\ 0 & 0 & -1 & 0 & -a & -2b & 0 & -a^2 & -2ab & -3b^2 & -a^3 & -3ab^2 \\ 0 & 1 & 0 & 2a & b & 0 & 3a^2 & 2ab & b^2 & 0 & 3a^2b & b^3 \\ 1 & a & b & a^2 & ab & b^2 & a^3 & a^2b & ab^2 & b^3 & a^3b & ab^3 \end{bmatrix} \begin{bmatrix} \alpha_1 \\ \alpha_2 \\ \alpha_3 \\ \alpha_4 \\ \alpha_5 \\ \alpha_6 \\ \alpha_7 \\ \alpha_8 \\ \alpha_9 \\ \alpha_{10} \\ \alpha_{11} \\ \alpha_{12} \end{bmatrix} \qquad \boldsymbol{\delta}^e = \mathbf{A}^e\,\boldsymbol{\alpha} \qquad (10.73a,b)$$

where $\mathbf{A}^e$ is a 12×12 matrix arising from the 3 degrees of freedom in each of the 4 nodes. For node 1, where $x = 0$, $y = 0$, eq(10.73a) gives

$\theta_{x1} = -\alpha_3,\ \theta_{y1} = \alpha_2$ and $w_1 = \alpha_1$

and for node 2, where $x = 0,\ y = b$,

$\theta_{x2} = -(\alpha_3 + 2\alpha_6 b + 3\alpha_{10} b^2),\ \theta_{y2} = \alpha_2 + \alpha_5 b + \alpha_9 b^2 + \alpha_{12} b^3$ and $w_2 = \alpha_1 + \alpha_3 b + \alpha_6 b^2 + \alpha_{10} b^3$

The expressions w_1, w_2, θ_{x1} and θ_{x2} show that the four coefficients α_1, α_3, α_6 and α_{10}, can be found from knowing w and θ_x at nodes 1 and 2. This will preserve continuity in the deflected shape of adjacent elements along edges parallel to x. Knowing θ_y at nodes 1 and 2 is insufficient for determining the four coefficients α_2, α_5, α_9 and α_{12} appearing in the two θ_{y1} and θ_{y2} expressions. This results in a discontinuity in θ_y between adjacent elements along edges parallel to y. The function $w(x, y)$ is said therefore to be non-conforming. Though not ideal, the numerical error involved can be made small by averaging θ_y along these edges. The inverse of eq(10.73b) is found from Gaussian elimination:

$$\begin{bmatrix} \alpha_1 \\ \alpha_2 \\ \alpha_3 \\ \alpha_4 \\ \alpha_5 \\ \alpha_6 \\ \alpha_7 \\ \alpha_8 \\ \alpha_9 \\ \alpha_{10} \\ \alpha_{11} \\ \alpha_{12} \end{bmatrix} = \begin{bmatrix} 0 & 0 & 1 & 0 & 0 & 0 & 0 & 0 & 0 & 0 & 0 & 0 \\ 0 & 1 & 0 & 0 & 0 & 0 & 0 & 0 & 0 & 0 & 0 & 0 \\ -1 & 0 & 0 & 0 & 0 & 0 & 0 & 0 & 0 & 0 & 0 & 0 \\ 0 & -2/a & -3/a^2 & 0 & 0 & 0 & 0 & -1/a & 3/a^2 & 0 & 0 & 0 \\ 1/a & -1/b & -1/ab & 0 & 1/b & 1/a & -1/a & 0 & 1/ab & 0 & 0 & -1/a \\ 2/b & 0 & -3/b^2 & 1/b & 0 & 3/b^2 & 0 & 0 & 0 & 0 & 0 & 0 \\ 0 & 1/a^2 & 2/a^3 & 0 & 0 & 0 & 0 & 1/a^2 & -2/a^3 & 0 & 0 & 0 \\ 0 & 2/ab & 3/a^2b & 0 & -2/ab & -3/a^2b & 0 & 1/ab & -3/a^2b & 0 & -1/ab & 3/a^2b \\ -2/ab & 0 & 3/ab^2 & -1/ab & 0 & -3/ab^2 & 2/ab & 0 & -3/ab^2 & 1/ab & 0 & 3/ab^2 \\ -1/b^2 & 0 & 2/b^3 & -1/b^2 & 0 & -2/b^3 & 0 & 0 & 0 & 0 & 0 & 0 \\ 0 & -1/a^2b & -2/a^3b & 0 & 1/a^2b & 2/a^3b & 0 & -1/a^3b & 2/a^3b & 0 & 1/a^2b & -2/a^3b \\ 1/ab^2 & 0 & -2/ab^3 & 1/ab^2 & 0 & 2/ab^3 & -1/ab^2 & 0 & 2/ab^3 & -1/ab^2 & 0 & -2/ab^3 \end{bmatrix} \begin{bmatrix} \theta_{x1} \\ \theta_{y1} \\ w_1 \\ \theta_{x2} \\ \theta_{y2} \\ w_2 \\ \theta_{x3} \\ \theta_{y3} \\ w_3 \\ \theta_{x4} \\ \theta_{y4} \\ w_4 \end{bmatrix} \tag{10.74a}$$

Equation (3.115a) is written in the symbolic form:

$$\boldsymbol{\alpha} = \mathbf{A}^{e-1}\, \boldsymbol{\delta}^e \tag{10.74b}$$

where $\mathbf{A}^e\mathbf{A}^{e-1} = \mathbf{I}$. Eliminating $\boldsymbol{\alpha}$ between eqs(10.72b) and (10.74b) gives the general displacement vector $\boldsymbol{\delta} = [\theta_x\ \ \theta_y\ \ w]^{\mathrm{T}}$ as

$$\boldsymbol{\delta} = \mathbf{A}\, \mathbf{A}^{e-1}\, \boldsymbol{\delta}^e$$

10.8.2 Strain and Curvature

The strains arising from flexure depend upon the radii of the plate curvature. In addition to the negative radii R_x and R_y of point P in each plane of the sagging plate (see Fig. 10.27a,b), R_{xy} refers to the curvature accompanying the rotation of a fibre out of its original unstrained plane. The reciprocals of the radii are given as second order partial derivatives [11]:

$$1/R_x = -\,\partial^2 w/\partial x^2,\quad 1/R_y = -\,\partial^2 w/\partial y^2 \text{ and } 1/R_{xy} = \partial^2 w/\partial x\partial y \tag{10.75a,b,c}$$

The direct strains in the plate at a distance z from the unstrained neutral plane is $\varepsilon = z/R$. Thus, with z-negative in the lower half of the plate thickness and with R_x and R_y both negative, the maximum tensile strains in the bottom surface are $\varepsilon_x = t/(2R_x)$ and $\varepsilon_y = t/(2R_y)$. With positive z for the top surface, these strains are a maximum in compression. R_{xy} is used to determine the engineering shear strain $\gamma_{xy} = 2z/R_{xy}$, i.e. the total angular change in two originally perpendicular unstrained line elements. For any plane, it follows from eq(10.75a - c) that the three strain components are

$$\varepsilon_x = z/R_x = -z\,(\partial^2 w/\partial x^2),\ \varepsilon_y = z/R_y = -z\,(\partial^2 w/\partial y^2),\ \gamma_{xy} = 2z/R_{xy} = 2z\,(\partial^2 w/\partial x\partial y) \quad (10.76a,b,c)$$

where, from eq(10.71a)

$$\begin{aligned}
\partial^2 w/\partial x^2 &= 2\alpha_4 + 6\alpha_7 x + 2\alpha_8 y + 6\alpha_{11} x\,y \\
\partial^2 w/\partial y^2 &= 2\alpha_6 + 2\alpha_9 x + 6\alpha_{10} y + 6\alpha_{12} x\,y \\
\partial^2 w/\partial x\partial y &= \alpha_5 + 2\alpha_8 x + 2\alpha_9 y + 3\alpha_{11} x^2 + 3\alpha_{12} y^2
\end{aligned} \quad (10.77a,b,c)$$

Combining eqs(10.76) and (10.77), it follows that

$$\begin{bmatrix} \varepsilon_x \\ \varepsilon_y \\ \gamma_{xy} \end{bmatrix} = z \begin{bmatrix} 0 & 0 & 0 & -2 & 0 & 0 & -6x & -2y & 0 & 0 & -6xy & 0 \\ 0 & 0 & 0 & 0 & 0 & -2 & 0 & 0 & -2x & -6y & 0 & -6xy \\ 0 & 0 & 0 & 0 & 2 & 0 & 0 & 4x & 4y & 0 & 6x^2 & 6y^2 \end{bmatrix} \begin{bmatrix} \alpha_1 \\ \alpha_2 \\ \alpha_3 \\ \alpha_4 \\ \alpha_5 \\ \alpha_6 \\ \alpha_7 \\ \alpha_8 \\ \alpha_9 \\ \alpha_{10} \\ \alpha_{11} \\ \alpha_{12} \end{bmatrix} \qquad \boldsymbol{\varepsilon} = z\,\mathbf{C}\,\boldsymbol{\alpha} \quad (10.78a,b)$$

Combining eqs(10.78b) and (10.74b) gives the three strains $\boldsymbol{\varepsilon} = [\varepsilon_x\ \ \varepsilon_y\ \ \gamma_{xy}]^T$ generally in terms of the nodal displacements $\boldsymbol{\delta}^e = [\delta_1\ \ \delta_2\ \ \delta_3\ \ \delta_4]^T$:

$$\boldsymbol{\varepsilon} = z\,\mathbf{C}\,(\mathbf{A}^{e-1}\,\boldsymbol{\delta}^e) = z\,\mathbf{B}\,\boldsymbol{\delta}^e \quad (10.79a)$$

where matrix $\mathbf{B} = \mathbf{C}\mathbf{A}^{e-1}$ depends upon x and y. The nodal strains $\boldsymbol{\varepsilon}^e = [\boldsymbol{\varepsilon}_1\ \ \boldsymbol{\varepsilon}_2\ \ \boldsymbol{\varepsilon}_3\ \ \boldsymbol{\varepsilon}_4]^T$, in which $\boldsymbol{\varepsilon}_1 = [\varepsilon_{x1}\ \ \varepsilon_{y1}\ \ \gamma_{xy1}]^T$ etc, follow from the substitution of the nodal co-ordinates into eq(10.78a). This gives

$$\boldsymbol{\varepsilon}^e = z\,\mathbf{B}^e\,\boldsymbol{\delta}^e \quad (10.79b)$$

where $\mathbf{B}^e = \mathbf{C}^e\mathbf{A}^{e-1}$ in which the dimebnsions if $\mathbf{C}^e$ are 12×12.

10.8.3 Stress and Moments

The stress components follow from the constitutive relations as follows:

$$\sigma_x = E\,(\varepsilon_x + \nu\,\varepsilon_y)/(1-\nu^2) = -Ez\,(\partial^2 w/\partial x^2 + \nu\partial^2 w/\partial y^2)/(1-\nu^2)$$
$$\sigma_y = E\,(\varepsilon_y + \nu\,\varepsilon_x)/(1-\nu^2) = -\,Ez\,(\partial^2 w/\partial y^2 + \nu\partial^2 w/\partial x^2)/(1-\nu^2) \qquad (10.80a,b,c)$$
$$\tau_{xy} = G\gamma_{xy} = Ez\,(\partial^2 w/\partial x\partial y)/(1+\nu)$$

Equations (10.80a-c) appear in the matrix forms:

$$\begin{bmatrix}\sigma_x\\ \sigma_y\\ \tau_{xy}\end{bmatrix} = \frac{Ez}{(1-\nu^2)}\begin{bmatrix}1 & \nu & 0\\ \nu & 1 & 0\\ 0 & 0 & (1-\nu)/2\end{bmatrix}\begin{bmatrix}-\partial^2 w/\partial x^2\\ -\partial^2 w/\partial y^2\\ 2\partial^2 w/\partial x\partial y\end{bmatrix} \qquad \sigma = z(D/I)\mathbf{D}\,\boldsymbol{\varepsilon} \qquad (10.81a,b)$$

In eq(10.81a), $E/(1-\nu^2)$ appears as D/I, the ratio between the plate flexural coefficient: $D = Et^3/[12(1-\nu^2)$ and its second moment of area I (see Fig. 10.28).

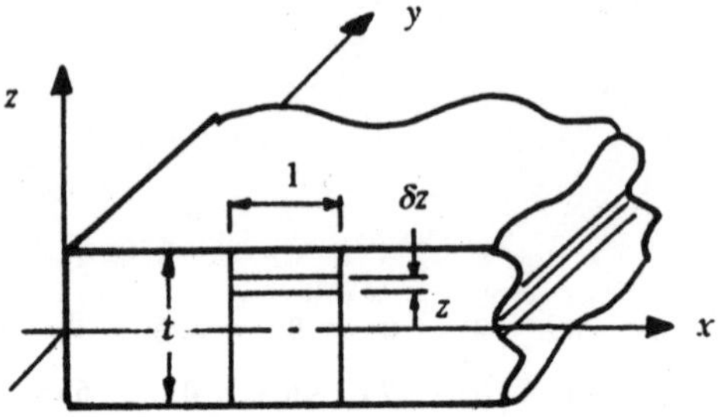

Figure 10.28 I for a plate in flexure

Now I applies to the plate cross-section for a unit length of its side:

$$I = \int_A z^2\,dA = \int_z z^2\,(1\times dz) = t^3/12 \qquad (10.82)$$

The moments m_x and m_y in Figs 10.25a,b follow from the engineering theories of bending and torsion as

$$m_x = \sigma_x\, I/z = -\,D(\,\partial^2 w/\partial x^2 + \nu\,\partial^2 w/\partial y^2)$$
$$m_y = \sigma_y\, I/z = -\,D(\,\partial^2 w/\partial y^2 + \nu\,\partial^2 w/\partial x^2) \qquad (10.83a,b,c)$$
$$m_{xy} = m_{yx} = (\tau_{xy} I/z) = (1-\nu)D\,\partial^2 w/\partial x\partial y$$

Equations (10.83a-c) are written as

$$\begin{bmatrix}\sigma_x\\ \sigma_y\\ \tau_{xy}\end{bmatrix} = \frac{z}{I}\begin{bmatrix}m_x\\ m_y\\ m_{xy}\end{bmatrix} \qquad \sigma = (z/I)\,\mathbf{m} \qquad (10.84a,b)$$

where

$$\begin{bmatrix} m_x \\ m_y \\ m_{xy} \end{bmatrix} = D \begin{bmatrix} 1 & \nu & 0 \\ \nu & 1 & 0 \\ 0 & 0 & (1-\nu)/2 \end{bmatrix} \begin{bmatrix} -\partial^2 w/\partial x^2 \\ -\partial^2 w/\partial y^2 \\ 2\partial^2 w/\partial x\,\partial y \end{bmatrix} \qquad \mathbf{m} = (D/z)\,\mathbf{D}\,\boldsymbol{\varepsilon} \qquad (10.85\text{a,b})$$

The general stress vector $\sigma = [\sigma_x\ \sigma_y\ \tau_{xy}]^{\mathrm{T}}$ is, from eqs(10.84b) and (10.85b)

$$\sigma = (z / I)(D/z)\,\mathbf{D}\,\boldsymbol{\varepsilon} = (D/I)\mathbf{D}\,\boldsymbol{\varepsilon} \qquad (10.86\text{a})$$

Substituting eq(10.79a) into eq(10.86a) gives

$$\sigma = z(D/I)\,\mathbf{D}\,\mathbf{B}\,\boldsymbol{\delta}^{e} \qquad (10.86\text{b})$$

The nodal point stresses $\sigma^{e} = [\ \sigma_1\ \ \sigma_2\ \ \sigma_3\ \ \sigma_4]^{\mathrm{T}}$ where $\sigma_1 = [\sigma_x\ \ \sigma_y\ \ \tau_{xy}]^{\mathrm{T}}$ etc, follow from eqs(10.86a,b) as

$$\sigma^{e} = (D/I)\,\mathbf{D}\,\boldsymbol{\varepsilon}^{e} = z\,(D/I)\,\mathbf{D}\,\mathbf{B}^{e}\boldsymbol{\delta}^{e} = z(D/I)\,\mathbf{H}^{e}\boldsymbol{\delta}^{e} \qquad (10.86\text{c})$$

where $\mathbf{H}^{e} = \mathbf{D}\mathbf{B}^{e} = \mathbf{D}\,(\,\mathbf{C}^{e}\mathbf{A}^{e\,-1})$.

10.8.4 Element Stiffness Matrix

Finally, we apply the PVW to find $\mathbf{K}^{e}$. Substituting eqs(10.79a) and (10.86b) into eq(10.5b),

$$\boldsymbol{\delta}^{evT}\,\mathbf{f}^{e} = \int_V \boldsymbol{\varepsilon}^{vT}\,\sigma\,\mathrm{d}V$$
$$= \int_x \int_y \int_z z\,(\mathbf{B}\,\boldsymbol{\delta}^{e})^{vT}\,z\,(D/I)\,\mathbf{D}\,\mathbf{B}\,\boldsymbol{\delta}^{e}\,(\mathrm{d}x\ \mathrm{d}y\ \mathrm{d}z) \qquad (10.87\text{a})$$

Equation (10.87a) is rearranged as follows:

$$\boldsymbol{\delta}^{evT}\,\mathbf{f}^{e} = \boldsymbol{\delta}^{evT}\,(D/I)\int_x \int_y \mathbf{B}^{\mathrm{T}}\mathbf{D}\,\mathbf{B}\,\mathrm{d}x\,\mathrm{d}y \int_z [z^2(1\times \mathrm{d}z)] \times \boldsymbol{\delta}^{e} \qquad (10.87\text{b})$$

Now from eq(10.82), the second integral defines I in Fig. 10.27. Cancelling $\boldsymbol{\delta}^{evT}$ and I gives

$$\mathbf{f}^{e} = D\int_x \int_y \mathbf{B}^{\mathrm{T}}\,\mathbf{D}\,\mathbf{B}\,\mathrm{d}x\,\mathrm{d}y \times \boldsymbol{\delta}^{e} \qquad (10.88\text{a})$$

from which the element stiffness matrix follows as

$$\mathbf{K}^{e} = D\int_x \int_y \mathbf{B}^{\mathrm{T}}\mathbf{D}\,\mathbf{B}\,\mathrm{d}x\,\mathrm{d}y \qquad (10.88\text{b})$$

Recall that as $\mathbf{B} = \mathbf{C}\,\mathbf{A}^{e\,-1}$ contains both x any y, the product $\mathbf{B}^{\mathrm{T}}\mathbf{D}\,\mathbf{B}$ will need to be found and then integrated term by term.

Example 10.9. Determine the single stiffness components $K_{11}{}^{e}$ and $K_{21}{}^{e}$ from eq(10.88b) without performing the full matrix multiplications.

Writing eq(10.88b) in its component form,

$$K_{ij}^{e} = D \int_x \int_y (B_{ik})^{T} D_{km} B_{mj} \, dx\, dy = D \int_x \int_y (B_{ki} D_{km} B_{mj}) \, dx\, dy \tag{i}$$

where i and $j = 1, 2...12$, k and $m = 1, 2$ and 3. Setting $i = j = 1$ in eq(i) gives K_{11}^{e} as

$$K_{11}^{e} = D \int_x \int_y (B_{k1} D_{km} B_{m1}) \, dx\, dy \tag{ii}$$

Summing eq(ii) over k and m, gives

$$\begin{aligned} K_{11}^{e} &= D \int_x \int_y (B_{11} D_{1m} B_{m1} + B_{21} D_{2m} B_{m1} + B_{31} D_{3m} B_{m1}) \, dx\, dy \\ &= D \int_x \int_y (B_{11} D_{11} B_{11} + B_{11} D_{12} B_{21} + B_{11} D_{13} B_{31} \\ &\quad + B_{21} D_{21} B_{11} + B_{21} D_{22} B_{21} + B_{21} D_{23} B_{31} \\ &\quad + B_{31} D_{31} B_{11} + B_{31} D_{32} B_{21} + B_{31} D_{33} B_{31}) \, dx\, dy \end{aligned} \tag{iii}$$

Setting, from eq(10.85a), $D_{11} = D_{22} = 1$, $D_{12} = D_{21} = \nu$, $D_{33} = ½(1 - \nu)$ and all other $D_{ij} = 0$ in eq(iii), gives

$$K_{11}^{e} = D \int_x \int_y (B_{11}^{2} + 2\nu B_{11} B_{21} + B_{21}^{2} + ½(1 - \nu) B_{31}^{2}) \, dx\, dy \tag{iv}$$

Alternatively, reverting eq(ii) to matrix form

$$K_{11}^{e} = D \int_x \int_y \begin{bmatrix} B_{11} & B_{21} & B_{31} \end{bmatrix} \begin{bmatrix} 1 & \nu & 0 \\ \nu & 1 & 0 \\ 0 & 0 & (1-\nu)/2 \end{bmatrix} \begin{bmatrix} B_{11} \\ B_{21} \\ B_{31} \end{bmatrix} dx\, dy$$

$$= D \int_x \int_y (B_{11}^{2} + 2\nu B_{11} B_{21} + B_{21}^{2} + ½(1 - \nu) B_{31}^{2}) \, dx\, dy \tag{iv}$$

Now in eq(i), B_{mj} follows from $\mathbf{B} = \mathbf{C}\,\mathbf{A}^{e\,-1}$ as

$$B_{mj} = C_{mp} (A_{pj}^{e})^{-1} \tag{v}$$

The B_{mj} terms required in eq(iv) are for $j = 1$ and $m = 1, 2, 3$. From eq(v)

$$B_{m1} = C_{mp} (A_{p1}^{e})^{-1} \tag{vi}$$

Setting $m = 1, 2, 3$ in turn and summing over $p = 1, 2..12$, eq(vi) gives three terms

$$\begin{aligned} B_{11} &= C_{11} (A_{11}^{e})^{-1} + C_{12} (A_{21}^{e})^{-1} + \ldots\ldots\; C_{1,12} (A_{12,1}^{e})^{-1} \\ B_{21} &= C_{21} (A_{11}^{e})^{-1} + C_{22} (A_{21}^{e})^{-1} + \ldots\ldots\; C_{2,12} (A_{12,1}^{e})^{-1} \\ B_{31} &= C_{31} (A_{11}^{e})^{-1} + C_{32} (A_{21}^{e})^{-1} + \ldots\ldots\; C_{3,12} (A_{12,1}^{e})^{-1} \end{aligned}$$

Substituting the appropriate components of $(\mathbf{A}^{e})^{-1}$ and $\mathbf{C}$ from eqs(10.74a) and (10.78a) respectively gives

$$\begin{aligned} B_{11} &= (0)(0) + (0)(0) + (0)(-1) + (-2)(0) + (0)(1/a) + (0)(2/b) + (-6x)(0) + (-2y)(0) \\ &\quad + (0)[-2/(ab)] + (0)(-1/b^{2}) + (-6xy)(0) + (0)[1/(ab^{2})] = 0 \end{aligned}$$

$$B_{21} = (0)(0) + (0)(0) + (0)(-1) + (0)(0) + (0)(1/a) + (-2)(2/b) + (0)(0) + (0)(0)$$
$$+ (-2x)[-2/(ab)] + (-6y)(-1/b^2) + (0)(0) + (-6xy)[1/(ab^2)]$$
$$= (-2/b)(2 - 2x/a - 3y/b + 3xy/(ab))]$$

$$B_{31} = (0)(0) + (0)(0) + (0)(-1) + (0)(0) + (2)(1/a) + (0)(2/b) + (0)(0) + (4x)(0)$$
$$+ (4y)[-2/(ab)] + (0)(-1/b^2) + (6x^2)(0) + (6y^2)[1/(ab^2)]$$
$$= (2/a)(1 - 4y/b + 3y^2/b^2)$$

Substituting these into eq(iv),

$$K_{11}^{e} = D\int_{x=0}^{a}\int_{y=0}^{b} \{(4/b^2)[2 - 2x/a - 3y/b + 3xy/ab)^2]$$
$$+ (2/a^2)(1 - \nu)(1 - 4y/b + 3y^2/b^2)^2\}\,dx\,dy$$

Integrating over x gives

$$K_{11}^{e} = (4D/b^2)\int_0^b (-4ay/b + 4a/3 + 3ay^2/b^2)\,dy$$
$$+ (2D/a)(1 - \nu)\int_0^b (1 - 8y/b + 22y^2/b^2 - 24y^3/b^3 + 9y^4/b^4)\,dy$$

and then integrating over y, we find

$$K_{11}^{e} = (4D/3)[(a/b) + (b/a)(1 - \nu)/5]$$

The off-diagonal component K_{12}^{e} follows from eq(i) as

$$K_{12}^{e} = D\int_x\int_y (B_{k1}D_{km}B_{m2})\,dx\,dy \qquad \text{(vii)}$$

Reverting eq(vii) to matrix form,

$$K_{12}^{e} = D\int_x\int_y \begin{bmatrix} B_{11} & B_{21} & B_{31} \end{bmatrix} \begin{bmatrix} 1 & \nu & 0 \\ \nu & 1 & 0 \\ 0 & 0 & (1-\nu)/2 \end{bmatrix} \begin{bmatrix} B_{12} \\ B_{22} \\ B_{32} \end{bmatrix} dx\,dy$$

$$= D\int_x\int_y (B_{11}B_{12} + \nu B_{11}B_{22} + \nu B_{12}B_{21} + B_{21}B_{22} + \tfrac{1}{2}(1 - \nu)B_{32}B_{31})\,dx\,dy \qquad \text{(viii)}$$

This integral again employs B_{11}, B_{21} and B_{31} and, from eq(vi),

$$B_{m2} = C_{mp}(A_{p2}^{e})^{-1}$$

which gives

$$B_{12} = C_{11}(A_{12}^{e})^{-1} + C_{12}(A_{22}^{e})^{-1} + \ldots\ldots\ C_{1,12}(A_{12,2}^{e})^{-1}$$
$$B_{22} = C_{21}(A_{12}^{e})^{-1} + C_{22}(A_{22}^{e})^{-1} + \ldots\ldots\ C_{2,12}(A_{12,2}^{e})^{-1}$$
$$B_{32} = C_{31}(A_{12}^{e})^{-1} + C_{32}(A_{22}^{e})^{-1} + \ldots\ldots\ C_{3,12}(A_{12,2}^{e})^{-1}$$

Substituting the components of $(\mathbf{A}^{e})^{-1}$ and $\mathbf{C}$ from eqs(10.74a) and (10.78a) and then integrating eq(viii) leads to $K_{12}^{e} = -\nu D$.

10.9 Concluding Remarks

The reader will now be familiar with the common procedure for formulating the stiffness matrix for any element by the displacement method of finite elements. A similar procedure is employed for the derivation of K for three-dimensional elements [1]. The 10 steps in this procedure are as follows:

1. Select a suitable element and co-ordinate system
2. Choose a suitable general displacement function $\boldsymbol{\delta} = \mathbf{A}\,\boldsymbol{\alpha}$
3. Establish the particular nodal displacements $\boldsymbol{\delta}^e = \mathbf{A}^e\,\boldsymbol{\alpha}$
4. Derive the general displacements from the nodal values $\boldsymbol{\delta} = \mathbf{A}\,(\mathbf{A}^e)^{-1}\boldsymbol{\delta}^e$
5. Derive the general strains $\boldsymbol{\varepsilon} = \mathbf{C}\,\boldsymbol{\alpha} = \mathbf{C}\,(\mathbf{A}^e)^{-1}\boldsymbol{\delta}^e = \mathbf{B}\,\boldsymbol{\delta}^e$
6. Establish the particular nodal strains $\boldsymbol{\varepsilon}^e = \mathbf{B}^e\,\boldsymbol{\delta}^e$
7. Determine the general stresses from the general strains $\boldsymbol{\sigma} = \mathbf{D}\,\boldsymbol{\varepsilon} = \mathbf{D}\,\mathbf{B}\,\boldsymbol{\delta}^e$
8. Establish the nodal stresses from the nodal strains $\boldsymbol{\sigma}^e = \mathbf{D}\,\boldsymbol{\varepsilon}^e = \mathbf{D}\,\mathbf{B}^e\,\boldsymbol{\delta}^e$
9. Employ the general stress and strains within PVW

$$\boldsymbol{\delta}^{evT}\,\mathbf{f}^e = \int_V \boldsymbol{\varepsilon}^{vT}\,\boldsymbol{\sigma}\,\mathrm{d}V = \int_V (\mathbf{B}\,\boldsymbol{\delta}^{ev})^T(\mathbf{D}\,\mathbf{B}\,\boldsymbol{\delta}^e)\,\mathrm{d}V$$

to give the element stiffness matrix $\mathbf{K}^e$ as

$$\mathbf{K}^e = \int_V \mathbf{B}^T\mathbf{D}\,\mathbf{B}\,\mathrm{d}V$$

10. Assemble the overall stiffness matrix $\mathbf{K}$ from $\mathbf{K}^e$ (e = I, II, III...), then invert $\mathbf{K}$ and solve for $\boldsymbol{\delta}^e = \mathbf{K}^{-1}\mathbf{f}^e$.

References

[1] Zienkiewicz O. C. and Taylor R. L., *The Finite Element Method, Vol 1: Basic Formulation and Linear Problems*, McGraw-Hill, U.K., 1989.
[2] Rees D. W. A., *Basic Solid Mechanics*, Macmillan, U.K., 1997.
[3] Boresi A. P. and Sidebottom O. M., *Advanced Mechanics of Materials*, John Wiley and Sons, 1985.
[4] Knight C. E., *The Finite Element Method in Mechanical Design*, PWS-Kent Pub Co, Boston, 1993.
[5] Fenner R. T., *Finite Element Methods for Engineers*, Macmillan, 1975.
[6] Ottosen N. and Peterson H., *Introduction to the Finite Element Method*, Prentice-Hall, 1992.
[7] Spencer W. J., *Fundamental Structural Analysis*, Macmillan, 1988.
[8] Fagan M. J., *Finite Element Analysis, Theory and Practice*, Longman, 1992.
[9] Tong P. and Rossettos, J. N., *Finite Element Method: Basic Technique and Implementation*, M.I.T. Press, 1977.
[10] Irons B. and Ahmad S., *Techniques of Finite Elements*, Ellis-Horwood, 1980.
[11] Timoshenko S. P and Goodier J. N., *Theory of Elasticity*, McGraw-Hill, 1970.
[12] Kleiber M. and Breitkopf P., *Finite Elements in Structural Mechanics*, Ellis-Horwood, 1993.

EXERCISES

10.1 Assemble the overall stiffness matrix for the framed structure given in Fig. 10.29. Determine the forces in bars I ...VII given that the areas and bar lengths are constant (included angles are all 60°).
Answer: (in kN) $p_I = 2.1$, $p_{II} = -1.1$, $p_{III} = 0.70$, $p_{IV} = -1.3$, $p_V = -4.8$, $p_{VI} = 11.6$ and $p_{VII} = 8.2$

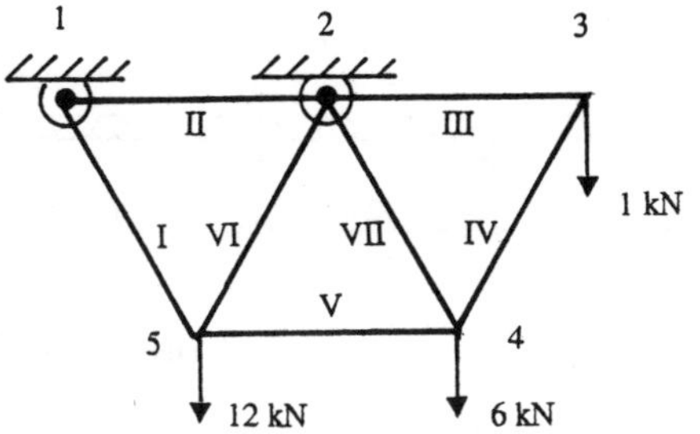

Figure 10.29

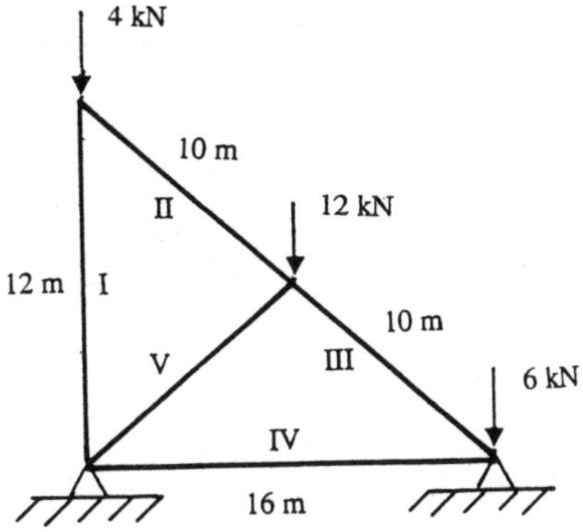

Figure 10.30

10.2 Determine the bar forces for the frame shown in Fig. 10.30 given that all bar areas are constant.
Answer: (in kN) $p_I = -4$, $p_{II} = 0$, $p_{III} = -10$, $p_{IV} = 8$ and $p_V = -10$

10.3 Derive the stiffness matrices in eqs(10.2), (10.3b) and (10.4b) using the principal of virtual displacements. Assume respective displacement functions for tension, torsion and bending as follows: (i) $u = \alpha_1 + \alpha_2 x$, (ii) $\theta = \alpha_1 + \alpha_2 z$ and (iii) $v = \alpha_1 + \alpha_2 z + \alpha_3 z^2 + \alpha_4 z^3$ where α are displacement coefficients to be determined and x, z are the length co-ordinates shown in Figs 10.1 and 10.2.

10.4 Deduce the stiffness matrix in the case of a uniform circular bar element subjected to combined axial force and torsion. Take each of the loadings to differ between the two nodes of the element.

10.5 The stepped shaft in Fig. 10.31 is subjected to a combination of axial nodal forces f_{x1}, f_{x2} ... f_{x5} and torques t_1, t_2 ..... t_5 at node junctions 1, 2 ... 5. Given that the nodal axial displacements and twists are (u_1, θ_1), (u_2, θ_2) .... .(u_5, θ_5), find the overall stiffness matrix.

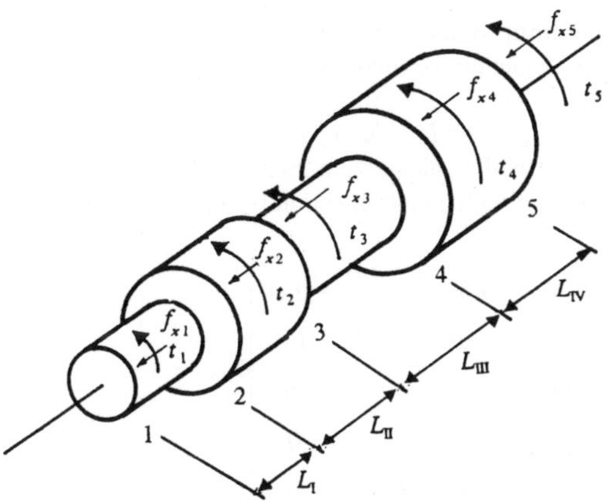

Figure 10.31

10.6 Examine whether it is possible to derive equations (10.20a) and (10.22) from the principle of virtual work.

10.7 Construct the net shear force and moment diagrams for the two-element beam shown in Fig. 10.32.
Answer: moments (kNm) -44.86, -8.71, 45.38, -8.87; shear forces (kN) -8.97, 6.03, 6.03, -13.56, 4.44

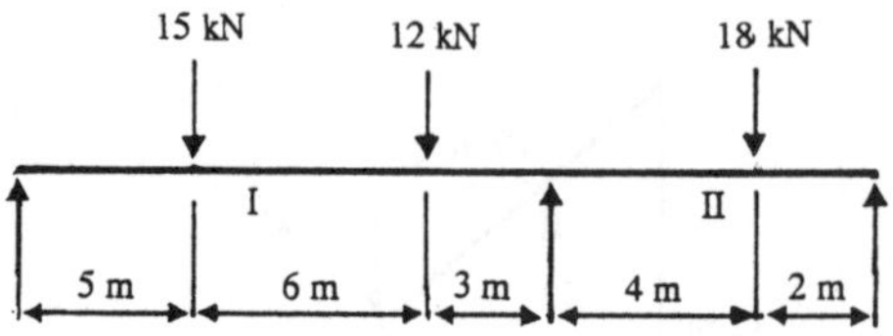

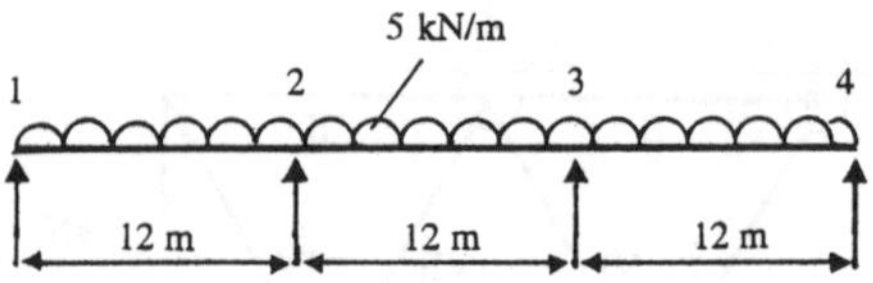

Figure 10.32

Figure 10.33

10.8 Determine the forces and moments at nodes 1, 2, 3 and 4 for the continuous 3 - element beam in Fig. 10.33.
Answer: $m_1 = m_4 = 0$, $m_2 = m_3 = 72$ kNm, $f_1 = f_4 = -24$ kN, $f_2 = f_3 = -66$ kN

10.9 The beam shown in Fig. 10.34 is composed of two elements I and II. Using the known slope and deflection at node 1 and the moment and shear force at node 3, determine the net moment and force at points 1 and 2.
Answer: $m_1 = 75.62$ kNm, $m_2 = 48.74$ kNm, $f_1 = -21.34$ kNm, $f_2 = 18.66$ kNm

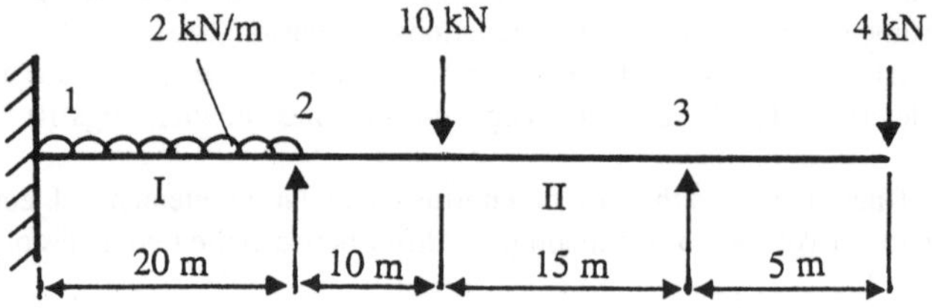

Figure 10.34

10.10 Determine the explicit forms of the matrix $\boldsymbol{\delta}$ that enable interior element displacements vector $\boldsymbol{\delta}_i = [\,u_i \;\; v_i\,]^T$ for a triangular element to be found when its nodal displacements $\boldsymbol{\delta}^e$ are known.

10.11 Determine the stiffness equation $\mathbf{f}^e = \mathbf{K}^e\,\boldsymbol{\delta}^e$ for the singular triangular element in Fig. 10.35 under conditions of plane stress with a thickness of 20 mm. Take $E = 210$ GPa and $\nu = 0.28$.
Answer: $\mathbf{K}^e$ in units of N and mm with a multiplication factor for each element of 10^6

$$\mathbf{K}^e = \begin{bmatrix} 3.50 & 1.46 & -2.85 & -0.82 & -0.66 & -0.64 \\ 1.46 & 2.85 & -0.64 & -1.03 & -0.82 & -1.82 \\ -2.85 & -0.64 & 2.85 & 0 & 0 & 0.64 \\ -0.82 & -1.03 & 0 & 1.03 & 0.82 & 0 \\ -0.66 & -0.82 & 0 & 0.82 & 0.66 & 0 \\ -0.64 & -1.82 & 0.64 & 0 & 0 & 1.82 \end{bmatrix}$$

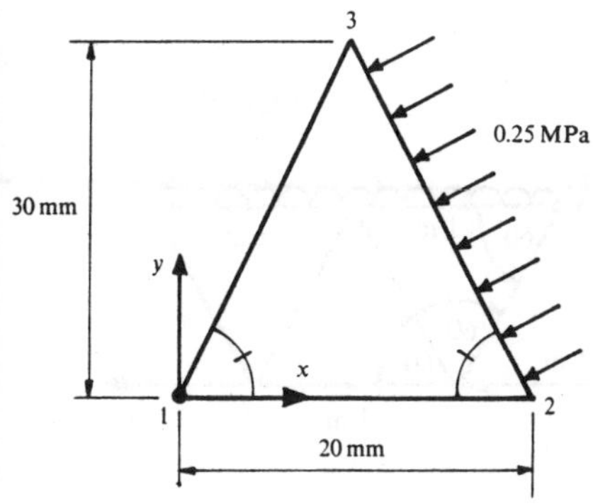

Figure 10.35

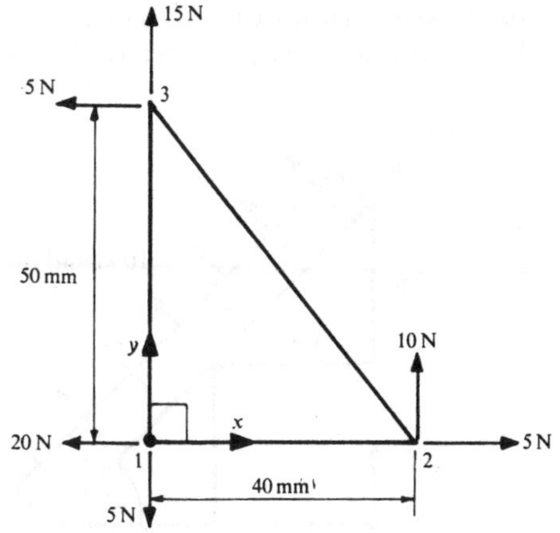

Figure 10.36

10.12 Determine the stiffness equation $\mathbf{f}^e = \mathbf{K}^e \delta^e$ for the singular triangular element in Fig. 10.36 under plane strain for a bar length of 200 mm. Take $E = 210$ GPa and $\nu = 0.28$.
Answers: to $\mathbf{K}^e$ in units of N and mm with a multiplication factor for each element of 10^6

$$\mathbf{K}^e = \begin{bmatrix} 40.12 & 18.64 & -33.56 & -8.20 & -6.56 & -10.44 \\ 18.64 & 31.73 & -10.44 & -10.25 & -8.20 & -21.48 \\ -33.56 & -10.44 & 33.56 & 0 & 0 & 10.44 \\ -8.20 & -10.25 & 0 & 10.25 & 8.20 & 0 \\ -6.56 & -8.20 & 0 & 8.20 & 6.56 & 0 \\ -10.44 & -21.48 & 10.44 & 0 & 0 & 21.48 \end{bmatrix}$$

10.13 Show that the stiffness equation (10.50a) for a plane rectangular element may also be derived from the principle of stationary potential energy.

10.14 Derive the components of matrix **B** in the general strain expression $\varepsilon = \mathbf{B}\,\delta^e$ for a plane rectangular element.

10.15 Derive the components of the rectangular element stiffness matrices $\mathbf{K}^e$ from eq(10.50a) (i) in the case of plane stress and (ii) in the case of plane strain. The dimensions of a single element are $a \times b \times t$ (see Fig. 10.19a,b).

10.16 Invert the matrix $\mathbf{A}^e$ in eq(10.48b) and hence determine the components of the general matrix equation $\delta = (\mathbf{A}\mathbf{A}^{e\,-1})\,\delta^e$ for a plane rectangular element.

10.17 Determine from eq(10.51b) the components of the element nodal stress matrix $\sigma^e = \mathbf{H}^e\,\delta^e$ for a plane rectangular element.

10.18 Show that the shape function for a plane rectangular element may be written as $u = N_i u_i$ and $v = N_i v_i$ ($i = 1 \ldots 4$). If the plate dimensions are $2a \times 2b$ and the origin for x, y lies at the plate centre, with x parallel to the side of length $2a$, show that $N_1 = (a - x)(b - y)/(4ab)$, $N_2 = (a + x)(b - y)/(4ab)$, $N_3 = (a + x)(b + y)/(4ab)$ and $N_4 = (a - x)(b + y)$. Nodes 3, 4, 1 and 2 lie in the first, second, third and fourth quadrants respectively.

10.19 Two interconnected, right-angled triangular elements are loaded along one side as shown in Fig. 10.37. Determine the equivalent nodal forces.

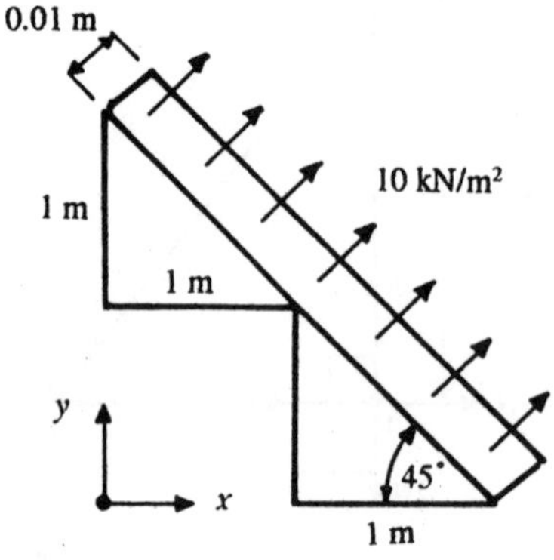

Figure 10.37

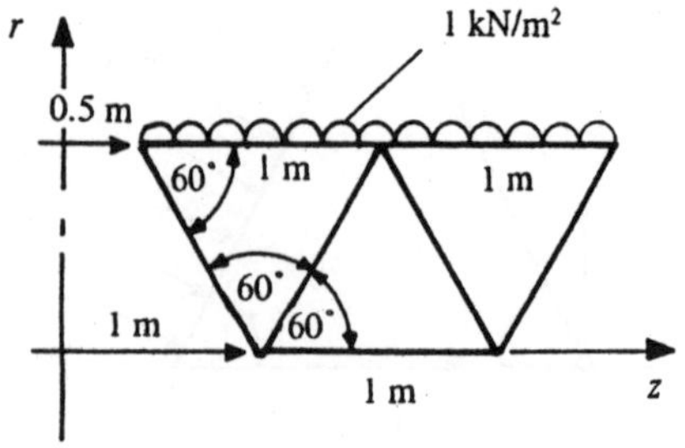

Figure 10.38

10.20 The cross-section of a steel ring of density 7600 kg/m^3 is represented by three elements as shown in Fig. 10.38. Determine the net nodal forces when the ring's horizontal top surface is subjected to distributed loading of 1 kN/m^2.

10.21 Examine the manner in which normal, uniformly distributed loading may be lumped at the nodes of a plane rectangular element when this loading is applied (i) along one horizontal side aligned with x and (ii) along one vertical side aligned with y.

10.22 Assemble the overall stiffness matrix **K** when the plane-stress cantilever beam in Fig.10.10 is divided into two rectangular plane stress elements with counter-clockwise node numbering 1342 and 3564 as shown.

10.23 Confirm the following two element stiffnesses for the toroidal element shown in Fig. 10.20:

$$K_{22}^{\,c} = [\pi/(2\Delta^2)] \int_r \int_z [D_{11}(r^3 - r^2) + D_{44}(z_2 - z_3)]\, dr\, dz$$

$$K_{12}^{\,c} = [\pi/(2\Delta^2)] \int_r \int_z \{D_{21}(r_3 - r_2)(z_2 - z_3) + D_{31}(r_3 - r_2)[(z_3 r_2 - z_2 r_3)/r$$

$$+ (z_2 - z_3) + (r_3 - r_2)\, z/r] + D_{44}(z_2 - z_3)(r_3 - r_2)\}dr\, dz$$

Then evaluate each double integrals using m and I as outlined in Section 10.7.3.

10.24 Confirm from eq(10.87b) the following diagonal stiffness components for a plate in flexure:

$$K_{11}^{\,c} = K_{44}^{\,c} = K_{77}^{\,c} = K_{10}^{\,c} = (4D/3)[a/b + (1 - \nu)\, b/(3a)]$$
$$K_{22}^{\,c} = K_{55}^{\,c} = K_{88}^{\,c} = K_{1111}^{\,c} = (4D/3)[b/a + a(1 - \nu)/(5b)]$$
$$K_{33}^{\,c} = K_{66}^{\,c} = K_{99}^{\,c} = K_{1212}^{\,c} = D[2b/a^3 + 2a/b^3 + 1/(ab) + 7(1 - \nu)/(5ab)]$$

CHAPTER 11

YIELD AND STRENGTH CRITERIA

Yielding in annealed polycrystalline materials is isotropic but the yield stress will become direction-dependent following a history of severe deformation processing, as with the cold-rolling of sheet metals. A prior-worked metal can also have different yield stresses in tension and compression similar to the fracture stresses found in brittle solids. Other combinations of stresses required to produce yield or fracture lie on a surface that bounds an elastic interior. This is known as the yield or fracture surface as appropriate to the material. There exist similar descriptions of the initial yield surface in metals and of the failure surface for brittle solids despite differences in their behaviour. For example, a steel yields before becoming plastic while cast iron remains elastic to the point of fracture. We refer respectively to the yield and the fracture stress as a limit to the elasticity in each material. In less brittle materials, such as wood and fibre-reinforced composites, a region of pseudo-inelasticity may follow yielding. This is due to the non-linear visco-elastic behaviour of the resin matrix. The fracture surface will envelop a more complex elastic, visco-elastic regime but this need not concern our strength estimates. The classical criteria of yield and fracture apply to each type of material without consideration of the preceding deformation. The constants they employ can be expressed in terms of a given material's uniaxial and shear strengths.

11.1 Yielding of Ductile Isotropic Metals

A ductile metallic material will begin to deform plastically under a uniaxial stress when the yield stress is reached. It has been seen that this situation can be avoided with a judicious choice of safety factor in deciding on an elastic safe working stress for the material. However, in practice, the stress state in many loaded structures is often biaxial or triaxial. Consider the principal triaxial stress system shown in Fig. 11.1a where $\sigma_1 > \sigma_2 > \sigma_3$

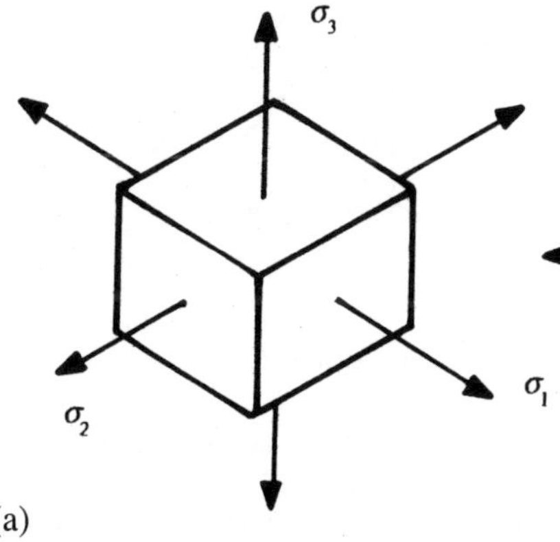

(a)

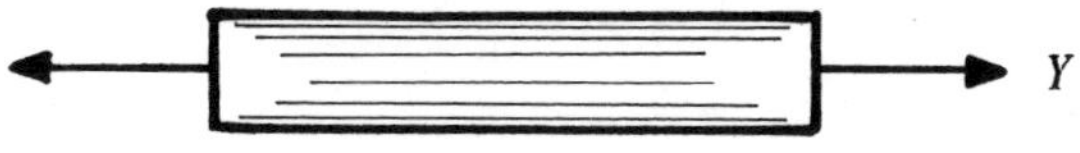

(b)

Figure 11.1 Principal and uniaxial stress systems

The question arises as to what magnitude of principal stresses will cause the onset of yielding. Where a material yields isotropically, the critical value of the chosen parameter at yield is independent of the orientation of the stress system. A number of yield criteria have been proposed over the past two centuries. Those given in Table 11.1 have been based upon critical values of stress, strain or strain energy being reached at yield in the principal system. The critical value is the corresponding quantity that prevails at yield in a simple tension test (Fig. 11.1b). This enables each criterion to be related to the constant uniaxial yield stress Y measured from a tension test for any orientation.

Table 11.1 Summary of yield criteria for metals

1. Maximum Principal Stress: $\sigma_1 = Y$ or $\sigma_3 = -Y$
2. Maximum Principal Strain: $[\sigma_1 - \nu(\sigma_2 + \sigma_3)] = Y$ or $[\sigma_3 - \nu(\sigma_1 + \sigma_2)] = Y$
3. Total Strain Energy: $(\sigma_1^2 + \sigma_2^2 + \sigma_3^2) - 2\nu(\sigma_1\sigma_2 + \sigma_2\sigma_3 + \sigma_1\sigma_3) = Y^2$
4. Maximum Shear Stress: $\sigma_1 - \sigma_3 = Y$
5. Shear Strain Energy: $(\sigma_1 - \sigma_2)^2 + (\sigma_2 - \sigma_3)^2 + (\sigma_1 - \sigma_3)^2 = Y^2$

For yielding under a plane principal stress state we set $\sigma_3 = 0$. The stress transformation equations, given in Chapter 1, enable these yield criteria to be expressed in any two or three dimensional combination of applied direct and shear stresses. It is now accepted that those attributed to von Mises [1] and Tresca [2] are the most representative of initial yield behaviour in metallic materials. It is instructive to consider the derivation and experimental verification of these two criteria and their confirmation using experimental data available in the literature.

11.1.1 Maximum Shear Stress Theory

Attributed to Tresca [2] (also, Coloumb and Guest) the maximum shear stress theory assumes that yielding, under the principal stress system in Fig. 11.1a, begins when the maximum shear stress reaches a critical value. The latter is taken as the maximum shear stress k at the yield point in simple tension or compression. That is, $k = Y/2$, which acts along planes at 45° to the tensile stress axis shown in Fig. 11.2a.

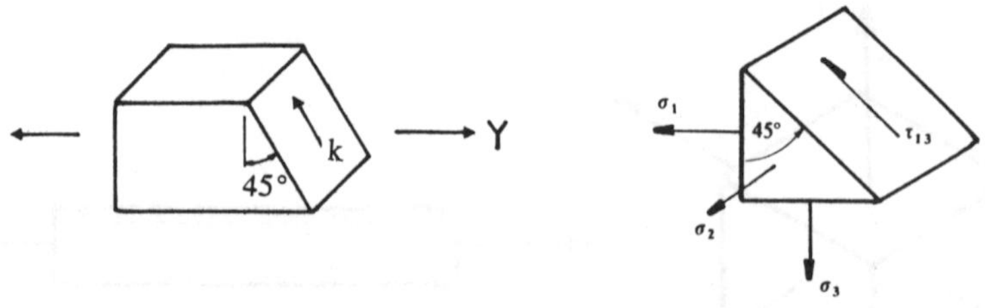

(a) (b)

Figure 11.2 Maximum shear stress under uniaxial and triaxial states

For the triaxial stress state in Fig. 11.1a, the greatest shear stress τ_{13} acts along the plane inclined at 45° to the 1 and 3 directions (Fig. 11.2b) with magnitude $\tau_{13} = (\sigma_1 - \sigma_3)/2$. Equating shear stresses for the uniaxial and triaxial cases $\tau_{13} = k$ gives the Tresca criterion:

$$\sigma_1 - \sigma_3 = Y \tag{11.1a}$$

Numerical values of σ_1 and σ_3 must be substituted in the left-hand side of eq(11.1a) with signs denoting tension or compression, e.g. if $\sigma_1 = 30$, $\sigma_2 = -15$ and $\sigma_3 = -20$ MPa then the left-hand side becomes 30 - (- 20) = 50 MPa, showing that the intermediate stress value is irrelevant. Often eq(11.1a) is stated in words as follows:

Greatest Principal Stress - Least Principal Stress = Tensile Yield Stress (11.1b)

In the case of plane stress, where (σ_x, σ_y, τ_{xy}) are non-zero, the principal stresses are

$$\sigma_1, \sigma_2 = ½(\sigma_x + \sigma_y) \pm ½\sqrt{[(\sigma_x - \sigma_y)^2 + 4\tau_{xy}^2]} \text{ and } \sigma_3 = 0 \tag{11.2}$$

where σ_1 is the greatest tensile and σ_2 is the least compressive stress. Substituting eq(11.2) into eq(11.1b) with $\sigma_y = 0$, the Tresca criterion reduces to

$$\sigma_x^2 + 4\tau_{xy}^2 = Y^2 \tag{11.3}$$

which defines the equation of a two-dimensional elliptical yield locus in axes of σ_x and τ_{xy}.

11.1.2 Shear Strain Energy Theory

Attributed to von Mises [1], this yield criterion has also been attributed to others (Maxwell, Huber and Hencky), who employed alternative derivations. They saw that the total strain energy was composed of dilatational (volumetric) and distortional (shear) components. The former depends upon the mean or hydrostatic component of the applied stress system while the latter is due to the remaining reduced (deviatoric) components of stress (as shown in Figs 1.11a - c). Clark Maxwell was among the first to realise that hydrostatic stress plays no part in yielding. He proposed that yielding under tension was due to the shear strain energy component of the total energy reaching a critical value (taken as the shear strain energy at the tensile yield point). This was later formalised by Richard von Mises and others in any one of the following three routes leading to the same yield criterion for ductile, initially isotropic, metallic materials. It will be shown later (see paragraph 11.1.4) that there is much experimental evidence to support this yield criterion.

(a) Shear strain energy

The total energy density for Fig. 2.4a is given by

$$u = \int_{\varepsilon_{ij}} \sigma_{ij}\, d\varepsilon_{ij} = \int (\sigma_1\, d\varepsilon_1 + \sigma_2\, d\varepsilon_2 + \sigma_3\, d\varepsilon_3)$$

Substituting the elastic constitutive relations from eq(2.15a-c) and integrating leads to

$$u = [1/(2E)](\sigma_1^2 + \sigma_2^2 + \sigma_3^2) - (2\nu/E)(\sigma_1\sigma_2 + \sigma_2\sigma_3 + \sigma_1\sigma_3) \tag{11.4}$$

The volumetric strain energy densitity is found from the hydrostatic component of stress (see Fig. 1.11b) as

$$\begin{aligned} u_v &= \int \sigma\, d\varepsilon = \int (\sigma_m\, d\varepsilon + \sigma_m\, d\varepsilon + \sigma_m\, d\varepsilon) \\ &= ½(\sigma_m\varepsilon + \sigma_m\varepsilon + \sigma_m\varepsilon) \end{aligned}$$

Substituting for the strain in any one direction, $\varepsilon = \sigma_m/(3K)$, from eqs(2.5b), leads to

$$u_v = \sigma_m^{\,2}/(2K) = 3(1 - 2\nu)\sigma_m^{\,2}/(2E)$$

and, since $\sigma_m = \sigma_{kk}/3$,

$$u_v = (1 - 2\nu)(\sigma_1 + \sigma_2 + \sigma_3)^2/(6E) \qquad (11.5)$$

Subtracting eq(11.5) from eq(11.4) leads to the shear strain energy associated with the deviatoric stress in Fig. 1.11c. This gives

$$\begin{aligned} u_S &= u - u_v \\ &= [1/(2E)](\sigma_1^{\,2} + \sigma_2^{\,2} + \sigma_3^{\,2}) - (2\nu/E)(\sigma_1\sigma_2 + \sigma_2\sigma_3 + \sigma_1\sigma_3) - (1 - 2\nu)(\sigma_1 + \sigma_2 + \sigma_3)^2/(6E) \\ &= (1 + \nu)[(\sigma_1 - \sigma_2)^2 + (\sigma_2 - \sigma_3)^2 + (\sigma_1 - \sigma_3)^2]/(6E) \qquad (11.6a) \end{aligned}$$

The value of u_S at the tensile yield point is found from putting $\sigma_1 = Y$, $\sigma_2 = \sigma_3 = 0$ in eq(11.6a). This gives

$$u_S = (1 + \nu)Y^2/(3E) \qquad (11.6b)$$

Equating (11.6a) and (11.6b) provides the principal stress form of the von Mises yield criterion:

$$(\sigma_1 - \sigma_2)^2 + (\sigma_2 - \sigma_3)^2 + (\sigma_1 - \sigma_3)^2 = 2Y^2 \qquad (11.7a)$$

If one principal stress, say σ_3, is zero, the biaxial form of eq(11.7a) becomes

$$\sigma_1^{\,2} - \sigma_1\sigma_2 + \sigma_2^{\,2} = Y^2 \qquad (11.7b)$$

Substitutions from eqs(11.2) for σ_1 (tensile) and σ_2 (compressive) provides the general biaxial form in terms of σ_x, σ_y and τ_{xy}. When $\sigma_y = 0$, a common plane form of the Mises criterion is found

$$\sigma_x^2 + 3\tau_{xy}^{\,2} = Y^2 \qquad (11.8)$$

Subscripts x and y are often omitted from eq(11.8) to account for yielding under combinations of direct stress and shear stress in any co-ordinate system.

(b) Octahedral shear stress

Alternatively, it may be proposed that yielding under the triaxial stress system in Fig. 11.1a commences when τ_o in eq(1.16b) reaches its critical value at the tensile yield point ($\sigma_1 = Y$, $\sigma_2 = \sigma_3 = 0$). We have

$$\tfrac{1}{3}\sqrt{[(\sigma_1 - \sigma_2)^2 + (\sigma_2 - \sigma_3)^2 + (\sigma_1 - \sigma_3)^2]} = \tfrac{1}{3}\sqrt{(2Y^2)}$$

It is seen that eq(11.7a) reappears because the normal stress for the octahedral planes, i.e. σ_o in eq(1.16a), is numerically equal to the mean stress $\sigma_m = \sigma_{kk}/3$ and causes dilitation only. The distortion, and hence yielding, occurs under the combination of principal stress differences that define τ_o. Clearly, energy considerations are unnecessary with this formulation of the von Mises criterion.

11.1.3 Determination of the Initial Yield Point

Where a material displays a sharp yield point, as shown in Fig. 11.3a, the division between elastic and elastic-plastic regions is clearly defined.

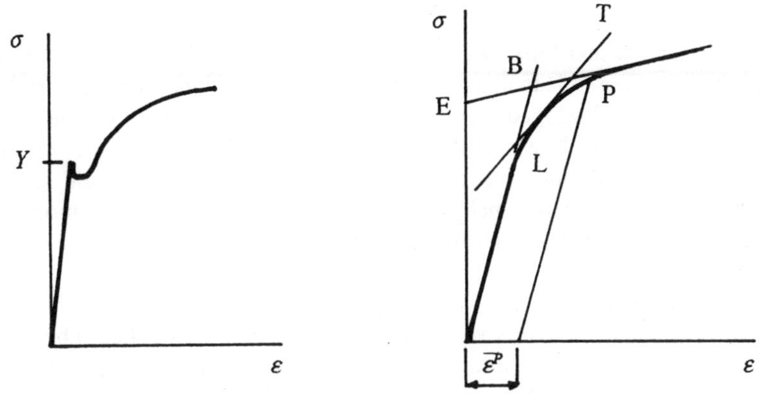

Figure 11.3 Definitions of yielding

On the other hand, there may be some doubt as to what the initial yield stress is for the many metallic materials that display a gradual transition between the two regions. To avoid such uncertainty in the determination of the yield stress, various definitions of yield have been employed. The most commonly used definition is the proof stress, i.e. the stress at P in Fig. 11.3b, corresponding to a plastic strain $\bar{\varepsilon}^P$ offset by a given small amount, e.g. 0.001% - 0.01%. Under combined stresses a similar equivalent plastic strain value is chosen to determine the amount that the component strains are to be offset. This method is often employed to determine the combined yield stresses necessary to construct a yield locus. One difficulty arising with this is that the usual Mises form of equivalent plastic strain expression $\bar{\varepsilon}^P = \sqrt{[(2/3)(\varepsilon_{ij}{}^P\varepsilon_{ij}{}^P)]}$, pre-supposes a Mises yield surface. If the yield points thus determined do not lie on a Mises locus the implication is that some other $\bar{\varepsilon}^P$ definition applies, corresponding to a locus that would contain the yield points. Unfortunately, we have no way of knowing in advance what the true yield function can be. All that can be said is the von Mises offset strain method provides a check on the validity of the Mises function. Many investigations on the shape of the yield surface have attempted to avoid this problem by employing low $\bar{\varepsilon}^P$ values [3]. The yield stresses found from offset strains employing different definitions of $\bar{\varepsilon}^P$ would then lie in a narrow region just beyond the limit of elastic proportionality, i.e. L in Fig.11.3b.

An alternative, back-extrapolation yield definition, shown in Fig. 11.3b, is much simpler in principle but its application requires large excursions into the plastic range. Point B, thus obtained, lies at the junction between the bi-linear approximation to a stress-strain curve, thereby providing only an estimation of large scale yielding. The method is more acceptable for those materials, e.g. alloy steels, that deform in an approximately similar manner. It is less applicable to materials with well-rounded plastic regions, e.g. copper and aluminium. The yield stress found from larger offset strains will approach that found from that back-extrapolation. As both methods involve considerable plastic strain, they require a new test-piece for every radial probe in order to avoid the effects of strain history.

Of all the yield definitions, the stress at the limit of proportionality, L, is the only one with physical significance. It divides the regions of elastic lattice distortion and inelastic slip. It is because the stress value at the limit of proportionality is sensitive to individual judgement that alternative methods have been devised. Mincho and Findley [4] have discussed results where, in Fig. 11.3b, point E and a tangent T, of predetermined slope, have defined yield.

11.1.4 Comparisons Between Tresca and von Mises Criteria

Appraisals of the Tresca and Mises yield criteria are usually made under two-dimensional stress states where it is possible to construct experimental yield loci. Comparisons between theory and experiment are best made for metallic materials that exhibit a well-defined yield point, such as low carbon steels, as shown in Fig. 11.3a. It is also instructive to examine the effect that yield point definition has on this comparison when initial yield stresses are determined at the limit of proportionality, by proof strain and back extrapolation methods. In this, the comments made in the preceding paragraph should be heeded.

(a) The σ_1 versus σ_2 plane

This principal biaxial stress state is achieved in the wall of a thin-walled tube when it is subjected to combined internal pressure and axial load. The radial stress σ_3 may be ignored provided the i.d./thickness ratio of the tube > 15. Yield loci have been determined either from proportional or step-wise loading paths applied mainly within the first quadrant. In the remaining quadrants, compressive buckling can precede yielding, particularly with yield definitions involving larger amounts of plastic strain. Table 11.2 sumarises the test conditions from recent published data [5-9] used in the construction of Fig. 11.4. The materials used in these investigations have received a processing method or subsequent heat treatment to ensure an initially isotropic condition. There was an obvious yield point (y.p.) in En 24 steel [5]. Other investigations [6-10] determined the yield stress either at the limit of proportionality (l.p.) or at the given proof strain value ($\bar{\varepsilon}^P$). All stress paths were radial with the exception of Moreton et al [7] who showed that it was possible to determine the full locus in alloy steels by applying a sequence of non-radial probes to a single test-piece. Their choice of an l.p. definition ensured a minimum of accumulated plastic strain from the probing technique. Most radial paths employ a new test-piece with a change to the stress ratio σ_2/σ_1. We have seen that this is good practice when defining yield for larger proof strains.

Table 11.2 Experimental yield point investigations in σ_1, σ_2 space

Material	Heat Treatment	Yield Definition	Reference	Symbol
En24 carbon steel	annealed	y.p.	5	×
SAE 1045 C-steel	hot-rolled	$\bar{\varepsilon}^P = 0.007$	6	■
2¼% Cr, 1% Mo steel	stress-relieved	l.p.	7	○
X-60 alloy steel	normalised	l.p.	7	∇
304 stainless steel	stress-relieved	l.p.	7	Δ
306 stainless steel	solution-treated	l.p.	7	□
M-63 brass	annealed	l.p.	8	◀
14S-T4 Al alloy	hot-rolled	$\bar{\varepsilon}^P = 0.002$	9	⬓
Ni-Cr-Mo steel	annealed	$\bar{\varepsilon}^P = 0.002$	10	●

The theoretical loci in Fig. 11.4 are plotted for axes normalised with the tensile yield stress Y. The von Mises prediction is, from eq(11.7b)

$$(\sigma_1/Y)^2 - (\sigma_1/Y)(\sigma_2/Y) + (\sigma_2/Y)^2 = 1$$

which defines an ellipse with 45° orientation. The Tresca locus is found by applying eq(11.1a) to each quadrant in turn. For example, in quadrant one, where $0 \le \sigma_1/Y < 1$, the greatest principal stress must be $\sigma_2/Y = 1$, (constant) when the least is $\sigma_3/Y = 0$. Similarly, where $0 \le \sigma_2/Y < 1$, the greatest principal stress is $\sigma_1/Y = 1$, with a least value of $\sigma_3/Y = 0$. The respective yield criteria $\sigma_2/Y = 1$ and $\sigma_1/Y = 1$ thus describe the horizontal and vertical sides of the Tresca hexagon for quadrant 1. In quadrant two, σ_2/Y is greatest in tension and σ_1/Y is least in compression. Hence, the left sloping side of the hexagon is expressed in the form $\sigma_2/Y - \sigma_1/Y = 1$. The completed Tresca hexagon is inscribed within the Mises ellipse. Taken overall, the superimposed test data in Fig. 11.4 lie closer to the Mises prediction, being apparently independent of the test conditions. The conclusion to be drawn from this is that an initial yield condition of Mises-isotropy applies to ductile polycrystalline materials. The data reveals that most points lie outside the hexagon, confirming that a Tresca prediction of yielding for these materials is design-safe. It is shown later that where the yield points for an individual material lie between or outside the two predictions, they may be represented by an invariant function (see Section 11.2). The uncertainties in yield point determination may however cast doubt on the appropriateness of an initial yield function that fits the test data precisely. This author's view is that the Mises condition is adequate provided the material has been heat-treated to a near isotropic condition. An examination of plastic strain paths can often provide a more definitive test of the yield function provided loading is radial.

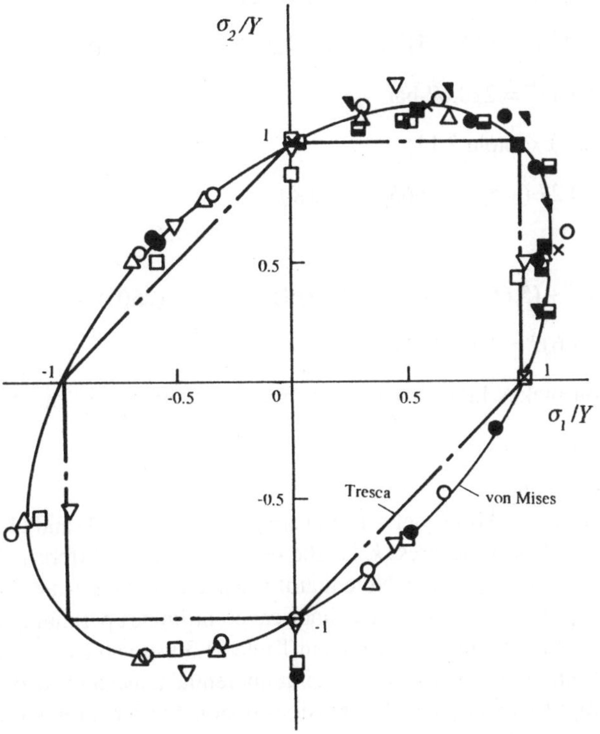

Figure 11.4 Tresca and von Mises yield loci in normalised σ_1, σ_2 space

Example 11.1 A thick-walled, closed-end cylinder, 100 mm inner diameter and 200 mm outer diameter, is subjected to both internal and external pressure. If the internal pressure is five times the external gauge pressure, find the pressure that would cause the bore to yield according to the Mises and Tresca criteria when the cylinder length (a) expands freely and (b) is prevented from expanding by rigid end closures. Take $\nu = 0.27$ and $Y = 250$ MPa.

Applying the boundary conditions $\sigma_r = -5p$ for $r = 50$ mm and $\sigma_r = -p$ for $r = 100$ mm to the Lamé eq(3.16a,b) gives

$$-5p = a - b/(50)^2 \quad \text{(i)}$$
$$-p = a - b/(100)^2 \quad \text{(ii)}$$

The solution to eqs(i) and (ii) is $a = p/3$ and $b = (4p/3) \times 10^4$. We find from eqs(3.16a,b) that the radial and hoop stresses at the bore are $\sigma_r = -5p$ (as expected) and $\sigma_\theta = 5.665p$. The solution also requires axial stress to be found as follows:

(a) free expansion (see eq(2.49)):

$$\sigma_z = \tfrac{1}{2}(\sigma_r + \sigma_\theta) = \tfrac{1}{2}(-5p + 5.665p) = p/3$$

Tresca: $\sigma_\theta - \sigma_r = Y$

$$5.665p - (-5p) = 250$$

$$p = 250/10.665 = 234.4 \text{ bar}$$

von Mises: $(\sigma_\theta - \sigma_z)^2 + (\sigma_\theta - \sigma_r)^2 + (\sigma_r - \sigma_z)^2 = 2Y^2$

$$p^2[(5.665 - 0.333)^2 + (5.665 + 5)^2 + (-5 - 0.333)^2] = 2 \times (250)^2$$

$$p = (2 \times 250^2/170.61)^{1/2} = 270.67 \text{ bar}$$

(b) plane strain (see example 2.11)

$$\sigma_z = \nu(\sigma_r + \sigma_\theta) = 0.27(-5p + 5.665p) = 0.18\,p$$

von Mises:

$$p^2[(5.665 - 0.18)^2 + (5.665 + 5)^2 + (-5 - 0.18)^2] = 2 \times (250)^2$$

$$p = (2 \times 250^2/170.66)^{1/2} = 270.64 \text{ bar}$$

The end condition makes little difference to the Mises pressure. The Tresca pressure is unchanged

(b) The σ versus τ plane

The respective Tresca and Mises yield criteria are given in eqs(11.3) and (11.8). An obvious difference is that the shear yield stress k, i.e. the semi-minor axis, differs as $k = Y/\sqrt{3}$ and $Y/2$ with Tresca being the smaller value. Note that it is usual to omit the subscripts x and y on σ and τ when applying these criteria in the cylindrical co-ordinates pertaining to the stresses in the wall of a tube. The experiments detailed in Table 11.3 were all conducted on thin-walled tubes subjected to torsion combined with (i) circumferential tension [5] and (ii) axial loading in tension and compression [11-16]. The tension-torsion combination includes the original experiments of Taylor and Quinney [16] who first employed the backward extrapolation (b.e.) technique. With the exception of the stepped stress probes employed by Ivey [12] and Phillips and Tang [13], all other investigations in Table 11.3 employed radial loading. Ellyin and Grass [15] applied multiple radial probes each emanating from the stress origin in a single test-piece. The final probe repeated their initial probe. The small difference between yield stresses for the initial and final probes showed that the effect of accumulated plastic strain was negligibly small for their chosen offset $\bar{\varepsilon}^P = 20$ microstrain.

Table 11.3 Experimental yield point investigations in σ, τ space

Material	Heat Treatment	Yield Definition	Reference	Symbol
En 24 carbon steel	annealed	y.p.	5	×
En 25 carbon steel	annealed	y.p.	11	○
Noral 19S Al alloy	stress relieved	l.p.	12	∇
1100-0 Al	annealed	l.p.	13	■
Brass	as-received	$\bar{\varepsilon}^P = 200 \times 10^{-6}$	14	◣
Ti-50A Ti-alloy	stress-relieved	$\bar{\varepsilon}^P = 20 \times 10^{-6}$	15	●
Copper (99.8%)	annealed	b.e.	16	Δ
Aluminium (99.7%)	annealed	b.e.	16	□

The comparison between theory and experiment in Fig. 11.5 reveals again that most of the experimental data in quadrants two and four lie closer to the Mises ellipse than to Tresca, irrespective of the chosen yield definition.

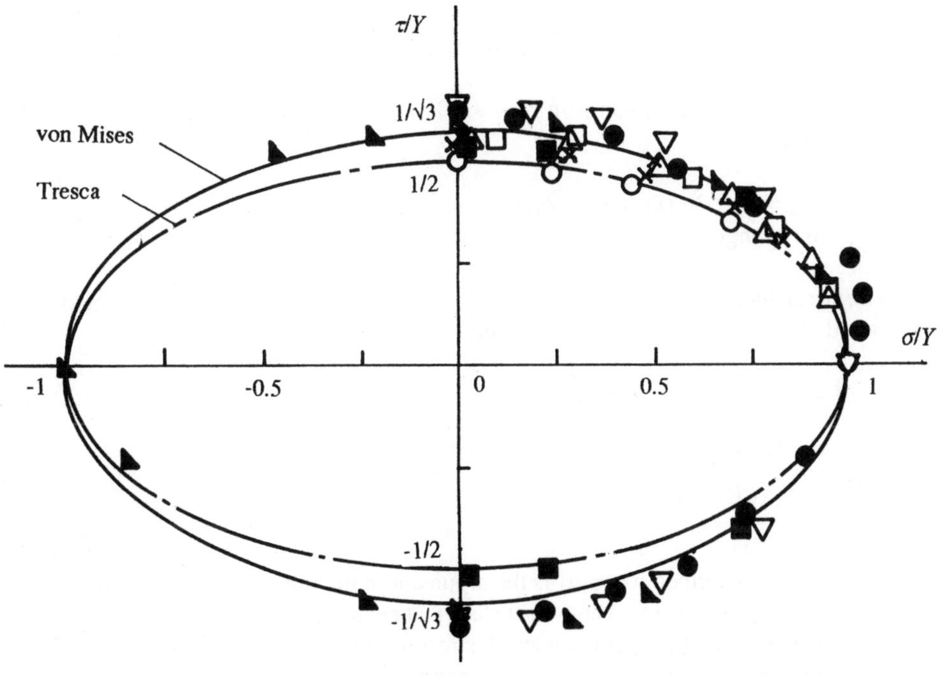

Figure 11.5 Tresca and Mises loci in σ, τ space

Figure 11.5 confirms our expectation that a Mises initial yield condition applies to ductile isotropic materials under σ, τ stress states, an observation first made by Taylor and Quinney from their own results for copper and aluminium. Similar comments made previously, for the definition of yielding under σ_1, σ_2 states again apply to Fig. 11.5. Again, the Tresca prediction will provides a conservative estimate of yielding for all definitions.

There are many practical instances where direct and shear stress states are combined. These include a circular shaft under direct shear or torsion combined with either an axial load or a bending moment. Equations(11.3) and (11.8) will also apply when all these loadings are applied simultaneously. A combination of direct stress and shear stress exists at points in the length of a loaded beam away from the neutral axis. Note that the hot-rolled [6,9] and as-received [14] materials given in Tables 11.2 and 11.3 did not receive any further heat treatment prior to test. It is possible that their departure from a Mises yield condition is due to the influence of initial anisotropy. The presence of anisotropy is detectable when initial yield stresses found for orthogonal tensile tests are dissimilar, but this was not reported in these investigations.

11.2 General Yield Function for Isotropic Metals

Because the mean or hydrostatic stress plays no part in yielding, it follows that the yield criterion is a general function of the deviatoric or reduced stress in Fig. 1.11c. The deviatoric stress tensor σ_{ij}' is the remaining part of the absolute tensor σ_{ij} after the mean or hydrostatic stress σ_m has been subtracted. This reduction applies only to the normal stress components σ_{11}, σ_{22} and σ_{33}, since shear stresses, σ_{12}, σ_{13} and σ_{23}, will cause no dilitation. Recall from eq(1.21a,b) that the introduction of the unit matrix **I** or the Kronecker delta δ_{ij} ensures the correct reductions. These are

$$\sigma_{ij}' = \sigma_{ij} - \delta_{ij}\sigma_m \quad \text{or} \quad \mathbf{S}^D = \mathbf{S} - \tfrac{1}{3}\,\mathbf{I}\,(\text{tr}\,\mathbf{S}) \qquad (11.9a,b)$$

where, δ_{ij} is unity for $i = j$ and zero for $i \neq j$. For example, with $i = 1$ and $j = 1$, 2 and 3, eq(11.9b) gives

$$\sigma_{11}' = \sigma_{11} - \sigma_m = \sigma_{11} - \tfrac{1}{3}(\sigma_{11} + \sigma_{22} + \sigma_{33})$$
$$\sigma_{12}' = \sigma_{12} \;\text{(no change)}$$

With further reductions it follows that the resulting stress deviator matrix $\mathbf{S}^D$ in eq(11.9b) is composed of σ_{11}', σ_{22}' and σ_{33}' and the original shear stresses σ_{12}, σ_{13} and σ_{23}.

$$\begin{bmatrix} \sigma'_{11} & \sigma_{12} & \sigma_{13} \\ \sigma_{21} & \sigma'_{22} & \sigma_{23} \\ \sigma_{31} & \sigma_{32} & \sigma'_{33} \end{bmatrix} = \begin{bmatrix} \sigma_{11} & \sigma_{12} & \sigma_{13} \\ \sigma_{21} & \sigma_{22} & \sigma_{23} \\ \sigma_{31} & \sigma_{32} & \sigma_{33} \end{bmatrix} - \frac{(\sigma_{11}+\sigma_{22}+\sigma_{33})}{3} \begin{bmatrix} 1 & 0 & 0 \\ 0 & 1 & 0 \\ 0 & 0 & 1 \end{bmatrix} \qquad (11.9c)$$

A yield criterion is formed when σ_{ij}' is the argument in the function $f(\sigma_{ij}')$ = constant. The yield function f defines a closed convex surface that encloses the elastic region. The surface governs the inception of plasticity under all stress states. Moreover, as yield behaviour is a property of the material itself f must be independent of the co-ordinates used to define σ_{ij}'. Yielding is thus a function of the deviatoric stress invariants J_1', J_2' and J_3'. These invariants may be obtained by subtracting σ_m from the absolute stress invariants, J_1, J_2 and J_3, the coefficients in the principal stress cubic eq(1.8b). This gives, in principal stress form,

$$J_1' = \sigma_1' + \sigma_2' + \sigma_3' = (\sigma_1 - \sigma_m) + (\sigma_2 - \sigma_m) + (\sigma_3 - \sigma_m)$$
$$= (\sigma_1 + \sigma_2 + \sigma_3) - 3\sigma_m = 0 \quad (11.10a)$$

$$-J_2' = \sigma_1'\sigma_2' + \sigma_2'\sigma_3' + \sigma_1'\sigma_3' = (\sigma_1 - \sigma_m)(\sigma_2 - \sigma_m) + (\sigma_2 - \sigma_m)(\sigma_3 - \sigma_m) + (\sigma_1 - \sigma_m)(\sigma_3 - \sigma_m)$$
$$= (\sigma_1\sigma_2 + \sigma_2\sigma_3 + \sigma_1\sigma_3) - 3\sigma_m^2$$
$$= ⅓[(\sigma_1\sigma_2 + \sigma_2\sigma_3 + \sigma_1\sigma_3) - (\sigma_1^2 + \sigma_2^2 + \sigma_3^2)]$$
$$= -(1/6)[(\sigma_1 - \sigma_2)^2 + (\sigma_2 - \sigma_3)^2 + (\sigma_1 - \sigma_3)^2] \quad (11.10b)$$

$$J_3' = \sigma_1'\sigma_2'\sigma_3' = (\sigma_1 - \sigma_m)(\sigma_2 - \sigma_m)(\sigma_3 - \sigma_m)$$
$$= \sigma_1\sigma_2\sigma_3 - \sigma_m(\sigma_1\sigma_2 + \sigma_2\sigma_3 + \sigma_1\sigma_3) + 2\sigma_m^3$$
$$= \sigma_1\sigma_2\sigma_3 - ⅓(\sigma_1 + \sigma_2 + \sigma_3)(\sigma_1\sigma_2 + \sigma_2\sigma_3 + \sigma_1\sigma_3) + 2(⅓)^3(\sigma_1 + \sigma_2 + \sigma_3)^3$$
$$= (⅓)^4[(2\sigma_1 - \sigma_2 - \sigma_3)^3 + (2\sigma_2 - \sigma_1 - \sigma_3)^3 + (2\sigma_3 - \sigma_1 - \sigma_2)^3] \quad (11.10c)$$

In changing the sign of the J_2' expression (11.10b) it becomes positive within its corresponding deviatoric stress cubic

$$(\sigma')^3 - J_2'(\sigma') - J_3' = 0$$

Note that the tensor component and matrix forms of the invariants in eqs(11.10a,b,c) are

$$J_1' = \sigma_{ii}' = \mathrm{tr}\, \mathbf{S}^{\mathrm{D}} = 0 \quad (11.11a)$$
$$J_2' = ½\, \sigma_{ij}'\, \sigma_{ji}' = ½\, (\mathrm{tr}\, \mathbf{S}^{\mathrm{D}})^2 \quad (11.11b)$$
$$J_3' = ⅓\, \sigma_{ij}'\sigma_{jk}'\sigma_{ki}' = ⅓\, (\mathrm{tr}\, \mathbf{S}^{\mathrm{D}})^3 \quad (11.11c)$$

Yielding begins when a function of the two non-zero deviatoric invariants J_2' and J_3' attains a critical constant value, C. That is,

$$f(J_2', J_3') = C \quad (11.12)$$

where C is normally defined from the reduction of eq(11.12) to yielding in simple tension or torsion. When J_3' is omitted in eq(11.12), f is simply equated to J_2', to give

$$J_2' = k^2 \quad (11.13a)$$

where $k^2 = Y^2/3$ is found from putting $\sigma_1 = Y$ (the tensile yield stress) with $\sigma_2 = \sigma_3 = 0$ in eq(11.10b). Then, from eqs(11.10b) and (11.13a) it follows that

$$(1/6)[(\sigma_1 - \sigma_2)^2 + (\sigma_2 - \sigma_3)^2 + (\sigma_1 - \sigma_3)^2] = Y^2/3 \quad (11.13b)$$

which again results in the von Mises yield criterion (eq 11.7a). Thus, according to von Mises, it is the second invariant of deviatoric stress which attains a critical value at the point of yield.

Example 11.2 A stirrer rotates within a chemical vessel at 600 rev/min under a compressive force of 4 kN. If the power absorbed by the stirrer is 400 W, determine the outer shaft diameter according to the von Mises and Tresca criteria using a safety factor of 3, given that the tensile yield stress is 180 MPa. How is the safety factor altered when, for each calculated diameter, the pressure within the vessel is increased to 120 bar?

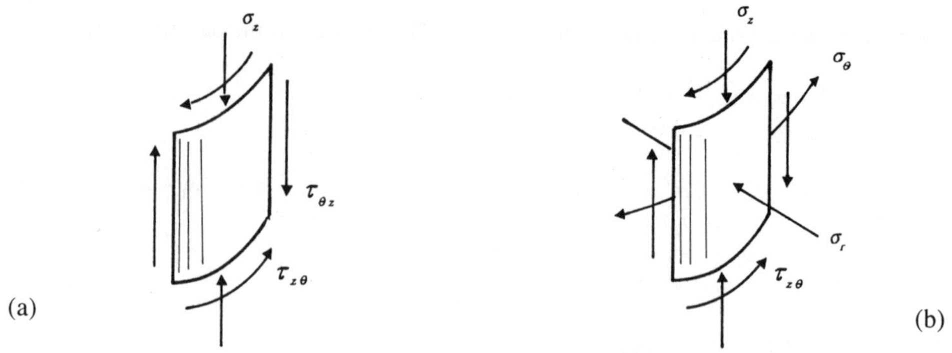

Figure 11.6 Surface elements of stirrer rod

In the absence of vessel pressure refer to the stress state shown in Fig. 11.6a Working in terms of the diameter d, the axial compressive stress σ_z is found from the thrust W

$$\sigma_z = W/A = -(4 \times 10^3)/(\pi d^2/4) = -5093/d^2 \text{ (MPa)} \tag{i}$$

The torque T in the shaft is found from the power P (Watt) and speed N (rev/min) as

$$T = 60P/(2\pi N) = (60 \times 400)/(2 \times \pi \times 600) = 6.366 \text{ Nm}$$

The shear stress $\tau_{\theta z}$ at the outer radius $r = d/2$ follows from torsion theory

$$\tau_{\theta z} = Tr/J = 16T/(\pi d^3) = (16 \times 6.366 \times 10^3)/(\pi d^3) = (32.42 \times 10^3)/d^3 \text{ (MPa)} \tag{ii}$$

where $J = \pi d^4/32$. The safety factor S provides a safe working stress of Y/S for which eqs (11.3) and (11.8) are re-written to provide a safe "working stress envelope". Tresca becomes

$$\sigma_z + 4\tau_{\theta z}^2 = (Y/S)^2$$
$$(-5093/d^2)^2 + 4[(32.42 \times 10^3)/d^3]2 = (180/3)^2$$

for which the trial solution is $d = 11.31$mm. The von Mises solution becomes

$$\sigma_z + 3\tau_{\theta z}^2 = (Y/S)^2$$
$$(-5093/d^2)^2 + 3[(32.42 \times 10^3)/d^3]2 = (180/3)^2$$

for which the solution is d = 10.97 mm. Clearly Tresca provides the safer prediction. When the stirrer shaft is under compressive pressure p from the vessel, the stress state at the rod surface is shown in Fig. 11.6b. The stress components become

$$\sigma_z = (-5093/d^2) - p$$
$$\sigma_\theta = p, \ \sigma_r = -p \tag{iii}$$
$$\tau_{\theta z} = 32.43 \times 10^3/d^3$$

where $p = 12$ MPa. The von Mises criterion must be account for the complete stress state in Fig. 11.6b. Where shear stress terms are present within the invariant eq (11.10b) this gives

$$(\sigma_\theta - \sigma_r)^2 - (\sigma_\theta - \sigma_z)^2 - (\sigma_r - \sigma_z)^2 + 6\tau_{\theta z}^2 = 2Y^2 \tag{iv}$$

Substituting eqs(iii) into eq(iv) leads to:

$$576 + (5093/d^2 + 24)^2 + (5093/d^2)^2 + 6(32.42 \times 10^3/d^3)^2 = 2(180/S)^2$$

and for $d = 10.97$ mm, S is reduced to 2.5. The Tresca criterion (11.1b) requires the greatest and least principal stresses. Within the plane of Fig. 11.6b the principal stresses are

$$\sigma_{1,2} = ½(\sigma_\theta + \sigma_z)^2 \pm ½\sqrt{[(\sigma_\theta - \sigma_z)^2 + 4\tau_{\theta z}^2]}$$

and substituting from eq(iii), with $p = 20$ MPa and $d = 11.31$mm , gives the principal stresses $\sigma_1 = 19.08$ MPa , $\sigma_2 = -58.9$ MPa. Comparing these to the third principal stress, $\sigma_r = p = -12$ MPa in Fig. 11.6b, it follows that σ_1 and σ_2 are the greatest and least numerical values required. From eq(11.1b)

$$19.08 - (-58.90) = (180/S)$$

from which S = 2.31 is a more conservative estimate.

11.2.1 Hydrostatic Stress

When the yield function is formulated from the stress deviator invariants it implies that initial yielding is unaffected by the magnitude of hydrostatic stress. We have employed this simplifying assumption allows the development of yield functions for both isotropic and anisotropic polycrystalline materials. Most experimental evidence in support of this assumption applies to materials in an apparently initially isotropic condition. High magnitudes of hydrostatic pressure, up to 3 kbar, when superimposed upon torsion [17-19], tension [20,21] and compression [22] have not altered significantly the yield stresses for mild steel, copper, aluminium and brass. For these materials we can assume a yield function of the general isotropic form $f(J_2', J_3')$.

Under certain conditions, an allowance for the influence of superimposed hydrostatic pressure may be required. For example, the upper, initial, shear yield stress $k = 200$ MPa for mild steel decreased by 5 - 6% under $p = 3$ kbar [17]. For prestrained brass, variations of -6% to 3% in k were found for a superimposed pressure range of 1 - 4 kbar [21]. To account for variations of this order, Hu [18] separated the influence of σ_m by including $J_1 = \sigma_{kk}$, along with J_2' and J_3', in the yield criterion:

$$f(\sigma_{ij}) = f(J_1, J_2', J_3') = c \qquad (11.14a)$$

in which J_1 was separated within two functions G and H as

$$f(\sigma_{ij}) = G(J_1)\, H(J_2', J_3') = c \qquad (11.14b)$$

Formulations of this kind are also required where the magnitude of σ_m is critical to the failure in brittle non-metals and plastic flow in porous compacts.

11.2.2 Influence of the Third Invariant

The fact that some initial yield points for isotropic material do not lie on the Mises loci, in Figs 11.4 and 11.5, suggests that initial yielding may conform to a more general function containing both deviatoric invariants. Firstly, note that Tresca's criterion may be expressed as the following complex function [23] in these invariants

$$f = ½(J_2')^3 - 3(J_3')^2 - 9k^2(J_2')^2 + 48k^4(J_3')^2 = 64\,k^6 \qquad (11.15a)$$

where the shear yield stress, $k = Y/2$ according to Tresca. Substituting eqs(11.10b,c) into (11.15a), and factorising gives

$$[(\sigma_1 - \sigma_2)^2 - 4k^2][(\sigma_2 - \sigma_3)^2 - 4k^2][(\sigma_1 - \sigma_3)^2 - 4k^2] = 0 \tag{11.15b}$$

The familiar Tresca criterion is the last of the following solutions to eq(11.15b): $(\sigma_1 - \sigma_2) = 2k$, $(\sigma_2 - \sigma_3) = 2k$ and $(\sigma_1 - \sigma_3) = 2k$. There are many other isotropic functions containing both invariants. Amongst these are the following homogenous stress functions f in eq(11.12)

$$f = (J_2')^3 - c(J_3')^2 = k^6 \tag{11.16a}$$

$$f = J_2' - b\,(J_3'/J_2')^2 = k^2 \tag{11.16b}$$

$$f = (J_2')^{3/2} - dJ_3' = k^3 \tag{11.16c}$$

In eqs(11.16a-c) c, b, and d are material constants and k is the shear yield stress. Unlike eq(11.15a), it is now possible to select a value for the constants in eqs(11.16a-c) to fit the initial yield data for most metallic materials [24]. Of these, the most well-known is eq(11.16a), proposed by Drucker[25], a homogenous function in stress to the sixth degree. The constant c lies in the range $-27/8 \le c \le 9/4$, in order to ensure convexity of the yield surface. Figure 11.7 indicates that, with $c = -2$ in eq(11.16a), the close representation of the initial yield points for stainless steel [7] is better than that which would be found from either Mises or Tresca.

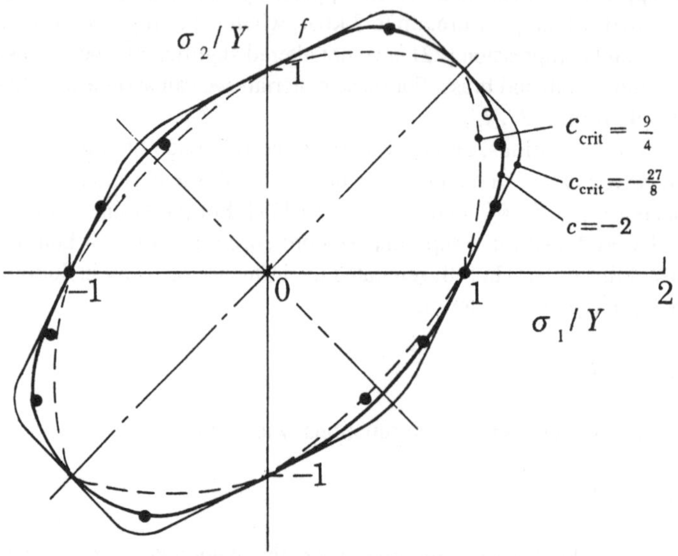

Figure 11.7 Drucker function applied to initial yielding in stainless steel

Also shown in Fig. 11.7 are the bounding loci from eq(11.16a) with the limiting values $c = -27/8$ and $9/4$, to ensure a closed surface [26]. Corners appear for the negative limit of c. The following chapter will show that, with the inclusion of J_3' in a homogenous yield function, the associated stress-plastic strain relations become more cumbersome. It is for this reason that the Mises yield function $f = J_2'$ is preferred.

A typical non-symmetrical function in the two deviatoric invariants J_2' and J_3' is formed from their linear combination in the following form [24]

$$f = (J_2')^n + (p/\sigma_t)^{(2n-1)} J_3' \tag{11.17}$$

where n and p are constants. The introduction of σ_t, the tensile yield stress, into the denominator of eq(11.17) will ensure homogeneity in stress. Functions of the form of eq(11.17) can account for so-called "second order" effects in plasticity [27]. These include (i) non-linear plastic strain paths under radial loading [24], (ii) the accumulation of axial strain under pure torsion [28] and (iii) a difference between initial tensile and compressive yield stresses. Explicit stress component forms of eq(11.17) are found by substituting from eq(11.10b and c). For example, taking $n = 1$ in eq(11.17) under a principal biaxial stress state ($\sigma_3 = 0$) gives the initial yield function

$$f = \tfrac{1}{3}(\sigma_1^2 - \sigma_1\sigma_2 + \sigma_2^2) + (\tfrac{1}{3})^4[(2\sigma_1 - \sigma_2)^3 + (2\sigma_2 - \sigma_1)^3 - (\sigma_1 + \sigma_2)^3](p/\sigma_t) \qquad (11.18)$$

The effect of (iii) above is revealed by respective substitutions for σ_1 of σ_t and $-\sigma_c$, each with $\sigma_2 = 0$, under uniaxial yielding in tension and compression. This gives

$$f = \sigma_t^2(1/3 + 2p/27) = \tfrac{1}{3}\sigma_c^2 - 2p\sigma_c^3/(27\sigma_t) \qquad (11.19a)$$

Introducing the ratio $\rho = \sigma_c/\sigma_t$ to eq(11.19a) leads to the relationship between p and ρ:

$$(2p/9)\rho^3 - \rho^2 + (2p/9 + 1) = 0 \qquad (11.19b)$$

We may normalise the function (11.18) in σ_t by equating to the first expression in eq(11.19a). This gives

$$(\sigma_1/\sigma_t)^2 - (\sigma_1/\sigma_t)(\sigma_2/\sigma_t) + (\sigma_2/\sigma_t)^2 + (p/27)[(2\sigma_1/\sigma_t - \sigma_2/\sigma_t)^3$$
$$+ (2\sigma_2/\sigma_t - \sigma_1/\sigma_t)^3 - (\sigma_1/\sigma_t + \sigma_2/\sigma_t)^3] = 1 + 2p/9 \qquad (11.20)$$

Figure 11.8 gives the loci found from eq(11.20) for p = 1.5, - 27/14 and - 3. Equation (11.19b) shows that these correspond to ρ = 2, 2/3 and 1/2 respectively.

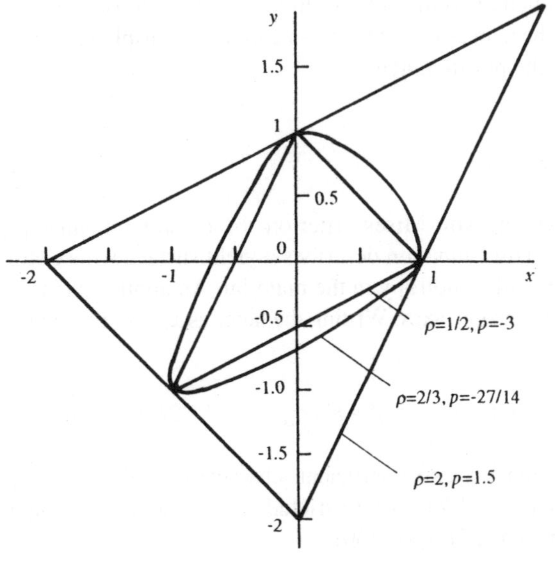

Figure 11.8 Non-symmetrical yield loci

It is seen that, for $\rho = 2$ and ½, the loci are bounded by three straight sides in contrast with Tresca's six-sided isotropic locus in Fig. 11.4. Convexity in the locus is ensured for $-3 < p < 3/2$. That is, the function is restricted to materials for which one uniaxial yield stress is not less than 50% of the other. For the intermediate value $p = -27/14$ shown, σ_t exceeds σ_c by 33%. Because stress deviator invariants appear in eq(11.17), the function preserves plastic incompressibility whilst accounting for second-order phenomena in quasi-isotropic ductile polycrystals. Equation (11.17) is not valid for brittle materials in which hydrostatic stress is responsible for the difference between σ_t and σ_c. For this, it will be shown that yield function must contain the first invariant of absolute stress $J_1 = \sigma_{ii}$.

In view of the uncertainty over the accuracy of initial yield data employing different yield point definitions, it could be argued that the influences of J_3' on the initial yield stress are inconclusive. It is also possible that initial anisotropy, and not the influence of J_3', is solely responsible for deviations from the Mises or Tresca conditions observed in Figs 11.4 and 11.5, where materials were not heat treated. It is therefore pertinent to consider next alternative, anisotropic yield functions.

11.3 Anisotropic Yielding

With the exception of eq(11.17), the yield criteria considered so far all assume that the yield stresses are of equal magnitude under the forward or reversed application of a given stress. That is, from the symmetry of Figs 11.4 - 11.6, the magnitude of the uniaxial tensile and compressive yield stresses are equal and the positive and negative shear yield stresses are also equal. This assumption is reasonably consistent with the observed, initial yield stress for annealed or stress-relieved metals and alloys. In the absence of heat treatment, the yield stress may vary depending upon the sense of the applied stress and its orientation. This so-called plastic anisotropy is associated with these directional variations in yield stress which do not appear within the foregoing isotropic functions. A common orthotropic form of initial anisotropy occurs in a sheet metal exhibiting different initial yield stresses for directions parallel and perpendicular to the rolling direction. As a consequence of their processing history, cold-rolled sheet and extruded bar can continue to display orientation-dependent flow behaviour well into the plastic range.

11.3.1 Orthotropic Sheets

Hill [29] generalised the von Mises criterion to account for anisotropic yielding in an orthotropic material. Hill's function describes a yield surface whose stress axes are aligned with the principal axes of orthotropy in the material. An applied stress σ_{ij} must therefore be resolved along the orthotropic axes. Writing the latter as x, y and z, the original function was given as

$$2f(\sigma_{ij}) = F(\sigma_{yy} - \sigma_{zz})^2 + G(\sigma_{xx} - \sigma_{zz})^2 + H(\sigma_{xx} - \sigma_{yy})^2 + 2L\tau_{yz}^2 + 2M\tau_{xz}^2 + 2N\tau_{xy}^2 = 1 \quad (11.21)$$

where F, G, H, L, M and N are six coefficients characterising the orthotropic symmetry in the yield stresses. In fact, eq(11.21) is a restricted form of a general quadratic yield function discussed by Edelman and Drucker [30]:

$$f(\sigma_{ij}) = ½\, C_{ijkl}\, \sigma_{ij}\, \sigma_{kl} \quad (11.22)$$

The fourth rank tensor C_{ijkl} contains 81 components. It is similar in form to that used for defining a strain energy function for anisotropic elasticity. The number of independent components reduces to 21 with imposed symmetry conditions: (i) symmetry in the Cauchy stress tensor $\sigma_{ij} = \sigma_{ji}$, giving $C_{ijkl} = C_{jikl} = C_{ijlk}$ and (ii) coincidence between the axes of stress and orthotropy, which gives $C_{ijkl} = C_{klij}$. With this number of coefficients, the function can account for the influence of hydrostatic stress on the yield behaviour of anisotropic materials. Otherwise, if yielding is to be independent of hydrostatic stress, the number of independent coefficients in eq(11.22) is reduced to 15. With further reductions to the number of independent components, C_{ijkl} will account for particular material symmetries. Amongst these are orthotropy (six coefficients), transverse isotropy (three coefficients) and a cubic form of anisotropy characterised by two coefficients [31]. The first of these corresponds with Hill's eq(11.21). The remaining two symmetry conditions follow from eq(11.21) by putting (i) $G = H$ with $M = N$ and (ii) $F = G = H$ with $L = M = N$ respectively. Wider application of the general forms of eqs(11.22) has been demonstrated in the absence of symmetry in [32,33]. Further applications of and developments to Hill's quadratic function have been made under particular conditions of plane stress and plane strain [34-36].

It is shown later that similar quadratic stress expressions reappear as failure criteria for intrinsically anisotropic non-metals. These apply to certain glass- and carbon-fibre reinforced composites that are insensitive to the sense of the fracture stress.

11.3.2 Other Anisotropic Yield Functions

Other yield criteria [37-44] have been proposed to account for the observed variations in yield stress for materials that may not conform to an orthotropic condition. They are summarised in Table 11.4.

Table 11.4 Yield criteria for an initially anisotropic condition

	Ref.	Anisotropic Yield Function
(i)	[37]	$f = C_{11}\sigma_{11}^2 + C_{12}\sigma_{11}\sigma_{22} + C_{13}\sigma_{11}\sigma_{12} + C_{22}\sigma_{22}^2 + C_{23}\sigma_{22}\sigma_{12} + C_{33}\sigma_{12}^2 = 1$
(ii)	[38]	$f = (3/2)[(\sigma_1/\sigma_{1y})^2 - (\sigma_1/\sigma_{1y})(\sigma_2/\sigma_{2y}) + (\sigma_2/\sigma_{2y})^2]^3$ $- (1/8)(\sigma_1/\sigma_{1y} + \sigma_2/\sigma_{2y})^2(\sigma_1/\sigma_{1y} - 2\sigma_2/\sigma_{2y})^2(\sigma_2/\sigma_{2y} - 2\sigma_1/\sigma_{1y})^2 = 1$
(iii)	[39]	$f = J_2'^2(J_2' + A_{ijkl}\sigma_{ij}'\sigma_{kl}') - C J_3'^2 = k^6$
(iv)	[40]	$f(\sigma_{ij}) = f(\sigma_2 - \sigma_3)^m + g(\sigma_1 - \sigma_3)^m + h(\sigma_1 - \sigma_2)^m$ $+ a(2\sigma_1 - \sigma_2 - \sigma_3)^m + b(2\sigma_2 - \sigma_3 - \sigma_1)^m + c(2\sigma_3 - \sigma_1 - \sigma_2)^m = \sigma^m$
(v)	[41]	$f = 3\sigma_x^3 - 6\sigma_x^2\sigma_y - 6\sigma_x\sigma_y^2 + 4\sigma_y^3 + (4\sigma_x + 21\sigma_y)\tau_{xy}^2$
(vi)	[42]	$f = \sum_{ijk} A_{ijk}\sigma_x^i\sigma_y^j\tau_{xy}^{2k},\ i + j + k \le 4$
(vii)	[43]	$f = a_1(\sigma_x - \sigma_y)^2 + a_2(\sigma_x - \sigma_z)^2 + a_3(\sigma_y - \sigma_z)^2 + a_4\tau_{yz}^2 + a_5\tau_{xy}^2 + a_6\tau_{xz}^2$ $+ a_7(\sigma_x - \sigma_y) + a_8(\sigma_x - \sigma_z) + a_9(\sigma_y - \sigma_z) + a_{10}\tau_{yz} + a_{11}\tau_{xy} + a_{12}\tau_{xz} = 1$
(viii)	[44]	$f = A\sigma_1^2 + B\sigma_2^2 + C\sigma_3^2 - D\sigma_1\sigma_2 - E\sigma_2\sigma_3 - F\sigma_1\sigma_3 + L\sigma_1 + M\sigma_2 + N\sigma_3 = 1$

In Table 11.4 the first 6 functions are homogenous in stress to various degrees. In the plane stress quadratic function (i), the cross-product terms in σ, τ introduce two additional constants (Hill has four). This form was claimed to improve descriptions of initial yielding in anisotropic sheet material. The function in (ii) is derived from the reduction to Drucker's eq(11.16a) for a plane principal stress state. The stresses σ_1 and σ_2 are normalised with respective yield stresses for the 1 and 2 directions, σ_{1y} and σ_{2y}, which are assumed to be different. In (iii), Drucker's isotropic function reappears in combination with a quadratic stress term of the orthotropic kind. This form accounts for deviations from an initial Mises isotropic condition due to the combined influences of J_3' and initial anisotropy. A non-quadratic function in (iv) was proposed by Hill [40] when principal stresses are aligned with the principal directions in an orthotropic material. The constants σ^m, a, b, c, f, g and h are all positive. It is seen that this form is a further generalisation of the isotropic invariants in eqs(11.10b,c). It also follows that his earlier quadratic function, eq(11.21), was a generalisation of J_2'. Moreover, the use of an exponent m, different from 2, within the range $1 \le m \le 2$, allows for a wider range of anisotropic behaviour. In particular, this function accounts for any ratio between transverse and thickness strain under uniaxial and equi-biaxial tension, including the common case of planar isotropy (i.e, $a = b = 0$, $f = g = 0$) in rolled polycrystalline sheets [45,46].

11.3.3 Earing

The question of whether the development of four or more ears from a cupping operation on an anisotropic sheet can be predicted successfully has been considered within higher order polynomial yield functions. The quadratic function, eq(11.21), can provide an explanation of two and, at most, four ears in certain cases. Bourne and Hill [41] showed that the the numerical coefficients within a cubic function (v) in Table 11.4 matched the six ears observed in brass cups. The quartic function (vi) was proposed by Gotoh [42] to represent up to a maximum of eight ears observed in soft aluminium and its alloy. The expanded plane stress form of eq(vi) is

$$f = [A_1\sigma_x^4 + A_2\,\sigma_x^3\,\sigma_y + A_3\sigma_x^2\,\sigma_y^2 + A_4\,\sigma_x\,\sigma_y^3 + A_5\,\sigma_y^4 + (A_6\,\sigma_x^2 + A_7\sigma_x\sigma_y + A_8\,\sigma_y^2)\tau_{xy}^2 + A_9\tau_{xy}^4]$$

where eight coefficients, together with $A_1 = 1$, are required to match eight ears. A further term $A_0(\sigma_x + \sigma_y)^2$ is added where an account of compressibility is required. The coefficients A_1 ... A_9 are found from the yield stresses under (a) uniaxial tension and either (b) through-thickness plane strain compression or (c) in-plane, equi-biaxial tension.

11.3.4 Inhomogenous Functions

Any yield function homogenous in stress to a degree of two or more assumes that the tensile and compressive yield stresses are equal for each principal material direction. This is reasonably near the truth for large offset strain and backward-extrapolation definitions of yield (see Fig. 11.3). However, a small offset strain definition of each yield point can reveal differences between them particularly in a material that has not received subsequent heat-treatment. An account of different positive and negative yield stresses appears with the inclusion of a linear stress term within the yield function. For example, two such functions are given in Table 11.4: eq(vii) for general 2D stress and eq(viii) for principal triaxial stress. Stassy-d'Alia [43] has examined both forms in some detail by placing restrictions on certain constants. A plane orthotropic function of linear plus quadratic terms applies when yielding remains independent of the magnitude of a superimposed hydrostatic stress [47]. With axes of orthotropy 1 and 2, coincident with the axes of a plane stress state σ_{11}, σ_{22} and σ_{12}, this is

$$f = L_1\sigma_{11} + L_2\sigma_{22} + L_3\sigma_{12} + Q_1\,\sigma_{11}^{\;2} + Q_2\sigma_{22}^{\;2} + Q_3\sigma_{11}\,\sigma_{22} + Q_4\sigma_{12}^{\;2} + Q_5\sigma_{11}\,\sigma_{12} + Q_6\,\sigma_{22}\,\sigma_{12} = 1 \quad (11.23a)$$

In a reduced space of σ_{11} and σ_{12}, eq(11.23a) becomes:

$$f = L_1\,\sigma_{11} + L_3\,\sigma_{12} + Q_1\,\sigma_{11}^{\;2} + Q_4\,\sigma_{12}^{\;2} + Q_5\,\sigma_{11}\,\sigma_{12} = 1 \quad (11.23b)$$

Figures 11.9a-h show several cases of anisotropy as represented by eq(11.23b).

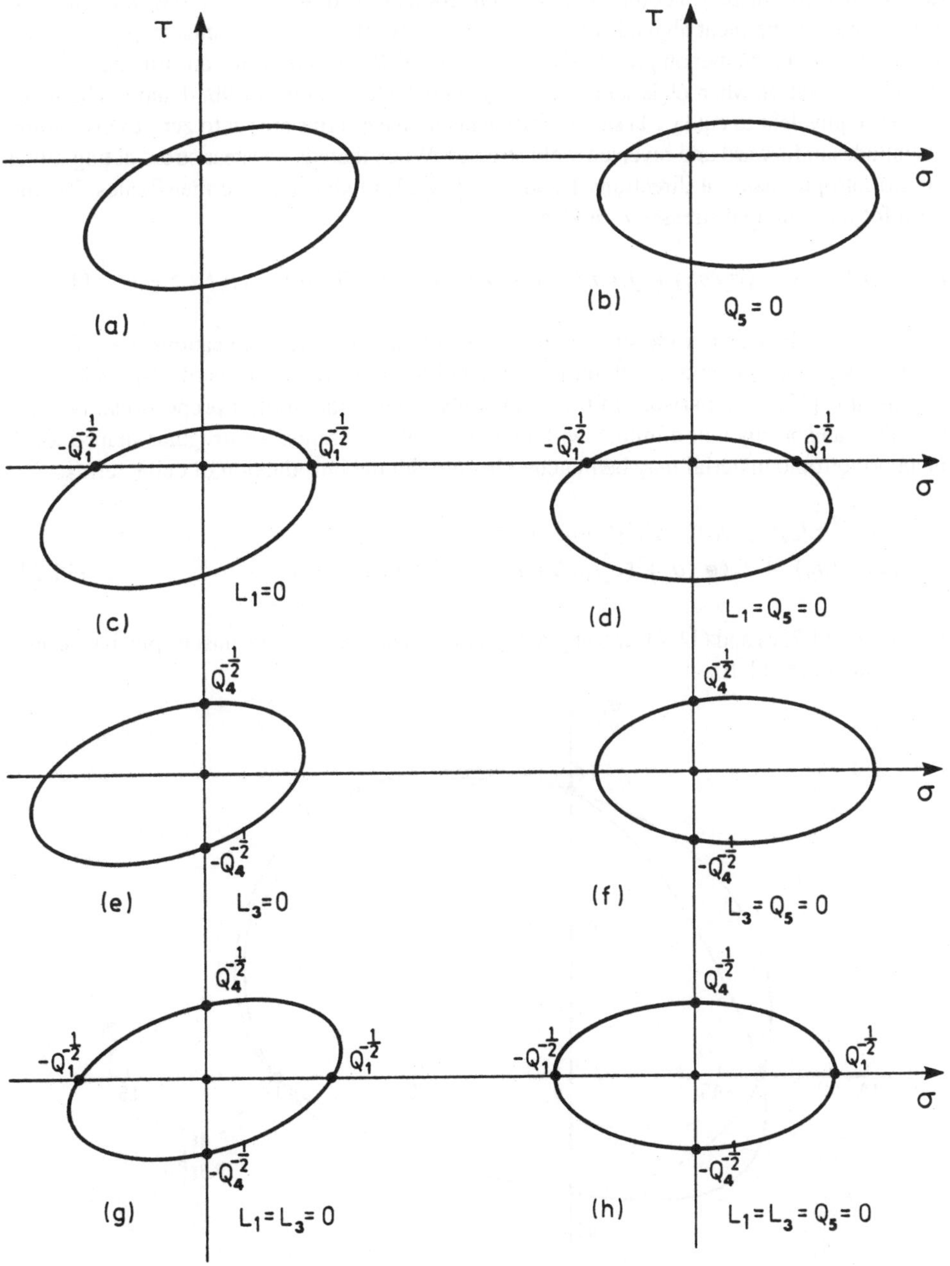

Figure 11.9 Yield loci in σ_{11}, σ_{12} space

The constants are readily matched to equal yield stresses as shown. Where tensile and compressive yield stresses differ, they become the roots of $Q_1 \sigma_{11}{}^2 + L_1\sigma_{11} - 1 = 0$ in a, b, e and f. Similarly, different shear yield stresses become the roots of $Q_4 \sigma_{12}{}^2 + L_3\sigma_{12} - 1 = 0$ in a, b, c and d. The product term $Q_5 \sigma_{11} \sigma_{12}$ is responsible for inclinations in the yield loci a, c, e and g. The most obvious test of an inclination within this stress space is to observe whether axial strain is produced under pure torsional loading. This follows from the direction of the outward normals at the two points of intersection between an inclined locus and the τ- axis. Each normal will display a component of axial strain aligned with the σ- axis in addition to a shear strain component aligned with the τ- axis. This effect is due to anisotropy and is not a second-order phenomenon previously associated with $J_3{}'$. Axial strain can still appear in the absence of rotation when Q_5 is set to zero, so giving the loci in Fig. 11.9b, d and f. The locus in (h) is equivalent to eq(11.21) since coefficients in linear terms are set to zero to give equal magnitudes in forward and reversed yield stresses. When there is no shear stress acting along the orthotropic material directions 1 and 2, eq(11.23a) reduces to a dimensionless biaxial form for the principal stresses σ_1 and σ_2:

$$f = L_1(\sigma_1/\sigma_{1t}) + L_2(\sigma_2/\sigma_{1t}) + Q_1(\sigma_1/\sigma_{1t})^2 + Q_2(\sigma_2/\sigma_{1t})^2 + Q_3(\sigma_1/\sigma_{1t})(\sigma_2/\sigma_{1t}) = 1 \quad (11.23c)$$

in which σ_{1t} is the tensile yield stress in the 1 - direction. Particular applications of eq(11.23c) have shown good agreement with initial yield loci for anisotropic sheets of Ti-Al alloy [27] and zircaloy [47]. Comparisons of this kind apply more to the limit of proportionality since the initial anisotropy will diminish with increasing offset strains. An irregular shaped locus in an orthotropic material may be matched better with the introduction of cubic terms:

$$\begin{aligned} &Q_1(\sigma_1/\sigma_{1t})^2 + Q_2(\sigma_2/\sigma_{1t})^2 + Q_3(\sigma_1/\sigma_{1t})(\sigma_2/\sigma_{1t}) + T_1(\sigma_1/\sigma_{1t})^3 \\ &\quad + T_2(\sigma_2/\sigma_{2t})^3 + T_3(\sigma_1/\sigma_{1t})^2(\sigma_2/\sigma_{1t}) + T_4(\sigma_1/\sigma_{1t})(\sigma_2/\sigma_{1t})^2 = 1 \end{aligned} \quad (11.24)$$

Equations (11.23c) and (11.24) are applied to initial yield locus for an anisotropic magnesium extrusion in Fig. 11.10.

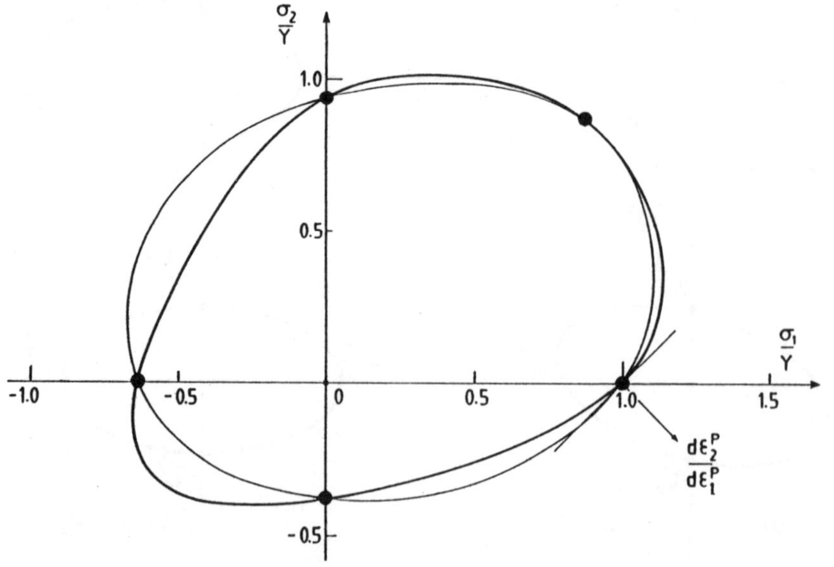

Figure 11.10 Yield locus for an orthotropic magnesium extrusion

The five constants, L and Q in eq(11.23c), are directly determined from the five experimental yield points shown. The determination of the seven constants in eq(11.24) requires an additional yield point and strain vector direction as indicated [48]. It is seen that both predictions represent the variations in yield stress along each axis of orthotropy. Both quadratic and cubic yield functions appear to describe the measured initial yield points equally well. However, eq(11.24) can provide a better account of the distorted locus in regions between yield points. It is seen from the comparison made in Fig. 11.10, that additional combined stress yield points are required to establish the degree of distortion. The literature reveals stronger evidence for distortion in the subsequent yield locus than in the initial locus. This becomes important to the direction of a plastic strain vector where it is assumed to be aligned with the position of the exterior normal.

11.4 Fracture Criteria for Brittle Materials

Brittle materials fail at stress levels just at or beyond their elastic limits with very little plasticity. We therefore use the term "fracture" and not "yield" to describe strength. However, the assumption of an isotropic fracture strength would not be consistent with observed behaviour for most brittle materials since these are inherently anisotropic. Amongst the materials displaying different strengths in tension and compression are glass, cast iron, rock, ceramic, concrete and soil. Pre-strained metals can show a similar anisotropy known as the Bauschinger effect. This refers to the reduction in reversed yield stress following forward deformation into the plastic range. Note that it is misleading to employ the term Bauschinger effect for a difference in brittle strengths where this is an inherent property and not one that is deformation induced.

11.4.1 Major Principal Stress Theory (MPS)

Rankine's principal stress criterion (Table 11.1) is easily modified to describe failure in brittle materials when their tensile and compressive strengths differ. Writing these as σ_t and σ_c respectively, Fig. 11.11 presents the failure locus in a principal biaxial stress plot. This allows for a compressive strength many times greater than the tensile strength. It is seen that, depending upon the quadrant in which the stress state lies, the MPS theory disregards the effect of one or other principal stress, since it is the numerically greater stress that is responsible for brittle fracture.

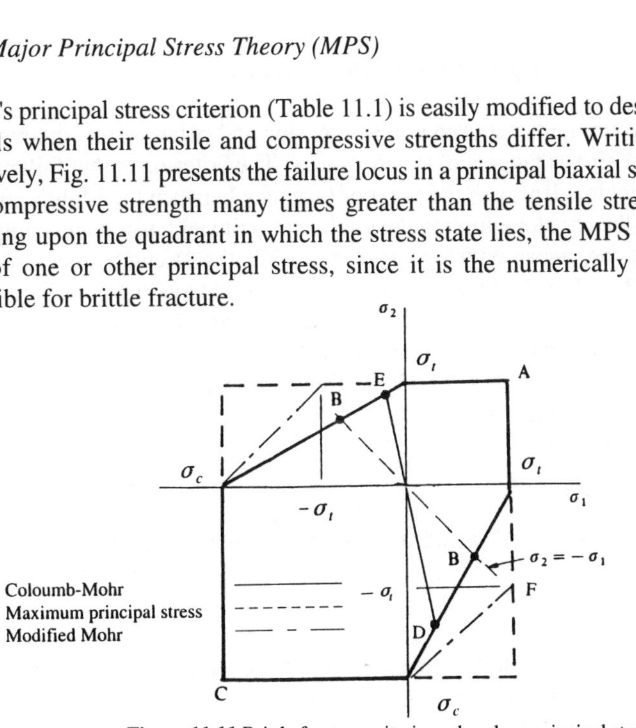

Figure 11.11 Brittle fracture criteria under plane principal stress

11.4.2 Internal Friction Theory

A further account of the failure locus in brittle materials is provided by the Couloumb-Mohr internal friction theory [49]. This is a combination of the major principal stress and maximum shear stress theories that also account for different tensile and compressive fracture strengths. Referring to Fig. 11.11, MPS failures occur in quadrants one and three and maximum shear failures in quadrants two and four. To account for failure under all stress states, an alternative presentation of Coloumb-Mohr's criterion is given in Fig. 11.12. Points on the Mohr's circles correspond with the particular combined stress points A, B, C and D given in Fig. 11.11.

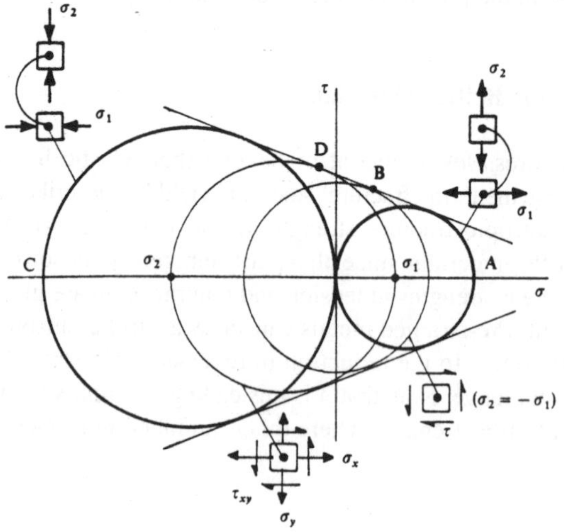

Figure 11.12 Coloumb-Mohr criterion

The different ultimate strengths between uniaxial tension and compression, are represented by points A and C on circles tangential to the τ- axis. The criterion is that failure under a biaxial stress state is coincident with the point of tangency between the Mohr's circle for a given stress state and the envelope linking the circles for uniaxial tension and compression. For example, within the linear envelope in Fig. 11.12, the strength under pure shear corresponds to point B. In Fig. 11.11 we can identify the corresponding intersection points B for pure shear failure where $\sigma_1 = - \sigma_2$. In the general case, for which tension (say) is combined with shear, the circle makes a tangent at D and the principal stresses at failure become the co-ordinates of point D in Fig. 11.11. These are expressed in the equation of the limb passing through D. That is

$$\sigma_2 = (\sigma_c / \sigma_t)\sigma_1 - \sigma_c \tag{11.25a}$$

In quadrant four, σ_1 is positive and σ_2 and σ_c are both negative, so eq(11.25a) is written as

$$\sigma_1 / \sigma_t + \sigma_2 / \sigma_c = 1 \tag{11.25b}$$

For other stress states lying within the remaining square envelopes of the failure locus, the major principal stress criterion applies.

11.4.3 Modified Mohr Theory

A further modification to the failure locus within its second and third quadrants in Fig. 11.11 was made by Mohr [49]. This was to achieve a closer fit to experimental data (see paragraph 11.4.5). Figure 11.11 shows that this modification places the failure envelope intermediate to others within quadrants two and four. All theories coincide in quadrants one and three. Now consider a state of pure torsion $\sigma_2/\sigma_1 = -1$ and $\sigma_1 = -\sigma_2 = \tau$, describing a line of slope -1, lying within quadrants two and four. The intersection between this line and the Coulomb-Mohr locus implies that the shear strength τ is less than the tensile strength. It is generally accepted that the two strengths are approximately equal for brittle materials, this being achieved with Mohr's modification. The modified Mohr theory is often applied graphically in this way to determine the combined stress states at fracture within quadrants two and four in Fig. 11.11. Alternatively, the line equation joining point F to σ_c in Fig. 11.11 is given by

$$\sigma_2 = \sigma_c - (\sigma_c + \sigma_t)\,\sigma_1/\sigma_t \tag{11.26a}$$

The modified theory is derived from eq(11.26a) in the form

$$\sigma_1/\sigma_t + (\sigma_1 + \sigma_2)/\sigma_c = 1 \tag{11.26b}$$

In eqs(11.26a,b), the signs are chosen to allow a positive substitution for σ_1 and negative substitutions for σ_2 and σ_c.

Example 11.3 A cast iron is subjected to a plane stress state as shown by the element inset in Fig. 11.12. Given that $\sigma_x = 100$ MPa tensile and $\tau_{xy} = 150$ MPa, determine the value of σ_y that according to Rankine and Mohr would cause brittle failure. The tensile and compressive strengths of this material are $\sigma_t = 400$ MPa and $\sigma_c = -1200$ MPa respectively.

Firstly, we determine the principal stress expressions in terms of the unknown σ_y. A Rankine tensile failure gives

$$\sigma_1 = \tfrac{1}{2}(\sigma_x + \sigma_y) + \tfrac{1}{2}[(\sigma_x - \sigma_y)^2 + 4\tau_{xy}^2]^{1/2} = \sigma_t$$

$$(0.25 + \sigma_y/400) + [(0.25 - \sigma_y/400)^2 + 0.5625]^{1/2} = 2 \tag{i}$$

from which $\sigma_y = 325$ MPa. A Rankine compressive failure gives

$$\sigma_2 = \tfrac{1}{2}(\sigma_x + \sigma_y) - \tfrac{1}{2}[(\sigma_x - \sigma_y)^2 + 4\tau_{xy}^2]^{1/2} = \sigma_c$$

$$(-0.0833 - \sigma_y/1200) - [(-0.0833 + \sigma_y/1200)^2 + 0.0625]^{1/2} = 2 \tag{ii}$$

from which $\sigma_y = -1182.7$ MPa. For a Coloumb-Mohr failure we substitute the left side of eqs(i) and (ii) into eq(11.25b) to give

$$(0.25 + \sigma_y/400) + [(0.25 - \sigma_y/400)^2 + 0.5625]^{1/2} + (-0.0833 - \sigma_y/1200)$$
$$- [(-0.0833 + \sigma_y/1200)^2 + 0.0625]^{1/2} = 2 \tag{iii}$$

Setting $x = \sigma_y/1200$ in eq(iii) leads to the cubic in x

$$x^3 - 0.0736x^2 - 0.3673x + 0.0867 = 0$$

for which the roots are $x = 0.463$, 0.2796 and -0.669. Correspondingly, $\sigma_y = 555.6$ MPa, 335.52 MPa and -802.8 MPa. The first value, being greater than σ_t, has no meaning. The second and third values of σ_y would produce tensile and compressive failure. In the graphical construction (Fig. 11.13) we fix point AB(100, -150) for the stress state on plane AB. Compressive failure BC lies on the circle in Fig. 11.13 with tangent at D. Tensile failure BC lies on another tangent circle, shifted to the right of the origin, with tangent at E also passing through AB. Points on circles D and E for plane BC (σ_y, 150) are then located as shown. However, since both principal stresses are tensile for circle E, we must use $\sigma_y = 325$ MPa at failure from the Rankine criterion.

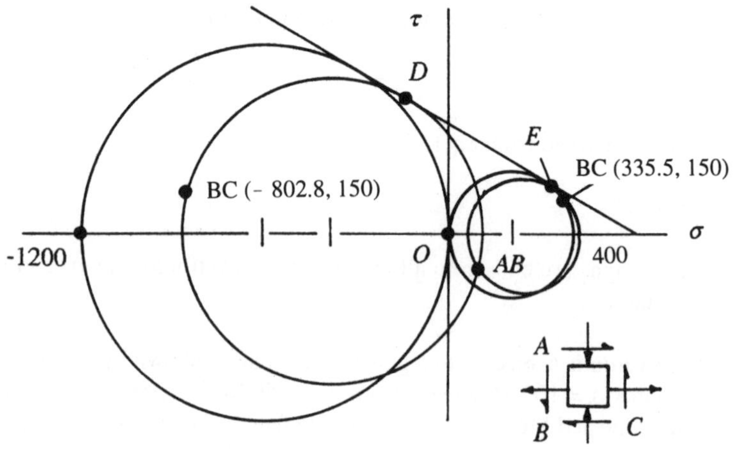

Figure 11.13 Coloumb-Mohr construction

11.4.4 Stassi-d'Alia Theory

Consider an unsymmetrical function (11.14a) combining the invariants J_1 and J_2' in the form

$$J_2' + \alpha J_1 = c \tag{11.27a}$$

where, from eqs (1.9a) and (11.10b),

$$(1/6)[(\sigma_1 - \sigma_2)^2 + (\sigma_2 - \sigma_3)^2 + (\sigma_1 - \sigma_3)^2] + \alpha(\sigma_1 + \sigma_2 + \sigma_3) = c \tag{11.27b}$$

Constant c is found by substituting the uniaxial condition (σ_1, 0, 0) in eq(11.27b). This gives

$$(1/6)(2\sigma_1^{\,2}) + \alpha\,\sigma_1 = c \tag{11.28a}$$

Since roots $\sigma_1 = \sigma_t$ and $\sigma_1 = -\sigma_c$ are required of eq(11.28a), it follows that

$$(\sigma_1 - \sigma_t)(\sigma_1 + \sigma_c) = \sigma_1^{\,2} + 3\alpha\,\sigma_1 - 3\,c$$

$$\sigma_1^{\,2} + \sigma_1(\sigma_c - \sigma_t) - \sigma_c\sigma_t = \sigma_1^{\,2} + 3\alpha\,\sigma_1 - 3\,c \tag{11.28b}$$

Equating coefficients in eq(11.28b) gives $\alpha = ⅓\,(\sigma_c - \sigma_t)$ and $c = ⅓\,\sigma_c\,\sigma_t$. Substituting into eq(11.27b) leads to

$$[(\sigma_1 - \sigma_2)^2 + (\sigma_2 - \sigma_3)^2 + (\sigma_1 - \sigma_3)^2] + 2\sigma_t(\sigma_c/\sigma_t - 1)(\sigma_1 + \sigma_2 + \sigma_3) = 2\sigma_c\,\sigma_t \qquad (11.29a)$$

Putting $\rho = \sigma_c/\sigma_t$ in eq(11.29a) leads to the criterion of Stassi-D'Alia [50]:

$$[(\sigma_1 - \sigma_2)^2 + (\sigma_2 - \sigma_3)^2 + (\sigma_1 - \sigma_3)^2] + 2\sigma_t(\rho - 1)(\sigma_1 + \sigma_2 + \sigma_3) = 2\rho\,\sigma_t^2 \qquad (11.29b)$$

Setting $\sigma_3 = 0$ in eq(11.29b), Stassi's normalised biaxial fracture criterion becomes

$$(\sigma_1/\sigma_t)^2 - (\sigma_1/\sigma_t)(\sigma_2/\sigma_t) + (\sigma_2/\sigma_t)^2 + (\rho - 1)(\sigma_1/\sigma_t + \sigma_2/\sigma_t) = \rho \qquad (11.29c)$$

The Stassi derivation [50] of eq(11.29c) did not involve the use of invariants but it is clear from the present derivation that hydrostatic stress is responsible for the difference between tensile and compressive strengths. When the derivation is extended to include shear stress, J_1 remains unchanged but additional terms in σ_{12}, σ_{23} and σ_{13} appear within J_2', to give

$$(\sigma_{11} - \sigma_{22})^2 + (\sigma_{22} - \sigma_{33})^2 + (\sigma_{11} - \sigma_{33})^2 + 6(\sigma_{12}^2 + \sigma_{13}^2 + \sigma_{23}^2)$$
$$+ 2\sigma_t(\rho - 1)(\sigma_{11} + \sigma_{22} + \sigma_{33}) = 2\rho\,\sigma_t^2 \qquad (11.30a)$$

Equation (11.30a) defines a locus in a reduced space σ_{11}, σ_{12}, as

$$\sigma_{11}^2 + 3\sigma_{12}^2 + \sigma_t(\rho - 1)\sigma_{11} = \rho\,\sigma_t^2 \qquad (11.30b)$$

We see that the major advantage of the invariant formulation is that it is reducible to a given stress state. When $\sigma_c = \sigma_t$, i.e. $\rho = 1$, a von Mises locus is recovered since hydrostatic stress then plays no part in failure.

Example 11.4 Derive an expression for the yield stress under a radial path $R = \sigma_{12}/\sigma_{11}$ ($\sigma_{22} = 0$) for a material where the ratio between compressive and tensile strengths is $\rho = \sigma_c/\sigma_t$. Find the yield stresses as a fraction of σ_t for (i) $\rho = 2$ and $R = 1$ and (ii) $\rho = 3$ and $R = 2$.

Normalising eq(11.30b) with respect to σ_t,

$$(\sigma_{11}/\sigma_t)^2 + 3(\sigma_{12}/\sigma_t)^2 + (\rho - 1)(\sigma_{11}/\sigma_t) = \rho \qquad \text{(i)}$$

Substituting $\sigma_{12} = R\sigma_{11}$ into eq(i) leads to the quadratic in σ_{11}/σ_t:

$$(1 + 3R^2)(\sigma_{11}/\sigma_t)^2 + (\rho - 1)(\sigma_{11}/\sigma_t) - \rho = 0 \qquad \text{(ii)}$$

The solution to eq(ii) is

$$\sigma_{11}/\sigma_t = \{-(\rho - 1) \pm \sqrt{[(\rho - 1)^2 + 4\rho(1 + 3R^2)]}\}/[2(1 + 3R^2)] \qquad \text{(iii)}$$

Setting $\rho = 2$ and $R = 1$ in eq(iii) gives $\sigma_{11}/\sigma_t = 0.593, -0.843$.

Setting $\rho = 3$ and $R = 2$ in eq(iii) gives $\sigma_{11}/\sigma_t = 0.41, -0.563$.

Figure 11.14 illustrates the solutions graphically for which the required stresses lie at the intersections with the loci for $\rho = 2$ and 3 and the linear path..

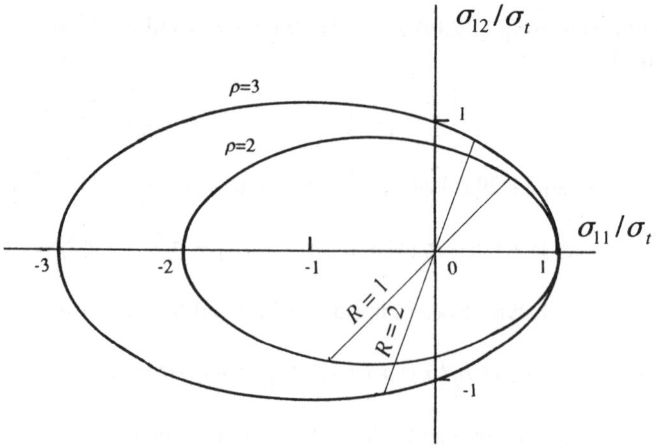

Figure 11.14 Failure loci in σ_{11}, σ_{12} space

11.4.5 Comparisons with Experiment

In Fig. 11.15, a comparison is made between the various strength criteria and experimental results [51-53] for fracture in grey cast iron. For the ratio $\rho = \sigma_c/\sigma_t = 3$, test data falls mainly within quadrants 1 and 4 for axes normalised with σ_t.

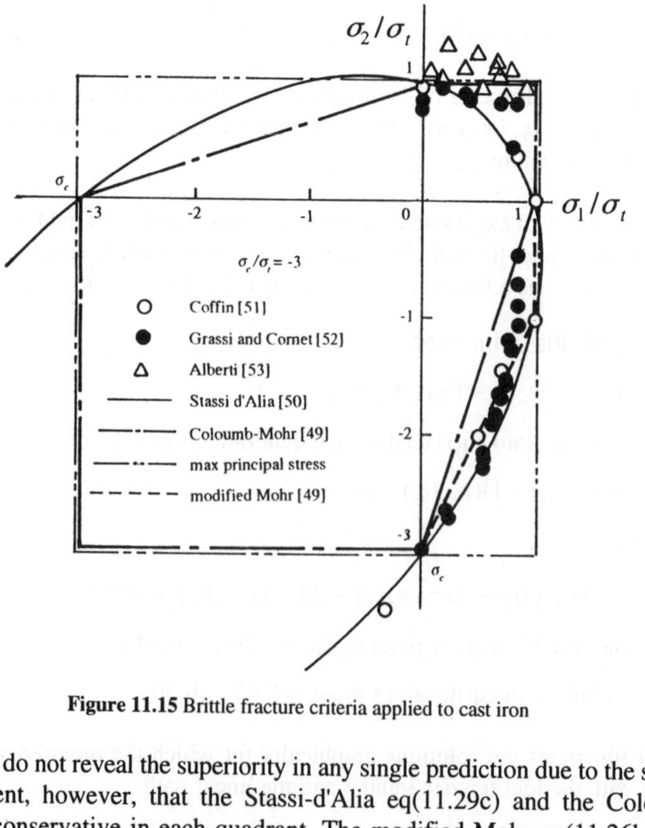

Figure 11.15 Brittle fracture criteria applied to cast iron

The experiments do not reveal the superiority in any single prediction due to the scatter in the data. It is evident, however, that the Stassi-d'Alia eq(11.29c) and the Coloumb-Mohr eq(11.25b) are conservative in each quadrant. The modified-Mohr eq(11.26b) represents failure adequately within quadrant four. Within the first quadrant, low-stress failures observed under biaxial tension, conform with theoretical predictions. The fact that fewer data are available under biaxial compression reflects the difficulty in conducting these experiments. However, some data lying within quadrant 3 suggests that Stassi-d'Alia's criterion best predicts the considerably enhanced compressive strengths observed.

11.5 Strength Criteria for Lamina

Consider next the failure of non-metallic laminae exhibiting quasi-brittle behaviour. Such materials include wood, slate and fibre-reinforced composites. Their properties are orientation dependent, strength, for example, being greater along the grain than across it. Many criteria have been proposed to account for strength variations in these materials. In what follows we classify them according to the applied stress system. The most versatile are those that admit failures from any combination of stresses inclined at any orientation to the grain or fibre direction. The general forms are more complex but they do admit reductions to simpler stress states.

11.5.1 Uniaxial Criteria

Early criteria dealt only with off-axis tensile and compressive failure. Those given in given in Table 11.5 [54-60], all predict the stress σ_{0f} at fracture when it is applied with an inclination θ to the fibre direction (i.e. direction - 1 in Fig. 11.16). The respective tensile and compressive strengths for the 1 - direction are σ_{1t} and σ_{1c} and those for the transverse 2 - direction are σ_{2t} and σ_{2c}. The shear strength for the 1-2 plane is denoted as σ_{12f}.

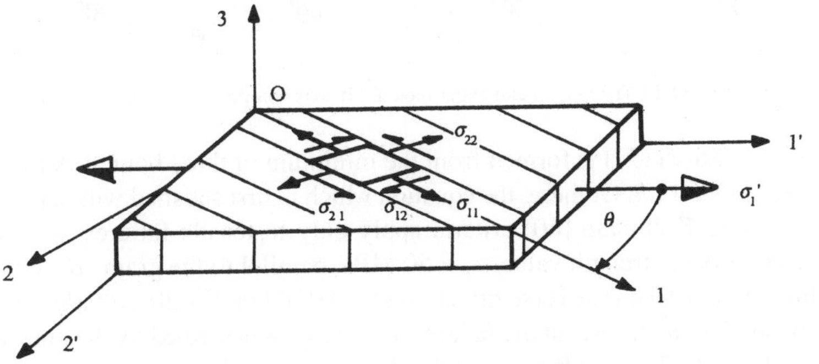

Figure 11.16 Off-axis tensile test on uni-directional composite

Table 11.5 Un iaxial failure criteria for directional materials

Jacoby	[54]	$\sigma_{0f} = \sigma_{1c}\cos^2\theta + \sigma_{2c}\sin^2\theta$
Howe	[55]	$\sigma_{0f} = \sigma_{2c} + (\sigma_{1c} - \sigma_{2c})(1 - \theta°/90°)^{5/2}$
Jenkins	[56]	$\sigma_{0f} = \sigma_{1t}/\cos^2\theta$ or $\sigma_{0f} = \sigma_{2t}/\sin^2\theta$ (smaller value applies)
Stussi	[57]	(a) $\sigma_{0f} = \sigma_{1c}/\cos^2\theta$, (b) $\sigma_{0c} = \sigma_{2t}/\sin^2\theta$, (c) $\sigma_{0f} = \sigma_{12f}/(\sin\theta\cos\theta)$
Hankinson	[58]	$\sigma_{0f} = \sigma_{1c}\sigma_{2c}/(\sigma_{1c}\sin^2\theta + \sigma_{2c}\cos^2\theta)$
Osgood	[59]	$\sigma_{0f} = \sigma_{1c}\sigma_{2c}/\{\sigma_{2c} + (\sigma_{1c} - \sigma_{2c})[a\sin^2\theta + (1-a)\sin^4\theta]\}$ a is a material constant
Kollmann	[60]	$\sigma_{0f} = \sigma_{1t}\sigma_{2t}/(\sigma_{1t}\sin^n\theta + \sigma_{2t}\cos^n\theta)$ $(1.5 \le n \le 2)$

Figure 11.17 compares four of these criteria with the compressive strength of Iroko wood. The Howe [55] and Hankinson [58] criteria best describe the observed fall in strength with an increasing orientation between the applied stress axis and the grain direction (0° inset).

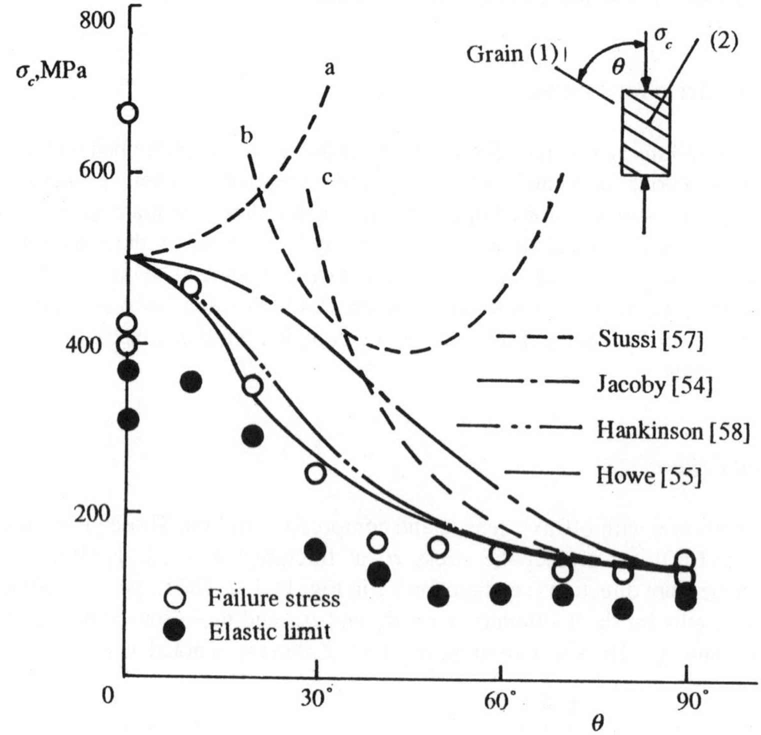

Figure 11.17 Theoretical and experimental off-axis compressive strength for wood

The criterion of Stussi [57] is formed from the minimum of three bounds defined by eqs(a), (b) and (c) in Table 11.5. Of these, the equation which is first satisfied will define failure. The Jenkins [56] and Kollmann [60] criteria apply only to tensile failures. It is seen that an average compressive strength value $\sigma_{1c} = 50$ MPa, parallel to the grain ($\theta = 0°$), is between 3 and 4 times that for the transverse direction ($\theta = 90°$). For $\theta \geq 30°$, a failure resulted from shearing parallel to the fibres but the failure stress was not described by the pure shear theory, curve (c) in Fig. 11.17. The difference in the stress ordinates between points established from the limit of proportionality and final fracture indicate the contribution to the fracture stress from inelasticity in this material. The greatest extent of inelasticity occurred for $\theta = 0°$, where the elastic and inelastic strain are of similar magnitudes $\approx$ 1% (see inset figure).

In contrast to its good description of the anisotropic strength for wood, Hankiinson's criterion cannot account for the purely brittle behaviour of Dinorwig slate (see Fig. 11.18). All other two-constant criteria in Table 11.5 are also unable to fit the slate data. With the constant a, in addition to σ_{1c}, σ_{2c}, Osgood's criterion [59] provides acceptable agreement with minimum strength observed at $\theta \approx 45°$. Within the orientations $30° \leq \theta \leq 60°$, the strength attains a minimum as failure occurs by shear along a single cleavage plane. Comparable maximum strengths are found for crushing failure, when $\theta = 90°$, and multiple splitting and buckling of individual layers for $\theta = 0°$.

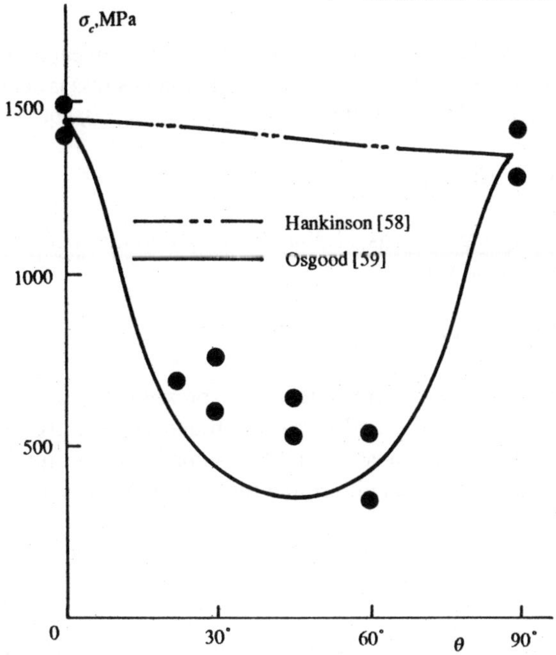

Figure 11.18 Theoretical and experimental off-axis compressive strength for slate

11.5.2 2D Criteria

A number of criteria have been proposed to predict the magnitudes of biaxial stress states at failure. The simplest plane-stress criterion is due to Jenkins [56]

$$\sigma_{11} = \sigma_{1t} \text{ and } \sigma_{22} = \sigma_{2t} \tag{11.31a,b}$$

where independent, direct tensile stresses σ_{11} and σ_{22} act along and perpendicular to fibres whose respective tensile strengths are σ_{1t} and σ_{2t}. The Stussi criterion [57] for compressive failure, is similar to eq(11.31a,b):

$$\sigma_{11} = \sigma_{1c}, \ \sigma_{22} = \sigma_{2c} \text{ and } \sigma_{12} = \sigma_{12f} \tag{11.32a,b,c}$$

in which additional account is given of shear failure under σ_{12f}. Failure occurs for whichever equality in eqs(11.31a,b) and (11.32a-c) is first achieved. In other biaxial-stress failure criteria, the plane stress components σ_{11}, σ_{22} and σ_{12} are combined. These criteria apply when the stress axes lie along, and transverse to, the fibre direction. Where the applied stresses are inclined they should be transformed into these directions according to eq(1.34a,b). Griffith and Baldwin [61] employed a similar derivation to the von Mises isotropic criterion. Failure was assumed to occur when the distortion energy reached a critical value independent of the hydrostatic stress $\sigma_m = \frac{1}{3}\sigma_{kk}$. The shear energy expression follows from taking an anisotropic plane-stress form of eq(11.6a) in which σ_m has been subtracted from σ_{11} and σ_{22} according to eq(1.21a). This gives

$$u_s = (\sigma_{11}^2/3)[C_{11} - (C_{12} + C_{13})/2] + (\sigma_{22}^2/3)[C_{22} - (C_{12} + C_{23})/2]$$
$$+ (\sigma_{11}\,\sigma_{22}/3)[2C_{12} - (C_{11} + C_{22} + C_{13} + C_{33})/2] + C_{44}\,\sigma_{12}^2 \tag{11.33a}$$

When C_{ij} in eq(11.33a) are constants the material behaves in a linear elastic manner to the point of fracture. The conversion of eq(11.33a) to a criterion in stress is achieved from taking u_{crit} in uniaxial stressing parallel to the fibres. Putting $\sigma_{22} = \sigma_{12} = 0$, eq(11.33a) becomes

$$u_{crit} = (\sigma_{1f}{}^2/3)[C_{11} - (C_{12} + C_{13})/2] \tag{11.33b}$$

Equating the right-hand sides of eqs(11.33a,b) and grouping coefficients leads to the simplified form of fracture criterion:

$$\sigma_{1f}{}^2 = \sigma_{11}{}^2 + a\sigma_{22}{}^2 + b\sigma_{11}\sigma_{22} + c\sigma_{12}{}^2 \tag{11.34}$$

in which the constants a, b and c are algebraic expressions in C_{ij}. Where non-linear deformation precedes fracture, a, b and c may be found empirically from the observed strengths in given directions. Amongst other Mises-type quadratic functions, Norris and McKinnon [62] proposed an empirical ellipsoidal function to describe the failures observed in plywood:

$$(\sigma_{11}/\sigma_{1f})^2 + (\sigma_{22}/\sigma_{2f})^2 + (\sigma_{12}/\sigma_{12f})^2 = 1 \tag{11.35}$$

Note that both eqs(11.34) and (11.35) are elliptical in form. Norris [63] later provided an interactive term, as with $b\sigma_{11}\sigma_{22}$ in eq(11.34), to give a biaxial failure criterion for the 1-2 plane as

$$(\sigma_{11}/\sigma_{1f})^2 - (\sigma_{11}\sigma_{22})/(\sigma_{1f}\sigma_{2f}) + (\sigma_{22}/\sigma_{2f})^2 + (\sigma_{12}/\sigma_{12f})^2 = 1 \tag{11.36a}$$

Two additional equations for the 1-3 and 2-3 planes follow with an appropriate change to the subscripts in eq(11.36a):

$$(\sigma_{11}/\sigma_{1f})^2 - (\sigma_{11}\sigma_{33})/(\sigma_{1f}\sigma_{3f}) + (\sigma_{33}/\sigma_{3f})^2 + (\sigma_{13}/\sigma_{13f})^2 = 1 \tag{11.36b}$$
$$(\sigma_{22}/\sigma_{2f})^2 - (\sigma_{22}\sigma_{33})/(\sigma_{2f}\sigma_{3f}) + (\sigma_{33}/\sigma_{3f})^2 + (\sigma_{23}/\sigma_{23f})^2 = 1 \tag{11.36c}$$

These three-plane fracture criteria apply particularly to cellular-type materials. In the absence of buckling, failure of their orthogonal walls may be considered from their in-plane stress states. Failure occurs when any one of the criteria (11.6a,b,c) is first reached. For example, a plane stress state σ_{11}, σ_{22} and σ_{12} shows that eq(11.36a) is accompanied by two reduced relationships from eqs(11.36b,c):

$$(\sigma_1/\sigma_{1f})^2 = 1 \text{ and } (\sigma_2/\sigma_{2f})^2 = 1$$

These restate the maximum principal stress criteria. Thus, in the absence of stress components σ_{33}, σ_{23} and σ_{13}, the lowest magnitude plane failure stresses are given by eq(11.36a).

11.5.3 General Strength Criteria

(a) Hill

Hill's homogenous quadratic function [29] eq(11.21) was originally proposed as a yield criterion for orthotropic metals. It has also been used to predict triaxial stress states that cause brittle fracture when written in the form

$$2f(\sigma_{ij}) = F(\sigma_{22} - \sigma_{33})^2 + G(\sigma_{11} - \sigma_{33})^2 + H(\sigma_{11} - \sigma_{22})^2 + 2(L\sigma_{23}^2 + M\sigma_{13}^2 + N\sigma_{12}^2) = 1 \quad (11.37a)$$

The constants are expressed in terms of known failure strengths as follows

$$2F = 1/\sigma_{2f}^2 + 1/\sigma_{3f}^2 - 1/\sigma_{1f}^2$$
$$2G = 1/\sigma_{3f}^2 + 1/\sigma_{1f}^2 - 1/\sigma_{2f}^2$$
$$2H = 1/\sigma_{1f}^2 + 1/\sigma_{2f}^2 - 1/\sigma_{3f}^2$$
$$2L = 1/\sigma_{23f}^2,\ 2M = 1/\sigma_{13f}^2 \text{ and } 2N = 1/\sigma_{12f}^2$$

Here σ_{12f}, σ_{23f} and σ_{13f} are the respective shear strengths within the planes 1 - 2, 2 - 3 and 1 - 3 and σ_{1f}, σ_{2f} and σ_{3f} are the respective direct strengths for the 1, 2 and 3 directions. The reduction to eq(11.37a) for plane stress applied to a material with transverse isotropy was considered by Azzi and Tsai [64]. Setting all stress components except σ_{11}, σ_{22} and σ_{12} to zero and with $\sigma_{2f} = \sigma_{3f}$ in eq(11.37a) gives

$$(\sigma_{11}^2 - \sigma_{11}\sigma_{22})/\sigma_{1f}^2 + \sigma_{22}^2/\sigma_{2f}^2 + \sigma_{12}^2/\sigma_{12f}^2 = 1 \quad (11.37b)$$

In the off-axis uniaxial test, with $0 < \theta < 90°$, (see Fig. 11.16) they showed from eq(11.37b) that the direction of the failure path was not necessarily perpendicular to the direction of the major principal stress. This influence of shear stress is absent in fracture criteria composed only of normal stresses. Note that in eqs(11.34) - (11.37), Griffith and Baldwin [61], Norris [62,63] and Hill [29] did not stipulate whether direct strengths σ_{1f}, σ_{2f} and σ_{3f} are tensile or compressive. These homogenous quadratic functions describe ellipses centred ast the origin and thereby provide equal tensile and compressive strengths. Where the two strengths differ, these criteria are inappropriate.

(b) Marin

In order to account for differenet tensile and compressive strengths Marin [65] proposed and empirical failure criterion. This includes both linear and quadratic terms in stress in the form

$$(\sigma_1 - a)^2 + (\sigma_2 - b)^2 + (\sigma_3 - c)^2 + q[(\sigma_1 - a)(\sigma_2 - b)$$
$$+ (\sigma_2 - b)(\sigma_3 - c) + (\sigma_3 - c)(\sigma_1 - a)] = \sigma^2 \quad (11.38a)$$

where a, b, c and σ are empirical constants determined from the observed uniaxial strengths in tension and compression for each orthogonal direction. Because the remaining constant q must be evaluated under a complex stress condition, Marin's function can be made to fit test data more reliably than other functions which rely solely on uniaxial data. The absence of shear stress in eq(11.38a) means that it is strictly restricted to where principal stresses σ_1, σ_2 and σ_3 are aligned with the material orthogonal axes. In order to gain wider applicability it could be assumed that shear stress has no effect on the failure condition and σ_{11}, σ_{22} and σ_{33} are the normal components of applied stress lying in the principal material co-ordinates. The plane stress reduction to eq(11.38a) takes the form

$$\sigma_{11}^2 + K_1\sigma_{11}\sigma_{22} + \sigma_{22}^2 + K_2\sigma_{11} + K_3\sigma_{22} = K_4 \quad (11.38b)$$

Substituting the known uniaxial strengths into eq(11.38b) leads to

$$K_2 = \sigma_{1c} - \sigma_{1t},\quad K_3 = \sigma_{2c}\,\sigma_{2t} \text{ and } K_4 = \sigma_{1c}\,\sigma_{1t} = \sigma_{2c}\,\sigma_{2t} = \sigma_{3c}\,\sigma_{3t}$$

where subscripts t and c denote the tensile and compressive strengths respectively. The constant K_1 may be found from either the failure stress σ_b under equi-biaxial tension ($\sigma_{11} = \sigma_{11} = \sigma_b$) or from pure shear ($\sigma_1 = -\sigma_2 = \sigma_{12f}$). These give K_1 respectively as

$$K_1 = \{\sigma_{1c}\,\sigma_{1t} + [(\sigma_{1t} + \sigma_{2t}) - (\sigma_{1c} + \sigma_{2c})]\sigma_b\}/\sigma_b^{\,2} - 2$$
$$= 2 - \{\sigma_{1c}\,\sigma_{1t} - [(\sigma_{1c} - \sigma_{2c}) - (\sigma_{1c} - \sigma_{2c})]\sigma_{12f}\}/\sigma_{12f}^{\,2}$$

Where shear stress is known to influence fracture, it has been introduced within the following "modified-Marin" form of plane stress criterion:

$$A_1\,\sigma_{11}^{\,2} + A_2\,\sigma_{11}\,\sigma_{22} + A_3\,\sigma_{22}^{\,2} + A_4\,\sigma_{11} + A_5\,\sigma_{22} + A_6\,\sigma_{12}^{\,2} = 1 \qquad (11.39a)$$

Constants A_1, A_3, A_4 and A_5 in eq(11.39a) are determined from the known strengths. A_2 becomes a "floating constant" and A_6 follows from failure under pure, in-plane shear. This gives eq(11.39a) in the form

$$(\sigma_{11} - A_2\,\sigma_{22})\sigma_{11}/(\sigma_{1c}\,\sigma_{1t}) + \sigma_{22}^{\,2}/(\sigma_{2c}\,\sigma_{2t}) + (\sigma_{1c} - \sigma_{1t})\sigma_{11}/(\sigma_{1c}\,\sigma_{1t})$$
$$+ (\sigma_{2c} - \sigma_{2t})\sigma_{22}/(\sigma_{2c}\,\sigma_{2t}) + \sigma_{12}^{\,2}/\sigma_{12f}^{\,2} = 1 \qquad (11.39b)$$

The effect of setting K_1 and A_2 in eqs(11.39a,b) different from unity is examined in the following section.

(c) Hoffman

A modification made to eq(11.37a) by Hoffmann [66] provides for a general orthotropic condition. The addition of linear terms in the components of direct stress gives

$$C_1(\sigma_{22} - \sigma_{33})^2 + C_2(\sigma_{33} - \sigma_{11})^2 + C_3(\sigma_{11} - \sigma_{22})^2 + C_4\,\sigma_{11}$$
$$+ C_5\,\sigma_{22} + C_6\,\sigma_{33} + C_7\,\sigma_{23}^{\,2} + C_8\,\sigma_{13}^{\,2} + C_9\,\sigma_{12}^{\,2} = 1 \qquad (11.40a)$$

The material constants C_i ($i = 1, 2 \ldots 6$) now become

$$2C_1 = 1/(\sigma_{2c}\,\sigma_{2t}) + 1/(\sigma_{3c}\,\sigma_{3t}) - 1/(\sigma_{1c}\,\sigma_{1t})$$
$$2C_2 = 1/(\sigma_{3c}\,\sigma_{3t}) + 1/(\sigma_{1c}\,\sigma_{1t}) - 1/(\sigma_{2c}\,\sigma_{2t})$$
$$2C_3 = 1/(\sigma_{1c}\,\sigma_{1t}) + 1/(\sigma_{2c}\,\sigma_{2t}) - 1/(\sigma_{3c}\,\sigma_{3t})$$
$$C_4 = 1/\sigma_{1t} - 1/\sigma_{1c}\,,\; C_5 = 1/\sigma_{2t} - 1/\sigma_{2c} \text{ and } C_6 = 1/\sigma_{3t} - 1/\sigma_{3c}$$

The remaining constants C_7, C_8 and C_9 relate to the shear strengths σ_{12f}, σ_{23f} and σ_{13f} and in the orthogonal planes as follows:

$$C_7 = 1/\sigma_{23f}^{\,2}\,,\; C_8 = 1/\sigma_{13f}^{\,2} \text{ and } C_9 = 1/\sigma_{12f}^{\,2}$$

The plane stress form of eq(11.40a) is

$$(\sigma_{11} - \sigma_{22})\sigma_{11}/(\sigma_{1c}\,\sigma_{1t}) + (\sigma_{22} - \sigma_{11})/(\sigma_{2c}\,\sigma_{2t}) + (\sigma_{1c} - \sigma_{1t})\sigma_{11}/(\sigma_{1c}\,\sigma_{1t})$$
$$+ (\sigma_{2c} - \sigma_{2t})\sigma_{22}/(\sigma_{2c}\,\sigma_{2t}) + \sigma_{12}^{\,2}/\sigma_{12f}^{\,2} = 1 \qquad (11.40b)$$

which is not identical to eq(11.39b). In this plane form, Hoffman described fracture surfaces for a uni-directional glass-fibre composite and a syntatic foam consisting of microspheres embedded in a resin matrix.

(d) Tsai and Wu

Equations (11.37 -11.40) predate the now most widely used fracture criterion of Tsai and Wu [67]. They combined linear and quadratic terms for a six component stress space as

$$F_{ij}\sigma_{ij} + F_{ijkl}\sigma_{ij}\sigma_{kl} = 1 \tag{11.41a}$$

where i, j, k and l take values from 1, 2 and 3. F_{ij} and F_{ijkl} total 12 strength coefficients which appear in the full expansion of eq(11.41a) as

$$F_{11}\sigma_{11} + F_{22}\,\sigma_{22} + F_{33}\,\sigma_{33} + F_{1111}\,\sigma_{11}^2 + F_{2222}\,\sigma_{22}^2 + F_{3333}\,\sigma_{33}^2 + 2\,(F_{2233}\,\sigma_{22}\,\sigma_{33} + F_{1133}\,\sigma_{11}\,\sigma_{33}$$
$$+ F_{1122}\,\sigma_{11}\,\sigma_{22}) + 4\,(F_{1212}\,\sigma_{12}^2 + F_{1313}\,\sigma_{13}^2 + F_{2323}\,\sigma_{23}^2) = 1 \tag{11.41b}$$

in which interactive stress terms; σ_{11} , σ_{12} etc, are ignored. Nine coefficients relate to the uniaxial strengths

$$F_{11} = 1/\sigma_{1t} - 1/\sigma_{1c}\,,\ F_{22} = 1/\sigma_{2t} - 1/\sigma_{2c}\,,\ F_{33} = 1/\sigma_{3t} - 1/\sigma_{3c}$$
$$F_{1111} = 1/(\sigma_{1c}\,\sigma_{1t}),\ \ F_{2222} = 1\,/(\sigma_{2c}\,\sigma_{2t}),\ \ F_{3333} = 1/(\sigma_{3c}\,\sigma_{3t})$$
$$F_{1212} = 1/\sigma_{12f}^{\,2}\,,\ F_{1313} = 1/\sigma_{13f}^{\,2}\,,\ F_{2323} = 1/\sigma_{23f}^{\,2}$$

from which the relationship to coefficients in earlier theories is apparent. The remaining coefficients in eq(11.41b) are found from biaxial stress tests.

(e) Waddoups

Up to now fracture criteria have been formulated in terms of the stress components. Waddoups [68] postulated an anisotropic form of St Venant's principal strain criterion. This simply states that failure occurs when one component of principal strain (ε_1, ε_2) reaches a critical value. Where plane principal stresses are aligned with the material directions:

$$\varepsilon_1 = \sigma_1/E_1 - \nu_{12}\,\sigma_2/E_1 \tag{11.42a}$$
$$\varepsilon_2 = \sigma_2/E_2 - \nu_{21}\,\sigma_1/E_2 \tag{11.42b}$$

where E_1 and ν_{12} are elastic constants for a direction aligned with the fibres and E_2 and ν_{12} are elastic constants for the transverse direction.

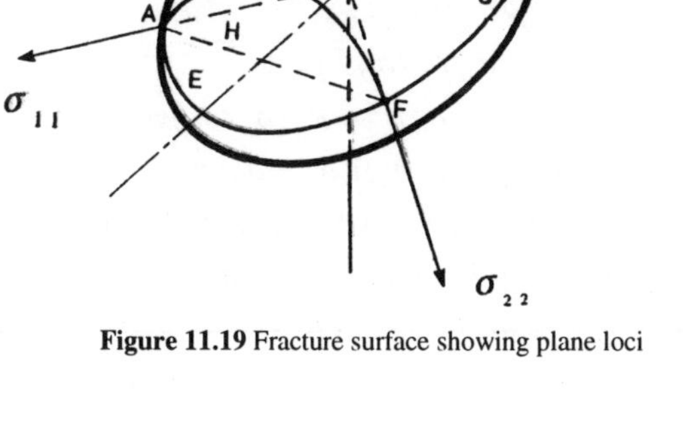

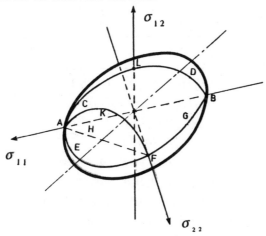

Figure 11.19 Fracture surface showing plane loci

11.6 Comparisons with Experiment

We have seen how fracture strength criteria for an orthotropic layer of composite material are formulated under various stress states. Very often the stress state in a multi-layered composite will be plane in nature, so it is instructive to compare appropriate 2D criteria with experimental fracture data. Where the lay-up uses different fibre orientations, the one orientation that offers least resisitance to the loading may be assumed to control the strength of the laminate.

The failure locus AKF in Fig. 11.19 corresponds to intersection of a failure surface in σ_{11}, σ_{22}, σ_{12} space and a plane defining off-axis uniaxial stressing.

When shear stress is absent along principal material directions 1 and 2, the normal stresses become principal stresses σ_1 and σ_2. The failure locus (AEFGB) is formed from the intersection between the plane $\sigma_{12} = 0$ and the surface in Fig. 11.19. Locus ACLDB lies in the σ_{11}, σ_{12} plane where $\sigma_{12} = 0$.

11.6.1 Off-Axis Compression

Resolution of an applied stress σ, inclined at θ to the principal fibre direction 1, gives the three components of plane stress for the material directions (see Fig. 11.16):

$$\sigma_{11} = \sigma\cos^2\theta,\quad \sigma_{22} = \sigma\sin^2\theta \ \text{ and } \ \sigma_{12} = (\sigma/2)\sin 2\theta \qquad (11.43\text{a,b,c})$$

Since these components are not independent, the experimental fracture locus may be derived within σ_{11}, σ_{22} space by eliminating θ and σ_{12}. This gives

$$\sigma_{11} + \sigma_{22} = \sigma \ \text{ and } \ \sigma_{12}^2 = \sigma_{11}\,\sigma_{22} \qquad (11.44\text{a,b})$$

In Table 11.6 the plane stress forms of the previous criteria are applied to the off-axis test. For this, the stress components are normalised with respect to failure strength giving variables x and y as indicated. Where the theoretical tensile and compressive strengths are assumed to be the same, σ_f is used for normalising the axes. Where these strengths are different, the compressive strength σ_c is used as shown.

Table 11.6 Offset-axis failure criteria

x	y	Function f	Ref
σ_{11}/σ_{1f}	σ_{22}/σ_{2f}	$x^2 + y^2 + (\sigma_{2f}/\sigma_{1f})[(\sigma_{1f}/\sigma_{12f})^2 - 1]xy = 1$	Hill [29]
σ_{11}/σ_{1f}	σ_{22}/σ_{2f}	$x^2 + (b/a)\,y^2(\sigma_{2f}/\sigma_{1f})^2 + (c/a + d/a)(\sigma_{2f}/\sigma_{1f})xy = 1$	Griffith [61]
σ_{11}/σ_{1f}	σ_{22}/σ_{2f}	$x^2 + y^2 + (\sigma_{1f}\sigma_{2f}/\sigma_{12f}^{\,2})x\,y = 1$	Norris [62]
σ_{11}/σ_{1f}	σ_{22}/σ_{2f}	$x^2 + y^2 - x\,y\,(1 - \sigma_{1c}\sigma_{2c}/\sigma_{12f}^{\,2}) = 1$	Norris [63]
σ_{11}/σ_{1c}	σ_{22}/σ_{2c}	$(\sigma_{1c}/\sigma_{1t})x^2 + (\sigma_{2c}^{\,2}/\sigma_{1c}\sigma_{1t})y^2 + K_1(\sigma_{2c}/\sigma_{1t})xy$ $+ (\sigma_{1c}/\sigma_{1t} - 1)x + [\sigma_{2c}/(\sigma_{1c}\sigma_{1t})](\sigma_{2c} - \sigma_{2t})\,y = 1$	Marin [65]
σ_{11}/σ_{1c}	σ_{22}/σ_{2c}	$(\sigma_{1c}/\sigma_{1t})x^2 + (\sigma_{2c}/\sigma_{2t})\,y^2 - (\sigma_{2c}/\sigma_{1t} + \sigma_{1c}/\sigma_{2t})xy$ $+ (\sigma_{1c}/\sigma_{1t} - 1)x + (\sigma_{2c}/\sigma_{2t} - 1)\,y + (\sigma_{12}/\sigma_{12f})^2 = 1$	Hoffmann [66]

Table 11.6 predictions apply to the first quadrant in x and y. In deriving the Hill function, transverse isotropy $\sigma_{2c} = \sigma_{3c}$ has been assumed. Taking again the off-axis compression results for Iroko wood (see Fig. 11.17), we have for the 0° and 90° directions $\sigma_{1c} = 50$ MPa and $\sigma_{2c} = 13.6$ MPa respectively. Equation (11.44b) is used to find σ_{12f}. The corresponding tensile strengths $\sigma_{1t} = 83$ MPa and $\sigma_{2t} = 15.3$ MPa were obtained for each direction from three-point bend tests. Fig. 11.20 compares the theoretical failure loci in Table 11.6 with the experimental data.

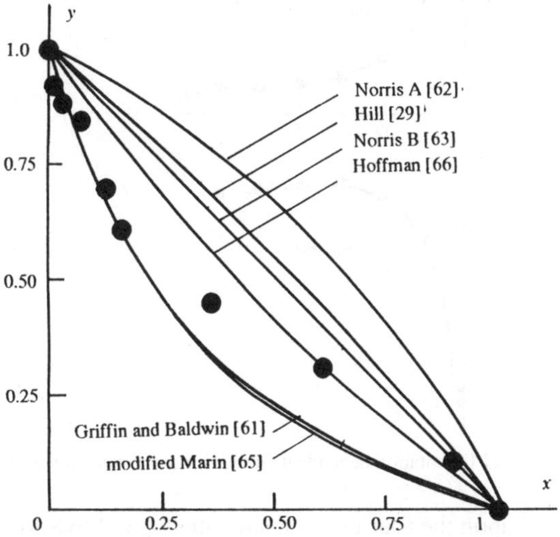

Figure 11.20 Predictions to the compressive failure in wood

These loci are simply views of AKF (Fig. 11.19) looking in the direction of σ_{12}. The comparison between theory and experiment shows that the test data falls between the Griffith and Baldwin [61] and Hoffmann [66] predictions. It is later shown that Hoffmann represents test data realistically over the four quadrants, admitting a difference between the tensile and compressive strengths. However, within the single quadrant in Fig. 11.20, a better description is found with the Griffith and Baldwin locus since the additional constant ($c/a + d/a$) allows it to pass through any intermediate point. The condition $\sigma_{1c}\,\sigma_{1t} = \sigma_{2c}\,\sigma_{2t}$ gives $K_1 = 1$ in Table 11.6. If this condition does not hold, as for this material, then K_1 is found from the failure stress either under biaxial tension or pure shear. It may also be found from any intermediate point. For example, Fig. 11.20 shows that K_1 may be chosen to coincide with the Griffith and Baldwin locus by choosing a common point. Both loci display different tensile and compressive strengths. Generally, criteria with a "floating constant" allow a more precise fit to the data. Thus, if we substitute eqs(11.43a,b,c) into Tsai-Wu criterion (11.41b) the off-axis strength is predicted as

$$[\cos^4\theta/(\sigma_{1c}\,\sigma_{1t}) + \sin^4\theta/(\sigma_{2t}\,\sigma_{2c}) + \tfrac{1}{4}(2F_{1122} + 4F_{1212})\sin^2 2\theta]\sigma^2$$
$$+ [\cos^2\theta(\sigma_{1c} - \sigma_{1t})/\sigma_{1c}\,\sigma_{1t} + \sin^2\theta(\sigma_{2c} - \sigma_{2t})/(\sigma_{2c}\,\sigma_{2t})]\sigma = 1 \qquad (11.45)$$

The two roots of this quadratic are the tensile and compressive strengths for a given orientation θ. For example, with $\theta = 0$, $\sigma = \sigma_{1t}$ and $\sigma = -\sigma_{1c}$ as expected. The comparison between eq(11.45) and the strengths of a 50% volume uni-directional glass fibre reinforced composite (gfrc) is shown in Fig. 11.21.

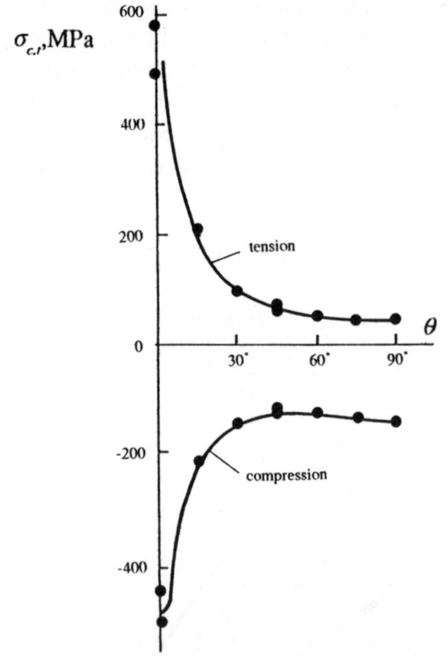

Figure 11.21 Orientation dependent strengths for uni-directional gfrc

The predictions are based upon the four uniaxial strengths $\sigma_{1c} = 476.8$ MPa, $\sigma_{2c} = 136.3$ MPa, $\sigma_{1t} = 529$ MPa and $\sigma_{2t} = 49.2$ MPa [69]. The "floating constant" $(2F_{1122} + 4F_{1212}) = 310.3$ mm^4/kN^2 is found by matching eq(11.45) to the tensile strength for the 45° direction. In this way eq(11.45) provides the strengths for all intermediate orientations with good accuracy.

11.6.2. Plane Principal Stresses

An experimental σ_1, σ_2 locus may be established from proportional and step-wise loading to the point of fracture. Such loadings may include combinations of internal pressure, pure shear and biaxial tension. Test procedures are similar to well-established methods employed for the determination of the yield surface for metals.

(a) Symmetrical functions

Consider failure criteria that assume equal tensile and compressive strengths, i.e. $\sigma_{1c} = -\sigma_{1t}$ and $\sigma_{2c} = -\sigma_{2t}$ but with $\sigma_{1c} \neq \sigma_{2c}$ and $\sigma_{1t} \neq \sigma_{2t}$. Simplifying this, we write σ_{1f} and σ_{2f} to denote respective strengths in the 1 and 2 directions. The normalised stress variables within each prediction in Table 11.7 are then $x = \sigma_1/\sigma_{1f}$ and $y = \sigma_2/\sigma_{2f}$.

Table 11.7 Symmetrical failure functions

$x^2 + y^2 - (\sigma_{2f}/\sigma_{1f})\,xy = 1$	Hill [29]
$x^2 + y^2 + c\,(\sigma_{2f}/\sigma_{1f})\,xy = 1$	Griffith [61]
$x^2 + y^2 = 1$	Norris [62]
$x^2 - xy + y^2 = 1$	Norris [63]
$y = (\sigma_{1f}/\sigma_{2f})(x-1)/\nu_{12}$ and $y = 1 + \nu_{12}(E_2/E_1)(\sigma_{1f}/\sigma_{2f})x$	Waddoups [68]

In Waddoups criterion the elastic constants in eqs(11.42a,b) are related: $\nu_{21}/\nu_{12} = E_2/E_1$. These equations then invert into the two normalised stress forms given in Table 11.7. One material with approximately equal tensile and compressive strengths is the woven glass-fibre, reinforced-epoxy matrix composite. The average strengths are: $\sigma_{1f} = 307$ MPa and $\sigma_{2f} = 264$ MPa and the elastic constants are $E_1 = 17.96$ GPa, $E_2 = 19.17$ GPa, $\nu_{12} = 0.142$ and $\nu_{21} = 0.152$ [69]. Comparison between predicted failure loci for this material appears in Fig. 11.22.

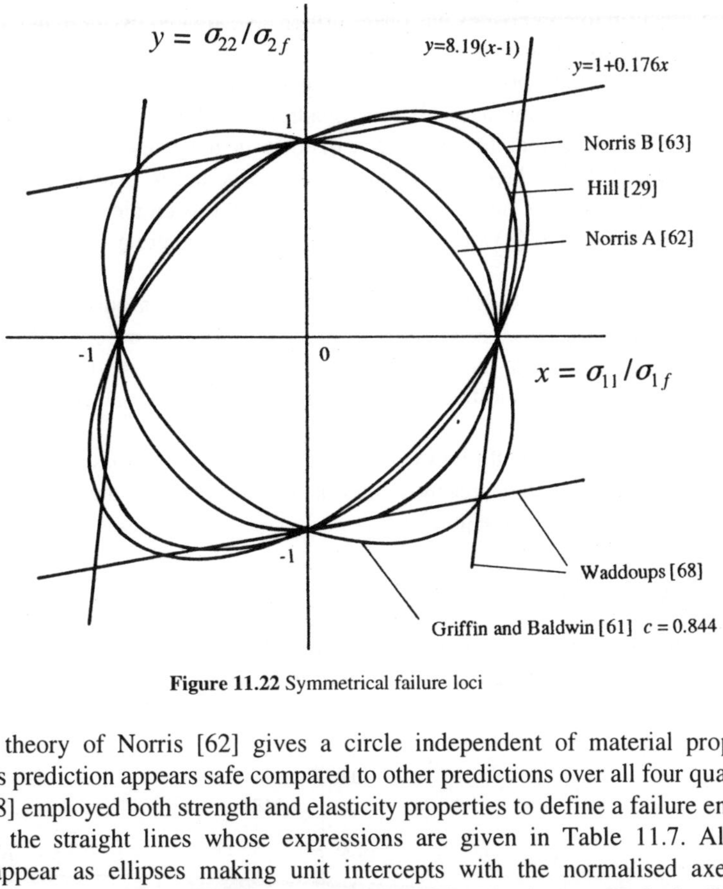

Figure 11.22 Symmetrical failure loci

The original theory of Norris [62] gives a circle independent of material properties. Moreover, this prediction appears safe compared to other predictions over all four quadrants. Waddoups [68] employed both strength and elasticity properties to define a failure envelope formed from the straight lines whose expressions are given in Table 11.7. All other predictions appear as ellipses making unit intercepts with the normalised axes. The transversely isotropic function of Hill [29] lies between the Norris A and B loci, all having the same orientation. One further material property is required to determine the constant c in the elliptical function of Griffin and Baldwin [61]. Assuming a shear strength of 250 MPa gives $c = 0.84$. This alters the orientation to give predicted strengths less than others for quadrants 1 and 3. When the compliances C_{ij} in eq(11.33a) are pre-determined from linear anisotropic elasticity the implication is that the material is brittle-elastic. This may limit its usefulness where there is evidence of inelasticity but in practice C_{ij} may be taken as empirical fracture constants, consistent with other theories.

(b) Unsymmetrical functions

The Marin [65] and Hoffman [66] criteria will account for differences between tensile and compressive strengths, i.e. where: $\sigma_{1c} \neq \sigma_{1t} \neq \sigma_{2c} \neq \sigma_{2t}$. The material is transversely isotropic if $\sigma_{2c} = \sigma_{3c}$. The original theory of Marin [65] was modified for when the required relationship $\sigma_{1c}\ \sigma_{1t} = \sigma_{2c}\ \sigma_{2t}$ does not apply. The modification (see eq(11.39b)) results in the plane, principal stress, failure criterion

$$(\sigma_1^2 - A_2\sigma_1\sigma_2)/(\sigma_{1c}\,\sigma_{1t}) + \sigma_2^2/(\sigma_{2c}\,\sigma_{2t}) + \sigma_1(\sigma_{1c} - \sigma_{1t})/(\sigma_{1c}\,\sigma_{1t}) + \sigma_2(\sigma_{2c} - \sigma_{2t})/(\sigma_{2c}\,\sigma_{2t}) = 1 \quad (11.46)$$

Again, the constant $A_2 \neq 1$ in eqs(11.46) may be determined from a further test conducted under pure shear, biaxial tension or any combined stress state. However, a check should be made that the A_2 value will result in a closed locus.

Kaminski and Lantz [70] showed that it was possible to fit symmetrical functions to unsymmetrical test data within individual quadrants. Similar applications of Hill's transversely isotropic function to composites has been further investigated by Azzi and Tsai [64]. They too removed the original restriction that tensile and compressive strengths are equal by admitting σ_c or σ_t appropriate to the given quadrant. An obvious advantage of Marin's modification is that it embraces all four quadrants within a single continuous function. Franklin [71] showed that Marin's locus distorts in the axes σ_1 and σ_2 when they are normalised with different directional strengths σ_{1t}, σ_{1c}, σ_{2t} and σ_{2c}. The reader should note that (i) a general summation of linear and quadratic stress terms employed by Tsai and Wu [67] will define an ellipse and (ii) all other functions of this kind are particular reductions to (i).

Example 11.5 Tensile and compressive tests conducted on a uni-directional glass-fibre, reinforced composite gave $\sigma_{1c} = 476.8$ MPa, $\sigma_{1t} = 529$ MPa, $\sigma_{2c} = 136.3$ MPa and $\sigma_{2t} = 49.2$ MPa. An additional, through-thickness, compression test gave $\sigma_{3c} = 167.7$ MPa. Compare the failure loci from Marin's modified criterion corresponding to $p = 1$, $p = -1$ and $p = 10$.

The through-thickness strength shows that it is not unreasonable to assume transverse isotropy $\sigma_{2c} = \sigma_{3c}$ when making the present comparisons. Substituting the strengths (MPa) into eq(11.46) gives the elliptical equation

$$(\sigma_1^2 - A_2\,\sigma_1\,\sigma_2) + 37.62\sigma_2^2 - 52.24\sigma_1 + 3275.1\sigma_2 = 25.22 \times 10^4 \quad \text{(i)}$$

which is composed of linear and quadratic terms. The given uniaxial strengths may be shecked from eq(i) by setting σ_1 and σ_2 to zero in turn. A comparison is made in Fig. 11.23 of the failure loci corresponding to eq(i) for $A_2 = 1, -1$ and 10.

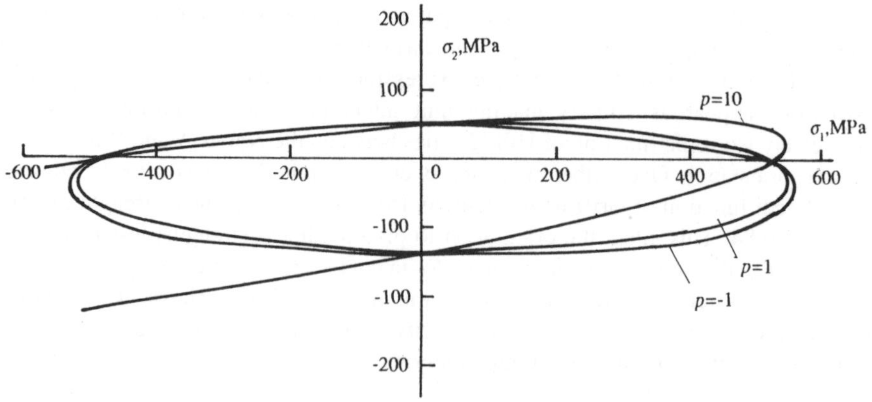

Figure 11.23 Unsymmetrical failure loci

It is seen that the change in sign from + 1 to − 1 does not alter appreciably the safe region enclosed within the resulting closed loci. In contrast, an invalid failure locus is found for $p = 10$ since it remains open on the compressive side.

11.7 Concluding Remarks

The classical von Mises and Tresca theories of yielding for metals may each be formulated as a function of the stress deviator invariants. This lies in the simple fact that yielding is a material phenomenon independent of the co-ordinates chosen for its description. Macro-mechanical failure criteria for brittle and orientated materials are mostly stress-based. All predictions appear as average stresses within a single layer and serve as a building block for multi-laminate designs. They offer little information about the mechanism of failure, e.g. buckling of the fibres and delamination. Micro-mechanical criteria offer a more precise description of the fracture process but are dependent upon more constituent data that are prone to statistical variation.

References

[1] Von Mises, R., *Nachr. Ges. Wiss. Gottingen*, (1913), 582.
[2] Tresca, H., *Memoire sur l'ecoulement des corps solides*, Memoirs Par Divers Savants, Paris, **18** (1968), 733 and **20** (1972), 75.
[3] Ikegami, K., *J.Soc.Mat.Sci.* **24**(261), (1975), 491 and **24**(263), (1975), 709 (*BISI* translation 14420, *The Metals Soc.* Sept 1976)
[4] Michno, M. J., and Findley, W. N., *Int. J. Non-Linear Mech.* **11** (1976), 59.
[5] Shahabi, S. N. and Shelton, A. J., *Mech. Eng. Sci.* **17** (1975), 82.
[6] Johnson, K. R. and Sidebottom, O. M., *Expl Mech.* **12** (1972), 264.
[7] Moreton, D. N., Moffat, D. G. and Parkinson, D. B., *J.Strain Anal.* **16** (1981), 127.
[8] Miastkowski, J and Szczepinski,W., *Int. J. Solids and Structures*, **1** (1965), 189.
[9] Marin, J., Hu, L.W. and Hamburg, J. F., *Proc. A.S.M.* **45** (1953), 686.
[10] Lessels, J. M. and MacGregor, C. W., *Jl Franklin Inst.* **230** (1940), 163.
[11] Rogan, J. and Shelton, A., *J.Strain Anal.* **4** (1969), 127.
[12] Ivey, H. J., *J. Mech. Eng. Sci.* **3** (1961), 15.
[13] Phillips, A. and Tang.J-L., *Int.J.Solids and Struct.* **8** (1972), 463.
[14] Shiratori, E., Ikegami, K. and Kaneko, K., *Trans Japan Soc. Mech. Engrs*, **39** (1973), 458.
[15] Ellyin, F. and Grass, J-P., *Trans Can. Soc for Mech, Engrs*, **3** (1975), 156.
[16] Taylor, G. I. and Quinney, H., *Phil Trans Roy Soc.* **230A** (1931), 323.
[17] Crossland, B. *Proc.I.Mech.E.* **168** (1954), 935.
[18] Hu. L. W., Proc: *Naval Structures*, (eds Lee, E. H. and Symonds, P. S.) Pergamon, Oxford, U.K., 1960, 924.
[19] Pugh, H. Ll. D. and Green, D., *Proc. I. Mech.E.* **179** (1964), 1.
[20] Ratner, S. I., *Zeitschrift fur Technische Physik*, **19** (1949), 408.
[21] Fung, P. K., Burns, D. J. and Lind, N. C., Proc: *Foundations of Plasticity*, (ed A.Sawczuk) Noordhoff, Warsaw, 1973, 287.
[22] Ros, M. and Eichinger, A., *Eidgenoess, Materialpruef. Versuchanstalt Ind. Bauw. Gewerbe, Zurich*, **34** (1929), 1.
[23] Shrivastava, H. P., Mroz, Z. and Dubey, R. N., *ZAMM*, **53** (1973), 625.
[24] Rees, D. W. A., *Proc. Roy. Soc. Lond.* **A383** (1982), 333.
[25] Drucker, D. C. *J.Appl.Mech.* **16** (1949), 349.
[26] Betten, J., *Acta Mechanica*, **25** (1976), 79
[27] Freudenthall, A. M. and Gou, P. F., *Acta Mechanica*, **8** (1969), 34.

[28] Rees, D. W. A., *J. Appl Mech.* **49** (1982), 663.
[29] Hill, R., *Proc. Roy. Soc.* **A193** (1948), 281.
[30] Edelman, F. and Drucker, D. C., *J.Franklin Inst.* **251** (1951), 581.
[31] Olsak, W. and Urbanowski, W., *Arch of Mech.* **8** (1956), 671.
[32] Sobotka, Z., *ZAMM*, **49** (1969), 25.
[33] Troost, A. and Betten, J., *Mech. Res. Comms*, **1** (1974), 73.
[34] Fava, F. J., *Annals of the C.I.R.P.* **15** (1967), 411.
[35] Hu, L. W., *J. Appl Mech.* **23** (1956), 444.
[36] Hazlett, T. H., Robinson, A. T. and Dorn, J. E., *Trans ASME*, **42** (1950), 1326.
[37] Jones, S. E. and Gillis, P. P., *Met Trans A*, **15A** (1984), 129.
[38] Hu, L. W. and Marin, J. J., *Applied Mech.* **22** (1955), 77.
[39] Takeda, T. and Nasu, Y., *J. Strain Analysis*, **26** (1991), 47.
[40] Hill, R., *Math. Proc. Camb. Phil. Soc.* **85** (1979), 179.
[41] Bourne, L. and Hill. R., *Phil Mag.* **41** (1950) 671.
[42] Gotoh, M., *Int. J. Mech. Sci.* **19** (1977), 505.
[43] Stassi-d'Alia, F., *Meccanica*, **4** (1969), 349.
[44] Harvey, S. J., Adkin, P. and Jeans, P. J., *Fatigue of Eng Mats and Struct.* **6** (1983), 89.
[45] Dodd, B., *Int. J. Mech. Sci.* **12** (1984), 587.
[46] Parmar, A. and Mellor, P.B., *Int. J. Mech. Sci.* **20** (1978), 385.
[47] Rees, D. W. A., *Acta Mechanica*, **43** (1982), 223.
[48] Rees, D. W. A., *Acta Mechanica*, **52** (1984), 15.
[49] Mohr, O., *VDI Z*, **44** (1900), 1524, 1572 and **45** (1901), 740.
[50] Stassi-d'Alia, F., *Meccanica*, **2** (1967), 178.
[51] Coffin, L. F., *J. Applied Mech.* **17** (1950), 233.
[52] Grassi, R. C. and Cornet, I., *J. Applied Mech.* **16** (1949), 178.
[53] Alberti, N., *La Ricerca Scientifica*, Dec. 1960.
[54] Jacoby, H. S., *Structural details or elements of design in heavy framing*, Wiley, 1909.
[55] Howe, M. A., *Eng. News* (1912).
[56] Jenkins, C. F., *Materials of construction used in aircraft and aircraft engines*, Aero Res Comm, U.K., 1920.
[57] Stussi, F., *Schw Bauzeitung*, **128** (1946), 251.
[58] Hankinson, R. L., *Air Ser Inf Cir.* U.S.A, **III** (1921), 259.
[59] Osgood, W. R., *Engng News Record*, **100** (1928), 243.
[60] Kollmann, F., *Der Bauingenieur*, **15** (1934), 198.
[61] Griffith, J. E. and Baldwin, W. M., *Developments in Theoretical and Applied Mechanics*, **1** (1962), 410.
[62] Norris, C. B. and McKinnon, P. F., *Forrest products Lab Rpt.* 1328, (1946).
[63] Norris, C. B., Forrest Products Lab Rpt., 1816, (1950).
[64] Azzi, V. D. and Tsai, S. W., *Expl Mech.* **5** (1965), 283.
[65] Marin, J., *J.Aero Sci.* **24** (1957), 265.
[66] Hoffman, O., *J. Composite Mats*, **1** (1967), 200.
[67] Tsai, S. W. and Wu, E. M., *J. Composite Mats*, **5** (1971), 58.
[68] Waddoups, M. E., *Fort Worth Div. Rpt.* FZM 4763, (1967).
[69] Rees, D. W. A. and Li, Y. K., *Advances in Engineering Materials*, Transtech, **99-100** (1995), 51.
[70] Kaminski, B. E. and Lantz, R. B., *ASTM*, STP **460** (1969), 160.
[71] Franklin, H. G., *Fibre Sci and Tech.* **1** (1968), 137.

Bibliography

Friedrich, K., (ed) *Application of Fracture Mechanics to Composite Materials*, Elsevier, 1989.
Jones, R. M., *Mechanics of Composite Materials*, McGraw-Hill, 1975.
Matthews, F. L., (ed) *The Failure of Reinforced Plastics*, Mech Eng Pub Ltd, London, 1990.
Pagano, N. J., (ed) *Interlaminar Response of Composite Materials*, Elsevier, 1989.
Shah, S.P., Swartz,S.E. and Wang, M. L., (eds) *Micro-Mechanics of Failure of Quasi-Brittle Materials*, Elsevier, 1990.
Tsai, W. S. and Hahn, H. T., *Introduction to Composite Materials*, Technomic Publishing Co, 1980.

EXERCISES

11.1 A thick-walled closed cylinder is 25 mm i.d. and 50 mm o.d has a constant axial force of 20 kN and an internal pressure p. Find p by von Mises and Tresca to produce bore yielding. Take $Y = 240$ MPa.
Answer: 1038.3 bar, 900 bar.

11.2 A hollow stirrer rod with diameter ratio $d_o/d_i = 2$, absorbs 5 kW when it rotates at 500 rev/min under a compressive force of 6 kN. Find the diameters, by Mises and Tresca, of the rod using a safety factor of 2 and a tensile yield stress of rod material 200 MPa. If a radial pressure of p also acts upon the rod set up an equation in terms of the rod diameter from von Mises yield criterion.

11.3 A solid circular shaft 125 mm diameter rotates in bearings at 30 rad/s. Lateral loading produce a bending moment of 10 kNm. Given the tensile yield stress of shaft material as 300 MPa, determine the power that the shaft can transmit according to the Tresca and von Mises yield criteria.

11.4 A solid mild steel bar, 120 mm diameter, withstands a torque of 3 kNm, an axial force of 5 kN and a bending moment of 4 kNm. Establish whether the bar remain elastic under these loadings according to von Mises and Tresca, given a tensile yield strength for the bar of 300 MPa.

11.5 A thin walled square tube of mean side length 120 mm and 5 mm thickness withstands a torque of 4.5 kNm and an axial tensile force of 40 kN. Find the safety factor, by Mises and Tresca, that has been used against yielding, given that the yield stress of the tube material is 310 MPa. If, when carrying these loadings, the tube is mounted as a 3 m long cantilever, determine the additional uniformly distributed load, according to von Mises, that would cause the material to reach its yield point.
Answer: 5.47, 4.79, 5.92 kN/m.

11.6 A thin pipe with mean radius to thickness of 10 withstands an axial torque of 1 kNm and an internal pressure of 25 bar. Using a safety factor of 2.5 and a material yield stress of 200 MPa, calculate the tube thickness according to the Mises and Tresca yield criteria.
Answer: 1.61, 1.535 mm.

11.7 Show that the relationship between k and Y defines the constants c, b and d in eqs(11.16a,b and c) as (i) $Y = k\,(1/27 - 4c/9^3)^{1/6}$, (ii) $Y = 9k/(27 - 4b)^{1/2}$ and (iii) $Y = k/(1/3^{3/2} - 2d/27)^{1/3}$.

11.8 Normalise the yield criterion (11.29a) of Stassi with the compressive fracture stress σ_c and plot the family of loci in σ_1, σ_2 space for $\rho = \sigma_c/\sigma_t = 2$, 3 and ∞.

11.9 Construct the family of yield loci in σ, τ space from the unsymmetrical yield function $f = J_2' + (p/Y)\,J_3'$ for $1.5 \le p \le -3$, where Y is the tensile yield stress.

11.10 Cast iron has tensile and compressive strengths of $\sigma_t = 300$ MPa and $\sigma_c = -900$ MPa respectively. Find the combination of combined shear and compressive stress that would cause this material to fail: (i) when they are applied in the respective ratio of 1 to 4 and (i) when a shear stress of 75 MPa remains constant whilst a compressive stress is steadily increased.

11.11 Cast iron has tensile and compressive strengths of $\sigma_t = 350$ MPa and $\sigma_c = -1000$ MPa respectively. Find the combinations of principal stresses (σ_1, σ_2) that cause failure by Rankine and Mohr when (i) $\sigma_1 = \sigma_t/3$ and (ii) the maximum shear stress $\frac{1}{2}(\sigma_1 - \sigma_2) = 400$ MPa.

11.12 Which of the criteria in Tables 11.6 and 11.7 are capable of matching a unidirectional composite in which $\sigma_{1c} = 476.8$ MPa, $\sigma_{2c} = 136.3$ MPa, $\sigma_{1t} = 529$ MPa, $\sigma_{2t} = 49.2$ MPa?

11.13 Examine which of the criteria in Tables 11.6 and 11.7 are capable of matching the following tensile and compressive strengths for a bi-directional composite: $\sigma_{1c} = 339$ MPa, $\sigma_{2c} = 284$ MPa, $\sigma_{1t} = 275$ MPa, $\sigma_{2t} = 243$ MPa?

11.14 Reduce the tensor function prediction (11.41b) to a principal biaxial stress state for transversely isotropic and orthotropic materials.

11.15 Compare the predictions from exercise 11.14 with the corresponding failure criteria of Marin (11.39b) and Hoffman (11.40b).

11.16 How would you predict failure in a brittle unidirectional fibrous material when a direct stress σ is combined with a shear stress τ, where (i) both σ and τ are aligned with the fibres and (ii) they are inclined at θ to the fibres? Note, the presence of complementary shear in each case.

11.17 Figure 11.24 shows an off-axis shear test on a unidirectional composite. The shear stress τ is applied at θ to the fibre - 1 direction. Determine the plane stress state components σ_{11}, σ_{22} and σ_{12} shown in terms of τ and examine from eq(11.37a) the dependence of the off-axis shear strength in terms of θ given σ_{12f} for the $\theta = 0°$ orientation.

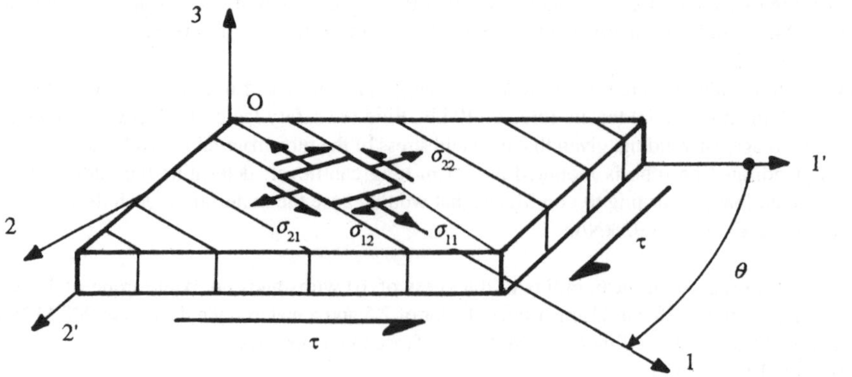

Figure 11.24

CHAPTER 12

PLASTICITY AND COLLAPSE

This chapter covers a number of aspects of deformation behaviour in metals that are not elastic. It includes the mechanics of plasticity and collapse for structures loaded beyond the elastic limit. Amongst these structures are beams, torsion bars, portal frames, the thick-walled cylinder and the rotating disc. The theory has been simplified by assuming an elastic-perfectly plastic material: an approximation for metallic materials that work harden. It is shown how uniaxial hardening is represented and incorporated into a flow rule for multiaxial stress states. This approach has been used to account for instability in pressure vessels.

12.1 Elastic-Plastic Bending of Beams

We can ensure a safe beam design when the working stress σ_w is derived from a safety factor S as follows:

$$S = Y/\sigma_w, \quad \therefore \quad \sigma_w = Y/S \tag{12.1}$$

The safe moment is then found from eqs(4.3) and (12.1):

$$M/I = \sigma_w/y, \quad \therefore \quad M = \sigma_w I / y = YI / (Sy) \tag{12.2}$$

When an applied moment reaches a limiting value M_Y, the most highly stressed fibres furthest from the neutral axis (n.a.) reach the yield stress while all other fibres in the cross-section remain elastic. As the moment is increased beyond M_Y the cross-section becomes partially plastic as the interior fibres approaching the n.a. successively reach the yield point. This state constitutes an elastic-plastic beam under an applied moment M_{ep}. When the plastic zone has penetrated the whole cross-section on the tensile and compressive sides of the n.a. This condition determines the ultimate moment M_P a given beam can withstand. Assuming that a beam of any section collapses under M_P and that the beam material behaves in an elastic-perfectly plastic manner, in which no increase in stress beyond Y occurs during plastic penetration, the following terms apply:
Shape Factor: $Q = M_P/M_Y > 1$, determined solely by the cross-section, i.e. independent of the applied loading.
Load Factor: $L = W_P/W$, the ratio between the corresponding collapse load W_P, and the safe elastic working load W. The latter depends upon the section, the applied loading and the manner of support.

12.1.1 Rectangular Section

Consider a beam with a rectangular section $b \times d$ in Fig. 12.1a under a hogging moment. The stress distributions for the transition from elastic to fully plastic behaviour are shown in Figs 12.1a - c.

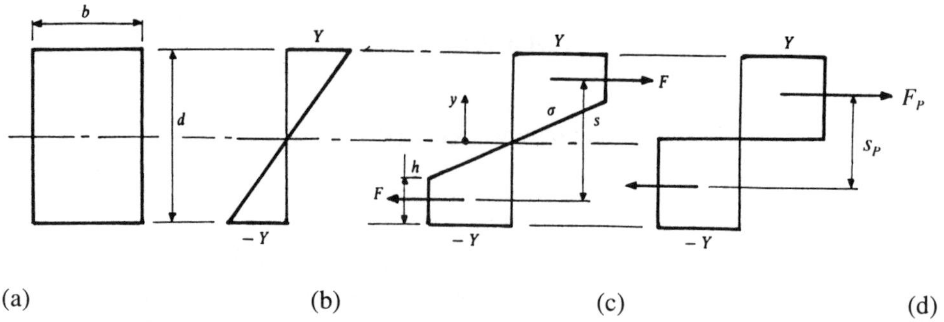

Figure 12.1 Penetration of plastic zone through a rectangular section

(a) Fully Elastic

Under M_Y in Fig. 12.1b the cross-section remains elastic and therefore the bending theory applies to the outer fibres where $\sigma = Y$, $y = d/2$. Equation (4.3) gives

$$M_Y = YI\,/\,y = Y\,(bd^3/12)/(d/2) = bd^2Y/6 \tag{12.3}$$
$$R_Y = Ey/\sigma = Ed/(2Y) \tag{12.4}$$

(b) Elastic-Plastic

Under M_{ep} (see Fig. 12.1c), where the plastic zone has penetrated by the amount h from top and bottom sides, the resistive moment exerted by the section is composed of elastic and plastic components such that

$$M_{ep} = M_e + M_p = YI_e/y + Fs \tag{12.5a}$$

where $s = d - h$ is the distance between the net force $F = Ybh$ in the plastic regions where $y = (d - 2h)/2$ defines the elastic-plastic boundary and $I_e = b(d - 2h)^3/12$ is the second moment of area of the elastic core. From eq(12.5a),

$$M_{ep} = 2Yb(d - 2h)^3/[12(d - 2h)] + (Ybh)(d - h)$$
$$= Y(bd^2/6)[1 + (2h/d)(1 - h/d)] \tag{12.5b}$$
$$M_{ep}/M_Y = 1 + (2h/d)(1 - h/d) \tag{12.5c}$$

The radius of curvature is found from applying bending theory to the edge of the elastic core, where $\sigma = Y$ for $y = (d - 2h)/2$. This gives

$$M_e/I_e = E/R_e = 2Y\,/\,(d - 2h)$$
$$R_e = E\,(d - 2h)/(2Y) \tag{12.6}$$

(c) Fully Plastic

Under M_P in Fig. 12.1d, the collapse moment of the fully plastic section is simply

$$M_P = F_P\,s_P = (bdY/2)(d/2) = bd^2Y/4 \tag{12.7}$$

The shape factor follows from eqs(12.3) and (12.7) as

$$Q = M_P/M_Y = (bd^2Y/4)/(bd^2Y/6) = 3/2 \tag{12.8}$$

12.1.2 Residual Bending Stresses

When the elastic-plastic beam (Fig. 12.1c) is fully unloaded from M_{ep}, a state of residual stress σ_R will remain in the cross-section after the elastic stresses σ_e have recovered. Assuming purely elastic springback in unloading from the elastic-plastic stress state, it follows that

$$\sigma_R = \sigma - \sigma_e \tag{12.9}$$

The elastic stress σ_e recovers over the whole depth $-d/2 \le y \le d/2$ in unloading from M_{ep} in eq(12.5a). Bending theory gives this as

$$\sigma_e = M_{ep}\, y/I = (M_{ep}/M_Y)(M_Y/Y)Y\, y/I$$

Substituting from eqs(12.3) and (12.5c),

$$\begin{aligned}\sigma_e &= (bd^2/6)[1 + (2h/d)(1 - h/d)][(12Y\, y/(bd^3)]\\ &= (2Yy/d)[1 + (2h/d)(1 - h/d)] \end{aligned}\tag{12.10}$$

Within the plastic zone: $\sigma = Y$ for y in the region $\pm(d/2 - h) \le y \le \pm d/2$. The residual stress in this zone follows from eqs(12.9 and 12.10):

$$\sigma_R = Y - (2Yy/d)[1 + (2h/d)(1 - h/d)] \tag{12.11}$$

Within the elastic zone for y in the region $0 \le y \le \pm(d/2 - h)$

$$\sigma = M_e\, y/I = (Yb/6)(d - 2h)^2 y \times 12/[b(d - 2h)^3] = 2Yy\,/\,(d - 2h) \tag{12.12}$$

which can be readily checked from similar triangles. From eqs(12.9, 12.10 and 12.12) the residual stress is

$$\sigma_R = 2Yy/(d - 2h) - (2Yy/d)[1 + (2h/d)(1 - h/d)] \tag{12.13}$$

Equations (12.11 and 12.13) reveal that σ_R varies linearly with y within each zone as shown in Fig. 12.2a. The construction in Fig. 12.2a is a graphical interpretation of eq(12.9) where σ_R is the difference between the horizontal ordinates. As M_{ep} is increased, σ_R approaches the stress distribution found by releasing a fully plastic moment. That is, $\sigma_R = Y(1 - 3y/d)$ is found by putting $h = d/2$ in eq(12.11). Figure 12.2b shows that the residual stress approaches the tensile yield stress Y at the centre ($y = 0$), and one half of Y at each edge ($y = \pm d/2$).

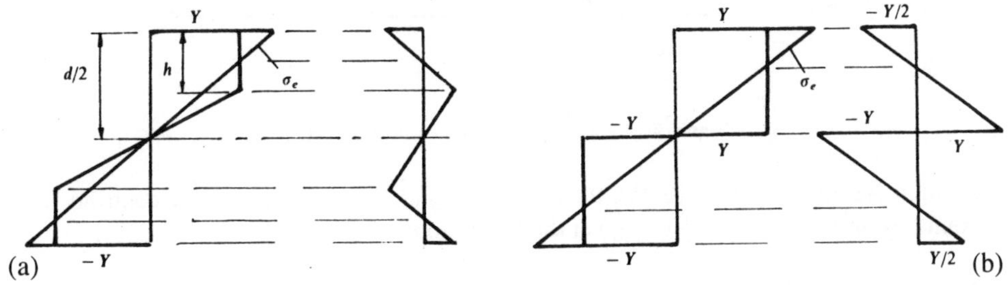

Figure 12.2 Residual stress distribution

12.1.3 Residual Curvature

The curvature remaining upon unloading is found from the residual stress eq(12.13) within the elastic core. The residual strain ε_R within any fibre $0 \le y \le (d/2 - h)$ in Fig. 12.2a can be written as

$$\varepsilon_R = y/R_R = \sigma_R/E$$

This gives

$$R_R = Ey/\sigma_R \tag{12.14a}$$

Since σ_R is proportional to y, we may set $y = (d - 2h)/2$ in eq(12.13) to give $\sigma_R = 2Y(h/d)^2 \times (3 - 2h/d)$. Substituting these into eq(12.14a) leads to

$$R_R = [Ed^3(1 - 2h/d)]/[4Yh^2(3 - 2h/d)] \tag{12.14b}$$

12.1.4 Non-Symmetric Sections

In a channel section (see Fig. 12.3a) the elastic n.a. does not lie at the central depth. Yielding under M_Y will begin first on the most severely stressed side (see Fig. 12.3b). Beyond this, as the bending moment (M_{ep}) is increased, the plastic zone will penetrate inwards from this side to a depth h as shown in Fig. 12.3c. In order that the horizontal tensile T, and compressive C, forces remain balanced, the stress distribution in Fig. 12.3c is accompanied by a shift in the n.a. to the new position $\bar{y}_{ep}$ such that $C = T$ under M_{ep}. As M_{ep} is further increased, the stress continues to redistribute, maintaining $C = T$ until full plasticity is reached under M_P in Fig. 12.3d when the n.a. divides the section area equally.

Example 12.1 Calculate M_Y and M_P for the inverted channel section in Fig. 12.4a, taking $Y = 235.5$ MPa. To what is the depth h has the plastic zone has penetrated the vertical webs when the flange surface first yields? Find the corresponding moment in this case.

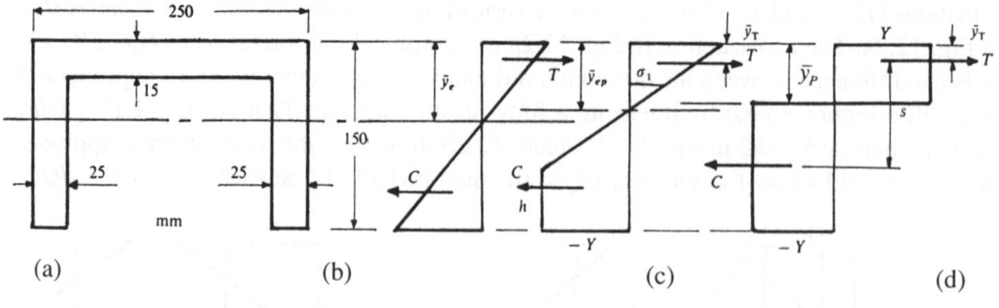

Figure 12.3 Plastic penetration in a channel section

For initial yielding (see Fig. 12.3b), the n.a. position ($\bar{y}_e$) passes through the centroid. This is found by applying $A\bar{y}_e = \sum(A_i\, y_i)$ at the flange top:

$$[(250 \times 15) + 2(135 \times 25)]\, \bar{y}_e = (250 \times 15 \times 7.5) + 2(135 \times 25 \times 82.5), \Rightarrow \bar{y}_e = 55.7 \text{ mm}$$

$\therefore I = \frac{1}{3}\,[250(55.7)^3 + 2 \times 25(94.3)^3 - 200(40.7)^3] = 23.88 \times 10^6\ \text{mm}^4$

Yielding commences at the web bottom where the corresponding moment is

$M_Y = YI / y = 235.5 \times 23.88 \times 10^6/(150 - 55.7) = 59.64$ kNm

For full plasticity in Fig. 12.4d, the n.a. ($\bar{y}_p$) divides the area:

$2 \times 25(150 - \bar{y}_p) = (250 \times 15) + 2\,(\bar{y}_p - 15)25, \Rightarrow \bar{y}_p = 45$ mm

The ultimate moment is found from $M_P = Cs = Ts$, where the equivalent force T acts at the centroid $\bar{y}_T$ of the tensile area in Fig. 12.3d. Taking moments about the top surface,

$[(250 \times 15) + (2 \times 30 \times 25)]\,\bar{y}_T = (250 \times 15)7.5 + 2\,(30 \times 25)30$

This gives $\bar{y}_T = 13.93$ mm. Similarly C acts at the centroid of the lower web giving the moment arm length s as

$s = (150 - 45)/2 + (45 - 13.93) = 83.57$ mm

$\therefore\ M_P = Cs = 25 \times 105 \times 2 \times 235.5 \times 83.57 = 103.32$ kNm

When the flange yields an elastic-plastic condition applies (see Fig. 12.4c). The n.a. position ($\bar{y}_{ep}$) is again found from $T = C$. Let σ_1 act at the flange bottom as shown. This gives

$(Y + \sigma_1)(250 \times 15)/2 + 2\sigma_1(\bar{y}_{ep} - 15)25/2 = (2 \times 25 \times Yh) + 2(150 - \bar{y}_{ep} - h)25Y/2$

Substituting $\sigma_1 = Y(1 - 15/\bar{y}_{ep})$ and $h = 150 - 2\,\bar{y}_{ep}$

$1875\,(2\bar{y}_{ep} - 15) + 25\,(\bar{y}_{ep}^2 - 30\bar{y}_{ep} + 225) = 50\,\bar{y}_{ep}(150 - 2\bar{y}_{ep}) + 25\bar{y}_{ep}^2$

$\bar{y}_{ep}^2 - 45\bar{y}_{ep} - 225 = 0, \Rightarrow \bar{y}_{ep} = 49.54$ mm

from which $h = 150 - 99.08 = 50.92$ mm. T acts at the centroid of the area above the n.a. in Fig. 12.3c. Let this lie at distance $\bar{y}_T$ from the top surface:

$[(250 \times 15) + (2 \times 34.54 \times 25)]\bar{y}_T = (250 \times 15)7.5 + 2(34.54 \times 25)(17.27 + 15)$

from which $\bar{y}_T = 15.31$ mm. Taking moments about the T force line, M_{ep} is the sum of two components: (i) due to the plastic stress distribution

$M_p = 2[235.5 \times 25 \times 50.92(50 - 15.31 - 50.92/2)] = 65.49$ kNm

and (ii) due to the elastic stress distribution

$M_e = 2 \times 235.5 \times 49.54 \times (25/2)(99.08 - 15.31 - 49.54/3) - (235.5/2)$
$\times [(250 \times 15) + 2(49.54 - 15)25](16.51 - 15.31) = 19.617 - 0.774 = 18.84$ kNm

$\therefore M_{ep} = M_e + M_p$
$= 65.49 + 18.84 = 84.33$ kNm

12.2 Plastic Collapse of Beams and Frames

12.2.1 Simply Supported Beams

(a) Central Concentrated Load

In Fig. 12.4 the maximum, central bending moment is $M = Wl/4$. Using eq(12.1), the safe working load in a rectangular section becomes

$$W = 4M/\,l = 4YI\,/\,(lS\,y) = 2Ybd^2\,/\,(3lS\,)$$

The ultimate load and the load factor are defined from eqs(2.2) and (12.7) as

$$W_P = 4M_P/l = bd^2Y/\,l$$
$$\therefore\; l = W_P/W = (bd^2Y\,/\,l\,)/\;[2Ybd^2/(3lS\,)] = QS$$

where $Q = 3/2$ is the shape factor. Thus, if a safety factor of $S = 2$ is employed, the central load is safe from collapse by a factor of $3/2 \times 2 = 3$.

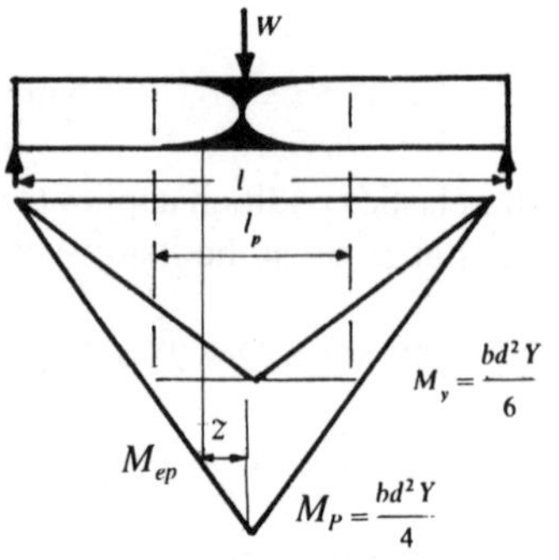

Figure 12.4 Plastic hinge under ultimate moment

At collapse, the plastic hinge length l_p is determined from the M - diagram (see Fig. 12.4). Beneath the load $M = M_P$ and at the extremities of the hinge $M = M_Y$. Where plasticity has penetrated to a depth h at a distance z from the centre, the moment at this position is elastic-plastic (M_{ep}). From similar triangles within the M - diagram,

$$2M_P/l = 2(M_P - M_Y)/l_p = (M_P - M_{ep})/\,z \qquad (12.15a)$$

The first and second expressions in eq(12.15a) reveal the hinge length

$$l_p = l\,(1 - M_Y/M_P) = l\,(1 - 1/Q) \qquad (12.15b)$$

Equation (12.15b) shows that for a beam with rectangular cross-section, with $Q = 3/2$, the central hinge extends over one third of the length ($l_p/\,l = 1/3$). The shape of the hinge follows from the first and third expressions in eq(12.15a):

$$z\,/\,l = \tfrac{1}{2}\,(1 - M_{ep}/M_P) \qquad (12.15c)$$

Substituting eqs(12.5b) and (12.8) into eq(12.15c), we find $z\,/\,l = (1/6)(1 - 2h/d)^2$, which shows that the hinge has a parabolic profile as shown.

(b) Uniformly Distributed Loading

Figure 12.5 shows the moment diagram for which the maximum moment is $-wl^2/8$. This beam will collapse under a uniformly distributed loading, $w_P = 8M_P/l^2$, so that if the section is rectangular we have from eq(12.7) $w_P = 2bd^2Y/l^2$.

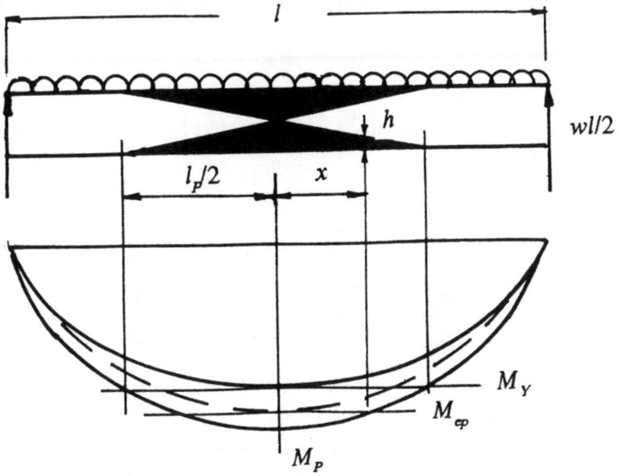

Figure 12.5 Hinge geometry for beam with uniformly distributed loading

The moment diagram shows the three ordinates, M_P, M_{ep} and M_Y, that define the shape and spread of the central plastic hinge. The moment expression is required in terms of a distance z from the centre

$$M = (w_p/2)(l/2 - z)(l/2 - z) - (w_p l/2)(l/2 - z)$$
$$= -(w_p/2)(l^2/4 - z^2) \qquad (12.16)$$

At the edges of the hinge, where $z = l_p/2$, eq(12.16) gives the fully elastic moment M_Y:

$$M_Y = -(w_p l^2/8)[1 - (l_p/l)^2] = M_P(1 - (l_p/l)^2]$$
$$l_p/l = \sqrt{(1 - M_Y/M_P)} = \sqrt{(1 - 1/Q)}$$

Taking $Q = 3/2$ gives $l_p = l/\sqrt{3}$, which is almost twice the length of the hinge for the simply supported beam in (a). Applying eq(12.16) within the hinge length $0 \le z \le l_p/2$, we may write

$$M_{ep} = -(w_p l^2/8)(1 - 4z^2/l^2) = M_P(1 - 4z^2/l^2)$$

Writing $M_{ep}/M_P = (M_{ep}/M_Y)(M_Y/M_P)$ and substituting from eqs(12.5c) and (12.8) for a rectangular section,

$$(2/3)[1 + (2h/d)(1 - h/d)] = 1 - 4z^2/l^2$$

from which z may be expressed in terms of h:

$$z/l = (1 - 2h/d)/(2\sqrt{3}) \qquad (12.17)$$

Equation (12.17) shows that the hinge profile is a straight line.

12.2.2 Encastre Beams

Three plastic hinges are required for the collapse of an encastre beam. This means that the bending moment diagram is no longer similar in shape to that for elastic working moments. When M_P is first reached in the most severely stressed section one hinge forms. Other hinges necessary for collapse form instantaneously as the stress redistributes to attain M_P in other less highly stressed sections.

(a) Single Concentrated Force
The encastre beam in Fig. 12.6a carries a concentrated load W that divides the length l into lengths a and b as shown.

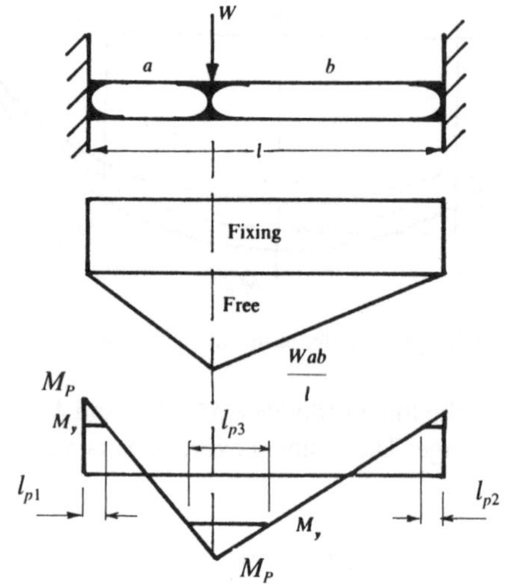

Figure 12.6 Collapse of an encastre beam

We wish to determine the collapse load and the geometry of the plastic hinges. The net moment diagram, in Fig. 12.6b, is constructed from Mohr's theorem (6.2b). Since there is no change in slope between the ends $\delta\theta = 0$ and so the areas of the free- and fixing-moment diagrams are equal. This results in differing net moments at the ends and centre. The mechanism necessary for plastic collapse depends upon M_P being reached together at each of these positions by stress redistribution. Equating moments at the end to the net moment at the load point in Fig. 12.6b gives:

M_P (ends) = [M (free) - M (fixing)] at W

$$\therefore M_P = Wab/l - M_P \tag{12.18}$$

From eq(12.18) the collapse load is $W = (2M_P l)/(ab)$, where M_P is determined separately for the given section. For example, a rectangular section has $M_P = bd^2Y/4$ giving a collapse load $W = bd^2Yl/(2ab)$. The length of the plastic hinges is found by marking off M_E at the ends and centre in Fig. 12.6c. At the left hand-end, by similar triangles:

$$(M_P - M_Y)/l_{p1} = 2M_P/a$$

$$\therefore l_{p1} = (a/2)(1 - M_Y/M_P) = (a/2)(1 - 1/Q) \tag{12.19a}$$

At the right-hand end:

$$(M_P - M_Y)/\,l_{p2} = 2M_P/b$$
$$l_{p2} = (b/2)(1 - M_Y/M_P) = (b/2)(1 - 1/Q) \qquad (12.19b)$$

At the centre:

$$l_{p3} = l_{p1} + l_{p2} = ½\,(a + b)(1 - M_Y/M_P) = (l/2)(1 - 1/Q) \qquad (12.19c)$$

We may now deduce from eqs(12.19a,b,c) the hinge lengths when the load is central. Setting $a = b = l/2$ for a central load and $Q = 3/2$ for a rectangular section (eq6.52), the hinge lengths become $l/12$ and $l/6$ at the ends and centre respectively.

(b) Uniformly Distributed Loading

Firstly, we find the uniformly distributed loading w corresponding to M_Y for the encastre rectangular section beam in Fig. 12.7a.

(a)

(b)

(c)

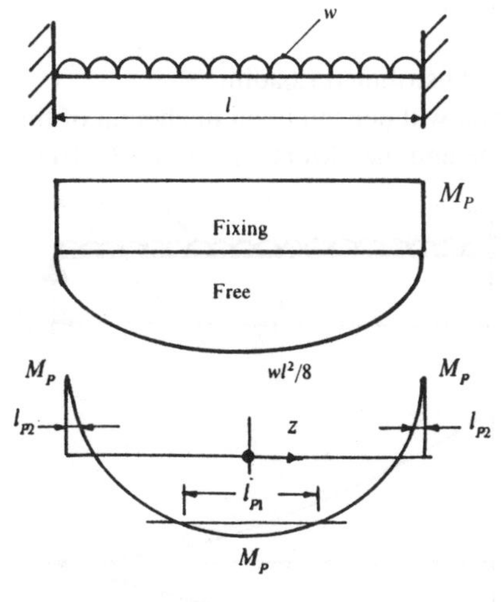

Figure 12.7 Encastre beam with uniformly distibuted loading

The net moment is greatest at the ends and there the initial yield moment will be reached first. Equating the areas of the free and fixing moment diagrams in Fig. 12.7b (see eq(6.2b)),

$$M_Y\, l = (2l/3)(wl^2/8)$$

which gives $M_Y = wl^2/12$. From eq(12.7), for a rectangular section,

$$M_Y = bd^2Y/6 = wl^2/12$$
$$\therefore\ w = 2bd^2Y/l^2$$

Collapse occurs as M_P is achieved simultaneously at the ends and the centre. From the free and fixing moment diagrams in Fig. 12.7b,

$$M(\text{ends}) = \text{net } M \text{ at centre}$$
$$M_P = wl^2/8 - M_P, \quad \Rightarrow \quad M_P = wl^2/16$$

The loading to collapse the beam follows from eq(12.7)

$M_P = bd^2Y/4 = w_p l^2/16, \Rightarrow \quad w_p = 4bd^2Y/l^2$

To estimate the hinge lengths at the centre and the ends take a z - co-ordinate from the centre. The net moment diagram at the point of collapse is expressed as

$$M = -M_P + 8M_P(z/l)^2 = -M_P[1 - 8(z/l)^2] \tag{12.20}$$

which gives $M = -M_P$ for $z = 0$ and $M = M_P$ for $z = \pm l/2$ as expected. Setting $M = M_Y$ and $z = l_{p1}/2$ in eq(12.20) gives the central hinge length

$$l_{p1}/l = \tfrac{1}{2}(1 - 1/Q) \tag{12.21a}$$

For the end hinge length set $M = M_Y$ and $z = l/2 - l_{p2}$

$$l_{p2}/l = \tfrac{1}{2}[1 - (1/\sqrt{2})(1 + 1/Q)^{1/2}] \tag{12.21b}$$

Setting $Q = 3/2$ in eqs(12.21a,b) shows that the hinge lengths for a rectangular cross-section are $l_{p1}/l = 1/6$ and $l_{p2}/l = 0.044$.

(c) Propped Cantilever with Distributed Loading

The maximum elastic moment will depend upon the height of the prop (see Fig. 12.8a). If it produces no deflection at the end then Mohr's theorem (6.2b) provides the prop reaction.

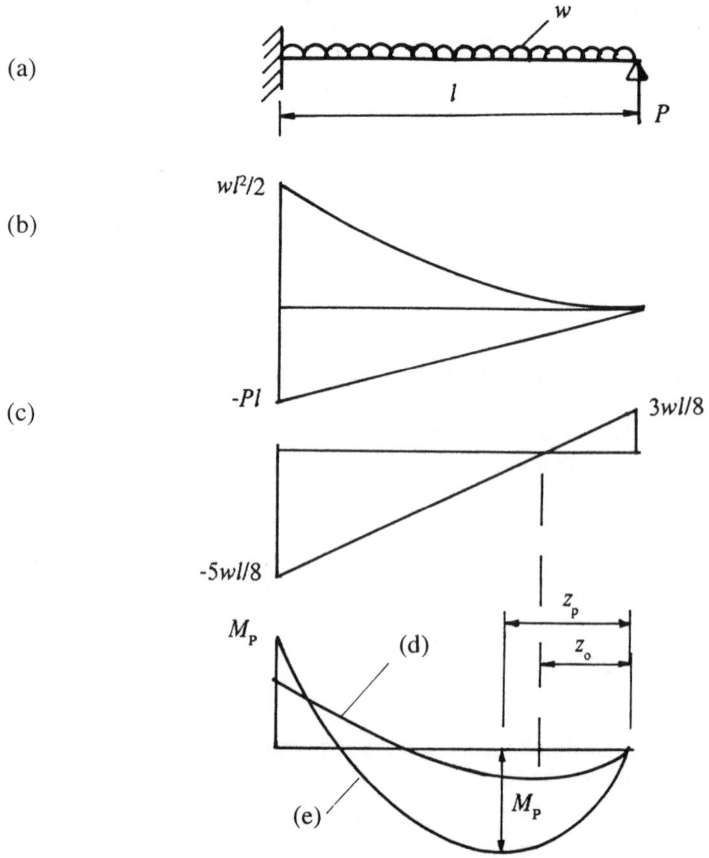

Figure 12.8 Force and moment diagrams for a propped cantilever

From Fig.12.8b:

Moment of M - Diag due to w about P - moment of M - Diag due to P about $P = 0$

$(wl^2/2)\,(l/3)\,(3l/4) - (Pl)\,(l/2)\,(2l/3) = 0$

from which $P = 3wl/8$. The maximum moment occurs at the fixed end

$M_{\max} = -\,Pl + wl^2/2 = wl^2/8$

When the beam is at the yield point, $M_{\max} = M_Y$ under a distributed loading $w = 8M_Y/\,l^2$. A local maximum occurs at a position z_o of zero shear force (see Fig. 12.8c). That is

$3wl/8 - z_o\, w = 0$

from which $z_o = 3l/8$ and in Fig. 12.8d the net moment at this position is

$M_o = -\,(3l/8)\,(3wl/8) + (3l/8)\,w\,(3l/16) = -\,9wl^2/128$

Figure 12.8d shows the net elastic moment diagram and Fig. 12.8e sdhows how the diagram is altered to accommodate two plastic hinges for collapse. A second hinge forms in the region where the elastic moment is a local maximum. At each hinge the fully plastic moment M_P is reached. Equating the numerical values of M_P at the z_o position to M_P at the fixed end,

$|\,M_P| = -\,3lP/8 + (3lw/8)(3l/16) = -\,(\,-\,Pl + wl^2/2)$

from which $P = 73wl/176$ and $M_P = \pm\, 15wl^2/176$. This gives collapse under a distributed loading of $w_p = 11.73M_P/\,l^2$. This solution is approximate. The true solution ensures that M_P occurs at a slightly modified position z_o' of zero shear force (see Fig. 12.8e). This gives

$$-\,P + z_o'\,w = 0 \qquad (12.22a)$$
$$|\,M_P| = -\,z_o'\,P + w\,(\,z_o'\,)^2/2 = -\,(-\,Pl + wl^2/2) \qquad (12.22b)$$

Combining eqs(12.22a.b) leads to a quadratic

$P^2 + 2Pwl - (wl)^2 = 0$

from which $P = (\sqrt{2} - 1)wl,\ z_o' = (\sqrt{2} - 1)\,l$ and, from eq(12.22b),

$M_P = \pm\,(-\,Pl + wl^2/2) = \pm\,(\sqrt{2} - 1.5)\,wl^2$

and therefore the true collapse loading is $w_p = 11.66M_P/\,l^2$.

12.2.3 Effect of Axial Loading Upon Collapse

(a) Rectangular Section $b \times d$

When an axial load is applied to rectangular section beam the moment carrying capacity is reduced. A tensile (or a compressive) force P will modify the fully plastic stress distribution to that shown in Fig. 12.9a.

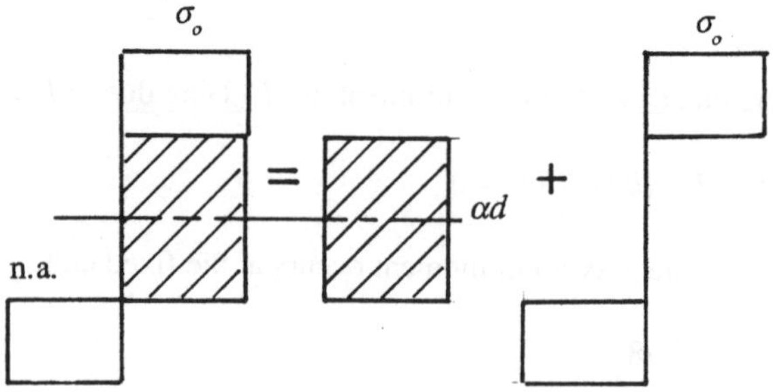

(a) (b) (c)

Figure 12.9 Stress distribution under combined tension and bending

It is convenient to split this into the sum of two components and use a multiplication factor α to locate the n.a. (see Figs 12.9b,c). From these moment- and force-carrying capacities are

$$M = Yb \times ½\,(d - \alpha d) \times ½\,(d + \alpha d) = (Ybd^2/4)(1 - \alpha^2) = M_P(1 - \alpha^2) \quad (12.23a)$$
$$P = Yb\alpha d = \alpha P_P \quad (12.23b)$$

where M_P (see eq(12.7)) and P_P refer to collapse under their separate actions. Equation 12.23b gives the shift in the neutral axis as $\alpha d/2 = (P/P_P)(d/2)$. Combining eqs 12.33a,b leads to an interaction equation

$$M/M_P = 1 - (P/P_P)^2 \quad (12.23c)$$

Equation (12.23c) provides a collapse moment M required fo collapse in the presence of any P value (tensile or compressive).

(b) I-Section

For the I beam in Fig. 12.10a the neutral axis may lie in the web or the flange depending upon the magnitide of P. The former case is shown for which the moment capacity arises from the flanges and some of the web (shaded areas in Fig.12.10a).

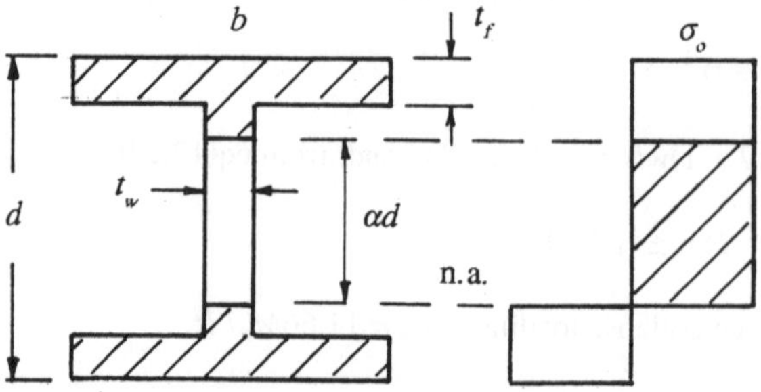

(a) (b)

Figure 12.10 Neutral axis in web

Taking this capacity as the sum of the moments from the flange and the web,

$$M = Y\,[bt_f\,(d - t_f) + t_w(d/2 - t_f - \alpha d/2)(d/2 - t_f + \alpha d/2)] \quad (12.24a)$$

The force capacity applies to the shaded rectangle in Fig. 12.10b,

$$P = \alpha d t_w Y \tag{12.24b}$$

from which the neutral axis shift is $\alpha d/2 = P/(2t_w Y)$. When P is absent $\alpha = 0$ and the fully plastic moment is

$$M_P = Y[bt_f(d - t_f) + (t_w/4)(d - 2t_f)^2] \tag{12.24c}$$

Combining eqs(12.24a,b,c) gives the interaction equation

$$M = M_P - P^2/(4Y t_w) \tag{12.24d}$$

Now let the neutral axis lie in the flange, the shift being αt_f from the bottom surface as shown in Fig. 12.11a.

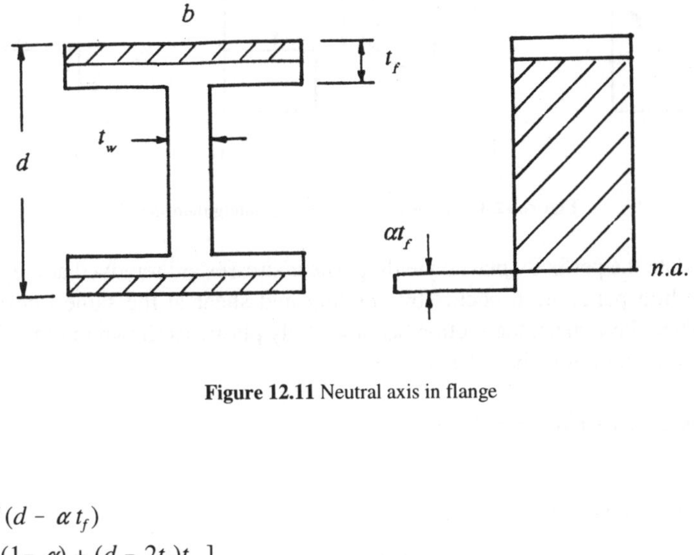

Figure 12.11 Neutral axis in flange

We have

$$M = b\alpha t_f Y (d - \alpha t_f) \tag{12.25a}$$
$$P = Y[2bt_f(1 - \alpha) + (d - 2t_f)t_w] \tag{12.25b}$$

From eq(12.25b) the position of the n.a. is

$$\alpha t_f = (P_P - P)/(2bY) \tag{12.25c}$$

where $P_P = [2bt_f + (d - 2t_f)\, t_w]Y$ is the fully plastic tensile force. Combining eqs(12.25a,b,c) leads to the interaction equation

$$M = (P_P - P)d/2 - (P_P - P)^2/(4bY) \tag{12.25d}$$

Equations (12.25d) may be applied to determine M in the presence of P provided eq(12.25c) shows that the position of the n.a. gives $\alpha < 1$. If it is found that $\alpha > 1$ then eq(12.24d) must be used.

12.2.4 Effect of Shear Force Upon Collapse

Up to now we have ignored the presence of shear force that may accompany bending of a beam. Depending upon the position in the length of a beam, there are two ways in which parabolic shear stress distribution across the section can interact with the linear bending stress distribution. While M is dominant in long beams, shear becomes important in shorter beams.

(a) Initial Yielding by Bending

In Fig. 12.12a plastic penetration has occurred in a rectangular section to depth $\alpha_o d/2$ by bending just as the shear yield stress k is reached at the neutral axis (see Fig. 12.12b).

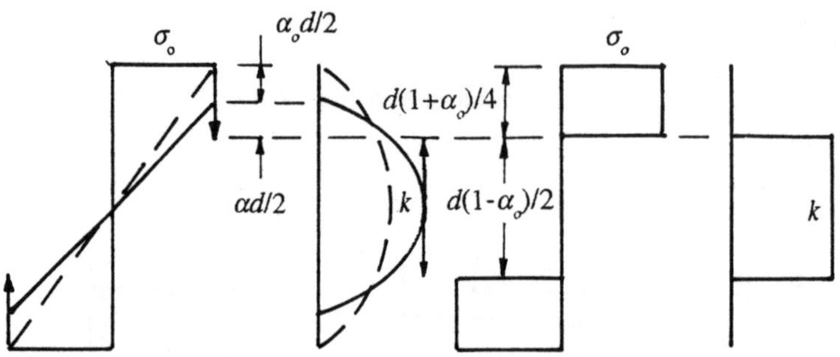

(a) (b) (c) (d)

Figure 12.12 Plastic penetration by bending then shear

To accommodate the perfectly plastic yielding, shear stresses are removed at the edges. With increased loading penetration occurs by bending and shear at the same rate in opposing directions. When these meet, the section becomes fully plastic as shown in Figs 12.12c,d. Let the subsequent penetrations be $\alpha d/2$, so that

$$2[(\alpha d/2) + (\alpha d/2) + (\alpha_o d/2)] = d$$
$$2\alpha + \alpha_o = 1 \tag{12.26a}$$

The plastic moment is, from Fig. 12.12c,

$$\begin{aligned} M &= Yb(d/4)(1 + \alpha_o)[(d/2)(1 - \alpha_o) + (d/4))(1 + \alpha_o)] \\ &= ¼\,(Ybd^2/4)(1 + \alpha_o)(3 - \alpha_o) \\ &= (M_P/4)(1 + \alpha_o)(3 - \alpha_o) \end{aligned} \tag{12.26b}$$

The plastic shear force is, from Fig. 12.12d,

$$\begin{aligned} F &= kb(d/2)(1 - \alpha_o) \\ &= (F_P/2)(1 - \alpha_o) \end{aligned} \tag{12.26c}$$

We find α_o from eq(12.26c) as

$$\alpha_o = 1 - 2F/F_P \tag{12.26d}$$

and eliminating α_o between eqs(12.26b,c) leads to the interaction equation

$$M/M_P = 1 - (F/F_P)^2 \tag{12.26e}$$

(b) Initial Yielding by Shear

In Fig. 12.13a plastic penetration has occurred to depth $\alpha_o d/2$ by shear on either side of the neutral axis before the axial yield stress Y is reached through bending at the edges (Fig. 12.13b).

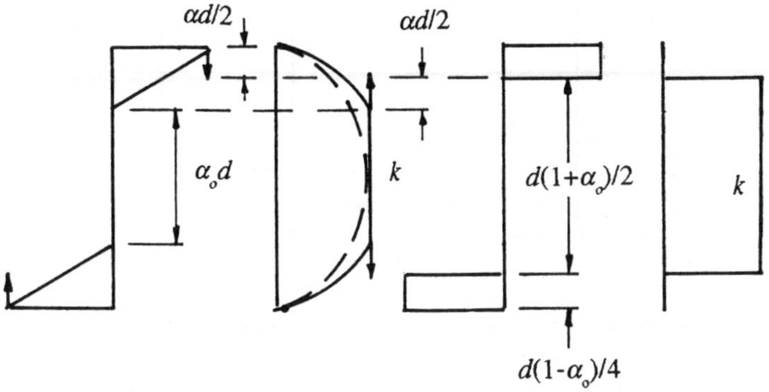

Figure 12.13 Plastic penetration by shear then bending

To accommodate initial yielding, the elastic bending stresses are removed within the penetration zone. With increased loading penetration occurs by bending and shear at the same rate in opposing directions. When these meet the section becomes fully plastic as shown in Fig. 12.13c,d. Let the subsequent penetrations be $\alpha d/2$ so that $2\alpha + \alpha_o = 1$. The plastic moment is, from Fig. 12.13c,d,

$$M = Yb\,(d/4)(1 - \alpha_o)[(d/2)(1 + \alpha_o) + (d/4)(1 - \alpha_o)]$$
$$= (M_P/4)(1 - \alpha_o)(3 + \alpha_o) \qquad (12.27a)$$

The plastic shear force is, from Fig. 12.13d,

$$F = kb(d/2)(1 + \alpha_o)$$
$$= (F_P/2)(1 + \alpha_o) \qquad (12.27b)$$

We find α_o from eq(12.27b) as

$$\alpha_o = 2F/F_P - 1 \qquad (12.27c)$$

and eliminating α_o between eqs(12.27a,b) leads to a similar interaction equation

$$M/M_P = 1 - (F/F_P)^2 \qquad (12.27d)$$

The spread of plasticity, by the simultaneous actions of bending and shear in each of Figs 12.12a,b and Figs 12.13a,b is confined to a rectangle breadth b and depth $(1 - \alpha_o)d$. Figures 12.14a,b show a partial spread of plasticity to a depth $\alpha d/2$ by bending and shear within this rectangle. Note that the beam depth is $(2\alpha + \beta + \alpha_o)d = d$ and therefore $\beta = 1 - \alpha_o - 2\alpha$. In the region defind by $\alpha d/2 \le y \le (1 - \alpha - \alpha_o)d/2$ there exists a linear bending stress in 12.14a and a parabolic shear stress in Fig. 12.14b. These are respectively

$$\sigma/Y = (2y/d - \alpha)/(1 - \alpha_o - 2\alpha) \qquad (12.28a)$$
$$\tau/k = [(1 - \alpha - \alpha_o)^2 - (2y/d)^2]/[(1 - \alpha_o)(1 - 2\alpha - \alpha_o)] \qquad (12.28b)$$

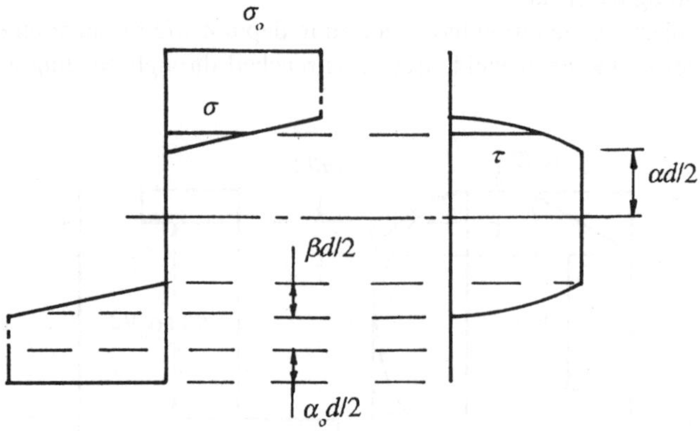

(a) (b)

Figure 12.14 Elastic-plastic section under bending and shear

The extent to which this region remains unyielded may be examined from the von Mises yield criterion (see eq(11.8)):

$$(\sigma / Y)^2 + (\tau / k)^2 = 1 \tag{12.28c}$$

where $Y^2 = 3k^2$. Substituting eqs(12.28a,b) into eq(12.28c), the left-hand side may be evaluated as an effective stress for given values of α and α_o. For example, if we take $\alpha_o = 0$ when initial yielding in bending and shear occurs simultaneously, then $0 \le \alpha \le 0.5$ during penetration. Figures 12.15a-c show the effective stress distribution for (a) initial yielding $\alpha = 0$, (b) an elastic-plastic section where $\alpha = 0.3$ and (c) the collapse condition when $\alpha = 0.5$.

(a) (b) (c)

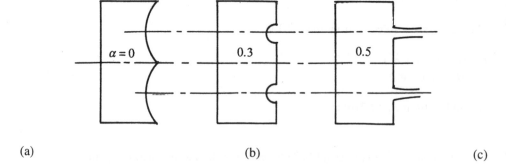

Figure 12.15 Effective stress in rectangular section

12.2.5 Effect of Hardening upon Collapse

The elastic-perfectly plastic analyses above may be refined further when an allowance is made for work-hardening in the plastic region [1]. The simplest linear hardening material has a constant rate of hardening $n = d\sigma/d\varepsilon$ which may be used to define the gradient of the penetration zones for both partial and full penetration of a rectangular beam cross-section $b \times d$ (see Figs 12.16a,b).

(a) 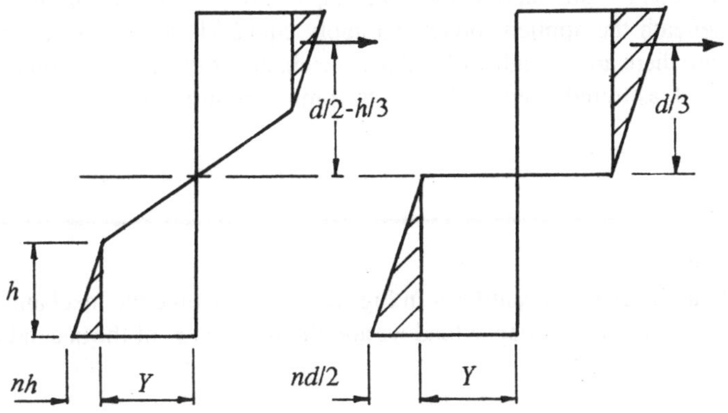 (b)

Figure 12.16 Account of hardening on plastic stress distribution

We see that an additional term arises from the shaded triangle in respect of (a) the elastic-plastic and (b) fully plastic moment. That is, we must add to eqs(12.5b) and (12.7) the respective moments

$$2(bh)(nh/2)(d/2 - h/3) = nbh^2(3d - 2h)/6$$
$$2(bd/2)(dn/4)(d/3) = nbd^3/12$$

It is appropriate to account for the hardening accompanying partial penetration of annealed metals. Alternatively, if only the collapse condition is required, a closer estimate of the fully plastic moment would be found when the ultimate strength replaces the yield stress in the foregoing equations.

12.3 Collapse of Structures

12.3.1 Principal of Virtual Work

A simple method to obtain the collapse loading of beams and plane frames is to treat collapse as a rigid mechanism under an equilibrium system of co-planar, non-concurrent forces. The equibrium conditions are

$$\sum F_i = 0 \text{ and } \sum M_i = 0 \qquad (12.29a,b)$$

where the moments must sum to zero at any two arbitrary points. Since the displacements and rotations occurring in a collapsing structure do not alter the equilibrium conditions (12.29a,b), we say that they are virtual, written as Δ^v and θ^v respectively. It follows that the equilibrium system of real forces and moments forces will do zero virtual work. That is,

$$\sum F_i \Delta_i^v = 0 \text{ and } \sum M_i \theta_i^v = 0 \qquad (12.30a,b)$$

Combining eqs(12.30a,b) leads to the useful forms

$$\sum F_i \Delta_i^v = \sum M_i \theta_i^v = 0 \qquad (12.31)$$

In eq(12.31) θ_i refers to the rotation under a collapse moment M_i at each hinge and Δ_i to the deflections beneath the applied forces. To apply eq(12.31) a collapse geometry must be assumed. Where there are a number of collapse mechanisms the true solution is that for which the collapse load, as found from eq(12.31), becomes a minimum.

12.3.2 Beams

(a) Propped Cantilever

Looking again at the propped cantilever in Fig. 12.8a, we assume the mechanism of collapse shown in Fig. 12.17a. The collapse load w_P and the position z_o of the second hinge is to be determined.

(a)

(b)

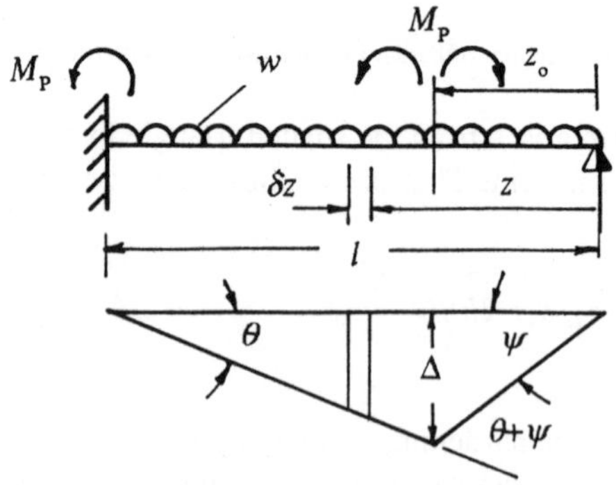

Figure 12.17 Collapse mechanism in a propped cantilever

Applying eq(12.31), omitting the superscript v for simplicity,

$$M_P\,\theta + M_P\,(\theta+\psi) = w_P \int v\,\mathrm{d}z \qquad (12.32)$$

where v is the deflection at position z and w dz is the elemental force at this position. The integral in eq(12.32) is clearly the enclosed area

$$\int v\,\mathrm{d}z = l\Delta/2 \qquad (12.33a)$$

The rotation ψ is found from simple geometry:

$$\Delta = (l - z_o)\theta = z_o\,\psi \qquad (12.33b)$$

which gives

$$\psi = (l - z_o)\theta / z_o \qquad (12.33c)$$

Substituting eqs(12.33a,b,c) into eq(12.32) leads to

$$w_P = 2M_P\,(l + z_o)/[\,l\,z_o(l - z_o)] \qquad (12.34)$$

To find the z_o value which minimises eq(12.34) we set $\mathrm{d}w_P/\mathrm{d}z_o = 0$. This gives

$$z_o^2 + 2\,z_o l - l^2 = 0$$

for which $z_o = (\sqrt{2} - 1)l = 0.4141l$. Then, from eq(12.34), $w_P = 11.66M_P/\,l^2$ as before.

(b) Continuous Beam

Figure12.18a shows a two-bay beam fixed at one end and resting upon two simple supports. The plastic collapse moment for the second bay is twice that of the first. If we wish to determine the collapse load W then all possible modes of failure must be considered (see Figs 12.18b - d).

(a)

(b)

(c)

(d)

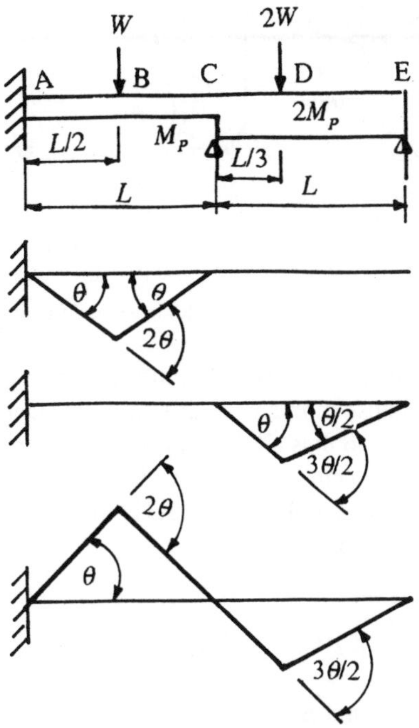

Figure 12.18 Collapse modes in a continuous beam

Apply eq(12.31) to each mode in turn. In Fig. 12.18b M_P is reached at A, B and C

$$M_P\,\theta + M_P(2\theta) + M_P\,\theta = \mathrm{W}\,(L/2)\theta \;\Rightarrow\; W = 8M_P/L \tag{12.35a}$$

In Fig. 12.18c M_P is reached in a hinge to the left of C and $2M_P$ must be reached for collapse at D. The rotation at E occurs in the absence of a collapse moment. This gives

$$M_P\theta + 2M_P\,(3\theta/2) = 2W\,(L/3)\theta \;\Rightarrow\; W = 6M_P/L \tag{12.35b}$$

Fig. 12.18d collapse moments M_P, M_P and $2M_P$ are reached at A, B and D respectively. Thus

$$M_P\,\theta + M_P\,(2\theta) + 2M_P\,(3\theta/2) = -\,W(L/2)\theta + 2W(L\theta/3) \;\Rightarrow\; W = 36M_P/L \tag{12.35c}$$

The least load from eqs(12.35a-c), $W = 6M_P/L$, is the collapse condition.

12.3.3 Portal Frames

The analysis of collapse in a portal frame involves the application of eq(12.31) to all possible failure modes [2]. Let the frame ABCDE in Fig. 12.19 carry vertical and horizontal loading when the supports A and E are hinged. This means that no moments are carried at these supports. Collapse can occur in the beam (Fig. 12.19b), by sway in the stanchions (Fig. 12.19c) and by a combination of collapse and sway (Fig. 12.19d)

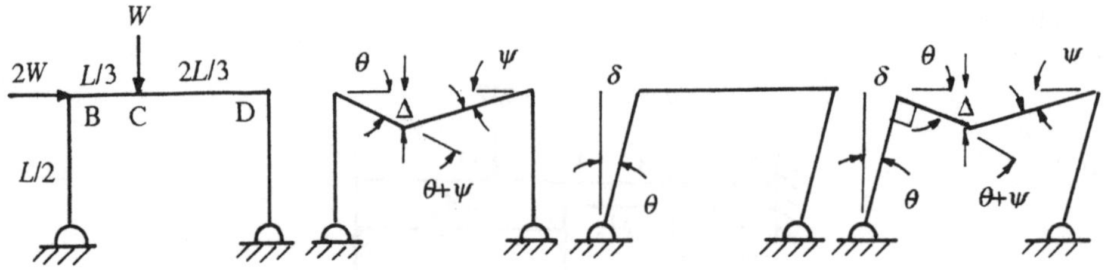

(a) (b) (c) (d)

Figure 12.19 Collapse modes in a portal frame

We must allow for different sections between the limbs, so that collapse at a corner occurs in the limb with the lower collapse moment. In this case the stanchions have a 50% greater moment capacity than the beam and thus all the hinges form on the beam side of the corners under M_P as shown. Collapse (b) occurs with hinges at B, C and D, (c) with hinges at B and D and (d) from hinges at C and D. Applying eq(12.31) to each mode in Figs 12.19b,c,d,

$M_P(\theta + \psi) = W\Delta$ (see Fig. 12.19b) (12.36a)

$2M_P\theta = 2W\delta$ (see Fig. 12.19c) (12.36b)

$M_P\theta + M_P\psi + M_P(\theta + \psi) = W\Delta + 2W\delta$ (see Fig. 12.19d) (12.36c)

Substituting $\psi = \theta/2$, $\delta = L\theta/2$ and $\Delta = L\theta/3$ into eqs(12.36a,b,c) leads to the respective collapse loads $9M_P/(2L)$, $2M_P/L$ and $9M_P/(4L)$, showing that a sway collapse would occur. Note that in these calculations the collapse moment $1.5M_P$ in the stanchions is not reached and therefore it would be more economical to select a uniform section of collapse moment M_P for the whole structure.

In the case of encastre fixings, further hinges form at at A and E under modes (c) and (d). Equations (12.36b,c) are modified to

$$2M_P\theta + 2(3M_P/2)\theta = 2W\delta \qquad (12.37a)$$

$$M_P\theta + M_P\psi + M_P(\theta + \psi) + 2(3M_P/2)\theta = W\Delta + 2W\delta \qquad (12.37b)$$

from which the respective collapse loads are $5M_P/L$ and $9M_P/(2L)$. This shows that failure could occur either from a combined mode or from collapse of the horizontal beam. The structure can be designed more economically if the sway and combined modes are satisfied simultaneously. Let $M_P' < M_P$ be the collapse moment of the encastre stanchions. Equations (12.37a,b) are modified to

$$2M_P'\theta + 2M_P'\theta = 2W\delta \qquad (12.38a)$$

$$M_P'\theta + M_P'\psi + M_P(\theta + \psi) + 2M_P'\theta = W\Delta + 2W\delta \qquad (12.38b)$$

Equation (12.38a) gives $M_P' = WL/4$. Substituting this into eq(12.38b) gives $M_P = 11WL/36$. Therefore $W = 36M_P/(11L)$ and $M_P' = 9M_P/11$, which shows that the stanchions are again overdesigned.

12.4 Plastic Torsion of Circular Sections

12.4.1 Solid Bar

Consider a solid circular bar of outer radius r_o subjected to an increasing torque T. While the cross-section is elastic, torsion theory applies to the ensuing deformation. When the outer fibres reach the shear yield stress k, the corresponding torque T_k and twist for the limiting fully elastic section is supplied from eq(5.6) as

$$T_k = Jk / r_o = \pi r_o^4 k /(2r_o) = \pi r_o^3 k /2 \qquad (12.39a)$$
$$\theta_k = kL/(Gr_o) = T_k L/(GJ) \qquad (12.39b)$$

where $J = \pi r_o^4/2$. Increasing the torque beyond T_k results in the penetration of a plastic zone inwards from the outer radius. This results in the elastic-plastic section shown in Fig. 12.20a in which an elastic core is surrounded by a plastic annulus with a common interface radius r_{ep} . Assuming that the material is elastic-perfectly plastic, i.e. it does not work-harden (Fig.12.20b), then k is constant for $r_{ep} \le r \le r_o$.

(a)

(b)

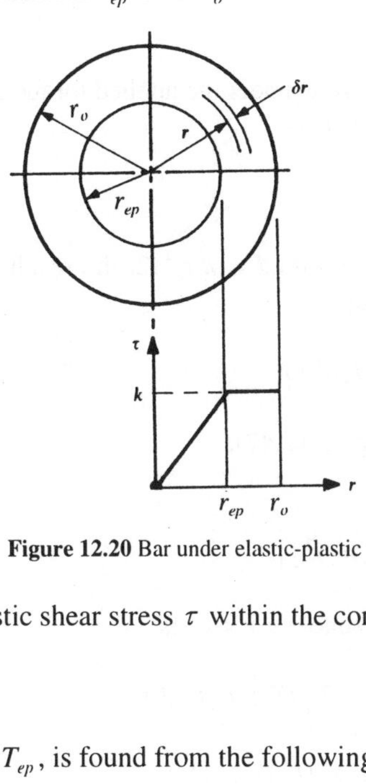

Figure 12.20 Bar under elastic-plastic torsion

The linear variation in elastic shear stress τ within the core $0 \le r \le r_{ep}$ is given by

$$\tau_e = k\,(r/r_{ep}) \qquad (12.40)$$

The corresponding torque T_{ep}, is found from the following equilibrium condition

$$T_{ep} = 2\pi \int_{r_{ep}}^{r_o} k\, r^2 \mathrm{d}r + 2\pi \int_0^{r_{ep}} \tau_e\, r^2 \mathrm{d}r \qquad (12.41a)$$
$$= 2\pi k \int_{r_{ep}}^{r_o} r^2 \mathrm{d}r + 2\pi (k / r_{ep}) \int_0^{r_{ep}} r^3\, \mathrm{d}r$$
$$= 2\pi (k/3) \left|\, r^3 \right|_{r_{ep}}^{r_o} + [2\pi k /(4r_{ep})] \left|\, r^4 \right|_0^{r_{ep}}$$
$$= (\pi k r_o^3 /2)\{(4/3)[1 - (r_{ep}/r_o)^3] + (r_{ep}/r_o)^3\} \qquad (12.41b)$$

and substituting from eq(12.39a),

$$T_{ep}/T_k = (4/3)[1 - \tfrac{1}{4}(r_{ep}/r_o)^3] \qquad (12.42)$$

The angle of twist is found by applying eq(5.6) to the elastic core $0 \le r \le r_{ep}$ where $\tau = k$ for $r = r_{ep}$. This gives

$$\theta_{ep} = kL/(Gr_{ep}) = T_e L/(GJ_e) \qquad (12.43a,b)$$

where $J_e = \pi r_{ep}^4/2$ and T_e is the torque carried by the elastic core, i.e. the second integral in eq(12.41a), not eq(12.39a). The section becomes fully plastic under a torque T_p that causes full plastic penetration to the bar centre. Putting $r_{ep} = 0$ in eq(12.42) and substituting from eq(12.39) gives the fully plastic torque

$$T_p/T_k = 4/3 \;\Rightarrow\; T_p = (2/3)(\pi r_o^3 k) \qquad (12.44a,b)$$

When an elastic-plastic bar is fully unloaded, elastic stresses recover to leave a residual stress distribution τ_R, found from

$$\tau_R = \tau - \tau_e \qquad (12.45a)$$

where τ is the shear stress in each zone under the applied torque T_{ep} and τ_e is the elastic stress that recovers on removal of T_{ep}. This is,

$$\tau_e = T_{ep}\, r/J \qquad (12.45b)$$

It follows from eqs(12.45a,b) that with $J = \pi r_o^4/2$, the residual stress distribution in the plastic zone $r_{ep} \le r \le r_o$ is given by

$$\tau_R = k - T_{ep} r/J = k - (T_{ep}/T_k) \times (T_k/J)\, r$$

Substituting $T_k/J = k/r_o$ and using eq(12.42),

$$\tau_R = k\,[1 - (T_{ep}/T_k)(r/r_o)] \qquad (12.46a)$$
$$= k\{1 - (4/3)(r/r_o)[1 - \tfrac{1}{4}(r_{ep}/r_o)^3]\} \qquad (12.46b)$$

The residual stress in the elastic zone $0 \le r \le r_{ep}$ is

$$\tau_R = k\,(r/r_{ep}) - T_{ep} r/J = k\,(r/r_{ep}) - (T_{ep}/T_k) \times (T_k/J)\, r$$
$$\tau_R = k\,(r/r_{ep})[1 - (T_{ep}/T_k)(r_{ep}/r_o)] \qquad (12.46c)$$
$$= k\,(r/r_{ep})\{1 - (4/3)(r_{ep}/r_o)[1 - \tfrac{1}{4}(r_{ep}/r_o)^3]\} \qquad (12.46d)$$

The distributions of eqs(12.46b,d) are represented graphically in Fig. 12.21. The residual stress τ_R is the difference between the ordinates τ and τ_e within each zone, the signs being determined from eq(12.45a). It is seen that τ_R has the largest values at the outer and interface radii, the greater value depending upon the depth of penetration.

The residual twist in the elastic core is found from eq(12.46c) using the elastic relationship:

$$\theta_R = \tau_R L/(Gr) = (\tau_R/k)[kL/(Gr)]$$
$$= [kL/(Gr_{ep})][1 - (T_{ep}/T_k)(r_{ep}/r_o)] \qquad (12.47)$$

Alternatively, using eqs(12.39b) and (12.43a) we may confirm eq(12.47):

$$\begin{aligned}\theta_R &= \theta_{ep} - \theta_e \\ &= kL/(Gr_{ep}) - T_{ep}L/(GJ) = kL/(Gr_{ep}) - (T_{ep}/T_k)[\,kL/(r_oG)] \\ &= [kL/(Gr_{ep})][1 - (T_{ep}/T_k)(r_{ep}/r_o)]\end{aligned}$$

(a)

(b)

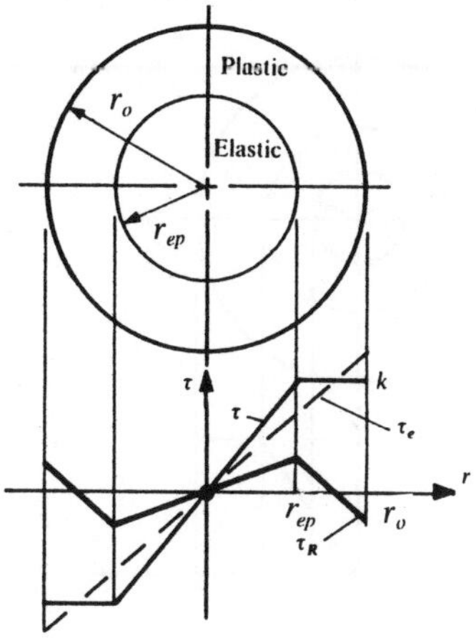

Figure 12.21 Residual stress in a partially plastic solid bar

12.4.2 Sandhill Analogy

A useful analogy can be employed to find the fully plastic torque of solid non-circular sections [1]. If we compare the expression for the volume V of the cone of sand that would rest on the end of a circular bar (Table 12.1) with eq(12.44b) then $T_P = 2V$ where $k = h/r_o$ defines the sloping side of the cone. For the rectangular and triangular sections T_P can be derived in a similar manner, from knowing (i) the volume of a pyramid of height h that each section would support and (ii) the sloping side of each pyramid has a gradient k.

Table 12.1 Fully plastic torques for solid sections

Section	V	Slope	$T_P = 2V$
circle (radius r_o)	$\pi r_o^2 h/3$	$k = h/r_o$	$2\pi r_o^3 k/3$
rectangle ($b \times t$)	$(th/6)(3b - t)$	$k = 2h/t$	$kt^2(3b - t)/6$
triangle (side a)	$a^2 h/(4\surd 3)$	$k = 2\surd 3h/a$	$ka^3/12$

We may take the sum of T_P for rectangles comprising I, T, U sections. The T_P for a rectangle also applies to thin-walled curved sections where b becomes the perimeter length and t is the constant thickness. For example, in an open circular tube $b = 2\pi r_m$ so that $T_P \simeq \pi k t^2 r_m$ where r_m is the mean wall radius. For tubular sections we may take the difference between T_P for the inner "solid" and T_P for the outer "solid".

12.4.3 Hollow Cylinder

Figure 12.22a shows a partially yielded hollow cylinder with inner and outer radii r_i and r_o in which the plastic zone has penetrated to radius r_{ep}, under torque T_{ep}. The corresponding distribution of shear stress is given in Fig. 12.22b. Again, an elastically-perfectly plastic material is assumed in which the plastic zone spreads under a constant yield stress value k.

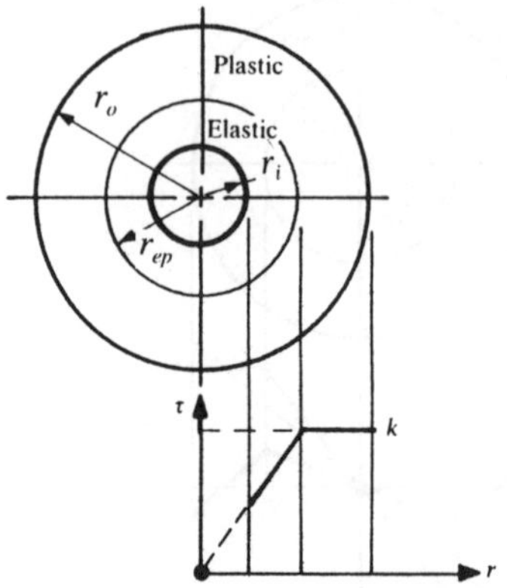

Figure 12.22 Elastic-plastic hollow cylinder

Torque equilibrium is expressed in

$$T_{ep} = 2\pi k \int_{r_{ep}}^{r_o} r^2\,\mathrm{d}r + 2\pi (k/r_{ep}) \int_{r_i}^{r_{ep}} r^3\,\mathrm{d}r \tag{12.48a}$$

$$= (2\pi k/3)(r_o^{\,3} - r_{ep}^{\,3}) + [\pi k/(2r_{ep})](r_{ep}^{\,4} - r_i^{\,4}) \tag{12.48b}$$

The following three torques may be identified from eq(12.48b):

(i) the fully elastic torque T_k, for $r_{ep} = r_o$

$$T_k = [\pi k / (2r_o)](r_o^{\,4} - r_i^{\,4}) \tag{12.49a}$$

(ii) a dimensionless elastic-plastic torque T_{ep}, ($r_i < r_{ep} < r_o$). Combining eqs(12.48b and 12.49a)

$$T_{ep}/T_k = (4r_o/3)(r_o^{\,3} - r_{ep}^{\,3})/(r_o^{\,4} - r_i^{\,4}) + (r_o/r_{ep})(r_{ep}^{\,4} - r_i^{\,4})/(r_o^{\,4} - r_i^{\,4})$$

$$= \left\{(4/3)[1 - \tfrac{1}{4}(r_{ep}/r_o)^3] - (r_i/r_{ep})(r_i/r_o)^3\right\} / [1 - (r_i/r_o)^4] \tag{12.49b}$$

which reduces to eq(12.42) for $r_i = 0$ in a solid cylinder.

(iii) the dimensionless fully plastic torque $r_{ep} = r_i$ in eq(12.49b)

$$T_p/T_k = (4/3)[1 - (r_i/r_o)^3] / [1 - (r_i/r_o)^4] \tag{12.49c}$$

Equations (12.46a,c) again supply the residual shear stresses within each zone where now the ratio T_{ep}/T_k is given by eq(12.49b). Equations (12.43a,b) provide the twist in an elastic-plastic cylinder when based upon the inner elastic annulus. In eq(12.43b) $J_e = (\pi/2)(r_{ep}^{\,4} - r_i^{\,4})$ and T_e is identified with the second integral in eq(12.48a). The residual twist within the elastic core $r_i < r < r_{ep}$ again follows from eqs(12.47) when T_{ep}/T_k is given by eq(12.49b).

Example 12.2 Find the torque required to cause yielding to (a) the inner radius, (b) the mean radius and (c) the outer radius in a hollow steel cylinder with 200 mm outer and 100 mm inner diameters (see Fig. 12.23a) for a shear yield stress of 180 MPa. Determine the distribution of residual shear stress in the section and the twist remaining in a 3 m length after the torque in (b) has been removed. Take G = 80 GPa.

(a)

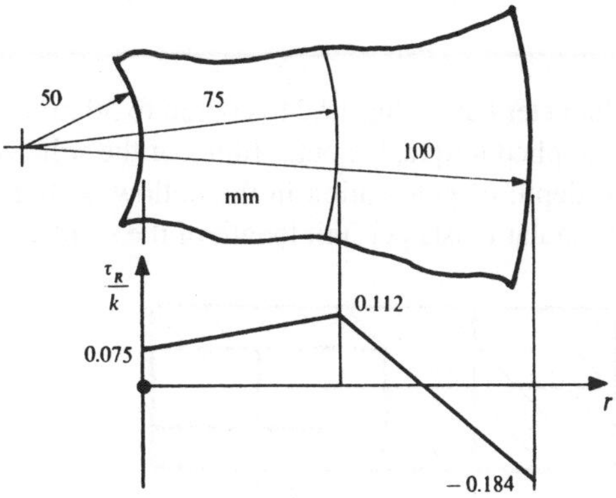

(b)

Figure 12.23 Residuals in hollow steel cylinder

(a) Setting r_o = 100 mm and r_i = 50 mm in eq(12.49a) gives the fully elastic torque

$$T_k = [(\pi \times 180)/(2 \times 100)]\,(100^4 - 50^4) = 265.07 \text{ kNm}$$

(b) With r_{ep} = 75 mm, the torque ratio in eq(12.49b) is

$$T_{ep}/T_k = \left\{(4/3)[1 - \tfrac{1}{4}(75/100)^3] - (50/75)(50/100)^3\right\} / \,[1 - (50/100)^4] = 1.1833$$

from which the elastic-plastic torque is

$$T_{ep} = 1.1833 \times 265.07 = 313.61 \text{ kNm}$$

(c) the torque ratio in eq(12.49b) is

$$T_P/T_k = (4/3)[1 - (50/100)^3] / \,[1 - (50/100)^4] = 1.244$$

from which the fully plastic torque is

$$T_P = 1.244 \times 265.07 = 329.87 \text{ kNm}$$

Using eq(12.46a) the residuals in the plastic annulus $75 \le r \le 100$ (mm) are

$$\begin{aligned}\tau_R / k &= 1 - (T_{ep}/T_k)(r/r_o)\\ &= 1 - 1.1833(r / 100)\\ &= 1 - 1.1833 \times 10^{-2}\, r \qquad \text{(i)}\end{aligned}$$

Using eq(12.46c) the residuals in the elastic annulus $50 \le r \le 75$ (mm) are

$$\begin{aligned}\tau_R / k &= [1 - (T_{ep}/T_k)(r_{ep}/r_o)](r/r_{ep})\\ &= [1 - 1.1833\,(75/100)](r/75)\\ &= 1.5 \times 10^{-3}\, r\end{aligned}$$

Clearly the distributions represented by eqs(i) and (ii) are both linear in the radius r. Fig. 12.23b shows that the greatest value, $\tau_R/k = 0.1837$, occurs at the outer radius. The residual twist is found from eq(12.47):

$$\theta_R = [\, kL/(Gr_{ep})\,[1 - (T_{ep}/T_k)(r_{ep}/r_o)] \qquad \text{(iii)}$$
$$= [(180 \times 3 \times 10^3)/(80 \times 10^3 \times 75)][1 - (1.1833 \times 75/100)]$$
$$= 0.0101 \text{ rad} = 0.58°$$

Example 12.3 The 100 mm diameter bar in Fig. 12.24 is bored to 50 mm diameter over half its length. If, under an axially applied torque, the outer fibres of the solid section reach their shear yield stress find (i) the depth of penetration in the hollow section and (ii) the ratio between the solid and hollow angular twists per unit length of the solid and hollow sections.

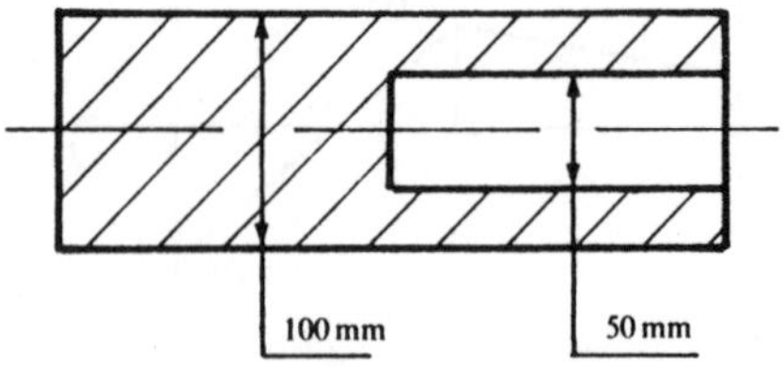

Figure 12.24 Torsion bar

For the solid section eq(12.39a) gives the fully elastic torque

$$T_k = \pi r_o^3 k/2 = (\pi/2)50^3 k \qquad \text{(i)}$$

Equation (12.48b) applies to the elastic-plastic hollow section, giving

$$T_{ep} = (2\pi k/3)(r_o^3 - r_{ep}^3) + [\pi k/(2r_{ep})](r_{ep}^4 - r_i^4)$$
$$= (\pi k/2)(4r_o^3/3 - r_{ep}^3/3 - r_i^4/r_{ep}) \qquad \text{(ii)}$$

Since $T_k = T_{ep}$ then equating (i) and (ii) with $r_o = 50$ mm, $r_i = 25$ mm gives the following equation for the elastic-plastic interface radius r_{ep}:

$$50^3 = (4/3)(50)^3 - r_{ep}^3/3 - 25^4/r_{ep}$$
$$r_{ep}^4 - (50)^3 r_{ep} + 3(25)^4 = 0$$

This quartic is satisfied by $r_{ep} = 46.25$ mm. The radial penetration depth for the hollow cylinder is therefore 3.75 mm in from the outer radius. In the solid cylinder, eq(12.39b) supplies the rate of twist

$$(\theta/L)_s = k/(Gr_o) = k/(50G)$$

For the hollow cylinder, applying eq(12.43a) to the elastic annulus gives its rate of twist as

$$(\theta/L)_h = k/(Gr_{ep}) = k/(46.25G)$$

and, therefore, from eqs(iii) and (iv), the unit twist ratio is

$$(\theta/L)_h/(\theta/L)_s = 50/46.25 = 1.08$$

12.4.4 Composite Shaft

The fully elastic torque capacity T_k for a composite shaft (see Fig. 12.25) is determined from the material which is the first to reach the shear yield stress at its outer fibres. With shear yield stresses k_1 and k_2 for the inner (1) and outer (2) components respectively, it follows from eqs(5.12a,b) that T_k is the lesser of

$$T_k = (k_1/r_1)[J_1 + (G_2/G_1)J_2] \qquad (12.50a)$$
$$T_k = (k_2/r_2)[J_2 + (G_1/G_2)J_1] \qquad (12.50b)$$

Assuming that the lesser T_k is determined from eq(12.50b) with k_2 first being attained in the outer fibres of the annulus 2, then the core remains elastic.

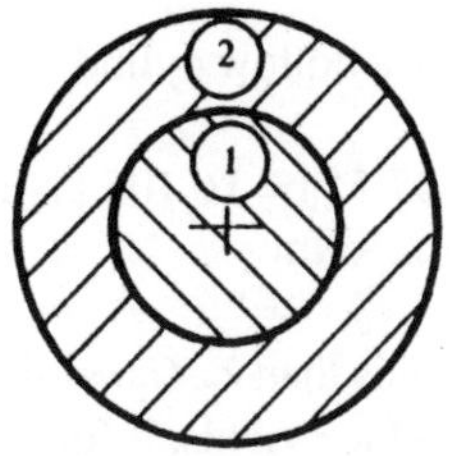

Figure 12.25 Composite section of torsion bar

The maximum shear stress $\tau_1 < k_1$, at the outer diameter of the core, follows from compatibility in that both materials suffer the same shear strain at the interface. That is,

$$(\tau/G)_1 = (\tau/G)_2$$
$$\therefore\ \tau_1 = k_2(r_1/r_2)(G_1/G_2) = (T_2 r_1/J_2)(G_1/G_2) \qquad (12.50a)$$

Alternatively, if the lesser T_k is found from eq(12.50a) then k_1 is reached in the core and the annulus remains elastic. The maximum stress $\tau_2 < k_2$, at the outer diameter of the annulus, is again found from equality of shear strain at the interface. That is,

$$(\tau_2/G_2)(r_1/r_2) = k_1/G_1,\ \Rightarrow\ \tau_2 = k_1(r_2/r_1)(G_2/G_1)$$
$$\therefore\ \tau_2 = (T_1 r_1/J_1)(r_2/r_1)(G_2/G_1) = (T_1 r_2/J_1)(G_2/G_1) \qquad (12.50b)$$

With increasing torque beyond T_k, the plastic zone inwardly penetrates either the shaft or the annulus independently until the remaining component becomes plastic. As plastic penetration then occurs in both components simultaneously, the foregoing elastic-plastic analyses again apply in which the total torque is the sum of the component torques.

Example 12.4 The composite bar in Fig. 12.25 is composed of a central brass core 75 mm diameter, firmly surrounded by a steel annulus 100 mm outer diameter. Determine the maximum elastic torque the section can sustain. Which material is understressed and by how much? Find also the torque needed to produce a fully plastic steel annulus. Take for brass(1) $G_1 = 41$ GPa, $k_1 = 60$ MPa and for steel(2) $G_2 = 82$ GPa, $k_2 = 120$ MPa.

$$J_1 = \pi(75^4)/32 = 3.106 \times 10^6 \text{ mm}^4$$
$$J_2 = \pi(100^4 - 75^4)/32 = 6.711 \times 10^6 \text{ mm}^4$$

Then, from eqs(12.50a,b),

$T_k = (60/37.5)[3.106 + (2 \times 6.711)]10^6 = 26.445 \times 10^6$ Nmm = 26.445 kNm
$T_k = (120/50)[6.711 + (0.5 \times 3.106)]10^6 = 19.834 \times 10^6$ Nmm = 19.834 kNm

The lesser T_k determines the maximum elastic torque. Thus the steel is at the point of yield at its outer diameter while the brass is understressed. The maximum stress in the brass is, from eq(12.51a),

$$\tau_1 = (G_1/G_2)(r_1/r_2)\,k_2 = (41/82)(37.5/50)120 = 45 \text{ MPa}$$

When the steel becomes fully plastic eq(12.49c) is applied to give

$$(T_p/T_k)_2 = (4/3)[1 - (37.5/50)^3] / [1 - (37.5/50)^4] = 1.1276$$

in which eq(12.49a) supplies T_k for a hollow cylinder

$$(T_k)_2 = [\pi \times 120/(2 \times 50)](50^4 - 37.5^4) = 16.107 \text{ kNm}$$
$$\therefore (T_p)_2 = 1.1276 \times 16.107 = 18.162 \text{ kNm}$$

The shear strain in the steel interface is then $k_2/G_2 = 1.463 \times 10^{-3}$. Since this is also the shear strain k_1/G_1 at which the brass interface yields, it follows that the brass is at its fully elastic condition. From eq(12.39a), for a solid bar,

$$(T_k)_1 = \pi \times 37.5^3 \times 60/2 = 4.97 \times 10^6 \text{ Nmm} = 4.97 \text{ kNm}$$

The net torque is then

$$(T_p)_2 + (T_k)_1 = 18.162 + 4.97 = 23.132 \text{ kNm}$$

These problems are complicated by the mismatch in shear strain arising at the interface when $k_1/G_1 \neq k_2/G_2$ for non-hardening component materials. The amount of simultaneous penetration in each material must then be determined from these limiting shear strain values.

12.5 Multiaxial Plasticity

In the case of a non-hardening material (elastic-perfectly plastic) $\sigma = Y =$ constant. The non-hardening plastic behaviour of beams in bending and solid shafts under torsion, was solved by the application of equilibrium alone. A yield criterion is required when the stress state is multiaxial. The plastic stresses can be found by combining the equilibrium condition with the yield criterion without recourse to the plastic strain behaviour. That is, the problem remains statically determinate. To illustrate this, principal plastic stress distributions will be established for structures with axial symmetry. These include a thick-walled pressurized cylinder with different end conditions and a rotating disc.

12.5.1 Thick-Walled Cylinder

Here we wish to derive expressions for the internal pressure required to penetrate a plastic zone to a radial depth r_{ep} in a non-hardening thick-walled cylinder. The inner and outer radii are r_i and r_o, the ends are closed and the material obeys a von Mises yield criterion.

(a) Elastic-plastic stress states
Since the outer annulus $r_{ep} \le r \le r_o$ is elastic, the radial and hoop stresses are supplied by Lamé (see eqs (2.48a,b) and (2.49)). The boundary conditions are $\sigma_r = -p_{ep}$ for the interface radius r_{ep}, and $\sigma_r = 0$ for the outer radius r_o. These give (see example 2.11)

$$\sigma_\theta = p_{ep} r_{ep}^2 (1 + r_o^2/r^2)/(r_o^2 - r_{ep}^2) \quad (12.52a)$$
$$\sigma_r = p_{ep} r_{ep}^2 (1 - r_o^2/r^2)/(r_o^2 - r_{ep}^2) \quad (12.52b)$$
$$\sigma_z = p_{ep} r_{ep}^2/(r_o^2 - r_{ep}^2) = (\sigma_\theta + \sigma_r)/2 \quad (12.52c)$$

At the interface a state of yield exists. Substituting eqs(12.52a-c) into eq(11.7a) with $r = r_{ep}$ and θ, r and z replacing 1, 2 and 3 leads to

$$p_{ep} = (Y/\surd 3)(1 - r_{ep}^2/r_o^2)$$

and, therefore, from eqs(12.52a-c) the stresses in the elastic zone are

$$\sigma_\theta = [Y r_{ep}^2/(\surd 3 r_o^2)](1 + r_o^2/r^2) \quad (12.53a)$$
$$\sigma_r = [Y r_{ep}^2/(\surd 3 r_o^2)](1 - r_o^2/r^2) \quad (12.53b)$$
$$\sigma_z = Y r_{ep}^2/(\surd 3 r_o^2) \quad (12.53c)$$

The radial equilibrium eq(3.13) applies to the inner plastic annulus $r_i \le r \le r_{ep}$

$$\sigma_\theta - \sigma_r = r\, d\sigma_r/dr \quad (12.54a)$$

and again assuming eq(12.52c) supplies the axial stress, the Mises criterion simply becomes

$$\sigma_\theta - \sigma_r = 2Y/\surd 3 \quad (12.54b)$$

Combining eqs(12.54a,b), and integrating gives

$$\sigma_r = (2Y/\surd 3) \ln r + K$$

in which K is found from the condition that $\sigma_r = -p_{ep}$ for $r = r_{ep}$. This gives

$$K = -p_{ep} - (2Y/\surd 3) \ln r_{ep} = -(Y/\surd 3)(1 - r_{ep}^2/r_o^2) - (2Y/\surd 3) \ln r_{ep}$$

and, therefore, the stresses in the plastic zone are

$$\sigma_r = (2Y/\surd 3) \ln (r/r_{ep}) - (Y/\surd 3)(1 - r_{ep}^2/r_o^2) \quad (12.55a)$$
$$\sigma_\theta = (2Y/\surd 3) \ln (r/r_{ep}) + (Y/\surd 3)(1 + r_{ep}^2/r_o^2) \quad (12.55b)$$
$$\sigma_z = (2Y/\surd 3) \ln (r/r_{ep}) + (Y/\surd 3)(r_{ep}^2/r_o^2) \quad (12.55c)$$

(b) Ultimate and mean wall-pressures
Now as $\sigma_r = -p_i$ for $r = r_i$ eq(12.55a) gives the required internal pressure

$$p_i = (2Y/\surd 3) \ln (r_{ep}/r_i) + (Y/\surd 3)(1 - r_{ep}^2/r_o^2) \quad (12.56a)$$

The pressure p_{ult} required to produce a fully plastic cylinder is found from putting $r_{ep} = r_o$ in eq(12.55a)

$$p_{ult} = (2Y/\surd 3) \ln (r_o/r_i) \quad (12.56b)$$

Figure 12.26a shows the distribution of radial, hoop and axial stress when $r_{ep} = (r_i + r_o)/2$. The plots combine the elastic (eqs12.53a-c) with the plastic (eqs 12.55a-c) stress distributions under pressure p_i.

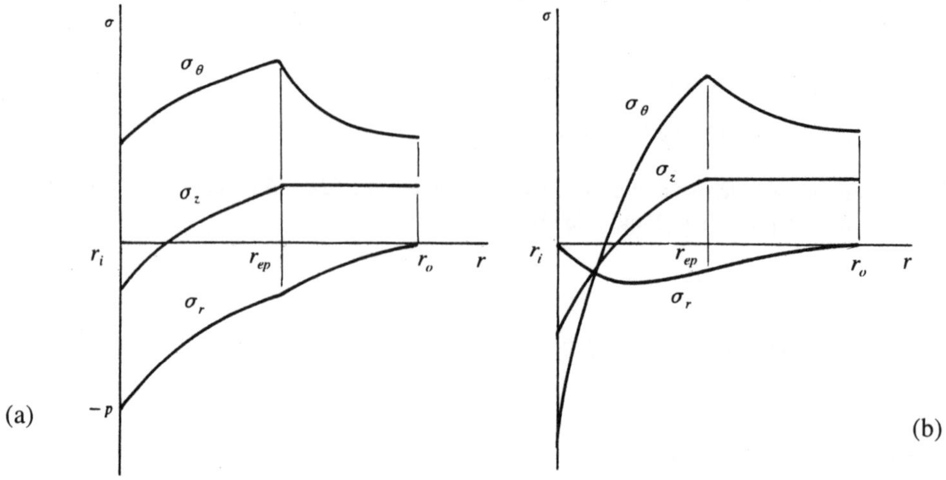

Figure 12.26 Elastic and plastic stress distributions

(c) Residual Stresses

When the pressure p_i is released an elastic stress σ_E recovers, leaving a multiaxial residual stress σ_{iR} ($i = \theta, r, z$). That is,

$$\sigma_{iR} = \sigma_i - \sigma_{iE} \tag{12.57}$$

where σ_{iE} take the Lamé form with subscript i replacing ep in eqs(12.52a-c) for $r_i \le r \le r_o$. Subtracting these from the corresponding stresses σ_i under pressure in each zone leads to the residuals. For the outer zone $r_{ep} \le r \le r_o$ eq(12.57) is written

$$\sigma_{\theta R} = \sigma_\theta - \sigma_{\theta E}$$
$$[Y r_{ep}^2/(\surd 3 r_o^2)](1 + r_o^2/r^2) - p_i r_i^2(1 + r_o^2/r^2)/(r_o^2 - r_i^2) \tag{12.58a}$$

$$\sigma_{rR} = \sigma_r - \sigma_{rE}$$
$$[Y r_{ep}^2/(\surd 3 r_o^2)](1 - r_o^2/r^2) - p_i r_i^2(1 - r_o^2/r^2)/(r_o^2 - r_i^2) \tag{12.58b}$$

$$\sigma_{zR} = \sigma_z - \sigma_{zE} = (\sigma_{\theta R} + \sigma_{rR})/2$$
$$= Y r_{ep}^2/(\surd 3 r_o^2) - p_i r_i^2/(r_o^2 - r_i^2) \tag{12.58c}$$

For the inner zone $r_i \le r \le r_o$ eq(12.57) becomes

$$\sigma_{\theta R} = \sigma_\theta - \sigma_{\theta E}$$
$$= (2Y/\surd 3)\ln(r/r_{ep}) + (Y/\surd 3)(1 + r_{ep}^2/r_o^2) - p_i r_i^2(1 + r_o^2/r^2)/(r_o^2 - r_i^2) \tag{12.59a}$$

$$\sigma_{rR} = \sigma_r - \sigma_{rE}$$
$$= (2Y/\surd 3)\ln(r/r_{ep}) - (Y/\surd 3)(1 - r_{ep}^2/r_o^2) - p_i r_i^2(1 - r_o^2/r^2)/(r_o^2 - r_i^2) \tag{12.59b}$$

$$\sigma_{zR} = \sigma_z - \sigma_{zE} = (\sigma_{\theta R} + \sigma_{rR})/2$$
$$= (2Y/\surd 3)\ln(r/r_{ep}) + Y r_{ep}^2/(\surd 3 r_o^2) - p_i r_i^2/(r_o^2 - r_i^2) \tag{12.59c}$$

These are distributed as shown in Fig. 12.26b. The compressive stresses remaining within the inner core of the cylinder are particularly beneficial to improving fatigue strength when they oppose and reduce subsequently applied cyclic tensile stresses (σ_θ and σ_z) due to fluctuating internal pressure. The process of pre-pressurizing thick cylinders in this way is known as autofrettage. Had the Tresca yield criterion $\sigma_\theta - \sigma_r = Y$ been preferred to that of von Mises it follows from eq(12.5b) that all the foregoing Mises stresses, including the residuals σ_R, need only be multiplied by $\sqrt{3}/2$. The resulting σ_θ and σ_r expressions will also apply to an open ended cylinder or thin disc since $\sigma_z = 0$ still remains the intermediate stress and does not enter Tresca's yield criterion. The Mises criterion is more difficult to combine with the equilibrium condition for an open cylinder. The following Section illustrates Nadai's parametric approach [1] to this problem.

12.5.2 Thin Annular Disc Under Internal Pressure

The following analysis provides the internal pressure required to (a) initiate yielding in the bore fibres, (b) produce full plasticity and (c) partially yield to an elastc-plastic interface radius r_{ep}. A non-hardening von Mises material is assumed.

(a) Initial Yielding
The initial yield pressure in an open-ended, thick-walled cylinder is required. Substituting σ_θ and σ_r from eqs(2.48a,b), together with $\sigma_z = 0$, into the von Mises yield criterion (11.7a) provides the initial yield pressure as

$$p = Y(K^2 - 1)/\sqrt{(3K^4 + 1)} \tag{12.60}$$

where $K = r_o/r_i$.

(b) Full Plasticity
The Mises yield criterion (11.7b) applies to the plastic material within the wall

$$\sigma_\theta^2 - \sigma_\theta\sigma_r + \sigma_r^2 = Y^2 \tag{12.61a}$$

Equation (12.61a) may be writen as a simpler elliptical equation:

$$\sigma^2/a^2 + \sigma'^2/b^2 = 1 \tag{12.61b}$$

where $a = \sqrt{2}Y$ and $b = \sqrt{(2/3)}Y$ with the following substitutions:

$$\sigma = (\sigma_\theta + \sigma_r)/\sqrt{2} \text{ and } \sigma' = (\sigma_\theta - \sigma_r)/\sqrt{2}) \tag{12.62a,b}$$

Figure 12.27 shows that eq(12.61b) is satisfied by introducing a parameter θ into the co-ordinates

$$\sigma = a\sin\theta = \sqrt{2}Y\sin\theta,\ \sigma' = b\cos\theta = \sqrt{(2/3)}Y\cos\theta \tag{12.63a,b}$$

Combining eqs(12.62a,b) and (12.63a,b) gives σ_θ and σ_r

$$\sigma_\theta = (1/\sqrt{2})(\sigma + \sigma') = Y[\sin\theta + (1/\sqrt{3})\cos\theta] \equiv (2Y/\sqrt{3})\sin(\theta + \pi/6) \tag{12.64a}$$
$$\sigma_r = (1/\sqrt{2})(\sigma - \sigma') = Y[\sin\theta - (1/\sqrt{3})\cos\theta] \equiv (2Y/\sqrt{3})\sin(\theta - \pi/6) \tag{12.64b}$$

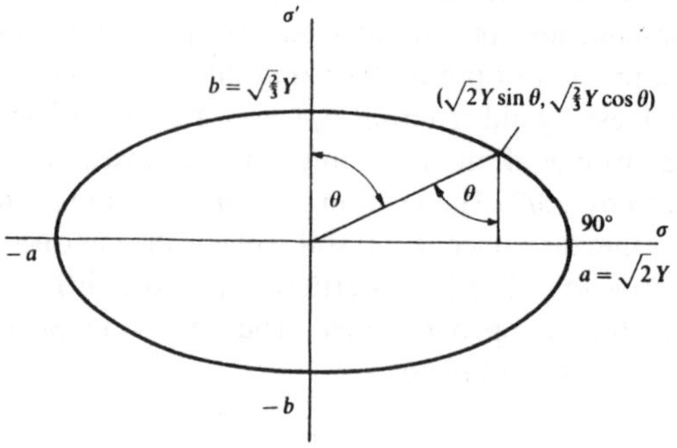

Figure 12.27 von Mises ellipse in parametric form

The equilibrium equation (12.53a) again applies. Substituting from eqs(12.64a,b),

$$r\,\mathrm{d}[\sin(\theta - \pi/6)]/\mathrm{d}r = [\sin(\theta + \pi/6) - \sin(\theta - \pi/6)] = \cos\theta$$

Putting $y = \cos\theta$ enables the variables to be separated into

$$\mathrm{d}y/\surd(1 - y^2) + \mathrm{d}y/\surd 3 = -(2/\surd 3)\,\mathrm{d}r/r$$

and integrating gives

$$A^2/r^2 = \cos\theta\,\exp(-\surd 3\theta) \qquad (12.65)$$

where A is an integration constant, found from the condition that $\sigma_r = 0$ for $r = r_o$. Hence from eq(12.64b), $\theta_o = \pi/6$ and eq(12.65) gives

$$A^2 = (\surd 3/2)\,r_o^2 \exp(-\surd 3\pi/6)$$
$$\therefore r_o^2/r^2 = (2/\surd 3)\cos\theta\,\exp[\surd 3(\pi/6 - \theta)] \qquad (12.66)$$

This equation enables θ to be found for any radius $r_i \le r \le r_o$. Then σ_θ and σ_r follow from eqs(12.64a,b). In particular, as $\sigma_r = -p_i$ for $r = r_i$ in eq(12.64b) the fully plastic pressure is

$$p_i = (2Y/\surd 3)\sin(\pi/6 - \theta_i) \qquad (12.67a)$$

where, correspondingly, θ_i is found from eq(12.66)

$$(r_o/r_i)^2 = (2/\surd 3)\cos\theta_i\,\exp[\surd 3(\pi/6 - \theta_i)] \qquad (12.67b)$$

The radial stress remains compressive under pressure for the valid range $-\pi/3 < \theta < \pi/6$.

(c) Partial plasticity

Let the internal pressure p_i produce plasticity to a radius r_{ep}. Equations (12.52a,b) again give the Lamé hoop and radial stresses in the outer elastic zone $r_{ep} \le r \le r_o$ (note $\sigma_z = 0$). The interface pressure p_{ep} is given from eq(12.60) for $K = r_o/r_{ep}$. The stresses in the plastic zone $r_i \le r \le r_{ep}$ are supplied by eqs(12.64a,b). However, the constant A in eq(12.65) is now found from the condition that σ_r is common to both zones at $r = r_{ep}$. That is, from eqs(12.60) and (12.64b),

$$Y(1 - r_o^2 / r_{ep}^2) / \surd(3r_o^4 / r_{ep}^4 + 1) = (2Y/\surd 3) \sin(\theta_{ep} - \pi/6)$$
$$(1 - r_o^2 / r_{ep}^2) / \surd(r_o^4 / r_{ep}^4 + ⅓) = 2 \sin(\theta_{ep} - \pi/6) \qquad (12.68a)$$

Having found θ_{ep}, to satisfy eq(12.68a), it follows from eq(12.65) that

$$A^2 = r_{ep}^2 \cos\theta_{ep} \exp(-\surd 3\theta_{ep})$$
$$r_{ep}^2 / r^2 = (\cos\theta / \cos\theta_{ep}) \exp[\surd 3(\theta_c - \theta)] \qquad (12.68b)$$

Equation (12.68b) enables θ to be found at any radius $r_i \le r \le r_{ep}$. In particular, when $r = r_i$, $\theta = \theta_i$, from which the internal pressure p_i required to produce an elastic-plastic cylinder is again found from eq(12.67a).

Example 12.5 An annular disc is made from material with a yield stress $Y = 500$ MPa and radii $r_i = 25$ mm, $r_o = 62.5$ mm. Find the internal yield pressures for the disc material at its (a) inner, (b) outer and (c) mean wall radii. Show the stress distribution for (b) and (c) and determine the maximum residual hoop stress for (c).

(a) Equation (12.60) gives the initial yield pressure. Setting $K = 62.5/25 = 2.5$ gives

$p = 500\,(6.25 - 1)/\surd(117.19 + 1) = 241.46$ MPa (2.415 kbar)

(b) The disc becomes fully plastic (Fig. 12.28), when $r_o / r_i = 2.5$ in eq(12.67b),

$(2.5)^2 = (2/\surd 3) \cos\theta_i \exp[\surd 3(\pi/6 - \theta_i)]$

A trial solution gives $\theta_i = -30.9° = -0.539$ rad. From eq(12.67a),

$p_i = (2 \times 500/\surd 3) \sin(\pi/6 + 0.539) = 504.5$ MPa

Equations (12.64a,b) give the fully plastic σ_θ and σ_r distributions. These appear as the broken lines in Fig. 12.28.

(c) For the partially plastic disc,

$$r_{ep} / r_i = (25 + 62.5)/(2 \times 25) = 1.75$$
$$\therefore r_o / r_{ep} = (r_o / r_i) \times (r_i / r_{ep}) = 2.5/1.75 = 1.4286$$

Substituting into eq(12.68a),

$$-0.4907 = 2 \sin(\theta_{ep} - \pi/6)$$

A trial solution gives $\theta_{ep} = 15.8° = 0.2758$ rad. From eq(12.68b), with $r = r_i$ and $\theta = \theta_i$,

$$1.75^2 = 1.0393 \cos\theta_i \exp[\surd 3(0.2758 - \theta_i)]$$

The trial solution yields $\theta_i = -22.55° = -0.3936$ rad. Hence from eq(12.67a),

$$p_i = (2Y/\surd 3) \sin(\pi/6 + 0.3936) = 458.4 \text{ MPa}$$

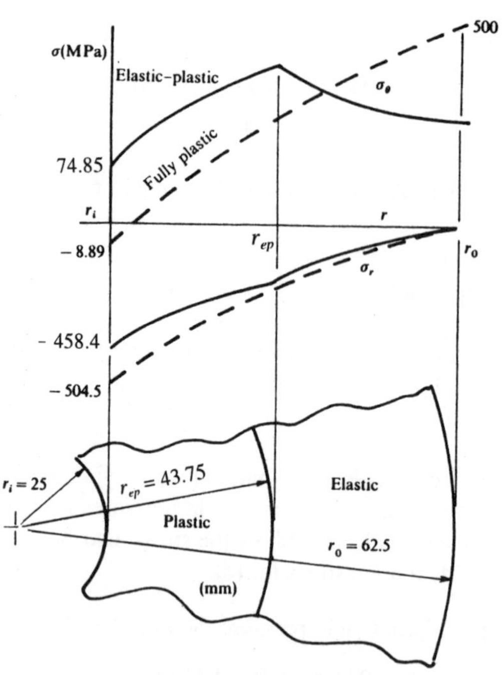

Figure 12.28 Yielding in a thin disc

Figure 12.28 shows the σ_r and σ_θ distributions for the elastic-plastic disc as found from eqs(12.64a,b) and (12.68b). For example, the maximum bore hoop stress is found from eq(12.64a) for $\theta = \theta_i$

$$\sigma_\theta = (2Y/\sqrt{3})\sin(\pi/6 - 0.3936) = 74.85 \text{ MPa}$$

Upon release of p_i the elastic hoop stress (12.58a) recovers

$$\sigma_{\theta E} = p_i(1 + r_o^2/r_i^2)/(r_o^2/r_i^2 - 1)$$
$$= 458.4(1 + 2.5^2)/(2.5^2 - 1) = 633.03 \text{ Mpa}$$

and, therefore the residual compressive hoop stress is

$$\sigma_{\theta R} = \sigma_\theta - \sigma_{\theta E} = 74.85 - 635.03 = -558.18 \text{ MPa}$$

indicating that reversed yielding occurs at the bore since Y is exceeded. A lesser pre-pressure is required if the disc is to benefit from autofrettage.

12.5.3 Thin Rotating Disc

As the speed of a thin rotating disc is raised, a plastic zone spreads towards the outer radius. The Tresca yield criterion enables the speed to be found for a given penetration in a non-hardening material.

(a) Initial Yielding

The elastic stresses in a solid rotating disc are given by eqs (3.28a-c). These are distributed in the manner of Fig. 12.29a.

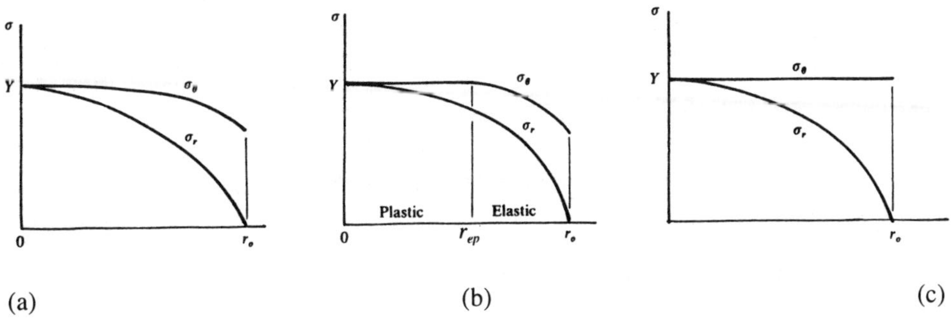

Figure 12.29 Stress distributions in a rotating disc

For a given radius $r > 0$, both σ_θ and σ_r are tensile and $\sigma_\theta > \sigma_r$. The Tresca criterion (11.1b) employs the greater stress together with the zero axial stress σ_z. This gives simply

$$\sigma_\theta = Y \tag{12.69}$$

Clearly, yielding will commence at the centre where σ_θ is a maximum. Substituting eq(3.28b,c) into (12.69) for $r = 0$ supplies the speed ω_Y to initiate yielding

$$\rho\,\omega_Y^{\,2}(3 + \nu)\,r_o^{\,2}/8 - 0 = Y$$
$$\therefore\ \omega_Y^{\,2} = 8Y/[\rho\,(3 + \nu)\,r_o^{\,2}] \tag{12.70}$$

(b) Partial plasticity

In the elastic-plastic case the equilibrium eq(3.21) holds for the inner plastic core $0 \le r \le r_{ep}$. Combining this with eq(12.69) leads to

$$(r\sigma_r)/\mathrm{d}r = Y - \rho\,\omega^2 r^2$$
$$r\,\sigma_r = Y r - \rho\,\omega^2 r^3/3 + A$$
$$\sigma_r = Y - \rho\,\omega^2 r^2/3 + A/r \tag{12.71}$$

Now as $\sigma_\theta = \sigma_r = Y$ for $r = 0$ then $A = 0$ in eq(12.71). For the outer elastic annulus ($r_{ep} \le r \le r_o$) eqs(3.24a,b) supply the stresses. The following boundary conditions enable the constants a and b to be found: $\sigma_r = 0$ for $r = r_o$ in eq(3.24a),

$$0 = a - b/r_o^{\,2} - (3 + \nu)\rho\,\omega^2\, r_o^{\,2}/8 \tag{12.72a}$$

and $\sigma_\theta = Y$ for $r = r_{ep}$ in eq(3.24b), giving

$$Y = a + b/r_{ep}^{\,2} - (1 + 3\nu)\rho\,\omega^2\, r_{ep}^{\,2}/8 \tag{12.72b}$$

Solving eqs(12.72a,b) leads to

$$a = \left\{Y - (\rho\omega^2/8)[(3 + \nu)\,r_o^{\,2} - (1 + 3\nu)\,r_{ep}^{\,2}]\right\} r_{ep}^{\,2}/(r_o^{\,2} + r_{ep}^{\,2}) + (3 + \nu)\,\rho\omega^2 r_o^{\,2}/8 \tag{12.73a}$$
$$b = \left\{Y - (\rho\omega^2/8)[(3 + \nu)\,r_o^{\,2} - (1 + 3\nu)\,r_{ep}^{\,2}]\right\} r_o^{\,2} r_{ep}^{\,2}/(r_o^{\,2} + r_{ep}^{\,2}) \tag{12.73b}$$

Equations (12.71) and (3.24a,b) provide the radial and hoop stress distribution shown in Fig. 12.29b.

The angular speed ω_{ep}^2 is found from the condition that σ_r is common to both zones at the interface radius r_{ep}. In the elastic zone eq(3.24a) gives

$$\sigma_r = a - b/r_{ep}^2 - (3 + \nu)\rho\,\omega_{ep}^2 r_{ep}^2/8 \tag{12.74a}$$

Substituting eqs(12.73a,b) into eq(12.74a) leads to

$$\sigma_r = \{Y - (\rho\,\omega_{ep}^2/8)[(3 + \nu)\,r_o^2 - (1 + 3\nu)\,r_{ep}^2]\}(r_{ep}^2 - r_o^2)/(r_{ep}^2 + r_o^2) + (\rho\,\omega_{ep}^2/8)(3 + \nu)(r_o^2 - r_{ep}^2) \tag{12.74b}$$

In the plastic zone eq(12.71) gives for $r = r_{ep}$

$$\sigma_r = Y - \rho\,\omega_{ep}^2 r_{ep}^2/3 \tag{12.74c}$$

Equating (12.74b,c) provides the speed

$$\omega_{ep}^2 = 24\,(Y/\rho)\,r_o^2/\ \{4(r_o^2 + r_{ep}^2)\,r_{ep}^2 + 3(r_o^2 - r_{ep}^2)[r_o^2(3 + \nu) + r_{ep}^2(1 - \nu)]\} \tag{12.75}$$

(c) Full Plasticity

When the disc becomes fully plastic, eq(12.71) applies for $0 \le r \le r_o$. Again $\sigma_r = Y$ for $r = 0$, giving $A = 0$. Equations (12.71) and $\sigma_\theta = Y$ are the stress distributions shown in Fig. 12.29c. Also, as $\sigma_r = 0$ for $r = r_o$,

$$0 = Y - \rho\,\omega^2 r_o^2/3$$

The fully plastic speed ω_P is therefore

$$\omega_P^2 = 3Y/(\rho\,r_o^2) \tag{12.76}$$

Alternatively eq(12.76) is found by setting $r_{ep} = r_o$ in eq(12.75).

Example 12.6 Determine the angular velocity ratios ω_{ep}/ω_Y and ω_P/ω_Y for a uniformly thin solid disc of radius r_o where ω_Y, ω_{ep} and ω_P are respectively the speeds for which the elastic-plastic interface radii are $r_{ep} = 0$, $r_{ep} = r_o/2$ and $r_{ep} = r_o$. Assume a Tresca yield criterion and take $\nu = ⅓$.

Firstly, the speed required for the elastic-plastic radius to reach the mean radius is found. Substituting $r_{ep} = r_o/2$ in eq(12.75) gives

$$\omega_{ep}^2 = 384Y/[\rho\,r_o^2(137 + 27\nu)] \tag{i}$$

Dividing eq(i) by eq(12.70), the speed ratio is

$$(\omega_{ep}/\omega_Y)^2 = 48(3 + \nu)/(137 + 27\nu) \tag{ii}$$

Setting $\nu = ⅓$ in eq(ii) gives $\omega_{ep}/\omega_Y = 1.047$. Taking the ratio between eqs(12.76) and (12.70) gives the ratio between the speeds for full and initial plasticity as

$$\omega_P/\omega_Y = 3(3 + \nu)/8 \tag{iii}$$

Setting $\nu = 1/3$ in eq(iii) gives $\omega_P/\omega_Y = 1.118$. The % increases in speed required for each spread of plasticity are 4.7% and 11.8% respectively.

12.6 Plasticity with Hardening

12.6.1 Tensile Instability

When metallic materials harden with the spread of plasticity some account needs to be made of the associated increase in flow stress. Firstly, it is instructive to examine the hardening behaviour in the stress-strain curve for uniaxial tensioh. The lower curve in Fig. 12.30a is typical of tension test results for ferrous and non-ferrous materials when the engineering stress and strain, σ_o and ε_o, are conventionally defined.

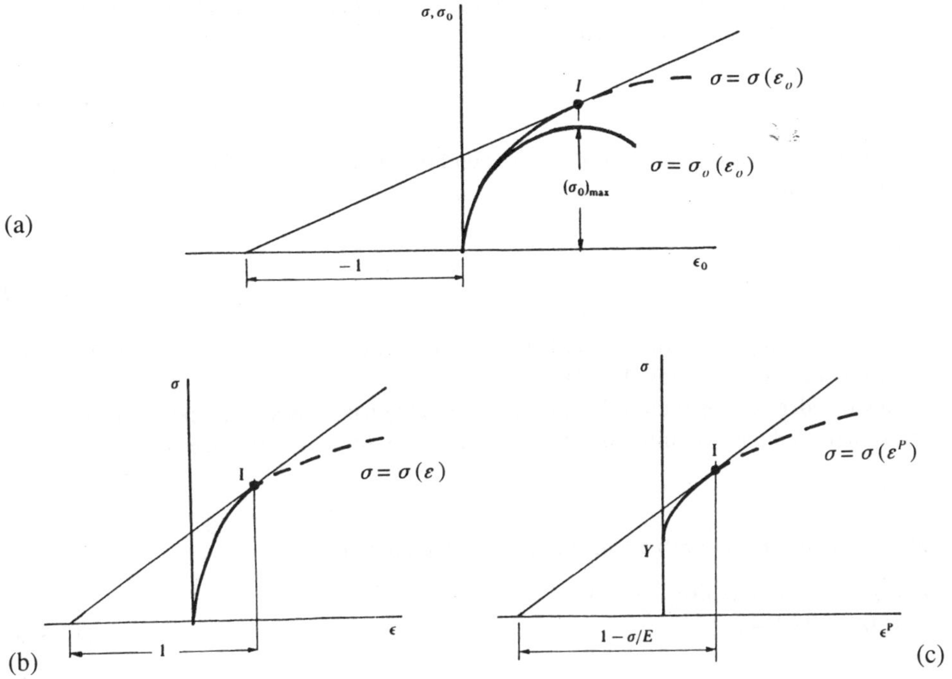

Figure 12.30 Hardening in nominal and true stress-strain axes

That is, if A_o and l_o are the original area and length respectively

$$\sigma_o = W/A_o \text{ and } \varepsilon_o = (l - l_o)/l_o$$

Since these expressions refer to the original testpiece section area and length, errors will arise in stress and strain calculations when large changes in testpiece dimensions occur within the plastic range. For example, the u.t.s. for a material is not a true stress value since no account has been made of the reduction in area to the point of necking. Moreover, the stress in a material will actually continue to rise, not fall, as the neck develops before final fracture. It is possible to correct the nominal tensile stress-strain curve to reflect this behaviour by employing true definitions for σ and ε:

$$\sigma = W/A \tag{12.77}$$
$$\varepsilon = \mathrm{d}l/l \tag{12.78}$$

where A and l are the current area and length. Unlike elastic behaviour, the plastic deformation produced by stressing beyond the elastic limit occurs without change in volume. The incompressibility or constant volume condition gives

$$Al = A_o\, l_o \tag{12.79a}$$
$$\therefore A = A_o l_o / l = A_o/(1 + \varepsilon_o) \tag{12.79b}$$

Substituting eq(12.79b) into eqs(12.77) leads to the true stress

$$\sigma = (W/A_o)(1 + \varepsilon_o) = \sigma_o(1 + \varepsilon_o) \tag{12.80}$$

Integrating eq(12.78) provides a measure of the true or logarithmic total strain between the original and current limits of length

$$\varepsilon = \ln\,(l / l_o) = \ln\,(1 + \varepsilon_o) \tag{12.81a}$$

The true plastic component of strain ε^P is found by subtracting the elastic strain $\varepsilon^E = \sigma/E$, from eq(12.81a),

$$\varepsilon^P = \varepsilon - \varepsilon^E$$
$$\varepsilon^P = \ln\,(1 + \varepsilon_o) - \sigma/E \tag{12.81b}$$

The corresponding corrections afforded to a nominal curve $\sigma_o = \sigma_o(\varepsilon_o)$ by eqs(12.80) and (12.81a,b) are shown in Figs. 12.30a-c.

In Considere's construction [3] in Fig. 12.30a the true stress at the point of instability corresponds to the maximum ordinate for σ_o. This condition is expressed in

$$\mathrm{d}\sigma_o/\mathrm{d}\varepsilon_o = 0$$

Substituting from eq(12.80) and differentiating the quotient,

$$\mathrm{d}\sigma_o/\mathrm{d}\varepsilon_o = (1 + \varepsilon_o)\,\mathrm{d}\sigma/\mathrm{d}\varepsilon_o - \sigma = 0$$
$$\therefore \mathrm{d}\sigma/\mathrm{d}\varepsilon_o = \sigma/(1 + \varepsilon_o) \tag{12.82a}$$

That is, the tangent to the true stress curve intersects the engineering strain axis at $\varepsilon_o = -1$ (see Fig. 12.30a). Now from eq(12.81a),

$$\mathrm{d}\varepsilon/\mathrm{d}\varepsilon_o = 1/(1 + \varepsilon_o)$$

Hence this instability condition, with respect to a true flow curve $\sigma = \sigma(\varepsilon)$, is given by

$$\mathrm{d}\sigma/\mathrm{d}\varepsilon = (\mathrm{d}\varepsilon_o/\mathrm{d}\varepsilon) \times (\mathrm{d}\sigma/\mathrm{d}\varepsilon_o)$$
$$= (1 + \varepsilon_o) \times \sigma/(1 + \varepsilon_o)$$
$$\therefore \mathrm{d}\sigma/\mathrm{d}\varepsilon = \sigma \tag{12.82b}$$

for which a sub-tangent value of unity appears along the true total strain axis in Fig. 17.1b. Note that if eq(12.81b) had been used instead of eq(12.81a) in this derivation, then instability on axes of true stress versuss true plastic strain would be expressed in

$$d\sigma/d\varepsilon^P = \sigma/(1-\sigma/E) \tag{12.82c}$$

where here the curve $\sigma = \sigma(\varepsilon^P)$ in Fig. 12.30c commences at the yield stress Y, not at the origin. Since there is no appreciable difference between eqs(12.82b and c), a unit sub-tangent is employed for the construction in Fig. 12.30c. A number of expressions $\sigma = \sigma(\varepsilon)$ are available. The simplest of these is the Hollomon [4] power law:

$$\sigma = A\varepsilon^n \tag{12.83a}$$

where A is the strength coefficient and $0.1 \le n \le 0.55$ is the strain hardening exponent. Since this equation represents a curve passing through the origin it is normally applied to a corrected true curve in Fig. 12.30b. A Ludwik law [5] represents the curve that commences at Y in Fig. 12.30c:

$$\sigma = Y + A(\varepsilon^P)^n \tag{12.83b}$$

The values of the constants A and n in eqs(12.83a,b) are quite different, as the following examples show.

Example 12.7 Determine the true stress and true strain at the point of tensile instability for the Hollomon and Ludwik functions.

From eq(12.83a),

$$d\sigma/d\varepsilon = (nA)\varepsilon^{n-1}$$

and from eq(12.82b), at the point of instability,

$$(nA)\varepsilon^{n-1} = \sigma = A\varepsilon^n$$
$$\therefore \varepsilon = n \text{ and } \sigma = A\, n^n$$

This shows that the strain at which necking begins equals the exponent value n for a given material. The Ludwik law (12.83b) gives

$$d\sigma/d\varepsilon^P = nA\,(\varepsilon^P)^{n-1}$$

and combining with eq(12.82c) we find the instability strain

$$\varepsilon^P \simeq n\,(1 - Y/\sigma)$$

Combining this with eq(12.82b) leads to the instability stress

$$\sigma = A\, n^n\,(1 - Y/\sigma)^{n-1}$$

Example 12.8 At the point of tensile plastic instability, the engineering stress and strain are 340 MPa and 30% respectively. Determine the constants A and n in the Hollomon and Ludwik laws. Take $Y = 250$ MPa and $E = 200$ GPa where appropriate.

Substituting $\sigma_o = 340$ MPa and $\varepsilon_o = 0.3$ in eqs(12.80) and (12.81a),

$\sigma = \sigma_o (1 + \varepsilon_o) = 340(1 + 0.3) = 442$ MPa
$\varepsilon = \ln (1 + \varepsilon_o) = \ln (1 + 0.3) = 0.2624$
$\varepsilon^P = \varepsilon - \sigma/E = 0.2624 - 442/(200 \times 10^3) = 0.26$

From example 12.8, the Hollomon law gives $n = \varepsilon = 0.2624$ and therefore

$442 = A\,(0.2624)^{0.2624}$
$\therefore A = 627.9$

In the Ludwick law,

$n = \varepsilon^P/(1 - Y/\sigma) = 0.26/(1 - 250/442) = 0.598$
$A = (442 - 250)/(0.26)^{0.598} = 191.6$

12.6.2 Tensile Necking

The use of true stress and strain is adequate to describe the uniform state of plastic deformation in a tensile testpiece up to the start of necking. However, during neck formation in ductile materials the stress state is neither uniform nor uniaxial within this region. The state of stress in the neck was analysed by Bridgeman [6] who showed that radial σ_r and tangential σ_θ stresses are induced under the applied axial stress σ_z. Consider the smallest neck diameter of $2a$ and a radius of curvature R in Fig. 12.31.

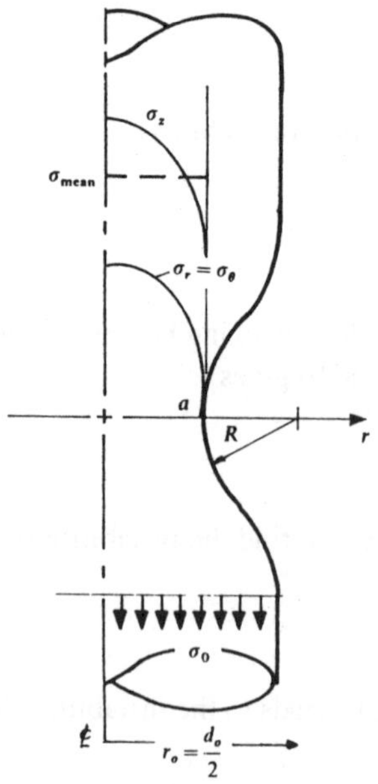

Figure 12.31 Tensile neck

Bridgeman proposed the following triaxial stress state for radius $0 \le r \le a$ within the neck:

$$\sigma_r = \sigma_\theta = \bar{\sigma} \ln [(a^2 + 2aR - r^2)/(2aR)] \tag{12.84a}$$
$$\sigma_z = \bar{\sigma} [1 + \ln (a^2 + 2aR - r^2)/(2aR)] \tag{12.84b}$$

in which $\bar{\sigma}$ is the true equivalent stress for the neck. The distribution of σ_r and σ_z is shown in Fig. 12.31. For an applied force W the mean axial stress σ_{mean} in the neck is found from

$$W = \pi a^2 \sigma_{mean} = \int_0^a 2\pi r\, \sigma_z\, dr$$

Substituting from eq(12.84b) gives

$$\pi a^2 \sigma_{mean} = 2\pi\bar{\sigma} \int_0^a \{r + r \ln [(a^2 + 2aR - r^2)/(2aR)]\}\, dr$$
$$= 2\pi\bar{\sigma}\, \{(aR + a^2/2)[\ln (a^2 + 2aR) - \ln (2aR)]\}$$
$$= \pi\bar{\sigma}\, \{(a^2 + 2aR) \ln [1 + a/(2R)]\}$$
$$\therefore \bar{\sigma}/\sigma_{mean} = \{(1 + 2R/a) \ln [1 + a/(2R)]\}^{-1} \tag{12.85a}$$
$$= [(1 + 1/X) \ln (1 + X)]^{-1} \tag{12.85b}$$

where $X = a/(2R)$ in eq(12.85b) and a/r_o was plotted against $\bar{\sigma}/\sigma_{mean}$ in the form of Bridgeman correction curves for particular materials. Hence, if the smallest diameter of the neck $2a$ is measured at intervals during loading W beyond the point of instability, the true equivalent stress at those intervals may be calculated. The true strain can also be calculated from the neck diameter $d = 2a$ using the following constant volume condition from eq(12.79a):

$$(\pi/4)\, d^2 l = (\pi/4)\, d_o^2 l_o$$

when, from eq(12.78),

$$\varepsilon = \ln (l / l_o) = 2 \ln (d_o / d) \tag{12.86}$$

Applying Bridgeman's correction to Figs. 12.30a-c show the corresponding extension to the flow curves during necking.

Example 12.9 The minimum radius of a tensile neck is 5.613 mm under an applied force of 59.63 kN. Given that $R = 70.16$ mm for a testpiece of original 12.5 mm diameter, determine the true stress in the neck.

Here $a/(2R) = 5.613/(2 \times 70.16) = 0.04$. Then from eq(12.85a),

$$\bar{\sigma}/\sigma_{mean} = [(1 + 25) \ln (1 + 0.04)]^{-1} = 0.9806$$

where $\sigma_{mean} = (59.63 \times 10^3)/[\pi(5.613)^2] = 602.45$ MPa

$$\therefore \bar{\sigma} = 0.9806 \times 602.45 = 590.78 \text{ MPa}$$

One of the problems with tensile testing is that the range of uniform strain is limited by the formation of neck. Larger strain can be achieved from the compression test on a short cylinder but unless the ends are well lubricated, non-uniform deformation arises from barreling of the testpiece's central region. Here the true strain may be calculated from either expression in eq(12.86) depending upon whether the current height l or diameter d of the cylinder is measured. For perfect lubrication of an isotropic material the true stress-strain curve for compression will coincide with that for tension.

12.6.3 Torsion

When a cylindrical bar of material hardens owing to torsional plasticity the distribution of shear stress through the section differs from that shown in Fig. 12.20. Nadai's method [1] allows the shear stress-strain curve, τ_o versus γ_o for the outer diameter to be derived from the torque-twist curve, T versus θ, in Fig. 12.32).

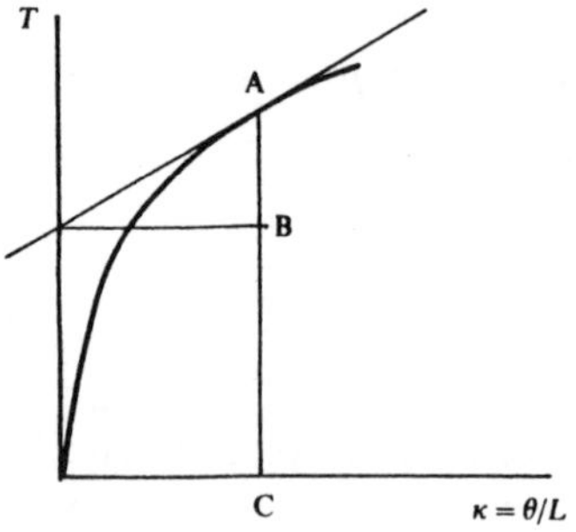

Figure 12.32 Nadai's construction

A solid bar, outer radius r_o is in equilibrium with the applied torque T when

$$T = 2\pi \int_0^a \tau r^2 \, \mathrm{d}r$$

Substituting from eq(5.3), $\gamma = r\kappa$ ($\kappa = \theta/L$ is the twist/unit length),

$$T = (2\pi/\kappa^3) \int_0^a \tau(\gamma)\gamma^2 \mathrm{d}\gamma$$

in which $\tau = \tau(\gamma)$ applies to radius r. Differentiating with respect to $\gamma_o = r_o\kappa$ gives

$$[1/(2\pi)]\ \mathrm{d}(T\kappa^3)/\mathrm{d}\gamma_o = \tau_o(r_o\kappa)^2$$

Since $\mathrm{d}\gamma_o = r_o \mathrm{d}\kappa$,

$$[1/(2\pi)]\ \mathrm{d}(T\kappa^3)/\mathrm{d}\kappa = \tau_o r_o^3 \kappa^2$$
$$\tau_o = [1/(2\pi r_o^3\kappa^2)]\ \mathrm{d}(T\kappa^3)/\mathrm{d}\kappa$$
$$\tau_o = [1/(2\pi r_o^3)](3T + \kappa \mathrm{d}T/\mathrm{d}\kappa) \tag{12.87a}$$

The geometrical interpretation of eq(12.87a), as given by the Nadai construction in Fig. 12.32, leads to

$$\tau_o = [1/(2\pi r_o^3)](3AC + AB) \tag{12.87b}$$

The Nadai method usefully reveals the actual shear stress-strain behaviour to fracture for a hardening material despite the presence of a stress gradient. Torsion tests conducted on thin-walled tubes avoid a stress gradient but these are prone to buckling well before before fracture, thereby limiting the range of strain.

Example 12.10 The torque-unit twist diagram for a solid bar can be expressed as $T = B\kappa + A\kappa^n$. Derive the corresponding shear stress-strain relationship for the outer radius r_o. Hence determine the torque and shear stress required to produce $\gamma_o = 2\%$ in a 20 mm diameter alloy bar for which $n = 1/3$, $A = 10^6$ MPa and $B = 25 \times 10^6$ (units of N and mm).

The linear term in the T expression represents the elastic line. With $n < 1$ the parabolic term dominates in the plastic range. The slope to the T versus κ curve is $dT/d\kappa = B + nA\kappa^{n-1}$. Substituting into eq(12.87a),

$$\tau_o = [1/(2\pi r_o^3)]\,[3\,(B\kappa + A\kappa^n) + \kappa(B + nA\kappa^{n-1})]$$
$$= [1/(2\pi r_o^3)][A\kappa^n(3 + n) + 4\,B\kappa]$$

Putting $\kappa = \gamma_o / r_o$ for the outer radius gives

$$\tau_o = [2B/(\pi r_o^4)]\{\,\gamma_o + (3 + n)[A\,r_o/(4B)](\gamma_o/r_o)^n\,\}$$
$$= (B/J)\{\,\gamma_o + [A\,(3 + n)\,r_o^{1-n}/\,(4B)]\,\gamma_o^n\,\} \qquad \text{(i)}$$

where $J = \pi r_o^4/2$, $\kappa = \gamma_o/r_o = 0.02/10 = 0.002$ and the torque is

$$T = (25 \times 10^6 \times 0.002) + 10^6(0.002)^{1/3} = (0.05 + 0.126)10^6$$
$$= 0.176 \times 10^6 \text{ Nmm} = 176 \text{ Nm}$$

Now $J = \pi(10)^4/2 = 1.571 \times 10^4$ mm^4 and eq(i) gives

$$\tau_o = (25 \times 10^2/1.571)\{0.02 + [(3.333 \times 10^6 \times 10^{2/3})/(4 \times 25 \times 10^6)](0.02)^{1/3}\,\}$$
$$= 1591.3\,(0.02 + 0.042) = 98.66 \text{ MPa}$$

12.6.4 Equivalent Stress and Strain

In the analysis of the triaxial stress state in the neck of a tensile testpiece, the concept of a stress equivalent to a true uniaxial stress was introduced. In the von Mises criterion (11.7a), yielding under σ_1, σ_2 and σ_3 commences in proportion to the root mean square of these stresses. For a hardening material this equivalent or effective stress continues to increase beyond the initial yield stress Y within the plastic range. The magnitude of the equivalent uniaxial stress is found by replacing Y by $\bar{\sigma}$ in eq(11.7a). This gives

$$\bar{\sigma} = (1/\surd 2)\,\surd[(\sigma_1 - \sigma_2)^2 + (\sigma_2 - \sigma_3)^2 + (\sigma_1 - \sigma_3)^2] \qquad (12.88)$$

which implies that the initial yield surface expands uniformly to contain the current stress point within the plastic range (isotropic hardening). Clearly, a uniaxial stress σ_1 becomes the equivalent stress $\sigma_1 = \bar{\sigma}$ when two of the stresses in eq(12.88) are set to zero. A similar expression holds for the plastic strain under a triaxial stress state. If $d\varepsilon_1^P$, $d\varepsilon_2^P$ and $d\varepsilon_3^P$ are the incremental plastic strains following a change $d\sigma_1$, $d\sigma_2$ and $d\sigma_3$ in the stress state, the equivalent plastic strain increment is defined as

$$d\bar{\varepsilon}^P = (\surd 2/3)\,\surd[(d\varepsilon_1^P - d\varepsilon_2^P)^2 + (d\varepsilon_2^P - d\varepsilon_3^P)^2 + (d\varepsilon_1^P - d\varepsilon_3^P)^2] \qquad (12.89)$$

The numerical factor √2/3 ensures that eq(12.89) correctly defines the equivalent strain under uniaxial stress while plasticity remains incompressible. That is, if the sides of a unit cube permanently change in length by $d\varepsilon_1{}^P$, $d\varepsilon_2{}^P$ and $d\varepsilon_3{}^P$ without changing its volume, then

$$(1 + d\varepsilon_1{}^P)(1 + d\varepsilon_2{}^P)(1 + d\varepsilon_3{}^P) - 1 = 0$$

Neglecting infinitesimal products, the constant volume condition is

$$d\varepsilon_1{}^P + d\varepsilon_2{}^P + d\varepsilon_3{}^P = 0 \qquad (12.90)$$

In the case of tension, say, in which the axial strain is $d\varepsilon_1{}^P$, the lateral strains are found from eq(12.90) to be $d\varepsilon_2{}^P = d\varepsilon_3{}^P = -\,d\varepsilon_1{}^P/2$. Then substituting into eq(12.89) gives $d\bar{\varepsilon}^P = d\varepsilon_1{}^P$. Note that the octahedral shear stress and strain (see table 1.1), which have also been used for equivalence, do not readily reduce to a simple stress state because the numerical factors differ, i.e. for similar substitutions under tension these are $\tau_o = \sqrt{2}\sigma_1/3 = \sqrt{2}\bar{\sigma}/3$ and $d\gamma_o{}^P = \sqrt{2}d\varepsilon_1{}^P = \sqrt{2}d\bar{\varepsilon}^P$. With the equivalent definitions in eqs(12.88) and (12.89) the plot of $\bar{\sigma}$ versus $d\bar{\varepsilon}^P$ will correlate hardening behaviour under all stress states in the single curve $\bar{\sigma} = \bar{\sigma}(\int d\bar{\varepsilon}^P) = \bar{\sigma}(\bar{\varepsilon}^P)$. The latter function is conveniently found from simple tension and compression tests for a limited range of plastic strain or from torsion and balanced biaxial tension (the bulge test) for a larger range of plastic strain. In practice, because materials are initially anisotropic to some degree, it is usually found that points from different tests lie within a narrow scatter band on $\bar{\sigma}$ versus $\bar{\varepsilon}^P$ axes.

The principal plastic strain increments in eq(12.89) and (12.90) require the removal of elastic strain increments from the total strains as follows:

$$\begin{aligned} d\varepsilon_1{}^P &= d\varepsilon_1 - d\varepsilon_1{}^E \\ &= d\varepsilon_1 - (1/E)[d\sigma_1 - \nu\,(d\sigma_2 + d\sigma_1)] \text{ etc} \end{aligned} \qquad (12.91)$$

When six component strains are involved in the deformation, the plastic components are found from a similar removal of elastic strain within a contracted tensor notation:

$$d\varepsilon_{ij}{}^P = d\varepsilon_{ij} - [d\sigma_{ij}/(2G) + \delta_{ij}d\sigma_{kk}/(9K)]$$

in which Kronecker δ_{ij} ensures that the final hydrostatic strain component is removed from the normal but not the shear strains.

12.6.5 The Flow Rule

Levy [7] and von Mises [8] proposed a flow rule of plasticity in which the incremental plastic strain tensor $d\varepsilon_{ij}{}^P$ is linearly dependent upon the deviatoric stress tensor σ_{ij}' in eqs(1.21a-c). A geometrical interpretation of this is that during loading the plastic strain increment vector lies in the position of the exterior normal to a flow surface that expands isotropically in stress space to contain the current stress point. With the surface defined by the von Mises equivalent stress function (eq12.88), the Levy-Mises flow rule is

$$d\varepsilon_{ij}{}^P = d\lambda\,\sigma_{ij}' \qquad (12.92)$$

where $d\lambda$ is a scalar factor of proportionality that changes with the progress of deformation under a given stress state. This is defined by simple tension ($\bar{\sigma}$, $\bar{\varepsilon}^P$) where, with a mean stress

$\sigma_m = \bar{\sigma}/3$, eq(13.24) gives $d\varepsilon_1^P = d\bar{\varepsilon}^P$, $\sigma_1' = 2\bar{\sigma}/3$, $\sigma_2' = \sigma_3' = -\bar{\sigma}/3$. Hence from eq(12.92),

$$d\lambda = 3\, d\bar{\varepsilon}^P / (2\bar{\sigma}) \tag{12.93}$$

which is inversely proportional to the incremental plastic modulus $d\bar{\varepsilon}^P/\bar{\sigma}$. For a principal triaxial stress system σ_1, σ_2 and σ_3, eq(12.92) expands to give, with eq(12.93), the principal plastic strain increments:

$$d\varepsilon_1^P = (d\bar{\varepsilon}^P/\bar{\sigma})[\sigma_1 - \tfrac{1}{2}(\sigma_2 + \sigma_3)] \tag{12.94a}$$
$$d\varepsilon_2^P = (d\bar{\varepsilon}^P/\bar{\sigma})[\sigma_2 - \tfrac{1}{2}(\sigma_1 + \sigma_3)] \tag{12.94b}$$
$$d\varepsilon_3^P = (d\bar{\varepsilon}^P/\bar{\sigma})[\sigma_3 - \tfrac{1}{2}(\sigma_1 + \sigma_2)] \tag{12.94c}$$

in which eq(12.90) holds. Often a comparison is made with the elastic increment in eq(12.91) where $\nu = \tfrac{1}{2}$ for incompressible plasticity and $1/E$ is replaced by the incremental compliance $d\bar{\varepsilon}^P/\bar{\sigma}$. For non-linear hardening, where the uniaxial function is $\bar{\sigma} = \bar{\sigma}(\int d\bar{\varepsilon}^P)$, the compliance is $d\bar{\varepsilon}^P/\bar{\sigma} = d\bar{\sigma}/(\bar{\sigma}\,\sigma')$ where $\sigma' = d\bar{\sigma}/d\bar{\varepsilon}^P$. The right sides of eqs(12.94a-c) can then be expressed in stress and integrated. Numerical methods are required to solve eqs(12.88-12.93) except in the case of the simplest uniform biaxial and triaxial stress states, where the stress components increase proportionately or follow a stepped path during loading.

Example 12.11 A pressure vessel steel has a stress-strain curve represented by the Swift law $\sigma = \sigma_o(1 + B\bar{\varepsilon}^P)^n$. At what internal pressure would tensile plastic instability occur in (a) a thin-walled sphere and (b) a closed cylinder? Given $B = 200$ and $n = 0.1$, determine the equivalent and maximum strain reached at the point of instability in each case. (IC)

(a) The hoop and meridional stresses are (see eq 3.4a,b) $\sigma_\theta = \sigma_\phi = pr/(2t)$.

$\therefore \ln \sigma_\theta = \ln(p/2) + \ln r - \ln t$
$d\sigma_\theta/\sigma_\theta = dp/p + dr/r - dt/t$

Since $dp/p = 0$ at the point of instability and $d\varepsilon_\theta^P = dr/r$, $d\varepsilon_r^P = dt/t$, then

$$d\sigma_\theta/\sigma_\theta = d\varepsilon_\theta^P - d\varepsilon_r^P \tag{i}$$

Now from eq(12.88), with $\sigma_1 = \sigma_\theta$, $\sigma_2 = \sigma_\phi = \sigma_\theta$ and $\sigma_3 = \sigma_r = 0$, the equivalent stress is

$$\bar{\sigma} = (1/\sqrt{2})\sqrt{[0 + \sigma_\theta^2 + \sigma_\theta^2]} = \sigma_\theta \tag{ii}$$

Replacing subscripts 1, 2 and 3 by θ, ϕ and r respectively in eqs(12.90 and 12.92),

$$d\varepsilon_\theta^P = d\lambda[\sigma_\theta - \tfrac{1}{3}(\sigma_\theta + \sigma_\phi + \sigma_r)] = (d\lambda/3)\sigma_\theta$$
$$d\varepsilon_\phi^P = d\lambda[\sigma_\phi - \tfrac{1}{3}(\sigma_\phi + \sigma_\theta + \sigma_r)] = (d\lambda/3)\sigma_\theta$$
$$d\varepsilon_r^P = -(d\varepsilon_\theta^P + d\varepsilon_\phi^P) = -2\,d\varepsilon_\theta^P$$
$$\therefore d\varepsilon_\theta^P = d\varepsilon_\phi^P = -d\varepsilon_r^P/2 \tag{iii}$$

when from eq(12.89) the equivalent plastic strain is

$$d\bar{\varepsilon}^P = \sqrt{(2/3)}\sqrt{[(3d\varepsilon_\theta^P)^2 + (3\,d\varepsilon_\theta^P)^2]} = 2\,d\varepsilon_\theta^P \tag{iv}$$

From eqs(ii) and (iv), $\mathrm{d}\bar{\sigma}/\mathrm{d}\bar{\varepsilon}^P = \mathrm{d}\sigma_\theta/(2\mathrm{d}\varepsilon_\theta{}^P)$. Then, from eqs(i) - (iii),

$$\mathrm{d}\bar{\sigma}/\mathrm{d}\bar{\varepsilon}^P = ½\,(1 - \mathrm{d}\varepsilon_r{}^P/\mathrm{d}\varepsilon_\theta{}^P)\sigma_\theta = 3\sigma_\theta/2$$
$$\mathrm{d}\bar{\sigma}/\mathrm{d}\bar{\varepsilon}^P = 3\bar{\sigma}/2$$

Thus, in an equivalent plot to Fig. 12.30b, the sub-tangent value is 2/3. The equivalent plastic instability strain is found from

$$\mathrm{d}\bar{\sigma}/\mathrm{d}\bar{\varepsilon}^P = Bn\bar{\sigma}/(1 + B\bar{\varepsilon}^P) = 3\bar{\sigma}/2$$
$$\therefore\ \bar{\varepsilon}^P = (1/B)(2Bn/3 - 1)$$
$$= (1/200)(2 \times 200 \times 0.1/3 - 1) = 0.062$$
$$\therefore\ \varepsilon_\theta{}^P = \bar{\varepsilon}^P/2 = 0.062/2 = 0.031$$

The equivalent stress at instability is $\bar{\sigma} = \sigma_o(2nB/3)^n$, which gives the critical pressure as

$$p = (2t\sigma_o/r)(2nB/3)^n$$

(b) For a thin closed cylinder the stresses are $\sigma_r = 0$ and, from eq(3.7a,b), $\sigma_\theta = pr/t = 2\sigma_z$ and $\sigma_r = 0$. Eq(i) again applies to the point of instability. The equivalent stress (eq 12.88) is

$$\bar{\sigma} = (1/\sqrt{2})\,\sqrt{[(\sigma_\theta - \sigma_z)^2 + \sigma_\theta{}^2 + \sigma_z{}^2]}$$
$$= (1/\sqrt{2})\,\sqrt{[(\sigma_\theta/2)^2 + \sigma_\theta{}^2 + (\sigma_\theta/2)^2]}$$
$$\bar{\sigma} = \sqrt{3}\sigma_\theta/2 \qquad \text{(v)}$$

and, from eqs(12.90) and (12.92),

$$\mathrm{d}\varepsilon_\theta{}^P = \mathrm{d}\lambda\,[\sigma_\theta - ⅓\,(\sigma_\theta + \sigma_r + \sigma_z)] = (\mathrm{d}\lambda/2)\sigma_\theta$$
$$\mathrm{d}\varepsilon_r{}^P = \mathrm{d}\lambda\,[\sigma_r - ⅓\,(\sigma_\theta + \sigma_r + \sigma_z)] = -\,(\mathrm{d}\lambda/2)\sigma_\theta$$
$$\therefore\ \mathrm{d}\varepsilon_r{}^P = -\,\mathrm{d}\varepsilon_\theta{}^P \text{ and } \mathrm{d}\varepsilon_z{}^P = 0 \qquad \text{(vi)}$$

The equivalent plastic strain follows from eq(12.89),

$$\mathrm{d}\bar{\varepsilon}^P = (\sqrt{2/3})\sqrt{[(\mathrm{d}\varepsilon_\theta{}^P)^2 + (2\mathrm{d}\varepsilon_\theta{}^P)^2 + (\mathrm{d}\varepsilon_\theta{}^P)^2]} = (2/\sqrt{3})\mathrm{d}\varepsilon_\theta{}^P \qquad \text{(vii)}$$

Combining eqs(v) and (vii),

$$\mathrm{d}\bar{\sigma}/\mathrm{d}\bar{\varepsilon}^P = 3\mathrm{d}\sigma_\theta/(4\mathrm{d}\varepsilon_\theta{}^P)$$

where from eqs(i) and (vi),

$$\mathrm{d}\bar{\sigma}/\mathrm{d}\bar{\varepsilon}^P = ¾\,(1 - \mathrm{d}\varepsilon_r{}^P/\mathrm{d}\varepsilon_\theta{}^P)\sigma_\theta = 3\sigma_\theta/2$$
$$\therefore\ \mathrm{d}\bar{\sigma}/\mathrm{d}\bar{\varepsilon}^P = \sqrt{3}\bar{\sigma}$$

i.e. a sub-tangent of $1/\sqrt{3}$ in this case. Hence, from the given law,

$$\mathrm{d}\bar{\sigma}/\mathrm{d}\bar{\varepsilon}^P = Bn\bar{\sigma}/(1 + B\bar{\varepsilon}^P) = \sqrt{3}\bar{\sigma}$$
$$\therefore\ \bar{\varepsilon}^P = (1/B)(Bn/\sqrt{3} - 1)$$
$$= (1/200)(200 \times 0.1/\sqrt{3} - 1) = 0.0527$$

$\therefore \varepsilon_\theta{}^P = (\surd 3/2) \times 0.0527 = 0.0457$

The corresponding equivalent stress is $\bar{\sigma} = \sigma_o (Bn/\surd 3)^n$ from which the critical pressure is

$p = [2\, t\sigma_o/(\surd 3r)](Bn/\surd 3)^n$

References

[1] Nadai A., *Theory of Flow and Fracture of Solids,* McGraw-Hill, New York, 1950.
[2] Bhatt P. and Nelson, H. M., *Structures*, 3rd edition, Longman, 1990.
[3] Considere A., *Ann ponts et Chaussees*, **9**(6), (1885) 574.
[4] Hollomon J. H., *Trans A.I.M.E.* **162**, (1945), 268.
[5] Ludwik P., *Elements der Technologischen Mechanik,* Verlag Julius Springer, 1909.
[6] Bridgeman, P. W., *Studies in Large Plastic Flow and Fracture,* McGraw-Hill, New York, 1952, p. 9.
[7] Levy M., *C.r. hebd. Seanc. Acad. Sci., Paris,* **70**, (1870), 1323.
[8] Von Mises R., *Nader Ges Wiss Gottingen,* 1913, 582.

EXERCISES

Elastic-Plastic Bending of Beams

12.1. Derive expressions for the moments M_Y, M_P and hence determine $Q = M_P/M_Y$ for each of the axi-symmetric beam cross-sections in Fig. 12.33.
Answer: $Q = 1.7,\ 16r^2(r_2{}^3 - r_1{}^3)/[3\pi(r_2^4 - r_1^4)],\ 2,\ 3[1-(bd^2)/(BD^2)]/\{2[1-(bd^3)/(BD^3)]\}$.

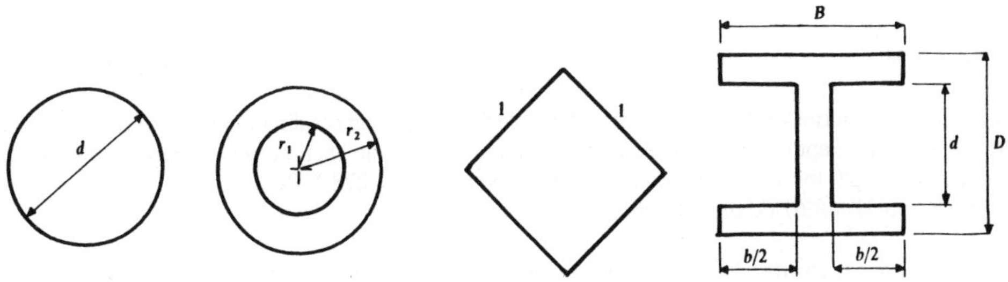

(a) (b) (c) (d)

Figure 12.33

12.2 Find the bending moment that will just initiate yielding in a circular steel rod 25 mm diameter given that its tensile yield stress is 300 MPa.

12.3 A simply supported beam 1.25 m long carries a central concentrated load W. If the cross-section is 25 mm wide × 75 mm deep, with a yield stress $Y = 278$ MPa, find W when (a) yielding first occurs and (b) yielding penetrates to a depth of 15 mm from top and bottom surfaces at mid span. Find the length over which yielding has occurred in (b) and the residual stress distribution for the unloaded beam.
Answer: 2195 kg, 2896 kg, 290 mm.

12.4 Find M_P for the inverted T - section in Fig. 12.34.
Answer: $M_P = Yah(a + h)/2$.

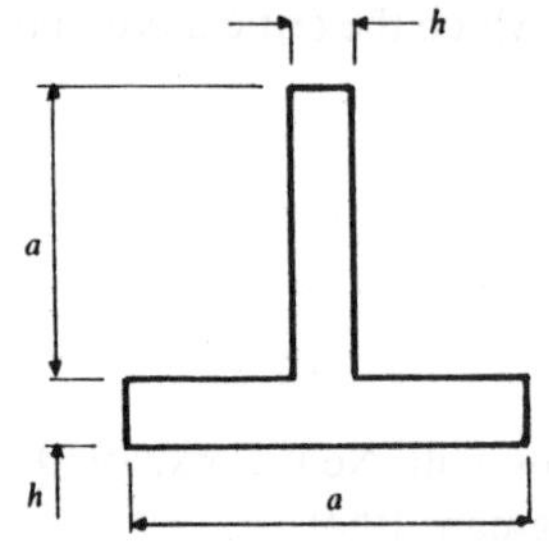

Figure 12.34

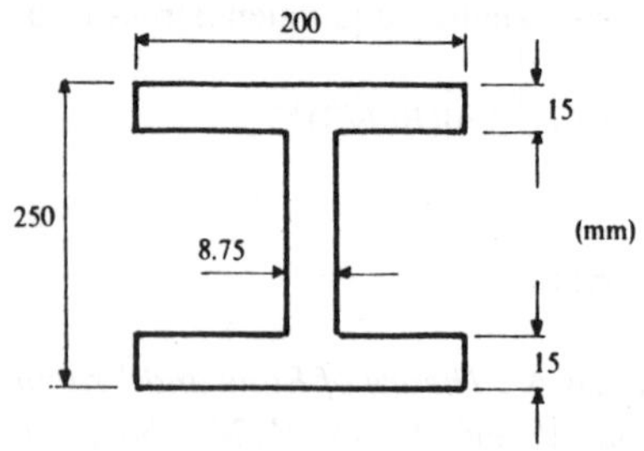

Figure 12.35

12.5 Calculate, for the I - section in Fig. 12.35, M_Y, the shape factor and load factor, for $Y = 324$ MPa, a single plastic hinge and a safety factor of 2.
Answer: 1.11, 2.22, 247 kNm.

12.6 Find the moment of resistance in the fully elastic and plastic conditions for the I-section of beam in Fig. 12.36, when $Y = 280$ MPa.

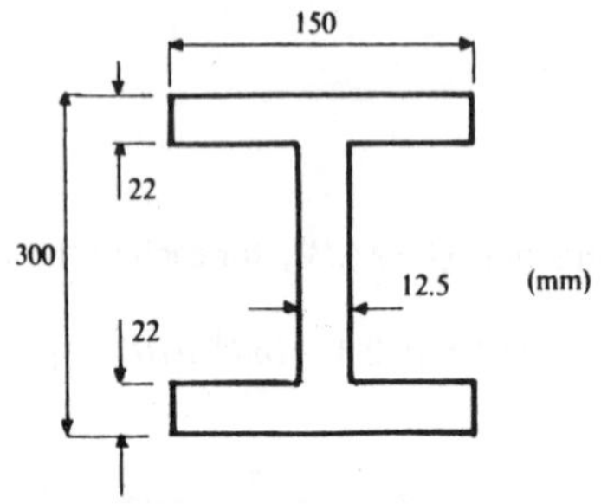

Figure 12.36

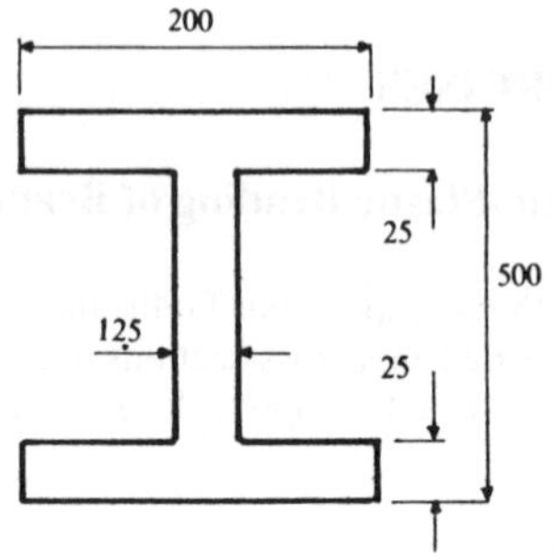

Figure 12.37

12.7 Express in terms of Y, for the I-section beam in Fig. 12.37, the bending moment to cause yielding throughout the depth of the flanges and throughout the whole section. Also establish the plastic penetration depth for an applied moment of 850 kNm with $Y = 278$ MPa.
Answer: $45.3Y$, $48.6Y$ (kNm), 100 mm.

12.8 Determine, for the notched section in Fig. 12.38, the moment to produce 15 mm of plastic penetration from each edge and the corresponding radius of curvature. Establish the residual stress distribution and radius of curvature upon removal of this moment.

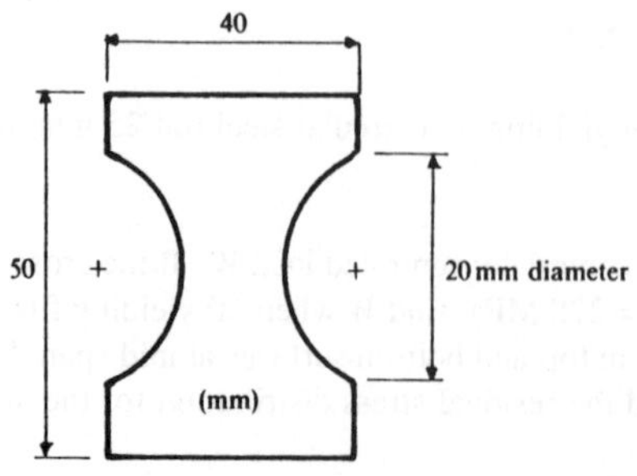

Figure 12.38

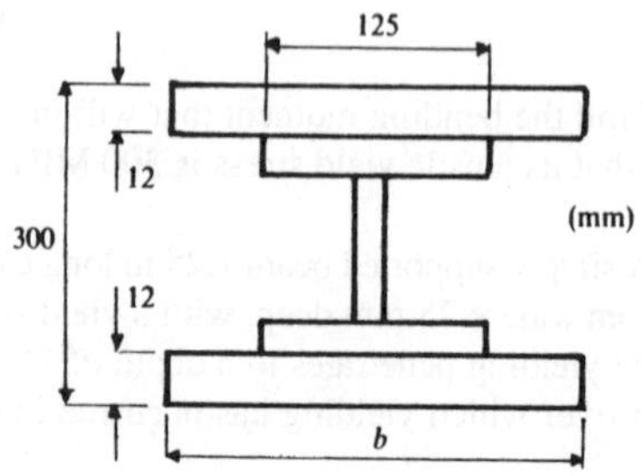

Figure 12.39

12.9 The steel beam I-section in Fig.12.39 is simply supported over a length of 5 m and carries a uniformly distributed load of 114 kN/m. Steel reinforcing plates 12 mm thick are welded to each flange. Calculate the plate width b such that they are fully plastic at mid-span and the extent of yielding in their length. Find the horizontal shearing stress at the interface for sections where the outer surfaces have just reached the yield point. Take the yield stress $Y = 300$ MPa and for the unreinforced section $I_g = 80 \times 10^{-6}$ m^4 (CEI)

12.10 Yielding occurs over the bottom 50 mm of the web for the T-section in Fig. 12.40 under an elastic-plastic bending moment. If the yield stress remains constant at 278 MPa find this applied moment, the stress at the flange top and the position of the neutral axis. Repeat for the fully plastic case. (Ans 71 kNm, 210 MPa, 135.5 mm, 155 mm)

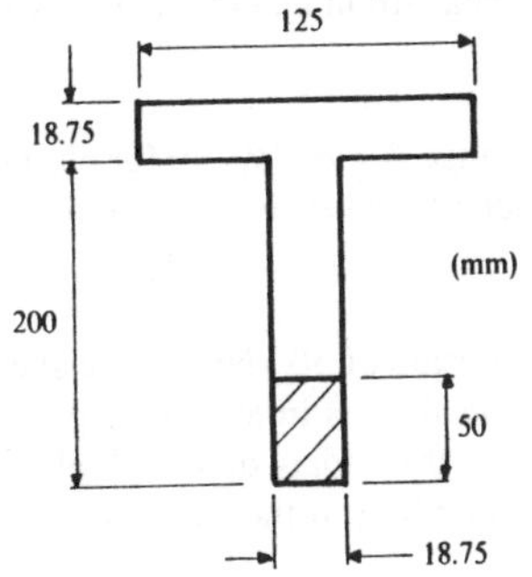

Figure 12.40

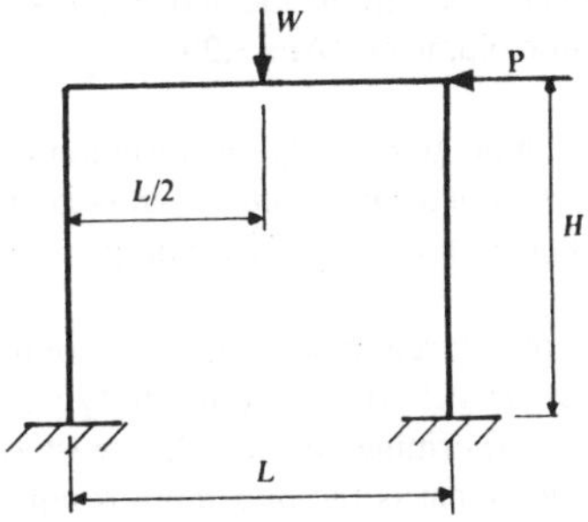

Figure 12.41

12.11 Derive, from energy considerations, expressions for the collapse load of the fixed base portal frame in Fig. 12.41 in the case of (a) a central vertical load W, (b) a horizontal load P and (c) combined W and P when $P = W/3$.

12.12 A 9 m length beam with 75 mm square section is to carry a uniformly distributed load w. Compare values of w that would cause the beam to collapse corresponding to support over two and three equal length bays when $Y = 300$ MPa.

12.13 A uniform beam of length l with rectangular cross-section ($b \times d$) is built in at both ends and carries a vertical load W at quarter span. Find the extent to which the plasticity spreads in the length when M_P is first achieved.

12.14 Determine expressions for the fully elastic and fully plastic concentrated load for a rectangular section, simply supported beam of breadth b and depth d, when the load W divides the length l into a and b. What is the length of the plastic hinge?
Answer: $W_Y = bd^2Yl/(6ab)$, $W_P = bd^2Yl/(4ab)$.

12.15 Determine the fully elastic and fully plastic moments for a cantilever beam of length l carrying a central concentrated load W when a prop prevents deflection at the free end. Confirm the fully plastic moment expression using the principle of virtual work.
Answer: $M_Y = 3W_Y l/16$, $M_P = W_P l/6$.

12.16 A span of 6 m is covered by a beam built in at both ends. The beam carries three equal concentrated loads W (kN) 1.5 m apart with equal 1.5 m lengths remaining at each end. The central 3 m of the beam is reinforced top and bottom to increase the moment of resistance by 80%. Given that $M_P = 7.5$ kNm for the unreinforced section, find the loading to cause collapse at the centre and at 1.5 m from the ends.

12.17 A 3 m length beam, carrying a central concentrated load W is built in at one end and simply supported on the same level at the other end. If the elastic section modulus $z = I/y = 15.63 \times 10^3$ mm³ and $Y = 235$ MPa, find the maximum elastic value of W using a safety factor of 2 under elastic conditions. Also find the value of W for a load factor of 2 against full plastic collapse.

Plastic Torsion

12.18 Find the torque that will just initiate yielding in a 25 mm diameter solid bar given that the shear yield stress is 175 MPa.

12.19 Determine the ratio T_{ep}/T_Y for a solid circular shaft when the plastic zone has penetrated through the section to one half the radial depth. If $k = 150$ MPa determine the shear strain and the rate of twist on a 25 mm diameter. (Ans 1.29)

12.20 Examine graphically the variation in T_{ep}/T_Y with θ_{ep}/θ_Y in the range $4 \leq \theta_{ep}/\theta_Y \leq 0$ for elastic-plastic torsion of a solid circular bar. Note that here θ is the twist/unit length and θ_Y is its value for the outer diameter at the start of yielding.

12.21 A solid, circular steel shaft 100 mm diameter is subjected to a torque of 30 kNm. If the shear yield stress is 125 MPa, find the fully plastic torque and show that the shaft is in an elastic-plastic condition. Determine the angular twist over a 3 m length and find the radial depth of plastic penetration. What is the maximum residual stress in the shaft corresponding to the fully unloaded condition? Take $G = 82$ GPa.

12.22 Determine the torque required to produce an elastic-plastic interface at the mean radius of a hollow tube with 25 mm inner and 100 mm outer diameters. Find the corresponding residual angle of twist/unit length and the stress distribution that remains upon removal of the torque. Take $k = 225$ MPa and $G = 78$ GPa.

12.23 A circular tube of mean radius 100 mm has a wall thickness of 10 mm. Determine the fully elastic and fully plastic torques given that the shear yield stress is 180 MPa. How are the torques altered when the tube contains a longitudinal split?.

12.24 A solid circular shaft 120 mm diameter is drilled to 60 mm diameter for one half its length. Under an applied torque a plastic zone has penetrated to a depth of 10 mm in the outer fibres of the solid shaft. Find the depth of yielding in the tubular section and the ratio between the angular twists per unit length of each section.

12.25 A solid, cylindrical composite shaft 1 m long consists of a solid copper core of 50 mm diameter surrounded by a well-fitting steel sleeve of 65 mm external diameter. If the steel is elastic-perfectly plastic determine the torque that will cause yielding to penetrate through to the common diameter. Calculate the residual shear stress at the outer steel diameter after removal of the torque. Neglect interference stresses and assume no slipping at the interface. For steel, take $G = 82.7$ GPa and $k = 124$ MPa, and for copper, take $G = 44.8$ GPa and $k = 75.8$ MPa. (CEI)

12.26 A solid shaft of non-hardening material is 0.25 m long and tapers uniformly between end diameters of 45 mm and 48 mm. Find the torque required to produce yielding at the outer surface of the smaller diameter. If this torque is then increased by 15% determine the length over which yielding has occurred at the outer surface and the radial depth of yielding at the smaller end.

12.27 The diameter of a tapered circular steel bar increases uniformly from 50 mm at one end to 55 mm at the other end over a length of 500 mm. Determine the torque which (a) initiates yielding in the bar and (b) which produces full plasticity at the smaller end. What is the angular twist in each case and the length of surface yielding?

12.28 The dimensions of a bonded-composite tubular section are brass: 50 mm i.d., 75 mm o.d.; steel: 75 mm i.d., 100 mm o.d. Find the maximum torque that may be applied without causing yielding. What then are the maximum shear stresses? What magnitude of applied torque would cause the steel annulus to become fully plastic? Take, for the brass and steel, shear yield stresses of 60 and 120 GPa, and rigidity moduli of 41 and 82 GPa respectively.

12.29 Give the principal reason for pre-straining of torsion bars and state why care has to be taken when this process is applied to the torsion bar suspension systems of motor vehicles. A cylindrical torsion bar measuring 20 mm in diameter and 0.85 m long is pre-strained by being twisted through an angle of 1.50 radians before being released. Assuming that strain-hardening effects may be neglected, (i) sketch, but do not calculate, the residual stress distribution across the cross-section of the bar and (ii) calculate the depth to which yielding takes place and the maximum torque required during pre-straining, given that the shear yield stress is 625 N/mm^2 and the modulus of rigidity is 80 kN/mm^2.

Cylinders and Discs

12.30 Find the internal pressure for which the hoop stress at the bore of a closed thick-walled cylinder is zero when it is fully plastic. Assume a non-hardening Mises material.

12.31 Show that the pressure required to penetrate a plastic zone to a radial depth r_{ep} in a non-hardening thick-walled cylinder of inner and outer radii r_i and r_o respectively is independent of the end condition according to the Tresca yield criterion. If $r_o/r_i = 3.5$ and $Y = 375$ MPa determine from Tresca the pressure to produce full plasticity and the maximum hoop and radial stresses when the ends are open.

12.32 Determine the internal pressure required to cause a plastic zone to penetrate to the mean radius in a thick-walled closed cylinder with respective inner and outer diameters of 25 and 100 mm. Plot the corresponding distribution of σ_θ, σ_r and σ_z throughout the wall when this pressure is held and when it is removed. Assume a von Mises yield criterion and take $Y = 350$ MPa and $E = 207$ GPa.

12.33 What internal pressure will result in an equivalent residual compressive stress (from eq 12.88) of magnitude equal to Y (the yield stress) in a closed, thick-walled, non-hardening von Mises cylinder of inner and outer radii r_i and r_o respectively?
Answer: $(2Y/\sqrt{3})(1 - r_i^2/r_o^2)$.

12.34 Derive the expression for the internal pressure required to produce a fully plastic thick-walled spherical shell. Plot the distribution of radial and meridional stress through the wall at this pressure.

12.35 Show, from the Nadai approach (section 12.5.2), that it is not possible to fully yield a thin annular disc of non-hardening Mises material under internal pressure when the radius ratio exceeds 2.963 and show also that the pressure then remains constant at $2Y/\sqrt{3}$.

12.36 A thin solid disc of uniform thickness and outside radius R rotates at an angular velocity ω rad/s. Determine the ratio between the speed necessary to initiate plastic flow and that for the flow to extend to 90% and 100% of the volume of the disc. Assume a Tresca yield criterion for which the constant flow stress is Y. The density is ρ and Poisson's ratio $\nu = 0.3$. Establish the stationary residual stress distributions corresponding to each flow speed (IC)
Answer: 1.112, 1.114.

12.37 Determine the speeds necessary to (i) initiate yielding, (ii) produce plastic flow to the mean radius and (iii) produce full plasticity in a thin annulus of inner and outer diameters of 50 and 300 mm respectively. Sketch the distributions of radial and hoop stresses in each case. Take $Y = 310$ MPa = constant, $\rho = 7750\ kg/m^3$ and $\nu = 0.28$.

Plasticity with Hardening

12.38 Why does fracture in the parallel length of a circular section tensile testpiece initiate from a point on its axis? How does a superimposed hydrostatic pressure influence the nominal axial strain required to cause tensile necking? Explain with reference to the Bridgeman analysis.

12.39 Determine the true fracture strain of a ductile material in terms of (i) the % elongation at fracture and (ii) the % reduction of area at fracture in a tensile test.

12.40 Convert the following nominal load-extension data to true stress and natural strain up to the instability point and thus determine the strain-hardening coefficient and exponent in the law $\sigma = A\varepsilon^n$. The testpiece is 12.5 mm diameter and the gauge length is 50 mm.

Load (kN)	0	5.56	11.13	16.45	22.25	27.80	33.38	38.94	44.50	53.40
Ext (mm)	0	0.01	0.020	0.030	0.040	0.050	0.060	0.070	0.080	0.102
Load (kN)	58.30	64.97	70.09	76.10	81.44	85.00	88.33	91.67	93.90	
Ext (mm)	0.112	0.127	0.142	0.163	0.183	0.203	0.224	0.254	0.279	

12.41 If the true stress-strain curve for a material is defined by $\sigma = A\varepsilon^n$ determine the tensile strength. Given A = 800 MPa and n = 0.2, find the u.t.s. and the true stress at instability.
Answer: An^n/e^n, 475 MPa, 580 MPa.

12.42 Show that when the Vocé exponential law $\sigma = \sigma_\infty - (\sigma_\infty - \sigma_0)\exp(-\varepsilon/\mu)$ describes true stress-strain behaviour then the stress and strain co-ordinates at the point of instability are $\sigma_u = \sigma_\infty/(1+\mu)$ and $\varepsilon_u = \mu\ln[(1+\mu)(\sigma_\infty - \sigma_0)/(\mu\,\sigma_\infty)]$, where μ, σ_0 and σ_∞ are constants.

12.43 Show that the stress and strain at the point of instability $\sigma_u = Kn^n$ and $\varepsilon_u = n - \varepsilon_0$ respectively, derive from the Swift law $\sigma = K(\varepsilon_0 + \varepsilon)^n$, where K, ε_0 and n are constants.

12.44 Use the Bridgeman correction factor $\bar{\sigma}/\sigma_{mean} = 0.4(d/d_o) + 0.62$, with the incompressibility condition (12.8b), to determine the true stress-strain behaviour beyond the point of tensile instability for the results referred to in Exercise 12.40.

12.45 Given that $T = T_0(1 + A\kappa)^n$ describes the non-linear portion of the torque versus unit-twist curve for a solid bar, determine the corresponding τ versus γ relationship for the bar material. If n = ¼ and A = 4 determine the torque required to produce a shear stress of 400 MPa at the 15 mm outer radius of a solid bar. What is the corresponding shear strain?

12.46 The following torque-twist/unit length data applies to a solid 25 mm diameter bar with 75 mm parallel length. Establish the shear stress-shear strain behaviour for the bar material.

T/Nm	514	582	729	797	910	983.4	1068	1124.7	1158.6	1192.5
θ°/mm	0.131	0.197	0.394	0.525	0.787	1.05	1.58	2.10	2.49	2.89
T/Nm	1221.9	1245.6	1265.9	1284	1295	1402				
θ°/mm	3.28	3.68	4.07	4.46	4.72	8.79				

12.47 Determine the equivalent stress and strain at the point of instability for a thin-walled, open-ended pressurized cylinder of mean diameter d and thickness t, if the material hardens according to the Swift law $\bar{\sigma} = \sigma_0(1 + B\bar{\varepsilon}^P)^n$. What is the pressure that causes this instability?

Answer: $\bar{\sigma} = \sigma_0(2nB/3)^n$, $\bar{\varepsilon}^P = (1/B)(2nB/3 - 1)$, $p = (4t\sigma_0/d)(2nB/3)^n$

CHAPTER 13

CREEP AND VISCO-ELASTICITY

13.1 The Creep Curve

Creep is the time-dependent deformation occurring in a material subjected to steady loading over a prolonged period. While some metallic materials such as lead, copper and mild steel creep at room temperature under sufficiently high loads, the phenomenon is normally associated with the load capacity at higher temperatures, e.g. stainless steel in steam and chemical plant operating in excess of 450°C and nickel base alloys in gas turbines at 750°C and above. The understanding of the mechanics of creep begins with a macroscopic representation of the accumulation of strain ε_c with time t under a given stress σ and absolute temperature T. When σ and T are constant under uniaxial stress conditions, the three stage creep curve in Fig. 13.1a is normally observed.

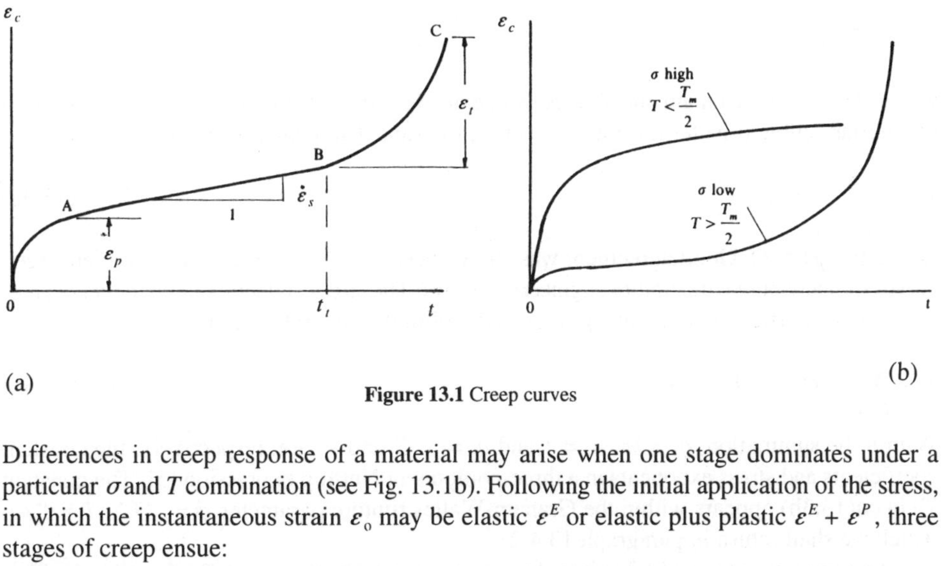

(a) (b)

Figure 13.1 Creep curves

Differences in creep response of a material may arise when one stage dominates under a particular σ and T combination (see Fig. 13.1b). Following the initial application of the stress, in which the instantaneous strain ε_0 may be elastic ε^E or elastic plus plastic $\varepsilon^E + \varepsilon^P$, three stages of creep ensue:

Primary Creep OA is a period of work-hardening in which the creep rate $d\varepsilon_c/dt$ decreases with time (also known as transient creep). As a result the material becomes harder to deform as an internal stress develops with increasing dislocation density.

Secondary Creep AB in which there is a balance between work-hardening and thermal softening. The latter is a recovery process activated by the energy within the dislocation structure. The result is that in the region AB the creep rate is constant and the material neither becomes harder nor softer. When AB is absent, the balance is achieved in a single point of inflection at which the creep rate is the minimum value for the curve. The secondary region normally forms the basis for engineering design, where a given creep strain can be tolerated, e.g. 0.01% in 10^4 hr (≈ 1½ years) or 0.1% in 10^5 hr (11½ years).

Tertiary Creep BC results from necking, cracking or metallurgical instability. It is characterised by an increasing creep rate culminating in fracture at point C.

13.1.1 Time Dependence of Creep Strain

A number of empirical functions $\varepsilon_c = \varepsilon_c(t)$ have been proposed to describe one or all of these stages within a specfied σ and T range. In the lower range of homologous temperatures $T/T_m < 0.5$, where T_m is the absolute melt temperature, primary creep dominates (see Fig. 13.1b). When the stress is low a limited amount of logarithmic primary creep is observed

$$\varepsilon_c = \alpha \ln t$$

where $\alpha = \alpha(\sigma, T)$. As σ increases, primary creep is better described by a Bailey power law [1]

$$\varepsilon_c = \alpha t^m \tag{13.1}$$

where $0 < m < 1$. For $m = ⅓$, eq(13.1) appears within Andrade's law [2] as follows

$$\varepsilon_c = (1 + \alpha t^{1/3}) \exp(\beta t) - 1 \approx \alpha t^{1/3}$$

since $\exp(\beta t) \rightarrow 1$ as $t \rightarrow 0$ in primary. With increasing time Andrade's β component of flow appears. This is now identified with the component of secondary creep (Fig. 13.1b) particularly for the higher temperature range $T/T_m > 0.5$. An equivalent description of high temperature creep is also achieved with the addition of a linear term to eq(13.1)

$$\varepsilon_c = \alpha t^m + \beta t \tag{13.2}$$

where $\beta = \beta(\sigma, T)$. Graham and Walles [3,4] extended eq(13.2) for $m = ⅓$, with the addition of a tertiary cubic time term for a wider range of σ and T in which the material will fail:

$$\varepsilon_c = \alpha t^{1/3} + \beta t + \gamma t^3 \tag{13.3a}$$

where $\gamma = \gamma(\sigma, T)$. Good agreement was shown between eq(13.3a) and the complete creep curves of many high temperature engineering alloys. The stress and temperature dependencies of the time coefficients α, β and γ appeared within the general formulation:

$$\varepsilon_c = \sum_{i=1}^{3} C_i \sigma^{m_i} t^{q_i} (T' - T)^{-20 q_i} \tag{13.3b}$$

Within the summation, $q_1 = ⅓$, $q_2 = 1$ and $q_3 = 3$. C_i and m_i are, respectively, the constant coefficients and stress exponents in a three-component Nutting expression [5]. The constant T' in eq(13.3b) appears within the Graham-Walles rupture parameter $\phi = t_f(T' - T)^{-20}$, to which we shall return in paragraph 13.4.2.

An alternative to eq(13.3a) is a three component expression of Davis et al [6]. This employs Garofalo's exponential-primary term [7] added to linear-secondary and exponential-tertiary terms in one of the following two forms

$$\varepsilon_c = \varepsilon_p[1 - \exp(-mt)] + \dot{\varepsilon}_s t + \varepsilon_L \exp[p(t - t_t)] \tag{13.4a}$$

$$\varepsilon_c = \varepsilon_p[1 - \exp(-mt)] + \dot{\varepsilon}_s t + \varepsilon_L[\exp(pt) - 1] \tag{13.4b}$$

where, with reference to Fig. 13.1a, ε_p in eq(13.4a,b) is the maximum amount of primary creep strain, $\dot{\varepsilon}_s$ is the secondary creep rate, and t_t is the time at the start of tertiary creep. Other constants α, β, γ, m, ε_L and p in eqs (13.1-13.4) are determined empirically for the given σ and T conditions. Note that $\varepsilon_o + \varepsilon_c$ is the net strain under σ.

13.1.2 Absent Secondary Creep

Where a creep curve (Fig. 13.1b) does not show a region of secondary creep, an inflexion occurs between the primary and secondary stages. At this point the creep rate is a minimum and it is possible to reproduce this with the omission of the linear term in either of eqs(13.4ab). The form employed [8] to predict the long-time creep deformation in a Cr, Mo, V ferritic steel is written in the form

$$\varepsilon_c = \theta_1 [1 - \exp(-\theta_2 t)] + \theta_3 [\exp(\theta_4 t) - 1] \qquad (13.4c)$$

Much work has been done on the determination of eq(13.4c) parameters θ_i (i = 1, 2, 3) when projecting from short to long service lives, typically in the region of 30 years [4]. Alternatively, the following two-term equation was proposed by de Lacombe [9] to fit the primary and tertiary creep strain regions

$$\varepsilon_c = A_1 t^p + A_2 t^q$$

where $0 < p < 1$ and $q > 1$ are constants and A_1 and A_2 are stress- and temperature-dependent functions. For a given σ and T, the exponents p and q are also constants that can be found by curve-fitting [10]. Numerical proceedures for this have been outlined by Conway [11].

13.2 Secondary Creep Rate

The secondary or minimum creep rate, $\dot{\varepsilon}_s$ in Fig. 13.1a, is the basis for most creep design. The rate, whilst being independent of time, often for long periods, will depend upon stress and temperature. Much attention has been paid to developing and validating suitable expressions for these dependencies in a range of high temperature alloys.

13.2.1 Stress Dependence

For a constant temperature T the family of creep curves in Fig. 13.2 applies to stress levels $\sigma_1 > \sigma_2 > \sigma_3 > \sigma_4$. As the curves display different amounts and rates of primary, secondary and tertiary creep, it follows that the strain function $\varepsilon_c = \varepsilon_c(t, \sigma)$ is complex.

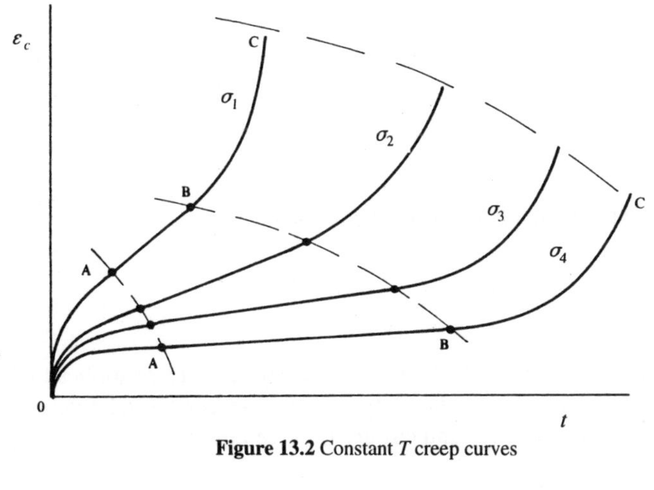

Figure 13.2 Constant T creep curves

However, in designing for creep the function is normally restricted to the secondary region where the creep rate ε_s depends solely upon σ. For lower stresses the linear law applies

$$\dot{\varepsilon}_s = A\sigma$$

where $A = A(T)$. For a medium stress range, the widely used Norton power law [12] applies

$$\dot{\varepsilon}_s = A\sigma^n \tag{13.5}$$

where $3 \le n \le 8$ is a material constant found from a double-log plot between ε_s and σ. Typical values of n, for common metals and alloys, are given in Table 13.1 for $T/T_m \ge 0.5$.

Table 13.1 Values of n and Q ($T/T_m \ge 0.5$)

Material	n	Q (kcal/mole)	T_m/K
Aluminium	5	34-36	930
Al-alloy	4	35	810
Al + Al_2O_3 dispersoids	10	150-400	800
Copper	5.5-6	47-55	1355
Ferrite	5	62	1810
0.23% C-steel	3-6	74	1750
0.56% C-steel	3-6	103	1750
Cr-V steel	12	143	1800
Ni-Cr steel	5	72	1800
Stainless (800 H)	6-8	75	1800
Lead	3	19-26	600
Nickel	5	66	1720
Magnesium	4	25-31	1590
Tin	4	21-36	505
Zinc	4-5	21	6802
Nimonic (80A)	8	98	1589

As the stress is increased, n often becomes dependent upon σ. An exponential law may then be preferable

$$\dot{\varepsilon}_s = A\exp(B\sigma)$$

where B is a material constant. Garofalo [7] proposed a hyperbolic sine law over the complete range of stress

$$\dot{\varepsilon}_s = A[\sinh(C\sigma)]^p \tag{13.6}$$

where C and p are experimentally determined constants.

13.2.2 Temperature Dependence

For a constant stress σ a further family of creep curves (see Fig. 13.3) applies to the absolute temperatures $T_1 > T_2 > T_3 > T_4$. As the temperature increases the relative proportions of lifetimes expended in secondary t_{AB} and tertiary t_{BC} creep increase.

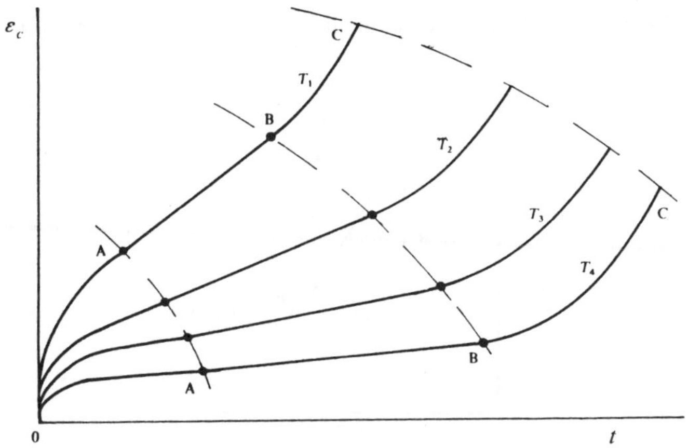

Figure 13.3 Constant σ creep curves

These trends would become cumbersome if expressed in a single function $\sigma_c = \sigma_c(t, T)$. However, the dependence of the secondary creep rate upon temperature $\dot{\varepsilon}_s = \dot{\varepsilon}_s(T)$ is uniquely defined by the thermal activation energy necessary for the dislocation recovery processes of climb and cross slip. This gives Arrhenius's diffusion expression for creep [13]:

$$\varepsilon_s = D \exp[-Q/(RT)] \tag{13.7}$$

where D is a constant, $R = 1.98$ cal/mole is the characteristic gas constant and Q is the creep activation energy, which may be assumed equal to the energy of self-diffusion of the metal when $T/T_m \geq ½$. Typical values of Q (kcal/mole) and T_m (K), for some metallics are listed in Table 13.1. At lower temperatures where creep is less dependent upon self diffusion, the Q values are lower, e.g. at $T/T_m = 0.25$, the value for aluminium is $Q = 25$ kcal/mole and for copper $Q = 23$ kcal/mole. The Q-value may be found experimentally under the given conditions from an $\dot{\varepsilon}_s$ versus $1/T$ plot as illustrated in the following example.

Example 13.1 The secondary creep rates given were obtained from a series of tensile creep tests at different temperatures all under the same load.

T, K	290	300	310	320	330
$\dot{\varepsilon}_s$, hr^{-1}	4.8×10^{-6}	2.74×10^{-5}	1.4×10^{-4}	6.44×10^{-4}	2.7×10^{-3}

Determine the activation energy Q in eq(13.7) and the constant D for secondary creep and so identify the material.

Note that identifying constant load with constant stress implies that the change in section area is negligible up to and during secondary creep. A true constant stress test reduces the load to account for loss in section. Taking the logarithm of eq(13.7) leads to

$$\ln \dot{\varepsilon}_s = -Q/(RT) + \ln D$$

Hence a plot of $\ln \dot{\varepsilon}_s$ versus $1/T$ in Fig. 13.4 results in a straight line of slope $-Q/R$. That is,

$$-Q/R = (\ln \dot{\varepsilon}_{s2} - \ln \dot{\varepsilon}_{s1})/(1/T_2 - 1/T_1)$$
$$= [\ln(1.5 \times 10^{-3}) - \ln(3 \times 10^{-6})]/(3.07 - 3.48)10^{-3} = -15158$$

$\therefore Q = 1.98 \times 15158 = 30012$ cal/mole, so identifying aluminium or one of its alloys.

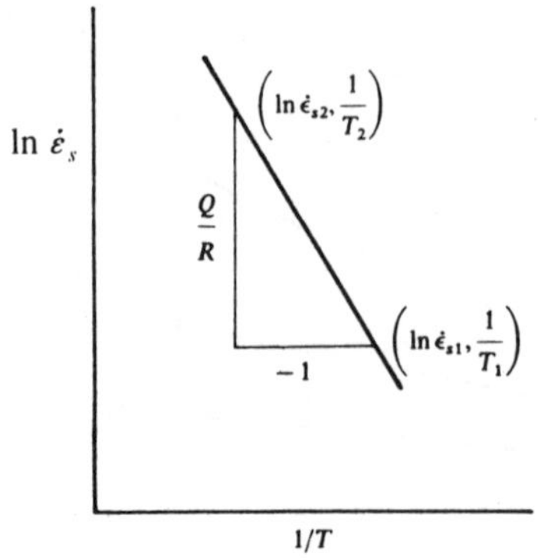

Figure 13.4 Arrhenius plot

Constant D is found from:

$$\ln D = \ln \varepsilon_s + Q/(RT) = \ln(1.5 \times 10^{-3}) + 30012/(1.98 \times 325.73)$$

$$= -6.502 + 46.534 = 40.03$$

$$\therefore D = 243 \times 10^{15}$$

13.2.3 Stress and Temperature Dependence

(a) Secondary Creep Rate

The separate equations describing the stress and temperature dependence of $\dot{\varepsilon}_s$ may be combined in a single function $\dot{\varepsilon}_s = (\sigma, T)$. For example, when $A = A(T)$ in eq(13.5) is identified with eq(13.7), this gives $\dot{\varepsilon}_s$ for the medium σ range

$$\dot{\varepsilon}_s = D\sigma^n \exp[-Q/(RT)] \tag{13.8}$$

Alternatively, when the function $A = A(T)$ in eq(13.6) is identified with eq(13.7) over the whole σ range

$$\dot{\varepsilon}_s = D[\sinh(C\sigma)]^p \exp[-Q/(RT)] \tag{13.9}$$

(b) Primary and Secondary Creep Strain

Depending upon the time dependence, there are various forms for the general creep strain function $\varepsilon_c = \varepsilon_c(\sigma, T, t)$. When primary creep strain is negligible the secondary creep strain is $\varepsilon_s = \dot{\varepsilon}_s t$, with $\dot{\varepsilon}_s$ given in eq(13.8) or (13.9). A commonly used function for primary creep strain combines eqs(13.1) and (13.8). This gives

$$\varepsilon_c = D\,\sigma^n t^m \exp[-Q/(RT)] \tag{13.10a}$$

If both primary and secondary creep are required, eq(13.2) indicates that a linear time term may be added to eq(13.10a). This gives

$$\varepsilon_c = D\sigma^n t^m \exp[-Q/(RT)] + D'\sigma^n t \exp[-Q/(RT)]$$
$$= (D\,t^m + D'\,t)\{\sigma^n \exp[-Q/(RT)]\} \tag{13.10b}$$

where D and D' are material constants. Equation (13.10b) shows that the stress and temperature dependence remains unaltered for each stage. In fact, the activation energy varies with primary creep strain to attain a constant value during secondary deformation [7]. Moreover, the respective stress dependencies of primary and secondary strains can differ respectively, as follows:

$\varepsilon_c = a\sigma + b\sigma^q$ and $\varepsilon_c = A\sigma^n$

where A, a and b are time and temperature dependent.

13.3 The Equation of State

The state equation assumes that for a given creep time or strain, the creep strain rate $\dot{\varepsilon}_c$ is a function of the current stress and temperature [14]. The implication is that the history of strain does not inflence $\dot{\varepsilon}_c$. Two state equations, strain and time hardening, result from this. In their general forms these are respectively

$$\dot{\varepsilon}_c = \dot{\varepsilon}_c(\sigma, T, t) \tag{13.11a}$$
$$\dot{\varepsilon}_c = \dot{\varepsilon}_c(\sigma, T, \varepsilon_c) \tag{13.11b}$$

which are related through the creep strain function $\varepsilon_c = \varepsilon_c(\sigma, T, t)$. Usually the state equations appear with separate function of σ, T, t and ε_c (see eqs 13.8 - 13.10). The differences between eqs(13.11a and b) are illustrated in Fig.13.5 for a uniaxial stepped stress.

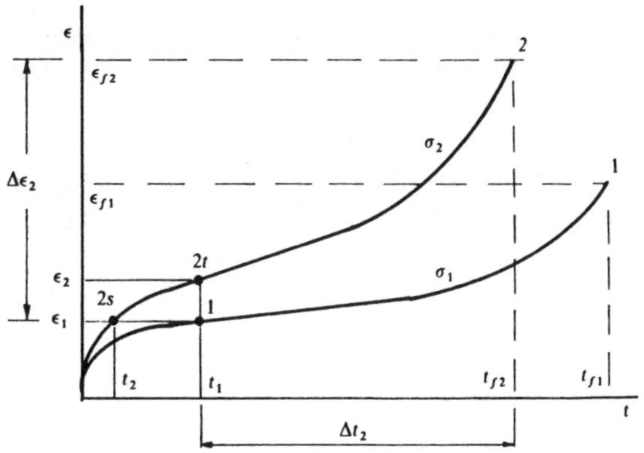

Figure 13.5 Time and strain hardening

This shows that after time t_1 the stress is increased from σ_1 to σ_2 at a constant temperature. According to eqs(13.11a,b), $\dot{\varepsilon}_c$ depends upon either the elapsed time or the accumulated strain. With time-hardening the new strain rate ε_c depends upon σ_2 at t_1 (path 1 - 2t) while for strain-hardening ε_c depends upon σ_2 at ε_1 (path 1 - 2s). For T = constant, eq(13.10a) describes the primary creep strain as

$$\varepsilon_c = A\,\sigma^n\,t^m \tag{13.12a}$$

The respective time- and strain-hardening rates follow from eqs (13.11a,b) and (13.12a) as

$$\dot{\varepsilon}_c = \dot{\varepsilon}_c(\sigma, t) = m\,A\,\sigma^n\,t^{m-1} \tag{13.12b}$$
$$\dot{\varepsilon}_c = \dot{\varepsilon}_c(\sigma, \varepsilon_c) = (m\,A^{1/m})\,\sigma^{n/m}\,\varepsilon_c^{(m-1)/m} \tag{13.12c}$$

Experiment shows that strain hardening is the more realistic although time-hardening may be used reliably for gradual stress changes. If the respective times and strains to fracture under σ_1 and σ_2 are (t_{f1}, ε_{f1}) and (t_{f2}, ε_{f2}), as shown, eqs(13.11a and b) supply the remaining strain ε_2 and lifetime t_2 according to the time and strain fraction rules [15]

$$\sum(t / t_f) = t_1 / t_{f1} + t_2 / t_{f2} = 1 \tag{13.13a}$$
$$\sum(\varepsilon / \varepsilon_f) = \varepsilon_1 / \varepsilon_{f1} + \varepsilon_2 / \varepsilon_{f2} = 1 \tag{13.13b}$$

Experiment has shown that the time fraction eq(13.13a) is more reliable than a fracture strain criterion. However, creep strain prior to fracture appears implicitly in the Monkman-Grant [16] criterion for rupture

$$\dot{\varepsilon}_s t_f^{\,k} = \text{constant} \tag{13.13c}$$

Normally as $k \approx 1$, the constant is equivalent to the secondary component of the fracture strain. This is often referred to as the true creep strain which is found to remain constant over a wide range of stress and temperature.

13.3.1 Relaxation

One useful application of the time-hardening eq(13.12a) is to predict stress relaxation behaviour. Relaxation refers to the stress decay following a stepped strain input. This is the opposite effect to creep, where strain accumulates following a stepped stress input. Consider, for example, the strain input to a bolt on initial tightening. As the initial stress σ_o will relax with time it becomes necessary to know when the bolts should be re-tightened to prevent the stress in them from falling below a critical level. If σ and ε_c are the stress and strain after time t then the net strain is

$$\sigma / E + \varepsilon_c = \text{constant} \tag{13.14a}$$

Differentiating with respect to time

$$(1/E)\, \mathrm{d}\sigma/\mathrm{d}t + \dot{\varepsilon}_c = 0 \tag{13.14b}$$

Substituting eq(13.12b) into eq(13.14b) leads to

$$(1/E)\, \mathrm{d}\sigma/\mathrm{d}t + mA\sigma^n t^{m-1} = 0$$

$$\int_0^t t^{m-1}\,\mathrm{d}t = -\,[1/(AEm)] \int_{\sigma_o}^{\sigma} \mathrm{d}\sigma/\sigma^n$$

$$(1/m)\,|\, t^m \Big|_0^t = -\,\{1/[(AEm)(1-n)]\}\,|\, \sigma_0^{\,1-n} - \sigma^{1-n} \Big|_{\sigma_o}^{\sigma}$$

$$t^m = \{1/[AE(n-1)]\}[1/\sigma^{n-1} - 1/\sigma_0^{\,n-1}] \tag{13.15a}$$

Had eq(13.11b) been used then from eq(13.14a),

$$(1/E)\, d\sigma/d\varepsilon_c + 1 = 0$$
$$\sigma = -\, E\varepsilon_c + C$$

Now as $\sigma = \sigma_0$ for $\varepsilon_c = 0$, then $C = \sigma_0$ and eq(13.12a) gives the relaxation time

$$t = [(\sigma_0 - \sigma)/(AE\sigma^n)]^{1/m} \tag{13.15b}$$

Example 13.2 In a cylindrical vessel a pressure of 1 MPa acts on a circular cover plate of area 0.13 m^2. The plate is held in position by twenty, 25 mm diameter steel bolts, equi-spaced around its rigid rim. What should be the initial tightening stress in the bolts in order that a safety factor of 2 is maintained after 10000 hr of creep relaxation at 455°C? After what time should the bolts be re-tightened to prevent leakage around the plate? Ignore primary creep but account for elasticity with $E = 172$ GPa and a secondary creep rate law of the form $\dot{\varepsilon}_c = 44.3 \times 10^{-16}\sigma^4$ for $\dot{\varepsilon}_c$ in hr^{-1} and σ in MPa. (CEI)

The force on the plate is 1 × 0.13 = 0.13 MN. Hence the minimum working stress in the bolts is given by

$$\sigma_w = (0.13 \times 4 \times 10^6)/(20 \times \pi \times 25^2) = 13.24 \text{ MPa}$$

The allowable relaxed stress is $\sigma = 2 \times \sigma_w = 26.48$ MPa. From eq(13.15a), with $m = 1$ for secondary creep, the initial tightening stress σ_o is

$$\begin{aligned}\sigma_0 &= 1/[1/\sigma^{n-1} - AEt\,(n-1)]^{1/(n-1)}\\ &= 1/[1/26.48^3 - 44.3 \times 10^{-16} \times 172 \times 10^3 \times 10000 \times 3]^{1/3}\\ &= 1/[(5.386 - 2.286)10^{-5}]^{1/3}\\ \sigma_0 &= 31.83 \text{ MPa}\end{aligned}$$

Leakage begins when the stress relaxes to its minimum working value $\sigma = 13.24$ MPa. Again from eq(13.15a) the corresponding time is

$$\begin{aligned}t &= [1/\sigma^{n-1} - 1/\sigma_0^{\,n-1}]/[AE\,(n-1)]\\ &= [1/13.24^3 - 1/31.83^3]/(3 \times 172 \times 10^3 \times 44.3 \times 10^{-16})\\ &= 174920 \text{ hr}\end{aligned}$$

Compare these time-hardening solutions with the less reliable strain hardening predictions from eq(13.15b) of $\sigma_0 = 30.23$ MPa and $t = 725620$ h.

13.3.2 Recovered and Anelastic Strains

Full unloading from within the primary region instantaneously recovers the initial elastic loading strain ε^E in Fig. 13.6a. The plastic strain component ε^P of the initial loading strain is permanent and does not recover. However, depending upon the σ and T, some creep strain ε_r may continue to recover so that a negative creep rate applies to time t_r during the recovery period.

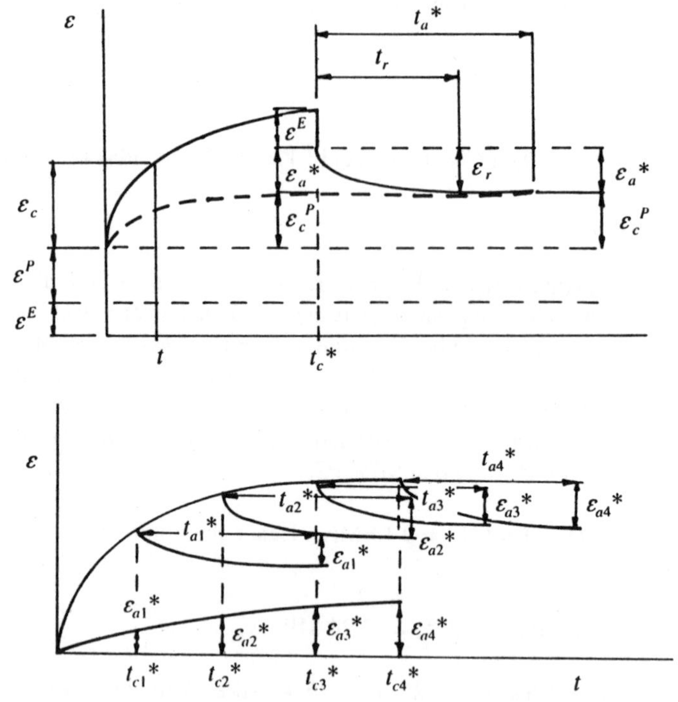

Figure 13.6 Recovery of primary creep strain

Figure 13.6a shows the case where the recovery rate diminishes to zero. That is, sufficient time t_a* has been allowed for all the anelastic strain ε_a* to recover following stress removal. A common interpretation of this behaviour is that the forward creep strain ε_c at time t_c* is composed of an anelastic component ε_a* and a permanent plastic strain component $\varepsilon_c^{\,P}$. The latter is the asymptotic strain value in Fig. 13.6a for a reversed creep curve following unloading at a time t_c*. The total permanent strain remaining after a full recovery period is $\varepsilon^P + \varepsilon_c^{\,P}$. We have from Fig. 13.6a two strain contributions to the forward creep strain

$$\varepsilon_c = \varepsilon_a* + \varepsilon_c^{\,P} \tag{13.16a}$$

At any time t on the forward creep curve the anelastic strain component can be be established from the difference betwen two measured strains: $\varepsilon_a* = \varepsilon_c - \varepsilon_c^{\,P}$, within a load-unload sequence (see Fig. 13.6b). This shows that for primary creep tmes t_{c1}*, t_{c2}*, t_{c3}* etc, anelastic strains ε_{c1}*, ε_{c2}*, ε_{c3}* etc, apply to recovery times t_{a1}*, t_{a2}*, t_{a3}* etc. Experiment [17] has shown that: (i) t_a* and ε_a* are each power functions of t_c*, (ii) ε_a* is the same order as the elastic loading strain $\varepsilon^E = \sigma/E$ and (iii) $\varepsilon_c^{\,P} >> \varepsilon_a$* unless the stress is very low. The anelastic strain ε_a* obeys a power law in stress and an exponential law in temperature. Taken with (i) the general form for ε_a* appears as a time-hardening state equation:

$$\varepsilon_a* = C\exp(T/T_0)\sigma^p t^m \tag{13.16b}$$

where C and T_0 are material constants. The exponents in eq(13.16b) depend upon the material, e.g for copper $p < 1$ and $m \approx 1/3$. The permanent plastic component follows from subtracting eq(13.16b) from eq(13.10a).

13.4 Brittle Creep Rupture

13.4.1 Damage in Tertiary Creep

In eqs(13.13a and b) the fractions of ductility ($\varepsilon/\varepsilon_f$) and life ($t/t_f$) expended in creep may be taken as measures of creep damage. The concept of creep damage was extended to brittle creep rupture by Katchanov [18] and Rabotnov [19]. This applies to low stress and long life in which a dominant tertiary stage results from early grain boundary cavitation and cracking. The loss of continuity is expressed in a structural damage parameter ω that takes extreme values $\omega = 0$ at the start of tertiary time ($t = 0$ for simplicity) and $\omega = 1$ at fracture ($t = t_f$). During tertiary creep, the strain rate $\dot{\varepsilon}$ and the damage rate $\dot{\omega}$ are assumed to depend upon σ and ω at a given temperature, in the forms:

$$\dot{\varepsilon} = A\sigma^n (1 - \omega)^{-q} \tag{13.17a}$$
$$\dot{\omega} = B\sigma^k (1 - \omega)^{-r} \tag{13.17b}$$

where A, B, k, n, q and r are positive material constants. The coupled eqs(13.17a,b) may be integrated to give rupture time and an equation to the tertiary creep curve. From eq(13.17b),

$$\int_0^\omega (1 - \omega)^r \,\mathrm{d}\omega = B\sigma^k \int_0^t \mathrm{d}t \tag{13.18a}$$

which gives the time $t = t_f$ to rupture for $\omega = 1$

$$t_f = 1/[B(1 + r)\sigma^k] \tag{13.18b}$$

With the upper limit of eq(13.18a) as $0 < \omega < 1$ it integrates to

$$1 - (1 - \omega)^{1+r} = B(1 + r)\sigma^k t$$

where from eq(13.18b),

$$(1 - \omega)^{1+r} = 1 - t/t_f \tag{13.18c}$$

Substituting eq(13.18c) into (13.17a) gives the tertiary creep strain

$$\varepsilon = A\sigma^n \int_0^t (1 - t/t_f)^{-q/(1+r)} \,\mathrm{d}t$$

$$= A\sigma^n t_f \lambda [1 - (1 - t/t_f)^{1/\lambda}]$$

$$\varepsilon/\varepsilon_f = 1 - (1 - t/t_f)^{1/\lambda} \tag{13.19a}$$

where $\lambda = (1 + r)/(1 + r - q)$ and the rupture strain is $\varepsilon_f = A\sigma^n t_f \lambda$. Equation (13.19) supplies a creep curve that is concave upwards (see Fig. 13.7), consistent with an increasing creep rate in the tertiary region. The ductility ratio λ is also given as

$$\lambda = \varepsilon_f / (A\sigma^n t_f) = \varepsilon_f / (\dot{\varepsilon}_s t_f) \tag{13.19b}$$

where from eq(13.13c), $\dot{\varepsilon}_s t_f =$ constant for $k \approx 1$. Thus λ in eq(13.19a,b) may be calculated from the measured quantities: minimum (or secondary) creep rate $\dot{\varepsilon}_s$, the rupture strain ε_f and the rupture time t_f as shown in Fig. 13.7.

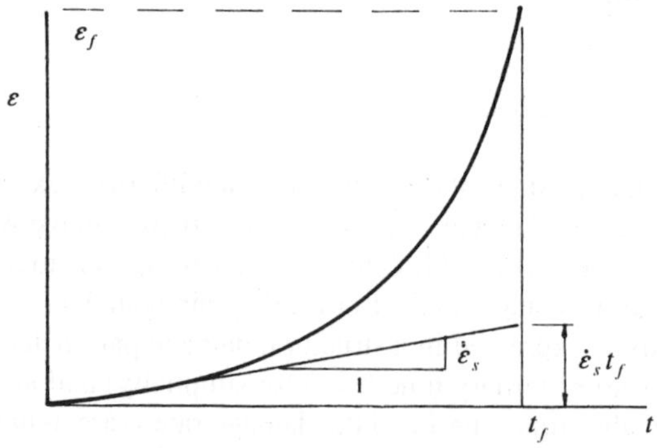

Figure 13.7 Tertiary creep curve

13.4.2 Rupture Life

The rupture life t_f in eq(13.18b) applies to low stress brittle intergranular creep fracture. It implies that there is no loss of section due to time-dependent (viscous) plastic flow. Equation (13.18b) will give a straight line of slope $-1/k$ on a double-log plot of σ_o versus t_f. Figure 13.8 illustrates this from an application to the tensile rupture behaviour of lead at 21°C.

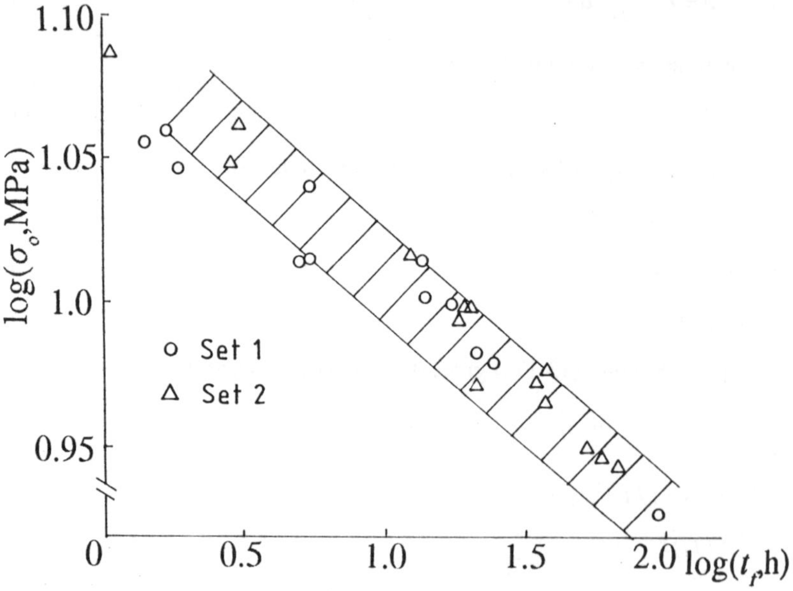

Figure 13.8 Creep rupture in lead

Despite some scatter in each set of testpieces, linear behaviour is observed in the lower stress region. The longer lives for this region are associated with brittle behaviour where rupture conforms to eq(13.18b) for $k = 11.5$ and $[B(1+r)]^{-1} = 10.23 \times 10^{12}$.

Empirical fits to the rupture curve have employed exponential and hyperbolic sine functions of stress [20]. Rupture data, extending to very long durations in commercial alloys, appear to show that the stress rupture plot is better approximated with a number of piecewise linear segments. The proper representation of the rupture curve becomes important to consider when extrapolating short-term data to longer lives. The segmented rupture curve is replicated within the Graham and Walles [3,4] extrapolation procedure. They assumed that the rupture strain was constant and that the tertiary term in eq(13.3a) was dominant. Applying this to eq(13.3b) provides their stress-time relationship for rupture.

$$\varepsilon_f = (C_1' \sigma^{m1} + C_2' \sigma^{m2} + C_3' \sigma^{m3})\, t_f^{\,q} = \text{constant} \tag{13.20a}$$

C_i' (i = 1, 2, 3) in eq(13.20a) are given by the temperature-dependent functions:

$$C_i' = C_i (T' - T)^{-20q} \tag{13.20b}$$

which attain constant values at a given temperature. The choice of a tertiary time exponent value of either $q = 3$ or $q = 9$ in eqs(13.20a,b) depended upon the best fit to a given material. The stress exponent m_i (i = 1, 2, 3) in eq(13.20a) took appropriate values from the geometric progression m_i/q =1, 2, 4, 8 ... A steep tertiary gradient contains the rupture times within a narrow band despite any variation that might be present in rupture strain. Taking logarithms and differentiating eq(13.20a) shows that the slopes of segments in Fig. 13.9 take values from $-q/m_i = -1, -1/2, -1/4, -1/8$....

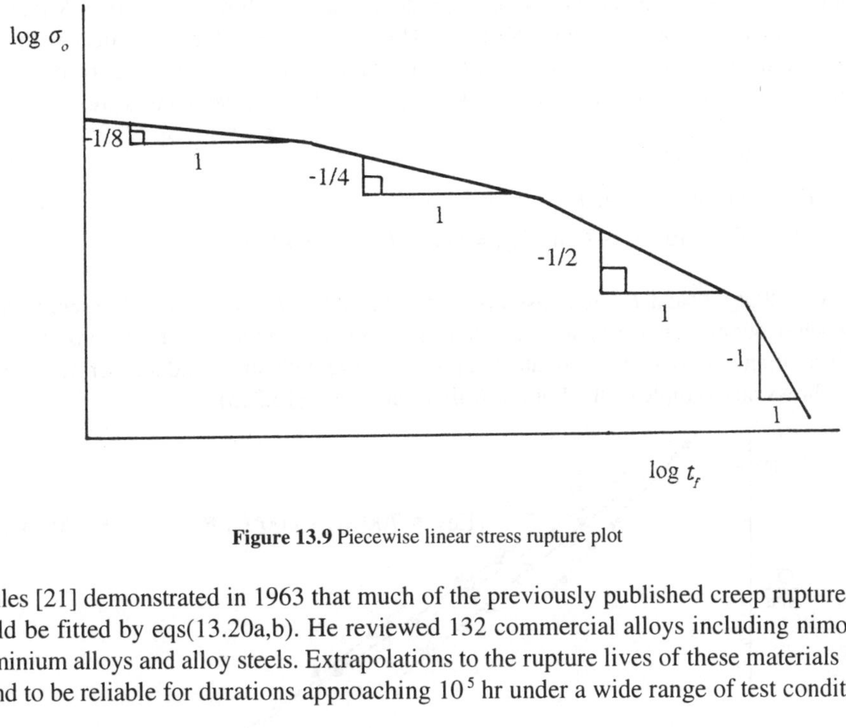

Figure 13.9 Piecewise linear stress rupture plot

Walles [21] demonstrated in 1963 that much of the previously published creep rupture data could be fitted by eqs(13.20a,b). He reviewed 132 commercial alloys including nimonics, aluminium alloys and alloy steels. Extrapolations to the rupture lives of these materials were found to be reliable for durations approaching 10^5 hr under a wide range of test conditions.

13.5 Correlation of Uniaxial Creep Data

State-equation parameters have been proposed for correlating both the creep deformation and rupture life of a material under given steady stress and temperature conditions. The Dorn [13] parameter $\theta = \theta(t, T)$ appears in the strain function $\varepsilon_c = \varepsilon_c(\sigma, \theta)$. When σ and θ are separated, ε_c may be plotted against θ, for a constant σ, to produce a single curve over a range of T (see Fig. 13.10a).

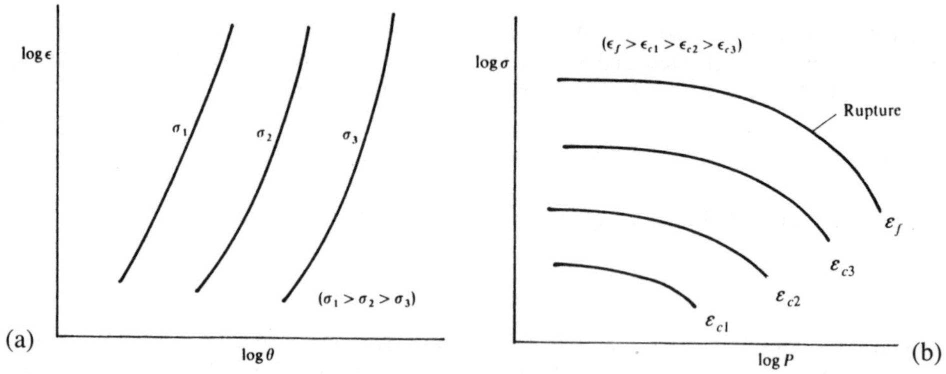

Figure 13.10 Creep parameters

Dorn introduced a parameter to correlate uniaxial creep data for pure metals.

$$\theta = t \exp\left[-Q/(RT)\right] \tag{13.21}$$

Equation (13.21) was found to be less appropriate for metallic alloys. The Larson-Miller [22], Manson-Haferd [23] and Graham-Walles [3,4] parameters $P = P(t, T)$, appear with creep strain ε_c in the stress function $\sigma = \sigma(\varepsilon_c, P)$. Thus, for a constant ε_c, the plot of σ versus P produces a single curve for a range of T (see Fig. 13.10b). The parameters are

$$P_{LM} = T\,(C + \log_{10} t\,) \tag{13.22a}$$

$$P_{MH} = (T - T_a)/(\log_{10} t - \log_{10} t_a) \tag{13.22b}$$

$$P_{GW} = t\,(T' - T\,)^{-20} \text{ for } T' > T \text{ and } P_{GW} = t\,(T - T'\,)^{20} \text{ for } T > T' \tag{13.22c}$$

where $C \approx 20$, t_a, T_a and T' are constants. Each of eqs(13.22a-c) was found to represent the creep behaviour of engineering alloys within an acceptable scatter band. Equations (13.22a-c) are more often employed to correlate creep rupture life with stress and temperature. Figure 13.11 shows an example of the Larson-Miller parameter (13.22a).

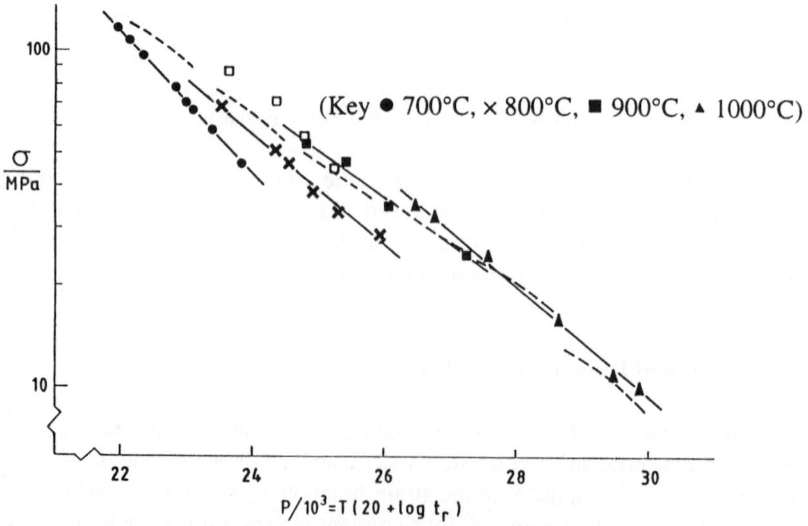

Figure 13.11 Effect of σ and T on the rupture life of Alloy 800H

The rupture data apply to an austenitic stainless steel (Alloy 800H) under uniaxial stress in the range 10 - 200 MPa for temperatures in the range 650 - 1000°C. Manufacturer's data (broken lines) apply to lower stresses and longer rupture lives at each temperature. Data for shorter rupture lives are taken from published sources for this alloy [24]. Figure 13.11 shows that all these data contract approximately on to a single master curve with an ordinate of stress and abscissa defined by eq(13.22a) for $C = 20$. The scatter band reveals that errors can arise from extrapolating shorter time data to longer lives at a given stress and temperature.

Example 13.3 Use Fig. 13.11 to estimate the creep rupture life of an austenitic stainless steel under 11 MPa at 925°C. Compare with the life estimate for similar conditions when the Larson-Miller parameter is based upon two tensile creep rupture lives: $t_f = 10^4$ hr under $\sigma = 121$ MPa at 650°C and $t_f = 5 \times 10^4$ hr under $\sigma = 28$ MPa at 815°C. Take $C = 20$.

Reading from the manufacturer's data (lower broken line) in Fig. 13.11b gives $P_{LM} = 29.28 \times 10^3$. Substituting into eq(13.22a) gives

$$29.28 \times 10^3 = (925 + 273)(20 + \log t)$$

from which the life is $t = 27588$ h. Substituting the two test conditions in eq(13.22a) gives

$$P_{LM} = (650 + 273)(20 + \log 10^4) = 923(20 + 4) = 22152$$
$$P_{LM} = (815 + 273)[20 + \log(5 \times 10^4)] = 1088(20 + 4.699) = 26872.5$$

If it is assumed that P_{LM} and σ plot linearly on double log axes within 1 logarithmic decade of stress, then $\sigma = N\,(P_{LM})^m$ so that

$$\log \sigma = m \log P_{LM} + \log N$$
$$\log 121 = m \log 22152 + \log N$$
$$\log 28 = m \log 26872.5 + \log N$$

from which $m = -\ 7.5757$, $\log N = 35$. Then for $\sigma = 11$ MPa, $P_{LM} = 30379$,

$$\log t_f = P_{LM}/T - 20 = 30379/(925 + 273) - 20 = 5.358$$

$\therefore\ t_f = 227{,}986$ h which is clearly an overestimation. The example illustrates the danger in extrapolating data without a reference curve to embrace all the test conditions.

13.6 Creep in Structures

The creep strain function $\varepsilon_c = \varepsilon_c\,(\sigma, T, t)$ may be applied to structures that creep. The following analyses apply to creep within beams, torsion bars and thin-walled pressure vessels. Where a stress distribution exists, i.e. for bending and torsion, the analysis of creep is based upon the elastic stress state existing at the skeletal point. The stress at this point is not altered by the stress redistribution that occurs during creep. A similar approach is applied to structures with multiaxial stress states that vary throughout the volume. This will be demonstrated for a thick-walled pressurised cylinder under steady state creep.

13.6.1 Beam in Bending

Let the power law (eq 13.12a) be written with a time function $\varepsilon_c = A\sigma^n f(t)$. Applying this to a beam at constant temperature, the moment equilibrium relation (4.2a) to be satisfied is

$$M = \int_A \sigma y \, \mathrm{d}A \tag{13.23}$$

The condition (4.2c) of zero axial force $F = \int_A \sigma \, \mathrm{d}A = 0$, implies that the neutral axis passes through the centroid of the section. In addition, plane sections remain plane when, from eq(4.1), $\varepsilon = y/R$. Thus, when instantaneous elastic and plastic strains are ignored, a simplified analysis gives the creep strain as

$$\varepsilon_c = y/R = A\sigma^n f(t) \tag{13.24}$$

Substituting eq(13.24) into (13.23)

$$M = [ARf(t)]^{-1/n} \int_A y^{(1+1/n)} \, \mathrm{d}A = I_c[ARf(t)]^{-1/n} \tag{13.25a}$$

where the moment of "inelastic" area is

$$I_c = \int_A y^{(1+1/n)} \, \mathrm{d}A \tag{13.25b}$$

For a rectangular section $b \times d$ with $\delta A = b\delta y$, eq(13.25b) gives

$$I_c = 2b \int_0^{d/2} y^{1+1/n} \, \mathrm{d}y = [2bn/(1+2n)](d/2)^{(1+2n)/n} \tag{13.25c}$$

Combining eqs(13.24) and (13.25a) completes the comparison with elastic bending theory (i.e. for $n = 1$ the creep theory reduces to eq(4.3)

$$M/I_c = 1/[ARf(t)]^{1/n} = \sigma/y^{1/n} \tag{13.26}$$

The skeletal position y_{sk} in the section, for which the elastic and creep stresses would be the same under the given M, is found by equating (4.3) and (13.26)

$$\sigma = M y_{sk}/I = M y_{sk}^{1/n}/I_c \tag{13.27a}$$

This gives

$$y_{sk} = (I/I_c)^{n/(n-1)} \tag{13.27b}$$

Thus, as an elastic stress calculation applies to this point, it is often used to simplify design procedures involving creep.

Provided creep deflections are small, the flexure equation may be combined with eq(13.26) to give

$$\mathrm{d}^2v/\mathrm{d}z^2 = 1/R = Af(t)[M(z)/I_c]^n \tag{13.28}$$

where $M(z)$ means the moment M is a function of length z. When elastic, plastic and thermal strains also contribute to the net strain, eq(13.24) becomes

$y/R = \sigma/E + \varepsilon^P + \alpha T + \varepsilon_c$
$\therefore\ \sigma = E\,(y/R - \varepsilon^P - \alpha T - \varepsilon_c)$

in which α is the expansion coefficient. The σ expression must be substituted into eq(13.23) and integrated numerically.

Example 13.4 A 0.5 m long cantilever with rectangular section 50 mm wide × 25 mm deep carries an end load of 45 N at 300°C. Neglecting elasticity and self-weight, determine the deflection at the centre and at the end following 10 hr of secondary creep at a rate given by $\dot{\varepsilon}_c = 3.5 \times 10^{-6}\sigma^2$ with $\dot{\varepsilon}_c$ in hr^{-1} and σ in MPa. What is the maximum stress in the beam and where is the skeletal point? (IC)

Here $M(z) = 45z$ (Nmm) with the origin of z at the free end. The secondary creep strain is simply $\varepsilon_c = 3.5 \times 10^{-6}\sigma^2 t$. Equation(13.25c) gives

$I_c = (2 \times 50 \times 2/5)(25/2)^{2.5} = 22097.09$

Hence eq(13.28) becomes

$d^2v/dz^2 = (3.5 \times 10^{-6})\,t\,(45z / 22097.09)^2 = (1.452 \times 10^{-11} t)\,z^2$
$dv/dz = (1.452 \times 10^{-11} t)\,z^3/3 + C_1$
$v = (1.452 \times 10^{-11} t)\,z^4/12 + C_1 z + C_2$

The boundary conditions are

$dv/dz = 0$ when $z = 500$ mm, $\Rightarrow\ C_1 = -\,0.605 \times 10^{-3} t$
$v = 0$ when $z = 500$ mm, $\Rightarrow\ C_2 = 0.2269\,t$
$\therefore\ v = (1.452 \times 10^{-11} t)\,z^4/12 - (0.605 \times 10^{-3} t)\,z + 0.2269\,t$

After $t = 10$ hr at $z = 0$ mm: $v = 2.269$ mm and after $t = 10$ hr at $z = 250$ mm

$v = 0.04726 - 1.5125 + 2.269 = 0.8038$ mm

The maximum stress occurs at the fixed end where $M = 45 \times 500 = 22500$ Nmm and $y = 12.5$ mm. From eq(13.26),

$\sigma = M\,y^{1/n}/I_c$
$= 22500 \times (12.5)^{0.5}/22097.09 = 3.6$ MPa

which is independent of time and distributed throughout the section as shown in Fig. 13.12. The position of the skeletal point is also shown at the intersection between the creep and elastic stress distributions. From eq(13.27b),

$y_{sk} = (50 \times 25^3)/(12 \times 22097.09)^2 = 8.68$ mm

Note that the initial elastic stress under M redistributes during creep to a more uniform condition through the depth.

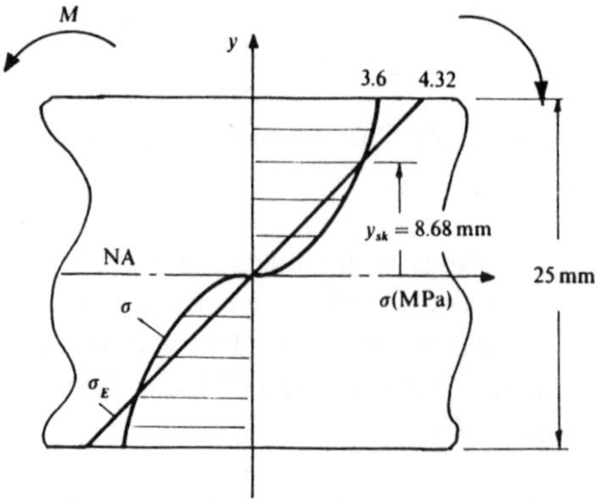

Figure 13.12 Skeletal point in beam section

13.6.2 Circular Shaft in Torsion

In a thick-walled tube of inner and outer radii r_i and r_o under torque T, the shear strain γ at radius r, after time t, is given by

$$\gamma = r\theta/\, l = A[\tau(r)]^n f(t) \tag{13.29a}$$

in which $\tau(r)$ indicates that τ varies with r. Any elastic and plastic shear strain under T are ignored. Equation(13.29a) is derived from $\varepsilon_c = A\sigma^n f(t)$ using equivalence relations (12.88) and (12.89) for torsion, $\varepsilon = \gamma/\sqrt{3}$ and $\sigma = \sqrt{3}\,\tau$. The shear stress is found from eq(13.29a)

$$\tau(r) = \{\theta/\,[A\,l f(t)]\}^{1/n} r^{1/n} = K\,r^{1/n} \tag{13.29b}$$

Now, the equilibrium condition (5.4) is

$$T = 2\pi \int_{r_i}^{r_o} \tau(r)\, r^2\, \mathrm{d}r \tag{13.30a}$$

Substituting eq(13.29b) into (13.30a) and integrating leads to

$$T = 2\pi K n\,(r_o^{3+1/n} - r_i^{3+1/n})\,/\,(1 + 3n) \tag{13.30b}$$

Eliminating K between eqs(13.29b) and (13.30b) gives

$$\tau(r) = (1 + 3n)T\,r^{1/n}/\,[2\pi n\,(r_o^{3+1/n} - r_i^{3+1/n})] \tag{13.31}$$

and combining eqs(13.29b) and (13.31) to complete the comparison with elastic torsion theory (for $n = 1$, see eq 5.6), we have, for creep

$$T/J_c = \{\theta/\,[\,A l f(t)]\}^{1/n} = \tau/r^{1/n} \tag{13.32a}$$

where, for a hollow shaft,

$$J_c = 2\pi n\,(r_o^{3+1/n} - r_i^{3+1/n})/(1 + 3n) \tag{13.32b}$$

For a solid shaft $r_i = 0$,

$$J_c = 2\pi n\, r_o^{3+1/n} / (1 + 3n) \tag{13.32c}$$

The skeletal radius r_{sk} at which the elastic and creep shear stresses are the same under the same T is found from

$$\tau = T r_{sk}/J = T\, r_{sk}^{1/n}/J_c \tag{13.33a}$$

For a tube, eq(13.33a) gives

$$r_{sk} = \left(\frac{J}{J_c}\right)^{n/(n-1)} = \left[\frac{(1+3n)(r_o^4 - r_i^4)}{4n\,[r_o^{(3n+1)/n} - r_i^{(3n+1)/n}}\right]^{n/(n-1)} \tag{13.33b}$$

For a solid shaft, $r_i = 0$ and eq(13.33b) simplifies to

$$r_{sk} = [(1 + 3n)/(4n)]^{n/(n-1)}\, r_o \tag{13.33c}$$

We can now determine an elastic reference stress at the skeletal radius from eq(13.33a,b) and use this stress to calculate the creep rate for the bar material.

Example 13.5 Determine the position of the skeletal radius for a tube inner radius 5 mm and outer radius 25 mm when it is subjected to a constant torque of 1 kNm under secondary creep. Compare elastic and steady state creep shear stress distributions for a stress index of $n = 5$.

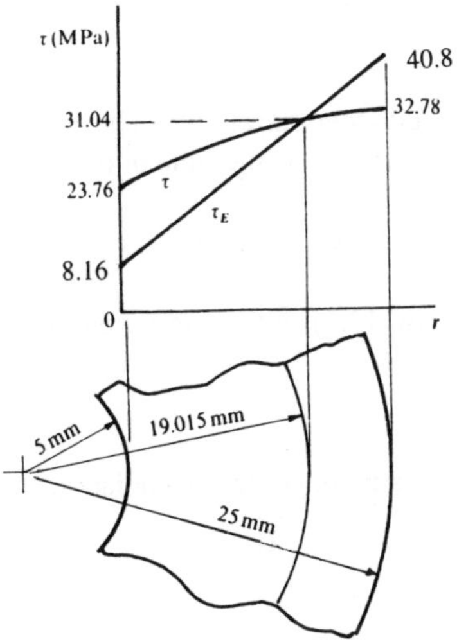

Figure 13.13 Stress distributions

Substituting $r_i = 5$ mm, $r_o = 25$ mm and $n = 5$ in eq(13.33b),

$$r_{sk} = \{[16\,(25^4 - 5^4)]/[20(25^{3.2} - 5^{3.2})]\}^{1.25} = 19.015$$

and from eq(13.31) the creep shear stress distribution is given by

$$\tau_c = (16 \times 1 \times 10^6 \times r^{0.2})/\,(10\pi \times 29572.13) = 17.22\, r^{0.2}\ \text{MPa} \qquad \text{(i)}$$

The elastic shear stress is found from eq(5.6)

$$\tau_E = Tr\,/\,J = (1 \times 10^6 \times r \times 2)/\,[\pi(25^4 - 5^4)] = 1.632\, r\ \text{MPa} \qquad \text{(ii)}$$

The elastic and creep stress distributions from eqs(i) and (ii) are shown in Fig. 13.13. It is evident that the initial, instantaneous elastic stress under T redistributes during primary creep to attain the more uniform condition shown under secondary creep.

13.7 Multiaxial Stress

Assume that creep is the result of dislocation slip mechanisms, as with low temperature plasticity. The time derivatives of eqs(12.93a-c) will supply the creep rates under multiaxial principal stress states. The principal creep rates are written as

$$\dot{\varepsilon}_1 = (\dot{\bar{\varepsilon}}_c/\bar{\sigma})[\sigma_1 - \tfrac{1}{2}(\sigma_2 + \sigma_3)] \qquad \text{(13.34a)}$$
$$\dot{\varepsilon}_2 = (\dot{\bar{\varepsilon}}_c/\bar{\sigma})[\sigma_2 - \tfrac{1}{2}(\sigma_1 + \sigma_3)] \qquad \text{(13.34b)}$$
$$\dot{\varepsilon}_3 = (\dot{\bar{\varepsilon}}_c/\bar{\sigma})[\sigma_3 - \tfrac{1}{2}(\sigma_1 + \sigma_2)] \qquad \text{(13.34c)}$$

where $\dot{\varepsilon}_1 = d\varepsilon_1^{\,P}/dt$ is the major principal creep strain rate, $\dot{\varepsilon}_3 = d\varepsilon_3^{\,P}/dt$ is the minor principal strain rate and $\dot{\varepsilon}_2 = d\varepsilon_2^{\,P}/dt$ is an intermediate principal strain rate. That is $\dot{\varepsilon}_3 > \dot{\varepsilon}_2 > \dot{\varepsilon}_3$ when $\sigma_1 > \sigma_2 > \sigma_3$ (numerically). Eqs(13.34a-c) shows that the sum of these three rates is zero:

$$\dot{\varepsilon}_1 + \dot{\varepsilon}_2 + \dot{\varepsilon}_3 = 0 \qquad \text{(13.35)}$$

which implies that creep from slip is incompressible. The equivalent stress $\bar{\sigma}$ is defined in eq(12.88) and the equivalent creep rate $\dot{\bar{\varepsilon}}_c$ is the time derivative of eq(12.89). This is

$$\dot{\bar{\varepsilon}}_c = (\sqrt{2}/3)\sqrt{[(\varepsilon_1 - \varepsilon_2)^2 + (\varepsilon_2 - \varepsilon_3)^2 + (\varepsilon_1 - \varepsilon_3)^2]} \qquad \text{(13.36)}$$

Now $\bar{\varepsilon}_c$ and $\bar{\sigma}$ obey the Norton law(13.5) so the coefficient in eqs(13.34a-c) follows from

$$\dot{\bar{\varepsilon}}_c = A\,\bar{\sigma}^{\,n}, \quad \dot{\bar{\varepsilon}}_c/\bar{\sigma} = A\,\bar{\sigma}^{\,n-1} \qquad \text{(13.37a,b)}$$

Equations (13.34a-c) appear within a more general formulation of the creep strain rate tensor. From eq(12.92),

$$\dot{\varepsilon}_{ij} = \dot{\lambda}\,\sigma_{ij}'\,f'(t) \qquad \text{(13.38a)}$$

where $\dot{\lambda}$ is a creep modulus, $f'(t)$ is a time derivative and σ_{ij}' is the deviatoric stress tensor from eq(1.21a)

$$\dot{\lambda} = (3\dot{\bar{\varepsilon}}_c)/(2\bar{\sigma}) \text{ and } \sigma_{ij}' = \sigma_{ij} - \tfrac{1}{3}\,\delta_{ij}\,\sigma_{ii} \qquad \text{(13.38b,c)}$$

Since $i = 1, 2, 3$ and $j = 1, 2, 3$, eq(13.38a) supplies six independent strain rates: three normal and three shear within the stage of creep defined by the time derivative $f'(t)$. For example, when the three stages of creep can be described by each term in eq(13.3a), we have

(i) in primary $f(t) \approx \alpha t^{1/3}$, $\Rightarrow f'(t) \propto (\alpha/3)\, t^{-2/3}$,

(ii) in secondary $f(t) = \beta t$, $\Rightarrow f'(t) = \beta$

(iii) in tertiary $f(t) \approx \gamma t^3$, $\Rightarrow f'(t) \propto 3\gamma t^2$

While (iii) will fit the tertiary curve reasonably well, it ignores the structural degradation that occurs. An alternative approach is to describe the increasing tertiary creep rates from the loss in load-bearing area that occurs. Thus, in certain cavity-forming alloys, e.g. stainless steels and nimonics, the Katchanov-Rabotnov damage parameter ω in (13.17a) will modify eqs(13.37b) and (13.38b) to

$$\dot{\bar{\varepsilon}}_c / \bar{\sigma} = A\, \bar{\sigma}^{\,n-1} (1 - \omega)^{-q} \tag{13.39a}$$

$$\dot{\lambda} = (3\, \dot{\bar{\varepsilon}}_c)/(2\bar{\sigma}) = (3/2) A \bar{\sigma}^{\,n-1} (1 - \omega)^{-q} \tag{13.39b}$$

in which ω is supplied from eq(13.17b). Substituting eq(13.39b) into eq(13.38a) gives the tertiary creep rate tensor (13.38a) as

$$\dot{\varepsilon}_{ij} = (3/2)\, A\, \bar{\sigma}^{\,n-1} (1 - \omega)^{-q} \sigma_{ij}' \tag{13.40}$$

Equation (13.40) also applies to secondary creep for the undamaged material when $\omega = 0$.

13.7.1 Thin-Walled Pressurised Vessels

Creep becomes a problem in pressure vessels when the wall material experiences stress and temperature within the creep regime through the fluid medium. We may make the usual assumptions that radial stress may be ignored in both cylindrical and spherical vessels so that the wall is under plane principal stress. The stresses may be assumed constant and so no stress redistribution occurs. The following example shows how eqs(13.34) and (13.37) will provide the steady creep rates in a given direction.

Example 13.6 A thin-walled pressure vessel of diameter d and wall thickness t is to be fitted with hemispherical ends. In addition, the vessel is to be designed to withstand an internal pressure p when operating under creep conditions. Assuming that the most satisfactory design life is obtained when the circumferential strain rates at the cylinder-hemisphere interface are matched, calculate the thickness of the ends if the material obeys a uniaxial Norton law $\dot{\bar{\varepsilon}}_c = A\sigma^4$. Neglect elastic strains. (IC)

With zero radial stress everywhere, the circumferential creep rate for the cylinder (subscript c) is, from eq(13.34a),

$$\dot{\varepsilon}_\theta = A\, \bar{\sigma}^{\,n-1} (\sigma_\theta - \sigma_z/2) \tag{i}$$

where, for biaxial stresses, the equivalent stress (eq 12.88) becomes

$$\bar{\sigma}_c = \sqrt{(\sigma_\theta^{\,2} - \sigma_\theta \sigma_z + \sigma_z^{\,2})} \tag{ii}$$

Substituting eq(ii) in (i) with $\sigma_\theta = pd/(2t_c)$, $\sigma_z = pd/(4t_c)$ leads to

$$\dot{\varepsilon}_\theta = A\,\bar{\sigma}_c^{\,n-1}\,[3pd/(8t_c)]$$
$$= A\,[3p^2d^2/(16t_c^2)]^{(n-1)/2}\,[3pd/(8t_c)] \qquad \text{(iii)}$$

For the hemispherical ends (subscript s), eqs(12.88) and (13.34a) reduce to

$$\dot{\varepsilon}_\theta = A\,\bar{\sigma}_s^{\,n-1}(\sigma_\theta - \tfrac{1}{2}\sigma_\phi) \qquad \text{(iv)}$$
$$\bar{\sigma}_s = \sqrt{(\sigma_\theta^2 - \sigma_\theta\sigma_\phi + \sigma_\phi^2)} \qquad \text{(v)}$$

Substituting eq(v) into (iv) with $\sigma_\theta = \sigma_\phi = pd/(4t_s)$ gives

$$\dot{\varepsilon}_\theta = A\,\bar{\sigma}_s^{\,n-1}\,[pd/(8t_s)]$$
$$= A\,[p^2d^2/(16t_s^2)]^{(n-1)/2}\,[pd/(8t_s)] \qquad \text{(vi)}$$

Equating (iii) and (vi) gives the thickness ratio:

$$t_c/t_s = [3^{(n+1)/2}]^{1/n} = 3^{(n+1)/2n} = 3^{5/8} = 1.987$$

13.7.2 Thick-Walled Cylinder

In a cylinder with closed or constrained ends the axial creep rate is zero. Using eq(13.35):

$$\dot{\varepsilon}_z = 0,\ \dot{\varepsilon}_r = -\dot{\varepsilon}_\theta \qquad \text{(13.41a,b)}$$

Let u be the radial velocity at a radius r in the wall, so that $\dot{\varepsilon}_\theta = u/r$ and $\dot{\varepsilon}_r = \mathrm{d}u/\mathrm{d}r$. Clearly if the two strain rates are to depend upon a single velocity they must obey compatibility:

$$r\,\mathrm{d}\dot{\varepsilon}_\theta/\mathrm{d}r = \dot{\varepsilon}_r - \dot{\varepsilon}_\theta \qquad \text{(13.42)}$$

Combining eqs(13.41a) and (13.42) and integrating leads to

$$\dot{\varepsilon}_\theta = C/r^2 \qquad \text{(13.43)}$$

where C is an integration constant. Now from eqs(13.41a,b) the equivalent creep rate (eq 13.36) reduces to

$$\dot{\bar{\varepsilon}}_c = 2\varepsilon_\theta/\sqrt{3} = (2/\sqrt{3})(C/r^2) \qquad \text{(13.44)}$$

Also, from eqs(13.41a) and (13.34c), the axial stress is

$$\sigma_z = \tfrac{1}{2}(\sigma_\theta + \sigma_r) \qquad \text{(13.45)}$$

Substituting eq(13.45) into the equivalent stress expression (12.88) leads to

$$\bar{\sigma} = (\sqrt{3}/2)(\sigma_\theta - \sigma_r) \qquad \text{(13.46)}$$

We may combine eqs(13.44) and (13.46) within the Norton law (13.37a) to give

$$(2/\sqrt{3})(C/r^2) = A[(\sqrt{3}/2)(\sigma_\theta - \sigma_r)]^n$$
$$\sigma_\theta - \sigma_r = C_1/r^{2/n} \qquad \text{(13.47)}$$

where $C_1 = (C/A)^{1/n}(2/\sqrt{3})^{(n+1)/n}$. We have previously derived equilibrium eq(3.13) for when both hoop and radial stresses vary with radius r in the cylinder wall. Substituting eq(13.47) into eq(3.13) and integrating gives the radial stress expression

$$\sigma_r = C_1 \int r^{-(1+2/n)}\, \mathrm{d}r$$
$$= C_2 - C_1 n/(2\, r^{2/n}) \tag{13.48a}$$

and from eqs(3.13), (13.45) and (13.48a),

$$\sigma_\theta = \sigma_r + r\, \mathrm{d}\sigma_r/\mathrm{d}r$$
$$= C_2 + C_1(1 - n/2)\, /\, r^{2/n} \tag{13.48b}$$
$$\sigma_z = C_2 + C_1(1 - n)\, /\, (2r^{2/n}) \tag{13.48c}$$

At this stage C_1 and C_2 are found from the boundary conditions. Let us take a cylinder where (i) $\sigma_r = 0$ when $r = r_o$ and (ii) $\sigma_r = -\,p$ when $r = r_i$. Substituting (i) and (ii) into eq(13.48a) gives two simultaneous equations, from which

$$C_1 = -\,2pr_o^{\,2/n}/\{n[1 - (r_o/r_i)^{2/n}]\} \tag{13.49a}$$
$$C_2 = -\,p\,/[1 - (r_o/r_i)^{2/n}] \tag{13.49b}$$

Substituting eqs(13.49a,b) into eqs(13.48a-c) gives the three principal stresses in the wall as

$$\sigma_r = -\,p\,[(r_o/r)^{2/n} - 1]/\,[(r_o/r_i)^{2/n} - 1] \tag{13.50a}$$
$$\sigma_\theta = p\,[(2/n - 1)(r_o/r)^{2/n} + 1]/\,[(r_o/r_i)^{2/n} - 1] \tag{13.50b}$$
$$\sigma_z = p\,[(1/n - 1)(r_o/r)^{2/n} + 1]/\,[(r_o/r_i)^{2/n} - 1] \tag{13.50c}$$

From eq(13.50c), the axial force acting on the cylinder ends is

$$\int_0^{r_i} 2\pi r\, \mathrm{d}r \sigma_z = \pi r_i^2 p$$

and this confirms our initial assumption that $\varepsilon_z = 0$. Figure 13.14 shows the variation in the three stresses (13.50a-c) from inner to outer radius for a cylinder with $r_o/r_i = 2$ and $n = 5$.

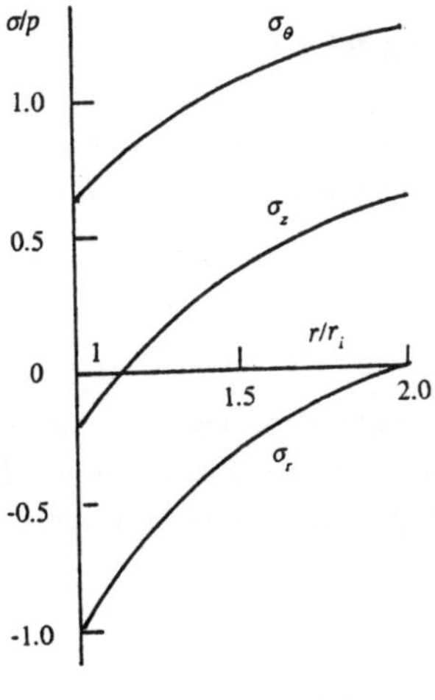

Figure 13.14 Steady state multixial stress distribution

If we compare Fig. 13.14 with the Lamé initial elastic stresses, i.e. for $n = 1$ in eqs(13.50a-c) (see Fig. 2.18), there has clearly been a redistribution during primary creep to attain the steady state condition shown. The skeletal position may be defined from the intersection between the elastic and steady state equivalent stress distributions. We can reduce the Lamé stresses (see eqs(iii) and (iv) in example 2.11) to a single Mises equivalent stress by substituting into eq(13.46). This gives

$$\bar{\sigma}/p = \sqrt{3}K^2(r_i/r)^2/(K^2 - 1) \tag{13.51}$$

where $K = r_o/r_i$, the radius ratio. Equation(13.51) plots as the continuous line in Fig. 13.15 for $K = 2$.

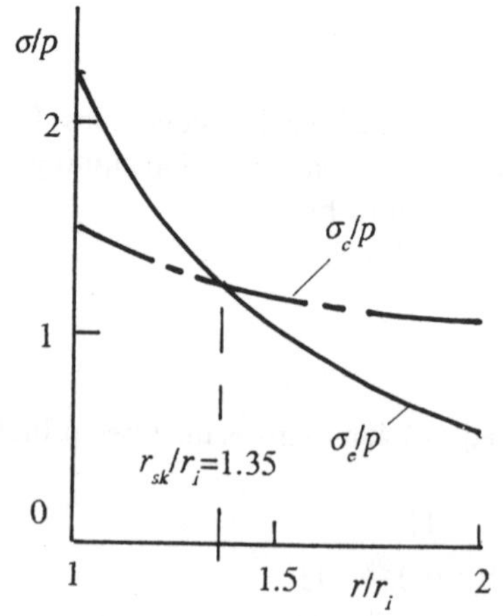

Figure 13.15 Skeletal radius in a thick-walled pressurised tube

Substituting eqs(13.50a,b,c) into eq(13.46), gives an equivalent steady-state creep stress

$$\bar{\sigma}/p = \sqrt{3}K^{2/n}(r/r_i)^{-2/n}/[n(K^{2/n} - 1)] \tag{13.52}$$

which appears as the chain line in Fig. 13.45 for $K = 2$ and $n = 5$. The intersection lies at the skeletal radius r_{sk}. Equating (13.51) and (13.52) gives

$$r_{sk}/r_i = K\{(K^2 - 1)/[n(K^{2/n} - 1)]^{n/2(1-n)}\} \tag{13.53}$$

Thus, the Lamé stress values at the skeletal radius may be used to determine the creep rates. In thinner-walled vessels it is reasonable to approximate r_{sk} with the mean wall radius [25].

13.8 Visco-elasticity

The traditional method for representing the visco-elastic strain response of non-metallic materials to stress is with rheological models. Firstly, an ideal elastic solid is modelled with a Hookean spring associated with proportionality between stress and strain

$$\sigma = E\varepsilon_s \tag{13.54a}$$

where E is the elastic modulus.

For a shear mode of deformation eq(13.54a) would be written $\gamma = G\tau_s$. Secondly, for a perfectly viscous fluid a Newtonian dashpot is associated with the linear relationship between stress and rate of strain

$$\sigma = \mu \, \mathrm{d}\varepsilon_d/\mathrm{d}t \qquad (13.54b)$$

where μ is the coefficient of viscosity. Where the observed behaviour is neither ideally elastic nor perfectly viscous, a spring-dashpot combination is employed to match particular visco-elastic effects. To describe creep, relaxation and recovery phenomena, for example, a series and parallel combination of a single spring and a dashpot are first examined.

13.8.1 Two Element Maxwell Model

In this series combination (see Fig. 13.16a) the net strain ε under stress σ is the sum of the strains in the spring and dashpot. Also the stress in each element is the same. That is

$$\varepsilon = \varepsilon_s + \varepsilon_d \qquad (13.55a)$$

Differentiating eq(13.55a) w.r.t. time t and substituting from eqs(13.54a,b) provides the governing constitutive equation

$$\mathrm{d}\varepsilon/\mathrm{d}t = (1/E)\,\mathrm{d}\sigma/\mathrm{d}t + \sigma/\mu \qquad (13.55b)$$

Now for *creep*, where $\sigma = \sigma_o$ = constant, eq(13.55b) becomes

$$\mathrm{d}\varepsilon/\mathrm{d}t = \sigma_o/\mu$$
$$\therefore \ \varepsilon = (\sigma_o/\mu)\, t + A_1 \qquad (13.56a)$$

For $t = 0$, $\varepsilon = \sigma_o/E$, indicating that the instantaneous strain is elastic only. This gives the constant $A_1 = \sigma_o/E$, whereupon eq(13.56a) becomes

$$\varepsilon = \sigma_o(t/\mu + 1/E) \qquad (13.56b)$$

This linear relationship between ε and t misrepresents creep strain that accumulates at a decreasing or an increasing rate (Fig. 13.16b).

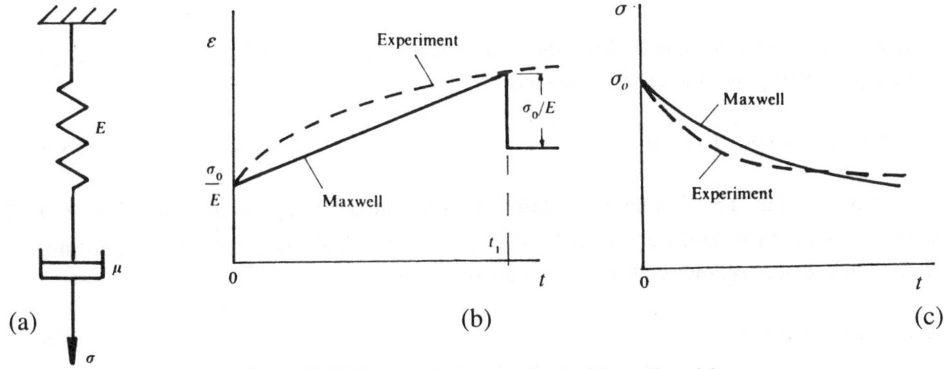

Figure 13.16 Creep and relaxation for the Maxwell model

When, in Fig. 13.16b, the stress σ_0 is completely removed after time t_1, an immediate *recovery* of the elastic spring strain, $\varepsilon_s = \sigma/E$, occurs. However, as eq(13.55b) gives a strain rate $\mathrm{d}\varepsilon/\mathrm{d}t = 0$, then no viscous dashpot strain is recovered regardless of the elapsed time. Again, this model does not reproduce observed behaviour.

In the case of *relaxation*, $\varepsilon = \varepsilon_o$ = constant and eq(13.55b) becomes

$$\mathrm{d}\sigma/\sigma = -\,(E/\mu)\,\mathrm{d}t$$

which integrates to

$$\ln\sigma = -\,(E/\mu)\,t + A_2 \tag{13.57a}$$

For $t = 0$, $\sigma = \sigma_0$ giving $A_2 = \ln\sigma_0$. Then eq(13.57a) gives the relaxed stress

$$\sigma = \sigma_0 \exp\,(-Et/\mu\,) \tag{13.57b}$$

in which $\sigma \to 0$ as $t \to \infty$. This exponential stress decay (Fig. 13.16c) is consistent with observed behaviour.

13.8.2 Two Element Voigt or Kelvin Model

For the parallel combination in Fig. 13.17a the net stress σ is the sum of the stresses for the spring and dashpot. Also the strain within each element is the same. That is

$$\sigma = \sigma_s + \sigma_d \tag{13.58a}$$

Substituting eqs(13.54 a,b) into eq(13.58a) leads to the governing constitutive equation

$$\sigma = E\varepsilon + \mu\,\mathrm{d}\varepsilon/\mathrm{d}t \tag{13.58b}$$

For *creep* $\sigma = \sigma_o$ and eq(13.58b) results in the differential equation

$$\mathrm{d}\varepsilon/\mathrm{d}t + (E/\mu)\varepsilon = \sigma_0/\mu \tag{13.59}$$

Integrating eq(13.59) after separating the variables leads to

$$\int \mathrm{d}\varepsilon/(\sigma_0 - E\varepsilon) = (1/\mu)\int \mathrm{d}t$$

$$-\,(1/E)\ln\,(\sigma_0 - E\varepsilon) = t/\mu + A_1 \tag{13.60a}$$

Here $\varepsilon = 0$ for $t = 0$ since the dashpot prevents instantaneous strain. Hence $A_1 = -\,(1/E)\ln\sigma_0$ and eq(13.60a) gives the creep strain as

$$\varepsilon = (\sigma_0/E)[1 - \exp(-\,Et/\mu\,)] \tag{13.60b}$$

It is seen from Fig. 13.17b that eq(13.60b) gives a primary creep curve in which $\varepsilon \to \sigma_0/E$ as $t = \infty$. Here, an instantaneous elastic strain is absent. With the removal of σ_0 at time t_1, eq(13.59) predicts *recovery* of strain ε after time $t' = t - t_1$ as

$$\mathrm{d}\varepsilon/\mathrm{d}t' + (E/\mu)\varepsilon = 0 \tag{13.61a}$$

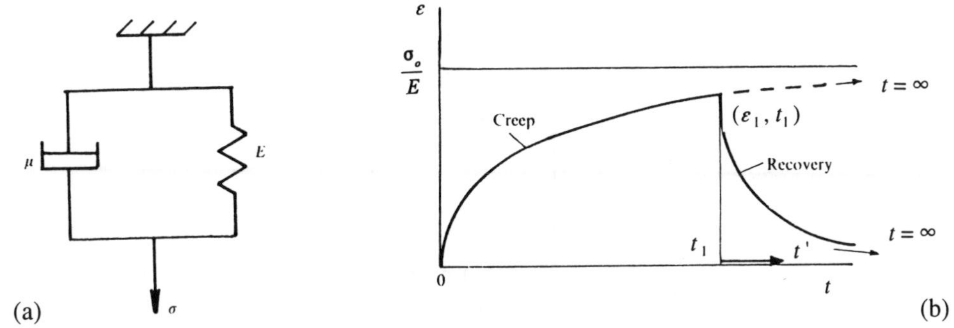

Figure 13.17 Creep and recovery for the Voigt model

Since $\varepsilon = \varepsilon_1$ (the creep strain at t_1) for $t' = 0$, eq(13.61a) integrates to

$$\varepsilon = \varepsilon_1 \exp(-Et'/\mu) = \varepsilon_1 \exp[-E(t - t_1)/\mu] \tag{13.61b}$$

in which ε_1, is given by eq(13.60b) for $t = t_1$. Equation (13.61b) becomes

$$\varepsilon = (\sigma_0/E)[1 - \exp(-Et_1/\mu)] \exp[-E(t - t_1)/\mu]$$
$$\varepsilon = (\sigma_0/E) \exp(-Et/\mu)[\exp(Et_1/\mu) - 1] \tag{13.61c}$$

It is seen from Fig. 13.17b that $\varepsilon = \varepsilon_1$ for $t = t_1$ and $\varepsilon \to 0$ as $t \to \infty$. Apart from the absence of instantaneous elastic recovery, the model is consistent with observed behaviour.

Under *relaxation* conditions where $\varepsilon = \varepsilon_0 =$ constant eq(13.58b) gives $\sigma = E\varepsilon$. The model therefore fails to admit relaxation of the initial stress.

13.8.3 Standard Linear Solids (SLS)

This refers to a combination of two springs and a dashpot. The more useful arrangement is when the Voigt model is connected in series with a second Hookean spring of different stiffness (Fig. 13.18a). It can be expected that the result will provide a better model of visco-elastic behaviour. Here there is equality in stress between the Voigt element (subscripts v and 2) and the single spring (subscript 1). This gives

$$\sigma = E_1\varepsilon_1 = E_2\varepsilon_v + \mu\, d\varepsilon_v/dt \tag{13.62}$$

Moreover, the total strain ε under an applied stress σ is

$$\varepsilon = \varepsilon_1 + \varepsilon_v$$
$$\therefore \varepsilon_v = \varepsilon - \sigma/E_1 \tag{13.63}$$

Substituting eq(13.63) into (13.62) leads to the constitutive equation

$$d\varepsilon/dt + (E_2/\mu)\varepsilon = (\sigma/\mu)(1 + E_2/E_1) + (1/E_1)\, d\sigma/dt \tag{13.64}$$

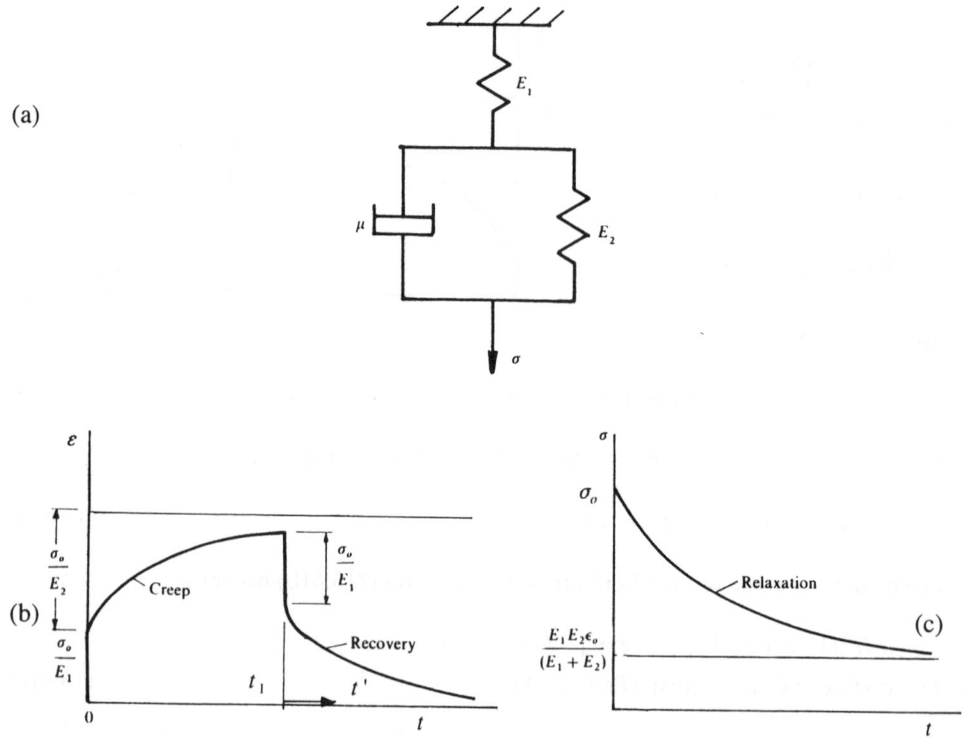

Figure 13.18 Creep, recovery and relaxation from an SLS

For *creep* $\sigma = \sigma_0$, $\mathrm{d}\sigma_0/\mathrm{d}\,t = 0$ and eq(13.64) becomes

$$\mathrm{d}\varepsilon/\mathrm{d}t + (E_2/\mu)\varepsilon = (\sigma_0/\mu)(1 + E_2/E_1) \tag{13.65a}$$

Separating the variables leads to

$$\mu \int \mathrm{d}\varepsilon/[\sigma_0 + E_2(\sigma_0/E_1 - \varepsilon)] = \int \mathrm{d}t$$

The integration, with $\varepsilon = \sigma_0/E_1$ for $t = 0$, gives the net strain as

$$\varepsilon = \sigma_0/E_1 + (\sigma_0/E_2)[1 - \exp(-E_2 t/\mu)] \tag{13.65b}$$

Equation (13.65b) displays an instantaneous strain σ_0/E_1 followed by time-dependent creep strain that approaches σ_0/E_2 with decreasing rate as $t \to \infty$ (Fig. 13.18b). Upon removal of σ_0 after time t_1, the elastic strain σ_0/E_1 is immediately recovered. Thereafter, for $t > t_1$ in Fig. 13.18b, it follows from eq(13.61c) that the *recovery* of creep strain conforms to

$$\varepsilon = (\sigma_0/E_2)\exp(-E_2 t/\mu)[\exp(E_2 t_1/\mu) - 1] \tag{13.65c}$$

That is, $\varepsilon \to 0$ as $t \to \infty$. The SLS model thus provides a good representation of observed creep and recovery behaviour in polymers.

For *relaxation* where $\varepsilon = \varepsilon_0$ and $\mathrm{d}\varepsilon_0/\mathrm{d}t = 0$, eq(13.64) becomes

$$\mathrm{d}\sigma/\mathrm{d}t + (E_1 + E_2)\sigma/\mu = E_1 E_2 \varepsilon_0/\mu \tag{13.66a}$$

With the condition that $\sigma = \sigma_0 = E_1\varepsilon_0$ for $t = 0$ the solution to eq(13.66a) is

$$\sigma = E_1\varepsilon_0\{E_1 \exp[-(E_1 + E_2)\,t/\mu] + E_2\}/(E_1 + E_2) \qquad (13.66b)$$

The additional elastic constant E_1 appearing in eq(13.66b) enables observed relaxation behaviour to be closely followed. Note from Fig.13.18(c) that the stress does not relax to zero as $t \to \infty$, but to a stress asymptote $\sigma = E_1 E_2 \varepsilon_0/(E_1 + E_2)$.

13.8.4 Other Models

There are numerous other combinations of springs and dashpots. The SLS is the simplest possible model to exhibit both creep and relaxation. Parallel combinations of the Maxwell model and series combinations of the Voigt model have been used extensively with different relaxation $(\mu/E)_m$ and retardation $(\mu/E)_v$ times. By increasing the number of such models without limit, a discrete spectrum of such times enables observed relaxation and creep behaviour to be fitted to any required degree of accuracy [26]. Specific forms of known creep and relaxation behaviour may be more conveniently modelled within a limited combination. For example, a rheological model for concrete employs a Maxwell element (E_1, μ_1) and two Voigt elements (E_2, μ_2 and E_3, μ_3) all connected in series [27]. This gives the creep strain,

$$\varepsilon = \sigma_0\{1/E_1 + t/\mu_1 + (1/E_2)[1 - \exp(-E_2 t/\mu_2)] + (1/E_3)[1 - \exp(-E_3 t/\mu_3)]\}$$

A simplified form of this model was employed by Burger for a representation of the observed creep behaviour in fibre-reinforced resins and bituminous materials (see Example 13.7).

Example 13.7 Deduce the creep response to a stress σ_0 when a Maxwell model (E_1 and μ_1) is connected in series with a Voigt model (E_2 and μ_2) as shown in Fig. 13.19a. Show that as $t \to \infty$ then the creep rate $\varepsilon \to \sigma_0/\mu_1$ and that when the dashpot μ_1 is absent, the SLS prediction applies.

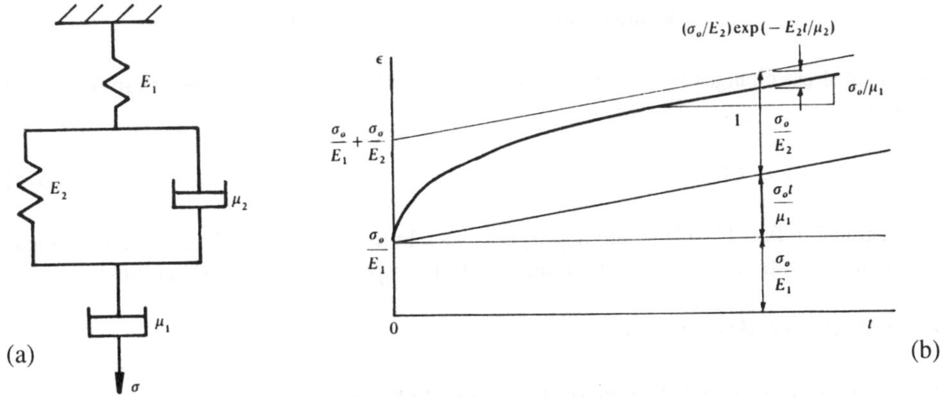

Figure 13.19 Burger's four parameter model

The creep strain is simply found by superimposing the separate Voigt and Maxwell predictions. That is, from the addition of eqs(13.56b and 13.60b),

$$\varepsilon = \varepsilon_v + \varepsilon_m$$
$$= (\sigma_0/E_2)[1 - \exp(-E_2 t/\mu_2)] + \sigma_0(t/\mu_1 + 1/E_1) \qquad \text{(i)}$$

It is seen that the net creep curve in Fig. 13.19b is the sum of those shown in Figs 13.16b and 13.17b. The curve shows an instantaneous elastic response σ_0/E_1, followed by primary creep at a decreasing rate. The latter is found by differentiating eq(i)

$$\varepsilon = (\sigma_0/\mu_1)[1 + (\mu_1/\mu_2)\exp(-E_2 t/\mu_2)] \qquad \text{(ii)}$$

It is apparent from eq(ii) that a steady rate $\varepsilon = \sigma_0/\mu_1$ is approached as $t \to \infty$. Clearly when the dashpot (μ_1) is absent, eq(i) reduces to the SLS prediction (13.65b).

13.8.5 Superposition Principle

Boltzman's [28] linear superposition or hereditary principle is useful when dealing with stress or strain inputs that vary with time. In the case of creep under a stepped stress input (see Fig. 13.20a), the principle asserts that the corresponding net strain response is the sum of the individual strain responses to each stress level. For a linear visco-elastic material, obeying $\varepsilon(t) = \sigma C(t)$, the creep strain is linear in stress for a given time. It follows that the compliance $C(t) = \varepsilon(t)/\sigma$ is a constant for that particular time.

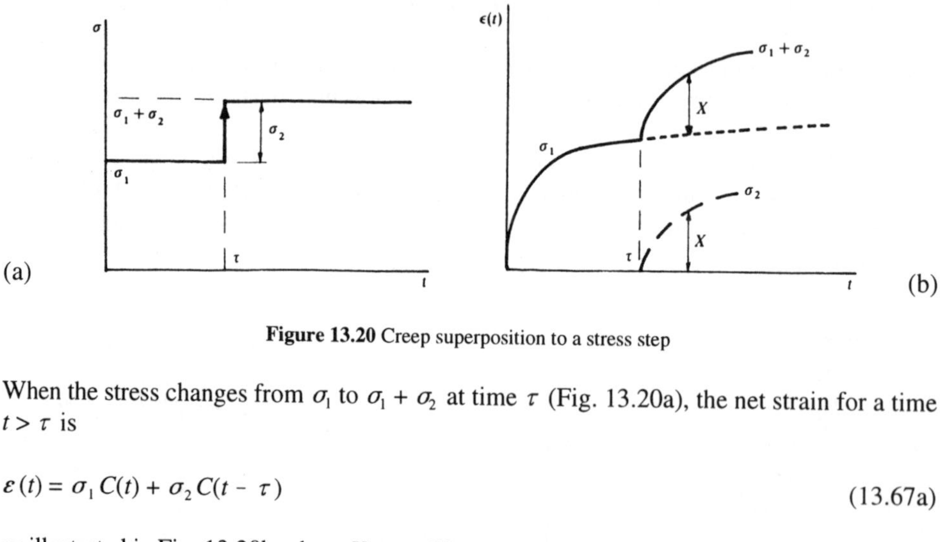

Figure 13.20 Creep superposition to a stress step

When the stress changes from σ_1 to $\sigma_1 + \sigma_2$ at time τ (Fig. 13.20a), the net strain for a time $t > \tau$ is

$$\varepsilon(t) = \sigma_1 C(t) + \sigma_2 C(t - \tau) \qquad \text{(13.67a)}$$

as illustrated in Fig. 13.20b where $X = \sigma_2 C(t - \tau)$. In the more general case of a non-linear visco-elastic material $\varepsilon(t) = \phi(\sigma, t)$, the net strain response to this step becomes

$$\varepsilon(t) = \phi(\sigma_1, t) + \phi(\sigma_2, t - \tau) \qquad \text{(13.67b)}$$

When the step corresponds to a removal of σ_1 after time τ (Fig. 13.21a), recovery occurs (see Fig. 13.21b) for $t > \tau$ with the substitution of $\sigma_2 = -\sigma_1$ in eqs(13.67a,b). For a linear visco-elastic material this substitution gives the net strain at $t > \tau$:

$$\varepsilon(t) = \sigma_1 C(t) - \sigma_1 C(t - \tau) \qquad \text{(13.67c)}$$

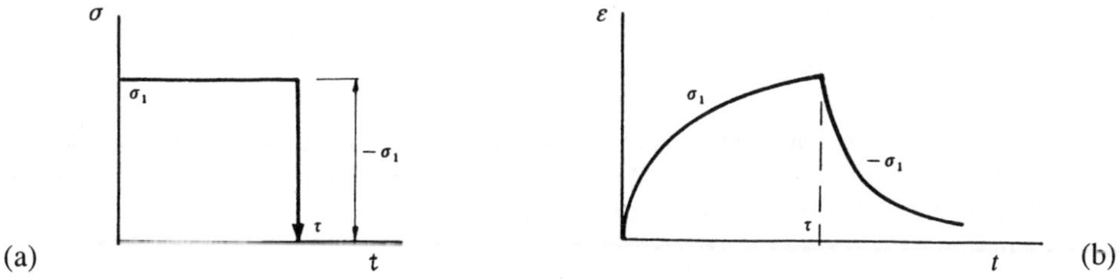

Figure 13.21 Recovery by superposition

If the creep strain reaches an asymptotic value $\varepsilon(\infty)$ (as with $t \to \infty$ in the Voigt model of Fig. 13.17b) then at time t' after the stress is changed, the net strain becomes

$$\varepsilon(t) = \sigma_1 C(\infty) + \sigma_2 C(t')$$

Where σ varies with time, this may be treated as a series of steps in an integral formulation. Let an increase in stress $\mathrm{d}\sigma$ apply to time τ. For a linear material, the increase in strain after time $t > \tau$ is given by

$$\mathrm{d}\varepsilon(t) = \mathrm{d}\sigma C(t - \tau) \tag{13.68a}$$

However, $\mathrm{d}\sigma = \mathrm{d}\tau(\mathrm{d}\sigma/\mathrm{d}\tau)$ since σ varies continuously with τ. The net strain at time t follows from integrating eq(13.68a)

$$\varepsilon(t) = \int_{-\infty}^{t} C(t - \tau)(\mathrm{d}\sigma/\mathrm{d}\tau)\,\mathrm{d}\tau \tag{13.68b}$$

in which the lower limit will account for any stress history prior to $t = 0$, as illustrated in Fig. 13.22a and b.

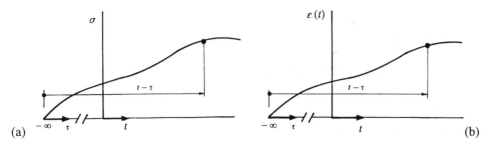

Figure 13.22 Continuously varying stress history

If there has been no prior history of stress then the lower limit may be taken as zero. Equation (13.68b) is known as the convolution integral. For creep in non-linear visco-elastic materials, eqs(13.68a and b) are replaced by the general formulation

$$\mathrm{d}\varepsilon(t) = (\partial\phi/\partial\sigma)\,\mathrm{d}\sigma \tag{13.69a}$$

$$\varepsilon(t) = \int_{-\infty}^{t} [\partial\phi(\sigma, t - \tau)/\partial\sigma](\mathrm{d}\sigma/\mathrm{d}\tau)\,\mathrm{d}\tau \tag{13.69b}$$

In one particular form of eq(13.69b), due to Leaderman [29], the variables σ and t are separable: $\phi(\sigma, t) \equiv C(t) f(\sigma)$. Equation (13.69b) becomes

$$\varepsilon(t) = \int_0^t C(t - \tau)\,(\partial f / \partial \tau)\,\mathrm{d}\tau$$

Under stress relaxation conditions, where a strain input varies continuously with time, we may deduce the $\sigma(t)$ expressions from eqs(13.68b) and (13.69b). They are, for linear $\sigma = \varepsilon M(t)$ and non-linear $\sigma = \theta(\varepsilon, t)$ visco-elasticity respectively:

$$\sigma(t) = \int_{-\infty}^{t} M(t - \tau)\,(\mathrm{d}\varepsilon / \mathrm{d}\tau)\,\mathrm{d}\tau$$
$$\sigma(t) = \int_{-\infty}^{t} [\partial\theta(\varepsilon, t - \tau) / \partial\varepsilon](\mathrm{d}\varepsilon / \mathrm{d}\tau)\,\mathrm{d}\tau$$

These are known as the Duhamel integrals. It is important to recognise that the relaxation modulus $M(t)$ for a linear visco-elastic material, and the function $\theta(\varepsilon, t)$, for non-linear visco-elasticity, are not, in general, the reciprocals of the corresponding creep functions. That is $M(t) \neq 1/C(t)$ and $\theta(\varepsilon, t) \neq 1/\phi(\sigma, t)$

Example 13.8 Isochronous data for the torsional creep of a linear visco-elastic polymer has established the following time dependence for the rigidity modulus:

t/min	10	10^2	10^3	10^4	10^5
$G(t)$/ GPa	11.25	10.62	10	9.1	7.75

A torque of 65 Nm is applied rapidly to a tube of this material 50 mm mean diameter, 1.5 mm thick and 175 mm long. If the torque is held constant for 10^3 min and then removed, determine the maximum angle of twist at the end of this period and 10^2 min after unloading.

Isothermal creep data for polymers is often presented in isochronous form. That is, stress-strain curves, each associated with a given creep time, are obtained from a family of creep curves. Since for a linear visco-elastic material $\gamma(t) = \tau C(t)$, a single modulus for each time may be identified with the slope of the stress-strain line. In the case of torsion that is

$$G(t) = \tau / \gamma(t) = 1/C(t)$$
$$\gamma(t) = \tau / G(t) = r\theta(t) / l \qquad \text{(i)}$$

Now from the Batho's torsion eq(5.64) the constant shear stress is

$$\tau = T / (2A\,t) = (65 \times 10^3) / [2 \times \pi (25)^2 \times 1.5] = 11.04 \text{ MPa}$$

and from eq(i), for $t = 10^3$ min,

$$\theta(10^3) = l\tau / [r\,G(10^3)] = (175 \times 11.04) / (25 \times 10 \times 10^3) = 77.3 \times 10^{-4} \text{ rad} = 0.443^\circ$$

Applying the superposition principle to unloading (see Fig. 13.21), the torsional equivalent of eq(13.67c) is

$$\gamma(t) = \tau C(t) - \tau C(t - \beta)$$
$$= \tau / G(t) - \tau / G(t - \beta) \qquad \text{(ii)}$$

Here $t = 10^3 + 10^2 = 1100$ min, $\beta = 10^3$ min, $t - \beta = 10^2$ min. Interpolating from the table $G(t) = 9.99$ GPa and $G(t - \beta) = 10.62$ GPa. Equation (ii) gives

$\gamma(t) = 11.04[(1/9.99) - (1/10.62)]10^{-3} = 0.0656 \times 10^{-3}$

and, from eq(i),

$$\theta(t) = l\gamma(t)/r = 175 \times 0.0656 \times 10^{-3}/25$$
$$= 0.459 \times 10^{-3} \text{ rad} = 0.0263°$$

Example 13.9 A linear visco-elastic material obeys the creep law $\varepsilon(t) = A\sigma t^m$. Determine the creep response to a stress input that increases linearly according to $\sigma = k\,\tau$ (k is a constant) for $0 \le \tau \le t_1$ and thereafter remains constant at $k\,t_1$ for $t_1 \le t \le 2t_1$ (Fig. 13.23a). If at time $2\,t_1$ the stress is abruptly removed, determine the creep strain at t_1 and $2\,t_1$ and the net strain at time $4\,t_1$ using Boltzman's superposition principle.

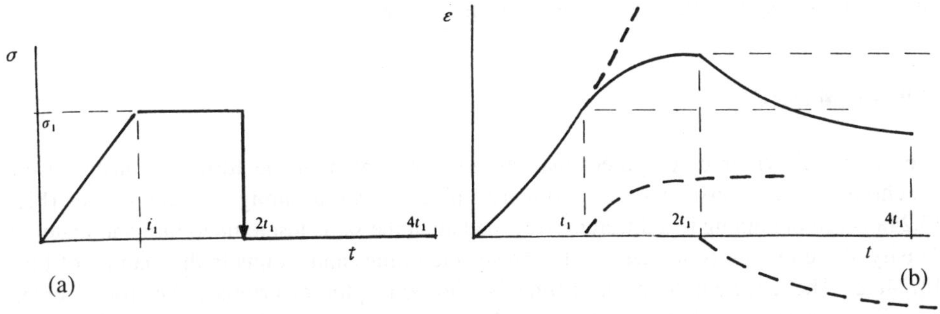

Figure 13.23 Stress and strain responses for linear visco-elasticity

During the linear increase in stress, where $d\sigma/d\tau = k$, eq(13.68b) becomes

$$\varepsilon(t) = kA \int_0^t (t-\tau)^m \, d\tau$$
$$= (kA)\, t^{m+1}/(m+1) \qquad \text{(i)}$$

Equation (i) supplies the creep strain at time t_1 as

$$\varepsilon(t_1) = (kA)\, t_1^{\;m+1}/(m+1) \qquad \text{(ii)}$$

During the period of constant stress the creep strain is

$$\varepsilon(t) = A\sigma(t - t_1)^m = (kA)\, t_1 (t - t_1)^m \qquad \text{(iii)}$$

Adding eqs(i) and (iii) gives the net strain for $t_1 \le t \le 2t_1$

$$\varepsilon(t) = (kA)\, t^{m+1}/(m+1) + (kA)\, t_1 (t - t_1)^m \qquad \text{(iv)}$$

Equation (iv) supplies the creep strain at time $t = 2t_1$ as

$$\varepsilon(2t_1) = (kA)\, t_1^{\;m+1}[1 + 2^{m+1}/(m+1)]$$

Applying eq(13.67c) for the net strain in the interval $2\,t_1 \le t \le 4t_1$,

$$\varepsilon(t) = (kA)\,t^{m+1}/(m+1) + (kA)\,t_1\,(t - t_1)^m - (kA)\,t_1\,(t - 2t_1)^m \qquad \text{(v)}$$

and for $t = 4\,t_1$ eq(v) gives the net strain

$$\varepsilon(4\,t_1) = (kA)\,t_1^{\;m+1}\,[4^{m+1}/(m+1) + 3^m - 2^m]$$

The separate strain responses are illustrated in Fig. 13.23b. They are:

$0 \le t \le t_1$, creep under a linear stress increase,
$t_1 \le t \le 2t_1$, creep under a constant stress,
$t \ge 2t_1$, recovery following stress removal.

The net strain response (heavy line) is the sum of these.

13.9 Design Data

The creep behaviour of polymers may also be presented in the form of constant time (isochronous) and constant strain (isometric) plots. Given a family of creep curves (Fig. 13.24a), each pertaining to a different stress and all at the same test temperature and material density, these two plots are derived for preselected times and strains in the manner of Figs 13.24b,c. The gradient to a short-time isochronous plot determines the 100 s creep compliance, $\varepsilon(t)/\sigma$, as a function of stress. The σ versus t plots apply to strain levels 0.01%, 0.05% ... 1% etc. These may be used to determine a creep modulus $\sigma/\varepsilon(t)$ versus time. If the strains are extended through to fracture, the stress to cause rupture in a given time may be found. The plots shown are typical of thermoplastic polymers when log time is the axis to plots 13.24a,c. Linear interpolation may be used for intermediate temperatures and densities, as shown in the following example, but extrapolations to large strains are not recommended.

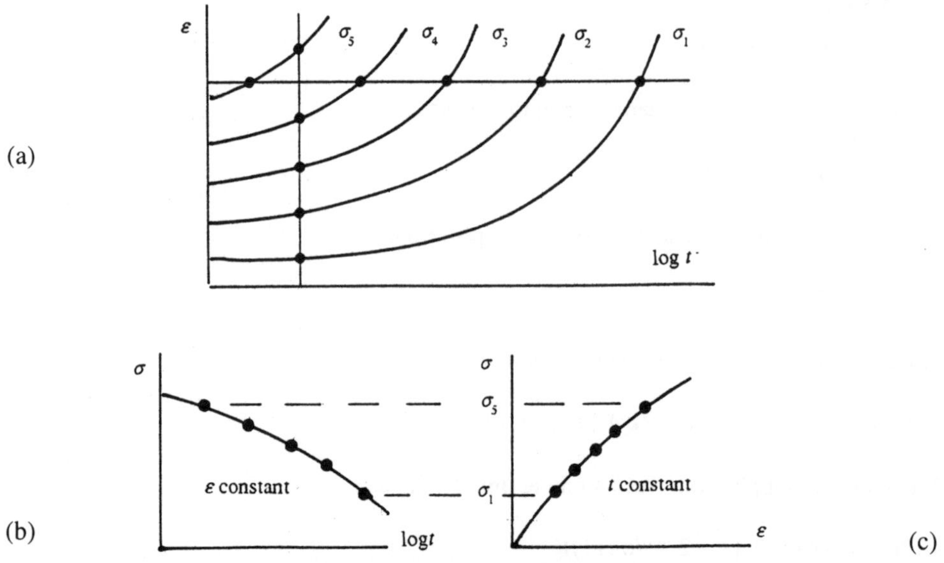

Figure 13.24 Isochronous and isometric plots for a thermoplastic

Example 13.10 A polypropylene component with density 905 kg/m^3 is to withstand constant uniaxial compressive loading. Determine the design stress when the strain is not to exceed 3% in three years at a maximum temperature of 50°C. Plots typical of Fig. 13.24b for a polypropylene of density 909 kg/m^3 gave a design stress of 8 MPa at 20°C and 5 MPa at 60°C. Assume for every 1 kg/m^3 change in density, the stress alters by 4% in the same sense.

Set $\sigma_1 = 8$ MPa, $T_1 = 293$K and $\sigma_2 = 5$ MPa, $T_2 = 333$ K. By linear interpolation we have for $t = 323$K (and $\rho_1 = \rho_2 = 909$ kg/m^3) a design stress:

$$\begin{aligned}\sigma &= \sigma_1 + (\sigma_2 - \sigma_1)(T - T_1)/(T_2 - T_1)\\ &= 8 + (5 - 8)(323 - 293)/(333 - 293)\\ &= 5.75 \text{ MPa}\end{aligned}$$

For $\rho = 905$ kg/m^3 the design stress reduces to

$$\begin{aligned}\sigma_D &= \sigma + 0.04\,\Delta\rho\,\sigma = \sigma(1 + 0.04\,\Delta\rho)\\ &= 5.75[1 + 0.04(905 - 909)]\\ &= 4.83 \text{ MPa}\end{aligned}$$

References

[1] Bailey R. W., *Proc. I. Mech. E.* **131** (1935), 131.
[2] Andrade N. da C., *Proc. Roy. Soc.* **A84** (1910), 1.
[3] Graham A. and Walles K. F. A., *J. Iron and Steel Inst.* **179** (1955), 105.
[4] Graham A. and Walles K. F. A., *Aero. Res. Cl.*, Paper No.**379**, HMSO, London, 1958.
[5] Nutting P. G., *Jl Franklin Inst.* **191** (1921), 679.
[6] Davis P. W., Evans W. J., Williams K. R. and Wilshire B., *Scripta Met.* **3** (1969), 671.
[7] Garofalo F., *Fundamentals of Creep and Creep Rupture in Metals*, MacMillan, New York, 1965.
[8] Evans R. W. Parker J. D. and Wilshire B., *Recent Advances in the Creep and Fracture of Engineering Materials and Structures*, Pineridge Press, 1982, 135.
[9] De Lacombe J., *Rev. Metall.* **36** (1939), 178.
[10] Cummings W. M. and Fairbairn J. A., *Numerical Procedure for Obtaining a Primary/Tertiary Creep Equation for a Range of Stress and Temperature*, NEL Rpt 562, East Kilbride, Glasgow, 1974.
[11] Conway J. B., *Numerical Methods for Creep and Rupture Analysis*, Gordon and Breach, 1969.
[12] Norton F. H., *Creep of Steel at High Temperatures*, McGraw-Hill, New York, 1929.
[13] Dorn J. E., *Creep and Recovery*, *A.S.M.* Cleveland, Ohio, 1957.
[14] Ludwik P., *Elements der Technologischen Mechanik*, Verlag Julius Springer, Berlin, 1909.
[15] Robinson E. L., *Trans A.S.M.E.* **100** (1938), 253.
[16] Monkman F. C. and Grant N. J., *Proc. A.S.T.M.* **56** (1956), 593.
[17] Berry D. A. and Anstee R. F. W., *Proc Roy Aero Soc.* **5** (1968), 24.
[18] Katchanov L. M., *Izv Akad Nauk SSSR, (Mech Eng)* **3** (1959), 84.
[19] Rabotnov Y. N., *Creep Problems in Structural Members,* North Holland, 1969.
[20] Penny R. K. and Marriot D. L., *Design for Creep*, McGraw-Hill, U.K, 1971.
[21] Walles K. F. A., *Summary of Creep Data Analyses in PlasticFlow Group*, NGTE Note S4077, 1963.

[22] Larson F. R. and Miller J., *Trans A.S.M.E.* **74** (1952), 765.
[23] Manson S. S. and Haferd, A. M., *NACA* (1953), Tech. Note 2890.
[24] Rees D. W. A., *Int Jl Press Vess and Piping*, **20** (1985), 191.
[25] Ohnahmi M and Motoie K., *Proc 9th Jap Cong Test Mat.* 1966, 39.
[26] Ferry J. D., *Visco-elastic Properties of Polymers,* Wiley, 1970.
[27] Illston J. M., *Creep of Engineering Materials*, Mech Eng Pub, 1978, 47.
[28] Boltzman L., *Pogg Ann Physik* **7** (1876), 624.
[29] Leaderman H., *Jl. Poly. Sci.* **16** (1955), 261.

EXERCISES

Creep

13.1 The following short-time transient creep behaviour of aluminium applies to two stress levels σ_1 and σ_2 at 100°C:

t/min	1/6	1/2	1	11/2	2	3	4	5	6	7	8	9	10	11	12	13	14	15
ε_1/%	0.31	0.34	0.36	0.37	0.38	0.39	0.40	0.41	0.42	0.426	0.435	0.44	0.445	0.45	0.455	0.46	0.465	0.47
ε_2/%	0.9	1.47	2.03	2.48	2.84	3.48	4.0	4.5	5.0	5.36	5.75	6.2	6.5	6.8	7.1	7.4	7.7	8

Show that the transient creep laws: $\varepsilon_1 = \varepsilon_o + b \ln t$ and $\varepsilon_2 = \varepsilon_o + at^m$ depend upon stress where ε_o is the instantaneous strain; a, b and m are constants.

13.2 A series of 100°C tensile creep tests on aluminium alloy gave the following secondary creep rates:

σ/MPa	100	150	200	250
$\dot{\varepsilon}_s$/hr^{-1}	9.2×10^{-6}	5.1×10^{-5}	1.55×10^{-4}	4.2×10^{-4}

Determine the law relating $\dot{\varepsilon}_s$ to σ
Answer: $\dot{\varepsilon}_s = 10^{-13}\sigma^4$.

13.3 Isochronous creep data at 400 hr obeys a linear stress-strain law $\varepsilon = M\sigma$ over a low range of stress for the following temperatures:

T/°C	650	550	500	450
$M/10^{-6}$	53.3	28	16	9.8

Determine the function $M = M(T)$ and hence find the stress that may be applied to give a creep strain of 0.4% in 400 hr.

13.4 The following results apply to the short-time creep of lead under σ = 11.5 MPa at room temperature. Apply the Graham-Walles and Davis representations to the full creep curve.

t/min	0.25	0.5	1	2	3	4	6	8	10	15	20	25	30	35	40	42	44	46
ε/ %	1.25	1.6	2.06	2.77	3.26	3.6	4.23	4.77	5.2	6.17	7.03	7.89	8.77	9.94	12.49	14.57	17.94	26.3

13.5 The following data applies to the creep deformation in a Ni-Cr alloy at 820°C:

t/hr	20	100	200	1000	2000	3000	4000	5000
ε/ %	1	1.6	2	3.2	4	4.8	6	8.5

Given that the minimum creep rate occurs at 200 hr, estimate the constants θ_1, θ_2, θ_3 and θ_4 in the prediction of eq(13.4c).

13.6 The following data applies to the tensile creep of lead at room temperature. Confirm that the respective Norton, Monkman-Grant and Katchanov-Rabotnov laws (13.5), (13.13c) and 13.18b) relate $\dot{\varepsilon}_s$ to σ, $\dot{\varepsilon}_s$ to t_f and σ to t_f. Determine the constants in each law.

σ/MPa	11.52	10.4	9.86	9.25	8.91	8.80
t_f/hr	3.05	10	18.5	37	52	67.1
$\dot{\varepsilon}_s$/hr^{-1}	0.0265	0.0075	0.0052	0.00263	0.00225	0.0019

Answer: $\dot{\varepsilon}_s = 37 \times 10^{-4}\sigma^{10}$, $\dot{\varepsilon}_s t_f = 0.1$, $t_f = 10^{13}\sigma^{-11.85}$)

13.7 In a constant temperature creep test the stress is increased in equal increments $\Delta\sigma$ at regular intervals Δt. Given that the creep law for the material is $\varepsilon_c = A\sigma^n t^m$, show, from the time- and strain-hardening laws, that the respective creep strain after $r = 1, 2 .. N$ steps is

$$\varepsilon_c = A(\Delta t)^m \sum_{r=1}^{N} \{r^m - (r-1)^m\}(r\Delta\sigma)^n \text{ and } \varepsilon_c = A(\sigma t)^m\{\sum_{r=1}^{N} (r\Delta\sigma)^{n/m}\}^m$$

13.8 A stainless steel pipeline 40 mm bore and 2 mm wall thickness is to carry fluid at pressure 0.6 bar and temperature 510°C for 9 years. Given a limiting creep strain of 1% and a secondary creep law of the form: $\dot{\varepsilon}_s = A\sigma^5 \exp[-Q/(RT)]$, comment upon the safety of the design when the following information is supplied for a stress of 200 MPa:

T/°C	618	640	660	683	707
$\dot{\varepsilon}_s/10^{-8}$ s^{-1}	10	17	43	77	200

13.9 At stress levels of 230 MPa and 155 MPa the Larson-Miller parameter for a nickel-base alloy has values of 25 and 26 respectively. Estimate the maximum operating temperature at a stress of 155 MPa for a life of 10^4 hr. If a component of this alloy spent 10% of its life at 230 MPa and the remaining life at 155 MPa, determine the time to failure from the life fraction rule. (IC)
Answer: 810°C, 5880 hr.

13.10 The uniaxial creep life of an alloy is 10^5 hr at 900 K and 120 MPa, whilst at 300 MPa and 800 K the life is 10^4 hr. Using the Larson-Miller and Manson-Haferd parameters (P), interpolate linearly on log P versus logσ axes and compare temperature predictions for which the life is 10^5 hr at 200 MPa.

13.11 Find the width of a cantilever beam 750 mm long × 75 mm deep carrying a 2.5 kN force at its free end, when the deflection there is restricted to 10 mm in 10^4 hr at 500°C. Take the secondary creep law $\dot{\varepsilon}_s = 3 \times 10^{-13}\sigma$ ($\dot{\varepsilon}_s$ in hr^{-1} and σ in MPa). Neglecting self-weight, what will be the maximum stress and strain in the beam after this time? (IC)

13.12 An alloy steel shaft 50 mm outer diameter, 38 mm inner diameter and 2.75 m long carries a steady torque of 500 Nm at 450°C. Given that the secondary creep rate $\dot{\varepsilon}_s = (30.5 \times 10^{-4})\sigma^3$, with $\dot{\varepsilon}_s$ in hr^{-1} and σ in MPa, applies to uniaxial tension at 450°C, determine the angular twist after 10^4 hr. (IC)

13.13 Compare the times taken for the stress in a bolt between two rigid flanges to relax to one half its initial value (σ_0) according to the time- and strain-hardening equations of state. Assume the creep law $\varepsilon_c = A\sigma^3 t^{1/2}$. (IC)
Answer: $t_{SH}/t_{TH} = 2.18$.

13.14 A thin-walled cylindrical vessel of inside diameter d with hemispherical ends operates under secondary creep conditions where $\dot{\varepsilon}_s = A\sigma^{4.8}$. Find the ratio between the hemisphere and cylinder thicknesses when each of the following conditions exists at the joint (i) the circumferential strain rates are the same, (ii) the longitudinal strain rates are the same and (iii) the hoop stresses are the same.

13.15 A cylindrical vessel is riveted with the single lap circumferential joint shown in Fig. 13.25. The 2 mm diameter rivets are formed to produce an initial axial stress of 160 MPa. If a pressure of 0.35 MPa must be exerted between the lapping surfaces to prevent leakage, determine the life of the joint at 500°C. Ignore primary creep, take $E = 70$ GPa and employ the following secondary creep data:

σ/MPa	80	100	120	140	160
$\dot{\varepsilon}_s/10^{-9}\,\text{hr}^{-1}$	50	150	380	800	1570

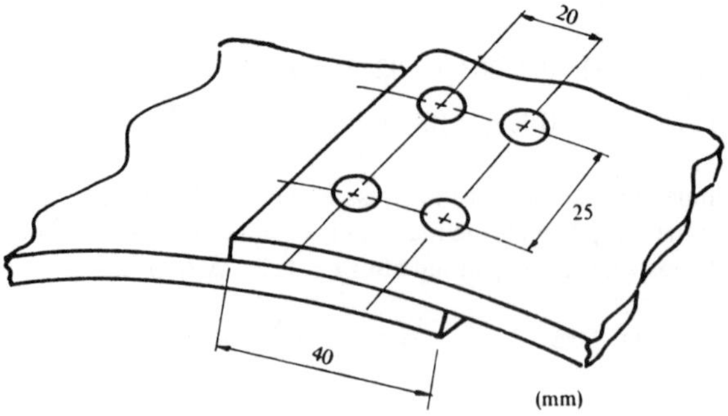

Figure 13.25

Visco-elasticity

13.16 Explain why the principal mechanical model representations of visco-elastic behaviour give only a qualitative picture.

13.17 State the creep compliance $C(t) = \varepsilon(t)/\sigma_o$ and relaxation moduli $M(t) = \sigma(t)/\varepsilon_0$ for the Maxwell, Voigt and SLS models.

13.18 Determine the strain function $\varepsilon(t)$ when the Maxwell, Voigt and SLS models are loaded at a constant stress rate $R = \mathrm{d}\sigma/\mathrm{d}t$. If, after time t_1, the stress is abruptly removed, find the recovery response $\varepsilon(t > t_1)$ for the Voigt and SLS models.
Answer: $\varepsilon_m(t) = R(1/E + t/\mu)$, $\varepsilon_v(t) = (R/E)\{t - \tau_o[1 - \exp(-t/\tau_o)]\}$

13.19 Determine the stress function $\sigma(t)$ when the Maxwell, Voigt and SLS models are loaded under a constant strain rate $\dot{\varepsilon} = \mathrm{d}\varepsilon/\mathrm{d}t$.
Answer: $\sigma_m(t) = \mu\dot{\varepsilon}\,[1 - \exp(-t/\tau_o)]$, $\sigma_v(t) = \dot{\varepsilon}\,(\mu + Et)]$

13.20 A spring of stiffness E_1 is connected in series with a dashpot of viscosity μ_1. The two are then connected in parallel with a second spring of stiffness E_2. Derive and sketch the stress relaxation response $\sigma(t)$ to a strain input ε_o. Given that $\sigma(t) = \varepsilon_0 M(t)$, determine the ratio $M(t)/E_1$ for $t = 20$ s when $E_1 = 5E_2$.

13.21 Show that when the Maxwell elements (E_1, μ_1) and Voigt elements (E_2, μ_2) are connected in parallel, the creep strain is given by:

$$\varepsilon(t) = (\sigma_o/E_1)[1/E_2 + t/\mu_2 - (1/E_2 + t/\mu_2)\exp(-E_1 t/\mu_1)]\,/$$
$$[(1/E_1 + 1/E_2) + t/\mu_2 - (1/E_1)\exp(-E_1 t/\mu_1)]$$

13.22 Sketch the strain-time curve represented by the following equation:

$$\varepsilon(t) = \sigma/E_1 + \sigma t/\mu_1 + (\sigma/E_2)[1 - \exp(-E_2 t/\mu_2)]$$

Determine the model from which this is derived and discuss whether this is realistic for creep and relaxation in polymeric materials. (CEI)

13.23 If for Example 13.8 the torque is increased from 65 Nm to 100 Nm after a time of 10^3 minutes, find the angular twist after 10^5 minutes. If the total torque is then removed, what angular twist remains after an equal recovery period of 10^5 minutes?

13.24 A viscous dashpot $\varepsilon = \mu\sigma$ is connected in series with a non-linear spring $\sigma = E\varepsilon - (a\varepsilon)^2$ where μ, a and E are constants. Determine the constitutive relation and deduce the relaxation behaviour for the model. (IC)
Answer: $\varepsilon = \sigma/\mu + (\sigma/E)[1 - (2a/E)^2\sigma]^{-1/2}$

13.25 A Voigt and Maxwell model are connected in series. Identify those elements of the combined model that represent the following components of strain: (i) instant elasticity, (ii) irrecoverable flow and (iii) retarded elasticity. Show where these appear in the corresponding creep curve. A certain polymer is modelled by this system where $E = 50$ kN/m^2, $\mu = 10^6$ Ns/m^2 for Voigt and $E = 10^6$ kN/m^2, $\mu = 100 \times 10^6$ Ns/m^2 for Maxwell. The polymer is subjected to a direct stress of 6 kN/m^2 for only 30 s. Determine the strain after times of 30 s, 60 s and 2000 s. (CEI.)

13.26 Derive and sketch the creep and relaxation responses for the SLS model in Fig. 13.26 given that the spring stiffnesses are E_1 and E_2 and that the dashpot compliance is $C(t)$. If $E_1 = 4E_2$ find the relaxation modulus $M(t)/E_1$ for $t = 10$ s given that the relaxation time is 100 s. (IC)

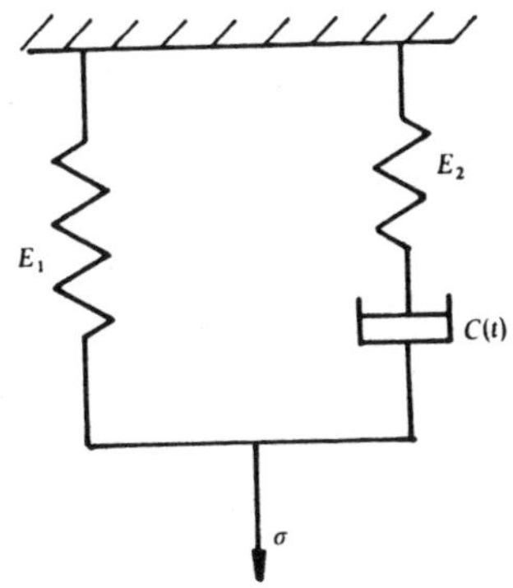

Figure 13.26

13.26 Stress rupture data for high density polyethylene are given in Fig. 13.27. Ignoring axial stress, estimate a suitable pipe thickness that is to sustain an internal pressure of 0.5 MPa at a temperature of 80°C for 12×10^6 hr.

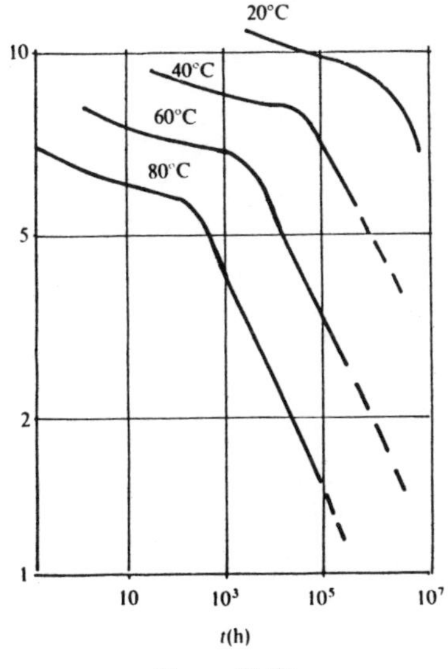

Figure 13.27

13.28 A linear visco-elastic material has a creep compliance function of the form $C(t) = t^m / E_o$. Derive the expression for constant stress rate loading and outline how the solution could be employed to describe saw tooth loading. (IC)

13.29 A uniform light cantilever with second moment of area I supports a uniformly distributed load w_1 from its centre to its free end. This load is applied suddenly at time t_1 and held constant until time t_2 when an additional distributed load w_2 is suddenly superimposed upon w_1 occupying the same length. If the cantilever material obeys the Voigt model creep behaviour, derive the deflection of the free end for time $t > t_2$, using the Boltzman superposition principle. (IC)

CHAPTER 14

HIGH AND LOW CYCLE FATIGUE

14.1 High Cycle Fatigue

So far in the analyses of elastic forces, moments and torques it has been assumed that these remain constantly applied to load-bearing components. In practice, however, loads are often cyclic in nature, e.g. random wind loading on an aircraft, reversed torsion of a watch spring, repeated bending of an axle and crankshaft in a road vehicle and pressure rippling in piping. High cycle fatigue is the name given to describe the failure resulting from cyclic elastic stresses in which the peak levels are less than the static yield stress. Fatigue occurs at low-medium temperatures and is characterised by its absence of gross macroscopic plastic deformation though the initiation and propagation of a crack. The process does involve plasticity but only at sites of high stress concentration on a microscopic scale. The long period of stage I crack initiation is associated with coarsening during slip band formation at a free surface along planes of maximum shear stress. Intrusions and extrusions can be observed within the roughening of the free surface. The shorter stage II crack propagation period is the result of progressive plastic sharpening and blunting of a crack front that advances in a direction normal to the major principal tensile stress.

The characteristic appearance of fatigue failures consists of two regions: the featureless stage I initiation site spreads across only a few grains in polycrystalline materials and is therefore invisible although it would normally lie within a narrow 45° shear lip on the surface and the stage II propagation area, in which the rate of advance is related to the spacing of characteristic fatigue striations or beach markings. One such marking is associated with a single application of cyclic stress. The fracture surface may be smooth initially as it emanates from the initiation site but often becomes rougher as the increased rate of crack propogation results in a wider spacing of fatigue markings. Finally, when the remaining uncracked area is unable to withstand peak cyclic stresses it fails catastrophically in either a brittle or ductile manner leaving a respective crystalline or fibrous appearance.

Fatigue is a common cause of failure, indicating that there is normally a limit to the number of elastic cycles a given material can withstand. There are many factors that influence this limit. Those concerned specifically with the cycle are: stress range, mean stress, changes in the stress amplitude, stress state and stress concentration, while those accounting for the effect of the environment include temperature and corrosion. Predictions to these influences have been examined experimentally from the standard fatigue test. This test, for example, may impose a constant amplitude reversed stress (Fig. 14.1a) under cyclic bending, torsion or tension-compression to a suitable testpiece until it fails in a number (N) of fatigue cycles. By testing geometrically identical testpieces to failure each at a different stress amplitude S, the usual S-N curve in Fig. 14.1b may be constructed. Alternatively, an S - log N plot may be preferred to distinguish more clearly the existence or otherwise of a fatigue limit. This limit is the asymptotic stress value S_L which the curve approaches in carbon steels, so indicating an infinite life in this material for amplitudes $S < S_L$. However, the S-N curve for most other materials shows that a fatigue limit does not exist since the curve falls continuously with increasing cycles.

(a)

(b)

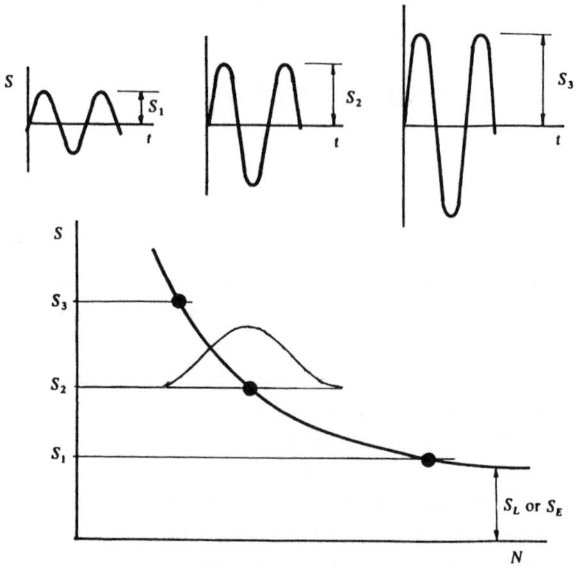

Figure 14.1 Reversed stress cycle and associated S-N curve

It then becomes necessary to employ an endurance limit S_E for a specified life, e.g. 10^9 or 10^{10} cycles. In general, the *S-N* curve reveals the fatigue life. That is the number of cycles to failure under a given stress amplitude (or semi-range of stress). Note that *N* indicates the probability of failure from an assumed distribution of life at each stress level. This is to account for the normally wide scatter observed in fatigue life. The Weibull distribution given in Fig. 14.1b shows the probability of fatigue at a given stress amplitude. We may construct *S-N* curves for, say 10% or 90% probabilities of failure as required. When S_L or S_E is regarded as a material property under specified test conditions, it may be expressed as a fraction of the ultimate tensile strength *U*. That is, an endurance ratio E_R is defined as

$$E_R = S_L/U \quad \text{or} \quad S_E/U \tag{14.1}$$

where $0.3 < E_R < 0.7$ for materials under the fully reversed stress cycle in Fig. 14.1a. Equation(14.1) provides a useful method for estimating fatigue strength, e.g. for mild steel with an $U = 455$ MPa, an average S_L value is $0.5 \times 455 = \pm\, 227.5$ MPa. The following factors within the stress cycle and the environment will influence the reversed fatigue strength.

14.2 The Stress Cycle

14.2.1 Mean Stress and Stress Range

For stress cycles that are not symmetrical about the time axis, a mean stress σ_M is defined as the average of the maximum and minimum peak stresses. The stress range is the difference between these peak values. They are

$$\sigma_M = (\sigma_{max} + \sigma_{min})/2 \quad (14.2a)$$
$$\Delta\sigma = \sigma_{max} - \sigma_{min} \quad (14.2b)$$

Four types of cycle, given in Fig. 14.2, may then be classified:

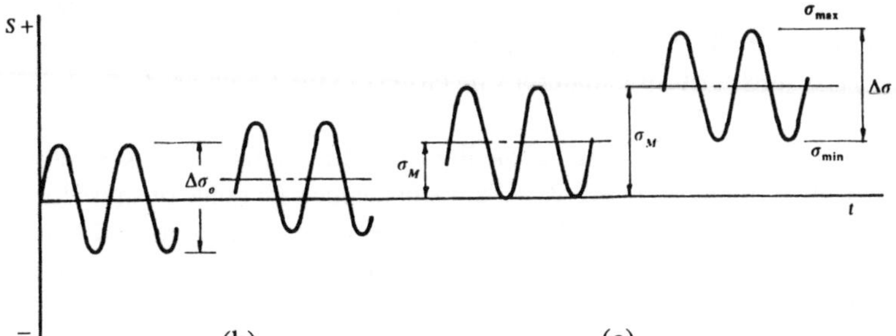

(a) (b) (c) (d)

Figure 14.2 Types of stress cycle

(a) Reversed cycle: zero mean stress with equal σ_{max} and σ_{min} magnitudes.
(b) Alternating cycle: σ_{max} is positive, σ_{min} is negative each with a different magnitude, so that σ_M may be either positive or negative.
(c) Repeated cycle: $\sigma_{min} = 0$, in which σ_{max} and $\sigma_M = \sigma_{max}/2$ lie either on the tensile or the compressive side.
(d) Fluctuating cycle: σ_{max}, σ_{min} and σ_M are either all positive or all negative.

Experiment has shown that as σ_M in (b), (c) and (d) increases so the safe range of cyclic stress $\Delta\sigma$, defining the fatigue or endurance limit, decreases. A number of design rules predict the reducing effect that σ_M has on the safe range of stress $\Delta\sigma_o$ under the reversed stress cycle (Fig. 14.2a). The limiting conditions are then

(i) when $\sigma_M = 0$, $\Delta\sigma_o = 2S_L$ or $\Delta\sigma_o = 2S_E$ for the reversed cycle, and
(ii) when $\Delta\sigma = 0$, $\sigma_M = Y$ or $\sigma_M = U$ for a static tension test.

These three limiting points, lying on the normalised axes of Fig. 14.3, have been connected with a parabola or the straight lines as shown.

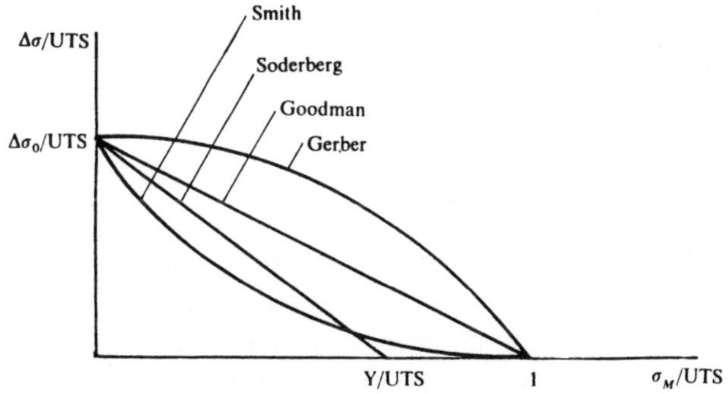

Figure 14.3 Effect of mean stress on the safe range of stress

The Gerber parabola [1], the Goodman line [2] and Smith's empirical equation [3] all employ $\sigma_M = U$ while Soderberg [4] employs $\sigma_M = Y$. These are respectively

$$\Delta\sigma = \Delta\sigma_o[1 - (\sigma_M/U)^2] \quad (14.3a)$$
$$\Delta\sigma = \Delta\sigma_o(1 - \sigma_M/U) \quad (14.3b)$$
$$\Delta\sigma = \Delta\sigma_o(1 - \sigma_M/U)/(1 + \sigma_M/U) \quad (14.3c)$$
$$\Delta\sigma = \Delta\sigma_o(1 - \sigma_M/Y) \quad (14.3d)$$

Equations (14.3a-d) enable the safe $\Delta\sigma$ to be estimated for a given σ_M. Test results for ductile materials are normally found to lie in the area bounded by the Soderberg and Gerber predictions. Results for brittle materials lie closest to the Smith prediction.

Example 14.1 The reversed stress fatigue limit for a low carbon steel is ± 230 MPa, its u.t.s. is 870 MPa and its yield stress is 380 MPa. Estimate the safe range of stress for (i) a repeated cycle (Fig. 4.2c) based upon the Gerber and Goodman predictions and (ii) an alternating tensile cycle (Fig. 4.2b) with mean stress 185 MPa based upon the Soderberg prediction. What are the peak stresses for (ii)?

(i) Here $\Delta\sigma_o = 2S_L = 460$ MPa and $\sigma_M = \Delta\sigma/2$ for a repeated cycle. Substituting into eq(14.3a) the safe Gerber range is found from

$$\Delta\sigma = 460[1 - \Delta\sigma/(2 \times 870)^2]$$
$$(\Delta\sigma)^2 + 6581.74\,(\Delta\sigma) - (3027.6 \times 10^3) = 0$$
$$\therefore \Delta\sigma = 432 \text{ MPa}$$

Substituting into eq(14.3b), the Goodman prediction becomes

$$\Delta\sigma = 460[1 - \Delta\sigma/(2 \times 870)]$$
$$\Delta\sigma = 364 \text{ MPa}$$

(ii) Here $\sigma_M = 185$ MPa, $\Delta\sigma_o = 460$ MPa and $Y = 380$ MPa. Equation (14.3d) gives the safe stress range for the alternating cycle as

$$\Delta\sigma = 460[1 - 185/380] = 236 \text{ MPa}$$

Combining eqs(14.2a,b), the peak stresses are

$$\sigma_{max} = \sigma_M - \Delta\sigma/2 = 185 - 236/2 = 67 \text{ MPa}$$
$$\sigma_{min} = \sigma_M + \Delta\sigma/2 = 185 + 236/2 = 303 \text{ MPa}$$

14.2.2 Cumulative HCF Damage

Often the stress amplitude may vary, thus complicating the prediction of fatigue life. A large number of investigations have examined the effect of block pattern cycling and of random changes to the stress amplitude. The lives found may be compared with predictions from a hypothesis due to Miner [5]. Consider the three blocks of reversed stress cycles in Fig. 14.4a where n_1, n_2 and n_3 cycles are applied at respective semi-stress ranges S_1, S_2 and S_3. If N_1, N_2 and N_3 are the corresponding fatigue lives (Fig. 14.4b) then the life expended, or the damage incurred, at each S is given by the cycle ratios n_1/N_1, n_2/N_2 and n_3/N_3. According to Miner's linear damage rule, failure occurs when the sum of the cycle ratios equals unity

$$\sum(n/N) = 1 \quad (14.4)$$

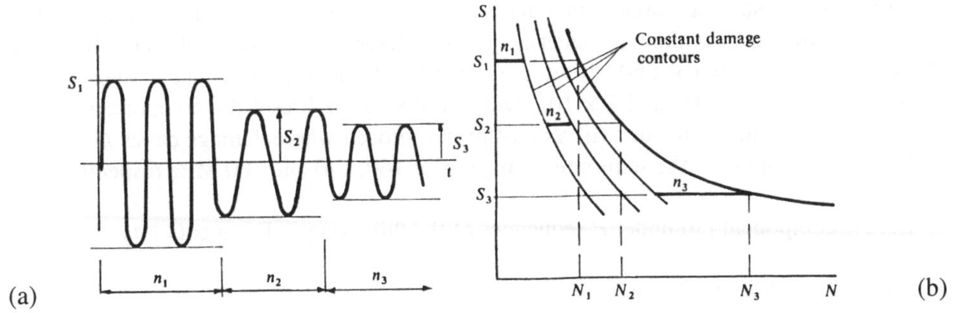

Figure 14.4 Damage under a variable amplitude cycle

The right hand side of eq(14.4) has often been replaced with a constant accounting for the observed stress dependence of the cumulative cycle ratio. Normally with a high to low block stress sequence $\sum(n/N) < 1$, while with a low to high sequence $\sum(n/N) > 1$. Alternatively, if D is the damage under cycle ratio $R = n/N$, these observations on stress dependence will appear when a given non-linear function $D = D(R)$ is assigned to each semi-stress range. A simple function proposed by Marco and Starkey [6] is the power law

$$D = (R)^x \tag{14.5a}$$

where $x = x(S)$ is a positive constant. Corten and Dolan [7] proposed the alternative function:

$$D = a\,n^b \tag{14.5b}$$

where $a = a\,(S)$ and b is a constant independent of S. In these non-linear rules a different D-R curve may be identified with each S level as illustrated in Fig. 14.5. For the high to low stress sequence in Fig. 14.4a the damage D_1 incurred by the cycle ratio $R = n_1/N_1$ is governed either by curve $D_1 = (R_1)^{x_1}$ or $D_1 = a_1 n_1^{\,b}$ as defined by eqs(14.5a and b). The change to stress level S_2, where the curve $D_2 = (R_2)^{x_2}$ or $D_2 = a_2 n_2^{\,b}$ applies, occurs at a constant damage value. Similarly for the change to S_3, another D-R curve describes the accumulation of damage as shown. Though failure again occurs when D attains a value of unity, it is seen from the relative disposition of the D-R curves in this approach that the cumulative cycle ratio $R_1 + R_2 + R_3$ is less than 1, in accordance with experiment. Moreover, for a low to high stress sequence S_3 - S_2 - S_1 these curves will also concord with $\sum R > 1$. Miner's rule (14.4) implies that the line $D = R$ curve applies to all stress levels. This simplistic, stress-independent approach accounts for its widespread use, particularly in the case of random cyclic loading where the application of non-linear rules is impracticable.

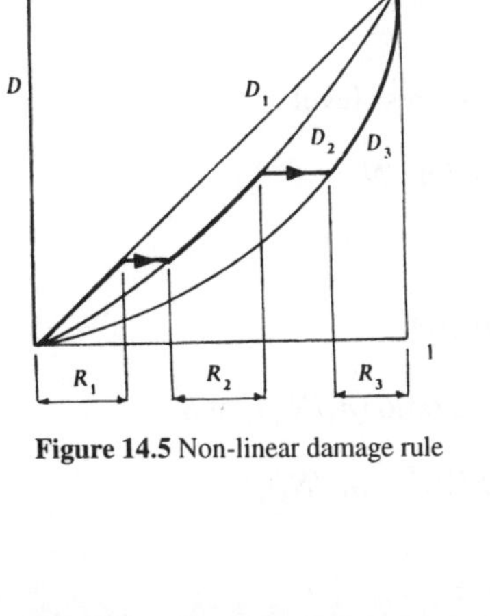

Figure 14.5 Non-linear damage rule

Example 14.2 A steel component in a machine tool is subjected to reversed cyclic loading at 100 cycles/day in a continuous sequence involving three stages: 200 cycles at ± 160 MPa, 200 cycles at ± 140 MPa and 600 cycles at ± 100 MPa. If the respective fatigue lives at these stress levels are 10^4, 10^5 and 2×10^5 cycles, estimate the life of the component in days according to (i) Miner's rule and (ii) a non-linear approach where damage cuves $D = \surd R$, $D = R$ and $D = R^2$ apply to the semi-stress ranges S of 160, 140 and 100 MPa respectively.

(i) Let the component fail under P sequences of the three stages. From eq(14.4),

$$P\,(n_1 / N_1 + n_2 / N_2 + n_3 / N_3) = 1$$
$$P\,[200/10^4 + 200/10^5 + 600/(2 \times 10^5)] = 1$$

$\therefore P = 40$ and for 1000 cycles/sequence at a rate of 100 cycles/day, the life is $40 \times 1000/100 = 400$ days.

(a)

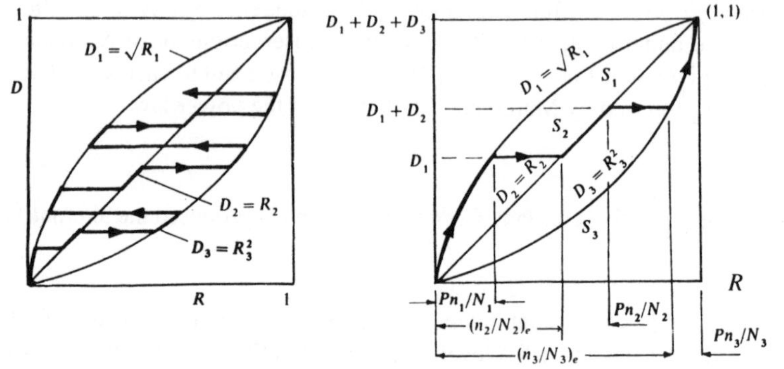

(b)

Figure 14.6 Cyclic loading sequence

(ii) With reference to Fig. 14.6a, the proper application of the non-linear approach clearly requires the laborious calculation of the damage value for every change in stress. However, by assuming, conservatively, that stresses of the same peak level are all applied before the stress is increased, then Fig. 14.6b may be applied to this problem. Thus for Pn_1/N_1 cycles at S_1 the damage is

$$D_1 = \surd R_1 = \surd(P \times 200/10^4) = 0.1414\surd P$$

The equivalent cycle ratio $(n_2/N_2)_e$ that would cause the same damage at S_2 is found from

$$D_1 = 0.1414\surd P = (n_2/N_2)_e \qquad \text{(i)}$$

and for Pn_2/N_2 cycles at this stress level

$$\therefore D_1 + D_2 = R_2 = (n_2/N_2)_e + Pn_2/N_2$$

where, from eq(i),

$$D_1 + D_2 = 0.1414\surd P + 200P/10^5$$

This is equivalent to the cycle ratio $(n_3/N_3)_e$ at S_3

$$D_1 + D_2 = 0.1414\surd P + 200P/10^5 = (n_3/N_3)_e^{\,2}$$

$\therefore (n_3/N_3)_e = \surd[0.1414\surd P + 200P/10^5]$ (ii)

and for Pn_3/N_3 cycles to failure

$D_1 + D_2 + D_3 = R_3^{\,2} = 1$
$[(n_3/N_3)_e + Pn_3/N_3]^2 = 1$

Substituting from eq(ii),

$\{\surd[0.1414\surd P + 200P/10^5] + 600P/(2 \times 10^5)\}^2 = 1$
$\surd\,[0.1414\surd P + 200P/10^5] + 300P/10^5 = 1$

The solution is $P = 30$ sequences, for which the life is $30 \times 1000/100 = 300$ days. This indicates that Miner's prediction is an overestimation for the high to low order of stressing.

14.2.3 Stress State

Fatigue under cyclic biaxial and triaxial elastic stress states may be treated in a similar manner to yielding under similar stress states (see Chapter 11). The uniaxial yield stress (Y) is replaced with a uniaxial fatigue strength (S_o) typically obtained under reversed stress conditions from a push-pull or reversed bending test. Since fatigue cracks initiate at a free surface, where one principal stress is zero, then biaxial fatigue criteria apply. For example, when principal biaxial stress ranges $\Delta\sigma_1$ and $\Delta\sigma_2$ are applied in a reversed cycle it follows from a von Mises type criterion (eq11.7b) that the combination of stress ranges required for fatigue failure in a ductile material is given by

$$\Delta\sigma_1^{\,2} - \Delta\sigma_1\Delta\sigma_2 + \Delta\sigma_2^{\,2} = \Delta\sigma_o^{\,2} \tag{14.6a}$$

where $\Delta\sigma_o = 2S_o$ In the presence of a mean stress σ_M, eq(14.6a) becomes

$$\Delta\sigma_1^{\,2} - \Delta\sigma_1\Delta\sigma_2 + \Delta\sigma_2^{\,2} = \Delta\sigma^2 \tag{14.6b}$$

where $\Delta\sigma$ applies to the given σ_M. When only the reversed fatigue strength is available, $\Delta\sigma$ is found from one of eqs(14.3a-d). A more conservative Tresca criterion of fatigue failure (eq11.1b) is often preferred:

$$\text{Greatest principal } \Delta\sigma - \text{Least principal } \Delta\sigma = \Delta\sigma_o \tag{14.7}$$

in which the least $\Delta\sigma$ may be negative for a wholly compressive cycle but would normally be zero (the third principal stress range) when biaxial stress ranges $\Delta\sigma_1$ and $\Delta\sigma_2$ are both tensile.

In the case of combined cyclic bending and torsion, where respective ranges of reversed stress $\Delta\sigma$ and $\Delta\tau$ are applied, Gough and Pollard [8] expressed eq(11.8) in a fatigue criterion for ductile steels:

$$(\Delta\sigma/\Delta\sigma_o)^2 + (\Delta\sigma/\Delta\tau_o)^2 = 1 \tag{14.8}$$

where $\Delta\sigma_o$ and $\Delta\tau_o$ are the critical stress ranges under the separate application of reversed bending and torsion respectively, i.e. twice their reversed fatigue strengths. Note that the

Tresca and von Mises criteria give $\Delta\tau_0 = \Delta\sigma_0/2$ and $\Delta\tau_0 = \Delta\sigma_0/\surd 3$ respectively. Equation(14.8) was verified experimentally for twelve ductile steels. Brittle cast did not conform to eq(14.8). The following modified form of a maximum principal strain type criterion (see table 11.1) was found to be more appropriate:

$$(\Delta\sigma_0/\Delta\tau_0 - 1)(\Delta\sigma/\Delta\sigma_0)^2 + (2 - \Delta\sigma_0/\Delta\tau_0)(\Delta\sigma_0/\Delta\tau_0) + (\Delta\tau/\Delta\tau_0)^2 = 1$$

Findlay and Mathur [9] modified eqs(14.6a) and (14.8) to account for fatigue failure from each stress state in both ductile and brittle materials as follows:

$$(\Delta\sigma_1/\Delta\sigma_0)^2 - \alpha(\Delta\sigma_1/\Delta\sigma_0)(\Delta\sigma_2/\Delta\sigma_0) + (\Delta\sigma_2/\Delta\sigma_0)^2 = 1$$
$$(\Delta\sigma/\Delta\sigma_0)^a + (\Delta\tau/\Delta\tau_0)^2 = 1$$

where α is a constant and $a = \Delta\sigma_0/\Delta\tau_0$ is unity for brittle fatigue and equal to two for ductile fatigue. Note that these modifications are identical when $\alpha = (\Delta\sigma_0/\Delta\tau_0)^2 - 2$. Stullen and Cummings [10] gave an account of the normal stress influence upon the plane of maximum shear stress within a triaxial stress state. This modified the Tresca criterion into the form

$$\Delta\sigma_1 - \lambda\,\Delta\sigma_3 = \Delta\sigma_0$$

where $\Delta\sigma_1$ and $\Delta\sigma_3$ refer to the greatest and least (numerical) ranges of applied stress and λ is a material constant that can be found from reversed torsion fatigue. That is, setting $\Delta\sigma_1 = \Delta\tau_0$ and $\Delta\sigma_3 = -\Delta\tau_0$ gives $\lambda = \Delta\sigma_0/\Delta\tau_0 - 1$. It should be noted that where safe ranges of stress need to be applied in practice then $\Delta\sigma_0$ and $\Delta\tau_0$ in these equations incorporate suitable safety factors.

Example 14.3 A thin-walled steel cylinder ($d/t = 20$) with closed ends is subjected to a range of internal pressure $p/3$ about a mean value p. Given that the reversed fatigue limit is ± 175 MPa, and that the yield stress is 320 MPa, find p for an infinite number of pressure cycles. Employ the von Mises and Tresca criteria of equivalent fatigue strength. Use a Soderberg law and neglect radial stress.

Applying eq(14.3d) to find the uniaxial stress range in the presence of a mean stress $\sigma_M = p$,

$$\Delta\sigma = \Delta\sigma_0(1 - \sigma_M/Y)$$
$$= 350(1 - p/320) = 350 - 1.0938p \qquad \text{(i)}$$

The principal stress ranges are

$$\Delta\sigma_\theta = (\Delta p)\,d/(2t) = (p/3)\,20/2 = 10p/3 \qquad \text{(ii)}$$
$$\Delta\sigma_z = (\Delta p)\,d/(4t) = (p/3)\,20/4 = 5p/3 \qquad \text{(iii)}$$

Substituting from eqs(i - iii) into eq(14.6b) gives

$$(10p/3)^2 - (5p/3)(10p/3) + (5p/3)^2 = (350 - 1.0938p)^2$$

This leads to the following quadratic in p:

$$p^2 + 107.28p - 17164 = 0$$

for which the positive root is p = 87.93 MPa. The Tresca criterion eq(14.7) simply becomes

$$10p/3 - 0 = 350 - 1.0938p$$

giving a safer estimate p = 79.06 MPa.

14.2.4 Stress Concentrations

The fatigue strength of a material depends strongly upon concentrations of stress surrounding abrupt changes in geometry, e.g. at holes and fillets and from poor surface finish. The elastic stress concentration for these "notches" is defined as

$$K_t = \frac{\text{Maximum stress at notch tip}}{\text{Average stress across notch section}} \tag{14.9}$$

Elastic stress concentrations have now been well documented in graphical, tabular and empirical forms [11]. Some of the more common geometries are given in Appendix III including those shown in Fig. 14.7a - c.

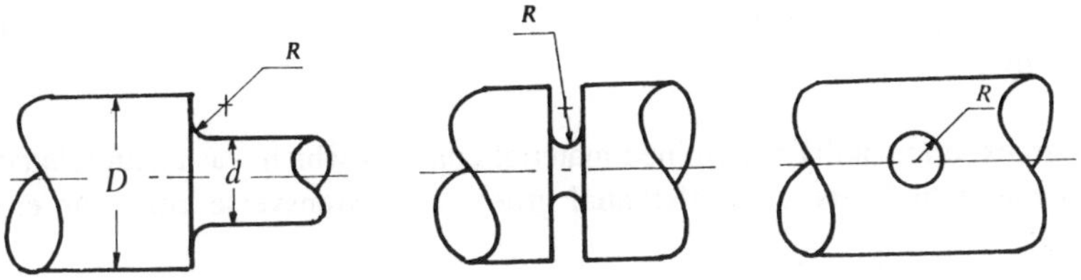

(a) (b) (c)

Figure 14.7 Representative geometries with stress concentration

Consider the stress concentration within the fillet radius that arises from applying an axial force to the stepped the bar in Fig. 14.7a. The following K_t values apply when D = 50 mm and d = 25 mm: $K_t = 2.6$ for R = 1.5 mm, $K_t = 1.9$ for R = 3 mm and $K_t = 1.4$ for R = 6 mm.

In general, the presence of a notch will reduce fatigue life at a given stress range, i.e. it will lower the *S-N* curve. The effect that a particular notch has on reducing fatigue strength for a given life may be assessed from a strength reduction factor $K_f \geq 1$, defined as

$$K_f = \frac{\text{Plain fatigue strength } (S)}{\text{Notched fatigue strength } (S_n)}. \tag{14.10}$$

Normally K_f is greater for long endurances. For example, eq(14.10) could give $K_f = 2$ for $N_f = 10^7$ and $K_f = 1.2$ for $N_f = 10^5$ cycles. Because materials display different K_f values for the same K_t, a notch sensitivity index $0 \leq q \leq 1$ combines K_t and K_f in the following form:

$$q = (K_f - 1)/(K_t - 1) \tag{14.11}$$

in which $q \to 1$ as $K_f \to K_t$ for a fully notched sensitive material, and $q \to 0$ as $K_f \to 1$ in a material that is insensitive to notches. We should use $K_f = K_t$ for the worst case. Note that where K_t appears very high for sharp notches, q for the material may appear unrealistically low. Neither K_f nor q in eq(14.11) is a material constant. Notch sensitivity increases with (i) geometry, since the likelihood of an internal defect will increase within a large volume, (ii) static strength of the material, (iii) alignment between the cyclic stress axis and material rolling direction, (iv) decreasing grain size and (v) cyclic loading in the presence of stress gradients. The application of eq(14.11) depends upon knowing K_t and K_f for a given geometry and material.

Heywood [12] examined the notched fatigue strength S_n (MPa) of ductile materials at long endurances. For the geometries given in Fig. 14.7a-c he gave an empirical formula:

$$S_n = (S/K_t)\{1 + 2[(K_t - 1)/K_t]\sqrt{(a/R)}\} \tag{14.12a}$$

where R (mm) are the radii indicated in Fig. 14.7a-c and a represents are the respective notch alleviation factors: (a) $a = (139/U)^2$, (b) $a = (104.3/U)^2$ and (c) $a = (173.8/U)^2$ where U is the ultimate tensile strength in MPa. Combining eqs(14.10) - (14.12a) gives K_f and q as

$$K_f = K_t/\{1 + 2[(K_t - 1)/K_t]\sqrt{(a/R)}\} \tag{14.12b}$$

$$q = \{(K_t - 1)[K_t - 2\sqrt{(a/R)}]\}/[K_t - 2\sqrt{(a/R)} + 2K_t\sqrt{(a/R)}] \tag{14.12c}$$

Neuber [13] proposed an alternative empirical strength reduction factor K_f in eq(14.10):

$$K_f = 1 + (K_t - 1)/[1 + \sqrt{(A/R)}] \tag{14.13a}$$

$$q = [1 + \sqrt{(A/R)}]^{-1} \tag{14.13b}$$

where A is an experimentally determined material constant which was again related to the u.t.s. For steels with fillets, circumferential grooves and transverse bores, an empirical relation holds,

$$\sqrt{A} = 1.0 - U/874 + (U/1703)^2 \tag{14.13c}$$

In eqs(14.13a-c) U is in MPa and A is in mm.

Example 14.4 A high tensile steel beam, which contains a 75 mm diameter cross bore (for which $K_t = 3$), is subjected to a repeated bending cycle. Using the Goodman relationship, compare q and the maximum bending stress predictions of Heywood and Neuber for infinite life given that its u.t.s. is 1100 MPa and its reversed bending fatigue strength is ± 460 MPa.

The repeated bending fatigue strength is found from Goodman's eq(14.3b) with $\Delta\sigma_o = 920$ MPa, $\sigma_M = \Delta\sigma/2$ and $U = 1100$ MPa. This gives,

$$\Delta\sigma = 920[1 - (\Delta\sigma/2) \times 1100]$$

$$\Delta\sigma = 920 - 0.4182\Delta\sigma \Rightarrow \Delta\sigma = 648.72 \text{ MPa}$$

Heywood: The notch alleviation factor for Fig. 14.7c is

$$\sqrt{(a/R)} = 173.8/(U \times \sqrt{R}) = 173.8/(1100 \times \sqrt{37.5}) = 0.0258$$

Substituting into the Heywood eqs(14.12b,c),

$$K_f = 3/\{1 + 2[(3 - 1)/3]0.0258\} = 2.9 \quad \text{and} \quad q = 0.95$$

Then from eq(14.10) $S_n = S/K_f$ and, since $S = \Delta\sigma$ expresses both the maximum stress and the range for the unnotched repeated cycle, the corresponding notched values are

$$S_n = \Delta\sigma/K_f = 648.72/2.9 = 227.74 \text{ MPa}$$

Neuber: The material constant A is found from eq(14.13c) as

$$\sqrt{A} = 1.0 - U/874 + (U/1703)^2 = 0.159$$

Substituting into eq(14.13a,b),

$$K_f = 1 + (3 - 1)/[1 + 0.159/\sqrt{37.5}] = 2.95 \quad \text{and} \quad q = 0.975$$

$$\therefore\ S_n = \Delta\sigma/K_f = 648.72/2.95 = 229.9 \text{ MPa}$$

which shows that the Heywood prediction is slightly more conservative.

14.2.5 Residual Stresses

The presence of residual surface stresses can alter the fatigue strength of a material. The general rule is that when the sense of the residual stress opposes that within a fatigue cycle, an improvement in fatigue strength will be found. A residual compressive stress, for example, would be beneficial to the fatigue strength under repeated or fluctuating tensile cycles (Figs 14.2c,d). This is because their effect is to reduce the mean stress which in turn raises the *S-N* curve. On the other hand, tensile residual stresses will effectively lower the *S-N* curve to the detriment of the fatigue strength under tensile cycling. It has long been known that favourable compressive residual stresses may be induced by mechanical deformation and thermal treatment processes. These include cold rolling, shot and hammer peening, honing, carburising and nitriding. In contrast, heavy grinding and welding can produce surface cracks in addition to tensile residual stresses, both of which will impair the fatigue strength under tensile cycling.

It has been shown earlier that partial plasticity of beams in bending, torsion bars and auofrettage of thick cylinders results in a residual stress distribution following unloading. The residual stresses may be used to enhance the fatigue life of these structures where they serve to cancel the stress within the applied cycle. Compressive residuals are usually beneficial since most fatigue failure arises from the tensile part of a cycle. Consider the residual stress distribution in the beam arising from applying an elastic-plastic bending moment (see Fig. 12.2a). In Fig. 14.8a we superimpose upon this the peak elastic stress distributions from a reversed bending cycle.

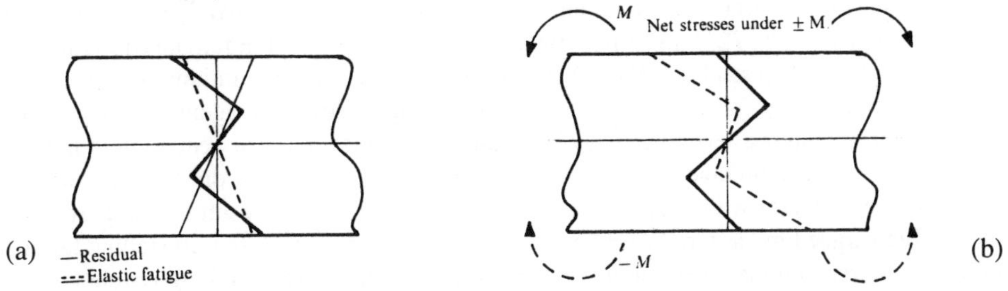

Figure 14.8 Interaction between residual and cyclic bending stresses

It is seen from the net stress distribution in Fig. 14.8b that the surface bending stresses are reduced with a hogging moment but increased with a sagging moment. The residuals are beneficial under the former moment but detrimental under the latter. A similar interaction occurs between residual stresses from pre-torsion (see Fig. 12.21b) and the peak elastic stresses from reversed torsion. In an autofrettaged thick cylinder the interaction between the residual stresses and the repeated application of Lamé's elastic pressure stresses is triaxial. For example, if we add the residual stress distributions of Figs 12.26b to the Lamé stresses (shown in Fig. 2.18) the net distributions in Fig. 14.9 are found.

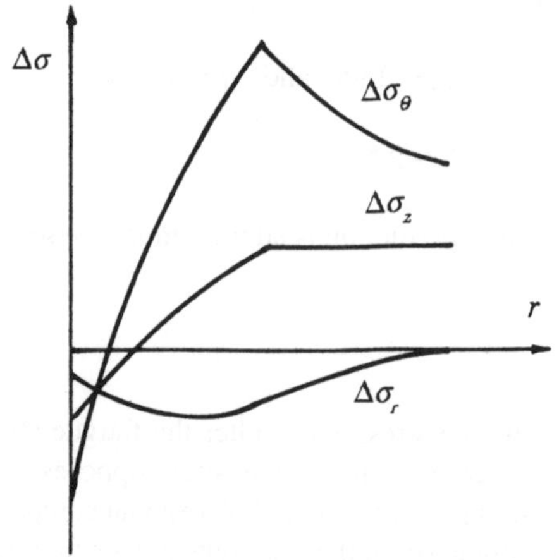

Figure 14.9 Net triaxial stress state in autofrettaged cylinder

Typically these residuals apply to the peak pressure in a closed-end cylinder that has been autofrettaged to the mean wall radius. The range of Lamé's elastic bore shear stress is $\Delta\tau = \frac{1}{2}\Delta(\sigma_\theta - \sigma_r)$ which is used as the measure of fatigue strength. An optimally autofrettaged cylinder will have infinite life when a net hydrostatic compression exists at the bore so that $\Delta\tau = 0$. In practice, slight differences between the three principal net bore stresses will be found (see Fig. 14.9) depending upon the peak pressure, but we can still expect an improvement in fatigue strength [14].

14.3 Environmental Effects

14.3.1 Temperature

The two stages of fatigue crack initiation and propagation will prevail with increasing temperature. Up to $T < T_m/2$, most materials show only a slight fall in fatigue strength. The exception is for carbon steels (at 350°C), cast iron (at 450°C) and aluminium-copper alloys (at 150°C) which show improved fatigue and static strengths due strain ageing. However, for increasing temperatures beyond $T > T_m/2$ all materials display a drastic reduction in fatigue strength due to the simultaneous presence of creep. The effect is most drastic under higher peak and mean stresses within repeated and alternating cycling. The result is that there is no discernible fatigue limit as the fracture path changes from the normal transcrystalline fatigue failure to intercrystalline creep failure. Figure 14.10 combines creep and high cycle fatigue behaviour under a high temperature alternating cycle with range $\Delta\sigma$ and mean σ_M.

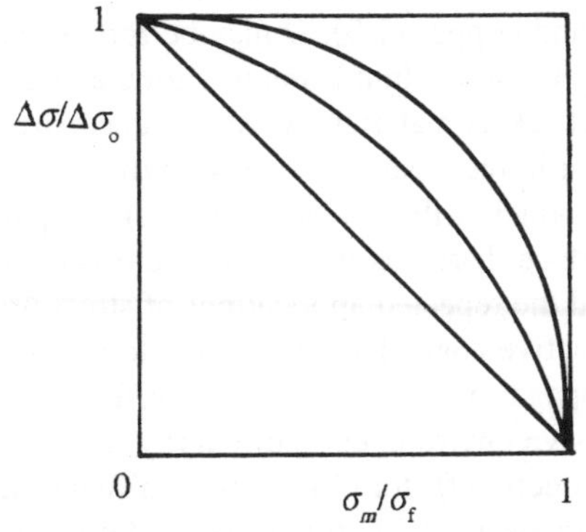

Figure 14.10 Stress range diagram for combined creep-fatigue

Figure 14.10 is similar to Fig. 14.3 where fatigue and creep strengths, instead of the u.t.s., are used to normalise $\Delta\sigma$ and σ_M respectively. The ordinate is the ratio $\Delta\sigma/\Delta\sigma_o$ where $\Delta\sigma_o$ is the reversed stress range for a given cyclic life (i.e. $\Delta\sigma_o = 2 \times S_E$). The abscissa is the stress ratio σ_M/σ_f where σ_f is the creep stress to give rupture in a given time. We may select S_E and σ_f according to our design requirements ($S_E = 10^5, 10^6, 10^7$ .... cycles: $\sigma_f = 10^2, 10^3, 10^4$ ... hr etc). The linear, parabolic and quadratic predictions shown apply to low, medium and high temperatures respectively. Experiment [15,16] has shown that the line applies to low-temperature fatigue

$$\Delta\sigma/\Delta\sigma_o + \sigma_M/\sigma_f = 1$$

and the quadratic applies to high-temperature dynamic creep

$$(\Delta\sigma/\Delta\sigma_o)^2 + (\sigma_M/\sigma_f)^2 = 1$$

Further investigations have been made to establish the relationship between dynamic and static creep and the influence of fatigue upon the creep curve [17,18]. These showed that a vibratory stress test with small amplitude superimposed upon a mean stress impaired the rupture life of alloy steels below 815°C. Also, when a vibratory stress was applied to copper at 400°C at a certain time along a creep curve, the creep rate increased and the rupture life decreased. The strain rate was relatively insensitive to the frequency of cycling employed within each investigation.

14.3.2 Corrosion

A corrosive environment which assists the formation of an initiation site, through pitting and scaling, will drastically reduce the normal fatigue strength for air. This is most marked for higher strength materials. In salt water, for example, there is a 50 - 80% loss in fatigue strength for steels, a 60% loss in aluminium alloys, while copper remains unaffected. Moreover, for high strength steels in particular, there is no evidence of a fatigue limit. These influences of corrosion may be expressed in a damage ratio D :

$$D = \frac{\text{Corrosion fatigue strength}}{\text{Normal fatigue strength}}$$

For a salt water corrodent, D depends upon the material: carbon steel ($D = 0.2$), stainless steel (0.5), aluminium alloys (0.4) and copper (1.0). D may be further reduced when (i) higher temperatures accentuate oxidising and sulphidising environments and (ii) low frequencies allow more time for corrosive attack. A material's resistance to corrosion fatigue will depend upon its resistance to corrosion in the absence of stress. When both corrosion and fatigue occur together, the chemical attack will increase the crack propagation rate. The main influence of corrosion is to produce sharp pits from which numerous cracks form. The pitting reduces the initiation phase and the repeated applications of stress prevent the formation of protective barriers within the active zone. The pits penetrate inwards and sharpen into the cracks responsible for the propagation phase. The notch effect of stress concentrations is accentuated further in a corrosive environment particularly in high strength materials. To account for this the strength reduction factor (14.10) is modified to $K_f' = K_f \times D$.

There are a number of methods available for corrosion-fatigue protection. The surface may be coated with a continuous metal that should also protect the base metal by electrochemical action if the coating is broken. This can be achieved by selecting a coat that is anodic to the base metal. For example, zinc, being anodic to steel, provides good protection. The surface of components is also coated with nickel, cadmium, chromium, copper and aluminium provided they are not cathodic to the base, e.g. as with copper on steel, since this can serve to reduce the fatigue resistance. Non-metallic coating formed by anodising, varnish, paints and epoxy resins have all been found to be effective under certain conditions [19]. Mechanical surface treatments such as shot peening, rolling and nitriding produce residual compressive stresses that serve to close cracks, thus preventing penetration by the corrosive medium. Chemical inhibitors such as sodium carbonate and emulsifying oils, when employed as adherent films, give protection to steels in particular. A sacrificial anode of aluminium or magnesium is often used to provide protection for steel structures immersed in sea water. This method is referred to as cathodic protection because the steel becomes the cathode in an electrolyte of brine. Care must be taken with this method to avoid over-protection where an excess of atomic hydrogen can penetrate and embrittle the steel.

Fretting corrosion is the name given to the failure resulted in the contact zone when two metals rub together usually from a vibration source. Initiation of fretting occurs by molecular plucking of surface particles. These particles oxidise to form a debris which can then act as an environment for corrosion fatigue attack. Fretting of dissimilar materials may be avoided by nitriding the surface or by electroplating. Lubrication and accentuation of the movement, when allowable, are also effective inhibitors.

14.4 Low Endurance Fatigue

Low endurances ($N \leq 10^4$ cycles in Fig. 14.1b) are not of great practical interest where long life is required under stress cycling. Within this regime the inherent ductility of a material becomes a controlling factor in withstanding the plastic deformation occurring in every cycle. Fig. 14.11a illustrates typical stress-strain hysterisis loops corresponding to stress or load cycling between positive and negative limits of equal magnitude. It is seen that the loops progress in the direction of positive (or negative) strain, indicating an accumulation of strain with each cycle. In contrast, under purely elastic strain cycling (Fig. 14.11b) within high cycle fatigue, the cyclic life is dependent upon the material's strength. Figure 14.11c shows two materials with contrasting fatigue behaviour. The more ductile material B has the greater resistance to low cycle fatigue failure, while the stronger material A has the greater resistance to high cycle fatigue failure. The low cycle fatigue behaviour of a material becomes more important to consider in practice when plasticity occurs between alternating limits of cyclic strain. Cyclic plasticity can arise either from stress or temperature cycling.

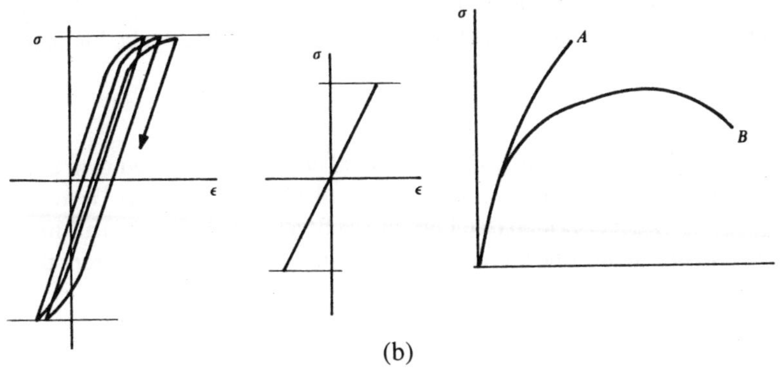

Figure 14.11 Stress cycling in low and high cycle fatigue

14.4.1 Strain Limited Cycling

The term low cycle fatigue (LCF) normally refers to this condition. Figures 14.12a and b illustrate typical stress-strain hysterisis loops corresponding to symmetrical strain cycles bounded by fixed limits of positive and negative strain of equal magnitude. Under the constant total strain range $\Delta\varepsilon^T$ the range of stress $\Delta\sigma$, during the first 10% of cyclic life, will initially increase (Fig. 14.12a) for a strain-hardening material or it may fall (Fig. 14.12b) for a strain softening material.

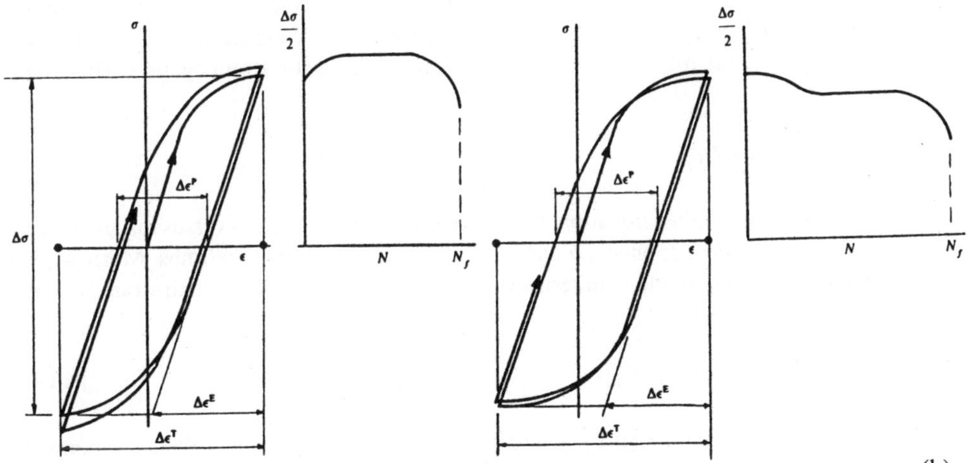

Figure 14.12 Strain cycling hardening and softening materials

As a rule annealed materials harden while prior worked materials with *Y*/*U* > 0.8 soften. Thereafter, a cyclically stable state is normally attained, where $\Delta\sigma$ remains constant. Cyclic stability accounts for 80 - 90% of life before a further continuous fall in $\Delta\sigma$ leads to rapid failure in N_f cycles as shown. The resistance of the material to fatigue failure under plastic strain cycling is governed by its ductility. To show this we conduct a number of LCF tests each at a different range of plastic strain $\Delta\varepsilon^P$. For lower endurances in LCF, where plastic strains dominate, the Coffin [20] - Manson [21] law expresses the cyclic life in the form

$$\Delta\varepsilon^P = C\,N_f^{\,-\beta} \tag{14.14}$$

where C and $\beta\,(0.4 \le \beta \le 0.7)$ are positive material constants identified with the slope and intercept in a $\log(\Delta\varepsilon^P)$ versus $\log(N_f)$ plot (see Fig. 14.13a). Normally $\Delta\varepsilon^P$ is identified with the width of a stable hysteresis loop (see Figs 14.12a,b) measured at cyclic half-life. Note that C may be identified with the true plastic fracture strain $\varepsilon_f^{\,P}$ in a tension test from the first load reversal.

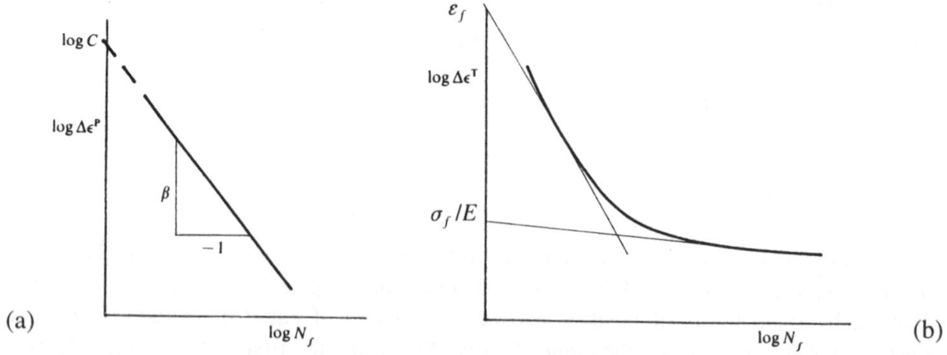

Figure 14.13 Dependence of strain range upon cyclic life

When, for greater LCF lives, the range of cyclic elastic $\Delta\varepsilon^E$ and plastic strains are of comparable magnitudes, then the life N_f is better related to the total strain range by the addition of a further term to account for the influence of cyclic elasticity on LCF life [22]. Here the strength of the material controls the resistance to elastic strain cycling. This gives

$$\begin{aligned}\Delta\varepsilon^T &= \Delta\varepsilon^E + \Delta\varepsilon^P \\ &= (B/E)(N_f)^b + C(N_f)^c \end{aligned} \tag{14.15a}$$

where E is Young's modulus. Equation(14.15a) describes the $\log(\Delta\varepsilon^T)$ versus $\log(N_f)$ curve in Fig. 14.13b when the exponents b and c are the gradients of its two limbs. Morrow [23] interpreted the constants B and C in terms of the true static strength (σ_f) and strain (ε_f) in a tension test as follows:

$$\Delta\varepsilon^T/2 = (\sigma_f/E)(2N_f)^b + \varepsilon_f(2N_f)^c \tag{14.15b}$$

where $2N_f$ refers to the number of reversals (2 per/cycle). Setting $N_f = ½$ corresponds to loading in tension under the first reversal. Equation (14.15b) gives the intercepts between each limb and the strain axis as σ_f/E and ε_f. Measured fracture strains may be only 50% of ε_f and so lives from eq(14.15b) will then be overestimated. In Manson's universal slopes relationship [24] b and c were taken as constants for all materials. Equation (14.15a) is written in the form

$$\Delta\varepsilon^T = 3.5(U/E)\,N_f^{\,-0.12} + D^{0.6}\,N_f^{\,-0.6} \tag{14.15c}$$

in which $D = \ln\,[100/(100 - R_A)]$, where R_A is the percentage reduction in area in static tension, for which E and the U take their usual meaning. Equations(14.15a-c) apply to the transcrystalline LCF failures found for many ductile materials at low-medium temperatures under high cyclic frequencies.

Example 14.5 A material conforms to the Coffin-Manson law $\Delta\varepsilon^P N_f^{\,0.4} = 0.25$. If it is subjected to $\Delta\varepsilon^P = 0.015$ for 250 cycles followed by $\Delta\varepsilon^P = 0.0075$ for the remainder of its life, how many cycles can it endure by Miner's rule?

The Coffin-Manson law (14.14) supplies the life N_f for a given plastic strain range in this material as

$$N_f = (0.25/\Delta\varepsilon^P)^{2.5} \tag{i}$$

Substituting $\Delta\varepsilon_1^{\,P} = 0.015$ and $\Delta\varepsilon_2^{\,P} = 0.0075$ in eq(i) gives $N_{f1} = 1134$ cycles and $N_{f2} = 6415$ cycles respectively. Then, from Miner's rule (eq14.4),

$$n_1 / N_{f1} + n_2 / N_{f2} = 1$$
$$250/1134 + n_2 /6415 = 1 \;\Rightarrow\; n_2 = 5000 \text{ cycles}$$

14.4.2 Frequency Modified Law

Further modifications to LCF life become necessary to account for the effects of creep at higher temperatures and lower frequencies of cycling. Very slow cycling can result in a predominance of intergranular oxidised cracking. The Coffin-Manson law (14.14) will not apply directly unless failures are wholly transcrystalline. Qualitatively, the effect of a different or interacting mode of failure is to alter the slopes of Figs 14.13a,b. In Coffin's [25] frequency modified LCF law the elastic range of strain is included within the total range of strain. The dependence of high temperature cyclic life upon frequency becomes

$$\Delta\varepsilon^T = (B/E)\, N_f^{\,-\alpha}\, \nu^{k'} + C\, N_f^{\,-\beta}\, \nu^{\,-\beta(k-1)} \tag{14.16a}$$

where α and β are the slopes and B, C, E, k and k' are all dependent upon temperature but not upon frequency. At low temperatures $k = 1$ and $k' = 0$ to restore the form of eq(14.15a). Equation (14.16a) shows that decreasing the frequency ν within constant strain cycles decreases the cyclic life N_f. Now k is related to the total time to failure ($t_f = N_f/\nu$) in

$$\nu^k\, t_f = K\,(\Delta\varepsilon^P) = \text{constant} \tag{14.16b}$$

where $k = 1$ for pure fatigue (rapid cycling) and $k = 0$ when creep becomes dominant with slow cycling at high temperature. The prediction (14.16b) also applies to constant period cycles with dwell, e.g. a fixed period of relaxation at peak strain or sreep at peak stress.

Where frequency variations occur between the positive and negative sides of a closed hysteresis loop, Coffin [26] separated the frequencies within the second inelastic strain term in eq(14.16a):

$$\Delta\varepsilon^P = C\, N_f^{\,-\beta}\, \nu_t^{\,-\beta(k-1)}\, (\nu_c/\nu_t)^q \tag{14.17}$$

where q is a further constant and under uniaxial stressing, t and c refer to tension and compression. Equation (14.17) has been shown to predict the cyclic life within a factor of 2 both for LCF cycles with varying frequency and with relaxation hold times at peak strain [27].

Frequency separation is less satisfactory with relaxation holds for off-peak strains and when periods of creep result in ratcheting.

Berling and Conway [28] proposed the method of characteristic slopes to account for the influence of frequency upon the high-temperature (650 - 850°C) LCF life of stainless steel. For this, short-time tensile data is required at the similar conditions of temperature and strain rate $\dot{\varepsilon}^T$. That is,

$$\Delta\varepsilon^T = 2\varepsilon_f^{\,E}(N_f/10)^{-n/2} + (D^2/\dot{\varepsilon}^T)(N_f/\nu)^{-1} \tag{14.18a}$$

where D is the reduction in area (see eq14.15c), $\varepsilon_f^{\,E} = \sigma_f/E$ is the true elastic fracture strain and n, the static strain-hardening exponent, is allowed to vary with the test conditions. Both $\varepsilon_f^{\,E}$ and n are derived from a true stress-strain curve in the high strain region. Assuming a triangular waveform, the frequency becomes $\nu = \dot{\varepsilon}^T/(2\Delta\varepsilon^T)$ so that eq(14.18a) can be solved for $\Delta\varepsilon^T$ if required for a given N_f

$$\Delta\varepsilon^T = \varepsilon_f^{\,E}(N_f/10)^{-n/2} \pm \sqrt{[(\varepsilon_f^{\,E})^2(10/N_f)^n + D^2/(2N_f)]} \tag{14.18b}$$

Example 14.6 A stainless steel is subjected to high-temperature LCF cycling under a constant strain rate $\dot{\varepsilon}^T = 5\times10^{-3}\ \mathrm{s}^{-1}$ within a total range of strain of 1%. Given that the material tensile properties are U = 280 MPa, R_A = 20.8%, E = 180 GPa and n = 0.3, compare the cyclic lives from the methods of universal and characteristic slopes with Coffin's frequency modification for k' = 0.2 and k = 0.7.

Universal Slopes: Referring to eq(14.15c),

$$U/E = 280/(180\times10^3) = 1.56\times10^{-3}$$
$$D = \ln[100/(100 - 20.8)] = 0.233$$
$$\begin{aligned}0.01 &= 3.5\,(1.56\times10^{-3})(N_f)^{-0.12} + (0.233)^{0.6}(N_f)^{-0.6}\\ &= (5.46\times10^{-3})(N_f)^{-0.12} + 0.417(N_f)^{-0.6}\end{aligned} \tag{i}$$

A trial solution shows that eq(i) is satisfied by N_f = 800 cycles. Whilst the high-temperature strength and ductility are accounted for the frequency is not.

Characteristic slopes: The true fracture strain and ductility are

$$\varepsilon_f^{\,E} \approx U/E = 280/(180\times10^3)] = 1.55\times10^{-3}$$

and the frequency is

$$\nu = \dot{\varepsilon}^T/(2\Delta\varepsilon^T) = (5\times10^{-3})/(2\times1\times10^{-2}) = 0.25\ \mathrm{Hz}$$

Substituting into eq(14.18a)

$$\begin{aligned}0.01 &= (2\times1.55\times10^{-3})(N_f/10)^{-0.15} + [(0.233)^2/(5\times10^{-3})](N_f/0.25)^{-1}\\ &= (4.379\times10^{-3})(N_f)^{-0.15} + 2.715(N_f)^{-1}\end{aligned} \tag{ii}$$

A trial solution shows that eq(ii) is satisfied by N_f = 332 cycles. The life is lowered by the influence of cycle frequency and slope values not conforming to the universal values.

Frequency modification: We shall identify the constants B and C in eq(14.16a) with σ_f and ε_f respectively and take the exponents as the characteristic slope values. This gives

$$\Delta\varepsilon^T = (\sigma_f /E)\,(N_f)^{-0.12}\,\nu^{0.2} + \varepsilon_f\,(N_f)^{-0.6}\,\nu^{0.18} \qquad \text{(iii)}$$

Recall from eq(12.83a) that n is the true axial strain at instability. So σ_f and ε_f can be estimated from eqs(12.80) and (12.81a). Take $\varepsilon_f > n$ (say 0.4). Thus

$$e_f = e^{\varepsilon_f} - 1 = \mathrm{e}^{0.4} - 1 = 0.492$$
$$\sigma_f = U\,(1 + e_f) = 280 \times 1.492 = 417.7 \text{ MPa}$$

Substituting the values of E, ν, ε_f and σ_f into eq(iii) gives

$$\Delta\varepsilon^T = (2.321 \times 10^{-3})\,(N_f)^{-0.12} + 0.311(N_f)^{-0.6} \qquad \text{(iv)}$$

A trial solution shows that eq(iv) is satisfied by $N_f = 358$ cycles. As this near the characteristic slope value, it confirms that universal slopes may be used with frequency modification.

14.4.3 Cumulative LCF Damage

We saw in Example 14.5 how Miner's rule was employed to determine LCF life where changes occurred to the inelastic strain range. Broadly, the same procedure is applied to assess damage when the amplitude of strain (and stress) varies in a random manner throughout LCF. However, a method is required to count the number of whole and half-cycles within a σ versus ε plot in order to assess cyclic damage. Usually the history is known in the form of an ε versus t and/or a σ versus t record so these must be converted into reversals. A rainflow count [29] employs the ε versus t record. Imagine rain following the vertically aligned ε versus t plot in Fig. 14.14a. Beginning at the inside of each peak it would fall vertically from the peaks at 2, 3, 4 etc. In the construction of the half and whole σ versus ε cycles in Fig. 14.14b every part of the ε, t trace must be included just once.

We shall refer to peaks 2, 4, 6 etc as maxima and peaks 3, 5, 7 etc as minima. The origin 1 is taken as a minimum with negative sense. To define a half-cycle originating from a maximum, the falling rain must stop when it lies opposite a more positive maximum. Similarly, with a half-cycle originating from a minimum the rain fall stops when it lies opposite a more negative minimum. Thus, rain flows from 1 - 2 - 2' - 4 and stops opposite 7 (ε_7 is more negative than ε_1). Starting at point 2 rain flows to 3 where it falls to stop opposite 4 (ε_4 is more positive than ε_2). Rain flow originating at 3 will terminate at 2' where it meets the flow from the "roof" above. The path 2 - 3 - 2' gives a closed loop which interrupts the half cycle 1 - 2 - 4 at point 2 (see Fig. 14.14b). From 4, a reversed half-cycle follows 4 - 5 - 5' - 7, falls from 7 to stop opposite 10 ($\varepsilon_{10} > \varepsilon_4$). A closed loop from path 5 - 6 - 5' is located at point 5 as shown. Paths 7 - 8 - 8' - 10 and 8 - 9 - 8' provide the remaining half and whole cycles shown. The rainflow count gives half cycles: 1 - 2 - 4, 4 - 5 - 7 and 7 - 8 - 10 and whole cycles 2 - 3 - 2', 5 - 6 - 5' and 8 - 9 - 8' within Fig. 14.14b. The damage fraction from each half and whole cycle (1 and 2 reversals respectively) is assessed as $1/(2N_f)$ and $2/(2N_f)$ where N_f is the number of reversals at each of the respective total strain ranges ($\Delta\varepsilon^T = \varepsilon_4 - \varepsilon_1$, $\varepsilon_4 - \varepsilon_7$, $\varepsilon_{10} - \varepsilon_7$, $\varepsilon_2 - \varepsilon_3$, $\varepsilon_6 - \varepsilon_5$ and $\varepsilon_8 - \varepsilon_9$). It is convenient to use eq(14.15b) to find $2N_f$ for a given $\Delta\varepsilon^T$. The damage fractions are then summed to predict failure when

$$\sum 2/(2N_f) + \sum 1/(2N_f) = 1.$$

(a)

(b)

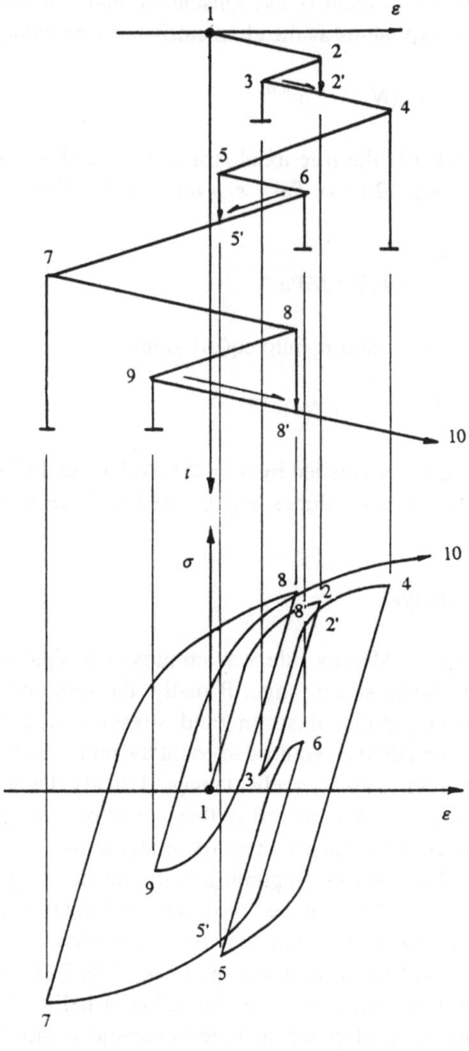

Figure 14.14 Rainflow counting method

If regular blocks of random cycles are applied the damage summation need only be applied to one block then multiplied by the number of blocks. Figure 14.14b shows that the cycles are often displaced from the origin and therefore an account of the mean stress will provide N_f more accurately [23]. When unbalanced loops occur on the compressive side σ_M is negative and the life is enhanced. For loops with a tensile bias, σ_M is positive and the life is reduced. Where stress amplitudes of balanced and unbalanced cycles are the same then the life N_a in the presence of mean stress is reduced, according to

$$N_a = N_f(1 - \sigma_M/\sigma_f)^{1/b} \tag{14.19a}$$

in which b and σ_f appear in a law, similar to eq(14.15b). This law now applies to $\Delta\varepsilon^T$ versus N_a in the presence of σ_M and may be combined with eq(14.19a) to give

$$\Delta\varepsilon^T/2 = (\sigma_f/E)(1 - \sigma_m/\sigma_f)(2N_f)^b + \varepsilon_f\,(1 - \sigma_M/\sigma_f)^{c/b}\,(2N_f)^c \tag{14.19b}$$

Alternatively, we may relate the peak stress $\sigma_p = (\sigma_a + \sigma_M)$ in a cycle with mean stress to a reversed cycle, of amplitude σ_A, within the following energy relation [30]:

$$\sigma_p \Delta\varepsilon^T/2 = \sigma_A \Delta\varepsilon^T/2 \quad \text{and} \quad \sigma_A = \sigma_f (2N_f)^b \tag{14.20a,b}$$

Combining eqs(14.15b) and eqs(14.20a,b) gives

$$\sigma_p(\Delta\varepsilon^T/2) = (\sigma_f^{\,2}/E)(2N_f)^{2b} + \sigma_f\,\varepsilon_f\,(2N_f)^{b+c} \tag{14.20c}$$

where the constants are determined for $\sigma_M = 0$ but N_f applies to all σ_m.

14.4.4 LCF Temperature Cycling

A fixed range of plastic strain will also be induced when a material is prevented from expanding naturally under temperature cycling. The general term "thermal fatigue" describes the failure resulting from both constrained and unconstrained expansions involving plastic deformation. The failure can occur from a single cycle (thermal shock) in a brittle material or from repeated cycles in a ductile material. Expansions are non-uniform where temperature cycling is localised and where the thermal mass varies. Consider a uniform encastre beam in Fig. 14.15a. Let the beam temperature be cycled in the range ΔT uniformly and rapidly over its length.

(a) 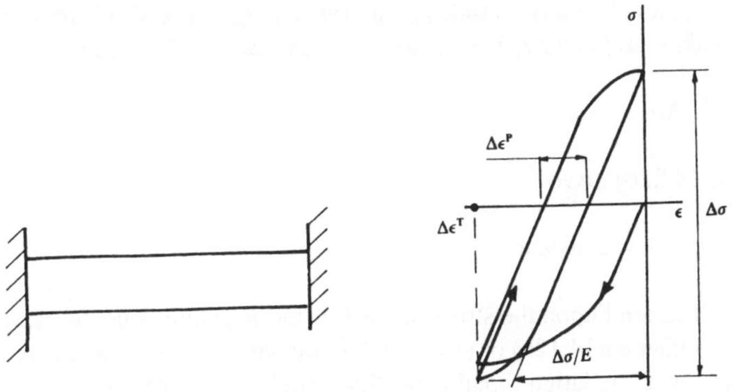 (b)

Figure 14.15 Thermal strain cycling of an encastre beam

The total strain induced by restraining the free expansion is

$$\Delta\varepsilon^T = \alpha\,\Delta T = \Delta\varepsilon^P + \Delta\varepsilon^E \tag{14.21a}$$

where α is the thermal expansion coefficient. During the initial cycling the stress range $\Delta\sigma$ alters to attain a steady state value. Thereafter, $\Delta\varepsilon^E = \Delta\sigma/E$ and the range of plastic strain within the stable loop (see Fig. 14.15b) follows from eq(14.21a) as

$$\Delta\varepsilon^P = \alpha\,\Delta T - \Delta\sigma/E \tag{14.21b}$$

Plots of $\Delta\varepsilon^P$ versus N_f obey the Coffin-Manson law. Thus, eqs(14.14) and (14.21) may be combined to estimate N_f. The asymmetry of the σ versus ε^P loop results in a lower coefficient C in eq(14.14) compared to stress cycling at a constant temperature of $\Delta T/2$ but the exponent

β is similar [30]. It follows from eq(14.21b) that, for a given ΔT, the parameters governing thermal fatigue strength are α, $\Delta\varepsilon^P$, $\Delta\sigma$ and E. A further property of the material, controlling its resistance to $\Delta\varepsilon^P$, is the thermal conductivity k. High k values are generally beneficial to reducing temperature gradients and induced strains. The Eichelberg quality factors provide a measure of a material's resistance to thermal strain cycling. These factors are $\Delta\sigma k/(\alpha E) \equiv 2Yk/(\alpha E)$, where $\Delta\sigma = 2Y$ in the absence of hardening, and $\Delta\varepsilon^P k/\alpha$. In each case high values are desirable.

14.4.5 Stress Concentrations

We have seen how the presence of a notch raises the local stress to reduce the high cycle fatigue life. If the cyclic stress is raised by the notch, to produce local reversals of plastic strain, a low cycle fatigue failure is likely. A modification to Neuber's rule [13] allows the local ranges of stress and strain to be approximated. This rule states that the geometric mean of the stress and strain concentration factors, K_σ and K_ε respectively, remains equal to K_t (see eq 14.9) during plasticity. The rule is modified for fatigue so that K_t is replaced by K_f (see eq(14.10) to give

$$K_f = \sqrt{(K_\sigma K_\varepsilon)} \tag{14.22a}$$

where $K_\sigma = \Delta\sigma/\Delta\sigma_0$ and $K_\varepsilon = \Delta\varepsilon/\Delta\varepsilon_0$ in which $\Delta\sigma$ (and $\Delta\varepsilon$) are the maximum stress (and strain) range at the notch and $\Delta\sigma_0$ (and$\Delta\varepsilon_0$) are the average stress (and strain) ranges at the notch section. Taking $\Delta\sigma_0$ and $\Delta\varepsilon_0$ to be elastic, we have $\Delta\sigma_0 = E\Delta\varepsilon_0$ and eq(14.22a) becomes

$$K_f = (E\Delta\sigma\Delta\varepsilon)^{1/2}/\Delta\sigma_0 \tag{14.22b}$$

Rearranging eq(14.22b) gives

$$\Delta\sigma\Delta\varepsilon = (\Delta\sigma_0 K_f)^2/E = \text{constant} \tag{14.23}$$

The constant will depend upon the stress range $\Delta\sigma_0$ and K_f, which may be estimated from a known K_t, using either eq(14.12b) or (14.13a). Say we wish to predict the cyclic life N_f from eq(14.14) under low cycle fatigue conditions. Since eq(14.14) applies to steady state cycling, the equation to a cyclic stress-strain curve allows $\Delta\sigma$ and $\Delta\varepsilon$ to be separated from within eq(14.23). The loci of the vertices of stable cycles shown in Fig. 14.16a is given by

$$\Delta\sigma = A'(\Delta\varepsilon)^{n'} \tag{14.24}$$

where A' and n' are material constants. In the axes shown $\varepsilon = \Delta\varepsilon/2$ and $\sigma = \Delta\sigma/2$ are half-ranges. The cyclic hardening curve lies beneath a monotonic stress-strain curve and therefore A' and n' are not the usual Hollomon constants (i.e. A and n in eq(12.83a)). A cyclic hardening exponent $n' \approx 1/7$ applies to all polycrystalline metals [10] remaining insensitive to the initial condition. In contrast, the strain-hardening exponent n can vary from 1/25 for cold-worked metals to ½ for annealed metals. If we now superimpose upon Fig. 14.16b a family of rectangular hyperbolae each for a different $\Delta\sigma_0$ value in eq(14.23), the intersection Q provides the stress and strain ranges at the notch root. The Coffin-Manson law (14.14) will supply the number of cycles to failure for a plastic strain range of $\Delta\varepsilon - \Delta\sigma/E$.

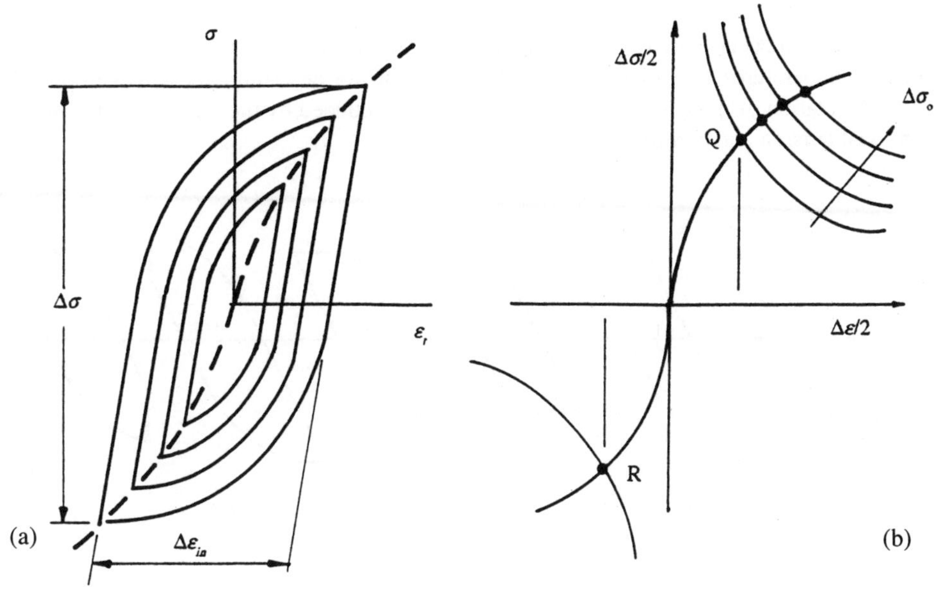

Figure 14.16 The cyclic hardening curve

Example 14.7 The high tensile steel cross-bored beam referred to in Example 14.4 is subjected to a reversed bending cycle for which the applied stress range is $\Delta\sigma_o = 1000$ MPa. Given that $A' = 4030.7$ and $n' = 0.133$ in eq(14.24) and that $C = 1.27$ and $\beta = 0.7$ in eq(14.14), find the cyclic life. Take $E = 207$ GPa.

We have previously shown that $K_f = 2.95$ from eq(14.13a). Substituting into the right-hand side of eq(14.23) gives

$$\Delta\sigma \times \Delta\varepsilon = (1000 \times 2.95)^2 / 207000$$
$$\Delta\sigma \times \Delta\varepsilon = 42.04 \text{ MPa} \qquad \text{(i)}$$

The cyclic stress strain curve is

$$\Delta\sigma = 4030.7(\Delta\varepsilon)^{0.133} \qquad \text{(ii)}$$

The Coffin-Manson law is

$$\Delta\varepsilon^P = 1.27\,(N_f)^{-0.7} \qquad \text{(iii)}$$

The solution to eqs(i) and (ii) gives $\Delta\sigma = 2358.9$ MPa and $\Delta\varepsilon = 0.0178$, i.e. two times each co-ordinate of point Q in Fig. 14.16b. The plastic component of this strain range is

$$\varepsilon^P = \varepsilon - \Delta\sigma / E$$
$$= 0.01078 - 2358.9 / 207000 = 6.404 \times 10^{-3}$$

Substituting into eq(iii), we find $N_f = 1914$ cycles.

14.5 Creep-Fatigue Interaction

An interaction occurs when periods of dwell are introduced into a LCF cycle, particularly at higher temperatures. Fig. 14.17a-f shows two notations for describing these waveforms. The NASA notation [31] applies to strain-limited cycles produced by the repeated application of uniaxial tension and compression.

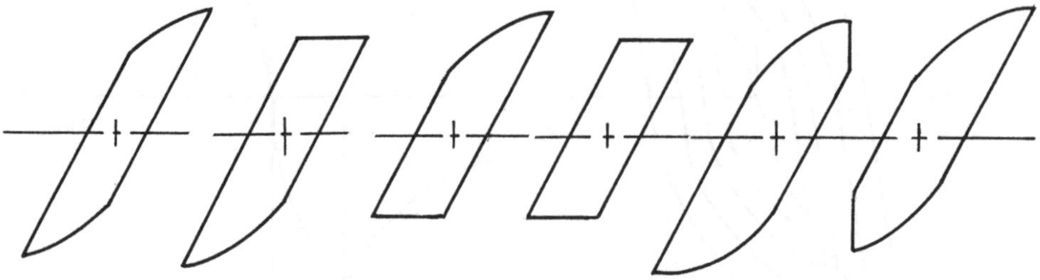

Figure 14.17 NASA cycle notations

These are designated as follows: (a) high rate strain-cycle (HRSC), (b) tensile cyclic-creep rupture (TCCR), (c) compressive cyclic-creep rupture (CCCR), (d) balanced cyclic-creep rupture (BCCR), (e) tensile hold strain-cycle (THSC) and (f) compressive hold strain-cycle (CHSC). The CCCR and CHSC cycles mirror the TCCR and THSC cycles. The NASA notation also admits cycles, similar in shape to (a)-(f), produced from alternating torsion. Practical examples include thermal fatigue in gas turbine blading and cyclic loading on steam casing bolts.

There are many methods available for predicting life under the combined action of creep and fatigue within these elevated temperature cycles [32]. All rely on baseline data from pure creep and continuous low-cycle fatigue tests conducted under uniaxial stress conditions at a similar temperature. The predictions differ in the manner they employ this baseline data. To illustrate this two of the more commom methods will be outlined.

14.5.1 Life Fraction Rule (Linear Damage Rule)

Taira [33] proposed that within creep-fatigue cycles, the fractions of damage due to creep ϕ_c and fatigue ϕ_f sum to unity at failure. This gives a linear damage rule:

$$\phi_c + \phi_f = 1 \tag{14.25a}$$

where ϕ is equated to the life fractions expended in each process. Applying eq(14.25a) to the strain-limited cycles in Figs 14.17b-d we write

$$\sum(\Delta t_c / t_f) + N / N_f = 1 \tag{14.25b}$$

where $\sum(\Delta t_c)$ is the total time expended in creep, and N is the cyclic life for the particular cycle. In the denominators of eq(14.25b), t_f is the monotonic creep rupture time at a stress level equal to the dwell stress in cycles (b-d). N_f is the life for continuous cycling (Fig. 14.17a) at the same inelastic strain range as the cycles in Fig. 14.17b - d. One or other life fraction will become dominant in the absence of an interaction between creep and fatigue. Equation (14.25b) reduces to a time-hardening criterion $\sum(\Delta t / t_f) = 1$ for a pure creep failure and to Miner's [5] hypothesis $\sum N / N_f = 1$ for a pure fatigue failure. Fatigue fracture under HRSC is associated with transgranular cracking. Creep fracture is associated with intergranular cracking in cycles with low frequencies and long hold times. When an interaction occurs, usually in cycles of shorter dwell times and higher frequencies for temperatures in the range $0.5 < T/T_m < 0.25$, mixed-mode cracking is observed [34].

14.5.2 Relaxation Dwell

In the presence of a relaxation period at peak strain (Figs 14.17e,f) the time fraction in eq(14.25b) is less clearly defined than for cycles with hold times in creep. Accounts of relaxation are usually made by evaluating an equivalent creep stress σ_c, which, if applied for the relaxation time t_r, would produce the same amount of inelastic strain $\Delta\varepsilon_r$ (see Figs 14.17e,f). We can then find t_f corresponding to σ_c from creep data to determine the life fraction/cycle (t/t_f) as before.

To find σ_c we must consider the interchange between elastic and creep strain as the stress relaxes from its peak value σ_p to σ_r (see Fig. 14.17e,f). During relaxation the total strain ε^T remains constant. Assuming that ε^T is composed only of elastic and secondary creep strain components, we can write

$$\sigma / E + A\sigma^n t = \varepsilon^T \tag{14.26a}$$

where E is Young's modulus and A, n are Norton's secondary creep constants. Differentiating eq(14.26a) with respect to time gives

$$(1/E)\, \mathrm{d}\sigma/\mathrm{d}t + A\sigma^n = 0 \tag{14.26b}$$

Separating the variables in eq(14.26b) leads to

$$\int \sigma^{-n}\, \mathrm{d}\sigma = -AE\int \mathrm{d}t \tag{14.26c}$$

Equation (14.26c) may be integrated to find (a) the relaxed stress σ after a time t and (b) the final relaxed stress σ_r after time t_r. Taking (a) and (b) above in turn as upper limits of the integration with the lower limit σ_p for $t = 0$ gives respectively

$$\sigma^{1-n} - \sigma_p^{1-n} = A(1-n)Et \tag{14.27a}$$

$$\sigma_r^{1-n} - \sigma_p^{1-n} = A(1-n)Et_r \tag{14.27b}$$

Dividing eqs(14.27a,b) gives

$$(\sigma/\sigma_p)^{1-n} = 1 + [(\sigma_r/\sigma_p)^{1-n} - 1](t/t_r) \tag{14.27c}$$

An equivalent stress σ_c for relaxation provides the same amount of creep strain. That is,

$$A\sigma_c^n t_r = A\int_0^{t_r} \sigma^n\, \mathrm{d}t \tag{14.28a}$$

Substituting eq(14.27c) into eq(14.28a) and integrating leads to

$$\sigma_c^{\,n} t_r = \frac{(1-n)\sigma_p^{\,n} t_r(\sigma_r/\sigma_p - 1)}{(\sigma_r/\sigma_p)^{1-n} - 1} \tag{14.28b}$$

from which, with $n > 1$ and $\sigma_p > \sigma_r$:

$$\left(\frac{\sigma_c}{\sigma_p}\right)^n = \frac{(n-1)(1-\sigma_r/\sigma_p)}{(\sigma_p/\sigma_r)^{n-1} - 1} \tag{14.28c}$$

14.5.3 Life-Fraction Codes

In practice, eq(14.25b) is prevented from overestimating life. The 1976 ASME code [35] recommends that safety factors S be used with monotonic creep and continuous low cycle fatigue lives. Factors of $S = 2$ were originally used for 304 and 316 stainless steels. For the extended use of eq(14.25b) it is modified to

$$\sum (t_c/t_d) + N/N_d = d \tag{14.29a}$$

where $t_d = t_f/S$ and $N_d = N_f/S$ are the design lives and d is a temperature-dependent material parameter. Campbell [36] measured a mean value $d = 0.6$ for 304 stainless steel with equal interaction between fatigue and creep. In the absence of such an interaction the life fractions from pure creep and fatigue failures will exceed unity. These will lie in regions of pure creep and pure fatigue in Fig. 14.18.

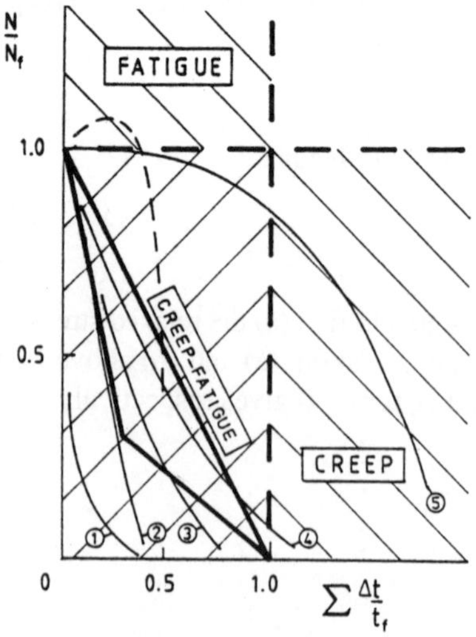

Figure 14.18 Life fraction codes

The code also employs a bilinear representation shown. Each line intersects its axes at unity and intersect with each other at co-ordinates (0.3, 0.3). The equations of the two lines are

$$(7/3)\sum(\Delta t/t_d) + N/N_d = 1 \quad \text{for } 0.3 \le N/N_d \le 1 \tag{14.29b}$$

$$\sum(\Delta t/t_d) + (7/3)(N/N_d) = 1 \quad \text{for } 0 \le N/N_d \le 0.3 \tag{14.29c}$$

Figure 14.17 compares eqs(14.25b) and (14.29b,c) with the experimental plots for five steels. This shows how the predictions can be made safe for the majority of service conditions where creep-fatigue interactions occur. However, material 5 shows little interaction, failure occurring either by one mode or the other.

14.5.4 Strain Range Partitioning (SRP)

A shorthand notation [37] is used in which p refers to plasticity and c to both creep and relaxation. The cycles in Fig. 14.17a-f are denoted as:

(a) *pp* forward plasticity reversed by plasticity
(b) *cp* forward creep reversed by plasticity (also (e))
(c) *pc* forward plasticity reversed by creep (also (f))
(d) *cc* forward creep reversed by creep

Figures 14.19a-d show how to partition these four cycles into their component strain ranges. In a more complex loop (see Fig. 14.19d) failure is attributed to the combined effects of three inelastic strain range components.

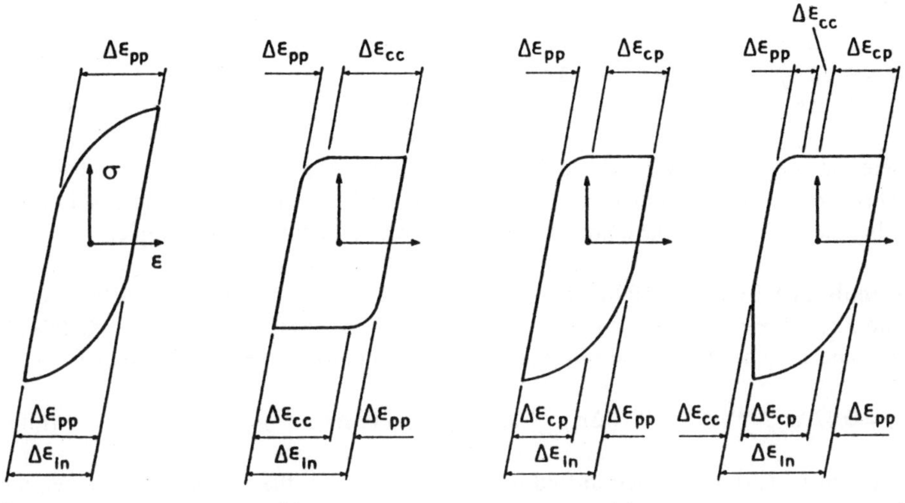

(a) (b) (c) (d)

Figure 14.19 SRP of a complex hysteresis loop

These are pure plastic strain ($\Delta\varepsilon_{pp}$), pure creep ($\Delta\varepsilon_{cc}$) and interactive creep and plasticity ($\Delta\varepsilon_{cp}$). The three strain ranges are found by balancing the strain in positive and negative directions. This ensures that fast forward plasticity $\Delta\varepsilon_{pp}$ is completely balanced by fast reversed plasticity and that forward creep strain $\Delta\varepsilon_{cc}$ is reversed by creep (relaxation in this case). The remaining part of $\Delta\varepsilon_{in}$ is assigned to the interactive strain range component $\Delta\varepsilon_{cp}$. A balance exists in partitioned ranges on the forward and reversed sides of a complex loop so that the width of the loop defines an inelastic strain range $\Delta\varepsilon_{in} = \Delta\varepsilon_{pp} + \Delta\varepsilon_{cc} + \Delta\varepsilon_{cp}$. When F_{ij} denotes the strain fraction, $F_{pp} = \Delta\varepsilon_{pp}/\Delta\varepsilon_{in}$ etc, it follows that $F_{pp} + F_{cc} + F_{cp}$ (or F_{pc}) =1. Let cyclic lives N_{pp}, N_{cc}, N_{cp} (or N_{pc}), apply to pure cycling under each mode at the same range of inelastic strain as in the complex cycle. These lives are related to the cyclic life N for any complex cycle by the following rule:

$$F_{pp}/N_{pp} + F_{cc}/N_{cc} + F_{cp}/N_{cp} \text{ (or } F_{pc}/N_{pc}) = 1/N \tag{14.30}$$

The SRP method requires the lives under the separate action of pure *pp*, *cc* and *cp* cycles at their partitioned values. Because SRP assumes that creep and plasticity exert separate influences on fracture, pure *pp* (Fig. 14.19a) applies to HRSC at frequencies greater than 10^{-1} Hz to avoid creep [38].

It is rarely possible to achieve pure *cc* and pure *cp* (or *pc*) in practice. For a balanced *cc* loop (Fig. 14.19b) the *cp* component is absent but a *pp* component cannot be discounted when strain ranges are partitioned at half-life. When softening occurs a *pp* component appears with the progressive rounding of the loop corners (see Fig. 14.19b). From eq(14.30):

$$N_{cc} = F_{cc} / (1/N - F_{pp}/N_{pp}) \tag{14.31a}$$

Similarly, while the cycle in Fig. 14.19c may have begun as pure *cp*, softening has again introduces a *pp* component. Applying eq(14.30) to the *pp* and *cp* components,

$$N_{cp} = F_{cp} / (1/N - F_{pp}/N_{pp}) \tag{14.31b}$$

The lives N_{pp}, N_{cc} and N_{cp} in eqs(14.31a,b), are described by an inverse Coffin-Manson law:

$$N_{pp} = A_{pp}\, \Delta\varepsilon_{pp}^{\ \alpha_{pp}} \tag{14.32a}$$
$$N_{cp} = A_{cp}\, \Delta\varepsilon_{cp}^{\ \alpha_{cp}} \quad (\text{or} \quad N_{pc} = A_{pc}\, \Delta\varepsilon_{pc}^{\ \alpha_{pc}}) \tag{14.32b}$$
$$N_{cc} = A_{cc}\, \Delta\varepsilon_{cc}^{\ \alpha_{cc}} \tag{14.32c}$$

where the coefficients A_{pp} etc and exponents α_{pp} etc (usually between -1 and -3) are experimental constants. Equations (14.32a-c) are used in combination with eqs(14.30) to predict the cyclic life of the complex cycle as shown in the following example.

Example 14.8 Employ SRP to predict the cyclic life for the loop in Fig. 14.19d. The loop width $\Delta\varepsilon_{in} = 1.5\%$ partitions as follows: $\Delta\varepsilon_{pp} = 0.25\%$, $\Delta\varepsilon_{cc} = 0.40\%$ and $\Delta\varepsilon_{cp} = 0.85\%$, for a nimonic alloy at 850°C. The SRP life equations to conform to eqs(14.32a,b,c) as

$$N_{pp} = 0.42\Delta\varepsilon_{pp}^{\ -1.43},\ N_{cc} = 0.18\Delta\varepsilon_{cc}^{\ -1.1} \text{ and } N_{cp} = 0.003\Delta\varepsilon_{cp}^{\ -2.21} \tag{i,ii,iii}$$

Equations (i)-(iii) are applied to find N_{pp}, N_{cc} and N_{cp} for the complex cycle when their inelastic ranges $\Delta\varepsilon_{pp}$, $\Delta\varepsilon_{cc}$ and $\Delta\varepsilon_{cp}$ are set to equal $\Delta\varepsilon_{in}$. The notation can lead to confusion since the subscripted strain ranges in eqs(i)-(iii) are not the components of $\Delta\varepsilon_{in}$. The partitioned strains are used only to determine F_{ij} in eq(14.30). Further steps in predicting a life of $N = 30$ cycles from eq(14.30) are given in Table 14.1.

Table 14.1 Application of SRP

$\%\Delta\varepsilon_{in}$	$\%\Delta\varepsilon_{pp}$	$\%\Delta\varepsilon_{cc}$	$\%\Delta\varepsilon_{cp}$	F_{pp}	F_{cc}	F_{cp}	N_{pp}	N_{cc}	N_{cp}	N
1.50	0.25	0.40	0.85	0.167	0.267	0.567	170.4	18.3	32.2	30

It is seen from Table 14.1 that SRP does not account directly for the magnitude of the creep stress. Also, no account is provided of the effect on life of the initial deformation sequence. However, SRP life-estimates are usually quite accurate.

References

[1] Gerber H., *Ing-Verins*, **6** (1874), 101.
[2] Goodman J., *Mechanics Applied to Engineering*, Longmans, London, 1899.
[3] Smith J. O. and Collins W. L., *Proc. ASTM*, **42** (1942), 639.
[4] Soderberg C. R., *Trans ASME*, **52** (1930), 13.
[5] Miner A. M., *Jl. Appl. Mech.*, **12** (1945), A159.
[6] Marco S. M. and Starkey W. L., *Trans ASME*, **76** (1954), 627.
[7] Corten H. T. and Dolan T. J., *Proc. Int. Conf. on Fatigue*, I.Mech.E., London, 1956, p.235.
[8] Gough H. J. and Pollard H. V., *Proc. I. Mech. E.* **13** (1935), 3.
[9] Findlay W. N. and Mathur P. N., *Proc. SESA*, **XIV(1)** (1956), 35
[10] Stullen F. B. and Cummings H. N., *Proc. ASTM*, **54**, (1954), 822.
[11] Peterson R. E., *Stress Concentration Factors*, Wiley, 1974.
[12] Heywood R. B., *Coll. on Fatigue*, Springer-Verlag, 1956, p.92.
[13] Neuber H., *Theory of Notch Stresses*, J.W. Edwards, Michigan, 1946.
[14] Rees D. W. A., *Int. J. Pres. Ves. and Piping*, **30** (1987), 57.
[15] Meleka A. H., *Met Rev.*, **17** (1962), 43
[16] Vitovec F. H., *Proc ASTM*, **57** (1957), 977.
[17] Lazan B. J., *Proc ASTM*, **49** (1949), 757.
[18] Meleka A. H., and Evershed,A.V. *Jl Inst Met.* **88** (1959-60), 88.
[19] Forrest P. G., *Fatigue of Metals*, Pergamon, 1970.
[20] Coffin L. F., *Trans ASME,* **76** (1954), 931.
[21] Manson S. S., *NACA, Tech. Note.* **2933** (1953).
[22] Basquin D. H., *Proc. A.S.T.M.* **10** (1910), 625.
[23] Morrow J. D., *Proc ASTM*, STP **378** (1965), 45
[24] Manson S. S., *Expl. Mech.* **5** (1965), 193.
[25] Coffin L. F., *Jl Mat.*, **6** (1971), 388.
[26] Coffin L. F., *Proc I. Mech.E.* **189** (1974), 109.
[27] Bernstein H. L., *Proc ASTM*, STP **770** (1982), 105.
[28] Berling J. T. and Conway J. B., *Met Trans*, **1**, (1970), 805.
[29] Endo, T. and Morrow J., *Proc ASTM*, **4** (1969), 159.
[30] Topper T. H. and Sandor, B. I., *Proc ASTM*, STP **462** (1970), 93.
[31] NASA Tech Mem, X67838, 1971.
[32] Rees D. W. A., *Prog Nuc Energy*, **19** (1987), 211.
[33] Taira S., *Creep in Structures*, (ed. Hoff.N.J.) Academic Press, New York, 1962, 96.
[34] Feltner C. E. and Sinclair G. M., *Proc Jnt Int Conf on Creep*, I.Mech.E., London, 1963, **3**, 7.
[35] ASME Boiler and Pressure vessel Code: *Criteria for Design of Elevated Temperature Class 1 Components*, 1976.
[36] Campbell R. D., *Trans ASME* (B), **93** (1971), 887.
[37] Halford G. R., Hirschberg M. H. and Manson, S. S., *Proc ASTM*, STP **520** (1973), 658.
[38] Manson S. S., *Proc ASTM*, STP **520** (1973), 744.

EXERCISES

14.1. A carbon steel has an ultimate tensile strength of 475 MPa and a yield strength of 350 MPa with a fatigue limit under a fully reversed stress cycle of ± 180 MPa. Estimate the maximum stress that would give infinite life for a minimum cyclic stress limit of 110 MPa according to the Gerber, Goodman and Soderberg predictions.

14.2 For a particular steel the endurance limit is expressed as 0.6 × u.t.s. for reversed bending at 10^5 cycles. Construct a Goodman diagram and determine the permissible stress range to avoid failure at 10^5 cycles when the mean stress is increased to 0.25 × u.t.s. (CEI)

14.3 It is required to design a tie bolt. Fatigue data, obtained from reverse bending tests, are available for the particular steel to be used. Describe how such data can be used to determine a safe working stress, taking into account that (i) the bolt will sustain a constant tensile load P on which will be superimposed a cyclic load of amplitude $0.25P$, (ii) the bolt has a constant cross-section area A except at a small circumferential V-shaped groove and (iii) the amplitude is constant except for a 5% increase on two occasions in the bolt life. (CEI.)

14.4 A titanium alloy is to be subjected to a fluctuating tensile cycle in which the mean stress is 450 MPa. If an endurance of 10^8 cycles corresponds to a fully reversed stress cycle of amplitude 650 MPa, calculate the allowable stress amplitude to give the same endurance for the fluctuating cycle. The yield and ultimate tensile stresses for the alloy are 800 and 1400 MPa respectively.

14.5 The fatigue behaviour of an aluminium alloy, under reversed stress conditions, is given by $\Delta\sigma^b N_f = C$, where $\Delta\sigma$ is the stress range, N_f is the cyclic life and b and C are constants. Given that $N_f = 10^8$ cycles when $\Delta\sigma = 340$ MPa and $N_f = 10^5$ cycles when $\Delta\sigma = 520$ MPa, determine the values of b and C. It is required that a component, made from the same material, should not fail in less than 10^7 cycles. Calculate the permissible amplitude of the cyclic tensile stress if the component suffers a tensile mean stress of 20 MPa and the yield stress of the material is 400 MPa. (CEI)

14.6 Cyclic bending stresses of 75, 60 and 40 MPa are induced within a beam in a lifting machine. The respective parts of the beam's life spent at these stress levels are 30%, 50% and 20%. Estimate the working life in days when the machine operates continuously at 10 cycles/day. Take the fatigue lives as 10^3, 10^4 and 10^5 cycles respectively.
Answer: 284 days.

14.7 Estimate the working life in days of a torsional spring that is subjected to reversed cyclic shear stresses of 60, 50, 40 and 35 MPa for the following respective amounts of its life: 10%, 50%, 20% and 20%, when the spring operates continuously at 40 cycles/day. The fatigue lives at the given stresses are 10^3, 10^4, 10^5 and 10^6 cycles respectively.
Answer: 164 days.

14.8 A centre zero moving coil has a torsion spring of circular section 0.15 mm diameter with effective length 50 mm. It is used for inspection at a rate of 5 components/minute. A satisfactory component twists the wire through ± 20° but rejected components twist the wire through 30°. The latter comprise 10% of the total. Estimate the working life of the spring in years for a working year of 50 weeks of 5 days/week and 8 hr/day. If the twist limit for rejects were reduced to ± 25°, determine the percentage increase in life. Take $G = 48.25$ GPa and the following fatigue data under reversed torsion:

Life, (cycles)	10^5	10^6	10^7	10^8
Shear stress, (MPa)	42.7	29	22.8	20.7

14.9 An aircraft wing experiences the following block loading spectrum repeatedly in service. Calculate from Miner's rule the number of such spectra that may be applied, given the following fatigue lives N_f at each stress amplitude.

$\Delta\sigma$ (MPa)	345	290	240	215	195
$n \times 10^3$ (cycles)	20	43	36	30	27
$N_f \times 10^3$ (cycles)	65	300	950	3100	42000

14.10 The loading spectrum on an aluminium alloy aircraft component is defined by the number of cycles n occurring at various reversed stress levels in every 100×10^3 cycles as given. Also tabulated is the 90% survival fatigue life for each stress level. Determine an expected fatigue life for the component based upon Miner's hypothesis of damage. (CEI)

$\Delta\sigma$ (MPa)	344.5	289.4	241.2	213.6	192.9
$n \times 10^3$ (cycles)	2	8	17	28	45
$N_f \times 10^3$ (cycles)	70	290	960	3000	40000

14.11 In two-step fatigue tests, repeated stress amplitudes of 250 and 300 MPa are applied in a low-high and high-low sequence. If, for each sequence, the initial stress is applied for one half the cyclic life, calculate the remaining life under the final stress according to (i) Miner's rule and (ii) when the following non-linear damage laws are assumed for each stress level:

$D = (n/N)^2$ for 300 MPa, $N_f = 14 \times 10^4$ and $D = (n/N)^3$ for 250 MPa, $N_f = 26 \times 10^4$.

Taking failure for $D = 1$, establish the effect of the stress sequence according to the non-linear theory.

14.12 A cylindrical vessel of 760 mm internal diameter and 6.5 mm wall thickness is subjected to internal pressure fluctuating from $P/2$ to P. The fatigue limit of the material at 10^7 cycles under reversed direct stress is ± 230 MPa and the static yield stress in simple tension is 310 MPa. Assuming a shear strain energy criterion of equivalent fatigue strength under biaxial cyclic stress and using the Soderberg diagram, determine a nominal value for P to give an endurance of 10^7 cycles. Neglect the radial stress. (CEI)

14.13 How does Heywood's method of estimating fatigue strength of a notched component overcome some of the limitations of the notch sensitivity index method? Describe its use for an alloy steel having a plain fatigue strength in rotating bending of ± 770 MPa when made into a stepped shaft with diameters 150 and 75 mm, joined by a 4.75 mm fillet radius. (IC)

14.14 Show that the following LCF results for steel obey the Coffin-Manson law: $\Delta\varepsilon^P = CN_f^{\beta}$ where C and β are constants to be determined:

$\Delta\varepsilon^P$	0.04	0.021	0.016	0.0084
N_f	100	500	1000	5000

If a component made from the same steel is subjected to $\Delta\varepsilon^P = 2\%$ for 300 cycles, followed by $\Delta\varepsilon^P = 1\%$ for 500 cycles, find the cyclic life remaining corresponding to $\Delta\varepsilon^P = 0.75\%$, assuming that Miner's rule applies.

14.15 At a temperature of 800°C the ends of a steel bar are clamped rigidly. Thereafter the bar is subjected to a temperature cycle of repeated cooling and heating between the limits of 100°C and 800°C until failure occurs by thermally induced fatigue. Calculate the endurance, ignoring work-hardening, when a Coffin-Manson law $N^{1/2}\Delta\varepsilon^P = 0.04$ applies to the steel. Assuming that Miner's rule applies, how many thermal cycles between 800°C and 400°C remain when the bar had been previously cycled from 800°C to 500°C for 1000 cycles? Take $Y = 310$ MPa, $E = 207$ GPa and $\alpha = 10 \times 10^{-6}$/°C.

14.16 A fatigue crack is assumed initiated when a surface crack of length 10^{-6} m is created. The corresponding percentage of cyclic life consumed is given by: $1000N_i = \sqrt{(2.02)}N_f^{\sqrt{2.02}}$, where N_i and N_f are the number of cycles to initiation and failure respectively. Determine the cyclic lifetime of two specimens one having a N_i/N_f ratio of 0.01 corresponding to a stress range of $\Delta\sigma_1$ and the other having a N_i/N_f ratio of 0.99 corresponding to a stress range of $\Delta\sigma_2$. If the crack at failure is 1 mm deep, determine the mean crack propagation rate at $\Delta\sigma_1$ and the mean crack nucleation rate at $\Delta\sigma_2$. (CEI)

CHAPTER 15

FRACTURE MECHANICS

An understanding of the behaviour of cracks under static and cyclic loading can establish the safe working stresses for components operating in various environments. We use the term linear elastic fracture mechanics (LEFM) to describe a purely brittle fracture in high strength low ductility materials. For more ductile materials LEFM may be corrected where localised crack tip plasticity occurs, but if plasticity is widespread a general yielding fracture mechanics (GYFM) theory is more appropriate. This chapter outlines these theories and considers their applications to creep and fatigue cracking, especially where corrosion may also influence behaviour.

15.1 Blunt Cracks

We saw in Chapter 2 that a circular hole in a stressed plate raised the applied stress threefold at the edges of the hole. A blunt crack of tip radius ρ may be approximated as an elliptical hole. Inglis [1] solved the problem of the elastic stress distribution in a stressed plate containing an elliptical hole. Let the major axis of the ellipse with semi-major and minor axes lengths a and b respectively lie normal to a uniaxial stress σ (see Fig. 15.1a). The normal stress σ_y is distributed as shown where a maximum is reached at the hole boundary

$$(\sigma_y)_{max} = \sigma(1 + 2a/b) \qquad (15.1)$$

Alternatively we may write eq(15.1) in terms of the radius of curvature $\rho = b^2/a$ at the end of the major axis

$$(\sigma_y)_{max} = \sigma[1 + 2\sqrt{(a/\rho)}] \qquad (15.2a)$$

In this form eq(15.2) may be used for blunt cracks that do not conform exactly to an elliptical profile. Equation (15.1) gives a stress concentration $S = 1 + 2a/b$ which increases toward infinity as b diminishes to crack-like proportions. Then, $a >> \rho$ in eq(15.2a), gives

$$(\sigma_y)_{max} \approx 2\sigma\sqrt{(a/\rho)} \qquad (15.2b)$$

When $a = b$, $S = 3$ in agreement with that previously derived for a circular hole.

For a plate uniformly stressed in biaxial tension ($Q = 1$ in Fig. 15.1) the maximum σ_y is

$$(\sigma_y)_{max} = 2\sigma(a/b) \qquad (15.3)$$

A cracked plate under biaxial tensile stress ratio $Q = \sigma_2/\sigma_1$, where σ_1 acts normal to the major axis and σ_2 acts normal to the minor axis, has a stress concentration $S_y = \sigma_y/\sigma_1$ at the hole of:

$$(\sigma_y)_{max} = \sigma_1(1 + 2a/b) - \sigma_2 \;\Rightarrow\; S_y = (1 + 2a/b) - Q \qquad (15.4a,b)$$

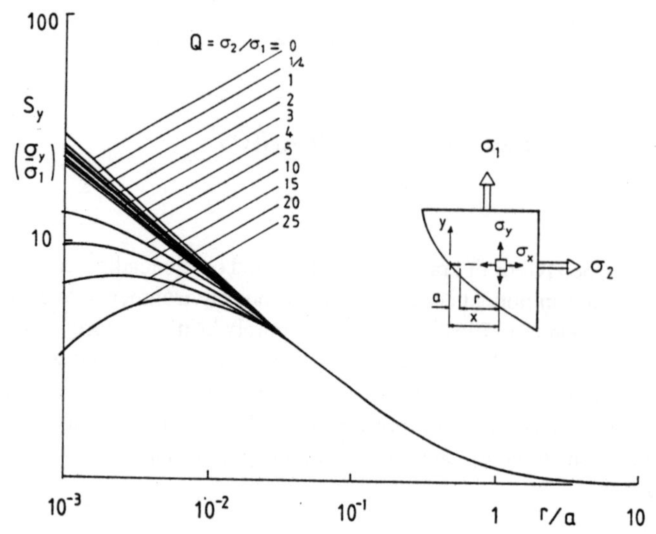

Figure 15.1 Stress concentrations around an elliptical hole

Figure 15.1 shows the dependence of S_y upon Q and distance r/a from the crack tip. Figures 15.2a,b compares theoretical S_y and S_x distributions with the measured distributions [2] for an elliptical crack, with $a = 12.5$ mm and $b = 1$mm, in an aluminium plate under stress ratios $Q = ⅓$ and 3. At the notch tip S_y reaches its maximum mum value given by eq(15.4b).

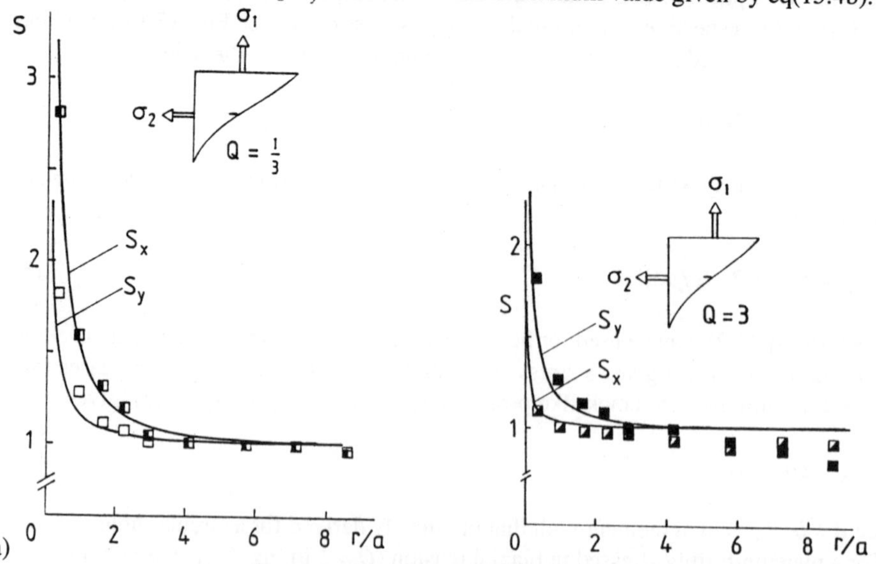

Figure 15.2 Elliptical crack under biaxial stress

In contrast, $S_x = \sigma_x/\sigma_2$ falls to zero at this position. Each normalised stress distribution S_x and S_y falls off rapidly to the nominal applied stresses at distances beyond one crack length away from the ends of the hole ($r/a > 1$). For each applied stress ratio, the theoretical stress distributions are confirmed by the test data. From this it is appeerent that highly localised concentrations of stress occur whenever crack-like defects are present in loaded bodies.

15.2 LEFM

A sharp crack may propagate in one of three modes (Figs 15.3a-c). The tension- opening mode I in (a) is most often referred to, but the two shear modes II and III in (b) and (c) describe other brittle fracture modes from sliding and anti-plane tearing.

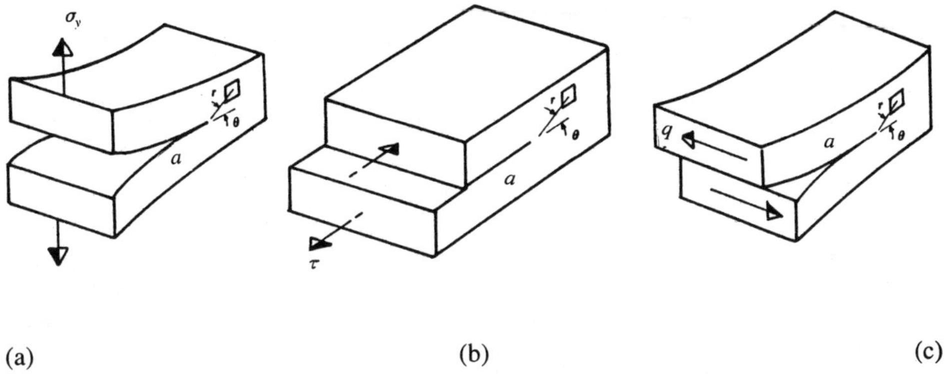

Figure 15.3 The three modes of brittle fracture

In order to determine the critical loading that would cause failure under each mode, a failure criterion must be specified. There are two criteria: one based upon the stress state around the crack tip and the other upon the release of elastic strain energy produced by crack extension. It is instructive to show firstly that the two approaches are compatible in the case of mode I opening of a crack of length $2a$ in a large (infinite) plate subjected to a normal boundary stress σ.

15.2.1 Stress Function Approach

Westergard [3] examined the stress state in the vicinity of the crack tip. Let the origin of axes x and y lie at the crack centre and x be aligned with the crack length (see Fig. 15.4a). The following Cartesian stress components σ_x, σ_y and τ_{xy} were derived for a point (r, θ) near the crack tip ($r << a$) in the infinite plate:

$$\sigma_y = \sigma[a/(2r)]^{1/2} \cos(\theta/2)[1 + \sin(\theta/2)\sin(3\theta/2)] \quad (15.5a)$$

$$\sigma_x = \sigma[a/(2r)]^{1/2} \cos(\theta/2)[1 - \sin(\theta/2)\sin(3\theta/2)] \quad (15.5b)$$

$$\tau_{xy} = \sigma[a/(2r)]^{1/2} \sin(\theta/2)\cos(\theta/2)\cos(3\theta/2) \quad (15.5c)$$

For a thin plate, plane stress prevails with $\sigma_z = 0$. In the case of a thick plate, when the thickness strain is zero, this gives the through thickness stress as

$$\sigma_z = \nu(\sigma_x + \sigma_y) = 2\nu\sigma[a/(2r)^{1/2}]\cos(\theta/2) \quad (15.5d)$$

When $\theta = 0$ for a point on the x - axis, close to the crack tip, eqs(15.5a-c) give $\tau_{xy} = 0$ and:

$$\sigma_y = \sigma_x = \sigma[a/(2r)]^{1/2}$$

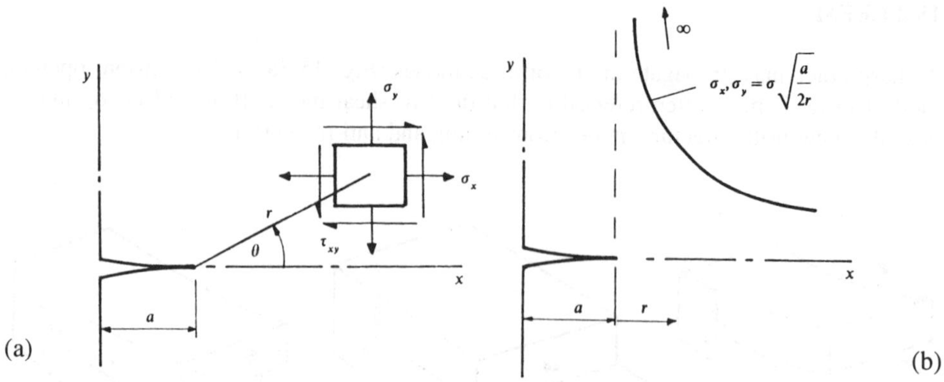

Figure 15.4 Mode I stress state in an infinite plate

If the crack length is $2a = 1$ mm, this shows that a boundary stress $\sigma = 20$ MPa is magnified fivefold in the components of σ_x and σ_y for when $r = 1/100$ mm, as shown in Fig. 15.4b. Of these stress components the magnitude of opening stress σ_y governs crack stability. Irwin [4] examined the elastic opening stress expressions for a number of loaded cracked plates with different geometries. He discovered that σ_y had the common form

$$\sigma_y = [K_I / (2\pi r)^{1/2}] \cos(\theta/2)[1 + \sin(\theta/2)\sin(3\theta/2)] \tag{15.6}$$

in which the mode I stress intensity factor K_I distinguished one plate from another. For example, K_I for an infinite plate is found by comparing eqs(15.5a) and (15.6)

$$K_I = \sigma\sqrt{(a\pi)} \tag{15.7}$$

If the crack is blunt, with tip radius ρ, (see Fig. 15.5) the stress distribution (15.5a - c) in modified with an additional term [5,6]

$$\sigma_y = \sigma[a/(2r)]^{1/2}\cos(\theta/2)\,\{1 + \sin(\theta/2)\sin(3\theta/2) + [\rho/(2r)]\cos(3\theta/2)\sec(\theta/2)\} \tag{15.8a}$$
$$\sigma_x = \sigma[a/(2r)]^{1/2}\cos(\theta/2)\{1 - \sin(\theta/2)\sin(3\theta/2) - [\rho/(2r)]\cos(3\theta/2)\sec(\theta/2)\} \tag{15.8b}$$
$$\tau_{xy} = \sigma[a/(2r)]^{1/2}\cos(\theta/2)\,\{\sin(\theta/2)\cos(3\theta/2) - [\rho/(2r)]\sin(3\theta/2)\sec(\theta/2)\} \tag{15.8c}$$

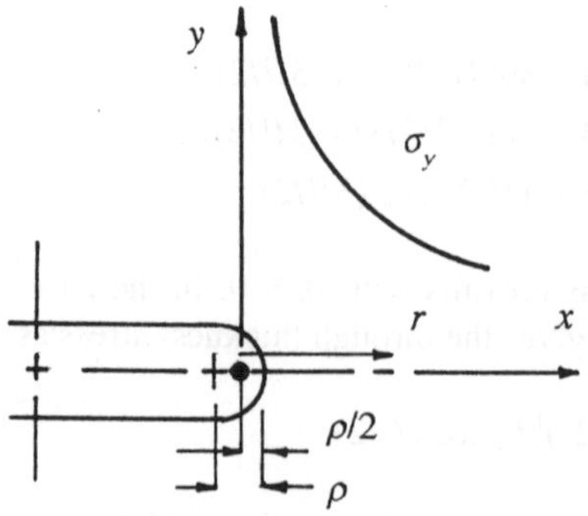

Figure 15.5 Mode I stress distribution for a blunt crack

where r is measured from a point distance $\rho/2$ behind the notch end. Equations (15.7) and (15.8a) show that when $\theta = 0$, the stress distribution ahead of the notch tip is

$$\sigma_y = \sigma[1 + \rho/(2r)][a/(2r)]^{1/2} = [K_I/(2\pi r)^{1/2}][1 + \rho/(2r)]$$

so that when $r = \rho/2$, eq(15.2b) is recovered.

The other plates considered were similarly characterised with sharp-crack stress intensity factors. They have the common form

$$K_I = Q\sigma\sqrt{a} \tag{15.9}$$

where Q is either a constant, for infinite plates, or a function of a finite plate geometry. Some of the more common crack geometries are given in Fig. 15.6a-f. The particular forms of eq(15.9) are the following:

(a) A semi-infinite plate containing an edge crack

$$K_I = 1.12\sigma\sqrt{(a\pi)}$$

(b) An infinite plate containing an embedded circular crack or a semi-circular surface crack, radius a, lying in a plane normal to σ

$$K_I = (2/\pi)\sigma\sqrt{(a\pi)}$$

(c) An infinite plate containing an embedded elliptical flaw $2a \times 2b$ or a semi-elliptical surface crack of width $2b$ in which the depth a is less than half the plate thickness, each lying normal to σ

$$K_I = (1.12/\phi)\,\sigma\sqrt{(a\pi)}$$

in which ϕ varies with the ratio a/b as follows:

a/b	0	0.2	0.4	0.6	0.8	1.0
ϕ	1.0	1.05	1.15	1.28	1.42	$\simeq \pi/2$

Clearly when $a/b = 1$ this approximates to the case in (b).

(d) An infinite plate containing a through thickness crack of length $2a$ with a concentrated opening force P applied normal to the crack surfaces, distance b from the crack centre:

$$K_I = P\sqrt{\{a/[\pi(a-b)(a+b)]\}} \tag{15.10}$$

When the force P is aligned with y, $b = 0$ and eq(15.10) becomes

$$K_I = P/\sqrt{(\pi a)}$$

(e) A plate of finite width w containing a central crack of length $2a$, where $a \le 0.3w$, with a normally applied remote stress σ

$K_I = \sigma\sqrt{[w \tan(a\pi/w)]} \approx \sigma\sqrt{(a\pi)}[1 + \pi^2 a^2/(6w^2) + \ldots]$

or, in the alternative form

$K_I = \sigma\sqrt{[(a\pi)\sec(a\pi/w)]}$

(f) A plate of finite width w containing symmetrical edge cracks each of depth a

$K_I = \sigma\sqrt{[w \tan(\pi a/w) + (0.1w)\sin(2\pi a/w)]}$ (approximately)

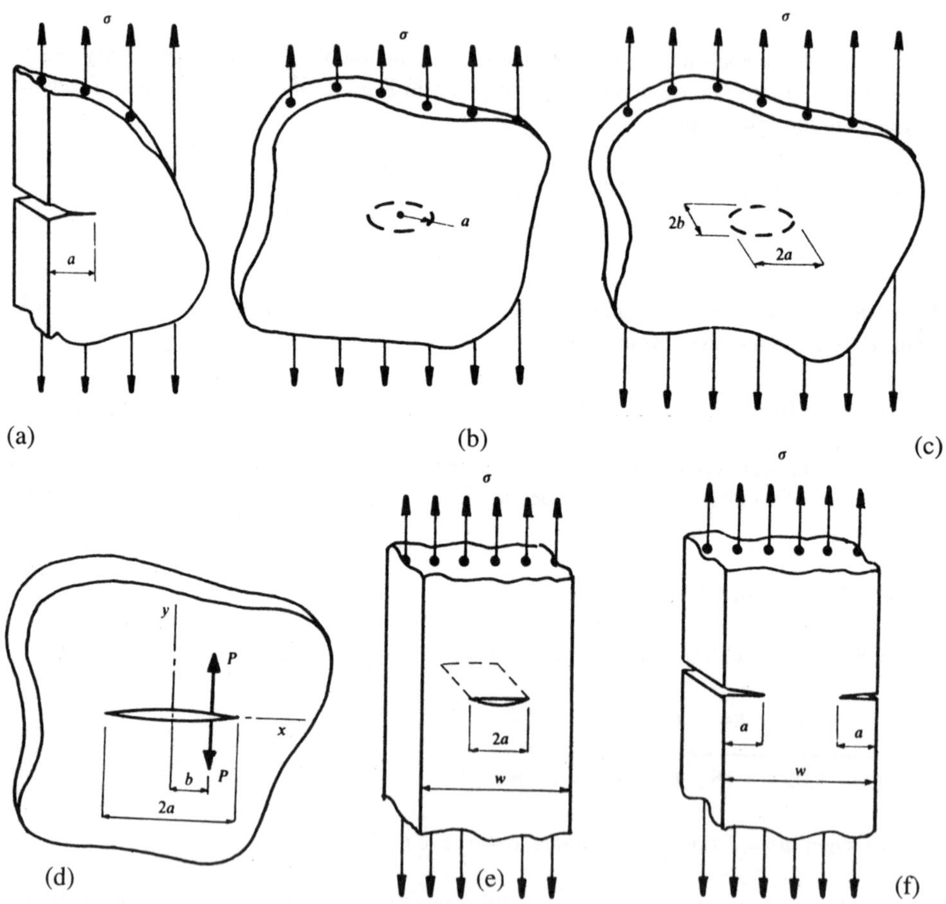

Figure 15.6 Common crack geometries

It is seen that as the stress intensity ahead of the crack can be expressed in the single parameter K_I then failure will occur when K_I reaches a critical value K_{Ic}. This material property is known as the fracture toughness (units: $MNm^{3/2} \equiv MPa\sqrt{m}$ or $Nmm^{3/2}$).

Example 15.1 A 5 mm long crack exists in an infinite steel plate of fracture toughness K_{Ic} = 105 $MN/m^{3/2}$. Calculate the maximum allowable design stress that could be applied around the boundary. If the yield stress for the plate material is 1500 MPa comment on the type of failure that would occur if this design stress was exceeded.

Replacing K_I in eq(15.7) with K_{Ic}, the maximum allowable stress for $a = 2.5$ mm is

$$\sigma = K_{Ic}/\sqrt{(a\pi)} = 105/\sqrt{(\pi \times 2.5 \times 10^{-3})} = 1185 \text{ MPa}$$

Since this is less than the yield stress, i.e. elastic, then failure in this high strength, low ductility material would be brittle. In the low to medium temperature range ($T/T_m < 0.5$) the fracture path is transcrystalline (transgranular). The crack path proceeds from grain to grain along crystallographic cleavage planes characterised by a smooth bright granular appearance.

15.2.2 Shear and Mixed Mode Loading

The stress intensity factors K_{II} and K_{III} may be derived in a similar manner for various geometries under the shear loading modes II and III. In an infinite plate, for example, containing a through thickness crack of length $2a$ under the II and III modes (Figs 15.3b,c), the respective stress intensities are

$$K_{II} = \tau\sqrt{(a\pi)} \text{ and } K_{III} = q\sqrt{(a\pi)}$$

where the shear stresses τ and q act as shown in each case. Critical values of K_{II} and K_{III} again define the fracture toughness of the material in each mode. Note that for a given geometry the stress intensities under each mode are additive. This is useful where loadings can be separated to give their contribution to each mode. For example, mixed mode I and II loadings arise when a uniaxial stress field is inclined at an angle β to the axis of a crack (see Fig. 15.7a).

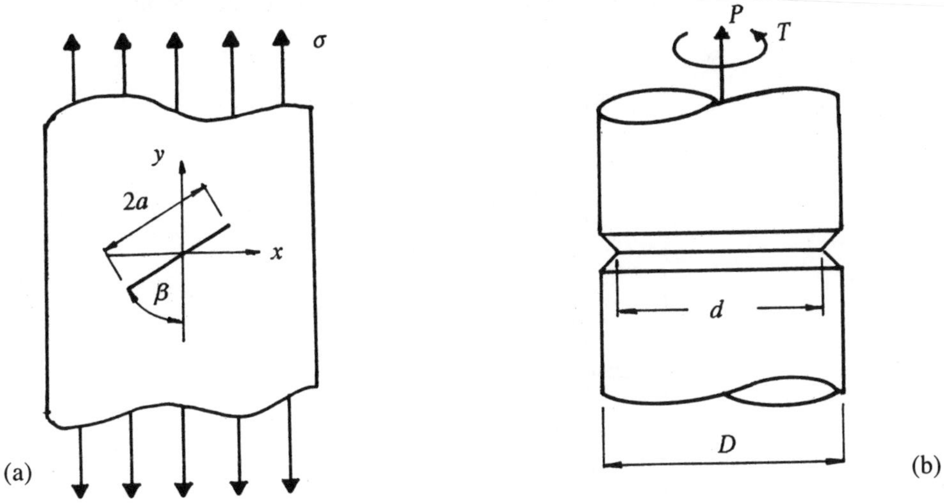

Figure 15.7 Mixed mode loadings

This gives:

$$K_I = \sigma\sqrt{(\pi a)}\sin^2\beta \text{ and } K_{II} = \sigma\sqrt{(\pi a)}\sin\beta\cos\beta$$

Mixed mode I and III loading arises when a notched bar of outer diameter D and root diameter d is subjected to a combined tensile force P and a torque T (see Fig. 15.7b). Then

$K_I = YP/D^{3/2}$ and $K_{III} = q/(\pi^2 d)$

where $Y = 1.72D/d - 1.27$ and $q = 16T/(\pi d^3)$. A circumferential crack is more severe and requires a compliance function to define K_I (see Exercise 15.15). Under these conditions experiment has shown that fracture may be taken to occur when

$$(K_I/K_{Ic})^2 + (K_{II}/K_{IIc})^2 = 1$$
$$(K_I/K_{Ic})^2 + (K_{III}/K_{IIIc})^2 = 1$$

The forms of these elliptical equations are similar to a normalised von Mises yield criterion.

15.3 Crack Tip Plasticity

In medium strength materials the local stresses in the vicinity of the crack tip will normally exceed the yield stress under plane stress conditions. The following correction to linear elastic fracture mechanics provides an equivalent crack length where there exists a region of localised crack tip plasticity.

15.3.1 Circular Plastic Zone

In Irwin's [5] estimate the σ_y distribution shown in Fig. 15.8a is cut off at an ordinate equal to the yield stress value Y. The elastic stress distribution does not extend within a radius r_y from the crack tip but the area A_1 is assumed to be available for further yielding.

(a)

(b)

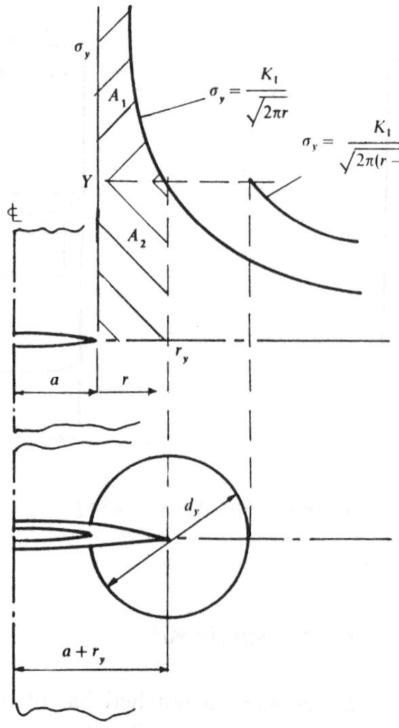

Figure 15.8 Crack tip plasticity

Now from eqs(15.6) and (15.7) at the crack tip in an infinite plate,

$$\sigma_y = \sigma[a/(2r)]^{1/2} = K_I/(2\pi r)^{1/2}$$

$$\therefore\ r_y = K_I^2/(2\pi Y^2) \qquad (15.11a)$$

The area beneath the region of the curve for $r_y \le r \le 0$ is given by

$$A_1 + A_2 = \int_0^{r_y} K_I \, \mathrm{d}r/(2\pi r)^{1/2} = (2/\pi)^{1/2} K_I \, r_y^{1/2}$$

and substituting from eq(15.11a),

$$A_1 + A_2 = K_I^2/(\pi Y) = 2Y r_y$$

but as $A_2 = Yr_y$ it follows that $A_1 = A_2$. Hence the plastic zone (Fig. 15.8b) is taken to be a circle of diameter $d_y = 2r_y$. This modifies the σ_y distribution as shown so that the fictitious crack tip of half length $a + r_y$ is assumed to lie at the centre of the plastic zone. Differences between the stresses σ_x, σ_y and σ_z (see eqs 15.5) are necessary for plasticity to occur. The presence of a through thickness stress σ_z in thick plates under plane strain conditions creates a central triaxial stress state close to the root of the crack. At mid-thickness the plastic zone size is estimated to be three times smaller than under plane stress where $\sigma_z = 0$. Thus for plane strain, eq(15.11a) becomes

$$r_y = K_I^2/(6\pi Y^2) \qquad (15.11b)$$

This radius increases from the plate centre to its edges where, with $\sigma_z = 0$, eq(15.11a) will again apply. The effect that a plastic zone has on the fracture toughness in each case may be assessed from the equivalent crack length. For an infinite plate, eq(15.8) is written as

$$K_{Ic}^* = \sigma\sqrt{[\pi(a + r_y)]} = \sigma\sqrt{(a\pi)}\,\sqrt{[1 + r_y/a]}$$

$$K_{Ic}^* = K_{Ic}\sqrt{[1 + (1/\beta)(\sigma/Y)^2]} \qquad (15.12)$$

where $\beta = 2$ for plane stress and $\beta = 6$ for plane strain.

15.3.2 Plastic Strip

Dugdale [7] presented an alternative analysis for the plastic zone size under a plane stress condition. The method employs the known solution (15.10) to the stress intensity under a normal force to the crack boundary. To find the closing force in a non-hardening plastic zone, the crack is assumed to extend elastically to length c (see Fig. 15.9).

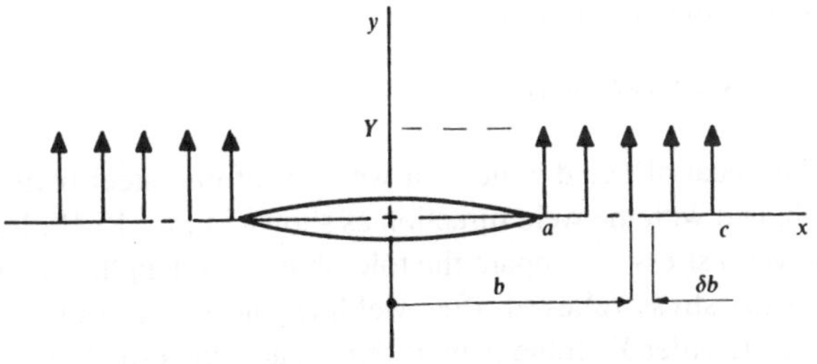

Figure 15.9 Dugdale's model of crack tip plasticity

For a plate of unit thickness a closing force $\delta P = Y \times \delta b$ is applied at a distance b ($a \le b \le c$) from the origin in a crack of semi-length c, as shown. Equation(15.10) becomes

$$K_{\mathrm{I}} = 2Y \int_a^c \sqrt{\{c/[\pi(c+b)(c-b)]\}}\,db$$
$$= 2Y(c/\pi)^{1/2}\cos^{-1}(a/c)$$

To obtain the same stress intensity for an elastic crack of length $2c$, it follows that:

$$\sigma\sqrt{(c\pi)} = 2Y(c/\pi)^{1/2}\cos^{-1}(a/c)$$
$$\therefore\ a/c = \cos[\pi\sigma/(2Y)]$$

Putting $d_y = c - a$ the cosine expansion reveals that the diameter of the zone under a limited spread of plasticity is

$$d_y = 2r_y = \pi K_{\mathrm{I}}^2/(8Y^2) \qquad (15.13a)$$

From eq(15.13a) we define $r_y = \pi K_{\mathrm{I}}^2/(8\beta Y^2)$ so that Dugdale's correction to eq(15.7) for the fracture toughness of an infinite plate becomes

$$K_{\mathrm{Ic}}{}^* = K_{\mathrm{Ic}}\sqrt{\{1 + [\pi^2/(8\beta)](\sigma/Y)^2\}} \qquad (15.13b)$$

where again $\beta = 2$ and $\beta = 6$ for plane stress and strain respectively.

15.3.3 Plastic Zones in Modes II and III

Bilby, Cottrell and Swindon [8] considered the equilibrium of screw dislocations within the yield zone arising from modes II and III. Referring to Figs 15.3b,c, they expressed the plastic zone within a ratio a/c where c is the length beyond a where plasticity is active (similar to length c in Fig. 15.9). The respective ratios are simply

$$a/c = \cos[\pi\tau/(2k)]$$
$$a/c = \cos[\pi q/(2k)]$$

where k is the shear yield stress. The corrected stress intensities $K_{\mathrm{IIc}}{}^*$ and $K_{\mathrm{IIIc}}{}^*$ become identical in form to eq(15.13b) in which τ/k and q/k replace σ/Y. The corresponding crack tip displacements were also derived. For mode II, u in the direction of x is

$$u = [4ak(1-\nu)/(G\pi)]\ln\{\sec[\pi\tau/(2k)]\}$$

and for mode III, w in the direction of z is

$$w = [4ak/(G\pi)]\ln\{\sec[\pi q/(2k)]\}$$

Example 15.2 The heat-affected zone in a welded infinite steel plate contains a through thickness crack aligned with the weld direction as shown in Fig. 15.10. If the boundary stress is to be 2/3 of the yield stress Y compare the tolerable crack lengths, assuming brittle failure when the plate is (a) stress relieved after welding and (b) as-welded, containing tensile residual stresses of the order Y. If the plate is thin what is the effect of crack tip plasticity on this comparison?

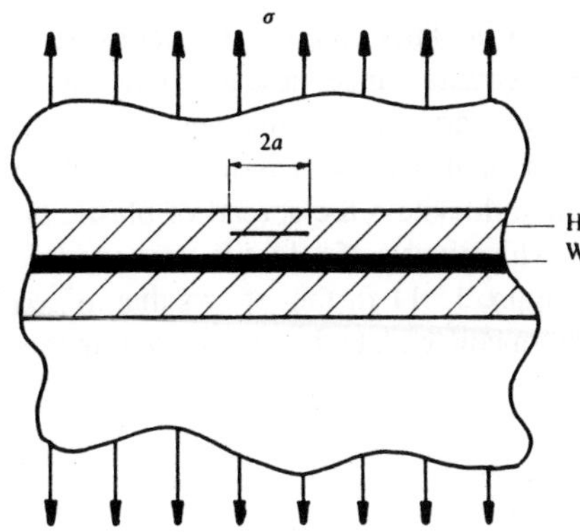

Figure 15.10 Welded plate

(a) *Stress relieved.* The tolerable crack length ($2a$) is found from eq(15.7) as

$$2a = (2/\pi)(K_{Ic}/\sigma)^2 \qquad \text{(i)}$$

and with $\sigma = 2Y/3$, eq(i) becomes

$$2a = [9/(2\pi)](K_{Ic}/Y)^2 \qquad \text{(ii)}$$

(b) *As-welded.* Here the tensile residual stress Y assists the boundary stress $2Y/3$ to produce an effective applied stress $\sigma = 5Y/3$ for the crack tip. Substituting into eq(i) leads to a tolerable crack length

$$2a = [18/(25\pi)](K_{Ic}/Y)^2 \qquad \text{(iii)}$$

Dividing (ii) by (iii) shows that stress relieving increases the permissible crack length by a factor of 25/4. Using eq(15.12) for an account of plane stress plasticity, eq(i) becomes

$$2a = 2K_{Ic}^{*2}/\{\pi\sigma^2[1 + \sigma^2/(2Y^2)]\} \qquad \text{(iv)}$$

Substituting $\sigma = 2Y/3$ and $\sigma = 5Y/3$ into eq(iv), the ratio between the tolerable crack lengths for stress relieved and welded conditions is

$$a_{sr}/a_w = [25(1 + 25/18)]/[4(1 + 2/9)] = 12.2$$

This shows that crack tip plasticity promotes a further twofold increase in the permissible crack length.

15.4 Fracture Toughness Measurement

15.4.1 Dependence of K_{Ic} upon Plate Thickness

The β factor in eqs(15.12) and (15.13b) show that K_{Ic}^* is dependent upon plate thickness. In general, since the toughness of a material decreases with decreasing plasticity it follows that the true fracture toughness for a material is the lower limiting plane strain value (Fig. 15.11).

The limiting toughness is particularly important to consider in high strength alloys and structural steels at lower temperatures since these are all at risk from brittle failure. When the plastic zone size is small compared to the thickness, the brittle transcrystalline cleavage fracture will be flat except for small shear lips, indicative of plane stress conditions at its free edges. With decreasing plate thickness the spread of plasticity increases until a full plane stress condition is reached. This gives a 45° ductile transcrystalline shear failure exibiting a dull fibrous appearance. Figure 15.11 further shows that mixed 45° shear and 0° cleavage fractures will appear for intermediate plate thicknesses within the transition from plane stress to plane strain.

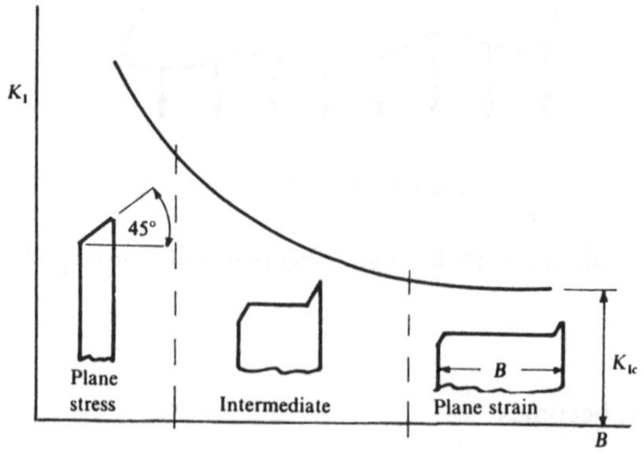

Figure 15.11 Dependence of K_{Ic} upon plate thickness

The assumption that the plastic zone is circular is more representative in the case of plane stress than in plane strain (see Fig. 15.12). It is possible to determine the true shape of the plastic zone assuming that a von Mises yield criterion applies to the elastic-plastic interface. Substituting eqs (15.5a - c) into eq(11.13a) with $\sigma_z = 0$ for plane stress and $\sigma_z = \nu(\sigma_x + \sigma_y)$ for plane strain and rearranging leads to the respective expressions for $r = r_y$ in terms of θ

$$r_y = [K_I^2/(4\pi Y^2)]\,[1 + (3/2)\sin^2\theta + \cos\theta] \tag{15.14a}$$

$$r_y = (K_I^2/(4\pi Y^2))\,[(3/2)\sin^2\theta + (1 - 2\nu)^2(1 + \cos\theta)] \tag{15.14b}$$

Equations (15.14a,b) define the profiles shown in Fig. 15.12. Setting $\theta = 0$ and $\nu = ¼$, reveals that the depth of each zone beyond the crack tip is respectively

$$r_y = K_I^2/(2\pi Y^2) \quad r_y = K_I^2/(8\pi Y^2) \tag{15.15a,b}$$

which shows that both Irwin's and Dugdale's estimates (15.11) and (15.12) of plastic advance are reasonable. The restricted spread of plasticity under plane strain lies within two lobes. The plastic zone shape has been confirmed by etch pit and finite element investigations [9].

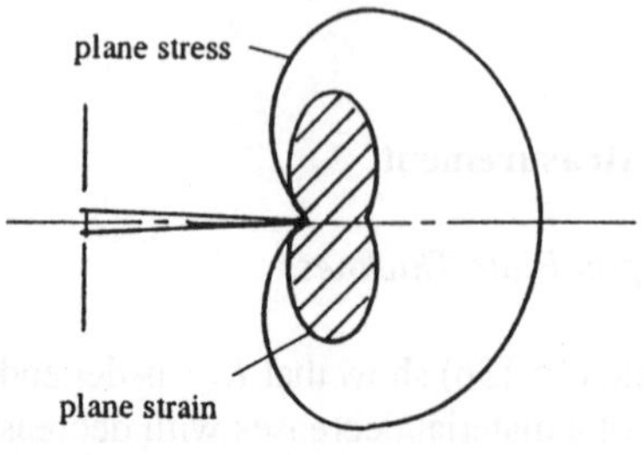

Figure 15.12 Plastic zones under plane stress and strain

At the mid-thickness of a thick plates the crack tip plasticity approximates to plane strain, while at the edges plane stress prevails. Hence, the size of the plastic zone increases from the centre to the edges as the central triaxial stress state becomes increasingly biaxial towards the free surfaces.

15.4.2 K_{Ic} Determination

In order to avoid the expense in determining K_{Ic} from testing infinite plates to fracture, a number of smaller test-pieces with finite dimensions have evolved. Their geometry is chosen to ensure a plane strain condition despite the nearness of the boundary to the crack surface. A boundary collocation analysis provides the appropriate K_{Ic} expression for the test.

A British Standard [10] outlines the procedures for testing to failure (a) three point bend, (b) compact tension and (c) single-edge notch test-pieces shown in Figs. 15.13a - c.

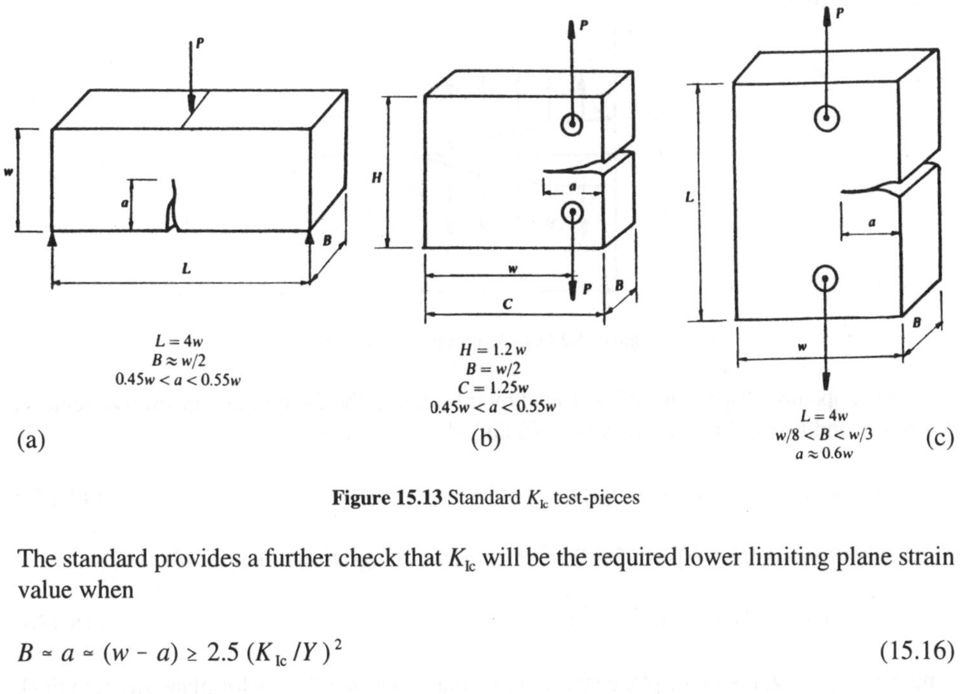

Figure 15.13 Standard K_{Ic} test-pieces

The standard provides a further check that K_{Ic} will be the required lower limiting plane strain value when

$$B \simeq a \simeq (w - a) \geq 2.5\,(K_{Ic}/Y)^2 \tag{15.16}$$

Once the dimensions of these standard test-pieces are chosen to satisfy eq(15.16) and the other conditions shown, the plane strain fracture toughness K_{Ic} is calculated directly from

$$K_{Ic} = P_c\, f(a/w)\,/\,(Bw^{1/2}) \tag{15.17}$$

where B and w are the respective thicknesses and widths shown, P_c is the critical load at fracture and the compliance function $f(a/w)$ accounts for the influence that the finite boundaries have on the stress intensity in each case. For Figs 15.13a-c, these are respectively

(a) $f(a/w) = 11.58(a/w)^{1/2} - 18.42(a/w)^{3/2} + 87.18(a/w)^{5/2} - 150.66(a/w)^{7/2} + 154.8(a/w)^{9/2}$

(b) $f(a/w) = 29.6(a/w)^{1/2} - 185.5(a/w)^{3/2} + 655.7(a/w)^{5/2} - 1017(a/w)^{7/2} + 638.9(a/w)^{9/2}$

(c) $f(a/w) = 1.99(a/w)^{1/2} - 0.41(a/w)^{3/2} + 18.7(a/w)^{5/2} - 38.48(a/w)^{7/2} + 53.85(a/w)^{9/2}$

15.5 Energy Balance Approach

An alternative criterion for brittle fracture is based upon a balance between the strain energy stored in a cracked body and the energy required to extend the crack. The solutions depend upon whether the plate response is elastic, giving a purely brittle fracture, or elastic-plastic, when the crack extension requires a greater amount of energy.

15.5.1 Elastic Fracture

Griffith [11] considered the balance between the loss in strain energy for a growing crack and the surface energy required to form the new crack surfaces. It was shown that the elastic work done in opening the upper crack surface in an infinite plate may be found more readily by applying a stress, with magnitude equal to the boundary stress, normal to the crack surfaces (see Fig. 15.14).

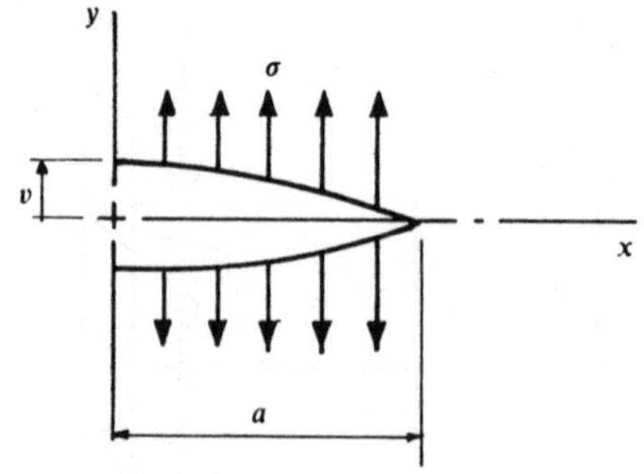

Figure 15.14 Crack-opening displacement

The crack opening displacement v is found by integrating the following constitutive relation for both conditions of plane stress ($\sigma_z = 0$) and plane strain ($\varepsilon_z = 0$):

$$\varepsilon_y = \partial v/\partial y = (1/E)[\sigma_y - \nu(\sigma_x + \sigma_z)] \tag{15.18a}$$

With substitutions from eqs(15.5a-c) for $x << a$, the integration of eq(15.18a) leads to

$$v = [\sigma/(2E)](1 + \nu)(1 + \alpha)(a^2 - x^2)^{1/2} \tag{15.18b}$$

where $\alpha = (3 - \nu)/(1 + \nu)$ for plane stress (a thin plate) and $\alpha = 3 - 4\nu$ for plane strain (a thick plate). Half the amount U of strain energy stored is equal to the work done in deforming the upper crack surface. It follows from eq(8.5a) U for this uniaxial stress system is given by:

$$U = \int_V \int_\varepsilon \sigma \,\mathrm{d}\varepsilon \,\mathrm{d}V$$

$$U = ½ \int_V \sigma \varepsilon_y \,\mathrm{d}V$$

$$= ½ \int_x \int_y \int_z \sigma(\partial v/\partial y)\,(\mathrm{d}x\,\mathrm{d}y\,\mathrm{d}z)$$

$$= (2 \times ½) \int_{z=0}^{B} \int_{x=0}^{a} \sigma v \,\mathrm{d}x\,\mathrm{d}z$$

$$= B \int_0^a \sigma v \,\mathrm{d}x \tag{15.19}$$

where B is the plate thickness. Substituting eq(15.18b) into (15.19):

$$U = [B\sigma^2/(2E)](1+\nu)(1+\alpha)\int_0^a (a^2 - x^2)^{1/2}\,dx$$
$$= a^2 B\pi(1+\nu)(1+\alpha)\sigma^2/(8E) \qquad (15.20)$$

Let S_e (J/m^2) be the elastic surface tension per unit area of crack surface (the specific surface energy). The total surface energy U_s of the upper crack is given by

$$U_s = 2aS_e B \qquad (15.21a)$$

Equation (15.21a) shows that the rate of strain energy released (with respect to the crack length) is given as

$$\delta U_s/\delta a = 2S_e B \qquad (15.21b)$$

Griffith postulated that an unstable crack growth δa occurs when the strain energy released equals or exceeds the energy required to extend the crack by this amount. That gives

$$\delta U/\delta a \geq \delta U_s/\delta a$$

where, from eqs(15.20) and (15.21b),

$$aB\pi(1+\nu)(1+\alpha)\sigma^2/(4E) \geq 2S_e B$$
$$\sigma \geq \sqrt{\{8S_e E/[a\pi(1+\nu)(1+\alpha)]\}} \qquad (15.22a)$$

For the plane stress condition $1+\alpha = 4/(1+\nu)$, when eq(15.22a) becomes

$$\sigma \geq \sqrt{[2S_e E/(a\pi)]} \qquad (15.22b)$$

whilst for plane strain $1+\alpha = 4(1-\nu)$, when eq(15.22a) becomes

$$\sigma \geq \sqrt{\{2S_e E/[a\pi(1-\nu^2)]\}} \qquad (15.22c)$$

It follows from eq(15.21b) that the critical energy release rate for a plate of unit thickness under mode I opening is

$$G_{Ic} = (1/B)(\delta U/\delta a) = 2S_e \qquad (15.23a)$$

G_{Ic} in eq(15.23a) is Griffith's fracture parameter which may also be interpreted as a critical crack extension force per unit width. The factor of 2 appears from the creation of two fracture surfaces. In Figs 15.15a and b the energy balances appropriate to plane stress and plane strain fracture are represented graphically to give in (a) the critical fracture stress for a given crack length and in (b) the critical crack length for a given applied stress.

Completely brittle materials containing sharp cracks, such as glass conform to eq(15.23a), i.e. the fracture energy will equal twice the elastic surface energy S_e. For most metallic materials the effect that crack tip plasticity has in blunting the crack tip invalidates eq(15.23a) since the fracture energy is far more than $2S_e$. However, with limited crack tip plasticity, both Irwin [4] and Orowan [12] recognised that fracture still occurred at a critical rate of energy released from the plate. Equation (15.23a) is modified to

$$G_{Ic} = (1/B)(\delta U/\delta a) = 2\,(S_e + S_p) \tag{15.23b}$$

in which an irreversible plastic work term $S_p >> S_e$ is added to the surface energy S_e. It follows from eqs(15.22b and c) that the fracture stress for an infinite plate under conditions of plane stress and plane strain are respectively

$$\sigma = \sqrt{[EG_{Ic}/(\pi a)]} \tag{15.24a}$$

$$\sigma = \sqrt{\{EG_{Ic}/[\pi a(1 - \nu^2)]\}} \tag{15.24b}$$

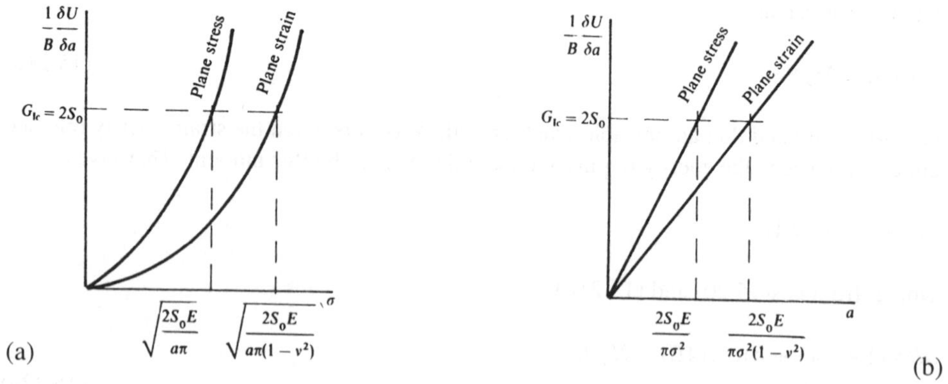

Figure 15.15 Energy for plane stress and plane strain fracture

Combining eqs(15.7) and eq(15.24a,b), the following relationships between K_{Ic} and G_{Ic} are found for plane stress and plane strain respectively:

$$G_{Ic} = K_{Ic}^2/E \tag{15.25a}$$

$$G_{Ic} = K_{Ic}^2(1 - \nu^2)/E \tag{15.25b}$$

In the case of Mode I plane stress, the crack-opening displacement (COD), v in eq(15.18b), can be written as

$$v = [8Ya/(\pi E)]\ \ln\{\sec[\pi\sigma/(2Y)]\}$$

from which

$$\exp[-\pi Ev/(8Ya)] = \cos(\pi\sigma/(2Y)) \tag{15.26a}$$

Provided σ/Y is small, we may take the first two terms in series exponential and cosine expansions to give

$$\sigma = \sqrt{[EYv/(\pi a)]} \tag{15.26b}$$

Comparing eq(15.26b) with eq(15.24a), we see that $G_{Ic} = Yv$. This shows that a brittle fracture occurs at a critical value of COD which can be derived from K_{Ic} and G_{Ic}. Alternatively, K_{Ic} is often found or checked from a COD measurement.

Example 15.3 A large thin-walled alloy steel tube has a uniaxial yield stress $Y = 1200$ MPa and an elastic modulus $E = 200$ GPa. A separate test on a cracked infinite plate showed that the critical rate of energy released at fracture was 24 kJ/m^2. The allowable circumferential stress in the tube is based upon Y with a factor of safety of 1.5. Calculate the maximum permissible size of longitudinal defect assuming (i) a purely brittle fracture and (ii) the presence of crack tip plasticity. (DU)

(i) The allowable circumferential stress:

$$\sigma = Y/1.5 = 1200/1.5 = 800 \text{ MPa}$$

In the brittle case, eq(15.8) gives the allowable defect size. Combining this with eq(15.25a) for plane stress, in units of N and m,

$$a = EG_{Ic}/(\pi\sigma^2)$$
$$= (200 \times 10^9 \times 24 \times 10^3)/[\pi \times (800 \times 10^6)^2] = 2.4 \times 10^{-3} \text{ m}$$
$$\therefore\ 2a = 4.8 \text{ mm}$$

(ii) Crack tip plasticity

Under plane stress conditions the radius of the plastic zone is estimated by combining eqs(15.15a) and (15.25a):

$$r_y = [1/(2\pi)](K_{Ic}^2/Y^2) = [1/(2\pi)(EG_{Ic}/Y^2)$$

Then from eq(15.12), assuming that the given G_{Ic}* is the critical rate of energy release for non-brittle fracture,

$$a + r_y = K_{Ic}^{*2}/(\pi\sigma^2) = EG_{Ic}^*/(\pi\sigma^2)$$
$$a = EG_{Ic}^*/(\pi\sigma^2) - r_y = EG_{Ic}^*/(\pi\sigma^2) - [1/(2\pi)](EG_{Ic}^*/Y^2)$$
$$= (EG_{Ic}^*/\pi\sigma^2)[1 - \sigma^2/(2Y^2)] = 2.4\,(1 - 2/9) = 1.8666 \text{ mm}$$
$$\therefore\ 2a = 3.73 \text{ mm}$$

Example 15.4 A compact tension test is conducted on a steel sample $B = 50$ mm thick with a crack length $a = 50$ mm and $w = 100$ mm (see Fig. 15.13b). If this test provides a fracture toughness value of $K_{Ic} = 110$ MN/m$^{3/2}$, estimate the fracture load and find the minimum yield strength for which the test is valid. Determine also the critical rate of strain energy released and the energy required to create the crack surfaces. Take $E = 220$ GPa and $\nu = 0.26$. (DU)

Here $a/w = 50/100 = 0.5$. Substituting into the compliance function $f(a/w)$ corresponding to Fig. 15.12b,

$$f(a/w) = 20.93 - 65.58 + 115.91 - 89.9 + 28.24 = 9.6$$

Then from eq(15.17),

$$P_c = K_{Ic}Bw^{1/2}/f(a/w)$$
$$= 110 \times 50 \times 10^{-3} \times (0.1)^{1/2}/9.6 = 0.181 \text{ MN} = 181 \text{ kN}$$

Equation (15.16) confirms the validity of this test since $B \geq 2.5(K_{Ic}/Y)^2$. The minimum yield strength is therefore

$$Y = \sqrt{(2.5K_{Ic}^2/B)}$$
$$= \sqrt{(2.5 \times 110^2/0.05)} = 777.8 \text{ MPa}$$

Under these plane strain conditions eq(15.25b) applies. Therefore, the critical rate of strain energy released per unit thickness is

$$G_{Ic} = (K_{Ic}^2/E)(1 - \nu^2)$$
$$= (110 \times 10^3)^2(1 - 0.26^2)/(220 \times 10^6)$$
$$= 51.28 \text{ kJ/m}^2$$

It follows from eq(15.23a) that this is also the specific surface energy required to create the two crack surfaces. The total energy consumed is

$$U = G_{Ic}(Ba)$$
$$= 51.28 \times 10^3(50 \times 50 \times 10^{-6})$$
$$= 128.2 \text{ J}$$

15.5.2 G_{Ic} Determination

The previous example illustrates how G_{Ic} may be found indirectly from K_{Ic}. Alternatively, a compliance technique enables G_{Ic} to be obtained from measurements of the deflection Δ beneath the applied force P up to the point of fracture. Figures 15.16a,b show two possible load-deflection responses.

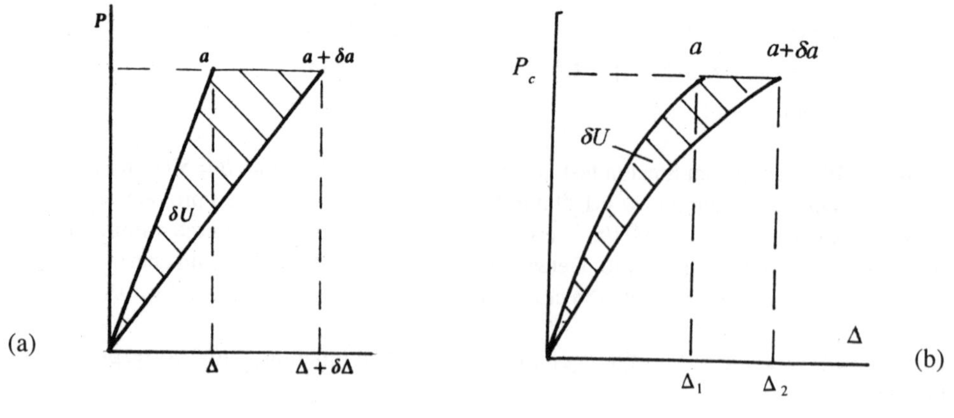

Figure 15.16 Load versus displacement diagrams

For a purely brittle fracture (Fig. 15.16a) P versus Δ will remain linear for plates with initial crack lengths of a and $a + \delta a$. For the longer crack there is a reduction in stiffness as seen in the shallower slope P/Δ, i.e. an increase in compliance $\lambda = \Delta/P$. When the crack extends by the amount δa under a constant load, it follows from eq(15.23a) that the energy absorbed in the crack formation is

$$\delta U = BG_{Ic}\,\delta a \tag{15.27}$$

The extension is assumed to occur under a constant force so that the change in compliance is $\delta\lambda = \delta\Delta / P$. The release of strain (potential) energy stored in the plate is the shaded area in Fig. 15.16a. This is

$$\delta U = P_c \delta\Delta / 2 = P_c^2 \delta\lambda / 2 \tag{15.28}$$

Equating (15.27) and (15.28) gives G_{Ic} as

$$G_{Ic} = [P_c^2/(2B)](\delta\lambda / \delta a) \tag{15.29}$$

Since eq(15.29) is independent of Δ, it would still apply had the crack extended without load point displacement. This arises from fixed grip loading where the load drops from A to B giving G_{Ic} = area OAB. The area ABC is of secondary order and may be neglected. The compliance calibration involves testing a number of identical test-piece geometries with different crack lengths to establish plots typical of Fig. 15.16a. A plot of λ versus a is found and, in applying eq(15.29), $\delta\lambda / \delta a$ is the slope to this plot at the value of fracture force P. G_{Ic} is readily converted to K_{Ic} for plane strain from eqs(15.25b) and (15.29):

$$K_{Ic} = \sqrt{\left[\{P_c^2 E/[2B(1 - \nu^2)]\}(\delta\lambda/\delta a)\right]} \tag{15.30}$$

The experimental compliance method relies only upon the measurement of P and Δ in a test-piece of given geometry. This method is often preferred to a theoretical derivation, the complexity of which is typified by the $f\,(a/w)$ expressions given with the standard test-pieces in Fig. 15.13a-c. However, a compliance test will not yield a valid G_{Ic} value if P versus Δ is non-linear (see Fig. 15.16b). In this case of general yielding fracture mechanics we must refer J to the crack extension force per unit width as

$$J_c = (1/B)(\delta U / \delta a) \tag{15.31a}$$

where δU is now the area enclosed between the two non-linear plots shown. To find δU, assume that the non-linear function $\Delta = \Delta(P)$ has an explicit form for each curve $\Delta_1 = A_1 P^n$ and $\Delta_2 = A_2 P^n$. This is equivalent to assuming a Hollomon hardening law in which $\varepsilon \propto \sigma^n$. The shaded area δU is given by the integral

$$\delta U = \int_0^{P_c} \delta P\,(\Delta_1 - \Delta_2)$$

$$= \int_0^{P_c} (A_1 - A_2)P^n \delta P$$

$$\delta U = (\Delta_1 - \Delta_2)P_c / (n + 1)$$

Writing $\delta\Delta = \Delta_1 - \Delta_2$ and dividing by δa, we find from eq(15.31a),

$$J_c = \{P_c/[B(n + 1)]\}(\delta\Delta / \delta a) \tag{15.31b}$$

In the fully elastic case where n = 1, $\Delta = \lambda P_c$, then $J_c = G_{Ic}$ and eq(15.31b) reduces to eq(15.28).

Example 15.5 Tests on an edge-cracked, high-tensile alloy steel plate model gave the following displacements beneath an applied force of 1 MN

a (mm)	0	5	10	15	20
$\Delta/10^{-3}$ (mm)	0.503	0.539	0.612	0.793	1.194

in which the crack was machined to the new length after each test. A tensile member 12.5 mm thick, made from the same steel, is to support an axial force of 5 MN. Estimate, from the model results, the maximum permissible crack length in the member to give a brittle fracture. Examine the possibility of an elastic-plastic plane stress fracture where $J_c \simeq G_c$. Take K_{Ic} = 123 MN/m$^{3/2}$, Y = 1540 MPa, n = 3, E = 207 GPa and ν = 0.27.

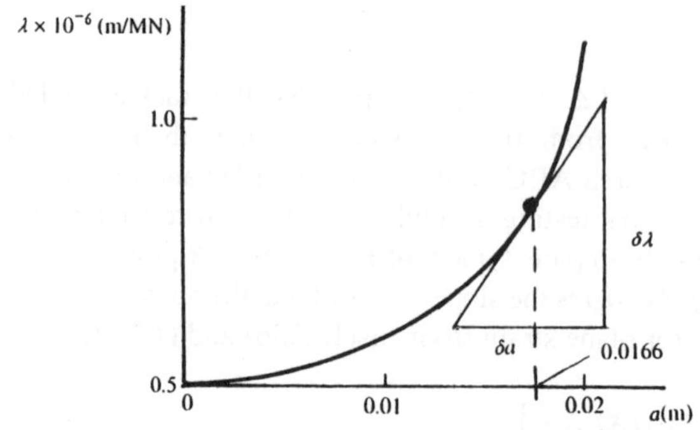

Figure 15.17 Compliance calibration

Firstly, check the plane strain condition from eq(15.15)

$$B \geq 2.5(K_{Ic}/Y)^2 = 2.5(123/1540)^2 = 0.016 \text{ m}$$

Since B = 0.0125 m, it appears that a mixed-mode fracture is likely.

(a) For linear elastic behaviour the direct measurement of load point displacement under a constant load of 1 MN gives a compliance conversion factor $\lambda = \Delta/P = \Delta \times 10^{-6}$ (m/MN), corresponding to each crack length a. This enables the plot of log λ versus log a in Fig. 15.17 to be constructed. For a plane-strain fracture, it follows from eq(15.30) that the critical value of $\delta\lambda/\delta a$, under a force of 5 MN, is

$$\delta\lambda/\delta a = [2 \times 0.0125(1 - 0.27^2) \times 123^2]/(5^2 \times 207 \times 10^3) = (67.76 \times 10^{-6})\text{ MN}^{-1}$$

Applying this gradient to Fig. 15.17 as shown, we find from the point of tangency a critical crack length of 16.6 mm. The plane stress condition, i.e. without the $(1 - \nu^2)$ factor in eq(15.30), gives a gradient $\delta\lambda/\delta a = 73.09 \times 10^{-6}$ and therefore the lower plane strain gradient provides the permissible crack length.

(b) If a plane stress condition is accompanied by elasto-plastic failure where $J_c \approx G_{Ic} = K_{Ic}^2/E$ = 73 MJ/m^2 we have from eq(15.31b)

$$\delta\Delta/\delta a = J_c B(n+1)/P_c = (73 \times 12.50 \times 10^{-3} \times 4)/5 = 0.73\text{ MN}^{-1}$$

This gradient is far ouside the range of Fig. 15.17, showing that the calibration applies to linear-elastic conditions.

15.5.3 The J - Integral

Equations(15.31a,b) apply to GYFM. The area J enclosed within Fig. 15.16b is an elasto-plastic fracture mechanics parameter also known as the J-integral [13]. To derive this integral any closed, counterclockwise contour Γ is taken whose path encloses the crack tip as shown in Fig. 15.18.

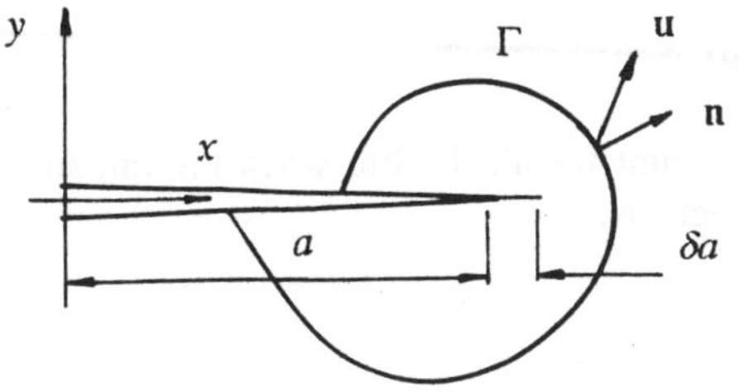

Figure 15.18 Crack tip contour

Let s be the distance along Γ so that δs is an element of Γ upon which a traction produces a displacement **u** with components u, v and w. Let the components of the outward normal vector **n** be n_i. The cartesian components T_j of the traction vector **T** are found from the stress state σ_{ij} as $T_j = \sigma_{ij}\, n_i$, where i and j refer to the co-ordinate directions x, y and z (see eqs(1.4a-c)). As the crack length increases by an amount δa, a point on the contour will displace by the amounts Δu, Δv and Δw in the x, y and z directions respectively. Consider a plane stress or plane strain problem where σ_{ij} is independent of z. The difference between the strain energy stored in the plate and the work of traction as the crack extends is.

$$- \Delta U = \int_{\Delta A} W \, \mathrm{d}A - \int_{\Delta s} \mathbf{T} \cdot \Delta \mathbf{u} \, \mathrm{d}s \qquad (15.32a)$$

where $W = \sigma_{ij}\varepsilon_{ij}$ is the strain energy density, ΔA is the change to the area enclosed by Γ and Δs is change to the length of Γ. For simplicity, we align x with a so that dA = dx dy. The potential energy release rate is found from eq(15.32a) as

$$- \partial U / \partial a = \int_x \int_y (\partial W / \partial a) \, \mathrm{d}x \, \mathrm{d}y - \int_{\Delta s} \mathbf{T} \cdot (\partial \mathbf{u} / \partial a) \, \mathrm{d}s \qquad (15.32b)$$

Since $\delta a = \delta x$ then $J = - \partial U / \partial a$ is identified with eq(15.32b) when written in following form

$$J = \int_\Gamma W \, \mathrm{d}y - \int_\Gamma \mathbf{T} \cdot (\partial \mathbf{u} / \partial x) \, \mathrm{d}s \qquad (15.32c)$$

Equation(15.32c) applies to both elastic and plastic loadings in which Γ is path independent provided it contains the crack tip. We have seen that a critical value J_{Ic} at fracture defines the area shaded in Fig. 15.15b. Methods for determining J_{Ic} in three-point bend and compact tension tests are outlined in BS 7448 (Part 1) 1991 [14].

15.6 Application of LEFM to Fatigue

The brittle nature of fatigue failure implies that fatigue strength, like fracture toughness, also depends upon the crack tip parameter K. During crack propagation under a repeated or fluctuating tensile stress cycle of constant stress range $\Delta\sigma$ (see Figs 14.2c,d), the range of stress intensity ΔK is written from eq(15.9) as:

$$\Delta K = Q\Delta\sigma\sqrt{a} \tag{15.33}$$

where $\Delta\sigma = \sigma_{max} - \sigma_{min}$. Equation (15.33) applies to all the plate shown geometries in Figs 15.6a-f. It is now well established that, following initiation, the cyclic rate of crack propagation da/dN in Fig. 15.19a is a function of the range of stress intensity ΔK in modes I, II and III [15]. That is,

$$da/dN = f(\Delta K) = g(\Delta\sigma, a) \tag{15.34}$$

Equation (15.34) will account for all the life when the initiation phase is absent in the presence of pre-existing defects.

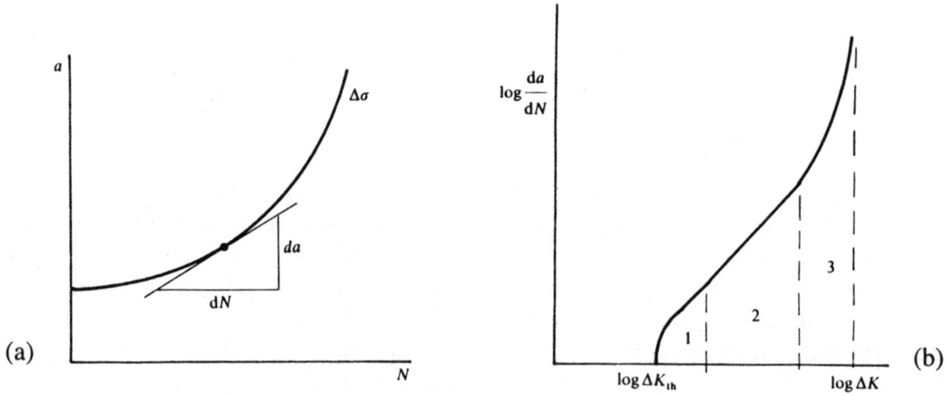

Figure 15.19 Fatigue crack growth

The logarithmic plot of da/dN versus ΔK in Fig. 15.19b correlates the growth rates for separate tests in air each with a different $\Delta\sigma$ value. The sigmoidal curve shape applies to non aggressive environments and indicates three regions of fatigue crack growth. Region 1 is associated with low and non-propagating rates of crack growth as ΔK lies near its threshold value (ΔK_{th}). For the mid-range of growth in region 2, a power law (15.34) applies:

$$da/dN = C(\Delta K)^n \tag{15.35}$$

where C and n are material constants. The exponent n normally lies in the range 2 - 7 depending upon the environment, cyclic frequency, mean stress and temperature. In region 3, unstable, accelerating crack growth precedes fracture as K_{max} approaches K_{Ic}. Most estimates of mode I fatigue life in structures are based upon combining eq(15.33) with eq(15.35). This gives

$$da/dN = C(Q\Delta\sigma/a)^n = CQ^n(\Delta\sigma)^n a^{n/2} \tag{15.36a}$$

For a crack of initial length a_0 which extends to a length a_f before the plate fails in N_f cycles, eq(15.36a) becomes

$$\int_{a_0}^{a_f} da/a^{n/2} = CQ^n(\Delta\sigma)^n \int_0^{N_f} dN$$

and provided Q does not vary with a, this integrates directly to give the cyclic life as:

$$N_f = \frac{2}{(n-2)CQ^n(\Delta\sigma)^n}\left[\frac{1}{a_o^{(n-2)/2}} - \frac{1}{a_f^{(n-2)/2}}\right] \tag{15.36b}$$

It is seen that eq(15.36b) will not apply to those plates in Fig. 15.6 where Q is trigonometric in a. In this case a numerical integration of eq(15.36a) will supply N_f.

Example 15.6 Calculate the number of cycles to failure when a cyclic repeated stress of range 175 MPa is applied to an infinite plate containing a 0.2 mm long crack. Take $K_{Ic} = 54$ MN/m $^{3/2}$ and the rate of crack growth as $da/dN = 40 \times 10^{-12}(\Delta K)^4$ m/cycle.

Substituting $Q = \sqrt{\pi}$ and $n = 4$ into eq(15.36b) gives

$$N_f = (1/a_o - 1/a_f)/[C\pi^2(\Delta\sigma)^4] \tag{i}$$

Now since $\Delta K_o^2 = \Delta\sigma^2(a_o\pi)$ and $\Delta K_{Ic}^2 = \Delta\sigma^2(\pi a_f)$, eq(i) becomes

$$N_f = [\pi\Delta\sigma^2/(\Delta K_o)^2 - \pi\Delta\sigma^2/(\Delta K_{Ic})^2]/(C\pi^2\Delta\sigma^4)$$
$$= [(\Delta K_{Ic}/\Delta K_o)^2 - 1]/[C\pi(\Delta\sigma\Delta K_{Ic})^2] \tag{ii}$$

This equation shows that $\Delta K_{Ic}/\Delta K_o$ and $\Delta\sigma$ govern the fatigue crack growth rate. Substituting into eq(ii): $\Delta K_o = 175\sqrt{(0.1 \times 10^{-3} \times \pi)} = 3.102$ MN/m$^{3/2}$, with the given values for ΔK_{Ic}, $\Delta\sigma$ and C,

$$N_f = [(54 / 3.102)^2 - 1]/[40 \times 10^{-12} \times \pi(175 \times 54)^2]$$
$$N_f = 26919 \text{ cycles}$$

15.6.1 Effect of Mean Stress

The compressive part of reversed and alternating cycles (see Fig. 14.2a,b) does not contribute to crack growth. Thus ΔK in eq(15.35) must be applied to the tensile part of these cycles. For wholly tensile cycles (Fig. 14.2c,d), the question arises as to how fracture mechanics can account for the effect that mean stress has in reducing cyclic life (see Fig. 14.3). Plots similar to Fig.15.19b, with different ratios of $\sigma_{max}/\sigma_{min}$, have shown that crack growth rates increase, mostly in region 3, with an increase in this stress ratio. Since the growth rate is also inversely proportional to toughness K_c (under modes I, II and III), Forman et al [16] modified eq(15.35) to include both effects:

$$da/dN = C\Delta K^n / [(1 - R)K_c - \Delta K]$$

where now $R = K_{max}/K_{min}$. Other investigators have re-defined ΔK in eq(15.35) with an effective stress intensity range $\Delta K_{eff} = K_{max} - K_{op}$ in Fig. 15.20, where K_{op} accounts for that part of the tensile cycle for which the crack may remain partially closed in the presence of high mean stresses.

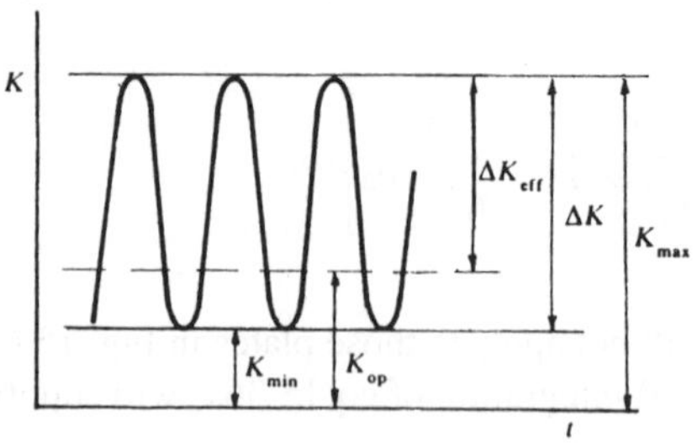

Figure 15.20 Effective opening range

15.6.2 Environmental Effects

Static loading of cracked bodies in an aggressive environment may cause failure at stress levels less than those calculated from K_{Ic}. This occurs where stress corrosion cracking, hydrogen and liquid metal embrittlement enable existing flaws to grow to the necessary critical size. When the environment assists cracking (EAC) under mode I loading, it becomes necessary to re-define a lower critical stress intensity K_{IEAC} if failure is not to occur by this means over long periods. Normally K_{IEAC} is found for the more common material and environment combinations in laboratory tests lasting at least 1000 hr on constant bolt-loaded compact tension test-pieces (Fig.15.12b). Ratios K_{IEAC}/K_{Ic} are approximately 1/3 but have been found to be as low as 1/10 for alloy steels in an NaCl solution. Intercrystalline failure occurs for stress intensities in the region $K_{IEAC} < K < K_{Ic}$. Here the logarithm of the crack growth rate da/dt varies with K_I for most materials in the manner shown in Fig. 15.21.

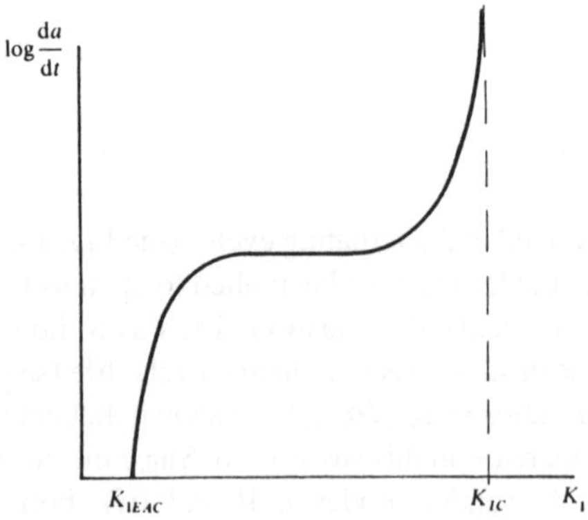

Figure 15.21 Environmental cracking

Normally the two regions of rapid crack growth rate are interrupted by one of relative insensitivity to K. The threshold K_I level is identified with K_{IEAC} in this plot. Fatigue crack growth rates are also sensitive to both the cyclic frequency and the temperature. In general, for a given ΔK the growth rates increase as the environmental attack is enhanced with increasing temperature and decreasing cyclic frequency. This particularly applies to square waveforms incorporating periods of dwell under the maximum and minimum stresses in the cycle. Equation(15.34) will account for this, provided the function $f(\Delta K)$ is determined experimentally under simulated conditions. To simplify life-prediction from eq(15.35), the effects of the environment on fatigue may only need to be considered in terms of a change in C and n for the region 2 in Fig. 15.19b. Alternatively, the separate effects of environment and frequency on fatigue crack growth may be superimposed when it is assumed that fatigue and environmentally assisted cracking act independently. The net crack growth rate is then the sum of the individual growth rates, as shown in the following example. When applying this method, the baseline fatigue crack growth data should be established in an inert atmosphere where cyclic frequency and temperature are known to have little effect on the growth rate.

Example 15.7 An aluminium alloy is subjected to cyclic tensile stressing at a frequency f = 1 Hz with short, constant dwell periods in the ratio $\Delta t_{max}/\Delta t_{min} = 3/2$ at the maximumum and minimum stresses (Fig. 15.22) where the corresponding stress intensities are $(K_I)_{max} = 15$ MPa√m and $(K_I)_{min} = 8$ MPa√m respectively. The crack growth rate in argon is given by the power law $da/dN = 13.2 \times 10^{-12} \Delta K^{2.9}$ m/cycle for this fluctuating cycle. Determine the crack growth rate in m/cycle for an aggressive environment where cracking occurs under the given maximum and minimum stress intensities at growth rates of $da/dt = 1.5 \times 10^{-8}$ m/s and $da/dt = 9 \times 10^{-10}$ m/s respectively. How does the net growth rate depend upon frequency?

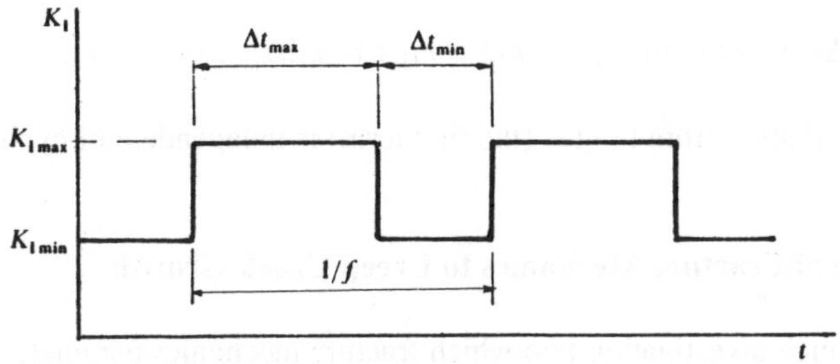

Figure 15.22 Square wave

In the fatigue cycle $\Delta K = 7$ MPa √m, and the Paris law gives the corresponding crack growth rate as

$$da/dN = 13.2 \times 10^{-12} (7)^{2.9}$$
$$= 3.73 \times 10^{-9} \text{ m/cycle}$$

For the square wave in Fig. 15.22 let Δt_{max} and Δt_{min} be the times spent under K_{Imax} and K_{Imin} where the cycle period is

$$\Delta t_{max} + \Delta t_{min} = 1/f \qquad \text{(i)}$$

The corresponding environmental crack rates/cycle are

$$(da/dN)_{max} = (da/dt)_{max} \Delta t_{max} \qquad \text{(ii)}$$
$$(da/dN)_{min} = (da/dt)_{min} \Delta t_{min} \qquad \text{(iii)}$$

Since $\Delta t_{max}/\Delta t_{min} = 1.5$ and $f = 1$ cycle/s, then from eq(i), $\Delta t_{max} = 0.6$s, $\Delta t_{min} = 0.4$s. Substituting into eqs(ii) and (iii),

$(\mathrm{d}a/\mathrm{d}N)_{max} = 1.5 \times 10^{-8} \times 0.6 = 9.0 \times 10^{-9}$ m/cycle
$(\mathrm{d}a/\mathrm{d}N)_{min} = 9 \times 10^{-10} \times 0.4 = 0.36 \times 10^{-9}$ m/cycle

The net crack growth rate/cycle is then

$$(\mathrm{d}a/\mathrm{d}N)_{net} = (3.73 + 9.0 + 0.36)10^{-9} = 13.09 \times 10^{-9} \text{ m/cycle}$$

It is seen from eq(i) that the growth rates in eqs (ii) and (iii) depend upon frequency. In general, when $r = \Delta t_{max}/\Delta t_{min}$ = constant, eq(i) gives

$$\Delta t_{max} = r/[f(1 + r)] \quad \text{(iv)}$$
$$\Delta t_{min} = 1/[f(1 + r)] \quad \text{(v)}$$

Substituting eqs(iv) and (v) into (ii) and (iii),

$$(\mathrm{d}a/\mathrm{d}N)_{max} = (\mathrm{d}a/\mathrm{d}t)_{max}\, r/[f(1 + r)]$$
$$(\mathrm{d}a/\mathrm{d}N)_{min} = (\mathrm{d}a/\mathrm{d}t)_{min}/[f(1 + r)]$$

The net growth rate/cycle then takes the general form:

$$(\mathrm{d}a/\mathrm{d}N)_{net} = C(\Delta K)^n + (\mathrm{d}a/\mathrm{d}t)_{max}\, r/[f(1 + r)] + (\mathrm{d}a/\mathrm{d}t)_{min}/[f(1 + r)]$$

in which the contribution from fatigue (the first term) is independent of cyclic frequency.

15.7 Application of Fracture Mechanics to Creep Crack Growth

There has been much investigation into which fracture mechanics parameter is appropriate to characterise creep crack growth. The latter applies to the growth of a single pre-existing flaw throughout life rather than the linkage of many cracks appearing within the tertiary regime of creep. The first creep crack growth studies were conducted on centre-notched (Fig. 15.6e), CTS and SEN test-pieces (Figs 15.13b,c) in stainless and alloy steels, under constant load at elevated temperatures [17,18]. They showed that the crack growth rate in these materials was given by

$$\mathrm{d}a/\mathrm{d}t = AK^n \quad (15.37)$$

where A and $5 \le n \le 30$ are constants. Equation (15.37) is a LEFM prediction that is most suitable for high strength, low ductility materials where little bulk creep occurs in the uncracked ligaments. With stable crack growth in a more ductile material, both bulk creep and crack growth occur simultaneously. It is then more appropriate for a net section stress or a reference stress to replace K in eq(15.37). In each case the n - value is similar to that used with the Norton secondary creep rate law.

There is some difficultly in evaluating K when it continually changes in these test-pieces. The double cantilever bend (DCB) testpiece (Fig. 15.23a) was designed with contoured edges to maintain a constant K with increasing crack length [19]. Crack growth in an aluminium alloy DCB test-piece in the temperature range 100 - 200°C appeared to violate eq(15.37) since $\mathrm{d}a/\mathrm{d}t \neq$ constant throughout life. The crack growth curve (Fig. 15.23b) shows a decreasing growth rate $\mathrm{d}a/\mathrm{d}t$ while K remains constant. Within this region it can be shown that the instantaneous value of J from eq(15.31b) also decreases with crack length under a constant K, so J was believed to control growth. When a crack grows with time we need the rate form of eq(15.31b), so that $\dot{\Delta} = d\Delta/dt$ is the COD rate and $\dot{J} = dJ/dt$ becomes the power release rate. The symbol C^* has replaced $\dot{J}$ in eq(15.32) when time derivatives of w and ε_{ij} appear in this integral. In this way C^* may be applied more generally to cracks that creep under mode I in elastic-plastic structures of various geometry [20,21].

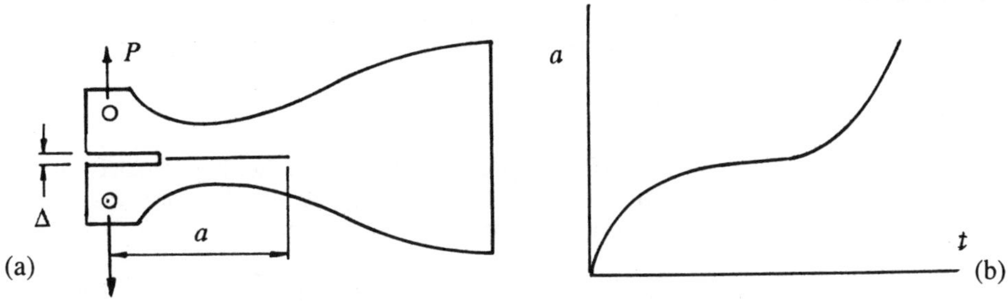

Figure 15.23 Creep crack growth in a DCB test-piece

References

[1] Inglis C. E., *Trans Inst Nav Arch.* **LV**(1) (1913), 219.
[2] Rees D. W. A., *Strain*, August (1997), 87.
[3] Westergard H. M., *Jl. Appl. Mech.* **61** (1939), A49.
[4] Irwin G. R., *Jl Appl. Mech.* **24** (1957), 109.
[5] Irwin G. R., *Appl. Mater. Res.* **3** (1964), 65.
[6] Craeger M. and Paris, P. C., *Int Jl Fract Mech*, **3** (1967), 247.
[7] Dugdale D. S., *Jl. Mech. Phy. Solids,* **8** (1960), 100.
[8] Bilby B. A., Cottrell A. H. and Swindon K. H., *Proc Roy Soc*, **A272** (1963), 304.
[9] Rice J. R. and Johnson M. A., *Inelastic Behaviour of Solids*, McGraw-Hill, New York, 1970, p.641.
[10] BS 5447: *Methods of Test for Plane Strain Fracture Toughness* (K_{Ic}) of Metallic Materials, 1977.
[11] Griffith A. A., *Phil. Trans. Roy. Soc.* **A221** (1921), 163.
[12] Orowan E., *Weld. Jl. Res. Suppl.* **20** (1955), 157s.
[13] Meguid A. S., *Engineering Fracture Mechanics*, Elsevier, 1989.
[14] BS 7448 (Pt.1): *Methods For the Determination of* K_{Ic}*, Critical CTOD and Critical J values of Metallic Materials*, 1991.
[15] Paris P. C., *Fatigue – An Interdisciplinary Approach*, Syracuse University Press, 1964, p.107.
[16] Forman R. G., Kearney V. E. and Engle, R. M., *Jl. Basic Eng, Trans. ASME,* **89** (1967), 459.
[17] Siverns M. J. and Price A. T., *Int Jl of Fracture*, **9**, (1973), 199.
[18] James L. A., *Int Jl of Fracture*, **8**, (1972), 347.
[19] Webster, G., *Mech and Phys of Fracture*, Inst Physics, **18**, (1975), 169.
[20] Van Leeuwen H. P., *Engng Frac Mech.* **9** (1977), 951.
[21] Fu L. S., *Engng Fract Mech*, **13**, (1980), 307.

EXERCISES

Fracture

15.1 A glass sheet of uniform thickness fails under a stress of 60 MPa. Estimate the size of the inherent defect in glass given that E = 72 GPa and S_o = 1 J/m^2. (CEI)

15.2 Determine the stress concentration factor at a point 0.01mm away from the tip of a sharp 0.5 mm edge crack in a semi-infinite plate (see Fig. 15.6a) when a stress of 20 MPa is applied around the boundary.
Answer: 5.

15.3 A large thick steel plate contains a single edge crack of length 1mm (see Fig. 15.6a). If the steel has a fracture toughness of K_{Ic} = 53 MN/m$^{3/2}$ and a yield strength of 950 MPa, find the stress at which fast fracture would occur and determine whether the fracture would be brittle or ductile.

15.4 A thin steel sheet 3.5mm thick by 305mm wide contains a semi-elliptical surface flaw 25 mm long × 0.5 mm deep (see Fig. 15.6c). If the plane strain fracture toughness is K_{Ic} = 66 MN/m$^{3/2}$ and the yield strength is 1930 MPa, estimate the remote stress that would cause failure when applied normal to the crack area. Do not neglect crack tip plasticity.
Answer: 1492 MPa.

15.5 A weld in an infinite plate contains residual stresses approximately equal to the yield stress Y = 800 MPa for the parent plate. The allowable applied stress is to be based upon Y with a safety factor of 2. If the fracture toughness value for the plate material is K_{Ic} = 60 MN/m$^{3/2}$, find the maximum permissible size of crack in the weld.

15.6 Find the critical defect size for a large diameter thick-walled steel cylinder, 500 mm o.d. and 375mm i.d., that is to sustain an internal pressure of 350 MPa if K_{Ic} = 100 MN/m$^{3/2}$ and Y = 1500 MPa. Check the validity of an assumed plane strain condition.

15.7 A single-edge notch tension test is conducted on a steel plate 15 mm thick by 100 mm wide with a sharp notch 50 mm deep (see Fig. 15.13c). If the fracture load is 400 kN, evaluate the fracture toughness and the critical rate of strain energy release. Is K_{Ic} valid? Take E = 210 GPa, ν = 0.27 and Y = 1235 MPa.
Answer: K_{Ic} = 809.5 MN/m$^{3/2}$, $\delta U/\delta a$ = 0.043 MJ/m.

15.8 A light alloy three-point bend specimen (Fig. 15.13a) is 50mm thick and spans 200mm between the supports with w = 100mm and a = 53mm. If the test-piece fails under a force of 53.2 kN, calculate (i) the fracture toughness giving a check on its validity for a yield stress of 450 MPa, (ii) the minimum value of yield stress for the calculated K_{Ic} that would ensure plane strain, (iii) the size of the plastic zone, (iv) the critical rate of strain energy release and (v) the specific fracture energy. Take ν = 0.31 and G = 75 MPa.
Answer: 39.42 MN/m$^{3/2}$, 278.7 MPa, 407μm, 0.937 kJ/m, 9.365 kJ/m^2.

15.9 A plate, 100 mm wide and 5 mm thick, containing a central crack 8 mm long (see Fig. 15.6e) fails under an applied tensile force of 200 kN. If the yield stress of the plate material is 500 MPa, evaluate the fracture toughness, assuming both brittle and ductile failure.
Answer: 45 MN/m$^{3/2}$, 51.87 MN/m$^{3/2}$.

15.10 The stress function $\phi = Ar^{3/2}\cos^3(\theta/2)$ prescribes the stress in the vicinity of a sharp crack when the origin for r and θ lies at its tip. Express A in terms of K when, for a remotely applied stress, there is no stress applied normally to the crack surfaces. Hence show that the two normal stress components are $\sigma_{x,y} = [K/(2\pi r)^{1/2}]\cos(\theta/2)[1 \pm \sin(\theta/2)]$ (IC)

15.11 A function of the form $C(\theta)\, r^q$ where $q > 1$ prescribes the stress state around the tip of an edge notch with a 90° included angle and origin for r, θ at the notch tip. Determine the value of the constant q. (IC)

15.12 A tie bar, 150 mm wide and 12.5 mm thick, is to withstand a tensile stress of 100 MPa. Given that edge cracks 2.5 mm deep and semi-elliptical surface cracks 10mm long by 1mm deep (see Fig. 15.6c,f) are present, select the most suitable steel from two that are available with fracture toughnesses and yield strengths values of 100 MN/m$^{3/2}$, 1350 MPa and 54.5 MN/m$^{3/2}$, 1700 MPa respectively. If the quoted K_{Ic} for the chosen steel is to be checked from three-point bending of a beam of width 50 mm, thickness 12.5mm, span 200mm and crack length 15mm, what should be the fracture load of the selected steel? (IC)

15.13 Show that the radius r_y of the plastic zone ahead of a blunt crack of tip radius ρ (see Fig. 15.5) is expressed by

$$r_y = (2/\pi)(K_I/Y)^2[1 + (\rho/r_y) + \surd(2\rho/r_y)]$$

given that the opening stress distribution is $\sigma_y = K_I(1 + \rho/r)/(2\pi r)^{1/2}$. Take the origin of co-ordinates at $\rho/2$ from the notch tip. (DU)

15.14 Two large steel plates, each 3 m wide, contain 2.5 mm central cracks through their thicknesses. If for one plate K_{Ic} = 115 MN/m$^{3/2}$, Y = 910 MPa and for the other K_{Ic} = 55 MN/m$^{3/2}$, Y = 1035 MPa, which plate could support a tensile force of 4.5 MN with the lesser weight? Assume similar densities for which the plate thickness is to be decided.

15.15 Inspection of a large bolt with outer and root diameters 100 mm and 92 mm respectively reveals a circumferential crack 1 mm deep around the root of one thread. If, for the bolt material, K_{Ic} = 50 MN/m$^{3/2}$ and Y = 300 MPa, estimate the axial tensile force P to cause failure, given that the stress intensity factor for a bar of radius R with a single circumferential crack of depth a is:

$$K = P \times f(a/R) / (\pi R^3)^{1/2}$$

where the compliance function is given as:

$$f(a/R) = [1.12(a/R)^{1/2} - 1.18(a/R)^{3/2} + 0.74\,(a/R)^{5/2}] / (1 - a/R)^{3/2}$$

Fatigue Crack Growth

15.16 Calculate the pressure that would cause a thin cylinder, 7.5 m i.d. and 7.5 mm thick, to fail when it contains a through-thickness crack 12.5 mm long aligned with the longitudinal direction. If this vessel is to be subjected to 4000 cycles of repeated internal pressure, determine the maximum cyclic pressure, based upon a safety factor of 2, given the crack growth rate

$$da/dN = 50 \times 10^{-12}\,(\Delta K)^4 \text{ m/cycle}$$

for ΔK in MN/m$^{3/2}$.

15.17 A steel plate is 350 mm wide and 35 mm thick. It contains an edge crack 9 mm deep. If K_{Ic} = 82 MN/m$^{3/2}$, find the cyclic life when a remotely applied tensile force varies between limits of 1.75 MN and 2.75 MN. Take the crack growth rate as

$$da/dN = 4.62 \times 10^{-12}\,(\Delta K)^{3.3} \text{ m/cycle.}$$

15.18 A thin-walled steel cylinder 8m inner diameter and 50 mm wall thickness is to withstand 3000 repeated cycles at a maximum pressure of 5.5 MPa. If the rate of growth of pre-existing cracks follows the law

$da/dN = 24 \times 10^{-15} (\Delta K)^4$ (m/cycle)

and the fracture toughness is $K_{Ic} = 210$ MN/m$^{3/2}$, find the permissible length of a longitudinal crack.

15.19 An infinite aluminium alloy plate containing a 10 mm crack is subjected to a remotely applied stress that fluctuates between 5 and 60 MPa. How many cycles must be applied to double the crack length? Takethe crack growth rate to be

$da/dN = 45.55 \times 10^{-12} (\Delta K)^3$ m/cycle.

15.20 An aluminium alloy panel is subjected to a constant amplitude repeated stress cycle in which the maximum stress value is $\sigma_{max} = 175$ MPa. If the panel contains a central crack 1.5 mm long, find the number of cycles to failure given that the critical crack length is 35 mm and the growth rate is

$da/dN = 15.2 \times 10^{-4} (\Delta K)^{2.77}$ mm/cycle

where $\Delta K = 11.1\ \sigma_{max} \sqrt{a}$.

APPENDIX I

PROPERTIES OF AREAS

Throughout this book a number of different properties of plane areas appear. As well as the area A, are the co-ordinates $(\bar{X}, \bar{Y})$ of its centroid, its first moments of area i_X and i_Y, referred to co-ordinate axes X and Y, and the second moments of area I_X, I_Y and I_{XY}. The origin of the reference axes may be taken anywhere but centroidal axes x and y have their origin at the centroid. The following definitions and theorems apply to these properties.

I.1 Centroid and Moments of Area

Moments of area are required for the calculation of internal stress when a plane area within a beam, a strut, a torsion bar etc, are to resist the externally applied loading. For a beam, the centroid locates the position of an unstressed neutral axis. Consider an element of area, $\mathrm{d}A = \mathrm{d}X \times \mathrm{d}Y$, in the section with respect to the general cartesian co-ordinates X, Y, for any arbitrary origin O (see Fig. I.1).

First Moments of Area:

$$i_X = \int_A Y\,\mathrm{d}A, \quad i_Y = \int_A X\,\mathrm{d}A,$$

Centroid Co-ordinates:

$$\bar{X} = i_Y / A, \quad \bar{Y} = i_X / A$$

Second Moments of Area:

$$I_X = \int_A Y^2\,\mathrm{d}A, \quad I_Y = \int_A X^2\,\mathrm{d}A, \quad I_{XY} = \int_A XY\,\mathrm{d}A \qquad \text{(I.1a,b,c)}$$

I.2 Parallel and Perpendicular Axes

When the axes X and Y originate from the centroid g, i.e. O is co-incident with g, then $\bar{X} = \bar{Y} = 0$ and $i_X = i_Y = 0$. When O does not co-incide with g, I_X may be transferred between the axes X,Y and the centroidal axes x, y (Fig. I.1) using the *parallel axis theorem*. From eq(I.1a)

$$I_X = \int_A Y^2\,\mathrm{d}A = \int_A (y + \bar{Y})^2\,\mathrm{d}A = \int_A (y^2 + 2\,y\bar{Y} + \bar{Y}^2)\,\mathrm{d}A$$

Substituting second moments referred to the centroidal axes,

$$I_x = \int_A y^2\,\mathrm{d}A, \; I_y = \int_A x^2\,\mathrm{d}A, \; I_{xy} = \int_A x\,y\,\mathrm{d}A \qquad \text{(I.2a,b,c)}$$

$$i_x = \int_A y\,\mathrm{d}A = 0, \;\; i_x = \int_A y\,\mathrm{d}A = 0$$

leads to the parallel axis theorem:

$$I_X = I_x + A\bar{Y}^2 \tag{1.3a}$$

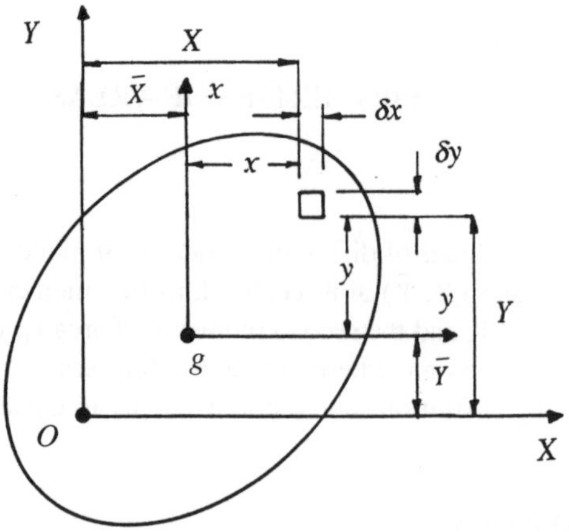

Figure I.1 Parallel axes

Similarly it can be shown from eqs(I.1b,c) and (I.2b,c) that

$$I_Y = I_y + A\bar{X}^2 \text{ and } I_{XY} = I_{xy} + A\,\bar{X}\,\bar{Y} \tag{1.3b,c}$$

in which the signs of $\bar{X}$ and $\bar{Y}$ are determined from the positive x, y co-ordinate directions defining the first quadrant. The *perpendicular axis theorem* enables the polar second moment of area I_z about a centroidal axis z, to be found from I_x and I_y.

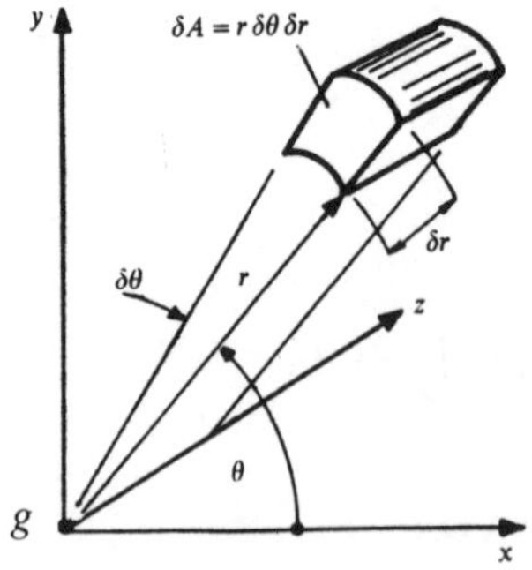

Figure I.2 Perpendicular axes

Representing dA in polar co-ordinates (Fig. I.2) it follows that

$$I_z = \int_A r^2\,dA = \int_A (x^2 + y^2)\,dA \tag{1.4a}$$

and with similar substitutions from eqs(I.2a,b) for I_x and I_y

$$I_z = I_x + I_y \tag{1.4b}$$

where, in this theorem, the subscripts x, y and z denote centroidal values.

I.3 Principal Second Moments of Area

The centroidal axes x and y are called principal axes for the section if $I_{xy} = 0$. In general this will apply when x and y are axes of symmetry for the plane area. When, for the centroidal axes, $I_{xy} \neq 0$ it is necessary to identify principal second moments of area, I_u and I_v, with a new set of orthogonal symmetry axes u and v in Fig. I.3. To find the magnitude of I_u and I_v and the inclination θ_P of principal axis u relative to x, a transformation is required. Let another pair of axes x', y' be inclined at θ as shown.

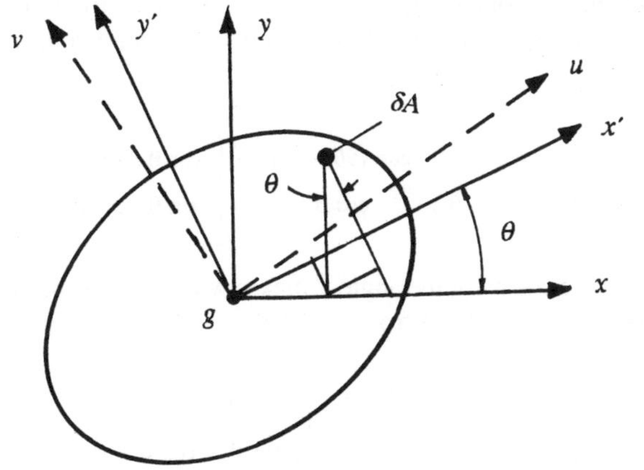

Figure I.3 Principal axes

This gives

$$I_{x'} = \int_A y'^2 \, dA, \quad I_{y'} = \int_A x'^2 \, dA, \quad I_{x'y'} = \int_A x' \, y' \, dA = 0 \qquad \text{(I.5a,b,c)}$$

where from the geometry given in Fig. I.3

$$x' = x \cos\theta + y \sin\theta \qquad \text{(I.6a)}$$
$$y' = y \cos\theta - x \sin\theta \qquad \text{(I.6b)}$$

Substituting eqs(I.6a,b) into eq(I.5a-c) leads to

$$I_{x'} = I_x \cos^2\theta + I_y \sin^2\theta - I_{xy} \sin 2\theta \qquad \text{(I.7a)}$$
$$I_{y'} = I_y \cos^2\theta + I_x \sin^2\theta + I_{xy} \sin 2\theta \qquad \text{(I.7b)}$$
$$I_{x'y'} = ½\,(I_x - I_y) \sin 2\theta + I_{xy} \cos 2\theta = 0 \qquad \text{(I.7c)}$$

When the axes x', y' co-incide with u and v it follows that $I_{x'y'} = 0$ in eq(I.7c). This gives the inclination θ_P of u, with respect to x,

$$\tan 2\theta_P = 2I_{xy} \, / \, (I_y - I_x) \qquad \text{(I.8)}$$

Now θ_P may be eliminated between eqs(I.7a,b) and (I.8) to give the principal values I_u, I_v in terms of I_x, I_y and I_{xy}

$$I_{u,\,v} = ½\,(I_x + I_y) \pm ½\,\sqrt{[(I_x - I_y)^2 + 4I_{xy}^{\,2}]} \qquad \text{(I.9)}$$

noting that: $I_u + I_v = I_x + I_y$

I.4 Matrix Representation

Equations (I.7a-c) are identical in form to the transformation equations for stress and strain (see Table 1.1). It follows that second moments of area I_x, I_y and I_{xy} for an symmetrical matrix that will transforms from axes x, y to axes x', y' as

$$\mathbf{I}' = \mathbf{L}\,\mathbf{I}\,\mathbf{L}^{\mathrm{T}} \tag{I.10}$$

where **I** is the matrix containing I_x, I_y and I_{xy}. **I'** is the matrix of transformed I values for any inclined axes, including u and v, whose directions are defined in **L**: a 2 × 2 matrix of direction cosines

$$\mathbf{L} = \begin{bmatrix} l_{x'x} & l_{x'y} \\ l_{y'x} & l_{y'y} \end{bmatrix} = \begin{bmatrix} \cos\theta & \cos(90-\theta) \\ \cos(90+\theta) & \cos\theta \end{bmatrix} = \begin{bmatrix} \cos\theta & \sin\theta \\ -\sin\theta & \cos\theta \end{bmatrix}$$

in which $l_{x'x}$ is the cosine of the angle x' makes with x, $l_{x'y}$ is the cosine of the angle x' makes with y etc (see Fig. I.3). Then from eq(I.10)

$$\begin{bmatrix} I_{x'} & I_{x'y'} \\ I_{x'y'} & I_{y'} \end{bmatrix} = \begin{bmatrix} \cos\theta & \sin\theta \\ -\sin\theta & \cos\theta \end{bmatrix} \begin{bmatrix} I_x & I_{xy} \\ I_{xy} & I_y \end{bmatrix} \begin{bmatrix} \cos\theta & -\sin\theta \\ \sin\theta & \cos\theta \end{bmatrix}$$

for which the principal values in eq(I.9) follow from the expansion of the determinant

$$\det \begin{vmatrix} I_x - I & I_{xy} \\ I_{xy} & I_y - I \end{vmatrix} = 0$$

I.5 Graphical Solution

Geometrically, the transformation eqs(I.7a-c) will describes a circle similar to Mohr's circle for stress and strain. The circle lies in axes (I_x, I_y) and I_{xy}, with centre co-ordinates given by [½ ($I_x + I_y$), 0] and a radius of $\sqrt{[\frac{1}{4}(I_x - I_y)^2 + I_{xy}^2]}$. The principal values I_u and I_v lie on opposite ends of the horizontal diameter, where $I_{xy} = 0$, as shown in Fig. I.4. In constructing the circle the calculated value of I_x and I_{xy} are plotted as the co-ordinates of point A. The sign of I_{xy} is changed when accompanying I_y to plot point B. The circle is then drawn with AB as its diameter. Figure I.4 applies to I_{xy} positive with $I_x > I_y$. This shows that $I_u < I_v$ so that the u - axis makes an acute angle θ with the x - axis. The directions x, y, u and v re-appear at the focus point F as shown. It follows that F is found by projecting the x - direction from A or the y - direction from B. Once F is located the directions of any pair of orthogonal axes x', y', including u and v, may be projected to cut the circle again at points that supply (i) the required co-ordinates $I_{x'}$, $I_{y'}$ and $I_{x'y'}$ and (ii) the orientation of the principal axes.

The following worked examples illustrate the application of the foregoing theory in respect of some elemental plane areas and for typical cross-sections found in structures appearing within this book.

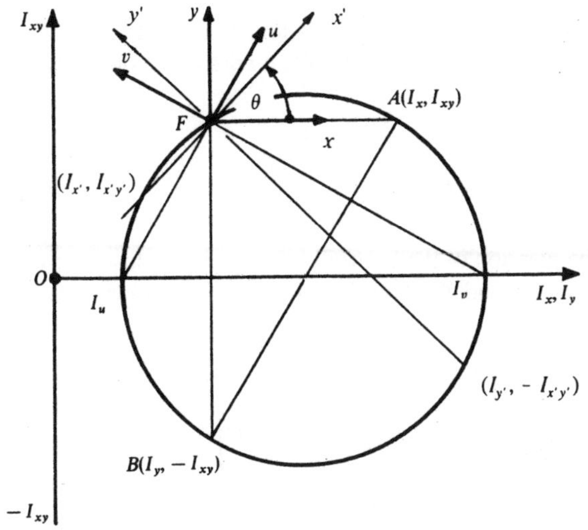

Figure I.4 Mohr's circle for principal second moments of area

Example I.1 Find I_x, I_y and I_z for the semi-circular section in Fig. I.5.

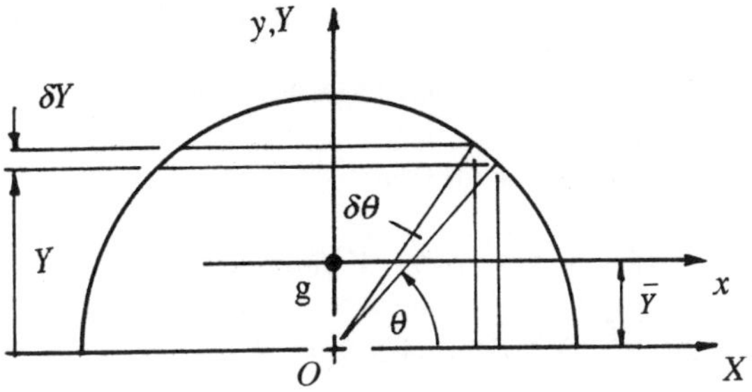

Figure I.5 Semi-circular section

Putting $y = r\sin\theta$:

$$i_X = \int_A Y\,\mathrm{d}A = \int_A Y\,(2X)\,\mathrm{d}Y = \int_0^{\pi/2} (r\sin\theta)(2r\cos\theta)(r\cos\theta\,\mathrm{d}\theta)$$
$$= 2r^3 \int_0^{\pi/2} \sin\theta\cos^2\theta\,\mathrm{d}\theta = -(2r^3/3)\left|\cos^3\theta\right|_0^{\pi/2} = 2r^3/3$$

$$\bar{Y} = i_X/A = (2r^3/3)/(\pi r^2/2) = 4r/(3\pi)$$

$$I_X = \int_A Y^2\,\mathrm{d}A = 2r^4 \int_0^{\pi/2} \sin^2\theta\cos^2\theta\,\mathrm{d}\theta$$
$$= (r^4/4)\left|\theta - \tfrac{1}{4}\sin 4\theta\right|_0^{\pi/2} = \pi r^4/8$$

$$I_x = I_X - A\bar{Y}^2 = \pi r^4/8 - (\pi r^2/2)[4r/(3\pi)]^2$$
$$= r^4[\pi/8 - 8/(9\pi)]$$

$$I_y = I_Y = \int_A X^2\,\mathrm{d}A = \int_A X^2 Y\,\mathrm{d}X = -2r^4 \int_{\pi/2}^{0} \sin^2\theta\cos^2\theta\,\mathrm{d}\theta$$
$$= (r^4/4)\left|\theta - \tfrac{1}{4}\sin 4\theta\right|_0^{\pi/2} = \pi r^4/8$$

$$I_z = I_x + I_y = r^4[\pi/4 - 8/(9\pi)]$$

Note for a solid circular section $I_x = I_y = \pi r^4/4$ and $I_z = I_x + I_y = \pi r^4/2$. The polar moment for a circular section is found independently from eq(I.4a). Noting that $\delta A = \delta r \times r\delta\theta$

$$I_z = \int_0^R \int_0^{2\pi} r^2\,(\mathrm{d}r \times r\,\mathrm{d}\theta)$$

$$= 2\pi \int_0^R r^3\,\mathrm{d}r = \pi R^4/2 = \pi D^4/32$$

Example I.2 Find the position of the centroid for the area enclosed between the positive X, Y axes, the parabola $Y^2 = 4aX$ and $X = b$ in Fig. I.6. Find the following second moments of area I_X, I_x, I_Y, I_y and I_{XY}.

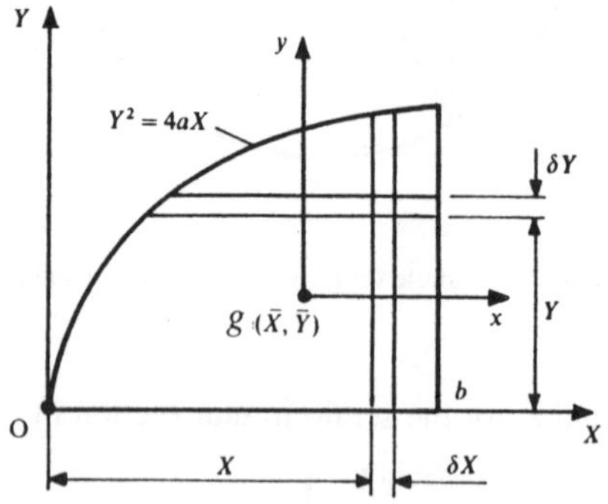

Figure I.6 Parabolic area

$$A = \int_0^b Y\,\mathrm{d}X = 2a^{1/2} \int_0^b X^{1/2}\,\mathrm{d}X = (4/3)(ab^3)^{1/2}$$

$$i_X = \int_0^{2\sqrt{(ab)}} Y\,(b - X)\,\mathrm{d}Y = \int_0^{2\sqrt{(ab)}} Y\,[b - Y^2/(4a)]\,\mathrm{d}Y = ab^2$$

$$i_Y = \int_0^b X\,(Y\,\mathrm{d}X) = 2a^{1/2} \int_0^b X^{3/2}\,\mathrm{d}X = (4/5)(ab^5)^{1/2}$$

$$\bar{X} = i_Y / A = 3b/5$$

$$\bar{Y} = i_X / A = ¾\,(ab)^{1/2}$$

$$I_X = \int_A Y^2\,\mathrm{d}A = \int_Y 4aX\,(b - Y)\,\mathrm{d}Y = \int_0^{2\sqrt{(ab)}} [Y^2 b - Y^4/(4a)]\mathrm{d}Y$$

$$= \Big|\; bY^3/3 - Y^5/(20a) \Big|_0^{2\sqrt{(ab)}} = (16/15)(a^3 b^5)^{1/2}$$

$$I_Y = \int_A X^2\,\mathrm{d}A = \int_A X^2 (Y\,\mathrm{d}X) = 2a^{1/2} \int_0^b X^{5/2}\,\mathrm{d}X$$

$$= (4/7)a^{1/2} \Big|\; X^{7/2} \Big|_0^b = (4/7)(ab^7)^{1/2}$$

$$I_{XY} = \int_X \int_Y XY\,\mathrm{d}X\,\mathrm{d}Y = \int_0^{2\sqrt{(ab)}} Y \Big(\int_{Y^2/4a}^{b} X\,\mathrm{d}X\Big)\,\mathrm{d}Y = (1/2) \int_0^{2\sqrt{(ab)}} Y\,[b^2 - Y^4/(16a^2)]\,\mathrm{d}Y$$

$$= (1/2) \Big|\; b^2 Y^2/2 - Y^6/(96a^2) \Big|_0^{2\sqrt{(ab)}} = (2/3)\,ab^3$$

$$I_x = I_X - A\,\bar{Y}^2 = (16/15)(a^3 b^5)^{1/2} - (4/3)(ab^3)^{1/2}[(3/4)(ab)^{1/2}]^2 = (19/60)(a^3 b^5)^{1/2}$$
$$I_y = I_Y - A\,\bar{X}^2 = (4/7)(ab^7)^{1/2} - (4/3)(ab^3)^{1/2}(3b/5)^2 = (16/175)(ab^7)^{1/2}$$
$$I_{xy} = I_{XY} - A\bar{X}\bar{Y} = (2/3)ab^3 - (3/5)ab^3 = ab^3/15$$

Example I.3 Find the position of the centroid and I_x, I_y, I_{xy}, I_u and I_v for a quadrant of a circle, radius r, lying in the second quadrant of X, Y in Fig. I.7.

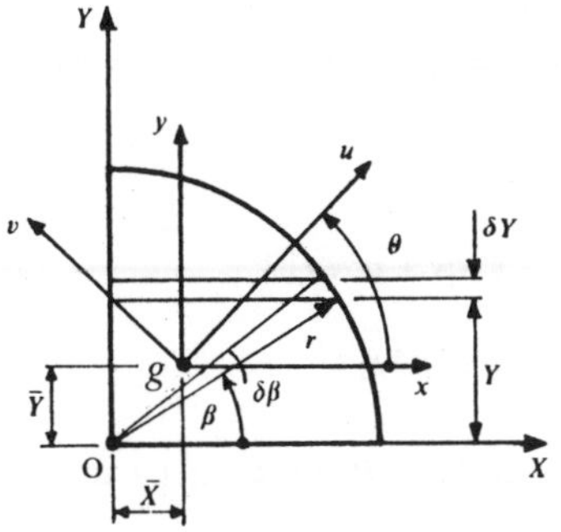

Figure I.7 Quadrant

Because of symmetry it is only necessary to find the centroid about one axis (say X)

$$i_X = \int_0^r Y X \, dY = \int_0^r Y (r^2 - Y^2)^{1/2} \, dY = r^3/3 \; (= i_Y)$$

$$\bar{Y} = i_X / A = (r^3/3)/[(\pi r^2/4) = 4r/(3\pi) \; (= \bar{X})$$

$$I_X = \int_0^r Y^2 (r^2 - Y^2)^{1/2} \, dY = (r^4/4) \int_0^{\pi/2} \sin^2 2\theta \, d\theta = \pi r^4/16 \; (= I_Y)$$

$$I_{XY} = - \int_0^r (X \, dY) YX/2 = - ½ \int_0^r (r^2 - Y^2) Y \, dY = - r^4/8$$

Then from parallel axes

$$I_x = I_X - A\bar{Y}^2 = (\pi r^4/16) - (\pi r^2/4)[4r/(3\pi)]^2 = [\pi/16 - 4/(9\pi)] r^4 \; (= I_y)$$

$$I_{xy} = I_{XY} - A\bar{X}\bar{Y} = - r^4/8 - (\pi r^2/4)[- 4r/(3\pi)][4r/(3\pi)] = - [(1/8 - 4/(9\pi)] \, r^4$$

when from eqs(1.4) and (1.5) $\theta = 45°$ and

$$I_u = r^4 [\pi/16 + 1/8 - 8/(9\pi)]$$

$$I_v = r^4 (\pi/16 - 1/8)$$

Example I.4 Find $\bar{Y}$, I_x, I_X and I_y for an isocelles triangular tubular of mean base length a, side lengths $2a$ and thickness t, with one side lying parallel to the y - direction (see Fig. I.8).

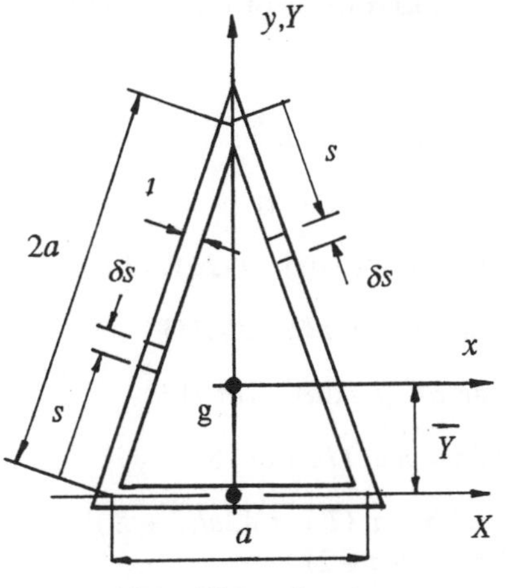

Figure I.8 Isocelles tube

Let s, be the distance along the sloping side as shown and the X-axis lie in the mid-thickness of the base. Then $dA = t\delta s$ and

$$i_X = \int_A Y\,dA = 2(\surd 15/4)\,t\int_0^{2a} s\,ds = (\surd 15/2)\,t\,\Big|\, s^2/2\,\Big|_0^{2a} = \surd 15\,ta^2$$

$$\bar{Y} = i_X/A = \surd 15ta^2/(5ta) = \surd(3/5)a$$

$$I_X = \int_A Y^2\,dA = 2\,(\surd 15/4)^2\,t\int_0^{2a} s^2\,ds + a^3t/12$$

$$(15/8)\,t\,\Big|\, s^3/3\,\Big|_0^{2a} + a^3t/12 = 5a^3t + at^3/12$$

$$I_x = I_X - A\bar{Y}^2 = 5a^3t + at^3/12 - 5at\,[\surd(3/5)a]^2 = 2a^3t + at^3/12$$

Since Y is an axis of symmetry, $\bar{X} = 0$ and

$$I_y = I_Y = \int_A X^2\,dA = ta^3/12 + 2\int_0^{2a} (s/4)^2\,(t\,ds) = 5a^3t/12$$

Example I.5 Find I_X and I_Y for a closed semi-circular tube (Fig. I.9) of mean radius a and thickness t, when the diameter coincides with the Y - axis.

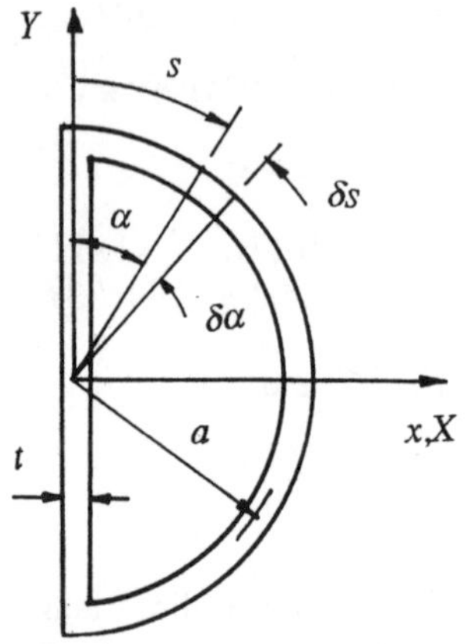

Figure I.9 Closed semi-circular tube

$$A = 2at + \pi at = at\,(2+\pi)$$

$$i_Y = \int_A X\,dA = 2\int_0^{\pi/2} (a\sin\alpha)(t\,a\,d\alpha) = 2a^2t$$

$$\bar{X} = i_Y/A = 2a^2t/[at\,(2+\pi)] = 2a/(2+\pi)$$

$$I_x = I_X = \int_A Y^2\,dA = 2\int (a\cos\alpha)^2(t\,ds) + t\,(2a)^3/12$$

$$= 2\int_0^{\pi/2} (a\cos\alpha)^2\,(t\,a\,d\alpha) = \pi a^3t/2 + 2\,ta^3/3$$

$$I_Y = \int_A X^2\,dA = 2\int_0^{\pi/2} (a\sin\alpha)^2\,(t\,a\,d\alpha) + 2at^3/12$$

$$= a^3t\int_0^{\pi/2} (1-\cos 2\alpha)\,d\alpha = \pi a^3t/2 + at^3/6$$

$$I_y = I_Y - A\bar{X}^2 = \pi a^3t/2 + at^3/6 - at\,(2+\pi)[2a/(2+\pi)]^2$$
$$= at^3/6 + (a^3t/2)(\pi-2)(\pi+4)/(\pi+2)$$

Example I.6 Find I_X for the single cell structure shown in Fig. I.10.

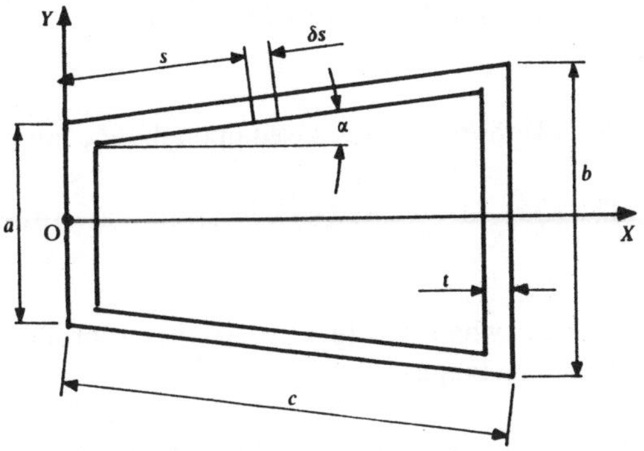

Figure I.10 Closed single cell

Taking the vertical webs and sloping sides separately

$$I_X = (t\,b^3/12 + t\,a^3/12)_{\text{webs}} + \int_{\text{sides}} Y^2\,\mathrm{dA}$$

where for the sides:

$$\int_{\text{sides}} Y^2\,\mathrm{d}A = 2\int_0^c (a/2 + s\sin\alpha)^2 (t\,\mathrm{d}s)$$

$$= t\,c\,[a^2/2 + ac\sin\alpha + (2c^2/3)\sin^2\alpha]$$

$$\therefore I_X = t\,[\,b^3/12 + a^3/12 + a^2c/2 + ac^2\sin\alpha + (2c^2/3)\sin^2\alpha\,]$$

Example I.7 Determine for the equal angle section given in Fig. I.11a the position of the centroid $(\bar{X}, \bar{Y})$ and I_x, I_y, I_{xy} for the centroidal axes x, y. Find analytically I_u and I_v and the inclination θ, of the principal axes u and v, with resect to x and y.

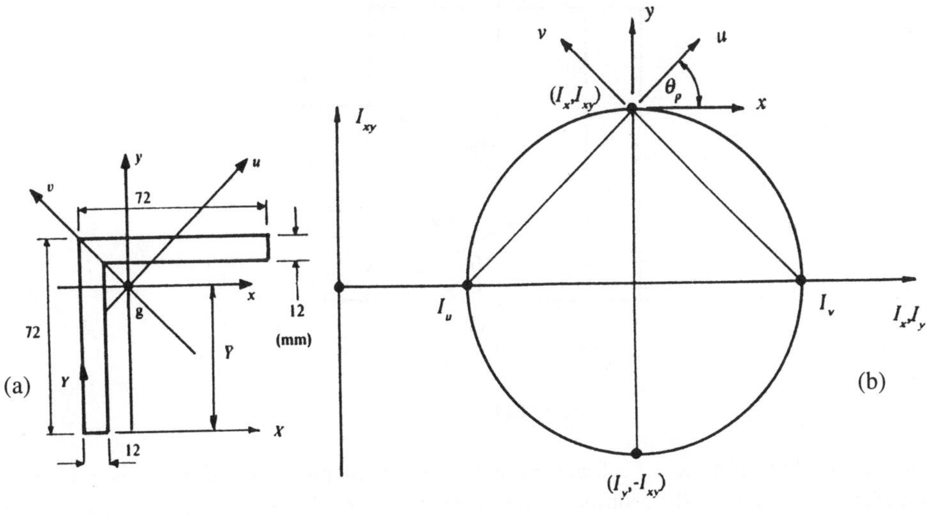

Figure I.11 Principal I - values for an equal angle section

First moments about the X - axis in Fig. I.11a gives the co-ordinates of the centroid $g(\bar{X}, \bar{Y})$:

$(72 \times 12 \times 36) + (60 \times 12 \times 66) = [(72 \times 12) + (60 \times 12)]\bar{Y}$
$\therefore\ \bar{Y} = \bar{X} = 49.64$ mm

Using the form $I = \sum(bd^3/3)$ for rectangles based upon these centroidal axes

$I_x = 72(22.36)^3/3 - 60(10.36)^3/3 + 12(49.64)^3/3$
$= 734.6 \times 10^3\ \text{mm}^4 = I_y$

Using the form $I_{xy} = \sum A\,\bar{x}\,\bar{y}$ where where $\bar{x}, \bar{y}$ are the co-ordinates between G and the individual centroids of each rectangle.

$I_{xy} = (72 \times 12)(+13.6)(+16.4) + (60 \times 12)(-19.6)(-16.4)$
$= 424.1 \times 10^3$ mm

Analytically, from eqs(I.8) and (I.9):

$\tan 2\theta = 2(424.1)/(734.6 - 734.6) = \infty$
$\therefore\ \theta = 45°$

$I_u, I_v = ½(734.6 + 734.6)\,10^3 \pm ½\sqrt{[(734.6 - 734.)^2 + 4(424.1)^2]} \times 10^3$
$\therefore\ I_u = 310.5\ \times 10^3\ \text{mm}^4$ and $I_v = 1158.7 \times 10^3\ \text{mm}^4$

The corresponding graphical solution is given in Fig. I.11b. Here $I_x = I_y$ and I_{xy} is positive, which gives $I_u < I_v$.

Example I.8 Determine analytically and graphically I_u, I_v and θ for the complex section given in Fig. I.12.

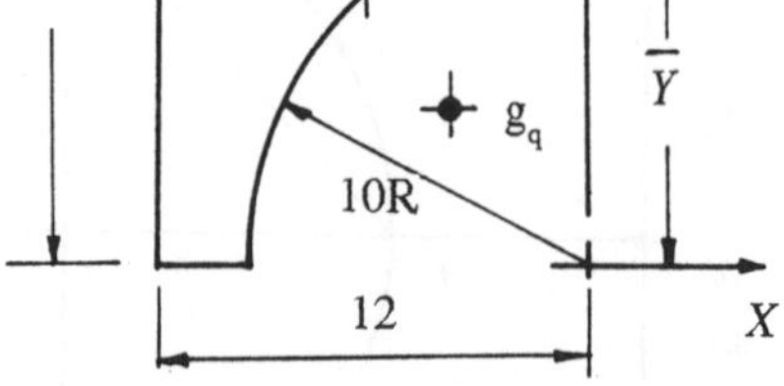

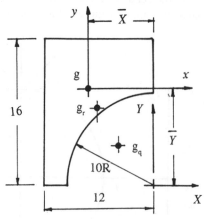

Figure I.12 Complex section

The centroidal position $(\bar{X}, \bar{Y})$ is found by taking first moments of area about axes X, Y. Note that example I.3 showed the centroid of a quadrant to be $4r/(3\pi)$ from its centre. This gives:

$(16 \times 12)6 - (\pi/4)(10)^2(4 \times 10)/(3\pi) = [(12 \times 16) - (\pi/4)(10)^2]\bar{X}$
$\therefore\ \bar{X} = 7.22$ mm

$(12 \times 16)8 + (\pi/4)(10)^2(4 \times 10)/(3\pi) = [(12 \times 16) - (\pi/4)(10)^2]\bar{Y}$
$\therefore\ \bar{Y} = 10.6$ mm

To find I_x and I_x, the parallel axis theorems (I.3a,b), transfer second moments between individual centroidal axes for the rectangle and the quadrant and the axes x, y:

$I_x = [12(10.6)^3/3 + 12(16 - 10.6)^3/3]_{\text{rect}} - [\pi(10)^4/16$
$- (\pi/4)(10)^2(4.25)^2 + (\pi/4)(10)^2(10.6 - 4.25)^2]_{\text{quad}} = 1676\ \text{mm}^4$

$I_y = [16\,(7.22)^3/3 + 16(12 - 7.22)^3/3]_{\text{rect}} - [\pi(10)^4/16$
$- (\pi/4)(10)^2(4.25)^2 + (\pi/4)(10)^2(7.22 - 4.25)^2]_{\text{quad}} = 1357\ \text{mm}^4$

To find I_{xy} for the rectangle, note that for its own centroid g_r the product moment is zero:

$I_{xy} = A\,\bar{x}_r\,\bar{y}_r = (12 \times 16)[-(10.6 - 8)](7.22 - 6) = -609.02\ \text{mm}^4$

where $\bar{x}_r, \bar{y}_r$ are the co-ordinates between g and g_r. For the quadrant, first refer $I_{XY} = -r^4/4$ (see Example I.3) to its own centroid using eq(I.3c):

$-r^4/4 = I_{g_q} + (\pi r^2/4)[-(4r/3\pi)](4r/3\pi)$

$I_{g_q} = r^4[4/(9\pi) - 1/8] = 164.71\ \text{mm}^4$

Next, transfer I_{g_q} to the centroid of the whole section, again using eq(I.3c)

$I_{xy} = 164.71 + (\pi/4)(10)^2[-(7.22 - 4.25)](10.6 - 4.25) = -1316.51\ \text{mm}^4$
$\sum I_{xy} = -609.02 - (-1316.51) = 707\ 49\ \text{mm}^4$

Finally, substituting I_x, I_y and I_{xy} into eqs(I.8) and (I.9) gives:

$I_{u,v} = ½(1676 + 1357) \pm ½\sqrt{[1676 - 1357)^2 + 4(707.49)^2]}$
$I_u = 791.25\ \text{mm}^4$ and $I_v = 2241.8\ \text{mm}^4$

$\tan 2\theta_P = (2 \times 707.49) / (1357 - 1676)$
$\theta_P = 51.35°$

Figure I.4 provides the graphical construction consistent with I_{xy} being positive and $I_x > I_y$. This shows that I_u is the lesser of the two principal inertias when axis u in inclined to x with positive θ_P.

EXERCISES

I.1 Show that for a rectangle, breadth b and depth d, that the second moments of area about the centroidal axes are $I_x = bd^3/12$, $I_y = db^3/12$ and $I_{xy} = 0$. When one corner lies at the origin of positive X,Y show that $I_X = bd^3/3$, $I_Y = db^3/3$ and $I_{XY} = b^2d^2/4$.

I.2 Determine the principal second moments of area I_u and I_v and their inclinations relative to the X-axis for the parabola in Fig. I.6.

I.3 Find $\bar{Y}$, I_x, I_X and I_Y for the isocelles triangular section, base b and height a in Fig. I.13.
Answer: $\bar{Y} = a/3$, $I_x = a^3b/36$, $I_X = a^3b/12$, $I_Y = b^3a/48$.

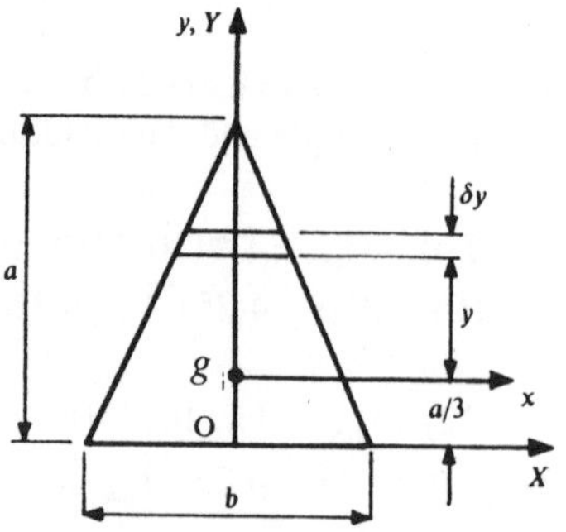

Figure I.13

I.4 Find I_x, I_y and I_z for an elliptical section, with major axis $2a$ and minor axis $2b$ in Fig. I.14.
Answer: $I_x = \pi ab^3/4$, $I_y = \pi ba^3/4$, $I_z = (\pi ab/4)(a^2 + b^2)$.

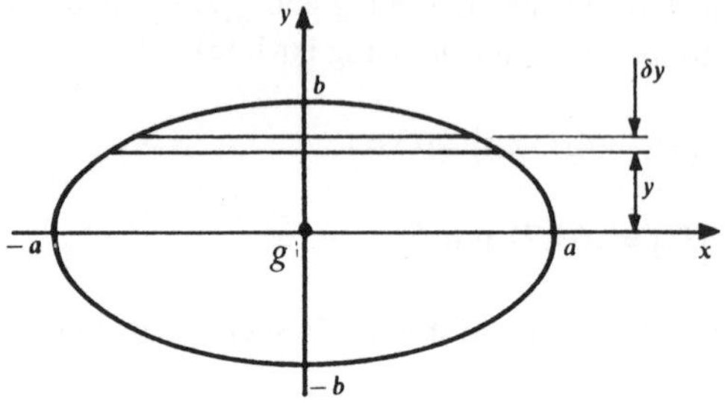

Figure I.14

I.5 Find I_X, I_Y, I_{XY}, I_x, I_y, I_{xy}, I_u and I_v and θ_P for a right triangle, base b, height a in Fig. I.15.
Answer: $I_X = a^3b/12$, $I_Y = ab^3/12$, $I_{XY} = a^2b^2/24$, $I_x = a^3b/36$, $I_y = ab^3/36$, $I_{xy} = -a^2b^2/72$,
$I_{u,v} = (ab/72)[(a^2 + b^2) \pm \sqrt{(a^4 + b^4 - a^2b^2)}$, $\theta_P = ½ \tan^{-1}[-ab/(b^2 - a^2)]$

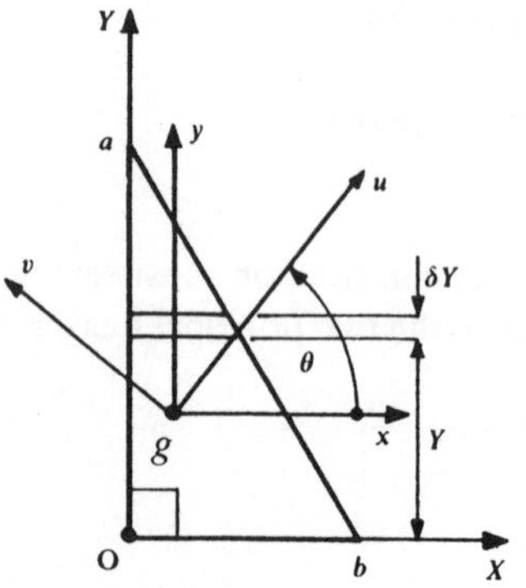

Figure I.15

I.6 Examine the effect on I_{XY}, I_{xy}, I_u, I_v and θ when the areas in Figs I.6 and I.7 are rotated about the Y-axis into their adjacent quadrants X, Y.

I.7 Show that for an equilateral triangular tube of side length a and thickness t, in Fig. I.16, the centroid co-ordinate $\bar{X} = a/\surd 3$, $I_x = a^3t/4$. Also find I_y, I_Y and I_z.

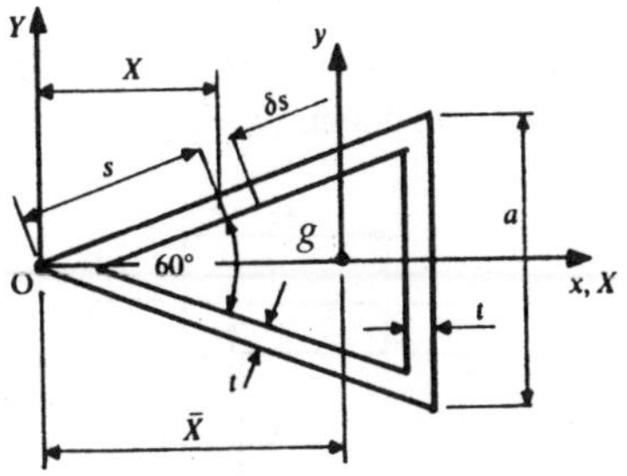

Figure I.16

I.8 Show that the horizontal position of the centroid for the open-tube in Fig. I.17 is given by $\bar{X} = 2a/\pi$. Find I_x, I_y and I_Y.

Answer: $I_x = a^3\,t/4$, $I_y = a^3\,t\,(\pi/2 - 4/\pi)$, $I_Y = \pi\,a^3\,t/2$

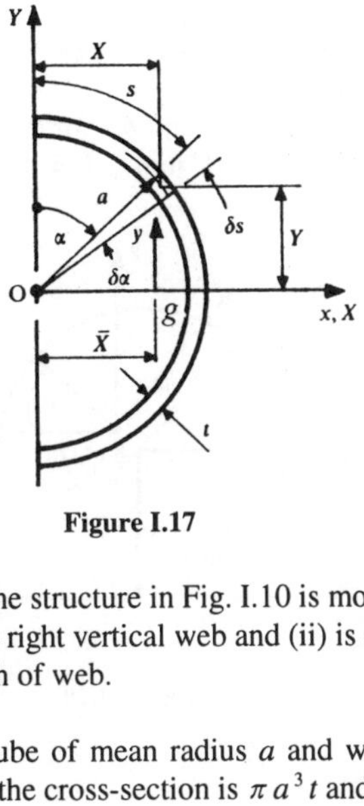

Figure I.17

I.9 Find I_x for the centroidal axis when the structure in Fig. I.10 is modified with a semi-circular tube of the same thickness that (i) replaces the right vertical web and (ii) is added to this web to form a two cell tube. The diameter equals the length of web.

I.10 Show for a thin-walled circular tube of mean radius a and wall thickness t, that the second moments of area about any diameter in the cross-section is $\pi\,a^3\,t$ and about its axis is $2\pi\,a^3\,t$.

I.11 Find the position of the centroid and the I_x value for each of the unsymmetrical sections shown in Figs I.18a-c. (All dimensions in mm)

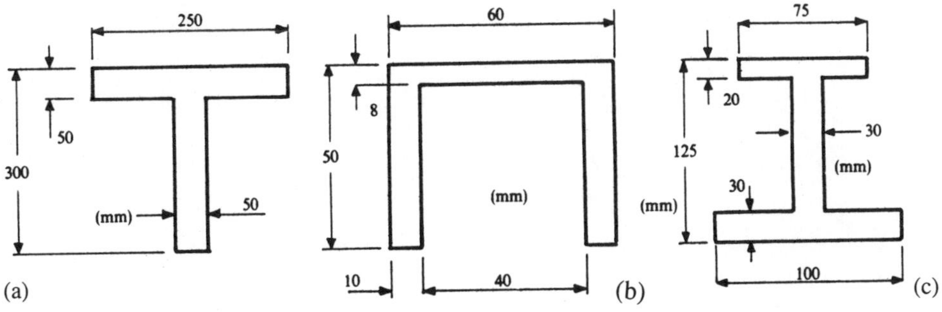

Figure I.18

I.12 Determine for the channel section shown in Fig. I.19 the position of the centroid and I_x, I_y and I_{xy}. Find by both analytical and graphical methods the principal second moments of area and the inclination of their axes with respect to x and y.

Answer: $\bar{X} = 9.45$ mm, $\bar{Y} = 27.22$ mm, $I_x = 36.31 \times 10^4$ mm^4, $I_y = 4.97 \times 10^4$ mm^4, $I_{xy} = 3.89 \times 10^4$ mm^4, $I_u = 4.47 \times 10^4$ mm^4, $I_v = 36.87 \times 10^4$ mm^4, $\theta_P = 83°$

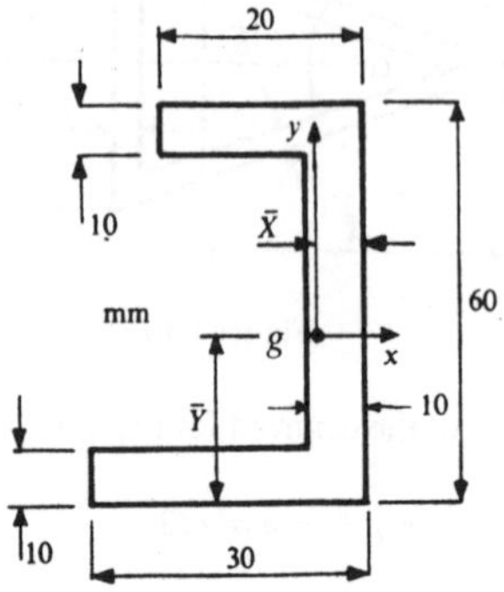

Figure I.19

I.13 Find analytically and graphically I_u, I_v and θ_P for the complex section shown in Fig. I.20.
Answer: $I_u = 15.3 \times 10^4$ mm^4, $I_v = 6.1 \times 10^4$ mm^4, $\theta_P = 41°$

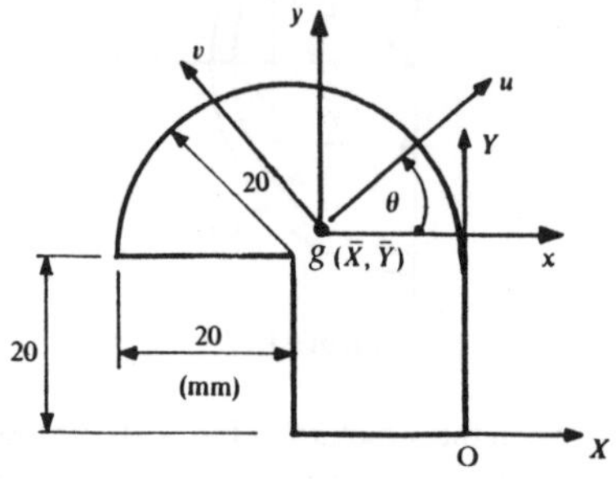

Figure I.20

APPENDIX II

MATRIX ALGEBRA

II.1 The Formation of a Matrix

It is convenient to represent n simultaneous equations in n unknowns with an abbreviated matrix notation

$$\sum a_{ij} y_j = x_i \tag{II.1}$$

where the summation applies for $i, j = 1, 2, 3 \ldots n$. For example, when $n = 5$ eq(II.1) represents the five equations:

$$\begin{aligned}
a_{11} y_1 + a_{12} y_2 + a_{13} y_3 + a_{14} y_4 + a_{15} y_5 &= x_1 \\
a_{21} y_1 + a_{22} y_2 + a_{23} y_3 + a_{24} y_4 + a_{25} y_5 &= x_2 \\
a_{31} y_1 + a_{32} y_2 + a_{33} y_3 + a_{34} y_4 + a_{35} y_5 &= x_3 \\
a_{41} y_1 + a_{42} y_2 + a_{43} y_3 + a_{44} y_4 + a_{45} y_5 &= x_4 \\
a_{51} y_1 + a_{52} y_2 + a_{53} y_3 + a_{54} y_4 + a_{55} y_5 &= x_5
\end{aligned} \tag{II.2a-d}$$

Often we need to solve these for the dependent variable y_j, knowing the coefficients a_{ij} and the independent variable x_i. The problem may be re-expressed by writing the coefficients within a square matrix $\mathbf{A} = a_{ij}$ and the variables as column matrices $\mathbf{x} = x_i$ and $\mathbf{y} = y_j$ so that

$$\mathbf{A}\,\mathbf{y} = \mathbf{x}$$

which, in full, becomes

$$\begin{bmatrix} a_{11} & a_{12} & a_{13} & a_{14} & a_{15} \\ a_{21} & a_{22} & a_{23} & a_{24} & a_{25} \\ a_{31} & a_{32} & a_{33} & a_{34} & a_{35} \\ a_{41} & a_{42} & a_{43} & a_{44} & a_{45} \\ a_{51} & a_{52} & a_{53} & a_{54} & a_{55} \end{bmatrix} \begin{bmatrix} y_1 \\ y_2 \\ y_3 \\ y_4 \\ y_5 \end{bmatrix} = \begin{bmatrix} x_1 \\ x_2 \\ x_3 \\ x_4 \\ x_5 \end{bmatrix} \tag{II.3}$$

II.2 Matrix Addition and Subtraction

Let **A**, **B**, **C**, **D** and **E** be matrices of similar dimension ($n \times m$) and p and q be scalars. Taking a_{jk}, b_{jk}, c_{jk}, d_{jk} and e_{jk} to be an element at the j th row and the k th column we may add and subtract the elements of a matrix as follows:

$c_{jk} = a_{jk} + b_{jk}$, $\mathbf{C} = \mathbf{A} + \mathbf{B}$

$d_{jk} = a_{jk} - b_{jk}$, $\mathbf{D} = \mathbf{A} - \mathbf{B}$

where

$$\mathbf{A} \pm \mathbf{B} = \begin{bmatrix} a_{11} \pm b_{11} & a_{12} \pm b_{12} & \cdots & \cdots & a_{15} \pm b_{15} \\ a_{21} \pm b_{21} & a_{22} \pm b_{22} & \cdots & \cdots & a_{25} \pm b_{25} \\ \vdots & \vdots & \cdots & \cdots & \vdots \\ \vdots & \vdots & \cdots & \cdots & \vdots \\ a_{51} \pm b_{51} & a_{52} \pm b_{52} & \cdots & \cdots & a_{55} \pm b_{55} \end{bmatrix}$$

II.3 Matrix Multiplication

If we multiply the elements of a matrix **A** by a scalar p then every elemet in **A** is multiplied by p. The result is a new matrix $\mathbf{E} = p\mathbf{A}$ whose elements are $e_{jk} = p\, a_{jk}$.

Two matrices may be multiplied. By multiplying the rows into columns in eq(II.3) we recover the five eqs(II.2a-d). In general, if **A** is an $m \times n$ matrix and **B** is an $n \times r$ matrix then the elements of each row in **A** may be multiplied by the elements of each column in **B** to give a matrix **C** with dimension $m \times r$. The multiplication involves the summation of the product of elements a_{jk} and b_{jk} in each row and column as follows:

$$\begin{bmatrix} a_{j1} & a_{j2} & a_{j3} & \cdots & a_{jn} \end{bmatrix} \begin{bmatrix} b_{1k} \\ b_{2k} \\ b_{3k} \\ \vdots \\ b_{nk} \end{bmatrix}$$

so that the element c_{jk} of **C** becomes

$$c_{jk} = a_{j1} b_{1k} + a_{j2} b_{2k} + a_{j3} b_{3k} + \ldots a_{jn} b_{nk} \tag{II.4}$$

Intrducing dummy subscript q, eq(II.4) may be contracted within the summation convention: $c_{jk} = a_{jq}\, b_{qk}$ where $q = 1, 2 \ldots n$. Multiplication of **A** and **B** is only possible if the number of columns in matrix **A** equals the number of rows in matrix **B**. Then **A** and **B** are conformable.

The multiplication of matrices is *associative* and *distributive*. This means

$(\mathbf{AB})\mathbf{C} = \mathbf{A}(\mathbf{BC}) = \mathbf{ABC}$

$(\mathbf{A} + \mathbf{B})\mathbf{C} = \mathbf{AC} + \mathbf{BC}$

$\mathbf{C}(\mathbf{A} + \mathbf{B}) = \mathbf{CA} + \mathbf{CB}$

$(\mathbf{A} + \mathbf{B})(\mathbf{C} + \mathbf{D}) = \mathbf{AC} + \mathbf{AD} + \mathbf{BC} + \mathbf{BD}$

However, multiplication is not *commutative*, i.e. $\mathbf{AB} \neq \mathbf{BA}$, except for the case where $\mathbf{B} = \mathbf{A}^{-1}$ when $\mathbf{AA}^{-1} = \mathbf{A}^{-1}\mathbf{A} = \mathbf{I}$, where **I** is a *unit matrix*.

II.4 Transpose of a Matrix

The transpose $\mathbf{A}^T$ of a matrix **A** is formed from writing the rows of **A** as columns, i.e the first row becomes the first column, the second row becomes the second column, and so on. Transposing matrix **A** in eq(II.3) gives the components as

$$\mathbf{A}^T = \begin{bmatrix} a_{11} & a_{21} & a_{31} & a_{41} & a_{51} \\ a_{12} & a_{22} & a_{32} & a_{42} & a_{52} \\ a_{13} & a_{23} & a_{33} & a_{43} & a_{53} \\ a_{14} & a_{24} & a_{34} & a_{44} & a_{54} \\ a_{15} & a_{25} & a_{35} & a_{45} & a_{55} \end{bmatrix}$$

II.5 The Inverse of a Matrix

We shall see later that by inverting the matrix **A**, the solution to y_j is found from $\mathbf{y} = \mathbf{A}^{-1}\,\mathbf{x}$. The inverse matrix $\mathbf{A}^{-1}$ only exists if its *determinant* is non zero. To find the inverse $\mathbf{A}^{-1}$ of a square $n \times n$ matrix **A** we must find the co-factor corresponding to each element a_{ik}:

$$c_{ik} = (-1)^{i+k}\, m_{ik} \tag{II.5}$$

in which m_{ik} is the minor for the element a_{ik}. The inverse becomes

$$\mathbf{A}^{-1} = (\det \mathbf{A})^{-1} \times (c_{ik})^T \tag{II.6a}$$

where

$$\det \mathbf{A} = a_{i1}\, c_{i1} + a_{i2}\, c_{i2} \ldots\ldots + a_{in}\, c_{in}\,. \tag{II.6b}$$

Example II.1 Find the inverse of the following matrix **A**:

$$\mathbf{A} = \begin{bmatrix} -3 & 6 & -11 \\ 3 & -4 & 6 \\ 4 & -8 & 13 \end{bmatrix}$$

The co-factors are found from eq(II.5). For example

$$c_{11} = (-1)^2 m_{11} = \begin{vmatrix} -4 & 6 \\ -8 & 13 \end{vmatrix} = -4$$

$$c_{12} = (-1)^3 m_{12} = \begin{vmatrix} 3 & 6 \\ 4 & 13 \end{vmatrix} = -15$$

$$c_{13} = (-1)^4 m_{13} = \begin{vmatrix} 3 & -4 \\ 4 & -8 \end{vmatrix} = -8$$

$$c_{21} = (-1)^3 m_{21} = \begin{vmatrix} -6 & -11 \\ -8 & 13 \end{vmatrix} = -10$$

The determinant, is from eq(II.6b)

$$\det \mathbf{A} = a_{11} c_{11} + a_{12} c_{12} + a_{13} c_{13}$$
$$= (-3)(-4) + (6)(-15) + (-11)(-8) = 10$$

Equation (II.6a) then gives

$$\mathbf{A}^{-1} = \frac{1}{10}\begin{bmatrix} -4 & 10 & -8 \\ -15 & 5 & -15 \\ -8 & 0 & -6 \end{bmatrix}$$

and to check the result use $\mathbf{A}\,\mathbf{A}^{-1} = \mathbf{I}$

$$\frac{1}{10}\begin{bmatrix} -3 & 6 & -11 \\ 3 & -4 & 6 \\ 4 & -8 & 13 \end{bmatrix}\begin{bmatrix} -4 & 10 & -8 \\ -15 & 5 & -15 \\ -8 & 0 & -6 \end{bmatrix} = \begin{bmatrix} 1 & 0 & 0 \\ 0 & 1 & 0 \\ 0 & 0 & 1 \end{bmatrix}$$

II.6 Matrix Types

Square matrix: one having an equal number of rows an columns

Diagonal Matrix: Is a square matrix whose elements above and below the principal diagonal are zero.

Unit Matrix (**I**): Is a diagonal matrix whose diagonal elements are all unity. That is, $a_{jk} = 0$ for $j \neq k$ and $a_{jk} = 1$ for $j = k$. This has the property: **IA** = **A**, **IB** = **B** etc

Zero Matrix: A matrix with all elements equal to zero. It is formed by subtracting a matrix from itself: **O** = **A** − **A**.

Scalar Matrix: is a diagonal matrix whose principal elements are all equal. This is formed from **I** as p **I**.

Orthogonal Matrix: Is a square matrix that obeys $\mathbf{A}^{-1} = \mathbf{A}^{T}$ i.e. $(a_{jk})^{-1} = a_{kj}$

Symmetric Matrix: Is a square matrix that obeys $\mathbf{R}^{T} = \mathbf{R}$, i.e. $r_{kj} = r_{jk}$.

Skew-Symmetric Matrix: Is a square matrix obeying $\mathbf{S}^{T} = -\mathbf{S}$, i.e. $s_{kj} = -s_{jk}$

Any square matrix **A** may be expressed as a sum:

$\mathbf{A} = \mathbf{R} + \mathbf{S}$

where **R** and **S** are respectively symmetric and skew-symmetric matrices derived from **A** as:

$\mathbf{R} = \frac{1}{2}(\mathbf{A} + \mathbf{A}^T)$

$\mathbf{S} = \frac{1}{2}(\mathbf{A} - \mathbf{A}^T)$

Upper Triangular Matrix: Is a square matrix with zero elements below the principal diagonal.

Lower Triangular Matrix: Is a square matrix with zero elements above the principal diagonal.

II.7 Matrix Operations

Matrix algebra will conform to

$$\mathbf{A} + \mathbf{B} = \mathbf{B} + \mathbf{A}$$
$$\mathbf{A} + (\mathbf{B} + \mathbf{C}) = (\mathbf{A} + \mathbf{B}) + \mathbf{C}$$
$$p\,(\mathbf{A} + \mathbf{B}) = p\,\mathbf{A} + p\,\mathbf{B}$$
$$p\,(q\mathbf{A}) = (pq)\,\mathbf{A}$$
$$(p\,\mathbf{A})^T = p\mathbf{A}^T$$
$$(\mathbf{A}^T)^T = \mathbf{A}$$
$$(\mathbf{A} + \mathbf{B})^T = \mathbf{A}^T + \mathbf{B}^T$$
$$(\mathbf{AB})^T = \mathbf{B}^T\mathbf{A}^T$$

EXERCISES

II.1 Given $\mathbf{p}^T = [p_1\ p_2\ p_3]$ and $\mathbf{q}^T = [q_1\ q_2\ q_3]$ find $\mathbf{pp}^T$, $\mathbf{q}^T\mathbf{q}$, $\mathbf{pq}^T$ and $\mathbf{p}^T\mathbf{q}$.

II.2 Calculate $\mathbf{F}^2 = \mathbf{FF}$, **FG**, **GF**, **FH**, **HF**, $\mathbf{FH}^T$, $\mathbf{H}^T\mathbf{F}$, **GH**, **HG**, $\mathbf{GH}^T$, $\mathbf{H}^T\mathbf{G}$, $\mathbf{H}^T\mathbf{H}$ and $\mathbf{HH}^T$ given that:

$$\mathbf{F} = \begin{bmatrix} 2 & 1 \\ 1 & 2 \end{bmatrix}, \quad \mathbf{G} = \begin{bmatrix} 3 & 2 \\ 2 & 3 \end{bmatrix}, \quad \mathbf{H} = \begin{bmatrix} 4 & 3 \\ 2 & 1 \end{bmatrix}$$

Would you expect symmetric matrices to be commutative in general? Would you expect the products $\mathbf{HH}^T$ and $\mathbf{H}^T\mathbf{H}$ to be (i) equal? (ii) symmetric?

II.3 Show that the matrices given in Exercise II.2 obey: (i) $(\mathbf{FG})\mathbf{H} = \mathbf{F}(\mathbf{GH})$, (ii) $\mathbf{F}(\mathbf{G} + \mathbf{H}) = \mathbf{FG} + \mathbf{FH}$ and (iii) $(\mathbf{F} + \mathbf{G})(\mathbf{G} + \mathbf{H}) = \mathbf{FG} + \mathbf{FH} + \mathbf{G}^2 + \mathbf{GH}$.

II.4 Calculate **FA**, **HA**, $\mathbf{A}^T\mathbf{F}$, $\mathbf{A}^T\mathbf{H}$, $\mathbf{AA}^T$ and $\mathbf{A}^T\mathbf{A}$ for matrices **F**, and **H** in Exercise II.2 and:

$$\mathbf{A} = \begin{bmatrix} 1 & 2 & 3 \\ 4 & 5 & 6 \end{bmatrix}$$

Does the product **AF** exist?

II.5 Verify that the matrices **F** and **H** in exercise II.2 conform to: $\mathbf{F}\mathbf{F}^2 = \mathbf{F}^2\mathbf{F} = \mathbf{F}^3$ and $\mathbf{H}\mathbf{H}^3 = \mathbf{H}^2\mathbf{H}^2 = \mathbf{H}^3\mathbf{H} = \mathbf{H}^4$.

II.6 Calculate $\mathbf{BB} = \mathbf{B}^2$, **AB** and $\mathbf{BA}^T$ taking **A** from Exercise II.4 and:

$$\mathbf{B} = \begin{bmatrix} 1 & 2 & 3 \\ 2 & 3 & 4 \\ 3 & 4 & 5 \end{bmatrix}$$

Which product is symmetric? Under what condition is $(\mathbf{AB})^T = \mathbf{BA}^T$?

II.7 Calculate $\mathbf{C}^2$, $\mathbf{CC}^T$, $\mathbf{C}^T\mathbf{D}^T$, **CD**, **DC**, $\mathbf{CD}^T$, $\mathbf{C}^T\mathbf{C}$, $\mathbf{C}^T\mathbf{D}$, $\mathbf{D}^T\mathbf{C}$, $\mathbf{DC}^T$, $\mathbf{D}^T\mathbf{C}^T$ given that:

$$\mathbf{C} = \begin{bmatrix} 1 & 2 & 3 \\ 1 & 1 & 2 \\ 1 & 1 & 1 \end{bmatrix}, \quad \mathbf{D} = \begin{bmatrix} 2 & 3 & 1 \\ 1 & 1 & 1 \\ 1 & 2 & 3 \end{bmatrix}$$

Hence show that (i) $\mathbf{C}^T\mathbf{C}$ is symmetric, (ii) $\mathbf{DC} = (\mathbf{C}^T\mathbf{D}^T)^T$, (iii) $\mathbf{CD}^T = (\mathbf{DC}^T)^T$, (iv) $\mathbf{C}^T\mathbf{D} = (\mathbf{D}^T\mathbf{C})^T$, and (v) $\mathbf{CD} = (\mathbf{D}^T\mathbf{C}^T)^T$.

II.8 Show that the **O** matrix is formed from the multiplication of two matrices **PQ** and **QP** where,

$$\mathbf{P} = \begin{bmatrix} 0 & 2 \\ 0 & 0 \end{bmatrix}, \quad \mathbf{Q} = \begin{bmatrix} 0 & 3 \\ 0 & 0 \end{bmatrix}$$

II.9 Find the products **Vy**, $\mathbf{y}^T\mathbf{V}$, $\mathbf{y}^T(\mathbf{Vy})$ and $(\mathbf{y}^T\mathbf{V})\mathbf{y}$ given that,

$$\mathbf{V} = \begin{bmatrix} 1 & 1 & 1 \\ 1 & 2 & 2 \\ 1 & 2 & 3 \end{bmatrix}, \quad \mathbf{y} = \begin{bmatrix} 1 \\ 2 \\ 1 \end{bmatrix}$$

Hence show that: $\mathbf{Vy} = (\mathbf{y}^T\mathbf{V})^T$ and $\mathbf{y}^T(\mathbf{Vy}) = (\mathbf{y}^T\mathbf{V})\mathbf{y}$.

II.10 Calculate **Ry**, $\mathbf{R}^T\mathbf{y}$, $\mathbf{y}^T\mathbf{R}$, $\mathbf{y}^T\mathbf{R}^T$, $\mathbf{RR}^T$, $\mathbf{R}^T\mathbf{R}$ given **y** above and

$$\mathbf{R} = \begin{bmatrix} 3 & 2 & 1 \\ 4 & 3 & 2 \\ 5 & 4 & 3 \end{bmatrix}$$

Verify that: $\mathbf{R}^T\mathbf{y} = (\mathbf{y}^T\mathbf{R})^T$, $\mathbf{Ry} = (\mathbf{y}^T\mathbf{R}^T)^T$, $(\mathbf{R}^T\mathbf{y})^T = \mathbf{y}^T\mathbf{R}$, $(\mathbf{Ry})^T = \mathbf{y}^T\mathbf{Q}^T$, $(\mathbf{y}^T\mathbf{R})(\mathbf{R}^T\mathbf{y}) = \mathbf{y}^T(\mathbf{RR}^T)\mathbf{y} = (\mathbf{y}^T\mathbf{R}^T)(\mathbf{Ry}) = \mathbf{y}^T(\mathbf{R}^T\mathbf{R})\mathbf{y}$. Would you expect these to hold if **y** were not symmetric?

II.11 Evaluate using **V** and **R** above the following: **EV**, **VE**, **ER**, **RE**, **EVE** and **ERE** given that What are the conditions that ensure $\mathbf{EV} = (\mathbf{VE})^T$?.

$$\mathbf{E} = \begin{bmatrix} 3 & 0 & 0 \\ 0 & 2 & 0 \\ 0 & 0 & 1 \end{bmatrix}$$

II.12 Calculate the following products of two lower triangular matrices: $(\mathbf{L}_1)^2$, $(\mathbf{L}_2)^2$, $\mathbf{L}_1\mathbf{L}_2$ and $\mathbf{L}_2\mathbf{L}_1$ given that

$$\mathbf{L}_1 = \begin{bmatrix} 1 & 0 & 0 \\ 1 & 1 & 0 \\ 1 & 1 & 1 \end{bmatrix}, \quad \mathbf{L}_2 = \begin{bmatrix} 1 & 0 & 0 \\ 2 & 1 & 0 \\ 3 & 2 & 1 \end{bmatrix}$$

Hence show that $\mathbf{L}_1^2 = \mathbf{L}^2$ and $\mathbf{L}_1\mathbf{L}_2 = \mathbf{L}_2\mathbf{L}_1$.

II.13 Two upper triangular matrices are formed transposing $\mathbf{L}_1$ and $\mathbf{L}_2$ above such that $\mathbf{U}_1 = \mathbf{L}_1^T$ and $\mathbf{U}_2 = \mathbf{L}_2^T$. Calculate $(\mathbf{U}_1)^2$, $(\mathbf{U}_2)^2$, $\mathbf{U}_1\mathbf{U}_2$ and $\mathbf{U}_2\mathbf{U}_1$ and show that $\mathbf{U}_1^2 = \mathbf{U}_2$, $\mathbf{U}_1\mathbf{U}_2 = \mathbf{U}_2\mathbf{U}_1$.

II.14 Verify with examples that the product of any matrix with its transpose is a symmetrical square matrix.

II.15 Confirm that $\mathbf{M}^{-1}$ is the inverse of **M**:

$$\mathbf{M} = \begin{bmatrix} 1 & -1 & 2 \\ 2 & 1 & -1 \\ -1 & 3 & 3 \end{bmatrix}, \quad \mathbf{M}^{-1} = \begin{bmatrix} 6/25 & 9/25 & -1/25 \\ -1/5 & 1/5 & 1/5 \\ 7/25 & -2/25 & 3/25 \end{bmatrix}$$

Selected Answers to Exercises:

II.1

$$\mathbf{pp}^T = \begin{bmatrix} p_1^2 & p_1p_2 & p_1p_3 \\ p_1p_2 & p_2^2 & p_2p_3 \\ p_1p_3 & p_2p_3 & p_3^3 \end{bmatrix}, \quad \mathbf{p}^T\mathbf{q} = p_1q_1 + p_2q_2 + p_3q_3$$

II.2

$$\mathbf{FG} = \begin{bmatrix} 8 & 7 \\ 7 & 8 \end{bmatrix}, \quad \mathbf{FH}^T = \begin{bmatrix} 11 & 5 \\ 10 & 4 \end{bmatrix}, \quad \mathbf{H}^T\mathbf{G} = \begin{bmatrix} 16 & 14 \\ 11 & 9 \end{bmatrix}$$

II.4

$$\mathbf{A}^T\mathbf{F} = \begin{bmatrix} 6 & 9 \\ 9 & 12 \\ 12 & 15 \end{bmatrix}, \quad \mathbf{HA} = \begin{bmatrix} 16 & 23 & 30 \\ 6 & 9 & 12 \end{bmatrix}$$

II.6

$$\mathbf{AB} = \begin{bmatrix} 14 & 20 & 26 \\ 32 & 47 & 62 \end{bmatrix}, \quad \mathbf{BA}^T = \begin{bmatrix} 14 & 32 \\ 20 & 47 \\ 26 & 62 \end{bmatrix}$$

II.7

$$\mathbf{CD}^T = \begin{bmatrix} 11 & 6 & 14 \\ 7 & 4 & 9 \\ 6 & 3 & 6 \end{bmatrix}, \quad \mathbf{CD} = \begin{bmatrix} 7 & 11 & 12 \\ 5 & 8 & 8 \\ 4 & 6 & 5 \end{bmatrix}$$

II.9

$$\mathbf{Vy} = \begin{bmatrix} 4 \\ 7 \\ 8 \end{bmatrix}, \quad \mathbf{y}^T\mathbf{Vy} = 26$$

II.10

$$\mathbf{R}^T\mathbf{y} = \begin{bmatrix} 16 \\ 12 \\ 8 \end{bmatrix}, \quad \mathbf{y}^T\mathbf{R} = \begin{bmatrix} 16 & 12 & 8 \end{bmatrix}, \quad \mathbf{RR}^T = \begin{bmatrix} 14 & 20 & 26 \\ 20 & 29 & 38 \\ 26 & 38 & 50 \end{bmatrix}$$

II.11

$$\mathbf{ER} = \begin{bmatrix} 9 & 6 & 3 \\ 8 & 6 & 4 \\ 5 & 4 & 3 \end{bmatrix}, \quad \mathbf{ERE} = \begin{bmatrix} 27 & 12 & 3 \\ 24 & 12 & 4 \\ 15 & 8 & 3 \end{bmatrix}$$

II.12

$$\mathbf{L}_2^2 = \begin{bmatrix} 1 & 0 & 0 \\ 4 & 1 & 0 \\ 10 & 4 & 1 \end{bmatrix}, \quad \mathbf{L}_1\mathbf{L}_2 = \begin{bmatrix} 1 & 0 & 0 \\ 3 & 1 & 0 \\ 6 & 3 & 1 \end{bmatrix}$$

II.13

$$\mathbf{U}_1^2 = \begin{bmatrix} 1 & 2 & 3 \\ 0 & 1 & 2 \\ 0 & 0 & 1 \end{bmatrix}, \quad \mathbf{U}_1\mathbf{U}_2 = \begin{bmatrix} 1 & 3 & 6 \\ 0 & 1 & 3 \\ 0 & 0 & 1 \end{bmatrix}$$

APPENDIX III

STRESS CONCENTRATIONS

III.1 Introduction

It was shown in Chapter 2 that a hole in an infinite plate under tension raised the stress at its periphery to three times the nominal value. Stress concentrations will arise wherever there is an abrupt change to the geometry of a load bearing component. When loading relatively simple geometries under tension, bending and torsion it is convenient to show the greatest stress concentration factor K in a graphical form employing convenient geometric ratios. They have been determined from photoelastic and finite element studies given in the Bibliography.

Graphs (Figs III.1 - III.8) are presented for several of the more common plate and bar geometries using a combination of linear and logarithmic scales. Two definitions K and K' are employed

$$K = \sigma_{max} / \sigma_{nom} \tag{III.1}$$

$$K' = \sigma_{max} / \sigma_{av} \tag{III.2}$$

where σ_{max} is the maximum stress at the notch tip, σ_{nom} is the nominal stress away from the notch and σ_{av} is the average stress within ligament area. We must define σ_{nom} and σ_{av} in eqs(III.1) and (III.2) appropriately for a tensile force P and a bending moment M. For this the dimension symbols given in the inset diagrams are employed. This will enable σ_{max} to be found when K or K' is read from the graphs. When a notch is subjected to combined tension and bending the maximum notch stress is found from adding the maximum stress under the separate actions of P and M.

In the case of an applied torque T, eqs(III.1) and (III.2) are written as

$$K = \tau_{max} / \tau_{nom} \tag{III.3}$$

$$K' = \tau_{max} / \tau_{av} \tag{III.4}$$

where τ_{nom} and τ_{av} are to be defined for each geometry. This will enable τ_{max} to be found when K or K' is read from the graphs.

III.2 Tension and Bending of a Flat Plate with a Central Hole

Referring to Fig. III.1A we have respective stresses away and at the edges of the hole

$$\sigma_{nom} = P/(wt), \quad \sigma_{max} = K\sigma_{nom}$$

where K follows from the d/w ratio. We see that K increases beyond 3 as d/w increases in a plate of finite width.

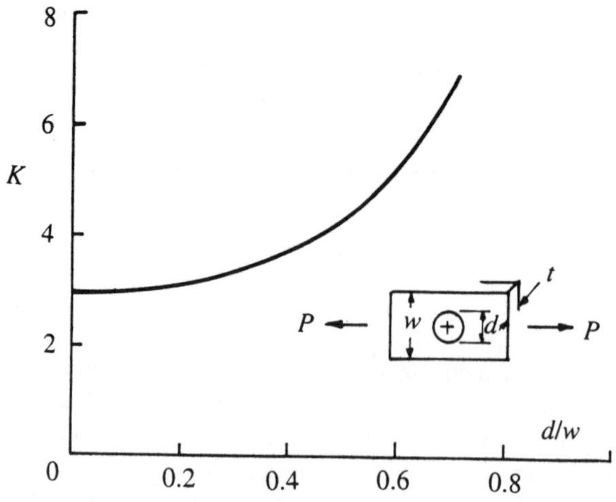

Figure III.1A Tension of a plate with a central hole

Referring to Fig. III.1B, bending theory gives

$$\sigma_{nom} = My/I = 6M/(w^2 t), \quad \sigma_{max} = K\sigma_{nom}$$

where $y = w/2$ and $I = wt^3/12$. Since the hole centre lies on the neutral axis $K \simeq 0$ when the hole is small or the plate width is large. In a plate of finite width K increases as the edges of the hole approach the free edges.

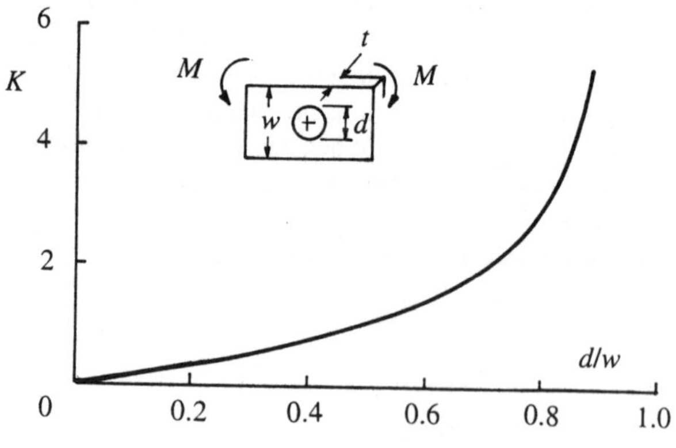

Figure III.1B Bending of a plate with a central hole

III.3 Tension and Bending of a Flat Plate with a Double U-Notch

The plots in Figs III.2A,B employ double logarithmic axes to accommodate a range of ratios $1 < w/w' \leq 2$ and $0.01 \leq r/w' \leq 0.2$. This accounts for notches of any depth, i.e. with parallel sides and root radius r. In tension, an average stress is defined for the ligament area giving

$$\sigma_{av} = P/(w't), \quad \sigma_{max} = K'\sigma_{av}$$

where K' in Fig. III.2A depends upon the ratios w/w' and r/w'.

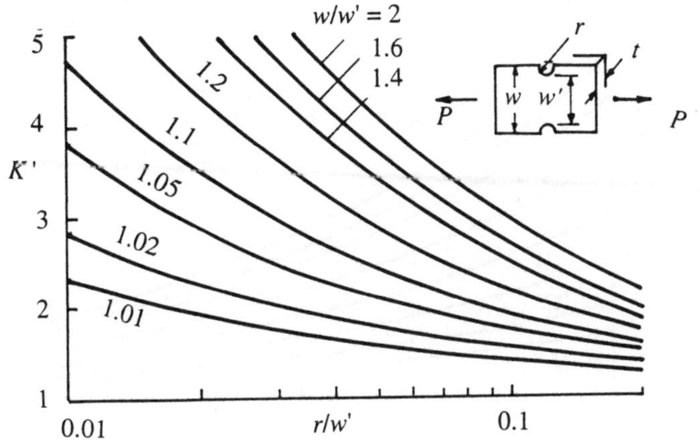

Figure III.2A Tension of a plate with a double U-notch

In bending an average bending stress is used to give:

$\sigma_{av} = 6M/(w'^2 t)$, $\sigma_{max} = K'\sigma_{av}$

where K' is read from Fig. III.2B for the given ratios w/w' and r/w'.

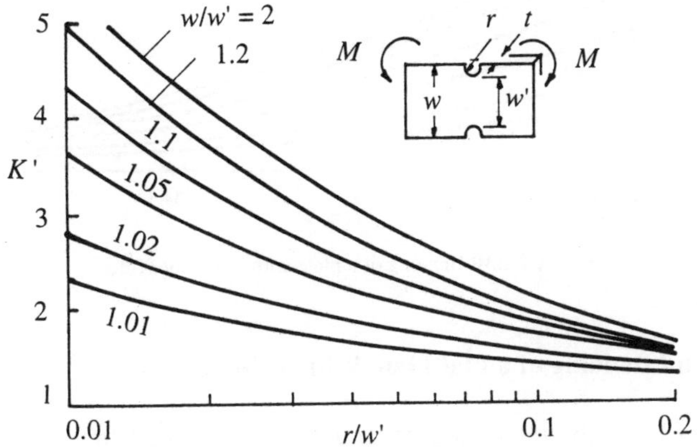

Figure III.2B Bending of a plate with a double U-notch

III.4 Tension of a Flat Plate with Shoulder Fillets

The average and maximum stresses are determined from Figs III.3A,B in a similar manner to III.2. In general, the reduced width w' is less than w by more than $2r$. Comparing Figs III.2A,B and III.3A,B for similar w/w' and r/w' ratios shows that the removal of material on one side of the notch serves to lower K' at the transition between the fillet root and the reduced section.

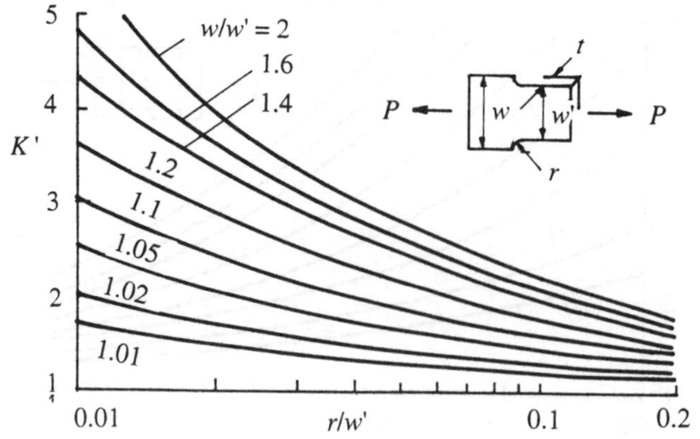

Figure III.3A Tension of a plate with a shoulder fillet

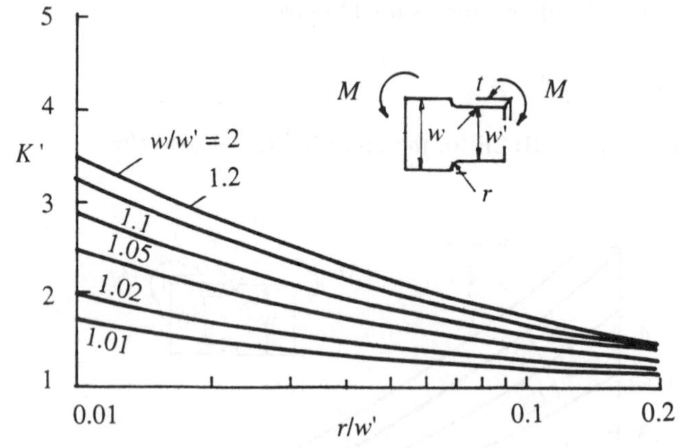

Figure III.3B Bending of a plate with a shoulder fillet

III.5 Tension and Bending of a Flat Plate With a U-notch

The plots in Figs III.4A,B employ log-linear axes to accommodate the following range of ratios: $1 < w/w' \leq 2$ and $0.01 \leq r/w' \leq 3$. The width w' will be less than w by more than $2r$ in a parallel sided notch. In tension, an average stress is defined for the ligament area giving

$$\sigma_{av} = P/(w'\,t\,), \quad \sigma_{max} = K'\sigma_{av}$$

where σ_{max} occurs at the bottom of the notch and K' is found from Fig. III.4A.

In bending we have from Fig. III.4B, with $0.01 \leq r/w' \leq 0.2$,

$$\sigma_{av} = 6M/(w'^2\,t\,),\ \sigma_{max} = K'\sigma_{av}$$

where σ_{max} occurs at the bottom of the notch.

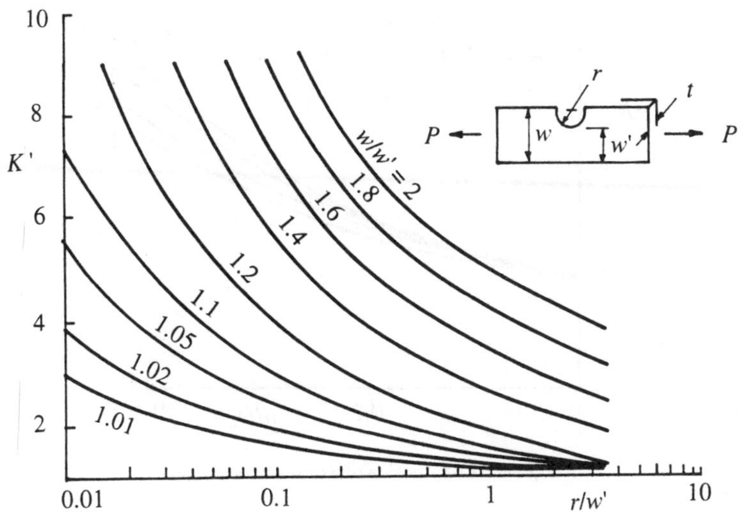

Figure III.4A Tension of a plate with a U-notch

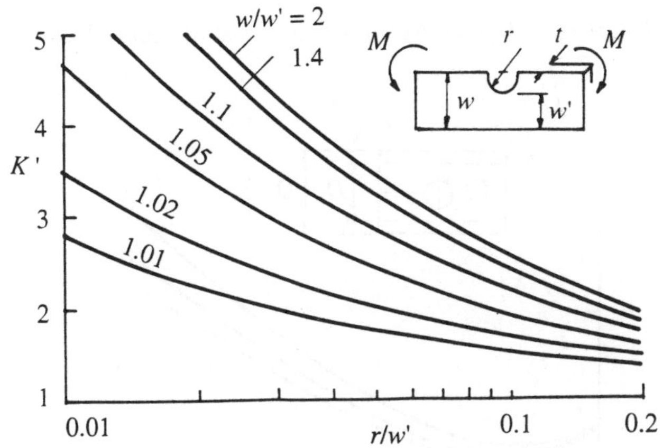

Figure III.4B Bending of a plate with a U-notch

III.6 Tension, Bending and Torsion of Circular Bars and Tubes with Cross-Bores

Figures III.5A,B,C employ linear scales for the hole and tube diameter ratios. The stress concentration factor K is now based upon a nominal stress value remote from the cross-bore. A solid bar applies to the particular case where the inner diameter $d_i = 0$. For tension (see Fig. III.5A)

$$\sigma_{\text{nom}} = 4P/[\pi(d_o^2 - d_i^2)], \quad \sigma_{\text{max}} = K\sigma_{\text{nom}}$$

where σ_{max} occurs at the edges of the bore, slightly beneath the outer surface.

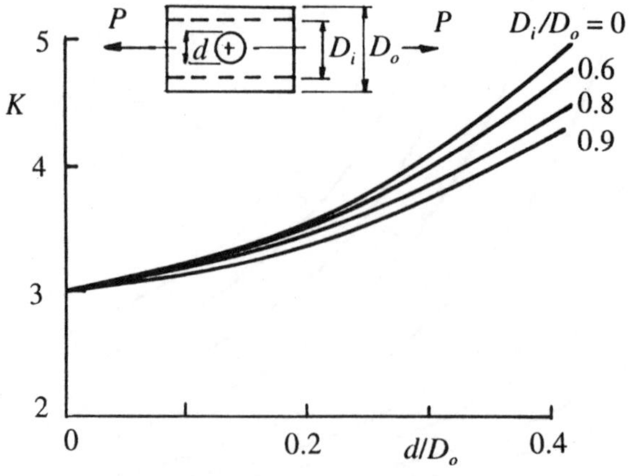

Figure III.5A Tension of a tube with a cross bore

In bending (see Fig. III.5B) the nominal stress is found for the outside diameter

$\sigma_{nom} = My/I = 32Md_o/[\pi(d_o^4 - d_i^4)]$, $\sigma_{max} = K\sigma_{nom}$

where $y = d_o/2$ and $I = \pi(d_o^4 - d_i^4)/64$. Again σ_{max} occurs at the hole edges slightly beneath the outer surface.

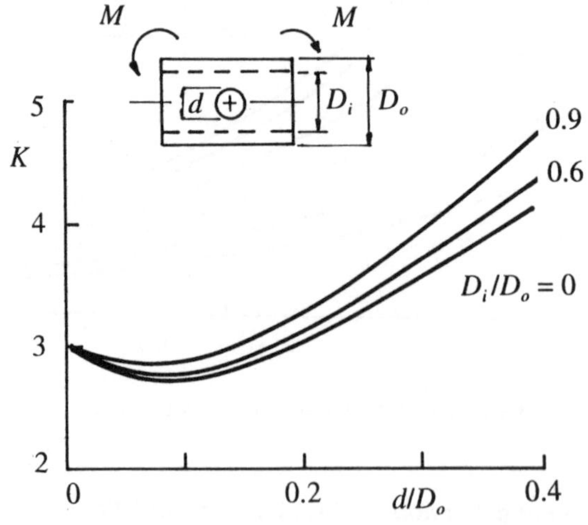

Figure III.5B Bending of a tube with a cross-bore

In torsion (see Fig. III.5C) the nominal stress is found for the outside diameter

$\tau_{nom} = Tr/J = 16Td_o/[\pi(d_o^4 - d_i^4)]$, $\sigma_{max} = K\sigma_{nom}$

where $r = d_o/2$, $I = \pi(d_o^4 - d_i^4)/32$ and τ_{max} occurs at the hole edges slightly beneath the outer surface.

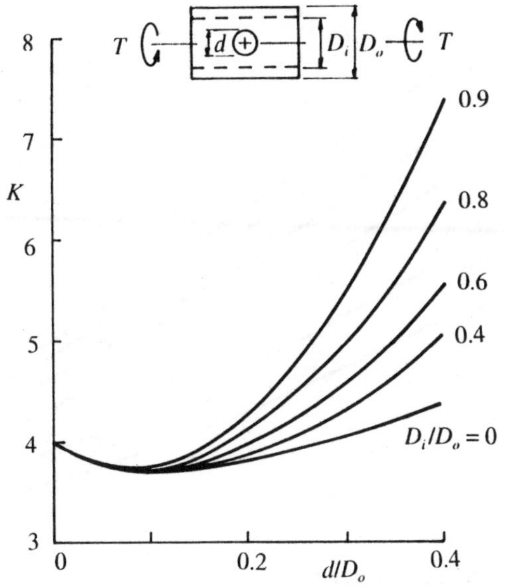

Figure III.5C Torsion of a tube with a cross-bore

III.7 Tension, Bending and Torsion of a Solid Bar With a Circumferential U-Notch

Within the logarithmic scale of Figs III.6A,B,C values of K' are plotted against r/d for the range $1 < D/d \leq 2$. The respective average stresses across the notch section are

$$\sigma_{av} = 4P/(\pi d^2)$$
$$\sigma_{av} = 32M/(\pi d^3)$$
$$\tau_{av} = 16T/(\pi d^3)$$

from which $\sigma_{max} = K'\sigma_{av}$ and $\tau_{max} = K'\tau_{av}$ apply to the bottom of the groove, which may lie at any depth.

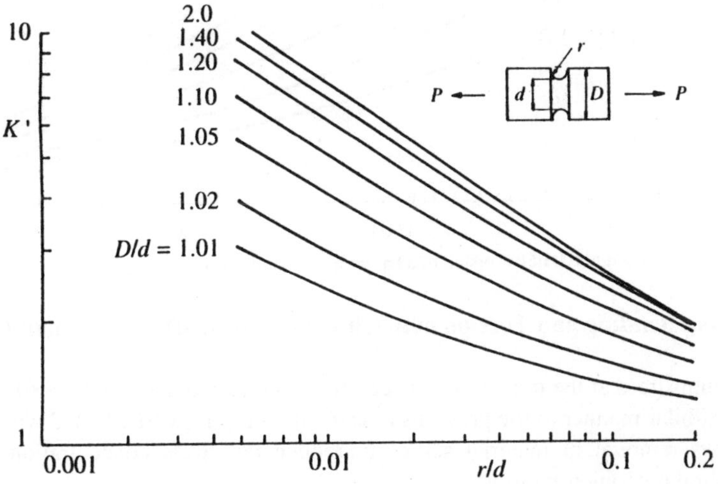

Figure III.6A Tension of a bar with a circumferential U-notch

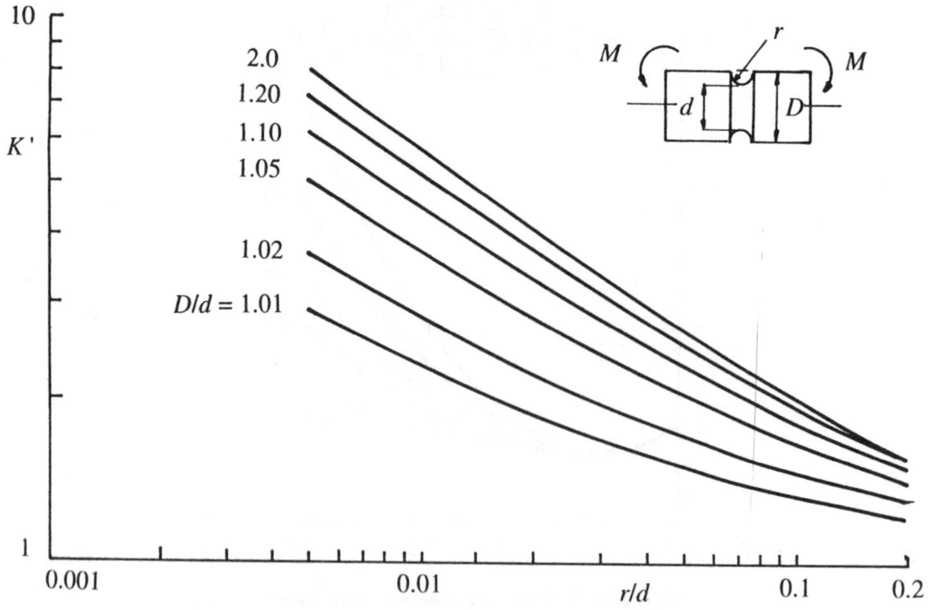

Figure III.6B Bending of a bar with a circumferential U-notch

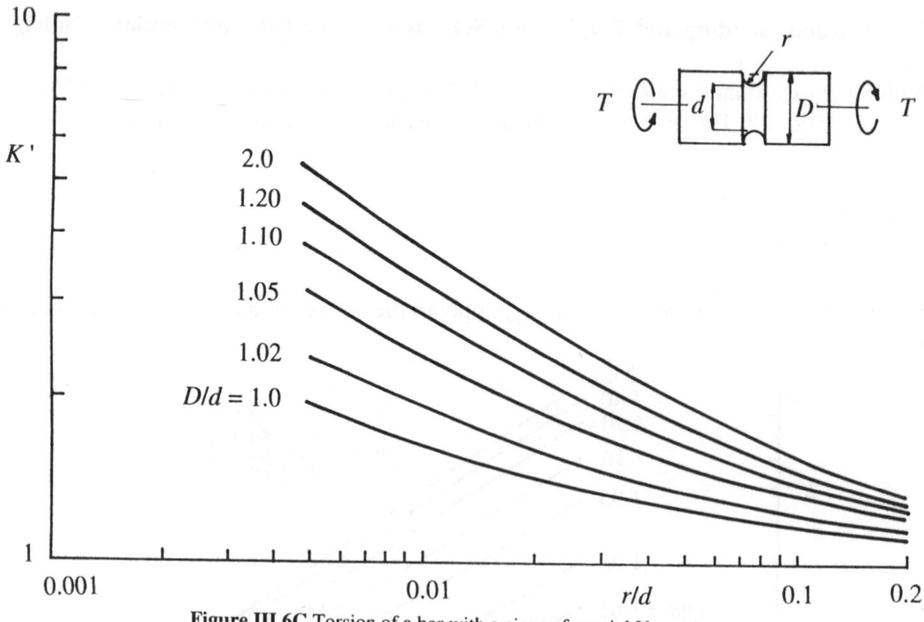

Figure III.6C Torsion of a bar with a circumferential U-notch

III.8 Tension, Bending and Torsion of a Solid Circular Shaft with a Shoulder Fillet

The maximum stress at the transition between the reduced diameter and the fillet radius is found in a similar manner to the previous case. Comparing Figs III.7A,B,C with Figs III.6 A,B,C shows removal of material serves to reduce the stress concentration factor for otherwise similar geometric ratios.

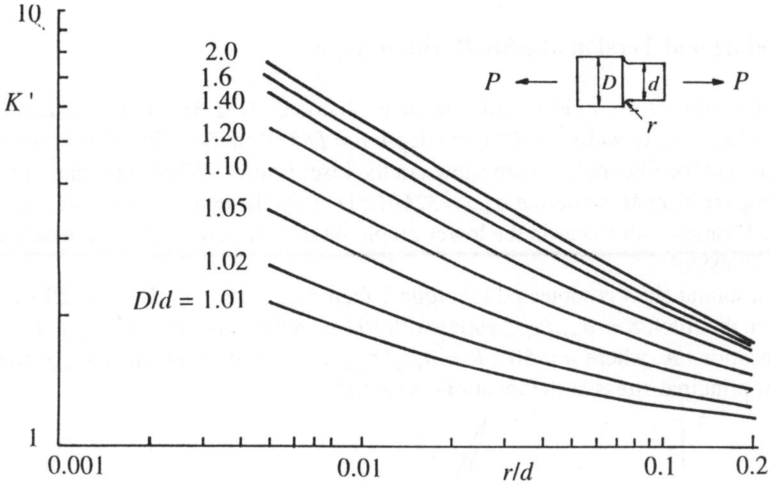

Figure III.7A Tension of a shaft with a shoulder fillet

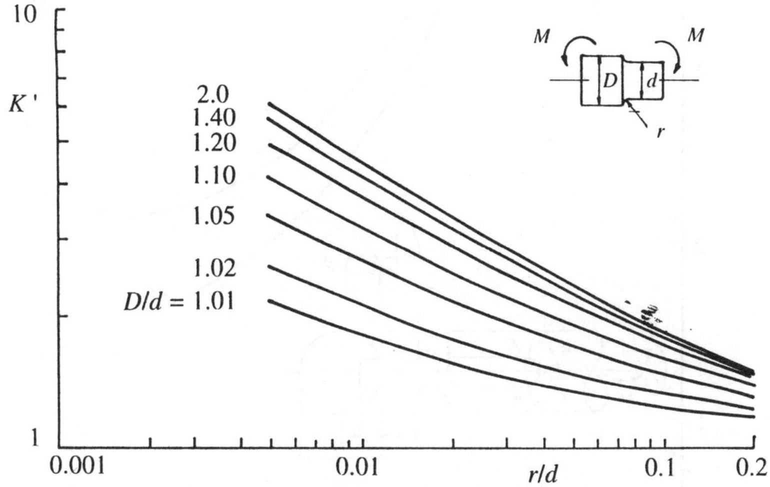

Figure III.7B Bending of a shaft with a shoulder fillet

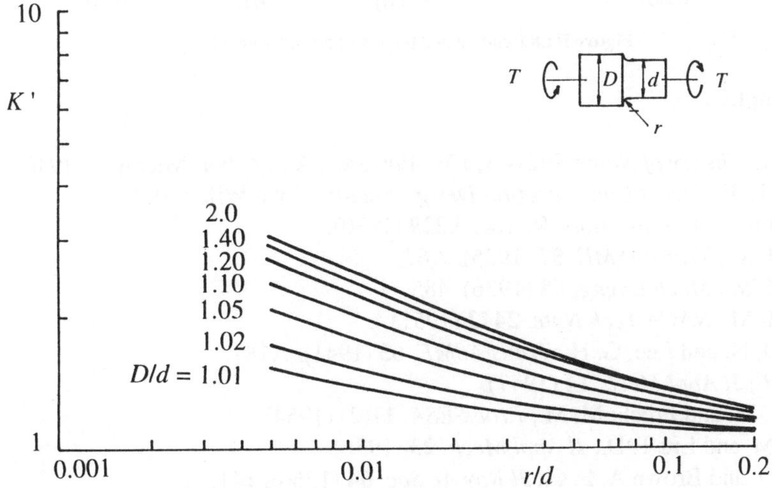

Figure III.7C Torsion of a shaft with a shoulder fillet

III.9 Bending and Torsion of a Shaft with a Keyway

Figure III.8 refers to a solid circular shaft of diameter $D \leq 165$ mm, machined with a longitudinal keyway to width $b = D/4$ and depth $h = D/8$. Details of the semi-circular end of the keyway and the fillet radius r are shown in the inset figures. When this shaft is subjected to a bending moment M we define $\sigma_{nom} = 32M/(\pi D^3)$. At the end of the keyway, location B ($\beta = 15°$), K varies with r/D as in the lower graph. At the surface location A, where $\alpha = 80°$, $K = 1.6 =$ constant.

When a similar shaft is subjected to torque T then $\tau_{nom} = 16T/(\pi D^3)$. At all locations B ($\beta = 15°$) in the fillet, $K = \sigma_{max}/\tau_{nom}$ varies with r/D as in the upper graph of Fig. III.8. At the surface locations A, where $\alpha = 50°$, $K = \sigma_{max}/\tau_{nom} = 3.4 =$ constant. Here σ_{max} refers to the maximum principal stress at the locations A and B.

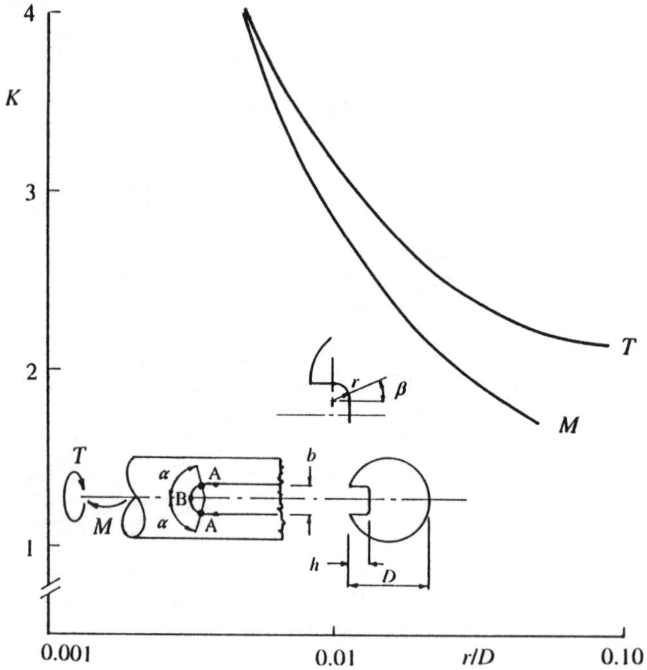

Figure III.8 Bending and torsion of a shaft with a keyway

Bibliography

Neuber H., *Theory of Notch Stresses*, J.W. Edwards, Ann Arbor, Michigan, 1946.
Peterson R. E., *Stress Concentration Design Factors*, John Wiley, 1974.
Howland R. C. J., *Phil Trans*. R. Soc. **A229** (1930).
Frocht M. M., *Trans ASME*, **57** (1935), A67.
Frocht M. M., *Mech Engng*, **58** (1936), 485.
Frocht M. M., *NACA Tech Note*, **2442** (1951).
Goodier J. N. and Lee, G. H., *Trans ASME*, **63** (1941), A187.
Ling C. B., *Jl Appl Mech*. **14** (1947).
Leven M. M. and Frocht M. M., *Proc SESA*, **11**(2) (1954).
Hetenyi M. and Liu T. D., *Jl Appl Mech*. **23** (1956).
Cole A. G. and Brown A. F. C., *Jl Roy Ae Soc*. **64** (1960), 141.
ESDU Items: 69020, March 1983 and 89048, December 1989.

INDEX